HANDBOOK OF OPTIMIZATION IN TELECOMMUNICATIONS

HANDBOOK OF OPTIMIZATION IN TELECOMMUNICATIONS

Edited by

MAURICIO G.C. RESENDE
AT&T Labs Research, Florham Park, New Jersey

PANOS M. PARDALOS
University of Florida, Gainesville, Florida

Library of Congress Control Number: 2005936165

ISBN-10: 0-387-30662-5   e-ISBN: 0-387-30165-8
ISBN-13: 978-0387-30662-9

Printed on acid-free paper.

AMS Subject Classifications: 90-00, 90B18, 90B10, 90B15, 90C35

© 2006 Springer Science+Business Media, Inc.

All rights reserved. This work may not be translated or copied in whole or in part without the written permission of the publisher (Springer Science+Business Media, Inc., 233 Spring Street, New York, NY 10013, USA), except for brief excerpts in connection with reviews or scholarly analysis. Use in connection with any form of information storage and retrieval, electronic adaptation, computer software, or by similar or dissimilar methodology now known or hereafter developed is forbidden.

The use in this publication of trade names, trademarks, service marks, and similar terms, even if they are not identified as such, is not to be taken as an expression of opinion as to whether or not they are subject to proprietary rights.

Printed in the United States of America.

9 8 7 6 5 4 3 2 1

springeronline.com

This book is dedicated to
Lucia and Rosemary.

Contents

Preface   xi

Contributing Authors   xv

Part I   Optimization algorithms

1
Interior point methods for large-scale linear programming   3
J.E. Mitchell, K. Farwell, and D. Ramsden

2
Nonlinear programming in telecommunications   27
A. Migdalas

3
Integer programming for telecommunications   67
E.K. Lee and D.P. Lewis

4
Metaheuristics and applications to optimization problems in telecommunications   103
S.L. Martins and C.C. Ribeiro

5
Lagrangian relax-and-cut algorithms   129
A. Lucena

6
Minimum cost network flow algorithms   147
J. L. Kennington and R. V. Helgason

7
Multicommodity network flow models and algorithms in telecommunications   163

M. Minoux

8
Shortest path algorithms 185
P. Festa

Part II Planning and design

9
Network planning 213
H.P.L. Luna

10
Multicommodity flow problems and decomposition in telecommunications networks 241
A. Lisser and Ph. Mahey

11
Telecommunications network design 269
A. Forsgren and M. Prytz

12
Ring network design 291
M. Henningsson, K. Holmberg, and D. Yuan

13
Telecommunications access network design 313
T. Carpenter and H. Luss

14
Optimization issues in distribution network design 341
G. R. Mateus and Z. K. G. Patrocínio Jr.

15
Design of survivable networks 367
B. Fortz and M. Labbé

16
Design of survivable networks based on p-cycles 391
W.D. Grover, J. Doucette, A. Kodian, D. Leung, A. Sack, M. Clouqueur, and G. Shen

17
Optimization issues in quality of service 435
J.G. Klincewicz

18
Steiner tree problems in telecommunications 459
S. Voß

19
On formulations and methods for the hop-constrained minimum spanning tree problem — 493
G. Dahl, L. Gouveia, and C. Requejo

20
Location problems in telecommunications — 517
D. Skorin-Kapov, J. Skorin-Kapov, and V. Boljunčić

21
Pricing and equilibrium in communication networks — 545
Q. Wang

Part III Routing

22
Optimization of Dynamic Routing Networks — 573
G. R. Ash

23
ILP formulations for the routing and wavelength assignment problem: Symmetric systems — 637
B. Jaumard, C. Meyer, and B. Thiongane

24
Route optimization in IP networks — 679
J. Rexford

25
Optimization problems in multicast tree construction — 701
C.A.S. Oliveira, P.M. Pardalos, and M.G.C. Resende

Part IV Reliability, restoration, and grooming

26
Network reliability optimization — 735
A. Konak and A.E. Smith

27
Stochastic optimization in telecommunications — 761
A. A. Gaivoronski

28
Network restoration — 801
D. Medhi

29
Telecommunication network grooming — 837
R.S. Barr, M.S. Kingsley, and R.A. Patterson

Part V Wireless

30
Graph domination, coloring and cliques in telecommunications 865
B. Balasundaram and S. Butenko

31
Optimization in wireless networks 891
M. Min and A. Chinchuluun

32
Optimization for planning cellular networks 917
E. Amaldi, A. Capone, F. Malucelli, and C. Mannino

33
Load balancing in cellular wireless networks 941
S. Borst, G. Hampel, I. Saniee, and P. Whiting

Part VI The web and beyond

34
Optimization issues in web search engines 981
Z. Liu and Ph. Nain

35
Optimization in e-commerce 1017
M. Kourgiantakis, I. Mandalianos, P.M. Pardalos, and A. Migdalas

36
Optimization issues in combinatorial auctions 1051
S. van Hoesel and R. Müller

37
Supernetworks 1073
A. Nagurney

Index 1121

Preface

Telecommunications has had a major impact in all aspects of life in the last century. There is little doubt that the transformation from the industrial age to the information age has been fundamentally influenced by advances in telecommunications. What sounded like science fiction just a few years ago is now reality. For example, in 1945, Arthur C. Clarke envisioned the integration of rockets and wireless communications in a system of orbiting space stations to relay radio signals around the world. Only twenty years later, Intelsat, the international satellite telecommunications organization, successfully placed the Early Bird satellite over the Atlantic Ocean.

Innovation and growth in telecommunications have been staggering. In 1927, AT&T introduced transatlantic telephone service, via radio, between the U.S. and London. The service had capacity for one call at a time and cost $75 for a three-minute call. Customer dialing for long distance domestic calls was introduced in 1951, and international calls only in 1970. The first fiber optic cable in a commercial communication system was put in place in 1977. Today, international calls cost consumers only a few cents a minute.

The International Telecommunication Union (ITC) estimates that the number of landlines worldwide grew from about 689 million in 1995 to over 1 billion in 2001. In the developing world, the growth was much greater. In China, for example, the number of lines quadrupled from 41 to 179 million in those six years. As recently as twenty years ago, personal wireless communication was limited to a handful of government and military officials. The first commercial cellular telephone system in the U.S. was opened in 1983. Now, it has spread all over the world. Again, in the period from 1995 to 2001, the number of mobile phones worldwide grew from about 91 million to almost a billion. In 64 developing countries, the number of mobile lines grew a hundred-fold in that period. In many countries, there are more wireless lines than wire lines. Wireless penetration in Europe is expected to reach 100% by 2007.

Broadband was unknown only ten years ago. In 2005, nearly half of the U.S. population had broadband Internet access. In some U.S. markets, nearly 70% of homes had broadband. In March 2005, according to the website internetworldstats.com, Sweden had the highest Internet penetration with about 74% of its population having access. Regionally, North America had the highest penetration (64%). However, only about 13% of the world's population could access the Internet (in Africa this value was only 1.5%), suggesting that there is still a long road ahead in the deployment of telecommunication systems around the world.

In the early days, telecommunication networks carried mainly voice traffic. With time, an increasing portion of traffic consisted of data. By 2000, the volume of data traffic on AT&T's network was greater than the volume of voice traffic. With voice over IP (VOIP), voice has become data and soon only data will be transported on telecommunication networks.

Optimization problems are abundant in the telecommunications industry. The successful solution of these problems has played an important role in the development of telecommunications and its widespread use. Optimization problems arise in the design of telecommunication systems, and in their operation.

This book brings together experts from around the world who use optimization to solve problems that arise in telecommunications. The editors made an effort to cover recent optimization developments that are frequently applied to telecommunications, and a spectrum of topics, such as planning and design of telecommunication networks, routing, network protection, grooming, restoration, wireless communications, network location and assignment problems, Internet protocol, world wide web, and stochastic issues in telecommunications. It is our objective to provide a reference tool for the increasing number of scientists and engineers in telecommunications who depend upon optimization in some way. Target readers will include students, researchers, and practitioners in engineering, computer science, statistics, operations research, and mathematics.

Each chapter in the handbook is of an expository, but also of a scholarly nature, and includes a brief overview of the state-of-the-art thinking relative to the topic, as well as pointers to the key references in the field. It is our expectation that specialists as well as nonspecialists will find the chapters stimulating and helpful.

The handbook is organized in six parts.

- Part I deals with basic optimization algorithms, including linear, integer, and nonlinear programming, network and multicommodity flow, and shortest path algorithms, metaheuristics, and Lagrangian relax-and-cut algorithms.

- Part II focuses on planning and design. This includes chapters on network planning, multicommodity flow and decomposition, network design, ring network design, access network design, distribution network design, survivable network design, location problems, Steiner tree problems, hop-constrained minimum spanning tree problem, quality of service, and pricing and equilibrium in communication networks.

- Part III addresses routing, with chapters on dynamic routing networks, routing and wavelength assignment, optimization in IP networks, and optimization of multicast trees.

- Part IV covers reliability, restoration, and grooming. This includes chapters on optimization of network reliability, stochastic optimization, network restoration, and network grooming.

- Part V focuses on issues arising in wireless telecommunications. This includes chapters on graph domination, coloring, and cliques, optimization in wireless networks, optimization for planning cellular networks, and dynamic load balancing in CDMA networks.

- Finally, Part VI deals with the web and beyond telecommunication networks, including optimization issues in web search engines, e-commerce, combinatorial auctions, and supernetworks.

Bibliographies are given at the end of each chapter. For ease of reference, the bibliographies at the ends of chapters have been compiled into a single HTML document which can be found online at http://www.springer.com/0-387-30662-5.

We would like to take this opportunity to thank the contributors, the reviewers, and the publisher for helping us to complete this handbook. We would also like to thank AT&T Labs Research and the National Science Foundation for partial support of this project.

Mauricio G. C. Resende and Panos M. Pardalos
Florham Park and Gainesville

Contributing Authors

Edoardo Amaldi received a "Diplôme" in Mathematical Engineering and a "Doctorat ès Sciences" (Ph.D.) from the Swiss Federal Institute of Technology at Lausanne (EPFL). After three years at the School of Industrial Engineering and Operations Research, Cornell University, USA, he joined in 1998 the Dipartimento di Elettronica e Informazione, Politecnico di Milano, Italy, where he is currently an Associate Professor in Operations Research. His main research interests include discrete optimization and algorithm design with applications in telecommunications, image processing and machine learning.

Gerald R. Ash is an AT&T Fellow and Senior Technical Consultant at AT&T Labs. He is the author of *Dynamic Routing in Telecommunications Networks* and authored many papers on dynamic routing, network design, and network optimization. Telecommunication networks have been central to the applied part of his research work. In addition to being an AT&T Fellow, he is also a Bell Labs Fellow, IEEE Fellow, recipient of the IEEE Alexander Graham Bell Medal, and was elected to the New Jersey Inventors Hall of Fame.

Balabhaskar Balasundaram is a doctoral student at the Department of Industrial Engineering at Texas A&M University. He received his bachelor's degree in Mechanical Engineering from Indian Institute of Technology – Madras, India. His research interests include mathematical programming approaches and algorithms for optimization problems on graphs, with emphasis on social, biological, and wireless networks.

Richard S. Barr is an Associate Professor and Chair of the Department of Engineering Management, Information, and Systems. He received his Ph.D. from the University of Texas at Austin.

Valter Boljunčić is Associate Professor of Mathematics at Faculty of Economics and Tourism "Dr. Mijo Mirković" in Pula, University of Rijeka. His field of research is efficiency and sensitivity analysis, and design of communication networks.

Sem Borst received the M.Sc. degree in applied mathematics from the University of Twente, The Netherlands, in 1990, and the Ph.D. degree from the University of Tilburg, The Netherlands, in 1994. In 1995 he joined Bell Labs, Lucent Technologies, as a Member of Technical Staff in the Mathematics of Networks and Systems Research department in Murray Hill, NJ, USA. Since the fall of 1998, he has also been with the Center for Mathematics and Computer Science (CWI) in Amsterdam. He also holds a part-time appointment as a professor of Stochastic Operations Research at Eindhoven University of Technology. Dr. Borst is a member of ACM Sigmetrics and IFIP Working Group 7.3, and serves or has served as a member of several program committees and editorial boards. His main research interests are in performance evaluation and resource allocation issues in communication networks and computer systems.

Sergiy Butenko is an Assistant Professor of Industrial Engineering at Texas A&M University. He received his master's degree in Mathematics from Kyiv Taras Shevchenko University in Ukraine, and Ph.D. degree in Industrial and Systems Engineering from the University of Florida. Dr. Butenko's primary research interests are in the areas of optimization, operations research, and mathematical programming and their applications. He has published more than 20 papers in refereed journals and edited books.

Antonio Capone received the Laurea and Ph.D. degrees in telecommunication engineering from the Politecnico di Milano, Italy, in 1994 and 1998, respectively. From November 1997 to June 1998, he was an Adjunct Professor at the University of Lecce, Italy. From June to October 2000, he visited the Computer Science Department, University of California, Los Angeles. He is now an Associate Professor in the Department of Elettronica e Informazione, Politecnico di Milano. His current research activities mainly include packet access in wireless cellular network, congestion control, and quality-of-service issues for Internet Protocol networks, network planning, and optimization.

Tamra Carpenter is Director of the Network Models and Algorithms Research Group at Telcordia Technologies. Her primary research interest lies in developing models and algorithms to optimize communication networks. Her particular emphasis has been in access network design, where she has published several related papers.

Altannar Chinchuluun is a Ph.D. student in the Department of Industrial and Systems Engineering at the University of Florida. His research interests include combinatorial optimization and global optimization.

Matthieu Clouqueur is a graduate from École Nationale Supérieure de l'Électronique et de ses Applications (National School of Electronics and Electrical Engineering), France. In 1998, he joined TRLabs at the University of Alberta, Edmonton, Canada to start his graduate studies in the Network Systems Group. At the end of 2003, he completed his Ph.D. work on the topic of "service availability in mesh-restorable transport networks." He is now a Research Scientist at Siemens, Corporate Technology in Munich, Germany. His research interests include all topics related to the optimal design of optical transport networks and network survivability.

Geir Dahl is Professor at the Centre of Mathematics for Applications at the University of Oslo, Norway. Professor Dahl's research interests are linear algebra, combinatorial optimization, and network optimization including applications of these areas.

John Doucette is an Assistant Professor in Engineering Management in the Department of Mechanical Engineering at the University of Alberta and an Adjunct Scientist in the Network Systems group at TRLabs. He received a B.Sc. in Mathematics from Dalhousie University in 1992, a B.Eng. in Industrial Engineering from the Technical University of Nova Scotia (TUNS, now Dalhousie University) in 1996, and a Ph.D. in Electrical and Computer Engineering at the University of Alberta in 2004. He has been with the Department of Mechanical Engineering since 2005, has held various positions within TRLabs since 1997, and he was an instructor in the Department of Electrical and Computer Engineering at the University of Alberta from 1998 to 2002, where he taught courses in probability and statistics and in telecommunication systems engineering. Dr. Doucette has published approximately 20 conference and journal papers, a book chapter, and approximately 15 technical reports and presentations. He has patents granted or pending on four topics, and is a P.Eng. in the province of Alberta. He has twice been one of three finalists for the Alberta Science and Technology (ASTech) Leadership Awards Foundation Leaders of Tomorrow Award and was awarded an Alberta Ingenuity Industrial Associateship in 2004. His current research interests include network restoration and protection, network planning and design, and network resource management and optimization.

Kris Farwell is a Doctoral Candidate in Mathematics at Rensselaer Polytechnic Institute. He received his M.S. in Mathematics at South Dakota State University. He developed what are known as *Farwell Points* while attending Houghton College. His research is in Integer Programming.

Paola Festa is Assistant Professor in Operations Research at the Mathematics and Applications Department of the University of Napoli FEDERICO II, Italy. She earned her Ph.D. in Operations Research and authored and co-authored many research papers in Network Flow Problems, Discrete Optimization, and Hard Combinatorial Optimization Problems.

Anders Forsgren is a Professor of Optimization and Systems Theory at the Department of Mathematics, Royal Institute of Technology (KTH), Stockholm, Sweden, from where he also earned his Ph.D.

Bernard Fortz is Professor of Operations Research at the Louvain School of Management (Université Catholique de Louvain). His main research interests are combinatorial optimization, network design problems, and the optimization of Internet resources using efficient routing protocols. He holds a Ph.D. in Operations Research from the Université Libre de Bruxelles. In 1997, he was awarded the AT&T Research Prize. He is currently coordinator of the European Network Optimization Group.

Alexei Gaivoronski is a Professor in the Department of Industrial Economics and Technology Management of NTNU – Norwegian University of Science and Technology in Trondheim, Norway. He obtained an M.Sc. from the Moscow Institute for Physics and Technology (1977) and a Ph.D. in applied mathematics from the V. M. Glushkov Institute of Cybernetics, Kiev (1979). Gaivoronski formerly worked in the public and the private sectors, including at the V. M. Glushkov Institute of Cybernetics in Kiev, at the International Institute for Applied Systems Analysis (IIASA) in Austria, and with Italtel, a world leading telecom supplier (part of the Siemens group). He held positions at the University of Milan. His research interests are in industrial optimization and quantitative decision support for planning of manufacturing and services. In particular, his focus has been in theory and algorithms for optimization under uncertainty, software systems for solution of problems related to industrial optimization, and specific decision support models for planning of industrial processes and services from different segments of industry, energy and finance.

Luis Gouveia is Associate Professor of the Department of Statistics and Operations Research, Faculty of Sciences, at the University of Lisbon, Portugal. He is currently the coordinator of the Operations Research Center at the University of Lisbon and his research interests are network optimization and combinatorial optimization. Telecommunication networks have been central to the applied part of his research work.

Wayne D. Grover obtained his B.Eng. from Carleton University, an M.Sc. from the University of Essex, and Ph.D. from the University of Alberta, all in Electrical Engineering. He had 10 years experience as scientific staff and management at BNR (now Nortel Networks) on fiber optics, switching systems, digital radio and other areas before joining TRLabs as its founding Technical VP in 1986. In this position he was responsible for the development of the TRLabs research program and contributing to development of the TRLabs sponsorship base and he saw TRLabs through its early growth as a start-up to over the 100-person level. He now functions as Chief Scientist – Network Systems, at TRLabs and as Professor, Electrical and Computer Engineering, at the University of Alberta. He has patents issued or pending on 26 topics to date and in has received two TRLabs Technology Commercialization Awards for the licensing of restoration and network-design related technologies to industry. He is a recipient of the IEEE Baker Prize Paper Award for his work on self-organizing networks, as well as an IEEE Canada Outstanding Engineer Award, an Alberta Science and Technology Leadership Award and the University of Alberta's Martha Cook-Piper Research Award. In 2001–2002 he is also holder of a prestigious NSERC E.R.W. Steacie Memorial Fellowship. He is a P.Eng. in the Province of Alberta and a member of SPIE and a Fellow of the IEEE.

Georg Hampel is a member of research staff in the Wireless Research Laboratory at Bell Labs in Murray Hill, New Jersey. He holds M.S. and Ph.D. degrees in physics from J.W. Goethe Universität in Frankfurt am Main, Germany. Hampel's current research efforts involve the dynamic optimization of third- and forth-generation wireless networks.

Richard V. Helgason is an Associate Professor in the School of Engineering at Southern Methodist University in Dallas, Texas. He is co-author of *Algorithms for Network Programming* and has published in the areas of network optimization and the identification of minimal hull generators. His recent work in telecommunications is focused on traffic engineering modeling in MPLS networks.

Mathias Henningsson is Associate Professor in Optimization, Department of Mathematics at Linköping University. A central part of his research is ring network design in telecommunication networks, using optimization based methods.

Kaj Holmberg is Professor in Optimization at the Department of Mathematics, Linköping Institute of Technology, Sweden. His research interests range from mathematical decomposition and relaxation methods to telecommunication optimization problems, with focus on network design, and in these areas he has more than 40 refereed publications in international journals. A current interest is optimization of telecommunication networks using OSPF.

Brigitte Jaumard is a Professor in the Department of Computer Science, Université de Montréal where she holds a Canada Research Chair on the Optimization of Communication Networks. She obtained a Ph.D. in Computer Engineering from the École Nationale Supérieure des Télécommunications, Paris in 1986. She is the author or coauthor of more than 150 research papers in mathematical programming and combinatorial optimization, part of them with a focus on the design or the management of various types of communication networks.

Jeffery L. Kennington is a Professor in the School of Engineering at Southern Methodist University in Dallas, Texas. He is co-author of *Algorithms for Network Pro-*

gramming and has published many manuscripts in the area of network optimization. His recent publications are in the area of telecommunication network design.

M. Scott Kingsley is founder and president of OptionTel LLC, a telecommunications consulting services and brokerage firm in Dallas, Texas. He received his Masters of Science in Telecommunications from Southern Methodist University, Dallas, Texas.

John G. Klincewicz is a Senior Technical Specialist in the Network Design and Performance Analysis Department of AT&T Labs. He has an S.B. in Mathematics from M.I.T. and a Ph.D. in Operations Research from Yale University. Since joining AT&T in 1979, he has worked on a variety of applications, including backbone network design, ring network planning, inventory control, transportation planning, warehouse location and capacity expansion. His research interests include telecommunication network design, facility location models, and heuristics for combinatorial problems. He has over twenty refereed publications and is a member of INFORMS.

Adil Kodian has been with TRLabs, Edmonton, Alberta, Canada since 2001, where he is currently completing his Ph.D. in Survivable Network Design. He received a B.E. in Electronics Engineering from National Institute of Technology, Surat, India. His current research interests include failure independent path protecting p-cycles, multiple quality of protection service classes in p-cycle networks and ring to p-cycle evolution.

Abdullah Konak is Assistant Professor of Information Sciences and Technology at Penn State Berks. He received his B.S. degree in Industrial Engineering from Yildiz Technical University, M.S. in Industrial Engineering from Bradley University, and Ph.D. in Industrial Engineering from University of Pittsburgh. Previous to this position, he was an instructor in the Department of Systems and Industrial Engineering at Auburn University. His current research interest is in the application of Operations Research techniques to complex problems, including such topics as telecommunication network design, network reliability analysis/optimization, facilities design, and data mining. He is a member of IIE and INFORMS.

Markos Kourgiantakis is a Ph.D. candidate in the Department of Economics, University of Crete. He holds a M.Sc. in Operational Research from the Department of Production Engineering and Management of the Technical University of Crete and a M.Sc. in Economics and Management from the Mediterranean Agronomic Institute of Chania. His Ph.D. thesis focuses on the economic and managerial analysis of business to business electronic marketplaces. He has publications in scientific journals and presentations in international conferences in several issues on e-business.

Martine Labbé is Professor of Operations Research at the Computer Science Department of the Université Libre de Bruxelles. Her main research area is combinatorial optimization, including graph theory and integer programming problems and with a particular emphasis on location and network design problems. Professor Labbé serves on the editorial board of several journals.

Eva Lee is Associate Professor in the School of Industrial and Systems Engineering at Georgia Institute of Technology, and Director of the Center for Operations Research in Medicine. Lee works in the area of mathematical modeling and optimization, including computational algorithms for optimal operations planning, resource allocation, and logistics. She primarily focuses on applications to medical and healthcare delivery problems. She has developed general-purpose linear and nonlinear mixed integer programming solvers; and decision support systems for improvement in healthcare delivery, disease prediction and diagnosis, optimal cancer treatment planning, large-scale scheduling, and transportation. She is also leading research in real-time resource allocation and operational and strategic planning for emergency responses to bioterrorism and infectious disease outbreaks.

Dion Leung received his B.Sc. degree in electrical engineering from the University of Alberta in 2000. He conducts research with the Network Systems Group at TRLabs and has recently finished his Ph.D. thesis, entitled *Capacity Planning and Management for Mesh-Survivable Networks under Demand Uncertainty*, at the University of Alberta. Besides his research interest in the area of transport network modeling and planning, he also received a project management certificate and has been a communications director for Hong Kong Canada Business Association, Edmonton Section (HKCBA). He is a member of IEEE, Project Management Institute (PMI), the Association of Professional Engineers, Geologists and Geophysicists of Alberta (APEGGA).

David Lewis is a Ph.D. student in the School of Industrial and Systems Engineering at Georgia Institute of Technology. His research focuses on optimization, in particular theory and computational approaches for large-scale network-type instances.

Abdel Lisser is University Professor in Computer Science in the University of Paris 11. He authored many research papers in network design problems, multicommodity flow problems, Discrete Optimization, Mathematical Programming and Operations Research. Telecommunication networks have been central to the major part of his research work.

Zhen Liu received the Ph.D. degree in Computer Science from the University of Orsay (Paris XI), France, in 1989. He was with the France Telecom R&D as an Associate Researcher from 1986 to 1988. He joined INRIA (the French national research center on information and automation) in 1988, first as a Researcher, then became a Research Director. Zhen Liu joined IBM T. J. Watson Research Center in 2000 as a Research

Staff Member, and has been the manager of the System Analysis and Optimization Department since 2001. His research interests are on control and performance analysis of communication networks, modeling of traffic and transport protocols, design and analysis of routing algorithms in wired and wireless networks, scheduling and performance evaluation of parallel and distributed systems. He is also interested in Petri nets and queueing networks. Zhen Liu was the program co-chair of the joint conference of *ACM Sigmetrics 2004* and *IFIP Performance Conference 2004*. He is a member of the *IFIP W.G. 7.3* on performance modeling.

Abilio Lucena is a Professor at the Business Administration Department of the Federal University of Rio de Janeiro. He received a Ph.D. degree from Imperial College, London, and, after that, has been a Visiting Professor at Erasmus University (Rotterdam), a Research Fellow at the Center for Operations Research and Econometrics (Louvain-la-Neuve), and a Senior Research Fellow at Imperial College. His main research interest is in combinatorial optimization, an area in which he authored many research papers. He has been a consultant for the airline and petroleum industries.

Henrique Pacca L. Luna is a Professor of the Department of Information Technology, at Federal University of Alagoas (UFAL), Maceió, Brazil, since 2003, and he is also a retired Professor of the Department of Computer Science, at Federal University of Minas Gerais (UFMG), Belo Horizonte, where he had taught since 1972. He earned his Docteur d'Etat title from Paul Sabatier University (UPS) in Toulouse, France, and he is currently vice-president of SOBRAPO, the Brazilian Society of Operations Research.

Hanan Luss is a Senior Scientist at Telcordia Technologies and an Adjunct Professor at Columbia University. From 1973 to 1998, he was at AT&T Bell Laboratories and AT&T Labs serving as the Technical Manager of the Operations Research Studies Group. Hanan published over 65 papers in major refereed journals on resource allocation, capacity planning, network design, and other related topics. His applied work has primarily focused on telecommunications networks and on logistics problems.

Philippe Mahey is a Professor in Computer Science in the University of Clermont-Ferrand, France. His research is centered on Decomposition Techniques in Mathematical Programming and he has published many contributions to the Network Design problem applied to modern broadband Telecommunication networks.

Federico Malucelli obtained a Laurea in Computer Science and a Ph.D. in Computer Science, both from Università di Pisa in 1988 and 1993, respectively. Since 2003, he is Professor of Operations Research at the Politecnico di Milano. His main research interests include: models and algorithms for combinatorial optimization problems, with applications in particular to telecommunications, transportations, logistics, and electronic circuit design. He has published more than 30 articles in international scientific journals.

Iraklis Mandalianos is a researcher in the Decision Support Systems Laboratory – ERGASYA of Technical University of Crete. He holds a M.Sc. in Operational Research from the Department of Production Engineering and Management of Technical University of Crete (2005). His research interests on optimization issues in e-commerce.

Carlo Mannino is Associate Professor in Operations Research at the University of Rome 1 – *La Sapienza*. He authored many research papers in Discrete Optimization, Mathematical Programming, and Operations Research. Telecommunication networks have been central to the applied part of his research work. He contributed to the development of the methodology adopted by the Italian Communications Authority to realize the Italian National Radio/TV Frequency plan, both for analog and digital technology.

Simone de Lima Martins graduated with an Electrical Engineering degree from PUC-Rio in 1984, and obtained a Master of Science in 1988 and a Ph.D. in 1999, both in Computer Science from PUC-Rio. She worked as a researcher at the Scientific Center of IBM-Brazil in 1988–1989 and from 1991–1993 as a computer systems analyst at IBM-Brazil, developing activities for voice and data integration in local area networks and multimedia systems. From 1999 to 2001, she worked as a visiting researcher in the Informatics Department of PUC-Rio and from 2001 to 2002 at LNCC (National Laboratory for Scientific Computation), both in Brazil. Since 2002, she is a Lecturer at the Universidade Federal Fluminense, where she develops activities in metaheuristics applications, bioinformatics, and parallel processing. She participates in several research projects financed by Brazilian government agencies in metaheuristics applications, parallel processing, and computational grids.

Geraldo Robson Mateus is a Professor in Computer Science at Federal University of Minas Gerais, Belo Horizonte, Brazil. He received his M.S. and Ph.D. in computer science from the Federal University of Rio de Janeiro, Brazil, in 1980 and 1986, respectively. He spent 1991 and 1992 at the University of Ottawa, Canada, as a visiting researcher. His research interests span network optimization, combinatorial optimization, algorithms and telecommunications. He has published over 200 scientific papers, 22 journal articles, 2 books, and 3 book chapters, and is a leader of several national and international projects. He has worked as a consultant for companies such as Telemig, Telemar, Usiminas, CVRD, MBR, France Telecom, and Embratel.

Deep Medhi is Professor in the Computer Science & Electrical Engineering Department at the University of Missouri–Kansas City. He is the co-author of *Routing, Flow and Capacity Design in Communication and Computer Networks*. His research spans protection and restoration design of multi-layered networks, dynamic QoS routing, network optimization, and network management.

Christophe Meyer is an Associate Researcher at the Department of Computer Science and Operations Research, Université de Montréal. He holds a Ph.D. in Applied Mathematics from the École Polytechnique de Montréal. He authored more than 20 research papers in the fields of discrete and global optimization.

Athanasios Migdalas is a Professor in the Department of Production Engineering and Management (DPEM) of the Technical University of Crete. He has published five international scientific books, and has edited several special issues of international scientific journals. He acts as referee for over ten international scientific journals and has been in the scientific board of several of them. He has published over sixty scientific articles in international scientific journals and special editions of books.

Manki Min is a Post Doctoral Associate in the Department of Industrial and Systems Engineering at the University of Florida. His research interests include wireless ad hoc networking, approximation algorithm design and analysis, and optimization problems in networks.

Michel Minoux is University Professor in Computer Science at the University of Paris 6. He is the author of *Mathematical Programming, Theory and Algorithms*, co-author of *Graphs and Algorithms* and authored many research papers in Discrete Optimization, Mathematical Programming, and Operations Research. Telecommunication networks have been central to the applied part of his research work.

John E. Mitchell is a Professor of Mathematical Sciences at Rensselaer Polytechnic Institute. His research interests include interior point methods (IPM) and integer programming (IP), especially the use of IPM's to solve IP problems. A current interest is the robustness of interdependent networks. He has published many papers in these areas.

Rudolf Müller is Professor of Quantitative Infonomics at Maastricht University. He received a Ph.D. in Mathematics from TU Berlin, and a habilitation in information systems from HU Berlin. His research focuses on the interplay of computational complexity, communication complexity and economic properties of distributed systems, with applications in auction design, scheduling and online decision support. His workshops on electronic market design have greatly stimulated the dialogue between computer scientists, economists and operations researchers.

Anna Nagurney is the John F. Smith Memorial Professor at the Isenberg School of Management at the University of Massachusetts at Amherst. She is the Founding

Director of the Virtual Center for Supernetworks (http://supernet.som.umass.edu) and the Supernetworks Laboratory for Computation and Visualization. Professor Nagurney also holds appointments in the Department of Civil and Environmental Engineering and the Department of Mechanical and Industrial Engineering at UMASS Amherst. She has devoted her career to research and education that combines management science / operations research, economics, and engineering. Her focus is the theoretical and applied aspects of network systems, particularly in the areas of transportation and logistics; economics and finance, and telecommunications, including the Internet. She is the editor of the book, *Innovations in Financial and Economic Networks*, published in 2003, and has authored or co-authored eight other books, including *Supernetworks: Decision-Making for the Information Age*, *Sustainable Transportation Networks*, and *Network Economics: A Variational Inequality Approach*. She has published over 100 refereed journal articles. Anna holds Ph.D., Sc.M., Sc.B. degrees (all in Applied Mathematics) and an A.B. degree (in Russian Language and Literature) from Brown University in Providence, Rhode Island. Among the awards she has received include: Fellow of the Radcliffe Institute for Advanced Study at Harvard University, a Bellagio Center Research Team Residency in Italy from the Rockefeller Foundation, a Distinguished Chaired Fulbright held at the University of Innsbruck, Austria, two AT&T Industrial Ecology Fellowships, an Eisenhower Faculty Fellowship, an NSF Visiting Professorship for Women held at MIT, an NSF Faculty Award for Women, and the Kempe Prize from the University of Umea, Sweden.

Philippe Nain received the Maîtrise Es-Sciences in Mathematics in 1978, the Diplôme d'Etudes Approfondies in Statistics in 1979, and the Doctorat de 3éme Cycle, specializing in Modeling of Computer Systems in 1981 from the University of Paris XI, Orsay, France. In 1987, he received the Doctorat d'Etat in Applied Mathematics from the University Pierre and Marie Curie, Paris, France. Since December 1981, he has been with INRIA where he is currently the head of the research project *Maestro* devoted to the modeling of computer systems and telecommunications networks. He has held visiting appointments at the University of Massachusetts (1993–94), at the University of Maryland (1987), and at North Carolina State University (1988). His research interests include modeling and performance evaluation of communication networks. He is an Associate Editor of Performance Evaluation and Operations Research Letters, and was an Associate Editor of *IEEE Transactions on Automatic Control*. He was a co-program chair of the *ACM Sigmetrics 2000* conference, the general chair of *Performance 2005*, and he is a member of *IFIP WG 7.3*.

Carlos A. S. Oliveira is Assistant Professor at the School of Industrial Engineering and Management in the Oklahoma State University. He obtained a Ph.D. in Industrial and Systems Engineering from the University of Florida, and a M.S. in Computer Science from the Federal University of Ceará, Brazil. He has been working on the development of efficient solution methods for NP-hard problems in diverse areas, such as Computer and Telecommunications Networks, Internet Modeling, and Biological Computing. He is the author of several papers in areas related to Optimization, Mathematical Programming and Computing. He is a member of the Editorial board of the Journal of Combinatorial Optimization.

Panos M. Pardalos is Distinguished Professor of Industrial and Systems Engineering at the University of Florida. He is also affiliated faculty member of the Computer Science Department, the Hellenic Studies Center, and the Biomedical Engineering Program. He is the Co-Director of the Center for Applied Optimization. He obtained a Ph.D. degree (1985) from the University of Minnesota in Computer and Information Sciences. He has held visiting appointments at Princeton University, DIMACS Center, Institute of Mathematics and Applications, FIELDS Institute, AT&T Labs Research, Trier University, Linköping Institute of Technology, and universities in Greece. He has received numerous awards including, University of Florida Research Foundation Professor, Foreign Member of the Royal Academy of Doctors (Spain), Foreign Member of the Lithuanian Academy of Sciences, Foreign Member of the National Academy of Sciences of Ukraine, and Foreign Member of the Petrovskaya Academy of Sciences and Arts (Russia). He is a Fellow of AAAS (American Association for the Advancement of Science), and in 2001 he was honored with the Greek National Award and Gold Medal for Operations Research. He is a world leading expert in global and combinatorial optimization. He is the editor-in-chief of the Journal of Global Optimization, managing editor of several book series, and a member of the editorial board of twenty international journals. He is the author of seven books and the editor of more than 50 books. He has written numerous articles and developed several well known software packages. His research is supported by National Science Foundation, NIH, and other government organizations. His recent research interests include network design problems, biomedical applications, datamining, optimization in telecommunications, e-commerce, and massive computing. He has been an invited lecturer at many universities and research institutes around the world. He has also organized several international conferences.

Zenilton K. G. Patrocínio Jr. received the B.S. and M.S. degrees in computer science from the Federal University of Minas Gerais, Belo Horizonte, Brazil, in 1990 and 1993, respectively. He is currently working toward the Ph.D. degree in computer science. In 2003, he joined the faculty of the Department of Computer Science of Pontifical Catholic University of Minas Gerais, Belo Horizonte, Brazil. He has published about 10 conference papers and his research interests include combinatorial optimization, network design and traffic grooming in WDM optical networks.

Raymond A. Patterson is a Canada Research Chair and Associate Professor of Management Information Systems at the University of Alberta, Alberta, Canada. He received his Ph.D. in Accounting and M.I.S. from the Ohio State University, Columbus, Ohio.

Mikael Prytz is an optimization expert at Ericsson Research, Stockholm, Sweden. He earned his Ph.D. from the Royal Institute of Technology (KTH), Stockholm, Sweden.

Daryn Ramsden is a graduate student in the Department of Mathematical Sciences at Rensselaer Polytechnic Institute. His research interests include Semidefinite Programming, Combinatorial Optimization, and Interior Point methods.

Cristina Requejo is Assistant Professor at the University of Aveiro, Portugal. She co-authored research papers in Discrete and Combinatorial Optimization. Telecommunication networks have been central to the applied part of her research work.

Mauricio G. C. Resende is a research scientist at the Algorithms and Optimization Research Department at the AT&T Shannon Laboratory of AT&T Labs Research. His undergraduate studies were in electrical engineering (systems concentration) at the Pontifícia Universidade Católica do Rio de Janeiro (PUC-Rio), Brazil (1978) and he earned a M.Sc. in operations research at the Georgia Institute of Technology (1979). He has been at AT&T Bell Labs and AT&T Labs since earning his Ph.D. in operations research at the University of California, Berkeley, in 1987. His research has focused on optimization, including interior point algorithms for linear programming, network optimization, and nonlinear programming, as well as heuristics for discrete optimization problems arising in telecommunications, scheduling, location, assignment, and graph theory. Most of his work with heuristics has focused on GRASP (greedy randomized adaptive search procedures), a metaheuristic that he and Thomas A. Feo developed in the late 1980s. He has developed several decision support systems (tools) for optimization problems arising in telecommunications. He has published over 100 papers. In addition to co-editing this handbook, he is co-editor of *Handbook of Applied Optimization, Handbook of Massive Datasets, Metaheuristics: Computer Decision-Making, Parallel Processing of Discrete Optimization Problems*, and the book series *Massive Computing*. He is on the editorial board of six journals. Besides working in the telecommunications industry, he has worked in the electrical power and semiconductor manufacturing industries.

Jennifer Rexford is a Professor in the Computer Science Department at Princeton University. From 1996–2004, she worked at AT&T Labs–Research, where she interacted closely with the designers and operators of AT&T's IP backbone network. Jennifer's research focuses on network measurement and network management, and she is the co-author of the book *Web Protocols and Practice*.

Celso Ribeiro is Full Professor at the Department of Computer Science of Universidade Federal Fluminense, Brazil. He chaired the departments of Electrical Engineering (1983–1987) and Computer Science (1993–1995) of the Catholic University of Rio de Janeiro. He has a bachelor degree in Electrical Engineering (Catholic University of Rio de Janeiro, 1976) and an M.Sc. degree in Systems Engineering (Federal

University of Rio de Janeiro, 1978). He obtained his doctorate in Computer Science at the École Nationale Supérieure des Télécommunications (Paris, France) in 1983. His research is supported by the Brazilian Council of Scientific and Technological Development (CNPq) and by the Rio de Janeiro State Foundation for Research Support (FAPERJ). Professor Ribeiro acted as President of the Brazilian Operations Research Society (SOBRAPO, 1989–1990) and of the Latin-American Association of Operations Research Societies (ALIO, 1992–1994), and as Vice-President of the International Federation of Operational Research Societies (IFORS, 1998–2000). He was a visiting researcher at AT&T Labs Research, International Computer Science Institute (ICSI, Berkeley), École Polytechnique de Montréal, and Université de Versailles (France). He is the editor of four books and the author of almost 100 papers in international journals. Professor Ribeiro supervised 11 doctorate dissertations and 26 master of science theses.

Anthony Sack completed his M.Sc. research at TRLabs in 2004, investigating the optimal design of mixed networks with APS, p-cycle, and SBPP techniques, along with studies on path length constraints for p-cycles and the use of Hamiltonian p-cycles in semi-homogeneous networks. In 2001, he received a B.E. in Electrical Engineering and a B.Sc. three-year in Computer Science from the University of Saskatchewan, followed by an M.Sc. in Electrical and Computer Engineering from the University of Alberta in 2004. He is now with TELUS, one of Canada's largest network operators. He is a member of the IEEE and an Engineer-in-Training in the Province of Alberta.

Iraj Saniee is the director of the Mathematics of Networks and Systems Research Department at Bell Laboratories, Lucent Technologies, Murray Hill, New Jersey, USA. Dr. Saniee has designed and developed many network design tools and optimization-based control mechanisms used in communication systems. He has also authored numerous articles in INFORMS and IEEE journals and proceedings. Dr. Saniee received his B.A. (Hon) and M.A. (Hon) in mathematics, and Ph.D. in operations research and control theory, all from the University of Cambridge.

Gangxiang Shen obtained his B.Eng. in 1997 from Zhejiang University, P. R. China, and M.Eng. from Nanyang Technological University, Singapore, in 1999. After that, he joined the laboratory of Network Technology Research Centre (NTRC) of Nanyang Technological University as a Research Associate, then Institute for Infocomm Research (I2R) of Singapore as a Senior Research Engineer. Now he is with Network Systems Group of TRLabs and University of Alberta, Canada, pursuing a Ph.D. His main research interests focus on all-optical networks and survivable networks. He has published around 30 papers in journals and conferences.

Darko Skorin-Kapov is Professor of Information Technology and Operations Management at the School of Business, Adelphi University, New York. He authored numerous research papers in management science and related areas. His research interests are in various aspects of network design, focusing on two topics: optimization and cost allocation in communication networks.

Jadranka Skorin-Kapov is Professor of Operations Management at the College of Business, State University of New York at Stony Brook. She authored many research papers in Operations Research. Her research interests include development of algorithms (heuristic search and learning, and polynomial algorithms for special cases) and applications of discrete optimization to location and layout, telecommunications, scheduling, manufacturing design, and network design.

Alice E. Smith is Professor and Chair of the Industrial and Systems Engineering Department at Auburn University. Previous to this position, she was on the faculty of the Department of Industrial Engineering at the University of Pittsburgh, which she joined in 1991 after ten years of industrial experience with Southwestern Bell Corporation. Dr. Smith has degrees in engineering and business from Rice University, Saint Louis University, and University of Missouri – Rolla. Her research in analysis, modeling and optimization of manufacturing processes and engineering design has been funded by NASA, the National Institute of Standards (NIST), Lockheed Martin, Adtranz (now Bombardier Transportation), the Ben Franklin Technology Center of Western Pennsylvania and the National Science Foundation (NSF), from which she was awarded a CAREER grant in 1995 and an ADVANCE Leadership grant in 2001. Her industrial partners on sponsored research projects have included DaimlerChrysler Electronics, Eljer Plumbingware, Extrude Hone, Ford Motor, PPG Industries and Crucible Compaction Metals. International research collaborations have been sponsored by the federal governments of Japan, Turkey, United Kingdom, and the U.S. Dr. Smith has served as a principal investigator on over $3 million of sponsored research. She was named a Philpott-WestPoint Stevens Distinguished Professor in 2001 by the Auburn University College of Engineering. For outstanding achievements in research and scholarly activity she received the annual Senior Research Award of the College of Engineering at Auburn University in 2001 and the University of Pittsburgh School of Engineering Board of Visitors annual Faculty Award in 1996. Dr. Smith holds one U.S. patent and several international patents and has authored over 50 publications in books and journals including articles in *IIE Transactions, IEEE Transactions on Reliability, INFORMS Journal on Computing, International Journal of Production Research, IEEE Transactions on Systems, Man, and Cybernetics, Journal of Manufacturing Systems, The Engineering Economist*, and *IEEE Transactions on Evolutionary Computation*. She won the E. L. Grant Best Paper Award in 1999 and the William A. J. Golomski Best Paper Award in 2002. Dr. Smith holds editorial positions on *INFORMS Journal on Computing, Computers & Operations Research, International Journal of General Systems*, IEEE *Transactions on Evolutionary Computation* and *IIE Transactions*. Five of her doctoral students have obtained tenure track positions at U.S. universities and two of these are NSF CAREER awardees. Dr. Smith is a fellow of IIE, a senior member of IEEE and SWE, a member of Tau Beta Pi, INFORMS and ASEE, and a Registered Professional Engineer in Industrial Engineering in Alabama and Pennsylvania. She serves on the Educational Foundation Board of the Institute of Industrial Engineers.

Babacar Thiongane is a postdoctoral fellow in the Department of Computer Science of UQAM – Université du Québec à Montréal. He was previously a postdoctoral fellow in the ORC research group in the Department of Computer Science and Operations Research, Université de Montréal. He has obtained his Ph.D. in Computer Science and Operations Research at Université de Paris 13, France.

Stan van Hoesel is Professor of Operations Research at Maastricht University. He received a Ph.D. in Operations Research from Erasmus University in Rotterdam. His research in Discrete Optimization concentrates on polyhedral methods for network optimization problems. His research in telecommunication focuses on frequency assignment, and network design and synthesis

Stefan Voß, born 1961 in Hamburg, is Professor and Director of the Institute of Information Systems at the University of Hamburg. Previous positions include Professor and Head of the Department of Business Administration, Information Systems and Information Management at the University of Technology Braunschweig (Germany) from 1995 up to 2002. He holds degrees in Mathematics (diploma) and Economics from the University of Hamburg, and a Ph.D. and the Habilitation from the University of Technology Darmstadt. His current research interests are in quantitative / information systems approaches to supply chain management, logistics, public mass transit, and telecommunications. He is author and co-author of several books and numerous papers in various journals. Stefan Voß serves on the editorial board of some journals including being Associate Editor of INFORMS Journal on Computing and Area Editor of Journal of Heuristics. He frequently organizes workshops and conferences. Furthermore, he consults with several companies.

Qiong Wang is a Member of Technical Staff with the Mathematical Sciences Research Center at Bell Labs. He does research on the interface of network engineering, management science, and economics. He has authored many papers on network pricing, revenue management, and capacity planning, and has been publishing on both engineering (e.g., IEEE Transactions, INFOCOM proceedings) and management journals (e.g., Operations Research, Journal of Marketing Research). In addition, he is recently working on issues related to technology transfer management and value chain coordination in telecommunications industry.

Phil Whiting received his B.A. degree in Mathematics from the University of Oxford, his M.Sc. from the University of London, and his Ph. D. was in queueing theory from the University of Strathclyde. After a post-doc at the University of Cambridge, Phil's interests centered on wireless. He was involved in the development of standards for UMTS as part of RACE. Subsequently, Phil participated in the Telstra trial of Qualcomm CDMA in South Eastern Australia. He then joined the Mobile research Centre at the University of South Australia Adelaide. Since 1997 he has been with Bell Labs where his main interests include the mathematics of wireless networks, particularly resource allocation in networks and the application of stochastic control to scheduling in wireless data networks.

Di Yuan is Associate Professor in Telecommunications at the Department of Science and Technology, Linköping Institute of Technology, Sweden. His research interests span over design, analysis, and resource optimization of telecommunication systems. He has authored or co-authored over 30 refereed articles in international journals and conference proceedings. His current research addresses network design and bandwidth allocation of UMTS systems, and resource management in ad hoc networks.

I Optimization algorithms

1 INTERIOR POINT METHODS FOR LARGE-SCALE LINEAR PROGRAMMING

John E. Mitchell[1], Kris Farwell[1], and Daryn Ramsden[1]

[1] Mathematical Sciences
Rensselaer Polytechnic Institute
Troy, NY 12180 USA
mitchj@rpi.edu
farwek@rpi.edu
ramsdd@rpi.edu

Abstract: We discuss interior point methods for large-scale linear programming, with an emphasis on methods that are useful for problems arising in telecommunications. We give the basic framework of a primal-dual interior point method, and consider the numerical issues involved in calculating the search direction in each iteration, including the use of factorization methods and/or preconditioned conjugate gradient methods. We also look at interior point column generation methods which can be used for very large scale linear programs or for problems where the data is generated only as needed.

Keywords: Interior point methods, preconditioned conjugate gradient methods, network flows, column generation.

1.1 INTRODUCTION

The performance of solvers for linear programming problems has improved dramatically in recent years. A user of linear programming packages now has available the option of using sophisticated interior point methods developed in the last twenty years. In addition, the development of interior point methods has spurred considerable successful research into efficient implementations of the simplex algorithm. It appears that different methods are better for different problems, with interior point methods possibly becoming a better choice as problem size grows. In addition to size, the structure of the linear programming problem is a major determinant as to which algorithm should be chosen. The simplex algorithm is discussed elsewhere in this book, so we focus on interior point methods.

The use of an interior point method is a good choice for general large-scale linear programming problems. We introduce interior point algorithms in Section 1.2, and discuss the computational issues that arise in Section 1.3.

For problems with well-defined structure, it is often true that an algorithm can be developed which will successfully exploit the structure. For example, the network simplex algorithm exploits the nature of basic feasible solutions to network flow problems. Interior point methods can also be refined to solve network flow problems efficiently, and we survey preconditioned conjugate gradient approaches to these problems in Section 1.4. The structure of multicommodity network flow problems can also be exploited in a carefully designed interior point method, as discussed in Section 1.5.

Linear programming problems often arise as subproblems of other problems. For example, integer programming problems can be solved using branch-and-cut, and multicommodity problems can be formulated with a huge number of variables and then attacked using a column generation approach. We survey the use of interior point methods in column generation and constraint generation settings in Section 1.6.

Conclusions are offered in Section 1.7.

1.2 PRIMAL-DUAL INTERIOR POINT METHODS

Several excellent textbooks on interior point methods were published in the 1990s, all of which discuss the material in this section in far greater detail. In particular, the reader is referred to Roos et al. (1997), Vanderbei (1996), Wright (1996), and Ye (1997). Several papers also surveyed primal-dual interior point methods, and our presentation is closest to that in Andersen et al. (1996).

We take the following to be our standard primal-dual LP pair:

$$\begin{aligned} \min \quad & c^T x \\ \text{subject to} \quad & Ax = b \\ & x + s = u \\ & x, s \geq 0 \end{aligned} \quad (P)$$

and

$$\begin{aligned} \max \quad & b^T y - u^T w \\ \text{subject to} \quad & A^T y + z - w = c \\ & z, w \geq 0 \end{aligned} \quad (D)$$

Here, A is an $m \times n$ matrix, c, x, z, and w are n-vectors, and b and y are m-vectors. We assume that the rank of A is m, that the feasible region of (P) is bounded, and that (P) and (D) each have a bounded set of optimal solutions.

An interior point method is an iterative scheme for solving (P) and (D) with each iterate strictly satisfying the nonnegativity restrictions. The primal-dual interior point barrier method can be motivated by setting up the subproblem

$$\begin{aligned} \min \quad & c^T x - \mu (\sum_{i=1}^n \ln(x_i) + \sum_{i=1}^n \ln(s_i)) \\ \text{subject to} \quad & Ax = b \\ & x + s = u \end{aligned} \quad (P(\mu))$$

where μ is a positive scalar. As μ is varied, the solution to $(P(\mu))$ traces out the *central path*. The limiting solution to $(P(\mu))$ as $\mu \to \infty$ is the *analytic center* of the feasible region of (P). The algorithm finds an approximate solution to $(P(\mu))$ for a particular μ, reduces μ, and repeats until a sufficiently accurate solution to (P) is found. As $\mu \to 0$, the solution to $(P(\mu))$ converges to the analytic center of the set of optimal solutions to (P).

The Karush-Kuhn-Tucker optimality conditions for $(P(\mu))$ can be written

$$Ax = b \tag{1.1}$$
$$x + s = u \tag{1.2}$$
$$A^T y + z - w = c \tag{1.3}$$
$$Zx = \mu e \tag{1.4}$$
$$Ws = \mu e \tag{1.5}$$
$$x, s, z, w \geq 0 \tag{1.6}$$

where Z and W denote diagonal matrices containing the entries of z and w respectively on the diagonal, and e denotes the vector of all ones of the appropriate dimension. The matrices S and X are defined similarly to Z and W.

Given an iterate $(\bar{x}, \bar{s}, \bar{y}, \bar{z}, \bar{w})$ strictly satisfying the nonnegativity constraint (1.6), one approach to Newton's method for finding a solution to the barrier problem is to solve the following system to find a direction $(\Delta x, \Delta s, \Delta y, \Delta z, \Delta w)$:

$$\begin{bmatrix} A & 0 & 0 & 0 & 0 \\ 0 & 0 & A^T & I & -I \\ I & I & 0 & 0 & 0 \\ Z & 0 & 0 & X & 0 \\ 0 & W & 0 & 0 & S \end{bmatrix} \begin{bmatrix} \Delta x \\ \Delta s \\ \Delta y \\ \Delta z \\ \Delta w \end{bmatrix} = \begin{bmatrix} r_b \\ r_c \\ r_u \\ \tau \mu e - XZe \\ \tau \mu e - SWe \end{bmatrix} \tag{1.7}$$

where τ is a scalar between 0 and 1, the diagonal matrices X, Z, S, and W are defined at the current iterate, and

$$r_b = b - A\bar{x} \tag{1.8}$$
$$r_c = c - A^T \bar{y} - \bar{z} + \bar{w} \tag{1.9}$$
$$r_u = u - \bar{x} - \bar{s} \tag{1.10}$$

Choosing different values of τ gives different directions. Taking $\tau = 0$ corresponds to a pure descent direction with no centering component, and this is known as the affine direction or the predictor direction. Choosing $\tau = 1$ and $\mu = \frac{1}{2n}(\bar{x}^T \bar{z} + \bar{s}^T \bar{w})$ is a pure centering direction, and it results in a direction that leaves the duality gap $x^T z + s^T w$ unchanged.

The system (1.7) can be converted into an equivalent system of equations with the making of the following eliminations:

$$\Delta z = X^{-1}(\tau \mu e - XZe - Z\Delta x) \tag{1.11}$$
$$\Delta s = u - \bar{x} - \bar{s} - \Delta x \tag{1.12}$$
$$\Delta w = S^{-1}(\tau \mu e - SWe - Wr_u + W\Delta x) \tag{1.13}$$

the system becomes

$$\begin{bmatrix} -D^{-2} & A^T \\ A & 0 \end{bmatrix} \begin{bmatrix} \Delta x \\ \Delta y \end{bmatrix} = \begin{bmatrix} r_c - X^{-1}(\tau\mu e - XZe) + S^{-1}(\tau\mu e - SWe - Wr_u) \\ r_b \end{bmatrix} \quad (1.14)$$

where

$$D^2 = (X^{-1}Z + S^{-1}W)^{-1}. \quad (1.15)$$

Equation (1.14) is known as the *augmented system* of equations. All interior point methods have to solve a system essentially of this form, possibly with a different diagonal matrix and with a different right hand side vector. The system can be reduced further to give the *normal equations*, through the elimination of Δx. In particular, multiplying the first set of equations by AD^2 and adding the second set gives:

$$\begin{aligned}(AD^2A^T)\Delta y &= r_b + AD^2(r_c - X^{-1}(\tau\mu e - XZe) \\ &\quad + S^{-1}(\tau\mu e - SWe - Wr_u)) =: g.\end{aligned} \quad (1.16)$$

Note that we define g to represent the right hand side of the normal equations. The *augmented system* and the *normal equations* lend their names to the two main approaches to solving for the vector of Newton directions. These approaches are discussed in the next section.

Once this vector of directions has been computed, a new feasible point can be found by scaling each of the steps down (to maintain feasibility) before adding them to the current feasible solution. At this point μ is updated and the procedure is repeated. Predictor-corrector algorithms are the current methods of choice. These methods alternate between a predictor step with $\tau = 0$ and a corrector step with $\tau = 1$. The algorithm can be implemented in such a way that the work of calculating both steps is not much greater than the work of calculating either step individually — see the next section.

1.3 COMPUTATIONAL CONSIDERATIONS FOR GENERAL LINEAR PROGRAMS

Computational considerations can intuitively be divided up into three categories, namely

- those related to initialization,
- those related to finding and scaling the vector of newton directions, and
- those related to termination.

As in the preceding section, we will adhere most closely to Andersen et al. (1996).

Initialization of an implementation consists of a presolving stage and finding an initial point. Presolving is the process by which the matrix A is examined to see if there are any straightforward measures that can be taken to make the problem easier to solve. For instance, the process may unearth duplicate rows or columns, in which case the corresponding dual or primal variables respectively can be combined, or it may find, with minimal effort, that certain variables should have fixed values or that some constraints are redundant. Also it is possible that inefficient formulation of the problem may result in zero columns, meaning we can simply push the corresponding variable to its upper or lower bound depending on whether or not its coefficient in the objective function is negative or positive. Measures such as these drastically reduce the workload in the body of the algorithm and overall it can be said that a preprocessing stage is well worth the work required. A comprehensive treatment of presolving has been done by Andersen and Andersen (1996).

Finding an initial point can be done by solving the quadratic program below, which has a solution given by an explicit formula:

$$\begin{aligned} \min \quad & c^T x - \tfrac{\sigma}{2}(x^T x + s^T s) \\ \text{subject to} \quad & Ax = b \\ & x + s = u \end{aligned}$$

The formula may result in values for x and s that may be negative and as such they can be pushed toward positivity. Parallel to this a solution (y, z, w) for the dual can be constructed with $y = 0$.

As mentioned before, there are two approaches to solving for the vector of Newton directions, namely the *augmented systems* approach and the *normal equations* approach. The choice of approach is based on the relative density of the matrix AD^2A^T when compared to the matrix A. If the former is relatively dense in comparison to the latter, formulation and usage of the *normal equations* is unnecessarily expensive. We can now look at the solving of the *normal equations*. If we let C be the coefficient matrix in equation (1.16) after a possible symmetric permutation of its rows and columns, under the assumption that A is of full row rank, we have that C is positive definite and we can use Cholesky factorization to find a lower triangular matrix L such that $C = PAD^2A^TP^T = LL^T$ where P is a permutation matrix. Systems of equations of the form $C\Delta y^* = r$ can then be solved (y^* representing an appropriate renumbering of the components of y to correspond with the permutations made to AD^2A^T), by solving the following systems in turn:

$$\begin{aligned} L\gamma &= \nu \\ L^T \Delta y^* &= \gamma \end{aligned}$$

The stability of the calculation is independent of the choice of permutation and thus much of the task at hand is the choice of an appropriate permutation which makes L reasonably sparse. There are two main ordering heuristics:

- the minimum degree, and
- minimum local fill-in heuristics (Wright, 1996).

Using the minimum degree ordering involves looking at the remaining matrix and choosing as the pivot element, the diagonal element that has the least number of nonzeroes (we call this its degree, d) in its row/column as the new pivot element at each step of the Cholesky algorithm. This is motivated by the fact that the algorithm generates d^2 nonzeroes in the update matrix and as such d should be kept as small as possible.

Minimum local fill-in is the main competing heuristic and is motivated by the fact that the minimum degree figures out how many elements are actually changed in the elimination steps as opposed to figuring out how many go from being zero to nonzero (a subset of the former category). Many times it is possible to choose a pivot element that results in a more sparse factorization at the expense of carrying out more analytical work beforehand. Choosing the right variant of minimum local fill-in can possibly lead to results that are competitive with that using the minimum degree heuristic.

The normal equations approach is particularly susceptible to weak performance in situations where A has dense columns and thus generates dense blocks in the AD^2A^T matrix. One simple possible remedy is to examine both of the factorizations, AA^T and A^TA. In situations that A has both columns and rows which are dense, this trick loses its effectiveness. At this point in the implementation more sophisticated methods can be employed, but it is often more appropriate to resort to working with the *augmented system*.

The augmented system approach utilizes the Bunch-Parlett factorization algorithm to get the following:

$$\begin{bmatrix} -D^{-2} & A^T \\ A & 0 \end{bmatrix} = L\Lambda L^T \tag{1.17}$$

where Λ is a block diagonal matrix. This formulation is more robust as far as coping with the effects of dense columns of A. Also while the matrix is prone to being ill-conditioned it is relatively easy to make judgments on the quality of the solution vector.

Once a solution vector has been found it is important to scale it before adding it to the current iterate. Conventionally, the maximum value of a multiplier of the solution vector that allows for maintained feasibility is scaled down by a predetermined constant between 0 and 1 to obtain the size of the step taken in the direction of the solution vector.

There are many possible predictor-corrector variants. One possibility is to first solve (1.7) with $\tau = 0$, then modify the right hand side of (1.7) as if the predictor step had been taken to the boundary of the feasible region, and then calculate the corrector step with $\tau = 1$ and with a guess for the new value of μ. If μ^0 denotes the old value of μ and μ^{aff} denotes the value obtained at the boundary of the feasible region after the affine step, Mehrotra (1992) suggests taking $(\frac{\mu^{aff}}{\mu^0})^3 \mu^0$ as the guess for the new value of μ in the calculation of the corrector step. The direction taken from the current iterate is the solution to this corrector system of equations. Note that the diagonal scaling matrices in the calculation of the corrector step are the same as those in the predictor step, so it is only necessary to factorize the matrix AD^2A^T once for each pair of a predictor and corrector step. This idea of calculating two vectors Δy with the same factorization of AD^2A^T can be extended naturally to finding multiple corrections to the affine direction, giving *higher order* methods. These higher order methods may reduce the number of iterations required to solve the problem slightly, with each iteration marginally more expensive, and they can lead to considerable overall reductions in computational time.

As far as termination criteria are concerned, we have only to monitor the duality gap and feasibility at each iterate until they are sufficiently small to meet the prescribed required precision. For a required precision of 8 digits, (a very common standard), the conditions are as follows:

$$\frac{||Ax-b||}{1+||b||} \leq 10^{-8} \text{ and } \frac{||x+s-u||}{1+||u||} \leq 10^{-8} \tag{1.18}$$

$$\frac{||A^T y + z - w - c||}{1+||c||} \leq 10^{-8} \tag{1.19}$$

$$\frac{|c^T x - (b^T y - u^T w)|}{1+|b^T y - u^T w|} \leq 10^{-8} \tag{1.20}$$

The last of these conditions is often the most stringent and difficult to satisfy and consequently in most cases it is adequate simply to check only its validity. Some implementations may even feature a termination phase in which an optimal vertex/basis is recovered. This is most notably implemented in CPLEX and is comparable to $O(m)$ simplex iterations in terms of complexity. A strongly polynomial algorithm for the recovery of a primal-dual optimal basis given optimal solutions to both the primal and dual problems was provided by Megiddo (1991).

1.4 PRECONDITIONED CONJUGATE GRADIENT METHODS FOR NETWORK FLOW PROBLEMS

A standard implementation of an interior point method, using a complete factorization of AD^2A^T, is not competitive with the network simplex method for solving network flow problems. An alternative is to use a preconditioned conjugate gradient algorithm to find Δy in (1.16). The structure of the problem allows the construction of preconditioners that work very well.

In the set of equations (1.16) for interior point methods

$$AD^2 A^T \Delta y = g \tag{1.21}$$

multiplying by the preconditioner M^{-1} gives

$$M^{-1} A D^2 A^T \Delta y = M^{-1} g \tag{1.22}$$

where M is a symmetric and positive definite matrix. The matrix M should be chosen appropriately so that equation (1.22) can be solved more easily than (1.21). It should also make the matrix $M^{-1} A D^2 A^T$ well-conditioned, with few extreme eigenvalues. These two requirements for M are somewhat in conflict. For example, taking $M = AD^2 A^T$ makes $M^{-1} A D^2 A^T = I$, so it minimizes the condition number of $M^{-1} AD^2 A^T$; however, determining the right hand side of (1.22) is as hard as solving the original system. It has been observed in practice that the number of conjugate gradient steps depends on the number of distinct eigenvalues, in fact typically on the number of distinct clusters of eigenvalues. Thus, it is desirable to pick M in order to cluster the eigenvalues of $M^{-1}AD^2A^T$. The Preconditioned Conjugate Gradient Algorithm for solving (1.21) is summarized in Algorithm 1.1. (Far more information on preconditioned conjugate gradient methods and related topics in numerical linear algebra can be found in the texts by Demmel (1997) and Trefethen and Bau (1997).)

procedure *PCG*
begin
    Take Δy_0 equal to some initial guess.;
    $r_0 \leftarrow g - AD^2A^T \Delta y_0$;
    $z_0 \leftarrow M^{-1} r_0$;
    $p_0 \leftarrow z_0$;
    $i \leftarrow 0$;
    while *some stopping criterion not met* **do**
        $q_i \leftarrow AD^2A^T p_i$;
        $\alpha_i \leftarrow z_i^T r_i / p_i^T q_i$;
        $\Delta y_{i+1} \leftarrow \Delta y_i + \alpha_i p_i$;
        $r_{i+1} \leftarrow r_i - \alpha_i q_i$;
        Solve $M z_{i+1} \leftarrow r_{i+1}$;
        $\beta_i \leftarrow z_{i+1}^T r_{i+1} / z_i^T r_i$;
        $p_{i+1} \leftarrow z_{i+1} + \beta_i p_i$;
        $i \leftarrow i+1$;
    end
    $\Delta y \leftarrow \Delta y_i$;
end

Algorithm 1.1: Preconditioned conjugate gradient algorithm for solving the system $AD^2A^T \Delta y = g$.

For the minimum cost network flow problem on a graph $G = (\mathcal{V}, \mathcal{A})$ with vertices $\mathcal{V}$ and arcs $\mathcal{A}$, A is the node-arc incidence matrix (after deleting one row for each component, to ensure A has full row rank). At an interior point, the matrix AD^2A^T has nonzeroes on the diagonal and in position (i,j) for each $(i,j) \in \mathcal{A}$, assuming no numerical cancellation. With a preconditioned conjugate gradient algorithm, it is typically sufficient to solve (1.16) approximately as part of an infeasible interior point algorithm. The desired accuracy of the conjugate gradient algorithm can be chosen to depend on the primal residual r_b, defined in (1.8).

At a nondegenerate basic feasible solution, the elements of D corresponding to nonbasic variables are zero, so we have

$$AD^2A^T = BD_B^2 B^T$$

where B denotes the columns of A corresponding to the basic variables and D_B the corresponding elements of D. At such a point, $M = BD_B^2 B^T$ would be a perfect preconditioner, in the sense that the conjugate gradient method would require just one step. Further, determining the right hand side of (1.22) is also straightforward: the basic feasible solutions are spanning trees, so finding $M^{-1}g$ requires only linear time. Therefore, once the interior point method gets close to an optimal solution, it is hoped that a preconditioner based on a nearby spanning tree would be a good choice, so preconditioners that use spanning trees are the most relevant. Some examples of the possible choices for preconditioners are the diagonal, the maximum spanning tree, or the diagonally compensated maximum spanning tree preconditioner.

The *Diagonal Preconditioner* was used by Resende and Veiga (1993b) in a dual affine algorithm for network flow problems. It is defined as:

$$M = \mathrm{diag}(AD^2A^T) \tag{1.23}$$

This particular preconditioner is commonly used in the first few iterations of the preconditioned conjugate gradient algorithms, when it is most effective, an empirical observation somewhat supported by theoretical results in Júdice et al. (2003). The diagonal preconditioner is known as the Jacobi preconditioner in numerical linear algebra, and it reduces the condition number of $M^{-1}AD^2A^T$ to within a factor of $|\mathcal{V}|$ of its minimum value.

The *Maximum Spanning Tree preconditioner* is determined by using the current solution to select weights for the arcs of the graph and then finding a maximum spanning tree. It is defined as

$$M = A_{\mathcal{T}} D_{\mathcal{T}}^2 A_{\mathcal{T}}^T \tag{1.24}$$

where $A_{\mathcal{T}}$ contains the columns of A corresponding to the edges of the maximum spanning tree of $G = (\mathcal{V}, \mathcal{A})$ with

$$D_{\mathcal{T}} = \mathrm{diag}(d_{t_1}, ..., d_{t_{m-1}}) \tag{1.25}$$

where $t_1, ..., t_{m-1}$ are the edge indices of the maximum spanning tree and D is defined in (1.15). The edge weights for the spanning tree problem need to be defined appropriately. Portugal et al. (2000) suggest using the edge weight vector:

$$w = D^2 e \tag{1.26}$$

where e is a vector of ones.

The *Diagonally Compensated Maximum Spanning Tree preconditioner* uses the matrix M defined as follows

$$M = A_{\mathcal{T}} D_{\mathcal{T}}^2 A_{\mathcal{T}}^T + \phi\, \mathrm{diag}(A_{\tilde{\mathcal{T}}} D_{\tilde{\mathcal{T}}}^2 A_{\tilde{\mathcal{T}}}^T) \tag{1.27}$$

where $G' = (\mathcal{V}, \mathcal{T})$ is maximum spanning tree of $G = (\mathcal{V}, \mathcal{A})$, $\tilde{\mathcal{T}} = \mathcal{A} - \mathcal{T}$ and ϕ is a nonnegative parameter. The diagonally compensated maximum spanning tree preconditioner is reduced to the maximum spanning tree preconditioner when $\phi = 0$. Some suggestions for ϕ used in Mehrotra and Wang (1996) are $\phi = 1$, $\phi = 0.1 \cdot \frac{\min(D_k)}{\max(D_k)}$, and $\phi = 10 \cdot \frac{\min(D_k)}{\max(D_k)}$.

Júdice et al. (2003) state and prove many theorems regarding the condition numbers of these preconditioners. They show that the condition number of AD^2A^T is bounded above by a function of the condition number of $D_{\mathcal{T}}^2$, where $D_{\mathcal{T}}$ consists of the elements of D corresponding to the edges of a maximum spanning tree for $G = (\mathcal{V}, \mathcal{A})$ with edge weights D. If the algorithm is converging to a primal nondegenerate basic feasible solution, then this gives a bound on the condition number of AD^2A^T. They are able to obtain stronger bounds on the condition number of $M^{-1}AD^2A^T$ with the use of the two tree-based preconditioners than the bound they obtain for either AD^2A^T itself or $M^{-1}AD^2A^T$ with the diagonal preconditioner. Further, they show that even in the presence of primal degeneracy of the optimal solution, the condition number of $M^{-1}AD^2A^T$ is uniformly bounded using the tree preconditioners, once the duality gap becomes small.

There are two stopping criteria for the Conjugate-Gradient Method suggested by Portugal et al. (2000). The first stopping criteria works well for the beginning iterations:

$$\| r_i \| \leq \beta_0 \| r_b \| \tag{1.28}$$

where r_i is the residual at the i-th iteration of Algorithm 1.1, $\| r_b \|$ is defined in (1.8) and β_0 is suggested to be 0.0999 in Portugal et al. (2000). This stopping criterion may be too conservative in later iterations, once $\| r_b \|$ is small, requiring too many conjugate gradient iterations.

The second stopping criteria exploits the fact that Δy is a direction, and that a dual steplength is still to be chosen. Therefore the magnitude of Δy is unimportant; what is important is that Δy point in approximately the right direction, that is, the angle θ between

$$AD^2A^T \Delta y_i \text{ and } g \tag{1.29}$$

should be small. This leads to the criterion:

$$| 1 - \cos\theta | < \varepsilon_{cos}^k \tag{1.30}$$

where ε_{cos}^k is the tolerance for the interior point iteration k. Also, ε_{cos}^k can be tightened by multiplying by $\Delta\varepsilon_{cos} < 1$ at each iteration. The angle θ can be computed

$$\cos\theta = \frac{| g^T (AD^2A^T) \Delta y_i |}{\| g \| \cdot \| (AD^2A^T) \Delta y_i \|} \tag{1.31}$$

or approximated (since calculating (1.31) is as expensive as a conjugate gradient iterate) by

$$\cos\theta \approx \frac{| g^T (g - r_i) |}{\| g \| \cdot \| (g - r_i) \|} \tag{1.32}$$

where r_i is the residual at the i-th iteration. This approximation works well when solving network linear programs.

The stopping criteria for the interior point method has two types of conditions. The first is the primal-basic (PB) stopping rule and the second the maximum flow (MF) stopping criterion; both are summarized in Portugal et al. (2000) and Resende and Veiga (1993a). Both criteria use basis identification techniques, with the (PB) method exploiting the spanning tree found in the calculation of the preconditioner and the (MF) method using a method from the interior point literature. For a survey of basis identification techniques in interior point methods, see El–Bakry et al. (1994).

The PB stopping rule uses the tree found in the preconditioner and takes the edges of the tree to be the basic variables. Let $\mathcal{T}$ be the index set of the edges of the maximum spanning tree used in the creation of (1.25). Let

$$\Omega^+ = \left\{ i \in \{1,2,...,n\} \backslash T : \frac{x_i}{z_i} > \frac{s_i}{w_i} \right\} \quad (1.33)$$

and set these nonbasic edges of the tree to their upper bounds. The basic variables must satisfy

$$A_T x_T^* = b - \sum_{i \in \Omega^+} u_i A_i \quad (1.34)$$

If this set of equations has a solution $0 \leq x_T^* \leq u$ then x_T^* is a basic feasible solution. Let

$$F = \{ i \in T : 0 < x_i^* < u_i \}. \quad (1.35)$$

be the set of edges where the dual slacks are zero. By orthogonally projecting the current dual solution onto the set F, we preserve complementary slackness by solving

$$\min_{y^* \in \mathbb{R}^m} \{ \| y^* - y^k \| : A_F^T y^* = c_F \} \quad (1.36)$$

which can be solved efficiently for network flow problems.

A feasible dual solution (y^*, z^*, w^*) can be found by changing the dual slacks

$$w_i^* = \begin{cases} -\delta_i & \text{if } \delta < 0 \\ 0 & \text{otherwise} \end{cases}$$

$$z_i^* = \begin{cases} 0 & \text{if } \delta < 0 \\ \delta_i & \text{otherwise} \end{cases}$$

where $\delta_i = c_i - A_{\cdot i}^T y^*$. We can stop when $c^T x - b^T y^* + u^T w^* = 0$ because (x^*, s^*) is a primal feasible solution and (y^*, w^*, z^*) is a dual feasible solution. The idea of projecting an interior point onto the boundary in this manner was proposed by Ye (1992) as a means to get finite convergence of interior point algorithms.

The other stopping criteria (MF) involves a different edge indicator, namely let edge i be inactive at its lower bound when

$$\frac{x_i}{z_i} < \xi \quad \text{and} \quad \frac{s_i}{w_i} > \xi^{-1}$$

and edge i be inactive at its upper bound when

$$\frac{x_i}{z_i} > \xi^{-1} \quad \text{and} \quad \frac{s_i}{w_i} < \xi$$

The other edges are then active. Using these active edges, we can form the maximum weighted spanning forest. Like before we can project to find y^* as in (1.36). However now we must also build a primal feasible x^*. The forest is extended slightly to $\overline{F} = \{ i \in \{1,2,...,n\} : |c_i - A_{\cdot i}^T y^*| < \varepsilon_r \}$ for some small tolerance ε_r (e.g., $\varepsilon_r = 10^{-8}$) and then the nonbasic variables are set:

$$x_i^* = \begin{cases} 0 & \text{if } i \in \Omega^- = \left\{ j \in \{1,2,...,n\} \backslash \overline{F} : c_j - A_{\cdot j}^T y^* > 0 \right\} \\ u_i & \text{if } i \in \Omega^+ = \left\{ j \in \{1,2,...,n\} \backslash \overline{F} : c_j - A_{\cdot j}^T y^* < 0 \right\} \end{cases}$$

Flow on the edges in $\overline{F}$ produces a *restricted network* which must satisfy

$$A_{\overline{F}} x_{\overline{F}} = b - \sum_{i \in \Omega^+} u_i A_i \tag{1.37}$$

and

$$0 \leq x_i \leq u_i, i \in \overline{F} \tag{1.38}$$

If a feasible primal solution exists on the restricted network then it is also complementary to y^*. As the algorithm proceeds, more edges will be declared inactive, resulting in a sparser spanning forest if the optimal solution is degenerate. In the limit, the algorithm will identify edges with flow strictly between the bounds, but some of these edges might not be included in the spanning forest if multiple primal optimal solutions exist. The expansion to $\overline{F}$ is necessary to recover these edges. Finding a feasible flow in the restricted network requires solving a maximum flow problem.

In summary, when using interior point methods to solve network flow problems, a preconditioned conjugate gradient algorithm can be used to approximately solve the direction finding subproblem at each iteration. It is recommended to start with the diagonal preconditioner, since it is easy to compute for beginning iterations, and then switch to a more complicated preconditioner based on spanning trees. The latter preconditioners have nice theoretical properties as the optimal solution is approached.

The primal-dual infeasible interior point method for network flow problems, PDNET, is available online at http://www.research.att.com/~mgcr/pdnet/. Much of this code is written in FORTRAN, and is described in detail in Patrício et al. (2004).

1.5 MULTICOMMODITY NETWORK FLOW PROBLEMS

In this section, we consider multicommodity problems. These problems are considered in more detail in two other chapters of this book. In a multicommodity problem, the resources of the network are shared among several different competing commodities. For example, multiple different calls must be routed in a telephone network, and each call can be regarded as a commodity. It is not appropriate to aggregate the calls from a particular customer, rather the calls between each pair of customers must be considered separately. Similarly, in a traffic network, the set of vehicles moving from a particular origin to a particular destination should be considered a commodity.

If there are no capacity restrictions on the nodes and arcs of the graph then the multicommodity problem separates into several single commodity problems which can be solved independently. We will assume in this section that each arc e has capacity u_e. For simplicity, we will not assume capacities on the nodes. Let $\mathcal{K}$ denote the set of commodities. Let b^k and c^k denote the demand and cost vectors, respectively, for commodity k. For simplicity, we assume that the node-arc incidence matrix A is identical for each commodity. The basic minimum cost multicommodity network flow problem can be formulated as follows, with variables x_e^k equal to the amount of commodity k

flowing along arc e:

$$\begin{array}{rl} \min & \sum_{k \in \mathcal{K}} c^{k^T} x^k \\ \text{subject to} & A x^k = b^k \quad \forall k \in \mathcal{K} \\ & \sum_{k \in \mathcal{K}} x_e^k \leq u_e \quad \forall e \in \mathcal{A} \\ & x^k \geq 0 \quad \forall k \in \mathcal{K} \end{array} \quad (1.39)$$

In contrast with the single commodity case, the constraint matrix in this formulation is not totally unimodular, so the optimal solution may not be integral. This formulation may have a huge number of variables for even a reasonably realistic multicommodity problem, since the number of variables is the product of the number of arcs and the number of commodities. Nonetheless, the constraint matrix has a structure that can be exploited to devise efficient interior point solvers, with better complexity bounds than with a naive invocation of an interior point method; see, for example, Kapoor and Vaidya (1988; 1996) and Kamath and Palmon (1995). As with single commodity problems, preconditioned conjugate gradient approaches have been used successfully for multicommodity problems. See, for example, Choi and Goldfarb (1990), Yamakawa et al. (1996), Júdice et al. (1997), and Castro (2000); Castro and Frangioni (2001). Resende and Veiga (2003) survey interior point approaches to multicommodity network flow problems.

We consider the algorithm of Castro (2000) in a little more detail, in order to give a flavor of the numerical linear algebra techniques that can be used to speed up solution of multicommodity network flow problems. The crucial issue is the solution of (1.16). For the formulation (1.39), (1.16) takes the block form

$$\begin{bmatrix} B & C \\ C^T & F \end{bmatrix} \begin{bmatrix} \Delta y_1 \\ \Delta y_2 \end{bmatrix} = \begin{bmatrix} g_1 \\ g_2 \end{bmatrix} \quad (1.40)$$

where B is a block diagonal matrix, F is a symmetric matrix, C is a matrix, and Δy_1, Δy_2, g_1 and g_2 are vectors. This system can be solved by forming the Schur complement

$$H := F - C^T B^{-1} C, \quad (1.41)$$

solving

$$H \Delta y_2 = \tilde{g}_2 \quad (1.42)$$

for an appropriate $\tilde{g}_2$, and then solving

$$B \Delta y_1 = \tilde{g}_1 \quad (1.43)$$

for an appropriate $\tilde{g}_1$. For a general network, H has the same nonzero structure as $A^T (AA^T)^{-1} A$, so it is dense and (1.42) is hard to solve directly. Thus, Castro proposes solving (1.42) using a preconditioned conjugate gradient approach, with preconditioners based on a power series expansion.

An alternative approach to multicommodity network flow problems is a column generation method involving a path decomposition. This is useful when there is a large number of commodities; typically, the direct approach is better when the number of commodities is limited (Castro, 2000). Without loss of generality, we assume that each commodity has a single source and a single sink (otherwise, the commodities can be disaggregated). Further, we assume that no two commodities have the same source and sink (otherwise the commodities can be aggregated). A set P_k of paths for each commodity $k \in \mathcal{K}$ are generated. Flow x_p is routed along these paths, and we let w_p be the cost of routing one unit along path p. The multicommodity network flow problem can then be written as

$$\begin{array}{rrcll} \min & \sum_{k \in \mathcal{K}} \sum_{p \in P_k} w_p x_p & & & \\ \text{subject to} & \sum_{p \in P_k} x_p & = & d_k & \forall k \in \mathcal{K} \\ & \sum_{p \in P_k, k \in \mathcal{K}: e \in p} x_p & \leq & u_e & \forall e \in \mathcal{A} \\ & x_p & \geq & 0 & \forall e \in \mathcal{A} \end{array} \qquad (1.44)$$

where d_k denotes the demand for commodity k. If the sets P_k contain all possible paths then this is an exact formulation, but of course this is typically not computationally feasible due to the number of paths. Thus, this formulation is used in a column generation approach, with paths generated as needed in a subproblem. The path generation subproblem is a shortest path problem, with edge weights coming from the dual problem. This column generation approach has been used with the linear programs (1.44) solved using simplex (for example, Chardaire and Lisser (2002)) and when the linear programs have been solved with an interior point method (for example, Chardaire and Lisser (2002) and Goffin et al. (1997)). We discuss interior point column generation methods in more detail in §1.6.

1.6 INTERIOR POINT COLUMN GENERATION METHODS

Solving very large linear programming problems directly is sometimes computationally intractable. In this case, a column generation approach may be attractive, and we describe interior point column generation methods in this section. Integer programming problems can be solved using cutting plane methods. Adding a cutting plane in the primal linear programming problem is equivalent to adding a column in the dual problem, so the methods discussed in this section are also applicable to cutting plane algorithms.

1.6.1 Motivation

As already mentioned in Section 1.5, path-based formulations of multicommodity network flow problems are typically solved using column generation methods. Goffin et al. (1997) and Chardaire and Lisser (2002) have both investigated interior point column generation methods for the solution of such problems. Gondzio and Kouwenberg (2001) have solved large problems in financial optimization using interior point column generation methods; these problems involve assets that flow from one class to another over time, and hence the problem formulation has similarities with network flow problems and hence with problems in telecommunications.

Other applications of interior point column generation approaches to problems on networks and to integer programming problems include the solution of crew scheduling problems by Bixby et al. (1992), max cut problems by Mitchell (2000) and linear ordering problems by Mitchell and Borchers (2000). Further, Goffin et al. have solved

Lagrangian relaxations of various problems, including integer programming problems, using interior point column generation methods; see, for example, Elhedhli and Goffin (2004).

In telecommunications, many integer programming and / or column generation formulations of network design problems have been posed in the literature. See, for example, Atamtürk (2004), Barahona (1996), Bienstock and Muratore (2000), Dahl and Stoer (1998), Grötschel et al. (1992) Grötschel et al. (1995), Myung et al. (1999), Raghavan and Magnanti (1997), and Sherali et al. (2000). Frequency assignment problems for cellular networks can also be cast as integer programming problems, as in, for example, Eisenblätter (2001).

The introduction of stochasticity into linear telecommunications problems leads to growth in the size of the linear programs, so again a column generation approach may well be necessary. The classical L-shaped method for stochastic programs is a column generation method, of course. See Birge and Louveaux (1997) and Kall and Wallace (1994) for more information on stochastic programming. Robust optimization is another approach to handling uncertainty, and many robust optimization models are constructed as second order cone programs or semidefinite programming problems, and then solved using interior point methods; see Ben-Tal and Nemirovski (2003) for example.

1.6.2 Mechanics

Interior point column generation methods differ from simplex-based column generation approaches in one crucial respect: the linear programming subproblems are not solved to optimality at each stage, but solved approximately. This leads to the generation of dual cuts that cut off a larger proportion of the dual space, at least theoretically (see Goffin and Vial (2002) and Mitchell (2003) for recent surveys of the theoretical performance of interior point column generation methods).

When columns are added, it is desirable to return to the interior of the feasible region in order to allow fast convergence to the next approximate solution. This can be done by using a direction originally proposed in Mitchell and Todd (1992), which can be motivated through consideration of Dikin ellipsoids.

The current relaxation can be written

$$\begin{array}{ll} \min & c^T x \\ \text{subject to} & Ax = b \\ & x \geq 0 \end{array} \qquad \begin{array}{ll} \max & b^T y \\ \text{subject to} & A^T y \leq c \end{array} \qquad (1.45)$$

The complete problem can be written

$$\begin{array}{ll} \min & c^T x + h^T z \\ \text{subject to} & Ax + Hz = b \\ & x, z \geq 0 \end{array} \qquad \begin{array}{ll} \max & b^T y \\ \text{subject to} & A^T y \leq c \\ & H^T y \leq h \end{array} \qquad (1.46)$$

The number of columns in H may be very large, possibly infinite. As the algorithm proceeds, the current relaxation is modified by adding columns from H to A (and dropping columns from A also, if the columns no longer appear to be useful). Any dual point y that is feasible in the complete problem (1.46) provides a lower bound on the optimal value. Any point x feasible in the current primal relaxation (1.45) is also feasible in (1.46) and so provides an upper bound on the optimal value. The column generation process can be stopped once these two bounds are close enough.

An interior point column generation algorithm approximately solves the current relaxation (1.45), calls an oracle to search for violated dual constraints, adds a subset of the primal columns / dual constraints found by the oracle, finds a new interior point to restart, and repeats the process. If the oracle is unable to find violated dual constraints, another interior point iteration is taken, and the oracle is called again.

There are two disadvantages to the use of interior point column generation methods. First, it is harder to exploit a warm start with an interior point method than with simplex. For smaller problems, when only a few columns are generated, the simplex algorithm can reoptimize far faster than an interior point method. However, this disadvantage is reduced for larger problems and when large numbers of columns are generated at once. In such a situation, the stronger columns generated by the interior point method can lead to faster convergence than when using the simplex method; see, in particular, Bixby et al. (1992) and Mitchell and Borchers (2000).

The second disadvantage is that the attempt to generate columns from more central points may fail to find violated dual constraints, so time is wasted in the calls to the oracle. This failure to find a violated dual constraint doesn't imply that the problem has been solved (as would be the case if the oracle was always accurate and if the current relaxation was solved to optimality). The remedy is to take another interior point iteration and try again. The required accuracy for the subproblem can be tightened as the algorithm proceeds in order to reduce the occurrence of this situation. Further, the failure of the oracle to find a violated dual constraint does provide some useful information: the dual solution must be feasible in the complete problem, so a bound on the optimal value is obtained, provided the oracle is correct in stating that there are no violated dual constraints.

It is of interest to note two ideas that appear both here and in Section 1.4, in totally different contexts. In particular, in both interior point column generation methods and in interior point algorithms for network flows, it suffices to solve the current problem approximately. Secondly, the accuracy to which the subproblem is solved is updated dynamically, depending on the progress of the algorithm, with more accurate solutions desired as the solution to the overall problem is approached.

More details of the implementation of interior point column generation methods can be found in Mitchell et al. (1998), as well as in the references given above.

1.7 CONCLUSIONS AND EXTENSIONS

Interior point algorithms have found broad applicability in large scale linear programs arising in telecommunications, including successful implementations of interior point methods using preconditioned conjugate gradient algorithms for network flow problems. The use of interior point methods in column generation settings allows the possibility of solving very large linear programming problems. Interior point methods have been extended to other optimization problems such as semidefinite programming problems, second order cone programming problems, and general convex programming problems and these will allow the solution of more sophisticated models of problems in telecommunications. For example, network design problems have been solved by Lisser and Rendl using a semidefinite programming approach (Lisser and Rendl, 2003), and robust optimization problems such as antenna array design can be modeled as second order cone programs (for example, Lobo et al. (1998) and Ben-Tal and Nemirovski (2003)). As a final note, interior point methods for nonlinear programming problems are surveyed by Forsgren et al. (2002).

Acknowledgments

Research supported in part by NSF grant numbers DMS-0317323 and CMS-0301661. The research of the second author was supported by the NSF through the VIGRE program, grant number DMS-9983646.

Bibliography

E. D. Andersen and K. D. Andersen. Presolving in linear programming. *Mathematical Programming*, 71:221–245, 1996.

E. D. Andersen, J. Gondzio, C. Mészáros, and X. Xu. Implementation of interior point methods for large scale linear programming. In T. Terlaky, editor, *Interior Point Methods in Mathematical Programming*, chapter 6, pages 189–252. Kluwer Academic Publishers, 1996.

A. Atamtürk. On capacitated network design cut-set polyhedra. *Mathematical Programming*, 92(3):425–452, 2004.

F. Barahona. Network design using cut inequalities. *SIAM Journal on Optimization*, 6:823–837, 1996.

A. Ben-Tal and A. Nemirovski. Robust optimization — methodology and applications. *Mathematical Programming*, 92(3):453–480, 2003.

D. Bienstock and G. Muratore. Strong inequalities for capacitated survivable network design problems. *Mathematical Programming*, 89(1):127–147, 2000.

J. R. Birge and F. Louveaux. *Introduction to Stochastic Programming*. Springer, New York, 1997.

R. E. Bixby, J. W. Gregory, I. J. Lustig, R. E. Marsten, and D. F. Shanno. Very large-scale linear programming: a case study in combining interior point and simplex methods. *Operations Research*, 40:885–897, 1992.

J. Castro. A specialized interior point algorithm for multicommodity flows. *SIAM Journal on Optimization*, 10(3):852–877, 2000.

J. Castro and A. Frangioni. A parallel implementation of an interior-point algorithm for multicommodity network flows. In *Vector and Parallel Processing VECPAR 2000*, volume 1981 of *Lecture Notes in Computer Science*, pages 301–315. Springer-Verlag, 2001.

P. Chardaire and A. Lisser. Simplex and interior point specialized algorithms for solving nonoriented multicommodity flow problems. *Operations Research*, 50(2):260–276, 2002.

I. C. Choi and D. Goldfarb. Solving multicommodity network flow problems by an interior point method. In T. F. Coleman and Y. Li, editors, *Large-Scale Numerical Optimization*, pages 58–69. SIAM, Philadelphia, PA, 1990.

G. Dahl and M. Stoer. A cutting plane algorithm for multicommodity survivable network design problems. *INFORMS Journal on Computing*, 10:1–11, 1998.

J. W. Demmel. *Applied Numerical Linear Algebra*. SIAM, Philadelphia, PA, 1997.

A. Eisenblätter. *Frequency Assignment in GSM Networks: Models, Heuristics, and Lower Bounds*. PhD thesis, TU-Berlin and Konrad-Zuse-Zentrum für Informationstechnik, Berlin, 2001.

A. S. El–Bakry, R. A. Tapia, and Y. Zhang. A study of indicators for identifying zero variables in interior–point methods. *SIAM Review*, 36:45–72, 1994.

S. Elhedhli and J.-L. Goffin. The integration of interior-point cutting plane methods within branch-and-price algorithms. *Mathematical Programming*, 100(2):267–294, 2004.

A. Forsgren, P. E. Gill, and M. H. Wright. Interior methods for nonlinear optimization. *SIAM Review*, 44(4):525–597, 2002.

J.-L. Goffin, J. Gondzio, R. Sarkissian, and J.-P. Vial. Solving nonlinear multicommodity network flow problems by the analytic center cutting plane method. *Mathematical Programming*, 76:131–154, 1997.

J.-L. Goffin and J.-P. Vial. Convex nondifferentiable optimization: a survey focussed on the analytic center cutting plane method. *Optimization Methods and Software*, 17(5):805–867, 2002.

J. Gondzio and R. Kouwenberg. High-performance computing for asset-liability management. *Operations Research*, 49(6):879–891, 2001.

M. Grötschel, C. L. Monma, and M. Stoer. Computational results with a cutting plane algorithm for designing communication networks with low-connectivity constraints. *Operations Research*, 40(2):309–330, 1992.

M. Grötschel, C. L. Monma, and M. Stoer. Polyhedral and computational investigations for designing communication networks with high survivability requirements. *Operations Research*, 43(6):1012–1024, 1995.

J. J. Júdice, J. M. Patrício, L. F. Portugal, M. G. C. Resende, and G. Veiga. A study of preconditioners for network interior point methods. *Computational Optimization and Applications*, 24:5–35, 2003.

J. J. Júdice, L. F. Portugal, M. G. C. Resende, and G. Veiga. A truncated interior point method for the solution of minimum cost flow problems on an undirected multi-commodity network. In *Proceedings of the First Portuguese National Telecommunications Conference*, pages 381–384, 1997. In Portuguese.

P. Kall and S. W. Wallace. *Stochastic Programming*. John Wiley, Chichester, UK, 1994. Available online from the authors' webpages.

A. P. Kamath and O. Palmon. Improved interior point algorithms for exact and approximate solution of multi-commodity flow problems. In *Proceedings of the Sixth Annual ACM-SIAM Symposium on Discrete Algorithms*, pages 502–511, January 1995.

S. Kapoor and P. Vaidya. Fast algorithms for convex programming and multicommodity flows. *Proceedings of the 18th annual ACM symposium on the theory of computing*, pages 147–159, 1988.

S. Kapoor and P. Vaidya. Speeding up Karmarkar's algorithm for multicommodity flows. *Mathematical Programming*, 73:111–127, 1996.

A. Lisser and F. Rendl. Graph partitioning using linear and semidefinite programming. *Mathematical Programming*, 95(1):91–101, 2003.

M. S. Lobo, L. Vandenberghe, S. Boyd, and H. Lebret. Applications of second-order cone programming. *Linear Algebra and its Applications*, 284(1–3):193–228, 1998.

N. Megiddo. On finding primal- and dual-optimal bases. *ORSA Journal on Computing*, 3:63–65, 1991.

S. Mehrotra. On the implementation of a primal-dual interior point method. *SIAM Journal on Optimization*, 2(4):575–601, 1992.

S. Mehrotra and J.-S. Wang. Conjugate gradient based implementation of interior point methods for network flow problems. In L. Adams and J. L. Nazareth, editors, *Linear and nonlinear conjugate gradient-related methods*, pages 124–142. AMS/SIAM, 1996.

J. E. Mitchell. Computational experience with an interior point cutting plane algorithm. *SIAM Journal on Optimization*, 10(4):1212–1227, 2000.

J. E. Mitchell. Polynomial interior point cutting plane methods. *Optimization Methods and Software*, 18(5):507–534, 2003.

J. E. Mitchell and B. Borchers. Solving linear ordering problems with a combined interior point/simplex cutting plane algorithm. In H. L. Frenk *et al.*, editor, *High Performance Optimization*, chapter 14, pages 349–366. Kluwer Academic Publishers, Dordrecht, The Netherlands, 2000.

J. E. Mitchell, P. M. Pardalos, and M. G. C. Resende. Interior point methods for combinatorial optimization. In D.-Z. Du and P. M. Pardalos, editors, *Handbook of*

Combinatorial Optimization, volume 1, pages 189–297. Kluwer Academic Publishers, 1998.

J. E. Mitchell and M. J. Todd. Solving combinatorial optimization problems using Karmarkar's algorithm. *Mathematical Programming*, 56:245–284, 1992.

Y.-S. Myung, H.-J. Kim, and D.-W. Tcha. Design of cummunication networks with survivability constraints. *Management Science*, 45(2):238–252, 1999.

J. Patrício, L. F. Portugal, M. G. C. Resende, G. Veiga, and J. J. Júdice. Fortran subroutines for network flow optimization using an interior point algorithm. Technical Report TD-5X2SLN, AT&T Labs, Florham Park, NJ, March 2004.

L. Portugal, M. Resende, G. Veiga, and J. Júdice. A truncated primal-infeasible dual-feasible network interior point method. *Networks*, 35:91–108, 2000.

S. Raghavan and T. L. Magnanti. Network connectivity. In M. Dell'Amico, F. Maffioli, and S. Martello, editors, *Annotated bibliographies in combinatorial optimization*, pages 335–354. John Wiley, Chichester, 1997.

M. G. C. Resende and G. Veiga. An efficient implementation of a network interior point method. In D.S. Johnson and C.C. McGeogh, editors, *Network Flows and Matching: First DIMACS Implementation Challenge,*, pages 299–348. American Mathematical Society, 1993a. DIMACS Series on Discrete Mathematics and Theoretical Computer Science, vol. 12.

M. G. C. Resende and G. Veiga. An implementation of the dual affine scaling algorithm for minimum cost flow on bipartite uncapacitated networks. *SIAM Journal on Optimization*, 3:516–537, 1993b.

M. G. C. Resende and G. Veiga. An annotated bibliography of network interior point methods. *Networks*, 42:114–121, 2003.

C. Roos, T. Terlaky, and J.-Ph. Vial. *Theory and Algorithms for Linear Optimization: An Interior Point Approach.* John Wiley, Chichester, 1997.

H. D. Sherali, J. C. Smith, and Y. Lee. Enhanced model representations for an intra-ring synchronous optical network design problem allowing demand splitting. *INFORMS Journal on Computing*, 12(4):284–298, 2000.

L. N. Trefethen and D. Bau. *Numerical Linear Algebra.* SIAM, Philadelphia, PA, 1997.

R. J. Vanderbei. *Linear Programming: Foundations and Extensions.* Kluwer Academic Publishers, Boston, 1996. Second Edition: 2001.

S. Wright. *Primal-dual interior point methods.* SIAM, Philadelphia, 1996.

E. Yamakawa, Y. Matsubara, and M. Fukushima. A parallel primal-dual interior point method for multicommodity flow problems with quadratic costs. *Journal of the Operations Research Society of Japan*, 39(4):566–591, 1996.

Y. Ye. On the finite convergence of interior-point algorithms for linear programming. *Mathematical Programming*, 57(2):325–335, 1992.

Y. Ye. *Interior Point Algorithms: Theory and Analysis*. John Wiley, New York, 1997.

2 NONLINEAR PROGRAMMING IN TELECOMMUNICATIONS

Athanasios Migdalas[1]

[1]DSS Laboratory ERGASYA
Production Engineering & Management Department
Technical University of Crete
Chania 73100 Greece
migdalas@ergasya.tuc.gr

Abstract: Telecommunications have always been the subject of application for advanced mathematical techniques. In this chapter, we review classical nonlinear programming approaches to modeling and solving certain problems in telecommunications. We emphasize the common aspects of telecommunications and road networks, and indicate that several lessons are to be learned from the field of transportation science, where game theoretic and equilibrium approaches have been studied for more than forty years. Several research directions are also stated.

Keywords: Nonlinear optimization, Frank-Wolfe-like algorithms, simplicial decomposition, team games, routing, equilibrium flows, capacity assignment, network design, bilevel programming, Nash equilibrium, Wardrop's principle, Stackelberg game.

2.1 INTRODUCTION

Telecommunications have always been the subject of application for advanced mathematical techniques such as queuing theory (Kleinrock, 1964; 1974). With the advent of computer networks, research on and applications of optimization modeling and algorithms to problems in the area of data communication networks intensified in 1970s (Schwartz, 1977). Following the rapid technological development in telecommunications and the derivation of new efficient optimization methods, research and applications of optimization techniques to communications kept an increasing pace in the 1980s (Seidler, 1983) and 1990s (Bertsekas and Gallager, 1992) culminating the recent years with the new opportunities offered by the explosive expansion of fiber, broadband, mobile, and wireless networking. Most analysis or design problems in communication systems can be stated in terms of optimizing, i.e., minimizing or maximizing, some performance or utility function in several variables, the values of which must satisfy a set of prespecified requirements. These optimization problems quite often turn to be of combinatorial nature, and thus belong to the sphere of combinatorial optimization. However, nonlinear optimization has also been successfully applied to several problems arisen in the context of communication systems, including the areas of wireless networks, high speed Internet, equalization to broadband access, network topology, and routing.

The general form of the problems studied in nonlinear optimization may be stated as follows:

$$\textbf{(NLP)} \quad \min \quad f(\mathbf{x}) \qquad (2.1)$$
$$\text{s.t.} \quad \mathbf{g}(\mathbf{x}) \leq \mathbf{0} \qquad (2.2)$$
$$\mathbf{h}(\mathbf{x}) = \mathbf{0} \qquad (2.3)$$
$$\mathbf{x} \in \mathcal{S}, \qquad (2.4)$$

where $\mathcal{S} \subseteq \mathbb{R}^n$, $f : \mathbf{x} \in \mathcal{S} \to f(\mathbf{x}) \in \mathbb{R}$, $\mathbf{g} : \mathbf{x} \in \mathcal{S} \to \mathbf{g}(\mathbf{x}) \in \mathbb{R}^\ell$, and $\mathbf{h} : \mathbf{x} \in \mathcal{S} \to \mathbf{h}(\mathbf{x}) \in \mathbb{R}^m$. The function $f(\cdot)$ is the objective function of the problem, (2.2)-(2.4) are the problem constraints which define the feasible region of the problem, i.e., the set of points $\mathbf{x} \in \mathbb{R}^n$ that are admissible candidate solutions to **NLP**. We assume that (2.4) models simple constraints such as bounds on the values of $\mathbf{x}$. Problem **NLP** is a nonlinear optimization problem as long as at least one of the involved functions is nonlinear.

In the present chapter we are concerned with classical nonlinear programming approaches to solving problems of type **NLP**. This means in particular that the methods and algorithms discussed in Section 2.4 are able to attack only special instances of **NLP**, requiring, for example, continuity and differentiability of the involved functions in order to produce local minima (c.f. Definition 2.1) and certain convexity (c.f. Definitions 2.3 and 2.10) conditions to be satisfied in order to guarantee global optimality (c.f. Definition 2.1) of the produced solution. Hence some important cases of **NLP**, such as mixed integer nonlinear optimization problems, stochastic optimization problems, and global optimization problems, will not be discussed. Thus, certain very interesting problems in telecommunications, including hub or switching network design (Hakimi, 1965; Helme and Magnanti, 1989; O'Kelly et al., 1995), traffic management, capacity expansion and design of ATM networks (Bienstock and Saniee, 1998; Bonatti and Gaivoronski, 1994; Boland et al., 1994; Prycker, 1995), and pricing and costs in e-services (Kourgiantakis et al., 2005), will not be discussed in this chapter and the interested reader should refer to other chapters in this book. Here we place the emphasis on routing (or flow assignment) and capacity assignment problems for several reasons:

- Similar problems have been studied for road traffic networks for several decades, however, the crossover between the two fields has been minimal;

- The corresponding road traffic network problems have been based on game theoretic concepts since the seminal works of Wardrop (1952) and Dafermos and Sparrow (1969), while only recently, due to the deregulation of telecommunications and the increased competition in Internet services, has this game theoretic approach gained the attention of the telecommunications community, showing however impressive activity (Altman et al., 2002);

- Although we are basically concerned with networks of fixed topology, there are lessons to be learned from the findings of several decades and bring them into the study of networks with dynamically changing topology; and

- Several important improvements on algorithms for road traffic flow assignments can be carried over to communications.

The exposition of the chapter is as follows: Section 2.2 is dedicated to a brief overview of the basic theory of nonlinear optimization. In Section 2.3, several classical problems of routing and capacity assignment, as well road traffic flow assignment and network design are presented and emphasis is based on the interrelations of the two areas. The derivation of the models are based on the concept of team games, and distinction is made between, on one hand, equilibrium or user optimum flows and, on the other hand, system optimum flows. Several other nonlinear optimization problems, such as energy conserving routing in wireless ad-hoc networks and the extension of lifetime of power constrained networks, are also briefly surveyed. In Section 2.4, nonlinear programming algorithms able to solve problems stated in Section 2.3 are discussed, several of them in depth. Emphasis is placed on the Frank-Wolfe algorithm (Frank and Wolfe, 1956), known also as the flow deviation method (Fratta et al., 1973) to the communications community, as several new methods have been proposed in recent years which improve the performance of the original algorithm. Other methods, not related to Frank-Wolfe are also surveyed. We conclude with Section 2.5 where several future research issues are presented.

2.2 BRIEF OVERVIEW OF THE BASIC THEORY

In this section, we briefly review a few basic concepts from the theory of nonlinear optimization needed in subsequent sections. There are several excellent books (Bazaraa and Shetty, 1979; Bertsekas, 1982a; Bertsekas and Tsitsiklis, 1989; Martos, 1975) on the subject and the reader should turn to them for further details. The few concepts of game theory needed are borrowed from Migdalas (2004), Migdalas (1995), and Migdalas et al. (1997). The presentation will be based partly on **NLP** and partly on the following statement of a nonlinear optimization problem:

$$(\mathbf{P}) \quad \min \ f(\mathbf{x})$$
$$\text{s.t.} \ \mathbf{x} \in \mathcal{X},$$

where $\mathcal{X} \subseteq \mathbb{R}^n$ is the feasible region, and $f(\cdot)$ is the objective function of the problem. A point $\mathbf{x} \in \mathbb{R}^n$ is called feasible point to **P** if $\mathbf{x} \in \mathcal{X}$, otherwise $\mathbf{x}$ is infeasible. There are two solution concepts associated with any nonlinear programming problem and these are introduced next:

Definition 2.1 *A point* $\mathbf{x}^*$ *is a local minimum point of* **P** *if*

1. $\mathbf{x}^* \in \mathcal{X}$, *i.e., it is feasible, and*

2. there exists a neighborhood $\mathcal{N}_{\varepsilon}(\mathbf{x}^*) = \{\mathbf{x} : \|\mathbf{x} - \mathbf{x}^*\| \leq \varepsilon\}$, $\varepsilon > 0$, such that $f(\mathbf{x}^*) \leq f(\mathbf{x})$, $\forall \mathbf{x} \in \mathcal{N}_{\varepsilon}(\mathbf{x}^*) \cap X$.

A point $\mathbf{x}^*$ is a global minimum point of **P** if

1. $\mathbf{x}^* \in X$, i.e., it is feasible, and
2. $f(\mathbf{x}^*) \leq f(\mathbf{x})$, $\forall \mathbf{x} \in X$.

The point $\mathbf{x}^*$ is often referred to as a local (or global) optimal point or optimal solution.

A global minimum is always a local minimum. However, the reverse of this statement is true only for very specific instances of **P**, see for example Theorem 2.2. Moreover, a nonlinear optimization problem may have no optimal solution unless it satisfies certain conditions, such as lower semicontinuity of the objective function and compactness of the feasible set.

Definition 2.2 *The function $f(\mathbf{x})$ defined on $X \subseteq \mathbb{R}^n$ is said to be lower semicontinuous in X if the sets $L(b) = \{\mathbf{x} \in X | f(\mathbf{x}) \leq b\}$ are closed relative to X for any $b \in \mathbb{R}$.*

A continuous function is also lower semicontinuous. The reverse of the statement is not true. Existence of a solution to **P** is guaranteed by the next theorem.

Theorem 2.1 *If X is a nonempty compact set and the function $f(\mathbf{x})$ is lower semicontinuous in X, then **P** has a solution $\mathbf{x}^* \in X$.*

To guarantee the global optimality of the solution, additional requirements must be posed on **P**. Convexity of the objective function and the feasible region are two such requirements and are formally introduced next.

Definition 2.3 *The set $X \subset \mathbb{R}^n$ is convex if for all $\mathbf{x}^1, \mathbf{x}^2 \in X$ it satisfies the inclusion*

$$\alpha \mathbf{x}^1 + (1-\alpha)\mathbf{x}^2 \in X$$

for all $\alpha \in [0,1]$, that is, the convex combination of any two points in X is also in X. The function $f(\mathbf{x})$ is convex in the convex set X if for all $\mathbf{x}^1, \mathbf{x}^2 \in X$ and $\alpha \in (0,1)$ it satisfies the inequality

$$f(\alpha \mathbf{x}^1 + (1-\alpha)\mathbf{x}^2) \leq \alpha f(\mathbf{x}^1) + (1-\alpha)f(\mathbf{x}^2).$$

The function is strictly convex if strict inequality holds in the above relation. Moreover, the function $-f(\mathbf{x})$ is (strictly) concave if $f(\mathbf{x})$ is (strictly) convex.

The next theorem guarantees the global optimality of any local optimum in **P** under convexity assumptions.

Theorem 2.2 *If X and $f(\mathbf{x})$ are convex in **P**, then any local minimum point is also global. Moreover, if $f(\mathbf{x})$ is strictly convex the minimum point is unique.*

The case when the feasible region of a nonlinear program is defined by linear inequalities is of particular interest. The next few definitions and theorems address this issue.

Definition 2.4 *A polyhedron X in $\mathbb{R}^n$ is the intersection of a finite number of closed halfspaces, i.e., $X = \{\mathbf{x} \in \mathbb{R}^n | \mathbf{a}_i^T \mathbf{x} \leq b_i, i = 1, \ldots, m\}$, where $\mathbf{a}_i \in \mathbb{R}^n$ are constant vectors and b_i are real numbers. A bounded polyhedron is called a polytope.*

The convex hull of a finite number of points is the collection of all convex combinations of these points. Such a convex hull is always a polytope. The simplest example of a polytope is the simplex. Polyhedra and polytopes are convex sets of special structure, in particular they posses certain finite characteristics.

Definition 2.5 *The point $\mathbf{x}$ is an extreme point or a vertex of the polyhedron $X \subset \mathbb{R}^n$ if $\mathbf{x} = \alpha \mathbf{x}^1 + (1-\alpha)\mathbf{x}^2$ with $\mathbf{x}^1, \mathbf{x}^2 \in X$ and $\alpha \in (0,1)$ implies that $\mathbf{x} = \mathbf{x}^1 = \mathbf{x}^2$. That is, an extreme point cannot be written as a convex combination of two distinct points in X.*

Definition 2.6 *The nonzero vector $\mathbf{d} \in \mathbb{R}^n$ is called a direction of the polyhedron X if for every $\mathbf{x} \in X$, $\mathbf{x} + \alpha \mathbf{d} \in X$ for all $\alpha \geq 0$. The direction $\mathbf{d}$ of X is an extreme direction of X if $\mathbf{d} = \alpha \mathbf{d}^1 + \beta \mathbf{d}^2$ for $\alpha, \beta > 0$ implies $\mathbf{d}^1 = \gamma \mathbf{d}^2$ for some $\gamma > 0$. That is, an extreme direction cannot be written as a positive linear combination of two distinct directions in X.*

Theorem 2.3 *The number of extreme points or vertices of a polyhedron is finite. The number of extreme directions of a polyhedron is finite.*

Caratheodory's theorem states that a point in the convex hull of a set can be represented as a convex combination of a finite number of points in the set. The theorem in the case of polyhedral sets is also known as the theorem of inner representation and states that any point in a polytope can be expressed as a convex combination of the extreme point of the polytope and that any point in a polyhedron can be expressed as a convex combination of its extreme points plus a linear combination of its extreme directions.

Theorem 2.4 *A polytope is the convex hull of its vertices, that is, if $\bar{\mathbf{x}}^1, \bar{\mathbf{x}}^2, \ldots, \bar{\mathbf{x}}^K$ are the extreme points of a polytope $X \subset \mathbb{R}^n$, then*

$$X = \left\{ \mathbf{x} \in \mathbb{R}^n | \mathbf{x} = \sum_{k=1}^{K} \alpha_k \bar{\mathbf{x}}^k, \sum_{k=1}^{K} \alpha_k = 1, \alpha_k \geq 0 \text{ for } k = 1, \ldots, K \right\}$$

In particular, since $X \subset \mathbb{R}^n$, then at most $n+1$ extreme points are needed in order to represent any point $\mathbf{x} \in X$. Moreover, if $\mathbf{X} \subset \mathbb{R}^n$ is an unbounded polyhedron with K extreme points $\bar{\mathbf{x}}^1, \bar{\mathbf{x}}^2, \ldots, \bar{\mathbf{x}}^K$ and L extreme directions $\bar{\mathbf{d}}^1, \bar{\mathbf{d}}^2, \ldots, \bar{\mathbf{d}}^L$, then $\mathbf{x} \in X$ if and only if

$$\mathbf{x} = \sum_{k=1}^{K} \alpha_k \bar{\mathbf{x}}^k + \sum_{\ell=1}^{L} \beta_\ell \bar{\mathbf{d}}^\ell$$

$$\sum_{k=1}^{K} \alpha_k = 1$$

$$\alpha_k \geq 0, \forall k$$

$$\beta_\ell \geq 0, \forall \ell$$

Several algorithms for the solution of **P** are based on the notion of feasible directions of descent. That is, given a point $\mathbf{x}^k \in X$, a new point $\mathbf{x}^{k+1} \in X$ is produced such that $f(\mathbf{x}^{k+1}) < f(\mathbf{x}^k)$, where $\mathbf{x}^{k+1} = \mathbf{x}^k + \alpha_k \mathbf{d}^k$ for some $\alpha_k > 0$ and some vector $\mathbf{d}^k \neq \mathbf{0}$, called a direction. Thus, there is a particular interest in the behavior of the objective function $f(\cdot)$ along such directions. The concept of directional derivative provides in general the necessary information.

Definition 2.7 *Let $f(\mathbf{x})$ be defined on the nonempty set $X \subseteq \mathbb{R}^n$ and suppose that $\bar{\mathbf{x}} + \alpha \mathbf{d} \in X$ for $\bar{\mathbf{x}} \in X$, $\mathbf{d} \neq \mathbf{0}$ a vector in $\mathbb{R}^n$, and $\alpha > 0$ sufficiently small. The directional derivative of $f(\mathbf{x})$ at $\bar{\mathbf{x}}$ in the direction $\mathbf{d}$ is given by the following limit if it exists:*

$$\lim_{\alpha \to 0^+} \frac{f(\bar{\mathbf{x}} + \alpha \mathbf{d}) - f(\bar{\mathbf{x}})}{\alpha}$$

The limit in the previous definition with the values $-\infty$ and ∞ permitted exists for convex, and consequently for concave, functions.

Theorem 2.5 *Let X be nonempty and convex set and assume that $f(\mathbf{x})$ is convex, then the limit in the definition of the directional derivative exists.*

If the gradient of the objective function exists, the directional derivative can be calculated as the scalar product of the gradient and the direction.

Theorem 2.6 *If $f(\mathbf{x})$ has the gradient $\nabla f(\bar{\mathbf{x}})$ at $\bar{\mathbf{x}} \in \mathbb{R}^n$ and $\mathbf{d} \neq \mathbf{0}$ is a vector in $\mathbb{R}^n$, then the directional derivative of $f(\mathbf{x})$ at $\bar{\mathbf{x}}$ in the direction $\mathbf{d}$ is $\nabla f(\bar{\mathbf{x}})^T \mathbf{d}$.*

Whenever this scalar product is negative, the direction is descent, that is, a small step along it will decrease the current function value.

Definition 2.8 *Let $f(\mathbf{x})$ have the gradient $\nabla f(\bar{\mathbf{x}})$ at $\bar{\mathbf{x}}$. Then the vector $\mathbf{d} \neq \mathbf{0}$ is descent direction of $f(\mathbf{x})$ at $\bar{\mathbf{x}}$ if $\nabla f(\bar{\mathbf{x}})^T \mathbf{d} < 0$. The set $\mathcal{D}(\bar{\mathbf{x}}) = \{\mathbf{d} \in \mathbb{R}^n | \nabla f(\bar{\mathbf{x}})^T \mathbf{d} < 0\}$ is the set of descent directions of $f(\cdot)$ at $\bar{\mathbf{x}}$.*

However, not all descent directions at a point are admissible.

Definition 2.9 *Let $\bar{\mathbf{x}}$ be feasible to **P**. Then the vector $\mathbf{d} \neq \mathbf{0}$ is a feasible direction from $\bar{\mathbf{x}}$ if there exists a $\bar{\alpha} > 0$ such that $\bar{\mathbf{x}} + \alpha \mathbf{d}$ is feasible to **P** for all α satisfying $0 \leq \alpha \leq \bar{\alpha}$. The set of all feasible directions from $\bar{\mathbf{x}}$ will be denoted by $\mathcal{F}(\bar{\mathbf{x}})$.*

Clearly, a point $\bar{\mathbf{x}}$ feasible to **P** for which $\mathcal{F}(\bar{\mathbf{x}}) \cap \mathcal{D}(\bar{\mathbf{x}}) \neq \emptyset$ cannot be a (local) minimum point to **P**. Indeed, the next theorem states that if $\mathcal{F}(\bar{\mathbf{x}}) \cap \mathcal{D}(\bar{\mathbf{x}}) \neq \emptyset$, then it is always possible to find another feasible point with lower objective function value.

Theorem 2.7 *If $f(\mathbf{x})$ is differentiable at $\bar{\mathbf{x}}$, then it is continuous and has the gradient $\nabla f(\bar{\mathbf{x}})$ at $\bar{\mathbf{x}}$. Moreover, if $f(\mathbf{x})$ is differentiable on the segment joining $\mathbf{x}^1$ and $\mathbf{x}^2$, then the function $\phi(\alpha) = f(\alpha \mathbf{x}^1 + (1-\alpha)\mathbf{x}^2)$, defined for $\alpha \in [0,1]$, is differentiable in $[0,1]$ and has the derivative $\phi'(\alpha) = (\mathbf{x}^1 - \mathbf{x}^2)^T \nabla f(\alpha \mathbf{x}^1 + (1-\alpha)\mathbf{x}^2)$, that is, $\phi'(\alpha)$ is the directional derivative of $f(\mathbf{x})$ at $\alpha \mathbf{x}^1 + (1-\alpha)\mathbf{x}^2$ in the direction $\mathbf{x}^1 - \mathbf{x}^2$. If $\phi'(0) < 0$, i.e., $(\mathbf{x}^1 - \mathbf{x}^2)^T \nabla f(\mathbf{x}^2) < 0$, then there is a $\bar{\alpha} \in (0,1)$ such that $\phi(\alpha) < \phi(0)$, or equivalently $f(\alpha \mathbf{x}^1 + (1-\alpha)\mathbf{x}^2) < f(\mathbf{x}^2)$, for all $\alpha \in (0, \bar{\alpha})$.*

Differentiable convex functions have two important properties in terms of their gradient: their linearization by the tangent hyperplane at a given point provides an underestimation as stated by Theorem 2.8 below; and there exist necessary and sufficient conditions for an optimal solution as stated by Theorem 2.9 in terms of variational inequalities.

Theorem 2.8 *Let $f(\mathbf{x})$ be a differentiable function on a nonempty open set $X \subseteq \mathbb{R}^n$. Then $f(\mathbf{x})$ is convex if and only if for any $\mathbf{y} \in X$ the inequality*

$$f(\mathbf{x}) \geq f(\mathbf{y}) + \nabla f(\mathbf{y})^T (\mathbf{x} - \mathbf{y}), \forall \mathbf{x} \in X$$

holds.

Theorem 2.9 *Let $f(\mathbf{x})$ be a differential convex function defined on the convex set X. Then $\mathbf{x}^*$ is a global optimal solution to $\mathbf{P}$ if and only if*

$$\nabla f(\mathbf{x}^*)^T (\mathbf{x} - \mathbf{x}^*) \geq 0, \forall \mathbf{x} \in X.$$

Moreover, if X is open then $\mathbf{x}^*$ is an optimal solution if and only if $\nabla f(\mathbf{x}^*) = \mathbf{0}$.

The properties of the above theorem are not shared only by convex functions but also by pseudoconvex functions, which are introduced in the next definition.

Definition 2.10 *The differentiable function $f(\mathbf{x})$ defined on a nonempty open set $X \subseteq \mathbb{R}^n$ is said to be pseudoconvex if*

$$\nabla f(\mathbf{x}^1)^T (\mathbf{x}^2 - \mathbf{x}^1) \geq 0 \Rightarrow f(\mathbf{x}^2) \geq f(\mathbf{x}^1), \text{ or equivalently,}$$
$$f(\mathbf{x}^2) < f(\mathbf{x}^1) \Rightarrow \nabla f(\mathbf{x}^1)^T (\mathbf{x}^2 - \mathbf{x}^1) < 0,$$

for all $\mathbf{x}^1, \mathbf{x}^2 \in X$. Moreover, the function $-f(\cdot)$ is then called pseudoconcave.

In the absence of (pseudo-) convexity, global optimality of the local minimum point $\mathbf{x}^*$ cannot be guaranteed in $\mathbf{P}$. In particular, if $f(\mathbf{x})$ in $\mathbf{P}$ is a differentiable concave function we have the following results:

Theorem 2.10 *Let $f(\mathbf{x})$ be a differential concave function defined on the convex set X. If $\bar{\mathbf{x}}$ is a local optimal solution to $\mathbf{P}$ then*

$$\nabla f(\bar{\mathbf{x}})^T (\mathbf{x} - \bar{\mathbf{x}}) \geq 0, \forall \mathbf{x} \in X. \tag{2.5}$$

Moreover, if X is a nonempty polytope, then there exists an extreme point $\bar{\mathbf{x}} \in X$ which is an optimal solution of $\mathbf{P}$.

Actually, (2.5) in the above theorem does not require the concavity of the objective function. If, in particular, $\nabla f(\bar{\mathbf{x}})^T (\mathbf{x}^k - \bar{\mathbf{x}}) < 0$ for some $\mathbf{x}^k \in X$, then $\mathbf{d}^k = \mathbf{x}^k - \bar{\mathbf{x}}$ is a feasible direction of descent, and $f(\cdot)$ can be decreased from its current value $f(\bar{\mathbf{x}})$ by moving to a new feasible point $\mathbf{x}^{k+1} = \bar{\mathbf{x}} + \alpha_k (\mathbf{x}^k - \bar{\mathbf{x}})$ for some suitably chosen $\alpha_k > 0$. It should also be clear that if $X = \mathbb{R}^n$, then conditions (2.5) take the form of the well-known first order necessary conditions $\nabla f(\bar{\mathbf{x}}) = \mathbf{0}$.

When X is given explicitly, the optimality conditions derived above can be expressed in a more manageable form. Suppose that $X = \{g_i(\mathbf{x}) \leq 0, i = 1, \ldots, \ell\}$ in **P**. Let $\bar{\mathbf{x}} \in X$ and define the sets $I(\bar{\mathbf{x}}) = \{i | g_i(\bar{\mathbf{x}}) = 0\}$ and $\mathcal{D}^L(\bar{\mathbf{x}}) = \{\mathbf{d} \in \mathbb{R}^n | \mathbf{d}^T \nabla g_i(\bar{\mathbf{x}}) \leq 0, \forall i \in I(\bar{\mathbf{x}})\}$. The set $I(\bar{\mathbf{x}})$ is the index set of the binding or active constraints at $\bar{\mathbf{x}}$, while the set $\mathcal{D}^L(\bar{\mathbf{x}})$ gives a linearized approximation of the feasible region in a neighborhood of the feasible point $\bar{\mathbf{x}}$.

Definition 2.11 *The Kuhn-Tucker constraint qualification holds at $\bar{\mathbf{x}}$ if $\mathcal{D}(\bar{\mathbf{x}}) = \mathcal{D}^L(\bar{\mathbf{x}})$.*

In general, it is almost impossible to verify the Kuhn-Tucker constraint qualification directly. Therefore, several more or less checkable sufficient conditions for the Kuhn-Tucker constraint qualification have been proposed in the literature. See e.g. Bazaraa and Shetty (1979) for a hierarchy of such constraint qualifications. Clearly, if $\mathcal{D}^L(\bar{\mathbf{x}}) \cap \mathcal{F}(\bar{\mathbf{x}}) \neq \emptyset$, then $\bar{\mathbf{x}}$ is not a (local) minimum point since there exits a feasible descent direction (c.f. Theorem 2.7) from $\bar{\mathbf{x}}$ in which $f(\cdot)$ strictly decreases. The next theorem provides the Karush-Kuhn-Tucker (KKT) necessary conditions for (local) optimality of $\bar{\mathbf{x}}$.

Theorem 2.11 *In **NLP**, let S be a nonempty set with at least one interior point, assume that $f, \mathbf{g}$ and $\mathbf{h}$ are continuously differentiable. If $\bar{\mathbf{x}}$ is a local minimum point and some constraint qualification holds at $\bar{\mathbf{x}}$, then there exist $\lambda \in \mathbb{R}^\ell$ and $\mu \in \mathbb{R}^m$ such that*

$$-\nabla f(\bar{\mathbf{x}}) = \lambda^T \mathbf{g}(\bar{\mathbf{x}}) + \mu^T \mathbf{h}(\bar{\mathbf{x}}) \quad (2.6)$$
$$\lambda^T \mathbf{g}(\bar{\mathbf{x}}) = 0 \quad (2.7)$$
$$\lambda \geq \mathbf{0} \quad (2.8)$$

Under certain convexity assumptions, the KKT necessary conditions are also sufficient for optimality. The next theorem states the Karush-Kuhn-Tucker (KKT) sufficient conditions for **NLP**.

Theorem 2.12 *Assume that the conditions of Theorem 2.11 on $S, \mathbf{g}$ and $\mathbf{h}$ are valid. Assume further that f and the component functions $g_i(\cdot)$ of $\mathbf{g}(\cdot)$ are pseudoconvex and that $\mathbf{h}$ is linear, that is $\mathbf{h}(\mathbf{x}) = \mathbf{Ax} - \mathbf{b}$. Then, any feasible point $\bar{\mathbf{x}}$ in **NLP** which satisfies the KKT-conditions (2.6)-(2.8) is a global minimum point to **NLP**.*

Both Theorem 2.11 and Theorem 2.12 can be stated under far more mild assumptions, see e.g. Bazaraa and Shetty (1979), Bertsekas (1982a), Shapiro (1979), and Martos (1975). However, the assumptions made are satisfied by the problems considered in this chapter.

Sufficient optimality conditions for **NLP** can also be stated in terms of the Lagrangian function. For the problem **NLP**, define the Lagrangian function

$$L(\mathbf{x}, \lambda, \mu) = f(\mathbf{x}) + \lambda^T \mathbf{g}(\mathbf{x}) + \mu^T \mathbf{h}(\mathbf{x}) \quad (2.9)$$

for $\mathbf{x} \in S$ and $\lambda \geq \mathbf{0}$, and consider the Lagrangian subproblem

$$\Theta(\lambda, \mu) = \min_{\mathbf{x} \in S} f(\mathbf{x}) + \lambda^T \mathbf{g}(\mathbf{x}) + \mu^T \mathbf{h}(\mathbf{x}), \quad (2.10)$$

defined for $\lambda \geq 0$.

Theorem 2.13 *Let $\bar{\mathbf{x}}$ be a solution to (2.10) for a given $(\bar{\lambda}, \bar{\mu})$. If $\bar{\mathbf{x}}$ is a feasible point to* **NLP** *and satisfies the conditions*

$$f(\bar{\mathbf{x}}) + \bar{\lambda}^T \mathbf{g}(\bar{\mathbf{x}}) + \bar{\mu}^T \mathbf{h}(\bar{\mathbf{x}}) = \Theta(\bar{\lambda}, \bar{\mu}) \quad (2.11)$$

$$\bar{\lambda}^T \mathbf{g}(\bar{\mathbf{x}}) = 0 \quad (2.12)$$

$$\bar{\lambda} \geq \mathbf{0}, \quad (2.13)$$

then $\bar{\mathbf{x}}$ is an optimal solution to **NLP**.

Conditions (2.11)-(2.13) are not necessary since for an optimal solution $\bar{\mathbf{x}}$ to **NLP** there may not be, in general, any corresponding $(\bar{\lambda}, \bar{\mu})$ with $\bar{\lambda} \geq \mathbf{0}$ that satisfies (2.11)-(2.13). Theorem 2.11 and Theorem 2.12 establish conditions under which (2.11)-(2.13) are necessary and sufficient for the global optimality of $\bar{\mathbf{x}}$; clearly by requiring certain convexity and constraint qualification assumptions to be satisfied.

We close this section by introducing the notion of noncooperative games and the concept of noncooperative equilibria. Let $X_i \subset \mathbb{R}^{n_i}$ for $i = 1, \ldots, n$ be compact convex sets and assume that the functions $f_i : \prod_{i=1}^n X_i \to \mathbb{R}$ are (pseudo-) convex, where $\mathbf{x} = [\mathbf{x}_1, \ldots, \mathbf{x}_i, \ldots, \mathbf{x}_n]$, $\mathbf{x}_i \in X_i$ for $i = 1, \ldots, n$, and $X = \prod_{i=1}^n X_i = X_1 \times X_2 \times \cdots \times X_n$. Let $\mathcal{N} = \{1, 2, \ldots, n\}$ denote a set of players or agents, $\mathbf{x}_i$ denote a strategy of player i, X_i the set of strategies of player i, and f_i the loss function of player i. Then the triple $\mathcal{G} = [\mathcal{N}, \{X\}_{i \in \mathcal{N}}, \{f_i\}_{i \in \mathcal{N}}]$ defines a game. The vector $\mathbf{x} = [\mathbf{x}_1, \ldots, \mathbf{x}_n]$ of strategies $\mathbf{x}_i \in X_i$ chosen by the players is a situation, and the set X is the set of feasible situations. A situation $\mathbf{x}^*$ in a noncooperative game $\mathcal{G}$ is called admissible for the player i if for any other strategy $\mathbf{x}_i \in X_i$ for this player we have $f_i(\mathbf{x}^*) \leq f_i(\mathbf{x}^*_{i-}, \mathbf{x}_i, \mathbf{x}^*_{i+})$, where $\mathbf{x}_{i-}$ denotes the partial vector $[\mathbf{x}_1, \ldots, \mathbf{x}_{i-1}]$, and $\mathbf{x}_{i+}$ denotes the partial vector $[\mathbf{x}_{i+1}, \ldots, \mathbf{x}_n]$. A situation $\mathbf{x}^*$ which is admissible for all players is called a Nash equilibrium situation:

Definition 2.12 *The situation $\mathbf{x}^*$ is a Nash equilibrium of the game*

$$\mathcal{G} = [\mathcal{N}, \{X\}_{i \in \mathcal{N}}, \{f_i\}_{i \in \mathcal{N}}]$$

if

$$f_i(\mathbf{x}^*) = \min_{\mathbf{x}_i \in X_i} f_i(\mathbf{x}^*_{i-}, \mathbf{x}_i, \mathbf{x}^*_{i+}), \; \forall i \in \mathcal{N} \quad (2.14)$$

The game $\mathcal{G}$ is a team game if $f_i(\cdot) = f(\cdot)$, $\forall i \in \mathcal{N}$, that is, if all players share the same loss function. In such a case, a Nash equilibrium of the game is obtainable by solving the following problem

$$\text{(CPP)} \quad \min \; f(\mathbf{x}_1, \mathbf{x}_2, \ldots, \mathbf{x}_n)$$
$$\text{s.t.} \quad \mathbf{x}_i \in X_i, \; \forall i \in \mathcal{N},$$

which is an optimization problem defined over a Cartesian product of sets.

Consider next an additional player 0 with loss function $g : \mathcal{Y} \times \prod_{i=1}^{n} \mathcal{X}_i \to \mathbb{R}$ and suppose that the loss functions of the other players $i \in \mathcal{N}$ are $f_i : \mathcal{Y} \times \prod_{i=1}^{n} \mathcal{X}_i \to \mathbb{R}$. Suppose that player 0 announces her decision first and that she is committed to implement it once announced. Player 0 will be called a leader. The other players $i \in \mathcal{N}$ react to such an announced leader's strategy by optimizing their losses f_i. These players are called followers. Clearly, the leader must anticipate the reaction of the followers before she announces her decision in order to optimize her loss g. The situation can be stated as the following problem:

(GBP) $\quad$ min $\quad g(\mathbf{y}, \mathbf{x}^*)$ $\hfill$ (2.15)

$\qquad\quad$ s.t. $\quad \mathbf{y} \in \mathcal{Y}$, $\hfill$ (2.16)

$\qquad\quad$ where $\mathbf{x}^*$ is such that

$$f_i(\mathbf{y}, \mathbf{x}^*) = \min_{\mathbf{x}_i \in \mathcal{X}_i} f_i(\mathbf{y}, \mathbf{x}_{i-}^*, \mathbf{x}_i, \mathbf{x}_{i+}^*), \ \forall i \in \mathcal{N}. \qquad (2.17)$$

If the followers are involved in a team game, then the problem can be rewritten as

(BP) $\quad$ min $\quad g(\mathbf{y}, \mathbf{x}^*)$ $\hfill$ (2.18)

$\qquad\quad$ s.t. $\quad \mathbf{y} \in \mathcal{Y}$, $\hfill$ (2.19)

$\qquad\quad$ where $\mathbf{x}^*$ is such that

$$f(\mathbf{y}, \mathbf{x}^*) = \min_{\mathbf{x} \in \mathcal{X}} f(\mathbf{y}, \mathbf{x}). \qquad (2.20)$$

Problems **GBP** and **BP** are called bilevel or two-level problems and they model a situation of a noncooperative game known as the Stackelberg game. The leader's problem (2.15)-(2.16) and (2.18)-(2.19), respectively, known as the first level problem, is the same in both cases. The second level problems (2.17) and (2.20) differ in the sense that in the first case the followers are involved in a Nash noncooperative game, while in the second case the followers are involved in a team game. Clearly, in **BP** the modeled situation would have been the same if there was only one follower. Under certain conditions, both **GBP** and **BP** can be stated in an equivalent **NLP** or **P** form by replacing the second level problems by their corresponding optimality conditions either in the form of variational inequalities (c.f. inequality (2.5)) or in the form of Karush-Kuhn-Tucker (c.f. Theorem 2.11 and Theorem 2.12).

2.3 MODELS

2.3.1 Traffic assignment and routing

Traffic equilibria models are descriptive in the sense that their aim is to predict flow patterns and travel times which are the results of the network users' choices with regard to routes from their origins to their destinations. The input to the model is a complete description of the proposed or existing transportation system. The models are based on the behavioral assumption that "the journey times on all the routes used are equal, and less than those which would be experienced by a single vehicle on any unused route" (Wardrop, 1952). This is Wardrop's first condition, also known as descriptive assignment or equal times journey principle. The traffic flows that satisfy this principle are usually referred to as user equilibrium or user optimum flows, a term attributed to Dafermos and Sparrow (1969), since the routes chosen by the network users are those which are individually perceived to be the shortest under the prevailing conditions. The result from such individual decisions is a condition in which no user can reduce her travel time by choosing unilaterally another route, i.e., it is an equilibrium situation, similar to the Nash equilibrium introduced in Section 2.2, in a noncooperative game where the players are associated with origin-destination pairs (see e.g. Migdalas (1995) for further references on the subject). Actually it will soon become apparent that, under specific assumptions, the game played is a team game and can therefore be stated as an optimization problem of type **CPP**.

By contrast, system optimum flows satisfy Wardrop's second condition which states that "the average journey time is minimum." These flow patterns are characterized by the fact that all routes used between an origin and a destination have equal marginal travel times, that is, the total travel time in the network is minimized, and this is considered as the system's understanding of optimal network utilization. However, the total travel time is generally not minimized by the user optimal flows, and, moreover, observed flows in real life are closer to the user flow patterns than system optimum. The only situation in which the two flow patterns are equal is in the absence of congestion; this is an ideal case of course. In both system and user optimum flows, the fundamental units are considered to be vehicles or, in the case of public transport, the individual travelers.

Traffic equilibrium problems are frequently divided into two modeling cases: fixed demand and elastic demand. The additional case of stochastic demand will not be discussed here (see e.g. Patriksson (1994)). In the fixed demand case, an origin-destination demand matrix $\mathbf{R} = [r_k]$, with r_k being the travel demand between the kth origin-destination pair, is assumed given. By contrast, in the elastic demand case, the demand r_k is modeled as a function of the least travel cost between the end points of the kth origin-destination pair. Thus, the user has a number of travel choices available and she is economically motivated in her decision of making or not the trip.

In telecommunication networks, and especially in data and computer networks, where the routed units are packets, the routing has historically been based on the minimization of total average time (Bertsekas and Gallager, 1992; Bertsekas and Tsitsiklis, 1989; Fratta et al., 1973; Gartner, 1977; Gerla, 1981; Schwartz and Cheung, 1976; Schwartz, 1977), typically under the assumption of Poisson arrivals and exponential message lengths, and the independence assumption of Kleinrock (1964; 1974) for the M/M/1 queues of packets. However, equilibrium flow patterns have increasingly drawn the attention of the researchers in recent years. Indeed, the equilibrium principle of Wardrop can be adapted to model the situation where the routing decisions are made by the nodes instead of the users of the network in order to minimize the per packet delay. Actually, the routers at the nodes attempt to minimize the per packet delay in terms of hops, i.e., nodes, to the destination. In ad-hoc networks, however, where both the users and the base stations are mobile incurring constantly changing delays, it has been argued (Gupta and Kumar, 1997) that actual delays instead of hops should be considered. In such cases, the Wardrop equilibrium principle can be utilized to describe the resulting flow patterns. Wardrop equilibria have also been used in telecommunication networks where users can individually determine their route in order to route entire sessions instead of packages (Korillis et al., 1999). Wardrop's equilibrium principle is also applicable to the case of distributed computer networks where the routed entities are entire jobs, and an individual job can be processed in any of the interconnected computers. In such a case, the routing decisions concerns the minimization of the expected communication and processing delay in the system (see e.g. Altman et al. (2002) for further references).

Let $G = (\mathcal{A}, \mathcal{N})$ be the underlying network with $\mathcal{N}$ the set of nodes and $\mathcal{A}$ the set of links, $\mathcal{K} \subseteq \mathcal{N} \times \mathcal{N}$ the set of origin-destination pairs, $\mathcal{P}_k$ the set of simple paths (routes) between the end nodes of the kth origin-destination pair, c_{pk} and h_{pk} the travel time and flow respectively on the $p$the route in $\mathcal{P}_k$.

According to Wardrop's equilibrium principle, if π_k denotes the shortest route travel time between the end nodes of the kth origin-destination pair, then

$$h_{pk} > 0 \;\Rightarrow\; c_{pk} = \pi_k, \; \forall p \in \mathcal{P}_k,$$
$$h_{pk} = 0 \;\Rightarrow\; c_{pk} \geq \pi_k, \; \forall p \in \mathcal{P}_k,$$

hold for all pairs in $\mathcal{K}$. Thus, including flow feasibility constraints, the user equilibrium conditions for fixed demand can be stated as follows:

$$h_{pk}(c_{pk} - \pi_k) = 0, \; \forall p \in \mathcal{P}_k, \tag{2.21}$$
$$c_{pk} - \pi_k \geq 0, \; \forall p \in \mathcal{P}_k, \tag{2.22}$$
$$\sum_{p \in \mathcal{P}_k} h_{pk} = r_k, \tag{2.23}$$
$$h_{pk} \geq 0, \; \forall p \in \mathcal{P}_k \tag{2.24}$$
$$\pi_k \geq 0, \tag{2.25}$$

for all origin-destination pairs k.

For every link $a \in \mathcal{A}$, let x_a denote the total link flow, and let $s_a(x_a)$ be the link travel cost encountered by a user traveling on link a with a total flow x_a. Define the link-route incidence matrix $\Delta = [\delta_{kap}]$, where δ_{kap} is 1 if the route p of the kth pair uses link a, and 0 otherwise. We then have the following result (Dafermos and Sparrow, 1969):

Theorem 2.14 *Assume that the network $G = (\mathcal{N}, \mathcal{A})$ is strongly connected with respect to the pairs in $\mathcal{K}$, that the demand matrix $\mathbf{R}$ is nonnegative, and that the travel time function $s_a(\cdot)$ is positive, strictly monotone increasing and continuously differentiable. Then, conditions (2.21)-(2.25) are the Karush-Kuhn-Tucker optimality conditions of the convex optimization problem*

$$\text{(FTAP)} \quad \min \sum_{a \in \mathcal{A}} \int_0^{x_a} s_a(t) dt, \tag{2.26}$$
$$\text{s.t.} \quad \sum_{p \in \mathcal{P}_k} h_{pk} = r_k, \; \forall k, \tag{2.27}$$
$$\sum_k \sum_{p \in \mathcal{P}_k} \delta_{kap} h_{pk} = x_a, \; \forall a \in \mathcal{A}, \tag{2.28}$$
$$h_{pk} \geq 0, \; \forall p \in \mathcal{P}_k, \; \forall k \tag{2.29}$$

By contrast, the system optimum seeking problem can be stated as follows:

$$\text{(SFTAP)} \quad \min \sum_{a \in \mathcal{A}} s_a(x_a) x_a,$$
$$\text{s.t.} \quad (2.27) - (2.29)$$

Note that (2.28) are definitional rather actual constraints. Indeed, they can be eliminated by replacing x_a in the objective function (2.26) by the left-hand side in (2.28). Hence, both **FTAP** and **SFTAP** are defined over a Cartesian product of $|\mathcal{K}|$ simplices, defined by (2.27) and (2.29), and are therefore special cases of **CPP**.

Typically, while in road networks the function $s_a(x_a)$ has the form

$$c_a + b_a \left(\frac{x_a}{u_a}\right)^{\nu},$$

where c_a, b_a are constants specific to each link, c_a is the travel time on a at mean free speed, u_a is the practical capacity of the link, and ν is some positive integer, in data communication networks, Kleinrock's performance function

$$s_a(x_a) = \frac{1}{u_a - x_a} + \mu t_a$$

is used. Here u_a is the capacity (in bps) on link a, t_a represents the processing delay, propagation delay or some other fixed delay in link a, $\frac{1}{\mu}$ is the average message length, assumed equal for all arcs, and x_a is the average flow (in bps). In both **FTAP** and **SFTAP** the objective functions are convex in route flows and strictly convex in link flows. This means that while the flow pattern in link flows is unique, that is, the total flow on each link has a unique optimal value, there may be a variety of optimal path flow patterns that all result in the unique link flow pattern. Thus, there are different optimal ways of of splitting flow from each origin to each destination among optimal routes.

To extend the user equilibrium model to the case of elastic demands, let $r_k = g_k(\pi)$, where $\pi = [\pi_1, \ldots, \pi_k \ldots]$, that is, the travel demand r_k between the end nodes of the kth origin-destination pair is a function of the vector of the cheapest route costs. Then Wardrop's user equilibrium principle for both route flows and demands are mathematically stated as follows:

$$\begin{aligned} h_{pk} > 0 &\Rightarrow c_{pk} = \pi_k, \forall p \in \mathcal{P}_k, \\ h_{pk} = 0 &\Rightarrow c_{pk} \geq \pi_k, \forall p \in \mathcal{P}_k, \\ r_k > 0 &\Rightarrow r_k = g_k(\pi), \\ r_k = 0 &\Rightarrow g_k(\pi) \leq 0, \end{aligned}$$

for all pairs k. Introducing the flow feasibility requirements, the above equations lead, under the additional requirement of nonnegative g_k on the nonnegative orthant, to the following conditions:

$$h_{pk}(c_{pk} - \pi_k) = 0, \forall p \in \mathcal{P}_k, \quad (2.30)$$
$$c_{pk} - \pi_k \geq 0, \forall p \in \mathcal{P}_k, \quad (2.31)$$
$$\sum_{p \in \mathcal{P}_k} h_{pk} = g_k(\pi), \quad (2.32)$$
$$h_{pk} \geq 0, \forall p \in \mathcal{P}_k, \quad (2.33)$$
$$\pi_k \geq 0, \quad (2.34)$$

for all origin-destination pairs k.

In their seminal work, Beckmann et al. (1956) recognized in (2.30)-(2.34) an optimization problem. Indeed, assume that g_k has the additional property of being continuous and strictly increasing. Then it is invertible, in which case $r_k = g_k(\pi_k) \Leftrightarrow \pi_k = g^{-1}(r_k)$, whenever $r_k > 0$. Thus, we have the following result:

Theorem 2.15 *Assume that the network $\mathcal{G} = (\mathcal{N}, \mathcal{A})$ is strongly connected with respect to the pairs in $\mathcal{K}$, and that the travel time function s_a is positive, monotone increasing and continuous differentiable. Then, conditions (2.30)-(2.34) are the Karush-Kuhn-Tucker optimality conditions of the following convex optimization problem:*

$$\textbf{(ETAP)} \quad \min \sum_{a \in \mathcal{A}} \int_0^{x_a} s_a(t)dt - \sum_k \int_0^{r_k} g_k^{-1}(t)dt, \quad (2.35)$$

$$\text{s.t.} \quad \sum_{p \in \mathcal{P}_k} h_{pk} = r_k, \forall k, \quad (2.36)$$

$$\sum_k \sum_{p \in \mathcal{P}_k} \delta_{kap} h_{pk} = x_a, \forall a \in \mathcal{A}, \quad (2.37)$$

$$h_{pk} \geq 0, \forall p \in \mathcal{P}_k, \forall k, \quad (2.38)$$

$$r_k \geq 0, \forall k. \quad (2.39)$$

The objective function of **ETAP** is convex for any increasing $s_a(\cdot)$ and any decreasing $g_k(\cdot)$. Thus, **ETAP** consists of minimizing a convex function over linear constraints. Again, the definitional equations (2.37) can be removed by replacing x_a in the objective function (2.35) by the left-hand side of (2.37). Hence, **ETAP** is a problem over the Cartesian product of the simplices (2.36), and therefore a special case of **CPP**.

The problems **FTAP**, **SFTAP**, and **ETAP** can be formulated in terms of only link flows instead of route flows. For this, let $o(k)$ and $d(k)$ denote the origin and the destination of each pair $k \in \mathcal{K}$, let $\mathcal{A}^+(i)$ be the set of links emanating from node $i \in \mathcal{N}$ and $\mathcal{A}^-(i)$ the set of links terminating at node $i \in \mathcal{N}$. Then, **FTAP** can be stated as:

$$\textbf{(AFTAP)} \quad \min \sum_{a \in \mathcal{A}} \int_0^{x_a} s_a(t)dt, \quad (2.40)$$

$$\text{s.t.} \quad \sum_{a \in \mathcal{A}^+(i)} x_a^k - \sum_{a \in \mathcal{A}^-(i)} x_a^k = r_i^k, \forall i \in \mathcal{N}, \forall k \in \mathcal{K}, \quad (2.41)$$

$$x_a = \sum_{k \in \mathcal{K}} x_a^k, \forall a \in \mathcal{A}, \quad (2.42)$$

$$x_a^k \geq 0, \forall a \in \mathcal{A}, \forall k \in \mathcal{K}, \quad (2.43)$$

where x_a^k denotes the portion of flow from origin $o(k)$ to destination $d(k)$ that streams through link a, and

$$r_i^k = \begin{cases} r_k, & \text{if } i = o(k), \\ -r_k, & \text{if } i = d(k), \\ 0, & \text{otherwise.} \end{cases}$$

The corresponding system optimum flow assignment problem can then be stated as

$$\textbf{(SAFTAP)} \quad \min \sum_{a \in \mathcal{A}} s_a(x_a) x_a,$$
$$\text{s.t.} \quad (2.41) - (2.43).$$

Similarly, **ETAP** can be restated as follows:

$$\textbf{(AETAP)} \quad \min \sum_{a \in \mathcal{A}} \int_0^{x_a} s_a(t) dt - \sum_k \int_0^{r_k} g_k^{-1}(t) dt,$$
$$\text{s.t.} \quad (2.41) - (2.43),$$

where

$$r_i^k = \begin{cases} 0 \le r_k, & \text{if } i = o(k), \\ -r_k, & \text{if } i = d(k), \\ 0, & \text{otherwise.} \end{cases}$$

Constraints (2.41) are the so-called flow balance equations which state that flow is generated at each origin, absorbed at each destination, and is not destroyed in intermediate nodes. There is a separate set of such constraints for each origin-destination pair k. Note also that all restated models are still special cases of **CPP**, where the Cartesian product is now with respect to the polyhedral sets defined by (2.41) and (2.43).

Clearly, in **ETAP** and **AETAP**, the demands of the network users are modeled so as to depend on the network state, which in turn depends on the users' choices of routes. Elastic demands have been studied extensively for road networks (see e.g. Murchland (1970), Evans (1976), Gartner (1980), and Patriksson (1994)) but also in single commodity setting for trade networks, e.g. Glassey (1978).

The differences between system and user optimization are the sources of several paradoxes (Dafermos and Nagurney, 1984) with most known that of Braess (1968), which have been studied extensively for road networks (Murchland, 1970) and have recently attracted the attention of the telecommunications community (Korillis et al., 1999). Braess' paradox is illustrated in the next section.

2.3.2 Capacity assignment and network design

The purpose of improving a given network by adding to its links more capacity or by adding to it entirely new links is to improve a certain traffic situation, that is, to decrease the delays in routing flows between remote locations. However, since routing according to the previous section can be done in two different and probably conflicting ways, i.e., from the users' perspective or the system's, and since the investment in the design or improvement of the network is undertaken by some central authority, one can suspect that conflicting situations may arise. In order to highlight this fact, consider the next example.

Example 2.1 Consider the network in Figure 2.1(a), where the link costs s_{ij} are linear increasing functions of the flow x_{ij} for all links (i,j) of the network, and assume that there are 6 units of flow to be routed from node 1 to node 2. The total delay on each link (i,j) for the user equilibrium model **FTAP** is then given by:

$$\begin{aligned} f_{13}(x_{13}) &= 5x_{32}^2 \\ f_{32}(x_{32}) &= 50x_{32} + 0.5x_{32}^2 \\ f_{14}(x_{14}) &= 50x_{14} + 0.5x_{14}^2 \\ f_{42}(x_{42}) &= 5x_{42}^2 \end{aligned}$$

By inspection, the equilibrium flow pattern is given by routing 3 units through the path $1 \to 3 \to 2$ and 3 units through the path $1 \to 4 \to 2$ because of the symmetry. This equilibrium flow pattern generates a total (system) delay $\sum_{(i,j)\in\mathcal{A}} s_{ij}(x_{ij}) = 498$, and the delay encountered by each unit of flow on each of the two paths is 83.

Suppose next that the network is being expanded by the addition of a new link $(3,4)$ as shown in Figure 2.1(b). Then the equilibrium flow pattern in the so-modified network is given by routing 2 units along the path $1 \to 4 \to 2$, 2 units along the path $1 \to 3 \to 2$, and 2 units along the new path $1 \to 3 \to 4 \to 2$. The new total (system) delay is then 552, that is, the intention of improving the network performance with the addition of a new link results in a worse situation than before since the total (system) delay has increased. Moreover, the delay of each flow unit along the two old paths has now increased to 92. □

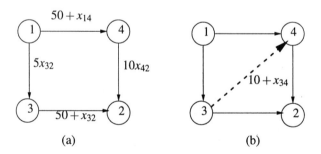

Figure 2.1 Illustration of Braess' paradox in network design

We may now conclude that in the context of network design/improvement, we cannot be sure that the addition (or even deletion) of a link to (from) an existing network will not increase (decrease) the total delay at equilibrium unless, somehow, the users' behavior has been taken into consideration in conjunction with the new network infrastructure. The phenomenon in which the equilibrium flow in an augmented network yields an increase in origin-destination travel times is called the Braess paradox. A direct consequence of its occurrence is that, if the equilibrium principle of Wardrop is adopted to represent the behavior of the network users, restriction or suppression of travel on some of the network links may reduce not only the total (system) delay, but also the delay of each individual flow unit. Possible occurrences of the paradox have been reported for road networks in Europe (Knödel, 1969) and in the USA (Berechman, 1984).

The network design problem from the system's perspective, i.e., ignoring user behavior, can be stated mathematically by slightly modifying the system flow assignment models of the previous section. Let y_a be the decision variable of the design/improvement of link a. We will assume that y_a mainly concerns assigning link a some capacity level and that capacity comes in arbitrarily divisible small units, that is y_a is a continuous and not a discrete variable. Moreover, let $\mathcal{Y}_a$ be the set from which y_a takes its values. Clearly a value $y_a = 0$ implies that the corresponding link is suppressed. We will further assume that the investment to a capacity level y_a for link a incurs a cost $g_a(y_a)$, where $g(\cdot)$ is some continuous function, and that the delays s_a are functions of the capacity level y_a as well as of the total flow x_a routed through link a. Then, a typical network problem can be stated as:

(SNDP) $\quad \min \quad \sum_{a \in \mathcal{A}} \{s_a(x_a, y_a) x_a + g_a(y_a)\}$

$\quad$ s.t. $\quad (2.27) - (2.29)$ or, equivalently, $(2.41) - (2.43)$

$\quad\quad\quad\quad y_a \in \mathcal{Y}_a, \ \forall a \in \mathcal{A}$

It should be noted that in data communication networks, explicit capacity constraints on the flow are often present (Gerla and Kleinrock, 1977; Bertsekas and Gallager, 1992), that is,

$$x_a \leq u_a + y_a, \ \forall a \in \mathcal{A}$$

should be included in the above formulation.

If the functions $g_a(y_a)$ are convex nonlinear or linear, then this is a problem which is not more difficult to solve than the problems of the previous section. However, typically these are not always very realistic assumptions (Steenbrink, 1974), and, in particular, convexity of the functions would in general result in small capacity improvements in almost all links. The presence of two blocks of variables encourages then the application of a so-called primal or Benders decomposition (Geoffrion, 1970; 1972; Minoux, 1984) to the problem in which, the $|\mathcal{A}|$ capacity assignment subproblems (Bertsekas and Gallager, 1992; Steenbrink, 1974; Schwartz, 1977)

$$f_a(x_a) = \min_{y_a} s_a(x_a, y_a) x_a + g_a(y_a)$$

$$\text{s.t. } y_a \in \mathcal{Y}$$

are solved and a so-called master problem, which is a flow assignment problem, is formed

(BMP) $\quad \min \quad \sum_{a \in \mathcal{A}} f_a(x_a)$

$\quad$ s.t. $\quad (2.27) - (2.29)$ or, equivalently, $(2.41) - (2.43)$

If all $g_a(y_a)$ have some particular form, then $f_a(x_a)$ could be obtained in closed form. The case of linear $g_a(y_a)$ is discussed in Gerla and Kleinrock (1977). However, this not the case in general. Moreover, in the presence of capacity constraints, $f_a(v_a)$ turn concave even if $g_a(y_a)$ are linear. Even worse, if $g_a(y_a)$ are not linear, $f_a(x_a)$ may not even be differentiable (Steenbrink, 1974).

Since in many situations the network manager or planner cannot impose the routing strategy of the network users, she must rather rely on some oracle to predict the user equilibrium flow patterns that will be the result of her decisions. Here lies the danger of the occurrence of Braess and other paradoxes. Thus, this case requires special modeling attention. The bilevel modeling of the Stackelberg game provides the way:

$$\textbf{(SNDP)} \quad \min_y \sum_{a \in \mathcal{A}} \{s_a(x_a^*, y_a)v_a + g_a(y_a)\}$$

$$\text{s.t.} \quad y_a \in \mathcal{Y}_a, \, \forall a \in \mathcal{A}$$

where $\mathbf{x}^* = [x_a^*]_{a \in \mathcal{A}}$ solves

$$\min_{\mathbf{x}} \sum_{a \in \mathcal{A}} \int_0^{x_a} s_a(t, y_a) dt$$

s.t. (2.27) – (2.29) or, equivalently, (2.41) – (2.43)

Clearly, the leader minimizes total system delay time and total investment subject to the restriction that the calculation of the total system delay is based on the equilibrium flow patterns and not on system routing flow patterns. **SNDP** can be restated as a nonlinear programming problem, if the second level problem is replaced by its optimality conditions corresponding to (2.21)-(2.25). However, the feasible region resulting from such a transformation is not a convex set and therefore the methods of Section 2.4 are, in general, unable to solve **SNDP**. See, for instance, Migdalas (1994) and Migdalas (1995), and the references therein for further analysis of the problem.

2.3.3 Other applications

We briefly survey a few telecommunications applications that have been formulated either as nonlinear optimization problems and/or have been based on game theoretic concepts similar to those encountered in road networks.

In the models of the previous sections, the route costs are additive, that is, the cost of any path is calculated as the sum of the costs on the links that form it, there are cases, however, where such an assumption does not apply. For instance, in the so-called power criterion, which is frequently used in flow control problems, the ratio between some power of the overall throughput to overall delay is used instead of link delays. In the case of road networks, equilibrium flow models for nonadditive costs are investigated in Gabriel and Bernstein (1997), Gabriel and Bernstein (1999), and Bernstein and Gabriel (1996), while Altman et al. (2000) discuss a simplified case of equilibrium routing in telecommunication networks.

The Braess paradox has been recently studied in the context of queuing networks (Beans et al., 1997; Calvert et al., 1997; Cohen and Jeffries, 1997), in capacity allocation to communication networks (Korillis et al., 1997; 1999), in distributed computing (Kameda et al., 2000), and in flow routing (Roughgarden and Tardos, 2002;

Roughgarden, 2002). Stackelberg strategies in telecommunications are considered in Roughgarden (2002), Yaiche et al. (2000), and Korillis et al. (1997).

In Kelly (1997), the issues of charging, rate control and routing for telecommunication networks, such as ATM networks, with elastic traffic are studied in a game theoretic setting and nonlinear optimization is employed.

Energy conserving routing is formulated as an optimization problem in Chang and Tassiulas (1997). Similarly, the problem of extending network life time in power constrained networks can be formulated and solved as a nonlinear optimization problem. The transmission rate assignment problem as well as other problems are formulated and solved as a nonlinear optimization problem in Berggren (2003)

2.4 ALGORITHMS

2.4.1 The Frank-Wolfe algorithm

Consider the problem of minimizing a differentiable function over a polyhedral set, i.e.,

$$(\textbf{P1}) \quad \min \ f(\mathbf{x})$$
$$\text{s.t.} \quad \mathbf{x} \in \mathcal{X},$$

where $\mathcal{X} = \{\mathbf{x} \in \mathbb{R}^n : \mathbf{Ax} = \mathbf{b}, \mathbf{x} \geq \mathbf{0}\}$ is a nonempty polyhedron.

The algorithm of Frank and Wolfe (Frank and Wolfe, 1956), originally stated for a quadratic objective function $f(\mathbf{x}) = \mathbf{q}^T \mathbf{x} + \mathbf{x}^T \mathbf{Q} \mathbf{x}$, is one of the most popular techniques for the solution of certain instances of the nonlinear program **P1**. The popularity of this technique is due in part to its ability to exploit special constraint structures such as network structures (Bienstock and Raskina, 2002; Cantor and Gerla, 1974; Daganzo, 1977a;b; Fratta et al., 1973; Gerla, 1981; Leblanc et al., 1975), and in part to the fact that it decomposes nonseparable problems over Cartesian product sets (cf. Larsson and Migdalas (1990), Migdalas (1994), and Migdalas (2004)). Also its implementation is in general quite simple. The algorithm was proposed as a routing technique in data networks under the name of flow deviation method by Fratta et al. (1973), and independently for the traffic assignment problem in road networks by Leblanc et al. (1975).

The algorithm is based on the linearization of the objective function. That is, given an iteration point $\mathbf{x}^k \in \mathcal{X}$, the algorithm approximates the objective with a first order Taylor expansion at $\mathbf{x}^k$, resulting in the linear programming subproblem

$$(\textbf{FW-SUB}^k) \quad \min \ \nabla f(\mathbf{x}^k)^T \mathbf{x}$$
$$\text{s.t.} \quad \mathbf{x} \in \mathcal{X},$$

where the constant terms have been dropped from the objective function. The solution $\bar{\mathbf{x}}^k$ of this subproblem is used in the construction of the search direction of descent $\mathbf{d}^k = \bar{\mathbf{x}}^k - \mathbf{x}^k$. A line search on the interval $[0, 1]$ furnishes the next iterate $\mathbf{x}^{k+1}$, that is, $\mathbf{x}^{k+1} = \mathbf{x}^k + \alpha_k \mathbf{d}^k$, where $\alpha_k \in \arg\min_{\alpha \in [0,1]} f(\mathbf{x}^k + \alpha \mathbf{d}^k)$, and the process is repeated.

If f in **P1** is convex, then an interesting aspect of the application of the Frank-Wolfe algorithm to **P1** is the generation of a lower bound on $f(\mathbf{x}^*)$ at each iteration point $\mathbf{x}^k$, namely $f(\mathbf{x}^k) + \nabla f(\mathbf{x}^k)^T (\bar{\mathbf{x}}^k - \mathbf{x}^k)$. This is a consequence of Theorem 2.8 and the minimization in **FW-SUB**^{k}. However, these lower bounds are not monotonically increasing. Hence, at iteration k, the current lower bound is defined as $\text{lbd}^k = \max\{\text{lbd}^{k-1}, f(\mathbf{x}^k) + \nabla f(\mathbf{x}^k)^T (\bar{\mathbf{x}}^k - \mathbf{x}^k)\}$, where lbd^{k-1} is the incumbent lower bound, initially set to $-\infty$. In practice, the algorithm can be terminated once $f(\mathbf{x}^k) - \text{lbd}^k \leq \varepsilon_1$ or $\frac{f(\mathbf{x}^k) - \text{lbd}^k}{f(\mathbf{x}^k)} \leq \varepsilon_2$ for suitably chosen $\varepsilon_1 > 0$ and $\varepsilon_2 > 0$.

We observe further that due to the linearity of the objective function in **FW-SUB**^{k}, the subproblem separates into n problems, one for each factor in the Cartesian product when the algorithm is applied to the team game **CPP**:

$$\begin{aligned} \min \quad & \nabla_i f(\mathbf{x}^k)^T \mathbf{x}_i \\ \text{s.t.} \quad & \mathbf{x}_i \in \mathcal{X}_i \end{aligned} \Bigg\} \; \forall i \in \mathcal{N}, \tag{2.44}$$

where $\nabla_i f(\mathbf{x})$ denotes the gradient of f with respect to the ith block of $\mathbf{x} = [\mathbf{x}_1, \ldots, \mathbf{x}_i, \ldots, \mathbf{x}_n]$ and it is itself the ith block component of f's gradient $\nabla f(\mathbf{x}) = [\nabla_1 f(\mathbf{x}), \ldots, \nabla_i f(\mathbf{x}), \ldots, \nabla_n f(\mathbf{x})]$. Clearly, the application of the Frank-Wolfe algorithm to **CPP** acts as a parallel decomposition scheme since the subproblems (2.44) are not interacting.

The application of the algorithm to the flow and traffic assignment problems **FTAP**, **AFTAP**, **STAP**, and **SAFTAP** in Section 2.3 results in subproblems **FW-SUB**^{k} or, equivalently, (2.44) which are so-called "all-or-nothing assignment" problems. That is, the shortest path p_κ between each origin-destination pair κ is calculated in terms of the linearized costs $\nabla_i f(\mathbf{x}^k)$ at the current flow situation $\mathbf{x}^k$, and the entire flow volume r_κ is assigned to this path, resulting into an extreme flow pattern $\bar{\mathbf{x}}^k$ which solves **FW-SUB**^{k} or, equivalently, (2.44). Thus, the character of the subproblems, and actually of the overall algorithmic approach, is independent of whether the problem is stated in terms of link or path flows. The application of the algorithm to problems with elastic demand, **ETAP** and **AETAP**, requires some additional precautions and modifications which are discussed in detail by Gartner (1977) and Gartner (1980). Evan's algorithm (Evans, 1976), although referred to as a partial linearization algorithm, is essentially more of a very specialized realization of the Frank-Wolfe algorithm than of the partial linearization discussed in Section 2.4.2

For concave $f(\cdot)$ in **P1**, the Frank-Wolfe algorithm without line search, i.e., $\alpha_k = 0$, $\forall k$, acts as a heuristic approach able to produce local solutions to **P1**, since by Theorem 2.10, the concave function $f(\cdot)$ attains a global minimum among the extreme points of the polytope $\mathcal{X}$, and the Frank-Wolfe subproblem produces an extreme point optimal to the linearization of $f(\cdot)$ at $\mathbf{x}^k$. This property of the method has been utilized in several applications, often within the scheme of a multi-start local search or some other metaheuristic, notably in misclassification minimization and data mining (Bennet, 1994; Bennet and Mangasarian, 1992; 1993; Blue and Bennett, 1996) and in flow routing with concave costs (Jr., 1971; Zadeh, 1973) as the result of the primal decomposition approach to the network design and capacity assignment problems of Section 2.3.

Several convergence results exist for the algorithm in the form given above as well as for several of its specializations (Canon and Cullum, 1968; Martos, 1975; Pshenichny and Danilin, 1978; Daganzo, 1977a;b; Hearn and Ribeira, 1981). For completeness of the exposition, we will demonstrate the global convergence of the approach to a global minimum point of **P1** under the following assumptions:

(A1) $f(\mathbf{x})$ is continuously differentiable on X,

(A2) $f(\mathbf{x})$ is pseudoconvex on X, and

(A3) X is closed and bounded (i.e. a polytope).

Assumption **A2** is essential for the verification of the global optimality of any accumulation point of the sequence $\{\mathbf{x}^k\}$. Assumption **A3** ensures that **P1** and **FW-SUB**^{k} have finite optima. It can be replaced by the coercivity assumption $\lim_{\|\mathbf{x}\| \to \infty} f(\mathbf{x}) = \infty$ on X, and the assumption that $\nabla f(\mathbf{x}^k)^T \mathbf{x}$ is bounded from below on X for all $\mathbf{x}^k \in X$.

Theorem 2.16 *Under the posed assumptions, the Frank-Wolfe algorithm either terminates finitely with an optimal solution of* **P1** *or it generates an infinite sequence* $\{\mathbf{x}^k\}$ *of feasible points in X such that any of its accumulation points is an optimal solution of* **P1**.

The algorithm would terminate at iteration k if the extreme point solution $\bar{\mathbf{x}}^k$ of **FW-SUB**^{k} is such that $(\bar{\mathbf{x}}^k - \mathbf{x}^k)^T \nabla f(\mathbf{x}^k) = 0$, implying $\nabla f(\mathbf{x}^k)^T \bar{\mathbf{x}}^k = \nabla f(\mathbf{x}^k)^T \mathbf{x}^k$. Indeed, the optimality of $\bar{\mathbf{x}}^k$ in **FW-SUB**^{k} implies $(\bar{\mathbf{x}}^k - \mathbf{x})^T \nabla f(\mathbf{x}^k) \leq 0$, $\forall \mathbf{x} \in X$, and therefore $(\mathbf{x} - \mathbf{x}^k)^T \nabla f(\mathbf{x}^k) \geq 0$, $\forall \mathbf{x} \in X$. Consequently, $\mathbf{x}^k$ is optimum in **P1**. We therefore assume subsequently that $(\bar{\mathbf{x}}^k - \mathbf{x}^k)^T \nabla f(\mathbf{x}^k) < 0$. Letting $\phi(\alpha) = f(\mathbf{x}^k + \alpha(\bar{\mathbf{x}}^k - \mathbf{x}^k))$, we observe that $\phi'(0) = (\bar{\mathbf{x}}^k - \mathbf{x}^k)^T \nabla f(\mathbf{x}^k)$, and therefore $\phi'(0) < 0$. Consequently, $\phi(\alpha) < \phi(0)$ for some $\alpha \in [0,1]$ implying $\phi(0) > \min_{\alpha \in [0,1]} \phi(\alpha)$, where the minimum exists by the continuity assumption. Thus, $\phi(\alpha_k) < \phi(0)$ implying $f(\mathbf{x}^{k+1}) < f(\mathbf{x}^k)$. Hence, $\{\mathbf{x}^k\}$ is a sequence of feasible points such that $\{f(\mathbf{x}^k)\}$ is a strictly decreasing sequence which is bounded from below due to the continuity of $f(\mathbf{x})$ and the compactness of X. Consequently, $\{f(\mathbf{x}^k)\}$ has a finite limit f^* and the sequence $\{\mathbf{x}^k\}$ has an accumulation point $\mathbf{x}^* \in X$ such that $f(\mathbf{x}^*) = f^*$. Assume that $\{\mathbf{x}^k\}$ is a subsequence such that $\lim_{k \to \infty} \mathbf{x}^k = \mathbf{x}^*$. Then it must have a subsubsequence $\{\mathbf{x}^k\}$ such that the corresponding $\bar{\mathbf{x}}^k$ is the same extreme point $\bar{\mathbf{x}}$ of X for all k. Such a subsubsequence exists since the number of extreme points is finite. Then, $(\bar{\mathbf{x}} - \mathbf{x})^T \nabla f(\mathbf{x}^k) \leq 0, \forall \mathbf{x} \in X, \forall k$, which for $k \to \infty$ implies

$$(\bar{\mathbf{x}} - \mathbf{x})^T \nabla f(\mathbf{x}^*) \leq 0, \forall \mathbf{x} \in X. \quad (2.45)$$

Since $\{f(\mathbf{x}^k)\}$ is decreasing, for $\ell > k$ we have $f(\mathbf{x}^\ell) \leq f(\mathbf{x}^{k+1}) < f(\mathbf{x}^k)$ which implies $f(\mathbf{x}^k + \alpha(\bar{\mathbf{x}} - \mathbf{x}^k)) \geq f(\mathbf{x}^\ell)$ for some $\alpha \in [0,1]$. Taking the limit as k and ℓ approach ∞ we have $\mathbf{x}^k \to \mathbf{x}^*$ and $\mathbf{x}^\ell \to \mathbf{x}^*$, and consequently

$$\frac{f(\mathbf{x}^* + \alpha(\bar{\mathbf{x}} - \mathbf{x}^*)) - f(\mathbf{x}^*)}{\alpha} \geq 0$$

for $\alpha \in [0,1]$. Letting $\alpha \to 0^+$ we obtain (c.f. Theorem 2.6)

$$(\bar{\mathbf{x}} - \mathbf{x}^*)^T \nabla f(\mathbf{x}^*) \geq 0. \quad (2.46)$$

Inequalities (2.45) and (2.46) imply $(\mathbf{x} - \mathbf{x}^*)^T \nabla f(\mathbf{x}^*) \geq 0$, $\forall \mathbf{x} \in X$ and consequently $\mathbf{x}^*$ is optimum in **P1**. $\square$

Although the most attractive features of the Frank-Wolfe algorithm are inherent in the subproblems **FW-SUB**^{k} as they are linear and decomposable, these same subproblems are the sources of the disadvantages associated with the approach. Due to their linearity, they always produce extreme point solutions, and the directions obtained may therefore depend more on the properties of the feasible region X than the properties of the objective function. It is also apparent from the convergence proof given above that the finiteness of the number of extreme points of X imply repeated re-generation of the same extreme points within an infinite sequence and consequently a zig-zagging approach towards the optimal solution $\mathbf{x}^*$ should be expected and jamming as $\mathbf{x}^*$ is being approached. An extreme such case is given by the example below.

Example 2.2 Consider the following instance of **CPP**:

$$\begin{aligned}
\min \quad & (x_1 - 2x_4)^2 + (3x_2 - x_3)^2 + (x_1 - 2)^2 + (x_3 - 2)^2 + (x_2 - 5)^2 + (x_4 - 3)^2 \\
\text{s.t.} \quad & 3x_1 + 2x_2 \leq 6 \\
& 5x_3 + 2x_4 \leq 10 \\
& x_j \geq 0, \ j = 1, 2, 3, 4
\end{aligned}$$

Its optimal solution is $(1.407, 0.890, 1.583, 1.042)$ and the corresponding optimal objective function value is 22.89.

If the algorithm is initialized with the origin, it must approach the optimal solution by following directions which are based on the alternate generation of the two extreme points $(\bar{x}_1^1, \bar{x}_2^1) = (2, 0)$ and $(\bar{x}_1^2, \bar{x}_2^2) = (0, 3)$ of the first simplex and the alternate generation of the two extreme points $(\bar{x}_3^1, \bar{x}_4^1) = (2, 0)$ and $(\bar{x}_3^2, \bar{x}_4^2) = (0, 5)$ of the second simplex, zig-zagging thus towards the optimal solution. This phenomenon is illustrated in Figure 2.2. The jamming starts as the steplengths α_k become smaller for each new iteration. Table 2.1 shows a few iterations of the algorithm. Note the rapid decrease of the objective function value in early iterations and the small decimal changes in later iterations. $\square$

A recent asymptotic analysis of the Frank-Wolfe algorithm for maximum concurrent flows is presented in Bienstock and Raskina (2002). It can be shown that the convergence rate of the Frank-Wolfe algorithm is indeed sublinear (Canon and Cullum, 1968; Dunn, 1979). Thus, several improvements to the original algorithm, based either on the concept of feasible direction of descent or on the concept of column generation, have been proposed in recent years and these are examined in the following subsections.

2.4.2 Feasible Direction Improvements

The basic idea of improving the Frank-Wolfe algorithm while retaining its feasible direction nature is to not let the generated directions be based so heavily on the extreme points of the feasible region. This can be done basically in two different ways; either by avoiding the complete linearization of the objective function or by enriching the

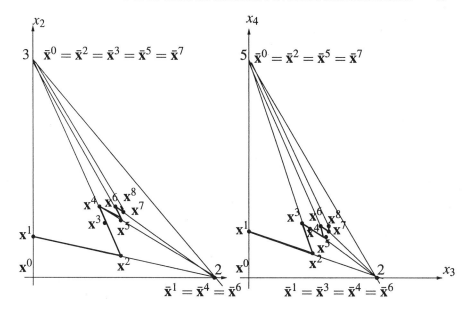

Figure 2.2 Movements of the Frank-Wolfe algorithm in the subproblem spaces

Table 2.1 Frank-Wolfe iterations

k	$\mathbf{x}^k$	$f(\mathbf{x}^k)$	$\bar{\mathbf{x}}_1^k$	$\bar{\mathbf{x}}_2^k$	lbd^k	$\mathbf{d}^k$	α_k
1	(0.000,0.000, 0.000,0.000)	42.000	(0.000,3.000)	(0.000,5.000)	-18.000	(0.000,3.000, 0.000,5.000)	0.139
2	(0.000,0.419, 0.000,0.695)	37.814	(2.000,0.000)	(2.000,0.000)	11.209	(2.000,-4.186, 2.000,-0.698)	0.432
3	(0.763,0.562, 0.763,0.936)	29.102					
33	(1.284,0.826, 1.428,1.015)	23.861	(0.000,3.000)	(2.000,0.000)	22.290	(-1.284,2.174, 0.572,-1.015)	0.0179
34	(1.261,0.865, 1.438,0.997)	23.847					

Frank-Wolfe subproblems with some nonlinear information. In both cases the original problem **P** is replaced iteratively by a sequence of easier (sub-) problems obtained by replacing the original objective function by a new one which may depend on the current iteration point.

Larsson and Migdalas (1990) proposed that the original function $f(\cdot)$ should only be partially linearized, that is, if $f(\mathbf{x}) = \sum_{i=1}^{n} f_i(x_i) + e(\mathbf{x})$, where $f_i(\cdot)$ are strictly convex functions and $e(\cdot)$ is not additively separable, then only $e(\cdot)$ needs to be linearized. Moreover, if $f(\cdot)$ does not have the necessary form, such a form can be enforced with the introduction of a second function $\varphi(\cdot)$, which may be assumed strictly convex and additively separable. Then, the original objective function $f(\cdot)$ is replaced by the equivalent $\varphi(\cdot) + [f(\cdot) - \varphi(\cdot)]$ and the "error" $e(\cdot) = f(\cdot) - \varphi(\cdot)$ is linearized. In the case of problems whose feasible region is the Cartesian product of polyhedral sets, i.e., of the form of the **CPP**, the partial linearization approach retains the parallel decomposition property of the original Frank-Wolfe problem, replacing (2.44) by the strictly convex subproblems

$$\left. \begin{array}{ll} \min & f_i(\mathbf{x}_j) + \nabla_i e(\mathbf{x}^k)^T \mathbf{x}_i \\ \text{s.t.} & \mathbf{x}_i \in X_i \end{array} \right\} \forall i \in \mathcal{N}. \qquad (2.47)$$

Letting $\bar{\mathbf{x}}^k = [\bar{\mathbf{x}}_1^k, \ldots, \bar{\mathbf{x}}_n^k]$ denote the point obtained by solving these subproblems, a feasible direction of descent $\mathbf{d}^k = \bar{\mathbf{x}}^k - \mathbf{x}^k$ is formed and the next iterate $\mathbf{x}^{k+1}$ is furnished by a line search on the interval $[0, \bar{\alpha}_k]$, where $\bar{\alpha}_k = \max\{\alpha \geq 0 | \mathbf{x}^k + \alpha \mathbf{d}^k \in X\}$, that is, $\mathbf{x}^{k+1} = \mathbf{x}^k + \alpha_k \mathbf{d}^k$, where $\alpha_k \in \arg\min_{\alpha \in [0, \bar{\alpha}_k]} f(\mathbf{x}^k + \alpha \mathbf{d}^k)$. The algorithm and its behavior is illustrated by the next example.

Example 2.3 Partial linearization is applied to the problem instance of Example 2.2 by linearizing only the nonseparable part of the objective function, i.e., $-4x_1x_4 - 6x_2x_3$, we then obtain the two subproblems:

$$\begin{array}{lll} (\text{SUB}^1) & \min & 2(x_1)^2 + 10(x_2)^2 - (4 + 4\bar{x}_4)x_1 - (10 + 6\bar{x}_3)x_2 \\ & \text{s.t.} & 3x_1 + 2x_2 \leq 6 \\ & & x_1, x_2 \geq 0 \end{array}$$

and

$$\begin{array}{lll} (\text{SUB}^2) & \min & 2(x_3)^2 + 5(x_4)^2 - (4 + 6\bar{x}_2)x_3 - (6 + 4\bar{x}_1)x_4 \\ & \text{s.t.} & 5x_3 + 2x_4 \leq 10 \\ & & x_3, x_4 \geq 0. \end{array}$$

Table 2.2 lists three iteration of the partial linearization algorithm, while Figure 2.3 illustrates the movements of the algorithm in the subproblems' simplices. Comparing to the Frank-Wolfe algorithm in Example 2.2, we observe that the point obtained by the latter in the 34th iteration is further away from the optimal point than the first point produced by the partial linearization. Also from the figures it is clear that the partial linearization algorithm is less dependent on the shape of the feasible region since it retains more information about the original objective function in its subproblems. □

One important property of the partial linearization algorithm, not shared by the Frank-Wolfe algorithm, is that the subproblems (2.47) are able to generate the optimal solution to the original problem.

Table 2.2 Iterations of the partial linearization algorithm

k	$\mathbf{x}^k$	$f(\mathbf{x}^k)$	$\bar{\mathbf{x}}_1^k$	$\bar{\mathbf{x}}_2^k$	$\mathbf{d}^k$	α_k
0	(0.000,0.000, 0.000,0.000)	42.000	(1.000,0.500)	(1.000,0.600)	(1.000,0.500, 1.000,0.600)	1.500
1	(1.500,0.750, 1.500,0.900)	23.625	(1.410,0.885)	(1.556,1.109)	(-0.089,0.135, 0.056,0.209)	1.000
2	(1.410,0.885, 1.556,1.109)	22.913	(1.417,0.875)	(1.582,1.045)	(0.007,-0.010, 0.026,-0.064)	0.880
3	(1.416,0.876, 1.578,1.053)	22.894				

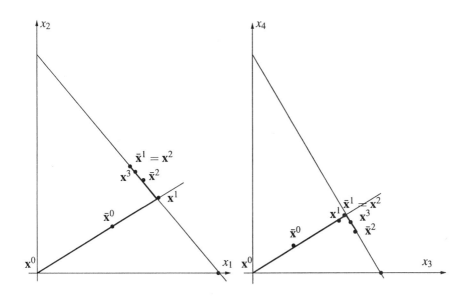

Figure 2.3 Movements of the partial linearization algorithm in the subproblem spaces

In Migdalas (1994), the concept of regularization of the Frank-Wolfe subproblems was introduced. To large extend, this regularized Frank-Wolfe algorithm is based on the observation that although the linearization of $f(\cdot)$ at $\mathbf{x}^k$ performed in (2.44) is mathematically good in some small neighborhood around $\mathbf{x}^k$, subproblems (2.44) lack anything that would enforce such a restriction. Thus, (Migdalas, 1994) introduces a regularization function $\phi : X \times X \to \mathbb{R}$ satisfying the following properties:

1. ϕ is continuously differentiable on $X \times X$,

2. ϕ is nonnegative and convex on $X \times X$,

3. $\phi(\mathbf{x}, \mathbf{y})$ is strictly convex for every fixed $\mathbf{y} \in X$,

4. $\phi(\mathbf{x}, \mathbf{y})$ is strictly convex for every fixed $\mathbf{x} \in X$, and

5. $\nabla \phi(\mathbf{x}, \mathbf{y}) = [\nabla_{\mathbf{x}} \phi(\mathbf{x}, \mathbf{y}), \nabla_{\mathbf{y}} \phi(\mathbf{x}, \mathbf{y})] = [\mathbf{0}, \mathbf{0}]$ if and only if $\mathbf{x} = \mathbf{y}$.

Because of the last property, such functions can be thought of as "distance" functions which, however, need not satisfy symmetry and/or triangle inequality. Examples of functions that satisfy the stated requirements include the proximal point function $\phi(\mathbf{x}, \mathbf{y}) = \frac{1}{2}\|\mathbf{x} - \mathbf{y}\|^2$, the projection function $\phi(\mathbf{x}, \mathbf{y}) = \frac{1}{2}(\mathbf{x} - \mathbf{y})^T \mathbf{D}(\mathbf{x} - \mathbf{y})$, where $\mathbf{D}$ is a positive diagonal matrix, and the entropy function

$$\phi(\mathbf{x}, \mathbf{y}) = \sum_{i=1}^{n} \{(x_i + \varepsilon) \ln\left(\frac{x_i + \varepsilon}{y_i + \varepsilon}\right) - (x_i - y_i)\},$$

where $\varepsilon > 0$ is a small constant.

In this regularized Frank-Wolfe approach, the subproblems (2.44) in the original algorithm are replaced by

$$\left. \begin{array}{l} \min \quad \nabla_i f(\mathbf{x}^k)^T \mathbf{x}_i + t_k \phi_i(\mathbf{x}_i, \mathbf{x}_i^k) \\ \text{s.t.} \quad \mathbf{x}_i \in X_i \end{array} \right\} \forall i \in \mathcal{N}, \quad (2.48)$$

where $t_k > 0$ is some positive constant and $\phi(\mathbf{x}, \mathbf{x}^k) = \sum_{i=1}^{n} \phi_i(\mathbf{x}_i, \mathbf{x}_i^k)$. The next iterate $\mathbf{x}^{k+1}$ is computed in the same manner as in the partial linearization algorithm. The algorithm shares with the partial linearization approach the property that the subproblems are able to generate the optimal solution of the original problem. Moreover, it can be shown that both algorithms converge globally to an optimal solution of **P1** or **CPP** under the conditions stated for the Frank-Wolfe algorithm. The proofs of convergence resemble that given above for the Frank-Wolfe algorithm (see Migdalas (1994) and Larsson and Migdalas (1990) for the details). It is shown in Migdalas (1994) that the regularized Frank-Wolfe algorithm unifies under the same umbrella several known nonlinear programming algorithms, such as the Goldstein-Levitin-Polyak projection algorithm (Levitin and Polyak, 1966; Bertsekas, 1976; 1982a), the Newton method (Bertsekas, 1982b;a; Klincewicz, 1983), and it also introduces several new methods, such as the partially linearized proximal point algorithm. Both partial linearization and regularized Frank-Wolfe are suitable for distributed computations (Patriksson, 1997), particularly as they convergence, under suitable assumptions (Karakitsiou et al., 2005), for fixed step lengths α_k. The subproblems in both algorithms can be solved in several different ways; approximate solutions based on the original Frank-Wolfe algorithm are discussed in Migdalas (1994), a Lagrangian relaxation approach along the lines of (2.9)-(2.10), which is in itself a fully distributed computation scheme (Bertsekas and Tsitsiklis, 1989), has been proposed in Larsson et al. (1993) and computational experience from its application to flow assignment problems are reported there, finally approaches similar to Klincewicz (1983) or the more recent Boland et al. (1991b), Boland et al. (1991a), and Boland et al. (1994) can be adapted to solve the subproblems.

In the sequential implementation of the Frank-Wolfe algorithm, the subproblems (2.44) must be solved sequentially under some prespecified order, typically by increasing i. It is thus obvious that under such a realization of the Frank-Wolfe algorithm, although new information is generated during the loop of subproblem solving, the algorithm fails to take advantage of it during the duration of the loop. Thus, it is suggested in Migdalas (2004), under the name of cyclic linearization, that new information generated during the Frank-Wolfe subproblem solving loop (2.44) should be utilized during the duration of the loop. Hence, the loop (2.44) in the original algorithm is replaced by the following scheme:

$$\left. \begin{array}{rcl} \bar{\mathbf{x}}_i^k & \in & \arg\min_{\mathbf{x}_i \in \mathcal{X}_i} \nabla_i f(\mathbf{x}_{i-}^{k+1}, \mathbf{x}_i^k, \mathbf{x}_{i+}^k)^T \mathbf{x}_i \\ \alpha_i^k & \in & \arg\min_{\alpha \in [0,1]} f(\mathbf{x}_{i-}^{k+1}, \mathbf{x}_i^k + \alpha(\bar{\mathbf{x}}_i^k - \mathbf{x}_i^k), \mathbf{x}_{i+}^k) \\ \mathbf{x}_i^{k+1} & = & \mathbf{x}_i^k + \alpha_i^k(\bar{\mathbf{x}}_i^k - \mathbf{x}_i^k) \end{array} \right\} \forall i \in \mathcal{N} \quad (2.49)$$

Finally, several other improvements have been proposed by several authors (Leblanc et al., 1985; Patriksson, 1994; Weintraub et al., 1985). The PARTAN or parallel tangents approach has been suggested in particular in order to provide an acceleration scheme for the Frank-Wolfe algorithm that overcomes to certain extend the zig-zagging behavior. Thus, given two consecutive Frank-Wolfe iterates, $\mathbf{x}^k$ and $\mathbf{x}^{k+1}$, a line search is performed along the descent direction $\bar{\mathbf{d}}^{k+1} = \mathbf{x}^{k+1} - \mathbf{x}^k$ and the new point $\bar{\mathbf{x}}^{k+1}$ obtained as $f(\bar{\mathbf{x}}^{k+1}) = f(\mathbf{x}^k + \alpha_k \bar{\mathbf{d}}^{k+1}) = \min_{\alpha \in [0, \bar{\alpha}_k]} f(\mathbf{x}^k + \alpha \bar{\mathbf{d}}^{k+1})$ replaces $\mathbf{x}^{k+1}$ in the next Frank-Wolfe iteration. It can be shown (Luenberger, 1984) that under certain conditions PARTAN has the behavior of the conjugate gradients method.

2.4.3 Simplicial Decomposition and Column Generation

Consider problem **P1**. Since $\mathcal{X}$ is a polyhedral set, by Caratheodory's Theorem 2.4, any point in $\mathcal{X}$ can be expressed as a convex combination of its extreme points plus a linear combination of its extreme directions. For simplicity of the exposition, let us assume that $\mathcal{X}$ is a polytope and only use the first part of Theorem 2.4, otherwise, $\mathcal{X}$ may be unbounded, in which case its extreme directions should be taken under consideration according to the second part of Theorem 2.4. Then **P1** can be restated in the equivalent form:

$$\textbf{(MP1)} \quad \min_\alpha \quad f(\sum_{k=1}^{K} \alpha_k \bar{\mathbf{x}}^k),$$

$$\text{s.t.} \quad \sum_{k=1}^{K} \alpha_k = 1,$$

$$\alpha_k \geq 0, \ k = 1, \ldots, K,$$

where $\bar{\mathbf{x}}^k$ are the extreme points of $\mathcal{X}$. If α^* is an optimal solution to **MP1**, the optimal solution to **P1** is calculated as $\mathbf{x}^* = \sum_{k=1}^{K} \alpha_k^* \bar{\mathbf{x}}^k$. Recall that since $\mathcal{X} \subset \mathbb{R}^n$, by Caratheodory's theorem, at most $n+1$ extreme points are needed to represent the optimum point $\mathbf{x}^*$.

If $\mathcal{X}$ is defined as the Cartesian product of polytopes, as in the case of the problem **CPP**, the above transformation still applies. However, another possibility is to apply

Caratheodory's theorem to each polytope separately. In such a case, **CPP** can be restated as follows:

$$(\textbf{MP2}) \quad \min_\alpha \quad f(\sum_{k=1}^{K_1} \alpha_k^1 \bar{x}_1^k, \ldots, \sum_{k=1}^{K_n} \alpha_k^n \bar{x}_n^k),$$

$$\text{s.t.} \quad \sum_{k=1}^{K_i} \alpha_k^i = 1, \ i = 1, \ldots, n,$$

$$\alpha_k^i \geq 0, \ k = 1, \ldots, K_i, \ i = 1, \ldots, n,$$

where $\bar{x}_i^k$ are the extreme points of the ith polytope X in the Cartesian product $X = X_1 \times \cdots \times X_n$. Clearly, **MP2** includes a convexity constraint for each polytope in the product and is therefore itself a minimization problem over a Cartesian product of simplices.

The simplicial decomposition approach is based on solving a restricted version of **MP1** or **MP2**, called the restricted master problem. The restriction concerns the number of extreme points used, that is, a number $\bar{K}$ of extreme points much lesser than the actual number K is used. If the produced solution is not optimal for the original problem then new extreme points are added and the new restricted problem is solved again. In order to generate the needed extreme point, the Frank-Wolfe subproblems, **FW-SUB**^{k} or (2.44), are solved in each iteration. Thus, simplicial decomposition can be thought of as an improved version of the original Frank-Wolfe algorithm, where the line search has been replaced by a restricted master of type **MP1** or **MP2**, or as a column generation scheme, where columns (i.e., extreme points) are generated as needed. The algorithm has been proposed and analyzed in its general setting as well as in traffic and flow assignment problems by several authors (Holloway, 1974; Cantor and Gerla, 1974; von Hohenbalken, 1977; Hearn et al., 1985; 1987; Larsson and Patriksson, 1992) and is shown to be finite by Hearn et al. (1985). However, the solution of the restricted master problems does require the application of, in general, infinitely converging algorithms, several of which have been proposed in the literature, including projected second-order methods such as Newton methods (Bertsekas, 1982b; Bertsekas and Gafni, 1983; Bertsekas et al., 1984), reduced gradient methods (Larsson and Patriksson, 1992), and modified Frank-Wolfe algorithms. In Damberg and Migdalas (1997a) and Karakitsiou et al. (2005), the restricted master problems are solved by applying a regularized Frank-Wolfe approach of Section 2.4.2. Computational results in Karakitsiou et al. (2005) show that the overall approach is about 50% more efficient when regularized Frank-Wolfe is used to solve the master problem in DSD instead of Wolfe's reduced gradient algorithm (Wolfe, 1970; Bazaraa and Shetty, 1979) employed by Larsson and Patriksson (1992).

The application of simplicial decomposition to the flow problems of Section 2.3 provides several different opportunities; the master problem can be of the form **MP1**, i.e., having only one convexity constraint incorporating all origin-destination pairs, or of the form **MP2** in which there is one convexity constraint for each origin-destination pair. The latter case is known as the disaggregate simplicial decomposition (DSD), a term coined by Larsson and Patriksson (1992). In between there are some other possibilities, such as to have one convexity constraint per origin or per destination, or different other levels of aggregation. It is observed in Larsson and Patriksson (1992) that the disaggregate case is the more efficient and offers several advantages in the case of traffic assignment problems in road networks. The extreme points generated by the Frank-Wolfe subproblems correspond again to route flows, that is the subproblems are still of the type "all-or-nothing assignment." Cantor and Gerla (1974) seems to be first in applying the simplicial decomposition algorithm to optimal routing in packet-switched networks. A distributed implementation of the algorithm is discussed in Damberg and Migdalas (1997a), while a corresponding sequential, very efficient implementation is presented by Karakitsiou et al. (2005).

Similar column generation approaches to the flow problems of Section 2.3 have been proposed in the case of traffic assignment in road networks already by Dafermos and Sparrow (1969); Leventhal et al. (1973). The method is based on solving a restricted version of the path flow formulations **FTAP** or **SFTAP**, that is, similarly to the simplicial decomposition approach, "all-or-nothing assignment" problems are solved in order to generate path flows to be included in (2.40)-(2.43). The differences from the simplicial approach are therefore basically conceptual rather than substantial (Damberg and Migdalas, 1997b). Efficient implementation of the method in this setting are offered by Bertsekas (1982b); Bertsekas and Gafni (1983); Bertsekas et al. (1984); Bertsekas and Tsitsiklis (1989), where the Goldstein-Levitin-Polyak gradient projection method (Bertsekas, 1976; Levitin and Polyak, 1966; Bertsekas, 1982a), and various of its extensions incorporating second order information are used in solving the restricted master problem, i.e., the restricted versions of **FTAP** and **SFTAP**.

2.4.4 Other solution techniques

Rosen's gradient projection algorithm (Rosen, 1960; Bazaraa and Shetty, 1979) has been suggested for solving the routing problem in data networks (Schwartz and Cheung, 1976; Schwartz, 1977). They present results for small networks in order to demonstrate the superiority of their approach over the Frank-Wolfe algorithm, however, it is questionable whether gradient projection in the form considered by the authors provides a viable alternative to Frank-Wolfe and Frank-Wolfe-like algorithms for large scale network. Indeed, no particular effort is placed in specializing the algorithm for the underlying network structure, only the Cartesian product is fully utilized.

, Cyclic decomposition, which can be interpreted in game theoretic terms as a fictitious play approach to the equilibrium, is particularly suited as a sequential approach to problems of type **CPP** and consequently to flow assignment problems. The method can be stated as follows for the **CPP**:

$$\mathbf{x}_i^{k+1} \in \arg\min_{\mathbf{x}_i \in \mathcal{X}} f(\mathbf{x}_{i-}^{k+1}, \mathbf{x}_i, \mathbf{x}_{i+}^k), \ \forall i \in \mathcal{N}, \qquad (2.50)$$

that is, starting with $\mathbf{x}^k$, the next point $\mathbf{x}^{k+1}$ is obtained by cycling through the factors in the Cartesian product. This block coordinate descent approach dates back to Dafermos and Sparrow (1969) and Murchland (1970). It does require some specific method to solve the single commodity flow problems in (2.50). Nguyen (1974) used the convex simplex method (Bazaraa and Shetty, 1979; Zangwill, 1967), specialized to the underlying network, and reported that the resulting overall approach is about

40% faster than the Frank-Wolfe algorithm. It is demonstrated in Migdalas (2004) that the subproblems in (2.50) need not be solved to optimality and can be solved rather inaccurately. A few Frank-Wolfe iterations are sufficient. If only one Frank-Wolfe iteration is used, then the cyclic linearization approach of Section 2.4.4 is obtained. It was suggested in Petersen (1975) that the cyclic order should be replaced by a so-called Gauss-Southwell order (Luenberger, 1984) according to which the subproblem in (2.50) is selected which is furthest from optimality and is then solved. In Petersen (1975), the subproblem to be selected is determined by solving auxiliary "all-or-nothing assignment" problems, then the selected problem is solved by piecewise linearization (Bazaraa and Shetty, 1979) of its objective function. It is shown in Migdalas (2004) that if the subproblems (2.50) are solved approximately using a few Frank-Wolfe iterations, then the subproblem to be solved next is identified without the need to solve auxiliary problems if the properties of the so-called gap function (Hearn, 1982), which is immediately available from the Frank-Wolfe subproblems, are appropriately utilized.

Embedding the regularized Frank-Wolfe approach into the disaggregate simplicial decomposition algorithm is suggested by Larsson et al. (1997) and computational results are presented for a number of large scale network problems including traffic assignment.

2.5 CONCLUSIONS

We have reviewed the role of nonlinear optimization in modeling and solving routing and network design problems in data and road networks with fixed topology and we have briefly mentioned new trends in telecommunications where nonlinear programming still plays a role.

We have, in particular, emphasized the role game theoretic and equilibrium models have played for decades in transportation science and indicated lessons to be learned from that field and to be brought into the field of telecommunications. We expect Wardrop's equilibrium principle, Nash games and Stackelberg bilevel optimization to play an increasingly important role in telecommunications as globalization, new services and new technology introduce ever increasing competitiveness between firms and more freedom of choice among the users of telecommunication systems.

Further research on routing strategies based on different equilibrium concepts is certain to be a fruitful direction. Elastic demand models provide a suitable instrument in order to motivate or discourage users in their choices of services through appropriate price setting and/or routing mechanisms. We strongly believe that bilevel models based on elastic demand equilibria should be the route to take for setting price and service levels. Network configurations, be it of fixed or ad-hoc topology, must take into consideration the users reaction as the Braess and other paradoxes studied in transportation science have reveled.

Acknowledgments

I thank Dr. Athanasia Karakitsiou for her help in preparing this chapter. Unpublished manuscripts and papers in the bibliography are available through CiteSeer, the Autonomous Citation Indexing and Scientific Literature Digital Library, at http://citeseer.ist.psu.edu.

Bibliography

E. Altman, R. El Azouzi, and V. Abramov. Non-cooperative routing in loss networks. *Performance Evaluation*, 49:43–55, 2002.

E. Altman, T. Başar, T. Jiménez, and N. Shikin. Competitive routing in networks with polynomial costs. Technical report, INRIA, B.P. 93, 06902 Sophia Antipolis Cedex, France, 2000.

M.S. Bazaraa and C.M. Shetty. *Nonlinear programming – Theory and algorithms*. John Wiley and Sons, New York, 1979.

N.G. Beans, F.P. Kelly, and P.G. Taylor. Braess's paradox in a loss network. *Journal of Applied Prob.*, 34:155–159, 1997.

M. Beckmann, C.B. McGuire, and C.B. Winsten. *Studies in Economics of Transportation*. Yale University Press, 1956.

K.P. Bennet. Global tree optimization: A non-greedy decision tree algorithm. *Computing Science and Statistics*, 26:156–160, 1994.

K.P. Bennet and O.L. Mangasarian. Robust linear programming discrimination of two linearly inseparable sets. *Optimization Methods and Software*, 1:23–34, 1992.

K.P. Bennet and O.L. Mangasarian. Bilinear separation of two sets in n-space. *Computational Optimization and Applications*, 2:207–227, 1993.

J. Berechman. Highway-capacity utilization and investment in transportation corridors. *Environment and Planning*, 16A:1475–1488, 1984.

F. Berggren. *Power control and adaptive resource allocation in DS-CDMA systems*. PhD thesis, Royal Institute of Technology, 2003.

D. Bernstein and S.A. Gabriel. Solving the nonadditive traffic equilibrium problem. Technical Report SOR-96-14, Statistics and Operations Research Series, Princeton University, 1996.

D.P. Bertsekas. On the Goldstein-Levitin-Polyak gradient projection method. *IEEE Trans. Automat. Control*, AC-21:174–184, 1976.

D.P. Bertsekas. *Constrained optimization and Lagrange multiplier methods*. Academic Press, New York, 1982a.

D.P. Bertsekas. Projected Newton methods for optimization problems with simple constraints. *SIAM Journal on Control and Optimization*, 20:221–246, 1982b.

D.P. Bertsekas and E.M. Gafni. Projected Newton methods and optimization of multicommodity flows. *IEEE Trans. Automat. Control*, AC-28:1090–1096, 1983.

D.P. Bertsekas, E.M. Gafni, and R.G. Gallager. Second derivative algorithms for minimum delay distributed routing in networks. *IEEE Trans. Comm.*, COM-32:911–919, 1984.

D.P. Bertsekas and R. Gallager. *Data networks*. Prentice-Hall, Englewood Cliffs, NJ, 1992.

D.P. Bertsekas and J.N. Tsitsiklis. *Parallel and distributed computation – Numerical methods*. Prentice-Hall, 1989.

D. Bienstock and O. Raskina. Aymptotic analysis of the flow deviation method for the maximum concurrent flow problem. *Mathematical Programming*, 91:479–492, 2002.

D. Bienstock and I. Saniee. ATM network design: Traffic models and optimization-based heuristics. Technical Report 98-20, DIMACS, 1998.

J.A. Blue and K.P. Bennett. Hybrid extreme point tabu search. Technical Report 240, Dept of Mathematical Sciences, Rensselaer Polytechnic Institute, Troy, NY, 1996.

N.L. Boland, A.T. Ernst, C.J. Goh, and A.I. Mees. A faster version of the ASG algorithm. *Applied Mathematics Letters*, 7:23–27, 1994.

N.L. Boland, C.J. Goh, and A.I. Mees. An algorithm for non-linear network programming: Implementation, results and comparisons. *Journal of Operational Research Society*, 42:979–992, 1991a.

N.L. Boland, C.J. Goh, and A.I. Mees. An algorithm for solving quadratic network flow problems. *Applied Mathematics Letters*, 4:61–64, 1991b.

M. Bonatti and A. Gaivoronski. Guaranteed approximation of Markov chains with applications to multiplexer engineering in ATM networks. *Annals of Operations Research*, 49:111–136, 1994.

D. Braess. Über ein paradoxen der werkehrsplannung. *Unternehmenforschung*, 12:256–268, 1968.

B. Calvert, W. Solomon, and I. Ziedins. Braess's paradox in a queueing network with state-dependent routing. *Journal of Applied Prob.*, 34:134–154, 1997.

M.D. Canon and C.D. Cullum. A tight upper bound on the rate of convergence of the Frank-Wolfe algorithm. *SIAM J. Control*, 6:509–516, 1968.

D.C. Cantor and M. Gerla. Optimal routing in a packet-switched computer network. *IEEE Transactions on Computers*, C-23:1062–1069, 1974.

J.-H. Chang and L. Tassiulas. Energy conserving routing in wireless ad-hoc networks. Technical report, Dept of Electrical and Computer Engineering, University of Maryland, College Park, MD 20742, 1997.

J.E. Cohen and C. Jeffries. Congestion resulting from increased capacity in single-server queueing networks. *IEEE/ACM Transactions on Networking*, 5:1220–1225, 1997.

S.C. Dafermos and A. Nagurney. On some traffic equilibrium theory paradoxes. *Transportation Research*, 18B:101–110, 1984.

S.C. Dafermos and F.T. Sparrow. The traffic assignment problem for a general network. *Journal of RNBS*, 73B:91–118, 1969.

C.F. Daganzo. On the traffic assignment problem with flow dependent costs – I. *Transportation Research*, 11:433–437, 1977a.

C.F. Daganzo. On the traffic assignment problem with flow dependent costs – II. *Transportation Research*, 11:439–441, 1977b.

O. Damberg and A. Migdalas. Distributed disaggregate simplicial decomposition – A parallel algorithm for traffic assignment. In D. Hearn et al., editor, *Network optimization*, number 450 in Lecture Notes in Economics and Mathematical Systems, pages 172–193. Springer-Verlag, 1997a.

O. Damberg and A. Migdalas. Parallel algorithms for network problems. In A. Migdalas, P.M. Pardalos, and S. Storøy, editors, *Parallel computing in optimization*, pages 183–238. Kluwer Academic Publishers, Dordrecht, 1997b.

J.C. Dunn. Rate of convergence of conditional gradient algorithms near singular and nonsingular extremals. *SIAM Journal on Control and Optimization*, 17:187–211, 1979.

S.P. Evans. Derivation and analysis of some models for combining trip distribution and assignment. *Transportation Research*, 10:35–57, 1976.

M. Frank and P. Wolfe. An algorithm for quadratic programming. *Naval Research Logistics Quarterly*, 3:95–110, 1956.

L. Fratta, M. Gerla, and L. Kleinrock. The flow deviation method. An approach to store-and-forward communication network design. *Networks*, 3:97–133, 1973.

S.A. Gabriel and D. Bernstein. The traffic equilibrium problem with nonadditive costs. *Transportation Science*, 31:337–348, 1997.

S.A. Gabriel and D. Bernstein. Nonadditive shortest paths. Technical report, New Jersey TIDE Center, New Jersey Institute of Technology, Newark, NJ, 1999.

N.H. Gartner. Analysis and control of transportation networks by Frank-Wolfe decomposition. In T. Sasaki and T. Yamaoka, editors, *Proceedings of the 7th International Symposium on Transportation and Traffic Flow*, pages 591–623, Tokyo, 1977.

N.H. Gartner. Optimal traffic assignment with elastic demands: A review. Part I: Analysis framework, Part II: Algorithmic approaches. *Transportation Science*, 14: 174–208, 1980.

A.M. Geoffrion. Elements of large-scale mathematical programming. *Management Science*, 16:652–691, 1970.

A.M. Geoffrion. Generalized Benders decomposition. *Journal of Optimization Theory and Applications*, 10:237–260, 1972.

M. Gerla. Routing and flow control. In F.F. Kuo, editor, *Protocols and techniques for data communication networks*, chapter 4, pages 122–173. Prentice-Hall, 1981.

M. Gerla and L. Kleinrock. Topological design of distributed computer networks. *IEEE Trans. Comm.*, COM-25:48–60, 1977.

C.R. Glassey. A quadratic network optimization model for equilibrium single commodity trade flows. *Mathematical Programming*, 14:98–107, 1978.

P. Gupta and P.R. Kumar. A system and traffic dependent adaptive routing algorithm for ad hoc networks. In *Proceedings of the 36th IEEE Conference on Decision and Control*, pages 2375–2380, San Diego, 1997.

S.L. Hakimi. Optimum distribution of switching centers in a communication network and some related graph theoretic problems. *Operations Research*, 13:462–475, 1965.

D.W. Hearn. The gap function of a convex program. *Operations Research Letter*, 1: 67–71, 1982.

D.W. Hearn, S. Lawphongpanich, and J.A. Ventura. Finiteness in restricted simplicial decomposition. *Operations Research Letters*, 4:125–130, 1985.

D.W. Hearn, S. Lawphongpanich, and J.A. Ventura. Restricted simplicial decomposition: Computation and extensions. *Mathematical Programming Study*, 31:99–118, 1987.

D.W. Hearn and J. Ribeira. Convergence of the Frank-Wolfe method for certain bounded variable traffic assignment problems. *Transportation Research*, 15B:437–442, 1981.

M.P. Helme and T.L. Magnanti. Designing satellite communication networks by zero-one quadratic programming. *Networks*, 19:427–450, 1989.

C.A. Holloway. An extension of the Frank and Wolfe method of feasible directions. *Mathematical Programming*, 6:14–27, 1974.

B. Yaged Jr. Minimum cost routing for static network models. *Networks*, 1:139–172, 1971.

H. Kameda, E. Altman, T. Kozawa, and Y. Hosokawa. Braess-like paradoxes in distributed computer systems, 2000. Submitted to IEEE Transactions on Automatic Control.

A. Karakitsiou, A. Mavrommati, and A. Migdalas. Efficient minimization over products of simplices and its application to nonlinear multicommodity network problems. *Operational Research International Journal*, 4(2):99–118, 2005.

F.P. Kelly. Charging and rate control for elastic traffic. *European Transactions on Telecommunications*, 8:33–37, 1997.

L. Kleinrock. *Communication nets: Stochastic message flow and delay*. McGraw-Hill, New York, 1964.

L. Kleinrock. *Queueing systems, Volume II: Computer applications*. John Wiley and Sons, New York, 1974.

J.G. Klincewicz. Newton method for convex separable network flow problems. *Networks*, 13:427–442, 1983.

W. Knödel. *Graphentheoretische Methoden und ihre Anwendungen*. Springer-Verlag, Berlin, 1969.

Y.A. Korillis, A.A. Lazar, and A. Orda. Capacity allocation under non-cooperative routing. *IEEE/ACM Transactions on Networking*, 5:309–173, 1997.

Y.A. Korillis, A.A. Lazar, and A. Orda. Avoiding the Braess paradox in non-cooperative networks. *Journal of Applied Probability*, 36:211–222, 1999.

M. Kourgiantakis, I. Mandalianos, A. Migdalas, and P. Pardalos. Optimization in e-commerce. In P.M. Pardalos and M.G.C. Resende, editors, *Handbook of Optimization in Telecommunications*. Springer, 2005. In this volume.

T. Larsson and A. Migdalas. An algorithm for nonlinear programs over Cartesian product sets. *Optimization*, 21:535–542, 1990.

T. Larsson, A. Migdalas, and M. Patriksson. The application of partial linearization algorithm to the traffic assignment problem. *Optimization*, 28:47–61, 1993.

T. Larsson and M. Patriksson. Simplicial decomposition with disaggregated representation for traffic assignment problem. *Transportation Science*, 26:445–462, 1992.

T. Larsson, M. Patriksson, and C. Rydergren. Applications of the simplicial decomposition with nonlinear column generation to nonlinear network flows. In D. Hearn et al., editor, *Network optimization*, volume 450 of *Lecture Notes in Economics and Mathematical Systems*, pages 346–373. Springer-Verlag, 1997.

L.J. Leblanc, R.V. Helgason, and D.E. Boyce. Improved efficiency of the Frank-Wolfe algorithm for convex network. *Transportation Science*, 19:445–462, 1985.

L.J. Leblanc, E.K. Morlok, and W.P. Pierskalla. An efficient approach to solving the road network equilibrium traffic assignment problem. *Transportation Research*, 9: 308–318, 1975.

T. Leventhal, G. Nemhauser, and L. Trotter Jr. A column generation algorithm for optimal traffic assignment. *Transportation Science*, 7:168–176, 1973.

E.S. Levitin and B.T. Polyak. Constrained minimization methods. *USSR Computational Mathematics and Mathematical Physics*, 6:1–50, 1966.

D.G. Luenberger. *Linear and nonlinear programming*. Addison-Wesley, Reading, Mass., second edition, 1984.

B. Martos. *Nonlinear programming: Theory and methods*. North-Holland, Amsterdam, 1975.

A. Migdalas. A regularization of the Frank-Wolfe method and unification of certain nonlinear programming methods. *Mathematical Programming*, 65:331–345, 1994.

A. Migdalas. Bilevel programming in traffic planning: Models, methods and challenge. *Journal of Global Optimization*, 7:381–405, 1995.

A. Migdalas. Cyclic linearization and decomposition of team game models. In S. Butenko, R. Murphey, and P. Pardalos, editors, *Recent developments in cooperative control and optimization*, pages 333–348. Kluwer Academic Publishers, Boston, 2004.

A. Migdalas, P.M. Pardalos, and P. Värbrand, editors. *Multilevel optimization – Algorithms and applications*. Kluwer Academic Publishers, 1997.

M. Minoux. Subgradient optimization and Benders decomposition for large scale programming. In R.W. Cottle et al., editor, *Mathematical programming*, pages 271–288. North-Holland, Amsterdam, 1984.

J.D. Murchland. Braess's paradox of traffic flow. *Transportation Research*, 4:391–394, 1970.

S. Nguyen. An algorithm for the traffic assignment problem. *Transportation Science*, 8:203–216, 1974.

M.E. O'Kelly, D. Skorin-Kapov, and J. Skorin-Kapov. Lower bounds for the hub location problem. *Management Science*, 41:713–721, 1995.

M. Patriksson. *The traffic assignment problem: Models and methods*. VSP, Utrecht, 1994.

M. Patriksson. Parallel cost approximation algorithms for differentiable optimization. In A. Migdalas, P.M. Pardalos, and S. Storøy, editors, *Parallel computing in optimization*, pages 295–341. Kluwer Academic Publishers, Dordrecht, 1997.

E.R. Petersen. A primal-dual traffic assignment algorithm. *Management Science*, 22: 87–95, 1975.

M. De Prycker. *Asynchronous transfer mode solution for broadband ISDN*. Prentice-Hall, Englewood Cliffs, NJ, 1995.

B.N. Pshenichny and Y.M. Danilin. *Numerical methods in extremal problems*. Mir Publishers, Moscow, 1978.

J.B. Rosen. The gradient projection method for nonlinear programming. Part I: Linear constraints. *SIAM Journal on Applied Mathematics*, 8:181–217, 1960.

T. Roughgarden. *Selfish routing*. PhD thesis, Cornell University, 2002.

T. Roughgarden and E. Tardos. How bad is selfish routing? *Journal of the ACM*, 49: 239–259, 2002.

M. Schwartz. *Computer-communication network design and analysis*. Prentice-Hall, Englewood Cliffs, NJ, 1977.

M. Schwartz and C.K. Cheung. The gradient projection algorithm for multiple routing in message-switched networks. *IEEE Trans. Comm.*, COM-24:449–456, 1976.

J. Seidler. *Principles of computer communication network design*. Ellis Horwood Ltd, Chichester, 1983.

J.F. Shapiro. *Mathematical programming – Structures and algorithms*. John Wiley and Sons, New York, 1979.

P.A. Steenbrink. *Optimization of transport networks*. John Wiley and Sons, London, 1974.

B. von Hohenbalken. Simplicial decomposition in nonlinear programming algorithms. *Mathematical Programming*, 13:49–68, 1977.

J.G. Wardrop. Some theoretical aspects of road traffic research. In *Proceedings of the Institute of Civil Engineers – Part II*, pages 325–378, 1952.

A. Weintraub, C. Ortiz, and J. Conzalez. Accelerating convergence of the Frank-Wolfe algorithm. *Transportation Research*, 19B:113–122, 1985.

P. Wolfe. Convergence theory in nonlinear programming. In J. Abadie, editor, *Integer and nonlinear programming*, pages 1–36. North-Holland, Amsterdam, 1970.

H. Yaiche, R. Mazumdar, and C. Rosenberg. A game theoretic framework for bandwidth allocation and pricing of elastic connections in broadband networks: Theory and algorithms. *IEEE/ACM Transaction on Networking*, 8:667–678, 2000.

N. Zadeh. On building minimum cost communication networks. *Networks*, 3:315–331, 1973.

W.I. Zangwill. The convex simplex method. *Management Science*, 14:221–283, 1967.

3 INTEGER PROGRAMMING FOR TELECOMMUNICATIONS

Eva K. Lee[1] and David P. Lewis[1]

[1] School of Industrial and Systems Engineering
Georgia Institute of Technology
Atlanta, GA 30332 USA
evakylee@isye.gatech.edu
dlewis@isye.gatech.edu

Abstract: This chapter presents an overview of integer programming in the field of telecommunications. Various integer programming models are described, and computational strategies for solving the integer programming instances are summarized. Techniques such as branching variable selection and node selection schemes are discussed; and the concepts of problem preprocessing and reformulation, heuristics, and continuous reduced-cost fixing are outlined. These latter techniques have been shown to be very effective when embedded within a branch-and-bound algorithm. The use of an interior point method as a subproblem solver is also described. Finally, Lagrangian relaxation in the context of solving specific telecommunication instances is analyzed as an alternative relaxation for use within the branch-and-bound tree search environment.

Keywords: Integer programming, preprocessing, heuristics, branching, reduced-cost fixing, branch-and-bound, interior point methods, Lagrangian relaxation, Benders' decomposition, column generation, branch-and-cut.

3.1 INTRODUCTION

Integer programs (IPs) and mixed integer programs (MIPs) provide a flexible and mathematically precise way of formulating many real-world problems. However, solving an IP or MIP instance to optimality can be computationally taxing, as general classes of these problems are NP-hard. In particular, large-scale instances and models exhibiting a high degree of symmetry are often computationally intractable. Even for a moderately-sized problem, care must be taken to formulate it in a way that lends it to be solved with sophisticated algorithmic methods. Indeed, given two distinct valid IP formulations of the same problem, one may be readily solvable with existing methods, while the other may require tremendous computational effort.

The purpose of this chapter is to review well-known IP formulations in the field of telecommunications and the corresponding methods used for solving them. Particular attention will be paid to those methods that enable large-scale instances to be solved. Also, the scope of this chapter will be restricted to methods that, though they may contain heuristic subroutines, are not heuristic themselves. We assume that the reader is familiar with linear programming concepts, including duality theory and complementary slackness, the simplex method and the dual ascent algorithm.

This chapter is organized as follows. The first section provides an overview of the four main MIP formulations that have been used extensively to model telecommunications problems. The next section cites examples of MIP problems formulated in each of these four ways and discusses the most common IP solution methods used for each type of formulation. Rather than providing an exhaustive reference list of telecommunications problems formulated and solved with integer programming methods, papers have been chosen that represent a cross-section of this wide field of study. The final section contains detailed descriptions about strategies for solving general integer programming problems; this section serves as a reference for readers unfamiliar with these techniques.

3.2 OVERVIEW OF INTEGER PROGRAMMING MODELS FOR TELECOMMUNICATION

Telecommunication problems to which integer programming has been applied can be roughly split into two types. The first consists of network synthesis problems, where the objective is to find an optimal allocation of capacity to edges of a complete graph such that all flow (transmission) demands, and any other requirements of the problem, are met. Examples of "other" requirements to be met include the following:

- specifying a particular topology of the edges with nonzero capacity,
- ensuring that transmission demands are met even in the event of the failure of any node or edge of the graph,
- preventing data transmissions of similar wavelengths to use the same edge (link) in a graph (network) so as to avoid crosstalk.

The second type of telecommunications problems consist of a wide variety of *assignment* problems. Examples of assignment problems that arise in telecommunications include:

- the clustering of nodes in a graph to determine an optimal design of synchronous optical network (SONET) rings,
- the assignment of frequencies to base stations to minimize crosstalk,
- and the assignment of customers to nearby facilities.

These two types of telecommunications problems have a simple distinction in their IP models. An IP model of a network synthesis problem must contain a set of constraints that conforms any feasible solution of the problem to the structure of the graph upon which the network is being built. We refer to the monograph by Wynants (2001) for three general formulations of network synthesis problems and include these formulations here after introducing some notation.

Let $G = (N,E)$ be an undirected connected graph with a set of nodes N and edges E. Let K be the set of all possible pairs of nodes of the graph G; that is, $K = \{(i,j) : i, j \in N\}$. If $k \in K$, denote the corresponding node pair as (o^k, d^k). For generality, let Γ be an index set of states that the graph G may be in. Each state $\gamma \in \Gamma$ is characterized by a subgraph $G^\gamma = (N^\gamma, E^\gamma)$ of G and a set of multicommodity flow requirements $K^\gamma = \{k \in K : r^{k\gamma} > 0\}$ to be routed on G^γ. Here, $r^{k\gamma}$ is the amount to be routed between o^k and d^k.

The node-arc formulation expresses the flow balance constraints of the network in terms of a node-arc incidence matrix. An IP model that contains such a constraint set is said to have a *network substructure*:

$$\text{minimize} \quad \sum_{\{i,j\} \in E} c_{ij} y_{ij}$$

$$\text{subject to} \quad \sum_{j:\{i,j\} \in E^\gamma} (f_{ij}^{k\gamma} - f_{ji}^{k\gamma}) = \begin{cases} r^{k\gamma} & \text{if } i = o^k \\ -r^{k\gamma} & \text{if } i = d^k \\ 0 & \text{otherwise} \end{cases} \quad \forall \gamma \in \Gamma, k \in K^\gamma, i \in N^\gamma$$

$$\sum_{k \in K^\gamma} (f_{ij}^{k\gamma} + f_{ji}^{k\gamma}) \leq y_{ij} \quad \forall \gamma \in \Gamma, \{i,j\} \in E^\gamma$$

$$f_{ij}^{k\gamma}, f_{ji}^{k\gamma} \geq 0 \text{ integer} \quad \forall \gamma \in \Gamma, k \in K^\gamma, \{i,j\} \in E^\gamma$$

$$y_{ij} \geq 0 \text{ integer} \quad \forall \{i,j\} \in E^\gamma.$$

Here, $f_{ij}^{k\gamma}$ is the amount of flow of commodity k routed on edge $\{i,j\}$ in E^γ when the graph is in state γ, y_{ij} is the amount of capacity assigned to edge i, j, and c_{ij} is the cost per unit capacity installed on edge $\{i,j\}$. Identifying network substructures in an IP formulation is important because special-purpose algorithms exist that are more efficient than the simplex method to solve these kinds of IP models.

Rather than expressing the graph G by its node-arc incidence matrix, a variable can be created for every path between any two nodes of G. Let $P^{k,\gamma}$ be the set of paths p from o^k to d^k in the subgraph G^γ. The edge-path formulation is as follows:

$$\text{minimize} \quad \sum_{e \in E} c_e y_e$$

$$\text{subject to}$$

$$\sum_{p \in P^{k,\gamma}} f_p = r^{k\gamma} \quad \forall \gamma \in \Gamma, k \in K^\gamma$$

$$\sum_{k \in K^\gamma} \sum_{p \in P^{k,\gamma}, p \ni e} f_p \leq y_e \quad \forall \gamma \in \Gamma, e \in E$$

$$f_p \geq 0 \text{ integer} \quad \forall \gamma \in \Gamma, k \in K^\gamma, p \in P^{k,\gamma}$$

$$y_e \geq 0 \text{ integer} \quad \forall e \in E$$

Note that the number of variables is exponential in the size of the graph for this model.

The third alternative is a cutset formulation, which ensures that the total capacity installed on the edges of any $o^\gamma - d^\gamma$ cut of G^γ must be at least as large as the flow requirement r^γ:

minimize $\sum_{e \in E} c_e y_e$
subject to
$$\sum_{e \in s} y_e \geq r^\gamma \quad \forall \gamma \in \Gamma, \text{ for all } o^\gamma - d^\gamma \text{ cuts } s \text{ of } G^\gamma$$
$$y_e \geq 0 \text{ integer} \quad \forall e \in E$$

The above model can only be used when $|K| = 1$, since the max-flow min-cut theorem does not hold for multicommodity flow problems.

Those telecommunications problems that are assignment problems can either be expressed as, or can be shown to contain, the following general assignment problem formulation (Nemhauser and Wolsey, 1988):

maximize $\sum_{i,j} c_{ij} x_{ij}$
subject to
$$\sum_j x_{ij} \leq 1 \quad \forall i \in M,$$
$$\sum_i l_i x_{ij} \leq b_j \quad \forall j \in N,$$
$$x \in B^{mn}$$

Here, M and N are index sets of items, x_{ij} is an indicator variable that is 1 when item $j \in N$ is assigned to item $i \in M$, c_{ij} is the cost of assigning item $j \in N$ to item $i \in M$, l_i is the weight of item $i \in M$, and b_j is the total amount of weight that can be assigned to item $j \in N$.

3.3 EXAMPLES AND SOLUTION METHODS

3.3.1 IP Models with Network Substructure

As stated earlier, the presence of network substructure in an IP formulation can be exploited to solve the problem to optimality more efficiently. Lagrangian relaxation is a popular method for solving problems with a network substructure (see Section 3.4.8 for an explanation of Lagrangian relaxation). Rosenwein and Wong (1995) use a Lagrangian relaxation approach to decompose an IP that constructs a constrained Steiner tree into two subproblems. The original IP consists of flow conservation constraints in node-arc form and a knapsack constraint on the capacity variables that ensures the edges with nonzero capacity do not exceed a resource constraint. This IP is decomposed into a standard node-arc network synthesis subproblem (without the knapsack constraint) and a second subproblem with just the knapsack constraint. This is accomplished by creating duplicates of each capacity variable that are set equal to the original variables, substituting these duplicate variables into the knapsack constraint, and dualizing the equality constraints that couple the two types of capacity variables. Though the network substructure is successfully isolated using this approach, empirical results show that it does not have a distinct advantage over a Lagrangian relaxation of the knapsack constraint itself for this particular problem.

Balakrishnan and Altinkemer (1992), Kawatra (2002), Kawatra and Bricker (2000; 2004) consider a different Lagrangian decomposition of IP models similarly structured to that of Rosenwein et al. Each chooses to separate the original IP into subproblems such that each subproblem contains only flow variables or only capacity variables. Such decompositions can be accomplished by dualizing all constraints that couple two different kinds of variables together. The resulting subproblems are shortest path problems (solved by Dijkstra's algorithm) and minimum weight arborescence problems (solved greedily). The solution to the Lagrangian decomposed problem serve as a lower bound to the optimal solution of the original. Upper bounds are created with heuristics that construct a feasible solution to the original IP based on the Lagrangian solution after each subgradient optimization step to renew the Lagrangian multipliers. After a terminating condition on the number of iterations of the subgradient problem is met, the size of the duality gap produced between the Lagrangian solutions and the heuristic upper bounds is recorded. The size of this gap is found to have a dependence on problem-specific parameter values, (e.g. a direct dependence on the number of nodes in the graph.)

Amiri and Pirkul (1996) also use a similar decomposition strategy (dualizing coupling constraints) to find two edge-disjoint paths between any communicating node pairs. This results in many subproblems: $(2+|E-1|)|E|$ continuous knapsack problems that are greedily solvable and $2|M|$ shortest path problems solvable with Dijkstra's algorithm. Here, $|E|$ is the number of edges in the graph and $|M|$ is the number of communicating node pairs.

An alternative to dualizing coupling constraints would be to dualize the flow conservation (node-arc) constraints themselves and solve the resulting Lagrangian relaxed problem. Often the Lagrangian relaxed problem is not integral, and thus the lower bound it produces is at least as strong as that of the LP relaxation of the original problem. These stronger bounds can then be used in a branch-and-bound context to keep the size of the branch-and-bound tree small. Magnanti et al. (1995), and Holmberg and Yuan (1998; 2000) decompose their IP models in such a fashion, yielding a fixed number of knapsack problems for each edge of the graph. Magnanti et al. go on to compare, both theoretically and computationally, the strengths of the lower bounds produced by this method to that of a branch-and-cut procedure, which depends on the parameters of the problem itself. In general, the Lagrangian approach proves to be a better method in a computational sense for large-scale instances. Holmberg and Yuan make the point in Holmberg and Yuan (2000) that dualizing the flow constraints may yield subproblems that are even easier to compute than those produced by dualizing coupling constraints. This is an important consideration in the event that dualizing coupling constraints yields an integral IP with a lower bound no better than the one that the LP relaxation produces.

In general, it is not straightforward to determine, without some empirical evidence, which set of constraints to dualize so as to achieve the maximum computational benefit. Gouveia (1996) and Randazzo and Luna (2001) consider both the dualized flow constraints and the dualized coupling constraints for several instances of their problems and weigh their performances.

Randazzo and Luna also consider another decomposition method, Benders decomposition, to be used in conjunction with their Lagrangian relaxation with dualized flow constraints. For clarity, it should be mentioned that the node-arc formulation embedded in the formulation of Randazzo and Luna differs slightly from the one shown earlier. Rather than integer capacity variables, y_{ij}, they use binary topological variables x_{ij} with a coupling constraint of the form

$$f_{ij}^{k\gamma} \leq d_{k\gamma} x_{ij} \qquad \forall \gamma \in \Gamma, \{i,j\} \in E^\gamma.$$

It is these topological variables that are fixed to some arborescence for the master problem. This method is shown to be robust for problem instances that have a gap between the optimal values of their LP relaxed and integer problems.

Benders decomposition has also been used by Sridhar and Park (2000) as an alternative to Lagrangian relaxation, since choosing to dualize any constraint set of their model yielded lower bounds no better than the LP relaxation. Benders cuts prove to be very effective to improve the LP bound of problem instances with large amounts of traffic. Finally, the use of Benders decomposition is well established for stochastic integer programming as well, such as Riis and Andersen's capacitated network problem with uncertain demand (Riis and Andersen, 2002).

Though it does not exploit node-arc structure directly, some authors have identified valid and strong valid inequalities to either strengthen an LP relaxed problem (Balakrishnan et al., 1994; Grover and Doucette, 2001; Jaillet et al., 1996) or to generate them as needed within a branch-and-cut procedure (Balakrishnan et al., 2002; Lee et al., 2003a; van Hoesel et al., 2002) (see Section 3.4.11 for an explanation of branch-and-cut).

Many other models are too complex to solve in a reasonable amount of time with a solution sufficiently close to optimality with the simplex method alone. Typically, these MIP formulations are intractable because they are designed to solve two or more complicated problems at once in an integrated fashion (e.g. installing capacity on edges while routing and assigning wavelengths simultaneously). These problems are often solved with standard heuristics (Tabu search) or heuristics based on graph-theoretic properties and LP duality (Bianco et al., 2003; Cao et al., 2003; Frantzeskaskis and Luss, 1999; Gençata and Mukherjee, 2003; Ho et al., 2003; Holmberg and Yuan, 2004; Kiran and Murthy, 2004; Kumar and Kumar, 2002; Lee et al., 2003b; Orincsay et al., 2003; Puech et al., 2002; Ramaswami and Sivarajan, 1996; Yan et al., 2003). Another approach is to simply split the complex MIP formulation into its component problems and solve each of these to optimality or near-optimality sequentially (Laborczi and Cinkler, 2002; Zhang et al., 1992).

3.3.2 IP Models with Edge-Path Formulations

Because the number of variables in this kind of formulation is exponential in the size of the graph, a column generation approach is necessary to solve large-scale instances of these models to optimality (see Section 3.4.10 for an explanation of column generation.) Holmberg and Yuan (2003) use column generation to solve a multicommodity network-flow problem with side constraints on the paths. Their column generation subproblem, a constrained shortest path problem, is solved in polynomial time with a multilabeling algorithm. The columns that are generated have the desirable characteristic of satisfying the side constraints of the original formulation, thereby simplifying the structure of the master problem.

Incorporating column generation into a branch-and-bound scheme so that columns may be generated at any node of the tree is a process called branch and price. (See Section 3.4.10 for an explanation of branch and price.) Lee et al. (2000b), Barnhart et al. (2000), and Kang et al. (2004) each solve a shortest path problem to generate columns that price out favorably. To ensure that new columns are priced out properly at the child nodes of the branch-and-bound tree, both Lee et al. and Kang et al. solve a k-shortest path problem where the value of k is incremented every time the optimal path returned by the k-shortest path problem corresponds to a path variable that has been fixed to zero. Neglecting to do this would mean that a column would be priced out that is already present in the reduced model. Barnhart et al., on the other hand, propose a different branching strategy that causes the shortest path problem to price out columns successfully throughout. This is accomplished by branching on sets of edge variables present in their model (not path variables) in such a way that the structure of the pricing problem is not destroyed.

Branch-and-cut methods have been used on several variations of problems (Alevras et al., 1998a;b; Myung and Kim, 2004; Riis and Lodahl, 2002) where capacity is allocated to links to make a survivable network. To increase the effectiveness of their algorithms, Barnhart et al. and Kang et al. also introduce cutting planes to the nodes of their branch and bound tree while generating columns (a *branch and price and cut* procedure). Generally, this is difficult to implement when the cutting planes that are being added contain columns that are being generated by the pricing subproblem. Kang et al. avoid this by using cutting planes that do not contain path variables. Barnhart et al. address this issue by converting lifted cover inequalities composed of edge variables to equivalent inequalities with path variables. Their pricing subproblem is altered accordingly for each inequality added by adjusting the cost of one of the arcs in the underlying graph.

When column generation is not considered, many practitioners turn to heuristic methods (Fumagalli et al., 2003; Hu, 2002; Iraschko and Grover, 2000; Jæger and Tipper, 2003; Lee et al., 2000a; 2002; Li et al., 2003; Luo and Ansari, 2002; Sridharan et al., 2002; Zhu and Mukherjee, 2002) or Lagrangian decomposition (Amiri, 1998; Balakrishnan et al., 1995; Dutta, 1994) while others solve such problems on very small graphs (usually six nodes) to optimality or near-optimality with standard MIP solvers (Doucette and Grover, 2000; Iraschko et al., 1998; Lo and Chuang, 2003; MacGregor et al., 1997; Miyao and Saito, 1998; Tong et al., 2000; Van Caenegem et al., 1998; Xiong, 1998; Zheng et al., 1997). One promising result that has not yet been adapted to work without knowledge of all possible origin-destination paths is in a paper by Ozdaglar and Bertsekas (2003). They show a formulation whose LP relaxation often returns integral optimal solutions in practice. The key to this desirable property is a link capacity installation cost function that is piecewise linear with breakpoints at integer values.

3.3.3 *IP Models with Cutset Formulations*

Because there are exponentially many cutset inequalities in the constraint set of the model, it is natural to consider a branch and cut approach for problems with these formulations. First, an LP relaxation of the model without the cutset inequalities can be solved for a lower bound. If the optimal solution to the relaxed problem is found to violate any of the cutset inequalities, they can be added to the relaxed problem and solved again. If no violated cutset inequalities are found and the optimal solution is fractional, then a fractional variable or set of fractional variables can be fixed to integral values in a branching scheme and the process continues on the child nodes of the tree.

Myung et al. (1999) calculate lower bounds for their problem in a similar manner as described above. They choose to solve a simpler form of the model — one that includes only those cutset inequalities with exactly one node — to optimality. The minimum cuts of the resulting solution are then enumerated and checked against all applicable cutset inequalities. The lower bound is further strengthened by the addition of violated inequalities that are redundant for the integer problem. These inequalities are added using the same heuristic that Grötschel et al. (1992) use to find violated partition inequalities.

Cutset formulations prove to be useful for, but not limited to, models that are designed to create survivable telecommunications networks. "Survivable" means that transmission demands of the network are met even in the event of the failure of any edge or node. By extending the class of cutset inequalities to include cutsets of graphs with some number of edges or nodes missing, a solution is feasible only when it satisfies the transmission demands under all possible conditions of network failure considered.

As expressed in the formulation in the previous section, the cutset inequalities are formed by considering all $o^\gamma - d^\gamma$ cuts of G^γ. These inequalities are actually a special case of a more general class of inequalities called *metric inequalities* (Bienstock et al., 1998). Several authors (Alevras et al., 1998a; Brunetta et al., 2000; Dahl et al., 1999; Fortz et al., 2000; Gabrel et al., 1999; Grötschel et al., 1995; Lee et al., 1998) make their LP relaxations tighter by introducing other metric inequalities as cutting planes (e.g., partition inequalities with three or more partite sets) (Kerivin and Mahjoub, 2002).

It is a common occurrence in telecommunication network design for the objective function to be a discontinuous step increasing cost function. As a result, some authors (Alevras et al., 1998a; Gabrel et al., 1999; Stoer and Dahl, 1994) design their models specifically for this case. Let T_e be the number of breakpoints in the cost function, m_e^t be the incremental capacity of edge $e \in E$ for each breakpoint t and k_e^t be the incremental cost of edge e for each breakpoint t, $0 \leq t \leq T_e$. Such an objective function can be accounted for by using the following constraint set in the formulation:

$$\text{minimize} \quad \sum_{e \in E} \sum_{t=1}^{T_e} k_e^t x_e^t$$

subject to

$$0 \leq x_e^{T_e} \leq \ldots \leq x_e^1 \leq x_e^0 = 1 \quad \forall e \in E$$
$$x_e^t \in \{0, 1\} \quad \forall e \in E, \forall t = 1, \ldots, T_e$$
$$y_e = \sum_{t=0}^{T_e} m_e^t x_e^t \quad \forall e \in E$$
$$\sum_{e \in s} y_e \geq r^\gamma \quad \forall \gamma \in \Gamma, \forall o^\gamma - d^\gamma \text{ cuts } s \text{ of } G^\gamma$$
$$y_e \geq 0 \text{ integer} \quad \forall e \in E.$$

We refer to the first set of constraints as the ordering constraints. Stoer and Dahl (1994) define a class of inequalities called band inequalities to strengthen the ordering constraints in the LP relaxation. These constraints are shown to be facets of the integer hull under certain conditions. Both Stoer and Dahl and Alevras et al. (1998a) use this class of valid inequalities (among others) on a single problem instance. Gabrel et al.

(1999) show an alternate way of approaching the problem that may work better on a wider variety of problem instances. They devise heuristic methods for generating multiple violated metric inequalities quickly. If their procedure happens to not return any violated inequalities, a single most violated inequality is computed using a less efficient subgradient algorithm.

Finally, we cite a paper by Prytz and Forsgren (2002) to illustrate an example where decomposition using Lagrangian relaxation is effective. They use a technique similar to the one used by Rosenwein et al. (described in the node-arc subsection). A copy y_e of each of the x_e variables is made by introducing an equality constraint to couple each of the $|E|$ y_e variables with the x_e variables. The y_e variables are then used as a substitute in a complicating knapsack constraint. By dualizing the new equality constraint that coupled the x_e and y_e variables, the problem decomposes to a set of Steiner tree subproblems in terms of the x_e variables and capacity-level subproblems in terms of the y_e variables. Unlike the case of Rosenwein et al.'s model, this method of Lagrangian decomposition (through copying variables) actually yields tighter lower bounds than both the LP relaxation or straightforward Lagrangian relaxation for this particular model.

3.3.4 IP Models with Assignment Formulations

Many kinds of assignment problems arise in telecommunications such as channel assignment, cell assignment to switches or location registration areas, traffic assignment of satellite networks to timeslots in a schedule, and the optimal placement of facilities, add/drop multiplexers, and/or base stations to create a network with a desired topological architecture or performance guarantee. As this formulation is quite general, there is no particular solution technique that dominates this category of problems. The efficacy of any technique is strongly dependent on the complicating constraints of the particular model being considered and the data values of the problem instances. Examples in the literature include Lagrangian relaxation for tighter bounds (Chardaire et al., 1999; Lo and Kershenbaum, 1989; Mazzini et al., 2003; Tragantalerngsak et al., 2000) as well as its use for decomposition into smaller subproblems (Barbas and Marin, 2004; Holmberg et al., 1999; Sherali et al., 2000), as well as branch-and-bound approaches (Belvaux et al., 1998; Giortzis et al., 2000; Mathar and Schmeink, 2001) with cuts (Feremans et al., 2004; Gabrel et al., 1999; Hardin et al., 1998; Melkote and Daskin, 2001; Montemanni et al., 2004; Park et al., 1996; Sherali and Park, 2000; Smith, 2004), and/or column generation (Lee and Park, 2001; Park et al., 1996; Sutter et al., 1998). For large-scale problem instances, it may be useful to employ dual ascent techniques rather than the more computationally intensive simplex method to solve any linearly relaxed subproblems that may arise (Kim et al., 1995; Klincewicz, 1996; Kumar et al., 2002; Sung and Jin, 2001).

3.4 SOLUTION METHODS FOR INTEGER PROGRAMS

In previous sections we briefly mentioned the concepts of branch-and-bound and branch-and-cut, and well as other computational methods for solving telecommunications problems. In this section we formally describe computational strategies that are broadly applicable to solving integer programming problems. Specific techniques pertinent to solving telecommunication instances are highlighted when appropriate.

Recall that an integer programming problem (IP) is an optimization problem in which some or all of the variables are restricted to take on only integer values. The exposition presented in this section will focus on the case in which the objective and constraints of the optimization problem are defined via linear functions. In addition, for simplicity, it will be assumed that all of the variables are restricted to be non-negative integer valued. Thus, the mathematical formulation of the problem under consideration can be stated as:

$$\begin{aligned} \text{maximize} \quad & c^T x \\ \text{subject to} \quad & Ax \leq b \\ & x \in Z_+^n \end{aligned} \quad \text{(IP)}$$

where $A \in \Re^{m \times n}$, $b \in \Re^m$ and $c \in \Re^n$. For notational convenience, let S denote the constraint set of problem (IP); i.e.,

$$S := \{x \in Z_+^n : Ax \leq b\}.$$

The classical approach to solving integer programs is branch-and-bound (Land and Doig, 1960). The branch-and-bound method is based on the idea of iteratively partitioning the set S (branching) to form subproblems of the original integer program. Each subproblem is solved — either exactly or approximately — to obtain an upper bound on the subproblem objective value. The driving force behind the branch-and-bound approach lies in the fact that if an upper bound for the objective value of a given subproblem is less than the objective value of a known integer feasible solution (e.g., obtained by solving some other subproblem) then the optimal solution of the original integer program cannot lie in the subset of S associated with the given subproblem. Hence, the upper bounds on subproblem objective values are, in essence, used to construct a proof of optimality without exhaustive search.

One concept that is fundamental to obtaining upper bounds on subproblem objective values is that of problem relaxation. A *relaxation* of the optimization problem

$$\max\{c^T x : x \in S\}$$

is an optimization problem

$$\max\{r(x) : x \in S_R\},$$

where $S \subseteq S_R$ and $c^T x \leq r(x)$ for all $x \in S$. Clearly, solving a problem relaxation provides an upper bound on the objective value of the underlying problem. Perhaps the most common relaxation of problem (IP) is the linear programming relaxation formed by relaxing the integer restrictions and enforcing appropriate bound conditions on the variables; i.e., $r(x) = c^T x$ and $S_R = \{x \in \Re_+^n : Ax \leq b, \ l \leq x \leq u\}$, where l and u are the lower and upper integer bounds for x.

A formal statement of a general branch-and-bound algorithm (Nemhauser and Wolsey, 1988) is presented below. The notation L is used to denote the list of active subproblems $\{\text{IP}^i\}$, where $\text{IP}^0 = \text{IP}$ denotes the original integer program. The notation $\bar{z}_i$ denotes an upper bound on the optimal objective value of IP^i, and $\underline{z}_{ip}$ denotes the incumbent objective value (i.e., the objective value corresponding to the current best integral feasible solution to IP).

General Branch-and-Bound Algorithm

1. *(Initialization)*: Set $L = \{\text{IP}^0\}$, $\bar{z}_0 = +\infty$, and $\underline{z}_{ip} = -\infty$.

2. *(Termination)*: If $L = \emptyset$, then the solution x^* which yielded the incumbent objective value $\underline{z}_{ip}$ is optimal. If no such x^* exists (i.e., $\underline{z}_{ip} = -\infty$) then IP is infeasible.

3. *(Problem selection and relaxation)*: Select and delete a problem IP^i from L. Solve a relaxation of IP^i. Let z_i^R denote the optimal objective value of the relaxation, and let x^{iR} be an optimal solution if one exists. (Thus, $z_i^R = c^T x^{iR}$, or $z_i^R = -\infty$.)

4. *(Fathoming and Pruning)*:

    (a) If $z_i^R \leq \underline{z}_{ip}$ go to Step 2.

    (b) If $z_i^R > \underline{z}_{ip}$ and x^{iR} is integral feasible, update $\underline{z}_{ip} = z_i^R$. Delete from L all problems with $\bar{z}_i \leq \underline{z}_{ip}$. Go to Step 2.

5. *(Partitioning)*: Let $\{S^{ij}\}_{j=1}^{j=k}$ be a partition of the constraint set S^i of problem IP^i. Add problems $\{\text{IP}^{ij}\}_{j=1}^{j=k}$ to L, where IP^{ij} is IP^i with feasible region restricted to S^{ij} and $\bar{z}_{ij} = z_i^R$ for $j = 1, \ldots, k$. Go to Step 2.

The actual implementation of a branch-and-bound algorithm is typically viewed as a tree search, where the problem at the root node of the tree is the original IP. The tree is constructed in an iterative fashion with new nodes formed by branching on an existing node for which the optimal solution of the relaxation is fractional (i.e., some of the integer restricted variables have fractional values). Typically, two child nodes are formed by selecting a fractional valued variable and adding appropriate constraints in each child subproblem to ensure that the associated constraint sets do not include solutions for which this chosen branching variable assumes the same fractional value.

The phrase "fathoming a node" is used in reference to criteria that imply that a node need not be explored further. As indicated in Step 4, these criteria include:

(a) the objective value of the subproblem relaxation at the node is less than or equal to the incumbent objective value; and

(b) the solution for the subproblem relaxation is integer valued.

Note that (a) includes the case when the relaxation is infeasible, since in that case its objective value is $-\infty$. Condition (b) provides an opportunity to prune the tree; effectively fathoming nodes for which the objective value of the relaxation is less than or equal to the updated incumbent objective value. The tree search ends when all nodes are fathomed.

Since the branch-and-bound algorithm is an enumeration procedure, the algorithm terminates with a proven optimal solution for (IP), if one exists, or concludes that the integer programming problem is infeasible. A variety of strategies have been proposed for intelligently selecting branching variables, for problem partitioning, and for selecting nodes to process. However, no single collection of strategies stands out as being best in all cases. In the remainder of this chapter, some of the strategies that have been implemented or proposed are summarized. Some of the related computational strategies – preprocessing and reformulation, heuristic procedures, and the concept of reduced-cost fixing – which have proven to be highly effective in branch-and-bound implementations are considered. In addition, there is a discussion of recent linear programming based branch-and-bound algorithms that have employed interior-point methods for the subproblem relaxation solver, which is in contrast to using the more traditional simplex-based solvers. Besides the linear programming relaxation, the notion of Lagrangian relaxation is also presented.

Though branch-and-bound is a classic approach for solving integer programs, there are practical limitations to its success in applications. Often integer feasible solutions are not readily available, and node pruning becomes impossible. In this case, branch-and-bound fails to find an optimal solution due to memory explosion as a result of excessive accumulation of active nodes. In fact, general integer programs are NP-hard; and consequently, as of this writing, there exists no known polynomial-time algorithm for solving general integer programs (Garey and Johnson, 1979).

Most commercial integer programming solvers use a branch-and-bound algorithm with linear programming relaxations. Unless otherwise mentioned, the descriptions of the strategies discussed herein are based on using the linear programming relaxation.

Readers are referred to the books by Schrijver (1986) and Nemhauser and Wolsey (1988) for detailed theoretical and algorithmic expositions on integer programming and the branch-and-bound algorithm, and for further bibliographic references. The text by Parker and Rardin (1988) also includes useful material about branch-and-bound.

3.4.1 Partitioning Strategies

When linear programming relaxation is employed, partitioning is done via the addition of linear constraints. Typically, two new nodes are formed on each division. Suppose x^R is an optimal solution to the relaxation of a branch-and-bound node. Common partitioning strategies include:

- *Variable Dichotomy* (Dakin, 1965). If x_j^R is fractional, then two new nodes are created, one with the simple bound $x_j \leq \lfloor x_j^R \rfloor$ and the other with $x_j \geq \lceil x_j^R \rceil$; where $\lfloor \cdot \rfloor$ and $\lceil \cdot \rceil$ denote the floor and the ceiling of a real number. In particular, if x_j is restricted to be binary, then the branching reduces to fixing $x_j = 0$ and $x_j = 1$, respectively. One advantage of simple bounds is that they maintain the size of the basis among branch-and-bound nodes, since the simplex method can be implemented to handle both upper and lower bounds on variables without explicitly increasing the dimensions of the basis.

- *Generalized-Upper-Bound (GUB) Dichotomy* (Beale and Tomlin, 1970). If the constraint $\sum_{j \in Q} x_j = 1$ is present in the original integer program, and x_i^R, $i \in Q$, are fractional, one can partition $Q = Q_1 \cup Q_2$ such that $\sum_{j \in Q_1} x_j^R$ and $\sum_{j \in Q_2} x_j^R$

are approximately of equal value. Then two branches can be formed by setting $\sum_{j \in Q_1} x_j = 0$ and $\sum_{j \in Q_2} x_j = 0$, respectively.

- *Multiple branches for bounded integer variable.* If x_j^R is fractional, and $x_j \in \{0, \ldots, l\}$, then one can create $l+1$ new nodes, with $x_j = k$ for node k, $k = 0, \ldots, l$. This idea was proposed in the first branch-and-bound algorithm by Land and Doig (1960), but currently is not commonly used.

3.4.2 Branching Variable Selection

During the partitioning process, branching variables must be selected to help create the children nodes. Clearly the choice of a branching variable affects the running time of the algorithm. Many different approaches have been developed and tested on different types of integer programs. Some common approaches are listed below. For telecommunication problems, popular approaches are to choose the most infeasible integer variable for branching or to set branching priorities on sets of variables before solving.

- *Most/Least Infeasible Integer Variable.* In this approach, the integer variable whose fractional value is farthest from (closest to) an integral value is chosen as the branching variable. See Brunetta et al. (2000); Lee et al. (1998) for examples where the most infeasible variable is chosen for branching.

- *Driebeck-Tomlin Penalties* (Driebeek, 1966; Tomlin, 1971). Penalties give a lower bound on the degradation of the objective value for branching each direction from a given variable. The penalties are the cost of the dual pivot needed to remove the fractional variable from the basis. If many pivots are required to restore primal feasibility, these penalties are not very informative. The *up penalty*, when forcing the value of the kth basic variable up, is

$$u_k = \min_{j: a_{kj} < 0} \frac{(1 - f_k)\bar{c}_j}{-a_{kj}}$$

where f_k is the fractional part of x_k, $\bar{c}_j$ is the reduced cost of variable x_j, and the a_{kj} are the transformed matrix coefficients from the kth row of the optimal dictionary for the LP relaxation. The *down penalty* d_k is calculated as

$$d_k = \min_{j: a_{kj} > 0} \frac{f_k \bar{c}_j}{a_{kj}}.$$

Once the penalties have been computed, a variety of rules can be used to select the branching variable (e.g., $\max_k \max(u_k, d_k)$, or $\max_k \min(u_k, d_k)$). A penalty can be used to eliminate a branch if the LP objective value for the parent node minus the penalty is worse than the incumbent integer solution. Penalties are out of favor because their cost is considered too high for their benefit.

- *Pseudo-Cost Estimate.* Pseudo-costs provide a way to estimate the degradation to the objective value by forcing a fractional variable to an integral value. The technique was introduced by Benichou et al. (1971). Pseudo-costs attempt to reflect the total cost, not just the cost of the first pivot, as with penalties. Once a variable x_k is labeled as a candidate branching variable, the pseudo-costs are computed as:

$$U_k = \frac{\bar{z}_k - z_k^u}{1 - f_k} \text{ and } D_k = \frac{\bar{z}_k - z_k^d}{f_k},$$

where $\bar{z}_k$ is the objective value of the parent, z_k^u is the objective value resulting from forcing up, and z_k^d is the objective value from forcing down. (If the subproblem is infeasible, the associated pseudo-cost is not calculated.) If a variable has been branched upon repeatedly, an average may be used.

The branching variable is chosen as that with the maximum degradation, where the degradation is computed as: $D_k f_k + U_k(1 - f_k)$. Pseudo-costs are not considered to be beneficial on problems where there is a large percentage of integer variables.

- *Pseudo-Shadow Prices* (Fenelon, 1991; Land and Powell, 1979). Similar to pseudo-costs, pseudo-shadow prices estimate the total cost to force a variable to an integral value. Given the up and down pseudo-shadow prices (π_i^+ and π_i^-) for each constraint and pseudo-shadow prices (q_j^+ and q_j^-) for each integer variable (these values can be specified by the user), the degradation in the objective function for forcing an integer variable x_k up to an integral value is

$$S_k^+ = (1 - f_k)[q_k^+ + \sum_i h(\pi_i^+, \pi_i^-)]$$

and for forcing down, it is

$$S_k^- = f_k[q_k^- - \sum_i h(\pi_i^+, \pi_i^-)]$$

where

$$h(\pi_i^+, \pi_i^-) = \max(a_{ik}\pi_i^+, -a_{ik}\pi_i^-, a_{ik}r_i).$$

Here, a_{ik} is the matrix coefficient in row i and column k, and r_i is the dual value for row i. The branching variable is chosen using criteria similar to penalties and pseudo-costs.

- *Strong Branching*. This branching strategy arose in connection with research on solving difficult instances of the traveling salesman problem and general mixed 0/1 integer programming problems (Applegate et al., 1995; Bixby et al., 1995). Applied to 0/1 integer programs within a simplex-based branch-and-cut setting, strong branching works as follows. Let N and K be positive integers. Given the solution of some linear programming relaxation, make a list of N binary variables that are fractional and closest to 0.5 (if there are fewer than N fractional variables, take all fractional variables). Suppose that I is the index set of this list. Then, for each $i \in I$, fix x_i first to 0 and then to 1 and perform K iterations (starting with the optimal basis for the LP relaxation of the current node) of the dual simplex method with steepest-edge pricing. Let $L_i, U_i, i \in I$, be the objective values that result from these simplex runs, where L_i corresponds to fixing x_i to 0 and U_i to fixing x_i to 1. A branching variable can be selected based on the best weighted-sum of these two values.

- *Priorities Selection.* Variables are selected based on their priorities. Priorities can be user-assigned, or based on objective function coefficients, or on pseudo-costs. Smith (2004) compares the effectiveness of four different priority branching schemes in assigning telecommunication traffic to a set of SONET rings. Each integer variable corresponds to a unique pairing of a SONET ring to a traffic demand between two nodes. The priorities are determined based on the values of the traffic demand and the cost of using the SONET ring.

3.4.3 Node Selection

Given a list of active problems, one has to decide which subproblem should be selected to be examined next. This in turn will affect the possibilities of improving the incumbent, the chance of node fathoming, and the total number of problems needed to be solved before optimality is achieved. Below, various strategies given in Beale (1979), Benichou et al. (1971), Benichou et al. (1977), Breu and Burdet (1974), Fenelon (1991), Forrest et al. (1974), Gauthier and Ribiere (1977), and Mitra (1973) are presented.

- *Depth-First-Search with Backtracking.* Choose a child of the previous node as the next node; if it is pruned, choose the other child. If this node is also pruned, choose the most recently created unexplored node, which will be the other child node of the last successful node.

- Best-Bound. Among all unexplored nodes, choose the one which has the best LP objective value. In the case of maximization, the node with the largest LP objective value will be chosen. The rationale is that since nodes can only be pruned when the relaxation objective value is less than the current incumbent objective value, the node with largest LP objective value cannot be pruned, since the best objective value corresponding to an integer feasible solution cannot exceed this largest value.

- *Sum of Integer Infeasibilities.* The sum of infeasibilities at a node is calculated as
$$s = \sum_j \min(f_j, 1 - f_j).$$
Choose the node with either maximum or minimum sum of integer infeasibilities.

- *Best-Estimate using Pseudo-Costs.* This technique was introduced (Benichou et al., 1971) along with the idea of using pseudo-costs to select a branching variable. The individual pseudo-costs can be used to estimate the resulting integer objective value attainable from node k:
$$\varepsilon_k = \bar{z}_k - \sum_i \min(D_i f_i, U_i(1 - f_i))$$

where $\bar{z}_k$ is the value of the LP relaxation at node k. The node with the best estimate is chosen.

- *Best-Estimate using Pseudo-Shadow Prices.* Pseudo-shadow prices can also be used to provide an estimate of the resulting integer objective value attainable from the node, and the node with the best estimate can then be chosen.

- *Best Projection* (Forrest et al., 1974; Mitra, 1973). Choose the node among all unexplored nodes which has the best projection. The projection is an estimate of the objective function value associated with an integer solution obtained by following the subtree starting at this node. It takes into account both the current objective function value and a measure of the integer infeasibility. In particular, the projection p_k associated with node k is defined as

$$p_k = \bar{z}_k - \frac{s_k(\bar{z}_0 - z_{ip})}{s_0},$$

where $\bar{z}_0$ denotes the objective value of the LP at the root node, z_{ip} denotes an estimate of the optimal integer solution, and s_k denotes the sum of the integer infeasibilities at node k. The projection is a weighting between the objective function and the sum of infeasibilities. The weight $(\bar{z}_0 - z_{ip})/s_0$ corresponds to the slope of the line between node 0 and the node producing the optimal integer solution. It can be thought of as the cost to remove one unit of infeasibility. Let n_k be the number of integer infeasibilities at node k. A more general projection formula is to let $w_k = \mu n_k + (1-\mu) s_k$, where $\mu \in [0,1]$, and define

$$p_k = \bar{z}_k - \frac{w_k(\bar{z}_0 - z_{ip})}{w_0}.$$

3.4.4 Preprocessing and Reformulation

Problem preprocessing and reformulation has been shown to be a very effective way of improving integer programming formulations prior to and during branch-and-bound (Bixby and Lee, 1993; Bixby and Wagner, 1987; Bordoefer, 1997; Bradley et al., 1975; Brearley et al., 1975; Crowder et al., 1983; Dietrich and Escudero, 1990; Guignard and Spielberg, 1981; Hoffman and Padberg, 1991; Savelsbergh, 1994). Below, some commonly employed preprocessing techniques are listed. For more details on these procedures, see the references.

1. Removal of empty (all zeros) rows and columns. Detection of implicit bounds and implicit slack variables.

2. Removal of rows dominated by multiples of other rows, including pairs of rows for which the support of one is a subset of the support of the other.

3. Strengthen the bounds within rows by comparing individual variables and coefficients to the right-hand-side. Additional strengthening may be possible for integral variables using integer rounding.

4. Use variable bounds to determine upper and lower bounds for the left-hand side of a constraint, and compare these bounds to the right-hand side. Where possible, conclude that a constraint is inconsistent, redundant, or forces the fixing of some or all variables in its support. Several of these row-driven operations can be dualized to columns.

5. Aggregation: Given an equality constraint where the bound on some variable is implied by the satisfaction of the bounds on the other variables, this variable can be substituted out, and the constraint deleted. Note that free variables always satisfy this condition. Note also that in order to control fill-in (and coefficient growth), not all such substitutions may be desirable. For integral variables, there is the added restriction that they can be eliminated only if their integrality is implied by the integrality of the remaining variables. For integer programming problems, an added advantage of aggregation relative to LP's, is that the reduction in the number of equality constraints increases the relative dimension of the underlying polytope.

6. Coefficient reduction: Consider a constraint $\sum_{j \in K} a_j x_j \geq b$ in which all $a_j \geq 0$ and all $x_j \geq 0$. If x_j is a 0/1 variable and $a_j > b$, for some $j \in K$, replace a_j by b. A stronger version of this procedure is possible when the problem formulation involves other constraints of appropriate structure.

7. Conflict graph, logical implications and probing: Conflict graph, logical implications and probing work by investigating the implications among other variables when a certain variable is fixed to 0 or 1. For example, this analysis may yield logical implications such as $x_k = 1$ implies $x_j = 0$, or $x_k = 1$ implies $x_j = 1$, for some other variable x_j. The implied equality is then added as an explicit constraint. Logical implications can be performed recursively, as well as in conjunction with some of the preprocessing techniques described above. Computational issues and numerical experiments regarding these techniques can be found in Bixby and Lee (1993), Bordoefer (1997), Guignard and Spielberg (1981), and Savelsbergh (1994).

3.4.5 Heuristics

Heuristic procedures provide a means for obtaining integer feasible solutions quickly, and can be used repeatedly within the branch-and-bound search tree. A good heuristic — one that produces good integer feasible solutions — is a crucial component in the branch-and-bound algorithm since it provides an upper bound for reduced-cost fixing (discussed later) at the root, and thus allows reduction in the size of the linear program that must be solved. This in turn may reduce the time required to solve subsequent linear programs at nodes within the search tree. In addition, a good upper bound increases the likelihood of being able to fathom active nodes, which is extremely important when solving large-scale integer programs as they tend to create many active nodes leading to memory explosion.

Broadly speaking, five ideas are commonly used in developing heuristics. The first idea is that of greediness. Greedy algorithms work by successively choosing variables based on best improvement in the objective value. Kruskal's algorithm, which is an exact algorithm for finding the minimum-weight spanning tree in a graph, is one of the most well-known greedy algorithms. Greedy algorithms have been applied to a variety of problems, including 0/1 knapsack problems, uncapacitated facility location problems, set covering problems, and the traveling salesman problem. Two examples of greedy algorithms applied to telecommunications problems appear in Balakrishnan and Altinkemer (1992) and Amiri and Pirkul (1996) where subproblems derived by decomposition strategies were solved greedily.

A second idea is that of local search, which involves searching in a local neighborhood of a given integer feasible solution for a feasible solution with a better objective value. The k-interchange heuristic is a classic example of a local search heuristic. Simulated annealing is another example, but with a variation. It allows, with a certain probability, updated solutions with less favorable objective values in order to increase the likelihood of escaping from a local optimum. See Chardaire et al. (1999) and Feremans et al. (2004) for examples applications of simulated annealing within a branch-and-bound scheme to telecommunications problems. The readers should refer to other chapters of this handbook for detailed descriptions of local search and meta-heuristics.

Randomized Enumeration is a third idea that is used to obtain integer feasible solutions. One such method is that of genetic algorithms, where the randomness is modeled on the biological mechanisms of evolution and natural selection (Goldberg, 1989). Recent work on applying a genetic algorithm to the set covering problem can be found in Beasley and Chu (1996).

The term primal heuristics refers to certain LP-based procedures for constructing integral feasible solutions from points that are in some sense good, but fail to satisfy integrality. Typically, these non-integral points are obtained as optimal solutions of LP relaxations. Primal heuristic procedures involve successive variable fixing and rounding (according to rules usually governed by problem structure) and subsequent re-solves of the modified primal LP (Baldick, 1992; Bixby et al., 1995; Bixby and Lee, 1993; Hoffman and Padberg, 1991). Dahl et al. (1999) use a primal heuristic for solving a routing problem within a branch-and-cut scheme. If an integer solution is found that violates one or many cut inequalities, only the most violated cuts are added to the model. The number of times the heuristic is called at a node is dependent on its success for finding feasible integer solutions. This prevents the model from being too large.

The fifth general principle is that of *exploiting the interplay between primal and dual solutions*. For example, an optimal or heuristic solution to the dual of an LP relaxation may be used to construct a heuristic solution for the primal IP. Problem dependent criteria based on the generated primal-dual pair may suggest seeking an alternative heuristic solution to the dual, which would then be used to construct a new heuristic solution to the primal. Iterating back-and-forth between primal and dual heuristic solutions would continue until an appropriate termination condition is satisfied. See Alevras et al. (1998b) for an example related to network synthesis.

It is not uncommon that a heuristic involves more than one of these ideas. For example, pivot-and-complement is a simplex-based heuristic in which binary variables in the basis are pivoted out and replaced by slack variables. When a feasible integer solution is obtained, the algorithm performs a local search in an attempt to obtain a better integer feasible solution (Balas and Martin, 1980). Obviously, for a specific branch-and-bound implementation, the structure of the problems that the implementation is targeted at influences the design of an effective heuristic (Applegate et al., 1995; Bixby et al., 1995; Bixby and Lee, 1993; Crowder et al., 1983; Erlenkotter, 1978; Hoffman and Padberg, 1991; Lee and Mitchell, 1997).

3.4.6 Continuous Reduced Cost Implications

Reduced cost fixing is a well-known and important idea in the literature of integer programming (Crowder et al., 1983). Given an optimal solution to an LP relaxation, the reduced costs $\bar{c}_j$ are nonpositive for all nonbasic variables x_j at lower bound, and nonnegative for all nonbasic variables at their upper bounds. Let x_j be a nonbasic variable in a continuous optimal solution having objective value z_{LP}, and let $\underline{z}_{ip}$ be the objective value associated with an integer feasible solution to (IP). The following facts specify circumstances when integer variables can be fixed without affecting the optimal objective value:

(a) If x_j is at its lower bound in the continuous solution and $z_{LP} - \underline{z}_{ip} \leq -\bar{c}_j$, then there exists an optimal solution to the integer program with x_j at its lower bound.

(b) If x_j is at its upper bound in the continuous solution and $z_{LP} - \underline{z}_{ip} \leq \bar{c}_j$, then there exists an optimal solution to the integer program with x_j at its upper bound.

When reduced-cost fixing is applied to the root node of a branch-and-bound tree, variables which are fixed can be removed from the problem, resulting in a reduction in the size of the integer program. A variety of studies have examined the effectiveness of reduced-cost fixing within the branch-and-bound tree search (Bixby et al., 1995; Bixby and Lee, 1993; Crowder et al., 1983; Hoffman and Padberg, 1991; Padberg and Rinaldi, 1989).

3.4.7 Subproblem Solver

When linear programs are employed as the relaxations within a branch-and-bound algorithm, it is common to use a simplex-based algorithm to solve each subproblem, using dual simplex to reoptimize from the optimal basis of the parent node. This technique of advanced basis has been shown to reduce the number of simplex iterations to solve the child node to optimality, and thus speed up the overall computational effort. Recently, with the advancement in computational technology, the increase in the size of integer programs, and the success of interior point methods in solving large-scale linear programs (Andersen et al., 1996; Lustig et al., 1994; Zhang et al., 1992), there have been some research efforts devoted to exploring branch-and-bound implementations that employ interior point algorithms as the linear programming solver (Borchers and Mitchell, 1991; de Silva and Abramson, 1998; Lee and Mitchell, 1996; 1997). In such an implementation, the notion of advanced basis is no longer applicable, and care is needed to take advantage of warmstart vectors for the interior point solver so as to facilitate effective computational results. In Lee and Mitchell (1996) and Lee and Mitchell (1997), the notion of advanced warmstart is discussed and computational results are presented.

3.4.8 Lagrangian Relaxation

Linear programming relaxation has been widely used within branch-and-bound implementations. However, other relaxations can also be used, some of which may lead to an improvement (compared to linear programming relaxation) to the upper bound of the optimal objective value to problem (IP). One such relaxation is known as Lagrangian relaxation . There were a number of forays prior to 1970 into the use of Lagrangian methods in discrete optimization (see Fisher (1981) and Nemhauser and Wolsey (1988) for an extensive list of references). However, the spark that ignited a significant amount of research into the computational possibilities of the Lagrangian approach occurred in 1970 when Held and Karp (1970) used a Lagrangian problem based on minimum spanning trees to devise a remarkably successful algorithm for the traveling salesman problem. Encouraged by this success, Lagrangian methods were applied in the early 1970s to scheduling problems, and the general integer programming problem. In 1974, Geoffrion (1974) coined the name "Lagrangian Relaxation" for this approach. Lagrangian relaxation works by partitioning the constraints into two sets. The relaxation is then obtained by "dualizing" one set of constraints. Consider an integer programming problem of the form

$$\begin{aligned} z_{ip} = \text{maximize} \quad & c^T x \\ \text{subject to} \quad & Ax = b \\ & Dx \geq f \\ & x \in Z^n_+ \end{aligned} \quad \text{(IP1)}$$

We consider two possible Lagrangian problems associated with (IP1). The first is obtained by dualizing the constraints $Ax = b$, and the second by dualizing the constraints $Dx \geq f$. Let

$$\begin{aligned} z_{LR1}(u) = \text{maximize} \quad & c^T x + u^T (Ax - b) \\ \text{subject to} \quad & Dx \geq f \\ & x \in Z^n_+ \end{aligned} \quad \text{(LR1)}$$

and

$$\begin{aligned} z_{LR2}(v) = \text{maximize} \quad & c^T x + v^T (Dx - f) \\ \text{subject to} \quad & Ax = b \\ & x \in Z^n_+ \end{aligned} \quad \text{(LR2)}$$

In (LR1), u is free, and in (LR2), $v \geq 0$. The following properties of the optimization problems (LR1) and (LR2) establish that they are indeed relaxations of (IP1).

(a) For all $u, z_{LR1}(u) \geq z_{ip}$.

(b) For all $v \geq 0, z_{LR2}(v) \geq z_{ip}$.

Either (LR1) or (LR2) can be used in place of linear programming relaxation to derive upper bounds in a branch-and-bound algorithm. Note that in contrast to linear programming relaxation, integrality restrictions are imposed in these Lagrangian relaxations. Hence, the Lagrangian relaxations have the potential to provide better upper bounds to the original integer program than the corresponding linear programming relaxation. Besides providing bounds, a Lagrangian relaxation can also be used as a medium for selecting branching variables and selecting nodes. In addition, good feasible integral solutions to (IP1) may be be obtained by perturbing nearly feasible solutions to (LR1) and (LR2) (Erlenkotter, 1978). Subgradient algorithms have been used to solve Lagrangian relaxations for various classes of combinatorial problems (Corneujols et al., 1977; Fisher, 1981; Geoffrion and McBride, 1978; Goffin, 1977; Held et al., 1974; Neebe and Rao, 1983; Shapiro, 1979). Although Lagrangian relaxation is not appropriate for all classes of integer programming problems, it is particularly suited for some; for example, the generalized assignment problem.

3.4.9 Benders' Decomposition

Consider a mixed integer programming problem of the form

$$z_{mip} = \text{maximize} \quad c^T x + h^T y$$
$$\text{subject to} \quad Ax + Gy \leq b \quad \text{(MIP1)}$$
$$x \in X \subseteq Z^n_+, y \in R^p_+$$

Benders (1962) published a technique in 1962 for reformulating mixed integer problems to problems in $X \times R$. Fixing the integer variables x yields a linear program (LP1)

$$z_{lp}(x) = \text{maximize} \quad h^T y$$
$$\text{subject to} \quad Gy \leq b - Ax \quad \text{(LP1)}$$
$$y \in R^p_+$$

with corresponding dual (LD1)

$$\text{minimize} \quad u(b - Ax)$$
$$\text{subject to} \quad uG \geq h \quad \text{(LD1)}$$
$$u \in R^m_+.$$

Let $\{u^k \in R^m_+ : k \in K\}$ be the set of extreme points of the dual feasible region and $\{v^j \in R^m_+ : j \in J\}$ be the set of extreme rays of $\{u \in R^m_+ : uG \geq 0\}$. (MIP1) can be reformulated as:

$$z = \text{maximize} \quad \eta$$
$$\text{subject to} \quad \eta \leq cx + u^k(b - Ax) \text{ for } k \in K \quad \text{(MIP')}$$
$$v^j(b - Ax) \geq 0 \text{ for } j \in J$$
$$x \in X, \eta \in R$$

Though the resulting formulation (MIP') has a huge number of constraints, only a small subset of these constraints need to be generated to find its optimal solution. Consider some relaxation of (MIP') that is missing some subset (possibly all) of the linear inequality constraints that correspond to index sets K and J. Let (η^*, x^*) be the optimal solution of this relaxed problem (if the relaxed problem is unbounded, choose some feasible pair with η^* larger than some fixed value.) Either (LP1) or (LD1) may be solved (using x^* as the fixed x values in these problems) to generate some violated inequality of (MIP') if such an inequality exists. See Nemhauser and Wolsey (1988) for details about these column generation subproblems and for references of further developments on Benders decomposition.

3.4.10 Column Generation

Column generation is a technique that enables MIPs with a large number of variables (columns) to be solved optimally. It is typical when solving such problems with the simplex method that most columns never enter the basis. Such columns need not be introduced to the formulation to solve the problem to optimality. This fundamental idea was first suggested by Ford and Fulkerson (1958) in the context of multicommodity flow problems. The technique was first used on a linear program by Dantzig and Wolfe (1960). Gilmore and Gomory (1961; 1963) made notable use of column generation as part of an efficient heuristic algorithm for solving the cutting stock problem.

Column generation requires a method of discovering variables with negative reduced costs without having to consider all columns of the original, or *master*, problem. This may be accomplished by finding the minimum reduced cost over the index set of all columns that are not yet included in the formulation. We refer to an MIP that is produced by removing some subset of the columns of the master problem as the *restricted* problem. In many cases, the formulation of the problem that identifies the minimum reduced cost (called the *column generation problem* or *pricing problem*) has a special structure that allows it to be solved quickly to optimality. If the value of the minimization problem is found to be greater than or equal to zero, then the optimal solution to the master problem has been found. Otherwise, the variable indicated to have the smallest value of reduced cost is introduced to the restricted problem and is resolved to optimality. Heuristic methods for generating columns are also common in practice.

The use of column generation to solve integer programs would become apparent when incorporated with a branch-and-bound framework. Desrosiers et al. (1984) demonstrated this possibility in 1984 to solve a vehicle routing problem with time window constraints. The technique of generating columns at several nodes of a branch-and-bound tree has come to be known as branch-and-price (Barnhart et al., 1998). This paper addresses the common difficulties associated with branch-and-price as identified by Johnson (1989), and identifies branching rules and formulations that help avoid these difficulties.

3.4.11 Branch-and-Cut

The optimal solution for a linear-programming (LP) relaxation of an IP need not be integer-valued. For such a fractional optimal solution, there exists a linear inequality that is violated by this solution while all the feasible solutions of the IP formulation satisfy it. Such inequalities are referred to as valid inequalities or cutting planes. If such an inequality is added to the LP relaxation, the once optimal solution is now infeasible, and the strengthened relaxation may be re-solved to optimality. As the feasible region of the strengthened LP relaxation is contained in the previous feasible region, the new optimal solution may provide a better lower bound. The incorporation of cutting planes to strengthen the subproblems encountered in a branch-and-bound tree is a technique called branch-and-cut.

The notion of cutting planes was introduced by Gomory as early as 1958 (Gomory, 1958). Unfortunately, the cutting planes he constructed in his general integer programming algorithms were too weak to be of practical use. In 1983, a breakthrough in the computational possibilities of branch-and-bound came as a result of the research by Crowder, Johnson, and Padberg. In their paper (Crowder et al., 1983), cutting planes were added at the root node to strengthen the LP formulation before branch-and-bound was called. In addition, features such as reduced-cost fixing, heuristics and preprocessing were added within the tree search algorithm to facilitate the solution process. Branch-and-cut would soon be found to be useful for several applications, but perhaps most notably for the traveling salesman problem (TSP) (D. Applegate and Cook, 1998; Padberg and Rinaldi, 1987; 1991). Other advances in branch-and-cut techniques include lift-and-project cutting planes for mixed 0-1 integer programming (E. Balas, 1996), using structure to obtain strong valid inequalities for integer programs (see Nemhauser and Wolsey (1988) for a reference) and more recently, the parallel implementation of branch-and-cut algorithms (Bixby et al., 1999; Lee, 2004).

Bibliography

D. Alevras, M. Grötschel, P. Jonas, U. Paul, and R. Wessäly. Survivable mobile phone network architectures: models and solution methods. *IEEE Communications Magazine*, 36(3):88–93, 1998a.

D. Alevras, M. Grötschel, and R. Wessäly. Cost-efficient network synthesis from leased lines. *Annals of Operations Research*, 76:1–20, 1998b.

A. Amiri. A system for the design of packet-switched communication networks with economic tradeoffs. *Computer Communications*, 21(18):1670–1680, 1998.

A. Amiri and H. Pirkul. Primary and secondary route selection in backbone communication networks. *European Journal of Operational Research*, 93(1):98–109, 1996.

E. D. Andersen, J. Gondzio, C. Mészáros, and X. Xu. Implementation of interior point methods for large scale linear programming. In T. Terlaky, editor, *Interior Point Methods in Mathematical Programming*, chapter 6. Kluwer Academic Publishers, 1996.

D. Applegate, R.E. Bixby, V. Chvátal, and W. Cook. Finding cuts in the TSP (a preliminary report). Technical Report 95-05, DIMACS, Rutgers University, New Brunswick, NJ 08903, 1995.

A. Balakrishnan and K. Altinkemer. Using a hop-constrained model to generate alternative communication network design. *ORSA Journal on Computing*, 4(2):192–205, 1992.

A. Balakrishnan, T.L. Magnanti, and P. Mirchandani. A dual-based algorithm for multi-level network design. *Management Science*, 40(5):567–581, 1994.

A. Balakrishnan, T.L. Magnanti, J.S. Sokol, and W. Yi. Spare-capacity assignment for line restoration using a single-facility type. *Operations Research*, 50(4):2120, 2002.

A. Balakrishnan, T.L. Magnanti, and R.T. Wong. A decomposition algorithm for local access telecommunications network expansion planning. *Operations Research*, 43(1):58–76, 1995.

E. Balas and C.H. Martin. Pivot and complement-a heuristic for 0/1 programming. *Management Science*, 26:86–96, 1980.

R. Baldick. A randomized heuristic for inequality-constrained mixed-integer programming. Technical report, Department of Electrical and Computer Engineering, Worcester Polytechnic Institute, Worcester, MA 01609, 1992.

J. Barbas and A. Marin. Maximal covering code multiplexing access telecommunication networks. *European Journal of Operational Research*, 159(1):219–238, 2004.

C. Barnhart, C.A. Hane, and P.H. Vance. Using branch-and-price-and-cut to solve origin-destination integer multicommodity flow problems. *Operations Research*, 48(2):318–326, 2000.

C. Barnhart, E.L. Johnson, G.L. Nemhauser, M.W.P. Savelsbergh, and P.H. Vance. Branch and price: column generation for solving huge integer programs. *Operations Research*, 46:316–329, 1998.

E.M.L. Beale. Branch and bound methods for mathematical programming systems. *Annals of Discrete Mathematics*, 5:201–219, 1979.

E.M.L. Beale and J.A. Tomlin. Special facilities in a general mathematical programming system for nonconvex problems using ordered sets of variables. *Proceedings of the Fifth International Conference on Operations Research, J. Lawerence, ed., Tavistock Publications*, pages 447–454, 1970.

J.E. Beasley and P.C. Chu. A genetic algorithm for the set covering problem. *European Journal of Operations Research*, 194:392–404, 1996.

G. Belvaux, N. Boissin, A. Sutter, and L.A. Wolsey. Optimal placement of add/drop multiplexers: static and dynamic models. *European Journal of Operational Research*, 108(1):26–35, 1998.

J.F. Benders. Partitioning procedures for solving mixed-variable programming problems. *Numerische Mathematik*, 4:238–252, 1962.

M. Benichou, J.M. Gauthier, P. Girodet, G. Hehntges, G. Ribiere, and O. Vincent. Experiments in mixed integer linear programming. *Mathematical Programming*, 1: 76–94, 1971.

M. Benichou, J.M. Gauthier, G. Hehntges, and G. Ribiere. The efficient solution of large-scale linear programming problems - some algorithmic techniques and computational results. *Mathematical Programming*, 13:280–322, 1977.

A. Bianco, M. Guido, and E. Leonardi. Incremental scheduling algorithms for WDM/TDM networks with arbitrary tuning latencies. *IEEE Transactions on Communications*, 51(3):464–475, 2003.

D. Bienstock, S. Chopra, O. Günlük, and C.-Y. Tsai. Minimum cost capacity installation for multicommodity network flows. *Mathematical Programming*, 81(2): 177–199, 1998.

R.E. Bixby, W. Cook, A. Cox, and E. K. Lee. Parallel mixed integer programming. Technical Report CRPC-TR95554, Center for Research on Parallel Computation, Rice University, Houston, Texas, 1995. Revised paper appeared in *Annals of Operations Research, Special Issue on Parallel Optimization, 1997*.

R.E. Bixby, W. Cook, A. Cox, and E. K. Lee. Computational experience with parallel mixed integer programming in a distributed environment. *Annals of Operations Research*, 90:19–43, 1999.

R.E. Bixby and E.K. Lee. Solving a truck dispatching scheduling problem using branch-and-cut. Technical Report TR93-37, Department of Computational and Applied Mathematics, Rice University, Houston, Texas, 1993. Companion paper appeared in *Operations Research* 46 (3) (1998), 355-367.

R.E. Bixby and D.K. Wagner. A note on detecting simple redundancies in linear systems. *Operations Research Letters*, 6:15–18, 1987.

B. Borchers and J.E. Mitchell. Using an interior point method in a branch and bound algorithm for integer programming. Technical Report 195, Mathematical Sciences, Rensselaer Polytechnic Institute, Troy, NY 12180, March 1991. Revised July 7, 1992.

R. Bordoefer. Aspects of set packing, partitioning, and covering. Technical report, Technischen Universitäte, Berlin, Germany, 1997.

G.H. Bradley, P.L. Hammer, and L. Wolsey. Coefficient reduction in 0-1 variables. *Mathematical Programming*, 7:263–282, 1975.

A.L. Brearley, G. Mitre, and H. P. Williams. Analysis of mathematical programming problems prior to applying the simplex method. *Mathematical Programming*, 5: 54–83, 1975.

R. Breu and C.A. Burdet. Branch and bound experiments in zero-one programming. *Mathematical Programming*, 2:1–50, 1974.

L. Brunetta, M. Conforti, and M. Fischetti. A polyhedral approach to an integer multicommodity flow problem. *Discrete Applied Mathematics*, 101(1-3):13–36, 2000.

X. Cao, V. Anand, X. Yizhi, and Q. Chunming. A study of waveband switching with multilayer multigranular optical cross-connects. *IEEE Journal on Selected Areas in Communications*, 21(7):1081–1095, 2003.

P. Chardaire, J.L. Lutton, and A. Sutter. Upper and lower bounds for the two-level simple plant location problem. *Annals of Operations Research*, 86:117–140, 1999.

G. Corneujols, M.L. Fisher, and G.L. Nemhauser. Location of bank accounts to optimize float: An analytic study of exact and approximate algorithms. *Management Science*, 23:789–810, 1977.

H. Crowder, E.L. Johnson, and M. Padberg. Solving large-scale zero-one linear programming problem. *Operations Research*, 31:803–834, 1983.

V. Chvátal D. Applegate, R. E. Bixby and W. Cook. On the solution of traveling salesman problems. *Documenta Mathematica Journal der Deutschen Mathematiker-Vereinigung ICM III*, pages 645–656, 1998.

G. Dahl, A. Martin, and M. Stoer. Routing through virtual paths in layered telecommunication networks. *Operations Research*, 47(5):693–702, 1999.

R.J. Dakin. A tree search algorithm for mixed integer programming problems. *Computer Journal*, 8:250–255, 1965.

G.B. Dantzig and P. Wolfe. Decomposition principle for linear programs. *Operations Research*, 8:101–111, 1960.

A. de Silva and D. Abramson. A parallel interior point method and its application to facility location problems. *Computational Optimization and Applications*, 9:249–273, 1998.

J. Desrosiers, F. Soumis, and M. Desrochers. Routing with time windows by column generation. *Networks*, 14:545–565, 1984.

B. Dietrich and L. Escudero. Coefficient reduction for knapsack-like constraints in 0/1 programs with variable upper bounds. *Operations Research Letters*, 9:9–14, 1990.

J. Doucette and W.D. Grover. Influence of modularity and economy-of-scale effects on design of mesh-restorable DWDM networks. *IEEE Journal on Selected Areas in Communications*, 18(10):1912–1923, 2000.

N.J. Driebeek. An algorithm for the solution of mixed integer programming problems. *Management Science*, 21:576–587, 1966.

A. Dutta. Capacity planning of private networks using DCS under multibusy-hour traffic. *IEEE Transactions on Communications*, 42(7):2371–2374, 1994.

G. Cornuéjols E. Balas, S. Ceria. Mixed 0-1 programming by lift-and-project in a branch-and-cut framework. *Management Science*, 42(9):1229–1246, Sept. 1996.

D. Erlenkotter. A dual-based procedure for uncapacitated facility location. *Operations Research*, 26:992–1009, 1978.

M. Fenelon. Branching strategies for MIP. *CPLEX*, 1991.

C. Feremans, M. Labbe, and G. Laporte. The generalized minimum spanning tree problem:polyhedral analysis and branch-and-cut algorithm. *Networks*, 43(2):71–86, 2004.

M.L. Fisher. The Lagrangian relaxation method for solving integer programming problems. *Management Science*, 27(1):1–18, 1981.

L.R. Ford and D.R. Fulkerson. A suggested computation for maximal multicommodity network flows. *Management Science*, 5:97–101, 1958.

J.J. Forrest, J.P.H. Hirst, and J.A. Tomlin. Practical solution of large mixed integer programming problems with umpire. *Management Science*, 20:736–773, 1974.

B. Fortz, M. Labbe, and F. Maffioli. Solving the two-connected network with bounded meshes problem. *Operations Research*, 48(6):866–877, 2000.

L.F. Frantzeskaskis and H. Luss. The network redesign problem for access telecommunications networks. *Naval Research Logistics*, 46(5):487–506, 1999.

A. Fumagalli, I. Cerutti, and M. Tacca. Optimal design of survivable mesh networks based on line switched WDM self-healing rings. *IEEE/ACM Transactions on Networking*, 11(3):501–512, 2003.

V. Gabrel, A. Knippel, and M. Minoux. Exact solution of multicommodity network optimization problems with general step cost functions. *Operations Research Letters*, 25(1):15–23, 1999.

M.R. Garey and D.S. Johnson. *Computers and Intractability - A Guide to the Theory of NP-Completeness*. W.H. Freeman and Company, Oxford, England, 1979.

J.M. Gauthier and G. Ribiere. Experiments in mixed integer programming using pseudo-costs. *Mathematical Programming*, 12:26–47, 1977.

A. Gençata and B. Mukherjee. Virtual-topology adaptation for WDM mesh networks under dynamic traffic. *IEEE/ACM Transactions on Networking*, 11(2):236–247, 2003.

A.M. Geoffrion. Lagrangian relaxation and its uses in integer programming. *Mathematical Programming*, 2:82–114, 1974.

A.M. Geoffrion and R. McBride. Lagrangian relaxation applied to capacitated facility location problems. *AIIE Transaction*, 10:40–47, 1978.

P.C. Gilmore and R.E. Gomory. A linear programming approach to the cutting-stock problem. *Operations Research*, 9:849–859, 1961.

P.C. Gilmore and R.E. Gomory. A linear programming approach to the cutting-stock problem – part ii. *Operations Research*, 11:863–888, 1963.

A.I. Giortzis, L.F. Turner, and J.A. Barria. Decomposition technique for fixed channel assignment problems in mobile radio networks. *IEE Proceedings: Communications*, 147(3):187–194, 2000.

J.L. Goffin. On the convergence rates of subgradient optimization methods. *Mathematical Programming*, 13:329–347, 1977.

D.E. Goldberg. *Genetic algorithms in search, optimization, and machine learning.* Addison-Wesley, 1989.

R.E. Gomory. Outline of an algorithm for integer solution to linear programs. *Bulletin American Mathematical Society*, 64:275–278, 1958.

L. Gouveia. Multicommodity flow models for spanning trees with hop constraints. *European Journal of Operational Research*, 95(1):178–190, 1996.

M. Grötschel, C.L. Monma, and M. Stoer. Computational results with a cutting plane algorithm for designing communication networks with low-connectivity constraints. *Operations Research*, 40(2):309–330, 1992.

M. Grötschel, C.L. Monma, and M. Stoer. Polyhedral and computational investigations for designing communication networks with high survivability requirements. *Operations Research*, 43(6):1012–1024, 1995.

W.D. Grover and J. Doucette. Topological design of survivable mesh-based transport networks. *Annals of Operations Research*, 106:79–125, 2001.

M. Guignard and K. Spielberg. Logical reduction methods in zero-one programming. *Operations Research*, 29:49–74, 1981.

J. Hardin, J. Lee, and J. Leung. On the boolean-quadric packing uncapacitated facility-location polytope. *Annals of Operations Research*, 83:77–94, 1998.

M. Held and R.M. Karp. The traveling salesman problem and minimum spanning trees. *Operations Research*, 18:1138–1162, 1970.

M. Held, P. Wolfe, and H.D. Crowder. Validation of subgradient optimization. *Mathematical Programming*, 6:61–88, 1974.

P.-H. Ho, H.T. Mouftah, and J. Wu. A scalable design of multigranularity optical cross-connects for the next-generation optical internet. *IEEE Journal on Selected Areas in Communications*, 21(7):1133–1142, 2003.

K.L. Hoffman and M. Padberg. Improving LP-representations of zero-one linear programs for branch-and-cut. *ORSA Journal on Computing*, 3:121–134, 1991.

K. Holmberg, M. Rönnqvist, and D. Yuan. An exact algorithm for the capacitated facility location problems with single sourcing. *European Journal of Operational Research*, 113(3):544–559, 1999.

K. Holmberg and D. Yuan. A Lagrangian approach to network design problems. *International Transactions in Operational Research*, 5(6):529–539, 1998.

K. Holmberg and D. Yuan. A Lagrangian heuristic based branch-and-bound approach for the capacitated network design problem. *Operations Research*, 48(3):461–481, 2000.

K. Holmberg and D. Yuan. A multicommodity network-flow problem with side constraints on paths solved by column generation. *INFORMS Journal on Computing*, 15(1):42–57, 2003.

K. Holmberg and D. Yuan. Optimization of internet protocol network design and routing. *Networks*, 43(1):39–53, 2004.

J.-Q. Hu. Traffic grooming in wavelength-division-multiplexing ring networks: a linear programming solution. *Journal of Optical Networking*, 1(11):397–408, 2002.

R.R Iraschko and W.D. Grover. A highly efficient path-restoration protocol for management of optical network transport integrity. *IEEE Journal on Selected Areas in Communications*, 18(5):779–794, 2000.

R.R. Iraschko, M.H. MacGregor, and W.D. Grover. Optimal capacity placement for path restoration in STM or ATM mesh-survivable networks. *IEEE/ACM Transactions on Networking*, 6(3):325–336, 1998.

B. Jæger and D. Tipper. Prioritized traffic restoration in connection oriented QoS based networks. *Computer Communications*, 26(18):2025–2036, 2003.

P. Jaillet, G. Song, and G. Yu. Airline network design and hub location problems. *Location Science*, 4(3):195–211, 1996.

E.L. Johnson. Modeling and strong linear programs for mixed integer programming. In S.W. Wallace, editor, *Algorithms and Model Formulations in Mathematical Programming*, pages 1–41. NATO ASI Series 51, 1989.

J. Kang, K. Park, and S. Park. ATM VP-based network design. *European Journal of Operational Research*, 158(3):555–569, 2004.

R. Kawatra. A multiperiod degree constrained minimal spanning tree problem. *European Journal of Operational Research*, 143(1):43–53, 2002.

R. Kawatra and D. Bricker. A multiperiod planning model for the capacitated minimal spanning tree problem. *European Journal of Operational Research*, 121(2):412–419, 2000.

R. Kawatra and D. Bricker. Design of a degree-constrained minimal spanning tree with unreliable links and node outage costs. *European Journal of Operational Research*, 156(1):73–82, 2004.

H. Kerivin and A.R. Mahjoub. Separation of partition inequalities for the (1,2)-survivable network design problem. *Operations Research Letters*, 30:265–268, 2002.

H.-J. Kim, S.-H. Chung, and D.-W. Tcha. Optimal design of the two-level distributed network with dual homing local connections. *IIE Transactions*, 27(5):555–563, 1995.

G. Raghu Kiran and C. Siva Ram Murthy. QoS based survivable logical topology design in WDM optical networks. *Photonic Network Communications*, 7(2):193–206, 2004.

J.G. Klincewicz. A dual algorithm for the uncapacitated hub location problem. *Location Science*, 4(3):173–184, 1996.

A. Kumar, R. Rastogi, A. Silberschatz, and B. Yener. Algorithms for provisioning virtual private networks in the hose model. *IEEE/ACM Transactions on Networking*, 10(4):565–578, 2002.

M.S. Kumar and P.S. Kumar. Static lightpath establishment in WDM networks-new ILP formulations and heuristic algorithms. *Computer Communications*, 25(1):109–114, 2002.

P. Laborczi and T. Cinkler. IP over WDM configuration with shared protection. *Optical Networks Magazine*, 3(5):21–33, 2002.

A.H. Land and A.G. Doig. An automatic method for solving discrete programming problems. *Econometrica*, 28:497–520, 1960.

A.H. Land and S. Powell. Computer codes for problems of integer programming. *Annals of Discrete Mathematics*, 5:221–269, 1979.

C.-M. Lee, C.-C. R. Hui, F. F.-K. Tong, and P. T.-S. Yum. Network dimensioning in WDM-based all-optical networks. *Photonic Network Communications*, 2(3):215–225, 2000a.

E. K. Lee. Generating cutting planes for mixed integer programming problems in a parallel computing environment. *INFORMS Journal on Computing*, 16(1):3–26, Winter 2004.

E.K. Lee and J.E. Mitchell. Computational experience in nonlinear mixed integer programming. In *The Operations Research Proceedings 1996*, pages 95–100. Springer-Verlag, 1996.

E.K. Lee and J.E. Mitchell. Computational experience of an interior-point SQP algorithm in a parallel branch-and-bound framework. In *Proceedings of High Performance Optimization Techniques 1997*. Springer-Verlag, 1997.

K. Lee, K. Park, S. Park, and H. Lee. Economic spare capacity planning for DCS mesh-restorable networks. *European Journal of Operational Research*, 110(1):63–75, 1998.

M. Lee, J. Yu, Y. Kim, C.-H. Kang, and J. Park. Design of hierarchical crossconnect WDM networks employing a two-stage multiplexing scheme of waveband and wavelength. *IEEE Journal on Selected Areas in Communications*, 20(1):166–171, 2002.

T. Lee, K. Lee, and S. Park. Optimal routing and wavelength assignment in WDM ring networks. *IEEE Journal on Selected Areas in Communications*, 18(10):2146–2154, 2000b.

T. Lee and S. Park. An integer programming approach to the time slot assignment problem in SS/TDMA systems with intersatellite links. *European Journal of Operational Research*, 135(1):57–66, 2001.

Y. Lee, J. Han, and K. Kang. A fiber routing problem in designing optical transport networks with wavelength division multiplexed systems. *Photonic Network Communications*, 5(3):247–258, 2003a.

Y. Lee, S. Kim, S. Lee, and K. Kang. A location-routing problem in designing optical internet access with WDM systems. *Photonic Network Communications*, 6(2):151–160, 2003b.

D. Li, Z. Sun, X. Jia, and K. Makki. Traffic grooming on general topology WDM networks. *IEE Proceedings-Communications*, 150(3):197–201, 2003.

C.-C. Lo and B.-W. Chuang. A novel approach of backup path reservation for survivable high-speed networks. *IEEE Communications Magazine*, 41(3):146–152, 2003.

C.-C. Lo and A. Kershenbaum. A two-phase algorithm and performance bounds for the star-star concentrator location problem. *IEEE Transactions on Communications*, 37(11):1151–1163, 1989.

Y. Luo and N. Ansari. Restoration with wavelength conversion in WDM networks. *Electronics Letters*, 38(16):900–901, 2002.

I. J. Lustig, R. E. Marsten, and D. F. Shanno. Interior point methods for linear programming: Computational state of the art. *ORSA Journal on Computing*, 6(1):1–14, 1994.

M.H. MacGregor, W.D. Grover, and K. Ryhorchuk. Optimal spare capacity preconfiguration for faster restoration of mesh networks. *Journal of Network and Systems Management*, 5(2):159–171, 1997.

T.L. Magnanti, P. Mirchandani, and R. Vachani. Modeling and solving the two-facility capacitated network loading problem. *Operations Research*, 43(1):142–157, 1995.

R. Mathar and M. Schmeink. Optimal base station positioning and channel assignment for 3G mobile networks by integer programming. *Annals of Operations Research*, 107:225–236, 2001.

F.F. Mazzini, G.R. Mateus, and J.M. Smith. Lagrangean based methods for solving large-scale cellular networks design problems. *Wireless Networks*, 9(6):659–672, 2003.

S. Melkote and M.S. Daskin. Capacitated facility location/network design problems. *European Journal of Operational Research*, 129(3):481–495, 2001.

G. Mitra. Investigations of some branch and bound strategies for the solution of mixed integer linear programs. *Mathematical Programming*, 4:155–170, 1973.

Y. Miyao and H. Saito. Optimal design and evaluation of survivable WDM transport networks. *IEEE Journal on Selected Areas in Communications*, 16(7):1190–1198, 1998.

R. Montemanni, D.H. Smith, and S.M. Allen. An improved algorithm to determine lower bounds for the fixed spectrum frequency assignment problem. *European Journal of Operational Research*, 156(3):736–751, 2004.

Y.-S. Myung and H.-J. Kim. A cutting plane algorithm for computing k-edge survivability of a network. *European Journal of Operational Research*, 156(3):579–589, 2004.

Y.-S. Myung, H.-J. Kim, and D.-W. Tcha. Design of communication networks with survivability constraints. *Management Science*, 45(2):238–252, 1999.

A.W. Neebe and M.R. Rao. An algorithm for the fixed charge assignment of users to sources problem. *Journal of the Operational Research Society*, 34:1107–1115, 1983.

G.L. Nemhauser and L.A. Wolsey. *Integer and Combinatorial Optimization*. Wiley, New York, 1988.

D. Orincsay, B. Szviatovszki, and G. Böhm. Prompt partial path optimization in mpls networks. *Computer Networks: The International Journal of Computer and Telecommunications Networking*, 43(5):557–572, 2003.

A.E. Ozdaglar and D.P. Bertsekas. Routing and wavelength assignment in optical networks. *IEEE/ACM Transactions on Networking*, 11(2):259–272, 2003.

M. Padberg and G. Rinaldi. Optimization of a 537-city tsp by branch-and-cut. *OR letters*, 6:1–8, 1987.

M. Padberg and G. Rinaldi. A branch-and-cut approach to a traveling salesman problem with side constraints. *Management Science*, 35:1393–1412, 1989.

M. Padberg and G. Rinaldi. A branch-and-cut algorithm for the resolution of large-scale symmetric traveling salesman problems. *SIAM Review*, 33:60–100, 1991.

K. Park, S. Kang, and S. Park. An integer programming approach to the bandwidth packing problem. *Management Science*, 42(9):1277–1291, 1996.

R.G. Parker and R.L. Rardin. *Discrete Optimization*. Academic Press, San Diego, 1988.

M. Prytz and A. Forsgren. Dimensioning multicast-enabled communications networks. *Networks*, 39(4):216–231, 2002.

N. Puech, J. Kuri, and M. Gagnaire. Topological design and lightpath routing in WDM mesh networks: a combined approach. *Photonic Network Communications*, 4(3-4): 443–456, 2002.

R. Ramaswami and K.N. Sivarajan. Design of logical topologies for wavelength-routed optical networks. *IEEE Journal on Selected Areas in Communications*, 14 (5):840–851, 1996.

C.D. Randazzo and H.P.L. Luna. A comparison of optimal methods for local access uncapacitated network design. *Annals of Operations Research*, 106:263–286, 2001.

M. Riis and K.A. Andersen. Capacitated network design with uncertain demand. *INFORMS Journal on Computing*, 14(3):247–260, 2002.

M. Riis and J. Lodahl. A bicriteria stochastic programming model for capacity expansion in telecommunications. *Mathematical Methods of Operations Research*, 56(1): 83–100, 2002.

M.B. Rosenwein and R.T. Wong. A constrained steiner tree problem. *European Journal of Operational Research*, 81(2):430–439, 1995.

M.W.P. Savelsbergh. Preprocessing and probing for mixed integer programming problems. *ORSA Journal on Computing*, 6:445–454, 1994.

A. Schrijver. *Theory of Linear and Integer Programming*. Wiley, New York, 1986.

J.F. Shapiro. A survey of Lagrangian techniques for discrete optimization. *Annals of Operations Research*, 5:113–138, 1979.

H.D. Sherali, Y. Lee, and T. Park. New modeling approaches for the design of local access transport area networks. *European Journal of Operational Research*, 127(1): 94–108, 2000.

H.D. Sherali and T. Park. Discrete equal-capacity p-median problem. *Naval Research Logistics*, 47(2):166–183, 2000.

J.C. Smith. Algorithms for distributing telecommunication traffic on a multiple-ring sonet-based network. *European Journal of Operational Research*, 154(3):659–672, 2004.

V. Sridhar and J.S. Park. Benders-and-cut algorithm for fixed-charge capacitated network design problem. *European Journal of Operational Research*, 125(3):622–632, 2000.

M. Sridharan, M.V. Salapaka, and A.K. Somani. A practical approach to operating survivable WDM networks. *IEEE Journal on Selected Areas in Communications*, 20(1):34–46, 2002.

M. Stoer and G. Dahl. A polyhedral approach to multicommodity survivable network design. *Numerische Mathematik*, 68(1):149–167, 1994.

C.S. Sung and H.W. Jin. Dual-based approach for a hub network design problem under non-restrictive policy. *European Journal of Operational Research*, 132(1):88–105, 2001.

A. Sutter, F. Vanderbeck, and L. Wolsey. Optimal placement of add/drop multiplexers: heuristic and exact algorithms. *Operations Research*, 46(5):719–728, 1998.

J.A. Tomlin. An improved branch and bound method for integer programming. *Operations Research*, 19:1070–1075, 1971.

F. Tong, T.-S. Yum, and C.-C. Hui. Supervisory management and lightpath restoration for wavelength routing networks. *Journal of Lightwave Technology*, 18(9):1181–1186, 2000.

S. Tragantalerngsak, J. Holt, and M. Rönnqvist. An exact method for the two-echelon, single-source, capacitated facility location problem. *European Journal of Operational Research*, 123(3):473–489, 2000.

S.P.M. van Hoesel, A.M.C.A. Koster, R.L.M.J. van de Leensel, and M.W.P. Savelsbergh. Polyhedral results for the edge capacity polytope. *Mathematical Programming*, Ser. A 92:335–358, 2002.

B. Van Caenegem, W. Van Parys, F. De Turck, and P.M. Demeester. Dimensioning of survivable WDM networks. *IEEE Journal on Selected Areas in Communications*, 16(7):1146–1157, 1998.

C. Wynants. *Network Synthesis Problems*. Kluwer Academic Publishers, 2001.

Y. Xiong. Optimal design of restorable ATM mesh networks. *IEEE ATM Workshop, Proceedings*, pages 394–399, 1998.

S. Yan, J.S. Deogun, and M. Ali. Routing in sparse splitting optical networks with multicast traffic. *Computer Networks: The International Journal of Computer and Telecommunications Networking*, 41(1):89–113, 2003.

Y. Zhang, R. Tapia, and J. Dennis Jr. On the superlinear and quadratic convergence of primal-dual interior-point linear programming algorithms. *SIAM Journal on Optimization*, 2:304–324, 1992.

Y. Zheng, W.D. Grover, and M.H. MacGregor. Dependence of network capacity requirements on the allowable flow convergence overloads in ATM backup VP restoration. *Electronics Letters*, 33(5):362–363, 1997.

K. Zhu and B. Mukherjee. Traffic grooming in an optical WDM mesh network. *IEEE Journal on Selected Areas in Communications*, 20(1):122–133, 2002.

4 METAHEURISTICS AND APPLICATIONS TO OPTIMIZATION PROBLEMS IN TELECOMMUNICATIONS

Simone L. Martins[1] and Celso C. Ribeiro[1]

[1] Department of Computer Science
Universidade Federal Fluminense
Niterói, Rio de Janeiro 22410-240, Brazil
simone@ic.uff.br
celso@ic.uff.br

Abstract: Recent years have witnessed huge advances in computer technology and communication networks, entailing hard optimization problems in areas such as network design and routing. Metaheuristics are general high-level procedures that coordinate simple heuristics and rules to find good approximate solutions to computationally difficult combinatorial optimization problems. Among them, we find simulated annealing, tabu search, GRASP, VNS, genetic algorithms, and others. They are some of the most effective solution strategies for solving optimization problems in practice and have been applied to a very large variety of problems in telecommunications. In this chapter, we review the main components that are common to different metaheuristics. We also describe the main principles associated with several metaheuristic and we give templates for basic implementations of them. Finally, we present an account of some successful applications of metaheuristics to optimization problems in telecommunications.

Keywords: Metaheuristics, telecommunications, networks, network design, network routing.

4.1 INTRODUCTION

Recent years have witnessed huge advances in computer technology and communication networks, entailing hard optimization problems in areas such as network design and routing (Martins et al., 2004). They often concern the minimization of the costs involved in the design of the networks or the optimization of their performance.

Combinatorial optimization problems in telecommunications and other areas involve finding optimal solutions from a discrete set of feasible solutions. However, even with the advent of new computer technologies and parallel processing, many of these problems cannot be solved to optimality in reasonable computation times, due to their inner nature or to their size. Moreover, reaching optimal solutions is meaningless in many practical situations, since we are often dealing with rough simplifications of reality and the available data is not precise. The goal of approximate algorithms (or heuristics) is to quickly produce good approximate solutions, without necessarily providing any guarantee of solution quality.

Metaheuristics are general high-level procedures that coordinate simple heuristics and rules to find good (often optimal) approximate solutions to computationally difficult combinatorial optimization problems. Among them, we find simulated annealing, tabu search, GRASP, genetic algorithms, scatter search, VNS, ant colonies, and others. They are based on distinct paradigms and offer different mechanisms to escape from locally optimal solutions, contrarily to greedy algorithms or local search methods. Metaheuristics are among the most effective solution strategies for solving combinatorial optimization problems in practice and they have been applied to a very large variety of areas and situations. The customization (or instantiation) of some metaheuristic to a given problem yields a heuristic to the latter.

In this chapter, we consider the combinatorial optimization problem of

$$\text{minimizing } f(S) \text{ subject to } S \in X,$$

defined by a ground set $E = \{e_1, \ldots, e_n\}$, a set of feasible solutions $X \subseteq 2^E$, and an objective function $f: 2^E \to \mathbb{R}$. We seek an optimal solution $S^* \in X$ such that $f(S^*) \leq f(S)$, $\forall S \in X$. The ground set E, the objective function f, and the feasible set X are specific to each problem.

Principles and building blocks that are common to different metaheuristics are reviewed in the next section. The main metaheuristics and templates of their basic implementations are described in Section 4.3. We also comment on hybridizations combining components from different metaheuristics, that are currently among the most effective algorithms for solving real-life problems. Applications of metaheuristics to several problems in telecommunications and network design and routing are reviewed in Section 4.4.

4.2 PRINCIPLES AND BUILDING BLOCKS

Several components are common to different metaheuristics. They are often blended using different strategies and additional features that distinguish one metaheuristic from the other.

4.2.1 Greedy algorithms

In a greedy algorithm, solutions are progressively built from scratch. At each iteration, a new element from the ground set E is incorporated into the partial solution under construction, until a complete feasible solution is obtained. The selection of the next element to be incorporated is determined by the evaluation of all candidate elements according to a greedy evaluation function. This greedy function usually represents the incremental increase in the cost function due to the incorporation of this element into the partial solution under construction. The greediness criterion establishes that the element with the smallest incremental increase is selected, with ties being arbitrarily broken. Figure 4.1 provides a template for a greedy algorithm for a minimization problem.

---

procedure GreedyAlgorithm()
1. $S \leftarrow \emptyset$;
2.   Evaluate the incremental cost of each element $e \in E$;
3.   **while** S is not a complete feasible solution **do**
4.   Select the element $s \in E$ with the smallest incremental cost;
5. $S \leftarrow S \cup \{s\}$;
6.   Update the incremental cost of each element $e \in E \setminus S$;
7.   **end_while**;
8.   **return** S;
end.

Figure 4.1 Greedy algorithm for minimization.

The solutions obtained by greedy algorithms are not necessarily optimal. Greedy algorithms are often used to build initial solutions to be explored by local search or metaheuristics.

4.2.2 Randomization and greedy randomized algorithms

Randomization plays a very important role in algorithm design. Metaheuristics such as simulated annealing, GRASP, and genetic algorithms rely on randomization to sample the search space. Randomization can also be used to break ties, so as that different trajectories can be followed from the same initial solution in multistart methods or to sample fractions of large neighborhoods. One particularly important use of randomization appears in the context of greedy algorithms.

---

procedure GreedyRandomizedAlgorithm(Seed)
1. $S \leftarrow \emptyset$;
2.   Evaluate the incremental costs of each element $e \in E$;
3.   **while** S is not a complete solution **do**
4.   Build the restricted candidate list (RCL);
5.   Select an element s from the RCL at random;
6. $S \leftarrow S \cup \{s\}$;
7.   Update the incremental cost of each element $e \in E \setminus S$;
8.   **end**;
9.   **return** S;
end.

Figure 4.2 Greedy randomized algorithm for minimization.

Greedy randomized algorithms are based on the same principle of pure greedy algorithms, but make use of randomization to build different solutions at different runs. Figure 4.2 illustrates the pseudo-code of a greedy randomized algorithm for minimization. At each iteration, the set of candidate elements is formed by all elements that can be incorporated into the partial solution under construction without destroying feasibility. As before, the selection of the next element is determined by the evaluation of all candidate elements according to a greedy evaluation function. The evaluation of the elements by this function leads to the creation of a restricted candidate list (RCL) formed by the best elements, i.e. those whose incorporation into the current partial solution results in the smallest incremental costs. The element to be incorporated into the partial solution is randomly selected from those in the RCL. Once the selected element has been incorporated into the partial solution, the incremental costs are reevaluated.

Greedy randomized algorithms are used e.g. in the construction phase of GRASP heuristics or to create initial solutions to population metaheuristics such as genetic algorithms or scatter search; see Section 4.3. Randomization is also a major component of metaheuristics such as simulated annealing and VNS, in which a solution in the neighborhood of the current one is randomly generated at each iteration; see also Section 4.3.

4.2.3 Neighborhoods

A neighborhood of a solution S is a set $N(S) \subseteq X$. Each solution $S' \in N(S)$ is reached from S by an operation called a *move*. Normally, two neighbor solutions S and $S' \in N(S)$ differ by only a few elements. Neighborhoods may also eventually contain infeasible solutions not in X.

A solution S^* is a local optimum with respect to a given neighborhood $N()$ if $f(S^*) \leq f(S), \forall S \in N(S^*)$. Local search methods are based on the exploration of solution neighborhoods, searching for improving solutions until a local optimum is found.

The definition of a neighborhood is not unique. Metaheuristics such as VNS make use of multiple neighborhood structures. A metaheuristic may also modify the neighborhood, by excluding some of the possible moves and introducing others. Such modifications might also lead to the need of changes in the nature of solution evaluation. The strategic oscillation approach (Glover, 1996; 2000; Glover and Laguna, 1997) illustrates this intimate relationship between changes in neighborhood and changes in evaluation.

4.2.4 Local search

Solutions generated by greedy algorithms are not necessarily optimal, even with respect to simple neighborhoods. A local search technique attempts to improve solutions in an iterative fashion, by successively replacing the current solution by a better solution in a neighborhood of the current solution. It terminates when no better solution is found in the neighborhood. The pseudo-code of a basic local search algorithm for a minimization problem is given in Figure 4.3. It starts from a solution S and makes use of a neighborhood structure $N()$.

```
procedure LocalSearch(S)
1.   while S is not locally optimal do
2.       Find S' ∈ N(S) with f(S') < f(S);
3.       S ← S';
4.   end;
5.   return S;
end.
```

Figure 4.3 Local search algorithm for minimization.

The effectiveness of a local search procedure depends on several aspects, such as the neighborhood structure, the neighborhood search technique, the speed of evaluation of the cost function, and the starting solution. The neighborhood search may be implemented using either a *best-improving* or a *first-improving* strategy. In the case of a best-improving strategy, all neighbors are investigated and the current solution is replaced by the best neighbor. In the case of a first-improving strategy, the current solution moves to the first neighbor whose cost function value is smaller than that of the current solution. In practice, we observe that quite often both strategies lead to the same final solution, but in smaller computation times when the first-improving strategy is used. We also observe that premature convergence to a non-global local minimum is more likely to occur with a best-improving strategy.

4.2.5 Restricted neighborhoods and candidate lists

Glover and Laguna (1997) point out that the use of strategies to restrict neighborhoods and to create candidate lists is essential to restrict the number of solutions examined in a given iteration, in situations where the neighborhoods are very large or their elements are expensive to evaluate.

Their goal consists in attempting to isolate regions of the neighborhood containing desirable features and inserting them into a list of candidates for close examination. The efficiency of candidate list strategies can be enhanced by the use of memory structures for efficient updates of move evaluations from one iteration to another. The effectiveness of a candidate list strategy should be evaluated in terms of the quality of the best solution found in some specified amount of computation time. Strategies such as aspiration plus, elite candidate list, successive filter strategy, sequential fan candidate list, and bounded change candidate list are reviewed in Glover and Laguna (1997).

Ribeiro and Souza (2000) used a candidate list strategy to significantly speedup the search for the best neighbor in their tabu search heuristic for the Steiner problem in graphs, based on quickly computed estimates of move values. Moves with bad estimates were discarded. Restricted neighborhoods based on filtering unpromising solutions with high evaluations are discussed, for example, in Martins et al. (1999) and Resende and Ribeiro (2003c).

4.2.6 Adaptive memory

The core idea of metaheuristics such as tabu search is the use of an *adaptive memory*, contrarily to memoryless approaches such as simulated annealing and rigid memory designs typical of branch-and-bound strategies. Such memory structures operate by reference to four principal dimensions: recency, frequency, quality, and influence. Memory can be explicit (full solutions are recorded, typically consisting of elite solutions visited during the search) or attributive (attributes that change in moving from one solution to another are recorded).

Fleurent and Glover (1999) and Fernandes and Ribeiro (2005) have successfully used adaptive memory strategies to improve the solutions constructed by multistart procedures.

4.2.7 Intensification and diversification

Two important components of metaheuristics are *intensification* and *diversification*. Intensification strategies encourage move combinations and solution features historically found good or return to explore attractive regions of the solution space to visit them more thoroughly. The implementation of intensification strategies enforces the investigation of neighborhoods of elite solutions and makes use of explicit memory to do so.

Diversification strategies encourage the search to examine unvisited solutions of the solution space or to generate solutions that significantly differ from those previously visited. Penalty and incentive functions are often used in this context.

4.2.8 Path-relinking

Path-relinking was originally proposed by Glover (1996) as an intensification strategy exploring trajectories connecting elite solutions obtained by tabu search or scatter search. Starting from one or more elite solutions, paths in the solution space leading toward other elite solutions are generated and explored in the search for better solutions. To generate paths, moves are selected to introduce attributes in the current solution that are present in the elite guiding solution. Path-relinking may be viewed as a strategy that seeks to incorporate attributes of high quality solutions, by favoring these attributes in the selected moves.

The algorithm in Figure 4.4 illustrates the pseudo-code of the path-relinking procedure applied to a pair of solutions S_s (starting solution) and S_t (target solution). The procedure starts by computing the symmetric difference $\Delta(S_s, S_t)$ between the two solutions, i.e. the set of moves needed to reach S_t from S_s. A path of solutions is generated linking S_s to S_t. The best solution $\bar{S}$ in this path is returned by the algorithm. At each step, the procedure examines all moves $m \in \Delta(S, S_t)$ from the current solution S and selects the one which results in the least cost solution, i.e. the one which minimizes $f(S \oplus m)$, where $S \oplus m$ is the solution resulting from applying move m to solution S. The best move m^* is made, producing solution $S \oplus m^*$. The set of available moves is updated. If necessary, the best solution $\bar{S}$ is updated. The procedure terminates when S_t is reached, i.e. when $\Delta(S, S_t) = \emptyset$.

```
procedure PathRelinking(S_s, S_t)
1.   Compute the symmetric difference Δ(S_s, S_t);
2.   f̄ ← min{f(S_s), f(S_t)};
3.   S̄ ← argmin{f(S_s), f(S_t)};
4.   S ← S_s;
5.   while Δ(S, S_t) ≠ ∅ do
6.       m* ← argmin{f(S ⊕ m) : m ∈ Δ(S, S_t)};
7.       Δ(S ⊕ m*, S_t) ← Δ(S, S_t) \ {m*};
8.       S ← S ⊕ m*;
9.       if f(S) < f̄ then
10.          f̄ ← f(S);
11.          S̄ ← S;
12.      end_if;
13.  end_while;
14.  return S̄;
end.
```

Figure 4.4 Path-relinking procedure for minimization.

Path-relinking may also be viewed as a constrained local search strategy applied to the initial solution S_s, in which only a limited set of moves can be performed and uphill moves are allowed. Several alternatives have been considered and combined in recent successful implementations of path-relinking in conjunction with tabu search, GRASP, and genetic algorithms (Aiex et al., 2003; 2005; Binato et al., 2001; Canuto et al., 2001; Festa et al., 2002; Martins et al., 2004; Reeves and Yamada, 1998; Resende and Ribeiro, 2003a; Ribeiro and Rosseti, 2002; Ribeiro et al., 2002; Ribeiro and Vianna, 2003; Souza et al., 2003): periodical, forward, backward, back and forward, mixed, randomized, truncated, and post-optimization relinking. Resende and Ribeiro (2003b) reviewed these alternatives, showing that they involve trade-offs between computation time and solution quality. In particular, Ribeiro et al. (2002) observed that exploring two different trajectories for each pair (S_s, S_t) takes approximately twice the time needed to explore only one of them, with very marginal improvements in solution quality. They have also observed that if only one trajectory is to be investigated, better solutions are found when the relinking procedure starts from the best among S_s and S_t. Since the neighborhood of the initial solution is much more carefully explored than that of the guiding one, starting from the best of them gives the algorithm a better chance to investigate in more detail the neighborhood of the most promising solution. For the same reason, the best solutions are usually found closer to the initial solution than to the guiding solution, allowing pruning the relinking trajectory before the latter is reached.

4.3 METAHEURISTICS AND TEMPLATES

In this section we review the main principles and ideas involved with five widely and successfully used metaheuristics and we give templates for their implementation: simulated annealing, tabu search, GRASP, VNS, and genetic algorithms.

4.3.1 Simulated annealing

Simulated annealing was introduced by Kirkpatrick et al. (1983) and its basic ideas come from an analogy with the physical annealing process. In physics, annealing is a thermal process for obtaining low energy states of a solid in a heat bath. Computer simulation methods based on Monte Carlo techniques are used to model this process. A solid in state i has an energy E_i and a new state j is generated by applying a perturbation mechanism that leads to the energy level E_j. If the difference $E_j - E_i$ is less than or equal to zero, the state of the solid is changed to j. Otherwise, the state transition is performed with probability $e^{(E_i - E_j)/(K_B T)}$, where T is the temperature of the heat bath and K_B is the Boltzmann constant.

The principle of the simulated annealing technique lies in the following analogy between the physical process and a combinatorial optimization problem: solutions of the combinatorial problem are equivalent to states of the physical problem and their cost to the energy of a state. Neighboring solutions correspond to states generated by the perturbation method.

```
procedure SimulatedAnnealing()
1.   Generate an initial solution S_0 and set S ← S_0;
2.   Compute the initial temperature T_0 and set T ← T_0;
3.   while stopping criterion is not reached do
4.       while thermal equilibrium is not reached do
5.           Obtain a neighbor solution S' ∈ N(S) at random;
6.           Compute ΔE = f(S') - f(S);
7.           if ΔE < 0 then S ← S';
8.           else if e^{-ΔE/(K_B T)} > random[0,1) then S ← S';
9.           end_if
10.      end_while
11.      Decrease T according with the annealing schedule;
12.  end_while;
13.  return S;
end.
```

Figure 4.5 Template of a simulated annealing heuristic for minimization.

The pseudo-code of a simulated annealing algorithm for a minimization problem is described in Figure 4.5. First, an initial solution S_0 is computed and the temperature is set at T_0. Iterations are performed until a certain *stopping criterion* is met. In most implementations, the latter corresponds to bringing the temperature to a very small value close to zero. For each temperature T, the inner loop is performed until *thermal equilibrium* is reached. The latter is often implemented as a fixed number of iterations which depends on the temperature. A neighbor solution S' and the variation ΔE in the objective value are computed at each iteration. The new solution S' replaces the incumbent if $\Delta E < 0$, i.e., if the new solution is better. The same happens with probability $e^{-\Delta E/(K_B T)}$ in case the new solution is worse than the current one. Once *thermal equilibrium* is reached, the temperature T is reduced according with the *annealing schedule*. In most implementations the new temperature is geometrically reduced, by the multiplication of the current temperature by a constant smaller than one.

Simulated annealing starts with large values for the temperature T, allowing bad solutions to replace the incumbent. As the algorithm executes, the temperature T is decreased and it becomes harder to accept bad solutions. No further deteriorations are accepted when T approaches to zero. Convergence relies on several implementation choices: (a) the initial value T_0 of the temperature, (b) the *annealing schedule* to reduce the temperature, (c) the *stopping criterion*, and (d) the conditions for reaching the *thermal equilibrium* at each temperature. Convergence and strategies for such implementation choice are discussed, for example, in Aarts and Korst (2002), Aarts and Korst (1989), and Henderson et al. (2003).

The main advantage of simulated annealing is its simplicity of implementation. Diversification is allowed by accepting bad solutions with decreasing probability. However, the speed of convergence and solution quality rely on several implementation choices that may not be easy to tune. Furthermore, the algorithm is memoryless and does not use cost information about complete neighborhoods.

4.3.2 Tabu search

The fundamental ideas of *tabu search* were independently proposed by Glover (1986) and Hansen (1986). The approach was later developed by Glover (1989) and Glover (1990). It may be viewed as a dynamic neighborhood method that makes use of memory to drive the search by escaping from local optima and avoiding cycling (Glover and Laguna, 1997). Contrarily to memoryless heuristics such as simulated annealing, and to methods that use rigid memory structures such as branch-and-bound, tabu search makes use of flexible and adaptive memory designs.

For any incumbent solution S, in the case of tabu search the neighborhood $N(S)$ over which local search is applied is not a static set. Instead, it can change according to the history of the search. At each local search iteration, tabu search looks for the neighbor solution that most improves the objective function. However, some neighbors in $N(S)$ are forbidden and discarded. The set of forbidden (tabu) neighbors is stored in a short-term memory formed by the `TabuTenure` last visited solutions, so as to preclude the search from returning to previously visited solutions. The list of forbidden solutions is usually implemented as a list of forbidden moves, that discard a broader subset of solutions. Contrarily to plain local search procedures, tabu search chooses the move which least deteriorates the objective function value in case there are no improving moves.

The basic steps of a simple tabu search heuristic for minimization are described in the pseudo-code of Figure 4.6. First, an initial solution S_0 is computed and the short-term memory is initialized. Iterations are performed until a certain *stopping criterion* is met. In most implementations, the latter corresponds to an upper limit on the number of consecutive moves without improvement in the best solution value. Procedure SelectBestNeighbor returns the best non-forbidden neighbor solution S'. The best known solution is eventually updated. If the tabu list is full, then the oldest forbidden solution is removed from T. The incumbent solution solution is inserted into the tabu list and is replaced by the best neighbor S'.

```
procedure TabuSearch();
1.   Generate an initial solution S_0 and set S ← S_0;
2.   S* ← S;
3.   T ← ∅;
4.   while stopping criterion is not reached do
5.        S' ← SelectBestNeighbor(N(S)\T);
6.        if f(S') < f(S*) then
7.             S* ← S';
8.        end_if;
9.        if |T| = TabuTenure then
10.            Remove from T the oldest solution;
11.       end_if;
12.       T ← T ∪ S;
13.       S ← S';
14.  end_while;
15.  return S*;
end.
```

Figure 4.6 Template of a short-term memory tabu search heuristic for minimization.

The tabu search scheme above described makes use of very simple ideas and can be further extended by the incorporation of more sophisticated features. Among them, we may cite the use of aspiration criteria to override the tabu status of unvisited solutions, the use of medium-term and long-term memories to implement intensification and diversification procedures, the use of candidate list strategies to speedup the search, and the use of hashing tables to speedup and filter the search.

Tabu search is certainly among the most effective approaches for solving hard combinatorial optimization problems. It has been successfully applied to a wide range of problems in many domains. However, implementations of tabu search often involve setting many parameters, that have to be appropriately tuned for achieving good performance in practice. The interested reader is referred to Glover and Laguna (1997) for a thorough study of tabu search and related ideas.

4.3.3 GRASP

GRASP (Greedy Randomized Adaptive Search Procedure) (Feo and Resende, 1989; 1995; Resende and Ribeiro, 2003b;c; Ribeiro, 2002) is a multistart metaheuristic, in which each iteration consists of two phases: construction and local search. The construction phase builds a feasible solution, whose neighborhood is investigated until a local minimum is found during the local search phase. The best overall solution is kept as the result. The pseudo-code in Figure 4.7 illustrates the main blocks of a GRASP procedure for minimization, in which MaxIterations iterations are performed and Seed is used as the initial seed for the pseudo-random number generator.

```
procedure GRASP(MaxIterations,Seed)
1.   Set f* ← ∞;
2.   for k = 1,...,MaxIterations do
3.       S ← GreedyRandomizedAlgorithm(Seed);
4.       S ← LocalSearch(S);
5.       if f(S) < f* then
6.           S* ← S;
7.           f* ← f(S);
8.       end_if;
9.   end_for;
10.  return S*;
end.
```

Figure 4.7 Template of a GRASP heuristic for minimization.

An especially appealing characteristic of GRASP is the ease with which it can be implemented. Few parameters need to be set and tuned, and therefore development can focus on implementing efficient data structures to assure quick iterations. Basic implementations of GRASP rely exclusively on two parameters: the number MaxIterations of iterations and the parameter used to limit the size of the restricted candidate list within the greedy randomized algorithm used by the construction phase. In spite of its simplicity and ease of implementation, GRASP is a very effective metaheuristic and produces the best known solutions for many problems (see also Festa and Resende (2002) for an extensive survey of applications).

GRASP as originally proposed is a memoryless procedure, in which each iteration does not make use of information gathered in previous iterations. Path-relinking is a major enhancement that is able to add memory to the basic GRASP procedure, leading to significant improvements in solution time and quality. The use of path-relinking within a GRASP procedure, as an intensification strategy applied to each locally optimal solution, was first proposed by Laguna and Martí (1999). It was followed by several extensions, improvements, and successful applications (Aiex et al., 2003; 2005; Canuto et al., 2001; Resende and Ribeiro, 2003a;b;c; Resende and Werneck, 2004; Ribeiro and Rosseti, 2002; Ribeiro et al., 2002; Souza et al., 2003).

In this context, path-relinking is applied to pairs of solutions, one of them being a locally optimal solution and the other randomly chosen from a pool with a limited number MaxElite of elite solutions found along the search. The pool of elite solutions is originally empty. Since we wish to maintain a pool of good but diverse solutions, each locally optimal solution obtained by local search is considered as a candidate to be inserted into the pool if it is sufficiently different from every other solution currently in the pool. If the pool already has MaxElite solutions and the candidate is better than the worst of them, then a simple strategy is to have the former replacing the latter.

The pseudo-code in Figure 4.8 illustrates the main steps of a GRASP procedure using path-relinking to implement a memory-based intensification strategy.

```
procedure GRASPwithPathRelinking(MaxIterations, Seed)
1.   Set f* ← ∞;
2.   Set Pool ← ∅;
3.   for k = 1,...,MaxIterations do
4.       S ← GreedyRandomizedAlgorithm(Seed);
5.       S ← LocalSearch(S);
6.       if k > 1 then
7.           Randomly select a solution S' ∈ Pool;
8.           S ← PathRelinking(S',S);
9.       endif;
10.      if f(S) < f* then
11.          S* ← S;
12.          f* ← f(S);
13.      end_if;
14.      Add S to Pool if it satisfies the membership conditions;
15.  end_for;
16.  return S*;
end.
```

Figure 4.8 Template of a GRASP with path-relinking heuristic for minimization.

4.3.4 Variable Neighborhood search

VNS(Variable Neighborhood Search) is a metaheuristic proposed by Hansen and Mladenović (1999; 2002; 2003); Mladenović and Hansen (1997) based on the systematic change of neighborhood. It makes use of a finite set of k_{max} pre-selected neighborhood structures identified as $N_1, N_2, \ldots, N_{k_{max}}$ that may be induced from one or more metric functions introduced in the solution space X. We denote by $N_k(S)$ the set of solutions in the kth neighborhood of S.

Figure 4.9 summarizes the main steps of the pseudo-code of a VNS heuristic. First, an initial solution S_0 is computed. Iterations of the outer loop are performed until a certain *stopping criterion* is met. The current neighborhood is set to $k = 1$ and neighborhoods N_1 to $N_{k_{max}}$ of the current solution are progressively scanned within the inner loop. The search starts from the lowest order neighborhood $N_1(S)$ of the current solution. The initial step of each iteration is a random perturbation move applied to generate a solution S' in the current neighborhood $N_k(S)$. Next, a locally optimal solution S'' is obtained by the application of local search to S'. If solution S'' is better than the incumbent S, then the former replaces the latter and the search resumes from neighborhood $N_1(S)$. Otherwise, the current solution S is not modified,

k is incremented by one, and the search resumes from a higher order neighborhood of S.

---

procedure VNS(k_{max}, Seed)
1.   Generate an initial solution S_0 and set $S \leftarrow S_0$;
2.   **while** *stopping criterion* is not reached **do**
3. $k \leftarrow 1$;
4.   **while** $k \leq k_{max}$ **do**
5.   Obtain a neighbor solution $S' \in N_k(S)$ at random;
6. $S'' \leftarrow \text{LocalSearch}(S')$;
7.   **if** $f(S'') < f(S)$ **then**
8. $S \leftarrow S''$;
9. $k \leftarrow 1$;
10.   **else**
11. $k \leftarrow k + 1$;
12.   **endif**;
13.   **end_while**;
14. **end_while**;
15. **return** S;
end.

Figure 4.9 Template of a VNS heuristic for minimization.

VNS is also a memoryless heuristic, being very easy to implement and relying on very few parameters: the *stopping criterion* and the number k_{max} of neighborhoods. The former may be the maximum number of iterations, the maximum number of iterations between two improvements, or the maximum number of times the highest order neighborhood $N_{k_{max}}$ is attained. Although this is not necessary, successive neighborhoods N_k are nested in most implementations of VNS, i.e., $N_1(S) \subset N_2(S) \subset \ldots \subset N_{k_{max}}(S)$. In this case, solutions in neighborhood $N_k(S)$ are progressively more distant from S as k increases. We refer to Aloise et al. (2005) for a successful application of VNS in which the neighborhoods are not nested.

4.3.5 Genetic algorithms

Genetic algorithms were first introduced and investigated by Holland (1975; 1992). They are population-based methods that use selection and recombination operators to generate new solutions in the search space, imitating the process of natural selection and evolution of species. Solutions are evaluated in terms of their fitness, which most often corresponds to the value of the objective function itself. In some implementations, the fitness of a solution may also account for penalties due to infeasibilities.

Contrarily to other metaheuristics reviewed in this chapter, genetic algorithms explore a population of solutions and not a single trajectory evolving from a unique initial solution. Given a population of solutions, the main operators involved in standard implementations of genetic algorithms are:

- `SelectMates`: a subset of solutions is selected, either randomly or using fitness information to privilege promising solutions.

- `RecombineParents`: pairs of solutions are combined and new solutions are generated (crossover operation).

- `ApplyMutation`: a few solutions are randomly selected and go through small modifications in their structure (mutation operation).

- `SelectBest`: the best solutions are selected, while the others are eliminated.

```
procedure GeneticAlgorithm()
1.   Generate the initial population P of solutions;
2.   k ← 0;
3.   EvaluateFitness(P);
4.   while stopping criterion is not reached do
5.       Parents ← SelectMates(P);
6.       Children ← RecombineParents(Parents);
7.       Children ← ApplyMutation(Children);
8.       EvaluateFitness(Children);
9.       P ← SelectBest(Children ∪ P);
10.      k ← k + 1;
11.  end_while
12.  return the best solution in P;
end.
```

Figure 4.10 Template of a genetic algorithm.

The pseudo-code of the basic steps of a genetic algorithm are described in Figure 4.10. The procedure starts by generating an initial population P of solutions, either entirely randomly or using a greedy randomized algorithm. The fitness of each solution S in the population is evaluated. The loop is performed until some *stopping criterion* is reached. Most applications use the number of iterations, the stabilization of the population, or the stabilization of the best solution in the population. Each iteration of this loop handles a generation of the population. The process of obtaining a new generation starts by the selection of a subset *Parents* of solutions from the current population. Solutions are pairwise combined by crossover operations and a set *Children* of off-springs results. A small subset of the latter go through mutations and the fitness of every newly generated solution is evaluated. Finally, the best among the old and the new solutions are selected to form a new generation of solutions and a new iteration starts.

Classical genetic algorithms do not incorporate any problem-specific knowledge and rely almost entirely on randomization for mate selection, crossover, mutation, and selection. Hybrid genetic algorithms, also called memetic algorithms, make use of specific knowledge available to solve the problem. One of such strategies consists in applying local search to some elements in the generation.

Genetic algorithms are very appealing and easy to implement. However, many implementation choices for each of the above operators are possible. A good implementation requires an appropriate solution encoding as finite length data strings and well tuned strategies for the generation of the initial population, mate selection, crossover, mutation, and selection; see e.g. Reeves (1993) and Reeves (2003).

Scatter search (Glover, 1996; 2000; Glover et al., 2003) and ant colony optimization (Dorigo and Stützle, 2003) are other population-based metaheuristics.

4.3.6 Hybridizations

Hybrid heuristics combining features and strategies borrowed from different metaheuristics are among the most efficient algorithms for finding good approximate solutions for combinatorial problems.

We illustrate this issue with some examples involving the hybridization of GRASP with other metaheuristics. Since GRASP and VNS are complementary in the sense that they make use of randomization at different stages, Festa et al. (2002) studied different variants and combinations of GRASP and VNS for the max-cut problem, showing that hybrid implementations outperform other algorithms in the literature by finding and improving the best known solutions for several benchmark instances.

GRASP has also been used in conjunction with genetic algorithms. The construction phase of a GRASP may be applied to generate the initial population for a genetic algorithm. The genetic algorithm of Ahuja et al. (2000) for the quadratic assignment problem makes use of the GRASP proposed by Li et al. (1994) to create the initial population. A similar approach was used in Armony et al. (2000).

The hybridization of GRASP with tabu search was first studied by Laguna and González-Velarde (1991). Delmaire et al. (1999) considered two approaches. In the first, GRASP is applied as a powerful diversification strategy in the context of a tabu search procedure. The second approach is an implementation of the Reactive GRASP algorithm (Prais and Ribeiro, 2000), in which the local search phase is strengthened by tabu search. Results reported for the capacitated location problem show that the hybrid approaches perform better than the isolated methods previously used.

4.4 APPLICATIONS IN TELECOMMUNICATIONS

The outbreak of new technologies in telecommunications networks, together with the demand for more computer intensive applications, leads to huge developments and needs in network design and routing. As most of these problems are NP-hard and exact methods can only solve small problems, approximate algorithms based on metaheuristics play a very important role in their solution.

Communication networks consist of nodes that can be computers, database repositories, instruments like tomography equipments or radio transmitters, connected by data transmission links such as copper cables, optical fibers, satellite and radio links. Their design involves making decisions on many issues like the number of nodes and their locations, routing paths, capacity installation, wavelength allocation, and frequency assignment. The main objective is often to obtain a least cost network configuration subject to constraints involving issues such as delay, throughput, reliability, link capacity, and crosstalk level.

Applications of metaheuristics to problems in telecommunications are plentiful in the literature. We give below a partial account of some of these applications, organized according to the main metaheuristic used to solve each problem.

Kim et al. (2000) worked out a simulated annealing algorithm to allocate nominal channels to the cells of a mobile radio system in such a way that the average blocking in the entire system is minimized. Experimental results show that their proposal gives the lowest overall blocking probability and the largest number of simultaneously usable channels, when compared with other three heuristics found in the literature.

Randall et al. (2002) showed results obtained by a simulated annealing algorithm developed to find paths in a network which minimize the cost of transporting required origin-destination flows subject to specified link, node capacity, node degree, and chain hop-limit constraints. They showed that the heuristics found good results for very large network design and in smaller times than another genetic algorithm also developed for this problem.

Amaldi et al. (2003) proposed two randomized greedy procedures and a tabu search algorithm for finding the location of universal mobile telecommunication system base stations to maximize the traffic covered and to minimize installation costs. The greedy function takes into account the fraction of traffic covered and the installation costs. The randomized greedy procedures provided good approximate solutions from medium- to large-size realistic instances, and the tabu search algorithm was able to further improve the solutions obtained by the greedy procedures.

Cox and Sanchez (2000) presented a new heuristic algorithm for designing least-cost telecommunications networks to carry traffic from cell sites to wireless switches while meeting survivability, capacity, and technical compatibility constraints. A short term tabu search metaheuristic was introduced (that do not use longer-term features, such as search intensification and diversification strategies), with embedded knapsack and network flow sub-problems. It has proved highly effective in designing such backhaul networks for carrying personal communications services traffic. It solved challenging problems for conventional branch-and-bound solvers in minutes instead of hours and found lower-cost solutions. Applied to real-world network design problems, the heuristic has successfully identified designs that save over 20% compared to the best previously known designs.

Fink et al. (1999) present a more general problem formulation for ring network design problems and associated methods that apply to a broad range of problems. They implemented a simulated annealing and a tabu search algorithm using a metaheuristics framework, so that no calibrations or specializations with respect to a specific problem type were performed. Reactive tabu search presented better results than simulated annealing.

Gendron et al. (2000) proposed a heuristic approach for a network loading problem that alternates construction and local search phases. Initially, a construction method provides a feasible solution, while at subsequent construction steps, a diversification approach is adopted for exploiting information gathered along previous iterations.

Two local search procedures are used: a pure descent search and a tabu search heuristic. An implementation of GRASP is also proposed.

Girard et al. (2001) examined the use of tabu search for the solution of an access network design problem that uses ADM equipments, which are both concentrators and multiplexers, so that demand can be sent on different SONET channels. They compared a simple tabu algorithm and a more elaborated one, using simulated and real instances. They showed that the latter found improved solutions in much less computation times. They also reported that the differences between the results obtained by tabu search and the exact values were quite small for some instances

Hao et al. (1998) presented a tabu search algorithm for assigning frequencies in mobile radio networks, while minimizing electromagnetic interference. Experimental results show that the tabu search algorithm largely outperforms other algorithms based on simulated annealing, constraint programming, graph coloring, and steepest descent, also developed for this problem. Castelino and Stephens (1999) developed a surrogate constraint tabu thresholding algorithm for the same problem. The aim of this work was to determine whether using surrogate constraints can improve the solutions found. The use of surrogate constraints consists in combining problem constraints to provide information for guiding the search, instead of using them in isolation. They compared their algorithm to a thresholding tabu search and showed that the surrogate thresholding tabu found better results within the same computation time.

Klincewicz (1992) proposed two heuristics based on tabu search and GRASP for the p-hub location problem. The objective is to overcome the difficulty that local search algorithms encounter. The local search procedure of the GRASP algorithm is based on a 2-exchange neighborhood. The same author (Klincewicz, 2002) also proposed heuristics based on GRASP and tabu search for locating hubs in a communication or transportation network. The greedy function of the GRASP heuristic takes into account the amount of originating and terminating traffic of each possible location, while the local search procedure uses a swap neighborhood.

Noronha and Ribeiro (2004) developed a tabu search approach to routing and wavelength assignment in all-optical networks. Their algorithm combines the computation of alternative routes for the light-paths with the solution of a partition coloring problem. The computational experiments showed that their approach outperforms the previously best known heuristic for the problem (Manohar et al., 2002).

Pamuk and Sepil (2001) proposed a tabu search algorithm to perform optimal location and allocation of hubs in switched networks. They developed three strategies to generate initial solutions and another three to search for an allocation scheme. Nine versions of tabu search heuristics were developed by combining these strategies. Experimental results showed that the three initial solution strategies perform in a similar way and the performance of allocation schemes depends on the type of instance.

Xu et al. (1997) employed tabu search for designing a least-cost telecommunications network where the alternate routing paths can be changed dynamically from hour to hour. They conducted computational experiments on data drawn from real and simulated problems and showed that the tabu search algorithm found better results than two other less computer intensive strategies. They concluded that the the most effective tabu search heuristic was the one which integrates tabu search memories, probabilistic rules, and a periodic solution recovery strategy.

Abello et al. (1999; 2002) developed GRASP heuristics to approximately solve the maximum clique and maximum quasi-clique in very large graphs. Their approach is used in data mining to extract communities of interest from telephone call detail graphs.

Canuto et al. (2001) developed a GRASP heuristic for the prize-collecting Steiner tree problem, with applications in the design of telecommunications access networks. They proposed a multi-start local search algorithm and path-relinking is used to improve the solutions found by local search. VNS is used as a post-optimization procedure. Their results show that the local search with perturbations approach found optimal solutions on nearly all of the instances tested, in much smaller computation than an exact branch-and-cut algorithm.

Gabrel et al. (2003) presented and compared approximate algorithms for discrete cost multicommodity network optimization problems. Firstly, extensions of classical greedy heuristics, based on link-rerouting and flow-rerouting heuristics, are presented in details. Secondly, a new approximate solution algorithm, which basically consists of a heuristic implementation of the exact Benders-type cutting plane generation method, is proposed. All these algorithms are extensively compared on randomly generated graphs up to 50 nodes and 90 links.

Li et al. (2000) addressed the problem of server replication and placement. In a multicast network, packets are forwarded from a source server to groups of receivers along a distribution tree, where the source is the root, the receivers are the leaves, and the multicast capable routers are the internal nodes. The problem consists in placing multiple replicated servers within the multicast-capable routers. Several heuristics are proposed, including a GRASP. The greedy function is the router cost function.

Prais and Ribeiro (2000) developed a reactive GRASP heuristic for traffic assignment in TDMA communication satellites. A geostationary communication satellite has a number of spot beam antennas covering geographically distributed areas. According to the slot switching configuration and on the on-board switch, the uplink traffic received at the satellite has to be immediately sent to ground areas through a set of transponders. The slot switching configurations are determined through the solution of a time slot assignment problem, which is equivalent to the decomposition of a nonnegative traffic matrix into the sum of a family of switching mode matrices.

Resende and Ribeiro (2003a) proposed a family of heuristics for routing private virtual circuits. The GRASP with path-relinking variant was able to significantly improve upon some simple heuristics currently used in traffic engineering, at the expense of additional computation time.

Srinivasan et al. (2000) presented an approach for efficient design of a signaling network for a network of software switches supporting Internet telephony. The optimal load balancing for given demand forecast is formulated as a quadratic assignment problem, which is solved with a GRASP.

Armony et al. (2000) implemented a genetic algorithm to determine how traffic should be routed on self-healing rings, a problem known as the stacked ring design problem. The objective is to optimize the trade-off between the cost of equipments to implement the rings and the cost of exchanging traffic among rings. The initial population of the proposed GA is generated by GRASP. They showed that their GA found good quality solutions using much less time than CPLEX, and also that it always improved the initial solutions generated by GRASP.

Buriol et al. (2005) presented a hybrid genetic algorithm for solving the OSPF weight setting problem, combining a genetic algorithm with a local search procedure applied to improve the solutions obtained by crossover. Experimental results showed that the hybrid heuristic found better solutions and led to a more robust implementation than the best known heuristic in the literature.

He and Mort (2000) elaborated a hybrid genetic algorithm, where the objective was to minimize the number of point-to-point transmission (hops) for all source-destination pairs and also to minimize the number of congested nodes and links for the routing table. The initial population was created using an heuristic based on a shortest path algorithm with a minimum hop metric. They showed that the hybrid genetic algorithm found solutions with a small number of congested links and nodes than the shortest path with hops heuristic.

Poon et al. (2000) described the GenOSys tool developed to optimize the design of secondary and distribution networks used on typical copper access network cabling to connect customers. Its objective is to determine the best locations for distribution points and to identify geographically advantageous tree-structured sub-networks to aggregate cables from customers. The tool allows the user to enter data about the network and provides information which can be used for ducting and cabling using the hybrid genetic algorithm. A practical problem on a network of 240 nodes was solved in less than 30 minutes on a Pentium 200 MHz.

Watanabe et al. (2001) proposed a new type of parallel genetic algorithm model for multi-objective optimization problems. It was applied to solve an antenna arrangement problem in mobile communications. Their algorithm showed a very good performance when compared to other methods.

Ant colony optimization was applied by Varela and Sinclair (1999) to the problem of routing and wavelength-allocation in a multi-wavelength all-optical virtual-wavelength-path routed transport network. Three variants were proposed and the best one provided results that approached those of an earlier problem-specific heuristic on small- and medium-sized networks.

Wittner et al. (2003) developed a swarm algorithm to find a path of resources from a client terminal to a service provider, such that all resources in the path conform with constraints and preferences of a request profile specified by the user. Given a network composed of users, terminals, and services that have individual profiles containing quality of service parameters, the objective is to search for resource paths for each peer-to-peer communication.

Some conclusions can be drawn from the above analysis of telecommunication applications. Tabu search is often applied and usually presents good performance when compared to other heuristics. The main difficulty when using this technique is the calibration of parameters. Simulated annealing also requires a great effort to tune implementation parameters.

GRASP is a memoryless metaheuristic that is quite simple to implement, because very few parameters have to be tuned. GRASP is largely and successfully used in telecommunication applications. It is often showed that the use of techniques which add memory to GRASP, such as path-relinking and reactive GRASP, always improves the results found by simpler implementations.

Genetic algorithms also present good performance, but quite often take large computation times. Most applications use hybrid algorithms, in which a heuristic is used to generate the initial population. All results show that hybrid genetic algorithms perform much better than pure ones.

Bibliography

E. Aarts and J. Korst. Selected topics in simulated annealing. In C.C. Ribeiro and P. Hansen, editors, *Essays and Surveys in Metaheuristics*, pages 1–37. Kluwer, 2002.

E.H.L. Aarts and J. Korst. *Simulated annealing and Boltzmann machines: A stochastic approach to combinatorial optimization and neural computing.* Wiley, 1989.

J. Abello, P.M. Pardalos, and M.G.C. Resende. On maximum clique problems in very large graphs. In J. Abello and J. Vitter, editors, *External memory algorithms and visualization*, volume 50 of *DIMACS Series on Discrete Mathematics and Theoretical Computer Science*, pages 119–130. American Mathematical Society, 1999.

J. Abello, M.G.C. Resende, and S. Sudarsky. Massive quasi-clique detection. *Lecture Notes in Computer Science*, 2286:598–612, 2002.

R.K. Ahuja, J.B. Orlin, and A. Tiwari. A greedy genetic algorithm for the quadratic assignment problem. *Computers and Operations Research*, 27:917–934, 2000.

R.M. Aiex, S. Binato, and M.G.C. Resende. Parallel GRASP with path-relinking for job shop scheduling. *Parallel Computing*, 29:393–430, 2003.

R.M. Aiex, M.G.C. Resende, P.M. Pardalos, and G. Toraldo. GRASP with path-relinking for three-index assignment. *INFORMS Journal on Computing*, 17:224–247, 2005.

D.J. Aloise, D. Aloise, C.T.M. Rocha, C.C. Ribeiro, J.C. Ribeiro Filho, and L.S.S. Moura. Scheduling workover rigs for onshore oil production. *Discrete Applied Mathematics*, 2005. To appear.

E. Amaldi, A. Capone, and F. Malucelli. Planning UMTS base station location: Optimization models with power control and algorithms. *IEEE Transactions on Wireless Communications*, 2:939–952, 2003.

M. Armony, J.C. Klincewicz, H. Luss, and M.B. Rosenwein. Design of stacked self-healing rings using a genetic algorithm. *Journal of Heuristics*, 6:85–105, 2000.

S. Binato, H. Faria Jr., and M.G.C. Resende. Greedy randomized adaptive path relinking. In J.P. Sousa, editor, *Proceedings of the IV Metaheuristics International Conference*, pages 393–397, 2001.

L.S. Buriol, M.G.C. Resende, C.C. Ribeiro, and M. Thorup. A hybrid genetic algorithm for the weight setting problem in OSPF/IS-IS routing. *Networks*, 46(1): 36–56, 2005.

S.A. Canuto, M.G.C. Resende, and C.C. Ribeiro. Local search with perturbations for the prize-collecting Steiner tree problem in graphs. *Networks*, 38:50–58, 2001.

D. Castelino and N. Stephens. A surrogate constraint tabu thresholding implementation for the frequency assignment problem. *Annals of Operations Research*, 86: 259–270, 1999.

L.A. Cox and J.R. Sanchez. Designing least-cost survivable wireless backhaul networks. *Journal of Heuristics*, 6:525–540, 2000.

H. Delmaire, J.A. Díaz, E. Fernández, and M. Ortega. Reactive GRASP and tabu search based heuristics for the single source capacitated plant location problem. *INFOR*, 37:194–225, 1999.

M. Dorigo and T. Stützle. The ant colony optimization metaheuristic: Algorithms, applications, and advances. In F. Glover and G. Kochenberger, editors, *Handbook of Metaheuristics*, pages 251–285. Kluwer, 2003.

T.A. Feo and M.G.C. Resende. A probabilistic heuristic for a computationally difficult set covering problem. *Operations Research Letters*, 8:67–71, 1989.

T.A. Feo and M.G.C. Resende. Greedy randomized adaptive search procedures. *Journal of Global Optimization*, 6:109–133, 1995.

E.L.R. Fernandes and C.C. Ribeiro. A multistart constructive heuristic for sequencing by hybridization using adaptive memory. *Electronic Notes in Discrete Mathematics*, 2005. In press.

P. Festa, P.M. Pardalos, M.G.C. Resende, and C.C. Ribeiro. Randomized heuristics for the max-cut problem. *Optimization Methods and Software*, 7, 2002.

P. Festa and M.G.C. Resende. GRASP: An annotated bibliography. In C.C. Ribeiro and P. Hansen, editors, *Essays and Surveys in Metaheuristics*, pages 325–367. Kluwer, 2002.

A. Fink, G. Schneidereit, and S. Voss. Solving general ring network design problems by meta-heuristics. In M. Laguna and J. L. González, editors, *Computing Tools for Modeling, Optimization and Simulation(Interfaces in Computer Science and Operations Research)*, pages 91–113. Kluwer Academic Publishers, 1999.

C. Fleurent and F. Glover. Improved constructive multistart strategies for the quadratic assignment problem using adaptive memory. *INFORMS Journal on Computing*, 11: 198–204, 1999.

V. Gabrel, , A. Knippel, and M. Minoux. A comparison of heuristics for the discrete cost multicommodity network optimization problem. *Journal of Heuristics*, 9:429–445, 2003.

B. Gendron, J.-Y. Potvin, and P. Soriano. Diversification strategies in local search for a nonbifurcated network loading problem. *European Journal of Operational Research*, 142:231–241, 2000.

A. Girard, B. Sansó, and L. Dadjo. A tabu search algorithm for access network design. *Annals of Operations Research*, 106:229–262, 2001.

F. Glover. Future paths for integer programming and links to artificial intelligence. *Computers and Operations Research*, 13:533–549, 1986.

F. Glover. Tabu search - Part I. *ORSA Journal on Computing*, 1:190–206, 1989.

F. Glover. Tabu search - Part II. *ORSA Journal on Computing*, 2:4–32, 1990.

F. Glover. Tabu search and adaptive memory programming – Advances, applications and challenges. In R.S. Barr, R.V. Helgason, and J.L. Kennington, editors, *Interfaces in Computer Science and Operations Research*, pages 1–75. Kluwer, 1996.

F. Glover. Multi-start and strategic oscillation methods – Principles to exploit adaptive memory. In M. Laguna and J.L. Gonzáles-Velarde, editors, *Computing Tools for Modeling, Optimization and Simulation: Interfaces in Computer Science and Operations Research*, pages 1–24. Kluwer, 2000.

F. Glover and M. Laguna. *Tabu Search*. Kluwer, 1997.

F. Glover, M. Laguna, and R. Martí. Scatter search and path relinking: Advances and applications. In F. Glover and G. Kochenberger, editors, *Handbook of Metaheuristics*, pages 1–35. Kluwer, 2003.

P. Hansen. The steepest ascent mildest descent heuristic for combinatorial programming. In *Congress on Numerical Methods in Combinatorial Optimization*, Capri, 1986.

P. Hansen and N. Mladenović. An introduction to variable neighbourhood search. In S. Voss, S. Martello, I.H. Osman, and C. Roucairol, editors, *Metaheuristics: Advances and trends in local search procedures for optimization*, pages 433–458. Kluwer, 1999.

P. Hansen and N. Mladenović. Developments of variable neighborhood search. In C.C. Ribeiro and P. Hansen, editors, *Essays and Surveys in Metaheuristics*, pages 415–439. Kluwer, 2002.

P. Hansen and N. Mladenović. Variable neighborhood search. In F. Glover and G. Kochenberger, editors, *Handbook of Metaheuristics*, pages 145–184. Kluwer, 2003.

J. Hao, R. Dorne, and P. Galinier. Tabu search for frequency assignment in mobile radio networks. *Journal of Heuristics*, 4:47–62, 1998.

L. He and N. Mort. Hybrid genetic algorithms for telecommunications network backup routeing. *BT Technol. J.*, 18:42–50, 2000.

D. Henderson, S.H. Jacobson, and A.W. Johnson. The theory and practice of simulated annealing. In F. Glover and G. Kochenberger, editors, *Handbook of Metaheuristics*, pages 287–319. Kluwer, 2003.

J.H. Holland. *Adaptation in natural and artificial systems*. University of Michigan Press, 1975.

J.H. Holland. Genetic algorithms. *Scientific American*, 267:44–50, 1992.

S-H. Kim, K-N Chang, and S. Kim. A channel allocation for cellular mobile radio systems using simulated annealing. *Telecommunication Systems*, 14:95–106, 2000.

S. Kirkpatrick, C.D. Gelatt Jr., and M.P. Vecchi. Optimization by simulated annealing. *Science*, 220:671–680, 1983.

J.G. Klincewicz. Avoiding local optima in the p-hub location problem using tabu search and grasp. *Annals of Operations Research*, 40:283–302, 1992.

J.G. Klincewicz. Enumeration and search procedures for a hub location problem with economies of scale. *Annals of Operations Research*, 110:107–122, 2002.

M. Laguna and J.L. González-Velarde. A search heuristic for just-in-time scheduling in parallel machines. *Journal of Intelligent Manufacturing*, 2:253–260, 1991.

M. Laguna and R. Martí. GRASP and path relinking for 2-layer straight line crossing minimization. *INFORMS Journal on Computing*, 11:44–52, 1999.

B. Li, F. Chen, and L. Yin. Server replication and its placement for reliable multicast. In *Proceedings of the IEEE ICCCN-00*, pages 396–401, 2000.

Y. Li, P.M. Pardalos, and M.G.C. Resende. A greedy randomized adaptive search procedure for the quadratic assignment problem. In P.M. Pardalos and H. Wolkowicz, editors, *Quadratic assignment and related problems*, volume 16 of *DIMACS Series on Discrete Mathematics and Theoretical Computer Science*, pages 237–261. American Mathematical Society, 1994.

P. Manohar, D. Manunath, and R.K. Shevgaonkar. Routing and wavelength assignment in optical networks from edge disjoint path algorithms. *IEEE Communication Letters*, 5:211–213, 2002.

S.L. Martins, P.M. Pardalos, M.G.C. Resende, and C.C. Ribeiro. Greedy randomized adaptive search procedures for the Steiner problem in graphs. In P.M. Pardalos, S. Rajasejaran, and J. Rolim, editors, *Randomization Methods in Algorithmic Design*, volume 43 of *DIMACS Series on Discrete Mathematics and Theoretical Computer Science*, pages 133–145. American Mathematical Society, 1999.

S.L. Martins, C.C. Ribeiro, and I. Rosseti. Applications and parallel implementations of metaheuristics in network design and routing. *Lecture Notes in Computer Science*, 3285:205–213, 2004.

N. Mladenović and P. Hansen. Variable neighbourhood search. *Computers and Operations Research*, 24:1097–1100, 1997.

T.F. Noronha and C.C. Ribeiro. Routing and wavelength assignment by partition coloring. Technical report, Department of Computer Science, Universidade Federal Fluminense, Niterói, Rio de Janeiro 22410-240, Brazil, 2004. To appear in *European Journal of Operational Research*.

F.S. Pamuk and C. Sepil. A solution to the hub center problem via a single-relocation algorithm with tabu search. *IIE Transactions*, 33:399–411, 2001.

K.F. Poon, A. Conway, G. Wardrop, and J. Mellis. Successful application of genetic algorithms to network design and planning. *BT Technol. J.*, 18:32–41, 2000.

M. Prais and C.C. Ribeiro. Reactive GRASP: An application to a matrix decomposition problem in TDMA traffic assignment. *INFORMS Journal on Computing*, 12: 164–176, 2000.

M. Randall, G. McMahon, and S. Sugden. A simulated annealing approach to communication network design. *J. of Combinatorial Optimization*, 6:55–65, 2002.

C. Reeves. Genetic algorithms. In F. Glover and G. Kochenberger, editors, *Handbook of Metaheuristics*, pages 65–82. Kluwer, 2003.

C.R. Reeves. Genetic algorithms. In C.R. Reeves, editor, *Modern heuristic techniques for combinatorial problems*, pages 151–196. Wiley, 1993.

C.R. Reeves and T. Yamada. Genetic algorithms, path relinking and the flowshop sequencing problem. *Evolutionary Computation*, 6:45–60, 1998.

M.G.C. Resende and C.C. Ribeiro. A GRASP with path-relinking for private virtual circuit routing. *Networks*, 41:104–114, 2003a.

M.G.C. Resende and C.C. Ribeiro. GRASP and path-relinking: Recent advances and applications. In T. Ibaraki and Y. Yoshitomi, editors, *Proceedings of the Fifth Metaheuristics International Conference*, pages T6–1 – T6–6, 2003b.

M.G.C. Resende and C.C. Ribeiro. Greedy randomized adaptive search procedures. In F. Glover and G. Kochenberger, editors, *Handbook of Metaheuristics*, pages 219–249. Kluwer, 2003c.

M.G.C. Resende and R.F. Werneck. A hybrid heuristic for the p-median problem. *Journal of Heuristics*, 10:59–88, 2004.

C.C. Ribeiro. GRASP: Une métaheuristique gloutone et probabiliste. In J. Teghem and M. Pirlot, editors, *Optimisation approchée en recherche opérationnelle*, pages 153–176. Hermès, 2002.

C.C. Ribeiro and I. Rosseti. A parallel GRASP heuristic for the 2-path network design problem. *Lecture Notes in Computer Science*, 2400:922–926, 2002.

C.C. Ribeiro and M.C. Souza. Tabu search for the Steiner problem in graphs. *Networks*, 36:138–146, 2000.

C.C. Ribeiro, E. Uchoa, and R.F. Werneck. A hybrid GRASP with perturbations for the Steiner problem in graphs. *INFORMS Journal on Computing*, 14:228–246, 2002.

C.C. Ribeiro and D.S. Vianna. A genetic algorithm for the phylogeny problem using an optimized crossover strategy based on path-relinking. *Revista Tecnologia da Informação*, 3(2):67–70, 2003.

M.C. Souza, C. Duhamel, and C.C. Ribeiro. A GRASP with path-relinking heuristic for the capacitated minimum spanning tree problem. In M.G.C. Resende and J. Souza, editors, *Metaheuristics: Computer Decision Making*, pages 627–658. Kluwer, 2003.

A. Srinivasan, K.G. Ramakrishnan, K. Kumaram, M. Aravamudam, and S. Naqvi. Optimal design of signaling networks for Internet telephony. In *IEEE INFOCOM 2000*, March 2000.

G.N. Varela and M.C. Sinclair. Ant colony optimisation for virtual-wavelength-path routing and wavelength allocation. In P.J. Angeline, Z. Michalewicz, M. Schoenauer, X. Yao, and A. Zalzala, editors, *Proceedings of the Congress on Evolutionary Computation*, volume 3, pages 1809–1816, Washington D.C., 1999. IEEE Press.

S. Watanabe, T. Hiroyasu, and M. Miki. Parallel evolutionary multi-criterion optimization for mobile telecommunication networks optimization. In *Proceedings of the EUROGEN 2001 Conference*, pages 167–172, Athens, 2001.

O. Wittner, P.E. Heegaard, and B. Helvik. Scalable distributed discovery of resource paths in telecommunication networks using cooperative ant-like agents. In *Proceedings of the 2003 Congress on Evolutionary Computation*, Canberra, 2003.

J. Xu, S.Y. Chiu, and F. Glover. Tabu search for dynamic routing communications network design. *Telecommunications Systems*, 8:55–77, 1997.

5 LAGRANGIAN RELAX-AND-CUT ALGORITHMS

Abilio Lucena[1]

[1]Departamento de Administração
Universidade Federal do Rio de Janeiro
Rio de Janeiro, Brazil
lucena@facc.ufrj.br

Abstract: Attempts to allow exponentially many inequalities to be candidates to Lagrangian dualization date from the early 1980s. In the literature, the term *Relax-and-Cut* is being used to denote the whole class of Lagrangian Relaxation algorithms where Lagrangian bounds are attempted to be improved by dynamically strengthening relaxations with the introduction of valid constraints (possibly selected from families with exponentially many constraints). In this chapter, Relax-and-Cut algorithms are reviewed in their two flavors. Additionally, a general framework to obtain feasible integral solutions (that benefit from Lagrangian bounds) is also presented. Finally, the use of Relax-and-Cut is demonstrated through an application to a *hard-to-solve* instance of the Knapsack Problem. For that application, Gomory cuts are used, for the first time, within a Lagrangian relaxation framework.

Keywords: Relax-and-cut, Lagrangian relaxation, cutting planes, knapsack problem.

5.1 INTRODUCTION

Attempts to allow exponentially many inequalities to be candidates to Lagrangian dualization date from the early 1980s. A short list of selected contributions in this area, in chronological order, is initiated with the *Restricted Lagrangian Approach* of Balas and Christofides (1981) for the Traveling Salesman Problem. Later on, Gavish (1985) suggested an *Augmented Lagrangian Approach* to solve a Centralized Network Design Problem. A contribution by Lucena (1992; 1993) follows with a scheme to dualize *violated* inequalities *on the fly*, as they become violated at the solution to a valid Lagrangian Relaxation of the problem being solved. Almost simultaneously, Escudero et al. (1994) proposed an algorithm to solve the Sequential Ordering Problem with Precedence Constraints. The authors called that algorithm *Relax-and-Cut*. In a more recent development, Barahona and Ladányi (2001) proposed the use of the Volume Algorithm Barahona and Anbil (2000) as an alternative to the use of the Simplex method (or Interior Point algorithms) in implementations of Branch-and-Cut algorithms (Padberg and Rinaldi, 1991). Finally, Ralphs and Galati (2003) describe a framework where various decomposition methods (including Lagrangian relaxation) and polyhedral cutting plane algorithms can be viewed from a common theoretical perspective.

Following Lucena (2004), we use the term Relax-and-Cut to refer not only to the algorithm in Escudero et al. (1994) but to a much larger class of algorithms which includes that one. According to that guideline, a Relax-and-Cut algorithm must be Lagrangian based. Furthermore, Lagrangian bounds must be attempted to be improved by dynamically strengthening relaxations with the introduction of valid constraints (possibly selected from families with exponentially many constraints). Finally, strengthening constraints may or may not be explicitly dualized.

The definition of Relax-and-Cut above is broad enough to include all algorithms in Balas and Christofides (1981), Gavish (1985), Lucena (1992), Lucena (1993), Escudero et al. (1994), and Barahona and Ladányi (2001). These algorithms, however, differ significantly among themselves. Differences are particularly pronounced in the way strengthening constraints are treated. For instance, in Gavish (1985) and Escudero et al. (1994), strengthening constraints are only used after Lagrangian Dual Problems are solved (and corresponding best possible Lagrangian bounds are obtained). Differently from that, strengthening constraints in Lucena (1992; 1993) are identified and used after every Lagrangian Relaxation Problem is solved.

As for any Lagrangian Relaxation algorithm, Relax-and-Cut is initiated with a relaxation of a given model where a set of *complicating constraints* is dualized while remaining constraints are *kept* (Guignard, 2004). Following the nomenclature introduced in Lucena (2004), *Delayed Relax-and-Cut* (DRC) algorithms (Lucena, 2004) then proceed by solving the corresponding Lagrangian Dual Problem. Valid constraints which violate the solution to that problem are then identified and may be either dualized or else *kept*. Either way, a new Lagrangian Dual Problem is formulated and solved and the procedure continues until a stopping criteria is reached. DRC was introduced in Escudero et al. (1994) and a number of DRC applications are discussed in Guignard (1998). These include the Asymmetric Traveling Salesman Problem, the Generalized Assignment Problem, and the Multiple Choice Knapsack Problem.

As mentioned above, Relax-and-Cut algorithms in Lucena (1992; 1993) do not *delay* the identification and use of violated constraints until the Lagrangian Dual Problem is solved. Differently from DRC, violated inequalities are attempted to be identified (and are dualized, in case of success) for every Lagrangian Relaxation Problem eventually solved. That variant of Relax-and-Cut is called *Non Delayed Relax-and-Cut* (NDRC) in Lucena (2004). NDRC was first proposed and successfully used for the Steiner Problem in Graphs in Lucena (1992; 1993). Later on, it was used for the Edge-Weighted Clique Problem (Hunting et al., 2001), the Quadratic Knapsack Problem (de Moraes Palmeira et al., 1999), the Traveling Salesman Problem (Belloni and Lucena, 2000), the Vehicle Routing Problem (Martinhon et al., 2004), the Linear Ordering Problem (Belloni and Lucena, 2003), the Rectangular Partition Problem (Calheiros et al., 2003), and the Capacitated Minimum Spanning Tree Problem (da Silva, 2002).

Assuming that the increasingly reinforced Lagrangian Dual Problems are solved to optimality, theoretical convergence of DRC algorithms is guaranteed (to a bound at least as good as that given by the Linear Programming (LP) relaxation of the original formulation, reinforced with the additional families of valid inequalities used within the Lagrangian framework). Recently, Belloni and Sagastizábal (2004) obtained a similar result, under not very restrictive conditions, for a Bundle Method implementation of NDRC.

Starting with the pioneering work of Everett III (1963) and Held and Karp (1970; 1971), a vast literature exists on Lagrangian relaxation. Among these, references such as Geoffrion (1974), Shapiro (1974; 1979), Fisher (1981), Beasley (1993), Lemaréchal (2001), and Guignard (2004) are highly relevant. On the other hand, Relax-and-Cut is a relatively recent development and surveys on it can be found in Guignard (1998), Guignard (2004), and Lucena (2004).

In this chapter, Relax-and-Cut algorithms are reviewed in their two variants, namely NDRC and DRC. In addition, the effectiveness of the approach is demonstrated in an application to a *hard-to-solve* instance of the Knapsack Problem (KP). That instance involves coefficients of the order of 10^{15}, which makes it considerably more challenging to solve than ordinary KP instances. Throughout that application, Gomory cuts (Gomory, 1963) are used, for the first time, within a Lagrangian relaxation framework.

NDRC and DRC algorithms are discussed, respectively, in Sections 5.2 and 5.3. A framework for generating primal integral solutions (which benefit from Lagrangian dual information) is presented in Section 5.4. The generation of Gomory cuts for KP follows in Section 5.5. The KP instance used as an example (for the application of NDRC and DRC) is also introduced in that section. In Section 5.6, the generic *dual* algorithms of Sections 5.2 and 5.3 and the generic *primal* algorithm of Section 5.4 are specialized to KP. The use of these algorithms is illustrated with a numerical example. Finally the chapter is closed in Section 5.7 with suggestions for future work.

5.2 NON DELAYED RELAX AND CUT

The NDRC algorithm in Lucena (1992; 1993) is based upon the use of SM and, throughout this chapter, we follow Lucena (1992; 1993; 2004) in using SM to describe and test NDRC. For the sake of clarity, the material presented in this section is focused on binary 0-1 problems. However, with the exception of Subsection 5.4, it generalizes to Mixed Integer Programming problems.

Assume that a formulation for a $\mathcal{NP}$-hard combinatorial optimization problem is given. Assume as well that exponentially many inequalities are included in it. Such a formulation can be generically described as

$$\max\{cx: \; Ax \leq b, \; x \in X\}, \tag{5.1}$$

where, for simplicity, x denotes binary $0-1$ variables (i.e. $x \in \mathbb{B}^n$, for positive integral values of n). Accordingly, for positive integral values of m, we have $c \in \mathbb{R}^n$, $b \in \mathbb{R}^m$, $A \in \mathbb{R}^{m \times n}$ and $X \subseteq \mathbb{B}^n$. Assume, as it is customary in Lagrangian relaxation, that

$$\max\{cx: \; x \in X\} \tag{5.2}$$

is an *easy* problem to solve. On the other hand, in what is unusual for the application of Lagrangian relaxation, assume that m is an exponential function of n, i.e. (5.1) contains exponentially many inequalities. Assume as well that one dualizes

$$\{a_i x \leq b_i : \quad i = 1, 2, \ldots, m\} \tag{5.3}$$

in a Lagrangian fashion (regardless of the difficulties associated with the dualization of exponentially many inequalities) and let $\lambda \in \mathbb{R}_+^m$ be the corresponding vector of Lagrangian multipliers. A valid upper bound on (5.1) is given by the solution to the *Lagrangian Relaxation Problem*

$$LRP(\lambda) = \max\{(c - \lambda A)x + \lambda b : \quad x \in X\}. \tag{5.4}$$

Subgradient Optimization (SO) could then be used to solve the corresponding *Lagrangian Dual Problem*

$$LDP = \min_{\lambda \in \mathbb{R}_+^m} \{LRP(\lambda)\} \tag{5.5}$$

and obtain the best possible Lagrangian bound on (5.1). Optimization is typically conducted here in an interactive way with multipliers being updated so that (5.5) is attained. For the sake of completeness, let us briefly review SM, as implemented in Fisher (1981). That implementation is precisely the one which is adapted in this paper to produce the computational results in the following sections.

5.2.1 A brief description of the Subgradient Method

At iteration k of SM, for a feasible vector λ^k of Lagrangian multipliers, let $\bar{x}$ be an optimal solution to $LRP(\lambda^k)$ and z_{lb} be a known lower bound on (5.1). Additionally, let $g^k \in \mathbb{R}^m$ be a subgradient associated with the relaxed constraints at $\bar{x}$. Corresponding, entries for g^k are

$$g_i^k = (b_i - a_i \bar{x}), \quad i = 1, 2, \ldots, m. \tag{5.6}$$

In the literature (see Fisher (1981), for instance), to update Lagrangian multipliers, one generates first a *step size*

$$\theta^k = \frac{\alpha [LRP(\lambda^k) - z_{lb}]}{\sum_{i=1,\ldots,m} (g_i^k)^2}, \tag{5.7}$$

where α is a real number assuming values in $(0, 2]$, and then computes

$$\lambda_i^{k+1} \equiv \max\{0; \lambda_i^k - \theta^k g_i^k\}, \quad i = 1, \ldots, m. \tag{5.8}$$

After Lagrangian multipliers are updated, one moves on to iteration $k+1$ of SM.

Under the conditions imposed here, the straightforward use of updating formulas (5.7)–(5.8) is not as simple as it might appear. The reason being the exceedingly large number of inequalities that one would have typically to deal with.

5.2.2 NDRC modifications to the Subgradient Method

Inequalities in (5.3), at iteration k of SM, may be classified into three sets. The first one contains inequalities that are violated by $\bar{x}$. The second set is for those inequalities that have nonzero multipliers currently associated with them. Notice that an inequality may be, simultaneously, in the two sets just defined. Finally, the third set consists of the remaining inequalities. Following Lucena (2004), we will refer to these sets of inequalities respectively as the *Currently Violated Active* set, the *Previously Violated Active* set, and the *Currently Inactive* set. Accordingly, they are respectively denoted by $CA(k)$, $PA(k)$, and $CI(k)$.

For the traditional use of Lagrangian relaxation, say when m is a polynomial function of n, Beasley (1993) reported *good practical convergence* of SM to (5.5), while, arbitrarily setting $g_i^k = 0$ whenever $g_i^k > 0$ and $\lambda_i^k = 0$, for $i \in \{1,\ldots,m\}$. In our context, all subgradient entries that are candidates to that modification belong to $CI(k)$.

In spite of the exponentially many inequalities one is faced with, we follow Beasley's advice, The reasoning behind the SM modifications suggested above comes from two observations. The first one is that, irrespective of the suggested changes, from (5.8), multipliers for $CI(k)$ inequalities would not change their present null values at the end of the current SM iteration. Clearly, $CI(k)$ inequalities do not directly contribute to Lagrangian costs (at the current SM iteration). On the other hand, they do play a decisive role in determining the value of θ^k and this fact brings us to the second observation. Typically, for the application being described, the number of strictly positive subgradient entries associated with $CI(k)$ inequalities, tends to be huge. If all these subgradient entries are explicitly used in (5.7), the value of θ^k would result extremely small, leaving multiplier values virtually unchanged from iteration to iteration and SM convergence problems should be expected.

By following Beasley's suggestion, we are capable of dealing adequately, within a SM framework, with the exceedingly large number inequalities in $CI(k)$. However, we may still face problems arising from a potentially *large* number inequalities in $(CA(k) \setminus PA(k))$, i.e. those inequalities that will become effectively dualized (i.e. have a nonzero multiplier associated with them) at iteration k of SM.

Assume that a *large* number of inequalities exist in $(CA(k) \setminus PA(k))$. These inequalities must therefore be violated at the solution to $LRP(\lambda^k)$ and have zero valued Lagrangian multipliers currently associated with them. Typically, such inequalities may be partitioned into subsets (associated, for instance, with a partitioning of the set of vertices in an associated graph, if that applies) and, according to some associated criteria, a maximal inequality would exist for each subset. In order to avoid repeatedly penalizing the same variables, again and again, we only dualize one maximal inequality per subset of inequalities. Remaining inequalities in $(CA(k) \setminus PA(k))$ have their subgradient entries arbitrarily set to 0, thus becoming, in effect, inequalities of $CI(k)$. Examples of applications where the situation described above prevails, are discussed in Lucena (2004). However, for the application considered in Section 5.6, only Gomory cuts are used and, at every SM iteration, there will always be at most one fractional variable to generate Gomory cuts from.

One should notice that, under the classification proposed above, inequalities may change groups from one SM iteration to another. It should also be noticed that the only multipliers that may directly contribute to Lagrangian costs $(c + \lambda^{k+1}A)$, at the end of iteration k of SM, are the ones associated with *active* inequalities, i.e. inequalities in $(CA(k) \cup PA(k))$.

An important step, in the scheme outlined above, is the identification of inequalities violated at $\bar{x}$. That problem must be solved at every iteration of SM and is equivalent to the separation problems found in Branch and Cut algorithms. However, NDRC separation problems typically involve a lower complexity than its Branch and Cut counterparts. This follows from the fact that LRP is normally chosen so that one will be separating over integral structures. For the application in Section 5.6, although one will not be working with integral structures, separation (of Gomory cuts) is quite straightforward.

According to Lucena (2004), an NDRC algorithm could be seem as a *Traditional Lagrangian Relaxation* (TLR) algorithm where exponentially many inequalities are dualized but subgradients are *projected*, at every iteration of SM, into a smaller space, i.e. the space implied by *active* inequalities (formed by inequalities in $(CA(k) \cup PA(k))$). In doing so, only a fraction of the exponentially many inequalities involved are explicitly considered, at every SM iteration, to update Lagrangian multipliers. An analog of that idea, in terms of LP based algorithms, is to find the LP relaxation of a formulation involving exponentially many inequalities. Clearly, only a *tiny* fraction of these inequalities are tight at a LP relaxation solution. Furthermore, in practice, only a *few* of these inequalities would be explicitly used in a cutting plane algorithm to attain LP bounds.

In another interpretation in Lucena (2004), NDRC may be seen as a Lagrangian relaxation algorithm where exponentially many candidate inequalities are dualized *on the fly*, as they become violated at an optimal solution to $LRP(\lambda^k)$. Since inequalities may be dualized for every LRP (and not only for LDP), an analog of the idea, in terms of LP based algorithms, generates cutting planes as the LP is being solved.

5.3 DELAYED RELAX AND CUT

Description of a DRC algorithm is quite straightforward. Assume that an initial LDP is solved and let $\bar{x}$ be the corresponding LDP optimal solution. Violated inequalities associated with $\bar{x}$ are then separated and are either kept or else dualized, thus giving rise to a new (strengthened) LDP (see Guignard (1998) for a discussion on efficient cuts in DRC). The new LDP is then solved and the procedure is repeated until some stopping criteria is met (for instance, until either optimality is proven or else the maximum number of allowed LDP solving rounds is reached). DRC could be seen as a Lagrangian relaxation analog of LP based cutting plane algorithms.

An important consideration in designing a DRC algorithm is the definition of an initial LDP to solve. In general, a trade off exists between LDP bound strength and the CPU time required to solve the problem.

An *easy to solve* LDP would typically return a *weak* bound, requiring, however, *few* SM (or Bundle Method, or Volume Algorithm) iterations to be solved. On the other hand, a *hard to solve* LDP would typically return a stronger bound requiring, however, a *large* number of SM (or Bundle Method, or Volume Algorithm) iterations to be solved. That issue is treated in more detail in Lucena (2004), where an application of

DRC to the Steiner Tree Problem is carried out. In the context of the KP application in Section 5.6, an *easy to solve* LDP is used. Such an LDP contains only one constraint, i.e. the knapsack inequality. An example of a *harder to solve* LDP would be, for instance, one containing one or more Lifted Minimum Cover Inequalities (see Wolsey (1998), for instance) associated with the LP relaxation of KP.

5.4 LAGRANGIAN HEURISTICS

Assume that a given basic heuristic, denoted here BH, is available for generating feasible solutions to (5.1). We typically call BH, for the first time, prior to initializing SM. Additional calls are made alongside some of the iterations of SM (or Bundle Method or Volume Algorithm). For the first call of BH, the original costs c are used. Remaining calls are performed either under the available Lagrangian modified costs $(c - \lambda^k A)$, or else under costs given by $(1 - \bar{x})c$ (where, $\bar{x}$, as before, is an optimal solution to $LRP(\lambda^k)$). Costs $(1 - \bar{x})c$ attempt to make it attractive to BH to select variables set to 1 in $\bar{x}$. After a feasible solution to (5.1) is generated, either under Lagrangian costs or else under costs $(1 - \bar{x})c$, the actual value for that feasible solution must be computed (under the original costs c). For the applications we have so far considered, the use of costs $(1 - \bar{x})c$ has proved, in most cases, more effective than the use of Lagrangian modified costs.

The motivation for the Lagrangian Heuristic (LH) sketched above, is the common sense belief that dual information must obviously be relevant to primal heuristics.

As it is the case for any non exact solution algorithm, feasible solutions generated by LH may also be attempted to be improved through local search procedures.

Ideally, it is desirable that BH be fast, to allow a large number of calls to be made to it. In our experience, in most cases, a simple greedy heuristic suffices to eventually return (alongside the application of SM) *good quality* feasible solutions to problem (5.1).

So far, LH have been specifically tailored to the Steiner Problem in Graphs (Lucena, 1992; 1993), the Traveling Salesman Problem (Belloni and Lucena, 2000), the Vehicle Routing Problem (Martinhon et al., 2004), the Linear Ordering Problem (Belloni and Lucena, 2003), the Rectangular Partition Problem (Calheiros et al., 2003), and the Capacitated Minimum Spanning Tree Problem (da Silva, 2002).

5.5 GOMORY CUTS FOR THE 0-1 KNAPSACK PROBLEM

The $0-1$ KP (see Martello and Toth (1997) and Pisinger (1997) for solution algorithms for the problem) is formulated as

$$z = \max \sum_{j=1,\ldots,n} c_j x_j \quad (5.9)$$

subject to

$$\sum_{j=1,\ldots,n} a_j x_j \leq b, \quad (5.10)$$

$$x \in \mathbb{B}^n, \quad (5.11)$$

where coefficients $\{a_j : j = 1,\ldots,n\}$ and b are positive integers. The analogy normally associated with the problem is that of filling a knapsack of capacity b with items selected from among n objects with capacities $\{a_i : i = 1,\ldots,n\}$. Items should be selected in order to maximize value (as expressed by the objective function) of knapsack load.

The LP relaxation of (5.9)–(5.11) is obtained by relaxing integrality enforcing constraints (5.11), thus resulting in

$$\bar{z} = \max \sum_{j=1,\ldots,n} c_j x_j \tag{5.12}$$

subject to

$$\sum_{j=1,\ldots,n} a_j x_j \leq b, \tag{5.13}$$

$$0 \leq x_j \leq 1, \quad j = 1,\ldots,n. \tag{5.14}$$

For the purposes of this chapter, it is more convenient to explicitly associate slack variables $\{x_{n+j} : j = 1,\ldots,n\}$ to the right most inequalities in (5.14), associate slack variable x_{2n+1} to inequality (5.13) and rewrite (5.12)–(5.14) as

$$z = \max \sum_{j=1,\ldots,n} c_j x_j \tag{5.15}$$

subject to

$$x_j + x_{n+j} = 1, \quad j = 1,\ldots,n, \tag{5.16}$$

$$\sum_{j=1,\ldots,n} a_j x_j + x_{2n+1} = b, \tag{5.17}$$

$$x_j \geq 0, \quad j = 1,\ldots,n. \tag{5.18}$$

Following Dantzig (1957), an optimal solution to (5.12)–(5.14) is straightforward to compute. Assume that the coefficients $\{c_j : j = 1,\ldots,n\}$ are ordered so that

$$c_1/a_1 \geq c_2/a_2 \geq \ldots \geq c_n/a_n \tag{5.19}$$

and that j^* is the largest integer for which

$$\sum_{j=1,\ldots,j^*} a_j \leq b. \tag{5.20}$$

An optimal solution to (5.12)–(5.14) is then given by

$$\bar{x}_j = 1, \quad j = 1,\ldots,j^*, \tag{5.21}$$

$$\bar{x}_j = 0, \quad j = j^* + 2,\ldots,n, \tag{5.22}$$

$$\bar{x}_{j^*+1} = (b - \sum_{j=1,\ldots,j^*} a_j)/a_{j^*+1}. \tag{5.23}$$

5.5.1 LP relaxation basis

Let us now translate solution (5.21)–(5.23) in terms of an optimal basis to (5.15)–(5.18). For each of the first j^* rows in (5.16), the corresponding basic variable is x_j, $j = 1, \ldots, j^*$. Accordingly, $x_{(n+j)}$, for $j = (j^* + 2), \ldots, n$, is basic for the last $n - (j^* + 1)$ rows in (5.16). If $\bar{x}_{(j^*+1)} > 0$ then the basic variable associated with the $(j^* + 1)$-th row in (5.16) is $x_{(j^*+1)}$ while the basic variable associated with row (5.17) is $x_{(n+j^*+1)}$. Otherwise, if $\bar{x}_{(j^*+1)} = 0$ then $x_{(n+j^*+1)}$ is the basic variable associated with the $(j^* + 1)$-th row in (5.16) and $x_{(2n+1)}$ is the basic variable associated with row (5.17).

Whenever $\bar{x}_{(j^*+1)} > 0$, a fractional Gomory cut may be generated from the $(j^* + 1)$-th row in (5.16). Furthermore, such a cut may be obtained with very few pivoting operations. More specifically, consider the rows in (5.16) for which a variable x_j, $j = 1, \ldots, n$, is basic. Each of these rows should be multiplied by their corresponding a_j and then be subtracted from row (5.17). One should then select x_{n+j^*} as the pivot for the updated row (5.17) and from that eliminate the variable from the $(j^* + 1)$-th row in (5.16). A fractional Gomory cut is then generated from the updated row j^* in (5.16). Such a cut may clearly involve slack variables in $\{x_{(n+j)} : j = 1, \ldots, n\}$ and one should then use constraints (5.16) and (5.17) to rewrite the cut only in terms of the original variables.

It is not difficult to verify that, due to the particular structure of (5.16)-(5.18), it is always possible to generate Gomory cuts where coefficients are integral valued.

5.5.2 Knapsack instance

Table 5.1 describes the coefficients $\{c_j : 1, \ldots, n\}$ and $\{a_j : 1, \ldots, n\}$ associated with a particular $0 - 1$ KP instance. For that instance, the RHS coefficient in (5.10) is $b = 12107067865319564$. One should notice that such an instance involves coefficients of the order of 10^{15}. Coefficients of such a magnitude preclude the use of Dynamic Programming (DP) recursions to attain optimality since the associated state space would be out of reach of currently available computer memory. Likewise, reinforcing Linear Programming (LP) relaxation (5.12)–(5.14) with violated valid inequalities, in an LP based cutting plane approach, would probably not be an option. That applies since, after generating a few, say Gomory cuts, one would, most likely, run into numerical problems while performing matrix inversions. Finally, a 0.000001% gap between upper and lower bounds tends to be, for ordinary applications, tight enough to provide an optimality certificate. However, for our KP instance such a gap corresponds to millions of units and is by no means a guarantee of optimality.

For the numerical results in Section 5.6, coding was carried out in FORTRAN. In a 32 bits computer, like the one we use, the largest integer variable one is capable of representing, may contain only 8 decimal digits. We have thus used floating point variables to represent the coefficients of the KP instance in Table 5.1. Great care was then taken to minimize the possibilities of incurring in rounding off errors. That involved an extensive use of FORTRAN compiler intrinsic functions that round down a real valued number to the real number which represents the resulting integer. Whenever one of the coefficients of the KP instance had to be explicitly considered, it was

Table 5.1 KP instance

j	c_j	a_j	j	c_j	a_j
1	5985887646675 12	1043758913 8747	26	796893775463105	459233075380326
2	838245868682862	205728669107128	27	270887374877930	492233395576478
3	127963647246 37	382125303149 23	28	945141553878785	502445280551911
4	427655428647996	498940432667 8	29	648177325725556	525000452995301
5	740525364875794	544768981635 58	30	145332604646683	535258352756501
6	578473687171937	629063695669 18	31	445022046566010	595017075538636
7	157311424612999	810336694121 37	32	275523483753205	605881750583649
8	313272386789322	198077023029328	33	636593878269196	647979378700257
9	368257552385331	200995206832886	34	267737418413163	666921436786652
10	877835592939 3769	217693403363228	35	141873389482499	670142650604249
11	885035455226899	271198809146882	36	196074530482293	677149176597596
12	217146083712578	281886577606202	37	150113984942437	694714903831482
13	272873789072037	299474477767945	38	204226523637772	704419076442719
14	522032976150513	300442874431611	39	559784591197968	723579525947571
15	673082530498505	314126789569855	40	442578673362732	728680551052094
16	556192934513093	327096879482270	41	195826128125191	744018435478211
17	994231045246125	340778887271882	42	692465543746949	749621152877808
18	235551774501801	354439347982407	43	567489504814149	781238436698914
19	864161729812623	376186668872834	44	942473530769349	805823385715485
20	464697897434235	377783030271531	45	444463163614274	887850999832154
21	453489422798157	392820507287980	46	518855929374695	893622457981110
22	193823680281640	431183815002442	47	460040390491486	924746572971345
23	748618841171265	440109968185425	48	475286751985550	935340881347657
24	180304899811745	444615721702576	49	649530351161957	937711596488953
25	468244463205338	450100183486939	50	594328463077546	988966226577759

first increased by a conveniently small tolerance and then rounded down as suggested above.

5.6 GOMORY BASED RELAX-AND-CUT ALGORITHMS FOR KP

Assuming that $p \geq 1$ Gomory cuts are currently dualized, the Lagrangian Relaxation Problem one will be faced with, either under NDRC or DRC, at any SM iteration, is given by

$$\text{LRP}(\lambda) = \max \sum_{j=1,\ldots,n} c'_j x_j + \text{const}(\lambda) \quad (5.24)$$

subject to

$$\sum_{j=1,\ldots,n} a_j x_j \leq b, \quad (5.25)$$

$$0 \leq x_j \leq 1, \quad j = 1,\ldots,n, \quad (5.26)$$

where $\lambda \in \mathbb{R}^p_+$ is a vector of Lagrangian multipliers associated with dualized Gomory cuts, const(λ) is a non negative constant associated with these multipliers and $\{c'_j : j = 1,\ldots,n\}$ are the corresponding Lagrangian modified objective function coefficients. Clearly, if no dualized Gomory cut exists, Lagrangian Relaxation Problem (5.24)–(5.26) corresponds to (5.12)–(5.14). One should also notice that a Lagrangian modified objective function coefficient, say c'_j, may eventually become negative throughout the application of SM. In that case, variable x_j is guaranteed to assume a value of 0 at an optimal solution $\bar{x}$ to (5.24)–(5.26).

5.6.1 A Lagrangian heuristic to KP

Following Section 5.4, a basic greedy heuristic BH was used to generate feasible solutions to the KP instance on Table 5.1. BH was called for every SM iteration and took, as an input, Lagrangian modified costs $\{c'_j : j = 1,\ldots,n\}$. Assume that an ordering

$$c'_{i_1}/a_{i_1} \geq c'_{i_2}/a_{i_2} \geq \ldots \geq c'_{i_n}/a_{i_n} \quad (5.27)$$

is obtained from these input costs. BH then considers indices in $\{i_1,\ldots,i_n\}$ (for corresponding item inclusion into the knapsack) in the order they appear in (5.27). Assume that iteration k of BH is being performed (i.e. that the item implied by index i_k is being considered for possible inclusion into the knapsack). Assume as well that b_k is the capacity left at the knapsack at iteration k of BH. The item implied by i_k should then be included into the knapsack if $a_{i_k} \leq b_k$. Iterations should be performed until either the knapsack is filled to capacity or else until it could be established that none of the items still to be investigated could be successfully introduced into the knapsack.

Once a feasible solution is returned by BH, the cost of that solution under $\{c_j : j = 1,\ldots,n\}$ must be computed (to obtain a valid KP lower bound). The solution should then be subjected to Local Search in an attempt to improve it. The search neighborhood we use is formed by those items that could feasibly replace an item currently included in the knapsack. Item replacement should be carried out for that pair of items (if any) leading to the largest KP lower bound increase. In case of success one should check if spare space still exists in the knapsack to include items currently left out of it (whenever applicable, items should be included by their cost/benefit ratios) and iterate.

Within a Lagrangian relaxation framework, repeated calls to BH followed by Local Search (as outlined above), give rise to a Lagrangian heuristic LH to KP (see Section 5.4).

5.6.2 Variable fixing tests

Whenever an optimal solution $\bar{x}$ is obtained for (5.24)–(5.26), a valid upper bound LRP(λ) is generated for KP. One may then use LP reduced costs, together with a feasible KP lower bound z_{lb}, in an attempt to price variables out of an optimal solution. Accordingly if $\bar{c}_j$ is the reduced cost associated with variable x_j in $\bar{x}$, the variable is guaranteed to be out of an optimal solution if

$$\text{LRP}(\lambda) + \bar{c}_j < z_{lb}. \tag{5.28}$$

Assume that variable x_j is such that $\bar{x}_j = 1$. One may attempt to price that variable into an optimal solution by solving (5.24)–(5.26) with the additional constraint $x_j = 0$. Clearly, if the solution value thus obtained is less than z_{lb}, variable x_j is guaranteed to be in an optimal solution to KP. The variable could then be fixed to 1.

5.6.3 A NDRC algorithm for KP

Whenever applicable (i.e. when $\bar{x}$ is fractional), a NDRC algorithm for KP would separate Gomory cuts, as suggested in Section 5.5. These cuts should be dynamically dualized, as proposed in Section 5.2. In our experiments with the KP instance in Table 5.1, a total of 1000 SM iterations were performed. Parameter α, initially set to 2.0, was halved after 50 consecutive SM iterations without an overall improvement on the best upper bound so far generated. LH, as described above, was called for every SM iteration. The same applies to the proposed variable fixing tests.

The computational results obtained are shown on Table 5.6.3. Best lower and upper bounds obtained up to the given fixed number of SM iterations are presented in the table. Twenty four variables were fixed into an optimal solution while ten variables were fixed out. Comparing the LP relaxation of the KP instance with the best upper bound generated, it is possible to see that the initial duality gap (i.e. LP relaxation value minus the best lower bound found) was closed by over 82% by the best NDRC upper bound obtained.

5.6.4 A DRC algorithm for KP

A DRC algorithm for KP would generate an initial Gomory cut from LP relaxation (5.12)–(5.14). Such a cut would then be dualized and a fixed number of SM iterations would be carried out in an attempt to solve the corresponding LDP. Assume that a few rounds of DRC Gomory cut separations have already been performed and consider the corresponding LDP. Let $\bar{x}$ be an optimal solution to the very last LRP(λ) solved while attempting to solve LDP with SM. Three outcomes are then possible. If $\bar{x}$ is integral, the DRC algorithm should be stopped since no Gomory cut could be generated from $\bar{x}$. If $\bar{x}$ is fractional but the newly separated Gomory cut is already currently

Table 5.2 Lower and upper bounds for NDRC

Iteration	Lower Bound	Upper Bound
1	18970083333551896	19110448305312768
100	18970083333551900	19068005714952528
200	19014186620712300	19049279232335140
300	19014186620712300	19042422190532812
400	19014186620712300	19041389490291912
500	19014186620712300	19038788425295852
600	19014186620712300	19036609424140228
700	19014186620712300	19035349741901788
800	19014186620712300	19033798315232872
900	19014186620712300	19032012697947240
1000	19014186620712300	19031191289895520

dualized, LDP should be, once again, stopped. Finally, if $\bar{x}$ is fractional and a Gomory cut different from the ones currently dualized is separated from $\bar{x}$, the cut should be dualized thus giving rise to a new LDP to be solved. In our experiments with the KP instance in Table 5.1, a total of 10 LDP rounds were performed. Each of these rounds involved 100 SM iterations where parameter α, initially set to 2.0, was halved after 5 consecutive SM iterations without an overall improvement on the best upper bound so far generated in the round.

LH, as described above, was called for every SM iteration. The same applies to the proposed variable fixing tests.

The computational results obtained are shown in Table 5.6.4. Best lower and upper bounds obtained up to the end of every LDP solving round are presented in the table (round 0 corresponds to the initial lower and upper bounds). Twenty four variables were fixed in an optimal solution while ten variables were fixed out. Comparing the LP relaxation of the KP instance with the best upper bound generated, it is possible to see that the initial duality gap (i.e. LP relaxation minus best lower bound) was closed by over 74% by the best upper bound obtained.

Comparing the best lower bounds generated respectively by NDRC and DRC, one should notice that a difference of only 4 units exists in favor of the NDRC bound. Such a small difference between feasible solution values involving 17 digit numbers, give a hint on the difficulty of finding proven optimal solutions to our KP instance.

The slightly better results obtained by NDRC are in accordance with results obtained in Lucena (2004) for the Steiner Tree Problem. In any case, duality gap reductions were substantial for both NDRC and DRC. One should then expect that a Branch-and-Bound algorithm based on either approach would perform well for the KP instance tested.

Table 5.3 Lower and upper bounds for DRC

Round	Lower Bound	Upper Bound
0	18970083333551896	19110448305312768
1	19014186620712292	19050199902159760
2	19014186620712292	19044749414273156
3	19014186620712292	19044745465383692
4	19014186620712292	19044744079996880
5	19014186620712300	19042060411170460
6	19014186620712300	19041048482141080
7	19014186620712300	19041026532351020
8	19014186620712304	19038583646728036
9	19014186620712304	19038583646728036
10	19014186620712304	19038583646728036

5.7 CONCLUSIONS AND SUGGESTIONS FOR FUTURE WORK

Relax-and-Cut is proving to be an attractive proposition for generating *good quality* dual bounds to Integer Programming Problems. The technique may be used on its own, as exemplified in this chapter, or be combined with LP based solution algorithms into a hybrid solution approach (see Calheiros et al. (2003), for details). Relax-and-Cut also appears very attractive for the development of Lagrangian heuristics.

Implementing NDRC under subgradient optimization methods different from SM, appears clearly relevant. One example of this appears in Belloni and Sagastizábal (2004) where NDRC was adapted to operate under a Bundle method (Bonnans et al., 1997). Investigating variants of SM to NDRC that retain *computational lightness* while improving accuracy also appears very attractive.

For the particular application focused in this chapter we plan to implement an exact solution Branch-and-Bound algorithm based on NDRC. Such an algorithm should include, in addition to the Gomory cuts studied here, Lifted Minimum Cover Inequalities.

Acknowledgments

This research was supported by CNPq (grant 300149/ 94-8) and FAPERJ (grant E26/ 71.906/ 00).

Bibliography

E. Balas and N. Christofides. A restricted Lagrangian approach to the traveling salesman problem. *Mathematical Programming*, 21:19–46, 1981.

F. Barahona and R. Anbil. The volume algorithm: producing primal solutions with the subgradient method. *Mathematical Programming*, 87:385–399, 2000.

F. Barahona and L. Ladányi. Branch and cut based on the volume algorithm: Steiner trees in graphs and Max-Cut. Technical report, IBM Watson Research Center, 2001.

J.E. Beasley. Lagrangean relaxation. In Collin Reeves, editor, *Modern Heuristic Techniques*, page Oxford. Blackwell Scientific Press, 1993.

A. Belloni and A. Lucena. A relax and cut algorithm for the traveling salesman problem, 2000. Talk given at the 17th International Symposium on Mathematical Programming.

A. Belloni and A. Lucena. A Lagrangian heuristic for the linear ordering problem. In M.G.C. Resende and J. Pinho de Sousa, editors, *Metaheuristics: Computer Decision-Making*, pages 123–151. Kluwer Academic Publishers, 2003.

A. Belloni and C. Sagastizábal. Dynamic Bundle Methods. Technical Report A 2004/296, Instituto de Matemática Pura e Aplicada, 2004.

J.F. Bonnans, J.Ch. Gilbert, C. Lemaréchal, and C. Sagastizábal. *Optimisation numérique: Aspects théoriques et pratiques*. Springer Verlag, 1997.

F. Calheiros, A. Lucena, and C. de Souza. Optimal Rectangular Partitions. *Networks*, 41:51–67, 2003.

J.B.C. da Silva. Uma heuristica lagrangeana para o problema da árvore capacitada de custo mínimo. Master's thesis, Programa de Engenharia de Sistemas e Computação, COPPE, Universidade Federal do Rio de Janeiro, 2002.

G.B. Dantzig. Discrete variable extremum problems. *Operations Research*, 4:266–277, 1957.

M. de Moraes Palmeira, A. Lucena, and O. Porto. A relax and cut algorithm for quadratic knapsack problem. Technical report, Departamento de Administração, Universidade Federal do Rio de Janeiro, 1999.

L. Escudero, M. Guignard, and K. Malik. A Lagrangian relax and cut approach for the sequential ordering with precedence constraints. *Annals of Operations Research*, 50:219–237, 1994.

H. Everett III. Generalized Lagrange multiplier method for solving problems of optimum allocation of resources. *Operations Research*, 11:399–417, 1963.

M.L. Fisher. The Lagrangian relaxation method for solving integer programming problems. *Management Science*, 27:1–18, 1981.

B. Gavish. Augmented Lagrangean based algorithms for centralized network design. *IEEE Trans. on Communications*, 33:1247–1257, 1985.

A.M. Geoffrion. Lagrangian relaxation for integer programming. *Mathematical Programming Study*, 2:82–114, 1974.

R.E. Gomory. An algorithm for integer solutions to linear programs. In R. Graves and P. Wolfe, editors, *Recent advances in mathematical programming*, pages 269–302. McGraw-Hill, 1963.

M. Guignard. Efficient cuts in Lagrangean relax-and-cut schemes. *European Journal of Operational Research*, 105:216–223, 1998.

M. Guignard. Lagrangean relaxation. *TOP*, 11:151–199, 2004.

M. Held and R.M. Karp. The traveling salesman problem and minimum spanning trees. *Operations Research*, 18:1138–1162, 1970.

M. Held and R.M. Karp. The traveling salesman problem and minimum spanning trees: Part II. *Mathematical Programming*, 1:6–25, 1971.

M. Hunting, U. Faigle, and W. Kern. A Lagrangian relaxation approach to the edge-weighted clique problem. *European Journal of Operational Research*, 131:119–131, 2001.

C. Lemaréchal. Lagrangian relaxation. In M. Jünger and D. Naddef, editors, *Computational Combinatorial Optimization*, pages 115–160. Springer Verlag, 2001.

A. Lucena. Steiner problem in graphs: Lagrangean relaxation and cutting planes. *COAL Bulletin*, 21:2–8, 1992.

A. Lucena. Steiner problem in graphs: Lagrangean relaxation and cutting planes. In *Proceedings of NETFLOW93*, pages 147–154. Univesitá degli Studi di Pisa, Dipartimento di Informatica, 1993. Technical report TR-21/93.

A. Lucena. Non Delayed Relax-and-Cut Algorithms. *Annals of Operations Research*, 2004. In press.

S. Martello and P. Toth. Upper bounds and algorithms for hard $0-1$ knapsack problems. *Operations Research*, 45:768–778, 1997.

C. Martinhon, A. Lucena, and N. Maculan. Stronger K-Tree relaxations for the vehicle routing problem. *European Journal of Operational Research*, 158:56–71, 2004.

M. Padberg and G. Rinaldi. A branch-and-cut algorithm for the resolution of large-scale symmetric traveling salesman problems. *SIAM Review*, 33:60–100, 1991.

D. Pisinger. A minimal algorithm for the $0-1$ knapsack problem. *Operations Research*, 45:758–767, 1997.

T.K. Ralphs and M.V. Galati. Decomposition and dynamic cut generation in integer linear programming. Technical report, Department of Industrial and Systems Engineering, Lehigh University, 2003.

J.F. Shapiro. A survey of Lagrangean techniques for discrete optimization. *Annals of Discrete Mathematics*, 5:113–138, 1974.

J.F. Shapiro. *Mathematical programming: Structures and algorithms*. John Wiley and Sons, 1979.

L.A. Wolsey. *Integer Programming*. John Wiley and Sons, 1998.

6 MINIMUM COST NETWORK FLOW ALGORITHMS

Jeffery L. Kennington[1] and Richard V. Helgason[1]

[1]Southern Methodist University
Dallas, Texas, 75275, USA
jlk@engr.smu.edu
helgason@engr.smu.edu

Abstract: The minimal cost network flow model is defined along with optimality criteria and three efficient procedures for obtaining an optimal solution. Primal and dual network simplex methods are specializations of well-known algorithms for linear programs. The primal procedure maintains primal feasibility at each iteration and seeks to simultaneously achieve dual feasibility, The dual procedure maintains dual feasibility and moves toward primal feasibility. All operations for both algorithms can be performed on a graphical structure called a tree. The scaling push-relabel method is designed exclusively for optimization problems on a network. Neither primal nor dual feasibility is achieved until the final iteration.

Keywords: Networks-graphs flow algorithms, integer programming algorithms, linear programming algorithms, linear programming simplex algorithms.

6.1 INTRODUCTION

The special structure found in the minimum cost network flow problem has been exploited in the design of highly efficient solution techniques. In the early 1950s it was known that this special structure permits radical simplifications of the simplex method (see Dantzig (1951)). Methods for solving this problem are among the most efficient known for solving large-scale optimization problems. The mathematical foundations for these highly successful procedures can be found in the following classic books that appeared in the 1960s: Ford and Fulkerson (1962), Dantzig (1963), and Charnes and Cooper (1967). Much of the history of this problem along with some thirteen distinct algorithms may be found in the award-winning manuscript Ahuja et al. (1993). Of the many algorithms found in the literature, the most computationally efficient appear to be the following: primal network simplex, dual network simplex, and scaling push-relabel. A pseudo-code for each of these algorithms is given in this presentation.

An important aspect of real-world network management is the rapid restoration of service in the event of fiber cuts and equipment failures. Our colleagues at MCI in Dallas, Texas have developed complex models for restoration that require the use of efficient solution software. Their first real-time restoration system involved solving a series of minimum cost network flow problems, each of which determined a least-cost restoration path for a prioritized list of customers. This particular application used a software implementation of the primal network simplex algorithm, as described in this presentation.

6.1.1 Notation

A *network* is a directed graph $[V,E]$ with node set $V = \{1,\ldots,\bar{v}\}$ and arc set $E = \{1,\ldots,\bar{e}\}$ of ordered pairs of nodes. We adopt the mild notational abuse of referring to an arc k of E only by its endpoints, e.g. $(i,j) \in E$. Ambiguity results when there are multiple arcs from node i to node j. Alternatively, we may refer to i as the *tail* and to j as the *head* of arc $k \in E$. An arc is said to be *incident* to its head and tail nodes and vice-versa. A arc is said to be *directed* from its tail node to its head node, corresponding to some mechanism being modeled which permits flow in that direction only. Arcs (i,j) and (j,i) are said to be *oppositely directed*. For a given $v \in V$, the sets $F^*(v) = \{(v,j) \in E\}$ and $B^*(v) = \{(i,v) \in E\}$ are known as the *forward star* and *backward star*, respectively, of node v.

A finite odd length sequence $P = \{v_1, e_1, v_2, e_2, \ldots, v_p, e_p, v_{p+1}\}$, whose odd elements are nodes of V and whose even elements are arcs of E, is called a *walk of length p* in $[V,E]$ if $e_k \in \{(v_k, v_{k+1}), (v_{k+1}, v_k)\}$. If the nodes of P are distinct with the exception that $v_1 = v_{p+1}$ is allowed, then P is said to be a *path* from v_1 to v_{p+1} and those nodes are said to be *linked* by P. A graph is said to be *connected* if any two of its nodes can be linked. Note that in a path of length $p > 2$ the arcs will be distinct. If P is a path of length $p \geq 2$ with $v_1 = v_{p+1}$ and the arcs of P are distinct, then P is called a *cycle*. The requirement that the arcs of a cycle be distinct is necessary to prevent an exceptional case from being a cycle when $p = 2$ (see example path $\bar{C}$ below). Given path or cycle P, the arc (v_k, v_{k+1}) in P is said to be traversed in the *normal direction* while the arc (v_{k+1}, v_k) in P is said to be traversed in the *reverse direction*. A graph from which no cycles can be formed is said to be *acyclic*. A connected acyclic graph is called a *tree*. A node in a tree having a single incident arc is known as a *leaf node*. A tree with at least one arc will have at least two leaf nodes. It is important to note that there is a unique path linking every pair of nodes in a tree. For the network illustrated in Figure 6.1, $V = \{1,2,3,4\}$, $E = \{(1,2),(1,3),(2,3),(2,4),(3,4)\}$, $F^*(2) = \{(2,3),(2,4)\}$, $B^*(2) = \{(1,2)\}$, $W = \{1,(1,3),3,(3,4),4,(3,4),3,(2,3),2\}$ is a walk of length 4 but not a path, $P = \{1,(1,3),3,(2,3),2,(2,4),4\}$ is a path from 1 to 4 with arcs $(1,3)$ and $(2,4)$ traversed in the normal direction and arc $(2,3)$ traversed in the reverse direction, $C = \{1,(1,3),3,(2,3),2,(1,2),1\}$ is a cycle, and $\bar{C} = \{1,(1,3),3,(1,3),1\}$ is a path but not a cycle.

A graph $[\bar{V}, \bar{E}]$ is said to be a *subgraph* of $[V,E]$ if $\bar{V} \subseteq V$ and $\bar{E} \subseteq E$ and when $\bar{V} = V$, the subgraph is said to *span* $[V,E]$ or to be a *spanning subgraph* of $[V,E]$. A *spanning tree* of $[V,E]$ is a tree that spans $[V,E]$. Some spanning trees for the example network in Figure 6.1 are illustrated in Figure 6.2.

NETWORK FLOW ALGORITHMS 149

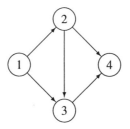

Figure 6.1 Example Network

6.1.2 The Problem

In network $[V, E]$, let $c(k)$ and $u(k)$ denote the *unit cost* and *arc capacity*, respectively, for arc $k \in E$ with corresponding $\bar{e}$-vectors c and u. In this presentation it is assumed that $0 \leq u(k) < \infty$. Let $r(v)$ denote the *requirement* for node $v \in V$ with corresponding $\bar{v}$-vector r. If $r(v) > 0$, then node v is said to be a *supply node* with supply $r(v)$. If $r(v) < 0$, then node v is said to be a *demand node* with demand $|r(v)|$. Note that some authors reverse this sign convention. If $r(v) = 0$, then node v is said to be a *transshipment node*.

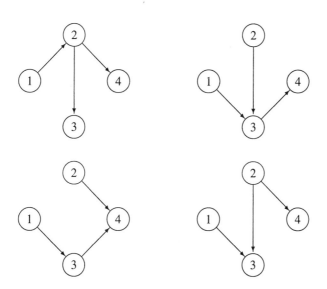

Figure 6.2 Spanning Trees

Let $x(k)$ denote the flow on arc $k \in E$ with corresponding $\bar{e}$-vector or *flow vector* x. Given problem data $D = (V, E, c, u, r)$, the set of *feasible flows* is $X = X^{fc} \cap X^{sb}$, where

$$X^{fc} = \left\{ x : \sum_{k \in F^*(v)} x(k) - \sum_{k \in B^*(v)} x(k) = r(v), \forall v \in V \right\}$$

are the *flow conservation equations* and

$$X^{sb} = \{ x : 0 \leq x(k) \leq u(k), \forall k \in E \}$$

are the *simple flow bounds*. Note that for each node $v \in V$, there is a flow conservation equation specifying that the difference of the total flow out of v and the total flow into v to be the requirement at v. Given D, the *minimal cost network flow problem* is to find a flow vector $\bar{x}$ such that

$$c\bar{x} = \min\{cx : x \in X\}.$$

Defining the $\bar{v} \times \bar{e}$ node-arc incidence matrix A associated with $[V,E]$ by

$$A_{n,k} = \begin{cases} +1, & \text{if the tail of arc k is node n} \\ -1, & \text{if the head of arc k is node n} \\ 0, & \text{otherwise}, \end{cases}$$

allows us to express this problem concisely as

$$\begin{aligned} \text{minimize} \quad & cx \\ \text{subject to} \quad & Ax = b \\ & 0 \leq x \leq u \end{aligned}$$

For the network illustrated in Figure 6.1 the node-arc incidence matrix is

$$A = \begin{bmatrix} 1 & 1 & & & \\ -1 & & 1 & & \\ & -1 & & 1 & \\ & & -1 & -1 & \end{bmatrix}$$

and the flow conservation equations are

$$\begin{aligned} x(1,2) + x(1,3) &&&&&&& = r(1) \\ -x(1,2) &+ x(2,3) + x(2,4) &&&& = r(2) \\ -x(1,3) &- x(2,3) &&+ x(3,4) &= r(3) \\ &&- x(2,4) &- x(3,4) &= r(4) \end{aligned}$$

Let $\bar{1}$ denote a $\bar{v}$-component row vector with all entries 1. Note that $\bar{1}A = 0$ since the only two nonzero entries in each column of A are $+1$ and -1. Hence, $\bar{1}A = \bar{1}r = \sum_{v \in V} r(v) = 0$, which implies that $X \neq \phi$ only if total supply equals total demand.

6.1.3 Specializations of the Network Flow Problem

The assignment problem, the semi-assignment problem, and the transportation problem are all special cases of the minimum cost network flow problem. The shortest path problem between a pair of nodes, say s and t, can be modeled by setting $r(s) = 1$, $r(t) = -1$, and $r(v) = 0, \forall v \in V \setminus \{s,t\}$. The cost $c(k)$ for $k \in E$ is the length of arc k and $u(k) = 1, \forall k \in E$. The maximal continuous flow problem between node 1 and node 4 in Figure 6.1 is modeled as illustrated in Figure 6.3, where $c(4,1) = -1$ and all other costs are zero. Physical links which permit an arbitrary direction of flow are referred to as *undirected links*. Models which use undirected links can be accommodated in the network structure presented here by replacing each undirected link with a pair of oppositely directed arcs.

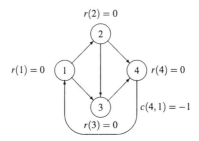

Figure 6.3 Maximum Flow From Nodes 1 to 4

6.1.4 Duality Theory and Optimality Conditions

The dual of the primal problem $\min\{cx : Ax = r, 0 \leq x \leq u\}$ is $\max\{r\pi - u\mu : \pi A - \mu \leq c, \mu \geq 0\}$. If $\bar{x} \in \{x : Ax = r, 0 \leq x \leq u\}$ and $(\bar{\pi},\bar{\mu}) \in \{(\pi,\mu) : \pi A - \mu \leq c, \mu \geq 0\}$, then $c\bar{x} \geq r\bar{\pi} - u\bar{\mu}$. If x^* is optimal for the primal and (π^*,μ^*) is optimal for the dual, then $cx^* = r\pi^* - u\mu^*$. For any $\bar{v}$-component $\bar{\pi}$ and any arc $k = (i,j)$, $\bar{c}(k) = c(k) - \bar{\pi}(i) + \bar{\pi}(j)$ is called the *reduced cost* for arc k. If $(\bar{x},\bar{\pi})$ satisfy the following conditions:

$$A\bar{x} = r \tag{6.1a}$$
$$\text{if } \bar{c}(k) = 0 \text{ , then } 0 \leq \bar{x}(k) \leq u(k) \tag{6.1b}$$
$$\text{if } \bar{c}(k) > 0 \text{ , then } \bar{x}(k) = 0 \tag{6.1c}$$
$$\text{if } \bar{c}(k) < 0 \text{ , then } \bar{x}(k) = u(k) \tag{6.1d}$$

then $\bar{x}$ is an optimum for the primal problem. Proofs of these results may be found in any standard textbook on linear programming.

6.1.5 Basic Solutions

Members of X of particular interest can be produced by partitioning E into three mutually disjoint sets (B,L,U) with $[V,B]$ a spanning tree for $[V,E]$. (Thus $B \cup L \cup U = E$

and $B \cap L = B \cap U = L \cap U = \phi$.) This induces partitionings of the flow vector, the arc capacity vector, and the flow conservation equations as $A^B x^B + A^L x^L + A^U x^U = r$. Let $\bar{x}(k) = 0, \forall k \in L$ and $\bar{x}(k) = u(k), \forall k \in U$. Since $[V,B]$ is a tree, it can easily be shown that $\bar{v} - 1$ is the rank of both A^B and the augmented matrix $\left[A^B \mid r - A^U \bar{x}^U\right]$, so that $A^B x^B = r - A^U \bar{x}^U$ can be solved uniquely. Let $\bar{x}^B$ be that unique solution. If $0 \leq \bar{x}^B \leq u^B$, then $0 \leq \bar{x} \leq u$ and $\bar{x}$ is called the *basic feasible solution* relative to the partitioning (B,L,U) with A^B often referred to as the *basis* (even though it is rank-deficient).

Since $[V,B]$ is a tree, the unique solution to $A^B x^B = r - A^U \bar{x}^U$ can be constructed one component at a time by the following simple method.

```
procedure Flows-On-A-Tree(V,B,b,x̄)
/* Flows On A Tree solves Nx = b. */
/* [V,B] is a tree with node-arc incidence matrix N. */
/* b is the right-hand side. */
/* x̄ is the solution. */
begin
    [V̄,B̄] ← [V,B] and b̄ ← b;
    while V̄ ≠ φ do
        ∃ a leaf node n of [V̄,B̄];
        if ∃(n,v) ∈ B̄ then
            B̄ ← B̄ \ {(n,v)}; V̄ ← V̄ \ {n};
            x̄(n,v) ← b̄(n); b̄(v) ← b̄(v) + b̄(n);
        else
            select (v,n) ∈ B̄;
            B̄ ← B̄ \ {(v,n)}; V̄ ← V̄ \ {n};
            x̄(v,n) ← -b̄(n); b̄(v) ← b̄(v) - b̄(n);
        end
    end
end
```

Algorithm 6.1: Flows on a tree

For the tree illustrated in Figure 6.4, $\bar{x}(1,6) = 2$, $\bar{x}(3,6) = -3$, $\bar{x}(4,2) = -3$, $\bar{x}(4,3) = 2$, $\bar{x}(5,3) = -6$, and $\bar{x}(7,6) = 1$.

It is also relatively easy to create a set of dual variables $\bar{\pi}$ such that the reduced costs are zero for all $k \in B$, i.e. $\bar{c}(k) = c(k) - \bar{\pi}(i) + \bar{\pi}(j) = 0$ for all $k = (i,j) \in B$. Note that $-\bar{\pi}(i) + \bar{\pi}(j) = -(\bar{\pi}(i) + \alpha) + (\bar{\pi}(j) + \alpha)$, which implies that adding any α to every component of a set of dual variables $\bar{\pi}$ yielding reduced costs of zero will produce another set of dual variables yielding reduced costs of zero. Since $[V,B]$ is a tree, a set of dual variables yielding zero reduced costs can be constructed by the following simple method.

For the tree illustrated in Figure 6.4, $\bar{\pi}(1) = -10$, $\bar{\pi}(2) = -4$, $\bar{\pi}(3) = 0$, $\bar{\pi}(4) = 2$, $\bar{\pi}(5) = -1$, $\bar{\pi}(6) = -3$, and $\bar{\pi}(7) = -3$.

6.2 THE PRIMAL NETWORK SIMPLEX ALGORITHM

This algorithm is a specialization of the primal simplex method for linear programming that exploits the underlying network structure. It constructs a series of basic feasible solutions $\bar{x}$ and associated partitionings (B,L,U) such that $A\bar{x} = r, 0 \leq \bar{x} \leq u$.

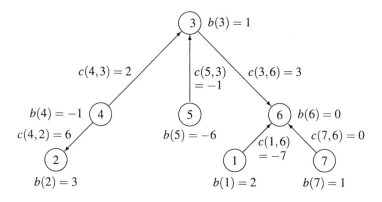

Figure 6.4 Example Tree

```
procedure Duals-On-A-Tree (V, B, c, π̄)
/* produces dual variables π̄ */
/* [V, B] is a tree */
/* c is the unit cost vector */
/* π̄ satisfies c̄(k) = c(k) − π̄(i) + π̄(j) = 0 ∀k = (i, j) ∈ B */
begin
    select a node v ∈ V; V̄ ← V \ {v}; π̄(v) ← 0;
    /* a arbitrary node has its dual set to 0 above */
    /* all other duals are set recursively as V̄ shrinks */
    while V̄ ≠ φ do
        if ∃(n, v) ∈ B with v ∈ V̄ and n ∈ V \ V̄ then
          | π̄(v) ← π̄(n) − c(n, v); V̄ ← V̄ \ {v};
        end
        if ∃(v, n) ∈ B with v ∈ V̄ and n ∈ V \ V̄ then
          | π̄(v) ← π̄(n) + c(n, v); V̄ ← V̄ \ {v};
        end
    end
end
```

Algorithm 6.2: Duals on a tree

As the basic feasible solutions progress, the dual variables $\bar{\pi}$ are updated to maintain reduced costs of zero for all $k \in B$. At optimality the basic feasible solution will also satisfy $\bar{c}(k) \geq 0 \; \forall k \in L$ and $\bar{c}(k) \leq 0 \; \forall k \in U$. The corresponding specialized procedure follows.

One simple strategy for obtaining the initial basic feasible solution is to append one additional node w and $\bar{v}$ additional arcs to the network. Before enlarging the network all original arcs are placed in L with zero flows and U is empty. Then as additional arcs are appended they are placed in B. For each $v \in V$ with $r(v) \geq 0$, append the arc $z = (v, w)$ with $c(z) = 0$, $u(z) = r(v)$, and $\bar{x}(z) = r(v)$. For each $v \in V$ with $r(v) < 0$, append the arc $z = (w, v)$ with $c(z) = \infty$ and $u(z) = \bar{x}(z) = -r(v)$. This strategy is illustrated in Figure 6.5. The procedure *Duals On A Tree* can be used to determine $\bar{\pi}$ at each iteration (see /* Dual Calculation */ in procedure *Primal Network Simplex*), although it is possible to update only a portion of $\bar{\pi}$ whenever B changes.

procedure *Primal-Network-Simplex*(V,E,c,u,r)
/* returns $\bar{x} \in X$ such that $\bar{x} \in \arg\min\{cx : x \in X\}$ */
/* $[V,E]$ is a network */
/* c is the unit cost vector */
/* u is the arc capacity vector */
/* r is the requirements vector */
/* assumption: $X = \{x : Ax = r, 0 \leq x \leq u\} \neq \phi$ */
/* assumption: $u(k) < \infty \; \forall k \in E$ */
begin
    Obtain (B,L,U) with $\bar{x}$ the corresponding basic feasible solution;
    repeat
        /* Dual Calculation */
        Obtain a set of duals $\bar{\pi}$ for which $\bar{c}(k) = 0 \; \forall k \in B$;
        /* Pricing */
        $N^- \leftarrow \{k \in L : \bar{c}(k) < 0\} \; ; N^+ \leftarrow \{k \in U : \bar{c}(k) > 0\}$;
        if $N^- \cup N^+ = \phi$ **then**
            return $\bar{x}$
        else
            select $k = (t,h) \in N^- \cup N^+$;
        end
        if $k \in L$ **then**
            $(o,d) \leftarrow (h,t)$;
        else
            $(o,d) \leftarrow (t,h)$;
        end
        /* Column Update */
        $P \leftarrow$ the path in $[V,B]$ from o to d; $Q \leftarrow \{k \in B : k \in P\}$;
        $Q^+ \leftarrow \{k \in Q : k$ is traversed in the normal direction in $P\}$;
        $Q^- \leftarrow Q \setminus Q^+$;
        /* Ratio Test */
        $\Delta^+ \leftarrow \min\{u(a) - \bar{x}(a) : a \in Q^+\}; \Delta^- \leftarrow \min\{\bar{x}(a) : a \in Q^-\}$;
        $\Delta \leftarrow \min\{u(k), \Delta^+, \Delta^-\}$;
        /* Flow Update */
        $\bar{x}(a) \leftarrow \bar{x}(a) + \Delta \; \forall a \in Q^+; \bar{x}(a) \leftarrow \bar{x}(a) - \Delta \; \forall a \in Q^-$;
        if $k \in L$ **then**
            $\bar{x}(k) \leftarrow \Delta$;
        else
            $\bar{x}(k) \leftarrow u(k) - \Delta$;
        end
        /* Basis Exchange */
        $L \leftarrow L \setminus \{k\}; U \leftarrow U \setminus \{k\}$;
        if $\Delta = u(k)$ **then**
            if $k \in L$ **then**
                $U \leftarrow U \cup \{k\}$;
            else
                $L \leftarrow L \cup \{k\}$;
            end
        else
            $R \leftarrow \{a \in Q^- : \bar{x}(a) = 0\} \cup \{a \in Q^+ : \bar{x}(a) = u(a)\}$;
            select $r \in R$; $B \leftarrow (B \setminus \{r\}) \cup \{k\}$;
            if $r \in Q^-$ **then**
                $L \leftarrow L \cup \{r\}$;
            else
                $U \leftarrow U \cup \{r\}$;
            end
        end
    until *exited*;
end

Algorithm 6.3: Primal network simplex method

Efficient software implementations of the primal simplex algorithm use special heuristics for the pricing operation and special data structures for maintaining and traversing the tree $[V,B]$. The data associated with the arcs is often indexed using a structure associated with the forward or backward stars for the nodes. Details may be found in Glover et al. (1974a), Glover et al. (1974b), Glover et al. (1974c), Bradley et al. (1977), Barr et al. (1979), Kennington and Helgason (1980), and Grigoriadis (1986).

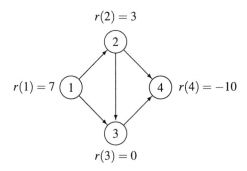

Figure 6.5a Example With 2 Source Nodes And 1 Demand Node

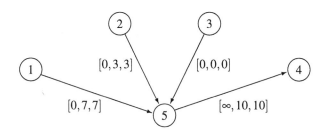

Legend: [cost, capacity, flow]

Figure 6.5b Initial Basic Feasible Solution

Figure 6.5 The Starting Solution

6.3 THE DUAL NETWORK SIMPLEX ALGORITHM

This algorithm is a specialization of the dual simplex method for linear programming that exploits the underlying network structure. It constructs a series of basic solutions $\bar{x}$ and associated partitionings (B,L,U) such that $A\bar{x} = r$ and $\bar{c}(k) = 0 \,\forall k \in B, \bar{c}(k) > 0 \,\forall k \in L$, and $\bar{c}(k) < 0 \,\forall k \in U$. Any basic solution that satisfies these conditions is said to be *dual feasible*. At optimality $0 \leq \bar{x}(k) \leq u(k) \, \forall k \in B$. That is, primal feasibility is only attained at the final step. The corresponding specialized procedure follows.

Under the assumption that all arc capacities are finite an initial dual feasible solution can easily be constructed. If this assumption does not hold, it is still possible to obtain such a solution using the two-phase approach described in Kennington and Mohamed (1997). The starting procedure (under finite arc capacities) follows.

6.4 THE SCALING PUSH-RELABEL ALGORITHM

The scaling push-relabel method extends the ideas of cost-scaling developed by Röck (1980) and Bland and Jensen (1992). The strategy is based on the concept of ε-optimality which can be found in Tardos (1985) and Bertsekas (1991). The presentation here follows that of Goldberg (1997) and Ahuja et al. (1993).

Let $\varepsilon \geq 0$ be a given scalar. If $(\bar{x}, \bar{\pi})$ satisfy the conditions:

$$A\bar{x} = r \tag{6.2a}$$
$$\text{if } |\bar{c}(k)| \leq \varepsilon, \text{ then } 0 \leq \bar{x}(k) \leq u(k) \tag{6.2b}$$
$$\text{if } \bar{c}(k) > \varepsilon, \text{ then } \bar{x}(k) = 0 \tag{6.2c}$$
$$\text{if } \bar{c}(k) < -\varepsilon, \text{ then } \bar{x}(k) = u(k) \tag{6.2d}$$

then $\bar{x}$ is said to be an ε–*optimum* for $\min\{cx : Ax = r, 0 \leq x \leq u\}$. For $\varepsilon = 0$, these conditions are the optimality conditions (6.1a)–(6.1d). If $\varepsilon < 1/\bar{v}$ and the cost vector is all integer, then an ε–optimum is also optimal for $\min\{cx : Ax = r, 0 \leq x \leq u\}$. Proofs are available in both Bertsekas (1991) and Ahuja et al. (1993). The method begins with a fairly large value for ε and a solution $(\bar{x}, \bar{c})$ that satisfies (6.2b)–(6.2d). After a series of flow and reduced cost adjustments, (6.2a) is also satisfied. Then ε is reduced and the process is repeated until $\varepsilon < 1/\bar{v}$.

For a given flow vector $\bar{x}$, the excess flow $e(v)$ at node v is given by

$$e(v) = r(v) - \sum_{a \in F^*(v)} \bar{x}(a) + \sum_{a \in B^*(v)} \bar{x}(a).$$

Hence, when $e(v) > 0$ the node v has excess supply that must be distributed. A *push* operation attempts to push flow away from v. This can be accomplished by increasing flow in some $a \in F^*(v)$ or decreasing flow in some $a \in B^*(v)$. If a push maintaining (6.2b)–(6.2d) is not possible, $\bar{\pi}(v)$ is modified so that a push will be permissible. The procedure implementing this approach follows.

A number of heuristic strategies for improving the computational performance of scaling push-relabel algorithms can be found in Goldberg (1997).

6.5 SOFTWARE IMPLEMENTATIONS AND EMPIRICAL EVALUATIONS

The first software implementations which clearly demonstrated the power of the primal network simplex algorithm were developed by Srinivasan and Thompson (1973)

procedure *Dual-Network-Simplex*(V, E, c, u, r)
/* returns $\bar{x} \in X$ such that $\bar{x} \in \arg\min\{cx : x \in X\}$ */
/* $[V, E]$ is a network */
/* c is the unit cost vector */
/* u is the arc capacity vector */
/* r is the requirements vector */
/* assumption: $X = \{x : Ax = r, 0 \leq x \leq u\} \neq \phi$ */
/* assumption: $u(k) < \infty \, \forall k \in E$ */
begin
    Obtain (B, L, U) with $\bar{x}$ the corresponding dual feasible solution;
    repeat
        /* Check Primal Feasibility */
        $I \leftarrow \{a : \bar{x}(a) < 0 \text{ or } \bar{x}(a) > u(a)\}$;
        if $I = \phi$ then
        | return $\bar{x}$;
        end
        /* Dual Calculation */
        Obtain a set of duals $\bar{\pi}$ for which $\bar{c}(k) = 0 \, \forall k \in B$;
        /* Select Leaving Variable */
        select $q \in I$;
        /* Select Entering Variable */
        $d(q) \leftarrow 1; d(a) \leftarrow 0 \, \forall a \in B \setminus \{q\}$;
        Obtain $\gamma(v) \, \forall v \in V$ so that $d(k) - \gamma(i) + \gamma(j) = 0 \, \forall k = (i, j) \in B$;
        /* the above values follow the dual calculation scheme */
        if $\bar{x}(q) > 0$ then
        | $w(i, j) \leftarrow (\gamma(j) - \gamma(i)) \, \forall (i, j) \in L \cup U$;
        else
        | $w(i, j) \leftarrow (\gamma(i) - \gamma(j)) \, \forall (i, j) \in L \cup U$;
        end
        $\Gamma \leftarrow \{a \in L : w(a) < 0\} \cup \{a \in U : w(a) > 0\}$;
        $\delta \leftarrow \max\left\{\dfrac{\bar{c}(a)}{w(a)} : a \in \Gamma\right\}, \Lambda \leftarrow \left\{a \in \Gamma : \dfrac{\bar{c}(a)}{w(a)} = \delta\right\}$;
        select $k = (t, h) \in \Lambda$;
        if $k \in L$ then
        | $(o, d) \leftarrow (h, t)$
        else
        | $(o, d) \leftarrow (t, h)$;
        end
        /* Column Update */
        $P \leftarrow$ the path in $[V, B]$ from o to d; $Q \leftarrow \{k \in B : k \in P\}$;
        $Q^+ \leftarrow \{k \in Q : k \text{ is traversed in the normal direction in } P\}$;
        $Q^- \leftarrow Q \setminus Q^+$;
        /* Flow Update */
        if $\bar{x}(q) > u(q)$ then
        | $\Delta \leftarrow \bar{x}(q) - u(q)$;
        else
        | $\Delta \leftarrow -\bar{x}(q)$;
        end
        $\bar{x}(a) \leftarrow \bar{x}(a) + \Delta \, \forall a \in Q^+$; $\bar{x}(a) \leftarrow \bar{x}(a) - \Delta \, \forall a \in Q^-$;
        if $k \in L$ then
        | $\bar{x}(k) \leftarrow \Delta$;
        else
        | $\bar{x}(k) \leftarrow u(k) - \Delta$;
        end
        /* Basis Exchange */
        $B \leftarrow (B \setminus \{q\}) \cup \{k\}$;
        if $\bar{x}(q) = u(q)$ then
        | $U \leftarrow U \cup \{q\}$;
        else
        | $L \leftarrow L \cup \{q\}$;
        end
        $L \leftarrow L \setminus \{k\}; U \leftarrow L \setminus \{k\}$;
    until *exited*;
end

Algorithm 6.4: Dual simplex network method

procedure *Dual-Feasible-Solution*(V,E,c,u,r)
/* returns (B,L,U) with $\bar{x}$ dual feasible */
/* $[V,E]$ is a network */
/* c is the unit cost vector */
/* u is the arc capacity vector */
/* r is the requirements vector */
/* assumption: $X = \{x : Ax = r, 0 \leq x \leq u\} \neq \phi$ */
/* assumption: $u(k) < \infty \, \forall k \in E$ */
begin
    Obtain B such that $[V,B]$ is a spanning tree for $[V,E]$;
    /* apply procedure *Duals On A Tree* */
    Obtain a set of duals $\bar{\pi}$ for which $\bar{c}(k) = 0 \, \forall k \in B$;
    $L \leftarrow \{a \in E : \bar{c}(a) > 0\}, \bar{x}(a) \leftarrow 0 \, \forall a \in L$;
    $U \leftarrow \{a \in E : \bar{c}(a) < 0\}, \bar{x}(a) \leftarrow u(a) \, \forall a \in U$;
    $b(v) \leftarrow r(v) - \sum_{(v,j) \in F^*(v)} u(v,j) + \sum_{(i,v) \in B^*(v)} u(i,v) \, \forall v \in V$;
    Apply **procedure** Flows-On-A-Tree to obtain $\bar{x}(a) \, \forall a \in B$;
end

Algorithm 6.5: Dual Feasible Solution

and by Glover and Klingman and their colleagues at the University of Texas at Austin (see Glover et al. (1974a), Glover et al. (1974b), and Glover et al. (1974c)). Other important contributions were made by Bradley et al. (1977) and Barr et al. (1979). Variations of the three methods presented in this exposition have been implemented in software and evaluated in a number of studies.

Goldberg (1997) reports on an experiment involving the five codes listed in Table 6.1. Kennington and Whitler (1998) report on an experiment with the six codes listed in Table 6.2. The scaling push-relabel code of Goldberg was found to be both fast and robust over a variety of problem structures. All the techniques and codes have their champions, but it is generally accepted that specialized network codes are at least two orders of magnitude faster than general linear programming software that can be used for these problems. Several solvers are available on-line at http://www-neos.mcs.anl.gov/neos/ or ftp://dimacs.rutgers.edu/pub/netflow/.

procedure *Scaling Push-Relabel*(V, E, c, u, r, α)
/* Returns $\bar{x} \in X$ such that $\bar{x} \in \arg\min\{cx : x \in X\}$ */
/* $[V, E]$ is a network */
/* c is the unit cost vector */
/* u is the arc capacity vector */
/* r is the requirements vector */
/* α is a reduction constant greater than 1 */
/* assumption: $X = \{x : Ax = r, 0 \leq x \leq u\} \neq \phi$ */
/* assumption: $u(k) < \infty \, \forall k \in E$ */
/* assumption: $c(k)$ is an integer $\forall k \in E$ */
begin
    $\varepsilon \leftarrow \max\{|c(a)| : a \in E\};$
    $\bar{x}(a) \leftarrow 0 \, \forall a \in E, \bar{c}(a) \leftarrow c(a) \, \forall a \in E;$
    repeat
        if $\varepsilon < 1/\bar{v}$ **then**
            return $\bar{x}$;
        end
        /* Scale */
        $\varepsilon \leftarrow \varepsilon/\alpha;$
        forall $a \in E$ **do**
            if $\bar{c}(a) > 0$ **then**
                $\bar{x}(a) \leftarrow 0;$
            end
            if $\bar{c}(a) < 0$ **then**
                $\bar{x}(a) \leftarrow u(a);$
            end
        end
        forall $v \in V$ **do**
            $e(v) = r(v) - \sum_{a \in F^*(v)} \bar{x}(a) + \sum_{a \in B^*(v)} \bar{x}(a);$
            while $\exists v \in V$ such that $e(v) > 0$ **do**
                /* Push */
                if $\exists a = (v, j) \in E$ such that $\bar{c}(a) < 0$ and $\bar{x}(a) < u(a)$ **then**
                    $\delta \leftarrow \min\{e(v), u(a) - \bar{x}(a)\};$
                    $\bar{x}(a) \leftarrow \bar{x}(a) + \delta;$
                    $e(v) \leftarrow e(v) - \delta;$
                    $e(j) \leftarrow e(j) + \delta;$
                else
                    if $\exists a = (i, v) \in E$ such that $\bar{c}(a) > 0$ and $\bar{x}(a) > 0$ **then**
                        $\delta \leftarrow \min\{e(v), \bar{x}(a)\};$
                        $\bar{x}(a) \leftarrow \bar{x}(a) - \delta;$
                        $e(i) \leftarrow e(i) + \delta;$
                        $e(v) \leftarrow e(v) - \delta;$
                    else
                      /* Relabel */
                      $\Gamma_1 \leftarrow \min\{\bar{c}(a) : a \in F^*(v) \text{ and } \bar{x}(a) < u(a)\};$
                      $\Gamma_2 \leftarrow \min\{-\bar{c}(a) : a \in B^*(v) \text{ and } \bar{x}(a) > 0\};$
                      $\Gamma \leftarrow \varepsilon + \min\{\Gamma_1, \Gamma_2\};$
                      $\bar{c}(a) \leftarrow \bar{c}(a) - \Gamma \, \forall a \in F^*(v);$
                      $\bar{c}(a) \leftarrow \bar{c}(a) + \Gamma \, \forall a \in B^*(v);$
                  end
              end
            end
        end
    until *exited*;
end

Algorithm 6.6: Scaling Push Relabel Algorithm

Table 6.1 Computer Codes in the Goldberg (1997) Study

Code Name	Algorithm Type	Reference
NETFLO	Primal Simplex	Kennington and Helgason (1980)
RNET Version 3.61	Primal Simplex	Grigoriadis (1986)
RELAX Version III	Scaling Push-Relabel	Bertsekas and Tseng (1990)
SPUR	Scaling Push-Relabel	Goldberg and Kharitonov (1993)
CS	Scaling Push-Relabel	Goldberg (1997)

Table 6.2 Computer Codes in the Kennington and Whitler (1998) Study

Code Name	Algorithm Type	Reference
NETFLO	Primal Simplex	Kennington and Helgason (1980)
CPLEX Version 4.0	Primal Simplex	CPLEX Callable Library
RELAX Version IV	Scaling Push-Relabel	Bertsekas and Tseng (1994)
CS2	Scaling Push-Relabel	Goldberg (1992)
NETFLO2 (Primal)	Primal Simplex	Kennington and Whitler (1998)
NETFLO2 (Dual)	Primal Simplex	Kennington and Whitler (1998)

Bibliography

R. Ahuja, T. Magnanti, and J. Orlin. *Network Flows: Theory, Algorithms, and Applications*. Prentice-Hall, Englewood Cliffs, NJ 07632, 1993.

R. Barr, F. Glover, and D. Klingman. Enhancement of spanning tree labeling procedures for network optimization. *INFOR*, 17:16–34, 1979.

D. Bertsekas. *Linear Network Optimization: Algorithms and Codes*. The MIT Press, Cambridge, MA, 1991.

D. Bertsekas and P. Tseng. RELAXT-III: A new and improved version of the RELAX code, lab. for information and decision systems report p-1990. Technical report, MIT, Cambridge, MA, 1990.

D. Bertsekas and P. Tseng. RELAX-IV: A faster version of the RELAX code for solving minimum cost flow problems. Technical report, Department of Electrical Engineering and Computer Science, MIT, Cambridge, MA, 1994.

R. Bland and D. Jensen. On the computational behavior of a polynomial-time network flow algorithm. *Mathematical Programming*, 54:1–41, 1992.

G. Bradley, G. Brown, and G. Graves. Design and implementation of large-scale primal transshipment algorithms. *Management Science*, 21:1–38, 1977.

A. Charnes and W. Cooper. *Management Models and Industrial Applications of Linear Programming: Volume I*. John Wiley and Sons, Inc., New York, NY, 1967.

G. Dantzig. Application of the simplex method to a transportation problem. In T. Koopmans, editor, *Activity Analysis of Production and Allocation*, pages 359–373. John Wiley and Sons, Inc., New York, NY, 1951.

G. Dantzig. *Linear Programming and Extensions*. Princeton University Press, Princeton, NJ, 1963.

L. Ford and D. Fulkerson. *Flows in Networks*. Princeton University Press, Princeton, NJ, 1962.

F. Glover, D. Karney, and D. Klingman. Implementation and computational comparisons of primal, dual, and primal-dual computer codes for minimum cost network flow problems. *Networks*, 4:191–212, 1974a.

F. Glover, D. Karney, D. Klingman, and A. Napier. A computational study on start procedures, basis change criteria, and solution algorithms for transportation problems. *Management Science*, 20:793–813, 1974b.

F. Glover, D. Klingman, and J. Stutz. Augmented threaded index method for network optimization. *INFOR*, 12:293–298, 1974c.

A. Goldberg. An efficient implementation of a scaling minimum cost flow algorithm,. Technical Report STAT-CS-92-1439, Computer Science Department, Stanford University, Stanford, CA, 1992.

A. Goldberg. An efficient implementation of a scaling minimum-cost flow algorithm. *Journal of Algorithms*, 22:1–29, 1997.

A. Goldberg and M. Kharitonov. On implementing scaling push-relabel algorithms for the minimum-cost flow problem. In D. Johnson and C. McGeoch, editors, *Network Flows and Matching: First DIMACS Implementation Challenge*, pages 157–198. AMS, Providence, RI, 1993.

M. Grigoriadis. An efficient implementation of the network simplex method. *Mathematical Programming Study*, 26:83–111, 1986.

J. Kennington and R. Helgason. *Algorithms for Network Programming*. John Wiley and Sons, Inc., New York, NY, 1980.

J. Kennington and R. Mohamed. An efficient dual simplex optimizer for generalized networks. In R. Barr, R. Helgason, and J. Kennington, editors, *Interfaces in Computer Science and Operations Research*, pages 153–182. Kluwer Academic Publishers, Norwell, MA 02061, 1997.

J. Kennington and J. Whitler. Simplex versus cost scaling algorithms for pure networks: An empirical analysis. Technical Report 96-CSE-8, Department of Computer Science and Engineering, Southern Methodist University, Dallas, TX, 1998.

H. Röck. Scaling techniques for minimal cost network flows. In U. Pape, editor, *Discrete Structures and Algorithms*, pages 181–191. Carl Hanser, Munich, 1980.

V. Srinivasan and G. Thompson. Benefit-cost analysis of coding techniques for the primal transportation algorithm. *Journal of the Association for Computing Machinery*, 20:194–213, 1973.

É. Tardos. A strongly polynomial minimum cost circulation algorithm. *Combinatorica*, 5:247–255, 1985.

7 MULTICOMMODITY NETWORK FLOW MODELS AND ALGORITHMS IN TELECOMMUNICATIONS

Michel Minoux[1]

[1]Laboratoire Lip6
Université Paris 6
Paris, France
Michel.Minoux@lip6.fr

Abstract: We present an overview of mathematical optimization models and solution algorithms related to optimal network design and dimensioning in telecommunications. All the models discussed are expressed in terms of minimum cost multicommodity network flows with appropriate choice of link (or node) cost functions. Various special cases related to practical applications are examined, including the linear case, the linear with fixed cost case, and the case of general discontinuous nondecreasing cost functions. It is also shown how the generic models presented can accommodate a variety of constraints frequently encountered in applications, such as, among others, constraints on the choice of routes, robustness and survivability constraints. Finally we provide an overview of recent developments in solution algorithms, emphasizing those approaches capable of finding provably exact solutions, which are essentially based on general mixed integer programming techniques, relaxation, cutting-plane generation techniques and branch & cut.

Keywords: Multicommodity flows, network design, mixed integer programming, relaxation, cutting-planes, constraint generation, branch & cut.

7.1 INTRODUCTION

We are concerned here with the development of models and exact solution methods for optimal design and dimensioning of Telecommunication networks.

Expressed in general terms, the basic problem addressed here may be subsumed as follows. Given a list of traffic nodes (sources or sinks), (telephone traffic, data, multimedia traffic) together with known anticipated values for the volume of traffic to be exchanged between nodes, it is required to build a network connecting sources and sinks and capable of handling all the requested traffic flows simultaneously. Achieving this involves the joint resolution of two types of problems:

- Defining a best network topology: determine which pairs of nodes should be connected by transmission links. Thus, the network topology is nothing else but the graph specifying how the traffic sources and sinks are interconnected.

- Defining link capacities (or dimensioning): determine the type and capacity of the transmission equipment to be installed on each link in order to construct the network according to the chosen topology.

A solution to the above network design problem (corresponding to a choice of a topology and link capacities) will be called *feasible* if the network thus constructed is actually capable of (simultaneously) flowing all the traffic requirements between all source-sink pairs. Let us point out here that, in order to check feasibility of a network (topology and dimensioning), we have to be sure that, for each source-sink requirement, we can find one or several paths to flow the corresponding traffic: this is the so-called *traffic routing problem*. Any network design problem therefore assumes proper handling (whether direct or indirect) of routing issues.

Clearly, a network design problem of the type described above will usually have a huge number of feasible solutions, but, among those, some will be less costly than others, where, depending on the context of application considered, cost will refer to various economic criteria to be minimized (investment costs, leasing costs) or maximized (payments received from subscribers). In the present chapter, the optimization criterion considered will be to minimize total network cost defined as the sum of the transmission equipment costs to be installed on the various links of the associated graph.

It will soon become apparent in the sequel that the practical difficulty in solving optimal network design problems depends, to a large extent, on the structure of the objective function, and on how closely it approximates actual costs of transmission equipment. For instance, if an approximation of actual costs by linear cost functions appears to be acceptable, then the problem may be reducible either to continuous linear programming or even to shortest path computations (for which many efficient algorithms are available). By contrast, if a much more accurate representation of reality is required, then step-increasing (discontinuous) cost functions have to be considered, giving rise to large scale integer linear programs, much more difficult to solve in practice. Other possible sources of difficulty in getting optimal solutions (with proof of optimality) are related to the presence of additional constraints derived either from technical restrictions in the use of equipment, or from various network management rules. Examples of this will be found in Sections 7.3.2 and 7.4.4.

7.2 BASIC MODELS: GRAPHS, FLOWS AND MULTICOMMODITY NETWORK FLOWS

The basic mathematical models used to formulate and solve optimal network design problems make use of graph-theoretic and/or linear programming-based models (see e.g. Berge (1970), Gondran and Minoux (1995), and Ahuja et al. (1993)).

The set of all possible topologies for the network to be constructed will typically be described by means of a given (undirected) graph $G = [\mathcal{N}, \mathcal{U}]$ where:

- the node set $\mathcal{N}$ represents the various traffic sources/sinks to be interconnected;
- the edge set $\mathcal{U}$ corresponds to the various pairs of nodes which may be physically connected by installing transmission links (cables, optical carriers, etc., ...).

It should be clear that the graph G defined above represents all the possible network topologies for interconnecting the $N = |\mathcal{N}|$ given nodes, which correspond to the decision of installing, or not installing, transmission equipment on each link $u \in \mathcal{U}$ of G. Thus any a priori possible network topology will correspond to a *partial graph* $G' = [\mathcal{N}, \mathcal{U}']$ of G (defined as a graph on the same node set $\mathcal{N}$, and edge set $\mathcal{U}' \subseteq \mathcal{U}$).

Now, a natural basic model to represent the way traffic between some source node and some sink nodes in G flows through the network, while using transmission resources installed on the links, is the so-called single-commodity network flow model (see Ford and Fulkerson (1962), Gondran and Minoux (1995), and Ahuja et al. (1993)). If $M = |\mathcal{U}|$ denotes the number of edges of G, a single-commodity flow between $s \in \mathcal{N}$ (source) and $t \in \mathcal{N}$ (sink), is a M-vector $\varphi = (\varphi_1, \varphi_2, \ldots, \varphi_M)^T$ such that:

- $|\varphi_u|$ represents the amount of transmission resource used on edge $u = (i, j)$;
- having chosen an arbitrary orientation on link $u = (i, j)$ (for instance assuming $i < j$, $\varphi_u > 0$ if the flow runs from i to j, $\varphi_u < 0$ if the flow runs from j to i), Kirchhoff's conservation law holds at each node $k \neq s$, $k \neq t$, in other words:

$$\forall k \in \mathcal{N}\setminus\{s,t\}:$$
$$\sum_{u \in \omega^+(k)} \varphi_u - \sum_{u \in \omega^-(k)} \varphi_u = 0 \tag{7.1}$$

and

$$\sum_{u \in \omega^+(s)} \varphi_u = \sum_{u \in \omega^-(t)} \varphi_u = v(\varphi) \tag{7.2}$$

In the above equations (7.1) and (7.2), $\omega^+(k)$ (resp: $\omega^-(k)$) denotes the subset of edges of the form (i,k) oriented from i to k (resp: the subset of edges of the form (k,j) oriented from k to j). The real value $v(\varphi)$ involved in equation (7.2) is referred to as the *value* of the flow φ. We recall that equations (7.1)-(7.2) may be rewritten in matrix-form as:

$$A \cdot \varphi = v(\varphi) \cdot b \tag{7.3}$$

where A is the so-called node-arc incidence matrix of the directed graph $\widetilde{G}$ deduced from G by choosing an arbitrary orientation on the edges as suggested above. b is a N-vector with one component b_i for each node $i \in \mathcal{N}$, defined as:

$$\begin{cases} b_i = 0, & \forall i \in \mathcal{N}\setminus\{s,t\} \\ b_s = +1 \\ b_t = -1 \end{cases}$$

Now, if the single-commodity flow model recalled above is well suited to representing how traffic flows through the network from one given source node to one given sink node, it is no longer appropriate to represent how to allocate transmission resources to several competing flows between distinct source-sink pairs. For this, we need an extended version of the single-commodity flow model, referred to as the *multicommodity network flow model*, defined as follows.

K single-commodity flows (numbered $k = 1, \ldots, K$) are given, where, for each k, $s(k)$ and $t(k)$ denote the source node and sink node respectively and d_k the requested value of the single commodity flow to be sent between $s(k)$ and $t(k)$ through the network. Then the corresponding multicommodity flow is represented as a M-vector $\psi = (\psi_u)_{u \in \mathcal{U}}$ satisfying the following linear system:

$$(MCF) \quad \begin{cases} \forall k = 1, \ldots, K: \\ \quad A\varphi^k = d_k b^k \\ \forall u \in \mathcal{U}: \\ \quad \psi_u = \sum_{k=1}^{K} |\varphi_u^k| \end{cases}$$

In the above, A denotes the node-arc incidence matrix of the directed graph $\widetilde{G}$ (deduced from G by giving each edge of G an arbitrary orientation); φ^k is the M-vector representing the kth single-commodity flow from $s(k)$ to $t(k)$ and b^k is the N-vector having all components 0 except $b^k_{s(k)} = +1$ and $b^k_{t(k)} = -1$.

7.3 MODELING NETWORK DESIGN PROBLEMS AS MINIMUM COST MULTICOMMODITY FLOWS

We discuss now a very general model, using the multicommodity flow concept, to formulate network design and dimensioning problems as optimization models.

7.3.1 A minimum cost multicommodity flow model

Let $G = [\mathcal{N}, \mathcal{U}]$ denote the graph representing all possible network topologies and suppose we are given a set of multicommodity flow requirements as a list of K source-sink pairs $\{s(k), t(k)\}$ ($k = 1, \ldots, K$) together with a requested flow value d_k between $s(k)$ and $t(k)$. For each edge $u = (i, j) \in \mathcal{U}$ on which one or several transmission equipment can be installed, we assume that we are given a cost function $\Phi_u : \mathbb{R}^+ \to \mathbb{R}^+$, where $\Phi_u(x_u)$ is the cost of installing the equipment necessary to accommodate a total traffic flow value x_u on link $u = (i, j)$. Note that, in most practical situations, the Φ_u functions will be nonnegative and nondecreasing on $[0, +\infty[$, or a subinterval in $[0, +\infty[$. Indeed, it is realistic to assume that each cost function Φ_u is defined only on an interval $[0, \beta_u]$ (β_u representing the maximum total transmission capacity which can possibly be installed on link $u = (i, j)$).

The problem of optimally designing a network capable of simultaneously handling all the requested flows at minimum cost can thus be formulated as the following min cost multicommodity flow problem:

$$\text{Minimize } \Phi(x) = \sum_{u \in \mathcal{U}} \Phi_u(x_u) \tag{7.4}$$

subject to:

$$(MCMCF) \quad \forall u \in \mathcal{U}: x_u = \sum_{k=1}^{K} |\varphi_u^k| \tag{7.5}$$

$$\forall k = 1, 2, \ldots, K: A\varphi^k = d_k b^k \tag{7.6}$$

$$\forall u \in \mathcal{U}: 0 \leq x_u \leq \beta_u \tag{7.7}$$

In this model, equations (7.5) and (7.6) define x as the vector representing the multicommodity flow, x_u being the total traffic flow through link u. Equations (7.6) express the single-commodity flow requirement constraints for each source-sink pair composing the multicommodity flow. Constraints (7.7) are bound constraints imposing that, on each link u, the total traffic flow through the link should not exceed the prescribed limit β_u. Finally (7.4) defines the objective function, to be minimized, as the sum of individual link cost functions.

Such a model involves $(K+1)M$ variables: M x_u variables and K unknown vectors $\varphi^1, \varphi^2, \ldots, \varphi^K$, each of dimension M. The total number of constraints, apart from the nonnegativity conditions on x, is $KN + 2M$.

The $(MCMCF)$ model is very generic in the sense that a huge variety of optimal network design problems can be cast into this form, depending mainly on the particular choice for the link cost functions Φ_u. The simplest situation corresponds to the case where the Φ_u functions are linear (cf. §7.4.1) but this may lead to poor approximations of reality. Therefore we will also investigate more accurate models: the linear with fixed cost case (§7.4.2), concave piecewise linear cost functions (see §7.4.3) or step-increasing cost functions (cf. §7.4.4).

7.3.2 Flexibility of the (MCMCF) model

The $(MCMCF)$ model introduced in the previous section can easily handle various constraints arising from practical concerns in network engineering. We illustrate here this flexibility on two typical examples:

(a) *restriction of possible choice of routes*

It frequently occurs that, for a given source-sink pair, routing has to be restricted to a given partial subgraph of G. This corresponds to the situation where, either for technical, managerial, or economical reasons, the flow under consideration is not allowed to use some links or nodes. This restriction can easily be taken into account by just replacing, for each such k ($1 \leq k \leq K$), equations (7.6) by: $A^k \varphi^k = d_k b^k$ where A^k denotes the node-arc incidence matrix of the partial subgraph allowed for flow k. Clearly the basic structure of the problem is essentially unchanged.

(b) *Handling node costs (switching costs)*

In many cases, the total network cost has to include, not only link costs (representing transmission equipment) but also node costs to represent the cost of switching equipment installed at the nodes. For any given node i, the cost of the necessary switching equipment essentially depends on y_i, the volume of traffic transiting through node i, which can easily be expressed in terms of x_u variables as:

$$y_i = \frac{1}{2}\left(\sum_{u \in \omega(i)} x_u - D_i\right)$$

where

$$D_i = \sum_{k/s(k)=i} d_k + \sum_{k/t(k)=i} d_k$$

is the total volume of traffic originating or terminating at i. Assuming that we are given, for each $i \in \mathcal{N}$, a cost function $\Delta_i(y_i)$ representing the cost of switching equipment as a function of y_i, the problem can be formulated as:

$$(MCMCF1)\begin{cases} \text{Minimize } \sum_{u \in \mathcal{U}} \Phi_u(x_u) + \sum_{i \in \mathcal{N}} \Delta_i(y_i) \\ \text{s.c:} \\ \forall u \in \mathcal{U}: \quad x_u = \sum_{k=1}^{K} |\varphi_u^k| \\ \forall i \in \mathcal{N}: \quad y_i = \frac{1}{2}\left(\sum_{u \in \omega(i)} x_u - D_i\right) \\ \forall k = 1,\ldots,K: \quad A\varphi^k = d_k b^k \\ \forall u \in \mathcal{U}: \quad 0 \leq x_u \leq \beta_u. \end{cases}$$

Indeed it is possible to show that $(MCMCF1)$ can be reduced to a minimum cost multicommodity flow model with link cost functions only, i.e. to the basic model $(MCMCF)$ (see e.g. Minoux (2001)) at the expense of increasing the number of nodes and edges. In view of this, it is seen that the two models are *essentially equivalent* and this helps in understanding why most published work on network optimization is mainly restricted to the case of link cost functions only.

A third example illustrating the flexibility of the $(MCMCF)$ model will be addressed in §7.5 in connection with the issue of survivability constraints.

7.4 EXACT SOLUTION METHODS FOR (MCMCF)

We provide in this section an overview of known exact solution approaches to the min cost multicommodity flow problem for the most frequently encountered types of cost functions.

7.4.1 Case of linear cost functions: problem (P1)

This is the special case where each link cost function is of the form:

$$\Phi(x_u) = \gamma_u x_u \text{ with } \gamma_u \geq 0$$

(so that Φ_u is non decreasing). Then $(MCMCF)$ can be simply expressed as an ordinary (continuous) linear program as follows.

Each vector φ^k having components unconstrained in sign, it can be rewritten as: $\varphi^{k+} - \varphi^{k-}$, where φ^{k+} and φ^{k-} are two nonnegative vectors with the additional condition that $\forall u$, φ_u^{k+} and φ_u^{k-} cannot be simultaneously nonzero. Then we have $|\varphi_u^k| = \varphi_u^{k+} + \varphi_u^{k-}$, and therefore the resulting problem reads:

$$(P1) \begin{cases} \text{Minimize } \sum_{u \in \mathcal{U}} \gamma_u x_u \\ \text{s.t:} \\ \forall u \in \mathcal{U}: \quad x_u = \sum_{k=1}^{K} \left(\varphi_u^{k+} + \varphi_u^{k-} \right) \\ \forall k = 1, \ldots, K: \quad A\varphi^{k+} - A\varphi^{k-} = d_k b^k \\ \qquad\qquad\qquad\qquad \varphi^{k+} \geq 0, \; \varphi^{k-} \geq 0, \\ \forall u \in \mathcal{U}: \quad 0 \leq x_u \leq \beta_u. \end{cases}$$

This gives rise to large scale (continuous) linear programs with special structure. Such problems have been extensively studied since the early 1970s. The first solution approaches have been based on the fact that $(P1)$ features the appropriate structure suited to the Dantzig-Wolfe decomposition technique (or "price decomposition"), see for instance Minoux (1995) and the survey papers Assad (1978) and Kennington (1978). Other approaches, using the decomposable structure of the problem have also been proposed: application of Lagrangean duality in conjunction with the use of subgradient algorithms (Kennington and Shalaby, 1977), resource decomposition for the maximum multicommodity flow problem (Sakarovitch, 1966), partitioning methods (Grigoriadis and White, 1972). More recently, extensive developments related to interior point methods in linear programming have motivated investigation of new approaches to multicommodity flow problems such as $(P1)$ (see e.g. Schultz and Meyer (1991), Choi and Goldfarb (1990), and Chardaire and Lisser (2002)). Also application of parallel computing techniques have been studied (Pinar and Zenios, 1992). It is worth pointing out here that a well-solved special case of $(P1)$ arises when the upper capacity constraints $x_u \leq \beta_u$ are not active (for instance, if, on each link, the β_u value is greater than the total sum of all requirements). In this case it is easily seen that an optimal solution is readily obtained by sending each flow k on a minimum cost chain (w.r.t. the γ_u valuations) between $s(k)$ and $t(k)$ in G. Thus solving the problem reduces to shortest path computations. This nice property is exploited in many approaches to capacitated multicommodity flow problems, in particular those using Dantzig-Wolfe decomposition and Lagrangean relaxation.

7.4.2 The linear with fixed cost case: problem (P2)

In this version of the problem, the cost function on link u is given by:

$$\Phi_u(x_u) = 0 \text{ if } x_u = 0$$
$$= \delta_u + \gamma_u x_u \text{ if } x_u > 0.$$

The corresponding $(MCMCF)$ problem, which we will denote $(P2)$, has been proposed in the literature as an appropriate model for taking into account infrastructure costs (such as digging trenches for cable installation, building towers for antennas, etc., ...). See, for instance, Yaged (1971) and Minoux (1989). Under this model, as soon as some traffic $x_u > 0$ has to be sent on a link $u = (i, j)$, then a fixed cost δ_u has to be paid, in addition to the linear cost $\gamma_u x_u$, proportional to the traffic flow value on link u. As will be seen in §7.4.3, this model is also relevant to approximate some more general cost functions, in particular nonlinear concave cost functions which are well suited to capture economy-of-scale phenomena in Telecommunications. Unfortunately, problem $(P2)$ belongs to the class of NP-hard problems. A possible proof of this uses the reduction of the Steiner tree problem in a graph, (already known to be NP-hard) to a single-commodity flow problem with fixed costs, a special case of $(P2)$ (see, e.g., Minoux (1989) for details).

For the case where the capacity constraints are not binding ($\beta_u = +\infty$, $\forall u \in \mathcal{U}$), an exact Branch & Bound algorithm has been proposed in Minoux (1976a); it makes use of lower bounds, obtained for each subset of edges $V \subset \mathcal{U}$ by carrying out the following two steps:

(a) Compute the minimum weight spanning tree with respect to the δ_u values on the partial graph $[\mathcal{N}, V]$. Let $\delta^*(V)$ denote the minimum value obtained.

(b) For each source-sink pair $k(1 \leq k \leq K)$ determine $\gamma_k^*(V)$, the length of a shortest chain between $s(k)$ and $t(k)$ with respect to the γ_u values, on the partial graph $[\mathcal{N}, V]$.

Then, for each $V \subset \mathcal{U}$, the resulting lower bound for the cost of an optimal network using only edges in V is

$$\mu(V) = \delta^*(V) + \sum_{k=1}^{K} d_k \gamma_k^*(V).$$

This bound can thus be obtained very efficiently, since it reduces to minimum spanning tree and shortest path computations. Exact solutions to problem $(P2)$ can be obtained with such an approach for medium-sized instances, with $|\mathcal{U}|$ typically not exceeding 40 to 50 edges (see Minoux (1976a)). For larger instances, heuristic solution methods providing only approximate solutions will have to be envisaged (see Minoux (1976a) and Minoux (1989)).

We note here that the special case of problem $(P2)$ where we have a set of single-commodity flow requirements (instead of multicommodity requirements) has been investigated by various authors, see Ortega and Wolsey (2003) and Rardin and Wolsey (1993).

7.4.3 The case of piecewise-linear concave cost functions: problem (P3)

If one is willing to consider models still closer to reality, an essential feature of telecommunications problems is the so-called *economy of scale* phenomenon: on each link of the network, the larger the installed capacity, the smaller the average cost (cost per unit capacity installed). Economic studies (see Ellis (1975), for instance) show that if we plot on a capacity/cost diagram the points representing the various transmission systems available to build transmission links (each one of these corresponding to an offered capacity c_u and associated cost γ_u) then these points lie, within a small tolerance, on a curve with equation $c_u = k_u(x_u)^\alpha$; here $k_u > 0$ is a coefficient depending on the link (and, in particular, on the link length) and α is a number close to 0.5, typically in the interval $[0.4, 0.6]$.

Minimum cost multicommodity flow problems with (concave) link cost functions of the form $\Phi_u(x_u) = k_u(x_u)^\alpha$ (where $0 < \alpha < 1$) are extremely difficult problems which, in view of the current state-of-the art, cannot be solved exactly (within acceptable computing times) except for very small sized networks (typically less than $N = 15$ nodes, $K = 20$ commodities and 10 possible paths per commodity). (Note that if approximate solutions are allowed, good heuristics for problems of this type have been described, e.g. those based on the so-called "accelerated greedy algorithm", see Minoux (1976b) and Minoux (1989)).

One possible way to bypass this difficulty is to approximate the cost function $k_u(x_u)^\alpha$ on each link by a piecewise linear concave function obtained, either via tangential approximation or barycentric approximation at a few points to be chosen. In either case, the approximation obtained can be viewed as the pointwise minimum of a finite number of linear with fixed cost functions. This version of the $(MCMCF)$ problem will be denoted $(P3)$. As an example, if we take the function $\Phi_u(x_u) = 100(x_u)^{0.5}$ to be approximated on the interval $[0, 900]$ then the tangential approximation at the points $x_u = 100$, $x_u = 400$, $x_u = 900$, is defined as:

$$\Phi_u^+(x_u) = \text{Min}\{500 + 5x_u\,;\, 1000 + \frac{5}{2}x_u\,;\, 1500 + \frac{5}{3}x_u\}.$$

The barycentric approximation at the same points is defined as:

$$\Phi_u^-(x_u) = \text{Min}\{10x_u\,;\, \frac{2000}{3} + \frac{10}{3}x_u\,;\, 1200 + 2x_u\}.$$

Note that $\Phi_u^+(x_u)$ and $\Phi_u^-(x_u)$ are upper and lower approximations to Φ_u respectively. Also, we note that in spite of the fairly reduced number of points defining Φ^+ and Φ^-, these constitute fairly good approximations (relative error $\leq 6\%$ on the range $[100, 900]$). However, on the range $[0.100]$ the relative error may be huge. For instance for: $x_u = 4$ $\Phi_u = 200$ whereas $\Phi_u^+ = 520$ and $\Phi_u^- = 40$ (the relative errors are 60 % and 80 % respectively). Thus, if more accurate approximations are required for small values of the capacity x_u, more points in the range $[0.100]$ will have to be used to build tangential or barycentric approximations.

Now we show that a problem of type $(P3)$ can be readily reduced to a problem of type $(P2)$ on a network with the same node set but an increased number of links. For each link $u \in \mathcal{U}$, let q_u denote the number of linear with fixed cost functions $\varphi_u^t(x_u)$ defining Φ_u by the formula

$$\Phi_u(x_u) = \text{Min}_{t=1,\ldots,q_u} \{\varphi_u^t(x_u)\} \qquad (7.8)$$

Let $\widehat{G}$ be the graph deduced from G by replacing each edge $u = (i,j) \in \mathcal{U}$ by q_u edges $u^1, u^2, \ldots, u^{q_u}$ between i and j. Denote $\widehat{\mathcal{U}}$ the edge set of $\widehat{G}$. Now it is easily seen that problem $(P3)$ readily reduces to the solution of $(P2)$ on $\widehat{G}$ where each edge $u^t \in \widehat{\mathcal{U}}$ has an associated linear with fixed cost function $\varphi_u^t(x_{u^t})$.

7.4.4 The case of step-increasing cost functions: Problem (P4)

In the previous section, an appropriate model for taking into account the so-called "economy-of-scale" phenomenon has been discussed. Here, while keeping this capability, we introduce a new model taking into account the essentially discontinuous character of allowed capacity augmentations on the transmission links. The resulting discontinuity of the link cost functions is due to what is commonly referred to as "modularity" of transmission equipment. For instance if we decide to install an additional optical carrier between two nodes i and j, then the capacity of edge (i, j) features an important increase, and the total cost of the network is simultaneously increased by the cost corresponding to this optical carrier. To represent this, we are therefore led to a new minimum cost multicommodity flow problem in which each link cost function is a step-increasing (discontinuous) function of installed capacity. We will denote $(P4)$ the corresponding problem.

A typical step-increasing cost function on link $u \in \mathcal{U}$ can be defined by specifying the finite set $V = \left\{v_u^0, v_u^1, \ldots, v_u^{q(u)}\right\}$ of capacities corresponding to the discontinuity points and the associated costs: $\gamma_u^0 = \Phi_u(v_u^0), \gamma_u^1 = \Phi_u(v_u^1), \ldots, \gamma_u^{q(u)} = \Phi_u\left(v_u^{q(u)}\right)$, with:

$$0 = v_u^0 < v_u^1 < \cdots < v_u^{q(u)} = \beta_u$$
$$0 = \gamma_u^0 < \gamma_u^1 < \cdots < \gamma_u^{q(u)}.$$

With the above notation, the link cost function Φ_u is defined by:

$$\Phi_u(x_u) = 0 \text{ if } x_u = 0$$
$$= \gamma_u^i \ \forall x_u \in]v_u^{i-1}, v_u^i], \ \forall i = 1, \ldots, q(u).$$

Note that Φ_u is defined for x_u in the range $[0, \beta_u]$, the $(P4)$ model therefore includes, by construction, upper bound constraints on variables.

Various special cases of the $(P4)$ model have been studied in the literature, which correspond to simplified forms of the step cost functions Φ_u. In the case where the v_u^i values (resp: the γ_u^i values) are integer multiples of a given capacity value c (resp: of a given cost value γ) we have the so-called "single facility" or "single module" case, studied e.g. in Magnanti and Mirchandani (1993) and Barahona (1996).

Also the "two-facility" case where two types of modules of capacity c_1 and c_2 respectively and corresponding costs γ_1 and γ_2 has been investigated by various authors (see Magnanti et al. (1995), Bienstock and Günlük (1996), and Günlük (1999)). We will provide an overview of previous work on exact solution methods for those special cases in §7.4.4.1 below. However, in practice, there is often significantly more than two types of equipment, and moreover, additional constraints have to be taken into account. A typical example of constraints to be taken into account is the following: if, for a given value x_u of the total traffic flow on link u, one or several modules of capacity c have been used, then for all traffic flow values $x'_u > x_u$, only modules with capacity at least c can be used.

Such a "non-regression" rule is a consequence of the usually rapid increase of traffic requirements over time. As an example, applied to a case with 3 modules of capacity c_1, c_2, c_3 and corresponding costs $\gamma_1, \gamma_2, \gamma_3$ with:

$$c_1 < c_2 < c_3$$
$$\frac{\gamma_1}{c_1} > \frac{\gamma_2}{c_2} > \frac{\gamma_3}{c_3}$$
$$\gamma_1 < \gamma_2 < 2\gamma_1$$
$$2\gamma_2 < \gamma_3 < 3\gamma_2$$

it leads to the following step-increasing cost function:

$$\begin{aligned}
\Phi_u(x_u) &= 0 \text{ if } x_u = 0 \\
&= \gamma_1 \text{ if } 0 < x_u \le c_1 \\
&= \gamma_2 \text{ if } c_1 < x_u \le c_2 \\
&= 2\gamma_2 \text{ if } c_2 < x_u \le 2c_2 \\
&= \gamma_3 \text{ if } 2c_2 < x_u \le c_3 \\
&= 2\gamma_3 \text{ if } c_3 < x_u \le 2c_3 \\
&\vdots \\
&= (k+1)\gamma_3 \text{ if } kc_3 < x_u \le (k+1)c_3
\end{aligned}$$

Since such step-increasing cost functions without well-identified special structure actually arise in practice, investigating solution methods capable of handling arbitrary step-increasing cost functions appears to be necessary. Exact solution methods applicable to this "general case" will be the subject of §7.4.4.2 below.

7.4.4.1 Exact solution methods for special cases. In this subsection, we provide a brief overview of existing exact solution methods applicable to the main two previously mentioned special cases, namely the single-facility case (see Magnanti and Mirchandani (1993), Barahona (1996), and Bienstock et al. (1998)) and the two-facility case (see Magnanti et al. (1995), Bienstock and Günlük (1996), and Günlük (1999)). The idea, common to all these references, is to formulate the problem as a mixed integer linear programming problem, and to apply a Branch & Cut approach (see e.g. Ortega and Wolsey (2003) and Padberg and Rinaldi (1991)). The basic ingredient of Branch & Cut approaches is to generate *valid inequalities* and to exploit them in the framework of Branch & Bound to strengthen the linear relaxation of the problem (the continuous linear program obtained by dropping the integrality constraints on the integer-constrained variables). Strengthening leads to improving the lower bounds, thus to reducing the number of generated nodes in the Branch & Bound tree.

Let us recall here that a *valid inequality* is an inequality necessarily satisfied by any integer solution which, when appended to the linear programming formulation, cuts off the polyhedron defining the set of fractional solutions. Among the simplest and most frequently used valid inequalities for strengthening LP relaxations in the context of 1-facility or 2-facility versions of problem ($P4$) we mention the so-called "cutset inequalities."

In the single-facility case, where there is only one module of capacity c and cost γ, such inequalities are obtained as follows.

On each edge $u \in \mathcal{U}$ of the network, let us denote $y_u \in \mathbb{N}$ the number of modules of capacity c installed on link u. For each cut separating the node set X into two complementary subsets $S \subset \mathcal{N}$, $\bar{S} = \mathcal{N} \setminus S$, let $\omega(S)$ denote the edge subset of the cut, i.e. the subset of edges having one endpoint in S and one endpoint in $\bar{S}$. Also, we denote $d(S,\bar{S})$ the sum of the d_k requirements taken on all k such that either $s(k) \in S$ and $t(k) \in \bar{S}$ or $s(k) \in \bar{S}$ and $t(k) \in S$. Then, for any such cut $(S,\bar{S})$, the inequality

$$\sum_{u \in \omega(S)} y_u \geq \left\lceil \frac{d(S,\bar{S})}{c} \right\rceil$$

is a valid inequality (where for any $\alpha \in \mathbb{R}$, $\lceil \alpha \rceil$ denotes the smallest integer greater than or equal to α).

Other types of valid inequalities which have been used in this context include the so-called 3-partition inequalities and arc-residual capacity constraints (see Magnanti et al. (1995) and Bienstock et al. (1998)), the rounded metric inequalities (see Bienstock et al. (1998), and in generalized form Minoux (2001)), flow-cut-set facets (Bienstock and Günlük, 1996), and mixed partition inequalities (Günlük, 1999). We refer to the survey in Minoux (2001) for a more detailed presentation of the above.

In the present state, according to the computational results reported by the various authors quoted, exact optimal solutions could only be obtained for problems of moderate size (at most 27 nodes and 51 links) and moreover the instances treated only concern fairly low density matrix requirements (the density of a flow requirement matrix is $2K/N(N-1)$, i.e. the ratio between the number K of source-sink requirements and the maximum possible number of source-sink pairs, i.e. $(N(N-1))/2$). It is worth pointing out here that the practical difficulty in getting exact optimal solution to ($MCMCF$) with step-increasing cost functions indeed appears to *increase very rapidly with this density parameter.*

7.4.4.2 Exact solution methods for the general case. The more general version of problem ($P4$) corresponding to arbitrary step-increasing cost functions has been studied in Dahl and Stoer (1998), Stoer and Dahl (1994), Gabrel et al. (1999), and Gabrel and Minoux (1997).

The approach in Dahl and Stoer (1998) and Stoer and Dahl (1994) basically makes use of the same kind of techniques as those mentioned in §7.4.4.1, namely the identification of valid inequalities leading to strengthening the formulation of the problem expressed as a (mixed) integer linear programming problem. The computational results presented in Dahl and Stoer (1998) involve a few instances with sizes up to 45 nodes and 53 links, but, in all cases, the requirement matrices have low density (less than 7 %). Only one of the instances is solved to exact optimality, in all other cases the integrality gaps (difference between optimal integer solution values and optimal continuous solution values) are quite important (ranging from 22 % to 66 %).

The approach proposed in Gabrel et al. (1999) is somewhat different and may be viewed as a specialization of the so-called Benders method (Benders, 1962). It leads to iteratively solving integer linear programming subproblems constructed along the iterations by appending additional constraints, according to a procedure referred to as *constraint generation*. At each iteration the integer linear programming subproblem is solved (exactly) by applying a standard Mixed Integer Linear programming solver (e.g. CPLEX or XPRESS). More precisely the approach works as follows.

At any current iteration k, the subproblem to be solved, denoted (R_k), is a *relaxation* of $(MCMCF)$ where the constraints (7.5)–(7.6) are replaced by a set of constraints (called *metric inequalities*), each of which expresses a necessary condition for feasibility of the vector $x = (x_u)_{u \in \mathcal{U}}$. Each such *metric inequality* corresponds to a choice of virtual lengths $\lambda_u \geq 0$ ($u \in \mathcal{U}$) assigned to the links of the network, and reads:

$$\sum_{u \in \mathcal{U}} \lambda_u x_u \geq \theta(\lambda) = \sum_{k=1}^{K} d_k \ell_k^*(\lambda) \qquad (7.9)$$

where, for each $k = 1, \ldots, K$, $\ell_k^*(\lambda)$ denotes the length of the shortest $s(k) - t(k)$ chain in the network with respect to the "lengths" λ_u ($u \in \mathcal{U}$). Moreover, it can be shown that $x = (x_u)_{u \in \mathcal{U}}$ is a feasible solution to the multicommodity flow problem if all inequalities (7.9) deduced for all possible $\lambda \geq 0$ are satisfied.

If we denote J^k the subset of indices of the metric inequalities generated up to iteration k, and λ^j the length vector corresponding to the jth inequality ($j \in J^k$), then the relaxed subproblem (R_k) reads:

$$(R_k) \quad \begin{cases} \text{Minimize } \sum_{u \in \mathcal{U}} \Phi_u(x_u) \\ \text{subject to:} \\ \sum_{u \in \mathcal{U}} \lambda_u^j x_u \geq \theta(\lambda^j) \quad \forall j \in J^k \\ x_u \in V_u, \quad \forall u \in \mathcal{U}. \end{cases}$$

As shown in Gabrel et al. (1999) and Minoux (2001), this problem can easily be reformulated as a pure 0-1 integer programming problem (with multidimensional knapsack-type structure). Let $\bar{x}$ be an (exact) optimal solution to (R_k). Then we know that it is an optimal solution to problem $(P4)$ if and only if all the metric inequalities (7.9) are satisfied. The algorithm therefore consists, at each iteration, in identifying metric inequalities violated by the current solution $\bar{x}$. If none can be (provably) found, then $\bar{x}$ is an exact optimal solution to $(P4)$ and the procedure terminates. On the contrary, if such violated inequalities can be identified, then these are appended to (R_k) to build up the new relaxed subproblem (R_{k+1}) at iteration $k+1$. This constraint-generation idea can be implemented in various ways. For instance, one can generate only one inequality at each step, the most violated one according to some chosen criterion ("single constraint generation"); or systematically generate several violated inequalities ("multiple constraint generation"). One can also consider the use of particular subclasses of metric inequalities, e.g. the so-called *bipartition inequalities* (these are particular metric inequalities obtained by setting $\lambda_u = 1$ for all edges of a cut and $\lambda_u = 0$ for the other edges).

In Gabrel et al. (1999), a multiple constraint generation procedure is described, based on the combined use of bipartition inequalities and general metric inequalities, and where violated inequalities are identified by maximizing the ratio

$$\rho = \frac{\theta(\lambda)}{\sum_{u \in \mathcal{U}} \lambda_u \bar{x}_u}$$

($\rho > 1$ corresponds to a metric inequality violated by the current solution $\bar{x}$).

Computational results are reported there on a series of 50 instances with up to 20 nodes, 37 links with cost functions featuring an average number of 6 steps, and *100 % dense requirement matrices*. A sample of these results (taken from Gabrel et al. (1999)) appears in Table 7.1. N is the number of nodes; M is the number of edges; Iter is the total number of main iterations necessary to reach exact optimality; NC is the total number of constraints ($=$ metricinequalities) generated in the process, and GMI is the number of general metric inequalities ($NC - GMI$ is thus the number of bipartition inequalities generated); T is the total computation time in seconds (using CPLEX 4.0 in MIP mode for the solution of the relaxed subproblems) including $T(CG)$, that part of total time devoted to generating constraints.

From these results it is seen that the number of main iterations is always quite limited (it does not exceed 12 on the instances treated) and increases only moderately fast with problem size. Also it is observed that general metric inequalities are very rarely needed to reach optimality, bipartition inequalities almost always suffice. Finally, the computation times needed for generating violated inequalities ($T(CG)$) represent only a very small fraction of total computation time.

Table 7.1 also illustrates the practical difficulty in obtaining exact (guaranteed) optimal solutions to problem ($P4$) with general step increasing cost functions and fully dense requirement matrices. Of course, network engineering applications often call for the solution of significantly larger instances, and in this case, approximate solution methods (heuristics) have to be envisaged. For problem ($P4$) many heuristic solution approaches are possible, including the use of standard meta-heuristics (simulated annealing, tabu search, genetic algorithms, see e.g. Reeves (1993)) in connection with the use of specialized local search operators such as link rerouting or flow rerouting. Recently it has been shown in Gabrel et al. (2003a) that the exact constraint-generation procedure of Gabrel et al. (1999) can also be used as the basis of an approximate solution procedure by solving the relaxed subproblems (R_k) *heuristically* rather than exactly. The computational results reported in Gabrel et al. (2003a) suggest that solutions thus obtained are much closer to exact optimality than those produced by several types of link-rerouting and flow-rerouting heuristics.

Table 7.1 Exact optimal solutions to problem $(P4)$ on a series of 10 to 20 node networks with cost functions having average 6-steps per link (taken from (Gabrel et al., 1999)).

N	M	Iter	GMI	NC	T (total)	T(CG)
10	16	7	0	25	26	1.1
10	17	9	2	53	171	2.4
10	18	6	0	30	48	1.1
12	21	11	0	68	1 471	4
12	20	7	0	37	150	2.4
12	20	12	0	68	361	4.2
15	26	9	0	79	1 621	8
15	27	12	0	132	10 911	11.2
15	26	8	0	69	984	7.2
20	36	12	0	183	18 795	35
20	35	9	0	103	2 139	25
20	35	12	0	147	12 476	34

7.5 HANDLING SURVIVABILITY CONSTRAINTS: THE (MCMCFS) MODEL

This section is devoted to an important extension to the $(MCMCF)$ model in which *survivability constraints* have to be taken into account. This will be denoted by $(MCMCFS)$. Such constraints frequently appear in practical network engineering applications, in view of the high capacities offered by modern transmission systems (e.g. optical carriers). Survivability constraints express the fact that the network to be set up, not only has to meet the flow requirements under regular operation, but should keep on meeting all the flow requirements (or a fixed minimum percentage α of these) in case of failure of one or several elements (transmission component, switching component). Since simultaneous failure of several transmission or switching components usually occurs with extremely low probability, most practical applications only require to make the network robust against single link failure and/or single node failure.

The $(MCMCFS)$ problem has been actively investigated in the special case of linear cost functions, which leads to solving large scale (continuous) linear programs, to which several types of decomposition techniques can be applied. The reader is referred to Dahl and Stoer (1998), Goemans and Bertsimas (1993), Minoux (1981), Minoux and Serreault (1984), Stoer and Dahl (1994), and to the survey papers Grötschel et al. (1995) and Minoux (1989). Also worth mentioning are published work where survivability constraints are indirectly and approximately taken into account by means of various types of connectivity constraints (e.g. k-connectedness constraints); the papers Grötschel and Monma (1990), Grötschel et al. (1992), and Monma and Shallcross (1989) are relevant references along this line.

We will discuss more thoroughly here recent work on handling survivability constraints in the case where the cost functions on the links are arbitrary step-increasing functions of total flow through the link (cf §7.4.4.2). In order to make the presentation of the model simpler, we will restrict to the case of *link failures* (though fairly straightforward, the extension to node failures would require more intricate notation). Furthermore, for the sake of simplicity, we will assume that when a given link $u = (i,j)$ fails, all the transmission systems corresponding to this link are in failure state. Of course, the $(MCMCFS)$ model presented below would easily extend if a more detailed representation of failure were required, for instance if, on any link (i,j), at most one of the installed transmission systems (modules) can break down at a time (for this, it would be enough to consider multiple copies of edge (i,j), one for each transmission system which can be installed between i and j, each being characterized by its capacity and cost).

To express robustness of the network w.r.t. any link failure, let us introduce, for each $v \in \mathcal{U}$, the *projection operator* π_v defined as follows. For any $x \in \mathbb{R}^M$ (such that $\forall u \in \mathcal{U}$, x_u represents the total flow through link u): $\pi_v(x) = x'$ where, $\forall u \neq v : x'_u = x_u$ and $x'_v = 0$. With this notation it is observed that $x \in \mathbb{R}^M$ satisfies the survivability constraint w.r.t. the failure of link $v \in \mathcal{U}$ if and only if $\pi_v(x) \in X$ where X denotes the set of solutions to (7.5)-(7.6). The $(MCMCF)$ model with survivability constraints can thus be stated as:

$$(MCMCFS) \quad \begin{cases} \text{Minimize:} \sum_{u \in \mathcal{U}} \Phi_u(x_u) \\ \text{s.t:} \\ \forall v \in \mathcal{U}: \quad \pi_v(x) \in X \\ \forall u \in \mathcal{U}: \quad 0 \leq x_u \leq \beta_u. \end{cases}$$

In the case where the cost functions Φ_u are step-increasing and where, on each link $u \in \mathcal{U}$, V_u denotes the set of discontinuity points of the Φ_u function, the constraint $0 \leq x_u \leq \beta_u$ has to be replaced by $x_u \in V_u$. The resulting optimization problem is a variant of problem $(P4)$ which we denote $(P4S)$.

Exact solution methods for $(P4S)$ have been investigated mainly by Dahl and Stoer (1998), Stoer and Dahl (1994), and Gabrel et al. (2003b), the proposed approaches may be viewed as generalizations of those described in §7.4.4.2.

In Dahl and Stoer (1998), computational results are reported on 23 instances involving networks up to 118 nodes and 134 edges, the link cost functions featuring up to 6 steps. For all the examples treated, the requirement matrices have *very low density* (less than 7 %). Exact optimal solutions are obtained for 8 instances over 23; for the other instances the approximate solutions obtained are sometimes close to exact optimality but the difference may be significant (up to 66 % off optimality), which suggests a huge variability in the practical difficulty of the problem instances.

Gabrel et al. (2003b) describes an extension to problem $(P4S)$ of the constraint-generation approach of Gabrel et al. (1999) (see §7.4.4.2 above). Several alternate strategies for generating constraints (metric inequalities) are described and tested on

30 instances involving 15 and 20 node networks with cost functions having average 6 steps per link and 100% dense requirement matrices. A sample of the results obtained is shown in Table 7.2 (MIP solver used: CPLEX 6.0 on a Sun UltraSparc 30 computer).

Table 7.2 Exact solutions to problem $(P4S)$ on a series of 15 node and 20 node networks with cost functions having average 6 steps on each link. (taken from Gabrel et al. (2003b)).

Instance	N, M	Iter	NC	Time (sec.)
15.1	15, 26	2	313	379
15.2	15, 26	6	461	5 422
15.3	15, 27	3	322	464
15.4	15, 26	3	306	108
15.5	15, 27	2	247	352
15.6	15, 26	4	305	237
20.1	20, 36	6	560	11 162
20.2	20, 35	4	597	3 457
20.3	20, 35	4	598	9 778
20.4	20, 36	8	609	41 752
20.5	20, 37	3	527	9 513
20.6	20, 35	3	550	209

The main conclusion which may be drawn from these experiments is that the exact solution of problem $(P4S)$ leads to computation times comparable to those obtained for problem $(P4)$ (without survivability constraints). The average computing times necessary to reach exact (guaranteed) optimality are significant, but it should be noticed that in all the instances in Table 7.2, *requirement matrices are 100% dense*. The comparison of the results with those presented in Dahl and Stoer (1998) seems to confirm that the practical difficulty of network optimization problems such as $(P4)$ and $(P4S)$ increases very rapidly with the density of requirement matrices. (Clearly, in view of practical application of the optimization models, it would not be realistic to restrict to low density requirement matrices).

For applications requiring solutions to $(P4S)$ on larger networks, approximate solution techniques will have to be envisaged. Among the many possible solution approaches, the one described in Gabrel et al. (2003a), which also readily applies to problem $(P4S)$, appears to be worth considering.

Bibliography

R. Ahuja, T. Magnanti, and J. Orlin. *Networks Flows, Algorithms and Applications.* Prentice Hall, 1993.

A.A. Assad. Multicommodity network flows – a survey. *Networks*, 8:37–91, 1978.

F. Barahona. Network design using cut inequalities. *SIAM Journal on Optimization*, 6(3):823–837, 1996.

J.F. Benders. Partitioning procedures for solving mixed-variable programming problems. *Num. Math.*, 4:238–252, 1962.

C. Berge. *Graphes et Hypergraphes.* Dunod, Paris, 1970.

D. Bienstock, S. Chopra, O. Günlük, and C.Y. Tsai. Minimum cost capacity installation for multicommodity network flows. *Mathematical Programming*, 81:177–199, 1998.

D. Bienstock and O. Günlük. Capacitated network design-polyhedral structure and computation. *INFORMS Journal on Computing*, pages 243–259, 1996.

P. Chardaire and A. Lisser. Simplex and interior-point specialized algorithms for solving non-oriented multicommodity flow problems. *Operations Research*, 50(2):260–276, 2002.

I.C. Choi and D. Goldfarb. Solving multicommodity network flow problems by an interior-point method. *SIAM Proc. in Appl. Math.*, 46:58–69, 1990.

G. Dahl and M. Stoer. A cutting-plane algorithm for multicommodity survivable network design problems. *Informs J. on Computing*, 10(1), 1998.

L.W. Ellis. La loi des volumes économiques appliquée aux télécommunications. *Rev. Telecom 1*, 50:4–20, 1975.

L.R. Ford and D.R. Fulkerson. *Flows in Networks.* Princeton University Press, Princeton, 1962.

V. Gabrel, A. Knippel, and M. Minoux. Exact solution of multicommodity network optimization problems with general step cost functions. *Operations Research Letters*, 25:15–23, 1999.

V. Gabrel, A. Knippel, and M. Minoux. A comparison of heuristics for the discrete cost multicommodity network optimization problem. *Journal of Heuristics*, 9:429–445, 2003a.

V. Gabrel, A. Knippel, and M. Minoux. Exact solution of survivable network problems with step-increasing cost functions. Unpublished research report, University Paris 6, 2003b.

V. Gabrel and M. Minoux. LP relaxations better than convexification for multicommodity network optimization problems with step-increasing cost functions. In *Acta Mathematica Vietnamica 22*, pages 128–145, 1997.

M. Goemans and D. Bertsimas. Survivable networks, linear programming relaxations and the parsimonious property. *Mathematical Programming*, 60(2):145–166, 1993.

M. Gondran and M. Minoux. *Graphs and Algorithms*. John Wiley & Sons, New York, 1995.

M.D. Grigoriadis and W.W. White. A partitioning algorithm for the multicommodity network flow problem. *Mathematical Programming*, 3(2):157–177, 1972.

M. Grötschel and C. Monma. Integer polyhedron associated with certain network design problems with connectivity constraints. *SIAM J. Discrete Math.*, 3:502–523, 1990.

M. Grötschel, C. Monma, and M. Stoer. Computational results with a cutting-plane algorithm for designing communication networks with low connectivity constraints. *Operations Research*, 40:309–330, 1992.

M. Grötschel, C. Monma, and M. Stoer. Design of survivable networks. *Handbook in OR and MS*, 27:617–672, 1995.

O. Günlük. A branch-and-cut algorithm for capacitated network design problems. *Math. Prog. A*, 86:17–39, 1999.

J. Kennington. Multicommodity flows: a state-of-the art survey of linear models and solution techniques. *Opns. Res.*, 26:209–236, 1978.

J. Kennington and M. Shalaby. An effective subgradient procedure for minimal cost multicommodity flow problems. *Management Science*, 23(9):994–1004, 1977.

T.L. Magnanti and P. Mirchandani. Shortest paths, single origin-destination network design and associated polyhedra. *Networks*, 23:103–121, 1993.

T.L. Magnanti, P. Mirchandani, and R. Vachani. Modeling and solving the two-facility network loading problem. *Operations Research*, 43(1):142–157, 1995.

M. Minoux. Multiflots de coûts minimal avec fonctions de coût concaves. *Annales des Télécommunications*, 31:77–92, 1976a.

M. Minoux. Optimisation et planification des réseaux de télécommunications. In G. Goos, J. Hartmanis, and J. Cea, editors, *Optimization Techniques*, pages 419–430. Springer-Verlag, 1976b. Lectures Notes in Computer Science 40.

M. Minoux. Optimum synthesis of a network with nonsimultaneous multicommodity flow requirements. *Studies on Graphs and Discrete Programming*, 11:269–277, 1981. P. Hansen, Editor. Annals of Discrete Mathematics.

M. Minoux. Network synthesis and optimum network design problems: models, solution methods and applications. *Networks*, 19:313–360, 1989.

M. Minoux. *Mathematical Programming. Theory and Algorithms*. John Wiley & Sons, 1995.

M. Minoux. Discrete cost multicommodity network optimization problems and exact solution methods. *Annals of Operations Research*, 106:19–46, 2001.

M. Minoux and J.Y. Serreault. Subgradient optimization and large scale programming: an application to network synthesis with security constraints. *RAIRO*, 15:185–203, 1984.

C. Monma and D. Shallcross. Methods for designing communication networks with certain two–connected survivability constraints. *Operations Research*, 37:531–541, 1989.

F. Ortega and L.A. Wolsey. A branch-and-cut algorithm for the single-commodity uncapacitated fixed-charge network flow problem. *Networks*, 41(3):143–158, 2003.

M. Padberg and G. Rinaldi. A branch-and-cut algorithm for the resolution of large-scale symmetric traveling salesman problems. *S.I.A.M. Review*, 33:60–100, 1991.

M.C. Pinar and S.A. Zenios. Parallel decomposition of multicommodity network flows using a linear quadratic penalty algorithm. *ORSA Journal on Computing*, 4(3):235–249, 1992.

R.L. Rardin and L.A. Wolsey. Valid inequalities and projecting the multicommodity extended formulation for uncapacitated fixed charge network flow problems. *European J. Operational Res.*, 71:95–109, 1993.

C.R. Reeves, editor. *Modern Heuristic Techniques for Combinatorial Problems*. Blackwell Scientific Publications, 1993.

M. Sakarovitch. The multicommodity maximum flow problem. ORC Report 65-25, University of California, Berkeley, 1966.

G.L. Schultz and R.R. Meyer. An interior-point method for block-angular optimization. *SIAM J. Optimization*, 1(4):583–602, 1991.

M. Stoer and G. Dahl. A polyhedral approach to multicommodity survivable network design. *Numerische Mathematik*, 68:149–167, 1994.

B. Yaged. Minimum cost routing for static network models. *Networks*, 1(2):139–172, 1971.

8 SHORTEST PATH ALGORITHMS

Paola Festa[1]

[1] Mathematics and Applications Department
University of Naples FEDERICO II
Naples, Italy
paola.festa@unina.it

Abstract: Shortest path problems are fundamental network optimization problems arising in many contexts and having a wide range of applications, including dynamic programming, project management, knapsack problems, routing in data networks, and transportation problems. The scope of this chapter is to provide an extensive treatment of shortest path algorithms covering both classical and recently proposed approaches.
Keywords: Shortest path problem, network flow problems, network optimization, combinatorial optimization.

8.1 INTRODUCTION

Shortest path problems are classical combinatorial problems that arise as subproblems when solving many optimization problems. As they are relatively easy to solve and at the same time they contain the most important ingredients of network flows, shortest path problems are a starting point for studying more complex network problems. In fact, they have been widely studied leading to a great number of algorithms adapted to find an optimal solution in various special conditions and/or constraint formulations (Ahuja et al., 1993; Gallo and Pallottino, 1986; 1988).

Shortest path problems have involved the interest of both researchers and practitioners, because they appear in a wide variety of contexts, whenever some material, or a telephone call, a computer data packet, etc. needs to be sent as cheaply or as quickly as possible.

The scope of this chapter is to provide an extensive treatment of shortest path algorithms. It starts providing the mathematical formulation of the problems in Section 8.2 and covering the classical approaches in Section 8.3. Then, it proceeds focusing on the most recent scientific literature in Section 8.4.

8.2 MATHEMATICAL FORMULATIONS

Throughout this chapter the following notation is used.
Let $G = (V, E, C)$ be a directed graph, then

- V is a set of nodes, numbered $1, 2, \ldots, n$;

- $E = \{(i, j) | \ i, j \in V\}$ is a set of m edges;

- $C : E \to \mathbb{R}$ is a function that assigns a length to any edge $(i, j) \in E$;

- A *simple path* P is a walk without any repetition of nodes. The set of edges connecting consecutive nodes of P can be partitioned into two groups: *forward edges* and *backward edges*. An edge (i, j) is forward if the path visits i before visiting node j, and it is backward otherwise;

- A *directed simple path* is a directed walk without any repetition of nodes, i.e. it has no backward edges;

- The length of any path P is defined as the sum of lengths of the edges connecting consecutive nodes in the path;

- A *cycle* C is a path $P = \{i_1, i_2, \ldots, i_k\}$ such that either $(i_1, i_k) \in E$ or $(i_k, i_1) \in E$;

- A *directed cycle* is a directed path $P = \{i_1, i_2, \ldots, i_k\}$ such that $(i_k, i_1) \in E$;

- G is an *acyclic graph* if it contains no directed cycle;

- A graph G is *connected* if, for each pair of nodes $i, j \in V$, there exists a path starting at i and ending at j; G is *strongly connected* if for each node $i \in V$, there exists at least one directed path from i to every other node;

- A *tree* T is a connected acyclic graph;

- A *spanning tree* T of a graph $G = (V, E, C)$ is a subgraph of G, which is a tree and includes all the nodes of G.

Shortest path problems can be posed in more than one way:

1. Finding a shortest path from a single node $s \in V$, called *source*, to a single node $d \in V$, called *destination*;

2. Finding shortest paths from a single source $s \in V$ to all other nodes;

3. Finding shortest paths from each of several sources to each of several destinations.

Problem type 1 will be referred as the *single source - single destination shortest path problem*, while problems type 2 and type 3 as *shortest path tree problems*. The single source - single destination shortest path problem has the following formulation

$$\min \sum_{(i,j) \in E} c(i,j) x(i,j)$$

s.t.
$$\sum_{\{j:\ (i,j) \in E\}} x(i,j) - \sum_{\{j:\ (j,i) \in E\}} x(j,i) = b_i, \quad i = 1, \ldots, n$$
$$b_s = 1,\ b_d = -1$$
$$b_i = 0,\ i \neq s, d$$
$$x(i,j) \in \{0,1\},\ \forall (i,j) \in E,$$

while the shortest path tree problem can be formulated as

$$\min \sum_{(i,j) \in E} c(i,j) x(i,j)$$

s.t.
$$\sum_{\{j:\ (i,j) \in E\}} x(i,j) - \sum_{\{j:\ (j,i) \in E\}} x(j,i) = b_i, \quad i = 1, \ldots, n$$
$$b_s = n - 1$$
$$b_i = -1,\ i \neq s$$
$$x(i,j) \in \{0,1\},\ \forall (i,j) \in E.$$

Since any type of shortest path problem consists of finding for every node $i \in V$ different from the source s the directed path having shortest length and connecting s to i, it can be easily viewed as a special network flow problem. In fact, for all $(i,j) \in E$, if $c(i,j)$ is interpreted as edge flow cost, a shortest path problem could be viewed as the problem of sending 1 unit of flow as cheaply as possible from s to d in the case of single source - single destination problem and to $V \setminus \{s\}$, in the case of shortest path tree problems.

As any minimization network flow problem, shortest path problems have associated a *dual* maximization problem and by solving one problem the other is solved as well. In the following, we report the mathematical formulation only of the more general shortest path tree problem.

$$\max\ (n-1)\pi(s) - \sum_{j \neq s} \pi(j)$$

s.t.
$$\pi(i) - \pi(j) \leq c(i,j),\ \forall (i,j) \in E,$$

where $\pi(i)$ is the dual variable associated with the node i and usually called *price* or *potential* of i.

In order to assure that a shortest path problem admits a solution, the following assumptions must be imposed.

1. *All edge lengths are integers.* This assumption is necessary only for algorithms whose complexity bound depends on the input data. However, it is not a restrictive assumption, because irrational numbers must to be converted to rational numbers to be represented on a computer and rational data can be always transformed to integer data by multiplying them by a suitably large number.

2. *The graph G is strongly connected.* This assumption can be removed by defining the distance between not connected nodes equal to $+\infty$.

3. *The graph G does not contain any directed cycle of negative length.* The shortest path problem defined on a graph with a negative cycle is $\mathcal{NP}$-complete. The linear programming formulations of **SSP** and **SPT** on a graph G containing a directed cycle C of negative length have an unbounded solution, because it is possible to send along C an infinite amount of flow.

8.3 CLASSICAL APPROACHES

Shortest path problems have a simple structure, particularly rich in intuition. Since the end of the 1950s plenty of intuitive algorithms have been developed for solving them and more than two thousand scientific works have been published in the literature. In spite of the huge number of proposed algorithms, there does not exist a *best* method that outperforms all the others. In fact, recent research lines tend to develop techniques designed *ad hoc* for solving special structured shortest path problems: either a special network topology or a special cost structure. Exhaustive surveys of the most interesting and efficient shortest path algorithms, important for their computational time complexity or for their practical efficiency, can be found among others in Ahuja et al. (1993), Bertsekas (1991b), Denardo and Fox (1979b), Denardo and Fox (1979a), Deo and Pang (1984), Festa (February 2000), Gallo and Pallottino (1986), and Shier and Witzgall (1981).

Shortest path problems come up in practice and arise as subproblems in many network optimization algorithms. Recently, as hardware becomes more powerful and more sophisticated algorithms need a shortest path subroutine, efficient shortest paths algorithms have growing importance. This is the case for other network optimization problems as well, motivating broad computational investigation of available algorithms.

The aim of the remainder of this section is to give an overview of shortest path techniques including both established and recently proposed methods. Most of these algorithms can be seen as a particular implementation of a unique, general, algorithmic schema, which is an adaptation of the primal network simplex approach (Gallo and Pallottino (1986)). All shortest path algorithms are iterative and essentially rely on the same important concept: *distance labels*. During the execution of any algorithm, a numerical value called distance label is associated with each node of the graph. If the label of any node $i \in V$ is infinite, a path joining the source node to i has still to be found; otherwise, it is the distance from the source s to that node along some path. We begin this section by describing optimality conditions that involve the distance labels, as they allow to assess when a set of distance labels are optimal, in the sense that they express the shortest distances from the given source node. Optimality conditions are important especially for a family of shortest path algorithms, since they provide a termination criterion, or optimality certificate, indicating when a feasible solution to the shortest path problem is also optimal.

8.3.1 Optimality conditions

Optimality conditions not only provide a termination criterion, but they also give helpful suggestions for solving optimization problems. In fact, when a tentative solution does not satisfy the optimality conditions, the conditions often suggest how to improve it in order to get another tentative solution, *closer* to the optimal solution, as measured by some underlying metric expressed through the distance labels associated with each node of the graph. At intermediate stages of an algorithm the distance label associated with any node $i \in V$ is an upper estimate of the shortest distance from s to i and is the shortest distance at the termination. Here, necessary and sufficient conditions are developed for a set of distance labels to represent shortest path distances.

Theorem 8.1 *For every node $i \in V \setminus \{s\}$, let $\phi(i)$ denote the distance label associated with i and let $\phi(s) = 0$. Then, $\{\phi(i)\}_{i \in V}$ represent shortest path distances if and only if*

$$\phi(j) \leq \phi(i) + c(i,j) \quad \forall (i,j) \in E. \tag{8.1}$$

Proof. It is easy to prove that conditions (8.1) are necessary for optimality. In fact, since the labels $\{\phi(i)\}_{i \in V}$ can be viewed as the length of some path P_i connecting the source node s to i, if $\phi(j) > \phi(i) + c(i,j)$, then the path obtained extending P_i by the edge (i,j) is shorter than P_j whose length is $\phi(j)$.

To prove that conditions (8.1) are also sufficient, let us consider any set of labels $\{\phi(i)\}_{i \in V}$ satisfying conditions (8.1) and let $P = \{s = j_0, j_1, \cdots, j_k = i\}$ be any directed path from the source s to the node i whose length will be denoted by $l(P)$. Then, from conditions (8.1) and assuming $\phi(j_0) = 0$, it follows that

$$\begin{aligned}
\phi(i) = \phi(j_k) &\leq \phi(j_{k-1}) + c(j_{k-1}, j_k) \\
&\leq \phi(j_{k-2}) + c(j_{k-2}, j_{k-1}) + c(j_{k-1}, j_k) \\
&\vdots \\
&\leq \phi(j_0) + \sum_{m=1}^{k} c(j_{m-1}, j_m) \\
&= \sum_{m=1}^{k} c(j_{m-1}, j_m) = l(P).
\end{aligned}$$

Hence, $\phi(i)$ is a lower bound on the length of any directed path from s to i. Moreover, since $\phi(i)$ is the length of some directed path from s to i, it expresses also an upper bound on the shortest path length. Therefore, $\phi(i)$ is the shortest path length as Theorem 8.1 claims. □

Definition 8.1 *The reduced edge length c_{ij}^d of an edge (i,j) with respect to the distance labels $\phi(\cdot)$ is defined as*

$$c_{ij}^d = c(i,j) + \phi(i) - \phi(j).$$

From Theorem 8.1, the following theorem holds.

Theorem 8.2 *The reduced edge lengths satisfy the following three properties.*

1. *For any directed cycle C,*

$$\sum_{(i,j) \in C} c_{ij}^d = \sum_{(i,j) \in C} c(i,j);$$

2. *For any directed path P from l to k, where $l, k \in V$,*

$$\sum_{(i,j) \in P} c_{ij}^d = \sum_{(i,j) \in P} c(i,j) + \phi(k) - \phi(l);$$

3. *If $\phi(\cdot)$ represents the shortest path distances array, then*

$$c_{ij}^d \geq 0 \quad \forall\, (i,j) \in E.$$

Corollary 8.1 *In the mathematical formulation of the shortest path problem pointed out in Section 8.2, the negative of the shortest path distances $\{-\phi(i)\}_{i \in V}$ define the optimal dual variables $\{\pi(i)\}_{i \in V}$ and conditions (8.1) are equivalent to require that the reduced lengths are nonnegative in the presence of an optimal solution. In the presence of a negative cycle, the primal problem is unbounded and hence the dual problem is infeasible.*

8.3.2 The generic shortest path algorithm

A prototype shortest path algorithm for the single source-all destinations problem maintains and adjusts a vector $\{\phi(1), \phi(2), \cdots, \phi(n)\}$ of *labels* that can be either scalars or ∞ and that satisfy the following theorem.

Theorem 8.3 *Let $\{\phi(1), \phi(2), \cdots, \phi(n)\}$ be scalars satisfying*

$$\phi(j) \leq \phi(i) + c(i,j) \quad \forall (i,j) \in E \tag{8.2}$$

and let P be a path starting from a node i_1 and ending at a node i_k. If

$$\phi(j) = \phi(i) + c(i,j) \quad \forall (i,j) \in P, \tag{8.3}$$

then P is a shortest path from i_1 to i_k.

Conditions (8.2) and (8.3) are also known as *Complementary Slackness Conditions* (CSC) from the connection of the shortest path problem with the minimum cost flow problem.

The generic shortest path algorithm starts with a vector of labels $\{\phi(i)\}_{i \in V}$ and successively selects edges (i,j) that violate condition (8.3). For each violating edge, it sets

$$\phi(j) = \phi(i) + c(i,j)$$

and stops when condition (8.3) is satisfied by all edges. The labels $\phi(i)$ can be intuitively viewed as the length of some path P_i from the source to the node i. Therefore, if $\phi(j) > \phi(i) + c(i,j)$, the path obtained by extending P_i by edge (i,j) is shorter than P_j whose length is $\phi(j)$. This can be iterated to find successively better paths from the source to various destinations.

Even if violating edges could be arbitrary chosen, a more efficient way is to establish an order of selecting nodes from a set Q, called *candidate list*, and checking violation of condition (8.3) for all their outgoing edges.

Let us suppose that the source node is node 1. Then, the pseudo-code of the prototype shortest path algorithm is shown in Algorithm 8.1.

procedure generic_SP
begin
    $Q := \{1\}$; $\phi(1) := 0$;
    $\phi(i) := \infty$; $\forall i \in V$; $i \neq 1$;
    while $Q \neq \emptyset$ **do**
        Select from Q a node i;
        foreach *outgoing edge* (i,j) **do**
            if $\phi(j) > \phi(i) + c(i,j)$ **then**
                $\phi(j) := \phi(i) + c(i,j)$;
                Add j to Q if it does not already belong to Q;
            end
        end
    end
end

Algorithm 8.1: Prototype shortest path algorithm

8.3.2.1 Implementations of the generic algorithm.
Many different implementations of the above described generic algorithm have been proposed in the literature. All of them are iterative approaches that at each step assign distance labels (i.e. upper bounds on the shortest path distances) to nodes. They differ only in the criterion of selection of the next node to be removed from the set Q and, consequently, in the different classes of shortest path problems they can solve. Traditionally, they are divided into two groups:

- **Label setting methods:** The algorithms belonging to this group remove from Q the node i corresponding to the minimum label and designate as optimal or *permanent* one label at each step.

  At each iteration, the minimum label over Q must be calculated and many implementations of this approach differ in the procedure they use to obtain that minimum.

  Label setting algorithms can be applied to solve only shortest path problems defined on acyclic graphs with arbitrary edge lengths and shortest path problems with nonnegative edge lengths. In fact, if the input data are nonnegative, it can be shown that each node will enter Q at most once and its label has permanent value the first time that node is extracted from Q.

- **Label correcting methods:** Label correcting algorithms consider all labels as temporary until the last iteration, when they become permanent.

  The choice of the node i to be removed from Q requires less calculations, but a node i can be involved more than once.

  The algorithms belonging to this group apply to all classes of shortest path problems, including those with negative edge lengths.

Worst case time complexity of label setting methods is better than that of label correcting ones, but on the other hand, label correcting methods are more general and offer more algorithmic flexibility.

8.3.3 Label setting algorithms

The first label setting algorithm is due to Dijkstra (1959) and independently to Dantzig (1960) and Hillier and Whiting (1960). Dijkstra proposed a method for solving the shortest path tree problem defined on graphs with nonnegative edge lengths. Successively, several different implementations of his algorithm have been developed giving birth to a family of more efficient algorithms.

8.3.3.1 Dijkstra's algorithm. Dijkstra's algorithm maintains and adjusts a vector of node distance labels $\{\phi(i)\}_{i \in V}$. During the computation, the set of nodes V is divided into two groups: the nodes designed as *permanently labeled* and those designed as *temporarily labeled*, which are stored in the candidate list Q. A permanent label associated with a node i represents the shortest distance from the source s to i, while a temporary label expresses an upper bound on that quantity.

Initially, the source node s has a permanent label equal to zero, while all other nodes a temporary label equal to ∞. The candidate list Q contains all nodes, except for the source. At each iteration, the algorithm removes from Q the node i corresponding to the minimum temporary label, designates as optimal or permanent its label and scans all outgoing edges from i in order to possibly update the temporary labels associated with the nodes adjacent to i.

Dijkstra's algorithm maintains a tree T rooted at the source node and spanning the nodes with finite labels. T is stored through predecessor indices $p(\cdot)$, such that if $(i,j) \in T$, then $p(j) = i$. For every tree edge $(i,j) \in T$, $\phi(j) = \phi(i) + c(i,j)$ with respect to the current distance labels $\phi(\cdot)$. When all temporary labels become permanent (i.e. when Q becomes empty), Dijkstra's algorithm terminates and T is a shortest path tree. The correctness can be easily proved following inductive arguments on the cardinality of the candidate list Q, as showed in Ahuja et al. (1993).

A pseudo-code for Dijkstra's algorithm is given in Algorithm 8.2.

8.3.3.2 Complexity of Dijkstra's algorithm. Dijkstra's algorithm performs two kinds of operations: node selections from the candidate list Q and distance label updates. It performs a total of n node selections. At each one, it scans the elements of Q in order to select and remove from Q the node corresponding to the minimum temporary label. Therefore, the total node selection time is $n + (n-1) + (n-2) + \cdots + 1 = O(n^2)$.

procedure dikjstra_SP
begin
$\quad Q := V \setminus \{s\}; \phi(s) := 0; p(s) := s;$
$\quad \phi(i) := \infty; p(i) := nil; \forall i \in V, i \neq s;$
$\quad$ **while** $Q \neq \emptyset$ **do**
$\quad\quad i = \arg\min_{j \in Q} \phi(j); Q = Q \setminus \{i\};$
$\quad\quad$ **foreach** *each outgoing edge* (i, j) **do**
$\quad\quad\quad$ **if** $\phi(j) > \phi(i) + c(i,j)$ **then**
$\quad\quad\quad\quad \phi(j) := \phi(i) + c(i,j); p(j) := i;$
$\quad\quad\quad\quad$ Add j to Q if it does not already belong to Q;
$\quad\quad\quad$ **end**
$\quad\quad$ **end**
$\quad$ **end**
end

Algorithm 8.2: Dijkstra's shortest path algorithm

It is easy to show that for updating all distance labels, $O(m)$ total time is needed, since the total number of distance label updates is $O(m)$ and each update is performed in time $O(1)$. Summing up, the running time of Dijkstra's algorithm in its simplest form is $O(n^2)$.

8.3.3.3 Reverse Dijkstra's algorithm. The (forward) Dijkstra's algorithm solves single source-single destination shortest path problems from node s to node d and shortest path tree problems from node s to every other node in $V \setminus \{s\}$. For solving shortest path problems from every node in $V \setminus \{d\}$ to d there is a slight different version of the algorithm: the Reverse Dijkstra's algorithm, which at each iteration performs operations that are opposite to those performed by the forward version. For every node $i \in V$, the Reverse Dijkstra's algorithm maintains and updates a distance label $\hat{\phi}(i)$, which is either temporary or permanent. As the forward algorithm, the reverse version divides the nodes into two groups: the nodes designed as permanently labeled and those designed as temporarily labeled, which are stored in the candidate list Q. A permanent label associated with a node i represents the shortest distance from i to d, while a temporary label expresses an upper bound on that quantity.

At each iteration, the Reverse Dijkstra's algorithm designs a node $j \in Q$ with minimum temporary label as permanent and then examines each incoming edge (i, j), possibly updating the temporary label associated with the node i. It stops when all distance labels are designated as permanent.

8.3.3.4 Combined Forward/Reverse Dijkstra's algorithm. The Reverse Dijkstra's algorithm can be helpful if *combined* with its forward counterpart. For the single source-single destination shortest path problems there is a further version of the algorithm which is particularly effective in practice, even though it does not improve the worst case complexity: the Combined Forward/Reverse Dijkstra's algorithm. It is simply a combination of the two versions that terminates when both

Forward and Reverse algorithms have permanently labeled the same node, say node k. At this point, let $P(k)$ be the shortest path from s to k found by forward algorithm and let $\hat{P}(k)$ be the shortest path from k to d found by reverse algorithm, then it is easy to prove that the shortest path from s to d is the path $P(k) \cup \hat{P}(k)$. The Combined Forward/Reverse Dijkstra's algorithm is more efficient, because it performs less node selections and consequently less distance label updates.

8.3.3.5 Efficient implementations of Dijkstra's algorithm. At each iteration, Dijkstra's algorithm removes from Q a node i such that

$$i = \arg\min_{j \in Q} \phi(j)$$

and updates the temporary distance labels of the nodes adjacent to i.

As the number of operations required for performing node selections and the number of operations required for performing distance label updates are not balanced, in the past decades research lines have tried to find implementation reflecting a compromise between practical efficiency and worst case complexity. In fact, several versions of Dijkstra's algorithm have been proposed that differ depending on the particular data structure representing the set Q and used to facilitate the removals and the insertions of nodes, as well as finding the node with the minimum label. As we will next point out, the choice of the most suitable data structure is crucial for good practical and theoretical performance.

8.3.3.6 Dial's implementation. Dial's implementation (Dial, 1969) assumes that all edge lengths are nonnegative. The candidate list Q is a vector whose size is equal to the maximum number of different distance label values and since every finite label is equal to the length of some path with no cycles, the possible label values range in

$$[0, (n-1) \max_{(i,j) \in E} c(i,j)].$$

The entry i of Q, called *bucket*, is a double-linked list containing each node whose label value is equal to i. The algorithm starts with the source node s in bucket $Q(0)$, while all other buckets are empty. At the first iteration, the algorithm examines each edge $(s,i) \in E$ and puts each node i in bucket $Q(c(s,i))$. It then proceeds examining bucket $Q(1)$. If $Q(1)$ is nonempty, it repeats the process by removing from Q each node with label 1 and moving other nodes to smaller numbered buckets; otherwise, it proceeds checking bucket $Q(2)$ and so on. Pseudo-code for Dial's implementation is shown in Algorithm 8.3.

The scope of Dial's implementation, as well as of each different implementation of Dijkstra's algorithm, is to perform more efficiently the operations of node selection and distance label updates. In fact, instead of scanning all temporarily labeled nodes, the node with minimum label is retrieved in constant time. Checking if a bucket is empty and inserting or removing a node from a bucket require $O(1)$ time; searching the minimum label node requires $O(n \max_{(i,j) \in E} c(i,j))$, while adjusting node labels and repositioning nodes between buckets require $O(m)$. Therefore, the running time is

procedure dial_SP
begin
  $Q(0) := \{s\}; z := 0;$
  $\phi(i) := \infty; p(i) := nil; \forall i \in V, i \neq s;$
  $\phi(s) := 0; p(s) := s;$
  while $Q \neq \emptyset$ **do**
    Move on Q until $Q(z) \neq \emptyset;$
    Set i equal to the first element of bucket $Q(z);$
    Remove i from $Q(z);$
    foreach *outgoing edge* (i,j) **do**
      if $\phi(j) > \phi(i) + c(i,j)$ **then**
        if $j \in Q$ **then**
          | Remove j from $Q(\phi(j))$
        end
        $\phi(j) := \phi(i) + c(i,j); p(j) := i;$
        Insert j into $Q(\phi(j));$
      end
    end
  end
end

Algorithm 8.3: Dial's implementation

pseudo-polynomial and is given by $O(m + n \cdot c_{max})$, where

$$c_{max} = \max_{(i,j) \in E} c(i,j).$$

The required memory space can be prohibitively large, because it uses $n \cdot c_{max} + 1$ buckets. Nevertheless, the number of buckets can be reduced to $c_{max} + 1$. In fact, since the distance labels designated as permanent by Dijkstra's algorithm are nondecreasing and since any distance label is bracketed from below by $\phi(i)$ and from above by $\phi(i) + c_{max}$, where i is the node permanently labeled at the current iteration, then it follows that buckets numbered $0, 1, \cdots, c_{max}$ are enough to store nodes with finite temporary distance labels. Node j is stored in bucket

$$\phi(j) \bmod (c_{max} + 1)$$

and in general bucket k contains nodes whose distance label is either k or $k + (c_{max} + 1)$, or $k + 2(c_{max} + 1)$, and so on.

8.3.3.7 Heap implementations. Other implementations propose a *heap* for representing the set Q. A heap is a collection of *objects*, each with an associated real number called *key*. It supports the following five operations.

create-heap(Q): create an empty heap;

find-min(i,Q): find the object with minimum key and store it in i;

procedure heap_SP
begin
    $\phi(i) := \infty;\ p(i) := nil;\ \forall i \in V,\ i \neq s;$
    $\phi(s) := 0;\ p(s) := s;$
    create-heap(Q);
    insert$(s, 0, Q)$;
    while $Q \neq \emptyset$ **do**
        find-min(i, Q);
        delete-min(i, Q);
        foreach *outgoing edge* (i, j) **do**
            $value = \phi(i) + c(i, j);$
            if $\phi(j) > value$ **then**
                $\phi(j) := value;\ p(j) := i;$
                if $j \notin Q$ **then**
                      insert$(j, value, Q)$;
                else
                      decrease-key$(value, j, Q)$;
                end
            end
        end
    end
end

Algorithm 8.4: Heap implementation of Dijkstra's algorithm

insert$(i, key(i), Q)$: insert object i with key $key(i)$;

decrease-key$(value, i, Q)$: decrease the key of the object i by *value*. Note that *value* must be smaller than the key it is modifying;

delete-min(i, Q): delete an object i with minimum key.

By using a heap, Dijkstra's algorithm can be implemented as shown in Algorithm 8.4. Operations **find-min**, **delete-min**, and **insert** are performed at most n times, while **decrease-key** is called at most m times.

Depending on the particular type of heap data structure, the running time can significantly change. In the following, some of the most popular Dijkstra's heap implementations are considered.

- **Binary heap.** A binary heap is a tree whose radix corresponds to the node having the minimum label and in which each node has label not greater than those of its two children. The removal of the node $Q(1)$ corresponding to the minimum label, the insertion of a new node in the last position $Q(last)$ and the correction of the label of a node already inserted in Q have complexity $O(\log q) \leq O(\log n)$, where q is the number of elements stored in Q.

  At each iteration, the radix of the heap Q is removed and some labels of nodes belonging to Q may decrease. Therefore, some nodes may have to be repositioned in Q, while some other nodes may enter for the first time in Q and have to be put at the right position. Each of the just mentioned operations takes $O(\log n)$ time. The total number of insertions is n as well as that of removals. Therefore, the number of operations needed to keep the heap Q ordered is $O((n+r)\log n)$, where r is the total number of repositioning operations. To get an upper bound on r, it is enough to observe that there is at most one repositioning for each edge, because each edge is involved at most once. Therefore, $r \leq m$ and the total operation count needed to maintain Q is $O(m \log n)$. Because this dominates the $O(m)$ operation count to examine each edge, the worst-case running time is $O(m \log n)$, even if experimental results indicate that it grows approximately like $O(m + n \log n)$, because usually r is a small multiple of n and considerably less than m.

- **d-heap.** In a d-heap each node has d children. *insert* and *decrease-key* operations require $O(\log_d n)$ time; *delete-min* requires $O(d \log_d n)$, while all the remaining operations require $O(1)$. Summing up, the computational time of this implementation is $O(m \log_d n + n d \log_d n)$. The optimal choice of d is obtained equating the two terms of the O-notation, giving $d = \max\{2, \lceil m/n \rceil\}$.

- **Fibonacci heap.** This implementation due to Fredman and Tarjan (1987) uses a Fibonacci heap, that allows the heap operations to be performed more efficiently than with a d-heap. In fact, every heap operation requires $O(1)$ amortized time, except for *delete-min*, which requires $O(\log n)$. The amortized complexity of an operation is the average worst-case complexity of performing that operation. For an exhaustive description of this kind of data structure and for a definition of amortized computational time the reader can refer to Cormen et al. (1990). The running time of this version of Dijkstra's algorithm is $O(m + n \log n)$.

- **Radix heap.** This version of Dijkstra's algorithm suggested by Ahuja et al. (1990) improves Dial's implementation. As with Dial's algorithm, it uses buckets to store the temporary node labels, but instead of storing in bucket k only labels whose value is exactly k, it stores labels in the range $[2^{k-1}, 2^k - 1]$. Therefore, only $\lceil \log(n \cdot c_{max}) \rceil + 1$ buckets are needed. The different temporary labels that can be stored in a bucket make up the *range* of the bucket, while the cardinality of the range is called its *width*. The ranges of the buckets are changed by the algorithm dynamically. At the beginning, they are such that

$$\text{range}(k) = [2^{k-1}, 2^k - 1].$$

Successively, whenever the algorithm finds that nodes with the minimum label are in a bucket i with width larger than 1, after selecting a node with minimum label, it redistributes the bucket ranges and shifts each node in the bucket i to lower-indexed buckets already empty.

Overall, the operation of moving nodes among buckets requires $O(m + n \cdot K)$ time, where $K = \lceil \log(n \cdot c_{max}) \rceil$. The term m comes from the number of label updates, while $n \cdot K$ arises because the algorithm can only move nodes in lower-indexed buckets, that can be at most $K + 1$. Therefore, $O(n \cdot K)$ is a bound on the total number of node movements.

The node selection operation requires $O(K)$ time at each iteration and $O(n \cdot K)$ time in total. Consequently, the running time of radix heap implementation is $O(m+n \cdot K)$. Since $K = \lceil \log(n \cdot c_{max}) \rceil$, this algorithm runs in $O(m+n \cdot \lceil \log(n \cdot c_{max}) \rceil)$.

As in Dial's algorithm, the number of buckets can be reduced to

$$\lceil \log c_{max} \rceil + 1,$$

giving an implementation running in $O(m + n \cdot \lceil \log c_{max} \rceil)$. Moreover, by using a Fibonacci heap data structure within the radix heap implementation, it is possible to further reduce the complexity to $O(m + n\sqrt{\log c_{max}})$.

8.3.4 Label correcting algorithms

Label correcting algorithms solve a larger class of shortest path problems with arbitrary edge lengths. In fact, they either identify a negative cycle, when one exists, or solve the given shortest path problem if the underlying graph contains no negative cycle.

Optimality conditions (8.1) play a central role in the development of label correcting algorithms, as they provide a termination condition and give helpful suggestions for solving optimization problems. Methods belonging to this family maintain a set of node distance labels stored in some data structure Q and iteratively update labels until they satisfy conditions (8.1). The selection of the next node to be removed from Q requires less calculations than in label setting methods, but each node can be involved more than once. All these methods use some sort of *queue* to store the set of distance labels and differ in the way the queue is structured and the choice of the queue position into which nodes are inserted.

Although the computational time of label correcting algorithms depends generally on the input data, they offer important advantages, such as flexibility and freedom in the choice of the techniques used for the selection of the edges violating optimality conditions (8.1).

In the following, some among the most interesting label correcting algorithms are described.

8.3.4.1 Bellman-Ford algorithm.

The Bellman-Ford method is related to the method proposed by Bellman (1957) and Ford (1956) based on dynamic programming ideas. It uses a FIFO strategy to maintain the queue Q: a node is removed only from the top and is inserted at the bottom of the queue. This method proceeds in cycles of iterations: the first cycle consists of iterating on the source node 1, while in each subsequent cycle the nodes entered Q during the previous cycle are removed from Q in the same order that they were inserted.

The Bellman-Ford algorithm solves the shortest path tree problem in the more general case in which the edge lengths can be negative. It fails if and only if there exists a path starting from the source and containing a negative cycle, whose presence is detected when the algorithm does not terminate after $n-1$ iteration cycles. In the case where all cycles in the graph have nonnegative length, the shortest distance of every node can be obtained after at most $n-1$ iteration cycles. Since in each iteration cycle each edge is involved at most once and each iteration cycle requires $O(m)$ operations, the running time of this method is $O(n \cdot m)$.

The pseudo-code of Bellman-Ford method is listed in Algorithm 8.5.

procedure Bell-Ford_SP
begin
$\quad Q := \{s\}; \phi(s) := 0; p(s) := s;$
$\quad \phi(i) := \infty; p(i) = nil; \forall i \in V, i \neq s;$
$\quad$ **while** $Q \neq \emptyset$ **do**
$\quad\quad$ Set i equal to the top element of the queue Q;
$\quad\quad$ Remove i from Q;
$\quad\quad$ **foreach** *outgoing edge* (i, j) **do**
$\quad\quad\quad$ **if** $\phi(j) > \phi(i) + c(i, j)$ **then**
$\quad\quad\quad\quad \phi(j) := \phi(i) + c(i, j); p(j) := i;$
$\quad\quad\quad\quad$ **if** $j \notin Q$ **then**
$\quad\quad\quad\quad\quad$ | Insert j at the bottom of Q;
$\quad\quad\quad\quad$ end
$\quad\quad\quad$ end
$\quad\quad$ end
$\quad$ end
end

Algorithm 8.5: The Bellman-Ford algorithm

The Bellman-Ford algorithm is the simplest label correcting approach. In the following, other three different implementations of label correcting methods will be considered that are more sophisticated than the Bellman-Ford algorithm. Their worst case complexity is no better than $O(n \cdot m)$, even if their practical performance is better on some instances.

8.3.4.2 D'Esopo-Pape algorithm. Like the Bellman-Ford algorithm, this method uses a queue to represent the candidate list Q. A node is always removed from the top of Q, but it is inserted at bottom if it has never been in Q before, otherwise it is put at the top. The choice of this insertion strategy comes observing that removal and updating of the label of a node i affect the labels of a subset N_i of neighbor nodes j with $(i, j) \in E$. Therefore, by placing the node i at the top of the queue, the labels of nodes belonging to N_i will be updated as quickly as possible. This singular strategy of manipulating Q works well on many instances, including those characterized by negative lengths of some edges, even if in the literature special examples have been constructed by Kershenbaum (1981), Shier and Witzgall (1981), and Bertsekas and Costanon (1993), where the D'Esopo-Pape algorithm behaves very poorly, as the number of entrances in Q of some nodes is not even polynomial.

Polynomial versions of this method have been proposed by Pallottino (1984) and Gallo and Pallottino (1988). Unfortunately, their practical performance does not change with respect to the original version.

8.3.4.3 SLF and LLL algorithms.
The *Small Label First* (SLF) and *Large Label Last* (LLL) algorithms adopt a particular policy of managing the queue Q that tries to place nodes with small label near the top of Q. In some way, they emulate Dijkstra's selection strategy of a node with smallest distance label, trying at the same time to reduce the computational overhead of Dijkstra's algorithm.

The SLF and LLL algorithms are particularly suitable for solving shortest path problems with nonnegative edge lengths. They also work on input graphs characterized by negative length of some edges, but in this case they do not behave better than every other label correcting algorithm.

- **SLF strategy.** The candidate list Q is a double ended queue. The node exiting Q is always the top node, while the insertion of nodes in Q respects the following rule. Whenever a node j enters Q, its label $\phi(j)$ is compared to the label of the top node i. If $\phi(j) \leq \phi(i)$, then j will occupy the top position, otherwise it will be entered at the bottom of Q.

- **LLL strategy.** At each iteration, LLL computes the average node label in Q, defined as
$$a = \frac{\sum_{j \in Q} \phi(j)}{|Q|}.$$
Then, starting from the top node and for each node $j \in Q$, it compares $\phi(j)$ to a. If $\phi(j) > a$, then j is moved to the bottom of Q; otherwise j is removed from the queue.

The SLF insertion and LLL extraction policies can be easily combined leading to the so called *combined SLF/LLL* method that on average requires a smaller number of iterations than either SLF or LLL. SLF, LLL, and combined SLF/LLL techniques behave better than Bellman-Ford and D'Esopo-Pape methods and can be easily implemented in parallel as shown by Bertsekas et al. (1996).

Chen and Powell (1997) showed that SLF and combined SLF/LLL have nonpolynomial worst case complexity, even if their non-polynomial behavior is rare in practice. Moreover, in the case of input graphs with nonnegative edge lengths, polynomial versions of these algorithms can be easily constructed. A first polynomial version of SLF and LLL is due to Chen and Powell (1997), who proposed an approach having complexity $O(n \cdot m^2)$ and based on the sorting by length of the outgoing edges of nodes removed from Q. Another approach to construct polynomial version of SLF and LLL, as well as of any other label correcting algorithm, is due to Bertsekas (1998), who describes a procedure that modifies SLF and/or SLF/LLL so that they have an $O(n \cdot m)$ worst case complexity.

8.3.4.4 Threshold algorithm.
Like the SLF/LLL methods, the underlying idea of *threshold algorithm* is to remove from the candidate list nodes with small distance labels. The candidate list Q is split into two queues Q_1 and Q_2 using a *threshold* parameter s, so that Q_1 contains nodes whose label is no larger than s, while Q_2 contains the remaining. At each iteration, a node is removed from Q_1 and any node j to be added will be inserted at the bottom of Q_1 or Q_2, depending on whether $\phi(j) \leq s$

or $\phi(j) > s$, respectively. When Q_1 becomes empty, the threshold s is adjusted and the queues Q_1 and Q_2 are recomputed, so that Q_1 will contain nodes whose label is no larger than the new computed threshold.

It is easy to show that in case of nonnegative edge lengths the threshold algorithm is identical to Dijkstra's algorithm if s is the minimum distance label, while for a generic value for s it can be viewed as a *block version of Dijkstra's algorithm*, since a whole subset of nodes becomes permanently labeled whenever Q_1 gets exhausted.

The complexity of the threshold algorithm is $O(n \cdot m)$ in the case of nonnegative edge lengths, even if its practical performance is strongly depending on the choice of s. Glover et al. (1985a) and Glover et al. (1985b) proposed some heuristics for selecting effective thresholds.

Bertsekas (1998) described an interesting approach that combines the good characteristics of the threshold and SLF/LLL methods. To assign permanent labels to the nodes in Q_1, the threshold approach uses essentially the Bellman-Ford algorithm restricted to the subgraph defined by Q_1. Instead, to reach this scope it could be possible to use different techniques, such as for example either the SLF or LLL strategy, obtaining a variant of both that computational experience has showed performing extremely well in practice. The average number of iterations it performs per node is only slightly larger than one, the minimum number of iterations per node performed by Dijkstra's algorithm. At the same time, it requires less overhead than Dijkstra's algorithm.

8.3.5 Negative-cycle detection algorithms

The negative cycle problem comes up both directly and as subproblem in algorithms for solving other network problems, as for example the minimum cost flow problem (see Klein (1967)).

All previously described algorithms assume that the input graph does not contain negative cycles. In fact, in the presence of a negative cycle no set of distance labels will satisfy optimality conditions (8.1) and any label correcting algorithm will keep decreasing distance labels indefinitely, never terminating. Nevertheless, any of the previously described algorithms can be easily modified in order to manage the presence of negative cycles. In fact,

$$LB = -\left(n \cdot \max_{(i,j) \in E} c(i,j)\right)$$

is a lower bound on any distance label. Therefore, if during the computation the distance label of some node k has fallen below LB, the algorithm can terminate, since it is involved in a negative cycle that can be obtained by tracing the predecessor indices starting at node k.

The above strategy for detecting negative cycles in a graph is one of the simplest. For a survey of more complex methods, the reader can refer to Cherkassky and Goldberg (1999), who studied various combinations of shortest path algorithms and cycle detection strategies and found the best combinations. One of their conclusions is that the cycle detection strategy proposed by Tarjan (1985) is one of the best combinations and greatly improves the practical performance of the state-of-art strategies. Tarjan's algorithm is a combination of the Bellman-Ford-Moore algorithm proposed by Bellman (1958), Ford and Fulkerson (1962), and Moore (1959) and a subtree-disassembly strategy for cycle detection. The Bellman-Ford-Moore algorithm maintains the set of distance labels in a first-in first-out (FIFO) queue. The next node to be processed is removed from the head of the queue, while a node that becomes temporarily labeled is added to the tail of the queue, if it not already in the queue. Assuming that there are no negative cycles, the Bellman-Ford-Moore algorithm runs in $O(n \cdot m)$ time in the worst case.

8.4 RECENT LITERATURE

A survey of the state-of-the-art shortest path algorithms until 1996 can be found in Cherkassky et al. (1996), where it is shown that in case of nonnegative edge lengths, implementations of Dijkstra's algorithm achieve the best computational time bounds.

Recent research lines tend to put their effort in two main directions. On one side, *ad hoc* techniques have been designed for solving special structured shortest path problems: either a special network topology or a special cost structure. On the other hand, several algorithms have been proposed implementing efficient classical algorithms, but using special data structures. One interesting approach in the first direction is due to Goldberg (2001), who proposed an algorithm with linear average time. By assuming the input lengths positive and uniformly distributed, the worst case running time of his approach is $O(m + n \log C)$, where C is the ratio of the largest and the smallest nonzero edge length.

8.4.1 Special data structures

In Goldberg and Silverstein (1997), a 2-level bucket implementation seemed to exhibit the best performance among the different implementations studied. It seemed to outperform also the classical 1-level bucket algorithm due to Dial (1969). Goldberg and Silverstein (1997) conducted a large set of computational experiments to show the major robustness of multi-level bucket with respect to the classical 1-level implementation.

A further investigation on efficient Dijkstra's implementations has been conducted by Cherkassky et al. (1999). The authors obtained a better complexity bound by proposing a new data structure, called heap-on-top priority queue (hot queue) that combines the multi-level bucket of Denardo and Fox (1979b) and a heap. They also discussed a practical implementation of their algorithm and showed that both multi-level bucket and heap implementations are less robust than their implementation.

8.4.2 Auction algorithms

An appealing family of algorithms recently proposed for shortest path problems is the auction family. The first auction algorithm was proposed by Bertsekas (1979) (see also Bertsekas (1985) and Bertsekas (1988)) for solving the assignment problem. It was then generalized for the transportation problem, the minimum cost flow problem (Bertsekas and Castanon, 1989a;b; 1991) and for shortest path problems (Bertsekas,

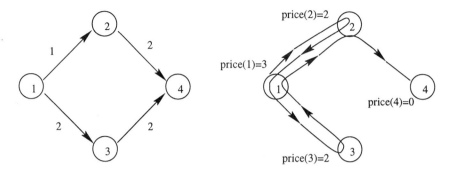

Figure 8.1 Example of the auction algorithm on a simple graph of 4 nodes: 1 is the source node, while 4 is the destination.

1991a). A complete survey of auction algorithms can be found in Chapter 4 of Bertsekas (1991b).

8.4.2.1 Standard auction approach. The standard forward auction algorithm follows a primal-dual approach and consists of the following three basic operations: path extension, path contraction, and dual price increase. For the single source and single destination case, the algorithm maintains a path P starting at the source and a set π of feasible values for the dual variables, also called *dual prices*. At each iteration, the candidate path P is either extended by adding a new node at the end of the path or contracted by deleting from P the last inserted node, called the terminal node. At any iteration, if no extensions or contractions are possible, the value of the dual variable corresponding to the terminal node of P is increased. The algorithm terminates when the candidate path P is extended by the destination node.

The algorithm maintains a pair (P, π) satisfying the following conditions:

$$\pi(i) \leq \pi(j) + c(i,j) \quad \forall \ (i,j) \in E, \tag{8.4}$$

$$\pi(i) = \pi(j) + c(i,j) \quad \forall \ (i,j) \in P. \tag{8.5}$$

It is easy to see that the above conditions are equivalent to (8.2) and (8.3), since the dual prices $\{\pi(i)\}_{i \in V}$ can be viewed as the negative of the labels $\{\phi(i)\}_{i \in V}$. Therefore, in the following we will also refer to conditions (8.4) and (8.5) as Complementary Slackness Conditions (CSC). The algorithm starts with a pair (P, π) satisfying the CSC (if $c(i,j) \geq 0, \ \forall \ (i,j) \in E$, P may consist initially only of s and π may be zero) and proceeds in iterations, transforming (P, π) into another pair satisfying CSC. At each iteration, a dual feasible solution and a primal (infeasible) solution are available for which CSC hold. While maintaining complementary slackness, the algorithm either constructs a new primal solution (not necessarily feasible) or a new dual feasible solution, until a primal feasible (and hence also optimal) is obtained. Figure 8.1 shows how the auction algorithm works on a simple graph of four nodes, while pseudo-code for a typical iteration of the standard auction algorithm is shown in Algorithm 8.6

The standard auction algorithm can be easily adapted to solve the shortest path tree problem. In fact, in this case it terminates when each node has been involved at least

```
procedure auction_iteration
begin
    Let i be the terminal node of P;
    min = min    {c(i,j) + π(j)};
          (i,j)∈E
    if π(i) < min then
    |   Go to 1;
    else
    |   Go to 2;
    end
1   /* CONTRACT PATH */
    π(i) := min;
    if i ≠ s then
    |   Contract P;
    |   Go to the next iteration;
    end
2   /* EXTEND PATH */
    Extend P by node j_i = arg    min      {c(i,j) + π(j)};
                              j∈V:(i,j)∈E
end
```

Algorithm 8.6: Typical iteration of an Auction Algorithm

once by the algorithm, i.e. when each node has been the terminal node of P at least once.

8.4.2.2 Reverse and combined forward/reverse auction algorithms.
The reverse version of the auction algorithm is particularly efficient for solving the single source-single destination shortest path problem. It is mathematically equivalent to the earlier described forward auction. The two approaches proceed in opposite directions. In fact, the reverse algorithm maintains a path R ending at the destination node d and a set π of feasible values for the dual variables. At each iteration, the candidate path R is either extended by adding a new node at the beginning of the path or contracted by deleting from R the last inserted node, called the initial node. At any iteration, if no extension or contraction is possible, the value of the dual variable corresponding to the initial node of R is decreased.

The algorithm starts with any pair (R, π) satisfying CSC, and proceeds in iterations, transforming (R, π) into another pair satisfying CSC. It terminates when the candidate path R is extended by the source node. This algorithm is most helpful when combined with its forward counterpart. The combined forward/reverse auction algorithm maintains a unique set π of feasible values for the dual variables and two distinct paths: a path P starting at the source and R ending at the destination. P and R are extended and contracted according to the rules of the forward and the reverse algorithms, respectively. The algorithm terminates when P and R have a common node. The correctness of the combined forward/reverse auction algorithm is guaranteed, since P, R, and π satisfy CSC at each iteration.

As shown by Bertsekas (1991a) and Bertsekas et al. (1995) through experiments on randomly generated graphs, combined forward/reverse auction algorithm outperforms its closest competitors for the single source-few destinations problems. The reason is that the selected few destinations are reached by the forward path faster than other nodes thanks to the mechanism of the reverse portion of the algorithm, that associates with each of them a feasible dual value.

8.4.2.3 Variants of the auction algorithm. The running time of the original standard auction algorithm can depend on the edge lengths. A first effort to turn it into a polynomial algorithm was made by Pallottino and Scutellá (1991), who derived conditions under which it is possible to "prune" the original graph. The authors showed that every time the standard forward auction algorithm reaches a node f, the optimality of the candidate path P enables the deletion from the original graph of all edges whose head is the node f, except for the edge (k, f), where k is its predecessor in P. The node set V becomes in this way partitioned into the set of the nodes never visited by the algorithm and those visited at least once and that have only one incoming edge. The algorithm is strongly polynomial with complexity $O(m^2)$ and does not that the cycle lengths be positive, and does not impose assumptions on the topology of the graph and input data.

Several other more sophisticated variants of the auction algorithm proposed in literature can be found in Bertsekas et al. (1995) and in Cerulli et al. (2001). Very recently, Cerulli et al. (2003) have proposed two algorithms having complexity $O(n^2)$ and in which the computational effort is reduced by fully exploiting the main feature of the auction algorithm: when it reaches for the first time a node i, the shortest path from the source to node i is found. Hence, all subtrees rooted at i can be computed applying the algorithm from a *virtual source* located on node i and that complete shortest path tree can be assembled joining pieces of *optimal* sub-paths so obtained.

Bibliography

R.K. Ahuja, T.L. Magnanti, and J.B. Orlin. *Network Flows: Theory, Algorithms and Applications*. Prentice-Hall Englewood Cliffs, 1993.

R.K. Ahuja, K. Mehlhorn, J.B. Orlin, and R.E. Tarjan. Faster algorithms for the shortest path problem. *Journal of ACM*, 37:213–223, 1990.

R.E. Bellman. *Dynamic Programming*. Princeton Univ. Press, 1957.

R.E. Bellman. On a routing problem. *Quart. Appl. Math.*, 16:87–90, 1958.

D.P. Bertsekas. A distributed algorithm for the assignment problem. Technical report, Lab. for Information and Decision Systems Working Paper, MIT, Boston, USA, 1979.

D.P. Bertsekas. A distributed asynchronous relaxation algorithm for the assignment problem. In *Proceedings of the 24th IEEE Conference on Decision and Control*, pages 1703–1704, Ft. Lauderdale, Fla., USA, 1985.

D.P. Bertsekas. The auction algorithm: A distributed relaxation method for the assignment problems. *Annals of Operation Research*, 14:105–123, 1988.

D.P. Bertsekas. An auction algorithm for shortest paths. *SIAM Journal on Optimization*, 1:425–447, 1991a.

D.P. Bertsekas. *Linear networks optimization: Algorithms and Codes*. MIT Press, 1991b.

D.P. Bertsekas. *Network Optimization: Continuous and Discrete Models*. Athena Scientific, 1998.

D.P. Bertsekas and D.A. Castanon. The auction algorithm for minimum cost network flow problem. Technical Report Report LIDS-P-1925, Lab. for Information and Decision Systems, MIT, Boston, USA, 1989a.

D.P. Bertsekas and D.A. Castanon. The auction algorithm for transportation problems. *Annals of Operation Research*, 20:67–96, 1989b.

D.P. Bertsekas and D.A. Castanon. A generic auction algorithm for the minimum cost network flow problem. Technical Report Report LIDS-P-2084, Lab. for Information and Decision Systems, MIT, Boston, USA, 1991.

D.P. Bertsekas and D.A. Costanon. Asynchronous hungarian methods for the assignment problem. *ORSA Journal on Computing*, 5:261–274, 1993.

D.P. Bertsekas, F. Guerriero, and R. Musmanno. Parallel asynchronous label correcting methods for shortest paths. *Journal of Optimization: Theory and Applications*, 88:297–320, 1996.

D.P. Bertsekas, S. Pallottino, and M.G. Scutellá. Polynomial auction algorithms for shortest paths. *Computational Optimization and Application*, 4:99–125, 1995.

R. Cerulli, P. Festa, and G. Raiconi. Graph collapsing in shortest path auction algorithms. *Computational Optimization and Applications*, 18:199–220, 2001.

R. Cerulli, P. Festa, and G. Raiconi. Shortest path auction algorithm without contractions using virtual source concept. *Computational Optimization and Applications*, 26(2):191–208, 2003.

Z.L. Chen and W.B. Powell. A note on bertsekas's small-label-first strategy. *Networks*, 29:111–116, 1997.

B.V. Cherkassky and A.V. Goldberg. Negative-cycle detection algorithms. *Mathematical Programming*, 85:277–311, 1999.

B.V. Cherkassky, A.V. Goldberg, and T. Radzik. Shortest path algorithms: Theory and experimental evaluation. *Mathematical Programming*, 73:129–174, 1996.

B.V. Cherkassky, A.V. Goldberg, and C. Silverstein. Buckets, heaps, lists, and monotone priority queues. *SIAM Journal of Computing*, 28:1326–1346, 1999.

T.H. Cormen, C.E. Leiserson, and R.L. Rivest. *Introduction to Algorithms*. MIT Press, 1990.

G.B. Dantzig. On the shortest route through a network. *Management Science*, 6: 187–190, 1960.

E.V. Denardo and B.L. Fox. Shortest route methods: 2. group knapsacks, expanded networks, and branch-and-bound. *Operational Research*, 27:548–566, 1979a.

E.V. Denardo and B.L. Fox. Shortest route methods: reaching pruning, and buckets. *Operational Research*, 27:161–186, 1979b.

N. Deo and C. Pang. Shortest path algorithms: taxonomy and annotation. *Networks*, 14:275–323, 1984.

R.B. Dial. Algorithm 360: Shortest path forest with topological ordering. *Comm. ACM*, 12:632–633, 1969.

E. Dijkstra. A note on two problems in connection with graphs. *Numer. Math.*, 1: 269–271, 1959.

P. Festa. *New Auction Algorithms for Shortest Path Problems*. PhD thesis, University of Naples *FEDERICO II*, Naples, Italy, February 2000.

L.R. Ford and D.R. Fulkerson. *Flows in Networks*. Princeton Univ. Press, 1962.

L.R. Jr. Ford. Network flow theory. Technical Report Report P-923, The Rand Corporation, Santa Monica, California, USA, 1956.

M.L. Fredman and R.E. Tarjan. Fibonacci heaps and their uses in improved network optimization algorithms. *Journal of ACM*, 34:596–615, 1987.

G. Gallo and S. Pallottino. Shortest path methods: A unified approach. *Math. Programming Study*, 26:38–64, 1986.

G. Gallo and S. Pallottino. Shortest path methods. *Ann. Oper. Res.*, 7:3–79, 1988.

F. Glover, D. Klingman, and N. Phillips. New polynomially bounded on finding shortest path trees. *Operations Research*, 33:65–73, 1985a.

F. Glover, D. Klingman, N. Phillips, and R.F. Schneider. *Network Models in Optimization and Their Applications in Practice*. Wiley, 1985b.

A.V. Goldberg. A simple shortest path algorithm with linear average time. Technical Report STAR-TR-01-03, InterTrust Technologies Corp., 2001.

A.V. Goldberg and C. Silverstein. Implementations of dijkstra's algorithm based on multi-level buckets. In P.M. Pardalos, D.W. Hearn, and W.W. Hager, editors, *Network Optimization*, pages 292–327. Springer Lecture Notes in Economics and Mathematical Systems, 1997.

J.A. Hillier and P.D. Whiting. A method for finding the shortest route through a road network. *Operation Research Quarterly*, 11:37–40, 1960.

A. Kershenbaum. A note on finding shortest path trees. *Networks*, 11:399–400, 1981.

M. Klein. A primal method for minimal cost flows with applications to the assignment and transportation problems. *Management Science*, 14:205–220, 1967.

E.F. Moore. The shortest path through a maze. In *Proceedings of the International Symposium on the Theory of Switching*, pages 285–292, 1959.

S. Pallottino. Shortest path methods: Complexity, interrelations and new propositions. *Networks*, 14:257–267, 1984.

S. Pallottino and M.G. Scutellá. Strongly polynomial auction algorithms for shortest path. *Ricerca Operativa*, 21:60–72, 1991.

D.R. Shier and C. Witzgall. Properties of labeling methods for determining shortest path trees. *Journals of Research of the National Bureau of Standards*, 86:317–330, 1981.

R.E. Tarjan. Shortest paths. In Y. Alavi et al., editor, *Graph Theory with Applications to Algorithms and Computer Science*, pages 753–759. John Wiley, 1985.

11 Planning and design

9 NETWORK PLANNING PROBLEMS IN TELECOMMUNICATIONS

Henrique Pacca L. Luna[1]

[1]Departamento de Tecnologia da Informação
Universidade Federal de Alagoas
57072-970 Maceio AL, Brazil
pacca@tci.ufal.br

Abstract: This chapter describes the main results of two decades of experience in studying and applying optimization models in the telecommunications sector. The Brazilian information technology policy has enabled during this time many fruitful contracts among some universities and companies of the sector, and the author with some colleagues have been engaged in this process with emphasis in the use of operations research (OR) methods. The author has also been benefited from the cooperation with French institutions that have similar engagements in that country, in such a way that the chapter is in part grounded in real case studies that have motivated research and development of different decision support systems. The chapter synthesize the framework that has been used to integrate models of local access and backbone networks, and then presents a past and future perspective on how is progressing the use of OR models to address the important questions of routing, planning and pricing services in telecommunication and computer networks. Most of the applications are related to the literature and have universal character, but we try to point out in this experience how different cultures, geography and politics impact and shape the way operations research is practiced around the world.
Keywords: Network design, routing, capacity assignment, pricing, network optimization.

9.1 INTRODUCTION

The actual telecommunication system is composed of intricate networks that enables the communication of hundreds of thousands of customers within a city and of millions of customers around the world. The hierarchical organization of this network is a matter of fact and plays a major role for operations research and management science (OR/MS) models, in as much as optimized levels of customer concentration enables the substantial economies of scale of increasing transmission bandwidth. Cost minimization is the objective of most of these operations research models. The main differences among the models concern the hierarchical level of network design, typically *backbone* versus *local access* network, and how are considered for such networks the relevant aspects of connectivity, survivability, modularity, origin-destination demands, routing, pricing and quality of service (Grötschel and Monma, 1990; Kelly, 1988). For each level of network design, a wide range of model formulations has been covered by the related literature, that may be classified by the *deterministic* versus *stochastic* axes, and also by the *continuous* versus *combinatorial* nature of the model.

9.1.1 Divide and conquer for OR/MS applications

A method where a large problem is broken up into several smaller problems is normally implicit in the literature concerning specific models of telecommunication network design (Gavish, 1983; 1991; Minoux, 1989). To provide a better description of different network levels, we have followed a practice of working explicitly with a hierarchical design organization (Balakrishnan et al., 1994; Mateus et al., 1994). The first level of this organization states that the *urban space* is partitioned in *local areas*. Each local area is served by a *switching center*. The communication among local areas is performed by a *backbone network*. The *switching center* is the element of linkage between the *backbone network* and the *local access network*. In our integrated models the *switching center* is a node representing a local area in the *backbone network* and it is also the *root* node of the correspondent *local access network*.

Nowadays, public telecommunication networks provide access for the customers assigned to each switching center with three levels of network design (see Figure 9.1):

1. At the first level, each local area is partitioned in *service sections*, and the *primary network* provides optical access of the service sections to the assigned switching center.

2. At the second level, each service section is partitioned in *terminal sections*, and the *secondary network* provides the access of these terminal sections.

3. At last, the *tertiary network* encompasses the *domestic networks* that assure the links of the customers to assigned terminal boxes.

9.2 PLANNING LOCAL ACCESS SYSTEMS

The most important component of the total cost of a public telecommunication system is normally related to the whole cost of these three classes of local access networks. This explains why our local access design systems have been implemented in first priority, together with the political fact that a developing country should provide a spatial infrastructure for telephones. In the early 1980s some Brazilian companies already used a facility location system to take decisions on the number, the places and sizes of switching centers. This OR/MS system also assigned clusters of customers to the switching centers, as guided by the minimization of the cost of these centers summed with the estimated cost of the local access networks. This traditional location system had been developed in a São Paulo state university, while we were engaged in the state of Minas Gerais with a first system dedicated to the above mentioned

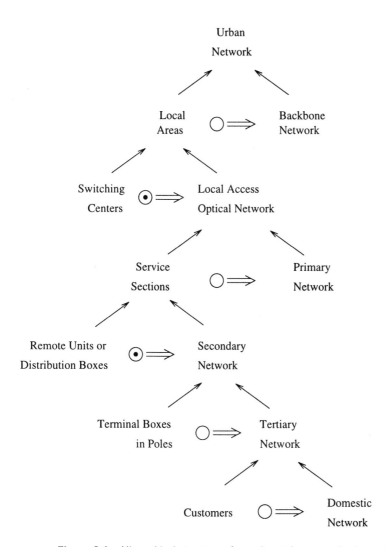

Figure 9.1 Hierarchical structure of an urban telecommunication network

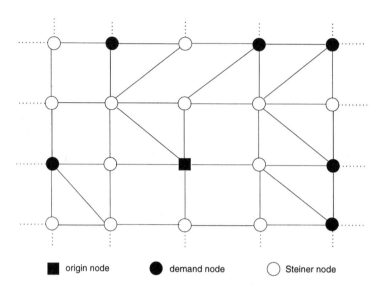

Figure 9.2 The local access network design problem.

primary networks. The life cycle of a series of these systems have finished in the early 1990s, with the possibility of using geographical information systems (GIS). The use of GIS technology has enabled a much better usability of the OR/MS models, and the spatially associated use of databases has permitted to go deeper in the automation of the local access design processes. In fact, the distribution network system where we have more recently worked concerns the above mentioned *secondary networks*, thus arriving much closer to the users local loops (Mateus et al., 2000).

Balakrishnan et al. (1991) discuss several local access network design formulations. The literature on local access network design problems covers a variety of settings, which raises issues of hierarchical organization, topological design, dimensioning and splicing cables (Balakrishnan et al., 1994; Magnanti et al., 1993; Bienstock and Günlük, 1995; Randazzo and Luna, 2001b; Mateus et al., 2000). Flow formulations for these problems concerns an important part of the literature on the subject. Many flow formulations of local access network design problems have been based in extensions of the minimum spanning tree problem (Gavish, 1983; 1991; Mirzaian, 1985; Hochbaum and Segev, 1989). We have preferred to work in Brazil with extensions of the Steiner tree problem on a graph (Luna et al., 1987; Randazzo and Luna, 2001a), as depicted in Figure 9.2. The problem always results to be NP-hard, but in the case of a minimum spanning tree one can find optimal solutions for such embedded sub-problems with the greedy algorithm of Kruskal (1956). Our experience suggests that the designer needs to express the real existence of intermediate or transshipment nodes. This is why we normally look for approximate solutions of the Steiner problem in graphs, thus facing the additional difficulty of working with an NP-hard embedded sub-problem (Hakimi, 1971; Aneja, 1980; Wong, 1984; Maculan, 1987; Beasley, 1989; Duin and Volgenant, 1989).

A complementary problem arising in local access design of computer and communication systems concerns the location of facilities. The facilities may be switching centers, as referred above, but may also be remote units, concentrators, distribution or terminal boxes (Minoux, 1989; Gavish, 1991). Normally a facility is the *root* (supply) node of a local access *tree* network, and it is one of many demand nodes of a higher level network, that may be a *tree*, a *ring* or a *multiconnected* network. We can be inspired by classical methods to solve facility location problems, either in capacitated versions (Sa, 1969; Beasley, 1988; Mateus and Luna, 1992) or in uncapacitated versions (Balakrishnan et al., 1989).

Distribution network planning consists of both locating the optical remote unit and selecting adequate cables to route the required flow to each terminal box. We assume in Mateus et al. (2000) that a previous phase has located the distribution boxes in the wider area of a switching center. The system then address the topology and dimensioning aspects of the distribution network, thus focusing the smaller area of one or some service sections. The model is a complex extension of a capacitated single commodity network design problem. We are given a network containing a set of sources with maximum available supply, a set of sinks with required demands, and a set of transshipment points. We need to install adequate capacities on the arcs to route the required flow to each sink (terminal node). Capacity can only be installed in discrete levels, i.e., cables are available only in certain standard capacities. Economies of scale induce the use of a unique higher capacity cable instead of an equivalent set of lower capacity cables to cover the flow requirements at any link. A path from a source to a terminal node requires a lower flow in the measure that we are closer to the terminal node, since many nodes in the path may be intermediate sinks. On the other hand, reduction of cable capacity levels across any path is inhibited by splicing costs. The objective is to minimize the total cost of the network, given by the sum of the arc capacity (cables) costs plus the splicing costs along the nodes. In addition to the limited supply and the node demand requirements, the model incorporates constraints on the number of cables installed on each edge and the maximum number of splices at each node.

The model for distribution network design is an NP-hard combinatorial optimization problem because it is an extension of the Steiner problem in graphs. The discrete levels of cable capacity and the need to consider splicing costs also increase the complexity of the problem, thus forcing the use of heuristic methods. Our Lagrangian relaxation satisfies the so-called integrality property. Hence, the best possible Lagrangian lower bound cannot exceed the optimal linear programming value, which in this case is weak. There are several strategies that might be devised for such a complicated model, but most of the approaches appear very difficult to be implemented. Researchers have long recognized that, for capacitated problems, optimal solution methods that use only the standard flow-based problem formulation similar to the one described in Mateus et al. (2000) are ineffective. The literature suggests that strong formulations, based on results from polyhedral combinatorics, appear to provide the best means to solve these problems. This is why we are changing our point of view with respect to these models, thus preferring to first look for global optimality of simplified models and then to try to accommodate many practical aspects of the problem. The point that we study now for some uncapacitated models is that, instead of working with heuristics based on weak single commodity flow formulations, we might have better results with strong multicommodity flow formulations. The combinatorial nature and the large scale induced by such redundant formulations can then be treated by Benders decomposition, as a successful alternative to the resolution of the very large scale linear programming relaxations (Randazzo and Luna, 2001b).

9.3 PLANNING AND ROUTING IN BACKBONE NETWORKS

In the mid-1980s the Brazilian telecommunication network had ten million telephones, after a period where the growth of the number of terminals in service had been significant in the last years. The volume of traffic carried by the Brazilian network at that time was estimated at a daily average of 5 million long distance calls and 40 million local calls. With the diversification of the development areas in the country, the number of long distance calls was growing at significant rates. This has asked for investment in the development and application of computer aided systems, which should help to take decisions on the design, expansion and maintenance of the network.

The first of these systems was designed to supply information for backbone traffic administration and engineering. It was in that time able to inform on the traffic profile in an arbitrary arc belonging to the routes between two arbitrary sets of nodes over an arbitrary observation time interval. As a consequence, the traditional origin-destination matrices could be derived for interurban network planning. The approach was based on an automaton that implemented the routing plan over the network topology. Each and every call was software-routed on the graph, and its effect was registered, in terms of circuit occupation in time, for every arc between origin and destination. Most of the telephone calls were directly gathered from the company billing file. At that time a part of the interurban calls was yet billed by pulse metered traffic, and so they where generated in consistence with the available data.

The way to the use of optimization methods to plan backbone networks has been paved by such traffic profile and auxiliary systems. An interactive, intelligent, topology-based editor of routing plans for telephone switching equipment was designed to ask as few questions as possible from its users. The network routing plan ought to be updated continuously, reflecting not only the changes in the topology, but also the inclusion of new terminals in a developing country with a relatively large rate of increase in population. Since that time traffic engineering was committed to keep track of all data relevant to network operation and to projection studies, and this includes the chronological sequence of topologies and their corresponding routing plans.

By the late 1980s was operative an optimization system to find the number of communication channels between each pair of switching centers, minimizing the total cost while maintaining the required quality of service (Mateus et al., 1990). The routing plan must be previously defined for this system. It assumes that the traffic from a node to another node can flow either directly by a *high-usage route* or by one or more *alternative routes*. The overflow traffic is successively offered to *alternative routes*, and so on until it is routed via a unique final route. A part of the traffic sent to the *final route* may eventually be blocked, and the *grade of service* of the network represents the capability to handle most of the traffic offered. We have then followed the widely accepted approach that associates a certain grade of service to a node-to-node blocking probability that do not exceed a given limit. The assumption of a Poissonian traffic with given mean offered to an integer number of channels permitted to calculate the blocking probability by the traditional Erlang B loss function. The crucial impact of non-voice services would then impose a serious limitation of that assumption on the traffic behavior.

Our feeling was that we had arrived at a quite clever model for the backbone problem, but that the traditional OR/MS methodology for telephone network planning was suffering from its natural limitations. Routing optimization and different classes of commodities for each origin-destination pair should be incorporated in a modern modeling approach. Also influenced by our French colleagues, since the mid- 1990s we are suggesting solutions in a certain sense borrowed from the traditional OR/MS methodology for computer network planning. We have then been driven to the use of multicommodity flow problems. These problems usually have a very large number of variables and constraints and arise in a great variety of applications. For instance, one of the most important problems in the design of packet-switched computer network consists in the determination of a set of routes on which packets have to be transmitted and which is minimal according to the average message delay.

We are now working with a relevant extension of this class of models, for which we have recently provided new bounds related to the global optimization of the problem of mixed routing and bandwidth allocation. The combinatorial nature of the problem, related to arc expansion decisions, is embedded in a continuous objective function that encompasses congestion and investment line costs. It results in a non-convex multicommodity flow problem, but we explore the separability of the objective function and the fact that each associated arc cost function is piecewise convex. The idea is to use the efficient algorithms for convex multicommodity flow problems, and we can calculate sharp bounds for the approximated solutions.

9.3.1 Discrete capacity allocation and routing

We present in this subsection an extension of an integrated formulation for the capacity and flow assignment problem (Luna and Mahey, 2000; Ferreira and Luna, 2003). The communication network is modeled as a graph $G = (V,A)$ where V is the set of network nodes ($|V| = n$) and A is the set of links (direct or indirect links)($|A| = m$). Any kind of traffic between a given pair of nodes $(O_k, D_k)_k$ is treated as a separate commodity k with a demand parameter d_k, where K is the number of commodities circulating in the network. Let P_{kh}, ($h = 1..., N_k$) be a set of N_k directed paths used by the product k to join O_k and D_k in G. This set may be a set of all simple directed paths between O_k and D_k or a restricted set of feasible paths. Let the variable x_{kh} be the amount of flow of commodity k through the path P_{kh} and let the parameter a^{kh} be an m−dimensional arc-path (link-path) incidence vector, with each component a_i^{kh} defined by:

$$a_i^{kh} = \begin{cases} 1 & \text{if arc } i \in P_{kh}, \\ 0 & \text{otherwise.} \end{cases}$$

We can state the network expansion problem as follows:

$$\text{minimize} \sum_{i=1}^{m} \tau_i(f_i) \tag{9.1}$$

$$\text{subject to} \sum_{k=1}^{K} \sum_{h=1}^{N_k} a_i^{kh} x_{kh} = f_i, \quad \forall i = 1,...,m \tag{9.2}$$

$$\sum_{h=1}^{N_k} x_{kh} = d_k, \quad k = 1,...,K \tag{9.3}$$

$$0 \leq f_i \leq c_i^{nc}, \quad \forall i = 1,...,m \tag{9.4}$$

$$x_{hk} \in \Re^+, \quad k = 1,...,K, h = 1,...,N_k \tag{9.5}$$

Each component f_i of the m-dimensional variable vector f denotes the total flow on link i. $\tau_i(f_i)$ is an increasing function in $[0, c_i^{nc})$, where c_i^{nc} is the value of the maximum available capacity on link i. Each $\tau_i(f_i)$ is assumed to be continuous, defined on R with values on the extended reals $R \cup \{+\infty\}$ and minored by at least one affine function.

Constraints (9.2) impose that the total flow on link i is equal to the sum of the individual path flows which use that link. Constraints (9.3) ensure that the demand d_k of each commodity k is satisfied. Constraints (9.4) guarantee that the total flow in each link i is smaller than the maximum available capacity for allocation.

The cost function (9.1) of the proposed model is assumed to be separable with respect to the links and is intended to integrate congestion and allocation costs. A congestion cost function is an increasing function which measures, for example, congestion on each link (Kleinrock, 1993), buffer overflow probability or call blocking probability (Kang and Tan, 1997). In fact, commodities are considered as competitive users of a limited resource. The choice of the cost function on the link does not intend to represent a specific traffic model, but should preserve the general characteristics of a large family of congestion measures and should be sufficiently simple to permit explicit computation.

The integrated cost function $\tau_i(f_i)$ on link i is defined by:

$$\tau_i(f_i) = \min\{T_i(f_i, c_i^0) + v_i^0 f_i + \pi_i^0, ..., T_i(f_i, c_i^j) + v_i^j f_i + \pi_i^j,$$
$$..., T_i(f_i, c_i^{nc}) + v_i^{nc} f_i + \pi_i^{nc}\}. \tag{9.6}$$

where $[c_i^0, c_i^1, ..., c_i^{nc}]$ is a set of commercially available line capacities ($c_i^{nc} >, c_i^{nc-1} > ..., > c_i^1 > c_i^0$), that are related, in terms of monetary value per unit of time (\$/$month$), to the following partial cost functions:

$T_i(f_i, c_i^j)$ congestion cost on the link i for line type j;

$v_i^j f_i$ variable operational cost, where the unitary cost v_i^j is evaluated for one $Kbps$ flowing in link i;

π_i^j capacity allocation cost of line type j on link i.

This new model is a unified way to work with a continuous optimization problem for both routing and discrete capacity assignment. The network capacity allocation and routing problem assumes that an initial graph is given, and that in each link it is installed a given capacity which is supposed to be able to support a standard traffic load. In other words, we are not concerned with topological aspects in the problem, in such a way that we consider here $\pi_i^0 = 0$, for all link $i \in A$.

For each link i and assigned capacity c_i^j the univariate congestion function $T_i(f_i, c_i^j)$ is an increasing proper convex, differentiable function. The derivatives $T_i'(f_i, c_i^j)$ are increasing. We also assume the following hypothesis $\forall i \in A$:

$$T_i(0, c_i^0) = T_i(0, c_i^j) = T_i(0, c_i^{nc}) = 0, \ \forall i = 1, ..., m \tag{9.7}$$

$$T_i(f_i, c_i^{nc}) < T_i(f_i, c_i^j) < T_i(f_i, c_i^0), \forall f_i \in (0, c_i^{nc}), \forall i = 1, ..., m \tag{9.8}$$

$$T_i'(f_i, c_i^{nc}) + v_i^{nc} < T_i'(f_i, c_i^0) + v_i^0, \forall f_i \in (0, c_i^{nc}), \forall i = 1, ..., m \tag{9.9}$$

$$T_i'(c_i^j, c_i^j) \geq M, \ \forall j \in (0, ..., n_c), \ \forall i = 1, ..., m. \tag{9.10}$$

Equation (9.7) trivially state that, for any link i, a null flow entails a zero congestion cost for any level of assigned capacity c_i^j. Relations (9.8) mean that, for a given flow f_i on link i, the congestion cost value decreases as far as the capacity of the link increases. Relations (9.9) mean that derivatives induce increasing marginal costs. Relations (9.10) mean that as the congestion cost function increases with the flow, we force the derivative at the saturation level to be greater than a given high value M.

Figure 9.3 illustrates the integrated cost function on the link, denoted in the figure by $\tau(f)$, omitting temporarily the link index i. The function is a pointwise infimum of a series of convex functions. The choice of the cost function is constructed on the basis of the congestion functions $T_i(f_i, c_i^0), T_i(f_i, c_i^1), ..., T_i(f_i, c_i^{nc})$, the variable operational costs $v_i^0, v_i^1, ..., v_i^{nc}$ and the fixed allocation costs $\pi_i^0, \pi_i^1, ..., \pi_i^{nc}$. We consider γ_i^j as being the percentage of the capacity c_i^j for which it becomes more economical to increase the link capacity from c_i^j to c_i^{j+1}. A consequence of this modeling approach is that the delay in any arc is bounded above.

The solution of this model is obtained by seeking path flows, link flows and implicit link capacities in order to minimize the total cost. The cost structure aims at distributing the load among all capacitated links to reduce the total cost of leasing lines and congestion expressed in monetary values. We remark that the capacity in any link may be possibly increased affording a given allocation cost π_i^j, in such a case reducing the marginal congestion cost.

9.3.2 Problem convexification

The objective function defined in (9.1) and (9.6) generates a nonconvex multicommodity flow problem. This nonconvex characteristic is inherent to the link capacity decision problem. Heuristic algorithms are usually proposed in the literature to solve problems of this class, but the algorithms proposed in Ferreira and Luna (2003) are

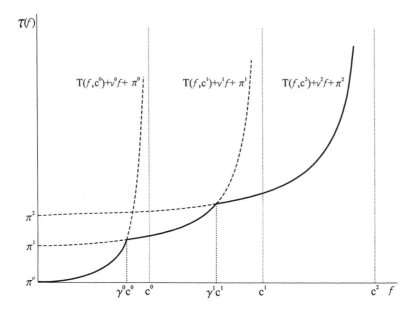

Figure 9.3 The integrated function $\tau(f)$ of congestion and allocation costs.

heuristic algorithms with performance guarantee, in the sense that worst-case errors are known with respect to global optimum decisions.

Let F be the nonempty convex set of link flow vectors for which we have feasible multicommodity flows, i.e., there exists $f \in F$ and corresponding path flows x_{kh} satisfying constraints (9.2) to (9.5). The closed convex hull of the function τ_i is denoted by $conv(\tau_i)$. The convex hull is defined by the greatest closed convex function majored by τ_i.

The following result in Luna and Mahey (2000) on convexification gaps between the convex hull and the original function justifies the methodology to calculate the convex hulls of the function τ_i. Suppose that each function τ_i is bounded below and that the problem (9.1)-(9.5) has an optimal solution f^* with optimal value ϕ^*. Then:
$$\ddot{\phi} = \inf_{f \in F} \{ \sum_{i=1}^{m} conv\, \tau_i(f_i) \}$$
is a lower bound of the optimal value, i.e. $\phi^* \geq \ddot{\phi}$. Moreover, it can be proved that

$$\phi^* - \ddot{\phi} \leq \sum_{i=1}^{m} \max_{f_i} [\tau_i(f_i) - conv\, \tau_i(f_i)] = \Delta. \qquad (9.11)$$

The maximal gap Δ associated with the above lower bound for the link cost function is, in general, greater than zero, as the convex hull of the sum of a set of functions is, in general, different from the sum of the convex hulls of each function. This is true even if the functions are separable because we have coupling constraints. The motivation to do this comes from the fact that the convex hull of certain one-dimensional functions is relatively easy to compute explicitly. Figure 9.4 illustrates the convex hull $conv(\tau_i)$ of the function τ_i, omitting temporarily the link index i.

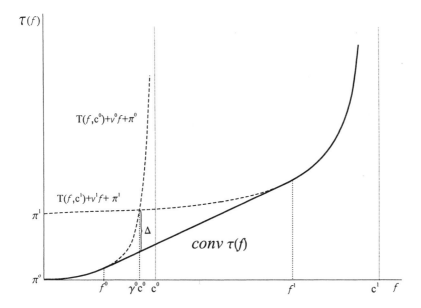

Figure 9.4 Convex hull of the integrated cost function $conv(\tau)$.

The multicommodity flow network problem obtained with the convex hull of each link cost function can be solved by any efficient algorithm addressed for a convex multicommodity flow problem (see Gerla (1973) and Ouorou et al. (2000), for example). Suppose we do not have any nonbifurcated requirement or any other kind of routing restriction for a commodity. In such case, solving the convex problem we get a consistent lower bound for the original capacity and flow assignment problem. On the other hand, an upper bound for the problem can be obtained using the Δ calculated by (9.11).

9.3.3 CFA algorithms with performance guarantee

We propose in Ferreira and Luna (2003) two different heuristics with guaranteed performance to find good solutions for the problem (9.1-9.5). These heuristic algorithms work by initially finding a feasible solution and then gradually reducing the objective value of the obtained solution until no better solution can be found. Each heuristic consists of two phases: in a common first phase, a convex approximation of the original nonconvex problem is solved and a lower bound is found; in the second phase, the obtained routing is used as starting point for swapping methods between capacity assignment and the application of a local search algorithm until no more improvement occurs.

9.4 PRICING TELECOMMUNICATION SERVICES

The need to define optimal prices for telephone services is traditional, and as the Internet makes the transition from the research testbed to commercial enterprise, the topic of pricing in computer networks has attracted great attention. Our suggestion is to use a model of multicommodity flows with price sensitive demands to address this important problem (Luna et al., 1997). The economic equilibrium problem is formulated by maximizing the net social benefit, that is measured by consumers welfare minus the operational and congestion costs associated with the total flow circulating on each link. The application may concern the study of the impact of different tariff policies in the network performance and in consumers welfare, and the impact of the introduction of competition in an opening market.

Much of the motivation for this work came from the above mentioned message routing problems, which were extended to cope with alternative performance measures for different classes of services and also with variable demands for these services. The demands are sensitive to prices that can vary continuously in order to maintain the demand for bandwidth closer to an efficient use of the telecommunication network. The idea is to optimize the use of the available network, always facing consumers willingness to pay with respect to both operation and congestion costs. The relevance of our OR/MS approach is to provide a modeling and an algorithmic framework to follow the tendency of working with discrete changes in prices, as far as we move from different commodities for users with different priorities and during different periods of time.

We stress the *competitive* nature of our modeling approach, thus providing a framework for the analysis of a regional or national telecommunication sector. The approach relies on the neoclassical economic theory, as a natural extension of the original work of Samuelson (1952) on spatial price equilibrium and linear programming. Under the theoretical convexity assumptions, a completely decentralized decision-making process – with many users and with different managers for different links – converges to a global optimum solution, as expressed by the competitive economic equilibrium conditions. Without convexity, the approach can yet be applied to a public monopoly that behaves as if it where placed in a situation of competitive market, thus fixing prices equal to marginal production costs (Curien and Gensollen, 1992). This situation was coherent with the Brazilian public telecommunications monopoly, but the institutional changes concerning the privatization of the sector have brought many questions to be investigated with respect to prices. We are also studying some possible applications concerning the role of the government to regulate the market, with taxes, subsides and minimal or maximal recommended tariffs.

9.4.1 Game theory literature in communications pricing

The objective of this chapter leads to introduce now some issues addressed by a related literature treating of pricing and cost allocation issues in telecommunication services. We do not follow the direction of a related literature that uses a *noncooperative* game theory framework to study the behavior of decentralized control decisions in large scale networks. The reader can see in Korilis et al. (1997) some aspects and references concerning the use of game theory to understand and to influence the behavior of such *noncooperative* networks.

The complementary use of game theory can also be addressed to study the *cooperative* nature of some relevant situations of the telecommunication industry (Skorin-Kapov and Beltran, 1994; Linhart et al., 1995). The *cost allocation* problem is concerned with the fair distribution of the cost of providing a telecommunication service among the customers. A referential example concerns the allocation of costs among the customers of an optimally designed capacitated network (Skorin-Kapov and Beltran, 1994). Before introducing some other *cooperative* situations, we remark that with marginal cost pricing the system is supposed to operate with the revenue greater than or equal to the total cost, instead of the equality normally searched by the cost allocation mechanisms related to the Shapley value of game theory.

The problem of allocation of value for jointly provided services, as exemplified by the caller identification service extended nationwide (Linhart et al., 1995), concerns another typical *cooperative* situation that arises in practice. The example refers to the caller identification commodity, whereby the telephone number of the calling party is visually displayed to the called party during ringing. Identification of a long-distance call requires, in a typical case, the participation of three companies: the local exchange carrier originating the call, the long distance carrier, and the local exchange carrier terminating the call. The paper addresses the question on how shall the revenues from the service be divided among the participating firms. The proposed schemes should be accommodated or else be alternatively treated in some studies of economic equilibrium for the telecommunication sector.

9.4.2 Maximizing net social benefit

We assume that we have a directed graph $G = (V, E)$ with m nodes and n arcs. In general, the nodes represent points of origin and/or destination of a number of K commodities, and the arcs represent links among some given pairs of nodes. A node may also represent a transshipment point, with neither origin nor destination of any commodity. In the telecommunication sector a node may represent a switching-center, a concentrator, a computer terminal, or else a radio station or a satellite. A link in telecommunication may represent a fiber optics cable, a copper cable, or else a radio channel in wireless systems.

A separate commodity k may be voice, file transfer, audio, video, or any kind of traffic between a given pair of nodes $(O_k, D_k)_k$. Note that many commodities can be considered between each communicating node pair. Each commodity k has a unitary operational cost t_k and is related to the scalar variables q_k and p_k, which represent, respectively, consumers demand and price for commodity k. Given a commodity k, we consider a given set of directed paths P_{kh}, $h = 1, \ldots, N_k$ joining the origin O_k and the destination D_k. This set may be the set of all simple directed paths – starting at node O_k and terminating at node D_k – or a restricted set of feasible paths, for instance with a limited number of hops. Let x_{kh} be the flow supplied to commodity k through the path P_{kh}, and a_{kh} its arc-path incidence vector defined by

$$a_{kh}^j = \begin{cases} 1 & \text{if arc } j \in P_{kh}, \\ 0 & \text{otherwise.} \end{cases}$$

The scalar variable x_{0j} denotes the total flow on arc j. Any arc j has a limited capacity c_j and may possibly be part of one or more paths used to flow some of the commodities. We assume given a congestion and operation cost function $f_j(x_{0j})$ on each arc j, that depends on the actual arc capacity c_j, that is increasing on $[0,c_j]$ and that is assumed to be convex and differentiable. It aims at distributing the load among all capacitated arcs to reduce the total cost of operating lines and congestion expressed in monetary values. We will not digress here neither on the question of the tradeoff between money and mean time delay, nor on the approximation in this case of any statistical congestion function like Kleinrock's delay function (see, for example, Bertsekas and Gallager (1987)).

The notation composes the column vectors p and q, which are formed, respectively, with the components p_k and q_k associated to the price and the demand for each commodity k. An invertible demand function is supposed to be known, in such a way that we can express the econometric interdependence between the vectors of prices and demands by the multi-valued inverse function $p = \phi(q)$. Figure 9.5 illustrates the inverse demand function for the single commodity case, when the demand q_k is assumed to be simply a function of the price p_k established for commodity k. In practice, any simplifying assumption of no demand interdependence among commodity prices would remove one of the most basic reasons why such models are important. A proper use of the demand function is crucial for the methodology. We introduce now the possibility of considering a multivariable *linear* interdependence between the vectors of prices and demands. We suppose that econometric studies have established a linear price dependent demand function for the telecommunication commodities

$$q = q^0 + Qp \quad (9.12)$$

where q^0 is a K-component positive vector and Q is a square matrix assumed to be symmetric and negative definite. This assures the existence of the inverse matrix Q^{-1}, so there is a one-to-one correspondence between q and p and the following inverse function can be defined:

$$p = p^0 + Rq \quad (9.13)$$

where $p^0 = -Q^{-1}q^0$ and $R = Q^{-1}$, implying that R is also a symmetric and negative definite matrix. The symmetry property permits to obtain the following integral for the function $p = \phi(q)$ expressed by (9.13):

$$\int_0^q \phi(\xi)\,d\xi = q'\left(p^0 + \frac{1}{2}Rq\right) \quad (9.14)$$

where the vector q' is the transpose of q.

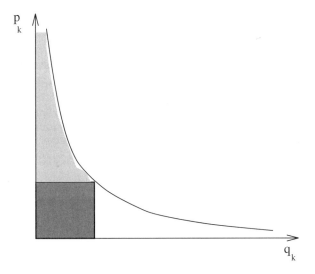

Figure 9.5 Relation between price and demand in a single commodity case.

In general, the Price Equilibrium Problem (PEP) can be formulated as

$$\min_{q,x \geq 0} - \int_0^q \phi(\xi) \, d\xi + \sum_{j=1}^n f_j(x_{0j}) + \sum_{k=1}^K t_k \sum_{h=1}^{N_k} x_{kh} \quad (9.15)$$

$$\text{subject to} \quad q_k - \sum_{h=1}^{N_k} x_{kh} \leq 0 \quad k = 1, \ldots, K \quad (9.16)$$

$$\sum_{k=1}^K \sum_{h=1}^{N_k} a_{kh}^j x_{kh} - x_{0j} \leq 0 \quad j = 1, \ldots, n \quad (9.17)$$

$$x_{0j} \leq c_j \quad j = 1, \ldots, n \quad (9.18)$$

We discuss now some computational aspects of the solution of problem (9.15-9.18), before discussing in the next section, in light of the optimality conditions, the physical and economic interpretation of the problem. The nonlinear convex model (9.15-9.18) could, in principle, be solved by general nonlinear programming techniques. However, the special structure of the problem makes decomposition methods more attractive in this case. A column generation procedure assures the correct selection of parameters a_{kh}^j associated to *optimal routes* when *all* possible routes are considered for *each* commodity. The explosive number of columns is only implicit in the formulation of the Price Equilibrium Problem (9.15-9.18). In practice, we need to memorize only a selected number of routes for each commodity. The selected routes are usually related to shortest paths on the graph G associated with marginal cost weights (Luna et al., 1997).

9.4.3 Economic Equilibrium Conditions

The mathematical representation for a competitive equilibrium can be described by the optimality conditions of the mathematical program (9.15-9.18). The problem searches for a minimization of a convex objective function subject to linear constraints, so the Kuhn-Tucker conditions are necessary and sufficient for a global optimum. At an optimal solution, the primal variables q (unrestricted) and x (non-negative) satisfy the feasibility constraints (9.16-9.18). Yet, there are non-negative dual variables p, β and γ such that the following relations are also verified and assure optimality together with (9.16-9.18).

$$p = \phi(q) \tag{9.19}$$

$$\left. \begin{array}{l} f'_j(x_{0j}) + \gamma_j \geq \beta_j \\[1em] [f'_j(x_{0j}) - \beta_j + \gamma_j]x_{0j} = 0 \end{array} \right\} \quad j = 1,...,n \tag{9.20}$$

$$\left. \begin{array}{l} t_k + \sum_{j=1}^{n} a^j_{kh}\beta_j \geq p_k \\[1em] [t_k + \sum_{j=1}^{n} a^j_{kh}\beta_j - p_k]x_{kh} = 0 \end{array} \right\} \quad k = 1,...,K,\ h = 1,...,N_k \tag{9.21}$$

The relationship among the gradient vectors of the objective function and the constraints is expressed by (9.19) and the first parts of (9.20) and (9.21). Note that the model (9.15)-(9.18) takes q as an unrestricted vector in order to meet the equality constraints (9.19). It is expected that in empirical applications the demand equation would bear positive q for every p contained within practical bounds, so restricting prices and quantities within ranges of economic sense. The second parts of (9.20) and (9.21) express the slack complementarity of the gradient relationship with respect to the non-negative primal variables x. The slack complementarity conditions associated with the dual variables p, β and γ and related, respectively, to the feasibility constraints (9.16), (9.17) and (9.18) are the following:

$$[q_k - \sum_{h=1}^{N_k} x_{kh}]p_k = 0 \quad k = 1,...,K \tag{9.22}$$

$$[\sum_{k=1}^{K}\sum_{h=1}^{N_k} a^j_{kh}x_{kh} - x_{0j}]\beta_j = 0 \quad j = 1,...,n \tag{9.23}$$

$$[x_{0j} - c_j]\gamma_j = 0 \tag{9.24}$$

Equation (9.19) becomes equation (9.13) in the linear case, reporting the econometric dependencies estimated among the prices and the consumers demands for the different commodities. Note that the non-negative vector p and the unrestricted vector q are unique, as proved in Luna (1979) for this type of formulation of sectorial economic equilibrium problems.

The set of constraints (9.16) assures no excess demand, while the slack complementarity condition (9.22) imposes supply equal to demand for every commodity with a positive price ($p_k > 0$). In practice, the price of every commodity is positive, and (9.16) is satisfied with strict equality for every commodity k. Constraints (9.16) then ensures that the demand for each commodity is fully supplied by the path flows activated for the commodity.

The inequalities (9.17) and (9.18) maintain all the activities within the production possibility set. The set of inequalities (9.17) relates the use of any arc j with the alternative routing and flow decisions x_{kh}, as imposed by the technology (arc-path) matrix of coefficients a_{kh}^j. The associated slack complementarity condition (9.23) provides a null slack for every link that is evaluated positively ($\beta_j > 0$). In such a case, the constraints (9.17) impose that the total flow on each arc is equal to the sum of the individual path flows which use that arc. The inequalities (9.18) maintain the use of any arc x_{0j} within the available capacity c_j; and the set of equations (9.24) say that a positive value ($\gamma_j > 0$) imply the full use of capacity in arc j, and it also imputes a null value ($\gamma_j = 0$) for additional capacities where slack is found.

Relations (9.20) express the fact that the marginal cost at any link j is never smaller than its marginal value β_j. Note that the marginal cost of the link j results from summing up the marginal operation and congestion cost $f_j'(x_{0j})$ plus the marginal value γ_j of the limited capacity c_j. If $x_{0j} > 0$ the effective use of link j entails an evaluation β_j that is *equal* to the marginal cost of the use of the link. Here β_j can be viewed, in a decentralized decision-making situation, as the equilibrium price to be paid by the service providers and telecommunication companies to the owner of link j. In most interpretations of the congestion notion, either in telecommunication or in transportation, the capacity limit c_j is never attained, in such a way that we have for any used arc $0 < x_{0j} < c_j$, $\gamma_j = 0$ and $\beta_j = f_j'(x_{0j})$. Note that for this value of β_j there is no incentive to increase the offer of traffic across j.

The relations (9.21) state that the marginal cost to produce a commodity k is never smaller than its price p_k. The condition states that no marginal production can make a profit in equilibrium because capital would otherwise move into the activity. The slack complementarity condition in (9.21) states that if a path is used to flow a positive value $x_{kh} > 0$ of commodity k then the price p_k is equal to the marginal cost of flowing an additional unit of commodity k across that path. The result is that all the paths that serve a commodity have the same marginal cost to produce (flow) that commodity. The price p_k to be paid by the consumers of k is formed by the operational cost t_k, to be directly afforded by the providers of the telecommunication service, and the addition of the prices β_j that should be paid to the owners of the links that compose any specific path where is flowing commodity k. In most applications each price is formed by the operational cost and the congestion costs across any path serving the commodity. A high congestion cost $\beta_j = f_j'(x_{0j})$ inhibits the inefficient use of the full capacity c_j because it implies a high price p_k that reduces the demand q_k for every commodity that is flowing through link j.

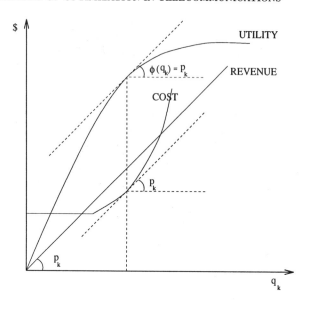

Figure 9.6 Price equilibrium for a single commodity case.

9.4.4 Remarks on price equilibrium models

Now some general remarks on this class of models together with some particular features of the present formulation deserve mention:

- The first point is the artificial character of the objective function. The idea is that it has not been proposed as a direct measure of social preference; instead, it is simply consistent with a mathematical programming framework whose optimality conditions correspond to a sectorial economic equilibrium. Although any interpretation of social welfare may be questioned, what follows is a attempt towards a better understanding of this methodology. In the special case where money provides an adequate measure of utility, the objective function (9.15) can be interpreted as the maximization of the *net social pay-off*, as suggested by Samuelson (1952). The interpretation for a telecommunication network is that the objective (9.15) maximizes the difference between the aggregate consumers utility and the total network operation and congestion costs. Note that the aggregate consumers utility is evaluated by integrating the inverse demand function from 0 until q. Figures 9.5 and 9.6 illustrate for the single commodity case how the consumers utility is evaluated and how a point that maximizes the net social benefit is found. The consumers utility corresponds to the shadowed area of Figure 9.5, and it is composed of the rectangular area measuring the commodity revenue (price p_k × quantity q_k) and the triangular area corresponding to the consumer's surplus. For each class of consumers, the surplus is given by the

- A second point concerns the general properties that should have a demand function, in particular when estimating multivariable *linear* relations between the vectors of prices and demands, as introduced in (9.12) and (9.13). A complete economic interpretation of the negative definiteness of matrix Q appears to be a little difficult, and there is also the question of whether or not this condition is met in the real world. Usually, in the economic literature, the comprehensive requirement that every product follows the law of downward-sloping demand – which means that the direct effect of a price change is negative, or that all the diagonal elements of Q are negative – has played an important role on sufficiency conditions for negative definiteness of Q. Another requirement has been that matrix Q be strongly quasi diagonalized, what roughly speaking means that there must exist some relation between direct and cross price effects such that the direct effects are stronger than some linear combination of cross effects. Moreover, there is no economic reason on which the assumption of a symmetric matrix Q is based. The only justification for it consists on working with an integrable inverse demand function, in order to obtain a mathematical program with the interesting property of having dual variables easily interpreted as the prices under study. Otherwise, we could not work with such a concise optimization model, or perhaps we should work with a variational inequality modeling and algorithmic framework (for instance, see Patriksson (1994)). A proper use of the demand function is crucial for the methodology, and the problem of reliable demand estimation is perhaps the major drawback to apply the model in empirical studies, in many cases overshadowing questions of symmetry or negative definiteness of the matrix Q.

- A third point is that the model is inserted within the general framework of the neoclassical equilibrium analysis, and as such it captures some common hypothesis. The model must be applied under a *ceteris paribus* hypothesis, by supposing that everything else in the environment of the studied sector remains constant, and yet this sector is relatively small so that it has no sensible influence on the overall equilibrium of the economy. It is also assumed that the *users* of the telecommunication services (commodities), the *providers* and the link *managers* take decisions under complete information about future market prices. Uncertainty has not been introduced in the model, and it is also supposed that there is no increasing returns of scale.

- Finally, we remark that this kind of study could be made either focusing the public telecommunication networks, as discussed above, or else focusing a corporate network with an internal pricing mechanism for the access of the enterprise divisions. The latter problem has been studied by Girard and St-Georges (1993), within the cooperative context of pricing and design of a corporate network. Their work incorporates routing decisions in stochastic models and uses Erlang loss functions. The problem is to define the optimal strategy for planning the corporate network and choose a pricing mechanism to allocate the cost of this optimally designed system to the enterprise divisions. The competitive environment refers to the alternative uses between the private and the public network. We do not cope here with the question of optimal planning of a private network, but we think that our formulation provides internal pricing mechanisms as a consequence of its general nature. Yet the question of using our approach in corporate networks deserves further analysis. We suggest that, in practice, the internal prices might eventually depart from the private network leasing cost, but they should coincide with the public transmission prices whenever the increased demands menaces the efficient use of the private network.

9.4.5 Intertemporal Dependence

Recall that the traffic between a given pair of nodes can be represented by many different commodities. A commodity may refer to a specific service in a certain hour between the pair of nodes, in such a way that a certain number of commodities can represent the same service in different periods of time. For instance, the telephone calls of an origin-destination pair may be represented by a number of commodities, each of them representing a time period of the week with a typical traffic pattern. For each of these time periods (charge bands) one of four possible charging rates may be assigned: expensive, normal, reduced, and very reduced. A multi-valued demand function for such origin-destination telephone calls can exhibit a natural interdependence within this kind of commodity requirements. Under these circumstances, the model can be used to study the impact of temporal charging policies in the network performance and in consumer welfare.

The remark above indicates that, for the telecommunication sector, some dynamical aspects can be accommodated by such a "static" formulation of the convex multicommodity flow problem. From the production (routing) point of view, the network should be repeated as many times as we need to represent different charge bands. The problem of pricing telecommunication services resembles the problem of pricing the electrical power sector and has also some analogies with the models of spatial and temporal prices in the agricultural sector (Geromel and Luna, 1981). But contrary to the traditional energy and agricultural sectors, where storage plays a major role, in telecommunication we do not need to express any linkage among different production periods. The real-time nature of telephone conversations and most Internet uses, and the high speed of information transmission make the role of storage very limited in telecommunication. In practice, coupling in telecommunication models may be limited to the cross-effects of some demand functions, while a convenient separability is maintained in the implicit production functions for the different periods. In summary, our "static" formulation serves to simulate the *dynamic* behavior of telephone users by calling during off-peak periods in such a way to benefit from reduced rates.

9.4.5.1 Example. We consider now an example to clarify the capability of the model to cope with demand interdependence and intertemporal aspects. Suppose we have a model of telephone services in a simple network with two nodes (one origin and one destination), two parallel arcs with equal capacity – of 2 Megabits per second (Mb/s) each – and two *non-simultaneous* commodities. The network is repeated once, because commodity 1 refers to telephonic calls during *day* period (more expensive), whereas commodity 2 refers to the cheaper *night* period. Any commodity can be routed across any of the two equal links, and for each commodity $k = 1, 2$ we need to determine the price p_k, in cents per call, and the demand q_k, in number of calls per period. Suppose that we can assign for each of the two links a nominal capacity of 7200 calls per period of 12 hours (day or night). Because of the symmetry, for each time period, the unique commodity is equally bifurcated across the two links, thus implying the same marginal cost in each link. Taking $t_1 = t_2 = 1$ cent/call as commodity operational cost and $f(x) = (1/3000)x^2$ as link operations and congestion cost function, we have to calculate the flow x_1 and the value β_1 associated with any link during day period and the flow x_2 and the value β_2 associated with any link during night period.

The underlying graph of the simple model is disconnected, but the interdependence between day and night periods can be expressed by inverse demand functions that look like $p_1 = \phi_1(q_1, q_2)$ and $p_2 = \phi(q_1, q_2)$ and that is here represented by the linear relation

$$\begin{pmatrix} p_1 \\ p_2 \end{pmatrix} = \begin{pmatrix} 60/7 \\ 30/7 \end{pmatrix} + \begin{pmatrix} -1/3500 & -1/21000 \\ -1/21000 & -1/1750 \end{pmatrix} \begin{pmatrix} q_1 \\ q_2 \end{pmatrix}$$

that must be satisfied together with the economic equilibrium conditions (9.20) and (9.21) expressed by

$$\frac{1}{1500} x_1 - \beta_1 = 0$$

$$\frac{1}{1500} x_2 - \beta_2 = 0$$

$$1 + \beta_1 = p_1$$

$$1 + \beta_2 = p_2$$

what results in prices $p_1 = 5$ and $p_2 = 2$ cents/call, demands $q_1 = 12000$ and $q_2 = 3000$ calls per period, link flows $x_1 = 6000$ and $x_2 = 1500$ calls per period and marginal link costs $\beta_1 = 4$ and $\beta_2 = 2$ cents/call for, respectively, day and night periods. Observe that $\gamma_1 = \gamma_2 = 0$ because the nominal capacities of the links are not attained in the optimal solution. The example illustrates the point of view that, given prices during one particular period, consumers would modify their demands and increase (or decrease) usage during cheaper (or more expensive) periods.
□

Apart the use of this model for the study of *pre-defined* charge bands we are also interested in the use of the model to accommodate a new tendency in the Internet and in Asynchronous Transfer Mode (ATM) networks, for which the prices might vary *continuously* in order to maintain the demand for bandwidth closer to its efficient supply (Karlsson, 1996). The idea is that a way to optimize an integrated services network is to involve the users more actively in the process. The users should be encouraged to accommodate their commodity flows having in mind that a network with a high load has more tendency to get into overloaded situations, which degrades quality of service. Feedback mechanisms with temporal charging could be used to smooth out the traffic by increasing charging at high loads and decreasing it at lower loads. Instead of being previously defined, changes in prices could be made in as much as times as them should be necessary to regulate an efficient use of the network.

9.5 TRENDS FOR OR/MS COMMUNICATION MODELS

Our conclusion is that decision support systems for engineering design in telecommunication require an hierarchical organization of optimization models. The rising complexity of integrated computer and communication systems imposes for the emerging problems a coherent divide-and-conquer solving strategy. Optimal and heuristic algorithms must be adapted to the different classes of OR/MS models, with consistent transfer of information and decision among the models. We have shown a used hierarchy of decision levels, and our experience indicates that the use of GIS technology in a developing country has effectively conducted the OR/MS models towards the lower levels of local access network design. Our guess is that the timing of our experience may have been somewhat specific with respect to local access design, because we were lead to the use of GIS in a time when most of the developed countries already had their local networks installed.

Our experience also suggests the importance of using flow formulations in all levels of telecommunication network design. Apart the economic study of the computer networks and telecommunications sector, the relevance of multicommodity flow models with variable demands will be to provide an algorithmic framework to cope with changes in prices of telecommunication services. These changes should be made as far as we move from alternative commodities for users with different priorities and during different periods of time. The idea is to optimize the use of the available network, always facing consumers willingness to pay with respect to congestion costs. Note that the idea fits the Internet challenge to meet more complex objectives, with users demanding specific service commitments during times of congestion, thus being tacitly assumed that will be offered higher throughput for users who pay more.

Another devised trend for OR/MS telecommunication models should include both the topology design and the bandwidth allocation in emerging technologies such as ATM nets, together with the mixed routing and pricing mechanisms here discussed. The motivation behind this modeling approach concerns the need to integrate the routing mechanisms (commodity flows) together with the control of bandwidth allocation (capacity assignment) in emerging technologies. The integrated problem is classical in network planning, but it has now a greater importance because now capacity assignment can also be done in real time, thus becoming an activity closely related to network monitoring and traffic measurement functions that must be carried out on line. Perhaps the network management in the future will remain automated with the current routing mechanisms, but this together with some kinds of pricing mechanisms as discussed in this chapter and with different kinds of decentralized bandwidth allocation.

Telephones have always transmitted voice, but the last twenty years encompass a period in which the way computer networks were used evolved from an academic curiosity to useful applications in large businesses, and then to the worldwide Internet, that has become a daily reality for millions of people. OR/MS models have been very useful to solve the evolving problems, and we see with optimism the future possibilities. Our view is that, in the more general cases, connectivity, survivability and modularity will be considered in a *combinatorial* part of the model, whether the aspects of multicommodity, routing, pricing and quality of service will be embedded in the large-scale *continuous* parts of the model. Heuristics will play the major role, but in many important situations we will find the possibility to solve real problems with exact decomposition algorithms.

Acknowledgments

The experience of the author in network planning has been benefited from the cooperation of a number of collaborators and co-authors, who are here represented by Adam Ouorou, Geraldo R. Mateus, Philippe Mahey and the doctoral students Cristina D. Randazzo, Gilberto Miranda Jr. and Ricardo P. M. Ferreira. This work has also been partially supported by two Brazilian scientific institutions, Conselho Nacional do Desenvolvimento Científico e Tecnológico (CNPq) and Fundação de Amparo à Pesquisa do Estado de Alagoas (FAPEAL).

Bibliography

Y.P. Aneja. An integer linear programming approach to Steiner problem in graphs. *Networks*, 10:167–178, 1980.

A. Balakrishnan, T.L. Magnanti, and P. Mirchandani. Modeling and heuristic worst-case performance analysis of two-level network design problem. *Management Science*, 40:846–867, 1994.

A. Balakrishnan, T.L. Magnanti, A. Shulman, and R.T. Wong. Models for planning capacity expansion in local access telecommunication networks. *Annals of Operations Research*, 33:239–284, 1991.

A. Balakrishnan, T.L. Magnanti, and R.T. Wong. A dual-ascent procedure for large-scale uncapacitated network design. *Operations Research*, 37:716–740, 1989.

J.E. Beasley. An algorithm for solving large capacitated warehouse location problems. *Journal of the Operational Research Society*, 33:314–325, 1988.

J.E. Beasley. An SST-based algorithm for the Steiner problem in graphs. *Networks*, 19:1–16, 1989.

D. Bertsekas and R. Gallager. *Data Networks*. Prentice Hall, New Jersey, 1987.

D. Bienstock and O. Günlük. Computational experience with a difficult mixed-integer multicommodity flow problem. *Mathematical Programming*, 68:213–237, 1995.

N. Curien and M. Gensollen. *Economie des Télécommunications: Ouverture et Réglementation*. Economica, Paris, 1992.

C.W. Duin and A. Volgenant. Reduction tests for the Steiner problem in graphs. *Networks*, 19:549–567, 1989.

R.P.M. Ferreira and H.P.L. Luna. Discrete capacity and flow assignment algorithms with performance guarantee. *Computer Communications*, 26:1056–1069, 2003.

B. Gavish. Formulations and algorithms for the capacitated minimal directed tree. *Journal of the ACM*, 30:118–132, 1983.

B. Gavish. Topological design of telecommunication networks - Local access design methods. *Annals of Operations Research*, 33:17–71, 1991.

M. Gerla. *The Design of Store-and-Forward (S/F) Networks for Computer Communications*. PhD thesis, UCLA, Los Angeles, 1973.

J.C. Geromel and H.P.L. Luna. Projection and duality techniques in economic equilibrium models. *IEEE Trans. on Systems, Man and Cybernetics*, 11:329–338, 1981.

A. Girard and S. St-Georges. Performance evaluation and routing optimization of ISDN networks with nonhierarchical and alternate routing. In *1th International Conference on Telecommunication Systems Modeling and Analysis*, pages 454–461, Nashville, USA, 1993. Vanderbilt University.

M. Grötschel and C. L. Monma. Integer polyhedra arising from certain network design problems with connectivity constraints. *SIAM J. Disc. Math.*, 3(4):502–523, 1990.

S.L. Hakimi. Steiner's problem in graphs and its implications. *Networks*, 1:113–133, 1971.

D.S. Hochbaum and A. Segev. Analysis of a flow problem with fixed charges. *Networks*, 19:291–312, 1989.

C.G. Kang and H.H. Tan. Combined channel allocation and routing algorithms in packet switched networks. *Computer Communications*, 20:1175–1190, 1997.

J.M. Karlsson. Some aspects of charging in ATM networks. In *4th International Conference on Telecommunication Systems Modeling and Analysis*, pages 567–573, Nashville, USA, 1996. Vanderbilt University.

F.P. Kelly. Routing in circuit-switched networks: Optimization, shadow prices and decentralization. *Adv. Appli. Prob.*, 20:112–144, 1988.

L. Kleinrock. On the modeling and analysis of computer networks. *Proceedings of the IEEE*, 81(8):1179–1191, 1993.

Y.A. Korilis, T.A. Varvarigou, and S.A. Ahuja. Optimal pricing strategies in noncooperative networks. In *5th International Conference on Telecommunication Systems Modeling and Analysis*, pages 110–123, Nashville, USA, 1997. Vanderbilt University.

J.B. Kruskal. On the shortest spanning tree of a graph and the traveling salesman problem. *Proc. Amer. Math. Society*, 7:48–50, 1956.

P. Linhart, R. Radner, K.G. Ramakrishnan, and R. Steinberg. The allocation of value for jointly provided services. *Telecommunication Systems*, 4:151–175, 1995.

H. P. L. Luna. Note on price-unicity in economic equilibrium models. *Socio-Economic Planning Sciences*, 13(4):223–225, 1979.

H.P.L. Luna and P. Mahey. Bounds for global optimization of capacity expansion and flow assignment problems. *Operations Research Letters*, 26:211–216, 2000.

H.P.L. Luna, P. Mahey, A. Ouorou, and J. Chifflet. Multicommodity flows with price sensitive demands. In *5th International Conference on Telecommunication Systems Modeling and Analysis*, pages 583–593, Nashville, USA, 1997. Vanderbilt University.

H.P.L. Luna, N. Ziviani, and R.M.B. Cabral. The telephonic switching centre network problem: Formalization and computational experience. *Discrete Applied Mathematics*, 18:199–210, 1987.

N. Maculan. The Steiner problem in graphs. *Annals of Discrete Mathematics*, 31:185–212, 1987.

T.L. Magnanti, P. Mirchandani, and R. Vachani. The convex hull of two core capacitated network design problems. *Mathematical Programming*, 60:233–250, 1993.

G.R. Mateus, F.R.B. Cruz, and H.P.L. Luna. Algorithm for hierarchical network design. *Location Science*, 2:149–164, 1994.

G.R. Mateus and H.P.L. Luna. Decentralized decision-making and capacitated facility location. *The Annals of Regional Science*, 26:361–377, 1992.

G.R. Mateus, H.P.L. Luna, and L.C.M. Lage. Optimal planning of telecommunication networks with logical and physical aspects. In *SBT/IEEE Telecommunications Congress*, pages 602–607, Rio de Janeiro, Brazil, 1990. SBT/IEEE.

G.R. Mateus, H.P.L. Luna, and A.B. Sirihal. Heuristics for distribution network design in telecommunication. *Journal of Heuristics*, 6:131–148, 2000.

M. Minoux. Network synthesis and optimum network design problems: Models, solution methods and applications. *Networks*, 19:313–360, 1989.

A. Mirzaian. Lagrangian relaxation for the star-star concentrator location problem: Approximation algorithm and bounds. *Networks*, 15:1–20, 1985.

A. Ouorou, P. Mahey, and J. Ph. Vial. A survey of algorithms for convex multicommodity flow problems. *Management Science*, 46:126–147, 2000.

M. Patriksson. *The Traffic Assignment Problem: Models and Methods*. Topics in Transportation, VSP, Utrecht, The Netherlands, 1994.

C.D. Randazzo and H.P. Luna. A comparison of optimal methods for local access uncapacitated network design. *Annals of Operations Research*, 106:263–286, 2001a.

C.D. Randazzo and H.P. Luna. Solving local access network design problems with two technologies. *Investigación Operacional*, 21:115–123, 2001b.

G. Sa. Branch-and-bound and approximate solutions to the capacitated plant location problem. *Operations Research*, 17:1005–1016, 1969.

P.A. Samuelson. Spatial price equilibrium and linear programming. *Amer. Econ. Rev.*, 42:284–303, 1952.

D. Skorin-Kapov and H.F. Beltran. An efficient characterization of some allocation solutions associated with capacitated network design problems. *Telecommunication Systems*, 3:91–107, 1994.

R.T. Wong. A dual ascent algorithm for the steiner problem in directed graphs. *Mathematical Programming*, 28:271–287, 1984.

10 MULTICOMMODITY FLOW PROBLEMS AND DECOMPOSITION IN TELECOMMUNICATIONS NETWORKS

Abdel Lisser[1] and Philippe Mahey[2]

[1]Université de Paris Sud, LRI - UMR 8623,
Bât 490, 91405 Orsay Cedex, France
lisser@lri.fr

[2]Université Blaise Pascal, LIMOS - UMR 6158,
B.P. 10125, 63173, Aubière Cedex, France
Philippe.Mahey@isima.fr

Abstract: The purpose of this chapter is to investigate multicommodity flow problems that appear in the network design and operation of modern broadband packet-switched networks. We present arc-node and arc-path models and analyze specialized formulations corresponding to hard to solve instances like the minimax congestion problem and the capacity assignment of data networks in the presence of failures. Decomposition methods are studied to cope with the coupling constraints which define interactions between commodities on critical arcs or the combinatorial choice between normal and spare capacities. We focus here mainly on continuous flow models with linear or convex costs.
Keywords: Multicommodity flow problems, network design, packet-switched networks, arc-node models, arc-path models, minimax congestion, capacity assignment, decomposition methods, continuous flow models.

10.1 INTRODUCTION

In the early 1980s, a new broadband technology was developed at Bellcore, called SONET (Synchronous Optical Network) (Advisory, 1990) and normalized for European countries under the name of SDH (Synchronous Digital Hierarchy). SDH and new optical fibers gave rise to the high speed transmission networks era. Due to SDH systems and equipment, difficult multiplexing and demultiplexing functionalities which characterize the previous technology become elementary operations and can be handled easily. The transmission capacities for telecommunications systems, grow exponentially and rates of Gigabit per second, Gb/s, are quickly achieved.

Transmission facilities gave rise to new multimedia services (voice, data, and video) with flexible bandwidth, and highly demanding in terms of bandwidth and quality of service. Virtual private networks, as well as Internet, came into sight with dazzling developments. Moreover, the competition brought about by deregulated telecommunication markets and customer requirements decreased in large scale the transmission costs and highlighted network survivability in order to provide a high quality of service.

SDH facilities use a pair of fibers per system and their high capacity simplifies network architectures. Therefore, the graphs underlying the networks are sparse and the traffic per cable is very high. This makes the network more sensitive to system or equipment failures and increases survivability issues.

Recent advances in wavelength technology, namely Wavelength Digital Technology WDM (Brackett, 1990), allow to handle several wavelengths, each carrying 2.5 Gb/s per fiber, raising the speed up to Terabit per second, Tb/s, through a long distance. The savings in the optical fibers as well as in the optical amplifiers (10 times SDH amplifying capabilities) are significant.

The fundamental task in telecommunication network design consists in transporting different services in order to connect different customers. This operation is called routing. In order to achieve this task, graph theory and optimization fields provide the multicommodity flow model.

It is well known that telecommunications networks can be modeled as graphs where the nodes correspond to telecommunication centers (local, central offices, ...) while the arcs represent the cables. Each arc, i.e. cable, has a given capacity according to the equipment it connects. This capacity is shared by individual demands. To route a given demand or commodity through the network, a cost should be paid according to different used arcs. This leads to a minimum cost multicommodity netflow problem, or MCNF for short. This problem has been widely studied in the literature both for theoretical (Schrijver, 2003) and practical purposes (Ahuja et al., 1993).

Recent applications in telecommunications in the last decade are: Traffic management of Asynchronous Transfer Mode (ATM) networks (Medova, 1998); Virtual path routing for survivable ATM networks (Murakami and Kim, 1996); Design of centralized telecommunication networks (Gupta and Pirkul, 1996); Design of telecommunication loss networks (Girard and Sanso, 1998; Medhi and Guptan, 1975); Design of fiber transport networks (Yoon et al., 1998); Routing and wavelength assignment in optical networks (Banerjee and Mukherjee, 1996); Terminal layout with hop constraints (Gouveia, 1996); Design of token ring LANs (LeBlanc et al., 1996); Routing in virtual circuit data networks with unreliable links (Sung and Lee, 1995); Telephone network traffic management (Chang et al., 1992; Chang, 1994); Ring network design (Arbib et al., 1992); Backbone network design in communications network (Bogdanowicz, 1993); Routing in packet-switched networks (Ribeiro and Elbaz, 1992); Virtual circuit routing in computer networks (Yee and Lin, 1992; Zhang and Hartmann, 1992); Fair integration of routing and flow-control in communication networks (Chang, 1992); Design of lightwave networks (Labourdette and Acampora, 1991); Design of survivable telecommunication networks (Lisser et al., 1998; 2000); Lightpath assignment for multifibers WDM networks with wavelength translators (Coudert and Rivano, 2002); A family of algorithms for network reliability problems (Shaio, 2002); A procedure for resource allocation in switchlet networks (Fonseca et al., 2002); A simple polynomial time framework for reduced-path decomposition in multi-path routing (Mirrokni et al., 2004); Optimized routing adaptation in IP networks utilizing OSPF and MPLS (Riedl, 2003); Design of a meta-mesh of chain subnetworks: Enhancing the attractiveness of mesh-restorable WDM networking on low connectivity graphs (Grover, 2002); and Simultaneous routing and resource allocation via dual decomposition (Xiao et al., 2004).

In this chapter, we present two formulations of MCNF. These formulations are used for the routing of different traffic demands in telecommunications. Several techniques have been applied to solve multicommodity flow problems and their extensions. These techniques exploit the special structure of MCNF to reduce the problem to a sequence of solutions of smaller problems. We will describe here different decomposition schemes illustrated on minimax congestion problems, routing with problems with QoS criterion and survivability problems. The chapter is organized as follows. In Section 10.2, we present different formulations of MCNF. We use the non-oriented multicommodity flow model for the flow formulation. Congestion problems and their equivalent counterpart, the maximum concurrent flow problem are presented in Section 10.3. In Section 10.4, we discuss and formulate survivability problems in telecommunications networks. Two decomposition techniques applied to spare capacity problem are presented in Section 10.4.3.

10.2 MCNF FORMULATIONS

In formulating MCNF, we can adopt either of two equivalent modeling approaches: the flow formulation, where flows are defined on arcs, and the path formulation, where flows are defined on paths (Ahuja et al., 1993).

Let $G = (V, E)$ be the graph representing the underlying telecommunications network with $|V| = n$ nodes and $|E| = m$ edges. Denote by x_{ij}^k the decision flow variable corresponding to the flow fraction of the commodity k on the arc (i, j) and $c_{ij}^k > 0$ be its routing cost. Let u_{ij} be the capacity of the edge (i, j).

The flow formulation can be expressed as follows:

$$\min \sum_{k \in \mathcal{K}} \sum_{(i,j) \in E} c_{ij}^k |x_{ij}^k| \qquad (10.1a)$$

$$\text{s.t.} \quad \mathcal{N} x^k = r^k, \qquad \forall k \in \mathcal{K}, \qquad (10.1b)$$

$$\sum_{k \in \mathcal{K}} \left| x_{ij}^k \right| \leq u_{ij}, \quad \forall (i,j) \in E. \qquad (10.1c)$$

where:

- $\mathcal{N}$ is the node-arc incidence matrix corresponding to an arbitrary orientation of the network graph (i.e. $\mathcal{N}_{ij} = +1$ if arc j is directed away from node i, $\mathcal{N}_{ij} = -1$ if arc j is directed towards node i, $\mathcal{N}_{ij} = 0$ otherwise),

- $\mathcal{K}$ is a set of commodities. k is a commodity characterized by a *demand* d^k, in number of traffic units, to be routed through the network between a given pair of nodes, s^k and t^k,

- r^k is the requirement vector for commodity k (i.e. $r_i^k d^k$ if $k = s^k$, $r_i^k = -d^k$ if $i = t^k$ and $r_i^k = 0$ otherwise),

- x^k is the flow vector for a given commodity k (i.e. x_{ij}^k is the flow on edge (i, j) with $x_{ij}^k > 0$ if the flow goes from i to j and $x_{ij}^k < 0$ if the flow goes from j to i).

An equivalent linear programming formulation is obtained by introducing nonnegative variables, x_{ij}^{k+} and x_{ij}^{k-}, such that $x_{ij}^k = x_{ij}^{k+} - x_{ij}^{k-}$ and $\left|x_{ij}^k\right| = x_{ij}^{k+} + x_{ij}^{k-}$, see Minoux (1986),

$$\min \sum_{k \in \mathcal{K}} \sum_{(i,j) \in E} c_{ij}^k (x_{ij}^{k+} + x_{ij}^{k-}) \qquad (10.2a)$$

$$\text{s.t.} \quad A\left(x^{k+} - x^{k-}\right) = r^k, \quad \forall k \in \mathcal{K}, \qquad (10.2b)$$

$$\sum_{k \in \mathcal{K}} \left(x_{ij}^{k+} + x_{ij}^{k-}\right) \leq u_{ij}, \quad \forall (i,j) \in E, \qquad (10.2c)$$

$$x_{ij}^{k+}, x_{ij}^{k-} \geq 0, \quad \forall k \in \mathcal{K}, \quad \forall (i,j) \in E. \qquad (10.2d)$$

Constraints (10.2b) express the flow conservation for each node $i \in V$ and for each $k \in \mathcal{K}$, whereas constraints (10.2c) express that for each $(i, j) \in E$ the flow routed on (i, j) must not exceed the capacity of edge (i, j). The standard form of this model has $M = |\mathcal{K}||V| + |E|$ constraints and $N = 2|\mathcal{K}||E|$ variables. This problem is large even for networks with a small number of nodes as shown in Table 10.1. It can be solved directly by LP packages or by specialized interior point and simplex based algorithms (Chardaire and A.Lisser, 2002a;b). The size of (10.2) can also be reduced by aggregating the commodities by origins or by destinations (Chardaire and A.Lisser, 2002b).

10.2.1 Path formulation

The path formulation starts with an enumeration of all paths p between any commodity pair of nodes. Denote by $\mathcal{P}^k$, the set of routing paths associated with commodity k, i.e. the set of all simple paths in G that connect s_k and t_k. For each path, $p \in \mathcal{P}^k$, denote by X_p the fraction of the demand, d_k, routed on p. For each edge, $e \in E$, denote by S_e the set of paths that contain e. Let $C_p = \sum_{e \in p} c_e$ be the cost of the path p. The path formulation problem can then be written as

Table 10.1 Size of *MCNF* problems

Problem	m	n	$\mathcal{K}$	Constraints	Variables
JLL-split	26	42	257	6981	34994
A-500-split	119	302	500	59802	302302
A-900-split	119	302	900	108302	758102
A-1000-split	119	302	1000	119302	604302
A-1885-split	119	302	1885	224617	1138842
B-5800-split	119	302	5800	835801	4240986
B-7021-split	119	302	7021	842822	5911984

$$\min \sum_{k \in \mathcal{K}} \sum_{p \in \mathcal{P}^k} C_p X_p \qquad (10.3a)$$

$$\text{s.t.} \sum_{p \in \mathcal{P}^k} X_p = d_k, \quad \forall k \in \mathcal{K}, \qquad (10.3b)$$

$$\sum_{p \in S_e} X_p \leq u_e, \quad \forall e \in E, \qquad (10.3c)$$

$$X_p \geq 0, \quad \forall p \in \mathcal{P}^k, \forall k \in \mathcal{K}. \qquad (10.3d)$$

Constraints (10.3b) specify that for each $k \in \mathcal{K}$ demand d_k has to be routed, whereas constraints (10.3c) express that for each $e \in E$, the flow routed on arc e must not exceed its capacity. The number of constraints, i.e. $|\mathcal{K}| + |E|$, is much smaller than in the flow formulation, but the number of variables is exponential in the number of nodes of G. Problem (10.3) can be solved using decomposition methods. Amongst all, Dantzig-Wolfe decomposition (Dantzig and Wolfe, 1961) was originally created for solving such problems. In Chardaire and A.Lisser (2002a), specialized algorithms for problem (10.3) based both on the simplex method and interior point methods are proposed and numerical results provided on large size instances. The simplex based method corresponds to Dantzig-Wolfe decomposition while the interior point methods are used within the Analytic Center Cutting Plane Method also called ACCPM (Chardaire and A.Lisser, 2002a; Goffin et al., 1992). It is also shown in (Chardaire and A.Lisser, 2002a) that the path formulation outperforms the node arc formulation and Dantzig-Wolfe decomposition outperforms ACCPM for the set of tested instances.

Numerical results for the flow and path formulations are displayed in Table 10.2, see Chardaire and A.Lisser (2002b). Concerning the flow formulation, columns "iter"

Table 10.2 Performance of flow and path formulations using specialized simplex algorithms

Problem	Flow Formulation (10.2)		Path Formulation (10.3)			
	Itr	CPU (secs)	Cols	MP Itr	Subprob Itr	CPU (secs)
JLL-split	23661	217.7	633	5	550	5.10
A-5000-split	207	16.68	513	3	518	5.10
A-900-split	7110	805.0	1423	4	1338	5.10
A-1000-split	10138	1285.11	1615	5	1487	5.10
A-1885-split	77892	17261.88	4379	6	4145	19.97
B-5800-split	181754	118716.29	15152	8	14634	208.13
B-7021-split	†	†	16522	8	20823	377.30

† Memory limit exceeded

and "CPU secs" report the number of iterations of specialized simplex algorithm and the computational time, respectively. The first column under the path formulation gives the number of generated paths, whereas the other columns give the number of iterations of the master program, the number of iterations of the subproblems and the computational time, respectively. It is easy to see that the path formulation requires less computation time than the flow formulation.

We will consider now a specialized version of the multicommodity flow problem, the optimal routing problem with a congestion cost function.

10.3 MINIMIZING CONGESTION ON THE NETWORK

10.3.1 Minimal congestion and the maximum concurrent flow problem

The minmax congestion network design problem consists in determining an optimal topology supporting a given traffic (with many origin-destination pairs) which minimizes the maximal congestion on the arcs of the network.

We will consider a network routing problem where we want to minimize the maximal relative congestion on the arcs of the network. The problem of minimizing the flow on the most congested link when we not only have to route some traffic in a network, but also have to decide the topology of the network, is of valuable interest for the design of data communication networks (see Bertsekas and Gallager (1987)).

The basic problem of minimizing the more congested arc in the routing of a multicommodity flow will be described below as:

$$(MINCONG) \quad \min \quad z$$
$$s.t. \quad \begin{cases} x_e - u_e z \leq 0 & \forall e \in E \\ \mathcal{N}(x^k = r^k, & \forall k \in \mathcal{K} \\ x \in X \end{cases}$$

where $x_e = \sum_{k \in \mathcal{K}} |x_e^k|$ and X represents eventual topology constraints. For instance, some authors have considered constraints on the degrees of the node in the topology (see Labourdette and Acampora (1991)), and they have stressed that even very small instances on an 8-node mesh and with in- and out-degrees forced to 2 are very hard to solve with the best available Branch-and-cut codes (Bienstock and Günlük, 1995).

We will focus first on the simplest case of (MINCONG) without topology constraints. Even if it is a linear multicommodity flow problem, thus an LP which can be solved by standard decomposition techniques exploiting the underlying flow structure, it is generally considered to be a hard problem.

The first reason why this occurs is that the objective function (in the min-max formulation) is convex piecewise linear and not separable with respect to arcs. The second reason is that it produces optimal solutions with a huge number of active paths, being in that sense equivalent to the Maximum Concurrent Flow problem as shown below. Bienstock (2003) reported a set of numerical experiments on very large maximum concurrent flow problems (with up to 4.10^5 rows and 2.10^6 columns) exhibiting abnormal cubic growth of the CPU time to solve them with the CPLEX dual code.

Observe that the minmax objective function implies that we can solve (MINCONG) without capacity constraints and test feasibility afterwards by comparing the optimal value of z with the value 1. (MINCONG) itself is feasible if and only if there exists at least one path linking each origin to its destinations. Let z^* be an optimal value for (MINCONG); there exists a feasible multicommodity flow satisfying the capacity constraints if and only if $z^* \leq 1$.

The path structure of an optimal solution of (MINCONG) is a very important issue for the traffic manager who needs to implement a given routing table in a given period. Too many paths will penalize the protocols which take charge of the packets at each switching equipment, and too few paths can make the solution too sensitive to failures.

We will call *active* a path that carries some positive flow and *critical* an arc e such that $x_e = u_e z$. An active path containing critical edges will be called a critical path. Suppose now that some commodity is routed on a critical path at the optimal solution and that there exists a second path supporting that commodity which is not critical. Both paths form a cycle, so that one can modify the solution deviating a small quantity from the critical path p to the non-critical one p'. A new basic optimal solution should be obtained when p' turns out to be critical. Critical edges appear thus as bottlenecks for some "critical" commodities:

Proposition 10.1 *There always exists an optimal solution of (MINCONG) such that, if any commodity is routed on a critical path, all paths used to route that commodity are critical.*

Analyzing the basic structure of the arc-path formulation of (MINCONG), an optimal solution will contain at most $|\mathcal{K}| + \sigma$ active paths (and at least $|\mathcal{K}|$), where σ is the number of critical edges.

As observed in Shahrokhi and Matula (1990), (MINCONG) is also strongly related to the Maximum Concurrent Flow problem (MAXCONCUR), where one wants to maximize the throughput of the network for a given set of capacities. The throughput of the network is a load factor that multiplies the traffic matrix.

(MAXCONCUR) $\quad\quad\quad$ max Z

$$s.t. \begin{cases} x_e & \leq u_e \quad \forall e \in E \\ \mathcal{N}x^k - Zr^k & = 0 \quad \forall k \in \mathcal{K} \end{cases}$$

One can easily verify that the optimal solutions x^* and $\tilde{x}^*$ of (MINCONG) and (MAXCONCUR), respectively, satisfy $x^* = z^* \tilde{x}^*$ and that $z^* = 1/Z^*$.

Both problems (MINCONG) and (MAXCONCUR) have received a lot of attention in the past decade, especially since the seminal paper by Shahrokhi and Matula (1990) who first proposed a fully polynomial approximation scheme to solve (MAXCONCUR) with uniform capacities. They showed that the minimization of a separable exponential penalty function on the arcs yields a flow with a nearly maximal throughput. They chose the following penalty function:

$$\phi_e(x_e) = \exp(\frac{2|E|^2}{\varepsilon} x_e)$$

where ε is a positive parameter which defines the approximation. They found a complexity bound of $O(|V||E|^7/\varepsilon^5)$ and showed that the number of active paths is $O(|E|^3/\varepsilon^2)$. Faster algorithms based on refinements of that exponential penalty function have been since proposed and extended to nonuniform capacities and to other packing and covering problems (see Garg and Könemann (1998); Grigoriadis and Khachiyan (1994); Leighton et al. (1995)).

An interesting link between the Flow Deviation algorithm of Section 10.3.2 and the resolution of (MINCONG) has been analyzed recently by Bienstock and Raskina (2002). They proposed an algorithm to solve (MINCONG) which alternates between a magnification step where the throughput is increased for a fixed congestion and a potential reduction step where congestion is minimized for a fixed throughput. That latter step is realized by performing flow deviation inner steps with the Kleinrock's congestion function $\phi_e(x_e) = x_e/(u_e - x_e)$ (proportional to the queueing delay under Poissonian hypotheses for the entering traffic routed on arc e). These are the crucial calculations as they will introduce new active paths in the solution. Bienstock and Raskina suggested to substitute the shortest-path calculations by min-cost flow subproblems. This comes naturally in the linearization procedure of Frank-Wolfe's algorithm if we add redundant capacity constraints for each commodity, i.e. $x_e^k \leq u_e, \forall e$. The higher cost of the subproblems is compensated by better bounds in the complexity estimates and that trick is also used in other papers (see Leighton et al. (1995) or Grigoriadis and Khachiyan (1994)). Recently, some authors modified the main step of the general algorithm to get back to shortest-path calculations without affecting the complexity bound (see Garg and Könemann (1998)).

Thus, the interesting idea behind these contributions brought by the "Approximation Algorithms" community is the possibility to approximate the non-separable congestion problem (MINCONG) by a separable multicommodity flow model with convex increasing costs on the arcs which can be resumed in the following problem (without topology constraints):

$$(MINDEL) \quad \min \sum_{e \in E} \phi_e(x_e)$$

$$\text{s.t.} \quad \begin{cases} x_e \leq u_e & \forall e \in E \\ \mathcal{N} x^k = r^k, & \forall k \in \mathcal{K} \end{cases}$$

We will analyze below two algorithmic approaches to solve (MINDEL) when the arc cost functions are given by the average delay suffered by a data packet on arc e. Indeed, independence assumptions with Poissonian arrival on the queueing network lead to Kleinrock's delay function which is proportional to $\Phi_e(x_e) = x_e/(u_e - x_e)$. The total cost is simply $\Phi(x) = \sum_e \Phi_e(x_e)$.

Observe that the objective function acts as a barrier function on the capacity constraints which can thus be ignored in the model.

We will begin with the Flow Deviation method which was used very early to solve capacity and flow assignment problems on packet-switched networks (see Fratta et al. (1973)). The second approach is the Proximal Decomposition method which has been tested on large routing problems for broadband data networks (see Ouorou et al. (2000)).

10.3.2 The flow deviation method

The Flow Deviation method (FD) is an adaptation of the classical linearization algorithm of Frank and Wolfe (1956) to solve multicommodity flow problems with convex costs. It was first proposed by Fratta et al. (1973) in the context of design of packet-switched networks and has been widely used by the community of transportation (see LeBlanc et al. (1985)) and in telecommunications networks (see Bertsekas and Gallager (1987)). We recall below the formal ideas behind Frank-Wolfe's algorithm for general nonlinear programs with linear constraints:

$$\text{Minimize} \quad \Phi(x)$$
$$\text{s.t.} \quad Ax = b$$
$$x \geq 0$$

The method proceeds by successive linearization for solving LP subproblems at each iteration t where the gradient $\nabla \Phi(x^t)$ has been computed. Let $\tilde{x}^t$ be the optimal solution of the subproblem

$$\tilde{x}^t = \text{Argmin}_{x \in P} \nabla \Phi(x^t).x$$

where $P = \{x \in \mathbb{R}^n \mid Ax = b, x \geq 0\}$ is the polyhedron of feasible solutions supposed bounded. Thus, we can suppose that $\tilde{x}^t$ is an extreme point of P. Then the new iterate is obtained by carrying out a line search on the segment $[x^t, \tilde{x}^t]$ with the nonlinear function Φ. Convergence results have been obtained in the strictly convex case (see Bertsekas (1995) for example) but the method suffers from very slow convergence tail which is evidenced by the following fact. The solution of the nonlinear program is in general not a vertex of the feasible polyhedron so that, as the sequence of feasible solutions tends to a point on a face of the polyhedron, the angle between successive descent directions tends to 180° (see Figure 10.1). It is shown in Bertsekas (1995) how typical sublinear convergence rate can be exhibited in very simple and low-dimensional situations.

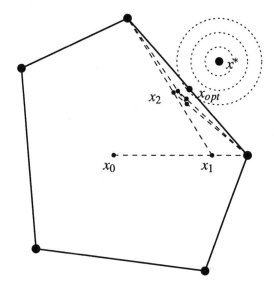

Figure 10.1 Zigzagging convergence of Frank and Wolfe's method

On the other hand, the positive aspect of the implementation of Frank-Wolfe's method to multicommodity flow problems is the simplicity of the subproblem resolution which reduces to shortest-path computations and the solution update which consists in deviating flows fairly from the active paths towards the new ones without explicitly computing the individual commodity flows as we can see below.

Now, we will apply the Flow Deviation method to (MINDEL) with the Kleinrock's delay function, a smooth convex function, separable with respect to arcs.

Assume that we get a strictly feasible multicommodity flow at iteration k, i.e. such that $x_e < u_e, \forall e$. Then, the linearization step at iteration t of (FD) reduces to separate shortest-path computations for each commodity k with arc lengths $\nabla \Phi_e(x_e^t)$. The solution $\tilde{x}^t$ corresponds to the situation where all demands d_k are routed separately on these new paths. Observe that $\tilde{x}^t$ may not satisfy the capacity constraints. The new solution x^{t+1} is obtained by carrying out a line search on the segment $[x^t, \tilde{x}^t]$. Observe again that the new solution will be forced strictly feasible by the barrier function.

The main point in the procedure above is the fact that all computations are performed on the total flow variables. However, in some situations, one may be interested in computing the optimal path flows. An arc-path formulation is then necessary and the (FD) iterations can be carried out on the path flow variables as explained in Bertsekas and Gallager (1987). The first step of the procedure is unchanged so that, if $\tilde{p}$ is the shortest path with the first-derivative arc lengths for a given commodity k, then we set $\tilde{X}_p = d_k$ and, denoting by P_t^k the subset of paths carrying the current solution, the flow deviation step can be computed in the following way:

$$X_p^{t+1} = X_p^t + \theta_t(\tilde{X}_p^t - X_p^t), \forall p \in P_{t+1}^k = P_t^k \cup \{\tilde{p}\}$$

where θ_t is the optimal step size which minimizes the objective function over all $\theta \in [0,1]$ (observe that the feasible direction whose components are the $\tilde{X}_p^t - X_p^t$ is a descent direction for Φ).

Thus an *equal* fraction of the flow on the non-shortest paths is shifted to the shortest path. One may observe that this update of path flows will tend to increase monotonically the number of active paths. Variants of the basic Flow Deviation method can be built by modifying one of the inner steps of the algorithm, i.e.

- either by modifying the search direction, for instance by substituting the shortest-path calculations by min-cost flow subproblems for each commodity,

- or by deviating *nonuniform* flow proportions on the newly generated paths.

To understand the first strategy, one can add the redundant constraints ($x_e^k \leq u_e, \forall e, k$) to the model (MINDEL). After the linearization step, the direction-finding subproblem of (FD) decomposes now in K minimum-cost flow problems. The computational overhead of these subproblems compared to the original shortest-path calculations is compensated by the choice of feasible paths (considering one commodity at a time). But the drawback is that more paths are likely to be generated which is exactly what we do not wish, and the path flow updates are not so straightforward as in the original method. Observe nevertheless that some authors have chosen to implement min-cost flow computations instead of shortest-path calculations to improve worst-case behavior of some approximation algorithms (see Plotkin et al. (1995)).

There are many different strategies to modify the search direction which result in nonuniform deviations from the current active paths to the shortest one. Second-order information can be used to yield Newton or Quasi-Newton directions as suggested in Bertsekas and Gallager (1987). Conjugate gradient strategies have also been tested in earlier works (see LeBlanc et al. (1985)). Again, the expected gain in convergence rate is overtaken by the extra work of performing each iteration and reconstituting the path support.

10.3.3 The Proximal Decomposition method for convex cost multicommodity flow problems

The Proximal decomposition method (PDM) is an adaptation of the Alternating Direction Method of Multipliers originally proposed by Glowinski and Marocco (1975) and further developed by Eckstein and Fukushima (1993) and Mahey et al. (1995). It can be viewed as a separable Augmented Lagrangian method and it is best understood when applied to the following generic separable convex program with coupling constraints:

$$\text{Minimize} \quad \Phi(x) = \sum_{i=1}^{p} \phi_i(x_i)$$
$$\text{s.t.} \quad \sum_{i=1}^{p} A_i x_i = b$$
$$x_i \in S_i, i = 1, \ldots, p$$

where the variables are partitioned into p blocks, all functions being convex on the compact convex sets S_i; suppose also that there exists $x_i \in S_i, i = 1, \ldots, p$ such that

$$\sum_{i=1}^{p} A_i x_i = b$$

so that the problem has an optimal solution with value v^*.

The key idea is to add primal allocation variables $y_i, i = 1, \ldots, p$ to get the equivalent formulation (where $\sum_i b_i = b$):

$$v^* = \min_{x,y} \{\Phi(x) | A_i x_i + y_i = b_i, x_i \in S_i, i = 1, \ldots, p, \sum_i y_i = 0\}$$

which is itself equivalent to:

$$v^* = \min_{x,y} \{\Phi(x) + \frac{\lambda}{2} \sum_i \|A_i x_i + y_i - b_i\|^2 | A_i x_i + y_i = b_i, x_i \in S_i, i = 1, \ldots, p, \sum_i y_i = 0\}$$

The Lagrangian dual with respect to the local constraints in (x_i, y_i) will induce the following subproblem:

$$v(u_1, \ldots, u_p) = \min_{x \in S, y \in Y} \{\Phi(x) + \frac{\lambda}{2} \sum_i \|A_i x_i + y_i - b_i\|^2 + \sum_i \langle u_i, A_i x_i + y_i - b_i \rangle\}$$

where $Y = \{(y_1, \ldots, y_p) | \sum_i y_i = 0\}$ is the primal subspace. To decompose that subproblem, the trick is to apply some kind of Gauss-Seidel scheme to alternate minimizations with respect to x and y. Indeed, the minimization w.r.t. y can be solved explicitly.

To sum up, we first solve the inner subproblem with respect to x for a fixed u^t and fixed allocations $y^{t-1} \in Y$; observe that the subproblem decomposes in p subproblems, where the i-th subproblem is:

$$\text{Minimize}_{x_i \in S_i} \ [\phi_i(x_i) + \frac{\lambda}{2} \|A_i x_i + y_i^{t-1} - b_i\|^2 + \langle u_i^t, A_i x_i + y_i^{t-1} - b_i \rangle]$$

Let x^t be the optimal solution. We must then solve the primal allocation subproblem, i.e. the following quadratic program with linear equality constraints:

$$\inf_{y \in Y} \sum_i [\frac{1}{2} \|A_i x_i^t + y_i - b_i\|^2 + \langle u_i^t, A_i x_i^t + y_i - b_i \rangle]$$

y^t is an optimal solution of that quadratic program if and only if there exists $v^t \in \mathbb{R}^m$ such that:

$$u_i^t + A_i x_i^t + y_i^t - b_i = v^t, \sum_i y_i^t = 0$$

We can easily solve the linear system in $\mathbb{R}^m$ to get

$$v^t = \frac{1}{p}\sum_i u_i^t + \frac{1}{p}r^t,$$

where $r^t = \sum_i(A_i x_i^t - b_i)$ is the residual of the coupling constraint, and substitute above to get y^t. Finally, the update of u^t will be simply $u_1^{t+1} = \cdots = u_p^{t+1} = v^t$ (i.e. $u^t \in Y^\perp$).

Observe that, even if the method may be interpreted as a separable Augmented Lagrangian technique, the parameter λ is a scaling parameter and not a penalty one, which must be estimated to drive the two primal and dual sequences at the same pace towards the optimal fixed point. Eckstein (1994) has shown the potential of that algorithm for massively distributed computing and different versions have been shown to be very efficient to solve multicommodity flow problems with linear or convex costs (Eckstein and Fukushima, 1993; Ouorou et al., 2000).

When applied to the convex multicommodity flow problem (MINDEL), the coupling constraints are not the capacity constraints as in the Dantzig-Wolfe's method (see Section 10.2 and 10.4.3.1) as they are already included in the delay function. Indeed, $\Phi(x)$ is finite whenever $x_e < u_e, \forall e$. We can apply the decomposition scheme to the multicommodity constraints

$$x_e - \sum_k x_e^k = 0, e \in E \qquad (10.4)$$

to split the problem into $\mathcal{K} + |E|$ subproblems, i.e. one quadratic subproblem for each commodity and one convex one-dimensional subproblem for each arc. It is shown in Ouorou et al. (2000) how quadratic flow problems can be avoided by including the $\mathcal{K}$ individual demand satisfaction constraints

$$\sum_p X_p = d_k, k \in \mathcal{K} \qquad (10.5)$$

in the coupling constraints to yield a completely distributed decomposition algorithm. Moreover, the arc-path formulation is used to induce the generation of supporting paths by successive shortest-paths calculations. These calculations may be performed by sweeping all possible destinations for a given origin, avoiding the explosion of the computational time when the requirement matrix is dense.

Numerical experiments with algorithm (PDM) have been performed by Ouorou et al. (2000) on medium-size networks but with dense requirement matrices (100 nodes, 900 arcs and 10,000 commodities). We limit ourselves, in Table 10.3, to only the comparisons made by the authors, who also illustrated the performance of ACCPM and a Projected Newton algorithm) on the Flow Deviation method versus the Proximal Decomposition algorithm. For a given tolerance on the final residuals of 10^{-4}, (PDM) performs better than (FD) on standard traffic demand matrices (where "scale" represents a load factor Z – the one used in model (MAXCONCUR), which varies from 1 to 2.5) and the average path dispersion is lower which makes its implementation on real routers much easier.

Table 10.3 Comparison of Proximal Decomposition with Flow Deviation

	Scale	1.0	1.5	2.0	2.5	3.0
	Delay	33.502997	52.292984	72.696367	94.984274	119.474206
CPU time (sec) PDM		76.93	83.88	156.49	161.54	712.04
Numb. of iter. PDM		24	36	84	192	494
Path dispersion for PDM		3	3	3	3	4
CPU time (sec) FD		355.8	362.7	359.4	347.6	361.3
Numb. of iter. FD		219	223	221	213	222

10.4 SURVIVABILITY ISSUES

Network survivability has always been a major preoccupation of telecommunication operating companies. The explosion of new services is attracting a new, more demanding customer. The intense competition between operating companies, due to market deregulation, demands that an ever increasing attention be given to the quality and cost of services.

The main concern of customers include the availability of network connections, which is the key component of quality based network services. Telecommunication companies are faced with the challenge of reducing the risk of an unpredictable failure while at the same time maximizing network efficiency.

Failures however cannot be entirely eliminated, whether it be a question of equipment (switching, transmission, ...) at a network node, or a physical break in a connecting link. In the event of such failures, the network operator should have planned an excess capacity in his network in order to reroute all demands which were being served by the failed network components. To ensure that sufficient capacity exists for any possible degree of failure, the current practice is to install an optimized spare capacity.

Basic network failures are of two types: arc and node failures. The former affects the traffic transiting by the failed arc while the latter affects all adjacent arcs traffic. The fundamental hypothesis in network survivability problems is that only one basic failure can be handled at once.

There exists two approaches to modeling and rerouting interrupted traffic demands: link restoration and path restoration. Link restoration considers that the rupture of a link creates, at its endpoints, a demand equal to the total flow which transited through the link. This demand must be rerouted by using the spare capacity. It can be described as a single commodity requirement. In the path restoration, the interrupted flow is analyzed and the fraction of demands in the nominal network affected by the failure is found. In this way, a set of demands is generated to be routed using the spare capacity. The demand requirement may now be described as one of multiple commodities. The

MULTICOMMODITY FLOW PROBLEMS   255

path restoration, although more difficult to put into effect, is economically preferable. It is the approach used in practice and the one we will study here.

The survivability problem consists in determining the excess capacities to be installed in the network and the required rerouting of demands in order to satisfy all demands and to minimize the total investment cost.

Under the hypotheses of linear capacity installation costs and divisible flows, the survivability problem can be formulated as a linear special case of MCNF problems using either flow formulation or path formulation.

10.4.1 Survivability model using flow formulation

In this section, we consider the global survivability model as we optimize network capacity for both routing and rerouting requirements. We do not consider any reuse of capacity released by any failure. Denote by $E_f \subseteq E$ the subset of edges with positive flows that can fail.

Suppose edge $e \in E_f$ fails. Let $(x_e^{k+})_{ij}$ and $(x_e^{k-})_{ij}$ denote the flows on edge (i, j) used to restore the demands affected by this failure, with corresponding vectors x_e^{k+} and x_e^{k+}. By definition, $(x_e^{k+})_e = (x_e^{k-})_e = 0$. i.e. when an edge fails both its working and spare capacities are unusable.

Therefore, at most $\mathcal{K}$ flows routed through edge (i, j) will be interrupted. Let $\mathcal{K}_e$ be the subset of such commodities. Let $y \in \mathbb{R}^{|E|}$ be the capacity to be installed and $c_{ij} > 0, \forall (i,j) \in E$ the installation unit cost. We can formulate the survivability model as follows:

$$\min \sum_{(i,j) \in E} c_{ij} \left(\sum_{k=1}^{\mathcal{K}} (x_{ij}^{k+} + x_{ij}^{k-}) + y_{ij} \right) \tag{10.6a}$$

$$\text{s.t.} \quad \mathcal{N}x^{k+} - \mathcal{N}x^{k-} = d^k, \quad \forall k \in \mathcal{K} \tag{10.6b}$$

$$\sum_{e \in E_f} \sum_{k \in \mathcal{K}_e} (x_e^{k+} + x_e^{k-}) \leq y, \tag{10.6c}$$

$$\mathcal{N}x_e^{k+} - \mathcal{N}x_e^{k-} = r^{ek}, \quad \forall k \in \mathcal{K}_e, \forall e \in E_f \tag{10.6d}$$

$$(x_e^{k+})_e = (x_e^{k-})_e = 0, \quad \forall k \in \mathcal{K}_e, \forall e \in E_f, \tag{10.6e}$$

$$x^{k+}, x^{k-} \geq 0, \quad \forall k \in \mathcal{K}, \tag{10.6f}$$

$$x_e^{k+}, x_e^{k-} \geq 0, \quad \forall k \in \mathcal{K}, \quad \forall e \in E_f, \tag{10.6g}$$

$$y_{ij} \geq 0, \quad \forall (i,j) \in E. \tag{10.6h}$$

where

- $\mathcal{N}$ is the $n \times m$-node-arc incidence matrix of the network as in Section 10.2.

- $r^{ek} = \sum_{k \in \mathcal{K}} x_{ij}^k, \forall (i, j) = e \in E_f$ is the flow interrupted by failure e to be rerouted.

The objective function (10.6a) concerns both working and spare capacities which have in this case the same cost. This formulation allows us also to assign different costs for the working and spare capacities (Lisser et al., 1998). Constraints (10.6c) ensure that sufficient capacity is installed on each edge to meet rerouting requirements over all edge failures. Constraints (10.6b) and (10.6d) are flow conservation constraints for the working and restoration flows respectively. Constraints (10.6e) prevent using the capacity of a failed edge for restoration needs. Some transmission and/or switching technologies require distinct working and spare capacities and thus consider two distinct networks, i.e. working network and reserve network (Lisser et al., 2000). In this case, problem (10.6) can be adapted easily by setting different capacity constraints for the working and protection networks and updating the objective function accordingly.

The main problem studied in the literature concerns only the spare capacity planning problem. In this case, the working flow and its routing are given. The problem consists in finding the minimum amount of spare capacity so that the network can survive the failure of an edge. This model can be deduced from (10.6) as follows :

$$\min \sum_{(i,j) \in E} c_{ij} y_{ij} \tag{10.7a}$$

$$\text{s.t.} \sum_{e \in E_f} \sum_{k \in \mathcal{K}_e} (x_e^{k+} + x_e^{k-}) \leq y, \tag{10.7b}$$

$$\mathcal{N} x_e^{k+} - \mathcal{N} x_e^{k-} = r^{ek}, \quad \forall k \in \mathcal{K}_e, \forall e \in E_f \tag{10.7c}$$

$$(x_e^{k+})_e = (x_e^{k-})_e = 0, \quad \forall k \in \mathcal{K}_e, \forall e \in E_f, \tag{10.7d}$$

$$x_e^{k+}, x_e^{k-} \geq 0, \quad \forall k \in \mathcal{K}, \quad \forall e \in E_f, \tag{10.7e}$$

$$y \geq 0. \tag{10.7f}$$

Models (10.6) and (10.7) consider the path restoration strategy. Link restoration models can be obtained by considering the amount of flows affected by an edge failure as a single commodity to be rerouted between the endpoints of the failed edge.

The integer version of (10.7) based on link restoration has been studied by Balakrishnan et al. (2001) while the integer path restoration variant of (10.7) was studied by Kennington and Lewis (2001).

10.4.2 Survivability model using path formulation

We consider the path formulation of problem (10.6). Each commodity $k \in \mathcal{K}$ is routed on a set of working paths $\mathcal{P}^{0k}$ between the source and the destination of the commodity. Each path p is characterized by a Boolean vector π_p on the set of edges with $(\pi_p)_{ij} = 1$ if path p intersects edge $(i, j) \in E$, and 0 otherwise. To each path p in the set $\mathcal{P}^{0k}$, we shall assign a variable X_p to represent the flow component of commodity k routed on p. The flow of demand k is split on different routes: its total value, $\sum_{p \in \mathcal{P}^{0k}} X_p$, must equal the demand d_k.

We shall additionally consider the set of restoration paths $\mathcal{P}^{ek}$ for the demand k when $e \in E_f$ fails. Each restoration path p is characterized by a Boolean vector π_p^e on the set of edges with $(\pi_p^e)_{ij} = 1$ if path p intersects arc $(i, j) \in E$, and 0 otherwise. If edge e fails, then at most $\mathcal{K}$ commodities will be interrupted on all paths that contain e. A variable X_p^e is introduced for each path for the flow component.

The problem can be written as

$$\min \sum_{(i,j)\in E} c_{ij}\left(\sum_{k\in \mathcal{K}}\sum_{p\in \mathcal{P}^{0k}} X_p(\pi_p)_{ij} + y_{ij}\right)$$

s.t.
$$\sum_{p\in \mathcal{P}^{0k}} X_p = r^k, \quad \forall k \in \mathcal{K}, \tag{10.8a}$$

$$\sum_{k\in \mathcal{K}^e}\sum_{p\in \mathcal{P}^{ek}} X_p^e(\pi_p^e)_{ij} \leq y_{ij}, \quad \forall (i,j) \in E, \forall e \in E_f, \tag{10.8b}$$

$$\sum_{p\in \mathcal{P}^{ek}} X_p^e = r^{ek}, \quad \forall k \in \mathcal{K}_e, \forall e \in E_f, \tag{10.8c}$$

$$X_p \geq 0, \quad \forall p \in \mathcal{P}^{0k}, \forall k \in \mathcal{K}, \tag{10.8d}$$

$$X_p^e \geq 0, \quad \forall p \in \mathcal{P}^{ek}, \forall k \in \mathcal{K}_e, \forall e \in E_f, \tag{10.8e}$$

$$y_{ij} \geq 0, \quad \forall (i,j) \in E. \tag{10.8f}$$

where

$$r^{ek} = \sum_{p\in \mathcal{P}^{0k}} X_p(\pi_p)_e, \forall k \in \mathcal{K}_e, \forall e \in E_f.$$

Problem (10.8) models the global survivability problem by using the path formulation. It is based on the path restoration. Constraints (10.8b) ensure that sufficient capacity is installed on each edge to meet rerouting requirements over all edge failures. Constraints (10.8a) and (10.8c) ensure that total flow routed or rerouted on several paths correspond to total flow of given commodity. As for the flow formulation, we can deduce the spare capacity planning from (10.8) as follows:

$$\min \sum_{(i,j)\in E} c_{ij} y_{ij}$$

$$\sum_{k\in \mathcal{K}_e}\sum_{p\in \mathcal{P}^{ek}} X_p^e(\pi_p^e)_{ij} \leq y_{ij}, \quad \forall (i,j) \in E, \forall e \in E_f, \tag{10.9a}$$

$$\sum_{p\in \mathcal{P}^{ek}} X_p^e = r^{ek}, \quad \forall k \in \mathcal{K}_e, \forall e \in E_f, \tag{10.9b}$$

$$X_p^e \geq 0, \quad \forall p \in \mathcal{P}^{ek}, \forall k \in \mathcal{K}_e, \forall e \in E_f, \tag{10.9c}$$

$$y_{ij} \geq 0, \quad \forall (i,j) \in E. \tag{10.9d}$$

This spare capacity planning problem is based on the path formulation. As for (10.7), we can consider link restoration by considering r^{ek} as a single commodity and reroute it as a whole, see Lisser et al. (1998) for more details.

Whatever the problem considered, routing or rerouting, *MCNF* and its extensions are so large scale linear programs, even for networks of moderate dimensions, that they challenge the capabilities of the most advanced commercial LP codes. The alternative is to turn to the principle of decomposition whose effect is to break down the huge initial problem into interconnected problems of much smaller dimensions.

10.4.3 Decomposition approaches for treating survivability models

There are at least two ways of implementing decomposition. The first one consists in separating the capacity design issue (the master program) and the non-simultaneous multicommodity flow requirements (the subproblems). The master program selects a tentative capacity design. The subproblems test whether this proposal meets the multicommodity flow requirements: If the proposal is not feasible, the subproblems return a cut, or constraint, to the master program. The merit of this first approach is that the master program is of moderate size. The difficulty lies in the subproblems, i.e., the constraint generation scheme, that requires the solution of a non-smooth optimization problem, see Minoux (1986). This approach is usually named Benders decomposition (Benders, 1962).

The other approach uses Lagrangian relaxation. An extensive formulation of the problem includes capacity constraints on each arc flow (one per arc and per failure configuration) and flow constraints (one set per commodity and per failure configuration). The idea is to dualize the capacity constraints and construct a Lagrangian in the space of the corresponding dual variables. The master program consists in maximizing the Lagrangian in the dual variables. The subproblems tests whether for a given set of dual variables there exists more profitable reroutings of the commodities. The information that is sent back to the master takes the form of a column generation scheme. In this decomposition mode, the subproblems are very simple: they are just shortest path problems. In contrast with Benders decomposition the master program can be very large. However this program is sparse and structured. It can be solved using appropriate techniques for exploiting sparsity.

As an illustration of those two decomposition principles, we apply Lagrangian relaxation and Benders decomposition techniques to the spare capacity planning problem (10.9). A positive upper bound $\bar{y}_{ij}$ is set on the edge capacity variable $y_{ij}, \forall (i,j) \in E$.

In Lisser et al. (2000; 1998), both Lagrangian relaxation and Benders decomposition are applied to problem (10.7) and also to problem (10.6) based on link restoration. The path restoration global survivability problem (10.8) is studied in Bonnans et al. (2000) using the proximal decomposition method described in Section 10.3.3.

10.4.3.1 Lagrangian Relaxation. Decomposition methods consist in converting the problem (10.9) into a smaller non-differentiable problem in convex optimization. This is achieved by partial dualization. In this section, we apply Lagrangian relaxation to problem (10.9) and show how to construct the elements of the sub-differential of the function thus obtained.

Consider the Lagrangian obtained by dualizing the coupling constraints (10.9a). The dual vector associated with (10.9a) is $v^e_{ij}, \forall (i,j) \in E, \forall e \in E_f$. The partial Lagrangian function is

$$L(X,y;v) = \sum_{(i,j) \in E} c_{ij} y_{ij} + \sum_{e \in E_f} \sum_{(i,j) \in E} v^e_{ij} \left(\sum_{k \in \mathcal{K}_e} \sum_{p \in \mathcal{P}^{ek}} X^e_p (\pi^e_p)_{ij} - y_{ij} \right). \quad (10.10)$$

Problem (10.9) can be formulated as the minmax problem

$$\min_{X \geq 0, y \leq \bar{y}} \{ \max_{v \geq 0} L(X,y;v) : \sum_{p \in \mathcal{P}^{ek}} X^e_p = r^{ek}, \quad \forall k \in \mathcal{K}_e, \forall e \in E_f \}. \quad (10.11)$$

By the minmax theorem, this problem has the same optimal value as

$$\max_{v \geq 0} L(v). \tag{10.12}$$

where

$$L(v) = \min_{X \geq 0, y \leq \bar{y}} \{L(X, y; v) : \sum_{p \in \mathcal{P}^{ek}} X_p^e = r^{ek}, \quad \forall k \in \mathcal{K}, \forall e \in E_f\}. \tag{10.13}$$

This problem is convex, non-differentiable, but of small size. To compute $L(v)$, we disaggregate it into a sum of elementary functions

$$L(v) = \gamma(v) + \varphi(v). \tag{10.14}$$

where

$$\gamma(v) = \min_X \sum_{e \in E_f} \sum_{k \in \mathcal{K}_e} \sum_{p \in \mathcal{P}^{ek}} \sum_{(i,j) \in E} v_{ij}^e X_p^e (\pi_p^e)_{ij}, \tag{10.15}$$

and

$$\varphi(v) = \min_{0 \leq y \leq \bar{y}} \{ \sum_{(i,j) \in E} c_{ij} y_{ij} - \sum_{e \in E_f} \sum_{(i,j) \in E} v_{ij}^e y_{ij} \}. \tag{10.16}$$

To decompose the function $\varphi(v)$, we extend the vector v with arbitrary components v_{ij}^e for each $(i, j) = e \notin E_f$ and introduce some new notations: $\delta_{ij}^e = 1$ if $(i, j) = e \in E_f$ and 0 otherwise.

Functions γ and φ can be decomposed into simple functions

$$\gamma_k(v) = \min_{X_p^e} \sum_{(i,j) \in E} v_{ij}^e X_p^e (\pi_p^e)_{ij}, \forall k \in \mathcal{K}_e, \forall e \in E_f. \tag{10.17}$$

and

$$\varphi_{ij}(v) = \min_{0 \leq y \leq \bar{y}_{ij}} y_{ij} (c_{ij} - \sum_{e \in E_f} \delta_{ij}^e v_{ij}^e). \tag{10.18}$$

Thus we have

$$\gamma(v) = \sum_{e \in E_f} \sum_{k \in \mathcal{K}_e} \gamma_k(v), \tag{10.19}$$

and

$$\varphi(v) = \sum_{(i,j) \in E} \varphi_{ij}(v). \tag{10.20}$$

The elementary functions $\gamma_k(v)$ and $\varphi_{ij}(v)$ each have the form of a minimum of functions linear in v. Hence, they are concave. In addition, their values at a given point v are easily calculated. Functions γ_k are the optimal values of simple flow problems corresponding to the shortest path problem over a graph with nonnegative costs. Functions φ_{ij} take either the value zero or $(c_{ij} - \sum_{e \in E_f} \delta_{ij}^e v_{ij}^e) \bar{y}_{ij}$ according to the sign of $(c_{ij} - \sum_{e \in E_f} \delta_{ij}^e v_{ij}^e)$. Thus $\bar{y}_{ij}$ essentially takes only two values, 0 and $\bar{y}_{ij}$.

To determine an element of the sub-differential problem, it suffices to express the dependence of the optimal values of γ and φ on v. Since this involves only simple linear expressions, the components of the sub-gradients are just the coefficients of v.

To be more precise, let $\hat{v}$ be a point at which we calculate $\gamma_k(\hat{v})$ and $\varphi_{ij}(\hat{v})$, and let $\hat{X}_p^e$ and $\hat{y}_{ij}$ be the values respectively where these functions attain their minima. Let v be an arbitrary point. Using the fact that, for a given v, γ_k is the minimum of $\sum_{(i,j)\in E} v_{ij}^e X_p^e(\pi_p^e)_{ij}$ for $X_p^e \in \mathcal{P}^{ek}$, we obtain for $k \in \mathcal{P}^{ek}$,

$$\begin{aligned} \gamma_k(v) &\leq \sum_{(i,j)\in E} v_{ij}^e \hat{X}_p^e(\pi_p^e)_{ij} \\ &= \gamma_k(\hat{v}) + \sum_{(i,j)\in E} \hat{X}_p^e(\pi_p^e)_{ij}(v_{ij}^e - \hat{v}_{ij}^e). \end{aligned} \quad (10.21)$$

Inequality (10.21) defines a support of the concave function γ_k, $k \in \mathcal{K}_e$ at the point $\hat{v}$. The coefficients $(v_{ij}^e - \hat{v}_{ij}^e)$ of $\hat{X}_p^e$ are the components of the sub-differential of the γ_k.

In the same way, we construct the sub-differential of φ_{ij} by the inequality

$$\begin{aligned} \varphi_{ij}(v) &\leq \hat{y}_{ij}\left(c_{ij} - \sum_{e\in E_f} \delta v_{ij}^{(e)}\right) \\ &= \varphi_{ij}(\hat{v}) - \sum_{e\in E_f} \delta \hat{y}_{ij}(v_{ij}^e - \hat{v}_{ij}^e). \end{aligned} \quad (10.22)$$

Inequality (10.22) defines a support of the concave function φ_{ij} at the point $\hat{v}$. The sub-gradient is therefore a null vector, excepting those components corresponding to edges in the failure network, which have value $\hat{y}_{ij}$.

10.4.3.2 Benders Decomposition.
The formulation of the survivability problem (10.9) suggests the following remark: an instantiation, y, of the design variables gives rise to elementary non-simultaneous multicommodity flow problems. We apply Benders decomposition on the variables y. Since there is no cost associated with the routings, the optimal value of the routing problems is either 0 or $+\infty$ depending on whether the point y belongs to the domain D defined by the set of capacities $y \in \mathbb{R}^{|E|}$. It is described by linear inequalities associated with the following degenerate linear program

$$\min 0$$

$$\sum_{k\in \mathcal{K}_e} \sum_{p\in \mathcal{P}^{ek}} X_p^e(\pi_p^e)_{ij} \leq y_{ij}, \forall (i,j) \in E, \quad (10.23a)$$

$$\sum_{p\in \mathcal{P}^{ek}} X_p^e = r^{ek}, \quad \forall k \in \mathcal{K}_e, \quad (10.23b)$$

$$X_p^e \geq 0, \quad \forall p \in \mathcal{P}^{ek}, \forall k \in \mathcal{K}_e, \quad (10.23c)$$

$$y_{i,j} \geq 0, \quad \forall (i,j) \in E. \quad (10.23d)$$

Let us consider the dual of (10.23) where v_{ij} and u_k are the dual variables associated with the constraints (10.23a) and (10.23b) respectively. This problem is written as follows:

$$\max \quad \sum_{(i,j) \in E} y_{ij} v_{ij} - \sum_{k \in \mathcal{K}_e} r^{ek} u_k \qquad (10.24a)$$

$$\sum_{(i,j) \in E} (\pi_p^e)_{ij} v_{ij} + u_k \leq 0, \forall p \in \mathcal{P}^{ek} \, \forall k \in \mathcal{K}_e, \qquad (10.24b)$$

$$u_k \geq 0, \quad \forall p \in \mathcal{P}^{ek}, \forall k \in \mathcal{K}_e. \qquad (10.24c)$$

Problem (10.24) is homogeneous and the trivial solution $(0,0)$ is feasible. It has an optimal value which takes either 0 or $+\infty$. By the strong duality theory, problem (10.23) has no feasible solution if and only if (10.24) is unbounded. In order to check unboundness, we solve problem (10.24) with the additional box constraints $v \leq \bar{v}$, where $\bar{v}$ is an arbitrary positive vector:

$$\max \quad \sum_{(i,j) \in E} y_{ij} v_{ij} - \sum_{k \in \mathcal{K}_e} r^{ek} u_k \qquad (10.25a)$$

$$\sum_{(i,j) \in E} (\pi_p^e)_{ij} v_{ij} + u_k \leq 0, \forall p \in \mathcal{P}^{ek} \, \forall k \in \mathcal{K}_e, \qquad (10.25b)$$

$$0 \leq v_{ij} \leq \bar{v}_{ij}, \quad \forall (i,j) \in E. \qquad (10.25c)$$

Since the constraints (10.25b) and (10.25c) are homogeneous, problem (10.24) has a bounded optimal value if and only if (10.25) has a positive optimal value. To check whether capacity y belongs to domain D, the feasibility cut is described by the following lemma (Lisser et al., 1998):

Lemma 10.1 *Let* $y \in \mathbb{R}^{|E|}$. *Then* $y \in \mathcal{D}$ *if and only if the optimal value of (10.25) equals 0. Consequently, if* $(u^\star, v^\star)$ *is a feasible solution to (10.24) with*

$$\sum_{k \in \mathcal{K}_e} r^{ek} u_k^\star - \sum_{(i,j) \in E} y_{ij} v_{ij}^\star > 0,$$

then $y \notin \mathbb{R}^{|E|}$. *Moreover, for all* $y \in \mathbb{R}^{|E|}$ *the following inequality holds*

$$\sum_{(i,j) \in E} y_{ij} v_{ij}^\star \geq \sum_{k \in \mathcal{K}_e} u_k^\star r^{ek}. \qquad (10.26)$$

10.5 CONCLUSION

Multicommodity flow formulations are very common in network design and routing problems. Unfortunately, they are much more difficult to solve than single commodity flow problems even in the continuous case. Their large scale combined with the underlying structure induces different decomposition schemes and they have greatly contributed to the development of that area since the pioneer works of Dantzig, Wolfe, and Benders. We have seen how path generation is a crucial issue to get efficient algorithms. This is the reason why the Flow Deviation algorithm has been very popular to solve practical flow and capacity assignment problems with congestion costs. More recently, the Proximal Decomposition method has been shown to be very competitive to solve convex cost multicommodity flow problems as it exploits the monotropic structure of the network model and allows path generation with fewer supporting paths than Flow Deviation.

Network survivability is a key component of the quality of based network services. We have given an illustration of network design problems with survivability constraints by extending different MCNF formulations. We have applied two different decomposition approaches to the spare capacity problem. Finally, survivability models are of higher degree of magnitude due to their huge size. However, they represent new exciting challenges for the mathematical programming community.

Bibliography

Bellcore Technical Advisory. Synchronous optical networks (sonet) fiber optic transmission systems requirements and objectives. Technical Report TA-NWT-000253, Bellcore, 1990.

R. Ahuja, T.L. Magnanti, and J.B. Orlin. *Network Flows : Theory and Algorithms.* Prentice-Hall, Englewood Cliffs, 1993.

C. Arbib, U. Mocci, and C. Scoglio. Methodological aspects of ring network design. *Lecture Notes in Control and Information Sciences*, 180:135–144, 1992.

A. Balakrishnan, T.L. Magnanti, J.S. Sokol, and Y. Wang. Telecommunication link restoration planning with multiple facility types. *Annals of Operations Research*, 106:127–154, 2001.

D. Banerjee and B. Mukherjee. A practical approach for routing and wavelength assignment in large wavelength-routed optical networks design. *IEEE journal on selected areas in telecommunications*, 14(5):903–908, 1996.

J.F. Benders. Partitioning procedures for solving mixed variables programming problems. *Numerische Mathematik*, 4:238–252, 1962.

D. Bertsekas and R. Gallager. *Data networks.* Prentice-Hall, Englewood Cliffs, 1987.

D. P. Bertsekas. *Nonlinear Programming.* Prentice-Hall, Englewood Cliffs, 1995.

D. Bienstock. *Potential Function Methods for Approximately Solving Linear Programming Problems : Theory and Practice.* Kluwer Publishers, 2003.

D. Bienstock and O. Günlük. Computational experience with a difficult mixed-integer multicommodity flow problem. *Mathematical Programming*, 68:213–237, 1995.

D. Bienstock and O. Raskina. Asymptotic analysis of the flow deviation method for the maximum concurrent flow problem. *Mathematical Programming*, 91:479–492, 2002.

Z.R. Bogdanowicz. A new optimal algorithm for backbone topology design in communications networks. *Mathematical and Computer Modelling*, 17(8):49–61, 1993.

J.F. Bonnans, M. Haddou, A. Lisser, and R.Rébaï. Proximal decomposition method for solving global survivability in telecommunication networks. Technical Report INRIA RR-4055, INRIA, Rocquencourt, 2000.

C.A. Brackett. Dense wavelength division networks: principles and applications. *IEEE Journal of Selected Areas in Communications*, 8:948–964, 1990.

S.G. Chang. Fair integration of routing and flow control in communication-networks. *IEEE Transactions on Communications*, COM-40(4):821–834, 1992.

S.G. Chang. An efficient approach to real-time traffic routing for telephone network management. *Journal of Operational Research Society*, 45(2):187–201, 1994.

Y.G. Chang, J.L.C. Wu, and H.J. Ho. Optimal virtual circuit routing in computer networks. *IEE Proceedings-I Communications Speech and Vision*, 139(6):625–632, 1992.

P. Chardaire and A.Lisser. Minimum cost multicommodity flows. In P.M. Pardalos and M.G.C. Resende, editors, *Handbook on Applied Optimization*, pages 404–422. Oxford University Press, 2002a.

P. Chardaire and A.Lisser. Simplex and interior point specialization algorithms for solving non-oriented multicommodity flow problems. *Operations Research*, 50(2): 260–276, 2002b.

D. Coudert and H. Rivano. Lightpath assignment for multifibers wdm networks with wavelength translators. *IEEE Global Telecommunications Conference*, 21(1):2693–2697, 2002.

G.B. Dantzig and P. Wolfe. The decomposition algorithm for linear programming. *Econometrica*, 29(4):767–778, 1961.

J. Eckstein. Parallel alternating direction multiplier decomposition of convex programs. *JOTA*, 80, 1994.

J. Eckstein and M. Fukushima. Some reformulations and applications of the alternating direction method of multipliers. In *Large-scale optimization : State of the art*. Kluwer Academic Publishers, 1993.

N.D. Fonseca, A.P. Castro, and A.T. Rios. A procedure for resource allocation in switchlet networks. *GLOBECOM 2002, IEEE Global Telecommunications Conference*, 21(1):1895–1898, 2002.

M. Frank and P. Wolfe. An algorithm for quadratic programming. *Naval Research Logistics Quarterly*, 3:95–110, 1956.

L. Fratta, M. Gerla, and L. Kleinrock. The flow deviation method: an approach to store-and-forward computer-communication network design. *Networks*, 3, 1973.

N. Garg and J. Könemann. Faster and simpler algorithms for multicommodity flow and other fractional packing problems. In *Proc. 39th Ann. Symp. on Foundations of Computer Science*, 1998.

A. Girard and B. Sanso. Multicommodity flow models, failure propagation, and reliable loss network design. *IEEE-ACM transactions on Networking*, 6(1):82–93, 1998.

R. Glowinski and A. Marocco. Sur l'approximation par éléments finis d'ordre 1 et la résolution par pénalisation-dualité d'une classe de problèmes de dirchlet non linéaires. *RAIRO*, 2, 1975.

J.-L. Goffin, A. Haurie, and J.-P. Vial. Cutting planes and column generation techniques with the projective algorithm. *Journal of Optimization Theory and Applications*, 65:409–429, 1992.

L. Gouveia. Multicommodity flow models for spanning trees with hop constraints. *European journal of Operational Research*, 95(1):178–190, 1996.

M. D. Grigoriadis and L. G. Khachiyan. Fast approximation schemes for convex programs with many blocks and coupling constraints. *SIAM Journal on Optimization*, 4(1):86–107, 1994.

W.D. Grover. Design of a meta-mesh of chain subnetworks: Enhancing the attractiveness of mesh-restorable wdm networking on low connectivity graphs. *IEEE Journal on Selected Areas in Communications*, 20(1):47–61, 2002.

S. Gupta and H. Pirkul. Design of centralized telecommunication networks. *International Transactions in Operational Research*, 3(1):1–21, 1996.

J. Kennington and M. Lewis. The path restoration version of the spare capacity allocation problem with modularity restrictions: Models, algorithms, and an empirical analysis. *INFORMS Journal on Computing*, 13(3):181–190, 2001.

J.F.P. Labourdette and A.S. Acampora. Logically rearrangeable multihop lightwave networks. *IEEE transactions on Communications*, COM-39(11):464–484, 1991.

L. B. J. LeBlanc, R. Helgason, and D.E. Boyce. Improved efficiency of the frank-wolfe algorithm for convex network programs. *Transportation Sci.*, 19:445–462, 1985.

L.J. LeBlanc, J.S. Park, V. Sridhar, and J. Kalvenes. Topology design and bridge-capacity assignment for interconnecting token ring lans: A simulated annealing approach. *Telecommunication Systems*, 6(1):21–43, 1996.

F. T. Leighton, F. Makedon, S. Plotkin, C. Stein, E. Tardos, and S. Tragoudas. Fast approximation algorithms for multicommodity flow problems. *Journal of Computer and System Sciences*, 50(2):228–243, 1995.

A. Lisser, R. Sarkissian, and J.P. Vial. Mid-range planning of survivable telecommunications networks: joint optimal synthesis of base and spare network capacities. Technical Report 14, France Telecom R&D, 1998.

A. Lisser, R. Sarkissian, and J.P. Vial. Solving lp relaxation for survivability problems in telecommunication networks. *Investigación Operativa*, 9(1-2-3):21–47, 2000.

P. Mahey, S. Oualibouch, and Pham D.T. Proximal decomposition on the graph of a maximal monotone operator. *SIAM J. Optimization*, 5:454–466, 1995.

D. Medhi and S. Guptan. Network dimensioning and performance of multiservice, multirate loss networks using dynamic routing. *IEEE-ACM transactions on Networking*, 5(6):994–957, 1975.

E.A. Medova. Chance constrained stochastic programming for integrated services network management. *Annals of Operations Research*, 81:213–229, 1998.

M. Minoux. *Mathematical Programming: Theory and Algorithms.* Wiley-Interscience, 1986.

V.S. Mirrokni, M. Thottan, H. Uzunalioglu, and S. Paul. A simple polynomial time framework for reduced-path decomposition in multi-path routing. *IEEE INFOCOM 2004, The Conference on Computer Communications*, 23(1):739–749, 2004.

K. Murakami and H.S. Kim. Virtual path routing for survivable ATM networks. *IEEE-ACM transactions on Networking*, 4(1):22–39, 1996.

A. Ouorou, P. Mahey, and J.P. Vial. A survey of algorithms for convex multicommodity flow problems. *Management Science*, 46:126–147, 2000.

Serge A. Plotkin, David B. Shmoys, and Eva Tardos. Fast approximation algorithms for fractional packing and covering problems. *Mathematics of Operations Research*, 20(2):257–301, 1995.

C. Ribeiro and D. Elbaz. A dual method for optimal routing in packet-switched networks. *Lecture Notes in Control and Information Sciences*, (180):199–208, 1992.

A. Riedl. Optimized routing adaptation in ip networks utilizing ospf and mpls. *ICC 2003 - IEEE International Conference on Communications*, 26(1):1754–1758, 2003.

A. Schrijver. *Combinatorial Optimization: Polyhedra and Efficiency.* Springer, 2003.

F. Shahrokhi and D. W. Matula. The maximum concurrent flow problem. *Journal of the ACM*, 37:318–334, 1990.

J. Shaio. A family of algorithms for network reliability problems. *ICC 2002 - IEEE International Conference on Communications*, 25(1):2167–2173, 2002.

C.S. Sung and K.B. Lee. A routing algorithm for virtual circuit data-networking with unreliable links. *Engineering Optimization*, 24(2):119–136, 1995.

L. Xiao, M. Johansson, and S.P. Boyd. A simultaneous routing and resource allocation via dual decomposition. *IEEE Transactions on Communications*, 52(7):1136–1144, 2004.

G.R. Yee and S.Y.S. Lin. A routing algorithm for virtual circuit data-networks with multiple sessions per o-d pair. *Networks*, 22(2):185–208, 1992.

M.G. Yoon, Y.H. Baek, and D.W. Tcha. Design of distributed fiber transport network with hubbing topology. *European journal of Operational Research*, 104(3):510–520, 1998.

Z. Zhang and H.L. Hartmann. On the non-bifurcated routing in virtual circuit communication networks. *European transactions on Telecommunications*, 3(1):45–53, 1992.

11 TELECOMMUNICATIONS NETWORK DESIGN

Anders Forsgren[1] and Mikael Prytz[2]

[1]Optimization and Systems Theory
Department of Mathematics
Royal Institute of Technology (KTH)
SE-100 44 Stockholm, Sweden
andersf@kth.se

[2]Wireless Access Networks
Ericsson Research
SE-164 80 Stockholm, Sweden
mikael.prytz@ericsson.com

Abstract: Telecommunications networks are fundamental in any telecommunications system. The network has to meet a number of criteria for the performance to be satisfactory. Hence, when designing the network, one may pose a number of optimization problems whose solutions give networks that are, in some sense, optimally designed. As the networks have become increasingly complex, the aid of optimization techniques has also become increasingly important. This is a vast area, and this chapter considers an overview of the issues that arise as well as a number of specific optimization models and problems. Often the problems may be formulated as mixed-integer linear programs. Due to problem size and problem structure, in many cases specially tailored solution techniques need to be used in order to solve, or approximately solve, the problems.

Keywords: Telecommunications optimization, network design, mixed-integer linear programming.

11.1 INTRODUCTION

Telecommunications network design is about creating a blueprint for a network. A network design is a plan for how the network should look like so that the involved parties—users, operators, regulators, etc.— will be happy with its performance and cost. Creating a network design is about choosing network structures, allocating resources, and configuring high-level parameters. It depends on the capabilities of the networking technologies, it needs a fair level of detail (but not too much since it is usually based on uncertain forecasts), and the results have to be carefully judged based on several contradicting merits.

Telecommunications networks are very complex today, in essence because they have to meet a wide range of requirements. Heterogeneous services and applications that coexist in the same network, mobility, interworking with a large number of other networks (including legacy networks), and deregulation are just a few factors that have driven complexity up. As a consequence, network design complexity has also been increased. The number of design options and possible design solutions are enormous.

The standard approach for handling the complexities of network design is to divide into subtasks and restrict possibilities, but even the subtasks can be very difficult decision problems. Many of the critical tasks relate to a practical, technical, and quantifiable problems in network design. Some examples are topology design, dimensioning (sizing/loading), configuration, routing, location of functionality, and resource allocation. Many of these problems can be naturally, and fruitfully, cast as optimization problems.

This chapter presents an overview of telecommunications network design from an optimization point-of-view, i.e., the focus is on discussing current problems in telecommunications network design where optimization models and algorithms are applicable and useful as tools. Business oriented aspects of network design, such as deciding on networking technology, and choosing targeted customers and services to offer, are not covered. Neither are technical problems that have no immediate optimization formulation (e.g., designing addressing plans).

The exposition is not complete, neither from a telecommunications network design perspective nor from an optimization application point of view. Network design in telecommunications encompasses many more aspects than what is covered here. There exists an extensive body of literature on the subject. The application of optimization to these problems also has a long history that is not described or surveyed in any detail. Nevertheless, it is our hope that this chapter can serve as a guide to the general network design issues.

11.2 BACKGROUND

To understand how optimization can be applied to telecommunications network design we begin by looking at the who, why, and how of network design: who is interested in doing a network design, why is it interesting, and what is the overall procedure.

Throughout this chapter, telecommunications network design is considered as a planning and configuration problem for a *network administrator*, which is someone, who is running, or planning to run, a telecommunications network. The network administrator can be a public operator, e.g., fixed telephony and mobile operators or Internet Service Providers (ISP). It can also be a corporation or larger company with a private network that connects its branch offices with each other and with external networks. There are also network administrators that are providers to other network administrators, e.g., access providers and leased line operators.

The planning and configuration problem is considered in the technical sense, i.e., the business-planning oriented activities of network design, such as selecting customer segments, deciding which services to offer, and determining the networking technology to use, etc., is assumed to be completed. The network is also assumed to be sufficiently large and sufficiently complex in order for there to actually be a planning problem of significance. This eliminates small home networks, which, if everything works, can be setup without planning or any detailed configuration.

Regarding network complexity there is a trend in telecommunications that strives to reduce the need for planning and configuration by developing low-cost, low-complexity, zero-configuration systems, such as ad-hoc networks. On the other hand there is an opposite trend to add new functionality and new services with increased performance requirements, all of which drives a tremendous increase in complexity of the systems. More services and applications with different inherent communication requirements, deregulation, more actors, legacy networks and technology, etc., are factors that all increase the complexity.

Costs for network components have steadily decreased. In addition, there are many unexploited potentially large markets around the world. This will give room for choices in the design, ranging from low complexity networks to smart, complex systems in optimized configuration. It is our belief that doing a good telecommunications network design will, in general, not be less difficult to do in the future than it is today.

The network design planning and configuration problem is not a static activity performed once in a predeployment phase. It is a continuous process with impacts on both the longer and shorter time scales.

There are many references on general telecommunications network design. An overview of network design issues for telecommunications networks can be found in Ericsson and Telia (1997; 1998). Long (2001) gives a similar overview directed towards Internet Protocol (IP) networks. Another general discussion covering detailed technical networking issues can be found in Stallings (1998).

The design of a network is made so as to accommodate certain traffic. The process leading to the design chosen is typically carried out in several steps. First, the network administrator business planning process yields basic demands and requirements for the network design. Examples of such requirements are served area, provided coverage, offered services, etc. We consider a long-term planning problem, where the time frame is typically many months or years up to the network's life expectancy. The life expectancy today is perhaps no more than five to ten years. The design problem is either a greenfield (design-from-scratch) problem or an expansion/modification problem in an existing network, perhaps where consideration also has to be taken to legacy equipment. The complex interaction and contradicting behavior of the design objectives discussed above, makes it very difficult to consider all aspects at the same time.

The network design process is usually separated into smaller subproblems so that each subproblem is easier to handle. An overall top-down approach is often used, beginning with a high-level design constructed to satisfy some carefully selected performance criteria. The high-level design is then refined into a detailed design, and the detailed design is then tested, for a proof-of-concept, in a lab environment (that mimics a real-life live network as closely as possible) so that all performance parameters and design objectives can be verified, see, e.g., Long (2001) or Kershenbaum (1993). It is often necessary to iterate through this process at various stages. Sometimes the high-level design has to be completely re-engineered after failing during lab verification. Many vendors now offer pre-verified and integrated solutions that minimize the effort spent in bringing the network into service.

11.3 NETWORKING

This section reviews some high-level basic technical networking concepts that are common to many different networking technologies. We describe the fundamental network components, nodes and links, and their basic properties, traffic models and demands, layered architectures and tiered network structures, protocols, and routing and forwarding. The material is brief and targeted for the optimization perspective.

A network is composed of *nodes* and *links*. The nodes represent user equipment, routers, switches, cross-connects, etc., while the links represent connections between the nodes, such as optical fibers, radio links, copper cables, etc. The distinction between nodes and links is usually apparent, but sometimes it may not be clear. In network design the nodes and links are often treated as *candidates* for where to locate nodes and links, i.e., in link topology design the nodes may exist and the full mesh of links can be candidates for where to locate links.

The purpose of the network is to shuffle *user traffic* between end users. In addition the network also transports *control traffic* for management and control of the network itself (examples are network availability and authentication information). Traffic is originated and terminated in nodes, and switched or forwarded through nodes. There may be different types or classes of traffic with different properties and requirements on the network. Traffic typically also varies over time, both over shorter and longer time scales.

11.3.1 Traffic models

To perform network capacity dimensioning it is necessary to characterize the network traffic generated by the applications and how much network resources that are needed to meet the performance requirements for the applications. We refer to such a characterization as a *traffic model*. A *traffic demand* (or just demand) is a requirement on the network design that is formulated using a traffic model. It prescribes the start- and endpoints for the traffic, often called an *origin-destination* pair for point-to-point traffic, and other parameters. The most important parameter is the *traffic volume*, i.e., how much traffic to send. It is usually given in units of some basic resource unit, or in rates of the unit (some examples are bits/s, 64 kbit/s channels, E1/T1s, fractional T1s, STM-1s, wavelengths, and OC-ns). A demand can also have requirements on the maximum tolerable *delay* for end-to-end delivery of the traffic.

The transportation of traffic between nodes is carried in the links. Links may connect two nodes (point-to-point) or more than two nodes, e.g. as in an Ethernet LAN segment or when using radio. Links can be unidirectional or bidirectional, and can have many more properties. For dimensioning, in particular link sizing and loading, the *capacity* of the link is important. It is measured in bits/s or some other resource unit, and it should be related to the traffic volume of the demands.

The network is often divided into different parts that are treated separately. The division may be motivated by different functionalities, different properties, and/or different topologies in terms of network structure, geography, operator organization, etc. It may also be a way to simplify the design, to limit design effort or, more importantly, to create a network that is easy to manage. The network parts, or subnets, are often arranged in hierarchical *tiers*. The first tier, which is closest to the end user, is usually called the *access network*, while the top tier is the *backbone*. Traffic in the backbone is usually aggregated from many users, and, hence, it may benefit from trunking gains. Hence, it is important to have appropriate traffic models for each tier. A traffic demand in an access network may be described differently than a traffic demand in a backbone network. Analogously, the networking functionality is often divided into *layers* to simplify management and hide details that are not necessary for the user, see, e.g. Stallings (1998), for more details. The layers are hierarchically ordered so that a certain layer provides *service* to the layer above, while using the services of the layer below.

In a network that does not reserve resources for individual traffic sessions, a delay design objective is often more appropriate than, e.g., blocking probability. Application sessions can be described using statistical models (birth-and-death processes) that capture when traffic is sent, the rate at which traffic is sent, and for how long the application transmits, see, e.g., Bertsekas and Gallager (1992). These models are often very complicated for "bursty" applications, such as FTP (File Transfer Protocol) or web-browsing traffic, see, e.g., Paxson and Floyd (1995). Given that a network exists, it is possible, in theory, to find the distribution of queue lengths and waiting times in the links and nodes. However, due to the complicated nature of the traffic models, it is often impossible to derive analytical expressions for these distributions. Simulations are an alternative way to compute numerical estimates of delays and queue lengths.

Extending the approach to network design problems is even more complicated, since many applications actually adapt their behavior to the available resources, i.e. the applications generate *elastic traffic*. This is the case for, e.g, Transmission Control Protocol (TCP) connections in an IP network which adapt their transmission rate to the current network conditions. Some results exist on dimensioning rules for elastic traffic, see, e.g., Berger and Kogan (2000) for a single bottleneck link model.

The planning time frame for network design problems affects the necessary accuracy of traffic models. Each parameter in a traffic model has to be estimated based on forecasts of future application traffic. The increased accuracy of a more detailed model may drown in the inaccuracy of forecasting errors. This motivates the use of approximate traffic models.

One approximation in capacity network design of data networks is to ignore the coupling between traffic rates and available capacity. This approximation is accurate if a fairly tight maximum utilization constraint is enforced in the design. Another approximation is to assume a constant transmission rate, see, e.g., Bertsekas and Gallager (1992,Sec. 5.4), which at first sight may seem very inaccurate. However, in a layered network where the backbone network design is considered, there may be a large number of application sources that have been aggregated into each origin-destination pair in the backbone network. If each source contributes a small amount of traffic, then the model can be sufficiently accurate. With constant rates, a design objective can be to minimize the total delay experienced by any origin-destination pair, or it can be to minimize the cost subject to a maximum utilization constraint. For traffic with significant variations in the sending rates, models exist that compute an effective sending rate (effective bandwidth) based on estimates of means and variances, see, e.g., Kelly (1996). The effective bandwidth can then be used in capacity dimensioning as a constant rate figure.

Network traffic often takes the form of *unicast traffic*, which means point-to-point traffic. This may be considered the "typical" traffic situation, for example when two persons are having a telephone conversation. However, today's networks also have the capability of handling *multicast traffic*, which is point-to-multipoint traffic. This could be the situation when a database is replicated to a group of receivers in the network, or when a movie is distributed to a number of subscribers. The senders and receivers in a multicast session are said to belong to a *multicast group*. When taking multicast traffic in to account, one may save capacity by connecting the members of a multicast group by a tree rather than by paths. A detailed discussion of multicasting in IP networks can be found in Williamson (2000). A recent review of multicast applications and implementation challenges is given in Quinn and Almeroth (2001). Further technical details can be found in Deering (1989).

11.3.2 Traffic routing and forwarding

A fundamental problem in the design of the network is the choice of the end-point to end-point paths for the traffic demands. Such paths are constructed by the *routing* process. In a network capacity dimensioning problem it is necessary to consider how the network constructs routing paths, and how the network uses resources when forwarding traffic. It may be the case, e.g., that certain combinations of routing paths are not simultaneously realizable, which may imply that more capacity is required for the design.

Routing can be static or adaptive based on current network conditions. Note that static here means preconfigured by the network administrator - it does not, e.g., exclude the possibility to have per-hop alternative paths. In an adaptive routing network, paths may be changed from end-point to end-point. Adaptive routing works on a time scale which typically is in the order of several minutes. It may take a fairly long time before new routing information has been propagated to all nodes in the network (convergence time). The actual on-line (if any) computations of routing paths are done by routing algorithms, see, e.g., Chen and Nahrstedt (1998) for a survey of algorithms for different routing cases.

The routing is implemented by *protocols*. A protocol is a set of rules, which are implemented in nodes. A protocol entity in a node has state information about itself. It also peers with similar entities in other nodes and gather information from them. Based on state and peer information there is predefined response. An important aspect is that the protocols have a *distributed* property, in that the nodes only have access to limited information about the network. A commonly used IGP (Interior Gateway Protocol) routing protocol in IP networks is OSPF (Open Shortest Path First), see, e.g., Moy (1998a;b). This adaptive link-state protocol uses a set of pre-configured positive integer link metrics (or weights/costs). The routing paths are found as the shortest paths, computed using Dijkstra's algorithm (Dijkstra, 1959), with respect to these link metrics (avoiding failed links). The metrics are often configured to be inversely proportional to the link bandwidth capacity, see, e.g., Jones and Mitchell (2001) and the Cisco OSPF Design Guide (Cis, 2001), but this rule-of-thumb may not always yield good routing patterns, see, e.g., Fortz and Thorup (2000).

The routing gives the paths along which the traffic is sent, or *forwarded*, from one end-point to another end-point. Hence, the forwarding is done on a much shorter time scale than the routing. Forwarding can be done by two general methods. These are often referred to as circuit-, and packet-switched delivery. In the first method the delivery is grouped into sessions (or calls, flows), which are initiated through a signaling phase in which the two end-point nodes contact each other by means of an addressing scheme (telephone numbers in PSTN). After contact is established, the network *reserves resources*, if possible, for the traffic flow from end-point to end-point. These reserved resources are maintained in the network for the duration of the session (or call), which requires the network nodes to maintain state information about all active traffic sessions. Since the network resources are reserved, it is relatively easy to issue performance guarantees. However, for an application that sends traffic at a highly varying rate, the network resource utilization can be very poor.

In the packet delivery method, traffic data is divided into small units, which are given an destination address tag, and are sent directly in the network without any signaling or resource reservation. Individual nodes in the network inspect the address tag and make their own forwarding decisions. Two immediate advantages with this method are the decreased need for nodes to maintain state information, and the possibilities for increased utilization. A disadvantage is that it is more difficult to issue performance guarantees. Traffic may arrive at a node that is very busy so that it may be forced to wait in buffers or may be discarded completely.

11.4 PERFORMANCE CRITERIA

The "what is a good network design" question is difficult to answer, since it depends on many, possibly conflicting, aspects. What is good from one perspective may not be as good from another point of view. A network design engineer typically has to evaluate tradeoffs between many different design parameters. To do this successfully requires experience and a thorough understanding of the design issues and the networking technologies. Many design objectives may not be so easy to quantify or compare with other objectives. Optimization models and methods may be used as tools to support the design process, but they do not produce the full answer.

Considering the aspects above, it is nevertheless possible to identify certain major design objectives. The focus here is on objectives that are quantifiable and that can be represented in an optimization model. We discuss the most important objectives:

- performance (throughput, delay, jitter, coverage, availability);
- redundancy, resiliency, survivability;

- cost.

In addition to these, there are other factors that have to be considered. One is how manageable the network design is. How easy is it to configure and reconfigure the network? How well can fault situations be handled (detected, isolated, and repaired), and how well does the network recover in a disaster situation? Can the network be adapted to changing requirements on performance? How good is the network design in terms of security? Is the network design flexible for accounting and charging of the users? The last two points are often treated together in systems for Authentication, Authorization, and Accounting (AAA).

The ability of a network to accommodate changing and growing traffic patterns is also an objective by which a network design can be measured. A good network design should be scalable so that it does not have to be completely redesigned to accommodate growth in users and traffic volumes.

11.4.1 Performance

One of the major design objectives for a telecommunications network is its *performance*, which can be measured in many different ways. An obvious measure is the *throughput*, i.e., the aggregate traffic handling capacity in bits per second across the network. The throughput capacity should be sufficiently large to accommodate the total traffic through the network.

Another very important measure is the *application response time*, i.e., the time an application experiences in letting the network deliver traffic from one endpoint to another endpoint. Some applications, such as telephony, are very sensitive to the response time (consider, e.g., the speech quality in early implementations of satellite- or Internet-routed phone calls). Requirements on the application response time often translates into requirements on *delay*, and on *jitter* (variations in delay), that can be tolerated in the network.

Another measure of network performance is its *availability*. How frequently is an application rejected network service due to failures or congestion situations? This may again be more or less important for different applications. Requirements on availability can be translated into requirements on, e.g., the percentage of data that has to be resent or the probability that an application session is being blocked due to congestion.

A complicating issue is that requirements on response times and availability may be different for different applications existing in the same network. Networking technologies have varying support for coexisting applications with different characteristics and requirements. Traditional PSTN (Public Switched Telephony Networks) have essentially only one service class, which is very well suited to the requirements of voice traffic. ATM (Asynchronous Transfer Mode) networks offer different traffic classes (constant or varying bitrate (CBR, VBR), available bitrate (ABR), etc.) with different performance guarantees. Frame Relay networks can also handle different traffic classes. IP (Internet Protocol) networks, such as the public Internet, have some support for traffic classes and traffic prioritization, but it has not been in use on the Internet. Differentiated Services (diffserv) is now being standardized and introduced in IP networks to improve the support for different traffic classes, see, e.g., Kilkki (1999).

11.4.2 Redundancy, Resiliency, and Survivability

Network *redundancy*, *resiliency*, and *survivability* are another group of design objectives. How well does the network cope under failure situations such as router/switch failures or network link failures? This design objective can translate into requirements on reserve network resources, both in terms of standby equipment and in availability of alternate paths to route traffic on.

11.4.3 Cost

There is a tradeoff between network availability/survivability discussed above and *cost*, which is another major design objective. The total network design cost includes both equipment and capacity leasing costs, as well as the cost of support and management of the network over its lifespan. For the backbone, or the Wide Area Network (WAN), the direct capacity costs, e.g., from leased lines, can be very large and often dominates the total cost.

11.5 OPTIMIZATION OF NETWORK DESIGN PROBLEMS

The purpose of this section is to give a brief introduction to optimization models for important classes of network design problems. The problems are all formulated using mathematical optimization models. Since these involve networks and optimization they are sometimes called *network optimization* problems. This should not be confused with another common interpretation of this term in the telecommunications industry, namely the activity of tuning the performance of an existing network by carefully tweaking parameters.

We mainly focus on problems associated with the high-level network design field. Following decisions on some fundamental design issues, such as selecting networking technologies and overall architecture, the problems arise as specific, quantifiable *decision problems* having many feasible solutions. Specifically, the problems concern *capacity dimensioning of network resources* in the presence of multicast and unicast traffic, including link topology selection and link sizing/loading, and *location selection* for shared tree multicast routing core/RP nodes. The capacity dimensioning problems concern the backbone network and a physical or logical transmission layer. The traffic model for these problems is described by fixed requirements, referred to as *demands*, on bandwidth capacities between different origin-destination pairs. The problems capture the cost, availability, and performance design objectives. Different solutions can behave very differently with respect to these objectives, e.g., two designs that satisfy the performance and availability requirements may have very different total costs.

An alternative to the optimization approach is to consider several possible cases and simulate the performance of each case. This approach allows for more detailed modeling of the problem, but it is not possible to characterize the best possible solutions and there is a risk that some important design choice is missing from all cases. The optimization problems normally have to focus on some specific areas of the network design process, and the results may not be what is ultimately implemented in the network. Changes and refinements might occur during the detailed network design phase or as a result of lab verifications. This does not, however, diminish the value of finding good solutions to the mathematical decision problems. Suppose, e.g., that an optimal solution (in some sense) has been found for the mathematical decision problem. This solution then constitutes a baseline against which any detailed network design changes or modifications can be compared with. The extra cost or the added delay or the increased availability of a proposed modification can be quantified against the optimal solution, which may assist in deciding if the change should be carried through.

A general overview of network optimization models and solution methods is given in the two handbooks of Ball et al. (1995a;b). A treatment of network flow models and algorithms is given in Ahuja et al. (1993). Several telecommunications network design models and algorithms are given in Kershenbaum (1993). Extensive surveys on network design and network optimization problems, models, and solution methods can be found in Magnanti and Wong (1984), Minoux (1989), and in the recent thesis by Yuan (2001). A survey of location problems in telecommunications is given in Gourdin et al. (2001).

11.5.1 Optimization Model Components

A telecommunications network can typically be represented by a directed graph $D = (V, A)$, where V is the set of nodes and A is the set of potential arcs connecting the nodes. If we consider *unicast* traffic, there is a set of demands, K, where demand $k \in K$ may be represented by its origin node s^k and its destination node t^k are the origin and destination nodes of each demand $k \in K$. If we, for simplicity, assume that demand k requires one unit to be sent from node s^k to node t^k, the requirement on the path of the demand may be represented by a flow f^k of unit capacity from node s^k to node t^k through D. Since link capacity is almost always bidirectional, the undirected graph $G = (V, E)$ is often used to represent the network, where each edge $e = [i, j] \in E$ corresponds to two anti-parallel directed arcs $(i, j) \in A$ and $(j, i) \in A$.

A flow may be modeled by *multicommodity flow* constraints

$$Nf^k = q^k, \quad k \in K, \tag{11.1}$$

where K is the set of demands, N is the node-arc incidence matrix of D, f^k is a vector of, continuous or binary, flow variables on A, and q^k is a vector where $q_i^k = 1$ for $i = s^k$, $q_i^k = -1$ for $i = t^k$, and $q_i^k = 0$ for $i \in V$, $i \neq s^k$, $i \neq t^k$. The choice of continuous or binary flow variables f depends on if the flows are allowed to be split or not in the network.

Although this is the "straightforward" modeling of flows through the network, the network connectivity constraints can also be represented using *cut inequalities*

$$\sum_{a \in \delta^+(S)} f_a^k \geq 1, \quad S \subset V, \ s^k \in S, \ t^k \in \bar{S}, \quad k \in K, \tag{11.2}$$

where the *cutset* $\delta^+(S) = \{a = (i, j) \in A : i \in S, j \in \bar{S}\}$ is the set of arcs leaving S (tail in S and head in $\bar{S}$). The cut inequalities given by (11.2) are exponentially many, in general. Hence, when using this formulation, the problem is not formulated *explicitly*, but rather inequalities from (11.2) are generated as needed through the iterative solution process, see Section 11.6.

The constraints on f^k, $k \in K$, given by (11.1) or (11.2) are *separable*, in that they apply to each demand individually. In the design process, the choice of flows through the network is normally restricted by common resources, such as link capacities, which give constraints that couple the different demands. Such coupling constraints may often be modeled by binary *decision variables*. To consider a specific situation where such variables arise, assume that the network resource constraints depend on the specific performance requirement. In the capacity dimensioning problems, the constraints relate the decision variables selecting network link capacity to the decision variables connecting the network demands. Assume that multiple capacities are available for each edge, let b_e^l denote the bandwidth capacity provided by level l on edge e, and let L denote the set of capacity levels. Then we may introduce binary decision variable x_e^l, $e \in E$, $l \in L$, with the interpretation that $x_e^l = 1$ if capacity level l is chosen for edge e, and $x_e^l = 0$ otherwise. Then the capacity constraints

$$\sum_{k \in K} d^k(f_{ij}^k + f_{ji}^k) \leq \sum_{l \in L} b_e^l x_e^l, \quad e = [i,j] \in E, \qquad (11.3)$$

can be used, where d^k is the demanded bandwidth capacity for demand $k \in K$. Additional constraints may also be present, e.g., constraints that capture restrictions on routing paths.

11.5.2 Important problem classes

As indicated by the discussion above, a wide range of problems within network design may be formulated as as *linear mixed integer programming* (MIP) problems, i.e., an optimization problem with linear objective function and linear constraint functions, subject to integrality requirements on some of the decision variables. We outline below a few fundamental problem classes and network requirements that may be modeled within this problem class.

11.5.2.1 Minimum cost multicommodity flow. The minimum cost multicommodity flow problem is an underlying basic optimization problem. The question is how to send a set of commodities through a network at minimal cost subject to capacity constraints on the arcs. Mathematically, the problem may be formulated as

$$\begin{aligned}
\text{minimize} \quad & \sum_{k \in K} \sum_{[i,j] \in A} c_{ij} d^k f_{ij}^k \\
\text{subject to} \quad & N f^k = q^k, \ f^k \geq 0, \quad k \in K, \\
& \sum_{k \in K} d^k f_{ij}^k \leq b_{ij}, \quad [i,j] \in A,
\end{aligned} \qquad (11.4)$$

where c_{ij} denotes the cost per unit for sending through arc $[i,j]$, b_{ij} is the capacity of arc $[i,j]$, and d^k is the bandwidth required by demand k.

11.5.2.2 Uncapacitated network design—fixed charge. A fundamental network design problem referred to as uncapacitated network design with fixed charge is obtained if the capacity constraints are replaced by a cost for utilizing the arc. We then get flow variables plus one design variable per arc, i.e., a problem of the form

$$\text{minimize} \quad \sum_{k \in K} \sum_{[i,j] \in A} c_{ij} d^k f_{ij}^k + \sum_{[i,j] \in A} g_{ij} x_{ij}$$
$$\text{subject to} \quad N f^k = q^k, \ f^k \geq 0, \quad k \in K, \quad (11.5)$$
$$f_{ij}^k \leq x_{ij}, \ x_{ij} \in \{0,1\}, \quad k \in K, [i,j] \in A,$$

where g_{ij} is the fixed cost. See, e.g., Hochbaum and Segev (1989).

11.5.2.3 Capacitated network design. The capacitated network design problem is obtained from the minimum cost multicommodity flow problem, by adding a fixed cost plus capacity constraints on the arcs. This gives

$$\text{minimize} \quad \sum_{k \in K} \sum_{[i,j] \in A} c_{ij} d^k f_{ij}^k + \sum_{[i,j] \in A} g_{ij} x_{ij}$$
$$\text{subject to} \quad N f^k = q^k, \ f^k \geq 0, \quad k \in K, \quad (11.6)$$
$$\sum_{k \in K} d^k f_{ij}^k \leq b_{ij} x_{ij}, \quad [i,j] \in A.$$

See, e.g., Balakrishnan et al. (1991); Barahona (1996); Bienstock et al. (1998); Holmberg and Yuan (1998); Günlük (1999); Holmberg and Yuan (2000).

11.5.2.4 Network loading problem. network loading problem
For the network loading problem, the topology of the network is given, i.e., it has been decided on beforehand which links that can be used. The question is to select capacity levels on the arcs from a give set of capacity steps. See, e.g., Mirchandani (2000).

11.5.2.5 Topology constraints. The topology of the networks is sometimes required to be of a certain type. For example, it may be of interest to use ring structures, tree structures, or some other type of particular structure. Such requirements may be included in the model, e.g., (Gavish, 1982; Holmberg and Yuan, 2000).

11.5.2.6 Routing constraints. The routing through the network may be constrained in a more complex way than through capacity levels. It may for example be required that only a restricted set of paths is used due to the fact that this has been preprocessed from a particular protocol, e.g., OSPF-routing. This may lead to more complicated problems of multilevel type. Similar complicating routing constraints may arise in networks based on restrictions on wavelengths or frequencies.

11.5.2.7 Multiperiod problems. As stated above, there is no timescale present in the problem, i.e., the demand is assumed to be static. We may consider models where the demand or capacity is forecasted for certain time intervals in the future. This may be included in the model by giving a time dimension to the design variables.

11.5.2.8 Hierarchical network design. The network design problem may be viewed on several levels. On the top level, a backbone network is designed, so as to meet different criteria. The lowest level may be a local network within a company, or even a private home. In between, there is a range of different levels of networks. By adding different levels into the model, the hierarchy of the network may be captured. See, e.g., Balakrishnan et al. (1994a;b).

11.5.2.9 Survivability. It is desirable to construct networks that are robust with respect to link or node failures. For example, if a link suddenly breaks, it is important that there are ways of rerouting the traffic so that the link failure does not lead to substantial disturbance in traffic. This may be done by selecting suitable network topology, and also by reserving spare capacity, see, e.g., Dahl (1994); Dahl and Stoer (1998).

11.6 SOLUTION STRATEGIES

The optimization models described in the previous section lead to *mixed-integer linear programs*, i.e., optimization problems on the form

$$\begin{aligned} \text{minimize} \quad & c^T x \\ \text{subject to} \quad & Ax \geq b, \\ & x \in Z_+^n, \end{aligned} \qquad (11.7)$$

where Z_+^n denotes the nonnegative integral points in $\Re^n$. (Not all components of x need to have integral requirement, but to simplify the exposition, we consider the form (11.7).) We denote by S the feasible set of (11.7), i.e., $S = \{x \in Z_+^n : Ax \geq b\}$. The decision problem is to find an optimal solution x^* (we will always assume that an optimal solution exists) for a given problem instance (11.7). If all variables are binary, then there can be at most 2^n solutions, where n is the number of variables. An optimal solution can always be found by generating all 2^n candidate solutions, verify which are feasible, and pick one of the feasible solutions with the lowest objective function value. This approach is clearly only viable for very small problems.

In terms of computational complexity, the general MIP problem and the general binary programming variant are so called NP-hard problems, see, e.g., Garey and Johnson (1979). Roughly this means that there are no known efficient, i.e. polynomial time, algorithms for the problem. The network capacity design problems considered here are typically special cases of MIP problems that are still NP-hard problems (Johnson et al., 1978). Hence, in general we cannot expect to find an efficient (polynomial time) algorithm to solve practical problem instances exactly.

There are several approaches that can be used to tackle the NP-hard optimization problem (11.7). The computational difficulty of solving the problem implies that there is no known approach which is guaranteed to be efficient for all instances. Typically, it is not sufficient to consider (11.7) as such, but a successful approach has to consider special problem structure, i.e., the particular problem formulation has to be considered, e.g., (11.4), (11.5) or (11.6).

11.6.1 Approximation Algorithms and Heuristics

One approach is to use *approximation algorithms*, see, e.g., Goemans and Williamson (1997) for an application to a general class of network design problems. An approximation algorithm sacrifices optimality for efficiency, but comes with a theoretical guarantee on the worst-case percentage away that the objective value of the approximate solution is from the actual optimum.

For some classes of problems and approximation algorithms, it is possible to derive tight worst case error bounds. If the worst case error bound is tight, then the approximate solution can be considered as satisfactory. The actual optimality bound may be much tighter for a specific instance, but there is no a priori bound other than the worst case error. It is also not possible to improve on the error if the bound is not deemed as sufficiently tight.

A *heuristic algorithm* is here considered as any algorithm where no formal performance guarantee is known. Many heuristic algorithms are based on a local, or neighborhood, search strategies, such as tabu search, see, e.g, Glover (1989); Glover et al. (1993), and/or including randomization, such as simulated annealing, see, e.g., Kirkpatrick et al. (1983) or Reeves (1993). Other type of heuristics may be based on an evolutionary strategy for finding improving solutions. Genetic algorithms belong to this class of heuristics. In spite of the lack of performance guarantees, these algorithms may nevertheless produce good feasible solutions in practice although any verification of the solution quality must be provided by other means. Heuristic algorithms may successfully be applied to wide ranges of telecommunications problems. However, we will focus or discussion towards optimization methods that provide bounds on the quality of the solutions given. Heuristic methods may be a useful complement to these methods for providing good feasible solutions, i.e., to give a good upper bound of the optimal value of the problem.

11.6.2 Relaxations

A general strategy for dealing with difficult integer programming problems is to consider relaxations of the problem. A relaxed problem is typically a simplified problem which provides a lower bound on the optimal objective function value z_{MIP}. This bound can be used together with an upper bound obtained for a feasible solution to estimate the quality of the feasible solution.

The problem

$$z_R = \begin{array}{l} \text{minimize} \quad z_R(x) \\ \text{subject to} \quad x \in S_R \end{array} \tag{11.8}$$

is said to be a *relaxation* of (11.7) if $S \subseteq S_R$ and $z_R(x) \leq c^T x$ for $x \in S$. For a general discussion of relaxations and duality for integer and mixed integer programming problems, see, e.g., Nemhauser and Wolsey (1988,Ch. II.3). The objective function value of an optimal solution, z_R is a lower bound on the optimal objective function value of (11.7), z_{MIP}.

The complexity of the relaxed problem (11.8) and the quality of the lower bound, depends on the choice of the objective function $z_R(x)$ and the set S_R. There is a similar tradeoff here between efficiency (or effectiveness) and bound quality as there is for approximation algorithms. For a MIP problem (11.7) a natural choice is to consider the *linear programming relaxation* where $z_R(x) = z_{\text{LP}}(x) = c^T x$ and $S_R = S_{\text{LP}} = \{x \in R^n : Ax \geq b\}$. In this case the relaxed problem (11.8) becomes a linear program (LP), which can be solved efficiently. The quality of the LP lower bound may be satisfactory for some problems, but may often be poor on network design problems with discrete capacity levels, see, e.g., Prytz and Forsgren (2002).

11.6.3 Lagrangian Relaxation

Another relaxation strategy is Lagrangian relaxation, which is appropriate when the constraint structure of the MIP problem (11.7) contains a set of "nice" constraints and a set of complicating constraints such that the problem is easy (or at least less difficult) to solve if the complicating constraints are dropped. In this case the constraint set S in (11.7) is of the form

$$S = \{x \in Z_+^n : A^1 x \geq b^1, A^2 x \geq b^2\}, \tag{11.9}$$

where $A^1 x \geq b^1$ are assumed to be the complicating constraints. The *Lagrangian relaxation* problem

$$\begin{aligned} z_{\text{LR}}(\lambda) = \quad & \text{minimize} \quad c^T x - \lambda^T(A^1 x - b^1) \\ & \text{subject to} \quad A^2 x \geq b^2, \\ & \qquad\qquad\quad x \in Z_+^n \end{aligned} \tag{11.10}$$

is a relaxation for any $\lambda \geq 0$. The relaxed problem (11.10) may be such that an efficient algorithm exists for solving it, or the relaxed problem may decompose into a set of smaller subproblems, which may lead to an overall improvement in effectiveness. The optimal objective function value $z_{\text{LR}}(\lambda)$ is a lower bound to z_{IP} for $\lambda \geq 0$. The greatest lower bound is given by the solution to the *Lagrangian dual problem*

$$\begin{aligned} z_{\text{LD}} = \quad & \text{maximize} \quad z_{\text{LR}}(\lambda) \\ & \text{subject to} \quad \lambda \geq 0. \end{aligned} \tag{11.11}$$

The function $z_{\text{LR}}(\lambda)$ is piecewise linear and concave in λ, but it is a nondifferentiable function in general. The Lagrangian dual problem (11.11) can be solved by methods for convex, nondifferentiable optimization problems such as iterative subgradient methods, see, e.g., Hiriart-Urruty and Lemaréchal (1993). The complexity of the relaxed problem (11.10) and the quality of the lower bound may depend on which of the constraints that are relaxed. The lower bound on the optimal objective function value obtained by a Lagrangian dual problem is always at least as good as the lower bound obtained by linear programming relaxation, and the bound obtained by Lagrangian relaxation may be significantly better.

In many network design problems there are constraints relating to each demand together with a set of connecting resource constraints. Relaxing the resource constraints yields relaxed problems (11.10) that decomposes into smaller subproblems for each demand. Other relaxations are also possible, e.g., *Lagrangian decomposition* (Nemhauser and Wolsey, 1988,Ch. II.3.6).

11.6.4 Relaxation Algorithms

Relaxations of (11.7) can be used to form implicit enumeration algorithms that embody a *divide-and-conquer* strategy. For a detailed treatment, see, e.g., Nemhauser and Wolsey (1988,Ch. II.4). These algorithms are exponential in the worst-case, but together with efficient derivations of lower and upper bounds they can be applied successfully on many instances of practical size. The performance of these methods depend critically on the quality of the relaxation lower bounds and the heuristic upper bounds, as well as on the details of how the divide-and-conquer strategy is implemented.

A general iterative *relaxation algorithm* considers an initial relaxation of (11.7) with objective function $z_R^1(x)$ and constraint set S_R^1, and finds an optimal solution x_R^1 to (11.8). If this solution satisfies $z_R(x_R^1) = c^T x_R^1$ and $x_R^1 \in S$, then x_R^1 is an optimal solution to (11.7). Otherwise the relaxed problem is refined by selecting a new constraint set S_R^2 and a new objective function $z_R^2(x)$, such that

$$S \subseteq S_R^2 \subseteq S_R^1, \text{ and } z_R^1(x) \leq z_R^2(x) \leq c^T x, \text{ for } x \in S. \tag{11.12}$$

The refinement should be strict, i.e. either $S_R^2 \neq S_R^1$ or $z_R^1(x) \neq z_R^2(x)$ for $x \in S$. The refined relaxed problem is solved and the new optimal solution x_R^2 is again checked if it is optimal to (11.7) or if a new refinement has to be made.

11.6.5 Cutting Plane, Branch-and-Bound, Branch-and-Cut

A *cutting plane algorithm* is a special relaxation algorithm based on the linear programming relaxation. Here $z_R^i(x) = c^T x$ in all iterations i and the initial constraint set is $S_R^1 = \{x \in R^n : Ax \geq b\}$. If the optimal LP solution x_R^1 also satisfies integrality, i.e. $x_R^1 \in S$, then it is an optimal solution to (11.7). Otherwise a *valid inequality* $a^1 x \geq b^1$ for S is found that separates x_R^1 from S, i.e. $a^1 x_R^1 < b^1$, and S_R^2 is selected as $S_R^2 = S_R^1 \cap \{x \in R^n : a^1 x \geq b^1\}$. The problem of how to find a valid inequality that separates x_R^1 from S, and perhaps also finding the inequality that is most violated in some sense, is called the *separation problem*.

Another relaxation approach is based on *dividing* the problem into smaller problems that are easier to manage. Suppose S is divided into sets S^j for $j = 1, \ldots, J$ such that $S = \bigcup_{j \in J} S^j$ and let

$$z_{\text{MIP}}^j = \underset{x \in R^n}{\text{minimize}} \left\{ c^T x : x \in S^j \right\}. \tag{11.13}$$

Then

$$z_{\text{MIP}} = \underset{j \in J}{\text{minimize}} \; z_{\text{MIP}}^j. \tag{11.14}$$

If all the sets S^j are mutually disjoint, then the division is a *partition* of S. The division of S can be done recursively by dividing or partitioning each S^j into smaller sets. This recursive division or partitioning is usually represented by a *search tree*, where the root node is the original problem and the sons, and sons of sons, etc. are the successive partitions. In a MIP problem with binary variables, a straightforward partition considers one binary variable and fixes the value of this variable to zero and one in two different subproblems, which leads to a binary search tree.

The recursive subdivisioning can be stopped for some j if either of the following three *pruning criteria* is fulfilled: (i) the subproblem is infeasible, (ii) an optimal solution to the subproblem is found, or (iii) the optimal objective function value to the

MIP problem (11.7), z_{MIP}, dominates the optimal objective function value of the subproblem (value dominance), i.e. $z_{MIP} \leq z_{MIP}^j$. Even after dividing or partitioning (11.7) the subproblems may be difficult to solve. The pruning criteria can also be difficult to apply, especially the third which assumes that z_{MIP} is known.

A natural strategy is to consider relaxations of the subproblems obtained on divisioning, which is the basis for *branch-and-bound*. The pruning criteria remain essentially the same. The second criteria can be applied if the optimal solution to the relaxed problem is an optimal solution to the subproblem, and the third criteria is modified so that z_{MIP} is replaced by $\bar{z}_{MIP}$, where $\bar{z}_{MIP}$ is an upper bound, which can be obtained, e.g., through heuristics. Branch-and-bound where relaxations are solved as subproblems can be viewed as a relaxation algorithm.

There are many aspects that affect the effectiveness of a branch-and-bound algorithm with relaxations as subproblems. In each step one subproblem node in the search tree is examined, but the overall performance of the algorithm depends on the order of the examined nodes. There may also be many ways to divide a subproblem into new subproblems, compare, e.g., the binary MIP problems where all of the remaining binary variables that have not been fixed so far are candidates for the next partition.

Branch-and-bound with linear programming relaxations is the most common branch-and-bound algorithm. It is also the basic algorithm used by most commercial MIP problem solvers. Branch-and-bound can also be used with other relaxations such as Lagrangian relaxation.

A *branch-and-cut* algorithm can be considered as a combination of branch-and-bound with linear programming relaxations and a cutting plane algorithm. Here the linear programming relaxation lower bounds are strengthened by adding valid inequalities to the problem. Note that it is possible to add inequalities that are valid only for a specific subproblem node in the branch-and-bound search tree, or that are valid for the original problem.

11.7 SUMMARY

Network design leads to many challenging optimization problems. The focus of the present chapter has been optimization problems that may be formulated as mixed-integer linear programming problems. In many cases, they are characterized by large size and cost- or constraint functions of staircase type, that may make linear programming relaxation poor. Hence, there has been significant work into specialized decomposition and relaxation methods directed towards telecommunications network design problems. The present chapter has highlighted some issues that arise within network design with respect to optimization—with an unavoidable bias towards the authors' research.

Bibliography

R.K. Ahuja, T.L. Magnanti, and J.B. Orlin. *Network Flows. Theory, Algorithms and Applications*. Prentice Hall, Englewood Cliffs, New Jersey, 1993.

A. Balakrishnan, T.L. Magnanti, and P. Mirchandani. A dual-based algorithm for multi-level network design. *Management Science*, 40:567–581, 1994a.

A. Balakrishnan, T.L. Magnanti, and P. Mirchandani. Modeling and heuristic worst-case performance analysis of the two-level network design problem. *Management Science*, 40:846–867, 1994b.

A. Balakrishnan, T.L. Magnanti, A. Shulman, and R.T. Wong. Models for planning capacity expansion in local access telecommunication networks. *Annals of Operations Research*, 33:239–284, 1991.

M.O. Ball, T.L. Magnanti, C.L. Monma, and G.L. Nemhauser, editors. *Network Models*, volume 7 of *Handbooks in Operations Research and Management Science*. Elsevier, 1995a.

M.O. Ball, T.L. Magnanti, C.L. Monma, and G.L. Nemhauser, editors. *Network Routing*, volume 8 of *Handbooks in Operations Research and Management Science*. Elsevier, 1995b.

F. Barahona. Network design using cut inequalities. *SIAM J. Optimization*, 6(3): 823–837, August 1996.

A.W. Berger and Y. Kogan. Dimensioning bandwidth for elastic traffic in high-speed data networks. *IEEE-ACM Transactions on Networking*, 8:643–654, 2000.

D. Bertsekas and R. Gallager. *Data Networks*. Prentice Hall, 1992.

D. Bienstock, S. Chopra, O. Günlük, and C.-Y. Tsai. Minimum cost capacity installation for multicommodity network flows. *Math. Program.*, 81(2):177–200, 1998.

S. Chen and K. Nahrstedt. An overview of quality of service routing for next-generation high-speed networks: Problems and solutions. *IEEE Network*, pages 64–79, November/December 1998.

Cisco Systems Inc., San Jose, CA, USA. *OSPF design guide*, 2001.

G. Dahl. The design of survivable directed networks. *Telecommunication Systems*, 2: 349–377, 1994.

G. Dahl and M. Stoer. A cutting plane algorithm for multicommodity survivable network design problems. *INFORMS Journal on Computing*, 10:1–11, 1998.

S. E. Deering. Host extensions for IP multicasting. Request for Comments 1112, Internet Engineering Task Force, August 1989.

E.W. Dijkstra. A note on two problems in connexion with graphs. *Numerische Mathematik*, 1:269–271, 1959.

Ericsson and Telia, editors. *Understanding telecoummunications, vol. 1*. Studentlitteratur, Lund, Sweden, 1997.

Ericsson and Telia, editors. *Understanding telecoummunications, vol. 2*. Studentlitteratur, Lund, Sweden, 1998.

B. Fortz and M. Thorup. Internet traffic engineering by optimizing OSPF weights. In *Proceedings of IEEE INFOCOM'2000*, pages 519–528. IEEE, 2000.

M.S. Garey and D.S. Johnson. *Computers and Intractability: A Guide to the Theory of NP-Completeness*. W.H. Freeman, New York, 1979.

B. Gavish. Topological design of centralized computer networks - formulations and algorithms. *Networks*, 12:355–377, 1982.

F. Glover. Tabu search, part I. *ORSA Journal on Computing*, 1:190–206, 1989.

F. Glover, E. Taillard, and D. de Werra. A user's guide to tabu search. *Annals of Operations Research*, 41:3–28, 1993.

M.X. Goemans and D.P. Williamson. A primal-dual method for approximation algorithms and its application to network design problems. In D.S. Hochbaum, editor, *Approximation Algorithms for NP-hard Problems*. PWS Publishing Company, 1997.

E. Gourdin, M. Labbé, and H. Yaman. Telecommunication and location. Technical report, Service de Mathematiques de la Gestion, Université Libre de Bruxelles, 2001.

O. Günlük. A branch-and-cut algorithm for capacitated network design problems. *Math. Program.*, 86:17–39, 1999.

J.-B. Hiriart-Urruty and C. Lemaréchal. *Convex Analysis and Minimization Algorithms II, Advanced Theory and Bundle Methods*. Springer-Verlag, 1993.

D.S. Hochbaum and A. Segev. Analysis of a flow problem with fixed charges. *Networks*, 19:291–312, 1989.

K. Holmberg and D. Yuan. A Lagrangean approach to network design problems. *International Transactions in Operational Research*, 5:529–539, 1998.

K. Holmberg and D. Yuan. A Lagrangean heuristic based branch-and-bound approach for the capacitated network design problem. *Operations Research*, 48:461–481, 2000.

D.S. Johnson, J.K. Lenstra, and A.H.G. Rinnooy Kan. The complexity of the network design problem. *Networks*, 8:279–285, 1978.

M. Jones and D. Mitchell. *JUNOS Internet Software Configuration Guide: Routing and Routing Protocols, Release 4.3*. Juniper Networks Inc., Sunnyvale, CA, USA, 2001.

F.P. Kelly. Notes on effective bandwidths. In *Stochastic Networks*, pages 141–168. Clarendon Press, Oxford, 1996.

A. Kershenbaum. *Telecommunications network design algorithms*. McGraw Hill, 1993.

K. Kilkki. *Differentiated services for the Internet*. Macmillan technical publishing, 1999.

S. Kirkpatrick, C.D. Gelatt, and M.P. Vecchi. Optimization by simulated annealing. *Science*, 220:671–680, 1983.

C. Long. *IP Network Design*. Osborne/McGraw-Hill, 2001.

T.L. Magnanti and R.T. Wong. Network design and transportation planning: models and algorithms. *Transportation Science*, 18:1–55, 1984.

M. Minoux. Network synthesis and optimum network design problems: models, solution methods and applications. *Networks*, 19:313–360, 1989.

P. Mirchandani. Projections of the capacitated network loading problem. *European J. Oper. Res*, 122:534–560, 2000.

J. Moy. *OSPF: Anatomy of an Internet routing protocol*. Addison Wesley, 1998a.

J. Moy. OSPF version 2. Request for Comments 2328, Internet Engineering Task Force, April 1998b.

G.L. Nemhauser and L.A. Wolsey. *Integer and Combinatorial Optimization*. Wiley, 1988.

V. Paxson and S. Floyd. Wide-area traffic: The failure of poisson modeling. *IEEE/ACM Transactions on Networking*, 3:226–244, 1995.

M. Prytz and A. Forsgren. Dimensioning multicast-enabled communications networks. *Networks*, 39:216–231, 2002.

B. Quinn and K. Almeroth. IP multicast applications: Challenges and solutions. Request for Comments 3170, Internet Engineering Task Force, September 2001.

C.R. Reeves. *Modern Heuristic Techniques for Combinatorial Problems*. Wiley, 1993.

W. Stallings. *High-Speed Networks, TCP/IP and ATM Design Principles*. Prentice Hall, 1998.

B. Williamson. *Developing IP Multicast Networks: The Definitive Guide to Designing and Deploying CISCO IP Multicast Networks*. Cisco Press, 2000.

D. Yuan. *Optimization models and methods for communication network design and routing*. PhD thesis, Division of Optimization, Department of Mathematics, Linköping University, Linköping, Sweden, 2001.

12 RING NETWORK DESIGN

Mathias Henningsson[1], Kaj Holmberg[1], and Di Yuan[2]

[1]Department of Mathematics
Linköping Institute of Technology
Linköping, Sweden
mahen@mai.liu.se
kahol@mai.liu.se

[2]Department of Science and Technology
Linköping Institute of Technology
Norrköping, Sweden
diyua@itn.liu.se

Abstract: Applying traditional methods of network design on modern telecommunication data often results in tree-like structures, due to the high capacities of the current optical fibers. However, the increasing importance of the traffic on telecommunication networks makes the issue of survivability more crucial. It is not acceptable that parts of the network are completely unable to communicate if a single link failure should occur. Therefore telecommunication networks must be designed so that certain survivability requirements are fulfilled. In this chapter we study the case where the survivability requirement is that the network should be composed of connected rings of links. In case of a failure in a ring, the traffic can simply be sent the other way around the ring. We describe a solution approach iteratively generating rings in a meaningful way, as the number of possible rings is very large. A model deciding the optimal usage of a given set of rings and a model generating valid rings are used together, to form a method of column generation-type. We also review work on other types of ring network design problems.

Keywords: Network design, rings, integer programming, column generation, heuristic.

12.1 INTRODUCTION

In a few decades telecommunication networks have been subject to rapid changes, and have been developed from being networks for telephones call, based on time slots, into optical fiber and package based networks carrying all kinds of information, such as text, voices, pictures, and moving pictures.

The optical fiber has been very important for the development of telecommunication networks and the fast progress in the telecommunication area. An optical fiber is not only more powerful then the older coaxial cable, it is also cheaper, which means that a higher capacity can be installed without the cost being higher. Also the packing of data, routing, network devices, and the physical structure of the networks have all become much more efficient in recent years, and have helped to develop the telecommunication networks.

The physical structure of telecommunication networks can be divided into a number of levels. For example, a network cluster may be divided into four levels. The first level represents the international level, where large routers are connected to each other. The cables have high capacity and may carry huge amounts of traffic. The next level represents the national level, connecting routers in main cities or regions. The third level is the city/region level, where routers, often called local stations, are connected. The last level is the network connecting the subscribers. The networks at lower levels often have a tree-like structure, which is cheap and has been shown to be efficient for telephone traffic. However, with increased traffic, these networks are very vulnerable.

Two major issues for telecommunication networks are to increase capacity and to speed up the network, for example shorten down-loading time. For most subscribers, these issues are of great importance. The increase in capacity and speed-up are important, but other issues, such as robustness, reliability, and safety are very important for the owner and the administrator of a network. A reliable network often costs more to build, but often costs less to run, and customers are in the long run more happy with a network they can rely on.

One aim when constructing a more reliable network is that it should be able to send most of the traffic, even if there is a failure in the network. It is common to require that there should be at least two disjoint paths in the network between every pair of nodes, i.e. there should exist at least two paths from a sender to a receiver. These paths can either have no links or no nodes in common. This means that if there is a failure in one of the paths, the traffic can be routed using the other path. Such requirements are called "two connectivity" requirements.

12.1.1 Ring Structure in Telecommunication Networks

A ring has one main advantage compared to other network structures, namely that traffic can always be sent in two directions, clockwise or counter clockwise. This means that the traffic can be routed in two different directions, thereby fulfilling the two connectivity requirement. Common rings used are so-called Self Healing Rings (SHR) where some capacity is unused until the event of a failure. The technology is based on the Synchronous Digital Hierarchy (SDH) standard for optical transmission, (the standard for the US is SONET, Synchronous Optical Network), see Hamel et al. (1996).

An SDH ring is classified into two types, unidirectional or bidirectional. When using a bidirectional ring, traffic can be sent in any of the two directions, using two or four fibers depending on the capacity arrangement. In the case of four fibers, one of the fibers in each direction is used for sending the traffic and the two others are used only for protection, i.e. only used in case of a failure. In the case of two fibers, half of the fiber capacity is used for sending the traffic and the other half for protection. This means that, for each package, one needs to determine in which direction in the ring this package should be sent. The goals are to decrease the fiber capacity in the ring and to balance the load. One must ensure that the capacity is large enough, so that, in case of a failure, all traffic can be sent. In this way the ring is used efficiently, but a routing problem, often quite difficult, must be solved for each ring. Research on routing on a single bidirectional ring can be found in Myung et al. (1997), Lee and Chang (1997), Dell'Amico et al. (1999), and Shepherd and Zhang (2001).

A unidirectional ring uses two fibers, one fiber to send all traffic in one direction and the other fiber, in the opposite direction, only for protection. If a failure occurs, the affected traffic will simply be sent on the other fiber in the opposite direction. The capacity required on each of the fibers is therefore equal to the total sum of the demand routed on the ring. This is called dedicated protection. It is more expensive due to higher capacity on most links in the ring, but this solution is simpler to administrate. Especially, there is no need to specify the routing in the ring.

When using a ring structure in a network, a number of different rings are connected in a network. Often, the physical underlying ducts without the rings are given. Demand is specified for pairs of nodes. Each ring starts in one node, passes a number of nodes, some or all of which are connected to the ring, before returning to the starting node. The main questions are in which ducts the rings should be placed, which nodes should be connected to which rings and how the traffic should be sent on these rings in order to minimize the total cost. The total cost includes a cost depending on the total ring length, the cost for routing, and the cost for installing devices in the connected nodes. Ring networks are basically used at the national and the city level. In the future, when fibers are even more common, they will probably be used all the way down to subscriber level. Rings connected to the same node can transfer traffic from one ring to another. However, in order to facilitate the administration of the network, it is often only feasible to change traffic from one ring to another using a few special nodes, called *transit* nodes.

Let us mainly study the case where the transit nodes are specified in advance. The number of transit nodes could vary, from one or two in local networks, to the case where all nodes may function as transit nodes. The latter is interesting when the ring network resembles a grid with many relatively small rings are connected "side by side" in a large number of nodes, see, for example, Stidsen (2002).

In an interesting real life case we have been working on, there are two specific transit nodes, so we will later pay special attention to this case. If a network with two transit nodes shall be connected, one must require that at least one of the rings passes both transit nodes, and that all rings must be connected to at least one of the transit nodes. In a ring network, there are always at least two link disjoint paths between every pair of nodes (if the network is connected). This means that the two connectivity requirement concerning the links is fulfilled. If there is a failure in a link, all traffic can still be sent, using the other path.

If there is a failure in any of the nodes (except the transit nodes), all traffic, apart from traffic with origin or demand in the failed node, can be sent in the same way as above. However, if there is a failure in either of the transit nodes, all traffic using this transit node can not be sent. If we add a two connectivity requirement concerning the nodes, i.e. require two node disjoint paths between every pair of nodes in the network, the traffic can be sent even if there is a failure in one of the transit nodes. This requirement is easy to fulfill in the case of two transit nodes, by simply requiring all rings to be connected to both transit nodes. Then, in case of a failure in one of the transit nodes, it is possible to use the other transit node. Of course, this requirement increases the cost of the network, but the survivability and redundancy increases.

The device used to connect a node to a ring is an OADM (Optical Add Drop Multiplexor), concentrating the demand into a larger flow, and adds it to the ring. It allows the main traffic of the ring to pass the node, and a part of the traffic to be added (or dropped) from the ring to the node.

The traffic is defined as a number of packages that should be sent through the network. We define a commodity as a certain number of packages with the same origin (sender) and destination (receiver). It is not feasible to split the demand, so a commodity must use the same path throughout the complete network.

12.2 A SURVEY OF RING NETWORK DESIGN

Design of ring networks has two major applications in communication networks. The first application is local area networks (LANs). Some LANs, such as token ring networks, use a ring as the basic topology. A ring can accommodate a limited number of computers. The LANs can be interconnected using bridge equipments.

A particular ring design problem in LANs concerns the interconnection of a number of token rings using bridges, with the objective of minimizing the total bridge cost. The problem can be viewed as finding a spanning subgraph in a complete graph, where the nodes are the LANs, and the edges are possible locations of bridges. This problem is studied in LeBlanc et al. (1996), in which several heuristics are presented. Later on, Sridhar et al. (2000) proposed and evaluated a cutting plane method and a Benders decomposition scheme for the problem. A related problem, which amounts to finding parameters for the bridges, such that the resulting topology spans over a set of LANs, was studied in Kaefer and Park (1998).

Altinkemer (1994) and Altinkemer and Kim (1995) presented two heuristics for solving a ring design problem in computer LANs. A set of nodes are to be connected using several rings. Each ring contains a bridge. The bridges are connected by a backbone ring. As the costs are associated with network links, the problem is similar to the multiple-depot vehicle routing problem. The bridge nodes in this case are, in fact, very similar to the transit nodes in our problem.

The second, and more important application of ring network design arises in optical fiber systems, which play a central role in today's telecommunication infrastructure. In particular, standardization of telecommunication systems has resulted in the SDH/SONET technology, which provides a uniform platform for building and interconnecting optical networks. For backbone networks, survivability is a vital performance measure. Ring networks are, by construction, (at least) two-connected. In addition, an SDH/SONET ring is self-healing, meaning that the process of network reconfiguration for rerouting traffic is automated. Self-healing SDH/SONET rings have therefore become an attractive candidate for organizing backbone networks.

A wide range of SDH/SONET ring design problems have been presented in the literature. A previous survey of these problems can be found in Soriano et al. (1999). Planning SDH/SONET ring networks is very complex in nature, and involves a number of decisions. Typically, design of SDH/SONET rings needs to consider the following issues.

- *Logical ring design* deals with the number of rings as well as the topology of each ring.

- *Physical ring design* translates the logical rings to optical fiber connections at the physical layer.

- *Capacity assignment* designs the capacity of the rings in order to serve the traffic demands.

- *Equipment location* concerns the location and capacity dimensioning of optical networking devices, such as add-drop multiplexers (ADMs).

- *Routing of traffic* deals with choosing the paths for traffic between the demand nodes.

Several cost components are relevant in ring design. The first component is the design cost, which is related to the links that are used to form rings. This cost often depends on the length of the ring, and, sometimes, also depends on the ring capacity. A second cost component, the equipment cost, stems from the networking devices that are used to connect demand nodes to rings, and to interconnect the rings. This cost may include a fixed part (for installing an equipment), and a variable part which depends on the traffic rate supported by the equipment. Thirdly, the routing cost is a variable cost that depends on the amount of traffic in rings and between the rings.

Some previous work in the area of SDH/SONET ring design are discussed below. Each of the references that we examine addresses a particular SDH/SONET ring design problem, involving some of the design issues and cost components that we have discussed.

An early reference on SONET ring network design is due to Wasem (1991). The author addressed the problem of finding a ring in an incomplete graph, such that a subset of the nodes are covered by the ring. The author proposed a construction heuristic for solving the problem. In addition to topology design, Wasem et al. (1994) presented a software that considers capacity dimensioning of SONET rings for multiple time periods. Much of the research effort devoted to software for SONET ring design is summarized in Cosares et al. (1995), in which a decision support system, SONET toolkit, is presented. The overall design problem is decomposed into a sequence of smaller problems, which are then solved using heuristics. The authors reported applications of the software to real planning scenarios, for which cost savings of up to 30% were obtained. The approach of decomposing SONET ring planning into several design steps is also used for developing another planning tool PHANET (Semal and Wirl, 1994). A similar decomposition approach was applied to a SONET planning scenario for the São Paulo metropolitan area in Bortolon et al. (1996).

Laguna (1994) applied a tabu search algorithm to a logical ring design problem, in which a set of demand nodes are to be connected using multiple SONET rings. A demand node can be connected to several rings, each of such connections requires an ADM. The objective is to minimize the sum of the equipment costs and the routing costs (the latter is due to inter-ring traffic). A similar ring design problem is solved using a genetic algorithm in White et al. (1999). In the problem studied in Lee et al. (2000), inter-ring traffic is not allowed (i.e., two nodes with demand between them must belong to the same ring). The authors presented a branch-and-cut technique for minimizing the total equipment cost.

The optimization model presented in Fortz et al. (2003) deals with not only the logical design of rings, but also the mapping between the logical and physical topologies. In addition, a ring must have at least two interconnection nodes to other rings (dual homing). The authors presented a tabu search algorithm for solving the problem. Lee and Koh (1997) also used tabu search for solving a ring design problem with dual homing requirement. The problem consists of finding a set of logical rings at minimum design cost, such that each ring contains two nodes that are connected to a given backbone ring. Another work that addresses a similar problem was presented in Proestaki and Sinclair (2000), in which the authors applied a multi-phase heuristic for minimizing the ring design cost and the inter-ring traffic.

In the problem addressed in Gawande et al. (1999), the access rings are connected using a backbone ring, in such a way that every access ring contains one node of the backbone ring (i.e., single homing). In addition, some nodes can be connected to the rings using a hub-and-spoke structure. The authors described a multi-phase heuristic for this problem. A similar problem was solved using heuristics in Shi and Fonseka (1994). In the problem studied by Rossi and Arbib (2000), the backbone system is not explicitly modeled. The design problem involves partitioning the demand nodes into disjoint sets (each forms an access ring), with the objective of minimizing the equipment costs for supporting intra-ring as well as inter-ring traffic.

One possible structure of an SDH/SONET network is multiple rings connected to a central transit node (a hub). In this structure, every ring passes the hub node. Chang and Chang (2000) presented a ring design problem with this network structure, and with the objective of minimizing the total cost for topology design and capacity dimensioning. The solution method is composed by a Lagrangean relaxation and a branch-and-bound scheme. Klincewicz et al. (1998) tackled a similar problem using greedy heuristics. A variant of this type of ring design problem, in which the rings are connected using one or two hub nodes, was solved using a column generation technique in Kang et al. (2000). The problem amounts to finding a number of rings to cover demand nodes with minimum design cost (which consists of a fixed charge per ring).

In another problem setting of SONET ring design, a set of demand nodes are to be connected to switching centers, which are attached to a backbone ring. A problem of this category was solved using a greedy heuristic in Chamberland et al. (1997). Later, Chamberland and Sansó (2000) presented a tabu search algorithm for solving en extension of the problem where the backbone consists of multiple rings. In a variant of the problem, the network has three levels of hierarchy (customer nodes, end office nodes, and hub nodes). A ring is used to connect the (hub) nodes at the highest level of the hierarchy. This problem was solved using a tabu search algorithm in Xu et al. (1999).

The network structure of a ring connecting switching nodes can also be used in cellular systems. In this case, the demand nodes (cells) are connected to hubs, which are connected using a backbone ring. The design issue is, however, not the ring topology, but the assignment of the cells to the hubs (Dutta and Kubat, 1999), and the assignment of the hubs to the ring (Cox, Jr. et al., 1996).

Some authors have focused on scenario analysis rather than developing solution algorithms for SONET ring design. These analyzes are intended to clarify various design objectives and trade-offs. Examples of work in this direction can be found in Flanagan (1990), Morley and Grover (1999), and Sutter and Fullsack (1994).

A relatively simple version of SONET ring design is to connect a set of demand nodes to a predefined set of rings using ADMs. The objective is to minimize the total number of ADMs subject to ring capacity constraints. This problem was studied in Sutter et al. (1998), in which solution methods based on linear and integer column generation are developed.

Kennington et al. (1999) addressed a design problem which involves generating multiple rings to cover a given set of nodes. A node must be included in at least one ring. The objective is to minimize the total ring cost. The problem was formulated using a set covering model, and solved using heuristics and a commercial optimization software. A similar problem was solved using a greedy heuristic in Slevinsky et al. (1993).

Instead of minimizing the total cost, an alternative objective function is to maximize the profit generated at the nodes that are covered by rings, minus the ring design cost. This trade-off was studied in Gendreau et al. (1995) and Fink et al. (1998).

Fortz and Labbé (2002) and Fortz et al. (2000) studied a very special optimization problem related to ring design. The problem can be described as finding a two-connected network in a graph, such that every edge in the network belongs to a ring with a limited length. The restriction in length is intended to limit the impact of rerouting in case of an edge failure. The authors presented solution methods based on heuristics and branch-and-cut.

Several optimization problems related to routing in a single ring have been studied in the literature. In the ring loading problem, for example, the objective is to minimize the maximum load of the links in the ring. A second example, which arises in optical rings using wavelength division multiplexing (WDM), is the problem of finding a minimum number of wavelengths needed to support any traffic matrix from an admissible set. As the focus of this chapter is on the design of ring networks, we will not discuss these problems in any detail, except referring to Chen and Modiano (2003), Dell'Amico et al. (1999), Karunanithi and Carpenter (1997), Lee and Chang (1997), Myung et al. (1997), Narula-Tam et al. (2002), Shepherd and Zhang (2001), Vachani et al. (1996), Wan and Yang (2000), and the references therein.

12.3 RING NETWORK DESIGN BY GENERATING RINGS

We now focus on ring networks where unidirectional rings with dedicated protection are used. The aim is to find a set of possible rings, and then use an optimization model in order to find out which of these rings to use and on which rings to route each commodity.

The number of possible rings is enormous for larger networks, up to a million possible rings for networks with 50-100 nodes. Even if there is a limit on the maximum number of nodes connected to a ring, the number of possible rings will still be large. If the network is sparse, which is often the case for telecommunication networks, it is often necessary for a ring to pass many nodes which are not connected to the ring, in order to reach certain nodes.

The main idea of this solution approach is to iteratively *generate* the rings needed, since it is impossible to evaluate all possible rings a priori.

A starting set of rings is generated, and an optimization model, the RING NETWORK DESIGN PROBLEM, RNDP, is used to optimally send the demand only using these rings. Then new rings are generated using more sophisticated optimization techniques. A reward is calculated for each node, based on the previous solution, and these rewards are then used in order to generate a new ring, which is beneficial to include in the set of possible rings.

The problem of generating a new ring is called the RING GENERATION PROBLEM, RGP. When collecting a reward, by including a node, a number of links need to be used, which generates a cost. The aim of the ring generation problem is to find a ring which maximizes the collected rewards and minimizes the cost of using the links. The rings may also have some restrictions, such as a maximal number of connected nodes to a ring.

This specific ring network design problem is quite new, see Henningsson (2003). Note that the rings are not given, but must be designed by the model. Also note that in most cases, a solution will not consist of only one ring, but of several connected rings.

The cost for a ring may depend on its length and capacity. The cost for the OADM (Optical Add-Drop Multiplexer) device occurs whenever a ring is connected to a node. In the most general case the cost structure for a ring is staircase formed, with a fixed cost for each fiber, i.e. each possible ring capacity. However, with a fast growing telecommunication market, the input data, in this case the costs, capacities and demand, are very uncertain. Therefore, it is probably better to have a cost structure with one large fixed cost for the first installation of a ring, but then approximate the steps corresponding to the different numbers of fibers by a linear cost. So our chosen model includes a fixed cost occurring if a ring is used, and a linear cost on the traffic on the ring. Furthermore, if the capacity of a ring is exceeded, one may simply add a new, identical ring, on top of the first one. This means that there is no capacity limit for a ring. The linear cost approximation will probably give a more stable solution (network topology) as the total traffic increases.

The general solution method used is the following (also described in Henningsson and Holmberg (2003b)). It starts with generating a small set of rings using a shortest path based heuristic. Given this set of rings, the LP-relaxation of a linear integer programming problem is solved. The solution to this problem chooses which rings to use and allocates the traffic onto the rings. In the next step the aim is to generate more rings using an approach based on Lagrangean relaxation. A reward connected to each node can be calculated using the dual variables from the Lagrangean relaxation. The details of these calculations are found in Henningsson and Holmberg (2003a).

The rewards calculated can be used in order to generate new rings with the help of the ring generation problem, which is a modification of the traveling salesman subtour problem with an additional constraint, TSSP+1. This problem is solved heuristically and the newly generated ring is added to the linear programming problem, which is reoptimized. The new ring contributes with both new columns and new rows.

At the end of the procedure an integer solution is identified. Let us now describe the problem and the method in more detail.

12.4 MATHEMATICAL FORMULATION

There is a given network $G = (N, A)$ with a set of nodes N and a set of undirected links A. Link $(i, j) \in A$ has length l_{ij}. There is a set of commodities C and each commodity, $k \in C$, has an origin $o(k)$, a destination $d(k)$ and a demand b_k.

The set of transit nodes, T, is a smaller set of nodes, $T \subset N$. In practice, a large quantity of the total demand may have a transit node as one of its demand nodes. In our applications, the most important case is $|T| = 2$.

A ring is a sequence of links starting and ending at the same node with the restrictions that a ring must not use any link more than once, and that the number of nodes connected to a ring must not exceed K. The number of rings grows exponentially with the size of the network, even if the network is sparse as in many telecommunication networks. Given a small, sparse network with 10 nodes, our test example has 400 rings, but when the number of nodes in the network increases to 15, the number of rings in our example is 19 000, and for networks with 50 nodes, the number of rings is over 10 million, see Henningsson et al. (2003a).

The topology of a ring is defined by the following coefficients:

$a_{ir} = 1$, if ring r passes node i, 0 otherwise,

$e_{ijr} = 1$, if link (i, j) is included in ring r, 0 otherwise.

Note that $a_{ir} = 1$ indicates that it is possible to connect node i to ring r, but not that the decision of whether or not to do this has been taken.

Each ring has a fixed cost, consisting of the sum of costs for the links included in the ring. The total fixed cost for a ring is thus

$$f_r = \sum_{(i,j) \in A} f^R_{ij} e_{ijr}.$$

The linear capacity costs also depend on the length of the ring as

$$c_r = \sum_{(i,j) \in A} c^R_{ij} e_{ijr}.$$

Note that the capacity of a ring is not limited, since a ring in our model could consist of any number of identical rings placed on top of each other.

In addition to the ring costs, there are fixed costs for the device (OADM) needed to connect a certain node to a certain ring. Several commodities can use the same device, if they have the same origin or destination. So one such cost occurs each time a certain ring is connected to a certain node (including the transit nodes). Finally there are also costs for all traffic leaving/entering a ring, which includes transit through a transit node. These costs are linear. Note, however, that since the total demand is assumed to be known, the total inflow/outflow at demand nodes is constant, and thus the linear cost for this is constant. This leaves only the transit flow as a variable factor for these costs.

12.4.1 A Ring Network Design Model

In Henningsson et al. (2002) and Henningsson et al. (2003a) an optimization model for deciding the best usage of a number of given rings is presented. We will here describe the Ring Network Design Problem (RNDP), but without giving all details. The model is based on a set of given rings, R, and uses the following main variables:

$y_r = 1$, if ring r is used, 0 otherwise,
$x_{rk} = 1$, if commodity k uses ring r, 0 otherwise,
$h_{ir} = 1$, if node i is connected to ring r, 0 otherwise.

Actually the x-variables are more detailed in order to distinguish between the different ways of using a ring (connecting only the origin, only the destination, both of them or none of them but using the ring for transit between two other rings). Furthermore there are variables for the amounts of flow through the transit nodes. However, the main variables defined above are sufficient for our level of detail.

The objective function minimizes the length of the rings, the linear cost for using a ring, the cost for installing an OADM device in the node, and the cost for changing between rings in the transit nodes.

There are many groups of constraints. The first of them ensures that each commodity is picked up once, and the second that each commodity is delivered once. Then there are constraints controlling the transition between rings, so that all commodities that are not picked up and delivered by the same ring passes appropriate transit nodes. (These constraints have the structure of node equilibrium constraints for a transshipment problem.)

Then there is a set of constraints effectively saying $x_{rk} \leq y_r$, which makes sure that unless the fixed cost is paid for a ring, it may not be used.

There are also constraint sets ensuring that a fixed cost for each OADM device is paid. There has to be an OADM device connecting the origin to a ring and one connecting the destination to a ring. Furthermore, if a commodity changes ring in one of the transit nodes, there has to be an OADM device in that transit node.

There is a constraint set limiting the number of nodes connected to a ring, and another enforcing that each ring should be connected to at least one transit node.

Finally there are constraints connecting the a-coefficients to the usage of the rings. It is only possible, for example, for ring r to pick up commodity k if ring r passes the origin, $o(k)$, i.e. $a_{o(k),r} = 1$ and ring r is paid for, i.e. $y_r = 1$.

The model has $8|R||C|+|N||R|+|R|$ variables and $9|R||C|+4|C|+|R|+|N||R|$ constraints for $|T|=2$. For example, for 50 nodes, 100 commodities and 1000 rings, there are 851000 variables and 951400 constraints. (However, please note that for that network size, the number of possible rings is much larger than 1000.)

If all rings are not known/included in model RNDP, one might consider the traditional column generation approach, where one is looking for "columns" not included, but with negative reduced cost. For this purpose it is important that the a-coefficients appear in the model as coefficients for y, since y represents the columns that we will generate. In other words, the items that can be generated are the rings.

Note that we are still planning to solve the model for fixed values of a and e. We then need a separate model for updating the values of a and e, with the help of dual information.

12.4.2 A Ring Generation Model

Assuming that all rings are known is not very realistic for large sized applications. Therefore one must be able to generate rings. Each ring should form a cycle, so if the nodes to be included were fixed, it would be a Traveling Salesman Problem (TSP). However, this is not the case. Instead there will be a reward for including a certain node, which yields a variant of the traveling salesman subtour problem, see Gensch (1978) and Mittenthal and Noon (1992). For further details about the traveling salesman subtour problem and related problems see Henningsson and Holmberg (2003c) and Westerlund (2002).

The variables in this problem are the coefficients a and e, in the previous section assumed to be fixed. Their definitions are repeated below for convenience. However, since we are generating one new ring, we do not need the ring index r. Telecommunication networks are usually sparse and since there is an additional constraint saying that only K nodes can be connected to a ring (using an OADM device), a variable z_i is needed to make sure that we only benefit from at most K rewards:

$$a_i = 1, \text{ if node } i \text{ is passed by the ring, 0 otherwise,}$$
$$z_i = 1, \text{ if node } i \text{ is connected to the ring, 0 otherwise,}$$
$$e_{ij} = 1, \text{ if link } (i,j) \text{ is included in the ring, 0 otherwise.}$$

The objective function consists of one part with costs for the links, $\hat{l}_{ij}$, depending on the length of the links, and one part with rewards for visiting nodes, $\hat{\pi}_i$ (< 0). We have the following model for the Ring Generation Problem, RGP.

$$\min \sum_{(i,j) \in A} \hat{l}_{ij} e_{ij} + \sum_{i \in N} \hat{\pi}_i z_i$$

$$\text{s.t.} \sum_{j:(i,j) \in A} e_{ij} + \sum_{j:(j,i) \in A} e_{ji} = 2 a_i \forall i \in N \tag{12.1}$$

$$\sum_{i,j \in S, \ (i,j) \in A} e_{ij} \leq \sum_{j \in S} a_j - a_k \forall S \subseteq N \setminus T, k \in S \tag{12.2}$$

$$z_i \leq a_i \forall i \in N \tag{12.3}$$

$$\sum_{i \in N} z_i \leq K \tag{12.4}$$

$$\sum_{t \in T} z_t \geq 1 \tag{12.5}$$

$$e_{ij} \in \{0,1\} \forall (i,j) \in A \tag{12.6}$$

$$a_i, z_i \in \{0,1\} \forall \ i \in N \tag{12.7}$$

Constraint set (12.1) ensures that if a node is passed by the ring, there must be two links connected to the node. Constraints set (12.2) is the subtour elimination constraints. The left-hand-side for constraint set (12.2) is the sum of the links in set S. The right-hand-side is the sum of the variables a, which is the same as the number of nodes passed by the ring minus one (k). There might be subsets S where the ring passes no one of the nodes. In these cases the a_k variable will also be zero, $k \in S$, and the constraint ensures that no links are used between nodes in that subset.

The subtour elimination constraints are constructed only for subtours which do not include any transit node, since a subtour including a transit node is feasible. The constraint set (12.3) ensures that we can only connect a node to a ring if the ring passes the node. The constraint set (12.4) makes sure that the number of nodes connected to the ring is less than K (the ring can still pass any number of nodes not connected). Finally, constraint (12.5) makes sure that at least one of the transit nodes is connected to the ring.

There are several possible solution approaches for this problem, see Section 12.5.2. If constraint (12.4) is relaxed, the problem becomes a variant of the prize collecting traveling salesman problem, see Dell'Amico et al. (1995). Alternatively one can use a heuristic for the traveling salesman subtour problem proposed by Mittenthal and Noon (1992). For further details about solving RGP, and how to generate rings connecting both transit nodes, see Henningsson and Holmberg (2003c).

12.5 RING GENERATION PROCEDURES

We will now discuss (heuristic) column generation approaches for the problem. However, we must first specify the information to be used for generating new columns. And what is a "column" that we may generate? How can we get reduced costs that tell us which column is best? One natural answer to the first question is that a column corresponds to a ring. However, it is also possible to define a column as a ring with certain functions, i.e. a certain ring with certain commodities connected in a certain way. Furthermore, there are two possible approaches for calculating reduced costs,

RING NETWORK DESIGN 303

based on LP-duality or Lagrangean duality. Finally, one might see a new ring as a modification of an existing ring, if this is useful.

A complication is the dependence between the different variables. If a certain ring is not used, there can be no connections between this ring and any node and there can be no transit to or from this ring. These variable dependencies are easier to handle in a Lagrangean approach, since there is a possibility of dualizing only some of the constraints, while other constraints may be kept.

Using Lagrangean relaxation on model RNDP, for details see Henningsson and Holmberg (2003a), we end up with the following reduced cost related to an existing ring r.

$$\bar{f}_r = \sum_{(i,j) \in A} \hat{l}_{ij} e_{ijr} + \sum_{i \in N} \hat{\pi}_{ir} a_{ir} + K_r$$

By *reduced cost*, we mean that we benefit from using the ring if $\bar{f}_r < 0$, and not if $\bar{f}_r \geq 0$. Here $\hat{l}_{ij} = f_{ij}^R + \theta_r c_{ij}^R$ and the parameter θ_r is the sum of all demand that will use the ring, i.e. an unknown parameter. One can in principle use a parametric programming approach centered around the total demand using ring r in the previous solution. $\hat{\pi}_{ir}$ is the reward for connecting node i to ring r. Finally, K_r is a constant, depending on the dual variables from the RNDP model.

Omitting the constants, we thus have an objective function for the ring generation problem, RGP, with one parameter, θ_r, for evaluating changes of ring r, obtained by searching for the most negative reduced cost.

The ring that seems best to include in the master problem is the one with the most negative reduced cost. One might, however, use any ring with negative reduced cost, which may shorten the time needed for finding such a ring.

Under certain circumstances, see Henningsson and Holmberg (2003b), the reduced costs will be independent of existing rings. (The $\hat{\pi}$-coefficients will not depend on r, and we can drop the index r from the expression above.) Furthermore, if the linear costs are zero, the parameter θ_r can be dropped from consideration. In this case, the column generation procedure is not a heuristic, but an exact method. It is no longer a matter of modifying an existing ring, but of generating a brand new one.

12.5.1 Solution Method

Solving the LP-relaxation of RNDP for most instances can be done with a general LP-code. For large instances other approaches are needed. An important output is the dual variables of some of the constraints. This can be exploited in a Lagrangean approach, where these constraints are relaxed. A subgradient optimization technique or the Danzig-Wolfe master problem can be used to find the best values of the Lagrange multipliers, which can then be used as dual solution for the column generation procedure.

We now present a solution method where we generate new rings based on the solution approach given in Henningsson and Holmberg (2003b). The calculation of the reduced costs for this specific ring network design problem is described in Henningsson and Holmberg (2003a). The method is as follows:

1. Generate a set of rings that enables a feasible solution to problem RNDP.

2. Solve the LP-relaxation of model RNDP, or solve the Lagrangean relaxation problem with a subgradient method, in order to get dual variable values.

3. Calculate $\hat{l}_{ij}$, $\hat{\pi}_{ir}$, based on the dual variable values.

4. Generate a new ring with heuristics.

5. If at least one new ring is found, add it and go to 2, otherwise go to 6.

6. Find a feasible integer solution, using the LP-solution and branch-and-bound.

12.5.2 Solving the Ring Generation Problem

The subproblem RGP, where the aim is to generate a new ring based on values from the dual variables, is a modification of a Traveling Salesman Subtour Problem with a additional constraint (TSSP+1), namely that at most K nodes may be connected. The TSSP+1 is known to be a difficult problem to solve to optimality, especially for larger instances, and there are even more difficulties in RGP, such as the estimation of the unknown θ_r parameter.

The first approach relaxes the constraint on the number of connected nodes. The problem is then a traveling salesman subtour problem, which can be reformulated to a standard traveling salesman problem with twice as many nodes, see Dell'Amico et al. (1995), and then be solved with a standard TSP code Carpaneto et al. (1995). We call this approach RTSP (reformulation to a TSP). Due to the relaxation of the constraint on the number of connected nodes, the solution may benefit from more than K connected nodes. This means that the suggested ring may not be as good as expected. However, if we only allow adding a small number of new nodes to the ring (for example 3), we will often fulfill the additional constraint on the number of connected nodes implicitly.

Another approach is to solve RGP heuristically by using a heuristic for the TSSP+1 problem. The heuristic is called TSSPBH (Traveling Salesman Subtour Problem Based Heuristic), and is described in detail in Henningsson and Holmberg (2003c). In short, the method includes three parts. First, if the network is sparse, the network is modified to a complete network by adding links. A new link represents a path passing other nodes not connected, with the cost equal to the cost of the total path. Next, the problem is solved using an implementation by Mittenthal and Noon (1992). Finally, the solution tour is converted into a ring in the original sparse network. There might be tours that can not be converted to rings, since two links in the solution that actually are paths in the original network, include the same node. This means that this node is included twice in the same ring, which is not feasible. This final step is solved heuristically by using shortest paths. The advantages with TSSPBH compared to RTSP is that TSSPBH is fast and that TSSPBH can benefit from a maximum of K rewards.

A third possibility is to modify a ring, i.e. increasing or decreasing the number of nodes in the ring, so that the modified ring gets a negative reduced cost. A modification of a ring is found by finding the shortest path from node i not connected to ring r with $\hat{\pi}_{ir} \ll 0$, to ring r and back again without using the same link or passing the same node twice. If $\bar{f}_r < 0$, include the new ring. One can also remove a node i with $\hat{\pi}_{ir}$ close to zero from ring r, and check $\bar{f}_r$. This method is called the Shortest Path Based Heuristic (SPBH).

12.5.3 Obtaining Integer Solutions

During the solution phase, RNDP often yield integer solutions (which are saved). However, there is no guarantee for this.

Theorem 12.1 *If the h variables all are fixed to integer values, then RNDP has an optimal solution where y and x are integer.*

The proof can be found in Henningsson et al. (2003b).

There are fewer y variables than h variables, and if the y variables are fixed to integers and x and h are allowed to be continuous, the solutions are often integral, so in order to find an integer solution fast, we let the CPLEX6.5 mixed integer programming solver first branch on the y variables, and then, if necessary, on the h variables.

The saved integer solutions are included as a feasible solution to speed up further the solution phase. Finding an integer solution for a given set of rings can often be made very fast.

In order to solve the problem more accurately, a branch-and-price algorithm may be used, i.e. a branch-and-bound scheme where new rings may be added in all nodes of the branch-and-bound tree. The branching should then be done on the h variables instead of the y variables, in order to enable the ring generation problem to obey the fixations.

If we branch on the y variables, a ring r that is forbidden ($y_r = 0$) can be generated again, which is not feasible. If we instead branch on the h variables, it is easy to forbid a specific node i from being included in the new ring r ($h_{ir} = 0$) or force a specific node to be included in the new ring ($h_{ir} = 1$).

12.6 COMPUTATIONAL TESTS

Computational tests have been done on 13 problems based on 7 different networks (one with real data and 6 randomly generated). The numbers of nodes lie between 10 and 100, the numbers of links lie between 42 and 534 and the numbers of commodities lie between 20 and 150. The number of transit nodes is 2.

The results first indicate that SPBH is too weak. It only improves the solution in 8 of the 13 cases, and the improvement is 2% or less. Using the other two heuristics for solving the RGP yields improvements from 4% up to approximately 25%. For most cases, the results are better using RTSP than TSSPBH. However, for the largest networks and the real data network, TSSPBH is approximately more than 8 times faster, and the solutions obtained are better or slightly worse than using RTSP. The solution time when using the TSSPBH is less than 3 minutes, while the solution time for RTSP is between 10 and 23 minutes for the networks with at least 50 nodes. In 40% of the cases, a better solution is found using both RTSP and TSSPBH together. However, the solutions are only in two cases significantly better than using the RTSP and TSSPBH separately, and the solution time increases by an average of 45% compared to the already long solution time when using RTSP. The solution time for the larger networks is from 15 minutes up to more than half an hour. However, given the importance of the underlying decisions, these solution times are probably acceptable.

Bibliography

K. Altinkemer. Topological design of ring networks. *Computers and Operations Research*, 21:421–431, 1994.

K. Altinkemer and B. Kim. Heuristics for ring network design. In *Third International Conference of Telecommunication Systems, Modelling and Analysis*, pages 429–442, 1995.

S. Bortolon, H.M.F. Tavares, R.V. Ribeiro, E. Quaglia, and M.A. Bergamaschi. A methodology to SDH networks design using optimization tools. In *Proceedings of IEEE GLOBECOM 1996*, pages 1867–1871, 1996.

G. Carpaneto, M. Dell'Amico, and P. Toth. Algorithm 750: CDT: A subroutine for the exact solution of large-scale, asymmetric traveling salesman problems. *ACM. Transactions on Mathematical Software*, 21(4):410–415, 1995.

S. Chamberland and B. Sansó. Topological expansion of multiple-ring metropolitan area networks. *Networks*, 36:210–224, 2000.

S. Chamberland, B. Sansó, and O. Marcotte. Heuristics for ring network design when several types of switches are available. In *Proceedings of IEEE International Conference on Communications 1997*, pages 570–574, 1997.

M.R. Chang and S.Y. Chang. A heuristic method for self-healing ring design in a single-homing cluster. *Telecommunication Systems*, 14:175–195, 2000.

L.-W. Chen and E. Modiano. Efficient routing and wavelength assignment for reconfigurable WDM networks with wavelength converters. In *Proceedings of IEEE INFOCOM 2003*, pages 1785–1794, 2003.

S. Cosares, D.N. Deutsch, I. Saniee, and O.J. Wasem. SONET toolkit: A decision support system for designing robust and cost-effective fiber-optic networks. *Interfaces*, 25:20–40, 1995.

L.A. Cox, Jr., L. Davis, L.L. Lu, D. Orvosh, X. Sun, and D. Sirovica. Reducing costs of backhaul networks for PCS networks using genetic algorithms. *Journal of Heuristics*, 2:201–216, 1996.

M. Dell'Amico, M. Labbé, and F. Maffioli. Exact solution of the SONET ring loading problem. *Operations Research Letters*, 25:119–129, 1999.

M. Dell'Amico, F. Maffioli, and P. Värbrand. On prize-collecting tours and the asymmetric traveling salesman problem. *International Transactions on Operational Research*, 81:289–305, 1995.

A. Dutta and P. Kubat. Design of partially survivable networks for cellular telecommunication systems. *European Journal of Operational Research*, 118:52–64, 1999.

A. Fink, G. Schneidereit, and S. Voss. Ring network design for metropolitan area networks. Technical report, Technische Universität Braunschweig, Germany, 1998.

T. Flanagan. Fiber network survivability. *IEEE Communications Magazine*, pages 46–53, June 1990.

B. Fortz and M. Labbé. Polyhedral results for two-connected networks with bounded rings. *Mathematical Programming*, 93:27–54, 2002.

B. Fortz, M. Labbé, and F. Maffioli. Solving the two-connected network with bounded meshes problem. *Operations Research*, 48(6):866–877, 2000.

B. Fortz, P. Soriano, and C. Wynants. A tabu search algorithm for self-healing ring network design. *European Journal of Operational Research*, 151:280–295, 2003.

M. Gawande, J.G. Klincewicz, and H. Luss. Design of SONET ring networks for local access. *Advances in Performance Analysis*, 2:159–173, 1999.

M. Gendreau, M. Labbé, and G. Laporte. Efficient heuristics for the design of ring networks. *Telecommunication Systems*, 4:177–188, 1995.

D. Gensch. An industrial application of the Traveling Salesman Subtour Problem. *AIIE Transactions*, 10(4):362–370, 1978.

A. Hamel, V. Tholey, A. Sutter, L. Blain, and F. Chatter. Increased capacity in an MS protection ring using WDM technique and OADM: the 'coloured section' ring. *Electronics Letters*, 32(3):234–235, 1996.

M. Henningsson. *Ring Network Design in Telecommunications: Optimization Based Solution Approaches*. Phd dissertation, Linköping University, Sweden, 2003. Linköping Studies in Science and Technology. Dissertation no. 829.

M. Henningsson and K. Holmberg. Calculating cost coefficients for generation of rings in network design. Research Report LiTH-MAT-R-2003-18, Department of Mathematics, Linköping Institute of Technology, Sweden, 2003a.

M. Henningsson and K. Holmberg. Lagrangean price directive ring generation for network design. Research Report LiTH-MAT-R-2003-17, Department of Mathematics, Linköping Institute of Technology, Sweden, 2003b.

M. Henningsson and K. Holmberg. A ring generation problem based on the traveling salesman sub tour problem. Research Report LiTH-MAT-R-2003-19, Department of Mathematics, Linköping Institute of Technology, Sweden, 2003c.

M. Henningsson, K. Holmberg, M. Rönnqvist, and P. Värbrand. Ring network design by Lagrangean based column generation. *Telecommunication Systems*, 21(2-4): 301–318, 2002.

M. Henningsson, K. Holmberg, M. Rönnqvist, and P. Värbrand. A ring network design problem and heuristics for generating a set of feasible rings. Research Report LiTH-MAT-R-2003-16, Department of Mathematics, Linköping Institute of Technology, Sweden, 2003a.

M. Henningsson, K. Holmberg, M. Rönnqvist, and P. Värbrand. A ring network design problem solved by a ring generation and allocation approach. Research Report LiTH-MAT-R-2003-20, Department of Mathematics, Linköping Institute of Technology, Sweden, 2003b.

F. Kaefer and J. S. Park. Interconnecting LANs and a FDDI backbone using transparent bridges: A model and solution algorithms. *INFORMS Journal on Computing*, 10:25–39, 1998.

D. Kang, K. Lee, S. Park, K. Park, and S.-B. Kim. Design of local networks using USHRs. *Telecommunication Systems*, 14:197–217, 2000.

N. Karunanithi and T. Carpenter. SONET ring sizing with genetic algorithms. *Computers and Operations Research*, 24:581–591, 1997.

J.L. Kennington, V.S.S. Nair, and M.H. Rahman. Optimization based algorithms for finding minimal cost ring covers in survivable networks. *Computational Optimization and Applications*, 14:219–230, 1999.

J.G. Klincewicz, H. Luss, and D.C.K. Yan. Designing tributary networks with multiple ring families. *Computers and Operations Research*, 25:1145–1157, 1998.

M. Laguna. Clustering for the design of SONET rings in interoffice telecommunications. *Management Science*, 40:1533–1541, 1994.

L.J. LeBlanc, J.S. Park, V. Sridhar, and J. Kalvenes. Topology design and bridge capacity assignment for interconnecting token ring LANs: A simulated annealing approach. *Telecommunication Systems*, 6:21–43, 1996.

C.Y. Lee and S.G. Chang. Balancing loads on SONET rings with integer demand splitting. *Computers and Operations Research*, 24:221–229, 1997.

C.Y. Lee and S.J. Koh. A design of the minimum cost ring-chain network with dual-homing survivability: A tabu search approach. *Computers and Operations Research*, 24:883–897, 1997.

Y. Lee, H.D. Sherali, J. Han, and S. Kim. A branch-and-cut algorithm for solving an intraring synchronous optical network design problem. *Networks*, 35:223–232, 2000.

J. Mittenthal and C.E. Noon. An insert/delete heuristic for the traveling salesman subset-tour problem with one additional constraint. *Journal of the Operational Research Society*, 43(3):277–283, 1992.

G.D. Morley and W.D. Grover. Current approaches in the design of ring-based optical networks. In *Proceedings of IEEE Canadian Conference on Electrical and Computer Engineering*, pages 220–225, 1999.

Y.-S. Myung, H.-G. Kim, and D.-W. Tcha. Optimal load balancing on SONET bidirectional rings. *Operations Research*, 45(1):148–152, 1997.

A. Narula-Tam, P.J. Lin, and E. Modiano. Efficient routing and wavelength assignment for reconfigurable WDM networks. *IEEE Journal on Selected Areas in Communications*, 20:75–88, 2002.

A. Proestaki and M.C. Sinclair. Design and dimensioning of dual-homing hierarchical multi-ring networks. *IEE Proceedings – Communications*, 147:96–104, 2000.

F. Rossi and C. Arbib. An optimization problem arising in the design of multiring systems. *European Journal of Operational Research*, 124:63–76, 2000.

P. Semal and K. Wirl. Optimal clustering and ring creation in the network planning system PHANET. In *Proceedings of the 6th International Network Planning Symposium - Networks '94*, pages 303–308, 1994.

B. Shepherd and L. Zhang. A cycle augmentation algorithm for minimum cost multicommodity flows on a ring. *Discrete Applied Mathematics*, 110:301–315, 2001.

J. Shi and J.P. Fonseka. Design of hierarchical self-healing ring networks. In *Proceedings of IEEE International Conference on Communications 1994*, volume 1, pages 478–482, 1994.

J.B. Slevinsky, W.D. Grover, and M.H. MacGregor. An algorithm for survivable network design employing multiple self-healing rings. In *Proceedings of IEEE GLOBECOM 1993*, pages 1568–1573, 1993.

P. Soriano, C. Wynants, R. Séguin, M. Labbé, M. Gendreau, and B. Fortz. Design and dimensioning of survivable SDH/SONET networks. In B. Sansó and P. Soriano, editors, *Telecommunications Network Planning*, pages 147–167. Kluwer Academic Publishers, 1999.

V. Sridhar, J.S. Park, and B. Gavish. LP-based heuristic algorithms for for interconnecting token rings via source routing bridges. *Journal of Heuristics*, 6:149–166, 2000.

T. Stidsen. *Optimisation problems in optical network design*. PhD thesis, Institute of Mathematical Modelling, Technical University of Denmark, 2002.

A. Sutter and J.L. Fullsack. SDH network planning in a changing environment. In *Proceedings of the 6th International Network Planning Symposium – Networks '94*, pages 149–154, 1994.

A. Sutter, F. Vanderbeck, and L. Wolsey. Optimal placement of add/drop multiplexers: Heuristic and exact algorithms. *Operations Research*, 46:719–728, 1998.

R. Vachani, A. Shulman, P. Kubat, and J. Ward. Multicommodity flows in ring networks. *INFORMS Journal on Computing*, 8:235–242, 1996.

P.-J. Wan and Y. Yang. Load-balanced routing in counter rotated SONET rings. *Networks*, 35:279–286, 2000.

O. J. Wasem. An algorithm for designing rings for survivable fiber networks. *IEEE Transactions on Reliability*, 40:428–432, 1991.

O.J. Wasem, T.-H. Wu, and R.H. Cardwell. Survivable SONET networks - design methodology. *IEEE Journal on Selected Areas in Communications*, 12:205–212, 1994.

A. Westerlund. Decomposition schemes for the traveling salesman subtour problem. Linköping Studies in Science and Technology. Theses No. 939, 2002. Licentiate thesis, LiU-TEK-LIC-2002:12.

A. R. P. White, J. W. Mann, and G. D. Smith. Genetic algorithms and network ring design. *Annals of Operations Research*, 86:347–371, 1999.

J. Xu, S.Y. Chiu, and F. Glover. Optimizing a ring-based private line telecommunication network using tabu search. *Management Science*, 45:330–345, 1999.

13 TELECOMMUNICATIONS ACCESS NETWORK DESIGN

Tamra Carpenter[1] and Hanan Luss[1]

[1] Telcordia Technologies
Piscataway, New Jersey 08854 USA
tcar@research.telcordia.com
hluss@telcordia.com

Abstract: A typical telecommunications network consists of a backbone network and multiple access networks. The investment in expanding and modernizing the access portion of the network is a significant part of the total. This chapter concentrates on describing representative models and solution approaches that are often found in access network design. Section 13.2 presents variations of concentrator location problems that play a major part in access network design. Section 13.3 focuses on current broadband access networks that deliver information at high speed, such as access networks that employ Digital Subscriber Line (DSL) and cable TV technologies. Finally, Section 13.4 describes the design of survivable access networks; in particular, access networks with dual homing and access networks that employ ring topologies.

Keywords: Access networks, network design, routing, facility location, concentrator location, broadband services, survivable networks.

13.1 PERSPECTIVE

This section presents a brief overview of the overall telecommunications network design problem and its relevance to other application areas, specific issues encountered in access networks, and models and solution approaches.

13.1.1 Telecommunications network design

Telecommunications networks move information among constituent locations. The information carried may include different types of traffic arising from voice, data and video services. At the center of a network is the backbone or core network. It consists of edge nodes, also called hubs or gateways, internal backbone nodes, and links that interconnect these nodes. Each gateway serves as an entry and exit point for an access

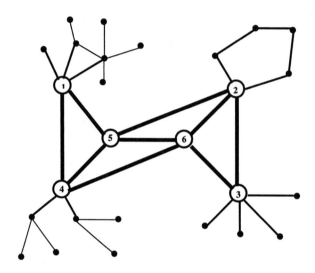

Figure 13.1 Illustration of a backbone network interconnecting access networks

network. An access network connects multiple demand nodes to the edge node that serves as its gateway. Each demand node represents a location where demand for information originates and/or terminates. Hence, the demand originating at some node of an access network is routed to its corresponding gateway. From there, it is routed, if needed, through the backbone network to a gateway of a different access network. Finally, it is routed through the latter access network to the destination.

Figure 13.1 illustrates a simple network. The backbone network has six nodes, where nodes 1–4 serve as gateways to access networks. Note that the access networks may have different topologies, including star networks, trees and rings. The backbone network links have a large capacity, capable of carrying a large volume of traffic at a high speed. In contrast, access network links are of lower capacity, moving traffic at a lower speed. Typically, the bottlenecks in a telecommunications network occur in the access portion of the network. The backbone network facilities route large traffic volumes among all pairs of edge nodes, and thus take advantage of economies of scale. The traffic originated at each demand node and routed from there to the gateway is relatively small, which limits the economy-of-scale savings in the access network. As a result, the investment in expanding and modernizing access networks is a significant percentage of the total investment in telecommunications networks.

The discussion above applies more generally to other application areas. Consider, for example, a transportation network that connects multiple large cities. At the heart of the network is a system of multi-lane highways. This is the backbone highway network. However, in order to reach a city center, we often have to exit the highway at the city outskirts and continue on an access network that consists of narrow roads with high congestion. These access roads are often the bottleneck of the entire transportation system. Similarly, consider air transportation. Often, the flying time is less than the time we spend on traveling to and from airports. The latter can be viewed as time spent on accessing the *backbone*. Furthermore, the domestic networks of most U.S. airlines have hub-and-spoke topologies. Typically, such a network consists of a few hub airports, where each of the other airports is served by a single hub. The network that interconnects the hubs is the airline's backbone network, whereas all other locations assigned to a hub constitute an access network. Taking advantage of the large volume of people that are transported among the hubs, the airplanes that transport these passengers are significantly larger than those serving the access networks. Another example is the shipment of products. The backbone logistics system ships large quantities from manufacturing facilities to major warehouse locations. These shipments can be done efficiently, using economies of scale. However, eventually, the products must be shipped in small quantities from regional distribution centers to individual demand locations.

A specific telecommunications network design considers a variety of issues, including topological constraints, existing embedded facilities, equipment and transport costs, capacity constraints, restoration requirements, and performance requirements. The resulting optimization problem is, in general, very difficult. A common approach, especially for large problems, has been to decompose the problem into more manageable sub-problems, which typically include:

- Finding a near-optimal set of nodes that serve as gateways for the backbone network and assigning every demand node to a gateway;
- Designing a near-optimal backbone network that routes traffic among the gateway nodes;
- Designing for each gateway node a near-optimal access network that routes traffic between the assigned demand nodes and the gateway.

The resulting network design problems, whether for backbone or access networks, may again contain tasks that include:

- Designing a near-optimal network infrastructure, which includes deciding the locations and sizes of links and node equipment;
- Determining optimal routes for given demands from sources to destinations.

Each of these two modules must satisfy a variety of constraints that are problem specific. Furthermore, the solutions of the two modules are not typically independent. Thus, an excellent infrastructure design based upon poor routing choices may result in a poor network design solution. Moreover, an excellent design for an optimized routing may still yield poor performance for different routing choices. Thus, it is sometimes preferable for the routing module to embed assumptions that mimic those of realistic protocols and practices.

Variations of decomposition schemes have been proposed in the literature. Klincewicz (1998) presents an excellent review of hub location problems in networks that consist of a backbone network that interconnects multiple access networks as described above. Several authors have also addressed network design problems using more integrated approaches. Representative publications include Gavish (1992); Bienstock and Saniee (2001); Rosenberg (2001).

13.1.2 Access network design

As shown in Figure 13.1, access networks may have different topologies with different degrees of complexity. Whereas backbone networks are designed to route traffic among numerous pairs of nodes, access networks are designed to route traffic from many demand nodes to a single node (the gateway) and from a single node to many demand nodes. Balakrishnan et al. (1991) and Gavish (1991) present excellent surveys on local access network design models and solution methods. Access networks may also serve internal demands that do not reach the gateway node. In this chapter, we do not address such intra-access network demand.

Consider the routing problem in a backbone network with many point-to-point demands that need to be routed over existing facilities. In the operations research literature, each point-to-point demand is viewed as a *commodity* and the problem of determining optimal routes for the numerous commodities is known as the multi-commodity network flow problem. The resulting problems are formulated as very large linear programming problems. Now consider an access network with specified demands that need to be routed from the demand nodes to the gateway. The routing problem from many nodes to a single one is a single commodity network flow problem, which is a significantly simpler problem. (See, for example, Ahuja et al. (1993).)

A commonly-used access network topology has a tree structure. In such a network, there is a single route from any demand node to the root of the tree, which serves as the gateway to the backbone network. Many access network technologies use concentrators at a subset of demand nodes for efficient resource utilization. The demands coming into a concentrator are multiplexed and routed from there at higher speed towards the gateway. The simple concentrator location problem selects a subset of demand nodes as concentration nodes, where a concentrator has a limited capacity. Each demand node is assigned directly to a single concentrator. From the concentration node, the demand is routed directly to the gateway. In this problem, the access network has a simple tree structure. However, even for this simplified problem, selection of the optimal locations for concentrators is a challenging problem. In Section 13.2, we will examine a sample of concentrator location problems.

In recent years, broadband access networks that deliver information at high speeds to residential subscribers are in growing demand. Telephone carriers offer Digital Subscriber Line (DSL) services, but face strong competition from cable companies that offer broadband access through their cable TV networks. We present several recent models for wireline broadband access networks in Section 13.3. As will be shown, such models may be viewed as variants of the concentrator location problems.

As the volume of information carried increases, so does the need for reliability of transport. Many corporate customers demand reliable movement of information with almost instantaneous restoration capability. Loss of traffic due to cable cuts and other network failures is unacceptable as it results in huge monetary losses. To that end, carriers pay significant attention to designing survivable networks. Obviously, access networks with pure tree topologies are not survivable. To that end, today's carriers provide survivable options that include access rings and dual homing. In Section 13.4, we examine a sample of relevant models.

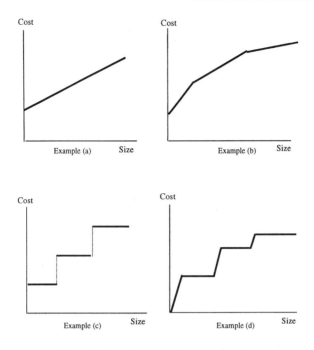

Figure 13.2 Representative cost functions

13.1.3 Models and solution approaches

Telecommunications network design involves large, complex mathematical programming models that are typically very difficult to solve. Often the resulting formulations are mixed integer programming (MIP) models. The integer variables represent decisions about which facilities should be installed and where. For example, which nodes should serve as concentrator locations, which links should be included in the network topology, and so forth. Figure 13.2 exhibits different cost functions associated with establishing a link or placing node equipment such as concentrators. In (a), we show a fixed charge cost function, where there is a fixed charge for establishing a facility and a variable cost that increases linearly with the facility size. In (b), we show a piecewise concave cost function, which is used to model fixed charge costs with decreasing variable costs for larger facilities. In (c), we show a step cost function, which represents installation of multiple identical facilities at the same location (this can be extended to functions for multiple facilities that are not identical). Finally, in (d) we present a more general piecewise-linear cost function that is neither convex nor concave. Incremental costs that are often considered include routing costs and processing costs. These are usually represented by a cost per unit of traffic.

In general, only relatively small network design problems can be solved to optimality using commercial packages for MIP models. Often, the problem is decomposed into modules, some of which are then solved optimally while others are solved using

318   HANDBOOK OF OPTIMIZATION IN TELECOMMUNICATIONS

heuristic methods. Some modules are formulated as facility location problems. Daskin (1995) presents many variations of facility location models and algorithms to solve these models. Optimal solutions are often derived using branch and bound schemes, where lower bounds are obtained using Lagrangian relaxation, dual ascent, or linear programming methods. Heuristic methods include combinations of add, drop, and exchange heuristics. More sophisticated heuristics include global methods such as simulated annealing, tabu search, and genetic algorithms. Reeves (1993) presents an overview of various global methods. Girard et al. (2001) describe a tabu search model for access network design. Other heuristic methods translate fixed charge costs to linear cost per capacity unit; the resulting problems are relatively easy to solve and the costs are updated after every iteration. Modules that address problems related to routing like shortest paths, maximum flows, minimum cost flows, and minimum spanning trees are equally important components for solving network design problems. These topics are covered in many books, for example, in Ahuja et al. (1993). Bodin et al. (1983) present an extensive review of vehicle routing problems, many of which are also applicable to routing issues in telecommunications networks.

This chapter is not a comprehensive survey. Instead, we concentrate on describing representative models that are often found in access network design and on providing examples from current technology. Several topics are not discussed in this chapter. Cellular wireless networks can be viewed as access networks as they provide connections from mobile stations to the base transceiver stations, which in turn are connected to backbone networks. However, the underlying problems are quite different and deserve a separate treatment. For instance, Aardal et al. (2001) present a survey of models and algorithms for the frequency assignment problem, which is one of the fundamental problems in multiple cellular wireless technologies. We also do not cover the topic of capacity planning over time, as the issues addressed in capacity planning models are not unique to access networks. Luss (1982) presents a survey on capacity planning.

13.2 CONCENTRATOR LOCATION PROBLEMS

An important recurring issue in the design of access networks consists of deciding where to locate concentrators. A concentrator at a node implies that a high-capacity link (a link with a high transmission rate) emanates from that node. In this section, we present a representative sample of concentrator location models and discuss some effective solution methods for these models.

13.2.1 The simple concentrator location problem

Consider the access network shown in Figure 13.3 with six demand nodes and a gateway (node 0). The originating demands at nodes 1–6, represented by the arrows, are given in units of DS1 (1.544 Mbps). A concentrator routes the demand on its outgoing link at a higher speed, in our example, in DS3 units (a DS3 has a capacity of 28 DS1's). In other words, the concentrator has a capacity constraint of 28 DS1. In Figure 13.3, concentrators are located at nodes 3 and 5. Thus 24 DS1's are routed from node 3 to the gateway and 25 units are routed from node 5 to the gateway. The links from nodes

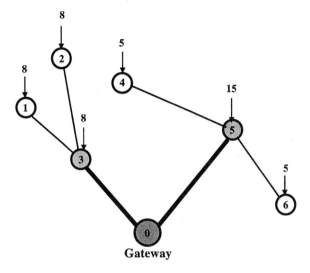

Figure 13.3 Illustration of the simple concentrator location problem

3 and 5 to node 0 have DS3 plug-ins at their end-points and possibly some equipment along the links. Therefore, concentrators installed at different locations may incur a different cost even if their capacity is the same. Such an access network topology is often referred to as a star/star network topology. In other words, the network of the concentrators and the gateway form a star topology with the gateway as its root. Likewise, the network formed by a single concentrator and the demand nodes assigned to that concentrator also form a star network with the concentrator as its root. The entire access network has a tree topology.

Consider an access network with demand nodes $i = 1, 2, \ldots, N$ and gateway node 0. Concentrators can be located at a subset J of the demand nodes. The originating demand at node i is h_i, the cost of installing a concentrator at location j is F_j, and the capacity of a concentrator at location j is b_j. This cost is represented by the cost function in Figure 13.2(a). The cost of assigning demand location i to a concentrator at j is c_{ij} per demand unit, and the processing cost at a concentrator at location j is p_j per demand unit. A demand node must be assigned to a single concentrator at some location $j \in J$. The decision variables are $y_j = 1$ if a concentrator is installed at location j and 0 otherwise, and $x_{ij} = 1$ if demand location i is assigned to a concentrator at location j and 0 otherwise. The problem is formulated as follows:

$$\text{Minimize} \sum_{j \in J} F_j y_j + \sum_{i=1}^{N} \sum_{j \in J} h_i (c_{ij} + p_j) x_{ij} \qquad (13.1)$$

so that:

$$\sum_{i=1}^{N} h_i x_{ij} \leq b_j, \text{ for } j \in J, \qquad (13.2)$$

$$\sum_{j \in J} x_{ij} = 1, \text{ for } i = 1, 2, \ldots, N, \qquad (13.3)$$

$$x_{ij} \leq y_j, \text{ for } j \in J \text{ and } i = 1, 2, \ldots, N, \qquad (13.4)$$

$$y_j = \{0, 1\}, \text{ for } j \in J, \qquad (13.5)$$

$$x_{ij} = \{0, 1\}, \text{ for } j \in J \text{ and } i = 1, 2, \ldots, N. \qquad (13.6)$$

Objective function (13.1) minimizes the fixed cost of the concentrators plus the costs of assigning the demand locations and processing the demand. A concentrator fixed cost depends on the location, since the cost depends on the distance from the gateway. Constraints (13.2) are capacity constraints. In this formulation, the concentrator capacity is node-dependent. Constraints (13.3) ensure that demand locations are assigned to concentrators, and combined with (13.6), they assure that each demand location is assigned to a single concentrator. Constraints (13.4) ensure assignments to locations that have a concentrator. Constraints (13.5) represent concentrator installation decisions.

Problem (13.1)–(13.6) is a version of a capacitated facility location problem (CFLP) with single source assignments. Klincewicz (1998) provides multiple references for this problem. Klincewicz and Luss (1986) and Pirkul (1987) use Lagrangian relaxation methods to find lower bounds and simple heuristics to find upper bound for problem (13.1)–(13.6). The former relax the capacity constraints, thus repeatedly solving uncapacitated facility location problems. The latter relaxes the assignment constraints, thus solving multiple knapsack problems at each iteration. Obviously, if the resulting gaps between the lower and upper bounds are viewed as too large, the methods can be incorporated into a branch-and-bound scheme.

Various extensions to the problem above have been published. For example, Lee (1993) considers a model that determines not only optimal concentrator locations, but also what type of concentrator should be installed at a selected location. Each concentrator type has different fixed cost and capacity. The author presents an algorithm that incorporates Benders decomposition and Lagrangian relaxation into a single framework that provides lower and upper bounds for this model.

Consider again Figure 13.3. Suppose the demand at node 4 is 10 instead of 5. Then, if we choose to install concentrators only at nodes 3 and 5, node 4 must be reassigned to the concentrator at node 3, while node 1 or node 2 would be reassigned to the concentrator at node 5. The resulting reassignments may increase the costs. However, if bifurcation of demand is allowed, we can assign some of the demand of node 4 to node 3 and some to node 5, for example, 4 units to node 3 and 6 units to node 5. Note that the resulting access network is not a tree anymore. The problem formulation is as in (13.1)–(13.6), where constraints (13.6) are replaced by:

$$x_{ij} \geq 0, \text{for } j \in J \text{ and } i = 1, 2, \ldots, N. \tag{13.7}$$

The problem is now formulated as the classical CFLP, see, for example, Daskin (1995, Chapter 7). This model can also be solved by the two Lagrangian relaxation methods mentioned above. Problem (13.1)–(13.6), with (13.7) replacing (13.6), is easier to solve than the original problem, as we replace 0-1 variables by continuous variables. The concentrators and the gateway node still form a star network. Furthermore, all nodes served by the same concentrator also form a star network with the concentrator serving as the root. However, since multiple concentrators may serve a demand node, the entire access network may not be a tree.

Suppose that we can install multiple, identical concentrators at a location. Now, variables y_j represent the number of concentrators installed at location j. The fixed cost of concentrators at a location is now depicted by part (c) of Figure 13.2. Allowing for bifurcated demands, the model formulation includes objective function (13.1) and constraints (13.3) and (13.7). Constraints (13.2) and (13.4) are replaced by:

$$\sum_{i=1}^{N} h_i x_{ij} \leq b_j y_j, \text{for } j \in J, \tag{13.8}$$

and constraints (13.5) are replaced by:

$$y_j = \{0, 1, 2, \ldots\}, \text{for } j \in J. \tag{13.9}$$

This problem represents a more challenging version of the CFLP. Frantzeskakis and Luss (1999) present a heuristic that repeatedly solves single-commodity minimum cost network flow problems. The link connecting a node in J to node 0 is represented in the network flow model by two links. Link $(j, 1)$ represents concentrators used at full capacity with a cost F_j/b_j per unit flow. Link $(j, 2)$ represents a single partially full concentrator with capacity b_j. The capacity of link $(j, 1)$ and the cost per unit flow on link $(j, 2)$ are adjusted at each iteration, based on the sum of flows on links $(j, 1)$ and $(j, 2)$ at the previous iteration. As shown in the paper, it may be desired to add some post-optimization schemes to further improve upon the solution.

The one-terminal TELPAK problem extends the problem above and considers a discrete set of facilities of different capacities that can be installed on the links. The link cost function is thus represented by a piecewise-linear function as shown in Figure 13.2(d). The TELPAK problem selects the links that should be included in the access network, the facilities installed on each selected link, and the routes used to route traffic from the demand nodes to the gateway. Rothfarb and Goldstein (1971) are the first to examine this problem. Gavish (1991) provides additional references to that problem.

13.2.2 A concentrator location problem with multi-level trees

Consider the access network shown in Figure 13.4. The network has 14 demand nodes routing traffic to the gateway (node 0). The problem is to determine concentrator locations and assign demand nodes to concentrators. A demand node is assigned to a

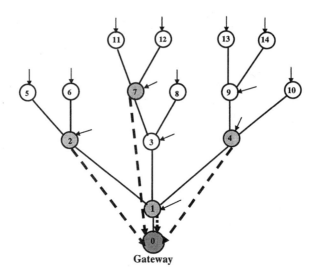

Figure 13.4 Illustration of a concentrator location problem for a multi-level tree

single concentrator. Demand is routed to a concentrator through the links of the tree; hence, the route is unique. We impose contiguity constraints: If node i is assigned to a concentrator at node j, then all demand nodes on the path from i to j, including node j, are also assigned to node j. We also assume that a concentrator has a dedicated connection to the gateway. An equivalent assumption is that the concentrators consume a negligible fraction of the link capacities. In Figure 13.4, we located concentrators at nodes 1, 2, 4 and 7. The dashed arrows represent the direct connections to the gateway. Nodes 2, 5 and 6 are assigned to the concentrator at node 2, nodes 7, 11 and 12 to the concentrator at node 7, nodes 4, 9, 10, 13 and 14 to the concentrator at node 4, and node 1 to the concentrator at node 1. Note that by the contiguity constraint, nodes 8 and 3 must be assigned to the same concentrator, either at node 1 or at node 7 (we allow *backward* assignments). Suppose we assign nodes 8 and 3 to node 7. Since the concentrators are directly connected to the gateway (the dashed links in Figure 13.4), the resulting network topology of all the concentrators is a star with the gateway serving as the root. The network comprised of a single concentrator and all nodes assigned to that concentrator form a tree topology. The topology of the entire network is often referred to as a star/tree topology.

Suppose the originating demand at node i is h_i, the cost of installing a concentrator at location j is F_j, and the fixed cost of routing demand over a link that connects nodes i and j is G_{ij}. The processing cost at a concentrator at location j is p_j per unit demand and the variable routing cost over link (i,k) is v_{ik} per unit. The 0-1 decision variables are: $y_j = 1$ if a concentrator is installed at location j and 0 otherwise, $x_{ij} = 1$ if demand location i is assigned to a concentrator at location j and 0 otherwise, and $z_{ik} = 1$ if link (i,k) is used and 0 otherwise. Finally, s_{ik} denotes the demand carried on link (i,k). Let T denote the set of links of the tree. The objective is then to minimize the sum of fixed and variable concentrator and link costs,

$$\text{Minimize} \sum_{j \in J} F_j y_j + \sum_{i=1}^{N} \sum_{j \in J} h_i p_j x_{ij} + \sum_{(i,k) \in T} G_{ik} z_{ik} + \sum_{(i,k) \in T} v_{ik} s_{ik}.$$

The constraints include assignment constraints that ensure that each demand node is assigned to one concentrator, constraints that coordinate assignments and concentrator installations, constraints that coordinate cable installations, assignment contiguity constraints, definitional constraints for s_{ik}, and integrality and non-negativity constraints. Balakrishnan et al. (1995) present this model. They showed how all the fixed and variable costs can be replaced by an assignment cost of demand node i to a concentrator at node j. Taking advantage of the contiguity constraints, they develop a dynamic programming model that finds an optimal solution. In their model they also consider existing spare capacity on the links and use Lagrangian relaxation to handle these capacities. The model does not include capacity constraints on the concentrators and on link expansions.

13.2.3 A sample of other related problems

13.2.3.1 Capacitated minimum spanning tree problems. Spanning trees are common structures in access networks. The Capacitated Minimum Spanning Tree (CMST) model attempts to find the minimum spanning tree that interconnects the gateway and all demand nodes, so that the total demand routed on a link must not exceed the capacity of that link. Gavish (1991) presents a review of various CMST models and solution approaches. Gouveia and Lopes (1997) present a CMST model, where bounds are imposed on the number of nodes directly connected to the gateway as well as on the number of nodes in the subtrees emanating from these *first-level* nodes.

13.2.3.2 Hierarchical concentrator problems. In the concentrator location problem discussed in Section 13.2, it was assumed that the concentrators are directly connected to the gateway. A more complicated topology assumes that the concentrators form a tree connected to the gateway, whereas any of the other demand nodes are assigned directly to one of the concentrators. Figure 13.5 illustrates this topology. The shaded nodes 6, 7, 10 and 11 represent concentrator locations and the link widths represent different capacities. Thus, for example, the concentrator at node 6 routes traffic to node 10 at a certain rate, while the concentrator at node 10 routes traffic to node 0 at a higher rate. The resulting network topology is often referred to as a tree/star topology. Pirkul and Narasimhan (1992) examine a version of this problem. Mateus and Franqueira (2000) and Andrews and Zhang (2002) present models for generalized access networks with modular link sizes. Their models consider multiple gateways, where each demand node is assigned to a single gateway.

13.2.3.3 Network redesign problems. Typically, existing networks have to be expanded over time in order to satisfy increasing demands. A common approach would focus on adding capacities to congested nodes and links. However, over time,

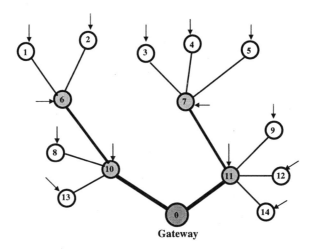

Figure 13.5 Illustration of an hierarchical concentrator access network

the incremental capacity expansions may result in a network that is far from optimal. Hence, models that expand, while simultaneously redesign, the existing network may lead to significantly better network design. Frantzeskakis and Luss (1999) present such a model for the concentrator location problem described in Section 13.2.1 with modular capacities and bifurcated demands. Existing demand assignments can be rearranged at a cost. The resulting model is a single-commodity network flow problem with continuous flow variables and integer variables on a subset of the links, representing modular capacity constraints. They present a heuristic method that repeatedly solves single-commodity minimum cost network flow problems and compare the results to those obtained by using software for mixed integer programs. Modeling and solving the network redesign problem for more general network topologies is a challenging topic with potentially significant gains.

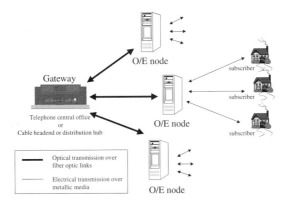

Figure 13.6 High-level view of a broadband access network

13.3 BROADBAND ACCESS NETWORKS

With the relatively recent push to deliver broadband services to residential and small business subscribers, the access network is becoming more frequently associated with the so-called *last mile* of a communication network. This is the portion of the network that connects the network provider directly to the customer and facilitates the provisioning of broadband services. These services may include data, voice and video, and they may be delivered to the customer by telephone, cable, or wireless access networks. At present, network providers are upgrading their infrastructure to support growing demand for broadband services by residential and small business subscribers.

In this section, we describe models for design of wireline broadband access networks, including the copper-based networks owned by telephone companies and the hybrid fiber coaxial (HFC) networks owned by cable companies. We will see that the design of both copper-based and HFC access networks involves many of the general issues described in the previous section, but each technology has some new details. At a high level, all of the networks discussed in this section have a similar logical structure, shown in Figure 13.6. They consist of a tree-shaped network that has subscribers at its leaves, a gateway node at its root, and selected intermediate nodes that contain equipment for multiplexing and for converting signals between the optical and electrical domains. More precisely, the resulting network topologies will be similar to star/tree topologies discussed in Balakrishnan et al. (1995). The intermediate nodes and the gateway form a star, while each intermediate node and the subscribers served by that node form a tree. Determining the size and location of these intermediate nodes lies at the heart of the corresponding design problems.

13.3.1 Copper-based access networks

The broadband access technologies being deployed by telephone companies typically leverage the existing copper infrastructure between customers and the central office. To enable broadband services, telephone companies are installing optical fiber between the central office and an intermediate point in the network, but they continue to use a twisted pair of copper wires between this intermediate node and each subscriber's premises. The use of fiber in the network allows higher bandwidth, while the use of copper into the premises makes these architectures attractive to telephone providers who have large investments in existing copper infrastructure. As a general rule, the shorter the length of copper between the intermediate node and the subscriber, the higher the bandwidth that can be delivered. Carpenter et al. (2001) use the term copper broadband access (CBA) to generically refer to these copper-based technologies which include DSL, fiber-to-the-curb (FTTC), and their many variants.

The intermediate node in such a network contains equipment that converts signals between the optical domain, used on the fiber, and the electrical domain, used on the copper. This equipment also typically performs other networking functions, such as signal multiplexing and demultiplexing. Following the convention in Carpenter et al. (2001), we refer to these nodes as optical/electrical (O/E) nodes, but they may be thought of as concentrator locations and the central office may be viewed as the gateway. The specific multiplexing equipment deployed at O/E locations depends on the details of the technology. For instance, DSL Access Multiplexers (DSLAMs) are deployed at O/E nodes in DSL networks and Optical Network Units (ONUs) are used in FTTC networks.

From a customer's perspective, the CBA technologies differ primarily in the amount of information that can be delivered. Each CBA technology is typically associated with a characteristic bandwidth, or line speed, that is assured by requiring that each customer be no more than a prescribed distance from the O/E node that connects it to the central office. This prescribed distance applies to every subscriber in the network and is, again, a characteristic of the particular technology. We refer to this characteristic distance as the range limit, or simply range, for the technology. Based on information in Abe (1997), the range limit for ADSL service is about 4,000 meters, while the range limit for FTTC may be on the order of 300 meters.

Although there are practical differences between the various CBA technologies, mathematically they manifest themselves in differences in the parameters of an otherwise identical problem. Recall the concentrator problem for a multi-level tree access network, presented in Balakrishnan et al. (1995). This network has a similar structure to that of the CBA networks. Referring back to Figure 13.4, in the CBA network context, nodes 1, 2, 4 and 7 are selected as the O/E nodes, and the dashed links from these nodes to node 0 are the fiber links that connect the O/E nodes to the gateway. The resulting network topology is a star/tree topology.

An initial version of the CBA network design problem is stated in Carpenter et al. (1996) and is extensively studied by Mazur (1999). This formulation captures the technical capabilities of the equipment but is later restricted (by Carpenter et al. (2001)) using a form of contiguity constraint to assure more practical solutions. The contiguity constraints assure that each O/E node serves a *connected region* within the network, thereby facilitating future maintenance and operations tasks.

We begin with the initial formulation and later describe how to restrict it. The problem is formulated on a network of nodes that forms a tree with the central office at its root and the customers at its leaves. This tree may be thought of as representing the physical topology of a copper network. Nodes that do not represent either customer locations or the central office would typically correspond to locations that provide convenient access to the copper wires, such as cross-connect locations. These nodes and the central office (but generally not the customer locations) are candidate O/E nodes. Each link in the tree has an associated length. In addition, we have specific equipment (DSLAMs or ONUs) that we can place at the candidate nodes to serve the customers. There may be several different equipment models that differ in their costs and the number of customers they can serve. Given the tree network, the equipment data, and the range limit, the CBA design problem is to place equipment at some subset of the candidate nodes so that customers can be feasibly served at the least cost.

Basic constraints require that each customer be served by a single O/E node within the prescribed range limit R, and that the number of customers served at an O/E node does not exceed the capacity of the equipment placed there. We use indices j, s and t to denote candidate O/E nodes, customer nodes and equipment types, respectively. J is the set of candidate O/E nodes, S is the set of customer nodes, J_s is the set of candidate O/E nodes that are within range limit R of customer node s, and T_j is the set of equipment types available at candidate O/E node j. The cost of equipment type t at candidate node j is c_{tj}, and the number of subscribers that can be served by equipment type t at candidate O/E node j is q_{tj}. The decision variables are $y_{tj} = 1$ if equipment type t is placed at candidate O/E node j and 0 otherwise, and $x_{sj} = 1$ if customer node s is assigned to O/E node j and 0 otherwise. The problem is formulated as follows:

$$\text{Minimize} \sum_{j \in J} \sum_{t \in T_j} c_{tj} y_{tj} \qquad (13.10)$$

so that:

$$\sum_{j \in J_s} x_{sj} = 1, \text{for } s \in S, \qquad (13.11)$$

$$\sum_{s \in S: j \in J_s} x_{sj} \leq \sum_{t \in T_j} q_{tj} y_{tj}, \text{for } j \in J, \qquad (13.12)$$

$$\sum_{t \in T_j} y_{tj} \leq 1, \text{for } j \in J, \qquad (13.13)$$

$$x_{sj} \in \{0,1\}, \text{for } s \in S \text{ and } j \in J_s, \qquad (13.14)$$

$$y_{tj} \in \{0,1\}, \text{for } j \in J \text{ and } t \in T_j. \qquad (13.15)$$

Objective function (13.10) minimizes the sum of equipment costs at the O/E nodes. Constraints (13.11) assure that each customer node is assigned to some O/E node for service, and in concert with (13.14), they assure that each customer is assigned to exactly one O/E node. Constraints (13.12) guarantee that the capacity placed at each O/E node is sufficient to serve the assigned subscribers. Constraints (13.13) assure that at most one type of equipment is placed at each node, and together with (13.15), they assure that if equipment is placed, it must be as a whole unit. We note that an equipment type may represent either a single piece of equipment or an allowable combination. Generally, we assume that it is possible to place concentrators in all possible combinations and multiples at any candidate site. Mazur (1999) shows that when all combinations and multiples are allowed, computing the equipment types is equivalent to solving an integer knapsack cover problem to obtain all non-dominated combinations. This computation appears practical for realistic-sized problems (Carpenter et al., 2001) and allows us to retain the binary-variable formulation. The range limits are not imposed by constraints, but rather, are assured by including the variable x_{sj} only if customer node s can reach candidate O/E node j.

The network design is very sensitive to distances because of the range limit for each customer, so the above formulation represents each subscriber explicitly. When multiple subscribers are in close proximity, so that their routes to an O/E node necessarily overlap, we may use a single node to represent them. Let h_s be the number of subscribers associated with customer node s. The set, J_s, of candidate nodes within reach is determined by the most restrictive customer represented by node s. Capacity constraints (13.12) are then modified as follows:

$$\sum_{s \in S: j \in J_s} h_s x_{sj} \leq \sum_{t \in T_j} q_{tj} y_{tj}, \text{for } j \in J. \tag{13.16}$$

Problem (13.10)–(13.15) can be viewed of as a variant of the capacitated concentrator location problem, but with some important differences from more general problems. For instance, the only cost considered is the equipment cost at the O/E nodes (e.g., no cost is associated with assigning a customer node to an O/E node). Furthermore, the underlying network is a tree, while the range limits narrow down the locations at which customer nodes may be served. If there is only a single type of concentrator to place at the candidate nodes, then the problem is polynomially solvable by a simple greedy algorithm proposed by Jaeger and Goldberg (1994) that extends the previous work of Kariv and Hakimi (1979). The survey by Magnanti and Wolsey (1995) discusses a wide variety of design problems on trees.

To make the formulation more practical, (Carpenter et al., 2001) add constraints that are analogous to the contiguity constraints used by Balakrishnan et al. (1995). These constraints state that if the copper wires for two customers meet en route to their serving locations, then these customers should ultimately be served at the same node. To add constraints on intersection let $P(s, j)$ denote the unique path in the tree that connects customer s to candidate node j. The path $P(s, j)$ can be thought of as being associated with variable x_{sj}. Adding the constraints

$$x_{si} + x_{rj} \leq 1, \text{for all } s \neq r \text{ and } i \neq j \text{ such that } P(s, i) \text{ and } P(r, j) \text{ intersect} \tag{13.17}$$

assures the desired behavior.

Adding the wire-intersection restriction allows the design problem to be solved by a dynamic programming approach described by Carpenter et al. (2001). This approach adapts to consider additional practical issues like the cost of installing new fiber on the links between the O/E node and the central office and also the desire to avoid investments in new copper. In order to deploy DSL quickly and cost-effectively, many

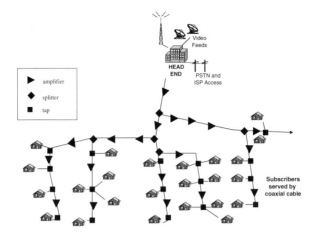

Figure 13.7 Traditional coaxial cable network

providers do not want to install any new copper, preferring to overlay DSL on their existing copper network. When this is the case, the set of candidate nodes is restricted to those candidates that are both within the reach limit and along the customer's path to the central office. These and other practical issues are also discussed in Behrens et al. (2000) and Behrens et al. (2001).

13.3.2 Hybrid fiber coaxial access networks

Just as telephone companies are trying to leverage their existing copper infrastructure to deliver new services, cable companies are similarly trying to leverage existing coaxial networks. Traditional cable networks have a tree-and-branch architecture, consisting of a head end node (or distribution hub) at the root that distributes content to customers at the leaves. These traditional networks contain trunk cables emanating from the head end that feed into feeder cables into neighborhoods. This feeder network distributes signals to taps that connect drop cables into individual customer premises. In traditional cable networks, trunk, feeder, and drop cables are coaxial cables carrying electrical signals to subscribers. Figure 13.7 provides a schematic illustration of a traditional tree-and-branch coaxial cable system.

The area served by a single head end may include 20,000–40,000 homes, a subset of which subscribes to cable services. The most distant customers may be as far as fifty miles from the head end. Signals injected at the head end weaken with distance traveled and because of losses that occur at taps and splitting devices along the cable route. To maintain signal strength to large numbers of customers over relatively long distances, amplifiers must be inserted to boost signal power. However, repeated amplification introduces noise and distortion. Signal quality issues, bandwidth limitations, and the lack of robustness in the tree-and-branch architecture have, therefore, led cable providers to modernize their facilities.

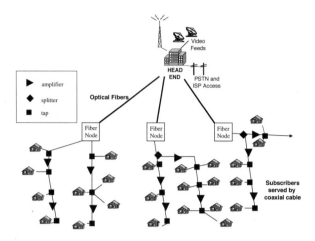

Figure 13.8 Hybrid fiber coaxial cable network

As in the case of DSL and other copper broadband access technologies, this modernization includes bringing fiber deeper into the network. More modern cable architectures feature ONUs placed at intermediate O/E nodes – often called fiber nodes – that are much closer to the customers they serve. Fiber nodes are directly connected to the head end by a fiber optic cable delivering optical signals and are connected to customers by the coaxial feeder network. Such networks are referred to as hybrid fiber coaxial (HFC) networks and exhibit the star/tree topology illustrated in Figure 13.8. As in CBA networks, the ONU at a fiber node performs functions like multiplexing and converting signals between the optical and the electrical domains. Details on HFC and the more traditional coaxial network architectures are provided in Abe (1997) and Ciciora et al. (1999).

The capacity and location of the fiber nodes in an HFC network are key design decisions in an HFC network. Gupta and Pirkul (2000) and Patterson and Rolland (2002) study this topological design problem for HFC networks. Unlike the design of copper-based networks described in Section 13.3.1, Gupta and Pirkul (2000) and Patterson and Rolland (2002) both pose the HFC design problem as if there were no existing infrastructure besides the head end. Thus, they locate both the ONUs and the coaxial links. Given a general network topology in which the head end and the customer locations are specified, they determine where to locate ONUs, the capacity of the ONUs located, and how to connect those ONUs to subscribers via coaxial links. Gupta and Pirkul (2000) develop a Lagrangian relaxation heuristic for the resulting network design problem, while Patterson and Rolland (2002) use an adaptive reasoning technique that is somewhat similar to tabu search methods. Both approaches are heuristics for minimizing total network cost, which includes both a connection cost to capture the cost of the coaxial network and the cost of the ONUs placed. The designs produced are tree-shaped networks directed from the subscribers to the head end.

This topological design problem for HFC networks contains both the concentrator location problem and the capacitated minimum spanning tree problem. Were it not for restrictions imposed by range limits, it would also contain the CBA network design problem as a special case. For cable companies that have an existing coaxial infrastructure and simply want to upgrade to HFC, the design problem may be simplified by exploiting the existing infrastructure as Carpenter et al. (2001) do for CBA networks. The corresponding network design problem would be quite similar to the problems presented in Balakrishnan et al. (1995) and in Carpenter et al. (2001). Moreover, with appropriate contiguity constraints, it seems that similar dynamic programming methods may be applicable.

Even after a topology is defined and ONUs are located, the coaxial portion of a cable network requires careful engineering to assure sufficient signal strength to each customer. Thus, once the star/tree topology is determined, there remains a feeder network engineering problem that locates amplifiers to assure sufficient signal strength to customers. The engineering issues involved are discussed in Carpenter et al. (1996) and Eiger (1996). The decisions in such an engineering problem can include choices on splitters, taps, and cable gauges, in addition to the choice and location of amplifiers, and the resulting problem is solvable by dynamic programming. We note that the corresponding problem does not exist in CBA network design because the copper wire from the O/E node to the subscriber is dedicated to a single subscriber, whose signal quality is assured through the range limit. Taylor et al. (1999) consider a feeder network design problem that includes both the choice of the coaxial links and the location of amplifiers. They refer to resulting problem as the spanning tree problem with amplification (SPTA) and present a randomized shortest-path tree heuristic as a solution approach.

13.4 SURVIVABLE ACCESS NETWORK DESIGN

The design of survivable networks has received significant attention in recent years. Since telecommunications networks move huge amounts of critical information among numerous locations, the loss of connection for even a short time duration may lead to significant monetary losses. Hence, companies require almost instantaneous restoration in the event that some network elements fail. Most of the literature presents models and algorithms for designing survivable networks in the event of a single link or node failure. The rationale is that failures can be detected and repaired relatively fast so that the probability of multiple failures at the same time is small. Although this assumption is usually reasonable (e.g., in the event of a link cut), some failures may in fact cause a chain of failures (e.g., a software failure), which may lead to severe service disruptions.

Although most of the attention in the literature has been devoted to designing survivable backbone networks, carriers are also demanding equally good restoration capabilities in access networks. So, unless carriers provide adequate restoration capabilities, customers will partition their demand among multiple access network service providers. Philpott et al. (2003) describe a practical method for designing survivable optical access networks in New Zealand.

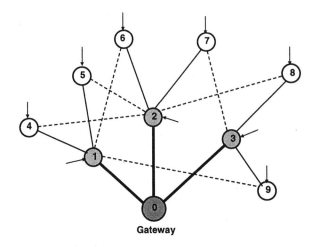

Figure 13.9 Illustration of an access network with dual homing

13.4.1 The concentrator location problem with dual homing

Access networks with tree structure are clearly not survivable under a link or node failure. For example, suppose node 3 in Figure 13.3 fails. In this case, the demands originating at nodes 1 and 2 cannot reach the gateway. Pirkul et al. (1988) extend the simple concentrator location problem to dual homing assignments. Each demand node is assigned to a primary and a secondary concentrator. Figure 13.9 exhibits such an access network, where the solid links represent primary connections and the dashed links represent secondary connections. If node 3 fails, then the concentrator at node 2 will serve the demand that originates at node 8, and the concentrator at node 1 will serve the demand of node 9. Formulation (13.1)–(13.6) can readily be extended to handle this network by having two sets of assignment variables, one for primary assignments and one for secondary assignments. The required processing resources of primary connections may exceed that of the secondary ones. The authors use Lagrangian relaxation to provide a lower bound to the model and a heuristic to derive an upper bound. Note that in order to ensure survivability in the event of a gateway failure, every concentrator should also be homed to two gateways.

13.4.2 Access networks with multiple ring families

With the development of synchronous (SONET/SDH) transmission technology and optical networking with wavelength division multiplexing (WDM) technology, detection of failures and traffic rerouting can be achieved in a fraction of a second. As a result, using rings in network design has become an attractive notion. Since two distinct routes connect every pair of nodes on a ring, the demand can be rerouted almost instantaneously in the event of a link or node failure. Ramaswami and Sivarajan (2002, Chapter 10) present an overview of SONET/SDH and optical rings in network design.

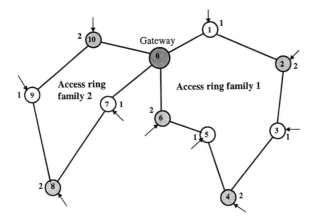

Figure 13.10 Illustration of an access network with multiple ring families

Although initially rings were employed primarily in backbone networks, improved technology and cost structure have recently made them also attractive for access networks.

Figure 13.10 shows an access network for ten demand nodes. Gateway 0 and nodes 1–6 constitute one access ring family, defined as a cycle of links of optical fiber cables. Gateway 0 and nodes 7–10 are on the second access ring family. A ring family consists of a stack of self-healing rings (SHRs), where each of the SHRs has the same capacity. A SHR is also constrained by the number of nodes it can serve. A demand node resides on one SHR. The gateway resides on all SHRs. Ring family 1 has two SHRs, where nodes 0, 1, 3 and 5 are on SHR 1 and nodes 0, 2, 4 and 6 are on SHR 2. Ring family 2 has also two SHRs, one with nodes 0, 7 and 9 and the other with nodes 0, 8 and 10 (the SHR number is adjacent to the node). The access rings serve traffic between the demand nodes and the gateway as well as traffic among the demand nodes. Note that traffic between nodes that reside on different SHRs is routed through the gateway.

Klincewicz et al. (1998) discuss this access network architecture, where each link used by a ring family incurs a fixed cost plus a variable cost per SHR associated with that family. The objective is to minimize the cost of constructing the ring families and the associated SHRs, while satisfying all the demands. The model can be viewed as a version of a single depot vehicle routing problem with multiple vehicles. Each vehicle forms a ring family. The nodes served by a ring family are analogous to locations served by a vehicle. The number of SHRs on a ring family may be viewed as the required vehicle size, with equal capacity reserves per SHR.

The proposed solution approach to the model consists of multiple heuristics to solve various components of the problem. The algorithm starts by creating many ring families. It then prepares a list of pairs of merger candidates. Ring families 1 and 2 in Figure 13.9 are good candidates for a merger if, for example, the traffic between node 6 and 7 is relatively large and the fixed cost of adding a link between these nodes is relatively small. The ring merging routine is quite involved. It requires designing the

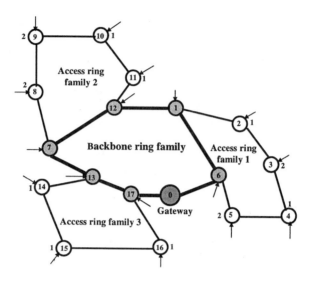

Figure 13.11 Illustration of a hierarchical access network

ring topology using a heuristic for the Traveling Salesman Problem and then designing the SHRs. For bi-directional SHRs, a load balancing routine is also needed.

In order to protect the access network from a gateway failure, the ring families must include two gateways. The model and solution method can readily accommodate this change.

13.4.3 A hierarchical access network design

Figure 13.11 exhibits an architecture designed to connect 17 demand nodes to the gateway. The demand nodes are partitioned into three clusters and each cluster forms an access ring family that may have multiple SHRs. Two nodes from each access ring family are designated as hubs (shaded nodes 1 and 6, 7 and 12, and 13 and 17). The hubs and the gateway node form a backbone ring that may also have multiple SHRs, usually with higher capacity. Each hub resides on all the SHRs of the associated access ring family and on a single SHR of the backbone ring. Each of the other demand nodes resides on a single SHR. A SHR has a capacity constraint, and lower and upper bounds on the number of nodes on a ring. Access ring families 1 and 2 have two SHRs, and access ring family 1 has one SHR.

Gawande et al. (1999) present a model and a heuristic solution for this architecture. The heuristic finds near-optimal solutions for $P = 2,3,4,\ldots$ access ring families and selects among these architectures the minimum cost solution.

For a specified P, the main module designs the network topology. Using a modified P-median heuristic, which considers lower and upper bounds on the number of nodes per cluster, the demand nodes are partitioned into P clusters. Two nodes, roughly at the center of each cluster, are initially designated as hubs. A modified Traveling

Salesman Problem heuristic is used to form access ring families and a backbone ring. The modification ensures that the two designated hubs on each access ring family are adjacent. Using approximate link costs (the link fixed cost is known, but the precise variable cost, which depends on the number of SHRs, is not known at this stage), the resulting solution is evaluated. This process is repeated multiple times with alternate hubs and the best solution is recorded. The entire design is done for different values of P and the minimum cost solution is selected as the desired topology.

Once the topology is determined, the stack of SHRs is designed for each access ring family and for the backbone ring. This problem is modeled as a bin-packing problem. The primary objective is to minimize the number of SHRs associated with a ring family, and a secondary objective is to balance the load among the SHRs in a ring family.

Bibliography

K. I. Aardal, S. P. M. van Hoesel, A. M. C. A. Koster, C. Mannino, and A. Sassano. Models and solution techniques for frequency assignment problems. Technical Report 01-40, Zentrum fur Informationstechnik Berlin, December 2001. Available on http://fap.zib.de/survey/index.html.

G. Abe. *Residential broadband.* Macmillan Technical Publishing, 1997.

R. K. Ahuja, T. L. Magnanti, and J. B. Orlin. *Network flows: Theory, algorithms, and applications*. Prentice Hall, 1993.

M. Andrews and L. Zhang. Approximation algorithms for access network design. *Algorithmica*, 34:197–215, 2002.

A. Balakrishnan, T. L. Magnanti, A. Shulman, and R. T. Wong. Models for planning capacity expansion in local access telecommunications networks. *Annals of Operations Research*, 33:239–284, 1991.

A. Balakrishnan, T. L. Magnanti, and R. T. Wong. A decomposition algorithm for local access telecommunications network expansion planning. *Operations Research*, 43: 58–76, 1995.

C. Behrens, T. Carpenter, M. Eiger, Y. Ho, H. Luss, G. Seymour, P. Seymour, and G. Truax. Network planning for xDSL. In *Proceedings of the 16th Annual National Fiber Optic Engineers Conference*, 2000.

C. Behrens, T. Carpenter, M. Eiger, Y. Ho, and P. Seymour. Enhanced xDSL planning. In *Proceedings of the 17th Annual National Fiber Optic Engineers Conference*, 2001.

D. Bienstock and I. Saniee. ATM network design: Traffic models and optimization-based heuristics. *Telecommunication Systems*, 16:399–421, 2001.

L. D. Bodin, B. L. Golden, A. A. Assad, and M. O. Ball. Routing and scheduling of vehicles and crews. *Computers and Operations Research*, 10(2):63–211, 1983.

T. Carpenter, M. Eiger, P. Seymour, and D. Shallcross. Automated design of fiber-to-the-curb and hybrid fiber-coax access networks. In *Proceedings of the 12th Annual National Fiber Optic Engineers Conference*, 1996.

T. Carpenter, M. Eiger, P. Seymour, and D. Shallcross. Node placement and sizing for copper broadband access networks. *Annals of Operations Research*, 106:199–228, 2001.

W. Ciciora, J. Farmer, and D. Large. *Modern cable television technology: Video, voice, and data communications.* Morgan Kaufmann, 1999.

M. S. Daskin. *Network and discrete location: Models, algorithms, and applications.* John Wiley, 1995.

M. Eiger. Coaxial network modeling and engineering. In *Proceedings of the International Conference on Telecommunications*, volume 1, pages 207–210, 1996.

L. F. Frantzeskakis and H. Luss. The network redesign problem for access telecommunications networks. *Naval Research Logistics*, 46:487–506, 1999.

B. Gavish. Topological design of telecommunications networks – Local access design methods. *Annals of Operations Research*, 33:17–71, 1991.

B. Gavish. Topological design of computer communication networks – The overall design problem. *European Journal of Operations Research*, 58:149–172, 1992.

M. Gawande, J. G. Klincewicz, and H. Luss. Design of SONET ring networks for local access. *Advances in Performance Analysis*, 2:159–173, 1999.

A. Girard, B. Sanso, and L. Dadjo. A tabu search algorithm for access network design. *Annals of Operations Research*, 106:229–262, 2001.

L. Gouveia and M. J. Lopes. Using generalized capacitated trees for designing the topology of local access networks. *Telecommunication Systems*, 7:315–337, 1997.

R. Gupta and H. Pirkul. Hybrid fiber co-axial CATV network design with variable capacity optical network units. *European Journal of Operational Research*, 123:73–85, 2000.

M. Jaeger and J. Goldberg. A polynomial algorithm for the equal capacity p-center problem on trees. *Transportation Science*, 28:167–175, 1994.

O. Kariv and S. Hakimi. An algorithmic approach to network location problems. I: The p-centers. *SIAM Journal of Applied Mathematics*, 37:513–538, 1979.

J. G. Klincewicz. Hub location in backbone/tributary network design: A review. *Location Science*, 6:307–335, 1998.

J. G. Klincewicz and H. Luss. A Lagrangian relaxation heuristic for capacitated facility location with single-source constraints. *Journal of the Operational Research Society*, 37:495–500, 1986.

J. G. Klincewicz, H. Luss, and D. C. K. Yan. Designing tributary networks with multiple ring families. *Computers and Operations Research*, 25:1145–1157, 1998.

C. Lee. An algorithm for the design of multitype concentrator networks. *Journal of the Operational Research Society*, 44:471–482, 1993.

H. Luss. Operations research and capacity expansion problems: A survey. *Operations Research*, 30:907–947, 1982.

T. Magnanti and L. Wolsey. Optimal trees. In M. Ball, T. Magnanti, C. Monma, and G. Nemhauser, editors, *Network Routing*, volume 7 of *Handbooks in Operations Research and Management Science*, chapter 9. North-Holland, 1995.

G. R. Mateus and R. V. L. Franqueira. Model and heuristic for a generalized access network design problem. *Telecommunication Systems*, 15:257–271, 2000.

D. Mazur. *Integer programming approaches to a multi-facility location problem*. PhD thesis, Johns Hopkins University, 1999.

R. A. Patterson and E. Rolland. Hybrid fiber coaxial network design. *Operations Research*, 50:538–551, 2002.

A. Philpott, A. Mason, and J. Davenport. 'FIDO': Telecom's best friend. *OR/MS Today*, 30(2):36–41, April 2003.

H. Pirkul. Efficient algorithms for the capacitated concentrator location problem. *Computers and Operations Research*, 14:197–208, 1987.

H. Pirkul and S. Narasimhan. Hierarchical concentrator location problem. *Computer Communications*, 15:185–191, 1992.

H. Pirkul, S. Narasimhan, and P. De. Locating concentrators for primary and secondary coverage in a computer communications network. *IEEE Transactions on Communications*, 36:450–458, 1988.

R. Ramaswami and K. N. Sivarajan. *Optical networks: A practical perspective*. Morgan Kaufmann Publishers, second edition, 2002.

C. R. Reeves, editor. *Modern heuristic techniques for combinatorial problems*. John Wiley, 1993.

E. Rosenberg. Dual ascent for uncapacitated telecommunications network design with access, backbone and switch Costs. *Telecommunication Systems*, 16:423–435, 2001.

B. Rothfarb and M. Goldstein. The one-terminal TELPAK problem. *Operations Research*, 19:156–169, 1971.

S. Taylor, N. Boland, and A. Philpott. Optimal spanning trees with attenuation and amplification. In *Proceedings of the 15th National Conference of the Australian Society for Operations Research*, pages 1225–1244, 1999.

14 OPTIMIZATION ISSUES IN DISTRIBUTION NETWORK DESIGN

Geraldo R. Mateus[1] and Zenilton K. G. Patrocínio Jr.[1]

[1] Department of Computer Science
Federal University of Minas Gerais
30123-970 Belo Horizonte, MG, Brazil
mateus@dcc.ufmg.br
zenilton@dcc.ufmg.br

Abstract: A distribution network design problem arises in a lower level of an hierarchical modeling approach for telecommunication network planning. Improvements of technologies used to deploy distribution networks have contributed to make distribution network planning more similar to other levels of access network. The major points that differentiate distribution network design problems are its huge dimensions and the several technological options that could be used to connect customers. Major technological trends to deploy distribution networks are discussed here. As an extension of the capacitated network design problem, it is a NP-hard combinatorial optimization problem. The need to install facilities and capacities in discrete levels and the incorporation of addition technology-related cost terms and constraints makes the exact solution of the mixed integer programming model even harder. There are several models and strategies that might be devised for solving those models, we present some of them.

Keywords: Distribution network, telecommunication system, capacitated network design.

14.1 INTRODUCTION

The urban telecommunication system is composed by intricate networks that enable the communication of hundreds of thousands of customers. The hierarchical organization of this network plays a major role, in as much as optimized levels of customers concentration enables the substantial economies of scale of increasing transmission bandwidth. The design of such networks with an hierarchical modeling approach is also a consequence of the fact that people solve complex problems by solving higher level problems first and then solving resulting lower level problems.

342 HANDBOOK OF OPTIMIZATION IN TELECOMMUNICATIONS

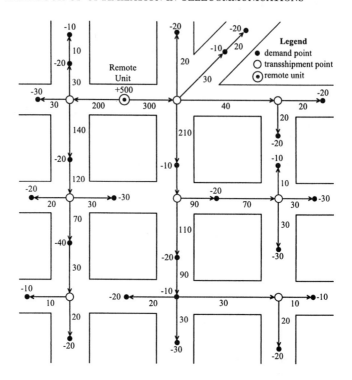

Figure 14.1 Spatial demands of a distribution network

A distribution network problem arises in a lower level of the hierarchical organization of the telecommunication system. The problem can be viewed as an extension of a capacitated single commodity network design problem, as illustrated in Figure 14.1. The commodity flow over an arc can be interpreted as the equivalent number of telecommunication channels that links a switching center or a concentrator to the customers served across that arc. The problem assumes demand points (sinks) as individual customer, a corporate customer or customers clustered and connected to terminal boxes, normally installed in electricity poles. The problem also assumes given the location and capacity of concentrators (sources) that can be optical remote units, which are supposed to be connected to the switching centers, distribution boxes or radio stations. We need to install adequate capacities on the arcs to route the required flow to each sink.

A method in which a large problem is broken up into several smaller problems is normally implicit in the literature concerning specific models of telecommunication network design (Gavish, 1982; 1983; 1991; Minoux, 1989). In order to provide a better description of the distribution network context we prefer to follow here the practice of working explicitly with an hierarchical design organization that can have different hierarchical levels depend on the technologies, services and customers (Balakrishnan et al., 1994; Mateus et al., 1994). Figure 14.2 depicts the classical hierarchical structure of an urban telecommunication network. The symbol $\bigcirc \Longrightarrow$ means that the element at left is a generic node of the network described at right, and the symbol $\odot \Longrightarrow$ means that the element at left is the *root* node of a *tree* network referred at right. The first level of the figure states that the *urban space* is partitioned in *local areas*. Each local area is served by a *switching center*. The communication among local areas is performed by a *backbone network*. The *switching center or concentrator* is the element of linkage between the *backbone network* and the *local access network*. Remark that a *switching center* is a node representing a local area in the *backbone network* and that it is also the *root* node of the correspondent *local access network*.

Our main focus here is on the second phase of the local access network design process. Figure 14.2 shows that, in order to provide access for the customers assigned to each switching center, we need to perform three levels of network design :

1. At the first level, each local area is partitioned in *service sections*, and the *primary network* provides optical access of the service sections to the assigned switching center.

2. At the second level, each service section is partitioned in *terminal sections*, and the *secondary network* provides the access of these terminal sections. An *optical remote unit*, a *distribution box* or a *radio station* is the element of linkage between the *primary* and the *secondary* networks.

3. At last, the *domestic network* encompasses the *tertiary networks* that assure the links of the customers to assigned terminal boxes.

The distribution network studied here concerns a particular local access network design problem, that in Figure 14.2 is associated with the secondary network. The literature on local access network design problems covers a variety of settings, which raises issues of dimensioning, topological design and routing. Balakrishnan et al. (1991) discuss several local access network design formulations, see also (Magnanti et al., 1993; Bienstock and Günlük, 1995). The local access design can be made either according to a Steiner tree (Luna et al., 1987; Balakrishnan et al., 1989) or else be based in an extension of the minimum spanning tree problem (Gavish, 1983; 1991; Hochbaum and Segev, 1989; Mirzaian, 1985). In any case the problem is NP-hard. In the first case we express the real existence of intermediate or transshipment nodes, but we face the difficulty of a NP-hard subproblem, thus looking for approximate solutions of the Steiner problem in graphs (Hakimi, 1971; Aneja, 1980; Wong, 1984; Maculan, 1987; Beasley, 1989; Duin and Volgenant, 1989; Agrawal et al., 1994). In the second case we deal with a restricted modeling approach, but we can take advantage of the fact that a greedy algorithm is able to find optimal solutions of the minimum spanning tree subproblem (Kruskal Jr., 1956).

A complementary problem arising in local access design of computer and communication systems concerns the location of facilities. The facilities may be switching centers, remote units, concentrators, radio stations, distribution or terminal boxes (Minoux, 1989; Gavish, 1991). Normally a facility is the *root* (supply) node of a local access *tree* network, and it is one of many demand nodes of a higher level network, that may be a *tree*, a *ring* or a *multiconnected* network. We can be inspired by classical methods to solve facility location problems, either in capacitated versions (Sa,

344 HANDBOOK OF OPTIMIZATION IN TELECOMMUNICATIONS

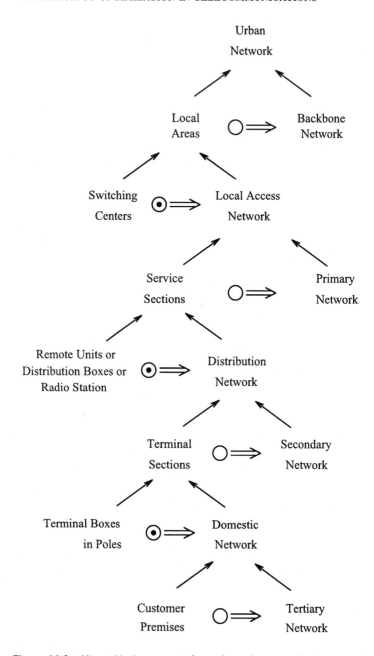

Figure 14.2 Hierarchical structure of an urban telecommunication network

1969; Beasley, 1988; Mateus and Luna, 1992) or in uncapacitated versions (Balakr-

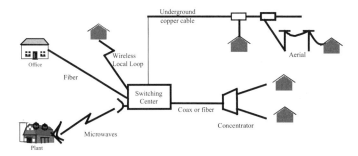

Figure 14.3 Technologies for distribution networks

ishnan et al., 1989). Distribution network planning consists of both locating optical remote units, concentrators or radio stations and selecting adequate cables or wireless channels to route the required flow to each demand point.

14.2 TECHNOLOGY FOR DISTRIBUTION NETWORK

The linkage of the urban communication customers is nowadays performed by fiberoptics for the backbone and for the primary networks. For the secondary (distribution) and the tertiary networks we can have wireless channels, fiber and/or copper cables (see Figure 14.3). For wireline networks, models should accommodate any possible configuration: *fiber-to-the-home* (FTTH), *fiber-to-the-curb* (FTTC) or traditional copper links. In this case the distribution network is composed by cables with standard capacity levels, normally given by a number of pairs of fiberoptics or copper wires. In the wireless network we need to cover all the urban area, assign frequencies and optimize the power control.

The use of copper-based and/or fiber-based systems is specially interesting for telecommunication operators that already have an installed infrastructure (switching centers, ducts, cables, access equipment, etc.), which can evolve in order to provide other services such as broadband access. Recently, new operators have joined the telecommunications market and usually they are not disposed to build a brand-new entire network and look for less expensive alternatives (such as, infrastructure renting). To these new operators, wireless technologies, either fixed or mobile, seem to represent an interesting way of creating an access/distribution network.

Wireless technologies are in their road to maturity and are supposed to be able to provide broadband services in the short-term. At the mobile side, the planning of *third generation* (3G) networks based on *wideband code division multiple access* (WCDMA) air interface, poses several differences and singularities when compared to the planning of *second generation* (2G) networks. The well known systems are the *universal mobile telecommunications system* (UMTS) or the *code division multiple access* (CDMA). In the case of fixed systems, the *local multipoint distribution service* (LMDS) is one of the most studied technologies.

Table 14.1 Technologies characteristics ([a]ATM-based passive optical network; [b]Ethernet-based passive optical network)

Type	Upstream Data Rate	Downstream Data Rate	Max Local Loop Reach (Km)	Life Line Voice
Analog Modems	14.4-33.6 Kbps	14.4-33.6 Kbps	N/A	Yes
56 Kbps Modems	33.6 Kbps	56 Kbps	N/A	Yes
Cellular Modems	14.4 Kbps	14.4 Kbps	N/A	N/A
ISDN	128 Kbps	128 Kbps	N/A	Yes
ASDL	176 Kbps 224-640 Kbps	1.54 Mbps 6-8 Mbps	5.4 3.6	Yes Yes
VSDL	640 Kbps 1.6-2.3 Mbps	13 Mbps 52 Mbps	1.4 0.3	Yes Yes
Cable Modems	0-768 Kbps	30 Mbps	N/A	No
LMDS	64 Kbps	34-40 Mbps	5.0	N/A
APON[a] FTTH/B	155 Mbps	622 Mbps	N/A	N/A
EPON[b] FTTH/B	10-1000 Mbps	10-1000 Mbps	N/A	N/A

In the following subsections, each of these technologies are briefly described and its major features are pointed out (see also Table 14.1). The focus here is not on the technical aspects of the systems, but rather on their socio and techno-economic dimensions.

14.2.1 Wired

14.2.1.1 Copper-based. In spite of antiquated technology and slow transmission rates, the *public switched telephone network* (PSTN) system is, on a world basis, the most reliable and inexpensive alternative for voice and low-rate data transmission. Moreover, it is this system through which the current Internet developed via broad access to modems of various speeds. It is also worth to mention that it is often quicker to carry out a variety of services by simply calling (via the PSTN system) rather than going through the procedure of starting a PC and logging onto a higher speed network.

However, the future of PSTN is not bright when it comes to services requiring high bandwidth. It is also doubtful that any compression technologies will be able to reduce the transmission requirements enough in order to guarantee transmission of even basic services over PSTN in the near future. So, in the context of the copper-based telephone network, on should consider other technologies for providing broadband access, such as *integrated services digital network* (ISDN) and the various *digital subscriber line* (xDSL) technologies.

ISDN. Employing the traditional telephone copper infrastructure and allowing for both voice and data transmission, ISDN is now widely available on a world basis.

Demand for this technology was quite high at the end of the last decade. The important factor to consider is its ability to provide faster Internet access as well as a support system for some business and LAN applications. However, ISDN must be defined as a medium range technology and, from the perspective of non-telecom operators, ISDN is rarely a solution to be consider in the development of future systems. Finally, ISDN represents a movement toward higher bandwidth. In the short-term, ISDN will continue being used, but in the future its limitations will likely mean that it will be superseded.

xDSL. The various xDSL technologies increase bandwidth to the subscriber by utilizing the copper wiring already installed in the PSTN infrastructure. xDSL comes in a seemingly endless number of versions that vary in the capacity carried and reached. In addition, while some version on the technology allows for symmetrical traffic others transmit two separate data streams with much more bandwidth devoted to the downstream leg to the customer than returning. One of the most popular versions is the *asymmetric* DSL (ADSL). It is effective because symmetric signals in many pairs within a cable (as occurs in cables coming out of the switching center) significantly limit the data rate and possible lime length. This variant increases the capacity of the copper system by 30 to 200 times when compared to PSTN modems. One of xDSL's drawbacks is that there is an inverse relationship between capacity and range. Versions of xDSL, such as *very high-bit-rate* DSL (VDSL), can operate at speeds over 50 Mbps but have a range of only a few hundred meters. The major advantage for xDSL is that it allows the further use of the copper-based telecom network. A second advantage is that the approach will allow for incremental development. Therefore, xDSL allows the operators of copper-based networks to offer broadband services at reasonable prices. For markets such as the *small office home office* (SOHO) users, xDSL offers advantages in those areas near to switching centers. As one moves farther into less densely settled areas the use of wireless technologies, such as cellular and LMDS, becomes a more realistic alternative.

14.2.1.2 Fiber-based.
Fiber to home has always been an attractive option. It has all the benefits of the fiber. It provides a future-proof network in that we do not have to go through the hassles of upgrading from ISDN to ADSL (and then, to xDSL, etc.). It does not have to contend with Electro Magnetic Interference (EMI) problems. No outside plant component implies highest reliability. It does not need electric powering and is immune to lightning and other transients. These properties of the fiber lead to lowest powering and operational costs (such as maintenance, provisioning and facilities planning).

The use of fiberoptics for the backbone and for the primary networks is not new, but only recently its use for the secondary (distribution) and the tertiary networks has started to be explored as a result of several new technologies that are leading to cost reduction such as *Passive Optical Networks* (PON).

Moreover, many protection schemes designed for fiber-based topologies rely on availability of another path between each remote unit and the associated switching center. So, establishment of local access rings has become more attractive as many operators have started to use fiberoptics.

FTTC. This scheme uses optical fibers to connect the switching center or central office (CO) to the *optical network units* (ONUs). The downstream traffic follows this path, namely, from the core to the CO and further on to the ONUs through fiber. The upstream traffic can also follow the same link. However, the ONUs are connected to the CO through a coaxial cable as well. The purpose of this cable is to send normal analog TV signal and to provide power to the ONUs. Each ONU is further connected in a star topology to a few (10-100) homes through a coaxial cable or a twisted pair. The cost of deploying FTTC must be equivalent to the cost of next generation *digital loop carrier* (DLC) — technology that makes use of digital techniques to bring a wide range of services to users via twisted-pair copper phone lines. It also provides a broadband-ready platform. And because the network is fiber to the serving pole or pedestal, maintenance savings are likely, too. The switched digital services FTTC architecture, as good as it is, fails to meet the needs and cost targets of a total service application. It costs about the same as a DLC system with a copper distribution for new build and rehabilitation projects. But it may not be cost-efficient for scattered demands of second lines and requirements for residential data, such as work at home, high-speed modems and LAN/WAN interfaces in existing neighborhoods.

FTTH/B. The optoelectronic equipment cost is a significant enough fraction of the total cost in the last mile that the future of FTTH/B is going to depend significantly on cost reduction in the optics. For example, one recent evaluation estimates 15% of the cost as lying in CO optoelectronics, 40% in the distribution network and its installation, and 45% in the customer premises optoelectronics and its installation. The same source breaks down the 40% for the infrastructure as follows: 53% in construction, 10% in engineering, 20% in couplers and splitters, 9% in splice closures, and only 8% in the cost of the fiber itself. A lot of optoelectronic cost reduction is already happening. While clever things are happening in the optical component world, equally inventive things are happening in the civil engineering of fiber installation. Machines can now cut narrow grooves in the pavements along city streets to lay in fiber bundles. Small trenching machines install plastic ducts one to two feet underground and remotely controlled directional drilling robots extend the path of the duct under highways. If the shorter ducts do not already contain the fiber, it can be later blown in by compressed air, a technique that allows upgrades and rehabs without further excavation. Thus, the costs, particularly the lifetime costs, of the all-glass solution are comparable to or less than those of any of the copper-based solutions, since the latter include along the right-of-way hot, finite-lifetime electronics with periodically spaced localized backup power sources. Some of the electronics cost is for data compression and decompression, unnecessary with fiber because of its huge offered bandwidth. Cost dominates everything in the last mile to a much greater extent than with the more traditional metro and long-haul situations.

14.2.2 Wireless

Telecom operators have long experience with copper access networks and will use them as long as possible. But, as we have mention before, new operators have recently joined the telecommunications market and usually they are not disposed to build a brand-new entire network and look for less expensive alternatives. To these new operators, wireless technologies (fixed or mobile) provide an interesting alternative for the deployment of an access/distribution network as shown in Figure 14.4.

14.2.2.1 Cellular. The deployment of a cellular system depends greatly upon the cellular network planning. Third generation mobile system for personal communication are becoming a reality faster than expected. Along with the new systems, a new set of services is intended to be offer to their user community. This new service set includes traditional voice and data access, network services, multimedia services and new services yet to be defined. All these services must be supplied with ubiquitous access, high performance and quality standard, must be able to run in a great variety of terminals and competitive price. Then, the network planning must consider radio propagation predictions, geographic and traffic parameters evaluation, radio network design optimization and network resource allocation. The radio station location at minimal cost aiming to cover the area under study. Link capacity is defined as the number of channels available in the wireless link. The capacity is limited by the reverse link rather than the forward link. Reverse link capacity depends on the interference received at the base station. Moreover, the link capacity varies with power control, cell coverage area, traffic load and radio path loss. The power control mechanisms allow each mobile unit to emit the minimum power needed to communicate with the target quality and hence generate the minimum possible interference at all others. So the network planning problem should consider the base station location and power control problems and the support for services based on higher transmission rates.

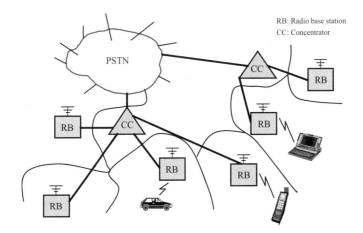

Figure 14.4 Wireless distribution network

14.2.2.2 LMDS. A fixed wireless cellular system, such as LMDS, is an alternative to the copper-based (xDSL), fiber-based (FTTx) and *hybrid fiber coax* (HFC) systems. Information broadcasting allows the high-speed downstream transmission, while service flexibility is obtained by an individual return channel whose performance is adequate to its end user (typically, residential or small/medium enterprises). LMDS has the advantage of low material and labor costs, and the ability to match construction to user demand avoiding cash-flow problems. Another advantage is the quick ability to install a system and lower maintenance because of less exposed equipment. The limitations for LMDS include the still unresolved constraints of the technology, lack of availability, the provisional nature of equipment, regulatory issues and unproven background when it comes to interactive services.

14.2.3 The cost structure

The available capacities of the distribution network come at different cost. The information about the costs of these capacities is rather accurate; for every particular type of communication link, the structure of the cost can be described by length and capacity dependent cost functions. These functions are structurally different and depend on which technologies are really used. The cost functions for wireline networks are well known, so in this section, we present typical cost functions for other two important types of links: leased lines and microwave links.

14.2.3.1 Leased lines. A new operator, such as a mobile-communication network operator, may rent part of its network from a leased line provider. As illustrative example, Figure 14.5 shows a typical cost structure for this case.

Figure 14.5 illustrates that the typical cost structure for a particular capacity, such as 30, 480, or 1920 channels, i.e., 2, 34, or 140 Mbps of *plesiochronous digital hierarchy* (PDH), is piecewise linear and monotonically increasing with the length of the link. The slope, however, decreases with the length of the link (e.g., it decreases at specified lengths such as l_1, l_2, and l_3 in Figure 14.5).

Changing the view from a particular capacity to available capacities on a particular link, as illustrated in Figure 14.6 for distinct digital signal levels of the PDH, it is easy to see that economies of scale are large. It depends on the distance between two end-nodes of a link, but as a rule of thumb, a capacity of six to eight times 30 channels is

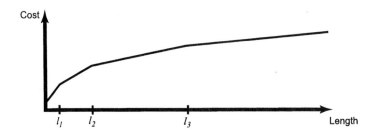

Figure 14.5 Typical cost structure for leased lines

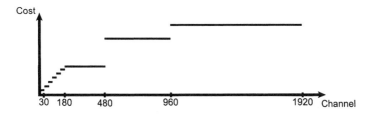

Figure 14.6 Typical cost structure for leased lines on a link

more expensive than a capacity of 480 channel, and three times 480 channels are more expensive than 1920 channel.

14.2.3.2 Microwave. The cost structure for microwaves is different since the maximum transmission distance through the air is limited. To guarantee a specified quality of the signal, it is necessary to periodically amplify the digital signal. So-called repeaters are needed after every interval of constant distance. For instance, if the maximum distance without amplification of the signal is 50 kilometers, two repeaters are necessary on a link of length 130 kilometers. The necessary amplification of the digital signals dominates the structure of the cost function for microwaves. Figure 14.7 illustrates a typical cost structure of microwaves for a particular capacity. The cost in dependence of the length is a staircase function with equal width intervals of constant cost. The width l_1 is the distance at which a repeater becomes necessary, and $l_i = i \cdot l_1$.

Figure 14.8 illustrates the cost structure of a microwave connection for a particular link. Similar to leased lines, it is a staircase function with considerable economies of scale.

The illustrated capacity and cost structures add significant complexity to the design of a distribution network. In the sense of complexity theory, the problem becomes difficult because of the discrete structure of the available capacities. It is not possible, for instance, to install 30.5 channels. Even if this is a required value, the network designer must choose between two discrete levels, e.g., 30 and 60 channels. The illustrated economies of scale cause further difficulties since it is not clear at which point it is appropriate to choose a 480 channel link instead of several 30 channel links. Of course, as shown in Figure 14.6, there exists a break-even point from which on it is

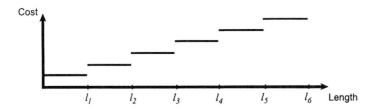

Figure 14.7 Typical cost structure for microwave connections

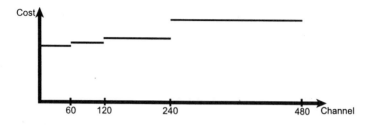

Figure 14.8 Typical cost structure for microwaves on a link

cheaper to use the higher capacity link, but it might pay to choose this higher capacity even below the break-even point because of the additional capacity. Using the larger capacity of 480 channels instead of six to eight times 30 channels, additional 240–300 channels are available at relatively small extra cost. Because of this additional capacity on one link it might be possible to decrease capacities on other links, and thus the overall network cost might decrease.

New local loop architectures (such as HFC, FTTC, FTTH/B, and LMDS) tend to have reduced economies of scale because of the new role played by variable costs associated with electronics and optronics (Pupillo and Conte, 1998). This means that it is very unlikely that a single uniform architecture will prevail in the way that twisted copper pair dominated the telephone network and coax cable dominated cable television in the past. Rather, the local loop of the future (and specially the distribution network) is more likely to be characterized by heterogeneous technologies. As a result, the inevitable competition will lead the successful operators to adopt most, if not all, of the forthcoming access/distribution technologies. Moreover, the competition in the local loop will increase due to a low probability of effective entry preemption. On the one hand, the decrease in economies of scale facilitates entry; on the other hand, the increase in economies of scope incentives to offer a new range of services through which entrants can differentiate themselves from dominant operators. Economies of scope mean that customers will be increasingly able to satisfy multimedia needs through any of a number of suppliers from formerly distinct industries (Pupillo and Conte, 1998).

14.3 OPTIMIZATION MODELS

In regard to network design problems, prior to the 1980s, location-allocation models were natural choice when designing telecommunication and computer networks (Boorstyn and Frank, 1977; Chou et al., 1978; Gerla and Kleinrock, 1977; McGregor and Shen, 1977). These models sought to determine the location of concentrators (or network access facilities) between a switching center and terminal units having known demands. The resulting network constituted an hierarchy of nodes between terminals and a switching center (or a gateway node) having a binary-tree-like structure such that whenever the flow passed to an upper level hierarchy, the flow would be combined or consolidated. The major cost savings from this design accrued from the fact that direct network connections from demand points to a switching center were obviated, and the installation permitted a more efficient use of link capacity from a concentrator to a switching center.

Given a capacitated network and point-to-point traffic demand, the objective function of more recently considered capacity expansion problems is to add capacity to the edges, in integral multiples of various modularities, and route traffic, so that the overall cost is minimized (Bienstock, 1993a;b; Bienstock and Günlük, 1996; Brockmüller et al., 1996; Cook, 1990; Dahl et al., 1995; Günlük, 1996; Mateus and Franqueira, 2000; Shaw, 1993; 1995). The underlying network has a directed tree rooted at a switching center or central office (CO). The arcs correspond to feeder sections along which signals can be sent, and the nodes represent geographic areas where demand for signals can occur. Each node, other than the CO, is entered by exactly one arc, and the CO is not entered by any arc. We are also given the demand at each node for equivalent DS0 signals (*Digital Signal Level 0*), which transmit at the rate of 64 Kbps. Furthermore, certain nodes are designated as remote terminals, where it is possible to install multiplexing equipment to convert DS3 signals (*Digital Signal Level 3*) to 28 DS1 signals (*Digital Signal Level 1*), and DS1 signals to 24 DS0 signals. For each arc, we are given its length, and the number of spare spaces available in the corresponding feeder section, that is, the number of cables that can be installed along the feeder section without undertaking the costly operation of constructing a new channel. For different types of signals, the space used by fiber cables is much less than that for copper cables. Hence, for large demand nodes, when we consider a capacity expansion to meet with such an increase in demand, the use of fiber optic cables is a more appropriate alternative. Naturally, there is a tradeoff between using cheaper copper cable installations or more expensive fiber optic cables, depending on the required feeder section capacity and the demand of customers. The objective function is to minimize the total cost of the copper and fiber cables and the multiplexers that must be installed in order to satisfy the demand.

As mention before, distribution networks are closely related to other levels of hierarchical organization of the telecommunication system especially to local access networks. Balakrishnan et al. (1991) demonstrated how to cast the general layered network framework for local access network planning problem as a fixed charge network design model. In order to do that the following assumption concerning the cost structure has to be made. The network design model ignores joint costs between various transmission media, and assumes a one-to-one correspondence between transmission rates and media, i.e., the planner preselects a preferred transmission medium for each transmission rate (Balakrishnan et al., 1991). Installation/expansion costs are also assumed to be piecewise linear, consisting of fixed and variable components.

In spite of that, a network design model can incorporate all features of the general layered network framework. In particular, it can handle general topologies, multiple service types, sectional and point-to-point media types, economies of scale in processor and transmission cost functions, and existing transmission and processing capacities. It also permits backfeed and bifurcated routing. Comprehensive surveys on the applications of network design models and their resolution by mathematical programming techniques can be found in (Magnanti and Wong, 1984; Minoux, 1989). If the cost functions are piecewise linear and concave, and if the network does not contain any existing capacities, the model reduces to an uncapacitated network design problem.

In general words, an uncapacitated network design problem is given by:

minimize *Design cost = Fixed cost + Variable cost*

subject to
    Flow balance constraints
    Linkage constraints

Fixed costs are related to installation/expansion of the facilities associated with arcs and/or nodes, while variable (or flow) costs are associated with routing customers demands through network topology. It is clear that, in order to minimize fixed costs, one should look for a solution that installs the minimum amount of facilities as possible, but this will probably rise variable costs as routing becomes harder. The same reasoning can be applied backwards. In order to minimize variable costs, more facilities have to be installed (to make routing easier and cheaper), but fixed costs will get higher.

So, the objective function above tries to minimize the summation of fixed costs and variable (or flow) costs. Flow balance constraints assure that installed (node/arc) facilities will meet all customers demands. Finally, linkage constraints establish a connection between decision and flow variables, i.e., if there is a flow unit passing through a facility (associated with an arc and/or a node), the existence of that facility should indicate by the corresponding decision variable, and vice-versa.

An uncapacitated network design solution that conserves flow at each node, and satisfies all demands at minimum total fixed plus flow costs corresponds to an optimum distribution network plan. It is possible to enrich this network design model in several ways. For instance, we can model economies of scale in processor and transmission costs if these economies can be approximated by piecewise linear concave cost functions.

Moreover, additional constraints can be used to set a limitation on the selected technology, e.g., one could limit the number of splices in a copper-based distribution network as described in (Mateus et al., 2000; Mateus and Franqueira, 2000); or the existence of overlapping cells in a LMDS system could be forbidden as in (Carlson and Authie, 2001). Similarly, additional terms (cost values) could also be incorporated to the objective function. Doing so, one could represent technology-related costs, e.g., splicing costs associated with copper-based infrastructure (Mateus et al., 2000), modularity (Mateus and Franqueira, 2000), or interference costs and the cost to connect the located base stations to the fixed networks related to wireless scenarios (Mazzini et al., 2003).

Existing resources and non-concave cost functions introduce arc capacities. Capacitated network design problem is much harder to solve than the uncapacitated version. One possible solution strategy is to dualize arc capacity constraints (i.e., multiply capacity constraints by Lagrange multipliers, and add these to the objective function). The resulting subproblem is an uncapacitated network design problem which is easier to solve. By iteratively modifying the Lagrange multipliers, e.g., using subgradient optimization, we can possibly originated good heuristic solutions and lower bounds for the original capacitated problem.

However, experiments with this approach for other related models (such as capacitated plant location and capacitated minimum spanning tree) suggest that the addition of arc capacities significantly increases the gaps between the upper and lower bounds. So, one should expect distribution network design problems with existing capacities and non-concave cost functions to be computationally more difficult to solve.

To cope with that, one could minimize the computational burden by simplifying the capacitated network design model. This can be done by ignoring some model aspects or by focusing on more relevant ones. For instance, if the location sites of node facilities (such as remote units, distribution boxes and radio stations) are predefined (or ignored), only arc facilities have to be selected along with the routes of customers traffic. In this case, Steiner tree problem (Luna et al., 1987; Balakrishnan et al., 1989; Agrawal et al., 1994; Mateus and Franqueira, 2000) is obtained, if there is intermediate or transshipment nodes, i.e., network elements which do not represent customers premises (e.g. concentrators). However, if every node has an associated demand value, which means that they all should be connect to some remote unit, distribution box or radio station, the generated problem is a minimum spanning tree problem (Gavish, 1983; 1991; Hochbaum and Segev, 1989; Mirzaian, 1985).

Considering the following aspects: node facilities, arc facilities and traffic routing, different simplified models can be obtained depending on which aspects are predefined or ignored. Table 14.2 summarizes some models that will be generated by this way. Although these problems are easier than the original one, many of them are still NP-hard. So, one might not except to obtain an exact procedure to some of them, and approximated solutions should be generated using good heuristics.

Many works that address those problems could be found in the literature, and it is not feasible to describe all of them here. In order to illustrated how those problems could be solved, some of those algorithms are revised in the next section.

14.4 ALGORITHMS

Because the underlying graph to be designed is a directed rooted spanning tree, the capacitated network model with point-to-point traffic demand described at the beginning of Section 14.3 can be formulated as a well structured integer linear program. Cook (1990) formulated and solved this problem using an algorithm based on the basis reduction technique of Lovász and Scarf (1990). Bienstock (1993a) formulated several subproblems related with Cook's problem, and developed various exact and ε-approximation algorithms.

After the success of cutting plane algorithms based on the facial structure of the convex hull of feasible solutions (so called branch-and-cut methods), recent research on telecommunication network design focuses on exploiting polyhedral descriptions to obtain tighter reformulations of the problem (Bienstock and Günlük, 1996; Brockmüller et al., 1996; Dahl et al., 1995; Grötschel and Monma, 1990; Grötschel et al., 1992a;b; 1994; Günlük, 1996; Stoer and Dahl, 1994). The capacitated network design problem described before generates a series of research problems related to expanding capacity on the communication network. Bienstock and Günlük (1996), Günlük (1996), and Dahl et al. (1995) consider installing more capacity on the edges and routing the traf-

Table 14.2 Distribution network optimization models

Model name	Aspects		Transshipment
	Predefined	*Focused*	*nodes exist ?*
Capacitated network design	—	Arc facilities Node facilities Traffic routing	yes
Steiner tree	Node facilities	Arc facilities Traffic routing	yes
Minimum spanning tree	Node facilities	Arc facilities Traffic routing	no
Minimum cost network flow	Node facilities Arc facilities	Traffic routing	yes
Transportation	Node facilities Arc facilities	Traffic routing	no
Capacitated facility location or Fixed-charge network flow	Arc facilities	Traffic routing Node facilities	yes/no
Capacity expansion or Minimum cost capacity installation	Node facilities Traffic routing	Arc facilities	yes/no
Tree covering or Discrete p-median	Arc facilities Traffic routing	Node facilities	yes/no

fic on the capacitated network, given a set of traffic demands between certain nodes of the network.

A second category of related problems is called the general network design problem, and is defined as follows. Given an undirected or directed graph $G = (V, E)$ where V and E are respectively the sets of vertices and edges of the graph G, and given the cost c_e, for corresponding each edge $e \in E$ (or for each arc in the directed case), the network design problem is to find a minimum cost subset $E' \subseteq E$ that meets some design criteria. Examples of the design criteria are Steiner tree and node- and edge-connectivity or reliability restrictions. In the node- and edge-connectivity problem, each node $s \in V$ has an associated nonnegative integer r_s, known as the type of s. We say that the graph G to be designed satisfies the node connectivity requirements, if, for each pair $s, t \in V$ of distinct nodes, G contains at least

$$r(s,t) = \min\{r_s, r_t\}$$

node-disjoint $[s,t]$-paths. Similarly, we say that G satisfies the edge connectivity requirements, if, for each pair s,t of distinct nodes, G contains at least $r(s,t)$ edge-disjoint $[s,t]$-paths. These conditions ensure that some communication path between s and t will survive a prespecified level of edge (or node) failures. Cutting plane algorithms are developed using the information on the facial structure of the convex hull of feasible solutions (Grötschel and Monma, 1990; Grötschel et al., 1992a;b; 1994). For $r_s \in \{0, 1, 2\}, \forall s \in V$, this problem is applied to Intra-LATA fiber network design problems faced by the regional Bell Operating Companies (BOC) and the reported computations show that their cutting plane algorithm with preprocessing solves problems having up to 116 nodes and 173 edges (Grötschel et al., 1991).

As a new alternative standardized network architecture, Synchronous Optical Network (SONET) technology with high-speed add/drop multiplexing technology is a recent innovation for protecting cable cuts or hub failures in the network. The Self-Healing Ring (SHR) structure has a ring structure such that any distinct pair of nodes s,t on this ring have two node-disjoint paths. For SONET ring architectures, the initial design problem is to determine the capacity of the ring and the clustering of the central offices into each ring (Wasem, 1991; Laguna, 1992). Another related problem is concerned with the load balancing issue (Shulman et al., 1991; Myung et al., 1997). This problem seeks to minimize the maximum aggregated link flow resulting from bidirectional link flows between all pairs of demand nodes on a ring.

The general location-allocation problem (LAP) is a class of mathematical programming problems that seeks the least cost method for simultaneously locating a set of service facilities and satisfying the demands of a given set of customers. A variant of this class of location-allocation problems is the p-median problem which is concerned with the location of p new facilities, called medians, on a network, in order to minimize the sum of weighted distances from each node to its nearest new facility (Handler and Mirchandani, 1979). If $p \geq 2$, this problem can be viewed as a location-allocation problem, because the location of the new facilities control the allocation of their service in order to best satisfy the demands. Hakimi (1964) has shown that for the p-median problem defined on a network, a set of optimal locations will always coincide with the vertices. Cavalier and Sherali (1986) have designed exact algorithms to solve the stochastic p-median problem on a chain graph and the 2-median problem on a tree graph, when the demand density functions are assumed to be piecewise uniform. Sherali and Nordai (1988) have developed certain localization results and algorithms for solving the capacitated p-median problem on a chain graph and the 2-median problem on a tree graph. Sherali (1991) later analyzed capacitated location-allocation problems on chain and tree graphs where the locational decisions need to be made sequentially, and Sherali and Rizzo (1991) addressed the case of unbalanced capacitated p-median problems having probabilistic link demands.

The fixed-charge or discrete location-allocation problem is a variant of the class of location-allocation problems that restricts the potential locations of the supply centers that might be constructed to certain preselected sites (Davis and Ray, 1969). Rardin and Choe (1979); Rardin (1982); Rardin and Wolsey (1993) have developed enhanced formulations and valid inequalities for uncapacitated fixed-charge network flow problems. Rardin and Wolsey (1993) introduced the class of dicut valid inequalities for an aggregated formulation, generating a tighter linear programming relaxation equivalent to the disaggregated formulation presented in Rardin and Choe (1979). Note that Steiner tree problems and uncapacitated facility location problems (UFL) are special cases of the class of uncapacitated fixed-charge problems. For the class of fixed-charge capacitated network flow problems, Herrmann et al. (1996) have generalized the dual ascent algorithm of Erlenkotter (1978). The related capacitated facility location problem (CFL) has been studied by Padberg et al. (1985); Leung and Magnanti (1989); Aardal et al. (1995); Aardal (1998), and several valid inequalities have been derived based on the facial structure of the problem. This problem seeks to construct any subset of a given number of fixed capacity facilities, so as to satisfy the specified demand at a minimal total construction plus distribution cost. Leung and Magnanti (1989) consider the case where at most one such facility having a capacity of U units can be constructed at each potential site, with no limit on the total number. Marathe (1992) considers a problem that permits any number of such facilities to be constructed at each site (hence, the net capacity is a multiple of U), with the total number of facilities located being p and presents a dynamic programming algorithm on a chain graph, and then solves a Lagrangian relaxation of a formulation on a general network via a conjugate subgradient optimization procedure.

The test problems considered in Marathe (1992) are solved exactly by a branch-and-bound algorithm that includes Lagrangian relaxation and Benders' decomposition strategies for computing lower bounds. The size of problems solved ranges from (number of network nodes, p) = (7, 5) to (20, 10) within about 1 CPU minute on an IBM 3090-300E supercomputer. For the capacitated facility location problem, Jacobsen (1983) has empirically compared various heuristics using a set of test problems derived from data published in Kuehn and Hamburger (1963). The size of problems solved involves 50 customers and 25 locations. More recently, Aardal (1998) has solved problems ranging up to 100 customers and 75 facilities using various strategies for generating valid inequalities. Although the initial linear programming (LP) relaxation generates tight lower bounds within about 1% of optimality, the effort to solve such problems to optimality ranges from 5.9 to 32.5 CPU hours on a SUN Sparc ELC computer. The LP relaxations of these problems produces significantly larger gaps, ranging up to about 70% of optimality, hence rendering them considerably more challenging to solve. For the fixed-charge capacitated network flow problems considered in Herrmann et al. (1996), test problems having the number of arcs ranging from 20 to 60 and the number of commodities ranging from 10 to 35 have been solved on a SUN Spark IPX Workstation. Problems having less than 30 arcs were solved to optimality via a branch-and-bound method and were used to evaluate the proposed dual ascent approach. The results indicated that the dual ascent procedure produced lower bounds lying typically within 12% of the optimal solution value.

14.5 CONCLUSIONS

Decision support systems for engineering design in telecommunication require an hierarchical organization of optimization models. The rising complexity of integrated computer and communication systems imposes for the emerging problems a coherent divide-and-conquer solving strategy. Optimal and heuristic algorithms must be adapted to the different classes of models, with consistent transference of information and decision among the models. We have showed a typical hierarchy of decision levels, and we have focused here the lower levels of local access network design.

Improvements of technologies used to deploy distribution networks have contributed to make distribution network planning more and more similar to other levels of access network, e.g., feeder network design. The major points that differentiate distribution network design problems are its huge dimensions (maybe involving hundreds of customers) and the several technological options that could be used to connect customers to the switching centers, distribution boxes or radio stations (maybe originating a distinct model for each distinct technology).

Capacitated network design problem is a referential model to design the lower levels of the local access networks. This makes distribution network design problem an extremely difficult one. As an extension of the capacitated network design problem, it is a NP-hard combinatorial optimization problem. The need to install facilities and capacities in discrete levels and the incorporation of additional technology-related cost terms and constraints makes the exact solution of the mixed integer programming model even harder. The computational results show that, from the theoretical point of view, it is very difficult to prove the optimality of the solutions. Heuristics is perhaps the only practical solution methodology for such a complex problem faced by telecommunications network providers. On the other hand, the computational results shows fast solution times, thus suggesting that heuristics can solve the large scale problems met in practice.

For this class of problems, duality gaps are substantial; however, when a more accurate lower bound procedure is used, the bounds found by many researchers were shown to be good. There are several strategies that might be devised for those models, but most of the approaches are not so easy to implement. Finally, researchers have long recognized that, for capacitated problems, optimal solution methods that use only the standard flow-based problem formulation are ineffective. The literature suggests that strong formulations, based on results from polyhedral combinatorics, appear to provide the best means to solve these problems.

Bibliography

K. Aardal. Capacitated facility location: Separation algorithms and computational experience. *Mathematical Programming*, 81:149–175, 1998.

K. Aardal, Y. Pochet, and L. A. Wolsey. Capacitated facility location: Valid inequalities and facets. *Mathematics of Operations Research*, 20:562–582, 1995.

A. Agrawal, P.N. Klein, and R. Ravi. When trees collide: An approximation algorithm for the generalized steiner problem on networks. Technical Report CS-90-32, Department of Computer Science, Brown University, Providence, Rhode Island 02912, 1994.

Y.P. Aneja. An integer linear programming approach to Steiner problem in graphs. *Networks*, 10:167–178, 1980.

A. Balakrishnan, T.L. Magnanti, and P. Mirchandani. Modeling and heuristic worst-case performance analysis of two-level network design problem. *Management Science*, 40:846–867, 1994.

A. Balakrishnan, T.L. Magnanti, A. Shulman, and R.T. Wong. Models for planning capacity expansion in local access telecommunication networks. *Annals of Operations Research*, 33:239–284, 1991.

A. Balakrishnan, T.L. Magnanti, and R.T. Wong. A dual-ascent procedure for large-scale uncapacitated network design. *Operations Research*, 37:716–740, 1989.

J.E. Beasley. An algorithm for solving large capacitated warehouse location problems. *Journal of the Operational Research Society*, 33:314–325, 1988.

J.E. Beasley. An sst-based algorithm for the Steiner problem in graphs. *Networks*, 19: 1–16, 1989.

D. Bienstock. Computational experience with an effective heuristic for some capacity expansion problems in local access networks. *Telecommunication Systems*, 1:379–400, 1993a.

D. Bienstock. A lot-sizing problem on trees, related to network design. *Mathematics of Operations Research*, 18:402–422, 1993b.

D. Bienstock and O. Günlük. Computational experience with a difficult mixed-integer multicommodity flow problem. *Mathematical Programming*, 68:213–237, 1995.

D. Bienstock and O. Günlük. Capacitated network design - polyhedral structure and computation. *INFORMS Journal on Computing*, 8:243–259, 1996.

R.R. Boorstyn and H. Frank. Large-scale network topological optimization. *IEEE Transactions Communications*, 25(1):29–47, 1977.

B. Brockmüller, O. Günlük, and L. A. Wolsey. Designing private line networks - polyhedral analysis and computation. Technical report, CORE Discussion Paper, 1996.

C.M.F. Carlson and Gerard Authie. Optimized design of LMDS cells: site location, equipment sizing and users homing. In *Proceeding 3rd IEEE International Conference on Mobile and Wireless Communications Networks (MWCN'2001)*, pages 111–118, Recife, Brazil, 2001.

T.M. Cavalier and H.D. Sherali. Network location problems with continuous link demands: p-medians on a chain and 2-medians on a tree. *European Journal of Operational Research*, 23:246–255, 1986.

W. Chou, F. Ferrante, and L. Gerke. An algorithm for optimally locating network access facilities. In *Proceedings of ICC '78*, pages 24.5.1–24.5.8, 1978.

W. Cook. Integer programming solution for capacity expansion of the local access network. Technical Report TM-ARH-017914, Bell Communications Research, 1990.

G. Dahl, A. Martin, and M. Stoer. Routing through virtual paths in layered telecommunication networks. Technical report, Telenor Research and Development, 1995.

P.S. Davis and T.L. Ray. A branch and bound algorithm for the capacitated facilities location problem. *Naval Research Logistic Quarterly*, 16(3):331–334, 1969.

C.W. Duin and A. Volgenant. Reduction tests for the Steiner problem in graphs. *Networks*, 19:549–567, 1989.

D. Erlenkotter. A dual-based procedure for uncapacitated facility location. *Operations Research*, 26(6):992–1009, 1978.

B. Gavish. Topological design of centralized computer networks - formulations and algorithms. *Networks*, 12:355–377, 1982.

B. Gavish. Formulations and algorithms for the capacitated minimal directed tree. *Journal of the ACM*, 30:118–132, 1983.

B. Gavish. Topological design of telecommunication networks - Local access design methods. *Annals of Operations Research*, 33:17–71, 1991.

M. Gerla and L. Kleinrock. On the topological design of distributed computer networks. *IEEE Transactions on Communications*, 25(1):48–60, 1977.

M. Grötschel and C. Monma. Integer polyhedra arising from certain network design problems with connectivity constraints. *SIAM Journal on Discrete Mathematics*, 3(4):502–523, 1990.

M. Grötschel, C. Monma, and M. Stoer. Polyhedral approaches to network survivability. In *DIMACS Series in Discete Mathematics and Theoretical Computer Science*, volume 5. American Mathematical Society, Providence, Rhode Island, 1991.

M. Grötschel, C. Monma, and M. Stoer. Computational results with a cutting plane algorithm for designing communication networks with low-connectivity constraints. *Operations Research*, 40(2):309–330, 1992a.

M. Grötschel, C. Monma, and M. Stoer. Facets for polyhedra arising in the design of communication networks with low-connectivity constraints. *SIAM Journal on Optimization*, 2(3):474–504, 1992b.

M. Grötschel, C. Monma, and M. Stoer. Design of survivable networks. In M. O. Ball, T. L. Magnanti, C. L. Monma, and G. L. Nemhauser, editors, *Handbooks in Operations Research and Management Science 7:Network Models*. North-Holland, 1994.

O. Günlük. A branch-and-cut algorithm for capacitated network design problem. Technical report, School of Operations Research and Industrial Engineering, Cornell University, 1996.

S.L. Hakimi. Optimal locations of switching centers and the absolute centers and median of a graph. *Operations Research*, 12:450–459, 1964.

S.L. Hakimi. Steiner's problem in graphs and its implications. *Networks*, 1:113–133, 1971.

G.Y. Handler and P.B. Mirchandani. *Location on Networks - Theory and Algorithms*. The MIT Press, Cambridge, Massachusetts, 1979.

J.W. Herrmann, G. Ioannou, I. Minis, and J. M. Proth. Fixed-charge capacitated network design problem. *European Journal of Operational Research*, 95:476–490, 1996.

D.S. Hochbaum and A. Segev. Analysis of a flow problem with fixed charges. *Networks*, 19:291–312, 1989.

S.K. Jacobsen. Heuristics for the capacitated plant location model. *European Journal of Operational Research*, 12:253–261, 1983.

J.B. Kruskal Jr. On the shortest spanning tree of a graph and the travelling salesman problem. *Proc. Amer. Math. Society*, 7:48–50, 1956.

A.A. Kuehn and M.J. Hamburger. A heuristic program for locating warehouses. *Management Science*, 9(4):643–666, 1963.

M. Laguna. Optimal design of sonet rings for interoffice telecommunication. Technical report, Graduate School of Business and Administration, University of Colorado, 1992.

J.M.Y. Leung and T.L. Magnanti. Valid inequalities and facets of the capacitated plant location problem. *Mathematical Programming*, 44:271–291, 1989.

L. Lovász and H. Scarf. The generalized basis reduction algorithm. Technical Report No. 946, Cowles Foundation Discussion Paper, Yale University, 1990.

H.P.L. Luna, N. Ziviani, and R.M.B. Cabral. The telephonic switching centre network problem: Formalization and computational experience. *Discrete Applied Mathematics*, 18:199–210, 1987.

N. Maculan. The Steiner problem in graphs. *Annals of Discrete Mathematics*, 31: 185–212, 1987.

T.L. Magnanti, P. Mirchandani, and R. Vachani. The convex hull of two core capacitated network design problems. *Mathematical Programming*, 60:233–250, 1993.

T.L. Magnanti and R.T. Wong. Network design and transportation planning: Models and algorithms. *Transactions Science*, 18(1):1–55, 1984.

V. Marathe. A discrete equal-capacity p-median problem. Master's thesis, Department of Industrial and Systems Engineering, Virginia Polytechnic Institute and State University, Blacksburg, Virginia, 1992.

G.R. Mateus, F.R.B. Cruz, and H.P.L. Luna. Algorithm for hierarchical network design. *Location Science*, 2:149–164, 1994.

G.R. Mateus and R.V.L. Franqueira. Model and heuristic for a generalized access network design problem. *Telecommunication Systems*, 15:257–271, 2000.

G.R. Mateus and H.P.L. Luna. Decentralized decision-making and capacitated facility location. *The Annals of Regional Science*, 26:361–377, 1992.

G.R. Mateus, H.P.L. Luna, and A.B. Sirihal. Heuristics for distribution network design in telecommunication. *Journal of Heuristics*, 6:131–148, 2000.

F.F. Mazzini, G.R. Mateus, and J.M. Smith. Lagrangean based methods for solving large-scale cellular network design problems. *Wireless Networks*, 9:659–672, 2003.

P.V. McGregor and D. Shen. Network design: an algorithm for the access facility location problem. *IEEE Transactions on Communications*, 25(1):61–73, 1977.

M. Minoux. Network synthesis and optimum network design problems: Models, solution methods and applications. *Networks*, 19:313–360, 1989.

A. Mirzaian. Lagrangian relaxation for the star-star concentrator location problem: Approximation algorithm and bounds. *Networks*, 15:1–20, 1985.

Y. Myung, H. Kim, , and D. Tcha. Optimal load balancing on sonet bidirectional rings. *Operations Research*, 45(1):148–152, 1997.

M.W. Padberg, T.J. Van Roy, and L.A. Wolsey. Valid linear inequalities for fixed charge problems. *Operations Research*, 33:842–861, 1985.

L. Pupillo and A. Conte. The economics of local loop architecture for multimedia services. *Information Economics and Policy*, 10:107–126, 1998.

R.L. Rardin. Tight relaxations of fixed charge network flow problems. Technical report, School of Industrial Engineering, Purdue University, West Lafayette, Indiana, 1982.

R.L. Rardin and U. Choe. Tighter relaxations of fixed charge network flow problems. Technical Report J-79-18, School of Industrial and Systems Engineering, Georgia Institute of Technology, Atlanta, Georgia, 1979.

R.L. Rardin and L.A. Wolsey. Valid inequalities and projecting the multicommodity extended formulation for uncapacitated fixed charge network flow problems. *European Journal of Operational Research*, 71:95–109, 1993.

G. Sa. Branch-and-bound and approximate solutions to the capacitated plant location problem. *Operations Research*, 17:1005–1016, 1969.

D.X. Shaw. Limited column generation technique for several telecommunication network design problems. Technical report, School of Industrial Engineering, Purdue University, West Lafayette, Indiana, 1993.

D.X. Shaw. Local access network design and its extensions. Technical report, School of Industrial Engineering, Purdue University, West Lafayette, Indiana, 1995.

H.D. Sherali. Capacitated, balanced, sequential location-allocation problems on chains and trees. *Mathematical Programming*, pages 381–396, 1991.

H.D. Sherali and F.L. Nordai. A capacitated balanced 2-median problem on a tree network with a continuum of link demands. *Transportation Science*, 22(1):70–73, 1988.

H.D. Sherali and T.P. Rizzo. Unbalanced, capacitated p-median problems on a chain graph with a continuum of link demands. *Networks*, 21:133–163, 1991.

A. Shulman, R. Vachani, J. Ward, and P. Kubat. Multicommodity flows in ring networks. Technical report, GTE Laboratories Inc., 1991.

M. Stoer and G. Dahl. A polyhedral approach to multicommodity survivable network design. *Numerische Mathematik*, 68:149–167, 1994.

O.J. Wasem. An algorithm for designing rings for survivable fiber network. *IEEE Transactions on Reliability*, 40:428–432, 1991.

R.T. Wong. A dual ascent algorithm for the steiner problem in directed graphs. *Mathematical Programming*, 28:271–287, 1984.

15 POLYHEDRAL APPROACHES TO THE DESIGN OF SURVIVABLE NETWORKS

Bernard Fortz[1] and Martine Labbé[2]

[1]Institut d'Administration et de Gestion
Université Catholique de Louvain
B-1348 Louvain-la-Neuve, Belgium
fortz@poms.ucl.ac.be

[2]Département d'Informatique
Université Libre de Bruxelles
1050 Bruxelles, Belgium
mlabbe@ulb.ac.be

Abstract: Long-term planning of backbone telephone networks has been an important area of application of combinatorial optimization over the last few years. In this chapter, we review polyhedral results for models related to these problems. In particular, we study classical survivability requirements in terms of k-connectivity of the network, then we extend the survivability model to include the notion of *bounded rings* that limit the length of the rerouting path in case of link failure.

Keywords: Network design, combinatorial optimization, branch-and-cut.

15.1 INTRODUCTION

Recently, the nature of services and the volume of demand in the telecommunication industry has changed drastically, with the replacement of analog transmission and traditional copper cables by digital technology and fiber optic transmission equipment. Moreover, we see an increasing competition among providers of telecommunication services, and the development of a broad range of new services for users, combining voice, data, graphics and video. Telecommunication network planning has thus become an important problem area for developing and applying optimization models.

Telephone companies have initiated extensive modeling and planning efforts to expand and upgrade their transmission facilities, which are, for most national telecommunication networks, divided in three main levels (see Balakrishnan et al. (1991)), namely,

- The *long-distance* or *backbone* network that typically connects city pairs through *gateway nodes*;

- The *inter-office* or *switching center* network within each city, that interconnects *switching centers* in different subdivisions (clusters of customers) and provides access to the gateway(s) node(s);

- The *local access* network that connects individual subscribers belonging to a cluster to the corresponding switching center.

These three levels differ in several ways including their design criteria. Ideally, the design of a telecommunication network should simultaneously account for these three levels. However, to simplify the planning task, the overall planning problem is decomposed by considering each level separately.

In this chapter, we study models and techniques for *long-term planning* in the first level of the hierarchy, i.e. the *backbone network*.

Planning in the backbone network is divided in two different stages : mid-term and long-term planning. Mid-term planning consists in dimensioning the network. More precisely, given a forecast of the demand matrix for this period and the current topology of the network, we have to compute how the expected demands will be routed as well as the necessary capacities of the cables. In some models, the addition of new edges is allowed. These problems involve, at the same time, survivable design criteria and routing constraints. A survey on these models can be found in De Jongh (1998).

Long-term planning involves a longer period of time so that demand data are not reliable enough, and we only deal with topological aspects. The goal is then to determine a set of cables connecting all nodes under some survivability criteria. In this context, the telephone network is seen as a given set of nodes and a set of possible fiber links that have to be placed between these nodes to achieve connectivity and survivability at minimum cost.

In traditional backbone networks, the limited capacity of copper cables resulted in highly diverse routing between offices. The developments in fiber-optic technology have led to components that are cheap and reliable, having an almost unlimited capacity. The introduction of such a technology has made hierarchical routing and bundling of traffic very attractive. This approach has resulted in sparse, even treelike network topologies with larger amounts of traffic carried by each link.

Two main issues appear in the planning process of fiber-optic networks: economy and survivability. Economy refers to the construction cost, which is expressed as the sum of the edge costs, while survivability refers to the restoration of services in the event of node or link failure. Trees satisfy the primary goal of minimizing the total cost while connecting all nodes. However, only one node or edge breakdown causes a tree network to fail in its main objective of enabling communication between all pairs of nodes.

This means that some survivability constraints have to be considered while building the network. Losing end-to-end customer service could lead to dramatic loss of revenue for commercial providers of telecommunication services. Constructing network topologies that provide protection against cable or office failures has become one of the most important problems in the field of telecommunications network design.

The most studied models deal with k-connectivity requirements, i.e. the ability to restore network service in the event of a failure of at most $k-1$ components of the network. Among them, the minimum-cost two-connected spanning network problem consists in finding a network with minimal total cost for which two node-disjoint paths are available between every pair of nodes. This means that two-connected networks are able to deal with a single link or node failure. two-connected networks have been found to provide a sufficient level of survivability in most cases, and a considerable amount of research has focused on so-called *low-connectivity constrained* network design problems, i.e. problems for which each node j is characterized by a requirement $r_j \in \{0,1,2\}$ and $\min\{r_i, r_j\}$ node-disjoint paths between every pair of nodes i,j are required. Section 15.3 presents a survey of the literature on these models.

Two-connectivity seems a sufficient level of survivability for most networks, since the probability of dealing with two simultaneous failures is very low. However, it turns out that the optimal solution of this problem is often a Hamiltonian cycle. Hence, any edge failure implies that the flow that passed through that edge must be rerouted, using all the edges of the network, an obviously undesirable feature. This led us to examine a new model for limiting the region of influence of the traffic which it is necessary to reroute : the *Two-Connected Network with Bounded Rings problem (2CNBR)*. This problem is studied in Section 15.4. In addition to the classical two-connectivity constraints, we require in this model that each edge belongs to at least one cycle (or *ring*) whose length is bounded by a given constant. It also finds its motivation in the emerging technology of *self-healing rings*. These are cycles in the network equipped in such a way that any link failure in the ring is automatically detected by the link end nodes and the traffic rerouted along the alternative path in the cycle. When such a strategy is chosen, rings must cover the network and their size is limited. These two requirements are fulfilled by our model.

In the case where edge lengths are equal to one, i.e. the edge (or node) cardinality of the rings is bounded, there exist more structural properties and polyhedral results. In particular, a lower bound on the number of edges in any feasible solution can be derived. These results are presented in Section 15.5. The chapter ends with a review of recent works on closely related models.

15.2 NOTATION AND DEFINITIONS

The aim in long-term planning of the backbone network is to determine a set of cables connecting given nodes and satisfying some survivability criteria that we will describe later. The given set of nodes and possible cable connections can be represented by an undirected graph $G = (V, E)$ where V is the set of *nodes* and E is the set of *edges* that represent the possible pairs of nodes between which a direct transmission link (cable) can be placed. The graph G may have parallel edges but should not contain loops. Graphs without parallel edges and without loops are called *simple*. If there exists an edge $e := \{i,j\}$ between two nodes i and j, these two nodes are called *adjacent*, and e is *incident* to i and j. Throughout this chapter, $n := |V|$ and $m := |E|$ will denote the number of nodes and edges of G.

Given the graph $G = (V, E)$ and $W \subset V$, the edge set

$$\delta(W) := \{\{i,j\} \in E \mid i \in W, j \in V \setminus W\}$$

is called the *cut* induced by W, and its size is given by $|\delta(W)|$. We write $\delta_G(W)$ to make clear – in case of possible ambiguities – with respect to which graph the cut induced by W is considered. For a single node $v \in V$, we denote $\delta(v) := \delta(\{v\})$. The *degree* of a node v is the cardinality of $\delta(v)$. The set

$$E(W) := \{\{i,j\} \in E \mid i \in W, j \in W\}$$

is the set of edges having both end nodes in W. We denote by $G(W) = (W, E(W))$ the subgraph induced by edges having both end nodes in W. If $E(W)$ is empty, W is an *independent set*. G/W is the graph obtained from G by contracting the nodes in W to a new node w (retaining parallel edges). Given two subsets of nodes W_1 and W_2, $W_1 \cap W_2 = \phi$, the subset of edges having one endpoint in each subset is denoted by

$$[W_1 : W_2] := \{\{i,j\} \in E \mid i \in W_1, j \in W_2\}.$$

We denote by $V - z := V \setminus \{z\}$ and $E - e := E \setminus \{e\}$ the subsets obtained by removing one node or one edge from the set of nodes or edges. $G - z$ denotes the graph $(V - z, E \setminus \delta(z))$, i.e. the graph obtained by removing a node z and its incident edges from G. This is extended to a subset $Z \subset V$ of nodes by the notation $G - Z := (V \setminus Z, E \setminus (\delta(Z) \cup E(Z)))$.

Each edge $e := \{i,j\} \in E$, has a *fixed cost* $c_e := c_{ij}$ representing the cost of establishing the direct link connection, and a *length* $d_e := d_{ij} := d(i,j)$. It is assumed throughout this work that these edge lengths satisfy the *triangle inequality*, i.e.

$$d(i,j) + d(j,k) \geq d(i,k) \quad \text{for all } i,j,k \in V.$$

The cost of a network $N = (V, F)$ where $F \subseteq E$ is a subset of possible edges is denoted by $c(F) := \sum_{e \in F} c_e$. The *distance* between two nodes i and j in this network is denoted by $d_F(i,j)$ and is given by the length of a shortest path linking these two nodes in F.

Without loss of generality, all costs are assumed to be nonnegative, because an edge e with a negative cost c_e will be contained in any optimum solution.

For any pair of distinct nodes $s, t \in V$, an $[s,t]$-*path* P is a sequence of nodes and edges $(v_0, e_1, v_1, e_2, \ldots, v_{l-1}, e_l, v_l)$, where each edge e_i is incident to the nodes v_{i-1} and v_i ($i = 1, \ldots, l$), where $v_0 = s$ and $v_l = t$, and where no node or edge appears more than once in P. A collection $P_1, P_2, \ldots, P_k$ of $[s,t]$-paths is called *edge-disjoint* if no edge appears in more than one path, and is called *node-disjoint* if no node (other than s and t) appears in more than one path. A *cycle* (containing s and t) is a set of two node-disjoint $[s,t]$-paths.

A *Hamiltonian cycle* is a cycle using each node of the network exactly once. The problem of determining if a graph contains a Hamiltonian cycle is NP-complete. The corresponding optimization problem – the *traveling salesman problem (TSP)* – has been well studied. We refer to Lawler et al. (1985) for an in depth treatment of this problem.

A graph $G = (V, E)$ is *k-edge-connected* (resp., *k-node-connected*) if, for each pair s, t of distinct nodes, G contains at least k edge-disjoint (resp., node-disjoint) $[s, t]$-paths.

When the type of connectivity is not mentioned, we assume node-connectivity. The *edge connectivity* (resp., *node-connectivity*) of a graph is the maximal k for which it is k-edge-connected (resp., k-node-connected). A 1-edge-connected network is also 1-node-connected, and we call it simply *connected*. A cycle-free graph is a *forest* and a connected forest is a *tree*. A *connected component* of a graph is a maximal connected subgraph. If $G - e$ has more connected components than G for some edge e, we call e a *bridge*. Similarly, if Z is a node set and $G - Z$ has more connected components than G, we call Z an *articulation set* of G. If a single node forms an articulation set, the node is called *articulation point*.

Node and edge-disjoint $[s, t]$-paths are related to cuts and articulation sets by Menger's theorem (Menger, 1927).

Theorem 15.1 (Menger)

1. *In a graph $G = (V, E)$, there is no cut of size $k - 1$ or less disconnecting two given nodes s and t, if and only if there exist at least k edge-disjoint $[s, t]$-paths in G.*

2. *Let s and t be two nonadjacent nodes in G. Then there is no articulation set Z of size $k - 1$ or less disconnecting s and t, if and only if there exist at least k node-disjoint $[s, t]$-paths in G.*

We will also use the following definitions arising from polyhedral theory (see e.g. Nemhauser and Wolsey (1988)). Given a polyhedron P, the *dimension* $\dim(P)$ of P is defined as the maximum number of affinely independent elements in P minus one. An inequality $a^T x \leq \alpha$ is *valid* with respect to P if $P \subseteq \{x : a^T x \leq \alpha\}$. The set $F_a := \{x \in P : a^T x = \alpha\}$ is called the *face* of P defined by $a^T x \leq \alpha$. If $dim(F_a) = \dim(P) - 1$ and $F_a \neq \phi$, then F_a is a *facet* of P and $a^T x \leq \alpha$ is called *facet-inducing* or *facet-defining*. A vector x is a *vertex* of P if it cannot be written as a non-trivial convex combination of points in P.

The convex hull of a set of points S will be denoted by conv(S). We also denote by e_i the i-th unit vector in $\mathbb{R}^n$.

15.3 LOW-CONNECTIVITY CONSTRAINED NETWORK DESIGN PROBLEMS

Throughout this chapter, a (backbone) telephone network is seen as a set of gateway nodes (or telephone offices) and fiber links that are placed between nodes. In this context, survivability refers to the restoration of services in the event of office or link failure, or, in other words, a network is survivable if there exists a prespecified number of node-disjoint or edge-disjoint paths between any two offices. The only costs considered are construction costs, like the cost of digging trenches and placing a fiber cable into service.

In this framework, a considerable amount of research has focused on low connectivity constrained network design problems. Following the terminology used by Monma and Shallcross (1989) and Stoer (1992), these models can be described informally as follows : we are given a set of telephone offices that have to be connected by a network. The offices may be classified according to importance, namely the

- *special offices*, for which a "high" degree of survivability has to be ensured in the network to be constructed;
- *ordinary offices*, which have to be simply connected to the network;
- *optional offices*, which may not be part of the network at all.

Given are also the pairs of offices between which a direct transmission link can be placed, and the associated cost of placing the fiber cable and putting it into service. The problem now consists in determining where to place fiber cables so that the construction cost, i.e. the sum of the fiber cable costs, is minimized and certain survivability constraints are ensured. For instance, we may require that

- the destruction of any single link may not disconnect any two special offices, or
- the destruction of any single office may not disconnect any two special offices.

These requirements are equivalent to ask that there exist

- at least two edge-disjoint paths, or
- at least two node-disjoint paths

between any two special offices.

Higher survivability levels may be imposed by requiring the existence of three or more paths between certain pairs of offices according to their importance class. However, up to now, low-connectivity requirements have been found to provide a sufficient level of survivability for telephone companies. For high-connectivity requirements, the reader is referred to Grötschel et al. (1995b); Stoer (1992).

In graph-theoretic language, the set of offices and possible link connections can be represented by an undirected graph $G = (V, E)$. The survivability requirement or importance of a node is modeled by node types. In particular, each node $s \in V$ has an associated nonnegative integer r_s, the *type* of s. Sometimes, we also write $r(s)$ instead of r_s. A network $N = (V, F)$, where $F \subseteq E$ is a subset of the possible links, is said to satisfy the *node-connectivity requirements*, if, for each pair $s, t \in V$ of distinct nodes, N contains at least

$$r(s,t) := \min\{r_s, r_t\}$$

node-disjoint $[s,t]$-paths.

Similarly, we say that N satisfies the *edge-connectivity requirements*, if, for each pair $s, t \in V$ of distinct nodes, N contains at least $r(s,t)$ edge-disjoint $[s,t]$-paths. If all node types have the same value k, it is equivalent to request that N is k-node-connected or k-edge-connected.

We restrict here to low-connectivity requirements, i.e. node types $r_s \in \{0, 1, 2\}$. Using our previous classification of nodes,

- special offices are represented by nodes of type 2,
- ordinary offices by nodes of type 1, and
- optional offices by nodes of type 0.

To shorten notation, we extend the type function r to sets by setting

$$\begin{aligned} r(W) &:= \max\{r_s \mid s \in W\} \text{ for all } W \subseteq V, \text{ and} \\ \text{con}(W) &:= \max\{r(s,t) \mid s \in W, t \in V \setminus W\} \\ &= \min\{r(W), r(V \setminus W)\} \\ &\quad \text{for all } W \subseteq V, \phi \neq W \neq V. \end{aligned}$$

We write $\text{con}_G(W)$ to make clear with respect to which graph $\text{con}(W)$ is considered.

In order to formulate network design problems as integer linear programs, we associate with every subset $F \subseteq E$ an *incidence vector* $\mathbf{x}^F = (x_e^F)_{e \in E} \in \{0,1\}^{|E|}$ by setting

$$x_e^F := \begin{cases} 1 & \text{if } e \in F, \\ 0 & \text{otherwise}. \end{cases}$$

Conversely, each vector $\mathbf{x} \in \{0,1\}^{|E|}$ induces a subset

$$F^{\mathbf{x}} := \{e \in E \mid x_e = 1\}$$

of the edge set E. For any subset of edges $F \subseteq E$ we define

$$x(F) := \sum_{e \in F} x_e.$$

We can now formulate the connectivity constrained network design problem as the following integer linear program:

$$\min \quad \sum_{e \in E} c_e x_e$$

s.t.

$$x(\delta(W)) \geq \text{con}(W) \quad W \subset V, \phi \neq W \neq V, \tag{15.1}$$

$$x(\delta_{G-z}(W)) \geq \text{con}_{G-z}(W) \quad z \in V, W \subset V \setminus \{z\},$$
$$\phi \neq W \neq V \setminus \{z\}, \tag{15.2}$$

$$x_e \in \{0, 1\} \quad e \in E. \tag{15.3}$$

It follows from Menger's Theorem that, for any feasible solution $\mathbf{x}$ of this program, the subgraph $N = (V, F^{\mathbf{x}})$ of G defines a network satisfying the node-connectivity requirements. Removing (15.2), we obtain an integer linear program for edge-connectivity requirements. Inequalities (15.1) are called *cut inequalities*, while inequalities (15.2) are called *node cut inequalities*.

The remainder of this section is devoted to a review of the work on these models, describing exact solution methods for more general or more specialized problems. Much of the material is taken from references cited in the surveys of Christofides and Whitlock (1981); Winter (1986b); Stoer (1992).

15.3.1 Structural properties and particular cases

A lot of research has focused on the survivability model presented in the previous section. The next sections survey these results. We begin by looking at the complexity of the problem, before considering some polynomially solvable cases. We also present work on particular cases, either with restricted connectivity requirements or restricted costs.

15.3.1.1 Complexity. The connectivity constrained network design problem is NP-hard in general. In particular :

- If $r_s \in \{0,1\}$, $\forall s \in V$, it reduces to the well-known NP-hard Steiner tree problem in networks. Winter (1987) made an in depth survey of these problems.

- If $r_s = 2$, $\forall s \in V$, it consists in determining a minimum cost two-connected network. This last problem is NP-hard even if the graph is complete and costs satisfy the triangle inequality, since with an algorithm for this problem, one could decide whether a graph has a Hamiltonian cycle by associating a cost equal to 1 to all graph edges and cost equal to 2 to all non-graph edges (see Eswaran and Tarjan (1976)).

However, for some particular connectivity requirements or costs, or when the underlying graph G is restricted, the problem may become polynomially solvable. We now review these cases.

15.3.1.2 Restricted connectivity requirements. By restricting the connectivity requirements r_s, the connectivity constrained network design problem reduces to some well-known polynomially solvable problems :

- If $r_s = 1$, $\forall s \in V$, the problem reduces to the minimum spanning tree problem. The most famous polynomial time algorithms for solving it are those from Kruskal (1956) and Prim (1957).

- If $r_s = 1$ for exactly two nodes of V and $r_s = 0$ for all the other nodes, the problem becomes a shortest path problem, solvable e.g. by the algorithms of Bellman (1958) or Dijkstra (1959).

- If $r_s = k$, $k \geq 2$, for exactly two nodes of V and $r_s = 0$ for all the other nodes, the problem becomes a k-shortest paths problem. This problem was studied by Suurballe (1974) and Suurballe and Tarjan (1984).

- If $r_s \in \{0,1\}$, $\forall s \in V$, the problem reduces to the Steiner tree problem in networks. This problem is NP-hard in general, but Lawler (1976) solved it in polynomial time in the case where either the number of nodes of type 0 or the number of nodes of type 1 is restricted.

15.3.1.3 Restricted costs. Under uniform or 0/1 costs, certain classes of connectivity constrained network design problems are polynomially solvable. We now examine these choices of costs.

Under uniform costs, the underlying graph G can be seen as a complete graph, and the problem turns into the construction of a sufficiently highly connected graph with a minimum number of edges. Chou and Frank (1970) solved this problem for edge connectivity requirements by producing a feasible graph where each node has degree r_s, except possibly for one node that has degree $r_s + 1$. Since these are the lowest possible degrees under the given connectivity requirements, the graph has the minimum number of edges. This proves the following lemma.

Lemma 15.1 *Given node types $r_s \geq 2$ for a set V of nodes, the minimum number of edges of a graph satisfying the edge-connectivity requirements given by r is*

$$\left\lceil \frac{1}{2} \sum_{s \in V} r_s \right\rceil.$$

The use of parallel edges is allowed in the construction.

Stoer (1992) describes a polynomial algorithm similar to that of Chou and Frank which also handles nodes of type 1. Frank and Chou (1970) also solved the problem when no parallel edges but extra nodes are allowed in the solution.

Unfortunately, to our knowledge, no general solution for the node connectivity version of the problem is available in the literature. But more can be said about uniform connectivity requirements $r_s = k$ for some $k \geq 2$.

An early work by Fulkerson and Shapley (1971) – written in 1961 but published ten years later – proved Lemma 15.1 for the edge-connectivity problem with uniform requirements, but without using parallel edges. Harary (1962) showed with the help of a polynomial algorithm that the same result holds for the node-connectivity problem with uniform requirements, leading to the following lemma.

Lemma 15.2 *Given $k \geq 2$ and $n \geq k+1$, the minimum number of edges in a k-node-connected graph on n nodes without parallel edges is*

$$\left\lceil \frac{kn}{2} \right\rceil.$$

Now, one may guess that Lemma 15.1 also holds for general node-connectivity requirements, but this conjecture is not true and a counter-example can be found in Stoer (1992).

We now turn to problems with 0/1 costs. These are known in the literature as augmentation problems, since these correspond to the problem of augmenting a graph $G = (V, E)$ by a minimum number of edges in $V \times V$, so that it meets connectivity requirements.

These augmentation problems were brought up by Eswaran and Tarjan (1976) for two-edge and two-node-connected graphs. Rosenthal and Goldner (1977) studied the augmentation to two-node-connected graphs. Their linear time algorithm contains an error that was corrected by Hsu and Ramachandran (1993), who also proposed a parallel implementation of their algorithm. Hsu and Ramachandran (1991) also developed a linear time algorithm for the augmentation to 3-node-connected networks.

The augmentation to k-edge-connected graphs was studied by Watanabe and Nakamura (1987), Ueno et al. (1988) and Cai and Sun (1989). The fastest known algorithm for this problem is the one by Naor et al. (1990). Frank (1992) solved the augmentation problem completely for general edge-connectivity requirements. All solution procedures allow the use of parallel edges, except those of Eswaran and Tarjan (1976) and Rosenthal and Goldner (1977). Again, the problem of augmentation to a node-connected graph is open in most cases.

15.3.1.4 Other polynomially solvable cases. Other cases of connectivity constrained network design problems are polynomially solvable if the underlying graph G is restricted to certain graph classes.

Among these, the class of *series-parallel graphs* has received a lot of attention. Series-parallel graphs are created from a single edge by two operations:

- addition of parallel edges,

- subdivision of edges by insertion of nodes.

Works on various connectivity requirements for these graphs can be found in the literature:

- $r_s \in \{0,1\}$, $\forall s \in V$ (Steiner tree problem):
  The problem was solved in linear time by Wald and Colbourn (1983). Goemans (1994) gave a complete description of the polytope associated with the solutions of the problem.

- $r_s = k$, $k \geq 2$, $\forall s \in V$, with edge connectivity requirements (k-edge-connected network problem):
  Mahjoub (1994) gave a complete description of the polytope associated with the solutions of the case when $k = 2$. This work was extended to any $k \geq 2$ by Didi Biha and Mahjoub (1996).

- $r_s \in \{0,2\}$, $\forall s \in V$ (two-connected Steiner subgraph problem):
  The problem was solved in linear time by Winter (1986a), both for edge and node-connectivity requirements. Coullard et al. (1991) gave a complete description of the polytope associated with the solutions of the node-connectivity case.

Winter has also developed linear-time algorithms for the case $r_s \in \{0,2\}$ in outerplanar (Winter, 1985b) and Halin graphs (Winter, 1985a). He also mentions in Winter (1987) that he solved the problem in linear time for $r_s \in \{0,3\}$ in Halin graphs. Coullard et al. solved the problem with $r_s \in \{0,2\}$ in W_4-free graphs (Coullard et al., 1993) and gave a complete description of the dominant of the corresponding polytope (Coullard et al., 1996).

Dominant of the polytopes of k-edge-connected networks where parallel edges are allowed were completely described by Cornuéjols et al. (1985) for k even and G series-parallel and by Chopra (1994) for k odd and G outerplanar.

15.3.2 Polyhedral studies and exact algorithms

Most Branch-and-Cut algorithms for these problems are based on the linear programming formulation (15.1)-(15.3). The first method for solving the problem exactly was developed by Christofides and Whitlock (1981). Their algorithm is based on the linear relaxation obtained by replacing integrality constraints (15.3) by $0 \leq x_e \leq 1$, and keeping only cut constraints (15.1) corresponding to subsets W such that $|W| = 1$. These particular cut constraints are called *degree constraints*. The starting linear program is thus

$$\min \sum_{e \in E} c_e x_e$$
$$\text{s.t.}$$
$$x(\delta(v)) \geq r(v) \quad v \in V,$$
$$0 \leq x_e \leq 1 \quad e \in E.$$

Given a solution to this LP, they impose the edge-connectivity requirements by adding violated cut constraints. These can be found in polynomial time by computing the minimum cut in the graph $G = (V, E)$ with edge capacities equal to the values of the corresponding variables in the solution of the current LP – using e.g. the Gomory-Hu algorithm (Gomory and Hu, 1961). When all cut constraints are satisfied, if some variables have fractional values, a branch-and-bound procedure is applied. Christofides and Whitlock (1981) mention that this algorithm is able to solve problems with "well over a hundred nodes" for edge-connectivity requirements.

If one wants to add node-connectivity requirements, they propose to check the node-connectivity each time an integer solution satisfying edge-connectivity requirements is found. If some node-connectivity requirements are violated, the corresponding node-cut constraints (15.2) are added to the LP.

Grötschel, Monma and Stoer studied in detail network design problems with connectivity constraints. A survey of their work can be found in (Grötschel et al., 1995a) and (Stoer, 1992).

In their earliest work on the subject, Grötschel and Monma (1990) introduced a general model mixing edge and node survivability requirements. They examined the dimension of the associated polytope and proved facet results for cut and node-cut inequalities.

They also described completely the polytope of the (1-)connected network problem, based on the work of Cornuéjols et al. (1985). This is done by the introduction of *partition inequalities*, that generalize cut inequalities (15.1). Given a partition $W_1, W_2, \ldots, W_p$ ($p \geq 2$) of V into p nonempty subsets, the inequality

$$\frac{1}{2} \sum_{i=1}^{p} x(\delta(W_i)) \geq p - 1$$

is valid for the polytope of connected networks.

Based on partition inequalities for connected networks, Grötschel and Monma introduced the *node-partition inequalities* for k-node-connected networks. These inequalities come from the fact that the deletion of $k - 1$ nodes from a k-node-connected network leaves a connected graph. Thus, if $Z \subseteq V$ is a node set with exactly $k-1$ nodes and $W_1, W_2, \ldots, W_p$ ($p \geq 2$) is a partition of $V \setminus Z$ into p nonempty subsets, the inequality

$$\frac{1}{2} \sum_{i=1}^{p} x(\delta_{G-Z}(W_i)) \geq p - 1$$

is valid for the polytope of k-node-connected networks. It is obvious that these node-partition inequalities are a generalization of node cut inequalities (15.2).

Grötschel, Monma and Stoer then attacked low-connectivity constrained problems (with $r_s \in \{0, 1, 2\}$), deriving new facets (Grötschel et al., 1992b) and implementing some of these into a Branch-and-Cut algorithm (Grötschel et al., 1992a). They generalized partition and node-partition inequalities, and introduced lifted 2-cover and comb inequalities. These results were extended to higher survivability requirements in Grötschel et al. (1995b).

More inequalities for two-edge-connected network problems were found by Boyd and Hao (1993) (complemented comb inequalities), and by Boyd and Zhang (1994) (clique tree inequalities). Baïou et al. (2000) and Kerivin and Mahjoub (2002) studied an extension of partition inequalities, the F-partition inequalities, first introduced by Mahjoub (1994), and showed these prove helpful for solving low survivability network design problems where edge-connectivity only is considered.

15.4 TWO-CONNECTED NETWORKS WITH BOUNDED RINGS

It turns out that the optimal solution of the two-connected network problem is often a Hamiltonian cycle. Hence, any edge failure implies that the flow that passed through that edge must be rerouted, using all the edges of the network, an obviously undesirable feature.

It is therefore necessary to add extra constraints to limit the region of influence of the traffic which is necessary to reroute if a connection is broken. Imposing a limit on the length of the rerouting can be done by limiting the length of the shortest cycle including each edge. Such a condition has also a direct implication in networks using the technology of *self-healing rings*. Self-healing rings are cycles in the network equipped in such a way that any link failure in the ring is automatically detected and the traffic rerouted by the alternative path in the cycle. It is natural to impose a limited length of these rings. This is equivalent to set a bound on the length of the shortest cycle including each edge.

The problem of designing a minimum *cost* network N with the following constraints:

1. The network N contains at least two node-disjoint paths between every pair of nodes (*2-connectivity constraints*), and

2. each edge of N belongs to at least one cycle whose *length* is bounded by a given constant K (*ring constraints*).

This problem is called the *Two-Connected Network with Bounded rings (2CNBR) problem*. It was first studied by Fortz et al. (2000). More polyhedral results can be found in (Fortz, 2000; Fortz and Labbé, 2002; 2004). Recently, Fortz et al. (2003a) studied the edge connectivity version of the problem.

A useful tool to analyze feasible solutions of 2CNBR is the *restriction of a graph to bounded rings*. Given a graph $G = (V, E)$ and a constant $K > 0$, we define for each subset of edges $F \subseteq E$ its restriction to bounded rings F_K as

$$F_K := \left\{ e \in F : \begin{array}{l} e \text{ belongs to at least one cycle} \\ \text{of length less than or equal to } K \text{ in } F \end{array} \right\}.$$

The subgraph $G_K = (V, E_K)$ is the *restriction of G to bounded rings*. Note that an edge $e \in E \backslash E_K$ will never belong to a feasible solution of 2CNBR.

Further we denote by $\mathcal{D}_{G,K}$ the set of incidence vectors x^F with $F \subseteq E$ such that

1. F is two-connected,

2. $F = F_K$.

Then, the 2CNBR problem consists in

$$\min \left\{ \sum_{e \in E} c_e x_e : x \in \mathcal{D}_{G,K} \right\}.$$

Checking that G_K is two-connected, i.e. that $\mathcal{D}_{G,K}$ is nonempty, can be done in polynomial time. We therefore assume in the remainder of this chapter that there always exists a feasible solution to the problem.

Since all costs c_e, $e \in E$ are assumed to be nonnegative, there always exists an optimal solution of 2CNBR whose induced graph is minimal with respect to inclusion. More precisely, if F_K is two-connected, as $F \supseteq F_K$, F is also two-connected and the cost of F is greater than or equal to the cost of F_K. We can thus relax the constraints and just require that F_K is two-connected for a set of edges F to be feasible. Hence, 2CNBR can be equivalently formulated as

$$\min \left\{ \sum_{e \in E} c_e x_e : x \in \{0,1\}^{|E|} \text{ and } F_K^x \text{ is two-connected} \right\}.$$

We denote by

$$\mathcal{P}_{G,K} := \text{conv}\{x \in \{0,1\}^{|E|} : F_K^x \text{ is two-connected}\}$$

the polyhedron associated to the 2CNBR problem.

Several formulations have been proposed for this problem. The first formulation using only design variables was proposed in Fortz and Labbé (2002). If a subset of edges $S \subseteq E$ is such that $(G - S)_K$ is not two-connected, then $G - S$ does not contain a feasible solution, and therefore each feasible solution contains at least one edge from S. As we are only interested in minimal feasible solutions, this is sufficient to formulate the 2CNBR problem as the following integer linear program :

$$\min \sum_{e \in E} c_e x_e$$

s.t.

$$x(S) \geq 1 \quad S \subseteq E, (G-S)_K \text{ is not two-connected}, \quad (15.4)$$
$$x_e \in \{0,1\} \quad e \in E. \quad (15.5)$$

Constraints (15.4) are called *subset constraints*.

15.4.1 Cut and ring-cut inequalities

Fortz and Labbé (2002) studied under which conditions cut constraints (15.1) are facet-defining for 2CNBR. Given a subset of nodes $W \subseteq V$, $\emptyset \neq W \neq V$, the cut constraint imposes that there are at least two edges leaving W, i.e.

$$x(\delta(W)) \geq 2.$$

To characterize which cut constraints define facets, it is useful to know, for any pair of edges $e, f \in \delta(W)$, if there exists a vector of $\mathcal{P}_{G,K}$ lying in the face $x(\delta(W)) = 2$ whose corresponding graph contains e and f. This is the case if and only if the incidence vector of

$$C_{e,f} := E(W) \cup E(V \setminus W) \cup \{e, f\}$$

belongs to $\mathcal{P}_{G,K}$, i.e. if $(C_{e,f})_K$ is two-connected. A useful tool to represent and analyze the vectors belonging to the face defined by a cut constraint is the *ring-cut graph* defined below.

Definition 15.1 (Ring-cut graph)
Let $G = (V, E)$ be a graph, $K > 0$ a given constant, and $W \subseteq V$ a subset of nodes, $\emptyset \neq W \neq V$.
The ring-cut graph $RCG_{W,K} := (\delta(W), RCE_{W,K})$ induced by W is the graph defined by associating one node to each edge in $\delta(W)$ and by the set of edges

$$RCE_{W,K} = \{\{e,f\} \subseteq \delta(W) : (C_{e,f})_K \text{ is two-connected}\}.$$

With the help of the ring-cut graph, we can characterize which cut constraints are facet-defining.

Theorem 15.2 *Let $G = (V, E)$ be a graph, $K > 0$ a given constant, and $W \subseteq V$ a subset of nodes, $\emptyset \neq W \neq V$. The inequality*

$$x(\delta(W)) \geq 2$$

defines a facet of $\mathcal{P}_{G,K}$ if and only if

1. *for all $e \in \delta(W)$, there exists $f \in \delta(W)$ such that $(C_{e,f})_K$ is two-connected;*

2. *in each connected component of $RCG_{W,K}$, there exists a cycle of odd cardinality;*

3. for all $e \in E(W) \cup E(V \setminus W)$, there exist $f, g \in \delta(W)$ such that $(C_{f,g} \setminus \{e\})_K$ is two-connected.

Moreover, Fortz and Labbé (2002) used the ring-cut graph to derive new valid inequalities. Let $G = (V, E)$ be a graph, $K > 0$ a given constant, $W \subseteq V$ a subset of nodes, $\emptyset \neq W \neq V$. If $S \subseteq \delta(W)$ is an independent subset in the ring-cut graph $RCG_{W,K}$, then

$$x(S) + 2x(\delta(W) \setminus S) \geq 3 \quad (15.6)$$

is a valid inequality for the 2CNBR problem. Inequalities (15.6) are called *ring-cut inequalities*. Fortz and Labbé (2002) also provide necessary conditions for these inequalities to be facet-defining.

15.4.2 Node-partition inequalities

In Section 15.3.2, we mentioned that node-partition inequalities (Grötschel and Monma, 1990) are valid for the two-connected network polytope.

Given a node $z \in V$ and a partition $W_1, W_2, \ldots, W_p$ ($p \geq 2$) of $V \setminus \{z\}$, the node-partition inequality for two-connected networks is

$$\frac{1}{2} \sum_{i=1}^{p} x(\delta_{G-z}(W_i)) \geq p - 1.$$

Since $\mathcal{P}_{G,K}$ is included in this polytope, node-partition inequalities are also valid for the 2CNBR problem. Fortz and Labbé (2002) give sufficient conditions for node-partition inequalities to define facets of $\mathcal{P}_{G,K}$.

15.5 RINGS OF BOUNDED CARDINALITY

An important application of ring constraints appears in topologies using the recent technology of *self-healing rings*. Self-healing rings are cycles in the network equipped in such a way that any link failure in the ring is automatically detected and the traffic rerouted by the alternative path in the cycle. Due to technological constraints, the length of self-healing rings must be limited. This is equivalent to set a bound on the length of the shortest cycle including each edge. In practice, the length of the ring is computed as the number of *hops*, i.e., the number of nodes that compose the ring. This corresponds to the particular case of 2CNBR that arises when a unit length is given to each edge. This model is only a first step in solving the self-healing ring network design problem, as it only ensures the presence of feasible rings in the network. The next step is dimensioning the rings, taking into account the demands and the additional cost for inter-ring transfer. A heuristic for the self-healing ring network design problem was proposed by Fortz et al. (2003b).

In this section, we present additional properties for this particular case, coming from Fortz et al. (2003a) and Fortz and Labbé (2004). We first describe a new class of valid inequalities, the cycle inequalities, that can be used to provide an alternative formulation of this special case. Another important result is a lower bound on the number of edges in any feasible solution of 2CNBR. This result is useful for showing that the problem is NP-complete for any fixed $K \geq 3$ and for deriving new valid inequalities.

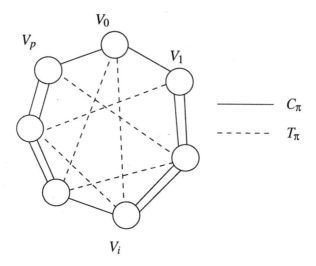

Figure 15.1 C_π and T_π

15.5.1 Cycle and metric inequalities

Let $G = (V, E)$ be a graph and $K \geq 3$. Let $\pi = (V_0, \ldots, V_p)$ be a partition of V such that $p \geq K$ and let $e \in [V_0, V_p]$. Moreover, let $C_\pi = \cup_{i=0}^{p-1}[V_i, V_{i+1}] \cup [V_0, V_p]$ and $T_\pi = \delta(V_0, \ldots, V_p) \setminus C_\pi$. Then, the inequality

$$x(T_\pi^e) \geq x_e \tag{15.7}$$

is valid for 2CNBR, with $T_\pi^e := T_\pi \cup ([V_0, V_p] \setminus \{e\})$, as illustrated in Figure 15.1. Inequalities (15.7) will be called *cycle inequalities*. Fortz et al. (2003a) showed that a formulation of 2CNBR is obtained by node-cut constraints, cycle inequalities and trivial inequalities.

Cycle inequalities are a special case of metric inequalities, that were studied by Fortz et al. (2000). Consider an edge $e := \{i, j\} \in E$ and a set of node potentials $(\alpha_k)_{k \in V}$ satisfying

$$\alpha_i - \alpha_j > K - 1.$$

Then

$$\sum_{f \in E - e} v_f x_f \geq x_e \tag{15.8}$$

is a valid inequality for $\mathcal{P}(G, K)$ where

$$v_f = \min\left(1, \max\left(0, \frac{|\alpha_l - \alpha_k| - 1}{\alpha_i - \alpha_j + 1 - K}\right)\right) \tag{15.9}$$

for all $f := \{k, l\} \in E - e$.

15.5.2 Cyclomatic inequalities

Theorem 15.3 *Let $G = (V, E)$ be a two-connected network with $n = |V|$ nodes and $m = |E|$ edges, such that there exists a covering of the network by cycles using at most K nodes. Then,*

$$m \geq M(n, K) := n + \min\left(\left\lceil\frac{n-K}{K-2}\right\rceil, \left\lceil\frac{n}{K-1}\right\rceil\right), \quad (15.10)$$

i.e., G contains at least $M(n, K)$ edges.

From this result, the complexity of the problem for K fixed can be established.

Problem 15.4 (R2CNBR) *Let $G = (V, E)$ be a graph, $K \geq 3$ a given constant and $B \geq 0$ an integer. To each edge $e \in E$ is associated a cost c_e and a unit length $d_e = 1$. Does there exists a subset $F \subseteq E$ of edges such that F_K is two-connected and $c(F) \leq B$?*

Theorem 15.5 *R2CNBR is NP-complete for any $K \geq 3$.*

Moreover, the result also applies to partitions of V, leading to a new class of valid inequalities:

Proposition 15.1 *Let $G = (V, E)$ be a graph with $n = |V|$ nodes, $K \geq 3$ a given constant, and $W_1, W_2, \ldots, W_p$ ($p \geq 2$) a partition of V. Then*

$$\frac{1}{2}\sum_{i=1}^{p} x(\delta(W_i)) \geq M(p, K) \quad (15.11)$$

is a valid inequality for $\mathcal{P}_{G,K}$.

Inequalities (15.11) are called *cyclomatic inequalities*. The inequality bounding the total number of edges (i.e., $p = n$) is facet-defining for complete graphs.

15.6 RELATED HOP-CONSTRAINED MODELS

Other network design problems with limits on the lengths of paths in the network have been studied. In most of these models, there must exist a path between any pair of nodes, or between a given root and any other node, using at most L links (hops). The hop-constrained minimum spanning tree problem was studied by Gouveia (1996); Gouveia and Magnanti (2003). Shortest paths with hop constraints have also received attention. The L-path polytope – the convex hull of incidence vectors of st-paths with no more than L edges – was first studied by Dahl (1999). Recently, Nguyen (2003) gave a complete description of this polytope. The directed version of the problem was studied by Dahl and Gouveia (2004); Dahl et al. (2004).

The hop-constrained network design problem (HCNDP) consists in finding at minimum cost a subgraph such that each pair of terminals is connected by at least K edge-disjoint paths using at most L links, where K and L are fixed constants. Balakrishnan and Altinkemer (1992) studied the problem for $K = 1$ within the framework of a more general model. The case $K = 1$ and $L = 2$ was considered by Dahl and Johannessen (2004). Huygens et al. (2004) consider a single pair of terminals with $K = 2$ and $L = 3$, and provide a complete description of the associated polytope.

Bibliography

M. Baïou, F. Barahona, and A.R Mahjoub. Separation of partition inequalities. *Mathematics of Operations Research*, 25:243–254, 2000.

A. Balakrishnan and K. Altinkemer. Using a hop-constrained model to generate alternative communication network design. *ORSA Journal on Computing*, 4(2):192–205, 1992.

A. Balakrishnan, T.L. Magnanti, A. Shulman, and R.T. Wong. Models for planning capacity expansion in local access telecommunication networks. *Annals of Operations Research*, 33:239–284, 1991.

R.E. Bellman. On a routing problem. *Q. Appl. Math.*, 16:87–90, 1958.

S.C. Boyd and T. Hao. An integer polytope related to the design of survivable communication networks. *SIAM J. Discrete Math.*, 6(4):612–630, 1993.

S.C. Boyd and F. Zhang. Transforming clique tree inequalities to induce facets for the 2-edge connected polytope. Technical Report TR-94-13, Department of Computer Science, University of Ottawa, 1994.

G.-R. Cai and Y.-G. Sun. The minimum augmentation of any graph to a k-edge-connected graph. *Networks*, 19:151–172, 1989.

S. Chopra. The k-edge-connected spanning subgraph polyhedron. *SIAM J. Discrete Math.*, 7(2):245–259, 1994.

W. Chou and H. Frank. Survivable communication networks and the terminal capacity matrix. *IEEE Transactions on Circuit Theory*, CT-17:192–197, 1970.

N. Christofides and C.A. Whitlock. Network synthesis with connectivity constraints — a survey. In J.P. Brans, editor, *Operational Research '81*, pages 705–723. North-Holland Publishing Company, 1981.

G. Cornuéjols, F. Fonlupt, and D. Naddef. The traveling salesman problem on a graph and some related integer polyhedra. *Mathematical Programming*, 33:1–27, 1985.

C.R. Coullard, A. Rais, R.R. Rardin, and D.K. Wagner. The 2-connected-Steiner-subgraph polytope for series-parallel graphs. Technical Report CC-91-32, Purdue University, 1991.

C.R. Coullard, A. Rais, R.R. Rardin, and D.K. Wagner. The dominant of the 2-connected-Steiner-subgraph polytope for W_4-free graphs. *Discrete Applied Mathematics*, 66:195–205, 1996.

C.R. Coullard, A. Rais, D.K. Wagner, and R.L. Rardin. Linear-time algorithms for the 2-connected Steiner subgraph problem on special classes of graphs. *Networks*, 23, 1993.

G. Dahl. Notes on polyhedra associated with hop-constrained paths. *Operations Research Letters*, 25:97–101, 1999.

G. Dahl and L Gouveia. On the directed hop-constrained shortest path problem. *Operations Research Letters*, 32:15–22, 2004.

G. Dahl and B. Johannessen. The 2-path network problem. *Networks*, 43:190–199, 2004.

G. Dahl, Foldnes N., and L Gouveia. A note on hop-constrained walk polytopes. *Operations Research Letters*, 32:345–349, 2004.

A. De Jongh. *Uncapacitated network design with bifurcated routing*. PhD thesis, Université Libre de Bruxelles, 1998.

M. Didi Biha and A.R. Mahjoub. k-edge connected polyhedra on series-parallel graphs. *Operations Research Letters*, 19:71–78, 1996.

E.W. Dijkstra. A note on two problems in connection with graphs. *Numer. Math.*, 1:269–271, 1959.

K.P. Eswaran and R.E. Tarjan. Augmentation problems. *SIAM Journal on Computing*, 5:653–665, 1976.

B. Fortz. *Design of Survivable Networks with Bounded Rings*, volume 2 of *Network Theory and Applications*. Kluwer Academic Publishers, 2000.

B. Fortz and M. Labbé. Polyhedral results for two-connected networks with bounded rings. *Mathematical Programming*, 93(1):27–54, 2002.

B. Fortz and M. Labbé. Two-connected networks with rings of bounded cardinality. *Computational Optimization and Applications*, 27(2):123–148, 2004.

B. Fortz, M. Labbé, and F. Maffioli. Solving the two-connected network with bounded meshes problem. *Operations Research*, 48(6):866–877, 2000.

B. Fortz, A.R. Mahjoub, S.T. Mc Cormick, and P. Pesneau. Two-edge connected subgraphs with bounded rings: Polyhedral results and branch-and-cut. IAG Working

Paper 98/03, Université Catholique de Louvain, 2003a. To appear in *Mathematical Programming*.

B. Fortz, P. Soriano, and C. Wynants. A tabu search algorithm for self-healing ring network design. *European Journal of Operational Research*, 151(2):280–295, 2003b.

A. Frank. Augmenting graphs to meet edge-connectivity requirements. *SIAM J. on Discrete Mathematics*, 5(1):22–53, 1992.

H. Frank and W. Chou. Connectivity considerations in the design of survivable networks. *IEEE Transactions on Circuit Theory*, CT-17:486–490, 1970.

D.R. Fulkerson and L.S. Shapley. Minimal k-arc connected graphs. *Networks*, 1: 91–98, 1971.

M.X. Goemans. Arborescence polytopes for series-parallel graphs. *Discrete Applied Mathematics*, 51:277–289, 1994.

R.E. Gomory and T.C. Hu. Multi-terminal network flows. *SIAM J. Appl. Math.*, 9: 551–570, 1961.

L Gouveia. Multicommodity flow models for spanning trees with hop constraints. *European Journal of Operational Research*, 95:178–190, 1996.

L. Gouveia and T.L. Magnanti. Network flow models for designing diameter-constrained minimum-spanning and steiner trees. *Networks*, 41(3):159–173, 2003.

M. Grötschel and C.L. Monma. Integer polyhedra arising from certain design problems with connectivity constraints. *SIAM J. Discrete Math.*, 3:502–523, 1990.

M. Grötschel, C.L. Monma, and M. Stoer. Computational results with a cutting plane algorithm for designing communication networks with low-connectivity constraints. *Operations Research*, 40(2):309–330, 1992a.

M. Grötschel, C.L. Monma, and M. Stoer. Facets for polyhedra arising in the design of communication networks with low-connectivity constraints. *SIAM J. Optimization*, 2(3):474–504, 1992b.

M. Grötschel, C.L. Monma, and M. Stoer. *Design of Survivable Networks*, volume 7 on Network models of *Handbooks in OR/MS*, chapter 10, pages 617–672. North-Holland, 1995a.

M. Grötschel, C.L. Monma, and M. Stoer. Polyhedral and computational investigations for designing communication networks with high survivability requirements. *Operations Research*, 43(6):1012–1024, 1995b.

F. Harary. The maximum connectivity of a graph. In *Proceedings of the National Academy of Sciences*, volume 48, pages 1142–1146, USA, 1962.

T.-S. Hsu and V. Ramachandran. A linear time algorithm for triconnectivity augmentation. In *Proc. 32nd Annual IEEE Symposium on Foundations of Computer Science*, pages 548–559, 1991.

T.-S. Hsu and V. Ramachandran. On finding a minimum augmentation to biconnect a graph. *SIAM Journal on Computing*, 22:889–891, 1993.

D. Huygens, A.R. Mahjoub, and P. Pesneau. Two edge-disjoint hop-constrained paths and polyhedra. *SIAM Journal on Discrete Mathematics*, 18(2):287–312, 2004.

H. Kerivin and A.R. Mahjoub. Separation of the partition inequalities for the (1,2)-survivable network design problem. *Operations Research Letters*, 30:265–268, 2002.

J.B. Kruskal. On the shortest spanning subtree of a graph and the traveling salesman problem. *Proc. Amer. Math. Soc.*, 7:48–50, 1956.

E.L. Lawler. *Combinatorial Optimization : Networks and Matroids*. Holt, Rinehart and Wilson, New-York, 1976.

E.L. Lawler, J.K. Lenstra, A.H.G. Rinnooy Kan, and D.B. Shmoys, editors. *The Traveling Salesman Problem*. John Wiley & Sons, New-York, 1985.

A.R. Mahjoub. Two-edge connected spanning subgraphs and polyhedra. *Mathematical Programming*, 64:199–208, 1994.

K. Menger. Zur allgemeinen kurventheorie. *Fundamenta Mathematicae*, 10:96–115, 1927.

C.L. Monma and D.F. Shallcross. Methods for designing communications networks with certain two-connected survivability constraints. *Operations Research*, 37(4):531–541, 1989.

D. Naor, D. Gusfield, and Ch. Martel. A fast algorithm for optimally increasing the edge-connectivity. In *Proceedings of the Foundation of Computer Science '90*, pages 698–707, St. Louis, 1990.

G.L. Nemhauser and L.A. Wolsey. *Integer and combinatorial optimization*. Wiley-Interscience series in discrete mathematics and optimization. Wiley, 1988.

V.-H Nguyen. A complete description for the k-path polyhedron. Technical report, LIP6, 2003.

R.C. Prim. Shortest connection networks and some generalizations. *Bell System Tech. J.*, 36:1389–1401, 1957.

A. Rosenthal and A. Goldner. Smallest augmentation to biconnect a graph. *SIAM Journal on Computing*, 6:55–66, 1977.

M. Stoer. *Design of Survivable Networks*, volume 1531 of *Lecture Notes in Mathematics*. Springer-Verlag, 1992.

J.W. Suurballe. Disjoint paths in a network. *Networks*, 4:125–145, 1974.

J.W. Suurballe and R.E. Tarjan. A quick method for finding shortest pairs of disjoint paths. *Networks*, 14:325–336, 1984.

S. Ueno, Y. Kajitani, and H. Wada. Minimum augmentation of a tree to a k-edge-connected graph. *Networks*, 18:19–25, 1988.

J.A. Wald and C.J. Colbourn. Steiner trees, partial 2-trees, and minimum ifi networks. *Networks*, 13:159–167, 1983.

T. Watanabe and A. Nakamura. Edge-connectivity augmentation problems. *Computer and System Sciences*, 35:96–144, 1987.

P. Winter. Generalized Steiner problem in halin networks. In *Proc. 12th International Symposium on Mathematical Programming*. MIT, 1985a.

P. Winter. Generalized Steiner problem in outerplanar networks. *BIT*, 25:485–496, 1985b.

P. Winter. Generalized Steiner problem in series-parallel networks. *Journal of Algorithms*, 7:549–566, 1986a.

P. Winter. Topological network synthesis. In B. Simeone, editor, *Combinatorial Optimization - Como 1986*, volume 1403 of *Lecture Notes in Mathematics*, pages 282–303. Springer-Verlag, 1986b.

P. Winter. Steiner problems in networks : a survey. *Networks*, 17:129–167, 1987.

16 DESIGN OF SURVIVABLE NETWORKS BASED ON P-CYCLES

Wayne D. Grover, John Doucette, Adil Kodian,
Dion Leung, Anthony Sack, Matthieu Clouqueur and Gangxiang Shen

TRLabs and University of Alberta
Edmonton, Alberta, Canada.
grover@trlabs.ca

Abstract: p-Cycles are a recently discovered and promising new paradigm for survivable networking. p-Cycles simultaneously provide the switching speed and simplicity of rings with the much greater capacity-efficiency and flexibility for reconfiguration of a mesh network. p-Cycles also permit shortest-path routing of working paths (as opposed to ring-constrained working path routing), which adds further to network capacity efficiency. Operationally p-cycles are similar to BLSRs in that, upon failure, switching actions are required at only two nodes and both those nodes are fully pre-planned as to the actions that are required for any failure detected at their sites. With the optimization models in this chapter, entire survivable transport networks can be easily designed with essentially the same spare to working capacity (redundancy) ratios as optimized span-restorable mesh networks. p-Cycles thus bridge the ring versus mesh debate that dominated work in survivable networks through the 1990s and provide the best of both worlds: *the efficiency of mesh with the speed of rings.*

Keywords: p-Cycles, survivable networking, optimization, capacity design, node protection, Multiple Quality of Protection.

16.1 INTRODUCTION

16.1.1 Span-Protecting p-Cycles

The operation of a p-cycle is portrayed in Figure 16.1. A single unit-capacity p-cycle, as in Figure 16.1(a), is a closed path composed of one spare channel on each span it crosses. When a failure occurs on a span covered by the cycle, the p-cycle provides one protection path for the failed span, as shown in Figure 16.1(b). In this aspect, p-cycles operate like a unit-capacity bi-directional line switched ring (BLSR). But p-cycles also protect so-called *straddling spans*, which are spans that have end nodes on the cycle but are not themselves on the cycle, as shown in Figure 16.1(c). The significance is that, because the *p*-cycle itself remains intact when a straddling span fails as in Figure 16.1(c), it provides *two* protection paths for each straddling span failure scenario, and straddling spans themselves require no spare capacity. This apparently minor difference actually has a great impact on the capacity requirements of *p*-cycle networks. *p*-Cycle network designs are in many cases exactly as efficient (or within a few percent) of the capacity efficiency of a span-restorable mesh network (Grover and Stamatelakis, 1998; Stamatelakis and Grover, 2000b).

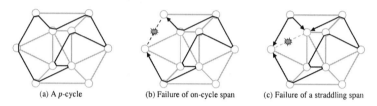

(a) A *p*-cycle (b) Failure of on-cycle span (c) Failure of a straddling span

Figure 16.1 An illustration of basic *p*-cycle concepts. (Grover and Stamatelakis (1998))

16.1.2 Node-Encircling p-Cycles

For a span failure, it is intrinsically possible (and often the goal) to achieve 100% restoration of affected demands by network re-routing. But when a *node* fails, no method of network-level re-configuration or re-routing can restore the originating or terminating demands at the failed node. Thus "node restoration" is really a misnomer. It is only pre-failure transiting flows through a node that can be restored by any type of network-level response. Source-sink flows at the failed node itself are inherently unrestorable by network-level re-routing. Therefore span restoration and protection of transiting flows upon a node failure have inherently different target recovery goals. Perhaps counter-intuitively as a result, 100% of all transiting paths can be protected against node failures within the spare capacity required for 100% span-failure restorability because fewer affected demands are actually considered in the restorability target.

Node-encircling *p*-cycles (Stamatelakis and Grover, 2000a) provide an efficient strategy for protecting such transiting flows at every node, especially when applied in the MPLS layer to protect against router failure. In this context a set of virtual MPLS-defined node-encircling *p*-cycles can complement an underlying set of physical-layer span-protecting *p*-cycles using the same physical capacity. A node-encircling *p*-cycle provides an alternate path amongst all of the nodes that are adjacent to the failed node. To do this a node-encircling *p*-cycle must contain all of the nodes that were adjacent to the failed node, but not the failed node itself. This constitutes a kind of "perimeter fence" which is assured to be intersected at ingress and egress by all transiting flows that may be affected by the given node's failure. Such a *p*-cycle must therefore be constructed within the sub-graph that results when the protected node is itself removed from the network, and it may or may not be possible to form a simple cycle within the resulting sub-network. There is, however, always a logically encircling cycle for each node in any two-connected (pre-failure) graph. Stamatelakis and Grover (2000a) provides details of the real time rerouting to and from node encircling *p*-cycles and how node failures are distinguished from span failures. Section 16.4 of this chapter treats node-encircling *p*-cycle network optimization.

16.1.3 Path-Segment or Flow-Protecting p-Cycles

A third type of *p*-cycle provides a path-oriented correspondent to basic *p*-cycles, which are span protecting. This is somewhat analogous to span-restorable and path-restorable mesh networks. So-called *path-segment* or *flow-protecting p*-cycles generalize straddling spans and on-cycle spans to the protection of contiguous flows of demand over path segments that may contain multiple-spans and nodes. The protected path segments do not have to be the whole path end-to-end as taken by a demand, but rather any portion of the entire path. Although they would yield even greater capacity efficiency, whole networks based on flow *p*-cycles would be fairly complex to operate. However, selective use of just one or a few flow *p*-cycles in conjunction with other protection schemes may be attractive. One such use is to support transparent optical transport of express flows through a regional network. Another is to use a flow *p*-cycle around the perimeter of an autonomous system domain to provide a single unified scheme for protecting all flows that completely transit the domain. The idea of using these two types of *p*-cycles in a network is presented in Shen and Grover (2003b;a) as well as the related design and operational theory for flow *p*-cycles, but we do not treat them further in this chapter.

16.1.4 Scope

In this chapter we emphasize the basic logical design problems for *p*-cycles. The extra complications on these models that pertain for optical networks in which precise wavelength assignment is explicitly modeled in the design problem is omitted. The models given thus pertain directly to a SONET-type transport network where full channel interchange is available at each node, or to an optical network with full wavelength conversion at each node (i.e., an OEO network), or to an optical network that has a sufficient number of wavelength conversion points so that the level of wavelength blocking is simply negligible. The fundamental problems of network architecture, design and evolution are issues of getting the capacity, topology, protection and working routing right. Given the right capacities and architecture in basic planning, the more operational problem of detailed wavelength assignment is essentially always solvable as a sub-problem with dozens of algorithms now available. Wavelength assignment without blocking due to "color mismatch" also fundamentally becomes easier as the number of channels available on each span increases (an implication of basic traffic theory). On the other hand, explicit representation of the decision variables for wavelength assignment would make most of the following ILP models virtually intractable above small sizes, unnecessarily removing one of the network planner and researchers most valuable tools (i.e. ILP). For these reasons, we treat the problems covered in this chapter as capacity allocation problems without individual wavelength assignment aspects.

For readers interested in aspects of *p*-cycles that are outside the current scope, Grover and Stamatelakis (1998) and Stamatelakis and Grover (2000b) are suggested as primary sources. In addition, Stamatelakis and Grover (2000a) presents the application of *p*-cycles in the IP or MPLS network layers and node-encircling *p*-cycles, and the use of *p*-cycles in WDM networks is addressed in Schupke et al. (2002). The chapter on *p*-cycles in Grover (2003) is also a comprehensive source. For an archive of all *p*-cycle literature known to us, we suggest our website A.Sack and W.D. Grover (2005).

16.1.5 Basic Notation and Models for p-Cycle Network Design

To define the basic *p*-cycle network design problem, we first introduce the following notation. This initial set of symbols and definitions applies for the remainder of the chapter, and will be supplemented with additional definitions as applicable for the more advanced design problems in subsequent sections.

- **Sets:**
  - S is the set of spans in the network, and is usually indexed by i for a failure span, and j for surviving spans.
  - M is the set of available modular transmission system capacities, and is indexed by m.
  - P is the set of (simple) cycles of the graph eligible for formation of *p*-cycles, and is usually indexed by k. This set of cycles is determined by a pre-processing method, and eligibility may be limited in some way (e.g., circumference).

- **Input Parameters:**
  - w_i is the number of working channels (or capacity units) on span i that require protection.
  - $x_{i,k} \in \{0,1,2\}$ encodes the number of protection relationships provided to span i by a unit-sized copy of *p*-cycle k. $x_{i,k} = 2$ if span i straddles cycle k, $x_{i,k} = 1$ if span i is on cycle k, and $x_{i,k} = 0$ in all other cases.
  - $\delta_{j,k} \in \{0,1\}$ encodes the spans on a *p*-cycle. $\delta_{j,k} = 1$ if cycle k includes span j (i.e., if $x_{i,k} = 1$), and $\delta_{j,k} = 0$ if it does not (i.e., $x_{i,k} \neq 1$).
  - C_j is the cost of a unit of capacity (i.e., a single channel) placed on span j (in the non-modular or integer-capacity design context). This generally includes considerations of the actual length of the span, the technology employed etc.
  - Z^m is the number of channels provided by a modular transmission system of type m.
  - C_j^m is the cost to install one transmission system or module of type m on span j (also includes the same considerations as C_j, except that now it is per module on each span).

- **Decision Variables:**

    - $s_j \geq 0$ is the integer number of spare channels assigned to span j in the design.
    - $n_k \geq 0$ is the integer number of unit-capacity p-cycles placed on cycle k by the design.
    - t_j^m is the integer number of modules of type m placed on span j in the design.

The basic p-cycle design problem is modeled as an integer linear programming (ILP) formulation for *spare capacity allocation* (SCA). Working demands are already routed (typically via shortest path routing or perhaps some flow-leveling approach) and the problem is to find the optimal set of p-cycles that can fully protect all working capacities, w_i, while minimizing the total investment in spare capacity needed to support those p-cycles. The ILP is formulated as follows:

$$\min \sum_{\forall j \in S} C_j s_j \tag{16.1}$$

$$\text{s.t.} \quad w_i \leq \sum_{\forall k \in P} x_{i,k} n_k \quad \forall i \in S \tag{16.2}$$

$$s_j \geq \sum_{\forall k \in P} \delta_{j,k} n_k \quad \forall j \in S \tag{16.3}$$

The objective we seek to minimize (16.1) is the total cost of spare capacity assignments. Constraint (16.2) asserts that the total number of protection relationships provided for span i by the set of p-cycles chosen must meet or exceed the number of working channels to be protected on span i. Constraints set (16.3) couples those requirements to the objective function by asserting that the spare capacity on each span must be sufficient to implement the set of p-cycles that provide this restorability to each span. Generically, these two constraint systems are often called the *restorability* and *spare capacity constraints*, respectively.

The SCA problem can easily be extended to recognize the modular and non-linear cost nature of the capacity increments of the actually available transmission systems, as in Doucette and Grover (2000). This is done by keeping constraints (16.2) and (16.3), changing the objective function to that in equation (16.4), and adding the new total modular capacity constraint in equation (16.5) as follows:

$$\min \sum_{\forall j \in S} \sum_{\forall m \in M} C_j^m \cdot t_j^m \tag{16.4}$$

$$\text{s.t.} \, w_j + s_j \leq \sum_{\forall m \in M} t_j^m \cdot Z^m \quad \forall j \in S \tag{16.5}$$

It is useful to note that the cost parameters associated with the available set of modular transmission capacities, C_j^m, can follow a significantly non-linear progression versus the corresponding capacity vales, Z^m, allowing economy-of-scale benefits to be captured in the network design (Doucette and Grover, 2000).

16.1.6 Efficiency of p-Cycles and p-Cycle Network Designs

In other approaches to survivable network design, where the decision variables typically pertain to restoration path choices or flow assignments to eligible routes, it is hard to identify general figures of merit for individual candidate backup paths or eligible restoration routes. In contrast, p-cycles lend themselves rather handily to compact measures of the potential (or actual) efficiency of individual p-cycles in different network contexts, and we can easily define several efficiency metrics associated with each p-cycle. Grover and Doucette (2002) defines the *topological score* (TS_k) of a candidate cycle k as the number of protection relationships it is capable of providing as a unit capacity p-cycle, and the *a priori efficiency* (AE_k) of a cycle as the ratio of TS_k to the total cost of constructing a unit-capacity copy of a p-cycle on cycle k:

$$TS_k = \sum_{\forall i \in S} x_{i,k} \qquad (16.6)$$

$$AE_k = \frac{TS_k}{\sum_{\forall i \in S: x_{i,k}=1} C_i} = \frac{\sum_{\forall i \in S} x_{i,k}}{\sum_{\forall i \in S: x_{i,k}=1} C_i} = \frac{\sum_{\forall i \in S} x_{i,k}}{\sum_{\forall j \in S} \delta_{j,k} C_j} \qquad (16.7)$$

We call AE_k an *a priori efficiency* because it reflects the best *potential* efficiency achievable by the cycle if all its straddlers are fully loaded with two working channels each and its on-cycle spans are loaded with one working channel. But this is calculated prior to obtaining any information about actual working loads, and this may not be the actual utilization efficiency of the p-cycle in any specific design. When the actual working channel quantities protected by the p-cycles are known, a closely related *working-weighted efficiency* measure can be defined (in Section 16.3.3).

The significance of p-cycles having such compact and highly characteristic efficiency measures is two-fold. First, they enhance intuitive understanding about what determines a good p-cycle and help to diagnose effects such as why an unconstrained optimum solution may tend to contain many large cycles, or even Hamiltonians. Second, these efficiency measures can fairly easily be built upon to realize some simple and effective heuristics, at least for the basic p-cycle design problem. In addition, we can use such measures to identify high merit cycles based on real insights about what makes a good p-cycles in the context of a given network design. The AE_k measure and concepts closely related to it are developed further in the discussion of algorithmic p-cycle design heuristics in Section 16.3.

The capacity-efficiency of 100% span-restorable p-cycle designs can be remarkably good in practice. In the special case of a particular class of semi-homogenous networks (probably most applicable for whole fiber-level protection), they can, by construction, actually *reach* the long-recognized lower bound on redundancy of $1/(\bar{d}-1)$ (see Doucette and Grover (2001)), where $\bar{d}$ is the network average nodal degree. A sample solution to SCA follows to illustrate the first claim. Figure 16.2 shows a 13-node, 23-span test network, with $\bar{d} = 3.54$. For simplicity, all span costs are $C_j = 1$, and the w_i values on each span in Figure 16.2(a) are from least-hop routing of a single demand unit between each node pair. In the specific design context of Figure 16.2(a) there are seven individual p-cycles constructed on five unique cycles, illustrated in

Figure 16.2(b)-(f). The design is 100% span-restorable with a logical (channel count) redundancy of 85/158 = 53.8%.

This is a highly efficient design solution that could be used for protection at the lightwave-channel level. But it is also possible with p-cycles to reach even greater efficiency. It was shown in Sack and Grover (2004) that if a single unit-capacity Hamiltonian cycle is used as a p-cycle in which all straddling spans are fully loaded, then we could achieve a 39.4% redundancy in that same network, which is exactly the $(1/(d-1))$ lower bound for generalized span-protection - $(1/(3.54-1) = 39.4\%)$. To see this, consider that in any such case we will have the ability to protect as many as

$$\sum_{\forall j \in S} w_j = |N| + 2(|S| - |N|) = |N| + 2\left(\frac{|N|\bar{d}}{2} - |N|\right) = |N|(\bar{d} - 1)$$

working channels (where N is the set of all nodes in the network), making the overall redundancy exactly equal to

$$\mathcal{R} = \sum_{\forall j \in S} s_j / \sum_{\forall j \in S} w_j = \frac{|N|}{|N|(\bar{d} - 1)} = \frac{1}{\bar{d} - 1}.$$

Clearly, these are motivating properties for a scheme that retains ring-like switching speed and structural simplicity. Heretofore, redundancies as low as 40% or so for 100% span survivability were only ever even approachable through end-to-end path restoration technologies that are far more complex to design and operate and much slower acting in practice.

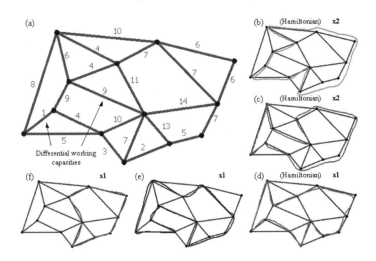

Figure 16.2 A fully restorable p-cycle network.(Sack and Grover (2004))

16.2 JOINT WORKING & SPARE CAPACITY DESIGN

To design a p-cycle network, one can first route the working demands via shortest paths (or by some other means) and then solve the minimum spare capacity allocation (SCA) problem. An alternate strategy is to optimize the choice of working routes in conjunction with the placement of p-cycles and spare capacity, with the goal of minimizing total (working plus spare) capacity. We call this the *joint* capacity allocation (JCA) problem.

The main advantage of joint optimization is that working routing is effectively allowed to deviate from shortest paths onto paths that allow a more efficient use of p-cycles. Allowing the optimization to consider alternate routings of working demands plays an important role in reducing total capacity requirements. One study in Iraschko et al. (1998) showed that total capacity requirements can be reduced by 4% to 27% in a span-restorable network and an average of 7% in a path-restorable network. Capacity savings in a p-cycle network are comparable; work in Grover and Doucette (2002) shows that for the COST 239 network (Schupke et al., 2002; Batchelor et al., 1999), capacity redundancy decreases by 25%, and translates to a 13% reduction in total capacity requirements. Studies on other networks show similar capacity reductions, attributable to JCA over SCA.

The SCA formulation in Section 16.1 can easily be modified to provide the JCA formulation. To do so, the prior w_i input parameters become decision variables, we modify the objective function, and we add new set, parameter, and variable notations as well as two new constraints to ensure the routing of working demands and adequate working capacity to support them. In addition to the notation from Section 16.1.5, we need the following additional notation for the joint problem:

- **Sets:**
  - D is the set of working lightpath demands, and is typically indexed by r.
  - Q^r is the set of all distinct pre-determined eligible routes available to carry the working paths for demand relation r, and is typically indexed by q.

- **Input Parameters:**
  - d^r is the number of lightpath demand units for demand relation r.
  - $\varsigma_i^{r,q} \in \{0,1\}$ is a parameter that encodes working routes. If $\varsigma_i^{r,q} = 1$, working route q used for demand relation r crosses span i. If $\varsigma_i^{r,q} = 0$, working route q used for demand relation r does not cross span i.

- **Decision Variables:**
  - $g^{r,q} \geq 0$ is the amount of working flow assigned to working route q used for demand relation r.
  - $w_i \geq 0$ is the integer amount of working capacity that is placed on span i (this is now a decision variable).

All previous notation from the SCA formulation of the problem in Section 16.1.5 still applies. In order to consider working capacities in addition to spare capacities, we also make the following change to the objective function:

$$\min \sum_{\forall j \in S} C_j(s_j + w_j) \tag{16.8}$$

Finally, the restorability and spare capacity constraint sets in constraints (16.2) and (16.3) also still apply, as do the following two new constraint sets:

$$\sum_{\forall q \in Q^r} g^{r,q} = d^r \quad \forall r \in D \tag{16.9}$$

$$\sum_{\forall r \in D} \sum_{\forall q \in Q^r} \varsigma_i^{r,q} g^{r,q} = w_i \quad \forall i \in S \tag{16.10}$$

Just as the SCA design problem chooses between eligible cycles, JCA also includes a choice between eligible routes for working demands. Constraint (16.9) is the working routing equivalent to equation (16.2). It ensures that the total working flow assigned to all eligible working routes for demand relation r is sufficient to fully route it. Note that as written, there is nothing in equation (16.9) that prevents a working flow bundle to be split (integrally) over multiple working routes. Equation (16.10) sizes the working capacities on each span in much the same way that equation (16.3) does so for spare capacity. The main structural difference between equations (16.10) and (16.3) is that in equation (16.10), the working flow for each demand relation is applied to each span *at the same time*, hence the double summation. In contrast, when sizing spare capacity, restoration flow is applied separately for each span failure scenario. Also, we use a strict equality here, rather than an inequality as in equation (16.3) for spare capacity. This is because working capacity assignment is equivalent to that required to carry all lightpath demands, which are all routed simultaneously, while spare capacity in equation (16.3) is sized to accommodate individual failure scenarios separately, some of which may require more or less spare capacity than others on any particular span. Note that the SCA design problem can actually be solved using this JCA formulation, by simply providing only a single eligible working route for each demand.

Analyses of working routing resulting from the JCA designs confirm that they are coordinating with p-cycle selections to provide a more balanced and efficient use of protection capacity. In the COST 239 network, working capacity costs increased by 5% in the JCA design relative to the SCA design, while spare capacity decreased by 43% (Grover and Doucette, 2002). This improvement was observed when we only provided all routes up to the 10th shortest eligible working route per O-D pair. Allowing working routes to deviate even further from their shortest paths rarely provides greater reductions in total capacity. Studies in Stidsen and Thomadsen (2004) show that joint optimization provided between 3% and 22% capacity savings versus the SCA designs on several test networks, and in Rajan and Atamturk (2002), capacity efficiency in JCA was found to increase by 30% versus SCA over a variety of test case networks.

16.3 HEURISTIC METHODS FOR THE P-CYCLE NETWORK DESIGN PROBLEM

The ILP methods developed above are quite effective at optimizing p-cycle network designs, and are an essential tool in many research activities, particularly when doing comparative analysis of new survivability mechanisms or design strategies. However, they generally require enumeration of a representative sample of all possible eligible cycles in a network graph (or the full set if strict optimality is required). This approach is suitable in practice for small or medium sized networks, but the number of possible cycles in a graph increases exponentially with the size of the graph. Providing a suitably large subset of eligible cycles for a large network, therefore, becomes difficult and time-consuming, and the number of eligible cycles directly increases the complexity of the ILP itself. We will therefore now address several techniques for dealing with these issues.

16.3.1 Eligible Cycle Pre-Selection Strategies

In Section 16.1.6, topological score, TS_k, and *a priori efficiency*, AE_k, were defined. In terms of these figures of merit, cycles with a large number of straddling spans relative to their size (or cost) are identified as having the highest potential efficiency as p-cycles. We can use this knowledge to pre-select high-merit cycles for population of the set of candidate cycles, P, when solving the SCA or JCA (or other) problems for large networks. Rather than providing all possible distinct cycles to the ILP problem, we can simply rank them by TS_k or AE_k and then provide only a limited number of the top-ranked cycles for representation in the model. Work in Grover and Doucette (2002) showed that this method of limiting eligible cycle sets in the COST 239 network was effective in reducing problem complexity and solution runtimes, while still reaching optimality in the resultant designs, using both SCA and JCA models. In the COST 239 network, there are a total of 3531 possible distinct cycles in the graph, but when given only the 250 top-ranked cycles by the AE_k metric, the SCA problem was able to be solved to within 1% of optimal in slightly more than 1/1000th of the time it took the same problem when all eligible cycles were provided. Similarly, when the JCA problem was provided with only 50 eligible cycles (again they were the top ranked by AE_k), it was solved to within 0.16% of optimal in 1/17th of the solution time of the complete problem. The cycle candidates provided by AE proved to be a more effective metric than TS, and allowed the eligible cycle set to be reduced significantly further while retaining optimality. The main drawback of this method is that it still requires enumeration of every cycle in the network prior to pre-selection in order to properly calculate AE_k or TS_k and rank them.

Work in Sack (2004) tested the hypothesis that using a candidate cycle set that "mimics" the statistical distribution of cycle lengths may permit even fewer cycles to be represented in the ILP problem for SCA or JCA. As an experiment to test this idea, following enumeration of the full set of cycles, the hop-length distribution observed for that set was scaled to reflect the number of cycles desired in the representative sample. Randomly selected cycles from the full set were then used to make up the sampled cycle set using a "bin-filling" approach - cycles of a certain length were only added if needed to achieve the necessary distribution. The random selection algorithm is expressed by the following pseudo-code:

```
StatisticalMimicry(NumCyclesDesired) {
    Find FullCycleSet (with TotalFullCycles number of elements)
    Find FullNumCyclesLength[l] for each length l in FullCycleSet
    If TotalFullCycles < NumCyclesDesired
        NumCyclesDesired = TotalFullCycles
    For each length l
        StatNumCyclesLength[l] = RoundUpToInteger(FullNumCyclesLength[l]
            * (NumCyclesDesired / TotalFullCycles))
    Do {
        Pick a cycle at random from FullCycleSet (length
        is l)
        If StatNumCyclesLength[l] > 0
            Add cycle to StatCycleSet
            Set length l of that cycle in FullCycleSet to 0
                (so it cannot be used if radomly chosen again)
    } while any value of StatNumCyclesLength[l] > 0
    Return StatCycleSet
}
```

Results showed that the average solution cost of several runs of the statistical sampling approach was lower than the cost with either AE_k or a set of short cycles, for three of four test networks where the same limited size of candidate cycle sets (as a percentage of the full set) was applied. Another test showed that the AE_k, shortest cycles, and statistical sampling methods were closer in performance at both very low and very high average nodal degree. Between these extremes, the average of the statistical sampling runs consistently produced lower-cost design solutions than either of the other two methods. Notably, in many cases, at least one of the statistically sampled test runs produced a design solution very close to the baseline solution cost with the full cycle set. Thus, even though the performance of a single run can be highly variable due to the random sampling approach, repeating the process several times can normally provide a very good result.

16.3.2 Other Candidate Cycle-Enumeration Strategies

The above methods are useful for reducing the eligible cycle set provided to the ILP model, but if the need for cycle-enumeration itself is a concern, there are methods to deal with that issue as well. The Straddling Link Algorithm (SLA) was proposed in Zhang and Yang (2002) as a means of enumerating only a very small subset of the possible cycles in a network. For each span in the network, SLA makes two calls to Dijkstra's shortest path algorithm to find the shortest route and the next shortest disjoint route connecting the span's end nodes, not using that span itself. When those two routes are combined end-to-end, the resultant *primary p-cycle* will have the span as a straddler. Although SLA is very fast (there are only $|S|$ such cycles to generate), the cycles produced tend to be quite inefficient and ill suited by themselves for overall capacitated network designs because they usually have only a single straddler and they are small cycles. While the fact that there are only at most $|S|$ primary p-cycles produced by SLA is the basis for its speed, such a small set of eligible cycles is far too limited for an optimization model to form efficient protection relationships.

The SLA procedure was therefore used as a basis for more advanced cycle-generation algorithms in Doucette et al. (2003), where primary *p*-cycles of a network are transformed into a larger set of higher-efficiency cycles. Three key algorithms are called *SP-Add*, *Expand* and *Grow*. The *SP-Add* algorithm starts with an existing cycle (say a primary *p*-cycle), and for each of its on-cycle spans, a new cycle is created by replacing the span with the shortest path connecting its end nodes such that the path is node-disjoint from the original cycle as a result. The span that is replaced becomes a straddling span in the new resultant cycle, which will usually have higher AE_k value than the original cycle. The *SP-Add* operation can be expressed by the following pseudo-code:

```
SP-Add(OriginalCycleSet) {
    Initialize NewCycleSet
    For each cycle p in OriginalCycleSet {
        Mark all spans and nodes on cycle p
        For each span i on cycle p {
            Dijkstra(i, unmarked spans/nodes) -> r
            If Dijkstra() returns a route r {
                Let cycle x = cycle p
                Add route r to cycle x
                Remove span i from cycle x
                Add cycle x to NewCycleSet
            }
        }
        Unmark all spans and nodes
    }
    Return NewCycleSet
}
Dijkstra() {
    Find the shortest route r between the end nodes of i using
    unmarked spans and nodes only
    Return route r
}
```

When the *SP-Add* algorithm was applied to the set of primary *p*-cycles (obtained from SLA), average AE_k values increased 8%-16% and their average on-cycle span coverage increased 36%-44% on the two test case networks in Doucette et al. (2003). The number of cycles generated by *SP-Add* increased from 29 in SLA to 90 for one test network and from 50 to 190 for the other, thereby producing larger and more efficient eligible cycle sets for the SCA and JCA models but in no case enumerating all possible cycles in the network graph.

The *Expand* and *Grow* algorithms are somewhat more complicated than *SP-Add*. In *Expand*, the *SP-Add* algorithm is modified so that after a span has been converted into a straddler, we move on to the next spans in the original cycle seeking yet further route replacements for spans to create ever-expanding candidate cycles. This is done until every span in the original cycle has been visited. Each time a span is replaced, the replacement route must not only be node-disjoint from the original cycle but also from each previous route already added. *Expand* is expressed by the following pseudo-code:

```
Expand(OriginalCycleSet) {
   Initialize NewCycleSet
   For each cycle p in OriginalCycleSet {
      Let cycle p' = cycle p
      Mark all spans and nodes on cycle p
      For each span i on cycle p {
         Dijkstra(i, unmarked spans/nodes) -> r
         If Dijkstra() returns a route r {
            Add route r to cycle p'
            Remove span i from cycle p'
            Mark all spans and nodes on route r
            Add cycle p' to NewCycleSet
         }
      }
      Unmark all spans and nodes
   }
   Return NewCycleSet
}
```

It was found in Doucette et al. (2003) that the new cycles resulting from the *Expand* algorithm are 10%-18% more efficient (by AE_k) than the original cycle set from SLA, and the number of cycles increased from 29 to 96 for one network and from 50 to 197 for the other. The *Grow* algorithm is a further modification of the *Expand* algorithm, where we effectively reset the span counter, i, each time we create a new cycle p'. This is essentially the same as forcing the *Expand* algorithm to recursively call itself each time a new cycle is found and added to the set. The *Grow* algorithm improves average AE_k values of the original cycle set from SLA by 25% and 41% in the two test case networks in Doucette et al. (2003), and the number of cycles was increased to 839 and 2407 respectively. The *Grow* algorithm can be expressed by the following pseudo-code:

```
Grow(OriginalCycleSetA) {
   Initialize SPAddCycleSetB
   SPAddCycleSetB = SP-Add(OriginalCycleSetA)
   Initialize NewCycleSet
   For each cycle p in SPAddCycleSetB {
      Let cycle p' = cycle p
      For each span i on cycle p' {
         Mark all spans and nodes on cycle p'
         Dijkstra(i, unmarked spans/nodes) -> r
         If Dijkstra() returns a route r {
            Add route r to cycle p'
            Remove span i from cycle p'
            Add cycle p' to NewCycleSet
            Restart count of i to first span in p'
         }
         Unmark all spans and nodes
      }
   }
   Add OriginalCycleSetA to NewCycleSet
   Add SPAddCycleSetB to NewCycleSet
}
```

The increases in the number of eligible cycles generated and their average AE_k values are only indications that the cycle sets have the potential to provide low-redundancy protection in a capacitated network. Results in Doucette et al. (2003) showed that for one test network, if the set of 2407 cycles produced by the *Grow* algorithm were provided to an SCA design model, the gap to optimality was 0.5% relative to when a conventional eligible cycle set enumeration method was used to provide the 15000 shortest distinct cycles in the network. Additional strategies for improving the eligible cycle sets provided by *Grow* include supplementing the cycle set with the set of all faces in the graph, using other methods for combining pairs of cycles (see Grover (2003)), or even adding a small set of the shortest distinct cycles (for which efficient depth-first search algorithms can be applied).

16.3.3 Greedy Heuristic Methods

Let us now further consider the *AE* metric and outline a greedy iterative heuristic algorithm to approximate SCA solutions. AE_k represents a candidate cycle k's *potential efficiency* as a p-cycle. Why is it only a potential, however? The reason is that in an actual network design, any given candidate cycle may not find working capacity present on every span it is capable of protecting. Only if sufficient working capacity exists on all of those spans will a p-cycle actually realize *AE* efficiency. Even in the most efficiently designed networks, inspection of the chosen p-cycles will usually show that they exhibit a variety of AE_k values. It is also easy to demonstrate that for a given working capacity arrangement, a cycle with relatively low AE_k might be preferable to one with a higher AE_k (e.g., if one span is very heavily loaded, then a very small cycle with that span as a straddler would suffice). So in the complete problem in the presence of capacitated working demand, finding cycles of high AE_k is only a first step towards finding a combination of p-cycles to collectively protect the network in an efficient manner. In light of these insights, AE_k can be modified to approximate a cycle's *actual efficiency* or *working-weighted efficiency* (E_k^W) based in part on the working capacities of the spans it protects as in Doucette et al. (2003):

$$E_k^W = \frac{\sum\limits_{\forall i \in S} w_i x_{i,k}}{\sum\limits_{\forall i \in S: x_{i,k}=1} C_i} = \frac{\sum\limits_{\forall i \in S} w_i x_{i,k}}{\sum\limits_{\forall j \in S} \delta_{j,k} C_j} \qquad (16.11)$$

An alternative definition, $E_k^{\prime W}$, also appears in Grover (2003):

$$E_k^{\prime W} = \frac{\sum\limits_{\forall i \in S} \min(w_i, x_{i,k})}{\sum\limits_{\forall i \in S: x_{i,k}=1} C_i} = \frac{\sum\limits_{\forall i \in S} \min(w_i, x_{i,k})}{\sum\limits_{\forall j \in S} \delta_{j,k} C_j} \qquad (16.12)$$

This new quantity gives us an indication of a p-cycle's actual suitability in a specific environment of working capacity still remaining to be assigned to a p-cycle for protection. Using E_k^W (or $E_k^{\prime W}$) as a basis for identifying efficient cycles, a greedy iterative p-cycle network design algorithm can be easily developed. Given a set of eligible cycles (either through complete enumeration, pre-selection, or some other method), the idea is to iteratively test the placement of eligible cycles to determine their suitability for placement in the network. We first calculate E_k^W for each candidate cycle using the network's initial working capacity values, select the cycle with highest E_k^W, and place one copy of it in the design. Since this newly placed p-cycle provides protection for one unit of working capacity on each on-cycle span and two units on each straddling span, we subtract those amounts from the working capacities on all spans protected by the cycle. The new working capacities represent the amount of protection still required on each span after placement of the first cycle. The process is then repeated by recalculating E_k^W for each eligible cycle, placing one copy of the cycle with the highest efficiency, and updating the working capacities on all spans protected by that cycle. The algorithm terminates when no further unprotected working capacity remains on any span in the network. We refer to the algorithm as the Capacitated Iterative Design Algorithm (CIDA), and it can be expressed by the following pseudo-code:

```
CIDA() {
    Initialize CycleSet, work[], and CycleUse[]
    CycleSet = EnumerateCycles()
    While work[i] > 0 for all spans i {
        BestCycle = 0
        For each cycle p in CycleSet {
            Calculate E_p^W
            If E_p^W > E_BestCycle^W {
                BestCycle = p
            }
        }
        CycleUse[BestCycle] = CycleUse[BestCycle] + 1
        For each on-cycle span i in BestCycle {
            work[i] = work[i] - 1
        }
        For each straddling span i in BestCycle {
            work[i] = work[i] - 2
        }
    }
    Return CycleSet and CycleUse
}
```

Analyses in Doucette et al. (2003) showed that CIDA produced p-cycle network designs that are within 5% or less of the optimal ILP solutions when given the same set of eligible cycles, but in a small fraction of the runtime of the ILP. Further studies suggest that various tactics such as placing multiple copies of the winning cycle at each iteration can increase the speed of CIDA even further with only minor impact on optimality.

16.3.4 Approaches with Little or No Cycle Enumeration

Another technique for solving the p-cycle design problem is the use of *column generation algorithms* as in Stidsen and Thomadsen (2004) and Rajan and Atamturk (2002) for JCA p-cycle design. Here, the algorithm allows solution of a relaxed ILP model while only *implicitly* also determining the p-cycles to use in the solution. In column generation, new constraints in the ILP model are only generated when needed to improve the objective function. At each iteration of the algorithm in Stidsen and Thomadsen (2004), the eligible working route with the lowest reduced cost is found for each O-D pair, as is the single eligible cycle with the lowest reduced cost. If any of the routes or the cycle has a negative reduced cost, those are added to the ILP. This continues until no such path or cycle with negative reduced costs can be added to the problem. Network design costs and runtimes reported in Stidsen and Thomadsen (2004) and Rajan and Atamturk (2002) with the column generation algorithms compare favorably to the ILP method presented above. Refer to Chvátal (1983) and Winston (1994) for more information on column generation techniques in general.

A fully distributed *p*-cycle forming process called the *distributed cycle preconfiguration* (DCPC) protocol is developed in Grover and Stamatelakis (1998) as a means of allowing a network to discover and organize its own set of *p*-cycles so that full protection is provided to every span. DCPC is essentially an adapted version of the SHN protocol Grover et al. (1991), and makes use of statelets that propagate through the network allowing nodes to identify potential cycles and collectively select the most efficient for configuration. The DCPC protocol was initially proposed as an in-situ on-line process to self-organize *p*-cycles in an existing network. However, it can also be simulated in which case it provides a polynomial time algorithm to find an efficient set of *p*-cycles.

An ILP model that does not require cycle enumeration is presented in Schupke (2004). Here, a transshipment-like set of constraints is developed so that *p*-cycles are actually computed within the ILP itself. The main drawback with this method is the increased complexity caused by the large number of variables, and its use is generally limited to networks of 25 nodes or less.

16.4 NODE-ENCIRCLING P-CYCLE DESIGN

Node-encircling *p*-cycles efficiently provide for rapid protection of transiting flows affected by a node failure. A node-encircling *p*-cycle must by its very nature, exhibit two important properties. First, it must cross through all nodes topologically adjacent to the node it is intended to protect; otherwise, it cannot capture all of the pre-failure flows that transit the failed node. Second, it cannot contain the node it is protecting; otherwise, the *p*-cycle itself will be disrupted when the node fails. Node-encircling *p*-cycles were initially designed for protection against node loss in the IP/MPLS layer, but can also offer protection against the loss of a logical link between a pair of adjacent routers, as well as node and span failures in an optical network. This is explained further in Grover (2003).In this section where we discuss the process of generating node-encircling *p*-cycles, then describe the node-encircling *p*-cycle network design problems for node and link failures. We close with an overview of ongoing research in node-encircling *p*-cycles and optical-layer *p*-cycles design for minimum over-subscription.

16.4.1 Types of Node-encircling p-Cycles

A node-encircling *p*-cycle is constructed within the subgraph that results when the protected node (and all its incident links) is removed from the network, as shown conceptually in Figure 16.3. In most cases, a *simple* cycle, one that crosses any node only once, can be formed within that resulting subgraph, as in Figure 16.3(a) and Figure 16.3(b). Such cycles visibly encircle the node they protect. In other cases this may not be possible. However, if the pre-failure network topology is at least two-connected, there will always be at least one *logically* encircling cycle possible if we allow *non-simple* node-encircling *p*-cycles that cross some node(s) and/or spans(s) more than once, as in Figure 16.3(c) and Figure 16.3(d). As shown in Figure 16.3(c), a cycle also does not necessarily have to physically encircle the protected node, so long as all nodes adjacent to that node are visited.

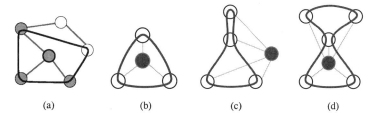

Figure 16.3 Illustrating (a) the basic (and simple) node-encircling p-cycle concept, (b) a simple node-encircling p-cycle, and (c)-(d) two non-simple node-encircling p-cycles. (Stamatelakis and Grover (2000a))

With a simple cycle, removal of the protected node does not disrupt the overall two-connectedness of the resulting network. In such cases the node encircling p-cycle is visually apparent and easy to find. In Figure 16.3(c) however, removal of the protected node results in a subgraph with a *stub node* (i.e., a degree-1 node) that can only be included in the node-encircling p-cycle through a segment that the cycle must cross twice. The node-encircling p-cycle in Figure 16.3(d) results when removal of the protected node creates a subgraph with a *bridge node* (i.e., a node whose removal would disconnect the graph entirely). Here a logically encircling p-cycle still exists but takes on the form of a figure eight, as it is forced to visit the bridge node twice. In the worst case, a network of N nodes would be fully protected against any single node outage by establishment of N node-encircling p-cycles. Each node would have a logically encircling p-cycle established for protection of its transiting flows through the set of its immediately adjacent nodes.

16.4.2 Generating Node-Encircling p-Cycles

A simple method of generating a set of node-encircling p-cycles is based on decomposition of the network into its bi-connected components following removal of the protected node and its incident edges. A preliminary step is to generate a file of all distinct simple cycles of the graph. If the network remains as a single bi-connected component, at least one simple node-encircling p-cycle exists and can be found by filtering the file of simple cycles to find the least-hop cycle that includes all of the potentially failed node's neighbors. Thus, if the failed node n has the set of neighbors $N_n = \{x, y, z, \ldots\}$, then the existence of a simple node-encircling p-cycle c'_n requires $N_n \in \{c'_n\} \in N - n$, after removing a specific node, where $\{c'_n\}$ is the set of nodes crossed by c'_n.

While removal of a node n cannot disconnect a bi-connected graph, G, it could result in multiple bi-connected components that are easily identified by an appropriate algorithm. In such a case, the bridge nodes or stub nodes are also identified. They will always be in N_n, and will connect the bi-connected components $G'_1, G'_2, \ldots$. Within each bi-connected component, some least-hop cycle p_i can be found for each component G'_i that will cross all nodes k such that $k \in N_n \cap G'_i$. The node-encircling p-cycle c'_n can then be easily constructed as $\{c'_n\} = \bigcup_{\forall i} p_i \cup N_n$. In other words, it is the cycle formed by merging the nodes $\{c'_n\}$ and bridge nodes and/or stub nodes p_i into a non-simple cycle.

This procedure can be easily implemented in an algorithm to determine at least one node-encircling p-cycle for each node in the network. These cycles can also be used in a version of CIDA that is modified to provide protection for node failures as well as span failures.

16.4.3 Designing Node-Encircling p-Cycle Networks

Although initially described in an IP-network context, the NEPC concept can easily be applied to optical networks, but there is reason for concern about cost if the NEPCs are formed as discrete wavelength channels separate and in addition to the span-protecting p-cycles. One observation, however, is that some span-protecting p-cycles will by chance also happen to have NEPC relationships with some nodes, in which case they can also be used to provide node-failure protection. However, the existence of such p-cycles will not be very widespread, so a deliberate design approach may be used to provide an assured level of node-failure protection. Therefore, a unified ILP design model to provide both span and node-failure protection is preferable. The main issue with such a model is that the ILP must contend with the many combinations of nodes, crossed with eligible NEPCs, crossed with working lightpaths, and then again with the spans on the NEPC so that we can efficiently reroute any lightpath one way or the other around the NEPC and minimize overall capacity requirements. In larger networks, this may result in a large number of potential input parameters, variables, and constraints, making the full ILP difficult to solve, so care must be taken when enumerating eligible cycle and route sets. In the future, algorithmic approaches similar to the CIDA algorithm may also be used to reduce the complexity of the problem when designing large networks. For present purposes, however, the joint working and spare capacity NEPC network design model is expressed as an ILP design model, which makes use of the following new notation:

- **Sets:**

    - N is the set of nodes in the network.

- **Input Parameters:**

    - $\phi_r^n \in \{0,1\}$ encodes the end-nodes of each demand. $\phi_r^n = 1$ if node n is an end-node of demand r, and $\phi_r^n = 0$ otherwise.
    - $Z_n^{r,q} \in \{0,1\}$ is a parameter that encodes working routes. If $Z_n^{r,q} = 1$, working route q used for demand relation r crosses node n. If $Z_n^{r,q} = 0$, working route q used for demand relation r does not cross node n.
    - $X_p^n \in \{0,1\}$ encodes which cycles can act as NEPCs for which nodes. If $X_p^n = 1$, p-cycle p can be an NEPC for node n, and if $X_p^n = 0$, it cannot.
    - $Z_{n,p}^{r,q} \in \{0,1\}$ further encodes working routes and their relationships to eligible cycles. $Z_{n,p}^{r,q} = 1$ if working route q used for demand relation r crosses node n, and can be protected by p-cycle p, and $Z_{n,p}^{r,q} = 0$ otherwise. $Z_{n,p}^{r,q}$ is explicitly defined as $Z_{n,p}^{r,q} = Z_n^{r,q} \cdot (1 - \phi_r^n) \cdot X_p^n$.

- $R1_{r,q}^{p,n,i} \in \{0,1\}$ defines one of the restoration routes around an NEPC. $R1_{r,q}^{p,n,i} = 1$ if span i is on cycle p and is crossed by the clockwise reroute of working route q for demand r in the event of the failure of node n, and $R1_{r,q}^{p,n,i} = 0$ otherwise.
- $R2_{r,q}^{p,n,i} \in \{0,1\}$ defines one of the restoration routes around an NEPC. $R2_{r,q}^{p,n,i} = 1$ if span i is on cycle p and is crossed by the counter-clockwise reroute of working route q for demand r in the event of failure of node n, and $R2_{r,q}^{p,n,i} = 0$ otherwise.

- **Decision Variables:**

  - $r1_{r,q}^{p,n} \geq 0$ is the amount of restoration flow in the clockwise direction over p-cycle p used to restore working route q for demand r in the event of failure of node n.
  - $r2_{r,q}^{p,n} \geq 0$ is the amount of restoration flow in the counter-clockwise direction over p-cycle p used to restore working route q for demand r in the event of failure of node n.

All previous notation from earlier formulations still applies, and we now define an ILP formulation that selects a set of span-protecting and node-encircling p-cycles so that all span and node failures are fully protected and total capacity requirements are minimized. The ILP design model is formulated as follows:

$$\min \sum_{\forall j \in S} C_j \cdot (s_j + w_j) \tag{16.13}$$

$$\text{s.t.} \quad \sum_{\forall q \in Q^r} g^{r,q} = d^r \quad \forall r \in D \tag{16.14}$$

$$\sum_{\forall r \in D} \sum_{\forall q \in Q^r} \zeta_i^{r,q} \cdot g^{r,q} = w_i \quad \forall i \in S \tag{16.15}$$

$$w_i \leq \sum_{\forall p \in P} x_{i,p} n_p \quad \forall i \in S \tag{16.16}$$

$$g^{r,q} = \sum_{\forall p \in P} \left(r1_{r,q}^{p,n} + r2_{r,q}^{p,n} \right) \quad \forall n \in N \quad \forall r \in D \quad \forall q \in Q^r \tag{16.17}$$

$$n_p \geq \sum_{\forall r \in D} \sum_{\forall q \in Q^r} \left(r1_{r,q}^{p,n} \cdot R1_{r,q}^{p,n,i} + r2_{r,q}^{p,n} \cdot R2_{r,q}^{p,n,i} \right) \quad \forall n \in N \quad \forall p \in P \quad \forall j \in S \tag{16.18}$$

$$s_j = \sum_{\forall p \in P} \delta_{j,p} \cdot n_p \quad \forall j \in S \tag{16.19}$$

The objective function in (16.13) minimizes total cost of working and spare capacity. The constraints in equations (16.14) and (16.15) are as already seen in equations (16.9) and (16.10), respectively, and ensure that there is sufficient working flow to fully route all working demands, and there is enough working capacity to accommodate all working flows. The constraints in equation (16.16) assert sufficient restoration flow for all span failures and is the same constraint set as seen previously in equation

(16.2). Equation (16.17) ensures that there is enough restoration flow in either or both directions around all relevant NEPCs so that all working flow on working route q for demand relation r affected by failure of node n is fully restorable. The constraints in equation places enough copies of NEPC p to accommodate the restoration flow simultaneously routed in either direction over any span j on the NEPC for all affected working routes q for all demand relations r. Finally, constraint set (16.20) is identical to the one in equation (16.3) and ensures there is enough spare capacity to accommodate all p-cycles placed in the design. Note that n_p now considers the fact that an eligible cycle can be used as an NEPC in addition to a span-protecting p-cycle.

A further modification can be made to the model so that only a specified level of node-failure restorability is required. To do so, we define $0 \leq L_r \leq 1$ as the proportion of lightpaths in demand relation r that must be restorable in the event of any node failure, and change equation (16.17) to the following:

$$g^{r,q} L_r = \sum_{\forall p \in P} \left(r1_{r,q}^{p,n} + r2_{r,q}^{p,n} \right) \quad \forall n \in N \forall r \in D \forall q \in Q^r \qquad (16.20)$$

Analyses in Doucette et al. (2005) and shown in Figure 16.4 illustrate that while providing full node-failure protection with NEPCs (the "NEPC 100%" curve) is quite costly relative to providing only span-failure protection (the "p-cycle" curve), limited node-failure protection can be quite cost-effective. For instance, providing node-failure protection to only 25% of the lightpaths for each demand relation requires an additional investment of only 2% to 8% additional capacity (the "NEPC 25%" curve). Note that in all cases shown in Figure 16.4, the test case network designs are 100% span-protected while at the same time providing the indicated levels of overall node failure protection. The test networks used are the 20-node network family from Doucette (2004).

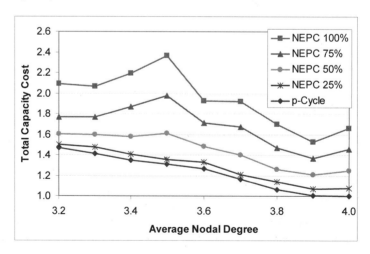

Figure 16.4 Normalized total capacity requirements for 25% to 100% node-failure restorable NEPC network design in the 20n40s1 network family.

16.4.4 Network Design using MPLS p-Cycles for Link Restoration

In a WDM-IP network, lightpaths in the WDM layer are used to set up logical connections between IP routers and/or label switched routers (LSRs). From the perspective of the IP network, these appear to be directional connections between routers, while in reality a single physical span may carry sections of multiple logical links protected between routers. Failure of a single physical span can then translate into the apparent failure of multiple links in the IP layer. This must be taken into consideration when designing a set of IP network p-cycles, so that any single physical span failure has a controlled or bounded maximum impact on the simultaneous failure of logical links with the same p-cycle.

In the IP network, restorability design also needs to consider the convergent flow effects arising from restoration. This is an aspect that does not exist for SONET or WDM where every working signal is either exactly replaced or not. In contrast, where packet or cell flows are being redirected upon restoration, one can take an over-subscription-based design to control the worst case simultaneously imposed flows on any link during any restoration scenario. Over-subscription is a comparative measure of actual capacity usage relative to pre-planned capacity allocations. Over-subscription-based capacity planning leads to assurance on the worst-case possible overloads, but it is a simpler and more practical planning framework than trying to plan capacity for protection while dealing directly with statistical traffic descriptions. Over-subscription under a restored or rerouted state in general is the ratio of the total of the pre-failure bandwidth allocations made to flows that now cross the link, relative to the actual capacity of the link.

In the MPLS context, it is technically meaningful to consider that the stochastic flows from more than one LSP may merge under protection rerouting to produce a combined load on a given link that somewhat exceeds the nominal utilization of that link, but still produces acceptable delay and cell loss probabilities in a restored-network state. In other words, we may plan to allow certain amounts of over-subscription of planned capacity during a failure. In the optical layer, however, over-subscription of capacity is clearly not an option because this layer requires exact matching of discrete working signals with corresponding signal paths for restoration. Over-subscription in the WDM layer would in effect correspond to simply not having enough protection lightpaths to cover a set of failed working lightpaths.

The over-subscription-based network design problem can be modeled as an ILP formulation, but we first define the following notation:

- **Set:**
  - S now refers to the set of IP links in the network. (This is generally more numerous than the set of physical layer spans.)

- **Input Parameters:**
  - M is the maximum number of p-cycles permitted in the design.
  - T_i is the total capacity allocated to link i.
  - w_i is the capacity allocation normally provided for the amount of working traffic on link i during normal operation.

- $\beta_{i,j,k} \in \{0, 0.5, 1\}$ is a pre-computed *imposed load ratio* corresponding to the $x_{i,k}$ coefficients for previous discrete-circuit p-cycles. Each $\beta_{i,j,k}$ is a constant that gives the fraction of the working flow from link i that is carried on link j if using cycle k for restoration. $\beta_{i,j,k} = 0$ if cycle k does not pass over link j, $\beta_{i,j,k} = 0.5$ if the p-cycle offers two restoration paths to link i and traverses link j (traffic is split), and $\beta_{i,j,k} = 1$ if p-cycle p offers only a single restoration path for link i and also crosses link j.

- **Decision Variables:**

    - $\delta_k \in \{0, 1\}$ is a binary decision variable, encoding selection of p-cycle k in the design. $\delta_k = 1$ if cycle k is formed into a p-cycle, and $\delta_k = 0$ if it is not.

    - $\alpha_{i,k} \in \{0, 1\}$ is a binary decision variable encoding whether a p-cycle k is assigned to carry restoration flow for failure of link i. $\alpha_{i,k} = 1$ if p-cycle k is assigned to the restoration of link i, and $\alpha_{i,k} = 0$ if it is not.

    - $X_{i,j} \geq 0$ is the over-subscription factor on link j during the restoration of link i (i.e., the total of the capacity allocations for normal flow that are now imposed on link j relative to the amount of its actual capacity.)

    - $\eta_M \geq 0$ is the maximum over-subscription ratio on any link during any restoration event.

All other notation from earlier formulations still applies. We can now define an ILP formulation that determines a set of IP p-cycles that minimizes the worst-case over-subscription factor on any link, over all failure scenarios. The design model for minimum peak oversubscription is formulated as follows:

$$\min \ \eta_M \tag{16.21}$$

$$\text{s.t.} \quad \sum_{\forall k \in P} \delta_k \leq M \tag{16.22}$$

$$\alpha_{i,k} \leq \delta_k \quad \forall i \in S, \ \forall k \in P \tag{16.23}$$

$$\sum_{\forall i \in S} \alpha_{i,k} \geq \delta_k \quad \forall k \in P \tag{16.24}$$

$$\sum_{\forall k \in P} \alpha_{i,k} = 1 \quad \forall i \in S \tag{16.25}$$

$$X_{i,j} = \frac{w_j + \sum_{\forall k \in P} \beta_{i,j,k} w_i \alpha_{i,k}}{T_j} \quad \forall i \in S \ \forall j \in S | i \neq j \tag{16.26}$$

$$\eta_M \geq X_{i,j} \quad \forall i \in S \ \forall j \in S | i \neq j \tag{16.27}$$

The constraints in equation (16.22) restricts the total number of p-cycles to be no more than M. Equation (16.23) allows link i to use a cycle p for its restoration (as designated by $\alpha_{i,k} = 1$) only if that cycle is selected for use in the overall design (i.e., $\delta_k = 1$). In equation (16.24), a cycle k is permitted to be selected in a design only if it is used for restoration of at least one link. The constraints in equation (16.25) assert the usage of one and only one cycle for each failure scenario. The over-subscription factor on link j for failure of span i is calculated in equation (16.26), as the ratio of the sum of the normal working flow on link j and all of the flows imposed on it by all p-cycles used to restore failure of link i relative to the total capacity on the link. (It is written this way to convey presence of the oversubscription ratio. Obviously T_j is a constant which can be moved to the left hand side.) Finally, equation (16.27) sets the maximum over-subscription ratio, η_M. Since the objective function in equation (16.21) seeks to minimize η_M, the interaction with the constraints in equation (16.27) will cause η_M to take the value of the largest $X_{i,j}$ on all links for all failure scenarios. In other words, η_M will be the worst-case over-subscription factor on all spans over all failure scenarios.

16.5 P-CYCLE DESIGN FOR ENHANCED AVAILABILITY

In this section, we present various formulations for capacity design of p-cycle networks in which single-failure restorability (referred to as R_1 demands or *Gold* quality of protection (QoP) class) is *not* the only protection option considered. We also consider an approach which improves the dual-failure restorability of selected demands with or without addition of any further capacity above that needed for the R_1 design above. These demands (of type *Gold-Plus*) are not strictly guaranteed dual-failure restorability but they will be more often restorable from dual failures and therefore will enjoy higher service availability. We then present an approach where selected demands (referred to as *Platinum* class) are provided with complete dual failure protection on all spans along their paths, a strategy that also provides extremely high availability. The following section then considers the simultaneous consideration of preemptible demands (type *Economy*), non-protected demands (*Bronze*), best-efforts restoration demands (*Silver*), single-failure restorable demands (*Gold*) and dual-failure restorable demands (*Platinum*).

Previous notation still applies, and we define the following new notation:

- **Sets:**
  - $D^g \subseteq D$ is the set of demand relations for the Gold protection class.
  - $D^{g+} \subseteq D$ is the set of demand relations for the Gold-*Plus* protection class.
  - $D^p \subseteq D$ is the set of demand relations for the Platinum protection class.

- **Input Parameters:**
  - $w_i^a \geq 0$ is the amount of working capacity allocated on span i for paths of protection class a, where $a \in \{g, g+, p\}$.

16.5.1 Capacity Placement for Selectively Enhanced Availability (p-Cycle SEACP)

The problem of Capacity Placement for Selectively Enhanced Availability jointly optimizes the routing of demands and the allocation of spare capacity to find the minimal cost capacity placement that allows us to serve all demands following the routing constraints described earlier and guarantee that each working channel is protected against single span failures. It is modeled as follows:

$$\min \sum_{\forall j \in S} \sum_{\forall m \in M} c_j^m t_j^m \tag{16.28}$$

$$\text{s.t.} \quad \sum_{\forall q \in Q^r} g^{r,q} = d^r \quad \forall r \in D^g \cup D^{g+} \tag{16.29}$$

$$w_i^g = \sum_{\forall r \in D^g} \sum_{\forall q \in Q^r} \varsigma_i^{r,q} g^{r,q} \quad \forall i \in S \tag{16.30}$$

$$w_i^{g+} = \sum_{\forall r \in D^{g+}} \sum_{\forall q \in Q^r} \varsigma_i^{r,q} g^{r,q} \quad \forall i \in S \tag{16.31}$$

$$w_i^g + w_i^{g+} \leq \sum_{\forall k \in P} x_{i,k} n_k \quad \forall i \in S \tag{16.32}$$

$$w_i^{g+} \leq \sum_{\forall k \in P} x_{i,k}(1 - \delta_{i,k}) n_k \quad \forall i \in S \tag{16.33}$$

$$s_j = \sum_{\forall k \in P} \delta_{j,k} n_k \quad \forall j \in S \tag{16.34}$$

$$w_j^g + w_j^{g+} + s_j \leq \sum_{\forall m \in M} t_j^m z^m \quad \forall j \in S \tag{16.35}$$

The objective is to minimize the total cost of required modular capacity. Constraint set (16.29) ensures that for every demand relation r in either protection class, there is enough flow over all eligible working routes to fully serve that demand. Constraint sets (16.30) and (16.31) ensure that the right amount of working capacity is allocated on each span k for the Gold and Gold-*Plus* protection classes respectively. Constraint set (16.32) guarantees that there are enough p-cycles in the solution to protect all working capacity units on all spans. By routing demands such that all Gold-*Plus* working channels are protected as straddlers, it is guaranteed that at least one of the protection paths will survive in the event of a dual span failure. To enforce the requirement that paths in the Gold-*Plus* class have to be routed on straddling spans only, constraint set (16.33) ensures that for each span i, the number of p-cycles that span i straddles is enough to protect all working capacity units in the Gold-Plus class. (Note that through constraint set (16.32), paths in the Gold class will be protected either by p-cycles they straddle or by p-cycles to which they have an on-cycle relationship.) Constraint set (16.34) makes sure that there is enough protection capacity allocated on all spans to support all p-cycles in the solution. Finally, constraint set (16.35) ensures that there is enough capacity placed on all spans to allow the allocation of working and spare capacity as imposed by (16.30), (16.31), and (16.34). Other approaches to improving the dual-failure restorability of service paths in p-cycle based networks are presented in Schupke et al. (2004).

16.5.2 *p-Cycle Multi-Restorability Capacity Placement (p-Cycle MRCP)*

The approach proposed in this section is based on an evolution of the basic p-cycle principle in which p-cycles can either be used to offer two backup paths protecting two working channels on straddling spans, as in the normal p-cycle scheme, or two protection *options* for a single platinum channel on any straddler. The latter option can be used for protection of service paths with ultra-high availability requirements. Based on studies in Clouqueur and Grover (2002) and Arci et al. (2003), it is known that offering two or more restoration options instead of just one leads to great improvements of the availability of service. This new class of service, that we call the *Platinum* class, is therefore expected to enjoy extremely high availability, and by virtue of design, full restorability to any dual span failures.

The multi restorability capacity placement (*p*-cycle MRCP) formulation is described below. It finds the optimal routing of demands and allocation of capacity that minimizes the required capacity placement subject to the routing constraints described earlier. The formulation is expressed as follows:

$$\min \sum_{\forall j \in S} \sum_{\forall m \in M} C_j^m t_j^m \qquad (16.36)$$

$$\text{s.t.} \sum_{q \in Q^r} g^{r,q} = d^r \quad \forall r \in D^g \cup D^p \qquad (16.37)$$

$$w_i^g = \sum_{\forall r \in D^g} \sum_{\forall q \in Q^r} \varsigma_i^{r,q} g^{r,q} \quad \forall i \in S \qquad (16.38)$$

$$w_i^p = \sum_{\forall r \in D^p} \sum_{\forall q \in Q^r} \varsigma_i^{r,q} g^{r,q} \quad \forall i \in S \qquad (16.39)$$

$$w_i^g + 2w_i^p \leq \sum_{\forall k \in P} x_{i,k} n_k \quad \forall i \in S \qquad (16.40)$$

$$2w_i^p \leq \sum_{\forall k \in P} x_{i,k}(1 - \delta_{i,k}) n_k \quad \forall i \in S \qquad (16.41)$$

$$s_j = \sum_{\forall k \in P} \delta_{j,k} n_k \quad \forall j \in S \qquad (16.42)$$

$$w_j^g + w_j^p + s_j \leq \sum_{\forall m \in M} t_j^m z^m \quad \forall j \in S \qquad (16.43)$$

The objective function is the same as before. Constraint sets (16.37) to (16.39) are similar to constraint sets (16.29) to (16.31), with class Gold-*Plus* being replaced by class Platinum. Constraint set (16.40) is similar to (16.32), but unlike for the Gold-*Plus* protection class, the Platinum protection class requires that a *p*-cycle provide two protection options for a single working link in that class, therefore a factor 2 is added in front of the number of working links in the Platinum class. Constraint set (16.41) ensures that Platinum-class working links on any span *i* can only be protected by *p*-cycles that span *i* straddles. Finally, constraint sets (16.42) and (16.43) are similar to (16.34) and (16.35) in the previous formulation.

16.6 MULTIPLE QUALITY OF PROTECTION P-CYCLE NETWORK DESIGN (MULTI-QOP *p*-CYCLE DESIGN)

In this section, we consider incorporating (among others) all the service classes discussed in the previous section in *p*-cycle networks. Additionally this is done without requiring any dynamic post failure reconfiguration of *p*-cycles to effect dual failure survivability. Reconfiguring *p*-cycles after a span failure to protect against any subsequent span failure has been studied in (Schupke et al., 2004). It is, however, interesting that with *p*-cycles, no reconfiguration is strictly necessary to effect guaranteed dual failure protection. Imposing the need for run-time reconfiguration requires that all or some of the nodes be full cross-connects capable of run-time switching at the line rate. Our interest here, however, is to design with a tradeoff in terms of increased capacity required to effect dual failure protection without assuming the flexibility of dynamic post failure reconfiguration in the network.

With *p*-cycles, to ensure that a particular platinum demand is protected end-to-end, it is sufficient that the two following conditions be met:

- (a) All platinum working capacity must be protected using straddling relationships only.

- (b) A whole unit of *p*-cycle capacity must be used to protect a unit of platinum capacity on the straddler. (As opposed to using half the *p*-cycle to protect a single unit of gold capacity on a straddler.)

Enforcing both conditions in the design ensures that a single *p*-cycle provides two disjoint routes (effectively 1+1+1 protection) to each unit of platinum capacity it protects. A second span failure along one of the protection routes (after failure of the straddling span) has no effect on survivability, as the end nodes of the failed span have only to switch over to the other still surviving and unused protection path. In the general case, it can also be shown that that *p*-cycles can be used to provide survivability solutions for an '*n*' failure protected network, with a theoretical minimum redundancy η, given by $\eta = n/(\bar{d}-n)$. Thus, as an example, a network with $\bar{d} = 4$ requires a minimum of 100% redundancy to offer platinum class protection (i.e., $n = 2$) to all demands.

Some additional notation used in the multi-QoP *p*-cycle design formulation is:

- **Sets:**

    - $D^{sb} \subseteq D$ is the set of demand relations for the Silver and Bronze protection classes.

    - $D^e \subseteq D$ is the set of demand relations for the Economy protection class.

- **Input Parameters:**

    - $c_{i,j}^k \in \{0,1\}$ encodes whether spans i and j are *both* on cycle k. $c_{i,j}^k = 1$ if spans i and j are both on the cycle k, and $c_{i,j}^k = 0$ if at least one is not. Mathematically, we set $c_{i,j}^k = \delta_{i,k}\delta_{j,k}$.

- **Decision Variables:**

    - $\eta_i^{k,g} \geq 0$ is the integer number of unit-capacity *p*-cycles on cycle k used to protect Gold class working capacity on span i.

    - $n_i^{k,p} \geq 0$ is the integer number of unit-capacity *p*-cycles on cycle k used to protect Platinum class working capacity on span i.

The multi-QoP p-cycle design formulation is expressed as follows:

$$\min \sum_{\forall j \in S} \sum_{\forall m \in M} C_j^m t_j^m \tag{16.44}$$

$$s.t. \quad \sum_{\forall q \in Q^r} g^{r,q} = d^r \quad \forall r \in D^p \cup D^g \cup D^{sb} \cup D^e \tag{16.45}$$

$$\sum_{\forall r \in D^p} \sum_{\forall q \in Q^r} \varsigma_i^{r,q} g^{r,q} = w_i^a \quad \forall i \in S, \forall a \in \{p, g, s, b, e\} \tag{16.46}$$

$$\sum_{\forall k \in P} x_{i,k} n_i^{k,j} \geq w_i^g \quad \forall i \in S \tag{16.47}$$

$$\sum_{\forall k \in P} x_{i,k} n_i^{k,p} \geq w_i^p \quad \forall i \in S \tag{16.48}$$

$$n^k \geq n_i^{k,g} + n_i^{k,p} \quad \forall i \in S \; \forall k \in P \tag{16.49}$$

$$n^k \geq 2 n_i^{k,p} \quad \forall i \in S \; \forall k \in P \tag{16.50}$$

$$n_i^{k,p} \leq \Delta(1 - c_{i,j}^k) \quad \forall i \in S \; \forall j \in S | i \neq j \; \forall k \in P \tag{16.51}$$

$$\sum_{\forall k \in P} (\delta_{i,k} n^k - w_i^e) \leq s_i \quad \forall i \in S \tag{16.52}$$

$$w_j^p + w_j^g + w_j^{sb} + w_j^e + s_j \leq \sum_{\forall m \in M} t_j^m Z^m \quad \forall j \in S \; \forall m \in M \tag{16.53}$$

The objective is to minimize the total cost of modular capacity of the design. Constraint (16.45) ensures that all demands (of all QoP classes) are routed. Constraint (16.46) to (16.48) ensures that there is sufficient working capacity placed on the network to accommodate them. The constraints in equations (16.49) to (16.51) place p-cycles in the network to fully protect all working capacity needing protection (Gold and Platinum). Constraints (16.50) and (16.51) ensure that Platinum capacity is protected using two disjoint protection paths over p-cycles, and that both rules (a) and (b) discussed earlier are obeyed. The constraint in (16.52) ensures that spare (and economy) capacity exists to build all the necessary p-cycles. Constraint set (16.53) introduces modularity (and economy of scale) into the problem. Initial results from Kodian et al. (2003) establish that, for certain networks, up to 30% of the total demand between every pair of nodes in a network can be offered platinum class protection without any extra capacity over the all-Gold design. (i.e. 30% of all services can have dual failure survivability in a network which has no more spare capacity in total than that required for a 100% single failure restorable network design.)

16.7 PROTECTION AGAINST SHARED RISK LINK GROUPS (SRLGS) WITH P-CYCLES

One special type of dual failures are called Shared Risk Link Groups (SRLGs). SRLGs are a special case where a single physical failure manifests itself as a logical dual span failure in the network (Doucette and Grover, 2002). (In general, SRLGs may also manifest as triple or quadruple simultaneous span failures, but currently we only consider up to dual span failures as a network manifestation of a single SRLG-type failure.) For instance the fiber optical cables leaving a network node forming two logically distinct and nominally disjoint spans, may actually cross a bridge together. They are thus logically distinct spans that share one common-cause failure possibility, i.e. the bridge that they cross. A variation of the multi-QoP model from the previous section is used to protect selected demands against stipulated SRLG-type failures. The following new parameters and variables are added to the previous models:

- **Input Parameters:**

    - $\phi_{i,j} \in \{0,1\}$ encodes span pairs affected by an SRLG. $\phi_{i,j} = 1$ if spans i and j can have some common cause of failure against which we want to design the network to be fully restorable.
    - $b_{i,j}^k \in \{0,1\}$ encodes whether SRLG spans i and j are a straddler and on-cycle pair of cycle k. We set $b_{i,j}^k = 1$ if span i straddles cycle k, span j is an on-cycle span of cycle k, and $\phi_{i,j} = 1$, and $b_{i,j}^k = 0$ otherwise. Mathematically, we can say $b_{i,j}^k = \left(\frac{x_{i,k}}{2}(1 - \delta_{i,k})\right)\delta_{j,k}\phi_{i,j}$.

The entire multi-QoP p-cycle design problem in Section (16.6) applies, and we add one new constraint:

$$n_k \geq (2n_i^{k,P})b_{i,j}^k \quad \forall i \in S \ \forall k \in P \ \forall j \in S | i \neq j \tag{16.54}$$

Constraint (16.54) ensures that the only dual failure cases that are considered are the potential SRLG span pairs. Work in Doucette and Grover (2002) shows that depending on the exact set of spans chosen as SRLG-pairs, protection of SRLGs in a span restorable mesh network can increase the capacity requirements from as little as a few percent more than that required for single-failure protection, to as high as that required for full dual-failure protection.

16.8 DESIGN OF P-CYCLE NETWORKS IN THE FACE OF DEMAND UNCERTAINTY

All the p-cycle design problems discussed above assume a specific demand forecast to which one optimizes the routing and transport capacity assignment for a single target planning view. In this section, we look at the p-cycle design problem when the demand forecast is uncertain. The demand uncertainty can be modeled by a set of possible future scenarios, with each assigned a probability estimate. In practice these might correspond to a set of "what-if" scenarios used by a planner to individually test the sensitivity of a nominal design. Here, however, we discuss a capacity design formulation that inherently considers a range of possible futures all at once. The new model can produce a capacity plan that leads to significantly lower expected total lifetime costs than traditional p-cycle designs that target a single forecast.

The problem is formulated as a two-part problem involving a bi-criteria objective function. The first part considers the budget X to be invested at present and the second part represents the total expected corrective or "recourse" cost Y to be incurred in the future when uncertainty unfolds. Compared to the traditional approach where a single demand forecast is assumed for a snapshot design, the two-part model better reflects the uncertainty arising from demand forecasting as well as the complete life-cycle investment costs associated with the capacity planning process. The two-part design also has the property of minimum expected total cost of the as-built current network plus future recourse actions as may be needed to cope with a range of different possible futures. Other work on capacity planning under demand uncertainty for span-restorable and/or path-protected mesh networks can be found in Kennington et al. (2003), Leung and Grover (2004b), and Leung and Grover (2004a).

16.8.1 Future-Proof p-Cycle Design

The goal of the future-proof p-cycle design (FPPC) problem is to minimize the total cost of the network that we commit to building in the present, as well as the expected value of future recourse actions to augment the design to serve possible future demand scenarios, where each scenario is associated with a probability estimate. To model the FPPC design problem, we must define new notation as follows:

- **Set:**

    - R_j is the recourse (or penalty) cost of placing an extra unit of capacity on span j to cope with future capacity additions. In the general case, these recourse costs are specific to each span and can reflect different practical considerations such as an abundance of dark fiber on some spans but not on others, the fact that one span may be nearing exhaust, the ability to lease capacity from another carrier, and so on. For comparative studies, we use a common recourse cost factor, R for all spans where $R_j = RC_j$.
    - $0 \leq P^u \leq 1$ is the probability estimate for demand scenario u.
    - $d^{r,u}$ is the number of lightpath demand units for relation r in demand scenario u.

- **Decision Variables:**

    - $y_j^u \geq 0$ is the integer number of additional working capacity units that would have to be placed on span j in the future to cope with scenario u now.(relative to the number required for the design solution to be built.)
    - $z_j^u \geq 0$ is the integer number of additional spare capacity units required on span j under future demand scenario u.
    - $g^{r,q,u} \geq 0$ is the amount of working flow assigned to working route q used for demand relation r in scenario u.
    - $n_k^u \geq 0$ is the integer number of unit-capacity p-cycles placed on cycle k in scenario u.

The FPPC problem is modeled as variation of the JCA problem in Section 16.2 where working routing and p-cycle design is performed jointly. The difference here is that this is done simultaneously for multiple demand scenarios so as to minimize the total of the actual cost incurred at present to build the network plus the expected future cost to deal with inaccuracies in the forecast. The model is expressed as follows:

$$\min \sum_{\forall j \in S} C_j(w_j + s_j) + \sum_{\forall j \in S} \sum_{\forall u \in U} P^u R_j(y_j^u + z_j^u) \quad (16.55)$$

$$\text{s.t.} \sum_{\forall q \in Q^r} g^{r,q,u} = d^{r,u} \quad \forall r \in D \; \forall u \in U \quad (16.56)$$

$$\sum_{\forall r \in D} \sum_{\forall q \in Q^r} \varsigma_j^{r,q} g^{r,q,u} = w_j + y_j^u \quad \forall j \in S \; \forall u \in U \quad (16.57)$$

$$w_i + y_i^u \leq \sum_{\forall k \in P} x_{i,k} n_k^u \quad \forall i \in S \; \forall u \in U \quad (16.58)$$

$$s_j + z_j^u \geq \sum_{\forall k \in P} \delta_{j,k} n_k^u \quad \forall j \in S \; \forall u \in U \quad (16.59)$$

$$y_j^u = 0 \quad u = 0 \; \forall j \in S \quad (16.60)$$

$$z_j^u = 0 \quad u = 0 \; \forall j \in S \quad (16.61)$$

For each scenario u, the constraints set in equation (16.56) allocates the demand flows $g^{r,q,u}$ of demand relation r onto working route q. Equation (16.57) determines the working capacity needed on each span to serve the demand flows. These two constraints sets are similar to those of equations (16.9) and (16.10) for the JCA formulation in Section 16.2, except that we now not only consider the *actual* working capacity, w_j, to be placed in the network, but also the *shortfall* working capacity $y_j^u \geq 0$ required to serve unexpected demands that may arise in future scenario u. Equations (16.58) and (16.59) correspond to the restorability and spare capacity constraint sets in equations (16.2) and (16.3), respectively, from the basic SCA design model. The final two constraint sets in equations (16.60) and (16.61) are needed to characterize a special scenario situation, $u = 0$, which represents any current existing demands that *must* be served and protected. These constraints ensure the condition by asserting that there can be no shortfall capacities for any demands in that scenario, forcing the design to contain adequate w_j and s_j capacities.

As a measure of the effectiveness of this model, tests were done in Leung and Grover (2004a). In that work we applied uniform random variations to an expected demand forecast to generate a set of plausible "what if" scenarios with total demand volumes ranging from 0.3 to 5.0 times the nominal forecasts. A recourse costs factor of $R = 5$ (i.e., $R_j = 5C_j$) allowed an FPPC network design with an expected cost reduction of 39% compared to a conventional single scenario JCA design on the COST 239 network. Note that in practice recourse costs could be either greater or less than unity (relative to present costs). It could be argued that technologically capacity is always less costly in the future. However recourse cost should include the total operational cost in the future of having to design, install and commission the capacity augmentation if it could have been included in the initial design.

16.9 PWCE DESIGN WITH P-CYCLES

The *protected working capacity envelope* (PWCE) is a new paradigm for provisioning dynamic survivable services, as described in Grover (2003), Grover (2004), Shen and Grover (2004). In the PWCE model, protection may be performed between the nodes adjacent to the failure, so span restoration and p-cycles are both amenable to the PWCE concept. In PWCE, it is the bearer capacity itself that is protected and so any service connection routed exclusively over such capacity in a PWCE is inherently protected as well, without any explicit designation of a backup route. In comparison with the conventional Shared Backup Path Protection (SBPP)-based provisioning method, PWCE shows advantages of simple operation, fast restoration, and good blocking performance. A PWCE can be simply formed within the working capacity determined from a conventional fully restorable p-cycle network design. But larger "envelopes" can also be designed so as to maximize the envelope within a specified capacity budget or some other criteria. Designing a maximal envelope is an opportunity for a PWCE-based network, so we now discuss various possible envelope design models based on different assumptions and capacity budgets for the p-cycle network.

The PWCE concept is illustrated in Figure (16.5), where a set of spare channels defines a *reserve network* of spare capacities corresponding to the s_j budget and the maximal PWCE of $\langle w_i^0 \rangle$ values the reserve network can accommodate (the $\langle w_i^0 \rangle$ values correspond to the solution of the above ILP problem). The PWCE is then able to dynamically serve multiple service paths as they arrive (three are shown). There are no per-path protection arrangements to make because the channels used for provisioning in the working layer are themselves protected by the reserve network and some embedded restoration or protection mechanism. In other words, as long as the $\langle w_i^0 \rangle$ quantities will support working routing of the demand, it is inherently protected end-to-end and no further action to explicitly deal with protection is required. Under PWCE, nodes need not track individual channel states as they do in SBPP because sharing relationships are not defined precisely between individual paths and individual backup channels. The nodes need only know that individual spans have one or more provisionable channels available, which requires no signaling for state dissemination. Consequently, this greatly simplifies the network operation and service provisioning.

16.9.1 Simple PWCE Volume Maximizing Model

There are various methods to construct a PWCE within a p-cycle network, one of which is to use conventional SCA or JCA designs given in Sections 16.1 and 16.2. However, those design methods do not provide an optimal envelope (at least in the volume-maximized sense), and so will not fully exploit the spare capacity from a PWCE standpoint. To achieve a better capacity efficiency, we need volume maximization models. The volume maximization model places more working demand flow on non-forcer spans such that more working capacity is protected without any increase in spare capacity allocation. (Forcers are spans which, in the design of a restorable mesh network, have working capacity quantities and surrounding topological circumstances such that they fully use the spare capacity available on one or more other spans in the design. The significance is that if a forcer span has additional demand routed over it, then the total network spare capacity must increase. On the other spans this will not be so in general. See (Shen and Grover (2003a)) The term "volume" refers to the total number of protected working channels the design supports. Three such models are developed in Shen and Grover (2003b;a), where the forcer structure of the conventional designs is used to identify extra working capacity to exploit.

The simplest PWCE maximizing model we can formulate is one where a spare capacity budget is known (say to correspond to a pre-determined maximal set of p-cycles), and we seek to place working capacity so as to maximize the protection envelope over all spans. We express this model as follows:

$$\max \sum_{\forall j \in S} C_j w_j \quad (16.62)$$

$$s.t. \sum_{\forall k \in P} x_{i,k} n_k \geq w_i \quad \forall i \in S \quad (16.63)$$

$$s_j \geq \sum_{\forall k \in P} \delta_{j,k} n_k \quad \forall j \in S \quad (16.64)$$

In this model, s_j is an input parameter rather than a decision variable as in past models. Such a spare capacity budget can be obtained from the conventional survivable network design, with which the working capacity of a forecasted demand matrix is protected (and of course, other options are possible for the budget selection). The objective of the model is then to maximize the volume of the envelope of protected working capacity within that budget. The constraints in equation (16.63) ensure that working capacity allocated to each span is fully protected, and the constraints in equation (16.64) guarantee that the spare capacity budget is sufficient to form the p-cycles required.

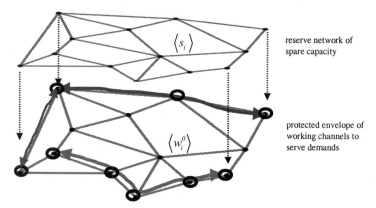

Figure 16.5 Three working paths routed within the PWCE of a p-cycle network (Grover (2003)).

16.9.2 Demand Target Pattern Matching

The above PWCE volume maximization model by itself simply creates the greatest total number of protected working channels over the entire network. However, they may not be in the best places in the network to be most effective. The problem can thus be modified to accommodate the idea of *demand target pattern matching* to improve the performance of an envelope design by structuring the distribution of the protected working channels on spans to be reflective of some basic pattern of *relative* intensities expected of future demand. Mathematically this pattern specification can be identical in form to a static demand matrix, but its interpretation and meaning is no longer that of an assumed exact forecast. Rather, it is taken simply as a relative shaping template to help structure the PWCE as it is simultaneously maximized in volume. The philosophy is that even though demand is random and the future pattern has uncertainty in forecasting, it is still worthwhile to provide the solver a best view of the relative future intensities of demand on each node pair. This serves as a guideline for structuring the PWCE to generally fit plausible future demand scenarios better than if pure volume maximization is effected with no such guiding hand at all. For instance, compared to giving no shaping guidance at all to a volume maximization problem, in practice it would always seem reasonable to at least enter a shaping template that reflects the relative number of purely topological shortest paths of the graph that cross each span. Given a plausible forecast of relative loads between node pairs (a *network load template*) we can identify which spans are traversed by high relative loads (or simply traversed by many shortest-routes between all node pairs) and hence should preferably have a lot of working capacity assigned to them in the envelope volume maximization. Given a total design budget cost, we can then design a PWCE with a maximized envelope volume and a form of structuring to the distribution of working channels that is reflective of the relative load template.

This modified model requires the following additional notations:

- **Input Parameters:**
    - l_j is the target predicted relative load on span j, which can be computed say by shortest path routing of some demand forecasts.

- **Decision Variables:**
    - λ is a shape-asserting factor which structures the PWCE relative to the target load distribution.
    - α is a bi-criterion trade-off factor between structure shaping and volume maximization of PWCE.

The demand target pattern matching problem is formulated as follows:

$$\max \quad \lambda + \alpha \sum_{\forall j \in S} w_j \tag{16.65}$$

$$\text{s.t.} \quad \sum_{\forall k \in P} x_{i,k} n_k \geq w_i \quad \forall i \in S \tag{16.66}$$

$$s_j \geq \sum_{\forall k \in P} \delta_{j,k} n_k \quad \forall j \in S \tag{16.67}$$

$$w_j \geq \lambda l_j \quad \forall j \in S. \tag{16.68}$$

The problem is now in the form of a bi-criteria optimization, where there is some user-defined tradeoff between PWCE volume maximization and the shape factor, λ.

424   HANDBOOK OF OPTIMIZATION IN TELECOMMUNICATIONS

This shape factor is calculated by the constraints in equation (16.68), which expresses utility in the similarity between the PWCE and the target network load distribution. Equations (16.66) and (16.67) are repeated from the previous model. Generally, α is chosen so that the total volume term in the objective in (16.65) does not reduce from its single criteria optimal maximal value, but subject to this λ is forced as high as it can be.

16.9.3 Other Capacity Budget Scenarios

In the model in Section 16.9.1, spare capacity on each span has been pre-determined and specified as the available budget within which to construct a volume-maximized PWCE, however, there are other contexts for the capacity budget. Budgets can be span-based where a specified maximum number of spare channels is allowed on each span, or network-wide where the limit is applied on the total spare capacity of the network as a whole. By its very nature, a network-wide budget normally has more freedom in the PWCE construction than a span-based budget, and so more efficient designs are possible. Capacity budgets can also be defined on the total of working and spare capacity. Therefore, there are four possible types of budget scenarios to consider.

1. *Span-based spare capacity budget* – There is a specified maximum spare capacity on each span individually.

2. *Span-based total capacity budget* – There is a specified maximum total capacity on each span individually. In this scenario, there are no bounds on how much of that capacity must be working capacity and how much is spare.

3. *Network-wide spare capacity budget* – There is a specified collective maximum amount of spare capacity over all spans combined. In this scenario, there are no requirements on how this budget is distributed among the spans.

4. *Network-wide total capacity budget* – There is a specified collective maximum amount of combined working and spare capacity over all spans combined. In this scenario, there are no requirements on how much of that capacity must be working capacity and how much is spare or on how this budget is distributed among the spans.

These capacity budget scenarios can all be applied to the ILP models in Sections 16.9.1 and 16.9.2. We first introduce new notation as follows:

- **Input Parameters:**

    - T_j is the total number of deployed (or deployable) channels on span j, among which some channels will be assigned as the working capacity and with the remainder assigned as spare capacity.
    - B_S is the collective network-wide *spare* capacity budget.
    - B_T is the collective network-wide *total* capacity budget.

The ILP models in Sections 16.9.1 and 16.9.2 correspond to the span-based spare capacity budget (capacity budget scenario 1). If we wish to implement the second capacity budget scenario where we have a total working plus spare capacity budget for each span individually, we let the spare capacity values, s_j, be decision variables, and add the new constraints in equation (16.69) to either the volume maximization problem in Section 16.9.1 or the demand target pattern matching problem in Section 16.9.2:

$$T_j \geq w_j + s_j \quad \forall j \in S \tag{16.69}$$

This scenario closely models the situation where there is an existing deployed network of transmission systems upon which we wish to construct an efficient p-cycle network under the PWCE paradigm. The problem is then to determine how best to assign the installed capacity as either working or spare capacity so as to produce an optimal protection envelope. The new constraint set in equation (16.69) ensures that the combined working and spare capacity does not exceed the allowable budget on each span.

The third capacity budget scenario where we have a single network-wide collective spare capacity budget can be implemented by the adding the following constraint set to either the volume maximization problem or the demand target pattern matching problem:

$$B_S \geq \sum_{\forall j \in S} s_j \tag{16.70}$$

Here, equation (16.70) limits the total spare capacity allowed in the network. Again, the spare capacity values, s_j, are decision variables, rather than input parameters. In general, resultant protection envelopes tend to be large for this scenario than the first two, since we essentially allow greater flexibility in where the budget is allocated.

Finally, the fourth capacity budget scenario with a network-wide total capacity budget can be modeled by adding the constraint set in equation (16.71) to either the volume maximization problem or the demand target pattern matching problem:

$$B_T \geq \sum_{\forall j \in S} (w_j + s_j) \tag{16.71}$$

Equation (16.71) ensures that the total combined working and spare capacity allocated throughout the entire network is within the specified budget limit. This scenario allows for the most optimal PWCE formation, since it allows the most flexibility of all capacity budget scenarios, and corresponds to a green-fields design problem, where the goal is to build a completely new network from the ground up, subject to a limited capacity (or financial) budget.

Work in Shen and Grover (2003b;a), showed that over five different test networks, the volume maximization model allows for an additional 16% to 78% increase in the amount of working capacity capable of being protected under the optimal spare capacity allocation from the SCA design model in Section 16.1. It should be emphasized that all of these working capacity increases are free in the sense that no spare capacity increase is needed to retain 100% restorability. For a network operator, this is a good source of extra revenue (or an opportunity for savings) and an attractive option for provisioning growing services. Comparisons to the SBPP-based provisioning method also show that PWCE with p-cycle protection provides lower probability of blocking, when both models are provided with the same capacity distribution.

16.10 RING-MINING TO P-CYCLES

Migration from existing ring-based networks towards a future mesh-based architecture and operation is of considerable interest to network operators. In one migration technique called "ring mining," the line capacity and high speed interfaces of existing ring transport systems are reclaimed to support further ongoing growth of demand by converting to mesh based routing and protection operating under the span capacities of the prior rings (Clouqueur et al., 2001). p-Cycles are an interesting and natural alternative target architecture for ring-mining because like rings, they are cycle-oriented, but at the same time, unlike rings, they are mesh-like in capacity efficiency.

16.10.1 Ring Mining to p-Cycles Without Capacity Addition

One simple model for converting a ring network to a p-cycle network is to find the largest common multiplier, λ', that can be applied to every member of the demand matrix, while still keeping the network restorable using p-cycles, without adding any new capacity at all. Like the JCA model in Section 16.2, this model jointly optimizes working and spare capacity for a p-cycle based network, but now we must respect the existing ring fiber capacity limits (we also call this the pure ring mining problem). We add the following new notation to that used for previous models:

- **Input Parameters:**
    - a_j^m is the number of modules of type m available on span j from an already existing ring network.

- **Variables:**
    - $\lambda' \geq 1$ is the uniform least common demand growth multiplier, and is the maximum value by which we can multiply all demands while still being able to construct a fully-restorable p-cycle network that respects the capacity limits on each span.

The problem is formulated as the following ILP:

$$\max \quad \lambda' \qquad (16.72)$$

$$s.t. \quad w_j \leq \sum_{\forall k \in P} x_{i,k} n_k \quad \forall i \in S \qquad (16.73)$$

$$s_j = \sum_{\forall k \in P} \delta_{j,k} n_k \qquad (16.74)$$

$$\sum_{\forall q \in Q^r} g^{r,q} = \lambda' d^r \quad \forall r \in D \qquad (16.75)$$

$$\sum_{\forall r \in D} \sum_{\forall q \in Q^r} \varsigma_i^{r,q} g^{r,q} = w_i \quad \forall i \in S \qquad (16.76)$$

$$w_j + s_j \leq \sum_{\forall m \in M} a_j^m Z^m \quad \forall j \in S \qquad (16.77)$$

The constraint sets in equations (16.73), (16.74), and (16.76) are identical to those for the JCA formulation in Section 16.2, while equation (16.75) replaces equation

(16.9). In equation (16.75), all working demands are routable even after the growth multiplier has been applied to each demand. Finally, equation (16.77) ensures that only the existing capacity mined from the prior ring network is used.

Work in Kodian et al. (2003) showed that for some networks, uniform demand growth multipliers as high as 2.75 were achievable, meaning without adding any new capacity, a ring network could be converted to a p-cycle network capable of fully routing and protecting all demands even if they are all increased to 275% of their original quantities.

From a practical view however, pure ring mining potential is just an indicator, and the use of a *uniform* demand growth multiplier can be very restrictive for the design since there are generally a few demands that act as upward forcers for the total network cost. Increasing these demands by a few units would soon exhaust capacity on some critical spans on all possible routes for the demand. Unused capacity may exist on other spans in the network, but it is effectively stranded because that capacity does not exist in a continuous fashion. These forcer demands then force the solver to choose a very low value for λ' so as to not go over the capacity bounds imposed by the prior ring networks, regardless of the fact that a good number of non-forcer demands may have supported higher growth multipliers individually.

There are two possible options at this stage. To start with, an accurate pattern of non-uniform growth multipliers could be specified per demand, which may result in a much better design. One possible solution is attempting to let the solver maximize all demands individually under the same capacity constraints. This would first require definition of new notation:

- **Variables:**

    - $\lambda'_r \geq 1$ is the maximum demand growth multiplier for demand r, and is the maximum value by which we multiply demand r in the new design while still being able to construct a fully-restorable p-cycle network that respects the capacity limits on each span.

The objective function then changes from the one in equation (16.72) to that in equation (16.78), below, and the constraints in equation (16.75) are replaced with those in equation (16.79).

$$\max \sum_{\forall r \in D} \lambda'_r \quad (16.78)$$

$$\text{s.t.} \sum_{\forall q \in Q^r} g^{r,q} = \lambda'_r d^r \quad \forall r \in D \quad (16.79)$$

All other constraints in equations (16.73), (16.74), (16.76), and (16.77) still apply. The result is that the solver first maximizes the demands between every adjacent node pair (subject to capacity availability), then the second adjacent node pair and so on. The second option is to add capacity selectively on spans that are very near exhaust, to effectively "unlock" the stranded capacity elsewhere in the network. We examine this in the next section.

428 HANDBOOK OF OPTIMIZATION IN TELECOMMUNICATIONS

16.10.2 Ring Mining with Selective Capacity Addition

In this model, we allow modular capacity to be selectively added while mining rings to p-cycles under an economy-of-scale module addition cost model. The uniform demand growth multiplier λ' now becomes a parameter to the optimization problem and is incremented in fixed steps, to calculate the minimal extra capacity-cost required to enable each step value of λ'. We define the following new notation:

- **Variables:**

  - $\xi_j^m \geq 0$ is the number of modules of type m strategically added to span j.

The objective function is now changed from that in equation (16.72) to the one in equation (16.80), which seeks to minimize the cost of additional capacity added to the network:

$$\min \sum_{\forall j \in S} \sum_{\forall m \in M} \xi_j^m C_j^m \qquad (16.80)$$

All of the constraints with constraints in equations (16.73), (16.74), (16.75), and (16.76), are kept and the constraints in equation (16.77) are replaced with those in (16.81).

$$w_j + s_j \leq \sum_{\forall m \in M} (a_j^m + \xi_j^m) Z^m \quad \forall j \in S \qquad (16.81)$$

This new constraint set limits the working and spare capacity total on each span to be under the span capacity available from the existing ring network, plus the added modular span capacity. Adding capacity increases the value of the objective, and thus ensures that the solver considers adding capacity only after fully exploiting the existing ring capacity. Minimizing the cost also ensures that the solver chooses the least costly construction of modular capacity. For example, adding 1 OC-48 module would be less expensive than 4 OC-12 modules on the same span. Again, if the solver has already added one module to a span, it will attempt to maximize the utilization of that module, even if some of the working/protection paths need to be longer, since using 1 or all 48 units of capacity on an added OC-48 module on a span does not affect the objective.

Work in Kodian et al. (2003) finds that the capacity addition profile for ring-mining to p-cycles is very close to the profile obtained for span restorable mesh. But at the same time there is no rule as to why a particular network requires a specific number of modules added for demand growth. Two interesting effects are also observed. First, there is a deferral of cost until significant demand growth occurs, and secondly, there is a relatively low amount of extra capacity required to support a high uniform demand growth multiplier for some test networks. To a network operator this means that, depending on the network details, substantial demand growth can be supported without much capital investment.

16.10.3 Ring Mining Without Changes to Existing Working Routes

In all the models discussed so far, working routes are jointly optimized with the placement of p-cycles. This is acceptable for experimental studies with growth demands and assumes that any implied rearrangement of existing paths may be acceptable if it is coordinated with customers and permitted by service contracts. But, based on industry feedback, this is not always an option. We now discuss a variation to the base model in Section 16.10.1 where we add a constraint restricting the rerouting of existing demand. We define the following new notation:

- **Input Parameters:**

    - $w_j^0 \geq 0$ is the previously existing working capacity on span j, obtained as part of the ring network data.

The objective function remains as that in equation (16.72), and we incorporate demand rerouting restrictions into the planning model by replacing the constraints in equations (16.75) and (16.77) by the constraints in equations (16.82) and (16.83), below, respectively:

$$\sum_{\forall q \in Q^r} g^{r,q} = (\lambda' - 1)d^r \quad \forall r \in D \tag{16.82}$$

$$w_j^0 + w_j + s_j \leq \sum_{\forall m \in M} a_j^m z^m \quad \forall j \in S \tag{16.83}$$

The other constraints in the base model in Section 16.10.1 still apply. Constraint set (16.82) ensures that only the new growth demands are considered for routing decisions that are jointly coordinated with the spare capacity placement decisions. The existing working capacity on each span is left untouched because of the constraints in equation (16.83), and so existing working routes are not disturbed. In Kodian et al. (2003) it was shown that the specific test network could handle a uniform demand growth multiplier of $\lambda' = 1.5$ when working routes are allowed to be re-optimized, compared to $\lambda' = 1.33$ with no changes to existing working routing.

16.10.4 Ring Mining with Selective Capacity Addition and no Change to Existing Working Routes

For similar reasons as discussed in Section 16.10.2, selective capacity addition is considered along with the constraint on altering existing working routing, with λ' increased in steps. We consider the same model as in Section 16.10.3 with the constraints in equation (16.83) replaced by equation (16.84), below.

$$w_j + s_j + w_j^0 \leq \sum_{\forall m \in M} (a_j^m + \xi_j^m) z^m \quad \forall j \in S \tag{16.84}$$

This new constraint set ensures that the existing working capacity is left untouched. The existing spare capacity is re-optimized for p-cycles, including the option to add new capacity in modular quantities. Of course, this assertion of fixed working routing may impose higher capacity requirements, as compared to the model in Section 16.10.2 where we do a global rerouting. But results in Kodian et al. (2003) for a real-world test network indicate that only a nominal amount of excess capacity is needed over the complete re-optimization design case.

16.11 ONGOING RESEARCH OUTLOOK AND ADDITIONAL RESOURCES

This chapter has recapped the p-cycle concept and collected together for reference a number of the now recognized problem models for design and further research in the topic area. Research in the area is still vigorous. As this chapter goes to press a couple of the most interesting ongoing developments include the extension of p-cycles into end-to-end path-protecting structures called Failure Independent Path Protecting (FIPP) p-cycles. (see Kodian and Grover (2005)). An important advantage of FIPP p-cycles is that, like span protecting p-cycles, the protection paths are fully pre-connected, end-to-end prior to failure. This gives added assurance of end-to-end optical transmission integrity of the protection paths compared to schemes of similar capacity efficiency for end-to-end protection in which optical backup paths have to be assembled "on the fly." This approach also inherently addresses node failures as well as span failures, and provides a simple end-node fault detection and protection activation paradigm. Another area under research is the combined or coordinated design of MPLS-layer node-encircling p-cycles and span-protecting optical layer p-cycles. Conventional span-protecting p-cycles with their fast protection switching and simple mechanism and operation are quite appealing in the optical layer but are not able to deal with node failures. On the other hand, oversubscription-based node-encircling p-cycles are capable of protecting node failures and potentially have high capacity efficiency because controlled oversubscription is allowed. This approach can also employ a "common pool" survivability concept to implement the node encircling MPLS layer p-cycles within the spare capacity used by span-protecting p-cycles, or at least re-using as much of that spare capacity as possible to deal with node failures (or even secondary span failures). Common pool survivability is based on reservation of resources in the optical layer to not only meet the needs of protection within that layer, but also as needed in the MPLS layer. The method allows saving capacity in the case of two-layer recovery procedures. Overall, p-cycles provide many new options for efficient survivable network architecture, as well as promising alternatives to dealing with new challenges such as highly differentiated availability guarantees to services in future networks, and the possibility of rapidly arriving and departing random transport path requirements and fundamental uncertainty in the future demand scenarios which a network design may have to face. To find out more about the ongoing research on p-cycles and the extent to which many aspects of p-cycle network design and operation have already been reduced to practice, see the p-cycles web site: A.Sack and W.D. Grover (2005). That web site will also serve as a forum for updates, addenda, or corrections as needed for this chapter.

Bibliography

D. Arci, G. Maier, A. Pattavina, D. Petecchi, and M. Tornatore. Availability models for protection techniques in WDM networks. In *Proceedings of the 4th International Workshop on Design of Reliable Communication Networks (DRCN 2003)*, pages 158–166, Banff, AB, October 2003.

A.Sack and http:netsys.edm.trlabs.cap-cycles W.D. Grover. Trlabs network systems group *p*-cycles website., June 2005. Online Reference.

P. Batchelor et al. *Ultra high capacity optical transmission networks, Final report of action COST 239*. Faculty of Electrical Engineering and Computing, Zagreb, Croatia, 1999.

V. Chvátal. *Linear Programming*. W. H. Freeman and Company, New York, NY, 1983.

M. Clouqueur and W. D. Grover. Availability analysis or span-restorable mesh networks. *IEEE Journal on Selected Areas in Communications*, 20(4):810–821, May 2002.

M. Clouqueur, W. D. Grover, D. Leung, and O. Shai. Mining the rings: Strategies for ring-to-mesh evolution. In *Proceedings 3rd International Workshop on the Design of Reliable Communication Networks (DRCN 2001)*, pages 113–120, Budapest, Hungary, October 2001.

J. Doucette. Advances on design and analysis of mesh-restorable networks. Master's thesis, University of Alberta, 2004.

J. Doucette, P. Giese, and W. D. Grover. Investigation of node protection strategies with node-encircling *p*-cycles. In *Design of Reliable Communication Networks(DRCN 2005)*, Ischia, Italy, October 2005.

J. Doucette and W. D. Grover. Influence of modularity and economy-of-scale effects on design of mesh-restorable DWDM networks. *IEEE Journal on Selected Areas in Communications*, 18(10):1912–1923, October 2000.

J. Doucette and W. D. Grover. Comparison of mesh protection and restoration schemes and the dependency on graph connectivity. In *Proceedings of 3rd International Workshop on Design of Reliable Communication Networks (DRCN 2001)*, pages 121–128, Budapest, Hungary, October 2001.

J. Doucette and W. D. Grover. Capacity design studies of span-restorable mesh networks with shared-risk link group (SRLG) effects. In *Optical Networking and Communications Conference (OptiComm 2002)*, pages 25–38, Boston, MA, July/August 2002.

J. Doucette, D. He, W. D. Grover, and O. Yang. Algorithmic approaches for efficient enumeration of candidate *p*-cycles and capacitated *p*-cycle network design. In *Proceedings of the 4th International Workshop on Design of Reliable Communication Networks (DRCN 2003)*, pages 212–220, October 2003.

W. D. Grover. *Mesh-Based Survivable Networks: Options and Strategies for Optical, MPLS, SONET, and ATM Networking*. W. H. Freeman and Company, Upper Saddle River, NJ, 2003.

W. D. Grover. The protected working capacity envelope concept: An alternative paradigm for automated service provisioning. *IEEE Communications Magazine*, 42(1):62–69, January 2004.

W. D. Grover and J. Doucette. Advances in optical network design with *p*-cycles: Joint optimization and pre-selection of candidate *p*-cycles. In *Proceedings of the IEEE-LEOS Summer Topical Meeting on All Optical Networking*, pages 49–50, Mont Tremblant, QC, July 2002.

W. D. Grover and D. Stamatelakis. Cycle-oriented distributed preconfiguration: Ring-like speed with mesh-like capacity for self-planning network restoration. In *Proceedings of IEEE International Conference on Communications (ICC 1998)*, pages 537–543, Atlanta, GA, June 1998.

W. D. Grover, B. D. Venables, M. H. MacGregor, and J. H. Sandham. Development and performance verification of a distributed asynchronous protocol for real-time network restoration. *IEEE Journal on Selected Areas in Communications*, 9(1): 112–125, January 1991.

R. R. Iraschko, M. H. MacGregor, and W. D. Grover. Optimal capacity placement for path restoration in STM or ATM mesh-survivable networks. *IEEE/ACM Transactions on Networking*, 6(3):325–336, June 1998.

J. Kennington, E. Olinick, K. Lewis, A. Ortynski, and G. Spiride. Robust solutions for the DWDM routing and provisioning problem: Models and algorithms. *Optical Networks Magazine*, 4(2):74–84, March/April 2003.

A. Kodian and W. D. Grover. Failure independent path protecting p-cycles: Efficient and simple fully pre-connected optical path protection, 2005. in press.

A. Kodian, W. D. Grover, J. Slevinsky, and D. Moore. Ring-mining to p-cycles as a target architecture: Riding demand growth into network efficiency. In *Proceedings of the 19th Annual National Fiber Optics Engineers Conference (NFOEC 2003)*, pages 1543–1552, Orlando, FL, September 2003.

D. Leung and W. D. Grover. Capacity planning of survivable mesh-based transport networks under demand uncertainty. *Journal of Photonic Network Communications*, June 2004a. Submitted.

D. Leung and W. D. Grover. Restorable mesh network design under demand uncertainty: Toward "future proof" transport investments. In *Optical Fiber Communication Conference (OFC 2004)*, Los Angeles, CA, February 2004b.

D. Rajan and A. Atamturk. Survivable network design: Routing of flows and slacks. In G. Anandalingam and S. Raghavan, editors, *Telecommunications Network Design and Management*, pages 65–82. Kluwer, 2002.

A. Sack. New techniques for p-cycle network design. Master's thesis, University of Alberta, 2004.

A. Sack and W. D. Grover. Hamiltonian p-cycles for fiber-level protection in homogeneous and semi-homogeneous optical networks. *IEEE Network, Special Issue on Protection, Restoration, and Disaster Recovery*, 18(2):49–56, March/April 2004.

D. A. Schupke. An ILP for optimal p-cycle selection without cycle enumeration. In *Proceedings of the 8th IFIP Working Conference on Optical Network Design and Modelling (ONDM)*, pages 2761–2765, Ghent, Belgium, February 2004.

D. A. Schupke, W. D. Grover, and M. Clouqueur. Strategies for enhanced dual-failure restorability with static or reconfigurable p-cycle networks. In *Proceedings of IEEE International Conference on Communications (ICC 2004)*, Paris, France, June 2004.

D. A. Schupke, C. G. Gruber, and A. Autenrieth. Optimal configuration of p-cycles in WDM networks. In *Proceedings of IEEE International Conference on Communications (ICC 2002)*, pages 2761–2765, New York, NY, April/May 2002.

G. Shen and W. D. Grover. Exploiting forcer structure to serve uncertain demands and minimize redundancy of p-cycle networks. In *Proceedings of 4th SPIE Optical Networking and Communications Conference (OptiComm 2003)*, pages 59–70, Dallas, TX, October 2003a.

G. Shen and W. D. Grover. Extending the p-cycle concept to path segment protection for span and node failure recovery. *IEEE Journal on Selected Areas in Communications, Optical Communications and Networking Series*, 21(8):1306–1319, October 2003b.

G. Shen and W.D. Grover. Design of protected working capacity envelopes based on p-cycles: An alternative framework for survivable automated lightpath provisioning. In A. Girard, B. Sanso, and F. Vazquez-Abad, editors, *Performance Evaluation* and *Planning Methods for the Next Generation Internet*, pages 65–82. Kluwer Academic Publishers, 2004.

D. Stamatelakis and W. D. Grover. IP layer restoration and network planning based on virtual protection cycles. *IEEE Journal on Selected Areas in Communications*, 18(10):1938–1949, October 2000a.

D. Stamatelakis and W. D. Grover. Theoretical underpinnings for the efficiency of restorable networks using pre-configured cycles ("p-cycles"). *IEEE Transactions on Communications*, 48(8):1262–1265, August 2000b.

T. Stidsen and T. Thomadsen. Joint optimization of working and p-cycle protection capacity. Technical Report IMM Technical Report 2004-8, Informatics and Mathematics Modelling, Technical University of Denmark, May 2004.

W. L. Winston. *Operations research applications and algorithms*. Duxbury Press, Belmont, CA, 3rd edition, 1994.

H. Zhang and O. Yang. Finding protection cycles in DWDM networks. In *Proceedings of IEEE International Conference on Communications (ICC 2002)*, volume 5, pages 2756–2760, New York City, NY, April/May 2002.

17 OPTIMIZATION ISSUES IN QUALITY OF SERVICE

John G. Klincewicz[1]

[1]AT&T Labs
Middletown, NJ 07748 USA
klincewicz@att.com

Abstract: The advent of the World Wide Web has fundamentally changed the nature of Internet traffic. New classes of applications, such as video conferencing, Internet telephony and various forms of e-commerce have arisen, for which so-called *best effort* service is no longer acceptable. These new applications represent delay-sensitive traffic with specific performance requirements. The term Quality of Service (QoS) is used to describe network features that are designed to provide the better than *best effort* performance that is required by such applications. In this chapter, we consider QoS in the context of network design. Specifically, we focus on network design or optimization problems that address link topology, link capacity, route assignment and/or router location, and that take various performance requirements into account. In particular, solutions to these network design problems ensure that sufficient resources (e.g., bandwidth) are made available so that certain specified performance requirements (e.g., delay requirements) will be explicitly met.

Keywords: Internet protocol, network design, quality of service, delay, routing, capacity, network topology

17.1 INTRODUCTION

Until recently, the Internet was dominated by applications such as file transfers and e-mail. Since such applications could tolerate considerable delays, so-called *best-effort* service, which does not provide any performance guarantees, was acceptable. The advent of the World Wide Web and Virtual Private Networks (VPNs), however, has fundamentally changed the nature of Internet traffic. New classes of applications, such as video conferencing, Internet telephony (i.e., *voice over IP*) and various forms of e-commerce have arisen. Many of these new applications have become critical for both business and government operations. They are typically delay-sensitive, with each one having its own set of specific performance requirements. In the case, for example, of Internet telephony, if delay or packet loss were to exceed certain levels, users would not be able to communicate effectively. In the case of e-commerce, delays in processing stock transactions could result in severe financial consequences for traders.

The term quality of service (QoS) is used to describe network features that are designed to provide applications with better than *best effort* performance. There are two broad categories of such QoS features: service differentiation and performance guarantees. (See, for example, Wang (2001).) In service differentiation, certain classes of Internet traffic are given priority or preferential treatment as they are processed and routed over the network. This does provide an improved quality of service for the priority traffic, at least relative to the other traffic, but does not necessarily ensure a specified level of performance. With performance guarantees, sufficient resources (e.g., bandwidth, processing time within the routers) are made available to the various classes of traffic so that certain specified performance requirements (e.g., delay requirements) will be explicitly met. It is this latter meaning that we will consider in this chapter.

In particular, we will consider QoS in the context of network design for Internet Protocol (IP) networks. We focus on network design or optimization problems that address one or more of the following issues, taking QoS considerations into account:

- **Link topology.** In an IP network, routers process packets of data and send them on to other routers, or to the end user. These routers are interconnected by transmission facilities. If one or more transmission facilities exist between a pair of routers, we say that there is a *link* between them. One common network design problem, then, is as follows. For a particular set of routers and their associated router-to-router traffic, how should the routers be interconnected? In other words, which links should be established?

- **Link capacity.** Transmission facilities typically come in certain modular sizes. For example, a T1 facility can carry 1.544 Mbps, an OC3 facility can carry 155.52 Mbps, and an OC48 facility can carry 2488.32 Mbps. Assigning capacity to a link corresponds to installing one or more modular transmission facilities between the corresponding pair of routers. The capacity assignment problem, then, is as follows. For a particular set of links, what level of capacity must be assigned to each of them?

- **Route assignment.** In an IP network, there is typically enough links to allow more than one path (or route) between each pair of routers. If a packet uses a particular path through the network, we say that it is *routed* on that path. In some cases, traffic can be split among multiple paths or routes. The route assignment problem, then, is as follows. How should a particular set of router-to-router traffic be routed through the network?

- **Router location.** An IP network is utilized by users on personal computers and workstations and by *servers* (workstations or computers that are accessed by users in order to conduct some transaction, such as to download a Web page or to obtain or modify data in a database). These users and servers are connected, in some way, to network routers, e.g. through Local Area Networks (LANs) or dial-up connections. The router location problem is then as follows: For a particular set of end users and servers, how many routers are needed, where should they be located, and which users and servers should be connected to each of them?

In reality, of course, all these problems are interrelated. For example, the required link sizing depends on the amount of traffic to be carried on the link, which depends, in turn, on the routing. The routings that are available depend on the link topology. The link topology, in turn, is influenced by the router-to-router traffic matrix, which depends on the location of the routers and how the users and servers are connected to them. In a similar vein, the required link sizing depends also on the delay that can be tolerated on that link. The delay that can be tolerated depends on which router-to-router traffic is routed on that link. The best routing of traffic, however, depends, in turn, upon the link sizing.

Determining how to solve all these interrelated network design problems simultaneously, taking QoS considerations into account, is an almost impossible undertaking. Fortunately, in the course of operating and planning an existing network, one typically does not have to consider all these problems at once. In planning for the short term, for example, one might consider changes to the routing of traffic in order to make best use of the existing capacity. Within a medium-range planning horizon, one might determine what additional capacity will be needed to handle anticipated growth in traffic. Over a longer-range planning horizon, one could consider modifying the set of links and/or router locations. Thus, in ongoing network planning, some features of the network are typically assumed to be fixed and given, while solutions are determined for the rest. Sometimes, in cases where multiple aspects do need to be determined simultaneously, iterative heuristic procedures can be developed that consider the various sub-problems sequentially.

The term Quality of Service (QoS) encompasses a wide variety of technologies and objectives, depending on the particular applications being considered (voice, video, real-time financial transactions, etc.) and the particular network being designed (private corporate enterprise network, public service provider network, local area network, etc.). Thus, the various models and methodologies that we discuss in this chapter will make different assumptions and differ in significant modeling details. The key issues that distinguish these models and methodologies include the following:

- **Routing discipline.** In IP networks, the most common routing mechanism is to use a shortest-path-based protocol, such as OSPF (Open Shortest Path First). In such a protocol, each link is assigned an administrative weight. Because routers in a network exchange topology and administrative weight information with each other, each router is able to compute the shortest paths to other routers (i.e., paths that minimize the sum of the administrative weights for the links in the path) and will forward packets accordingly. As a result, each packet will be routed from end to end in the network along a shortest path. In the event that there is more than one such shortest path emanating from a particular router, the router typically will attempt to balance the traffic among the multiple paths. For such IP networks, the solution to the route assignment problem is determined by the settings that are chosen for the administrative weights. Alternatively, MPLS allows the system administrator to explicitly set the routes between each pair of routers. This latter approach allows a greater degree of control over the traffic, but with significantly increased operational complexity.

- **QoS metrics.** Deciding how to best measure QoS for a particular network is not necessarily easy. Delay and link congestion are the two most common measures that appear in the literature. Some models concern themselves only with a constraint on the average delay of a packet in the network. This provides an overall performance guarantee, but may result in relatively poor performance for traffic between a particular pair of routers. A more stringent requirement (and one that, in general, results in a more difficult model to solve) is to set a constraint on the maximum end-to-end packet delay for each router-to-router pair. Other models avoid including explicit constraints on delay, but, instead, include a cost of delay or a cost of congestion, either on a link-by-link basis, or on the network as a whole. In practice, many network planners and administrators are concerned with measures such as packet loss rate and variation of delay (jitter). However, these metrics are more difficult to include in an optimization model focused on network design.

- **Components of delay.** Delay (i.e., the amount of time between when a packet enters the network and leaves the network) is the most common factor considered in QoS metrics. However, different models focus on different components of delay, depending on what is assumed to be the dominant factor in the network. The various possible components of delay include insertion delay, queueing delay, node processing delay and, propagation delay. Insertion (or transmission) delay refers to the time required to insert a packet of a given size (kilobits) on a transmission facility that serves packets at a given rate (kilobits per second). The queueing delay at the output buffer refers to the time that the packet has to wait to be served by the transmission facility. Node processing delay includes time required for the router to examine and route the packet and to perform other operations, such as encryption/decryption or data compression. The propagation delay refers to the time required to traverse the transmission facility (related to the length of the link and the speed of light). As routers become faster and transmission facilities become able to serve packets at increased rates of speed, the propagation delay (which depends on the physical route taken by the packet) becomes a more significant component of delay.

- **Classes of traffic.** When diverse applications share the same network, QoS requirements can either be met by engineering the network so that all traffic achieves the most stringent performance criteria, or, else, by differentiated treatment based on the class of traffic. Such differentiated treatment requires each packet to be tagged. This could be achieved, for example, using the Type of Service (TOS) field in the standard IP header. The increased need for QoS has motivated renewed interest in ways to utilize these bits in the header.

Differentiated treatment requires some form of priority queueing within the routers, and/or differentiated routing based on class. One such queueing algorithm is Cisco's

Custom Queueing (see Floyd and Jacobson (1995)). Under this bandwidth allocation scheme, a portion of the bandwidth is reserved for each specific class. The router serves a series of queues in a round robin order and provides each queue with its allocated portion before moving to the next queue. There is a first in, first out discipline within each class. If a particular class is not using its reserved bandwidth, it is given to the other classes.

Within this chapter, a limited number of models address the issue of differentiated performance requirements by class of traffic. Others apply a single specified performance requirement to all traffic, or assign different performance requirements (e.g., maximum delay) for each router-to-router pair.

- **Survivability and restoration.** Network components (e.g., routers, links) are subject to failure. In designing a network with particular performance characteristics, it is necessary to specify what, if any, performance characteristics are still guaranteed to be met in the case of a network failure.

- **Demand matrices.** For the most part, the types of network design problems that we consider in this chapter require some estimate of the traffic volume and traffic pattern that the network will be expected to carry. This generally takes the form of a router-to-router demand matrix. Creating such a matrix, either for current traffic or for forecast traffic, is not necessarily straightforward. First, today's systems for IP network measurement normally do not provide the inputs needed for the direct computation of such IP traffic matrices. Second, Internet traffic is highly variable over time; these variations make it difficult to describe using a single traffic matrix. Wasem et al. (1995) developed a model to create forecasts, based on demographic characteristics of various geographical areas. Zhang et al. (2003) developed an optimization-based approach, which they term tomo-gravity, to generate router-to-router matrices based on known quantities, such as maximum link loads, routing tables and volumes of terminating traffic. The optimization approach of Luss and Vakhutinsky (2001) utilizes a methodology based on resource allocation models. Typically, planners will utilize a single traffic matrix based on the demand pattern during a *peak* traffic time. However, some papers do attempt to engineer the network for multiple traffic matrices, corresponding to typical demand patterns at different hours of the day.

These and other key issues will be addressed further in the sections below.

As mentioned above, in this chapter, we focus on IP networks. Therefore, papers that specifically address features that are associated with other protocols, such as ATM or Frame Relay, are not included. Because the relevant body of literature is quite extensive, it would be virtually impossible to compile an exhaustive review. However, we have attempted, as far as possible to select papers that are representative of work in the area.

The remainder of this chapter is organized as follows. In Section 17.2, we briefly discuss some common modeling techniques used in optimization problems with QoS considerations. In Section 17.3, we consider papers that assume that all routes can be explicitly set, as is the case, for example, in an MPLS-enabled network. Section 17.4, on the other hand, discusses papers that can be applied to a network utilizing a shortest-path protocol. In each case, we distinguish among papers that utilize an average delay metric, those that utilize an end-to-end delay metric and those that use other metrics. Concluding remarks are given in Section 17.5.

17.2 SOME MODELING TECHNIQUES

Consideration of QoS in network design models introduces new features into both the problem formulations and the solution approaches that must be applied. Here, we illustrate some of the common modeling techniques that have been employed to take delay considerations into account. In Section 17.2.1, we present queueing models that are commonly employed in computing the average delay, either for a single link or for the overall network. In Section 17.2.2, we discuss a common heuristic solution approach that is applicable for problems that include constraints on the end-to-end delay along any given route.

17.2.1 Computing average delay

As noted above, several models consider average packet delay. A first step in computing the average packet delay over the network is to first compute the average packet delay over a single link i. The queueing plus transmission components of the delay have frequently been approximated using an M/M/1 model. Under this assumption of Poisson packet arrivals and exponential distribution of packet lengths, the average queueing plus transmission delay for a packet on a link i is given by

$$T_i = \frac{1}{\mu C_i - \lambda_i}, \qquad (17.1)$$

where $1/\mu$ is the average packet length (bits/packet), λ_i is the average packet arrival rate to link i (packets/second), and C_i is the capacity of link i (in bits/second).

This result can be extended to compute the average packet delay T over the entire network. For this expression, we require the following notation:

- P_i is the propagation delay on link i,
- K_i is the node processing delay entering link i,
- γ is total traffic in the network (packets/second), and
- f_i is the traffic flow on link i (bits/second),

where we can substitute $f_i = \lambda_i/\mu$. Given this, the formula for the average packet delay in the network T is as follows:

$$T = \frac{1}{\gamma} \sum_i f_i \left[\frac{1}{C_i - f_i} + \mu(P_i + K_i) \right]. \qquad (17.2)$$

Expression (17.2) for average packet delay in the network was initially derived by Kleinrock (1970). (See also Gerla and Kleinrock (1977).) It has appeared in a number of different contexts, as detailed in Sections 17.3 and 17.4 below.

17.2.2 Allocating a maximum delay to each link

Especially in models in which there are end-to-end delay constraints, it is often necessary to assign a maximum allowable delay to each link. These maximum delays must be chosen in such a way that, if the maximum delays are satisfied on each link, then the end-to-end delay constraints will be guaranteed to be satisfied over all the utilized routes. The process of assigning these maximum delays is known as delay allocation.

A typical delay allocation procedure is as follows. First, each route r is considered in turn, and an allowable delay is allocated to each link i in the route r, in such a way that $\sum_{i \in A_r} d_{ir} = D_r$, where D_r represents the maximum allowable delay on route r and A_r represents the set of links in route r. (Below, we discuss how one might choose to do this allocation.) Once this is done, we look at each link i and set the maximum delay on the link to be $d_i = \min_r \{d_{ir}\}$. This will yield a feasible solution, but may, in fact, be too stringent, in the sense that it may be possible to increase the allocation to certain links and still satisfy all the delay constraints. Therefore, one can examine all the routes to determine which are *tight* (i.e., $\sum_{i \in A_r} d_i = D_r$) and which are *slack* (i.e., $\sum_{i \in A_r} d_i < D_r$). The allocated delay for links on tight routes cannot be increased. For other links, the delays on slack routes that use them can then be re-allocated.

One could imagine a number of reasonable ways to choose the initial allocated values d_{ir} for each link i in each route r. For example, one could allocate D_r along each route in proportion to the length of the links, or in proportion to the traffic assigned to the links. However, a particular approach suggested by Klincewicz et al. (2002) bases the allocation on the solution to an optimization problem. First, let α_i denote a linear cost factor for link i. Then, consider the problem:

$$\text{Minimize } \frac{\alpha_i}{d_i}$$
$$\text{Subject to } \sum_{i \in A_r} d_i \leq D_r, \text{ for every route } r.$$

Since required capacity is roughly inversely related to delay, the objective function serves as a surrogate for capacity costs. If there is only a single route r, with links $i = i_1, \ldots, i_n$, then the solution to this problem (from an examination of the Kuhn-Tucker conditions) is given by

$$d_{i_1} = \frac{D_r}{\sum_j \sqrt{\alpha_{i_j}} / \sqrt{\alpha_{i_1}}},$$
$$d_{i_j} = \frac{\sqrt{\alpha_{i_j}}}{\sqrt{\alpha_{i_1}}} d_{i_1} \quad j = 2, \ldots, n.$$

Thus, this simple version of a delay allocation problem suggests that delays be allocated to each link i along a route in proportion to $\sqrt{\alpha_i}$. The values of α_i can be chosen, for example, based on the facility size that yields the least expensive cost per kilobit available on that link.

Note that several authors have considered optimal delay allocations for a set of demand pairs with a common origin or destination; see, for example, Monma et al. (1990); Shulman and Venkateswaran (1999), and Venkateswaran and Kodialam (1999).

However, none of these papers addresses the overall problem of determining link delay allocations for an arbitrary mesh network.

17.3 EXPLICIT ROUTING ON ASSIGNED PATHS

The vast majority of papers that take QoS considerations, such as delay, into account assume that routes between each router-to-router pair can be explicitly set, independent of a shortest-path weight setting scheme. In this section, we categorize these papers by the particular performance metrics that are taken into account.

17.3.1 Average delay

In this subsection, we consider papers that utilize average delay as a QoS metric. Table 17.1 summarizes the network features, assumptions and constraints that are addressed in these papers.

17.3.1.1 Flow assignment. The Flow Assignment (FA) problem assumes that the network topology is given and cannot be changed in the short term. Specifically, given the links, their current capacities and the traffic matrix, the problem is to route the traffic so as to minimize the average packet delay.

The FA problem assumes that routing of traffic can be modeled as a multi-commodity flow, where a commodity represents traffic between a particular origin-destination pair. This assumption implies that traffic for a given router pair can be split arbitrarily among multiple paths. The Flow Deviation (FD) algorithm of Fratta et al. (1973) is a well-known solution approach for this problem; see also Gerla and Kleinrock (1977). However, since the expression for average delay is a convex function, the FA problem is a convex multi-commodity flow problem and can be solved by any appropriate algorithm for such problems; see Ouorou et al. (2000).

17.3.1.2 Capacity and flow assignment. In the Capacity and Flow Assignment (CFA) Problem, it is necessary both to route the traffic and to choose the link capacities C_i for each link i. Installing capacity C_i incurs a cost $g_i(C_i)$. Fratta et al. (1973) consider a problem in which the objective is to minimize the average delay while keeping capacity costs within a specified budget constraint.

Other authors have considered the problem of minimizing the total cost $\sum_i g_i(C_i)$ of capacity, while satisfying a constraint on the average packet delay that can be tolerated (i.e., $T < T_{\max}$). For this problem, there are two cases to be considered. In one case, capacity is treated as a continuous variable, thus implicitly assuming that transmission facilities can be sized arbitrarily. In the other case, the model only allows capacity to be added in discrete increments. Except as noted below, work in this area assumes that traffic can be routed on the basis of a multi-commodity flow model.

For the continuous capacity case, Gerla and Kleinrock (1977) present adaptations of the Flow Deviation (FD) algorithm that can be applied to obtain local minima, when the cost of capacity is either a linear function or a concave function. Queiroz and Humes Jr. (2003) introduce a heuristic technique for the case of concave capacity costs that is based on the Tuy-Zwart algorithm for concave programming.

Table 17.1 Papers with explicit routes and average delay metric

Papers	Routes	Capacity	Links	Objective	Special constraints
Flow assignment					
Gerla & Kleinrock (1977) Fratta, Gerla & Kleinrock (1973)	MC Flow	Given	Given	Avg. delay	—
Capacity & flow assignment					
Fratta, Gerla & Kleinrock (1973)	MC Flow	Continuous	Given	Avg. delay	Max capacity cost
Gerla & Kleinrock (1977)	MC Flow	Continuous	Given	Capacity cost (linear/ concave)	Avg. delay
Queiroz & Humes (2003)	MC Flow	Continuous	Given	Capacity cost (concave)	Avg. delay
Ng & Hoang (1987)	MC Flow	Discrete	Given	Capacity cost	Avg. delay
Ng & Hoang (1987)	MC Flow	Discrete	Given	Capacity cost	Avg. delay (enhanced model)
Kang & Tan (1997)	MC Flow	Discrete	Given	Capacity cost	Avg. delay (restoration)
Betser et al. (1989)	MC Flow	Discrete	Given	Capacity cost	Avg. delay
Chari (1996)	Discrete	Discrete	Given	Capacity cost	Avg. delay (multi-period)
Topology, capacity & flow assignment					
Gerla & Kleinrock (1977)	MC Flow	Continuous	Binary choice	Capacity cost (concave)	Avg. delay
Gersht & Weihmayer (1990)	MC Flow	Continuous/ Discrete	Binary choice	Capacity cost (linear/concave)	Avg. delay
Stacey et al. (2000)	MC Flow	Discrete	Binary choice	Capacity cost (concave)	Avg. delay
Pierre & Legault (1998) Pierre & Elgibaoui (1997)	MC Flow	Discrete	Binary choice	Capacity cost	Avg. delay

For the discrete capacity case, Ng and Hoang (1987) extend the model of queueing delay to reflect more precisely the fact that, in the discrete case, each link consists of m parallel transmission facilities. They develop an FD-based optimization procedure for the extended model but assume that capacity costs are linear.

Few papers address issues of survivability. One that does is Kang and Tan (1997), which develops an iterative approach for the case where sufficient capacity must be available to meet the performance constraint even under link failure conditions. In addition to average packet delay, they allow an alternative performance constraint that limits the maximum buffer overflow probability. However, they assume that the costs for the discrete link capacity options are determined by a convex function.

Betser et al. (1989) present a novel application of the CFA problem to determine the size and number of servers needed for a computer network serving a number of remote PCs.

Unlike the above papers, which assume that traffic can be routed according to a multi-commodity flow, the paper by Chari (1996) considers a problem in which traffic is routed on one or more discrete routes. In particular, this work addresses a multi-hour version of the CFA problem in which there are multiple traffic matrices, corresponding to different times of the day. In this case, the problem is to choose a single set of capacities, as well as different routings for each time of day. The solution procedure iteratively solves a linear programming problem in order to find a heuristic solution.

17.3.1.3 Topology, capacity and flow assignment. Choosing which links to include in the network, simultaneously with the link capacity and the routing of the traffic, gives rise to the Topology, Capacity and Flow Assignment (TCFA) problem. As in the CFA problem above, there is a cost for installing capacity and a constraint on the maximum average delay. The addition of binary variables to indicate whether or not link i is included in the problem adds a significant degree of difficulty, especially when the number of nodes is large. However, all the papers reviewed in this subsection assume that the traffic routings can be based on multi-commodity flows.

A number of algorithms have been designed for the case where capacity costs are concave and capacity is treated as a continuous variable. The Concave Branch Elimination procedure, due to Gerla and Kleinrock (1977), starts with a fully connected topology and applies the FD algorithm for the CFA problem. With concave capacity costs, flow will be moved off uneconomical links, leading to a local minimum with a reduced link set. A greedy link reduction procedure, due to Gersht and Weihmayer (1990), starts with a set of candidate links in each successive iteration, to generate a sequence of topologies where each is reduced by one link, and then to select the least-cost reduced topology that satisfies the constraints. They indicate a methodology for extending the procedure for the continuous capacity problem to the discrete capacity case. The Concave Link Elimination (CLE) procedure for the discrete capacity case, due to Stacey et al. (2000), is based on formulating relaxed versions of the problems and utilizing them within a branch-and-bound framework.

Pierre and Legault (1998) develop a Genetic Algorithm for the discrete capacity problem that creates successive *populations* of link topologies. For each link topology, capacities and routings can be determined by solving a CFA problem. The lowest-cost solution, among all topologies in the various populations of topologies, is kept as the heuristic solution. Pierre and Elgibaoui (1997) applied a tabu search algorithm, which considers a succession of solutions, each obtained from the previous one by means of a *move* that removes, adds or substitutes links in the topology.

17.3.1.4 Router location. Kleinrock and Kamoun (1980) present a design methodology for a hierarchical data network. Backbone router and gateway locations are chosen based on hierarchical clustering. The average delay in such a hierarchical

Table 17.2 Papers with explicit routes and end-to-end delay metric

Papers	Routes	Capacity	Links	Objective	Special constraints
Route assignment					
Yen & Lin (2001a)	Discrete	Given	Given	Avg. delay	End-to-end delay
Capacity & route assignment					
Saksena (1989)	Discrete	Discrete	Given	Capacity cost	End-to-end delay
Topology, capacity & flow assignment					
Yen & Lin (2001b)	Discrete	Discrete	Binary choice	Capacity cost + switch cost	End-to-end delay

network is derived, expressed in terms of the average delays in each of the subnetworks (clusters).

17.3.2 End-to-end delay

Several authors have considered models in which M/M/1 formulas for queueing and insertion delay (as in Section 17.2.1) are used to compute the end-to-end delay along any chosen route from origin router to destination router. Constraints on this end-to-end delay, for each router pair, are incorporated into the problem formulation to insure that a specified delay requirement is met by all of the traffic. These models route traffic for each router-to-destination pair by choosing one or more paths from among a set of possible paths. By explicitly choosing discrete paths for routing traffic, end-to-end delays can be computed more readily. Table 17.2 summarizes the papers reviewed in Section 17.3.2.

17.3.2.1 Route assignment. Yen and Lin (2001b) consider a network with given link topology and given capacities. The problem is to choose a single route for each origin-destination pair, such that the capacity constraints are not violated and the end-to-end delay constraint on the chosen routes for each origin-destination pair are satisfied. The objective is to minimize the overall average delay in the network. The solution approach is to use Lagrangean Relaxation to incorporate constraints into the objective function and decompose the problem into a shortest path subproblem for each origin-destination pair and an independent, nonlinear subproblem for each link. Subgradient optimization is applied and primal heuristics are used to generate feasible solutions.

17.3.2.2 Capacity and route assignment. Saksena (1989) considers a problem with a given link topology, in which link capacity can be installed in increments of a single size. For each link, one must choose how many capacity increments to install.

For each origin-destination pair, one must choose a number of paths (bounded by some maximum number), and traffic between that pair of routers must be allocated among the chosen paths. There is a constraint on the average end-to-end delay for each origin-destination pair. The objective is to minimize the cost of the installed capacity. A heuristic algorithm is developed, based on successively dropping underutilized links.

17.3.2.3 Topology, capacity and route assignment. Yen and Lin (2001a) consider a problem to determine the topology, capacity and routing for a backbone network. There is a binary variable for each link that indicates whether or not the link is included in the solution. Capacity for each link must be chosen from among a discrete set of choices. It is necessary to route traffic on a single path for each origin-destination pair. In addition to an overall average network delay constraint, there is an end-to-end constraint on delay for each origin-destination pair. Additional constraints insure that the topology is two-connected. In addition, switches at the nodes must be sized to handle the number of incoming links and the traffic at the node. The objective is to minimize the total installation cost for the network, including cost of link capacity and cost of switch capacity. A solution procedure based on Lagrangean Relaxation is developed, in which five independent subproblems must be solved at each subgradient iteration.

17.3.3 Cost of delay

Various authors have taken delay into account by incorporating a *cost of delay* into the objective function. The cost of delay $T_i(f_i, y_{ik})$ for a particular link i is expressed as

$$T_i(f_i, y_{ik}) = \beta \frac{f_i}{\sum_k C_{ik} y_{ik} - f_i}, \qquad (17.3)$$

where β is a cost factor, f_i is the flow on link i, C_{ik} is the capacity associated with discrete size k for link i, and y_{ik} is a binary variable that indicates whether or not capacity size k for link i is chosen. Expression (17.3) is derived from the M/M/1 expression for queueing and insertion delay (see Section 17.2.1 above). By performing sensitivity analysis on the parameter β, one can obtain solutions with greater or less overall network delay. However, there is no explicit constraint to guarantee any particular level of performance for a given router-to-router pair.

Below, we consider two cases. In the first case, a single route for each router-to-router pair must be chosen from among pre-specified sets of possible routes. In the second, it is assumed that traffic can be routed like a multi-commodity flow. All the models discussed in Sections 17.3.4 and 17.3.4.1 assume that capacity must be chosen from among a discrete set of available capacities (or, else, that capacity is given). These models are summarized in Table 17.3.

17.3.4 Single routes for each router pair

Gavish and Neuman (1989) consider a type of capacity assignment and routing problem. The costs that they consider include a fixed cost for installing capacity, a cost per unit

Table 17.3 Papers with explicit routes and cost of delay metric

Papers	Routes	Capacity	Links	Objective	Special constraints
Capacity & route assignment					
Gavish & Neuman (1989) Gavish & Altinkemer (1990)	Discrete	Discrete	Given	Capacity cost + cost of delay	—
Neuman (1992)	Discrete	Discrete	Given	Capacity cost + cost of delay	Priority classes
Ferreira & Luna (2003)	MC flow	Discrete	Given	Capacity cost + cost of congestion	—
Link & route assignment					
Altinkemer & Yu (1992)	Discrete	Given	Binary choice	Link set-up cost + cost of delay	—
Router location, topology & capacity & flow assignment					
Gavish (1992)	Discrete	Discrete	Binary choice	Capacity cost + node cost + cost of delay	Node selection

of traffic that is routed on each link and a cost of delay on each link. Each of these costs depends both on the particular link and upon the choice of discrete capacity size for that link. The solution technique involves a Lagrangean Relaxation procedure. By incorporating capacity constraints into the objective function, the problem decomposes into a capacity subproblem and a routing subproblem. Subgradient optimization can then be applied.

Gavish and Altinkemer (1990) present a related algorithm in which routes can be generated as needed, rather than be limited by pre-specified sets of possible routes. Neuman (1992) extends this capacity assignment and routing problem to consider the case where two different priority classes must be routed. Altinkemer and Yu (1992) consider a related problem in which it is necessary to decide which links should be included in the network and how traffic should be routed over those links. In this case, each possible link has a predetermined capacity. Gavish (1992) presents an extended model in which router locations, assignment of end users to routers, inter-router links, link capacities and the routes for traffic must all be determined simultaneously. All of these papers incorporate cost of delay into the models and utilize Lagrangean Relaxation procedures to solve them.

17.3.4.1 Routes based on multi-commodity flow. Ferreira and Luna (2003) also consider a capacity assignment and routing problem that includes a *cost of congestion* on each link; this cost is a function of the traffic routed on the link. However, they assume that traffic can be routed like a multicommodity flow. The

Table 17.4 Papers consistent with shortest-path metrics

Papers	Routes	Capacity	Links	Objective	Special constraints
Capacity assignment					
Gerla & Kleinrock (1977)	Given	Continuous	Given	Capacity cost (linear/concave)	Average delay
Gerla (1973)	Given	Discrete	Given	Capacity cost	Average delay
Marayuma & Tang (1977) Levi & Ersoy (1994) Oommen & Roberts (2000, 2002)	Given	Discrete	Given	Capacity cost	Average delay priority classes
Route (weight) assignment					
Fortz & Thorup (2000) Ericsson, Resende & Pardalos (2002) Buriol, Resende, Ribeiro & Thorup (2005)	Shortest path	Given	Given	Cost of congestion	—
Fortz & Thorup (2002) Fortz, Rexford & Thorup (2002)	Shortest path	Given	Given	Cost of congestion	Limited changes, multi-period, or priority classes
Topology, capacity & flow assignment					
Klincewicz, Schmidt & Wong (2002)	Shortest path	Discrete	Binary choice	Capacity cost	End-to-end delay priority classes
Buriol, Resende & Thorup (2004)	Shortest path	Discrete	Binary choice	Capacity miles	End-to-end delay
Chattopadhyay, Morgan & Raghuram (1989)	Shortest path trees	Discrete	Binary choice	Capacity cost	Max link delay

heuristics that they develop, based on convexification of the problem, are guaranteed to produce a solution that is within a factor of the lower bound.

17.4 ROUTING CONSISTENT WITH SHORTEST PATH METRICS

Requiring the routes to be consistent with a shortest path metric imposes an additional constraint on the network design. Fewer papers have considered this more difficult case. In the subsections below, those papers are categorized by the particular performance metrics that are taken into account. In Table 17.4, we summarize the network features that are considered in the papers reviewed in this section.

17.4.1 Average delay: The capacity assignment problem

In the well-known Capacity Assignment (CA) Problem, the routing of the traffic for each origin-to-destination pair is assumed to be fixed. In particular, the routing could be implied by a set of administrative weights and a shortest-path routing protocol. Since the traffic demand matrix is given, the traffic flow on each link is therefore also fixed. In this problem, it is necessary to determine a set of capacities for the links, such that the average delay (computed using the same M/M/1 formulas as in Section 17.2.1) is less than or equal to a specified maximum (i.e., $T < T_{\max}$). The objective is to minimize the total cost of capacity. Increasing the capacity on a link will decrease the queueing and insertion delay, but will increase the cost.

17.4.1.1 Continuous capacity. For the case where capacity is treated as a continuous variable and the cost of capacity is linear, the problem can be solved analytically using Lagrange multipliers. For the case of continuous capacities and concave costs, an iterative procedure in which costs are linearized at each iteration leads, in general, to a local minimum; see Gerla and Kleinrock (1977).

17.4.1.2 Discrete capacity. For the perhaps more realistic case, where capacity must be added in discrete increments, solution techniques include dynamic programming and Lagrangian decomposition (Gerla, 1973).

In addition, various authors have extended the discrete CA problem to take into account additional requirements. For example, a number of authors have considered the problem in which there are multiple classes of traffic that have fixed routes over the network. Each class has its own priority level, where a lower priority level takes precedence. (That is, class j has priority over class k if $j < k$.) The standard expression (17.1) for average delay can be extended to determine the average delay T_k for each class k under the assumption of nonpreemptive FIFO priority queueing. Likewise, a constraint $T_k < T_{\max}^k$ is imposed on the allowable average delay for each class. With the additional consideration of priority classes, the problem of choosing capacities for each link becomes a more difficult, nonlinear, combinatorial problem. Maruyama and Tang (1977) developed some composite heuristics for choosing the capacities. Levi and Ersoy (1994) applied a Simulated Annealing approach. Oommen and Roberts (2000; 2002) developed Artificial Intelligence based approaches, utilizing Learning Automata.

17.4.2 End-to-end delay

17.4.2.1 Designing an enterprise network. Klincewicz et al. (2002) consider the specific case of designing an Enterprise Network. An Enterprise Network is a private Wide Area Network (WAN) serving a corporation, government entity or other enterprise. The objective is to minimize the cost of the network, taking into account both the cost of the WAN links themselves and the access costs to connect the Enterprise locations to the WAN provider points-of-presence. Since it is assumed that the WAN provider will guarantee link restoration and that other back-up technologies will be in place, failure scenarios are not explicitly considered.

The model features multiple classes of traffic, each with its own end-to-end delay requirements. Class-based OSPF weights are assumed, which allows each class of traffic to be routed differently. The routers are assumed to utilize a weighted round-robin queueing discipline, such as Cisco's Custom Queueing, that allocates a minimum bandwidth to each class of traffic. An M/G/1 expression for queueing delay is used.

Within this model, it is necessary to determine: which WAN links to provision; the capacity assigned to each link; how that capacity is allocated among the classes of traffic; and what OSPF administrative weight is assigned to each link for each class of traffic. These weights determine the routings. Thus, this is a type of Topology, Capacity and Route Assignment problem.

To begin the solution procedure, initial administrative weights, based on link costs, are assigned. The methodology first utilizes a *drop* heuristic to obtain an initial link topology. (That is, the heuristic starts with a fully meshed network. In an iterative fashion, traffic is routed and the most underutilized links are dropped.) Once an initial topology is found, another iterative procedure is performed. Within each of these iterations, the following steps are repeated:

- The allowable delay is allocated along the current routes.

- The bandwidth required on each link to meet the delay requirements for each class is determined. The links are sized to provide sufficient capacity to meet the sum of the bandwidth requirements for the classes.

- The administrative weights (and, hence, the routes) are modified, based on the cost gradients for each link.

- Underutilized links are dropped.

The iterations continue until the delay allocation and routings converge.

This sort of class-based differentiated service approach can result in considerable cost and performance advantages in the network design, when compared to single-class alternatives.

17.4.2.2 Survivable IP network design. Buriol et al. (2004) consider a problem where, for a given set of routers and a given demand matrix, it is necessary to choose both the OSPF weights and the number of circuits of a given size for each possible router-to-router link. The networks must have sufficient capacity to handle any single router or link failure, and the objective is to minimize the total circuit-miles required. Within their genetic algorithm, end-to-end delay constraints can be considered by means of a penalty function that disfavors solutions that violate these constraints.

17.4.3 Cost of congestion: Setting administrative weights

The Flow Assignment problem, which is discussed in Section 17.3.1.1, involves routing traffic in an MPLS-type environment, in which the system administrator can explicitly set the route for each router-to-router pair. In an OSPF-type environment, however, setting routes is a by-product of setting the administrative weights. In this section, we consider optimization approaches for setting administrative weights.

Default administrative weights for OSPF, as implemented by router manufacturers, are typically set to be inversely proportional to link capacity. That is, the greater the capacity of a link, the lower the corresponding administrative weight, and the more likely it is that traffic will end up being routed on that link. However, relying on the default settings can create problems. For example, suppose that a particular link were becoming congested and, in order to relieve that congestion, the system administrator increased the capacity of that link. Relying on default settings would then result in a smaller weight on that link. Because of the smaller weight, additional traffic might be routed on the link. If enough traffic were rerouted, congestion would once again result.

Therefore, to allow some greater degree of control over the routing, system administrators commonly override these default values and assign static administrative weights to the links. This is often as much of an art as a science, benefiting from the intuition and experience of the individual system administrator.

Various authors have considered optimization approaches for setting the administrative weights. In this body of work, QoS is addressed by means of an objective function that penalizes heavily congested (i.e. highly utilized) links. This objective is designed to reduce packet loss and increase throughput; however, it does not explicitly address any delay requirements.

Given a network, with a given link topology, link capacities and a router-to-router demand matrix, the problem is to choose administrative weights for each link i so as to minimize $\sum_i \phi_i(f_i)$, where f_i is the amount of traffic routed on link i and $\phi_i(\cdot)$ is a piece-wise linear, convex, increasing function. This function is typically designed to increase dramatically as the amount of traffic approaches the link capacity. The weights are chosen from among a discrete set of possible weights. Fortz and Thorup (2000) develop a local search heuristic for this problem. Ericsson et al. (2002) propose a genetic algorithm for setting the weights. Buriol et al. (2005) add a local search to a similar genetic algorithm.

Fortz and Thorup (2002) and Fortz et al. (2002) consider variations on the problem that address particular issues. For example, they modify their algorithm to limit the number of weight changes (given an incumbent set of weights), to find the single set of weights that is best for multiple demand matrices, and to set weights for two classes of service.

17.4.4 Maximum link delay

Chattopadhyay et al. (1989) considered a type of TCFA problem in which it is necessary to decide what the link topology should be (i.e., which links exist), what capacity should be installed on each link (chosen from among a discrete set of choices) and how traffic should be routed. There is a constraint, for each pair of routers i and j, on the maximum delay that can be tolerated on a link interconnecting routers i and j, if that link were chosen to be in the solution. Traffic must be routed in such a way that the routes for any given destination form a shortest-path tree. A branch-and-bound heuristic is developed.

17.5 CONCLUDING REMARKS

As can be seen from the variety of models considered in this chapter, the term QoS covers a breadth of topics. Given this variety and breadth, future research on QoS and network design is likely to be active in a number of different areas. In this final section, we discuss some possible areas for future research. These indicate both gaps in the existing literature and/or new topics that will be motivated by advances in technology. These possible areas include the following.

- **Real-time decisions.** In this chapter, we have focused on network design models that would typically be used in a medium-to-long-range network planning context, where the decisions suggested by the models would not necessarily be implemented immediately. Only a few optimization papers have addressed real-time QoS decisions. (Examples of these include QoS routing decisions for multimedia applications, as in Wang and Crowcroft (1996) and Vogel et al. (1996). Here, it is assumed that a multimedia data stream is about to enter the network and it is necessary to determine a route, for this particular data stream, that meets multiple QoS criteria, such as bandwidth, delay and packet loss probability.) Future advances in technology (e.g., ability to reconfigure network links and capacity through integration of the IP layer and optical layer) will likely make it ever more possible for networks to respond in real time to short-term changes in traffic demands. Optimization models that address various types of real-time decisions will be needed.

- **Probabilistic models of IP traffic.** A majority of the work reviewed in this chapter utilizes an M/M/1 based model of traffic on a link. (The papers by Ng and Hoang (1987) and Klincewicz et al. (2002) are exceptions.) The M/M/1 assumption is, of course, an overly simplified model of the behavior of IP traffic, especially in the case of multiple classes of traffic. There is need for network design procedures to utilize more realistic models and characterizations of traffic behavior both in the calculation of network delay and in the sizing of network links. Examples of such enhanced probabilistic models include Heyman (2003) and Heyman et al. (2003).

- **Router location.** Among papers in this chapter, only Gavish (1992) and Kleinrock and Kamoun (1980) explicitly consider the choice of backbone router location as part of the model. (Another paper that considers both router location and performance is Monma and Sheng (1986). However, this chapter addresses a specialized fully-interconnected backbone, and a performance measure, based on call setup blocking, that is not directly relevant to ordinary IP networks.) A QoS-conscious model that chooses router locations can be characterized as a type of *hub location model* with performance constraints (see Klincewicz (1998)). In general, this type of problem is difficult to solve. However, in the future, delay and similar performance metrics, measured from end user to end user, will become more pressing issues. It will be important to understand how such end-to-end performance is affected by the initial choice of the number and location of the backbone routers.

- **Shortest path metrics and delay.** Although shortest-path protocols are most commonly used in practice, a majority of the papers in this chapter assume that routes can be explicitly chosen. Additional work is needed on network design problems in which the routing must be based on a shortest-path metric. Models for setting administrative weights that explicitly consider end-to-end performance measures, such as delay, will be a great help in the operation of both new and existing networks.

- **Application delay.** The delay metrics considered in the literature have focused on the delay experienced per packet. However, in some Internet transactions, the QoS requirements can be more naturally expressed in terms of the time required to complete the entire application. Incorporating enhanced QoS metrics, such as application delay, into network design models will allow users to make a more direct connection between system requirements and model inputs.

- **Survivability and restoration.** In this chapter, only Kang and Tan (1997) explicitly consider survivability and restoration in their network designs. With IP networks being used to carry critical applications, survivability of IP services will become ever more important. Therefore, it is important to develop models that insure sufficient network resources (such as capacity) are available to meet QoS constraints, even in the event of the failure of a router or link. For the case of routing by shortest-path metrics, this involves modeling how traffic will be routed in each failure event.

Network design problems with QoS considerations are typically difficult combinatorial problems. These directions for research all point to the need for solving even more difficult and complex problems in the future. In order to do this, it will be necessary for researchers to take advantage of the latest developments in optimization methodology and computing technology. Success in these directions of research will enable network designers to solve practical problems, using models that realistically depict the behavior of IP networks and that measure QoS in appropriate ways.

Bibliography

K. Altinkemer and Z. Yu. Topological design of wide area communication networks. *Annals of Operations Research*, 3:365–38, 1992.

J. Betser, A. Avritzer, J.W. Carlyle, and W.J. Karplu. Configuration synthesis for a heterogeneous backbone cluster and a PC-interface network. In *Proceedings of IEEE INFOCOM 8*, pages 400–40, 1989.

L.S. Buriol, M.G.C. Resende, C.C. Ribeiro, and M. Thorup. A hybrid genetic algorithm for the weight setting problem in OSPF/IS-IS routing. *Networks*, 2005. In press.

L.S. Buriol, M.G.C. Resende, and M. Thorup. Survivable IP network design with OSPF routing. Technical Report TD-64KUAW, AT&T Labs Research, 2004.

K. Chari. Multi-hour design of computer backbone networks. *Telecommunications Systems*, 6:347–365, 1996.

N.G. Chattopadhyay, T.W. Morgan, and A. Raghuram. An innovative technique for backbone network design. *IEEE Transactions on Systems, Man and Cybernetics*, 19:1122–1132, 1989.

M. Ericsson, M.G.C. Resende, and P.M. Pardalos. A genetic algorithm for the weight setting problem in OSPF routing. *Journal of Combinatorial Optimization*, 6:299–333, 2002.

R.P.M. Ferreira and H.P.L. Luna. Discrete capacity and flow assignment algorithms with performance guarantee. *Computer Communications*, 26:1056–1069, 2003.

S. Floyd and V. Jacobson. Link sharing and resource management models for packet networks. *IEEE/ACM Transactions on Networking*, 3:365–386, 1995.

B. Fortz, J. Rexford, and M. Thorup. Traffic engineering with traditional IP routing protocols. *IEEE Communications Magazine*, 40:118–124, 2002.

B. Fortz and M. Thorup. Internet traffic engineering by optimizing OSPF weights. In *Proceedings of IEEE INFOCOM 2000*, pages 519–528, 2000.

B. Fortz and M. Thorup. Optimizing OSPF/IS-IS weights in a changing world. *IEEE Journal on Selected Areas in Communications*, 20:756–767, 2002.

L. Fratta, M. Gerla, and L. Kleinrock. The flow deviation method: An approach to store-and-forward computer communication network design. *Networks*, 3:97–133, 1973.

B. Gavish. Topological design of computer communication networks – The overall design problem. *European Journal of Operational Research*, 58:149–172, 1992.

B. Gavish and K. Altinkemer. Backbone network design tools with economic tradeoffs. *ORSA Journal on Computing*, 2:236–252, 1990.

B. Gavish and I. Neuman. A system for routing and capacity assignment in computer communication networks. *IEEE Transactions on Communications*, 37:360–366, 1989.

M. Gerla. *The design of store-and-forward networks for computer communications*. PhD thesis, School of Engineering and Applied Science, University of California, Los Angeles, 1973.

M. Gerla and L. Kleinrock. On the topological design of distributed computer networks. *IEEE Transactions on Communications*, COM-25:48–61, 1977.

A. Gersht and R. Weihmayer. Joint optimization of data network design and facility selection. *IEEE Journal on Selected Areas in Communications*, 8:1667–1681, 1990.

D.P. Heyman. Sizing Internet backbone links, 2003. Unpublished manuscript.

D.P. Heyman, T.V. Lakshman, and A.L. Neidhardt. A new method for analyzing feedback-based protocols with applications to engineering Web traffic over the Internet. *Computer Communications*, 26:785–803, 2003.

C.G. Kang and H.H. Tan. Design of packet switched network with unreliable links. *Computer Communications*, 20:544–553, 1997.

L. Kleinrock. Analytic and simulation methods in computer network design. In *Cong. Rec., Spring Joint Comput. Conf., AFIPS Conf Proc. 36*, pages 568–579. AFIPS Press, 1970.

L. Kleinrock and F. Kamoun. Optimal clustering structures for hierarchical topological design of large computer networks. *Networks*, 10:221–248, 1980.

J.G. Klincewicz. Hub location in backbone/tributary network design: A review. *Location Science*, 6:307–335, 1998.

J.G. Klincewicz, J.A. Schmitt, and R.T. Wong. Incorporating QoS into IP Enterprise Network Design. *Telecommunications Systems*, 20:81–106, 2002.

A. Levi and C. Ersoy. Discrete link capacity assignment in prioritized computer networks: Two approaches. In *Proceedings of the Ninth International Symposium on Computer and Information Sciences*, pages 408–415, 1994.

H. Luss and A. Vakhutinsky. A resource allocation approach for the generation of service-dependent demand matrices for communications networks. *Telecommunications Systems*, 17:411–433, 2001.

K. Maruyama and D.T. Tang. Discrete link capacity assignment in prioritized computer networks. *IBM Journal of Research and Development*, pages 254–263, 1977.

C.L. Monma, A. Schrijver, M.J. Todd, and V.K. Wei. Convex resource allocation problems on directed acyclic graphs: Duality, complexity, special cases and extensions. *Mathematics of Operations Research*, 15:736–748, 1990.

C.L. Monma and D.D. Sheng. Backbone Network Design and Performance Analysis: A methodology for packet switching networks. *IEEE Journal on Selected Areas in Communications*, SAC-4:946–965, 1986.

I. Neuman. A system for priority routing and capacity assignment in packet switched networks. *Annals of Operations Research*, 36:225–246, 1992.

T.M.J. Ng and D.B. Hoang. Joint optimization of capacity and flow assignment in a packet-switched communications network. *IEEE Transactions on Communications*, COM-35:202–209, 1987.

B.J. Oommen and T.D. Roberts. Continuous learning automata solutions to the capacity assignment problem. *IEEE Transactions on Computers*, 49:608–620, 2000.

B.J. Oommen and T.D. Roberts. Discretized learning automata solutions to the capacity assignment problem for prioritized networks. *IEEE Transactions on Systems, Man and Cybernetics – Part B: Cybernetics*, 32:821–831, 2002.

A. Ouorou, P. Mahey, and J-P. Vial. A survey of algorithms for convex multicommodity flow problems. *Management Science*, 46:126–147, 2000.

S. Pierre and A. Elgibaoui. A tabu-search approach for designing computer-network topologies with unreliable components. *IEEE Transactions on Reliability*, 46:350–359, 1997.

S. Pierre and G. Legault. A genetic algorithm for designing distributed computer network technologies. *IEEE Transactions on Systems, Man and Cybernetics – Part B: Cybernetics*, 28:249–258, 1998.

M. Queiroz and C. Humes Jr. A heuristic for the continuous capacity and flow assignment. *European Journal of Operational Research*, 146:444–459, 2003.

V.R. Saksena. Topological analysis of packet networks. *IEEE Journal on Selected Areas in Communications*, 7:1243–1252, 1989.

H.B. Shulman and V. Venkateswaran. Resource allocation in tree networks under delay constraints, 1999. Presented at INFORMS Fall 1999 Meeting, Philadelphia, PA.

C.H.E. Stacey, T. Eyers, and G.J. Anido. A concave link elimination (CLE) procedure and lower bound for concave topology, capacity and flow assignment network design problems. *Telecommunications Systems*, 13:351–372, 2000.

V. Venkateswaran and M. Kodialam. Bandwidth allocation in multicast trees with QoS constraints, 1999. Presented at INFORMS Spring 1999 Meeting. Cincinnati, OH.

R. Vogel, R.G. Herrtwich, W. Kalfa, H. Wittig, and L.C. Wolf. QoS-based routing of multimedia streams in computer networks. *IEEE Journal on Selected Areas in Communications*, 14:1235–1244, 1996.

Z. Wang. *Internet QoS: Architectures and Mechanisms for Quality of Service*. Morgan Kaufmann, 2001.

Z. Wang and J. Crowcroft. Quality-of-service routing for supporting multimedia applications. *IEEE Journal on Selected Areas in Communications*, 14:1228–1234, 1996.

O.J. Wasem, A.M. Gross, and G.A. Tiapa. Forecasting broadband demand between geographic areas. *IEEE Communications Magazine*, 33:50–57, 1995.

H-H. Yen and F.Y-S. Lin. Backbone network design with QoS requirements. *Lecture Notes in Computer Science*, 2094:148–157, 2001a.

H-H. Yen and F.Y-S. Lin. Near-optimal delay constrained routing in virtual circuit networks. In *Proceedings of IEEE INFOCOM 2001*, pages 750–756, 2001b.

Y. Zhang, M. Roughan, N. Duffield, and A. Greenberg. Fast accurate computation of large-scale IP traffic matrices from link loads. In *Proceedings of INFOCOMM 2003*, 2003.

18 STEINER TREE PROBLEMS IN TELECOMMUNICATIONS

Stefan Voß[1]

[1]Institute of Information Systems
University of Hamburg
20146 Hamburg, Germany
stefan.voss@uni-hamburg.de

Abstract: Connecting a given set of points at minimum cost may be rated as one of the most important problems in telecommunications network design. Related questions may be formulated in metric spaces as well as in graphs. Given a weighted graph, the *Steiner tree problem in graphs* asks to determine a minimum cost subgraph spanning a set of specified vertices. This problem may be viewed as *the* combinatorial optimization problem in telecommunications. In this chapter, we survey Steiner problems from a telecommunications perspective with a special emphasis on the problem in graphs.

Keywords: Steiner tree problems, telecommunications, network design, graphs.

18.1 INTRODUCTION

One of the oldest mathematical problems related to network design may be formulated as follows:

- *Given three points A, B and C in the plane, find a point P such that the sum of its distances to the three given points is minimal.*

Many famous mathematicians have contributed to solving this problem, including, e.g., Fermat and Torricelli around 1640, but also Steiner (1835). While this problem sounds somewhat simple, it provides the basic consideration of a most comprehensive treatment of various network design problems in the literature. Courant and Robbins (1941) (and, independently, Jarnik and Kössler (1934)) described a generalization where a least cost network had to be designed (in the plane) such that a given set of n points is connected at minimum cost. As they attributed this problem to the Swiss mathematician and geometer Jacob Steiner the name *Steiner tree problem* was termed.

Classifying the literature that has emerged beyond this version of the Steiner tree problem, three major areas may be considered:

- The Euclidean Steiner problem;
- The rectilinear Steiner problem;
- The Steiner problem in graphs.

Based on this, one may distinguish two major classes of Steiner tree problems. The first refers to Steiner's problem in metric spaces and the second to the Steiner problem in graphs. Given an edge-weighted graph, the Steiner problem in graphs is to determine a minimum cost subgraph spanning a set of specified vertices. The rectilinear case has attracted a separate consideration in the literature but also turns out to be a Steiner problem in special graphs.

This survey on Steiner tree problems in telecommunications provides an overview on various Steiner problems with a special emphasis on applications in telecommunications. While these applications mainly refer to the graph theoretic version of the problem, the problem in metric spaces plays a prominent role in providing important theoretical ideas and concepts that may be used in its graph pendent as well as in some applications. Based on this, we mainly focus on graph problems. We first provide a comprehensive treatment of the Steiner tree problem in undirected as well as in directed graphs (Section 18.2). Some specific pointers to the literature on Steiner problems in metric spaces are given in Section 18.3. Furthermore, we survey various modifications (Section 18.4) and generalizations (Section 18.5) that have been used in telecommunications applications. Whenever deemed practical we allow for some deviation to problems in metric spaces. The chapter closes with some final remarks.

Before going into detail we like to point the reader to an excellent reference providing a survey on Steiner problems even if it had been written more than a decade ago: Hwang et al. (1992). Additional books of interest had been written by Voß (1990a) and Prömel and Steger (2002). Comprehensive collections of papers on Steiner problems can be found in various journals including *Networks* as well as in the books edited by Du et al. (2000) and Du and Cheng (2001). Additional surveys on quite broad aspects of Steiner tree problems are provided by Winter (1987), Hwang and Richards (1992), and Voß (2000), as well as Du et al. (2001). Wu and Chao (2004) could be of interest regarding the design of algorithms for spanning tree problems.

Certainly, as this is meant to be a survey, we borrow many parts from previous publications cited throughout this chapter. Furthermore, we assume that the reader is familiar with some basic notation in graph theory and combinatorial optimization.

18.2 THE STEINER TREE PROBLEM IN GRAPHS

Consider an undirected connected graph $G = (V,E)$ with vertex set V, edge set E, and nonnegative weights associated with the edges. Given a set $Q \subseteq V$ of specified vertices (so-called terminals or basic vertices) *Steiner's problem in graphs* (SP) is to find a minimum cost subgraph of G such that there exists a path in the subgraph between every pair of basic vertices. In order to achieve this minimum cost subgraph, additional vertices from the set $S := V - Q$, so-called Steiner vertices, may be included. Since all edge weights are assumed to be nonnegative, there is an optimal solution which is a tree, called *Steiner tree*.

Correspondingly, *Steiner's problem in directed graphs* (SPD) is to find a minimum cost directed subgraph of a given graph that contains a directed path between a root node and every basic vertex. Obviously, any instance of the SP may easily be transformed into an instance of the SPD by replacing each undirected edge (i, j) with two arcs $[i, j]$ and $[j, i]$ directed opposite each other both having the same weight as the original edge.

Applications of the SP and the SPD are frequently found in many problems related to telecommunications as will be pointed out below. Beyond that, SP and SPD have equal importance also for the layout of connection structures in networks as, e.g., in topological network design, location science, and VLSI (very large scale integrated circuits) design. Stressing the notion of importance just a bit the SP may be referred to as having the same importance to telecommunications as the famous traveling salesman problem has to vehicle routing and transportation.

Only some well-known special cases of the SP are polynomially solvable. For instance, if $|Q| = 2$ the problem reduces to a shortest path problem and in the case where $Q = V$ the SP reduces to the *minimum spanning tree* (MST) problem. Restricting the number of Steiner vertices in a solution to a constant number (say 1) leads to polynomial solvability, too. On the other hand a large number of special cases turn out to be NP-hard, and therefore, in the general case the SP is an NP-hard problem (Garey and Johnson, 1979).

In this section, we consider the SP/SPD and provide a brief survey on exact and heuristic, as well as metaheuristic, algorithms and provide a wealth of pointers to the literature. Before doing that we review the efforts on providing problem instances and reduction techniques.

18.2.1 Data and applications

Benchmark instances for the SP had been devised by various authors until the 1990s. Based on that, it turned out that a specific set of 78 problem instances for the SP available from the OR-library (Beasley, 1990) had been considered by several authors (looking at the way they were treated one might refer to some sort of 'horse race' until they were all solved to optimality). These instances comprise data with the number of vertices ranging from 50 to 2500 and 63 up to 65000 edges.

An additional set of early benchmark instances for the SP (basically for the rectilinear version) that has received considerable attention in the literature was provided by Soukup and Chow (1973). They put together 46 instances by giving the coordinates of a number of basic vertices within the unit square. The instances arising are Steiner problems in the plane with the underlying metric being the Euclidean or the rectilinear metric.

Later many ideas related to developing some data generator were topics of interest, especially when it came to the definition of difficult instances. However, most of these instances were not taken from real applications.

The most comprehensive collection of Steiner tree problem instances to date can be found in the *SteinLib* library from Koch et al. (2001). They present a library of data sets for the SP and some generalizations which extends former libraries by many new interesting and difficult instances, some of them arising from real-world applications. Furthermore, they provide an overview regarding the difficulty of these problem instances by stating references to state-of-the-art software packages that were the first or are among the best to solve these instances.

Currently the most challenging instances are those that were developed to overcome the beneficial effects of certain reduction techniques (see Section 18.2.2) such as the so-called incidence data of Duin and Voß (1997). More recently, Rosseti et al. (2003) propose additional problem instances for the SP with the aim of common reduction techniques to fail on them. An interesting approach for data generation consists in developing data with known optimal solutions based on some mathematical programming formulations (see, e.g., Khoury et al. (1993)). However, state-of-the-art algorithms seem to be able to solve them quite easily. An additional field of interest refers to the study of random instances, e.g., data where a complete graph is assumed with edge weights drawn from a certain distribution and basic vertices chosen arbitrarily. Under the assumption that the number of nodes grows to infinity and the number of basic vertices $|Q| = O(|V|)$, Bollobás et al. (2004) provide some recent theoretical results related to an exponential distribution and a convergence to an optimal solution of a certain weight with high probability.

While applications of the SP are widespread, it is nevertheless difficult to really find detailed application descriptions (and respective data) provided in the literature. Usually, respective applications are related to certain modifications and generalizations such as those mentioned in Sections 18.4 and 18.5. One of the most important areas in that respect, where application descriptions arise, are multicast problems (see Section 18.4.1). As one explicit exception we mention Bachhiesl et al. (2004) who consider the access net domain (last mile) for building fiber optic networks. Although the investment volume in this domain exceeds the costs of core and distribution nets, they point out that network carriers hardly utilize intelligent planning tools for anticipatory investment decisions and cost optimization in this area. Based on that they describe strategic planning and optimization tools developed in cooperation with renowned European city carriers. The core is the utilization of geoinformation data within a hierarchical optimization approach.

18.2.2 Reduction techniques

In a large number of investigations, it has been shown that a very important step towards solving the SP is to perform some preprocessing routines (Balakrishnan and Patel, 1987; Duin and Volgenant, 1989a;c; Voß, 1990a; Winter and Smith, 1992; Duin, 2000). Among these routines are the following, just to mention some simple ones: Any vertex $v \in S$ of degree 1 can be removed and any vertex $v \in S$ of degree 2 can be replaced by a single edge. Any edge with its weight larger than a shortest path between its end vertices can be removed. If an edge between two basic vertices appears in an MST on the entire graph, then it belongs to an optimal solution and the graph can be contracted along this edge. This refers to the so-called bottleneck Steiner distances (see Duin and Volgenant (1989a) and Voß (1990a)) that have been shown to be most effective for reducing data of SP instances.

The most recent and comprehensive results in applying reduction techniques for the SP are reported in Duin (2000) and Polzin and Daneshmand (2001b). The results of these papers show that reduction techniques are in fact the most important ingredients in solving SP instances to optimality. One idea of interest in its own right is to use some heuristic reduction techniques that may be used to improve the running times of any algorithm considerably without discarding solution quality too much (see, e.g., Polzin and Daneshmand (2001b)).

18.2.3 Exact algorithms

The development of exact algorithms for the SP had been of major interest especially in the 1980s and 1990s and many good works have appeared in this respect; see, e.g., Chopra et al. (1992), Lucena (1992), Beasley (1989), Lucena and Beasley (1998), Polzin and Daneshmand (1997), and Koch and Martin (1998) for excellent results. Respective advances have been obtained as a mix of several features beyond the availability of more advanced computation infrastructure. Basically one may observe an intelligent interplay between various mathematical programming formulations providing lower bounds and the above mentioned reduction techniques as well as heuristics for upper bounds.

Surveys on mathematical programming formulations for the SP can be found in Goemans and Myung (1993) and Polzin and Daneshmand (2001a). While various formulations may be compared theoretically as well as empirically (regarding their respective relaxations within branch and bound or branch and cut) respective ideas may also be used to develop valid inequalities for different formulations that may proof useful for being used as cuts. The most important result of Polzin and Daneshmand (2001a) is a hierarchy of relaxations together with a wealth of polyhedral results putting most of the well known relaxations into perspective.

In Polzin and Daneshmand (2001b) the most important ingredients as mentioned above are put together to describe a state-of-the-art exact algorithm for the SP. As a formulation, the directed cut formulation of Aneja (1980) and Wong (1984) is used in a dual ascent fashion together with almost any known reduction technique. Furthermore, efficient heuristics as well as different lower bounds are also utilized within the reduction techniques. Moreover, if the proportion of basic vertices over the overall number of vertices is quite large, a modified formulation with a Lagrangean relaxation approach is used.

Additional works on exact algorithms beyond these concepts are quite rare. As an exception, Bahiense et al. (2003) describe the so-called volume algorithm.

Obviously, finding Steiner arborescences, i.e. solutions for the SPD, is a more general problem than the SP having additional real world applications. Nevertheless, it has not yet received the same amount of interest as its undirected version. Note that the directed version of a mathematical formulation has some advantages over its undirected counterpart (see, e.g., Chopra et al. (1992), as well as Voß (1990a)). Basically, however, state-of-the-art approaches like the one of Polzin and Daneshmand (2001b) may equally successfully been applied to the SPD.

18.2.4 Heuristics

Due to SP's complexity, one is interested in developing efficient heuristic algorithms to find good approximate solutions. Furthermore, as the SP reveals great importance with respect to its fundamental structural properties, algorithms are also investigated and developed for this already NP-hard 'special case' of more general network design problems. Voß (2000) provides a survey on heuristic search methods for the SP.

A central theme in Steiner tree heuristics is the use of some principles known from MST algorithms together with the computation of shortest paths. The general idea is to build up a feasible solution by insertion of shortest paths. Based on this observation, two main ideas may be distinguished (cf. Duin and Voß (1994) and Voß (1992)):

- **Single Component Extension**
  Start with a *partial solution* $T = (\{w\}, \emptyset)$ consisting of a single vertex w, the so-called root arbitrarily chosen from Q, if not stated otherwise. Extend T to a feasible solution by successive insertion of at most $|Q|$ shortest paths to basic vertices.

- **Component Connecting**
  Start with a *partial solution* $T = (Q, \emptyset)$ consisting of $|Q|$ singleton components. Expand T to a feasible solution by repeatedly selecting components which are connected by shortest paths.

These ideas may be related to the steps of two famous MST algorithms. The single component extension concept corresponds to the steps of the famous MST algorithm of Prim (1957). The MST algorithm of Kruskal (1956) is concerned if in each step of the component connecting approach two nearest components are connected.

Moreover, besides a specific partial solution to start with, different heuristics arise according to the number of built-up steps used to obtain a feasible solution T. In 1BASIC, we build by single component extension (e.g. cheapest insertion). In kBASIC, k given components (any basic vertex not yet part of a partial solution may be viewed as a component) are connected by use of shortest paths. Furthermore, heuristics differ in knowledge used in the computation of a feasible solution. Together with the specific implementation for extending or connecting components and including one of the above building blocks, we obtain different algorithms.

Let us revisit a unified approach based on the component connecting idea. Let $T_0, ..., T_\sigma$ denote the components of a partial solution T and $d(v, T_i)$ be the minimum of the shortest distances between v and any node of T_i in G. That is, with d_{ij} being the weight of a shortest path $P(i, j)$ between nodes i and j in G, $d(i, T) = \min\{d_{ij} | j \in V_T\}$ where V_T denotes the vertex set of T. A vertex of degree 1 in T is called leaf. Whenever ties occur, they are broken arbitrarily if not stated otherwise.

If the vertex set V_T of an optimal solution is known, the task is to polynomially compute an MST of the subgraph of G induced by V_T. The set $V_T - Q$ of Steiner vertices may be heuristically determined by means of a heuristic function f (see, e.g., Rayward-Smith (1983) and Rayward-Smith and Clare (1986)):

HEUM: *Heuristic Measure*
(1) Start with $T = (Q, \emptyset)$ comprising $|Q|$ basic vertices (subtrees of the final subgraph).
(2) **While** T is not connected,
   do choose a vertex v using a function f and unite the two components of T which are nearest to v by combining them with v via shortest paths (the vertices and edges of these paths are added to T).

The general idea of the heuristic measure relates to some artificial intelligence concepts (see Pearl (1984)). Usually f is minimized over all vertices where $f(v)$ is used to numerically express the attractiveness of v to be part of a solution. Intuitively $f(v)$ gives a measure of proximity of v to elements of a subset of vertices in T, or the least average distance of v to a set of vertices in T, respectively. In that sense we may easily modify Step 2 of HEUM by uniting more than two components, say three, at a time by means of kBASIC (Voß (1992)).

By choosing a suitable function f, e.g.

$$f_1(v) := \min_{1 \leq i \leq \sigma} \{d(v, T_0 \cap Q) + d(v, T_i \cap Q)\},$$

the distance network or shortest distance graph heuristic (SDISTG, see, e.g., Choukhmane (1978), Kou et al. (1981), and Plesnik (1981)) is obtained by HEUM. SDISTG has received considerable attention in the literature with respect to an efficient implementation even including a posterior improvement step. That is, in the original version of SDISTG a subsequent improvement procedure MST+P is applied which may be used in general after having performed any Steiner tree heuristic (see, e.g., Rayward-Smith and Clare (1986)):

MST+P: *Minimum spanning tree and pruning*
(1) Given a solution with vertex set V_T, construct an MST $T' = (V'_T, E'_T)$ of the subgraph of G induced by V_T.
(2) **While** there exists a leaf of T' that is a Steiner vertex,
   do delete that leaf and its incident edge.

When constructing the complete graph for vertex set $Q' \cup \{k\}$ for all non-basic vertices k where $|Q'| = 3$ we obtain a polynomial algorithm 3BASIC, which is useful as a subroutine in a component connecting approach: Determine with 3BASIC for every triple of components in the partial solution the best possible way to connect these three components. Execute the connection of minimum cost by corresponding path insertion (decreasing the number of components by two) and repeat these two steps until a feasible solution is obtained. (A somewhat different description of 3BASIC is given in Chen (1994)).

Likewise the cheapest insertion method (CHINS, Takahashi and Matsuyama (1980)) is obtained as a special case of HEUM by

$$f_2(v) := \min_{1 \leq i \leq \sigma} \{d(v, T_0) + d(v, T_i)\}$$

where, opposite to f_1, the non-basic vertices in T_0 are active. That is, both CHINS and SDISTG give special emphasis to the basic vertices. In addition, CHINS also uses the partial solution under development as its knowledge base. As a synthesis this clearly explains the restrictedness of the SDISTG approach when compared to CHINS and is clearly underlined in a large variety of numerical experiments (see, e.g., Rayward-Smith and Clare (1986) and Voß (1992)). A more general function incorporating a component connecting approach is

$$f_3(v) := \min_{1 \leq t \leq \sigma} \{\frac{1}{t}\sum_{i=0}^{t} d(v, T_i) | d(v, T_i) \leq d(v, T_j), 0 \leq i < j \leq \sigma\}$$

(cf. Duin and Voß (1994), Rayward-Smith and Clare (1986), and Voß (1992)).

One of the most interesting aspects of CHINS is the influence of how the root w is chosen. A repetitive approach results in the following modified algorithm called CHINS-V: *Cheapest insertion (all roots)*. CHINS is applied for every vertex (even possible Steiner vertices) defined as root w and the cheapest of all evaluated solutions gives the overall solution. Correspondingly, in CHINS-Q, CHINS is applied for every basic vertex as root w and in CHINS-2Q, CHINS is applied for every pair p, q of basic vertices where the root component is initialized as $P(p, q)$ and again the overall solution is the cheapest of all solutions obtained. Further modifications include repetitive applications of CHINS as well as the incorporation of 3BASIC into the framework of CHINS (Voß, 1992; Winter and Smith, 1992).

It should be noted that a common distinction within the area of heuristics is that of *single pass* versus *multiple pass* approaches depending on whether the proceeding underlying a specific heuristic calculates a feasible solution through a single pass through the given data or through multiple runs. Whereas CHINS itself belongs to the first category its modifications like CHINS-V belong to the second. With respect to HEUM in general and SDISTG there is no clear difference between these classes as the function evaluation is consecutively performed on the whole vertex set in each iteration. Considering the improvement part of SDISTG, we may interpret the method as a multiple pass approach.

A more versatile multiple pass approach may be described as follows (Duin, 1993). Consider as input a solution T given by a heuristic, say H, and let the *crucial set* of T be defined as the set $V_T^c := Q \cup \{i \in V_T - Q | \gamma_T(i) \geq 3\}$, where $\gamma_T(i)$ is the degree of vertex i in T. Then, for a number of iterations (e.g. as long as improved solutions are obtained) V_T^c could be taken as the new set of basic vertices for another pass of the heuristic retaining a new solution T' which repetitively replaces T. This multiple pass approach may also be implemented while allowing small changes of V_T^c in each iteration. While each Steiner tree heuristic may serve as a candidate for heuristic H, SDISTG and CHINS have been investigated within this framework by, e.g., Alexander and Robins (1994), Guitart and Basart (1998), and Minoux (1990)).

Another idea is to perturb the data such that the heuristic H is definitely finding a different solution and to project the results onto the original data (see e.g. Voß (1990a)).

Combining 3BASIC with SDISTG may lead to another heuristic which successively applies 3BASIC to every three-element subset of Q (or later a modified set of basic vertices) and then it effectively contracts the most promising of such subsets, to eventually obtain a new Steiner vertex after each contraction (see Zelikovsky (1993a) for the basic concept of this algorithm). The main idea is to compute, based on a specific heuristic measure function, a sequence of promising Steiner vertices that have to be included into the solution and then to apply SDISTG on this modified set of basic vertices.

With respect to versatile implementations, the running times of the above mentioned heuristics can be reduced compared to some standard implementation (see e.g. Mehlhorn (1988) and Zelikovsky (1993b) for SDISTG). One idea in this respect is to use efficient shortest path and MST computations as, e.g., presented by Gabow et al. (1986). With respect to the repetitive applications of CHINS, however, we can reduce the complexity in a different setting by simply rearranging some calculations. First, observe that only shortest paths between vertices of V and Q are necessary in CHINS. For CHINS-Q all those shortest paths can be calculated once in advance so that CHINS consumes only $O(|Q| \cdot |V|)$ time (cf. Duin and Voß (1997)). Therefore, for CHINS-Q (and also for the better variant CHINS-V) a time complexity equal to that of a single pass of CHINS follows.

Comparing different Steiner tree heuristics has been part of experimental as well as theoretical research for a long time. For a given instance, let Z_i and Z_j denote the objective function values derived by algorithms A_i and A_j, respectively. A_i and A_j are called *incomparable* if instances with $Z_i < Z_j$ as well as instances with $Z_i > Z_j$ exist. If A_i and A_j are incomparable independently of how ties are broken we call them *strongly incomparable*. The algorithms above, namely SDISTG and CHINS are strongly incomparable. However, this definition of incomparability does not give any idea on which algorithms could be preferred against others because even the simplest ones SPATH (i.e., computing the shortest path from a root w to any basic vertex) and MST+P on V (see above, computing a MST on G and then successively pruning any Steiner vertex with degree one) have to be added to this list of strongly incomparable algorithms (Voß, 1990a; Widmayer, 1986). For further results the reader is referred to, e.g., Plesnik (1992).

Alternatively, algorithms can be compared with respect to their worst case performance. Unfortunately, nearly all of the most well-known heuristics given in the literature (including CHINS and SDISTG) have a worst case bound of 2, i.e., $Z_T/Z_{opt} \leq 2(1 - 1/|Q|)$, where Z_T and Z_{opt} denote the objective function values of a feasible solution and an optimal solution, respectively (see e.g. Choukhmane (1978), Kou et al. (1981), Plesnik (1981), Plesnik (1991), Plesnik (1992), Voß (1992), and Waxman and Imase (1988)).

Zelikovsky (1993a) proved that SDISTG with 3BASIC has a worst case bound of $11/6$ instead of 2 which turned out to be a significant breakthrough in Steiner tree heuristics. At the same time, a slightly different approach had been considered by Berman and Ramaiyer (1994). Even earlier Minoux (1990) discussed a similar idea in a different context. Overall these approaches incorporate some sort of look ahead feature to overcome the possible myopic drawbacks of heuristics like SDISTG or CHINS. Some of these look ahead features may also be seen as preprocessing steps.

Subsequently, based on the 11/6-result of Zelikovsky (1993a), a number of researchers developed new heuristics with even better worst case bounds. Among them, Karpinski and Zelikovsky (1997) add a certain type of preprocessing which allows a reduction to approximately $1.644 < 1 + \ln 2$ instead of 11/6. Over time, further improvements have led to a bound of 1.55; see, e.g., Robins and Zelikovsky (2000).

Numerical results in various studies (see, e.g., Voß (1992), Winter and Smith (1992), and Duin and Voß (1997)) compare the quality of several heuristics for a great variety of problem instances. Overall the best way to describe the behavior of the different heuristics is that 'information helps' as well as that 'repetition pays.' As an example there are clear indications beyond the incomparability of SDISTG and CHINS that CHINS is able to utilize the connectivity information of given problem instances much better than SDISTG which may usually lead to better results. Furthermore, repetition helps in the sense that different choices of a root node for CHINS in a multipass setting may result in even better results.

For the SPD, Wong (1984) describes a dual ascent algorithm based on a multicommodity network flow formulation. Besides the implementations in Wong (1984) and Voß (1992), more recently very successful implementations based on improved data structures have come up by Polzin and Daneshmand (1997) and Polzin and Daneshmand (2001b). The dual ascent procedure produces as a side result reduced arc costs. The arcs with zero reduced costs define a *support graph*, a subgraph G' of G of promising arcs. Good upper bounds are obtained by running heuristics previously mentioned in this support graph (cf. Voß (1992)).

Subsequently different ideas including reduction techniques and modifications of concepts known from the SP have been investigated for the SPD (see Duin and Volgenant (1989b), Maculan et al. (1991), and Voß (1990a)). In Voß (1990a), Voß (1993), and Candia-Vejar and Bravo-Azlan (2004), the worst case performance of several heuristics for the SPD is investigated. With respect to this criterion, it is shown that none of the described algorithms for the SPD dominates any other. Contrary to the SP, even one of the simplest algorithms has the same worst case error ratio as the more sophisticated ones. Indeed, the best bound which can be given for any of the heuristics considered is $Z_T/Z_{opt} \leq |Q|$, which is quite discouraging. The proof idea is simply that the length of a shortest path from the root w to any basic vertex is at most Z_{opt}, i.e., $Z_T \leq |Q| \cdot Z_{opt}$. Furthermore, examples can be found to show that this bound is tight. Finally, it should be noted that all of these algorithms are strongly incomparable. Based on the work of Arora (1996) one may derive a much stronger result, i.e., that there cannot be any polynomial algorithm for the SPD with a constant error bound unless P = NP. For additional considerations see, e.g., Charikar et al. (1999).

Additional ideas have been developed and investigated for special cases of the SPD, e.g., when graphs are assumed to be acyclic (Rao et al., 1992; Voß and Duin, 1993; Zelikovsky, 1997). Furthermore, for the general case most ideas may be directly transfered from the SP to the SPD (see e.g. Osborne and Gillett (1991) and Voß (1990a)).

18.2.5 Metaheuristics

The main drawback of algorithms such as deterministic exchange procedures is their inability to continue the search upon becoming trapped in a local optimum. This suggests consideration of modern heuristic search techniques for guiding known strategies to overcome local optimality. Following this theme, one may investigate the application of local search based techniques such as the tabu search metastrategy.

18.2.5.1 The Pilot method.
Let us reconsider some possible drawbacks of SDISTG and CHINS. While SDISTG in each iteration only uses paths between basic vertices, CHINS takes into consideration all nodes inserted in previous iterations. Nevertheless, this is still myopic without any look ahead feature. That is, one of the weaker aspects of CHINS and similar heuristics is the fact that new paths are selected only based on shortest distances without evaluating the corresponding paths themselves. The algorithms fail to judge all intermediary nodes on these paths as junction nodes for future iterations. Only after inclusion, one hopes to exploit these connective qualities.

In this respect, HEUM with function f_3 is a first improvement. Another idea is to incorporate some tie breaking rule whenever more than one shortest path of the same length exists (cf. Duin and Voß (1997) and Khoury and Pardalos (1996)). However, it seems to be desirable that an evaluation function accounts for overlaps between paths and calculates in advance some connective consequences of including the paths associated with the evaluated node. This can be done by a look ahead metaheuristic, the so-called pilot method (see Duin and Voß (1999)).

In the pilot method the impact of the vertices is evaluated by additional calculations before actually inserting them. At any time within a component connecting approach, the partial solution $T = (V_T, E_T)$ induces an alternative smaller SP: assume the remaining components as basic vertices, while the non-basic vertex set is $V - V_T$. In this problem, we can tentatively run, as a heuristic measure, CHINS for various root nodes.

To formulate in general terms, the basic idea of the pilot method is as follows: By looking ahead with a heuristic H as a 'pilot heuristic,' one cautiously builds up a partial solution M, the master solution. Separately for each $e \notin M$, the pilot heuristic extends a copy of M by tentatively including e. Let $p(e)$ denote the objective value of the solution obtained by pilot H for $e \in E - M$, and let e_0 be the most promising element according to the pilot heuristic, i.e., $p(e_0) \leq p(e) \forall e \in E - M$. Element e_0 is included into the master solution M changing it in a minimal fashion. On the basis of the changed master solution M, new pilot calculations are started for each $e \notin M$, providing a new solution element e'_0, and so on. This process could proceed, e.g., until further pilot calculations do not lead to an improvement.

In various combinatorial contexts, a multiple pass method may be helpful, e.g., if providing the information on elements of a high quality solution can let a heuristic obtain better results. For instance, CHINS-V enhances the solution of a (single pass) heuristic by trial and error. The pilot method on the SP, applied with CHINS as pilot heuristic, starts similarly, but continues. At any time the master solution, a partial solution T, is a forest of trees: $T_1, T_2, ..., T_s$ covering the vertices of Q. In copies of this graph, with the components conceived of as basic vertices (as if contracted into a single vertex), one runs, separately for each vertex v CHINS with root v. Let vertex v^* produce the best objective function value. Then, two components of the master solution T are permanently joined in a rerun of CHINS from v^*, this time in the original graph for two iterations (or one if v^* is a basic node). Repetition of this process constitutes a pilot method for the SP denoted as PH_{CHINS}. Alternatively, one can interpret PH_{CHINS} as a synthesis of HEUM with f_3 and CHINS: the formula $f(v)$ is replaced by $p(v)$, the objective function value returned by subroutine CHINS with start v. Heuristic PH_{CHINS} can produce better solutions than CHINS-V, because it can find more than one (local) improvement.

Numerical results for the pilot method for the SP can be found in Duin and Voß (1999).

18.2.5.2 Local search. With respect to a proper neighborhood definition for SP two main ideas may be distinguished (see, e.g., Duin and Voß (1994)):

- Edge-oriented Transformation

    (1) Given a feasible solution $T = (V_T, E_T)$ randomly select an edge $(i,j) \in E_T$ and delete it from E. While there exists a leaf of T being a Steiner vertex do delete that leaf and its incident edge.

    (2) Let T_i and T_j be the two disconnected components of T. Find nearest vertices i^* and j^* belonging to T_i and T_j, respectively, and add the vertices and edges of the shortest path between i^* and j^* to T (eventually disallowing to use (i,j), i.e., defining a 'tabu status' for (i,j)).

- Node-oriented Transformation

    (1) Given a feasible solution $T = (V_T, E_T)$. Let $(x_1, \ldots, x_{|S|})$ be a binary vector with x_i denoting that i is included ($x_i = 1$) or excluded ($x_i = 0$) from the solution.

    (2) Choose a vertex $i \in S$ and define $x_i := 1 - x_i$.
    Recompute an MST of the graph induced by $Q \cup \{i \in S \mid x_i = 1\}$.

Within the local search literature usually an edge-oriented transformation is described (see, e.g., Verhoeven et al. (1996)). Moreover, Step (1) of this edge-oriented transformation has become known as the so-called key-path concept (i.e., a path between basic vertices or Steiner vertices with at least three incident edges is referred to as a key-path).

18.2.5.3 Local search and population based metaheuristics. While many different metaheuristics have been explored with respect to the SP it seems that mainly GRASP and tabu search approaches have turned out to be successful. Based on that we omit comments regarding neural networks and ant systems for the SP.

A *greedy randomized adaptive search procedure* (GRASP) is a multiple pass approach where each pass consists of two phases: constructing a feasible solution and then performing a local search. In the construction phase, e.g., feasible solutions may be built iteratively like in CHINS. However, to obtain diversity in each iteration the next element to be added is determined randomly choosing one of the best possible paths instead of the overall best one. Examples of GRASP implementations for the SP are provided by Martins et al. (2000) and Ribeiro et al. (2002).

Following a steepest descent mildest ascent approach, any transformation may either result in a best possible improvement or a least deterioration of the objective function value. To prevent the search from cycling between the same solutions endlessly, in *tabu search* some exchanges will be made tabu, i.e. forbidden for a specific number of iterations. In the literature, tabu search is based on the node-oriented neighborhood definition (cf. Duin and Voß (1994), Gendreau et al. (1999), and Voß (1990b) for the SP as well as Xu et al. (1996a) and Xu et al. (1996b) for some generalizations of the SP).

The first published reference on tabu search for the SP is Duin and Voß (1994) where the reverse elimination method (REM) is considered, however, with limited testing. Sondergeld and Voß (1996) investigate a specific intensification and diversification approach for the SP that strategically seeks a diversified collection of solutions by scattering multiple search trajectories over the solution space. The intensification strategy is based on strategic restarts of search strategies on so-called elite solutions previously encountered throughout the search. Depending on a periodic switching between diversification and intensification, multiple search levels may be distinguished.

Gendreau et al. (1999) present a static tabu search approach with radical diversification based on long term memory. An important contribution of the authors is their careful consideration of efficient data structures for manipulating given trees when adding or dropping nodes to or from a given solution/tree. As the computational effort for exploring a complete neighborhood may still be enormous, additional ideas for reducing it are given by Ribeiro and De Souza (2000). They describe transformation estimations, elimination tests, and neighborhood reduction techniques that may be helpful to speed up the local search, leading to a faster tabu search algorithm than that of Gendreau et al. (1999) without loosening too much the solution quality. For numerical results comparing different tabu search concepts see, e.g., Duin and Voß (1994) and Ribeiro and De Souza (2000).

Genetic algorithms for the SP are presented by Esbensen (1995) and Voß and Gutenschwager (1998).

18.3 STEINER PROBLEMS IN METRIC SPACES

Steiner problems in metric spaces play an expository role with respect to so-called *geometric network design problems*. In general one seeks some configuration or network of points and/or connections containing certain objects (points, connections), satisfying some predetermined requirements, and which optimizes a given objective function depending on one or more distance measures. A recent book on this topic is given by Cieslik (2005).

Complexity results are somewhat different as the calculation of square roots and respective comparisons may make it difficult to show that the Euclidean problem is in fact in NP. Following a result of Hanan (1966), the *rectilinear Steiner problem* may be formulated as a special case of the SP. For a given set Q of basic vertices, there exists an optimal solution which uses only vertices and edges of the graph induced by the set of basic vertices by drawing a horizontal and a vertical line through each basic vertex and retaining only the finite segments of these lines. The vertices of this so-called grid graph will be the set of all interconnections of any two lines.

Further extensive computations have been performed with respect to the rectilinear Steiner arborescence problem, i.e. the directed counterpart of the rectilinear Steiner problem. Despite some open complexity result for this problem for the case that all arcs are directed away from the origin with no cycles arising some heuristics were described in Voß and Duin (1993). The case of additional obstacles has been treated, too.

For the Euclidean problem, the state-of-the-art is the Geosteiner implementation of Warme et al. (2000). The core of the algorithm is the determination of so-called full Steiner trees. A full Steiner tree is a tree with no inner basic vertices, i.e., all basic vertices are leafs of the tree. For subsets of basic vertices full Steiner trees are determined one by one retaining the shortest. Then, these shortest full Steiner trees are checked and discarded or kept and those which are kept are in a next stage concatenated in any possible way to finally derive a solution for the overall problem instance. These ideas are successfully applied to the Euclidean and the rectilinear Steiner problem.

For a specific survey on the rectilinear Steiner problem see Zachariasen (2001). For older surveys on heuristics for the rectilinear Steiner problem the reader is referred to, e.g., Servit (1981) and Hwang et al. (1992).

A more general version of the Steiner problem assumes the problem to be considered on a sphere or on some general curved surfaces together with a related metric; see, e.g., Weng (2001).

18.4 PROBLEM MODIFICATIONS AND EXTENSIONS

An almost uncountable number of problem modifications and extensions are considered in the literature. The most important area of applications in this respect is multicast routing. Moreover, one may consider online optimization as an important telecommunications oriented modification. Also of interest, even if we do not discuss it in detail, may be the *k-restricted Steiner problem* which asks for a solution where the number of Steiner vertices in a solution is restricted to be at most k (with k being a non-negative number).

18.4.1 Multicast routing

Multicast routing has been amongst the most evolving areas in recent telecommunications applications of combinatorial optimization. In *multicast routing* the main objective is to send data from one or more sources to multiple destinations while minimizing the usage of resources (such as bandwidth, time, and connection costs). Recent surveys on multicast routing can be found in Novak et al. (2001) and Oliveira and Pardalos (2005), as well as Chapter 25 in this handbook. The wealth of problem settings in multicast routing is enormous while major objectives refer to concerns of, e.g., path delay, total cost of the chosen network and congestion. So-called quality of service is an important concern, e.g., for video-conferences. The path delay is an addictive delay function, corresponding to the sum of delays incurred from source to destination, for all destinations. Another example for an appropriate objective is to minimize the maximum network congestion where the congestion on a link is defined as the amount of capacity used. Here it is assumed that load-dependency comes into play, i.e., the more congestion is observed the harder it gets to handle failures in the network. Directly related to the SP is the minimization of the total cost of the routing tree. That is, the objective is to minimize the sum of costs for edges used in the routing tree.

With this, the SP indeed is one of the most important problems in multicast routing. In other words, some multicast problems are indeed just applications of the SP. For some additional thoughts the reader might consider Goel (1999) and Adelstein et al. (2003).

18.4.2 The online version

Interesting modifications of the Steiner problem arise from applications related to multicast routing as well as streaming applications. Various online versions are assuming that nodes and connections are changing over time, i.e., nodes and/or edges are added or deleted over time while a tree solution should be kept all the time.

One may consider an online version of the SP introduced by Imase and Waxman (1991), called by them the *dynamic Steiner tree problem* (DSP) and by others later the *online Steiner tree problem* (OSP). A natural classification of OSPs considers three different settings. First, it is assumed that additional nodes and edges arise over time allowing for later solution improvements but also necessitating the inclusion of new basic vertices into a tree. Second, data may change in the sense that nodes and edges are eliminated over time but a tree solution needs to be maintained all the time. Finally, a fully dynamic version assumes that any nodes and edges may arise and disappear again over time. In a sense, the SP may be referred to as a static problem while these online versions may be referred to as semi-dynamic (the first and second setting) and dynamic. A recent work on the online Steiner problem is Miller et al. (2005).

Related to multicast routing one may define online problems in various settings. For instance, Chakraborty et al. (2004) consider a heuristic approach for the problem where the staying duration of participants of, say a video session, is available at the time of joining. The approach is further tested assuming erroneous information given by the joining participants. Simulation results show that solution quality tends to be quite robust even in cases when the range of error is considerably high.

18.4.3 A stochastic version of the Steiner tree problem

One of the important real-world issues in telecommunications often neglected in combinatorial optimization is the consideration of infrastructure planning under uncertainty and related solution methods from stochastic optimization. For the SP not much has been done in this respect (one may see the consideration of online optimization as some exception; see Section 18.4.2). Exceptions are Gupta et al. (2004) and Ravi and Sinha (2004). The authors propose some approximation algorithms for a two stage proceeding. That is, in a first stage a partial solution is determined based on some cost function and some basic vertices 'possibly arising.' Based on the assumption of a probability distribution indicating the requirements (actually arising customers, i.e., basic vertices) in a second stage the partial solution might be extended, however, following a different cost function.

18.4.4 Steiner tree games

When defining game theoretic problems this may refer to important questions related to cost allocation, resource allocation and incentives. Famous early examples are the issue of cost allocation on spanning tree networks by Claus and Kleitman (1973). Regarding cost allocation and incentives issues a major avenue of research distinguishes centralized planning with a planner having complete information (so that in cooperative game theory each player is assumed to have full information) and cases where players have local or private information. A detailed survey on cost allocation on networks can be found, e.g., in Sharkey (1995).

A cooperative game may consist of a set $N = \{1, \ldots, n\}$ of players and a function $f : 2^N \to R_+$ which assigns a value to every coalition (i.e. subset) of players. Then, a solution of a game (N, f) consists of a set of vectors x which assign a payoff x_i to every player $i \in N$. The cooperative game approach to respective problems is to find a stable cost allocation which, to a certain extent, ensures that everyone is going to stay in the partnership (called coalition) and does not give any, at least monetary, incentive for any subset of users to secede and build their own competing subnetwork.

Assume a set $N = \{1, \ldots, n\}$ of customer nodes with an additional node w specified as the root node. For any $Q \subset N$ define $V_Q = Q \cup \{w\}$, and consider a graph $G_N = (V_N, E)$ with edge set E and symmetric nonnegative cost values c_{ij} for all $(i, j) \in E$ with $c_{ii} = 0$. Assume that any coalition Q of customers may be provided with a certain telecommunication service by a connected graph including the nodes from V_Q. That is, for any $Q \subset N$ let $T_Q = (V_Q, E_Q)$ represent a minimum cost connected graph such that $E_Q \subset E$. For nonnegative cost values there is an optimal solution that is a tree and the respective game is called the minimum cost spanning tree game for coalition Q of customers.

Here a spanning tree may only use nodes from V_S. The use of additional nodes leads to the monotone minimal cost spanning tree game and to the *Steiner tree game*. In the latter additional nodes from $V - Q$ (e.g. called transshipment or router nodes) may be used to achieve a tree T_Q connecting the nodes from Q which may allow additional cost reductions while making the problem more difficult to handle. A comprehensive collection of results regarding Steiner tree games can be found in Skorin-Kapov (2001). One may also consider a capacitated version of the Steiner tree game which can be adapted for our purposes. Here the set of nodes requires a certain service that can be provided by connecting them (eventually by using other nodes or switching points) to capacitated facilities, yet to be constructed.

18.5 GENERALIZATIONS

Many generalizations of the SP have been described in the literature. As an example let us define the *node-weighted Steiner problem* (NSP). For the NSP we assume the same data as for the SP with an additional cost function on the nodes. That is, every node is associated with a non-negative weight and the objective of the NSP is to find a minimum cost solution with respect to the sum of node and edge weights. One important problem with considerable subsequent interest had been investigated by Segev (1987) who assumes negative node weights and just one basic vertex. Cost minimization then refers to including nodes (customers with respect to a telecommunications service) only if it is beneficial for the overall objective.

Quite a few generalizations refer to some sort of capacity constraint such as constraints regarding a maximum number of incident edges to certain nodes or a certain hop limit. Others generalize with respect to protection against failures, etc. Here we provide some important pointers.

18.5.1 The Steiner problem with degree constraints

One important generalization of the SP refers to the case that the number of links from any node in the network is restricted to be at most a fixed value. This refers to the case that the number of copies of the same data is limited, e.g., when practical speed requirements may prohibit an unbounded number of copies of the received data. While the *degree-constrained Steiner problem* is considered, e.g., by Ravi et al. (2001), special multicast concerns of the problem are addressed by Bauer and Varma (1995).

18.5.2 The Steiner problem with hop constraints

The *Steiner problem with hop constraints* (HSP) is the same as the SP but with an additional capacity constraint. Beyond the definition of the SP we are given a positive integer H and wish to find a Steiner tree T of the graph with minimum total cost such that the unique path from a specified root node to any other node has no more than H hops (edges).

The HSP models the design of centralized telecommunication networks with quality of service constraints. The root node represents the site of a central processor or provider and the remaining nodes represent terminals that are required to be linked to the central processor. The hop constraints limit the number of hops (edges/nodes) between the root node and any other node and guarantee a certain level of service with respect to some performance constraints such as availability and reliability (see, e.g., Woolston and Albin (1988)). Availability refers to the probability that all the transmission lines in the path from the root node to the terminal are working, and reliability corresponds to the probability that a session will not be interrupted by a link failure. In general, these probabilities decrease with the number of links in the path implying that paths with fewer hops have a better performance with respect to availability and reliability. Centralized terminal networks are also usually implemented with multidrop lines for connecting the terminals with the center. In such networks, node processing times dominate over queuing delays and fewer hops mean, in general, smaller delays.

Lower bounding schemes for the HSP based on network-flow models have been suggested, e.g., by Gouveia (1998). Additional lower bounds can be found in Voß (1999). Interestingly enough, besides a cheapest insertion and the tabu search method described in Voß (1999) not much has been suggested for the HSP in terms of methods for obtaining good quality feasible solutions.

Related to the HSP is the SP with a constraint regarding the maximum diameter of a Steiner tree solution; see Gouveia et al. (2004).

18.5.3 The Steiner tree-star problem

We are given a set of nodes $V = Q \cup S$ with Q denoting a set of basic vertices to be spanned and S a set of possible Steiner vertices. For all possible connections between any two possible Steiner vertices let c_{ij} denote the cost of routing between i and j. If the basic vertex q can be served from the possible Steiner vertex i, then d_{iq} denotes the cost of this service, otherwise $d_{iq} = \infty$. The telecommunications problem that we are interested in is to find a tour, ring or routing through a subset of the possible Steiner vertices including a root node w so that all basic vertices are connected to at least one of the chosen nodes within the tour such that the total cost (routing and servicing) is minimized. Basically this problem is known for quite a long time as the so-called *traveling purchaser problem* which was treated in many earlier references (see, e.g., Voß (1990a) and the references therein).

Within the telecommunications context it was 're-invented' as the so-called *Steiner tree-star problem*. Xu et al. (1996b) consider a simple tabu search algorithm to solve this problem.

More generally, this problem refers to tree based *access network design*. That is, one may assume a backbone network (in the previous case the ring) and connect customers to the backbone nodes by means of trees (in the previous case single links or stars) building the access network. Standard solution procedures that can be found in the literature are based, e.g., on Lagrangean relaxations with the SP as an appropriate subproblem (see, e.g., Gavish (1992) and Yen and Lin (2005)). The latter reference uses some data from real networks with up to 26 nodes.

18.5.4 The generalized Steiner problem

The *generalized spanning tree* or *group Steiner problem* (GSP) is a generalization of the SP where one requires a tree spanning (at least) one vertex of each subset from a given family of vertex subsets, while minimizing the sum of the corresponding edge costs. In the literature, a large number of papers (more or less) prove the NP-hardness of the GSP (see, e.g., Dror et al. (2000)). Here we emphasize the strongest result which says that the GSP is NP-hard even on trees (see Ihler et al. (1999)). That is, with respect to common sense the GSP turns out to be somewhat harder than the SP.

The GSP has been 'reinvented' several times within the scientific literature throughout the last decade and subsequently been investigated in different somehow independent schools. Besides the above names one may also find the following: *Steiner tree problem with basic sets* and *class Steiner tree problem*. Therefore, we provide some historical remarks that try to put these different lines of research into perspective.

The problem was proposed in the 1960s by Cockayne and Melzak (1968) as follows: "Let $L_1, \ldots, L_n$ be n separate lakes which are to be interconnected by a network of canals of minimum possible length. Or, let $M_1, \ldots, M_n$ be n metropolitan areas which are to be joined by the shortest possible network of roads. Again, let $P_1, \ldots, P_n$ be n metal plates in the plane which are to be soldered together by the least possible amount of wire of fixed diameter. Assuming spatial homogeneity, we formulate an abstract problem which underlies the given examples. By a net N we shall understand a finite set of plane rectifiable arcs; the sum $L(n)$ of their lengths is the length of N. Let $n \geq 3$ and let $A = \{A_1,\ldots,A_n\}$ be a plane set with exactly n components A_i; it will be assumed throughout that each A_i is compact. Any such A will be called an n-terminal set and its n components will be called terminals. Our problem is: Given an n-terminal set A, to find the shortest net N, such set $A \cup N$ is a connected set."

The authors assume the Euclidean metric and provide some by now obvious theoretical results such as an upper bound on the number of Steiner nodes in an optimal solution. Most importantly they show the following theorem: "Let $A = \{A_1,\ldots,A_n\}$ be an n-terminal set and suppose that each A_i is arbitrarily well approximable by simple polygons. Then, the minimal connecting set N can be found by a finite sequence of Euclidean constructions to within arbitrary accuracy."

While many papers have considered the GSP it seems that the most important idea regarding its solution is a simple and yet effective transformation to the original SP as it had been proposed by Voß (1990b) and recently comprehensively studied by Duin et al. (2004). The latter paper shows that using known solution procedures (especially the pilot method) for the SP may solve benchmark instances specially defined in other GSP references and even outperform specialized approaches.

Independently, in the 1980s the GSP was described again for the Euclidean metric by Weng (1985). An approach for the GSP especially tailored to consider the rectilinear metric is provided by Zachariasen and Rohe (2003).

Generalizations of the GSP may be formulated, e.g., by requiring at least a certain number of vertices of any group or class to be included in a solution. That is, for each group there may be a different so-called requirement regarding the necessity of nodes 'covering' that group. The problem is also referred to as the *covering Steiner problem* (CSP). Many papers consider the CSP and discuss, e.g., approximability results on trees; see, e.g., Gupta and Srinivasan (2003).

A somewhat related but yet different problem is the following: Given an undirected graph, $G = (V,E)$ and subsets $Q_1, Q_2, \ldots, Q_k \subseteq V$, does there exist a family of subsets $E_1, E_2, \ldots, E_k \subseteq E$ such that $E_i \cap E_j = \emptyset$ for all $i \neq j$ and for each i, E_i forms a connected graph which includes every vertex in Q_i? This problem is another example for the fact that specific problems are described and considered independently; see, e.g., Du (1986) as well as Richey and Parker (1986).

18.5.5 Union of paths

Another problem that has become known as a *generalized Steiner problem* may be defined as follows. We are given a graph with non-negative edge weights and a set of pairs of vertices. The objective is to construct a minimum weight subgraph such that the two nodes of each pair are connected by a path. That is, a solution may be viewed as a union of paths (in fact any solution may be viewed as a (Steiner) forest). If we consider the special case of this problem where all pairs of some subset of vertices have to be connected then we derive the SP as a special case. Most efforts have been put into investigating approximation algorithms. Early references on this problem can be found in Agrawal et al. (1995) and Goemans and Williamson (1995).

More recently, an online version of this problem is considered, too; see, e.g., Awerbuch et al. (2004). Applications refer to building a network satisfying certain connectivity requirements, where new requirements appear over time. Moreover, it models some sort of communication aggregation where the cost of a communication protocol is measured by the number of edges used, rather than by the number of bits sent, a concern with long-term trunk reservation of telephone networks.

Awerbuch et al. (2004) also consider the network connectivity leasing problem where edges of the graph can be either bought or leased at different costs.

18.5.6 The prize-collecting Steiner problem

The SP seeks a minimum cost tree connecting a given set of basic nodes or terminals. In certain applications, however, even the basic nodes are not known beforehand (see, e.g., Johnson et al. (2000)). The *prize-collecting Steiner problem* (PCSP) is a generalization of the SP assuming a given connected graph $G = (V, E)$, positive edge weights c_{ij} for edges $(i, j) \in E$ as well as non-negative vertex profits or revenues r_i. The problem is to find a subtree $T = (V_T, E_T)$ of G such that the sum $\sum_{i \in V_T} r_i - \sum_{(i,j) \in E_T} c_{ij}$ is maximized. One may call vertices with strictly positive revenues basic vertices, and the SP is obtained once the positive revenues of the vertices to be spanned are sufficiently large while possible Steiner vertices have zero revenues. In a similar way the NSP may be related to the PCSP. A modified version of the problem asks to minimize the sum of the edge costs of E_T and the missing revenues for all vertices that are not included in V_T.

Telecommunications applications of the PCSP refer to, e.g., local access network design where one seeks a balance of potential revenues that can be obtained by providing services to customers and the cost to construct the network. While fixed costs for building fiber optic networks refer to edge weights, the node weights or profits are estimates of the potential revenues to be obtained by providing service to the customers on the specific nodes.

Canuto et al. (2002) present a local search heuristic for the PCSP with some perturbations related to the revenue data. It incorporates several ideas with respect to different neighborhoods and path relinking. Lucena and Resende (2004) propose a cutting plane algorithm for the PCSP based on the formulation of some generalized subtour elimination constraints. Klau et al. (2004) use an integer linear programming approach combined with a memetic algorithm and some preprocessing steps to obtain quite good results.

While reduction techniques for the SP have become one of the most important ingredients for successfully solving larger instances only a few studies have been undertaken to generalize and apply known reduction techniques to the PCSP. One such source is Uchoa (2005), who revisits the concept of bottleneck Steiner distances as well as some other preprocessing ideas in the context of the PCSP. His basic idea for obtaining good numerical results is to define the computation of bottleneck Steiner distances as a hard problem and then to resort to heuristics for approximating them.

18.5.7 The Steiner ring network design problem

A general discussion in telecommunications networks refers to the consideration of failures. Obviously, a ring network is able to overcome a link or node failure more easily than a tree network (if at all possible). The *ring network design problem (RNDP)*, as it has been considered by Gendreau et al. (1995), is a special case of the *general ring network design problem (GRNDP)* according to Fink et al. (2000). Given a graph G with vertex set $V = \{1,\ldots,n\}$ and edge set E with non-negative edge weights c_{ij} for all $(i,j) \in E$, the RNDP is defined as follows. A solution is a ring $R = (i_1,\ldots,i_k)$ with corresponding edges $((i_1,i_2),\ldots,(i_{k-1},i_k),(i_k,i_1))$. Whenever a ring contains a direct link between i and j, a cost c_{ij} is incurred. That is, only fixed construction costs are taken into account, and the sum of construction costs of R is given by $c(R) = \sum_{j=1}^{k-1} c_{i_j,i_{j+1}} + c_{i_k,i_1}$. A budget b limits the construction costs by $c(R) \leq b$. For every pair of nodes $i,j, i < j$, a revenue r_{ij} is obtained if and only if i and j both belong to the ring. The objective is to construct a feasible ring with maximum profit as defined by the obtained revenues minus the incurred construction costs.

Fink et al. (2000) describe a successful application of the pilot method to the RNDP. Starting from an empty solution (ring), they successively construct a larger ring by considering all possibilities for including some not yet included node at some position into the current ring until the budget constraint prevents any such extension. In each iteration of the pilot method all such options for extension are evaluated by means of the conventional (myopic) cheapest insertion procedure (i.e., by the respective objective function value eventually obtained). The pilot method is performed until an incomplete solution with a given number of nodes is reached; this solution is completed by continuing with a conventional cheapest insertion heuristic.

Closely related to this problem, Salazar-González (2003) considers the so-called *Steiner cycle problem* (SCP). Given an undirected graph G with a penalty associated to each node and a cost to each edge, and given a subset of nodes Q, the problem consists on finding a simple cycle in G visiting at least the nodes in Q and minimizing the sum of the costs of edges in the cycle plus the penalties of nodes not in the cycle. This problem finds applications in the optimal design of telecommunication systems, and generalizes the known traveling salesman problem (TSP) as well as the weighted Girth problem. The author analyzes the polyhedral structure associated to the SCP introducing two lifting procedures to extend facet-defining inequalities from the TSP polytope. The obtained results can be applied in a cutting-plane approach to solve the SCP.

18.5.8 Grade of service problem

The *grade of service Steiner minimum tree problem* GOSSP can be defined as follows: Let $Q = \{q_1, q_2, \ldots, q_n\}$ be a set of n terminal points or basic vertices in the Euclidean plane, where point q_i has a service request of grade $g(q_i) \in \{1,\ldots,r\}$ for some positive integer r. Let $0 < c(1) < c(2) < \cdots < c(r)$ be r real numbers. Each edge in the network is assigned a specific grade of service, which is a number in $\{1,\ldots,r\}$. We use $g(e)$ to denote the grade of service of edge e. The cost-per-unit-length for an edge with service grade u is $c(u)$. The GOSSP asks for a minimum cost network interconnecting the point set Q and possibly some Steiner points with service request of grade 0 such that between each pair of terminal points q_i and q_j there is a path whose minimum grade of service is at least as large as $min(g(q_i), g(q_j))$. The cost of an edge with service of grade u is the product of the Euclidean length of the edge with $c(u)$. The GOSSP is a generalization of the Euclidean Steiner problem, which is obtained when all terminal points have the same grade of service request. The GOSSP relates to the case of interconnecting many communication sites with the best choice of the connecting lines and the best allocation of the transmission capacities over these lines. Good solutions should provide paths with enough communication capacities between any two sites, with the least network construction costs. For a recent paper on the GOSSP see Kim et al. (2002), who propose a polynomial time approximation scheme.

Related to the GOSSP we may find various problem variations referring to applications, e.g., in multimedia distribution for users with different bit-rate requests. Different names and descriptions can be found, such as 'multi-tier Steiner tree problem' (Mirchandani (1996)). With respect to multimedia distribution one may also assume that each node possesses a rate and the cost of a link is not constant but depends both on the cost per unit of transmission bandwidth and the maximum rate routed through the link. Some references in this respect are Charikar et al. (2004) and Karpinski et al. (2005).

18.6 FINAL REMARKS

Steiner tree problems refer to important problem classes in graphs as well as in the Euclidean plane and more general in metric spaces. The Steiner tree problem in graphs may be called one of the most important combinatorial optimization problems not only in telecommunications. Almost any idea for solving optimization problems might have already been tried with respect to this problem. And still we see advances and enormous efforts in research.

While the 1990s have seen considerable improvements in solving the problem exactly as well as the investigation of many metaheuristics, the last few years have mainly been devoted to enhance the knowledge regarding important applications and generalizations of the problem mainly in the telecommunications domain such as, e.g., multicast routing. One has understood that the difficulty to solve problem instances need not only be judged by means of their size but also regarding their structure. In metaheuristics, solution quality need not only be judged by means of quality related to objective function calculations but also by means of structural difference while appropriate measures in this respect still need to be developed. This will shine through to generalized versions of the Steiner problem where insights on these generalizations will be a fruitful source for even further understanding while 'projected down' to the original version.

Additional modifications and generalizations of Steiner tree problems will certainly arise and become a core focus of research and telecommunications applications including additional online optimization problems as well as stochastic optimization approaches. Examples for possible generalizations may include, e.g., a weighted labeling Steiner tree problem with budget constraints (i.e., given a graph G with a label (color) assigned to each edge we look for a spanning tree with respect to a given subset of the nodes of G with the minimum number of different colors; furthermore, one may incorporate some weights on the edges and define a budget constraint on the sum of the weights of included edges while still minimizing the number of labels) or more versatile capacitated Steiner tree problems.

As a final comment we should note that the growth of references on modifications and generalizations of the SP in recent years is quite enormous. Based on that we had to select from available references. However, if a certain reference is missing it does not necessarily mean that it has no merit.

Bibliography

F. Adelstein, G.G. Richard, and L. Schwiebert. Distributed multicast tree generation with dynamic group membership. *Computer Communications*, 26:1105–1128, 2003.

A. Agrawal, P. Klein, and R. Ravi. When trees collide: An approximation algorithm for the generalized Steiner problem in networks. *SIAM Journal on Computing*, 24: 440–456, 1995.

M. J. Alexander and G. Robins. A new approach to FPGA routing based on multi-weighted graphs. In *Proceedings of the International Workshop on Field-Programmable Gate Arrays*, 1994.

Y.P. Aneja. An integer linear programming approach to the Steiner problem in graphs. *Networks*, 10:167–178, 1980.

S. Arora. Polynomial-time approximation scheme for Euclidean TSP and other geometric problems. In *Proceedings of the Symposium on Foundations of Computer Science*, pages 2–11, 1996.

B. Awerbuch, Y. Azar, and Y. Bartal. On-line generalized Steiner problem. *Theoretical Computer Science*, 324:313–324, 2004.

P. Bachhiesl, M. Prossegger, G. Paulus, J. Werner, and H. Stögner. Simulation and optimization of the implementation costs for the last mile of fiber optic networks. *Networks and Spatial Economics*, 3:467–482, 2004.

L. Bahiense, F. Barahona, and O. Porto. Solving Steiner tree problems in graphs with Lagrangian relaxation. *Journal of Combinatorial Optimization*, 7:259–282, 2003.

A. Balakrishnan and N.R. Patel. Problem reduction methods and a tree generation algorithm for the Steiner network problem. *Networks*, 17:65–85, 1987.

F. Bauer and A. Varma. Degree-constrained multicasting in point-to-point networks. In *Proceedings IEEE INFOCOM '95*, pages 369–376, 1995.

J. E. Beasley. An SST-based algorithm for the Steiner problem in graphs. *Networks*, 19:1–16, 1989.

J.E. Beasley. Or-library: distributing test problems by electronic mail. *Journal of the Operational Research Society*, 41:1069–1072, 1990.

P. Berman and V. Ramaiyer. Improved approximations for the Steiner tree problem. *Journal of Algorithms*, 17:381–408, 1994.

B. Bollobás, D. Gamarnik, O. Riordan, and B. Sudakov. On the value of a random minimum weigth Steiner tree. *Combinatorica*, 24:187–207, 2004.

A. Candia-Vejar and H. Bravo-Azlan. Performance analysis of algorithms for the Steiner problem in directed networks. *Electronic Notes in Discrete Mathematics*, 18:67–72, 2004.

S.A. Canuto, M.G.C. Resende, and C.C. Ribeiro. Local search with perturbations for the prize-collecting Steiner tree problem in graphs. *Networks*, 38:50–58, 2002.

D. Chakraborty, S.M.S. Zabir, A. Chayabejara, and G. Chakraborty. A distributed routing method for dynamic multicasting. *Telecommunication Systems*, 25:299–315, 2004.

M. Charikar, C. Chekuri, T. Cheung, Z. Dai, A. Goel, S. Guha, and M. Li. Approximation algorithms for directed Steiner problems. *Journal of Algorithms*, 33:73–91, 1999.

M. Charikar, J. Naor, and B. Schieber. Resource optimization in QoS multicast routing of real-time multimedia. *IEEE/ACM Transactions on Networking*, 12:340–348, 2004.

D. S. Chen. Constrained wirelength minimization of a Steiner tree. Technical report, Department of Electrical Engineering and Computer Science, Northwestern University, Evanston, Illinois, 1994.

S. Chopra, E.R. Gorres, and M.R. Rao. Solving the Steiner tree problem on a graph using branch and cut. *ORSA Journal on Computing*, 4:320–335, 1992.

E. A. Choukhmane. Une heuristique pour le probleme de l'arbre de Steiner. *R.A.I.R.O. Recherche Operationelle*, 12:207–212, 1978.

D. Cieslik. *Shortest Connectivity*. Springer, New York, 2005.

A. Claus and D.J. Kleitman. Cost allocation for a spanning tree. *Networks*, 3:289–304, 1973.

E. J. Cockayne and Z. A. Melzak. Steiner's problem for set-terminals. *Quarterly Applied Mathematics*, 26:213–218, 1968.

R. Courant and H. Robbins. *What is Mathematics?* Oxford University Press, New York, 1941.

M. Dror, M. Haouari, and J. Chaouachi. Generalized spanning trees. *European Journal of Operational Research*, 120:583–592, 2000.

D.-Z. Du. An optimization problem on graphs. *Discrete Applied Mathematics*, 14: 101–104, 1986.

D.-Z. Du and X. Cheng, editors. *Steiner Trees in Industries*. Kluwer, Boston, 2001.

D.-Z. Du, B. Lu, H. Ngo, and P.M. Pardalos. Steiner tree problems. In C.A. Floudas and P.M. Pardalos, editors, *Encyclopedia of Optimization*, volume 5, pages 227–290. Kluwer, Dordrecht, 2001.

D.-Z. Du, J. M. Smith, and J. H. Rubinstein, editors. *Advances in Steiner Trees*. Kluwer, Boston, 2000.

C. Duin. Preprocessing the Steiner problem in graphs. In D.-Z. Du, J. M. Smith, and J. H. Rubinstein, editors, *Advances in Steiner Trees*, pages 175–233. Kluwer, Boston, 2000.

C. W. Duin. *Steiner's Problem in Graphs: Approximation, Reduction, Estimation*. PhD thesis, Faculteit der Economische Wetenschappen en Econometrie, Universiteit van Amsterdam, 1993.

C.W. Duin and A. Volgenant. An edge elimination test for the Steiner problem in graphs. *Operations Research Letters*, 8:79–83, 1989a.

C.W. Duin and A. Volgenant. Reducing the hierarchical network design problem. *European Journal of Operational Research*, 39:332–344, 1989b.

C.W. Duin and A. Volgenant. Reduction tests for the Steiner problem in graphs. *Networks*, 19:549–567, 1989c.

C.W. Duin, A. Volgenant, and S. Voß. Solving group Steiner problems as Steiner problems. *European Journal of Operational Research*, 154:323–329, 2004.

C.W. Duin and S. Voß. Steiner tree heuristics - a survey. In H. Dyckhoff, U. Derigs, M. Salomon, and H.C. Tijms, editors, *Operations Research Proceedings 1993*, pages 485–496, Berlin, 1994. Springer.

C.W. Duin and S. Voß. Efficient path and vertex exchange in Steiner tree algorithms. *Networks*, 29:89–105, 1997.

C.W. Duin and S. Voß. The pilot method: A strategy for heuristic repetition with application to the Steiner problem in graphs. *Networks*, 34:181–191, 1999.

H. Esbensen. Computing near-optimal solutions to the Steiner problem in a graph using a genetic algorithm. *Networks*, 26:173–185, 1995.

A. Fink, G. Schneidereit, and S. Voß. Solving general ring network design problems by meta-heuristics. In M. Laguna and J.L. González Velarde, editors, *Computing Tools for Modeling, Optimization and Simulation*, pages 91–113. Kluwer, 2000.

H.N. Gabow, Z. Galil, T. Spencer, and R.E. Tarjan. Efficient algorithms for finding minimum spanning trees in undirected and directed graphs. *Combinatorica*, 6:109–122, 1986.

M. R. Garey and D. S. Johnson. *Computers and Intractability: A Guide to the Theory of NP–Completeness*. Freeman, San Francisco, 1979.

B. Gavish. Topological design of computer communication networks. *European Journal of Operational Research*, 58:149–172, 1992.

M. Gendreau, M. Labbé, and G. Laporte. Efficient heuristics for the design of ring networks. *Telecommunication Systems*, 4:177–188, 1995.

M. Gendreau, J.-F. Larochelle, and B. Sansò. A tabu search heuristic for the Steiner tree problem. *Networks*, 34:162–172, 1999.

A. Goel. *Algorithms for network routing, multicasting, switching, and design*. PhD thesis, Stanford University, Department of Computer Science, 1999.

M. Goemans and D. Williamson. A general approximation technique for constrained forest problems. *SIAM Journal on Computing*, 24:296–317, 1995.

M. X. Goemans and Y. S. Myung. A catalog of Steiner tree formulations. *Networks*, 23:19–28, 1993.

L. Gouveia. Using variable redefinition for computing lower bounds for minimum spanning and Steiner trees with hop constraints. *INFORMS Journal on Computing*, 10:180–187, 1998.

L. Gouveia, T.L. Magnanti, and C. Requejo. A 2-path approach for odd-diameter-constrained minimum spanning and Steiner trees. *Networks*, 44:254–265, 2004.

P. Guitart and J.M. Basart. *A high performance approximate algorithm for the Steiner problem in graphs*, pages 280–293. Springer, Berlin, 1998.

A. Gupta and A. Srinivasan. On the covering Steiner problem. In P.K. Pandya and J. Radhakrishnan, editors, *FSTTCS 2003*, volume 2914 of *Lecture Notes in Computer Science*, pages 244–251. Springer, Berlin, 2003.

A. Gupta, A. Srinivasan, and E. Tardos. Cost-sharing mechanisms for network design. In K. Jansen et al., editor, *APPROX and RANDOM 2004*, pages 139–150. Springer, Berlin, 2004.

M. Hanan. On Steiner's problem with rectilinear distance. *SIAM Journal on Applied Mathematics*, 14:255–265, 1966.

F. K. Hwang and D. S. Richards. Steiner tree problems. *Networks*, 22:55–89, 1992.

F. K. Hwang, D.S. Richards, and P. Winter. *The Steiner Tree Problem*. North-Holland, Amsterdam, 1992.

E. Ihler, G. Reich, and P. Widmayer. Class Steiner trees and VLSI-design. *Discrete Applied Mathematics*, 90:179–194, 1999.

M. Imase and B. Waxman. The dynamic Steiner tree problem. *SIAM Journal of Discrete Mathematics*, 4:369–384, 1991.

V. Jarnik and M. Kössler. O minimalnich grafech, obsahujicich n danych bodu. *Casopis pro Pestovani Matematiky a Fysiky*, pages 223–235, 1934.

D.S. Johnson, M. Minkoff, and S. Phillips. The prize collecting Steiner tree problem: Theory and practice. In *Proceedings of the Eleventh Annual ACM-SIAM Symposium on Discrete Algorithms*, pages 760–769. SIAM, 2000.

M. Karpinski, I.I. Mandoiu, A. Olshevsky, and A. Zelikovsky. Improved approximation algorithms for the quality of service multicast tree problem. *Algorithmica*, 42:109–120, 2005.

M. Karpinski and A. Zelikovsky. New approximation algorithms for the Steiner tree problems. *Journal of Combinatorial Optimization*, 1:47–65, 1997. Also in "Electronic Colloquium on Computational Complexity," TR95-003 (1995).

B. N. Khoury, P. M. Pardalos, and D. Z. Du. A test problem generator for the Steiner problem in graphs. *ACM Transactions on Mathematical Software*, 19:509–522, 1993.

B.N. Khoury and P.M. Pardalos. A heuristic for the Steiner problem in graphs. *Computational Optimization and Applications*, 6:5–14, 1996.

J. Kim, M. Cardei, I. Cardei, and X. Jia. A polynomial time approximation scheme for the grade of service Steiner minimum tree problem. *Journal of Global Optimization*, 24:427–448, 2002.

G.W. Klau, I. Ljubic, A. Moser, P. Mutzel, P. Neuner, U. Pferschy, G. Raidl, and R. Weiskircher. Combining a memetic algorithm with integer programming to solve the prize-collecting Steiner tree problem. Technical report, Vienna University of Technology, Vienna, 2004.

T. Koch and A. Martin. Solving Steiner tree problems in graphs to optimality. *Networks*, 32:207–232, 1998.

T. Koch, A. Martin, and S. Voß. SteinLib: An updated library on Steiner tree problems in graphs. In D.-Z. Du and X. Cheng, editors, *Steiner Trees in Industries*, pages 285–325. Kluwer, Boston, 2001.

L. Kou, G. Markowsky, and L. Berman. A fast algorithm for Steiner trees. *Acta Informatica*, 15:141–145, 1981.

J.B. Kruskal. On the shortest spanning subtree of a graph and the travelling salesman problem. *Proc. Amer. Math. Soc.*, 7:48–50, 1956.

A. Lucena. Steiner problem in graphs: Lagrangean relaxation and cutting-planes. *Bulletin of the Committee on Algorithms*, 21:2–7, 1992.

A. Lucena and J.E. Beasley. A branch and cut algorithm for the Steiner problem in graphs. *Networks*, 31:39–59, 1998.

A. Lucena and M. G. C. Resende. Strong lower bounds for the prize collecting Steiner problem in graphs. *Discrete Applied Mathematics*, 141:277–294, 2004.

N. Maculan, P. Souza, and A. Candia Vejar. An approach for the Steiner problem in directed graphs. *Annals of Operations Research*, 33:471–480, 1991.

S.L. Martins, M.G.C. Resende, C.C. Ribeiro, and P.M. Pardalos. A parallel GRASP for the Steiner tree problem in graphs using a hybrid local search strategy. *Journal of Global Optimization*, 17:267–283, 2000.

K. Mehlhorn. A faster approximation algorithm for the Steiner problem in graphs. *Information Processing Letters*, 27:125–128, 1988.

Z. Miller, D. Pritikin, M. Perkel, and I. H. Sudborough. The sequential sum problem and performance bounds on the greedy algorithm for the on-line Steiner problem. *Networks*, 45:143–164, 2005.

M. Minoux. Efficient greedy heuristics for Steiner tree problems using reoptimization and supermodularity. *INFOR*, 28:221–233, 1990.

P. Mirchandani. The multi-tier tree problem. *INFORMS Journal on Computing*, 8: 202–218, 1996.

R. Novak, J. Rugelj, and G. Kandus. Steiner tree based distributed multicast routing. In D.-Z. Du and X. Cheng, editors, *Steiner Trees in Industries*, pages 327–352. Kluwer, Boston, 2001.

C. A. S. Oliveira and P. M. Pardalos. A survey of combinatorial optimization problems in multicast routing. *Computers & Operations Research*, 32:1953–1981, 2005.

L.J. Osborne and B.E. Gillett. A comparison of two simulated annealing algorithms applied to the directed Steiner problem on networks. *ORSA Journal on Computing*, 3:213–225, 1991.

J. Pearl. *Heuristics: Intelligent Search Techniques for Computer Problem Solving*. Addison-Wesley, Reading, 1984.

J. Plesnik. A bound for the Steiner tree problem in graphs. *Math. Slovaca*, 31:155–163, 1981.

J. Plesnik. Worst-case relative performance of heuristics for the Steiner problem in graphs. *Acta Math. Univ. Comenianae*, 60:269–284, 1991.

J. Plesnik. Heuristics for the Steiner problem in graphs. *Discrete Applied Mathematics*, 37/38:451–463, 1992.

T. Polzin and S. Vahdati Daneshmand. Algorithmen für das Steiner-Problem. Diploma thesis, University of Dortmund, 1997.

T. Polzin and S. Vahdati Daneshmand. A comparison of Steiner tree relaxations. *Discrete Applied Mathematics*, 112:241–261, 2001a.

T. Polzin and S. Vahdati Daneshmand. Improved algorithms for the Steiner problem in networks. *Discrete Applied Mathematics*, 112:263–300, 2001b.

R.C. Prim. Shortest connection networks and some generalizations. *Bell Syst. Techn. J.*, 36:1389–1401, 1957.

H. J. Prömel and A. Steger. *The Steiner Tree Problem*. Vieweg, Wiesbaden, 2002.

S. K. Rao, P. Sadayappan, F. K. Hwang, and P. W. Shor. The rectilinear Steiner arborescence problem. *Algorithmica*, 7:277–288, 1992.

R. Ravi, M.V. Marathe, S.S. Ravi, D.J. Rosenkrantz, and H.B. Hunt III. Approximation algorithms for degree-constrained minimum-cost network-design problems. *Algorithmica*, 31:58–78, 2001.

R. Ravi and A. Sinha. Hedging uncertainty: Approximation algorithms for stochastic optimization problems. In D. Bienstock and G. Nemhauser, editors, *IPCO 2004*, pages 101–115. Springer, Berlin, 2004.

V. J. Rayward-Smith. The computation of nearly minimal Steiner trees in graphs. *Int. J. Math. Educ. Sci. Technol.*, 14:15–23, 1983.

V. J. Rayward-Smith and A. Clare. On finding Steiner vertices. *Networks*, 16:283–294, 1986.

C.C. Ribeiro and M.C. De Souza. Tabu search for the Steiner problem in graphs. *Networks*, 36:138–146, 2000.

C.C. Ribeiro, E. Uchoa, and R.F. Werneck. A hybrid GRASP with perturbations for the Steiner problem in graphs. *INFORMS Journal on Computing*, 14:228–246, 2002.

M. B. Richey and R. G. Parker. On multiple Steiner subgraph problems. *Networks*, 16:423–438, 1986.

G. Robins and A. Zelikovsky. Improved Steiner tree approximation in graphs. In *Proceedings of the Eleventh Annual ACM-SIAM Symposium on Discrete Algorithms*, pages 770–779, 2000.

I. Rosseti, M. Poggi de Aragao, C. Ribeiro, E. Uchoa, and R.F. Werneck. New benchmark instances for the Steiner problem in graphs. In M.G.C. Resende and J.P. de Sousa, editors, *Metaheuristics: Computer Decision-Making*, pages 601–614. Kluwer, Boston, 2003.

J.-J. Salazar-González. The Steiner cycle polytope. *European Journal of Operational Research*, pages 671–679, 2003.

A. Segev. The node-weighted Steiner tree problem. *Networks*, 17:1–17, 1987.

M. Servit. Heuristic algorithms for rectilinear Steiner trees. *Digital Processes*, 7: 21–32, 1981.

W.W. Sharkey. Network models in economics. In M.O. Ball, T.L. Magnanti, C.L. Monma, and G.L. Nemhauser, editors, *Network Routing*, pages 713–765. North-Holland, Amsterdam, 1995.

D. Skorin-Kapov. On cost allocation in Steiner tree networks. In D.-Z. Du and X. Cheng, editors, *Steiner Trees in Industries*, pages 353–376. Kluwer, Boston, 2001.

L. Sondergeld and S. Voß. A multi-level star-shaped intensification and diversification approach in tabu search for the Steiner tree problem in graphs. Technical report, TU Braunschweig, 1996.

J. Soukup and W.F. Chow. Set of test problems for the minimum length connection networks. *ACM / SIGMAP Newsletter*, 15:48–51, 1973.

J. Steiner. Aufgaben und Lehrsätze, erstere aufzulösen, letztere zu beweisen. *Journal für die reine und angewandte Mathematik*, 13:361–364, 1835.

H. Takahashi and A. Matsuyama. An approximate solution for the Steiner problem in graphs. *Math. Japonica*, 24:573–577, 1980.

E. Uchoa. Local search with perturbations for the prize-collecting Steiner tree problem in graphs. Technical report, Universidade Federal Fluminense, Niterói, Brazil, 2005.

M. G. A. Verhoeven, M. E. M. Severens, and E. H. L. Aarts. Local search for Steiner trees in graphs. In V.J. Rayward-Smith, I.H. Osman, C.R. Reeves, and G.D. Smith, editors, *Modern Heuristic Search Methods*, pages 117–129. Wiley, Chichester, 1996.

S. Voß and C.W. Duin. Heuristic methods for the rectilinear Steiner arborescence problem. *Engineering Optimization*, 21:121–145, 1993.

S. Voß. *Steiner-Probleme in Graphen*. Hain, Frankfurt/Main, 1990a.

S. Voß. A survey on some generalizations of Steiner's problem. In B. Papathanassiu and K. Giatas, editors, *1st Balkan Conference on Operational Research Proceedings*, pages 41–51. Hellenic Productivity Center, Thessaloniki, 1990b.

S. Voß. Steiner's problem in graphs: heuristic methods. *Discrete Applied Mathematics*, 40:45–72, 1992.

S. Voß. Worst case performance of some heuristics for Steiner's problem in directed graphs. *Information Processing Letters*, 48:99–105, 1993.

S. Voß. The Steiner tree problem with hop constraints. *Annals of Operations Research*, 86:321–345, 1999.

S. Voß. Modern heuristic search methods for the Steiner tree problem in graphs. In D.-Z. Du, J. M. Smith, and J. H. Rubinstein, editors, *Advances in Steiner Trees*, pages 283–323. Kluwer, Boston, 2000.

S. Voß and K. Gutenschwager. A chunking based genetic algorithm for the Steiner tree problem in graphs. In P.M. Pardalos and D.-Z. Du, editors, *Network Design: Connectivity and Facilities Location*, volume 40 of *DIMACS Series in Discrete Mathematics and Theoretical Computer Science*, pages 335–355. AMS, Princeton, 1998.

D.M. Warme, P. Winter, and M. Zachariasen. Exact algorithms for plane Steiner tree problems: A computational study. In D.-Z. Du, J. M. Smith, and J. H. Rubinstein, editors, *Advances in Steiner Trees*, pages 81–116. Kluwer, Boston, 2000.

B.M. Waxman and M. Imase. Worst-case performance of Rayward-Smith's Steiner tree heuristic. *Information Processing Letters*, 29:283–287, 1988.

J. F. Weng. Generalized Steiner problem and hexagonal coordinate system (in Chinese). *Acta Math. Appl. Sinica*, 8:383–397, 1985.

J.F. Weng. Steiner trees an curved surfaces. *Graphs and Combinatorics*, 17:353–363, 2001.

P. Widmayer. *Fast approximation algorithms for Steiner's problem in graphs*. Habilitation thesis, Institut für Angewandte Informatik und formale Beschreibungsverfahren, University Karlsruhe, 1986.

P. Winter. Steiner problem in networks: a survey. *Networks*, 17:129–167, 1987.

P. Winter and J. MacGregor Smith. Path-distance heuristics for the Steiner problem in undirected networks. *Algorithmica*, 7:309–327, 1992.

R. T. Wong. A dual ascent approach for Steiner tree problems on a directed graph. *Mathematical Programming*, 28:271–287, 1984.

K. Woolston and S. Albin. The design of centralized networks with reliability and availability constraints. *Computers & Operations Research*, 15:207–217, 1988.

B. Y. Wu and K.-M. Chao. *Spanning Trees and Optimization Problems*. Chapman & Hall / CRC, Boca Raton, 2004.

J. Xu, S.Y. Chiu, and F. Glover. A probabilistic tabu search for the telecommunications network design. *Combinatorial Optimization: Theory and Practice*, 1:69–94, 1996a.

J. Xu, S.Y. Chiu, and F. Glover. Using tabu search to solve Steiner tree-star problem in telecommunications network design. *Telecommunication Systems*, 6:117–125, 1996b.

H.-H. Yen and F.Y.-S. Lin. Near-optimal tree-based access network design. *Computer Communications*, 28:236–245, 2005.

M. Zachariasen. The rectilinear Steiner problem: A tutorial. In D.-Z. Du and X. Cheng, editors, *Steiner Trees in Industries*, pages 467–507. Kluwer, Boston, 2001.

M. Zachariasen and A. Rohe. Rectilinear group Steiner trees and applications in VLSI design. *Mathematical Programming*, 94:407–433, 2003.

A. Zelikovsky. A series of approximation algorithms for the acyclic directed Steiner tree problem. *Algorithmica*, 18:99–110, 1997.

A. Z. Zelikovsky. An 11/6-approximation algorithm for the network Steiner problem. *Algorithmica*, 9:463–470, 1993a.

A.Z. Zelikovsky. A faster approximation algorithm for the Steiner tree problem in graphs. *Information Processing Letters*, 46:79–83, 1993b.

19 ON FORMULATIONS AND METHODS FOR THE HOP-CONSTRAINED MINIMUM SPANNING TREE PROBLEM

Geir Dahl[1], Luis Gouveia[2], and Cristina Requejo[3]

[1]Department of Mathematics and Department of Informatics
University of Oslo
0316 Oslo, Norway
geird@ifi.uio.no

[2]DEIO – CIO
Faculdade de Ciências da Universidade de Lisboa
1749-016 Lisboa, Portugal
legouveia@fc.ul.pt

[3]University of Aveiro
1810-193 Aveiro, Portugal
crequejo@mat.ua.pt

Abstract: In this chapter we present a general framework for modeling the hop-constrained minimum spanning tree problem (HMST) which includes formulations already presented in the literature. We present and survey different ways of computing a lower bound on the optimal value. These include, Lagrangian relaxation, column generation and model reformulation. We also give computational results involving instances with 40 and 80 nodes in order to compare some of the ideas discussed in the chapter.
Keywords: Hop-constrained spanning trees, polyhedral characterizations, model reformulation.

19.1 INTRODUCTION

The Hop-constrained Minimum Spanning Tree Problem (HMST) hop-constrained spanning tree is defined as follows: Given a graph $G = (V,E)$ with node set $V = \{0,1,\ldots,n\}$ and edge set E as well as a cost c_e associated with each edge $e \in E$ and a natural number H, we wish to find a spanning tree T of the graph with minimum total cost such that the unique path from a specified root node, node 0, to any other node has no more than H hops (edges).

The HMST is NP-hard because it contains as a particular case (the case with $H = 2$) a NP-Hard version of the Simple Uncapacitated Facility Location problem (see Gouveia (1995) and Dahl (1998)). Manyem and Stallmann (1996) have shown that the HMST is not in APX, i.e., the class of problems for which it is possible to have polynomial time heuristics with a guaranteed approximation bound. Dahl (1998) has also studied the $H = 2$ case from a polyhedral perspective and compared related directed and undirected models. Lower bounding schemes for the HMST based on formulations using constrained versions of the well known Miller-Tucker-Zemlin constraints have been proposed in Gouveia (1995). The reported results indicate that these formulations have to be considerably strengthened in order to reduce the reported gaps. These results have, in a certain way, motivated the use of the network-flow based models as in Gouveia (1996; 1998). These models are surveyed in this chapter.

The HMST models the design of centralized telecommunication networks telecommunication networks with quality of service constraints. quality of service constraints The root node represents the site of a central processor (computer) and the remaining nodes represent terminals that are required to be linked to the central processor. The hop constraints limit the number of hops (arcs) between the root node and any other node and guarantee a certain level of service with respect to some performance constraints such as availability and reliability (see Woolston and Albin (1988)). Availability is the probability that all the transmission lines in the path from the root node to the terminal are working. Reliability is the probability that a session will not be interrupted by a link failure. In general, these probabilities decrease with the number of links in the path which means that paths with fewer hops have a better performance with respect to availability and reliability. Woolston and Albin (1988) have presented some computational results based on heuristic solutions which show that spanning trees design with no hop limit on the paths from the root node to the terminals behave poorly with respect to these measures. They have shown that by including hop limits it is possible to generate designs with a much better service and only with moderate increase on the total cost.

Centralized terminal networks are also usually implemented with multidrop lines for connecting the terminals with the center. In such networks, node processing times dominate over queuing delays and fewer hops mean, in general, lower delays. Hop constraints have also been discussed in the context of more general telecommunication network design problems. The reliability aspect of hop constraints have already been discussed by LeBlanc et al. (1999). These authors also pointed out that hop constraints can be used to avoid degradation of signal quality. Balakrishnan and Altinkemer (1992) have used different hop parameters in their model as a means of generating alternative solutions that span a wide range of quality of service levels. More recently, hop constraints have been used to guarantee quality of service constraints at the optical layer in the model described in Gouveia et al. (2003).

Note that we can formulate the HMST using directed edges (arcs). That is, we can direct all the edges of the tree paths from the root node to every other node in the tree. The directed formulation directed formulation replaces every edge $\{i,j\}$ in the graph by the two (directed) arcs (i,j) and (j,i) and associates to each of these two arcs the cost of the original edge. We let A denote the set of directed arcs in the directed model. As noted in several recent works (see, for instance, Goemans (1994), Magnanti and Wolsey (1996) and Dahl (1998)), the idea of directing the problem has proven to be quite useful in solving many special cases of network design network design problems. Thus, we shall focus our discussion on directed formulations for the HMST. Notice also that as our formulations will consider the trees directed away from the root node, node 0, each edge $\{0,j\}$ is only replaced by one single arc $(0,j)$.

Throughout the chapter, for any integer linear programming model P, we let P_L denote its linear programming relaxation, $F(P)$ denotes the set of feasible solutions in P, and let $v(P)$ denote the optimal value of P. We shall also need to distinguish between sequences of arcs defining a path between two given nodes and which may or may not repeat nodes. Thus, an $(s\text{-}t)$-walk is a sequence of arcs $\{(i_1,j_1),\ldots,(i_k,j_k)\}$ such that $i_1 = s$, $j_k = t$ and $j_p = i_{p+1}$ for $p = 1,\ldots,k-1$. An $(s\text{-}t)$-path is an $(s\text{-}t)$-walk which does not repeat nodes. Sometimes, we shall use the terms *path* or *walk* when one of the nodes s and t are obvious from the context. A path (walk) with at most H hops is a H-path (H-walk). hop-constrained path

The remainder of this chapter is organized as follows. Section 19.2 presents a general framework for constructing formulations for the HMST. We discuss a generic formulation for the HMST, show that this formulation includes previously known formulations as special cases and give some evidence that the corresponding linear programming bound is quite good. Different ways (including Lagrangian relaxation, column generation and model reformulation) of computing a good approximation of the linear programming bound given by the model of Section 19.2 are surveyed and discussed in Sections 19.3 to 19.8. Although this chapter is mostly a survey, the material of Sections 19.4 and 19.8 is new.

19.2 A GENERAL FORMULATION FOR THE HMST

In this section we describe a general multicommodity flow formulation multicommodity flow formulation for the HMST. This formulation uses two sets of variables. Variables $X_{ij}((i,j) \in A)$ indicate whether the spanning tree contains the arc (i,j) and directed flow variables $Y_{ijk}((i,j) \in A; k \in V \setminus \{0\}; k \neq i)$ specify whether the unique path from the root node to node k traverses the arc (i,j). Let F_k be the set of all incidence vectors of $(0\text{-}k)$-H-paths. Moreover, we let $Y_{.k}$ denote the vector with components Y_{ijk} for $(i,j) \in A$.

Formulation MCF-scheme.

$$\text{minimize} \sum_{(i,j)\in A} c_{ij} X_{ij} \tag{19.1}$$

$$\text{subject to:} \sum_{i\in V\setminus\{j\}} X_{ij} = 1, \quad j \in V \setminus \{0\} \tag{19.2}$$

$$Y_{\cdot k} \in F_k, \quad k \in V \setminus \{0\} \tag{19.3}$$

$$Y_{ijk} \leq X_{ij}, \quad (i,j) \in A; k \in V \setminus \{0\} \tag{19.4}$$

$$X_{ij} \in \{0,1\}, \quad (i,j) \in A \tag{19.5}$$

Constraints (19.3) state that the solution must contain a H-path to each node $k \in V \setminus \{0\}$. Constraints (19.4) are the coupling constraints relating flow variables with arc variables. They state that if an arc (i,j) is in an H-path to node k then arc (i,j) is in the solution. Constraints (19.3) together with constraints (19.4) guarantee that the arc set defined by the X variables is connected and contains an H−path from the root node 0 to each node $k \in V \setminus \{0\}$. This, together with (19.2), guarantees that the solution is a directed spanning tree rooted at node 0.

The MCF-Scheme gives a general way of building a MCF-like formulation for the HMST. A valid formulation for the HMST is obtained when 19.3 are replaced with a set of linear constraints guaranteeing that the condition in (19.3) holds. One example is given in Gouveia (1996). The sets F_k are modeled by

$$\sum_{i\in V\setminus\{k\}} Y_{ijk} - \sum_{i\in V\setminus\{0\}} Y_{jik} = \begin{cases} -1, & j=0 \\ 0, & j\neq 0,k \\ 1, & j=k \end{cases} \quad j \in V \setminus \{0,k\} \tag{19.6}$$

$$\sum_{(i,j)\in A} Y_{ijk} \leq H \tag{19.7}$$

$$Y_{ijk} \in \{0,1\}. \tag{19.8}$$

Constraints (19.6) are the usual flow conservation constraints. Constraints (19.7) state that no more than H arcs are in the path between the root node and any other node. These inequalities have been used in a more general network design problem with hop constraints by Balakrishnan and Altinkemer (1992). The solutions defined by (19.6)–(19.8) are sets of arcs containing $(0\text{-}k)$-H-paths. Thus, the set defined by (19.6)–(19.8) is a relaxation of F_k. However, under the presence of the other constraints involved in MCF-Scheme, the feasible sets that strictly contain feasible paths are ruled out and we obtain a valid formulation for the HMST. We denote by MCF the formulation obtained in this way. The corresponding linear programming relaxation, MCFL, is obtained by replacing constraints (19.5) and (19.8) with the corresponding lower bound constraints. Computational results given in Gouveia (1996; 1998) indicate that the LP bound given by $v(\text{MCFL})$ still needs to be improved for instances with H small.

Clearly, the LP bound given by a model built in this way depends on the set of linear constraints used to characterize the sets F_k. The best LP bound is obtained with the ideal model where constraints (19.3) are replaced by the convex hull of each set

F_k (denoted by conv(F_k) in the remainder of the chapter). Let MCF* denote this ideal formulation. G. and Gouveia (2003) have presented a linear description of conv(F_k) for $H = 2$ and 3 and the corresponding MCF* models are discussed in Section 19.3. Unfortunately, describing conv(F_k) for all values of H in the space of the Y variables is a difficult problem as the corresponding optimization problem for general arc costs is NP-Hard (for negative arc costs the problem becomes equivalent to the minimum Hamiltonian-path problem). In fact, as suggested by the more recent work by G. et al. (2003), finding a complete linear description of conv(F_k) with the hop parameter H as small as 4 and involving the Y variables alone is already too complicated. Thus, writing MCF* explicitly does not seem to be the best option for computing $v(\text{MCF}_L^*)$ when $H > 3$.

In the next sections we discuss several ways of computing $v(\text{MCF}_L^*)$ or good approximations of it. A Lagrangian relaxation is discussed in Section 19.4. A column generation scheme based on the fact that it is quite easy to give a complete and extended characterization of conv(F_k) involving an exponential sized set of extra variables is suggested in Section 19.5. A different idea is to replace each F_k by a suitable relaxation that is, a relaxation which is easier to analyze and such that the value $v(\text{MCF}_L^*)$ is not altered when the replacement is done, this is the topic of Section 19.6.

19.3 MCF* FOR H = 2,3

Consider a $k \in V \setminus \{0\}$. G. and Gouveia (2003) have shown that for $H = 2$, conv(F_k) becomes completely described by

$$\sum_{j \in V} Y_{0jk} = 1 \tag{19.9}$$

$$Y_{0jk} = Y_{jkk}, \quad j \in V \setminus \{0, k\} \tag{19.10}$$

$$\sum_{i \in V} Y_{ikk} = 1 \tag{19.11}$$

and $Y_{ijk} > 0$ for all $(i, j) \in A (i = 0$ or $j = k)$. The result follows simply from the fact that these equalities and inequalities define the network flow system in the sub-graph consisting of the arcs leaving node 0 and arcs entering node k. We can, now, use these systems for all $k \in V \setminus \{0\}$ to write the formulation MCF* for $H = 2$:

Formulation MCF*(H = 2). Minimize (19.1) subject to constraints (19.2), (19.9–19.11) for all $k \in V \setminus \{0\}$, (19.5), (19.8) and

$$Y_{0jk} \leq X_{0j}, \quad (0, j) \in A; \; k \in V \setminus \{0\} \tag{19.12}$$

$$Y_{ijj} \leq X_{ij}, \quad (i, j) \in A. \tag{19.13}$$

A few observations permit us to remove the flow variables Y_{ijk} from this formulation. First, we observe that one of the constraint sets (19.9) or (19.11) is redundant (we remove (19.9) for all k). Second, notice that constraints (19.2) and (19.11) imply that constraints (19.13) are satisfied as equalities. Thus, the variables Y_{ijj} can be eliminated from the formulation. Constraints (19.10) are modified accordingly and constraints (19.11) (which become equal to constraints (19.2)) and constraints (19.13) are eliminated from the model. Combining the modified constraints (19.10) with (19.12) permits us to remove the variables Y_{0jk} yielding the following formulation involving only the variables X_{ij}.

Formulation LOC. Minimize (19.1) subject to constraints (19.2), (19.5) and

$$X_{0j} \geq X_{ik}, \quad (j,k) \in A. \tag{19.14}$$

The above formulation was studied in Dahl (1998). It is quite similar to the so-called strong model for the uncapacitated location problem. Several experiments with this formulation have shown that the optimal linear programming solution is integer, in many of the cases tested. A similar experience is known for the strong model for the uncapacitated location problem. In fact, it is not easy to generate randomly instances for which the previous formulation gives an optimal linear programming solution with some fractional valued variables. The way we have derived LOC shows that

Proposition 19.1 *When $H = 2$, the projection of $F(\text{MCF}_L^*)$ into the space defined by the variables X_{ij} is given by $F(\text{LOC}_L)$.*

For $H = 3$, G. and Gouveia (2003) have shown that complete linear description polyhedral characterization $\text{conv}(F_k)$ becomes completely described by (19.6), $Y_{ijk} > 0$ for all $(i, j) \in A$ and

$$Y_{jkk} \geq \sum_{i \in V \setminus \{0\}} Y_{ijk}, \quad j \in V \setminus \{0, k\}. \tag{19.15}$$

Constraints (19.15) state that if a feasible path P contains an arc $(i, j)(i \neq 0, j \neq k)$ then P contains the arc (j, k). We note that the *symmetric* constraints

$$Y_{0ik} \geq \sum_{j \in V \setminus \{k\}} Y_{ijk}, \quad i \in V \setminus \{0, k\}$$

are implied by (19.15) and the conservation flow constraints (19.6). We can, now, write the formulation MCF* for $H = 3$:

Formulation MCF*(H = 3). Minimize (19.1) subject to constraints (19.2), (19.6), (19.15) for all $k \in V \setminus \{0\}$, (19.5), and (19.8).

The advantage of the previous formulation is that it is compact and thus, we can try to use it together with any available MIP package to solve HMST instances with $H = 3$. One interesting point of research is to find a formulation with an equivalent linear programming relaxation bound and involving only the X variables.

As suggested before, finding an explicit representation of MCF* for $H > 3$ seems to be quite difficult. We consider, next, several alternative ways of computing $v(\text{MCF}_L^*)$ for such values of H.

19.4 LAGRANGIAN RELAXATION

Consider the Lagrangian relaxation Lagrangian relaxation derived from MCF by associating nonnegative multipliers to each constraint (19.4), i.e. $Y_{ijk} < X_{ij}$ for all $(i, j) \in A$

and $k \in V \setminus \{0\}$, and dualizing them in the usual Lagrangian way. For a given set of multipliers the Lagrangian problem can be decomposed into a single inspection subproblem and n hop-constrained path subproblems. As noted before, for general costs these path problems are hard to solve. However, the corresponding arc costs are nonnegative (they are given by the value of nonnegative multipliers) and thus, the subproblems can be efficiently solved (using the algorithm suggested in Section 19.6). The convexity result of Geoffrion (1974) states that the Lagrangian bound given with the optimal set of multipliers is equal to the optimal value of the linear programming model given by
Minimize (19.1) subject to

$$Y_{.k} \in \text{conv}(F_k), \quad k \in V \setminus \{0\}$$
$$X_{ij} \in \text{conv}\{(19.2) \text{ and } (19.5)\}.$$

As the convex hull of the integer solutions defined by (19.2) and (19.5) is described by (19.2) and nonnegativity constraints associated to the X_{ij} variables, the previous model is MCF_L^*.

Computational results reported in Gouveia (1996; 1998) show that the bounds given by this relaxation are, for small values of H, much better than the LP bounds given by the MCF model. However, the results also show that for most of the cases, a large number of iterations of the associated subgradient optimization method may be needed for obtaining a lower bound which is close to the corresponding theoretical best bound. In fact, it was the inefficiency of this method that has, in a certain way, motivated the formulation described in Section 19.6.

19.5 COLUMN GENERATION

Consider a $k \in V \setminus \{0\}$, let U_p be the binary variable associated to the p-th $(0\text{-}k)\text{-}H$-path (we assume that the $(0\text{-}k)\text{-}H$-paths are ordered in F_k) and let P_k represent the set of $(0\text{-}k)\text{-}H$-paths. Consider the following column generation formulation, denoted by Path(k),

$$\sum_{p \in P_k} U_p = 1 \tag{19.16}$$

$$U_p \in \{0, 1\}, \quad p \in P_k \tag{19.17}$$

and its LP relaxation, Path$(k)_L$, obtained by replacing (19.17) with the corresponding lower bound inequalities. Constraint (19.16) states that one and only one path from P_k has to be selected in any feasible solution of Path(k). The path variables and the original variables are related in the following way:

$$\sum_{p \in P_{ijk}} U_p = Y_{ijk}, \quad (i, j) \in A \tag{19.18}$$

where P_{ijk} denotes the set of $(0\text{-}k)\text{-}H$-paths using arc (i, j). G. and Gouveia (2003) have shown that the projection of $F(\text{Path}(k)_L)$ into the space defined by the Y_{ijk} variables under the linear transformation (19.18) is given by conv(F_k). Thus, $F(\text{Path}(k)_L)$ gives an extended description of conv(F_k).

If we replace in MCF-Scheme, constraints (19.3) by the constraint set defining the model Path(k) for each $k \in V \setminus \{0\}$ (the U variables are re-indexed with a second index k) and if we replace Y_{ijk} in (19.4) by the right-hand side of (19.18) (again the path variables are re-indexed by k) we obtain the following formulation for the HMST.

Formulation path. Minimize (19.1) subject to constraints (19.2), (19.16) and (19.17) for all $k \in V \setminus \{0\}$, (19.5), and

$$\sum_{p \in P_{ijk}} U_{pk} \leq X_{ij}, \quad (i,j) \in A; k \in V \setminus \{0\}. \tag{19.19}$$

The result by G. and Gouveia (2003) implies that

Proposition 19.2 *The projection of $F(\text{Path}_L)$ into the space defined by the X_{ij} and Y_{ijk} variables under the linear transformation (19.18) (defined for all k) is given by $F(\text{MCF}_L^*)$.*

As a consequence we have $v(\text{MCF}_L^*) = v(\text{Path}_L)$ (and $v(\text{MCF}_L) < v(\text{Path}_L)$). The Path model also suggests a different scheme based on column generation for computing an approximation of $v(\text{MCF}_L^*)$. The whole idea of the method is to obtain the value $v(\text{Path}_L)$ with a restricted Path model involving, hopefully, much fewer path variables. It will be interesting to develop such a method and compare its performance with the performance of the other methods described in this paper.

It is also interesting to compare the fact that $v(\text{MCF}_L) < v(\text{Path}_L)$ with what happens when we consider the minimum spanning tree model without hop constraints. It is well known that the linear programming relaxation of the MCF model (without the hop constraints) is equivalent to the linear programming relaxation of a Path model where all paths between the root node and every other node k are considered (that is, no hop limit is given). This result follows from equality (19.18) (modified in order to consider all paths) and also holds for more general uncapacitated network design problems (see, for instance, Martin (1999)). That is, formulating a problem with a large number of path-commodity variables is equivalent to formulating the problem with a polynomial sized set of arc-commodity variables. The interesting issue is to know whether in the context of the HMST we can find a compact multicommodity flow type model or a generalization of it, whose LP relaxation is equivalent to the LP relaxation of the hop-constrained Path model. We have already shown that adding the cardinality constraints (19.7) to the MCF model is not sufficient for producing such a model. In the next section, we show that it is possible to derive a hop-index disaggregation of the MCF model whose LP relaxation is equivalent to the LP relaxation of the Path model. The arc-commodity variables are disaggregated into arc-hop-commodity variables and the flow conservation constraints that in MCF are defined for each node and each commodity, are now defined for each node, each hop and each commodity.

19.6 A SHORTEST-PATH REFORMULATION

Consider the relaxation, W_k, of F_k such that each element in W_k is a $(0\text{-}k)$-H-walk (i.e., a $(0\text{-}k)$-H-path which may repeat nodes). Let MCFW* be the model obtained from MCF* by replacing conv(F_k) with conv(W_k). We first argue that $v(\text{MCF}_L^*) =$

$v(\text{MCFW}_L^*)$. Assume that we use the Lagrangian relaxation described in Section 19.4 in the model MCFW*. shortest-path reformulation As the arc costs for each constrained walk subproblem are nonnegative (this follows from the fact that the multipliers are nonnegative), the corresponding optimal solutions are still $(0,k)$-H-paths. Thus, for any set of multipliers the two Lagrangian bounds are equal and the same happens for the optimal set of multipliers. By Geoffrion's result we obtain that $v(\text{MCF}_L^*) = v(\text{MCFW}_L^*)$.

The reason for considering these relaxed subsets is that the optimization problem over each W_k can be solved in polynomial time (see below). This fact suggests that finding a description of $\text{conv}(W_k)$ should be much easier than finding a description of $\text{conv}(F_k)$ (note, however, that $\text{conv}(F_k) = \text{conv}(W_k)$ when $H < 4$ as walks are paths for these values of H). The case $H = 4$ has been analyzed by G. et al. (2003). Their results show that a description of $\text{conv}(W_k)$ for $H = 4$ on the space of the Y variables (although much simpler than a description of $\text{conv}(F_k)$) contains an exponential sized set of inequalities. Thus, attempting to compute $v(\text{MCF}_L^*)$ for $H = 4$ by solving the linear programming relaxation of the explicit MCFW* model for $H = 4$ is far from practical (note, however, that this does not rule out a cutting plane approach using the MCFW* model obtained in this way). The results by G. et al. (2003) suggest that characterizing $\text{conv}(W_k)$ gets even more complicated when we consider bigger values for the parameter H.

At first sight, it seems that we did not gain much by replacing $\text{conv}(F_k)$ with $\text{conv}(W_k)$. However, the optimization problem over each W_k can be modeled as an unconstrained shortest path problem in an expanded acyclic graph. hop-constraints: expanded graph This expanded graph is built by levels with the nodes of the original graph replicated H times, one for each hop. Using the standard network flow formulation for the unconstrained shortest path problem on the expanded graph yields an exact formulation for the original problem (this follows from the fact that all the extreme points of the corresponding linear programming relaxation are integer (see, for instance, Ahuja et al. (1993)) and thus we obtain a complete description of $\text{conv}(W_k)$. This formulation involves variables associated to the arcs of the expanded graph and thus, it is written with a set of variables which is different from the original set with one variable associated to each arc of the original input graph. Thus, this formulation gives an extended description of $\text{conv}(W_k)$ and as the size of the expanded graph is a polynomial function of the size of the input graph, the extended formulation is compact (the number of variables and constraints is a polynomial function of the size of the graph). This extended description of $\text{conv}(W_k)$ can then, be adequately combined with the previous MCF-Scheme formulation in order to produce a compact formulation for the HMST with a LP value equal to $v(\text{MCF}_L^*)$. These ideas have been developed in Gouveia (1998) for the HMST and for completeness they will be briefly described in this chapter. We note that the framework for this kind of shortest path reformulation has been proposed by Martin (1987) (see also Martin (1999) for a more general approach).

As noted before, the expanded graph $G_E = (V_E, A_E)$ for modeling the hop constrained walk problem between node 0 and a given node $k \in V \setminus \{0\}$ is built by levels with the nodes of the original graph replicated H times, one for each hop. The node set V_E is defined as

$$V_E = (0,0) \cup \{(i,h) : 1 \leq h \leq H-1 \text{ and } V \setminus \{0\}\} \cup (k,H).$$

A node (i,h) is associated to node i being visited in position h in the original graph (i.e., node i is the h-th node in the path). The arc set A_E is defined as

$$A_E = \{((0,0),(j,1)) : j \in V \setminus \{0\}\} \bigcup$$
$$\{((i,h),(j,h+1)) : (i,j) \in A, i \neq 0, i \neq k \text{ and } 1 < h < H-2)\} \bigcup$$
$$\{((i,H-1),(k,H)) : i \in V \setminus \{0\} \text{ and } i \neq k\} \bigcup$$
$$\{((k,h),(k,h+1)) : 1 < h < H-1)\}.$$

It is not difficult to see that paths from node $(0,0)$ to node (k,H) in G_E correspond to $(0,k)$-H-walks in the original graph G. Notice, also, that if a path in G_E includes an arc of the form $((k,h),(k,h+1))$, then the corresponding walk in the original graph G includes a loop at node k (which is then discarded in terms of the original graph) and thus, contains less than H arcs.

By associating a binary variable Z_{ijh} to arc $((i,h-1),(j,h))$ in G_E and by writing down the usual network flow conservation constraints for the path problem we obtain an extended and compact formulation for the hop-constrained problem in the original graph (note that the size of G_E is a polynomial function of the size of the original graph G). In terms of the original problem, the hop variables Z_{ijh} can be interpreted as indicating whether arc (i,j) is the h-th arc in the path from the root node to node k. The definition of the expanded graph guarantees that $h = 1$ when $i = 0$ (i.e., an arc leaving the root is always in position 1) and $h > 2$ when $i \neq 0$. Variables Z_{kkh} modeling self loops at the target node k are also considered in the model. Using these variables we can formulate the hop-constrained walk problems in the following way:

HopPath(k), k $\in$ V $\setminus$ {0}.

$$Z_{0il} - \sum_{j \in V} Z_{ij2} = 0, \quad (0,i) \in A \tag{19.20}$$

$$\sum_{j \in V} Z_{jih} - \sum_{j \in V \setminus \{0\}} Z_{ij,h+1} = 0, \quad i \in V \setminus \{0\}; h = 2, \ldots, H-1 \tag{19.21}$$

$$\sum_{j \in V} Z_{jiH} = 1, \tag{19.22}$$

$$Z_{0j1} \in \{0,1\}, \quad (0,j) \in A \tag{19.23}$$

$$Z_{ijh} \in \{0,1\}, \quad (i,j) \in A; i \neq 0; h = 2, \ldots, H \text{ or } i = j = k. \tag{19.24}$$

Constraints (19.20) state that arc $(0,i)$ is the first arc in the path to node k if and only if one arc in position $h = 2$ emanates from node i. Constraints (19.21) state that an arc enters node i in position h if and only if another arc emanates from this node in position $h+1$. Constraints (19.22) state that one arc enters node k in position H. The remaining inequalities define the binary conditions on the variables Z_{ijh}. It is not difficult to see that the constraint set above implies that for each h, we should have one and only one variable Z_{ijh} equal to 1. When the Z_{ijh} variables with value equal to 1 represent a $(0\text{-}k)$-walk with less than H hops, we must have $Z_{kkh} = 1$ for some $h < H$ (i.e., self loops at node k) in order to obtain exactly H variables with value equal to 1. The linear programming relaxation of HopPath(k), HopPath$(k)_L$, is obtained by replacing (19.23)–(19.24) with the corresponding upper bound and lower bound inequalities. hop-indexed model The *shortest path system in expanded graph* argument used before guarantees that the polytope $F(\text{HopPath}(k)_L)$ satisfies the integrality property. The hop-index variables and the original variables are related in the following:

$$Z_{0j1} = Y_{0j}, \quad (0,j) \in A \tag{19.25}$$

$$\sum_{h=2}^{H} Z_{ijh} = Y_{ij}, \quad (i,j) \in A; i \neq 0 \tag{19.26}$$

The fact that $F(\text{HopPath}(k)_L)$ is integral permits us to show that for each k, the projection of $F(\text{HopPath}(k)_L)$ into the space defined by the Y_{ijk} variables under the linear transformation (19.25)–(19.26) is given by conv(W_k). That is, $F(\text{HopPath}(k)_L)$ gives an extended description of conv(W_k).

We can now obtain a new formulation for the HMST problem. For each k, the set of inequalities defined by F_k is replaced in MCF-Scheme by the set of inequalities defined by HopPath(k) (each Z variable is re-indexed with a fourth index k). Furthermore, we replace Y_{ijk} in each inequality (19.4) by the right-hand side of (19.25) or (19.26) (again the hop variables are re-indexed by k). The new formulation for the HMST is given by

Formulation HopMCF. Minimize (19.1) subject to constraints (19.2), (19.20)–(19.22) and (19.23)–(19.24) for all $k \in V \setminus \{0\}$, (19.5), and

$$Z_{0jlk} \leq X_{0j}, (0,j) \in A; k \in V \setminus \{0\} \tag{19.27}$$

$$\sum_{h=2}^{H} Z_{ijhk} \leq X_{ij}, (i,j) \in A; i \neq 0; k \in V \setminus \{0\}. \tag{19.28}$$

The fact that $F(\text{HopPath}(k)_L)$ gives an extended description of conv(W_k) guarantees that projection

Proposition 19.3 *The projection of $F(\text{HopMCF}_L)$ into the space defined by the X_{ij} and Y_{ijk} variables under the linear transformation (19.25)–(19.26) (defined for all k) is given by $F(\text{MCFW}_L^*)$.*

As $v(\text{MCF}_L^*)$ is not changed when conv(F_k) is replaced with conv(W_k), we obtain

$$v(\text{HopMCF}_L) = v(\text{Path}_L) = v(\text{MCF}_L^*) \geq v(\text{MCF}_L).$$

The equality $v(\text{HopMCF}_L) = v(\text{Path}_L)$ states that hop-limiting the paths in the Path model corresponds to a certain type of hop-index reformulation of the MCF model. Although we have established this equivalence for the HMST, it is not difficult to see that the same arguments can be used for more general problems. Several related problems also involving hop constraints and modeled with a similar Path formulation have been described in the literature (see, for instance, Alevras et al. (1996). In such cases, the LP relaxation was solved by column generation. For small values of H, the more compact hop-indexed model might be worth using.

We also note that the same kind of hop-indexed reformulation has been used in Pirkul and Soni (2003) and Gouveia et al. (2003) for more general telecommunication network design problems. The improvements reported in these two papers by using that reformulation are quite similar to the improvements obtained for the HMST (see Gouveia (1998) and Section 19.9).

The computational results given in Gouveia (1998) suggest that the LP bound given by the HopMCF model is quite often equal to the integer optimal value. The reported CPU times also suggest that it is worthwhile using the hop model when the value of H is small. On the other hand, we should also emphasize that if $v(\text{HopPath}_L)$ is not the optimal integer solution, then CPLEX might take a considerable amount of time to obtain the optimal integer solution, even if the LP bound is rather close to the optimal integer value.

Another disadvantage of using the linear programming relaxation of HopMCF is that one might get in trouble if more dense instances, instances with bigger values of H or instances with a bigger number of nodes are solved. In such cases, the size of the corresponding linear programming model might lead to huge computer storage requirements or to excessive CPU times. It is also important to state that for these instances, the Lagrangian relaxation described in Section 19.4 performs poorly (mainly due to the large number of multipliers). In the next section, we discuss a different Lagrangian relaxation method that is derived from the HopMCF model.

19.7 LAGRANGIAN RELAXATION REVISITED

There are two main alternatives in deriving a Lagrangian relaxation in a network flow based formulation. One of the alternatives has been described in Section 19.4. The coupling constraints are dualized and the network flow structure is maintained in the relaxed problem. In the second alternative, the flow conservation constraints are dualized and the relaxed problem is usually decomposed into $|A|$ subproblems, one for each arc. Although it can be proved that in general, both types of relaxations provide the same theoretical dual Lagrangian bound, computational results reported in the past for several problems (see, for instance, Beasley (1984) for the Steiner tree problem, and Gendron et al. (1999) for the capacitated network design problem) state that in general, the network flow based relaxation provides better results than the second relaxation. More precisely, for a fixed number of iterations of the subgradient optimization procedure, the first relaxation provides, quite often, a better bound than the second relaxation. These results do not necessarily imply that one should ignore the second type of relaxation. For instance, Holmberg and Yuan (1996) show that the second relaxation provides valuable information for a branch-and-bound procedure. Crainic et al. (2001) have shown that with an adequate *tuning* of the subgradient optimization procedure, the second relaxation is a viable alternative to the network flow-based relaxation.

With respect to the HopMCF formulation, the network flow-based relaxation is nothing else than the Lagrangian relaxation described in Section 19.4. The other relaxation has been described and tested in Gouveia and Requejo (2001). This relaxation is obtained by attaching Lagrangian multipliers to constraints (19.20)–(19.22) for all $k \in V \setminus \{0\}$, and dualizing them in the usual Lagrangian way. For further details we refer the reader to the original paper. The results of Gouveia and Requejo (2001) (and also confirmed in our experiment reported in Section 19.9) show that this relaxation provides a good alternative to the previous methods for the type of instances mentioned at the end of the previous section.

19.8 THE "JUMP" FORMULATION AND CONSTRAINT GENERATION

In Section 19.3 we have shown for $H = 2$, how to obtain a formulation involving only the X_{ij} variables and which has a linear programming bound equal to $v(\text{MCF}_L^*)$. In this section, we show how to obtain formulation for $H = 3$ satisfying the same properties. We also show that it is far from easy to obtain such a formulation for $H > 3$. The advantage of such a formulation is that it uses only one set of variables (although it involves an exponentially sized set of variables).

We suggest, next, a way of deriving such a formulation. Consider the relaxation D_k, of F_k (and of W_k) where $D_k = \text{conv}(F_k) + R_+^{|A|}$, that is D_k is the dominant of H-$(0\text{-}k)$-paths. Let MCFD* be the model obtained from MCF* by replacing $\text{conv}(F_k)$ by D_k for each k. Consider the optimization problem over D_k. For nonnegative arc costs, the corresponding optimal solution is a path. Thus, we can deduce that $v(\text{MCF}_L^*) = v(\text{MCFD}_L^*)$ in the same way that we have shown that $v(\text{MCF}_L^*) = v(\text{MCFW}_L^*)$. By providing a complete description of D_k and by substituting adequately in MCF-Scheme, we can obtain a new formulation for the HMST whose linear programming bound is equal to $v(\text{MCF}_L^*)$.

Unfortunately, obtaining such a description for all H appears to be difficult as well. Consider the well known cut inequalities cut inequalities

$$\sum_{(p,q) \in [V \setminus S, S]} Y_{pq}^k \geq 1, \quad S \subseteq V; 0 \notin S; k \in S \tag{19.29}$$

where $S_c = V \setminus S$. Inequalities (19.29) state that each $(0\text{-}k)$-path (or walk) must intersect any cut of the form $[V \setminus S, S]$ with $0 \in V \setminus S$ and $k \in S$. It is well known that (19.29) together with the trivial nonnegativity constraints give a complete description of the dominant of the incidence vectors of $(0\text{-}k)$-paths (see, for instance, Nemhauser and Wolsey (1988)). For the case of hop-constrained paths we need to consider a similar type of inequality, that is an inequality which involves, on the left-hand side, sets of arcs intersecting all feasible H-$(0\text{-}k)$-paths. Such an inequality has already been described by Dahl (1999) for the undirected case. Let $S_0, S_1, \ldots, S_{H+1}$ be node-disjoint sets defining a partition of the whole node set V such that each subset is nonempty, $S_0 = \{0\}$ and $S_{H+1} = \{k\}$. We then define $J = J(S_0, S_1, \ldots, S_{H+1}) = \bigcup_{[i+1 < j]} [S_i, S_j]$, where $[S_i, S_j]$ is the set of arcs $\{(u,v) \in A : u \in S_i, v \in S_j\}$. Observe that the sets $[S_i, S_j]$ are pairwise disjoint. We call such set J a $(0\text{-}k, H)$-jump. Let $J_{(0-k,H)}$ denote the set of all $(0\text{-}k, H)$-jumps. Consider, now, the following set of inequalities, the jump inequalities, jump inequalities

$$\sum_{(p,q)\in J} Y_{pq}^k \geq 1, \quad J \in J_{(0-k,H)}. \tag{19.30}$$

To see that (19.30) are valid for D_k and any H we consider the following argument taken from Dahl (1999). Assume we place all the nodes in $H+2$ consecutive layers where the i-th layer corresponds to the nodes in S_i. Assume also that a given $(0-k)$-H-path does not contain any arc in a given jump J. Then, the path would contain one arc in each of the (pairwise disjoint) arc sets $[S_i, S_{i+1}]$ and then it would contain at least $H+1$ arcs which is a contradiction. Thus, any $(0-n)$-H-path must have at least one arc in each jump J. Loosely speaking, any $(0-k)$-H-path needs to make a jump somewhere from a node set S_i to one of the sets $S_{i+2}, \ldots, S_{H+1}$.

The following result is a modification for the directed case of a result given in Dahl (1999):

Proposition 19.4 *Let F be a subset of arcs contained in A. Then F contains a path from 0 to k with at most H hops if and only if it intersects every 0-k cut and every $(0$-$k, H)$ jump.*

Let Jump(k) denote the set of solutions defined by (19.29), (19.30), and nonnegativity constraints on the variables. The previous proposition implies that Jump$(k) \supseteq D_k$.

The issue, now, is to know whether the inclusion is strict or not. G. et al. (2003) have noted that a strict equality holds when $H = 2$ or 3. This statement is a simple adaptation for the directed case of the results given by Dahl (1999). Unfortunately, G. et al. (2003) have also shown that describing D_k for $H > 3$ becomes much more difficult. They showed that for $H = 4$, a complete linear description of D_k contains exponential sized sets of constraints of the form

$$\sum_{(i,j)\in C} a_{ij} Y_{ij} \geq r \tag{19.31}$$

for adequate arc sets C and where r and a_{ij} are positive integers. The parameter r can be as big as $n - H + 1$ and the corresponding inequalities also have coefficients a_{ij} that cover all the integers between 1 and r.

Proposition 19.1 shows that if we replace in MCF-Scheme, F_k by Jump(k) we obtain a valid formulation for the HMST. The fact that Jump$(k) \supseteq D_k$ shows that this formulation, denoted by Jump-MCF, may provide a weaker lower bound when $H > 3$. The result by Dahl (1999) shows that Jump-MCF has the same linear programming bound as MCF* when $H = 3$. The interesting point is that we can obtain, from Jump-MCF, a formulation for the HMST which involves only the X_{ij} variables and whose linear programming bound is equal to the linear programming bound given by Jump-MCF (and by transitivity is equal to the LP bound of MCF* when $H = 3$). To see this, notice that the Y_{ijk} variables arise only in two types of inequalities: $\geq$ inequalities which are the cut inequalities (19.29) and the jump inequalities (19.30) and $\leq$ inequalities which are the linking constraints (19.4). Thus, by using Fourier-Motzkin elimination (see Martin (1999)), these variables can be eliminated and the resulting system is the linear programming feasible set of the following formulation:

Formulation Nat-Hop. Minimize (19.1) subject to (19.2), (19.5) and

$$\sum_{(p,q)\in [V\setminus S,S]} X_{pq} \geq 1, \quad S \subseteq V; 0 \notin S \qquad (19.32)$$

$$\sum_{(p,q)\in J} X_{pq} \geq 1, \quad k \in V \setminus \{0\}; J \in J_{(0-k,H)}. \qquad (19.33)$$

The previous arguments show that the projection of the set of feasible solutions of Jump-MCFL into the space of the X_{ij} variables is given by the set of feasible solutions of the Nat-Hop formulation. Proposition 19.29 and the arguments used to derive Nat-Hop show that

Proposition 19.5 $v(\text{MCF}_L^*) = v(\text{Nat-Hop}_L)$ when $H = 3$ and $v(\text{MCF}_L^*) \geq v(\text{Nat-Hop}_L)$ when $H > 3$.

The Nat-Hop model contains an exponential number of constraints. However, it can be seen that for $H = 2$ many of them are redundant and the model becomes equivalent to the LOC model described in Section 19.3 (for simplicity, we omit this equivalence proof). Albeit having a LP relaxation which is weaker than the LP relaxation of the MCF* formulation for $H > 3$, this natural formulation, Nat-Hop, may be worth investigating from a computational point of view, for instances where H is not small, because it contains only one variable for each arc in the input graph. This involves, of course, devising separation routines for the jump inequalities if we use an LP-based cutting plane approach.

Note also that as the convex hull of spanning trees without hop constraints rooted at node 0 is completely described by (19.2), (19.32), and nonnegativity constraint on the X_{ij} variables. Then, a Lagrangean relaxation scheme where the relaxed problem is a directed spanning tree and the constraints being penalized are the jump constraints (19.33), is also worth investigating. As the relaxed solutions are integer, separation of the jump inequalities can be done in polynomial time (although, at this time, it is not yet clear what is the best strategy for separating them). Note also that as there are too many jump constraints, an implicit generation scheme as devised by Gavish (1985) for the capacitated minimum spanning tree problem, must be implemented.

Note also that for $H > 3$, the linear programming relaxation of the Nat-Hop formulation can be improved by using some of the inequalities (19.31) rewritten with the X variables. These inequalities can be used either in a LP-based approach (needing, of course, a separation routine) or in a Lagrangean fashion as described in the previous paragraph.

19.9 COMPUTATIONAL RESULTS

In this section we compare some of the methods described in the previous sections. These comparisons have been reported before in different papers (with exception of the results given by the model of Section 19.3 for $H = 3$) but using different computers. With the results of this section, we present the results taken from these papers (and new ones, too) in the same computer platform, permitting a better comparison between the different methods.

Table 19.1 Size of reduced instances.

$|V|$	40	80
TC	60.6%	83.4%
TE	90.6%	96.7%
TR	72.2%	86.6%

We used three 40-node instances with roughly 400 edges and three 80-node instances with roughly 800 edges. Each group contains two Euclidean instances and one random instances, instances TR. Two locations for the root are considered for the Euclidean instances, one with the root located in the center of the grid, instances TC, and the other with the root located on a corner of the grid, instances TE. The hop parameter H was set to 3, 4 and 5 in every case.

To reduce the size of each instance, we have used the following simple arc elimination test (see Gouveia (1996)). If $c_{ij} > c_{0j}$, then any optimal solution does not use arc (i, j) and if $c_{ij} = c_{0j} (i \neq 0)$, then there is an optimal solution without arc (i, j). This means that arc (i, j) can be eliminated whenever $c_{ij} \geq c_{0j}$. This arc elimination test is applied to every instance before solving the Lagrangian or the linear programming relaxations. Table 19.9 shows, for each instance, the percentage of number of arcs still remaining in each instance, after the elimination test was performed. Note that the test is much more effective when applied to instances TC rather than to instances TE or TR and more effective to instances TR rather than to instances TE. This means that the reduced instances TE are larger than the reduced instances TC or TR suggesting that the TE instances will be much more difficult to solve than the remaining instances.

The computational results were obtained on a PC Pentium IV, 2.4 GHz with 768 Mb of RAM. We used the callable routine CPLEX 7.1 to obtain the linear programming bounds and the optimal integer solutions of the models tested. We first compare the performance of the models MCF (described in Section 19.2), MCF* (for $H = 3$ and described in Section 19.3) and HopMCF (described in Section 19.6). These results are shown in Table 19.1. The first column, denoted by Prob, identifies the type of instance (TC, TE or TR) and gives the number of nodes of the corresponding graph. The second column gives the value of H. In the third column, denoted by OPT we give the value of the corresponding optimal solution. The next three columns give the results obtained by the three models. For each value of H, we indicate the linear programming bound (on top) and two values a and b following the format $(a + b)$ which are given below the lower bound value. The value a indicates the time needed to solve the linear programming relaxation while the value b indicates the time needed to obtain the integer optimum.

Table 19.2 Comparing the linear and integer programming models. (m) means that the instance was not solved due to memory requirements. (a) indicates that the associated value is an upper bound.

Prob, $|V|$	H	OPT	MCF$_L$	MCF*3$_L$	HopMCF$_L$
TC, 40	3	566	559.1 (26+2949)	566.0 (1+0)	566.0 (1+0)
	4	519	509.2 (10+4384)		517.5 (6+4)
	5	496	488.1 (10+1329)		494.0 (13+164)
TE, 40	3	710	672.3 (183+ >2 days)	701.0 (23+1301)	701.0 (21+1217)
	4	625	602.7 (225+ >2 days)		619.5 (264+28494)
	5	581	568.5 (262+38196)		578.5 (1041+32038)
TR, 40	3	219	198.2 (45+2150)	219.0 (2+0)	219.0 (1+0)
	4	176	164.5 (39+2294)		176.0 (4+0)
	5	155	149.2 (32+3217)		155.0 (9+0)
TC, 80	3	1054	Not run	1054.0 (57+1)	1054.0 (27+1)
	4	967	Not run		964.5 (244+4390)
	5	918	Not run		913.1 (774+41229)
TE, 80	3	1808	Not run	1794.1 (390+29520)	1794.1 (210+75555)
	4	1568	Not run		1549.2 (5207+ >2 weeks)
	5	1751 (a)	Not run		1422.6 (29812+(m))
TR, 80	3	208	Not run	208.0 (3+1)	208.0 (3+1)
	4	180	Not run		180.0 (45+2)
	5	164	Not run		164.0 (112+3)

The reported results indicate the advantage of using HopMCF rather than MCF. Probably, the most surprising fact is that for the values of H tested, solving the linear programming relaxation of the disaggregated model HopMCF takes much less time than solving the linear programming relaxation of the models MCF and MCF* (for $H = 3$). In fact, for some of the instances tested, model HopMCF permits us to solve the integer problem in much less time than the linear programming relaxation of MCF is solved. With respect to the model MCF* for $H = 3$ the results are not conclusive. It appears that the performance of the less compact model is not superior to the performance of the hop-indexed model. The TE instances appear to be more difficult to solve. For these instances the LP bound given by the model HopMCF is not as close the integer optimum as it is for the two other sets of instances. The TE instance with 80 nodes and $H = 5$ is still unsolved. The same instance with $H = 4$ was very hard to solve. We note that this in this case, no upper cutoff with value less than 1600 have been generated by CPLEX have been generated after 10 days of computations. We believe that this problem would become easier to solve if somehow (perhaps, with the help of metaheuristic techniques) we could have provided a good upper bound value as a cutoff for the program.

The two Lagrangian relaxations described in Sections 19.4 and 19.7 (and denoted by Rel4 and Rel7 respectively) together with the corresponding subgradient optimization subgradient optimization method were implemented using the programming language FORTRAN. Different rules for initializing and updating the Lagrangean multipliers can be used for implementing the iterative subgradient optimization method proposed in Held et al. (1974). We have already referred that, usually, the best rules depend on the problem or instance being solved and for many cases it is difficult to find a set of rules that produces good lower bounds in a reasonable amount of time. Usually, the scalar used in the definition of the step size for updating the multipliers is recommended to be between 0 and 2 (see Held et al. (1974)). In our procedures we noticed that the rate of convergence could be significantly improved if higher values were tried. This scalar was halved whenever the lower bound was not improved after 15 iterations of the subgradient optimization method. The maximum number of iterations for the subgradient optimization methods were fixed to 5000.

The results comparing these relaxations are given in Table 19.2. The first three columns are as in Table 19.1. The fourth column gives the value of the linear programming relaxation of MCF and the fifth column gives the value of the linear programming relaxation of HopMCF (which give the theoretical best lower bound obtained by the Lagrangian relaxations). Each entry associated to an instance and to a relaxation contains three values: (i) the best lower bound (rounded up to the next integer when not integer); (ii) the number of iterations used by the associated subgradient optimization procedure to obtain the reported lower bound and (iii) the CPU time (in seconds) needed to obtain the best lower bound. The last column reports the lower bounds given by the LP relaxation of model HopMCF which are equal to the theoretical best bound given by the Lagrangian relaxations.

The reported results indicate that the lower bounds obtained by the Lagrangean relaxation Rel7 are a lot better than the lower bounds obtained by the Lagrangean relaxation Rel4. Notice that only $O(n2H)$ multipliers are involved in Rel7 while $O(n|A|)$

Table 19.3 Computational results for the Lagrangean problems.

Prob, $|V|$	H	OPT	Rel4 Value	Rel4 Iter	Rel4 Time	Rel7 Value	Rel7 Iter	Rel7 Time	$HMCF_L$
TC, 40	3	566	563	4598	66	566	241	8	566.0
	4	519	510	4401	72	517	4450	264	517.5
	5	496	488	4751	85	494	2542	225	494.0
TE, 40	3	710	690	4500	66	700	2995	111	701.0
	4	625	600	4951	82	616	4999	304	619.5
	5	581	556	4701	87	573	4744	520	578.5
TR, 40	3	219	213	4350	62	219	594	20	219.0
	4	176	172	4850	78	176	533	33	176.0
	5	155	152	3251	58	155	892	81	155.0
TC, 80	3	1054	1003	4950	1570	1051	3900	1488	1054.0
	4	967	896	4949	1638	960	3838	2157	964.5
	5	918	852	4801	1652	906	4794	3566	913.1
TE,80	3	1808	1665	4951	1582	1788	4950	1968	1794.1
	4	1568	1388	5000	1673	1508	4499	2682	1549.2
	5	1751 (a)	1276	4901	1708	1381	4397	3388	1422.6
TR,80	3	208	204	4744	1499	208	129	50	208.0
	4	180	171	4048	1336	180	590	339	180.0
	5	164	158	3500	1207	164	625	470	164.0

multipliers are involved in Rel4. However, the CPU times produced by Rel7 are, in general, greater than the CPU times produced by Rel4. The reason for this is that the CPU time used in each iteration of the relaxation Rel7 is substantially greater than the CPU time used in each iteration of the relaxation Rel4. At first sight this seems to be unexpected. However, our explanation for this is that variables with four indexes are involved in Rel7 and the computer programs implemented for solving the subproblems associated with Rel7 involve four nested cycles (notice that only three nested cycles are needed in the other relaxation).

19.10 CONCLUSIONS

In this chapter, we have presented a general framework for modeling the HMST which includes several previously known formulations as a special case. We have presented different ways of computing a theoretical lower bounding value associated to the previous formulation. We note that the ideas presented here can be used for variations of the HMST. For instance, Gouveia and Magnanti (2003) (see, also, Achuthan et al. (1992))

have noted that simple modifications of the HMST can be used to model the more general diameter constrained spanning tree problem where an hop limit is imposed on the number of edges in the tree between any node pair. The hop-indexed reformulation was also used with success in their work. As pointed out by Gouveia (1998) and Gouveia and Magnanti (2003) minor modifications to the models described before are needed to handle the Steiner tree version of the HMST (see Voss (1999)). For completeness, we note that Voss (1999) discusses several variations of a tabu search algorithm, for obtaining upper bounds for the Steiner version of the HMST.

Bibliography

N. R. Achuthan, L. Caccetta, P. Caccetta, and J. F. Geelen. Algorithms for the minimum weight spanning tree with bounded diameter problem. In et al. P. H. Phua, editor, *Optimization Techniques and Applications*, volume 1, pages 297–304. World Scientific, 1992.

R. Ahuja, T. Magnanti, , and J. Orlin. *Network flows: Theory, algorithms and applications*. Prentice Hall, 1993.

D. Alevras, M. Grötschel, and R. Wessäly. A network dimensioning tool. Technical Report SC 96–49, Konrad-Zuse-Zentrum für Informationstechnik, 1996.

A. Balakrishnan and K. Altinkemer. Using a hop-constrained model to generate alternative communication network design. *ORSA Journal on Computing*, 4:192–205, 1992.

J. Beasley. An algorithm for the Steiner problem in graphs. *Networks*, 14:147–159, 1984.

T. Crainic, A. Frangioni, and B. Gendron. Bundle-based relaxation methods for multicommodity capacitated fixed charge network design problems. *Discrete Applied Mathematics*, 112:73–99, 2001.

G. Dahl. "The 2-hop spanning tree problem. *Operations Research Letters*, 23:21–26, 1998.

G. Dahl. Notes on polyhedra associated with hop-constrained paths. *Operations Research Letters*, 25:97–100, 1999.

Dahl G., N. Foldnes, and L. Gouveia. A note on hop-constrained walk polytopes. Technical Report 1/2003, Centro de Investigação Operacional, Faculdade de Ciências da Universidade de Lisboa, 2003.

Dahl G. and L. Gouveia. On the directed hop-constrained shortest path problem. *Operations Research Letters*, 2003. To appear.

B. Gavish. Augmented Lagrangean based algorithms for centralized network design. *IEEE Transactions on Communications*, COM-33:1247–1257, 1985.

B. Gendron, T. Crainic, and A. Frangioni. Multicommodity capacitated network design. In B Sansò and P. Soriano, editors, *Telecommunications network planning*. Kluwer Academic Publishers, 1999.

A. Geoffrion. Lagrangian relaxation for integer programming. *Mathematical Programming Study*, 2:82–114, 1974.

M. Goemans. The Steiner polytope and related polyhedra. *Mathematical Programming*, 63:157–182, 1994.

L. Gouveia. Using the Miller-Tucker-Zemlin constraints to formulate a minimal spanning tree problem with hop constraints. *Computers and Operations Research*, 22: 959–970, 1995.

L. Gouveia. Multicommodity flow models for spanning trees with hop constraints. *European Journal of Operational Research*, 95:178–190, 1996.

L. Gouveia. Using variable redefinition for computing lower bounds for minimum spanning and Steiner trees with hop constraints. *INFORMS Journal on Computing*, 10:180–188, 1998.

L Gouveia and T. Magnanti. Network flow models for designing diameter constrained spanning and Steiner trees. *Networks*, 41:159–173, 2003.

L. Gouveia, P. Patrício, A. e Sousa, and Valadas R. MPLS over WDM network design with packet level QoS constraints based on ILP models. In *Proceedings of IEEE INFOCOM 2003*, April 2003.

L. Gouveia and C. Requejo. A new Lagrangian relaxation approach for the hop-constrained minimum spanning tree problem. *European Journal of Operational Research*, 132:539–552, 2001.

M. Held, P. Wolfe, and H. Crowder. Validation of subgradient optimization. *Mathematical Programming*, 6:62–88, 1974.

K. Holmberg and D. Yuan. A Lagrangean heuristic Based branch-and-Bound approach for the capacitated network design problem. Technical Report LiTH-MAT-R-1996-23, Department of Mathematics, Linköping Institute of Technology, 1996.

L. LeBlanc, J. Chifflet, and P. Mahey. Packet routing in telecommunication networks with path and flow restrictions. *INFORMS Journal on Computing*, 11:188–197, 1999.

T. Magnanti and L. Wolsey. Optimal trees. In *Network models*, volume 7 of *Handbooks in Operations Research and Management Science*, pages 503–615. Elsevier, 1996.

P. Manyem and M. Stallmann. Some approximation results in multicasting. Technical report, North Carolina State University, 1996.

R. K. Martin. Generating alternative mixed-integer programming models using variable redefinition. *Operations Research*, 35:820–831, 1987.

R. K. Martin. *Large scale linear and integer optimization: A unified approach*. Kluwer Academic Publishers, 1999.

G. L. Nemhauser and L.A. Wolsey. *Integer and combinatorial optimization*. John Wiley & Sons, 1988.

H. Pirkul and S. Soni. New formulations and solution procedures for the hop constrained network design problem. *European Journal of Operational Research*, 148: 126–140, 2003.

S. Voss. The Steiner tree problem with hop constraints. *Annals of Operations Research*, 86:271–294, 1999.

K. Woolston and S. Albin. The design of centralized networks with reliability and availability constraints. *Computers and Operations Research*, 15:207–217, 1988.

20 LOCATION PROBLEMS IN TELECOMMUNICATIONS

Darko Skorin-Kapov[1], Jadranka Skorin-Kapov[2], and Valter Boljunčić[3]

[1] School of Business
Adelphi University
Garden City, NY, 11530 USA
skorin@adelphi.edu

[2] College of Business
State University of New York at Stony Brook
Stony Brook, NY, 11794-3775 USA
jskorin@notes.cc.sunysb.edu

[3] Faculty of Economics and Tourism "Dr. Mijo Mirković"
Univesity of Rijeka
Pula, Croatia
vbolj@efpu.hr

Abstract: Location optimization is an important problem in design and usage of telecommunication networks. In different forms it appears in wire-based and in wireless networks, during unicast and multicast usage, in static, as well as in dynamic, networks. We provide an application oriented survey of recent results in location optimization pertaining to telecommunications networks, addressing problems such as converter, splitter, and amplifier placement in wavelength division multiplex (WDM) optical networks, concentrator location in designing access/backbone networks, base station location in Universal Mobile Telecommunication System (UMTS), and location management in mobile (cellular and ad hoc) networks.

Keywords: Location management, telecommunications, WDM optical networks, converter placement, amplifier placement, splitter placement, concentrator location, UMTS base station location, radio location.

20.1 INTRODUCTION

Telecommunication networks present rich application areas for different aspects of location optimization. In general, location problems in telecommunications are NP-hard. However, for some special cases it is possible to devise polynomial algorithms. This paper provides an application oriented survey of recent work addressing location issues in different types of telecommunication networks. Specifically, we address the following problems in wire-based networks: converter, amplifier, and splitter placement in wavelength division multiplex (WDM) optical networks, and concentrator location in designing access networks. Regarding wireless networks we discuss base station location in Universal Mobile Telecommunication System (UMTS), and some location management issues in mobile (cellular and ad hoc) networks.

The taxonomy of location problems presented in this survey is given in Table 20.1.

Table 20.1 Some location problems in telecommunications

Network type		Usage: unicast	Usage: multicast
Static	Wire-based	Converter placement(light paths) Amplifier placement Concentrator location	Converter placement(light trees) Splitter placement
	Wireless	UMTS base stations locations	—
Dynamic	Wire-based	Converter placement (with min blocking probability)	Converter placement (with min blocking probability)
	Wireless	Location position	Mobile multicast

The most paradigmatic location problem in telecommunications is probably the concentrator location problem, which has been studied extensively. (See the recent survey in Gourdin et al. (2002).) However, proliferation of telecommunication networks and new technologies led to new models and, consequently, to new location problems. On a general level, we can classify telecommunication networks as wire-based and wireless (mobile). Regarding wire-based networks there are optical networks utilizing fiber links and 'classical' telephone/cable networks using coaxial cables.

With respect to optical networks, optical fibers and WDM make it possible to superimpose a variety of virtual topologies on an existing physical topology. The use of different wavelengths in a fiber led to consideration of the converter placement problem: when and where in the network do we need to switch from one wavelength to another in order to optimize a certain criterion? Further, the need to maintain a consistent quality of a transported signal over long fiber links led to the problem of amplifier placement in an optical network.

The situation gets even more complicated when a group of users needs to communicate simultaneously, giving rise to multicast sessions. The problem, then, is to decide on placement of splitter nodes allowing for simultaneous transmission to all destination nodes.

Due to the cost of fiber links and due to the existing cable networking, the access part of the network is usually still based on non-optical technology (twisted pair or coaxial cable), while the backbone is fiber based. In such context arise problems related, but not identical, to the classical concentrator location problem.

Wireless networks create some specific location problems. In addition to locating base stations (which is related to concentrator location), cellular networks necessitate location area planning. Mobility of users in mobile networks presents a challenge for calculating precise location position and for delivering location-based services.

In this survey we will present some recent results in the above mentioned contexts.

20.2 LOCATION OPTIMIZATION IN WDM OPTICAL NETWORKS

Optical fibers propagate light and, therefore, are much faster transmission media than copper cables propagating electrons. An optical transmission system connects transmitters and receivers to the optical transmission medium. An electrical input signal arriving to a transmitter modulates a light source producing a beam of light that is transmitted over a fiber link, reaching a receiver whereby an optical detector converts it again to the electrical signal. (See Murthy and Gurusamy (2002, page 5)). Wavelength division multiplexing (WDM) is a method to efficiently use fiber's high bandwidth by sending simultaneously light beams of different wavelengths through the core of an optical fiber. In order to increase flexibility of an optical network, wavelength cross-connects (WXC) are optical devices used to switch a wavelength channel from one fiber to another. Wavelength converters are yet another type of optical devices that allow a wavelength conversion in an optical path.

There is a difference between optical fiber transmission links and optical networks. If point-to-point WDM optical links are used in connection with electro/optical devices, such a conversion creates bottlenecks due to differing speeds. Electro/optic bottleneck motivates the design of architectures providing optical connectivity between fiber links. All-optical networks employ switching and routing performed by optical devices such as wavelength cross-connects and wavelength converters. These devices allow totally optical connections called *lightpaths*. There are still other nodes (called edge nodes) providing the interface between the optical backbone and non-optical devices such as ATM switches, computers, or terminals. WDM optical network architectures can be either *broadcast-and-select* or *wavelength routed*. A network where the packet on its origin-destination path can be switched from one wavelength to another is called a *multi-hop network*, as opposed to a *single-hop network* where the optical connection between an origin-destination pair uses only one wavelength. In broadcast-and-select networks each node is equipped with tunable optical transmitters and receivers that allow simultaneous transmission of messages on different wavelengths. A logical (or virtual) connectivity between two nodes is realized when a transmitter of the origin node and a receiver of the destination node are tuned to the same wavelength. Such networks are well suited for multicasting and for covering smaller geographic areas, but do not scale well over longer distances due to loss of the transmitted power. Moreover, they do not allow for wavelength reuse in the network, hence requiring a large number of wavelengths to be available. Wavelength routed networks employ optical cross-connects as wavelength routing switches allowing an optical connection to cross intermediate nodes, hence becoming a *lightpath*. This relaxes constraints on possible physical networks, making wavelength-routed networks better suited for Wide Area Networking. Moreover, wavelength routed networks support wavelength reuse, as a possibility to employ the same wavelength for simultaneous transmission over fiber-link disjoint paths.

20.2.1 On Wavelength Converter Placement

The requirement that the same wavelength should be used on all links of the lightpath is known as the *wavelength continuity constraint*. This might be a considerable constraint necessitating a large number of wavelengths to be used. *Wavelength converters* relax this constraint by allowing optical wavelength conversion on various segments of the lightpath route. In effect, they allow for more efficient use of a fiber's bandwidth leading to better network throughput as a result of increased routing choices for a given lightpath. Wavelength conversion capability at a node can have different levels: full conversion (implying possible conversion between any two wavelengths employed in the network) or limited conversion (an input wavelength can be converted to a subset of usable wavelengths). Due to their cost and increased complexity of managing an optical network, the challenging question is to decide on the number and position of wavelength converters in an optical network. Since optical converters are expensive, it would be economically desirable to place the minimal number of converters, yet realizing good network performance. Hence, questions regarding the converter placement are as follows: (1) given a certain measure of network performance, what is the minimal number of converters to place, and (2) where (i.e. on which nodes) to place them?

20.2.1.1 On the minimal number of wavelength converters.
For a given optical network, let W denote the maximal number of allowable optical channels in a fiber link, i.e. the maximal number of different wavelengths that could be multiplexed in any of the fiber links used in the network (also called maximal link load). This is equal to the minimal number of wavelengths needed in order to use the optical bandwidth fully. In order to realize the traffic requirements, a much larger number of wavelengths might be needed, hence the need for optical converters to reduce the number of wavelengths employed (the relaxation of the wavelength continuity constraint). Finding the minimal number of converters is a challenging problem. For a given network and a corresponding traffic matrix, Wilfong and Winkler (1998) defined the so-called *sufficient set* S of nodes in the network as the set of converter placements such that the minimal number of wavelengths used equals W, i.e. the theoretical lower bound. Wilfong and Winkler (1998) showed that finding a sufficient set of minimum size is NP-complete even for special types of networks represented by planar graphs. Kleinberg and Kumar (1999) proposed a polynomial time 2-approximation algorithm for finding a sufficient set in an arbitrary directed graph. They also showed that any polynomial time c-approximation for this problem with $c < 2$ would imply a polynomial time c-approximation for the minimum vertex cover in an undirected graph. For

finding a sufficient set in a planar graph, they provided a polynomial approximation scheme. They were able to establish the above bounds by connecting the minimum sufficient set problem in bi-directed graphs with the minimum vertex cover problem in undirected graphs. Wan et al. (2003) proposed an adaptation of the results from Kleinberg and Kumar (1999) to special topologies of trees and trees of rings, as topologies especially relevant for telecommunications. Jia et al. (2002b) prove that for duplex communication channels an optimal solution to the minimal sufficient set can be obtained in polynomial time. We will not provide further details regarding the work on the minimal number of converters since our survey is mainly concerned with *location* problems arising in telecommunication networks, Hence, we wish to review the work on converter *placement*, given the number of converters to be placed. Since the actual placement of converters depends on the traffic matrix, one can distinguish between the static and the dynamic case.

20.2.1.2 On wavelength conversion, routing, and wavelength assignment in static networks.
In the case when the traffic matrix is given up-front, i.e. the network is static, Kennington and Olinick (2002) present a computational study regarding wavelength conversion. In fact, they present a formulation for the so-called *routing and wavelength assignment (RWA)* problem with and without wavelength conversion. The objective is to minimize the network construction cost (all capital expenses for hardware and software) for satisfying demand. They perform a comparison of solutions for twenty test problems ranging from 6 to 18 nodes. The two extreme cases (possible wavelength conversion at each node and no conversion at all) give rise to mixed integer formulations which they solve using CPLEX. With conversion at every node, CPLEX obtained optimal solution to all of the problems, while for instances with no conversion, optimal solutions were obtained only for the smallest problems. When only partial conversion is acceptable, the problem becomes more challenging and Kennington and Olinick (2002) develop a tabu search heuristic for this problem.

Regarding the routing and wavelength assignment problem (RWA), an interesting location problem arises when the virtual topology obtained by optical connections is restricted to a regular topology. In a regular topology all the nodes have the same nodal degree. A number of regular topologies (perfect shuffle, de Bruijn, Manhattan street network, GEMNET) exhibit good properties regarding reliability and management. Skorin-Kapov and Labourdette (1998) consider a broadcast-and-select optical network with wavelength conversion possible at each node and minimize the maximal flow on any link under the regular virtual topology constraint. The problem is to decide on placements of virtual topology nodes on physical nodes, and on optimal routing of traffic given the virtual topology. The first part of the problem gives rise to the *quadratic assignment problem* whereby an initial assignment of virtual topology nodes to physical nodes is obtained by maximizing the one-hop path traffic. Having an assignment of nodes to locations, the multicommodity network flow problem routes the flow so that the maximal flow on any link is minimized (i.e. the network throughput is maximized). Siu and Chang (2002) consider nodes assignment in WDM networks with regular topologies in order to minimize the average weighted hop distance. The hop distance between two nodes is the number of wavelengths used to send the traffic between them. Siu and Chang (2002) implement a generic regularity constraint via a distance matrix consisting of hop distances for all possible pairs of source and destinations. They formulate the problem as a quadratic assignment and solve it via a simulated Annealing algorithm.

We would next like to concentrate on the dynamic case when requests for connections come as the network is being used.

20.2.1.3 Converter placement problem: minimization of blocking probability. Converter placement problem is formally posed in Murthy and Gurusamy (2002, page 130), as: "Given K number of converters and a network with $N \geq K$ number of nodes, how do we choose K nodes such that placing converters at these nodes yields the best performance achievable using K converters?" The optimal placement of converters will depend on the objective function we wish to optimize and on the traffic pattern. For an arbitrary network this problem is NP-complete. Reflecting the dynamic nature of telecommunication networks, the usual optimization criterion is to minimize the blocking probability. The blocking probability is the probability that an origin-destination request for connection will be blocked due to the logical optical topology not supporting the requested lightpath. Without converters, the wavelength continuity constraint increases the blocking probability making the network less desirable. Minimization of the call-blocking probability increases the network performance. Intuitively, larger distances between converters increase the blocking probability. Furthermore, it seems that nodes with high amount of transit traffic and high amount of switching traffic present good choices for converter placement. Due to the complexity of the problem given arbitrary topology, it seems appropriate to consider simpler topologies and devise results that could be used in tackling arbitrary topologies.

Subramaniam et al. (1999) consider a problem of placing K number of converters on a path network topology in order to minimize the blocking probability. They prove that uniform spacing of converters is optimal given uniform and independent link loads. For non-uniform link loads, they propose a dynamic programming algorithm. The algorithm is then generalized to apply to bus and ring networks, running in $O(N^2K)$ time for the bus, and $O(N^3K)$ for the ring topology. Li and Somani (2002) reconsidered the optimal converter placement problem for K converters on a path. They develop a heuristic algorithm with the basic idea to divide the path into $K+1$ segments such that the blocking probability on each segment is equal. In cases when it is not possible to divide a path into segments with equal blocking probability (due to differing link loads), they propose three algorithms for dividing a path into segments having approximately equal blocking probabilities.

Jia et al. (2002a) in Chapter 6 consider the wavelength converter placement in order to minimize the overall blocking probability in an arbitrary network topology. They first express the overall blocking probability in a network as a polynomial function of 0/1 variables x_i indicating whether or not a converter is placed at a node i. They then employ a branch and bound algorithm to solve the problem of minimizing the blocking probability, subject to placing K converters. The complexity of the problem is concentrated in the calculation of the blocking probability. Considering a path P_{ij} between nodes v_i and v_j, and placing K converters, produces $K+1$ segments, each consisting of a number of links. A call is blocked in a segment if and only if there is no wavelength that is available on every link in the segment. A call will go successfully through the path P_{ij} only if it is not blocked on any of its segments. The probability of a successful connection depends on the number and locations of converters. Jia et al. (2002a) first consider the case with uniform link loads, i.e. they assume that the load on every link is ρ, representing in turn the probability that a wavelength on a link is occupied. Denoting the probability that a wavelength on a link is available as $q = 1 - \rho$, that a path consists of d links, and that there are w wavelengths available on each link, Jia et al. (2002a) state the following result (Lemma 6.1, page 144):

When d=1, the probability of successful path connection is given by $1 - (1 - q)^w$.
When $d = 2$, the probability of successful path connection is given by $(1 - (1 - q)^w)^{2x_2}(1 - (1 - q^2)^w)^{(1-x_2)}$.

The explicit expression is then provided for the case of $d > 2$, and subsequently generalized for the case of non-uniform traffic loads on links. The obtained polynomial function of binary variables $F(x_1, ... x_n)$ is then maximized subject to $\sum_{i=1}^{n} x_i = K$. This is done with a branch and bound algorithm.

Siregar et al. (2002) solve the converter placement problem for two realistic networks connecting cities in Japan. The number of converters is given and the task is to locate converters in order to minimize the blocking probability. They devise a genetic algorithm to obtain good suboptimal solutions.

20.2.2 Amplifier Placement

Transporting an optical signal over some distance creates a loss of power. Although the quality of optical fibers increases, in turn diminishing attenuation loss, it is still a problem to deal with. The attenuation is computed (in decibels per kilometer of fiber) as $10 log_{10}$(transmitted power/received power). Different wavelengths have different attenuations (see Murthy and Gurusamy (2002, page 6)). Amplifiers are devices that can simultaneously amplify optical signals using different wavelengths. Due to their cost, amplifiers should be used only where needed - which gives rise to the so-called *amplifier placement problem* whereby one decides what is the minimal number of amplifiers to be used, and where to place them in the network. Amplifiers can be used in three different places in a fiber link: power amplifiers can be located at the origin node and boost a signal before launching it, line amplifiers are located somewhere on the fiber line and restore a signal to its original power (dealing with fiber attenuation), and preamplifiers can be placed at the destination node to raise the signal level at the input of its optical receiver. (See Stern and Bala (2000, page 192).) In broadcast-and-select networks passive stars work by splitting the input signal and broadcasting it to output ports. The splitting of signals might result with power loss, so splitting nodes are natural locations for power amplifiers. We survey few works in the context of amplifier placement.

Ramamurthy et al. (1998) consider the problem of the minimum number of amplifiers needed and their placement for operation of multiwavelength optical local and metropolitan area networks. Signals arriving on different wavelengths can have different power levels, limiting in turn the amount of amplification available. In a previous work by the authors an integer linear programming formulation was used to model the simplified situation assuming equal power level for all the wavelengths in the network. In Ramamurthy et al. (1998), under the assumption of having wavelengths operating on different power levels, the problem is formulated as a nonlinear mixed integer problem. Due to difficulty in solving this formulation, the integrality constraints were relaxed and the problem of deciding on the minimal number of amplifiers was solved with a quadratic programming solver. The output from the solver was used to decide on the positions and gains of the amplifiers using the so-called Amplifier-Placement Module. The results showed that for small and medium size broadcast-and-select networks, consideration of different power levels for wavelengths resulted with smaller number of amplifiers needed.

Fumagalli et al. (1998) generalized the problem posed in Ramamurthy et al. (1998) allowing for star, tree and ring topologies, and taking into account the location dependent installation and maintenance amplifier cost. They assumed broadcast-and-select networks whereby each transmitting station has a distinct wavelength for transmission. Each station has a receiver tunable to any of the wavelengths used in the network. Fumagalli et al. (1998) propose a simulated annealing heuristic to get the minimal number of amplifiers in order to minimize the cost of amplifier placement.

Tao and Hu (2003) consider a path topology and take into account the fact that amplifiers introduce the spontaneous emission noise, degrading the optical signal-to-noise ratio, and add the constraint on the cumulative spontaneous emission noise. In addition, they assume different types of amplifiers. Two cases are considered: (1) the discrete case when possible locations for amplifier placements are given , and (2) the continuous case when amplifiers can be placed anywhere in the network. For both cases they propose integer programming formulations. Specifically, the discrete formulation uses the following input data: S is the number of types of amplifiers available; c_s is the cost of each type s amplifier; L is the total length of the path; E is the limit on the cumulative emission noise; $f_s(l)$ is the spontaneous emission noise produced by type s amplifier for a span with length l, N is the number of amplifiers to be placed ; M is the number of possible locations for amplifier placement; l_m is the fiber length between points $m-1$ and m, $m = 1,...M$; m_n is the index of the point where the n-th amplifier is placed, $n = 1...,N$; $L_n = \sum_{m=1+m_n}^{m_{n+1}} l_m$ is the length of the n-th fiber span, $n = 1,...,N$. The binary variable x_{ms} equals 1 if a type s amplifier is placed at point m, and 0 otherwise. The formulation for the discrete case is:

$$\min \sum_{m=1}^{M}\sum_{s=1}^{S} c_s x_{ms} \qquad (20.1)$$

subject to

$$\sum_{s=1}^{S} x_{ms} \leq 1, m=1,...M \qquad (20.2)$$

$$\sum_{n=1}^{N}\sum_{s=1}^{S} x_{m_n s} f_s(L_n) \leq E \qquad (20.3)$$

$$x_{ms} \in (0,1). \qquad (20.4)$$

For the continuous case, the variables are redefined as: x_{ns} equals 1 if the n-th amplifier placed is type s, 0 otherwise. The continuous case formulation is:

$$\min \sum_{n=1}^{N}\sum_{s=1}^{S} c_s x_{ns} \qquad (20.5)$$

subject to

$$\sum_{s=1}^{S} x_{ns} = 1, n=1,...,N \qquad (20.6)$$

$$\sum_{n=1}^{N} L_n = L \qquad (20.7)$$

$$\sum_{n=1}^{N}\sum_{s=1}^{S} x_{ns} f_s(L_n) \leq E \qquad (20.8)$$

$$x_{ns} \in (0,1). \qquad (20.9)$$

The proposed solution methods for the discrete case include a dynamic programming and a heuristic algorithm, and for the continuous case the solution procedure includes decomposition of the original optimization problem into two convex optimization problems. Future work could consider generalizing the approach to mesh networks having many lightpaths. This will complicate the underlying formulation since an amplifier can then be used in different lightpaths, each having its own cumulative spontaneous emission noise.

20.2.3 Splitter and Wavelength Converter Placement in Multicast Networks

A one-to-one transmission is known as *unicasting*. A one-to many communication (*multicasting*) whereby an origin communicates with a number of destinations is a preferable means of communications for a number of applications, e.g. video conferencing and distance learning. The trivial way of realizing multicast sessions by repeating unicast connectivity between the origin and each of the destination nodes is inefficient and multiplies lots of unnecessary traffic. A more efficient multicast communication is realized by network nodes having the capability to tap and split a signal. For a given multicast connection, the problem of finding a routing tree with the objective to minimize the overall network cost (the sum of the cost of all links in the network) can be posed as the classical Minimum Steiner Tree (MST) problem. In order to assure quality of service, a delay bound can be imposed on communications, giving rise to the so-called Constrained Minimum Steiner Tree Problem. A recent work in Skorin-Kapov and Kos (2003) proposes an efficient tabu search heuristic for constrained multicast routing.

A node with optical tapping capability (*drop and continue node (DaC)*) allows tapping of a small amount of optical signal that is dropped at a destination node. A node with splitting capability can make copies of traffic data and forward those copies to various destinations. When multicast communication is required, the virtual optical topology supporting it is realized via *light trees*. The notion of a light tree generalizes the notion of a lightpath by placing the multicast group users at the endpoints of this tree. A light tree allows multicasting performed in the optical layer, as opposed to performing multicasting at a layer requiring electro/optic conversion. As in the unicast case, the problem of placing converters in a network arises in the context of efficient network design and usage. In addition, multicasting environment creates a location problem of deciding on which cross-connect nodes to place power splitters, adding to the overall network cost.

The concept of a light tree on a WDM wavelength routed network was introduced in Sahasrabuddhe and Mukherjee (1999) as a way to increase the logical connectivity of the network and to decrease its hop distance. It is a point-to-multipoint all-optical channel possibly spanning multiple fiber links. Each branching node of the tree contains a power splitter that splits the optical signal into multiple signals forwarded to different destinations. The communication employing light tree requires more receivers than transmitters, but the resulting logical connectivity is increased. Since splitting light is easier than copying data in electronic buffer, optical multicasting is more desirable. For a given traffic matrix, i.e. a static network topology, Sahasrabuddhe and Mukherjee (1999) formulate the virtual topology design problem employing light trees as an optimization problem using two possible objective functions: (1) minimization of the average hop distance, and (2) minimization of the total number of transceivers in the network. They consider unicast and broadcast types of traffic. Broadcast traffic occurs when a node needs to send data to all the other nodes in the network.

As with unicast connections employing lightpaths, a multicast connection using light trees requires the solution to the routing and wavelength assignment problem. In this case, the routing problem involves constructing a tree rooted at the source node, and connecting all the destination nodes involved in multicasting. The wavelength assignment problem is then to assign wavelengths to light trees so that when two trees share a common link, they must use different wavelengths. Enforcing the light tree continuity constraint (i.e. the constraint that the same wavelength is used in the entire tree) would necessitate a large number of wavelengths. Placing wavelength converters relaxes this constraint and allows for more efficient logical connectivity. Using a static connectivity data, Yang and Liao (2003) formulate a complex mixed integer programming problem for the design of light trees based logical topology in WDM networks using power splitters and wavelength converters. Their problem consists of two parts.

First, a mixed integer linear problem is formulated in order to establish the optimal routing and wavelength assignment of light trees with an end-to-end delay bound and to obtain the optimal placement of power splitters and full range wavelength converters in the network (i.e. each wavelength on a channel can be converted to any other wavelength). The objective is to minimize the number of wavelengths used by all light trees, which in effect maximizes the number of light trees possibly created in the network. Next, the formulation is extended to design the logical topology based on light trees for multicast streams. Namely, Yang and Liao (2003) allow that a light tree carries data of multiple multicast streams, and that data of a multicast stream may use multiple light trees to reach its destinations. Hence, given a multicast stream, there is a need to determine the set of light trees used by it. The topology related to this problem can be represented by a *hypergraph* in which each link is a hyper-edge representing a light tree. The optimization criterion in designing the logical topology can either be minimization of the congestion, or minimization of the average hop distance. Yang and Liao (2003) present numerical results using CPLEX on the 14-node NSFNET backbone network. Further work in this area should concentrate on development of efficient heuristic approaches that will allow consideration of bigger problems.

In the context of multicast routing Ali and Deogun (2000) introduce the splitter placement problem in wavelength routed networks (SP-WRN) and show that it is NP-hard. The parameters of the problem include a network topology, a set of multicast sessions, and a fixed number of power splitters. The network does not use converters, hence the light tree continuity constraint is maintained by assuming that a given light tree uses only one wavelength. The blocking probability is used implicitly, incorporated in the objective function as a sum of profits for multicast sessions established in the network. (In a subsequent work, Ali (2002) develops an analytical model for the approximate blocking probability.) The task is to identify the best combination of tree/wavelength for each multicast session such that the overall profit of the established sessions is maximized. Each multicast session $\psi(s_i, D_i)$ is identified by a source s_i and a set of destinations D_i, and provides a profit ψ_i. For each multicast session a set of alternate trees, R_i, is computed by an approximation algorithm.

The variables used in the formulation are as follows: e_i equals 1 if the session ψ_i is established, and 0 otherwise; $\lambda_{i,j,w}$ equals 1 if session ψ_i is established using wavelength w and the j-th tree, and 0 otherwise; S_t equals 1 if node t contains a power splitter, and 0 otherwise. The additional input parameters are: Q is the set of multicast sessions; L is the set of links in the network, W is the set of wavelengths; M is the set of multicast capable cross-connects (possible splitter nodes); K is the number of power splitters; f_l is the number of physical fibers on link l; $B^{(i,j)}$ is the set of branching nodes in the j-th tree of i-th multicast session.

The integer programming formulation is given as :

$$\max Z = \sum_{\psi_i \in Q} \sigma_i e_i \qquad (20.10)$$

subject to

$$\sum_{\psi_i \in Q} \sum_{j \in R_i, T_{ij} \in V_l} \lambda_{i,j,w} \leq f_l, l \in L, w \in W \qquad (20.11)$$

$$|B^{(i,j)}| \sum_{w \in W} \lambda_{i,j,w} \leq \sum_{t \in B^{(i,j)}} S_t, \psi_i \in Q, 1 \leq j \leq |R_i| \qquad (20.12)$$

$$\lambda_{i,j,w} \in (0,1) \qquad (20.13)$$

$$e_i \in (0,1) \qquad (20.14)$$

$$S_t \in (0,1), t \in M \qquad (20.15)$$

where

$$e_i = \sum_{j \in K_i} \sum_{w \in W} \lambda_{i,j,w}, \psi_i \in Q \qquad (20.16)$$

$$\sum_{t \in M} S_t = K. \qquad (20.17)$$

The constraints ensure that that each wavelength w is used by at most f_l trees, that the trees which can not be realized due to lack of splitters are not considered in the solution, that at most one tree per session can be used, and that there are K splitters used in the network.

This problem is solved in two phases. First the multicast routing and wavelength assignment problem without the splitting capability is solved heuristically using a genetic algorithm. The objective is to establish the maximum number of sessions by selecting for each session a multicast tree and a lightwave used. In the next phase, K splitters are allocated by using either a simple heuristic based on the rule "most saturated node first" (MSNF), or a simulated annealing heuristic. Based on numerical experiments the authors conclude that simulated annealing outperforms MSNF, and that not more that 50% of the nodes in the network need to be equipped with power splitters in order to achieve good network performance. Due to complexity of the problem, development of additional heuristic approaches should be a fertile area of future research. In addition, the formulation could be modified to allow for wavelength converters.

20.2.4 On locating concentrators in access/backbone networks

As indicated in Anandalingam (2002), a design of a telecommunication network involves many aspects, including most suitable access and backbone network topologies and capacity considerations. A general formulation for designing a large-scale private telecommunications network is provided in Anandalingam (2002) as a complex mixed integer programming problem since the optimization in one aspect depends on other aspects of the network. The formulation can be heuristically solved by decomposing it into three subproblems: the access network design, the backbone network design, and the determination of link and node capacities.

Concentrator location is a classical location problem in telecommunications. It arises in different settings resulting with similar formulations. In order to increase the efficiency pertaining to network usage and management, the users are connected to concentrators and the traffic is consolidated. Gourdin et al. (2002) list the relevant questions regarding concentrator location: (1) needed number of concentrators; (2) concentrator locations in the network; (3) capacities of concentrators; (4) allocation of user nodes to concentrators. They survey the models for concentrator location and user allocation and classify them as uncapacitated, capacitated, and dynamic models. These types of models are not exclusive to telecommunication networks. Indeed, they are applicable to facility location and transportation networks, and are often known under the hub-and-spoke models. Since Gourdin et al. (2002) summarized the research concerned with the concentrator location problem, we will present here only some recent applications.

Rosenberg (2001) considers uncapacitated telecommunications networks with access, backbone, and switch costs. The problem is to decide, simultaneously, at which locations to place a switch, interconnecting all switches with backbone trunks, and to connect each location to some switch by an access circuit. Under no capacity constraints, the objective is to the minimize the sum of switch, backbone trunk, and access circuit costs. The problem gives rise to a mixed integer programming formulation which is heuristically solved by using a dual ascent method solving a sequence of dual uncapacitated facility location problems. This provides a lower bound to the optimal objective value. An upper bound is obtained via a Steiner tree based heuristic. The computational results are performed on 15 random problems with 100 locations, with the average duality gap of 2.0%.

The uncapacitated problem might not be realistic for the increased telecommunications traffic and new data-intensive applications. Hence, capacities should be taken into consideration. Raja and Han (2003) extend the capacitated concentrator location problem by simultaneously considering two capacity constraints (total connection ports and maximum data processing rate) on each concentrator. A star-star topology is assumed. The objective is to place concentrators to their potential locations, and to allocate each terminal node to exactly one concentrator, taking into account all processing demands and capacity constraints on concentrators, so that the total system cost is minimized. For illustrative purposes the formulation from Raja and Han (2003) will be presented. The notation is as follows: $I = \{1, ..., n\}$ is the set of terminal nodes; $J = \{1, ..., m\}$ is the set of potential concentrator sites; c_{ij} is the cost of link from terminal i to concentrator j; c_j is the cost of purchasing one unit of processing capacity for concentrator j; F_j is fixed cost of installation of concentrator j; K_j is the total number of connection ports of concentrator j; P_j is the effective processing limit of concentrator j; and p_{ij} is the processing demand of node i on concentrator j. The two classes of 0/1 variables are defined as: $x_{ij} = 1$ if terminal i is assigned to concentrator j, and 0 otherwise; $y_j = 1$ if concentrator at site j is installed, and 0 otherwise. The formulation of the extended capacitated concentrator location problem is then:

$$\min \sum_{i\in I}\sum_{j\in J} c_{ij}x_{ij} + \sum_{i\in I}\sum_{j\in J} c_j p_{ij} x_{ij} + \sum_{j\in J} F_j y_j \qquad (20.18)$$

subject to

$$\sum_{j\in J} x_{ij} = 1, \forall i \in I \qquad (20.19)$$

$$\sum_{i\in I} x_{ij} \leq K_j, \forall j \in J \qquad (20.20)$$

$$\sum_{i\in I} p_{ij} x_{ij} \leq P_j, \forall j \in J \qquad (20.21)$$

$$x_{ij} \leq y_j, \forall i \in I, \forall j \in J \qquad (20.22)$$

$$x_{ij}, y_j \in \{0,1\}, \forall i \in I, \forall j \in J \qquad (20.23)$$

Raja and Han (2003) develop a Lagrangian relaxation heuristic which makes use of a known algorithm for two-constraint knapsack problems. The method performed well on 100 randomly generated problems with sizes ranging from (30 nodes, 30 concentrators) to (150 nodes, 30 concentrators). Additional research could aim at development of other heuristic approaches for this formulation, as well as on modifying the formulation to take into account dynamic traffic patterns.

While backbone networks provide huge bandwidth enhanced by wavelength division multiplexing, access networks carrying data to and from individual users provide smaller bandwidth. Nevertheless, this bandwidth is being increased by advanced LAN technologies such as broadband access and cable modems, and by high-speed wireless systems such as universal mobile telecommunications system (UMTS). This is motivated by an increasing diversity of consumer applications requiring broadband network services, lead by requests for Internet access. Most cable companies presently use the Hybrid Fiber Coaxial (HFC) technology to design the access network, which is then connected to the fiber optic backbone. Patterson and Rolland (2002) consider the HFC network design problem implemented in a form of the capacitated tree-star topology. They propose a mixed integer programming formulation and a heuristic algorithm based on hierarchical decomposition of the problem. The access HFC network has a capacitated tree architecture and needs to be connected to a wide area network through optical network units (ONUs) which are fiber connected to a central office from which further fiber SONET links lead to the regional network (see Figure 1 in Patterson and Rolland (2002)). The solution to the problem involves determining the number and location of ONUs (viewed as concentrators), and determining directed spanning tree structures for connecting customers. The problem is decomposed to the capacitated minimum spanning tree (CMST) problem and to the capacitated star-star with concentrators (CSSC) problem. The solution procedure incorporates an Adaptive Reasoning Technique (ART) as a guiding memory-based heuristic framework over specialized heuristics for the subproblems. Computational results showing improvement over Lagrangian relaxation are reported for networks with up to 80 nodes.

In the context of Internet usage, Resende (2003) considers the so-called point-of-presence (PoP) placement problem faced by Internet access providers. For an Internet provider taking budgetary considerations into account, the task is to decide where to place modem pools (PoPs) accessible by customers' modem-phone calls, so as to maximize the customer based coverage. Resende (2003) formulates the problem as the maximum covering problem (MCP). In this case, the coverage is defined as the number of customers that can access at least one PoP through a local phone call. The set of potential locations for PoPs, $J = \{1,...,n\}$, is known upfront. For each potential PoP location i a set P_i is such that $I = \bigcup_{j \in J} P_j = \{1,...,m\}$ is the set of m exchanges that can be covered by the n potential PoPs. Each exchange i has an associated weight w_i. A cover $J^* \subseteq J$ with the weight $w(J^*) = \sum_{i \in I^*} w_i$ covers the exchanges in $I^* = \bigcup_{j \in J^*} P_j$. For a prescribed number p of PoPs to be placed, the objective is to find the set $J^* \subseteq J$ that maximizes $w(J^*)$ over all sets of cardinality p. Resende (2003) proposes a greedy adaptive search procedure (GRASP) for heuristic solvability of the PoP placement problem. The results are reported for a large PoP placement problem with $m = 18,419$ calling areas and $n = 27,521$ potential PoP location, and 27, 197,601 lines over the calling areas. The GRASP procedure compares favorably with the purely greedy and purely random heuristics.

20.3 LOCATION OPTIMIZATION IN WIRELESS NETWORKS

Cellular networks are defined by base stations and mobile users. The mobility of the network presents a new set of challenges related to location management. There is a number of location techniques based on cellular system signals such as GSM (second generation cellular mobile system) or UMTS (third generation cellular mobile system), including methods based on cell identification, signal strength, time delay, angle-of-arrival, signal pattern recognition, and database correlation. We will first present some recent work related to location of UMTS base stations, following with some recent developments in location research pertaining to cellular networks. Mobile wireless networks are classified into two broad categories (see Mukherjee et al. (2003)) as cellular (infrastructured) and ad hoc (infrastructureless). Cellular networks usually involve a single-hop wireless link to reach a mobile station. Ad hoc networks might need to traverse a multihop wireless path to reach a mobile station. Mobility can be viewed from the single mobile station, involving radio mobility consisting of the handover process, and from the network point of view, involving location updating and paging. Location management (updating, paging, and location area planning) arises in cellular networks, while routing is an issue in ad hoc mobile networks. In cellular networks, each cell is served by a base station (BS), responsible for communication with mobile terminals within its cell. The base stations are connected to a mobile switching center (MSC) via fixed links, then to the public phone system and the Internet. There are two multiple access techniques used in second generation mobile systems for using the spectrum in a cell. Time Division Multiple Access (TDMA) divides a given band into a number of time slots that correspond to a communication channel. In Code Division Multiple Access (CDMA) each signal is sent at the same time, but it is encoded differently (spread spectrum). The traffic increase has a very little influence when TDMA technique involves correct frequency planning process with respect to interference level. In CDMA environment the increase in traffic increases the interference level, degrading the quality of the telecommunication service. The consequence is that the cell area shrinks with the increased traffic, creating the so-called cell-breathing phenomenon. The capacity of each cell, measured as the number of connections, depends on the interference levels and is not limited by an a priori assignment as in TDMA systems. However, the CDMA environment allows more efficient and flexible use of radio resources.

20.3.1 UMTS Base Station Location

Universal Mobile Telecommunication System (UMTS) is the third generation mobile communication system using new frequency spectrum, in turn allowing the integration of voice and multimedia services on the same terminals and using the same network. The air interface of UMTS is based on wideband code division multiple access (W-CDMA) whereby the bandwidth is shared by all active connections. The service quality is usually measured as the signal-to-interference ratio (SIR) and it depends both on traffic distribution and on positions of base stations (BSs). Hence, in order to have good service coverage, the placement of a sufficient number of base stations is mandatory.

In a two-way communication systems links are established both from the base station to the mobile station (downlink) and from a mobile station to the base station (uplink). Downlinks are relevant for data service, e.g. web browsing. Uplinks are relevant for connections such as voice calls. The uplink and downlink models differ in intra-cell interference. In downlink model the level of interference between connections involving mobile stations assigned to the same base station is much lower than between connections involving mobiles assigned to different base stations. In general, interference levels depend on the mobile station positions. Due to limited power available for transmission, the mobile stations further away from the base station may experience an unacceptably low transmission quality (measured by the minimally acceptable SIR). Due to traffic distribution, areas belonging to different base stations may have different sizes. For a given traffic distribution and a given quality level, it becomes necessary to decide on base station locations. Amaldi et al. (2003a) consider optimal base station locations in cellular networks with respect to quality constraints for the uplink direction. In Amaldi et al. (2003b) they consider the same problem in cellular networks for the downlink direction. In both cases there are two common ways to model the power control mechanism: power-based and SIR-based. In the power-based power control mechanism, the emission power p_i required to connect a test point i, TS_i, is given implicitly. TS_i is viewed as a centroid where a given amount of traffic d_i is requested, a level of service measured in terms of SIR must be guaranteed, and there is a required number of simultaneously active connections u_i. (u_i is the function of the traffic demand, $u_i = \phi(d_i)$). In the case of power-based power control mechanism, the transmitted power is adjusted so that the power received equals a target value. In the SIR-based power mechanism each p_i is considered as an explicit variable and the transmitted power is adjusted so that the achieved SIR equals a target SIR value. This mechanism takes into account the interdependency of emitted powers of all cells, making it more complex.

Developing the model for uplink UMTS radio planning, Amaldi et al. (2003a) assume that each cell is assigned to a single base station and that the available number of spreading codes is higher than the number of connections assigned to any base station.

The task is to locate base stations and to allocate test points, given traffic demands, installation costs, and the signal quality requirements in terms of SIR. This problem is first formulated as the capacitated facility location problem. However, the basic model has to be enhanced by considering intercell interference explicitly and independently from intracell interference. This gives rise to a nonlinear constraint expressing the signal quality requirement. The following variables are used in the formulation: $y_j = 1$ if a base station is installed in location j, and 0 otherwise; $x_{ij} = 1$ if test point i is assigned to base station j, and 0 otherwise. The input parameters used are: $S = \{1,...,m\}$ is the set of available locations for base stations; $I = \{1,,,n\}$ is the set of test points; c_j is the installation cost for a base station at location j; $0 < g_{ij} < 1$ is the propagation factor of the radio link between test point i and base station j; $\lambda \geq 0$ is a tradeoff parameter between two parts of the objective function; P_{max} (resp., P_{target}) is the maximum (resp. target) emission power; SIR_{min} is the minimum signal-to-interference ratio required. The complete formulation in the case of power-based power control mechanism becomes:

$$\min \sum_{j=1}^{m} c_j y_j + \sum_{i=1}^{n} \sum_{j=1}^{m} u_i \frac{1}{g_{ij}} x_{ij} \quad (20.24)$$

$$\text{subject to} \quad (20.25)$$

$$\sum_{j=1}^{m} x_{ij} = 1, i \in I \quad (20.26)$$

$$x_{ij} \leq \min\left\{1, \frac{g_{ij} P_{max}}{P_{target}}\right\} y_j, i \in I, j \in S \quad (20.27)$$

$$y_j \left(\sum_{h=1}^{n} \sum_{t=1}^{m} u_h \frac{g_{hj}}{g_{ht}} x_{ht} - 1 \right) \leq \frac{1}{SIR_{min}}, j \in S \quad (20.28)$$

$$x_{ij}, y_j \in \{0,1\}, i \in I, j \in S \quad (20.29)$$

A related nonlinear mixed integer formulation is also developed for the SIR-based power control mechanism. Amaldi et al. (2003a) propose two randomized greedy procedures and a tabu search heuristic for solving the formulations. The computational results on some generated instances ranging from 22 to 200 possible locations for base stations, and from 95 to 750 test points indicate the effectiveness of tabu search. With respect to different power control mechanisms, the SIR-based control requires smaller number of base stations, hence more efficient use of radio resources. In a companion work, Amaldi et al. (2003b) consider the problem of location base stations in the presence of downlink directions (from base station to mobile). They propose a GRASP and a tabu search algorithms adapted to this situation.

A future relevant work could consider optimizing base station locations in cellular networks in which a cell could be assigned to more than one base station. This would imply that a soft-handover would need to be taken into account.

20.3.2 Ad Hoc Wireless Networks

In the traditional cellular network model mobile stations are connected to base stations, which are wire connected. Due to lack of wire structure in some geographic locations and in some situations (e.g. emergency relief operations or military communication), mobile stations would need to employ some routing in order to carry on the required communication. An ad hoc network is a temporary network that can operate without centralized administration. This can be achieved if every node has a capability of a mobile router equipped with a wireless transceiver. (Mukherjee et al., 2003) The problem associated with communication in an ad hoc network is the possibility that an intermediate node can break the communication path by either moving out of the range, or by switching off. Another problem is the limited energy resource due to using batteries, instead of constant power source supplied at wired networks. In the absence of base stations, an ad hoc network can have a cluster-based topology. In the cluster-based network, nodes are partitioned into clusters, and in each cluster one node is selected as the so-called cluster-head. Cluster-heads are responsible for local control such as handling inter-cluster traffic and delivering packets destined to the cluster, hence they experience high energy consumption. Clusters can have different sizes, depending on the cluster-head's transmission power. A good ad hoc network design should minimize cluster-head's energy consumption. An interesting problem is to determine the optimal cluster size and the optimal assignment of nodes to cluster-heads in order to maximize the network life-time by efficiently spending the energy. Chiasserini et al. (2003) present algorithms for maximizing the network life-time. The network life-time is defined as the time from the moment the network starts functioning until the moment when the first network node runs out of energy. Chiasserini et al. (2003) consider the situation when the traffic matrix is either static or changing slowly, and cluster-heads are given a priori. They devise an algorithm for maximizing the network life-time by optimally allocating nodes to cluster-heads which can dynamically adjust the size of the clusters through power control. The algorithm takes into account all sources of power consumption. A more complex network design would need to consider the scenario with rotating cluster-heads via an election procedure during which all nodes would need to be synchronized.

20.3.3 Location Area Planning in Cellular Networks

Mobility in cellular networks creates two procedures: (1) *registration* or updating initiated by a mobile station in order to update its location entry in the location database, and (2) *paging* or finding the mobile station's new position, initiated by the network. The paging process involves sending messages to cells where the mobile station could be located. A set of base stations in which a mobile is paged is called a location area (LA). The size of a location area is important since registration occurs when a mobile moves from one location area to another. Hence, the size of location areas determines the ratio of registration and paging costs: if each cell is a location area, the registration cost is maximized, if the service area of a mobile switching center is one location area, then the paging cost is maximized. In a cellular network such as GSM, several base stations (BSs) are wired to a base station controller (BSC), and several BSCs are wired to a mobile switching center (MSC). A location area (LA) contains the cells served by a single MSC. Mobile stations (MSs) are connected to BSs via radio signals.

Demirkol et al. (2001) consider the problem of determining location areas in a cellular network. They formulate the problem as a combinatorial optimization problem whereby the objective is to minimize the registration cost. The paging cost is considered implicitly as a constraint on the paging capacity of each base station controller (BSC). They design a simulated annealing procedure to find the assignment of base stations to location areas (LA-to-BS assignment matrix), the connections of base station controllers to mobile switching centers (BSC-to-MSC assignment), and the assignments of base stations to base station controllers (BS-to-BSC assignment).

Mandal et al. (2002) also consider the problem of grouping base stations into location areas, i.e. the assignment of cells to switches, in order to minimize the paging, updating, and physical infrastructure costs. They propose an algorithm employing the so-called Block Depth First Search (BDFS) that starts with the given number of cells and provides the number of location areas and their compositions. For a given runtime the algorithm performs better then the simulated annealing proposed for the same problem.

In the context of designing a personal communication services network (PCSN) overlaid on an existing wired network, Bhattacharjee et al. (2003) reconsider the problem of location area planning. Given the average speed of mobile terminals, the number, location, and call handling capacity of mobile switching centers, call arrival rate, and handoff cost between adjacent cells, the objective is to allocate cells to location areas and connect location areas to mobile switching centers in order to minimize the total system cost. The problem is decomposed into two subproblems. First, in subproblem I the total system recurring cost (i.e. location update and paging cost) is minimized by providing the optimum number of cells per location area for different call-to-mobility ratios. In the next phase, in subproblem II the handoff and the cable costs (cabling from the BS to the switch) are minimized by determining the LA compositions, i.e. by assigning cells to switches, or by identifying base stations which will form location areas. The simplifying assumption is that the cells have identical sizes with radius R, that the BS is located at the center of each cell, and that a cell can be assigned to one switch only. Subproblem I is posed as a nonlinear equation and solved using numerical analysis. Subproblem II is formulated as a 0/1 linear program having the following variables: $x_{ik} = 1$ if cell i is connected to switch k, and 0 otherwise; $y_{ij} = 1$ if cells i and j are connected via a common switch, and 0 otherwise. Denoting by C_{ik} the cable cost between cell i and switch k, by h_{ij} the handoff cost between cells i and j, and by M_k the call handling capacity of switch k, the formulation is as follows:

$$\min \sum_i \sum_k c_{ik} x_{ik} + \sum_i \sum_j h_{ij}(1 - y_{ij}) \tag{20.30}$$

subject to

$$\sum_k x_{ik} = 1, \forall i \tag{20.31}$$

$$\sum_i \lambda_i x_{ik} \leq M_k, \forall k \tag{20.32}$$

$$x_{ik}, y_{ij} \in \{0, 1\}, \forall i, j, k \tag{20.33}$$

Bhattacharjee et al. (2003) propose a greedy heuristic algorithm that constructs location areas by assigning cells in an iterative way. The assignments are carried on until all the cells are assigned, or the handling capacity of each mobile switching center is exhausted. The method is tested on a simulated problem whereby the output from subproblem I suggested 7 or 8 as the optimal number of cells per location area. A greedy heuristic performed well, compared with some previously proposed algorithms. Still, the authors noted, some enhanced heuristic approaches (such as simulated annealing and block depth first search) provided marginally better solution, albeit on the expense of longer computing time. Research on further heuristic approaches allowing for adjustment of LA sizes in context of different mobility models and call-to-mobility ratios would be desirable.

Bejerano and Cidon (2001) consider a mobility management scheme based on the traffic flow theory and concepts of location areas and location prediction. Traffic flow theory deals with the management of a national highway and street systems. The authors propose the so-called *moving location areas* as location areas defining a group of geographically concentrated mobile users that move in the same direction. Their scheme uses two complement sets of location areas overlapping each other: small location areas for locating mobile users in a quasi-static state, and the highways tracking traveling mobile users. Moving location areas are used for tracking mobile users on a highway. The authors provide optimal values for the scheme parameters for reducing the search and update costs at each cell along the highway.

20.3.3.1 Location discovery. One of the basic functions of a mobile telecommunication system is estimating the location of a mobile user. *Radiolocation* is an approach for measuring parameters of radio signals between a mobile station and a set of fixed transceivers in order to obtain the location estimate. A radiolocation system can be implemented in two different ways: (1) the mobile station uses signals transmitted by the base stations to calculate its own position, and (2) the base stations measure the signals transmitted by the mobile station and relay them to a central site for processing. Caffrey and Stuber (1998) provide an overview of radiolocation in code-division multiple access (CDMA) cellular systems. In general, radiolocation systems can be based on either signal strength, angle of arrival (AOA), or time of arrival (TOA). Location method based on signal strength uses a mathematical model describing the distance-based path loss attenuation. In AOA method, the mobile location is estimated by first measuring a signal from the mobile station at several base stations through the use of antenna arrays. The accuracy of this method diminishes with increasing distance between the mobile station and base stations. Time-based radiolocation systems can either measure time of arrival (TOA) of a signal from a mobile to multiple base stations, or time difference of arrival (TDOA) of a signal received at multiple pairs of base stations. The distance between a mobile and a base station is measured by the one-way propagation time between them. This provides a circle centered at a base station, on which a mobile lies. Considering three base stations, the TOA method locates a mobile on the intersection of the three circles. In the TDOA, differences in the TOAs are used creating hyperbolae, with foci at the base stations, on which the mobile lies. Some methods for TOA and TDOA are based on least squares minimization of the location error, some are based on the exact solutions to the hyperbolic equations (for TDOA), and some are based on Taylor-series expansion to linearize the equations that are subsequently used in an iterative algorithm. In the context of the second generation cellular systems (GSM), Caffrey and Stuber (1998) focus on TOA and TDOA algorithms for CDMA systems and their performance is compared favorably with AOA algorithms.

Addressing third generation cellular systems (UMTS), Ahonen (2002) presents location techniques inherited from GSM that are more accurate due to large bandwidth available in UMTS. However, TDOA method can be inaccurate due to non-line-of-sight in urban environments and multipath propagation. Furthermore, the weak signals from far-away base stations can be blocked by strong signals from near base stations. This so-called hearability problem can be diminished by using idle periods in which base stations turn their switches off. Another method in response to the above problems is the so-called Database Correlation Method (DCM), developed for the GSM network, and applied to the UMTS. In DCM, a database consisting of measured or predicted signal levels and some other location related data on the service area is collected a priori. By using the wide bandwidth of UMTS, the location information can be enhanced by power delay profiles (PDPs) containing amplitudes and delay on the multipath components, stored in the database. Due to this additional information, a location can be estimated by using only one base station. Ahonen and Eskelinen (2003) present an overview of UMTS mobile locations and evaluates TDOA and DCM through simulations in an urban microcell environment. The results show the effectiveness of DCM: DCM provided a location error of 25m for 67% of the calls, while respective TDOA accuracy was 215 m. The discrepancy is attributed to obstructions (e.g. buildings) in an urban environment, affecting the time measurements.

20.3.4 Location Management and Mobile Group Communication

Mobile and ad hoc networks (MANET) often need to support group communication, performed via multicast routing protocols. Each node in MANET can also serve as a router, supporting multi-hop path between nodes. The problem, resulting with large overhead, stems from mobility implying a dynamic network topology. Ji and Corson (2001) propose the so-called Differential Destination Multicast (DDM) protocol suitable for use with small multicast groups in ad hoc networks. In contrast to traditional multicast protocols that distribute membership control throughout the network, DDM is source-initiated and controlled. Namely, membership control is concentrated at the data sources via differentially-encoded, variable-length destination headers inserted in data packets. DDM protocol is suitable for multicasting to small groups in networks having unicast routing support because it queries the underlying unicast routing protocol and forwards data packets to multicast members.

Chen and Nahrstedt (2002) introduce another multicast scheme suitable for small multicast groups. Data delivery is based on packet encapsulation using packet distribution trees. The packet distribution tree is constructed so as to minimize the overall bandwidth cost of the tree. There are two types of trees: a location guided k-ary (LGK) tree and a location guided Steiner (LGS) tree. The trees are computed heuristically by using the geometric locations of the destination nodes. The network distance is measured in the number of hops and can be given only if the global topology of the network is known. Since the global topology of the network is not available in this case, assuming that bigger number of hops imply larger geometric distances, the geometric location information is used instead to construct location guided trees. A hybrid location update mechanism is used to disseminate location information among a group of nodes. This mechanism includes in-band and periodic updates. In an in-band update a node includes its geometric location with packets sent. If for an extended period no data is sent, a node sends a periodic update with its present location update. Chen and Nahrstedt (2002) present some simulation results indicating different efficiency depending on the location information of the nodes. If the location information is up-to-date, location guided Steiner tree approach works better (i.e. obtains smaller bandwidth cost). In case of out-dated location information, location guided k-ary tree approach is preferred due to its lower computational complexity. The constructed tree is an overlay multicast packet distribution tree on top of the underlying unicast routing protocol.

Multicast protocols can be either tree-based or mesh-based. Mesh-based protocols have multiple paths between any source and destination pair. When network topology changes frequently, mesh-based multicast protocols are more robust due to alternative available paths. Gui and Mohapatra (2003) present an efficient overlay multicast protocol in MANET environment whereby the constructed overlay mesh topology gradually adapts to the changes in underlying network topology. The adaptation occurs in a fully distributed way. To deal with data delivery, the authors propose the so-called *Source-based Steiner tree* algorithm to compute the delivery tree. The proposed multicast scheme allows for simple join and leave processes by which network topology and group membership change.

The future work on multicast schemes in MANET networks would need to emphasize even further issues such as Quality-of-Service (QoS), reliability, location awareness, and power efficient usage (de Morais Cordero et al., 2003). This might lead to research on efficient heuristic algorithms for constrained Steiner tree problems.

20.4 CONCLUDING REMARKS

Location problems in telecommunications are very diverse. In designing an access/ backbone network, some kind of concentrator location optimization is mandatory. However, new networking environments and new technologies create additional optimization problems involving locations and allocations. An example is provided by optical networks and their capability to superimpose a virtual topology on top of an existing physical topology.

Mobile networks can not function without location management schemes. Dynamic network topologies due to mobility pose a challenge for fast location optimization. Ad hoc mobile networks and the cluster-based approach lead to some kind of dynamic concentrator location (cluster-heads could be reelected).

The Internet contributed to group communication, which in turn leads to location problems associated with multicasting. The complexity of telecommunication networks will continue to present challenging location-based problems related to new topologies, technologies, applications, and the enormous and ever increasing traffic flow. The research area is very active and, unfortunately, it was not possible to list all the important papers in this limited survey. Nonetheless, we hope that we have addressed some diversity of relevant problems and approaches.

Bibliography

S. Ahonen. UMTS location for network planning and optimization in urban areas. In *Proc. LOBSTER Workshop*, Mykonos, Greece, October 2002.

S. Ahonen and P. Eskelinen. Mobile terminal location for UMTS. *IEEE Aerospace and Electronics Systems Magazine*, 18(2):23–27, 2003.

M. Ali. Optimization of splitting node placement in wavelength-routed optical networks. *IEEE Journal on Selected Areas in Communications*, 20(8):1571–1579, 2002.

M. Ali and J. Deogun. Allocation of splitting nodes in all-optical wavelength-routed networks. *Photonic Network Communications*, 2(3):247–265, 2000.

E. Amaldi, A. Capone, and F. Malucelli. Planning UMTS base station location: Optimization models with power control and algorithms. *IEEE Transactions on Wireless Communications*, 2(5):939–952, 2003a.

E. Amaldi, A. Capone, F. Malucelli, and F. Signori. Optimization models and algorithms for downlink UMTS radio planning. In *Proc. IEEE Wireless Communications and Networking Conference (WCNC-03)*, pages 827–831, 2003b.

G. Anandalingam. Optimization of telecommunication networks. In P.M. Pardalos and M.G.C. Resende, editors, *Handbook of Applied Optimization*, pages 826–841. Oxford University Press, 2002.

Y. Bejerano and I. Cidon. Efficient location management based on moving location areas. In *Proc. IEEE INFOCOM 2001*, volume 1, pages 3–12, 2001.

P.S. Bhattacharjee, D. Saha, and A. Mukherjee. An approach for location area planning in a personal communication services network (PCSN). Technical report, Indian Institute of Management (IIM), Calcutta, India, 2003.

J.J. Caffrey and G.L. Stuber. Overview of radiolocation in CDMA cellular systems. *IEEE Communication Magazine*, 36:38–45, 1998.

K. Chen and K. Nahrstedt. Effective location-guided tree construction algorithms for small group multicast in MANET. In *Proc. IEEE INFOCOM 2002*, volume 3, pages 1180–1189, 2002.

C.-F. Chiasserini, I. Chlamtac, P. Monti, and A. Nucci. Energy efficient design of wireless ad hoc networks. In E. Gregori, M. Conti, A.T. Campbell, G. Omidyar, and H. Zuckerman, editors, *NETWORKING 2002 – Networking Technologies, Services, and Protocols; Performance of Computer and Communication Networks; and Mobile and Wireless Communications*, volume 2345 of *Lecture Notes in Computer Science*, pages 376–386. Springer-Verlag, 2003.

C. de Morais Cordero, H. Gossain, and D. Agrawal. Multicast over wireless mobile and ad hoc networks: Present and future directions. *IEEE Network*, 17(1):2–9, 2003.

I. Demirkol, C. Ersoy, M.U. Caglayan, and H. Delic. Location area planning in cellular networks using simulated annealing. In *Proc. IEEE INFOCOM 2001*, pages 566–573, 2001.

A. Fumagalli, G. Balestra, and L. Valcarenghi. Optimal amplifier placement in multi-wavelength optical network based on simulated annealing. In *SPIE International Symposium on Voice, Video, and Data Communications*, volume 3531, pages 268–279, Boston, 1998.

E. Gourdin, M. Labbe, and H. Yaman. Telecommunication and location. In Z. Drezner and H.W. Hamacher, editors, *Facility Location: Application and Theory*, pages 275–305. Springer, 2002.

C. Gui and P. Mohapatra. Efficient overlay multicast for mobile and ad hoc networks. In *Proc. IEEE WCNC 2003*, volume 2, pages 1118–1123, 2003.

L. Ji and M. S. Corson. Differential destination multicast – A MANET multicast routing protocol for small groups. In *Proc. IEEE INFOCOM 2001*, pages 1192–1201, 2001.

X. Jia, X.-D. Hu, and D.-Z. Du. *Multiwavelength Optical Networks*. Kluwer Academic Publishers, 2002a.

X.-H. Jia, D.-Z. Du, X.-D. Hu, H.-J. Huang, and D.-Y. Li. Placement of wavelength converters for minimal wavelength usage in WDM networks. In *Proc. IEEE INFOCOM 2002*, volume 3, pages 1425–1431, 2002b.

J.L. Kennington and E.V. Olinick. Wavelength routing and assignment in a survivable WDM mesh network. Technical report, Department of Engineering Management and Information Systems, School of Engineering, Southern Methodist University, Dallas, 2002.

J. Kleinberg and A. Kumar. Wavelength conversion in optical network. In *Proc. 10th ACM-SIAM Symposium on Discrete Algorithms (SODA)*, pages 566–575, 1999.

L. Li and A.K. Somani. Efficient algorithms for wavelength converter placement. *Optimal Networks Magazine*, pages 1–9, March/April 2002.

S. Mandal, D. Saha, and A. Mahanti. A heuristic search for generalized cellular network planning. In *Proceedings of IEEE International Conference on Personal Wireless Communication (ICPWC)*, New Delhi, India, 2002.

A. Mukherjee, S. Bandyopadhyay, and D. Saha. *Location Management and Routing in Mobile Wireless Networks*. Artech House Mobile Communications Library, 2003.

C.S.R. Murthy and M. Gurusamy. *WDM Optical Networks: Concepts, Design, and Algorithms*. Prentice-Hall PTR, 2002.

R.A. Patterson and E. Rolland. Hybrid fiber coaxial network design. *Operations Research*, 50(3):538–551, 2002.

V.T. Raja and B.T. Han. An efficient heuristic for solving an extended capacitated concentrator location problem. *Telecommunication Systems*, 23(1–2):171–199, 2003.

B. Ramamurthy, J. Iness, and B. Mukherjee. Optimal amplifier placements in a multiwavelength optical LAN/MAN: The unequally powered wavelength case. *IEEE/ACM Transactions on Networking*, 6(6):757–767, 1998.

M.G.C. Resende. Combinatorial optimization in telecommunications. In P.M. Pardalos and V. Korotkich, editors, *Optimization and Industry: New Frontiers*, pages 59–112. Kluwer Academic Publishers, 2003.

E. Rosenberg. Dual ascent for uncapacitated telecommunications network design with access, backbone and switch costs. *Telecommunication Systems*, 16(3–4):423–435, 2001.

L.H. Sahasrabuddhe and B. Mukherjee. Light-trees: Optical multicasting for improved performance in wavelength-routed networks. *IEEE Communication Magazine*, 37(2):67–73, February 1999.

J.H. Siregar, H. Takagi, and Y. Zhang. Optimal wavelength converter placement in optical networks by genetic algorithm. *IEICE Trans. Commun.*, E85-B(6):1075–1082, 2002.

F. Siu and R.K.C. Chang. Effectiveness of optimal node assignments in wavelength division multiplexing networks with fixed regular virtual topologies. *Computer Networks*, 38:61–74, 2002.

J. Skorin-Kapov and J.-F. Labourdette. Rearrangeable multihop lightwave networks: Congestion minimization on regular topologies. *Telecommunication Systems*, 9:113–132, 1998.

N. Skorin-Kapov and M. Kos. The application of Steiner trees to delay constrained multicast routing: A tabu search approach. In *Proc. 7th International Conference on Telecommunications ConTEL*, 2003.

T.E. Stern and K. Bala. *Multiwavelength Optical Networks: A Layered Approach.* Prentice Hall PTR, 2000.

S. Subramaniam, M. Azizoglu, and A.K. Somani. On optimal converter placement in wavelength-routed networks. *IEEE/ACM Transaction on Networks*, 7:754–766, 1999.

Y. Tao and J.Q. Hu. Optimal amplifier placement for optical networks. Technical report, Department of Manufacturing Engineering, Boston University, Brookline, 2003.

P.J. Wan, O. Frieder, and L. Liu. Optimal placement of wavelength converters in trees, tree-connected rings, and tree of rings. *Computer Communications*, 26(1):718–722, 2003.

G. Wilfong and P. Winkler. Ring routing and wavelength translation. In *Proc. 9th ACM-SIAM Symposium on Discrete Algorithms*, pages 333–341, 1998.

D.-N. Yang and W. Liao. Design of light-tree based logical topologies for multicast streams in wavelength routed optical networks. In *Proc. IEEE INFOCOM 2003*, volume 1, pages 32–41, 2003.

21 PRICING AND EQUILIBRIUM IN COMMUNICATION NETWORKS

Qiong Wang[1]

[1]Industrial Mathematics and Business Analysis
Mathematical Sciences Research Center
Bell Labs, Lucent Technologies
Murray Hill, NJ 07974 USA
qwang@research.bell-labs.com

Abstract: Combine pricing and engineering in communication networks improves resources management to a great extent, and optimization models provide a bridge between the two disciplines. In this chapter, we survey a set of network pricing models. Managing network resources involves decision-making at many levels. Correspondingly, our survey takes a "divide and review" approach that classifies various pricing models into three categories: models that support on-line resources allocation; models that assist offline bandwidth provisioning; and models that complement long-term capacity planning. In discussing pricing schemes in each category, we highlight economic thinking behind model formulation, specific issues to be resolved by both pricing and engineering, and important insights obtained from the optimization results.

Keywords: Network pricing models, equilibrium, optimization.

21.1 INTRODUCTION

Network pricing is a topic of multi-disciplinary interest that has been studied from various perspectives. At times when the telecommunications industry was largely a regulated monopoly, pricing research aids regulators to develop effective governing policies. In this context, network services are viewed as goods that can be produced with economy of scale and economy of scope. In other words, the unit cost of providing a network service decreases when the volume of the service increases or the service is offered jointly with other services in one network. Demands for a service exhibit positive externality, i.e., the benefit of a service grows with the size of existing user base (Rohlfs, 1974; Gersho and Mitra, 1975). While these features should be exploited to improve economic efficiencies of network operation and enhance values of communication services, they also give rise to the monopoly power. Many studies were conducted to set prices according to the welfare of the society and align monopoly carriers' profit incentives with social and consumer benefits.

Another set of pricing research focuses on private networks such as intranets in big corporations. Network services are usually offered inside these organizations as free public goods the use of which is not mutually exclusive. Unless the system is sufficiently over-capacitated, users will impose congestion costs (negative externality) on each other. To reduce the damage of network congestion to herself, a user has the incentive to over-subscribe capacity, which can lead to a downward spiral of the infamous "tragedy of commons". One way to solve this problem is to install an internal transfer pricing scheme that encourages users to use the system only when the benefit they enjoy exceeds the congestion cost they create. Mendelson (1985), Dewan and Mendelson (1990), Whang and Mendelson (1990), and Westland (1992) derive optimal transfer prices and discuss their properties based on classical queueing models. In addition, Stidham (1992) and Rump and Stidham (1998) study uniqueness and stability of system equilibrium reached under the optimal transfer prices.

Recent work on network pricing is largely driven by the emergence of the Internet and deregulation of the telecommunications industry. As an open and distributed system, the Internet faces the same problem as the aforementioned private networks on the use of pricing mechanisms to regulate network usage. Deregulation provides incentives to explore the pricing mechanism for revenue and profit management. The two motives are increasingly interweaved as dominant carriers are adopting IP-based network architectures and services.

In the past decade, the topic of network pricing has attracted not only economists but also network engineers who perceive its potential to improve network resources management by integrating pricing into engineering processes. Many proposed pricing schemes are designed to work in tandem with engineering functions. Optimization models have played an important role of connecting the two mechanisms.

In this chapter, we review network pricing models with a focus on those that combine economics with network engineering. As pointed by Mitra (2002), managing network resources involves multiple levels of decision-making that apply to different time scales. Pricing complements engineering functions at each level. Figure 21.1 shows a three-level framework that characterizes different roles played by engineering and pricing. The figure also lays out the organization of the chapter. Pricing models that belong to the first, second, and third categories are reviewed in Sections 21.2, 21.3, and 21.4, respectively. Conclusions are given in Section 21.5.

Our discussions are concentrated on pricing for transmission services in wireline networks. Pricing of information storage and processing services, and pricing of wireless networks, though their importance is hardly to overstate, are not within the scope of our discussion. Also not included in our review are game theory models and empirical studies. Admittedly, the discussion is biased towards the author's own research.

Before we start, it is useful to define a few terms that will be used later. *Utility* refers to the value that a user derives from a network service. Assume rationality, a user's *willingness to pay* for a service equals her utility. The difference between a user's willingness to pay and the actual amount paid is defined as *surplus*. The sum

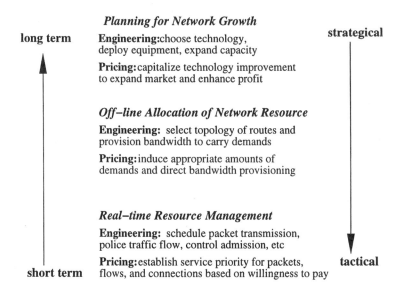

Figure 21.1 Combining pricing and network engineering at different levels

of utility of all users minus costs of providing network services is defined as *social welfare*, which equals the sum of the surplus of all users (defined as *consumer surplus*) plus carriers' *profit*. Users and carriers make decisions to maximize their surplus and profits, respectively, and the outcome is *socially optimal* if it maximizes social welfare.

21.2 ON-LINE PRICING AND RESOURCES MANAGEMENT

The benefit of integrating pricing into on-line network management was spotted quite early by the engineering community. The first volume of *IEEE/ACM Transactions on Networking* published a study by Cocci et al. (1993) who consider a network that offers transmission services to four types of applications: Telnet, FTP, email, and voice. Observing that different applications require different levels of service performance, they propose a scheme that gives performance-sensitive applications, such as Telnet and voice, higher transmission priorities. Performance-insensitive applications, such as email, receive lower service quality in exchange for a price discount. Simulation results show that the new scheme generates more benefits for *all* applications than the scheme that provides uniform service quality and charges a flat price.

The study proved by existence that it is highly desirable to differentiate users in resources allocation and use price incentives to encourage their cooperation. One may even argue that pricing should be the primary tool for traffic management. For instance, under the so-called "Paris Metro Pricing" (Odlyzko, 1999), the network is divided into several logic channels that differ from each other only in price. Even though traffic is served in the same manner in each channel, self-selection by users can lead to a de facto service differentiation where highly-priced channels are expected to experience less congestion than the lowly-priced ones. Nevertheless, in other schemes, pricing is developed as a complement rather than a replacement of engineering functions for managing network resources. The Internet takes the "divide and conquer" approach that places various engineering functions on different layers. Correspondingly, different pricing models have been proposed to work with engineering functions at different layers. In the following, we review these models layer by layer.

21.2.1 Pricing and resource management at physical layer

Pricing schemes in this category are based on the observation that in many cases, one network service can be provided by multiple combinations of resources. In economic terms, the set of all feasible combinations that provide the same service is defined as production possibility curve. Since the usage of network resources is usually unbalanced, the carrier should encourage users to move on the production possibility curve to the direction that substitutes heavily-used resources by lightly-used ones. Pricing provides an incentive mechanism to achieve this objective.

A good example is the combination of bandwidth and buffer, which can substitute each other to some extent (Krishnan and Mandjes, 2002). Low and Varaiya (1993) develop a pricing model in which the price is the weighted sum of rents of bandwidth and buffer space with the weights being the user's relative usage of each resource. The model allows dynamic adjustment of the rents based on total usage of bandwidth and buffer. Users change resource combinations to minimize their payments to the network, and the network changes prices to maximize social welfare.

21.2.2 Pricing and IP packet transmission

Mackie-Mason and Varian (1995) took a pioneering step to explore Internet pricing by introducing the concept of "smart market" into the scheduling of packet transmission. The scheme requires users to mark each IP packet with a bid that reflects their willingness to pay for its immediate delivery. Each router determines transmission priority of packets based on bid values and delay servicing low-bid packets when there is not enough capacity. A transmitted packet is *not* charged by its bid value, but rather the lowest bid value of all packets that are not delayed at the moment. The price is zero when there is sufficient bandwidth to transmit all packets and increases with the network congestion status.

This scheme is an application of the famous Vickery auction that is known to induce a "truth-telling" Nash equilibrium. In other words, a user's strategy that maximizes her surplus in all contingencies is to set the bid exactly equal to her willingness to pay, provided that all other users do the same. Since the transmission priority is established according to bid values, bidding truthfully by all users allows the network to allocate limited capacity to the most valuable packets, and thus produces a socially optimal outcome.

The key feature of the smart market scheme is its ability to induce users, out of self-interest, to provide truthful information to the network. Such kinds of "incentive compatible" mechanisms are highly desirable in a distributed environment. Consequently, the scheme has a lasting influence over many subsequent pricing models.

21.2.3 Pricing and flow control

The smart market mechanism assumes that a user's utility depends on instantaneous delivery of packets and prioritizes IP packet transmission based on willingness to pay. An alternative formulation is to model utility as a function of the end-to-end data rate and associate pricing with the control of the flow rate. An influential model in this category is developed by Kelly et al. (1998) who consider a network that use J types resources to serve R customers. Capacities of resources are fixed and denoted by the vector $\vec{C} = (c_1, c_2, .., c_J)$. A user r's ($r = 1, R$) utility, denoted by $U_r(x_r)$, is a continuous, increasing and concave function of end-to-end flow rate x_r. Total consumption of resources by all users are linear functions of their flow rates, expressed by $A\vec{x}$, where A is a J by R incidence matrix and $\vec{x}$ is the column vector of $x_r, r = 1, R$. The pricing problem is formulated as the following model for maximizing the total utility,

$$\max_{x_r \geq 0} \{ \sum_{r \in R} U_r(x_r) | A\vec{x} \leq \vec{c} \}. \tag{21.1}$$

Kelly et al. (1998) show that the optimal solution to the model can be obtained by solving two related subproblems. The first problem is for every user r to decide her willingness to pay per unit of time, denoted by w_r, to maximize the net utility (surplus), i.e.,

$$\max_{w_r \geq 0} U_r(w_r/\lambda_r) - w_r, \quad \forall r, \tag{21.2}$$

where λ_r is the charge for unit bandwidth and w_r/λ_r is the flow rate that is implicitly decided by user r through her choice of w_r. The network solves the second problem,

$$\max_{x_r \geq 0} \{ w_r \log(x_r) | A\vec{x} \leq \vec{C} \}. \tag{21.3}$$

It is shown that first, there exists a price vector $\vec{\lambda} = (\lambda_1, ..., \lambda_R)$ that when used as inputs, the joint optimization of models (21.2) and (21.3) yields a solution that achieves the global optimum of model (21.1). In fact, $\vec{\lambda} = A\vec{\mu}$, where $\vec{\mu}$ is the vector of Lagrange multipliers of capacity constraints $A\vec{x} \leq \vec{C}$.

Second, given w_r, (21.3) can be solved through two classes of decentralized adaptive algorithms, the primal form algorithm, which changes x_r, and the dual form algorithm, which changes the Lagrange multipliers of constraints $A\vec{x} \leq \vec{C}$.

Finally, if w_r are the same across all $r \in R$. Then

$$\sum_{r \in R} \frac{x_r - x_r^*}{x_r^*} \leq 0,$$

where x_r^* is the solution to (21.1) and x_r is any feasible solution. This implies proportional fairness, i.e., it is impossible to deviate from the optimal solution to give one user a percentage increase of her flow rate without causing a larger aggregated percentage drop of flow rates of all other users.

The work has inspired new studies to analyze network control functions from the utility maximization perspectives. For instance, Low (2003); Low et al. (2002) demonstrate that TCP flow controls can be viewed in general as implicit mechanisms for maximizing Equation (21.1), where different control schemes correspond to maximizing different utility functions. Stolyar (2005) shows that in many network resources allocation problems, a stylized convex optimization problem underlies various utility maximization strategies. He introduces a generalized Greedy Primal-Dual algorithm that is asymptotically optimal. Other studies (Lee et al., 2005) include maximizing utility by controlling many layers of network functions (not just flow rate) and maximizing utility functions that depend on more than one performance metrics.

21.2.4 Pricing and admission control

For the above pricing schemes to work, users have to be flexible to the network performance and willing to accept quality degradation during the service if given sufficiently large price incentive. The assumption does not fit applications that require guaranteed service quality, e.g., real-time voice and video. For these applications, pricing and performance control should be isolated from instantaneous network states. Otherwise, when there is a temporary congestion in the network, users may face an unpleasant choice between having their services terminated prematurely or being charged more than their willingness to pay. It will also be difficult for users to decide their willingness to pay if services they request are not guaranteed to be delivered properly.

One proposed remedy is to supplement the smart market with derivative instruments (Semret and Lazar, 1999). With this approach, a user pays a reservation fee to purchase an option contract before the service starts. The option gives the right to buy bandwidth at a fixed price within the expected service duration. During the service, the user can bid for capacity in the smart market or exercise her option to avoid paying high spot prices. Semret and Lazar (1999) calculate the fair price of an option based on an analogy to derivative pricing in finance literature.

In network engineering, admission control provides quality of service guarantee by keeping the number of users from exceeding some critical value. This threshold can be determined based on the notion of *effective bandwidth*, which is a function of traffic characteristics of all services in progress. In some cases, users may not be confined to any pre-specified service class. To enforce admission control, the network needs users to reveal their traffic parameters in advance as precisely as possible in order to accurately estimate the effective bandwidth. Kelly (1997) shows a pricing scheme that provides such incentives for a situation where traffic follows an on-off process with mean and peak rates chosen freely by users. Roughly speaking, the scheme charges each user a price of

$$P = f(R_d) + f'(R_d) * (R_a - R_d), \qquad (21.4)$$

where R_d is the estimated data rate declared by the user in advance, R_a is the actual data rate measured by the network during the service, and f is a differentiable, increasing, and concave function. It is easy to verify that because of the concavity of f, to pay the minimum amount for a given R_a, the user should declare R_d as close to R_a as possible, i.e., to reveal her best estimate of the actual data rate.

Alternatively, the network can limit users to a set of pre-specified service classes. Users choose the appropriate service class based on QoS requirement, traffic pattern, and expected service duration of their applications. The carrier applies pricing and admission control to each class. Pricing models for this situation are developed by

Jiang and Jordon (1996), Wang et al. (1997), Lanning et al. (1999a), and Keon and Anandalingam (2003).

For instance, in the model by Wang et al. (1997), there are N classes of services with guaranteed QoS and one best effort service with flexible QoS. All services are provided on a network link with capacity C. Users of class i guaranteed service are charged a price, $p_i(t)$, that applies for the duration of the service and varies according to the service starting time. User arrival of service i is assumed to be Poisson and the arrival rate, $\lambda_i(p_i(t),t)$, is time-dependent and controllable by price $p_i(t)$. The service duration is assumed to be exponentially distributed with mean $1/r_i$. Users of the best effort service is charged a time-varying per packet price $p_b(t)$. The aggregated packet arrival of the best effort service at time t is assumed to be Poisson, and the arrival rate, $\lambda_t(p_b(t),t)$, is also a function of time and price.

The pricing problem is formulated as the following optimal control model.

$$\max_{p_i,p_b \geq 0} \int_0^T \{\sum_{i=1,N}[1-\beta_i(t)]p_i(t)\lambda_i(p_i(t),t)/r_i + p_b\lambda_b(p_b,t)\}dt$$

$$\text{subject to} \quad \frac{dq_i}{dt} = [1-\beta_i(t)]\lambda_i(p_i(t),t) - r_i q_i(t),$$

$$A[q_i(t),..,q_N(t)|\beta_1(t),...,\beta_N(t)] \leq C,$$

$$\lambda_b(p_b,t) + s[q_1(t),...,q_N(t)] \leq C. \quad (21.5)$$

Here, $q_i(t)$ is the expected number of service i in progress at time t. Given $q_i(t)$, the amount of capacity needed to 1) satisfy QoS for all guaranteed services, and 2) meet blocking rate target $\beta_i(t)$, is denoted by $A[q_i(t),...,q_N(t)|\beta_1(t),...,\beta_N(t)]$. Values of $q_i(t)$ are determined by $\lambda_i(t)$ that are under control of prices, and price should be chosen to keep A less than capacity C. The last equation is the capacity constraint on average usage, in which $s[q_1(t),...,q_N(t)]$ is the average amount of bandwidth consumed by all guaranteed services at time t.

From this model, the optimal prices

$$p_i^*(t) = \frac{\varepsilon_i}{\varepsilon_i - 1}h_i(t), \; p_b^*(t) = \frac{\varepsilon_b}{\varepsilon_b - 1}l_2(t), \quad (21.6)$$

where

$$\varepsilon_i = \frac{p_i \partial \lambda_i}{\lambda_i \partial p_i} \text{ and } \varepsilon_b = \frac{p_b \partial \lambda_b}{\lambda_b \partial p_b}$$

are defined as price elasticities of demands of various classes of guaranteed services and the best effort service, respectively.

$$h_i(t) = \int_t^T [l_1(\tau)\frac{\partial A}{\partial q_i} + l_2(\tau)\frac{\partial s}{\partial q_i}]e^{r_i(t-\tau)}d\tau, \quad (21.7)$$

is the Hamiltonian associated with the state variable $q_i(t)$ and $l_1(\tau)$ and $l_2(\tau)$ are Lagrange multipliers to the two capacity constraints in (21.4) at time τ. One can interpret $l_1(\tau)$ and $l_2(\tau)$ as unit shadow prices for reserving and using bandwidth at time τ, and $h_i(t)$ as the opportunity cost of accepting a guaranteed service i at time t. Equation (21.7) shows that the opportunity cost equals the weighted sum of shadow costs (where the weights are marginal contributions of the service to capacity constraints) that are aggregated over time with application of the decay factor r_i. The cost is used as the basis for pricing the service, analogous to the way that the marginal cost is used to price a tangible good.

While the above analysis gives a general pricing principle, determining explicit values of the optimal prices remains a challenging mathematical problem. The difficulty is to characterize network states (such as the number of users in progress or the average usage of bandwidth) as functions of prices. This issue has to be addressed not only for the system defined in Equation (21.4), but also for many other network pricing problems. Besides the use of brute force (Wang, 1998) and standard Erlang formula (Keon and Anandalingam, 2003), two systematic approaches have been developed. The first approach is asymptotic analysis for systems with sufficiently large demands and capacities so that the law of large numbers and central limit theorem can be applied (Paschalidis and Tsitsiklis, 2000; Courcoubetis et al., 2001; Maglaras and Zeevi, 2005). As results, the optimal pricing strategy can be developed on the fluid scale where the original stochastic processes are approximated by deterministic ones, or on the diffusion scale where these processes are approximated by Brownian motions. Both deterministic and Brownian motion approximations make the problem tractable.

The other approach relies on the approximation of blocking rates by normal distributions. This line of thoughts has its root in (Jagerman, 1975) and has been extensively discussed in (Eick et al., 1993). Lanning et al. (1999b) apply this approach to solve pricing problems for a single class of network service with time-varying arrival rates. They start by relaxing the capacity constraint to create an offered-load model, and then use the tail of a normal distribution to approximate the probability that the number of users exceeds a threshold value. The latter probability is subsequently used as the approximation of the blocking rate in the original system. Hamshire et al. (2002) extend this approach to cases with multiple services, and develop controlling strategies that involve both pricing and bandwidth provisioning. They also show that the hazard rate function of a normal distribution is a better proxy for blocking rates than a tail value.

21.2.5 Discussions on network equilibrium

Having surveyed a number of specific pricing models, we now take our discussion to a more general level by examining the on-line pricing from the perspective of equilibrium analysis in economic theory.

One important feature of the Internet is that there is no central authority that monitors and controls the network to optimize its performance. Every network element, such as a router or a gateway, manages resources under its control without visibility of the entire network. They rely on a set of common protocols and proper exchange of local information. The overall network performance is a combined effect of decisions by individual elements.

This distributed control structure bears a close relevance to the general equilibrium model in microeconomics (Mas-Collel et al., 1995). In this model, there is a set of commodities possessed by a set of agents. An agent derives some utility by consuming certain amounts of commodities. Agents may also trade commodities among themselves based on a given set of exchange ratios, which are relative prices of commodities. Taking prices as inputs, the agent decide the amounts of commodities to consume and to trade for the maximization of its utility. The decision made by each agent is independent of those made by others.

For each commodity, the total quantity that all agents want to consume is defined as demand and the total amount that all agents plan to exchange for other commodities is defined as supply. The market equilibrium is reached when for every commodity, either the demand equals the supply, or the demand is less than the supply and the price is zero. The globally efficient equilibrium is the one that maximizes total utility of all agents. The general equilibrium theory indicates that the efficient equilibrium can be arrived in the absence of an omniscient global optimizer, and instead, through local optimizations by all agents under the regulation of the "invisible hand" of prices. Specifically, assume agents' utility functions are monotonically increasing and concave function in the amounts of goods consumed. Then there exists a set of prices that induces agents to make local decisions that result in a globally efficient equilibrium.

The network can be viewed as an instance of such system. Let the set of agents be the union of the set of all network elements (referred as element agents) and all users (referred as user agents). The set of commodities are network resources, which are initially owned by different element agents, and the money to pay for network services, which are initially owned by user agents. The utility function of an element agent equals the amount of the money it collects from user agents in exchange of the resources it provides. An user agent derives the utility from a network service, which is obtained by trading the commodity she owns (i.e., make a payment).

This view of network pricing leads to the question about how to arrive at the prices that induce the efficient equilibrium. Thomas et al. (2002) point out that many models, such as the ones developed by Jiang and Jordon (1996), Wang et al. (1997), Murphy et al. (1994), Kelly et al. (1998), Gupta et al. (1997), Low and Varaiya (1993), Veciana and Baldick (1998), and Thomas and Teneketzis (1997), implicitly lead to these prices by centralized optimization. They make this process explicit by replacing the role of centralized optimization by repeated auctioning. In this case, the equilibrium prices are not reached by optimization in one-shot, but through an iterative process which involves the auctioneer repeatedly adjusting prices based on current imbalance between demand and supply. At each iteration, agents take newly-adjusted prices as given and make decisions to maximize their utilities, resulting in new demands and supplies that are used as inputs by the auctioneer for the next round of price adjustment. Mathematically, the adjustment process is equivalent to searching the fixed point in the space of prices. In Thomas et al. (2002), the process is implemented based on Scarf's procedure (Scarf, 1982), which progressively divides the space into a set of increasingly smaller simplices and locates the simplex that contains the fixed point.

For the invisible hand to be effective, the price cannot be manipulated by a single agent. This condition is usually satisfied when there is a large number of independent agents so any uncoordinated action by an individual agent can only have a negligible influence on the price movement. However, if the number of agents is small, then the agents' incentives to game the system and their impacts cannot be ignored. The situation is normally formulated as a non-cooperative game, the outcome of which is known to be inferior to a system under the general equilibrium. The bounds on the difference of the total utility between the two models, sometimes referred as the "price of anarchy," has been estimated in recent studies (Hajek and Gopal, 2002; Johari and Tsitsiklis, 2004). These bounds quantify the impact of selfish user behavior on the economic efficiency of network systems.

21.3 PRICING AND OFF-LINE RESOURCES MANAGEMENT

Off-line resources management selects the topology of the explicit paths for different services and source-destination pairs and allocates link capacity. These functions are referred as *off-line* traffic engineering. The latter complements *on-line* traffic engineering that takes observed network states as inputs, employs real-time routing and admission control to balance realized loads over network links. A traditional approach to optimize off-line traffic engineering is to solve it as some types of multi-commodity flow (MCF) problems (Applegate and Thorup, 2003; Lagoa and Che, 2000; Mitra and Ramakrishnan, 2001; Suri et al., 2001). For instance,

$$\max_{X_r \geq 0} \{ \sum_{s,\sigma} P_{s\sigma} \sum_{r \in R(s,\sigma)} X_r | \sum_{r \in R(s,\sigma)} X_r \leq D_{s\sigma}, \sum_{l \in r} X_r \leq C_l \} \quad (21.8)$$

Here, the network has a set of links (L) with fixed capacity $C_l (l \in L)$. Demand $D_{s\sigma}$ is given as the aggregated traffic volume of service s between (source, destination) pair σ. The model selects route(s) r from a set of candidates $R(s,\sigma)$, defined as the admissible route set, to serve demand (s,σ), and decides X_r, the amount of bandwidth to be provisioned on the chosen route(s). Notice that $R(s,\sigma)$ is referred as the admissible route set in network engineering, the specification of which subjects to QoS and policy (e.g., security) considerations. As results, different services can differ in their admissible route sets even though they have the same source and destination. The choice of X_r should satisfy that a) the provisioned bandwidth does not exceed the demand; and 2) total provisioned bandwidth on a link does not exceed the capacity of the link. These conditions corresponds to two constraints in (21.8). The objective of the optimization is to maximize total revenue, which is obtained by multiplying the amount of provisioned bandwidth with the unit revenue $P_{s\sigma}$, summed over all demand indices (s,σ).

The above MCF model has recently been extended in two directions. One is to introduce price-demand relationship into the model and optimize pricing in conjunction with route selection and bandwidth provisioning. The other extension is to take randomness of demands into account and develop provisioning policies that manage profit risk caused by demand uncertainty. While price optimization does not feature directly in the second model, prices do play an important role in deciding optimal allocation of bandwidth among demands with various uncertainty levels. In the following, we review both models.

21.3.1 Joint optimization of pricing and traffic engineering

Empirical studies indicate that constant price-elasticity model, defined as

$$D_{s\sigma} = A_{s\sigma}P_{s\sigma}^{-\varepsilon_{s\sigma}}, \text{ where } \varepsilon_{s\sigma} = -\left(\frac{P_{s\sigma}}{D_{s\sigma}}\right)\frac{dD_{s\sigma}}{dP_{s\sigma}} \equiv \text{constant}, \qquad (21.9)$$

is adequate for modeling price-demand relationship for bandwidth (Lanning et al., 1999c). The model has been used "to describe the behavior of numerous product categories" and "derive widely accepted conclusions" (Radas and Shugan, 1999). The parameter $A_{s\sigma}$ is interpreted as demand potential.

Mitra and Wang (2002) integrates (21.9) into (21.8) and formulate the following optimization problem

$$\max_{D_{s\sigma},X_{sr}\geq 0}\left\{\sum_{s,\sigma}A_{s\sigma}^{1/\varepsilon_{s\sigma}}D_{s\sigma}^{(\varepsilon_{s\sigma}-1)/\varepsilon_{s\sigma}} \Big| \sum_{r\in R(s,\sigma)}X_{sr}=D_{s\sigma}, \sum_{s,\sigma}\sum_{r\in R(s,\sigma)}X_{s,r}\leq C_l\right\}. \qquad (21.10)$$

Notice that because demand is a function of price, the inequality in the first constraint of (21.8) is replaced by the equality. If there is a slack in the constraint for any (s,σ), a better solution can be generated by holding fixed all X_r so that the total carried bandwidth does not change, while decreasing $D_{s\sigma}$ to the point where the slack disappears, by increasing price $P_{s\sigma}$ appropriately. Furthermore, it is assumed that price elasticity $\varepsilon_{s\sigma} > 1$ for all (s,σ), which is normally the case. As a result, if every link is an admissible route for demands between its end points, the second constraint in (21.10) should also be held at equality. This is because the demand ($D_{s\sigma}$) and the carried bandwidth (X_r) on the direct link may always be increased equally by an amount that makes the unused capacity on the link vanish. Doing so always improves the revenue, which increases with $D_{s\sigma}$ when the elasticity is greater than unity.

The model of (21.10) is a typical instance of convex programming that can be solved efficiently. Applying the first-order necessary condition,

$$P_{s\sigma} = \frac{\varepsilon_{s\sigma}}{\varepsilon_{s\sigma}-1}\mu_{s\sigma}, \ D_{s\sigma} = A_{s\sigma}P_{s\sigma}^{-\varepsilon_{s\sigma}}, \text{ and } \sum_{l\in r}\lambda_l = \mu_{s\sigma} \text{ if } X_r > 0, \qquad (21.11)$$

where λ_l are Lagrange multipliers of link capacity constraints in (21.10), and

$$\mu_{s\sigma} = \min_{r\in R(s,\sigma)}\sum_{l\in r}\lambda_l$$

is interpreted as the *minimum route cost* for (s,σ), using λ_l as the cost metric. The condition establishes a "minimum cost provisioning" policy, i.e., traffic of each stream (s,σ) can only be carried on routes that belong to the *minimum cost admissible route set*. It also shows that the optimal price $P_{s\sigma}$ is determined by λ_l. In this sense, the link shadow cost provides a common basis for unifying pricing and route selection.

Assume that price elasticity depends on service type but not node-pair specific, i.e., $\varepsilon_{s\sigma} = \varepsilon_s$. Then the optimal shadow values of λ_l satisfy

$$\left(\frac{\tilde{A}_{s\min}}{C_{\max}}\right)^{1/\varepsilon_s} \leq \lambda_l \leq \max\left[\left(\frac{\sum_{s,\sigma}\tilde{A}_{s\sigma}}{C_{\min}}\right)^{1/\varepsilon_{\max}}, \left(\frac{\sum_{s,\sigma}\tilde{A}_{s\sigma}}{C_{\min}}\right)^{1/\varepsilon_{\min}}\right], \forall s,l, \qquad (21.12)$$

where $\tilde{A}_{s\sigma} = (\frac{\varepsilon_s - 1}{\varepsilon_s})^{\varepsilon_s} A_{s\sigma}$ and $\tilde{A}_{s\min} = \min_{\sigma} \tilde{A}_{s\sigma}$.

Equation (21.12) bears interesting insights that if the network capacity is large enough so that $C_{min} \geq \sum_{s\sigma} \tilde{A}_{s\sigma}$, then

- *Unbalanced demand growth.* Let $D_{s\sigma}(1)$ be the optimal values of demands when link capacities are at some base level, and $D_{s\sigma}(m)$ be the optimal demands when these capacities are scaled up uniformly by a factor of m. Then

$$\frac{D_{s\sigma}(m)}{D_{s\sigma}(1)} = O(m^{\varepsilon_s/\varepsilon_{\max}}), \tag{21.13}$$

where $\varepsilon_{\max} = \max_s \varepsilon_s$. In other words, when the carrier expands its network, the revenue-maximizing pricing policy should allow only demands with the highest price elasticity to grow at the same rate (linearly) with the capacity increase. Demands of other services grow at sub-linear rates determined by their price elasticity. Given that estimate of price elasticity of voice demand (approximately 1.05) is much lower than that of data service (in the range of 1.3 – 1.7), the above result indicates that there should no surprise that data traffic will surpass voice traffic as the network expands.

- *Dominance of direct traffic.* If a service has sufficiently high price elasticity, then the optimal demands between neighboring nodes will dominate total demands of that service. More specifically,

$$\lim_{\varepsilon_s \to \infty} D_s^d / \sum_{\sigma} D_{s\sigma} = 1, \tag{21.14}$$

where D_s^d is the total demand of service s carried on direct links, i.e., $D_s^d = \sum_{\sigma: M(s,\sigma) \subset L} D_{s\sigma}$.

- *Minimum-hop provisioning.* Let r_1 and r_2 be any two routes connecting a given node pair, with $H(r_1), H(r_2)$ being their respective hop counts. then

$$\exists \varepsilon_m > 0 \text{ if } \varepsilon_{\max} \geq \varepsilon_m, H(r_1) > H(r_2) \Rightarrow \sum_{l \in r_1} \lambda_l > \sum_{l \in r_2} \lambda_l. \tag{21.15}$$

Since bandwidth is always provisioned on minimum-cost routes, the above suggests that it is optimal to provision bandwidth only on minimum-hop routes for all services when price elasticity of at least one service is sufficiently high.

21.3.2 Risk-aware network profit management

Demand uncertainty bears profit risk, so carriers should take a balanced approach to allocate its capacity among multiple demands that are heterogeneous in uncertainty and revenue. The tradeoff is between increasing mean profit, which means to allocate bandwidth to demands that have high revenue potential but may not materialize, and containing profit risk, which means to provide capacity to more stable revenues. Demand uncertainty should also be considered in situations where a carrier engages in bandwidth sharing. The carrier can choose to take a risk in exchange for improvement in mean profit by buying bandwidth from other carriers to serve uncertain demands, or reduces both risk and mean profit by selling capacity to other carriers.

These essential features are captured in a risk-aware revenue management model by Mitra and Wang (2003b). The model starts with the formulation of a two-tier market. The first tier is a retail market where bandwidth is sold as services, such as voice, data, and video. The second tier is a wholesale market where bandwidth is sold as standardized commodity, e.g., DS3 circuits. In the wholesale market, the company can not only sell bandwidth for additional revenue, but also buy bandwidth to augment capacity in its network to serve retail demand. In this sense, the wholesale market provides a mechanism for many carriers to share resources and revenue, like the ones proposed in Anderson et al. (2005) and Linhart et al. (1995), and can be supported by an appropriate engineering architecture (Biswas et al., 2003). It is assumed that in the retail market, demands are random and associated with higher unit revenues. In the wholesale market, the carrier can buy and sell bandwidth at a lower price.

Turning to the formulation of the optimization model, let L_U be the set of links on which the carrier owns capacity and L_V be the set of links on which the carrier can buy capacity. The set of all links L is the union L_U and L_V, where L_U and L_V are not necessarily mutually exclusive. The amount of capacity the carriers owns on link $l \in L_U$, denoted by C_l, is an input parameter. The amount of capacity that the carrier buys on link $l \in L_V$, denoted by b_l, is a decision variable.

Assume that the price per unit of bandwidth on link $l \in L_V$ in the wholesale market is p_l, then $\sum_{l \in L_V} p_l b_l$ is the carrier's cost of buying bandwidth, and $C_l + b_l$ is the amount of bandwidth on link l at the carrier's disposal.

Let V_1 and V_2 be the collection of node pairs between which there are retail and wholesale demands, respectively. For each $v \in V_1$, the volume of retail demand is a random variable denoted by D_v. Each demand is associated with an admissible route set $R_1(v)$ for $v \in V_1$ and $R_2(v)$ for $v \in V_2$. Let ξ_r be the amount of bandwidth provisioned on route $r \in R_1(v)$ and ϕ_r be the amount of bandwidth provisioned on route $r \in R_2(v)$. Then $d_v = \sum_{r \in R_1(v)} \xi_r$ is total bandwidth provisioned to retail demand $v \in V_1$ and $y_v = \sum_{r \in R_2(v)} \phi_r$ is total bandwidth offered for wholesale between node pair $v \in V_2$. The choice of these variables should satisfy link capacity constraints

$$\sum_{r \in R_1(v): l \in r} \xi_r + \sum_{r \in R_2(v): l \in r} \phi_r \leq C_l + b_l \ (l \in L) \tag{21.16}$$

Denote π_v as the unit revenue of retail demand $v \in V_1$. Let e_v be the unit price of bandwidth between node pair $v \in V_2$ in the wholesale market, then $e_v = p_l$ where v is the end node pair of link $l \in L_V$. The carrier's profit

$$W = \sum_{v \in V_1} \pi_v \min(d_v, D_v) + \sum_{v \in V_2} e_v y_v - \sum_{l \in L_V} p_l b_l, \tag{21.17}$$

is a random variable. Let $m_v(d_v)$ and $s_v^2(d_v)$ be the mean and variance of $\min(d_v, D_v)$. Then

$$m_v(d_v) = \int_0^{d_v} x f_v(x) dx + d_v \bar{F}_v(d_v) = \int_0^{d_v} \bar{F}_v(x) dx \qquad (21.18)$$

and

$$s_v^2(d_v) = \int_0^{d_v} x^2 f_v(x) dx + d_v^2 \bar{F}_v(d_v) - m_v^2(d_v) = 2 \int_0^{d_v} x \bar{F}_v(x) dx - m_v^2(d_v), \qquad (21.19)$$

where $F_v(x)$ is the CDF of demand D_v and $\bar{F}_v(x) \equiv 1 - F_v(x)$.

In the presence of demand uncertainty, the objective function should take profit risk into proper consideration. To this end, one can borrow two popular risk-management frameworks from the finance literature:

- *Mean-risk analysis* addresses the issue of risk averseness by offering a optimization objective that is more general than the expected value. The approach starts with developing a risk index, which is a quantitative measure of the risk of profit shortfall, based on the profit distribution. It then maximizes the weighted combination of the mean profit and the risk index, i.e., *mean* $- \delta *$ (*risk index*), where $\delta \geq 0$ is a parameter. Different levels of risk averseness can be reflected by choosing different values for δ. A higher value of δ indicates greater willingness to sacrifice the mean profit to avoid risk.

- *Stochastic dominance theory* defines a partial ordering of random variables based on their probability distributions (Levy, 1998). Let W_1 and W_2 be two random variables, which represent profits under two different bandwidth management decisions. Then W_1 stochastically dominates W_2 to the first degree iff the former renders the carrier a better chance to exceed *any* profit target w, i.e.,

$$\forall w, \; Pr(W_1 \geq w) \geq Pr(W_2 \geq w). \qquad (21.20)$$

Furthermore, W_1 stochastically dominates W_2 to the second degree iff

$$\forall w, \; \int_w^\infty Pr(W_1 \geq \zeta) d\zeta \geq \int_w^\infty Pr(W_2 \geq \zeta) d\zeta. \qquad (21.21)$$

Clearly, one should make the provisioning decision that generates a stochastically efficient solution, i.e., the resulting profit distribution is not dominated in either degree,

Nevertheless, the two frameworks are not necessarily compatible. It is shown by Mitra and Wang (2003a) that for certain choices of the risk index, such as the variance of profit, an optimal solution to the mean-risk model can be stochastically inefficient. They also show that the optimal solution of the mean-risk model is asymptotically efficient if the standard deviation of profit,

$$s(W) = \sqrt{\sum_{v \in V_1} \pi_v^2 s_v^2(d_v)}.$$

is used as the risk index. In other words, in a network that has many independent demands, a solution that maximizes, $E(W) - \delta * s(W)$ is not stochastically dominated by other feasible solutions.

Using the standard deviation as the risk index, the goal of risk-aware network profit management model is to maximize

$$E(W) - \delta s(W) = \sum_{v \in V_1} \pi_v m_v(d_v) + \sum_{v \in V_2} e_v y_v - \sum_{l \in L_v} p_l b_l - \delta \sqrt{\sum_{v \in V_1} \pi_v^2 s_v^2(d_v)}. \quad (21.22)$$

$$\text{subject to} \sum_{r \in R_1(v)} \xi_r = d_v \ (v \in V_1), \ \sum_{r \in R_2(v)} \phi_r = y_v \ (v \in V_2), \quad (21.23)$$

$$\sum_{r \in R_1(v): l \in r} \xi_r + \sum_{r \in R_2(v): l \in r} \phi_r \leq C_l + b_l \ (l \in L). \quad (21.24)$$

It is shown (Mitra and Wang, 2003a; 2005) that while the non-linear objective function of this model is not concave everywhere, it is so in a region that contains the optimal solution. For problems with large number of random demands, the optimal solution is located in the intersection of a polyhedron and the region in which the objective function is concave. Consequently, this model can be solved as a convex programming problem.

21.4 PRICING AND CAPACITY PLANNING

The top block in Figure 21.1 refers to long-term, multi-period network capacity planning, which has been traditionally treated as a cost minimization problem. The model takes as inputs forecasted demands between origin-destination pairs in each period, and a technology roadmap that specifies a set of equipment for providing capacity, the time when each equipment becomes available, and costs of deploying and operating each equipment on every network link in different periods. The optimization model chooses capacity configurations over time to meet projected demands at the minimum net present value (NPV) of investment and operating costs within the planning horizon, with the application of a given discount rate. The main issues are a) whether to deploy larger equipment that requires higher up-front investment but saves incremental costs to accommodate future demand growth; b) whether to make an investment to replace existing equipment with newer, more efficient systems to save operating expenses; and c) whether to delay the adoption of new systems to benefit from future reductions of equipment costs.

From a broader perspective, capacity planning and pricing are tightly connected in carrier's overall strategy of profit maximization. Pricing determines demand, which is the basis for capacity planning. Reciprocally, the optimal price is a function of costs, which depend on capacity planning. Recent optimization models take this interdependency into consideration, making pricing and capacity decisions jointly to maximize profit instead of minimizing cost.

Lanning et al. (2000) develop a case study that illustrates just how significantly that the inclusion of pricing decisions in capacity planning optimization can influence the outcome. The model has the following ingredients:

- *Network.* The network has N nodes that are connected as a two-fiber unidirectional ring of $N+1$ links. For the sake of reliability, traffic originated from each node is sent along both directions. Therefore, each link has the same load that equals total demands between all node pairs.

- *Planning horizon.* The planning horizon is divided into T periods. The objective is to maximize NPV of cash flow (revenue minus cost), using a fixed discount rate ρ. The terminal value TV reflects the value of the network after the planning horizon and is assumed to be proportional to the cash flow in period T.

- *Price-demand relationship.* The same as Equation (21.9), the price-demand relationship is characterized by the constant price-elasticity model, parameterized by demand potential $A_{\sigma t}$ and price elasticity ε. The latter is uniform across all node pairs σ and periods t.

- *Equipment.* Network capacity is provided by installing different versions of wavelength division multiplexers (WDM), indexed by a sorted set K, i. e., If $k_1 < k_2 (k_1, k_2 \in K)$, then k_1 is an older version of WDM than k_2. The period in which version k WDM first becomes available is denoted by τ_k and its maximum capacity is denoted by κ_k.

- *Cost.* The investment cost of version k WDM in period t is I_{kt} per system and numbers of WDMs to install in each period are decision variables denoted by $b_{kt}(k \in K, t \in T)$. Given the symmetry of the single ring network, the same WDMs are installed on every node, so the total investment cost in period t equals $N\sum_{k \in K} I_{kt} b_{kt}$. The numbers of WDMs in operation in each period are also decision variables, denoted by u_{kt}. The operating expense is proportional to the total fiber-miles of the ring, which equals $2L\sum_k m_{kt} * u_{kt}$, where L is the perimeter of the ring and m_{kt} is the expense per mile for WDM of version k in period t.

- *Technology disruptiveness.* A newer version of WDM always improves old ones in both maximum capacity and unit investment cost, i.e.,

$$\kappa_k = \mu \kappa_{k-1},\ \mu > 1,\ \text{and}\ \frac{I_{k,\tau_k}}{\kappa_k} = (1-d)\frac{I_{k-1,\tau_{k-1}}}{\kappa_{k-1}}\ d < 1, \qquad (21.25)$$

where parameter d (d for disruptiveness) is defined as the reduction in initial investment cost per unit of capacity of kth system over $(k-1)^{th}$ system. The larger the value of d, the greater reduction in the per-unit investment cost, i.e., the greater disruptiveness.

The model is formulated as

$$\max_{\vec{d},\vec{u},\vec{b}\geq 0} NPV = \sum_{t=1}^{T} \rho^t [\sum_{\sigma} A_{\sigma t}^{1/\varepsilon} D_{\sigma t}^{1-1/\varepsilon}$$
$$- \sum_{k \in K} (NI_{kt}b_{kt} + 2L * m_{kt} * u_{kt})] + \rho^{T+1} TV$$

subject to
$$\sum_{\sigma} D_{\sigma t} \leq \sum_{k \in K} \kappa_k u_{kt}, \quad t = 1, \ldots T$$
$$u_{kt} \leq b_{kt} + u_{kt-1}, \quad k \in K, \ t = 2, \ldots T,$$
$$b_{kt} = 0, \quad t < \tau_k, \ k \in K, \ t = 1, \ldots T,$$
$$u_{k1} = b_{k1}, \quad k \in K. \tag{21.26}$$

The most insightful message from this optimization is conveyed by Figure 21.2, which shows the optimal capacity used ($\sum_{k \in K} \kappa_k u_{kt}$) in the first six periods for cases associated with different values of d and ε. When the unit cost of WDM decreases more steeply between generations, it makes sense to deploy a smaller number of the first version of WDM at the beginning but add capacity more rapidly later when newer systems become available. This explains why in the figure dashed lines always start below dotted lines but rise quickly to surpass the latter. More interestingly, the figure shows that given the same value of d and other parameters, there can still be huge differences in capacity expansion solely because different values of price elasticity is used. When price elasticity is high ($\varepsilon = 2.1$), capacity expands continuously and the total amount is much larger than the case when the elasticity is low ($\varepsilon = 1.3$). The deployment of new capacity in the latter case is also more sporadic, as shown by the plateaus on the curve. Intuitively, high elasticity means a slight drop in price stimulate a large increase in demands. So the carrier can be more profitable by decreasing the price and benefiting from large growth in demand period by period when unit cost are being reduced continuously by the emergence of newer WDMs. The opposite is true with the case of low price elasticity. When demands is less responsive to price change, then decreasing price may not benefit the carrier, and the demand growth and capacity expansion will remain stagnant. The discussion suggests that a profit-maximization planning model produces more than just an optimal capacity deployment schedule. It also suggests an appropriate business model – the carrier should provide service at a lower margin to a huge market if price elasticity is high, and do the opposite otherwise. In this sense, incorporating pricing into capacity planning elevates the optimization model from a tactical tool that enhances operations management to a modeling framework that supports strategic decision-making.

Optimizing real networks can be technically more complicated than the prototype model of (21.25). For example, a real network usually has too many nodes to be connected into a single ring, and the normal architecture is either a set of inter-connected rings or a mesh. In this case, routing is not obvious and needs to be optimized. To do that, one has to define a set of flow variables to represent amounts of each demand to be carried on different paths between the origin and the destination. The choice of flow variables should satisfy demand satisfaction and link capacity constraints, and sometimes meet additional reliability and protection requirements, e.g., flows can only be carried on disjoint paths in order to be able to backup each other when link fails.

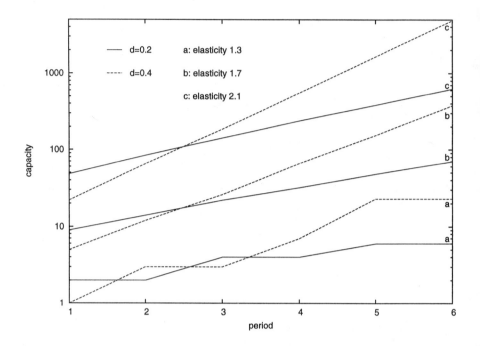

Figure 21.2 Capacity (shown on a log scale) over time.

As a result of these considerations, a large number of flow variables and associated constraints have to be included in the optimization model. Furthermore, more detailed capacity planning in optical networks involves more decisions than simply determine numbers of WDMs. A WDM typically hosts a variable number of wavelengths that can be added or dropped at the carrier's discretion. The number of wavelengths (for each link, each version of WDM, and each period) can form another set of variables to be optimized. After taking these factors into consideration, one usually ends up with a much larger optimization problems that contains hundreds thousands or even millions of variables and constraints for a reasonable size networks.

Bienstock et al. (2003) propose scalable techniques that solves (21.25) for large networks. In their approach, the problem is given in the following general format

$$\max_{\vec{D}, \vec{y} \leq 0} \{R(\vec{D}) - c^T \vec{y} | A\vec{D} + B\vec{y} \leq \vec{b}\}, \tag{21.27}$$

where $\vec{D}$ are demand variables and $\vec{y}$ are linear variables representing flows and numbers of equipment, and R is a concave and increasing in $\vec{D}$. The optimization is difficult because the number of y variables can be quite large. However, if these variables can be expressed as linear functions of $\vec{D}$, then the problems becomes a concave maximization model with far fewer variables, which is much easier to solve. The algorithm is built upon this observation.

Specifically, the Linear Programming indicates that the optimal values of $\vec{y}$ is a concave, piecewise linear function of $\vec{d}$. In other words, if F is the feasible region of $\vec{d}$, then $F = \cup_{j=1,M} F_j$ (M is a finite number). For any $d \in F_j$, the optimal value $\vec{y} = a_j^T \vec{d}$. Making use of this property, the optimization schedule by Bienstock et al. (2003) contains following steps.

1. Start with an initial solution $\vec{d} \in F_j$ and determine a_j.

2. Optimize $R(\vec{d}) - a_j^T \vec{d}$. Let $\vec{d}^*$ be the optimal solution.

3. Use a first-order method to find a better solution than $\vec{d}^*$. Either $\vec{d}^*$ cannot be improved, which means that the optimal solution has been found; or an improvement is found in another region $F_k \neq F_j$, in which case let $F_j = F_k$ and go to step 2.

Numerical experiments (Raskina, 2002) show that the above procedure can solve planing models for large networks quite efficiently.

21.5 CONCLUSION

While better engineering design makes communication networks more efficient, incorporating economic instruments such as pricing can create even more values. In this chapter, we review models of pricing optimization that combine economics with network engineering. As we have shown, there are different layers of resources management in the network, and pricing complements engineering functions at each layer. It provides incentives for users' cooperation, regulates demand volumes and improves revenues and profits, and translates cost reductions into market share and profit improvement.

Discussions in this chapter are mostly from a single carrier's perspective of maximizing a given objective function. An alternative paradigm is to formulate strategic interactions of multiple carriers and analyze the equilibrium states (Shakkottai and Srikant, 2005). In this case, optimization models play an important role as subroutines for calculating a carrier's basic moves that determine the final outcome. For instance, part of the motivation to build a stochastic traffic engineering models in (Mitra and Wang, 2003b;a; 2005) is to develop a framework for quantifying value propositions for a carrier in sharing capacity with its peers.

The common assumption under most models is that network resources are either constrained or have to be payed for. Consequently, the focus on the pricing design is about how to use limited resources more efficiently. Another interesting aspect of pricing study is about demand management. For instance, how to model substitution effects between different services and take them into consideration in pricing decisions. How to bundle various services into a single package. These questions become increasingly important in situations capacity constraints become a less concern to the carrier than the need to generate more revenues.

From a broader perspective, optimization models can be complemented by other ways of designing network pricing. For instance, one can start with a set of common principles that should be satisfied by all pricing designs, translate these principles into a set of axioms, and use these axioms to derive the pricing schedule. This approach has been used by Herzog et al. (1997) in their discussion on the pricing a multi-cast tree, and can be extended to other cases.

We end this chapter by noting that while design a network pricing schedule is a challenging task by itself, it is also important to address tradeoffs involved in pricing decisions between carrier's profit, consumer benefits, and social efficiency. As shown by Wang and Peha (2001), a pricing scheme that improves social welfare can be detrimental to consumer surplus. Therefore, it is not always desirable to design pricing models solely based on efficiency argument. In some cases, the best pricing may be a compromise of satisfying multiple objectives.

Bibliography

E. Anderson, F. Kelly, and R. Steinberg. A contract and balancing mechanism for sharing capacity in a comm unication network. Technical report, Australian Graduate School of Management, University of New South Wales, Australia, March 2005. Available online at http://www2.agsm.edu.au/agsm/web.nsf/Content/Faculty-FacultyDirectory-E%dwardAnderson-research.

D. Applegate and M. Thorup. Load optimal MPLS routing with $n+m$ labels. In *Proceedings of IEEE INFOCOM2003*, April 2003.

D. Bienstock, O. Raskina, I. Saniee, and Q. Wang. Combined network design and multi-period pricing: Modeling, solution techniques and computation. *to appear in Operations Research*, 2003.

S. Biswas, D. Saha, and N. Mandal. Intercarrier bandwidth exchange : An engineering framework. *IEEE Communications Magazine*, 41(1):130–138, 2003.

R. Cocci, S. Shenker, D. Estrin, and L. Zhang. Pricing in computer networks: Motivation, formulation, and example. *IEEE/ACM Transactions on Networking*, 1(6): 614–627, 1993.

C. A. Courcoubetis, A. Dimakis, and M. I. Reiman. Providing bandwidth guarantees over a best-effort network: Call admission and pricing. In *Proceedings of IEEE INFOCOM2001*, pages 459–467, April 2001.

S. Dewan and H. Mendelson. User delay cost and internal pricing for a service facility. *Management Science*, 36:1502–1517, 1990.

S. Eick, W. A. Massey, and W. Witt. The physics of the $m(t)/g/\infty$ queue. *Operations Research*, pages 400–408, 1993.

A. Gersho and D. Mitra. A simple growth model for the diffusion of a new communication service. *IEEE Transactions on Systems, Man, and Cybernetics*, 5(2): 209–216, March 1975.

A. Gupta, D. O. Stahl, and A. B. Whinston. A stochastic equilibrium model of Internet pricing. *Journal of Economic Dynamics and Control*, 21:697–722, 1997.

B. Hajek and G. Gopal. Do greedy autonomous systems make for a sensible internet? In *Conference on Stochastic Networks*, June 2002.

R. Hamshire, W. A. Massey, D. Mitra, and Q. Wang. Provisioning of bandwidth sharing and exchange. In G. Anandalingam and S. Ragahavan, editors, *Telecommunications network design and economics and management: Selected proceedings of the 6th INFORMS Telecommunications Conference*, pages 207–226. Kluwer Academic Publishers, Boston/Dordrecht/London, 2002.

S. Herzog, S. Shenker, and D. Estrin. Sharing the "cost" of multicast trees: An axiomatic analysis. *IEEE/ACM Transactions on Networking*, 5:847–860, December 1997.

D. L. Jagerman. Nonstationary blocking in telephone traffic. *Bell System Technical Journal*, pages 625–661, 1975.

H. Jiang and S. Jordon. A pricing model for high speed networks with guaranteed quality of service. In *Proceedings of IEEE INFOCOM*, pages 888–895, March 1996.

R. Johari and J. N. Tsitsiklis. Efficiency loss in network resource allocation game. *Mathematics of Operations Research*, 29(3):407–435, August 2004.

F. K. Kelly. Charging and accounting for bursty connections. In J. Bailey and L. McKnight, editors, *Internet Economics*, pages 253–278. MIT Press, 1997.

F. P. Kelly, A. K. Maullo, and D. K. H. Tang. Rate control in communication networks: Shadow prices, proportional fairness and stability. *Journal of the Operational Research Society*, 49:237–252, 1998.

N. Keon and G. Anandalingam. Optimal pricing for multiple services in telecommunications networks offering quality-of-service guarantee. *IEEE/ACM Transactions on Networking*, 11(1):66–80, February 2003.

K. Krishnan and M. Mandjes. The buffer-bandwidth tradeoff curve is convex. *Queueing Systems*, 38(4):471–483, 2002.

C. Lagoa and H. Che. Decentralized optimal traffic engineering in the internet. *Computer Communications Review*, 30:39–47, October 2000.

S. G. Lanning, W. A. Massey, B. Rider, and Q. Wang. Optimal pricing in queuing systems with quality of service constraints. In *Proceedings of the 16th International Teletraffic Congress-ITC16*, pages 747–756, June 1999a.

S. G. Lanning, W. A. Massey, B. Rider, and Q. Wang. Optimal pricing in queuing systems with quality of service constraints. In *Proceedings of the 16th International Teletraffic Congress*, pages 747–756, June 1999b.

S. G. Lanning, D. Mitra, Q. Wang, and M. H. Wright. Optimal planning for optical transport networks. *Phil. Transaction of Royal Society, London A.*, 358(1773): 2183–2196, August 2000.

S. G. Lanning, W. R. Neuman, and S. O'Donnell. A taxonomy of communications demand. In *Proceedings of 27th Telecommunications Policy Research Conference*, September 1999c.

J. W. Lee, M. Chiang, and R. A. Calderbank. Price-based distributed algorithm for optimal rate-reliability tradeoff in network utility maximization. *preprint*, 2005.

H. Levy. *Stochastic Dominance: Investment Decision Making Under Uncertainty*. Kluwer Academic Publishers, 1998.

P. B. Linhart, K. G. Ramakrishnan, R. Radner, and R. Steinberg. Allocation of value for jointly provided services. *Telecommunications Systems*, pages 151–175, 1995.

S. H. Low. A duality model of tcp and queue management algorithm. *IEEE/ACM Transactions on Networking*, 11(4):525–536, August 2003.

S. H. Low, L. Peterson, and L. Wang. Understanding tcp vegas: Theory and practice. *Journal of ACM*, 49(2):207–235, March 2002.

S. H. Low and P. P. Varaiya. A new approach to service provisioning in ATM networks. *IEEE/ACM Transactions on Networking*, 1(5):547–553, 1993.

J. K. Mackie-Mason and H. Varian. Pricing congestible resources. *IEEE Journal of Selected Areas in Communications*, 13(7):1141–1149, July 1995.

C. Maglaras and A. Zeevi. Pricing and design of differentiated services: Approximate analysis and structural insights. *Operations Research*, 53(2):242–262, March/April 2005.

A. Mas-Collel, M. Whinston, and J. Green. *Microeconomic Theory*. Oxford University Press, 1995.

H. Mendelson. Pricing computer services: Queuing effects. *Communications of ACM*, 28(3):312–321, 1985.

D. Mitra. Combine network economics and engineering over many scales. In *Plenary Speech at 6th INFORMS Telecommunications Conference*, March 2002.

D. Mitra and K. G. Ramakrishnan. Techniques for traffic engineering of multiservice multipriority networks. *Bell Labs Technical Journal*, 6(1):17–34, January 2001.

D. Mitra and Q. Wang. Generalized network engineering, optimal pricing and routing for multi-service networks. In *Proceedings of SPIE: Scalable Service Provisioning and Traffic Control in IP Networks*, volume 4868, August 2002.

D. Mitra and Q. Wang. Risk-aware network profit management in a two-tier market. In *Proceedings of 18th International Teletraffic Congress -ITC18*, August 2003a.

D. Mitra and Q. Wang. Stochastic traffic engineering, with applications to network revenue management. In *Proceedings of IEEE INFOCOM2003*, April 2003b.

D. Mitra and Q. Wang. Stochastic traffic engineering for demand uncertainty and risk-aware network revenue management. *IEEE/ACM Transactions on Networking*, 13(2):221–233, April 2005.

L. Murphy, J. Murphy, and J. K. Mackie-Mason. Bandwidth allocation by pricing ATM networks. In *Broadband Communications, II, Proceedings of the IFIP TC6 Second International Conference on Broadband Communications C-24*. Elsevier Science Publishers B. V., 1994.

A. Odlyzko. Paris metro pricing for the internet. In *Proceedings of the 1st ACM conference on Electronic Commerce*, pages 140–147, 1999.

I. Ch. Paschalidis and J. N. Tsitsiklis. Congestion-dependent pricing of network services. *IEEE/ACM Transactions on Networking*, 8(2):171–184, April 2000.

S. Radas and S. Shugan. Managing service demand: Shifting and bundling. *Journal of Service Research*, 1(1):47–64, August 1999.

O. Raskina. *Two large-scale network design problems*. PhD thesis, Columbia University, New York, NY, 2002.

J. Rohlfs. A theory of interdependent demand for a communications service. *Bell Journals of Economics*, 5:16–37, Spring 1974.

C. M. Rump and S. Stidham. Stability and chaos in input pricing for a service facility with adaptive customer response to congestion. *Management Science*, 44(2):246–261, 1998.

H. E. Scarf. The computation of equilibrium prices: An exposition. In K. J. Arrow and M. D. Intriligator, editors, *Handbook of Mathematical Economics*, volume 2, pages 1007–1061. North-Holland Publishing Company, 1982.

N. Semret and A. A. Lazar. Spot and derivative markets in admission control. In *Proceedings of the 16th International Teletraffic Congress-ITC16*, June 1999.

S. Shakkottai and R. Srikant. Economics of network pricing with multiple isps. In *Proceedings of IEEE INFOCOM 2005*, March 2005.

S. Stidham. Pricing and capacity decisions for a service facility: Stability and multiple local optima. *Management Science*, 38(8):1121–1139, 1992.

A. Stolyar. On the asymptotic optimality of the gradient scheduling algorithm for multiuser throughput allocation. *Operations Research*, 53(1):12–25, January/February 2005.

S. Suri, M. Waldvogel, and P. R. Warkhede. Profile-based routing: A new framework for MPLS traffic engineering. In *QofIS'01*, 2001.

P. Thomas and D. Teneketzis. An approach to service provisioning with quality of service requirements in ATM networks. *Journal of High Speed Networks*, 6(4): 263–291, 1997.

P. Thomas, D. Teneketzis, and J. K. Mackie-Mason. A market-based approach to optimal resource allocation in integrated-services connection-oriented networks. *Operations Research*, 50(4):603–616, 2002.

G. De Veciana and R. Baldick. Resource allocation in multi-service networks via pricing. *Computer Networks ISDN Systems*, 30(9-10):951–962, 1998.

Q. Wang. *Pricing of Integrated Service Networks*. PhD thesis, Carnegie Mellon University, Pittsburgh, PA, 1998.

Q. Wang and J. M. Peha. State-dependent pricing and its economic implications. *Telecommunications Systems: Modeling, Analysis, Design, and Management*, 18 (4):315–329, December 2001.

Q. Wang, J. M. Peha, and M. A. Sirbu. Optimal pricing for integrated-services networks. In J. Bailey and L. McKnight, editors, *Internet Economics*, pages 353–376. MIT Press, 1997.

J. C. Westland. Congestion and network externalities in the short run pricing of information services. *Management Science*, 38(7):992–1009, 1992.

S. Whang and H. Mendelson. Optimal incentive-compatible priority pricing for the M/M/1 queue. *Operations Research*, 38:870–883, 1990.

III Routing

22 OPTIMIZATION OF DYNAMIC ROUTING NETWORKS

Gerald R. Ash[1]

[1] AT&T Labs
Middletown, NJ 07748 USA
gash@att.com

Abstract: This chapter describes optimization methods useful for the design of dynamic routing protocols, as well as for the concomitant optimization of dynamic routing network design and dynamic routing performance. A variety of dynamic routing methods are used in practice, which include time-dependent, state-dependent, and event-dependent algorithms, and these are briefly outlined. We briefly review current practice in dynamic routing protocol design in voice, data, and integrated voice/data networks. Case studies are given for dynamic routing protocol design for a) real-time network routing, a state-dependent dynamic routing method used in practice in a very large-scale application, and b) integrated voice/data routing in an IP/MPLS network application. Methods for optimal min-cost network design and max-flow performance optimization include traffic-load-flow, discrete-event flow, and virtual-transport-flow optimization models. Examples are given in the application of these methods.

Keywords: Routing protocols, dynamic routing, concomitant optimization, telecommunications.

22.1 INTRODUCTION TO DYNAMIC ROUTING (DR) NETWORKS

Factors driving network routing evolution and optimization are performance quality improvements, new services introduction, and technological evolution. Performance quality improvement is achieved when network routing increases real-time adaptivity and robustness to load shifts and failures. New services are enabled, for example, when DR capability provides new high-priority services and dynamic bandwidth allocation. Technological evolution is supported when network routing capitalizes on new transmission, switching, and network management/design technologies to achieve simpler, more automated, and more efficient networks

Figure 22.1 illustrates a model for network routing and network management/design. The central box represents the network, which can have various architectures and configurations, and the traffic routing tables and transport routing tables within the network. Network configurations include metropolitan area networks, national intercity networks, and global international networks. Routing tables describe the route choices from an originating node to a terminating node, for a connection request for a particular service. Hierarchical and nonhierarchical traffic routing tables are possible, as are fixed routing tables and DR tables. Routing tables are used for a multiplicity of traffic and transport services on the telecommunications network.

Terminology used in the chapter is that a *link* connects two nodes, a *path* is a sequence of links connecting an origin and destination node, and a *route* is the set of different paths between the origin and destination that a connection might be routed on.

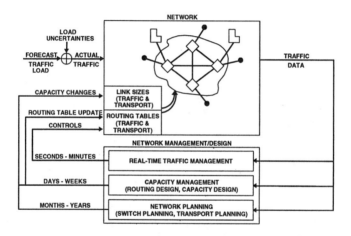

Figure 22.1 Network routing, management, & design

Network management/design functions include real-time traffic management, capacity management, and network planning. Real-time traffic management ensures that network performance is maximized under all conditions including load shifts and failures. Capacity management ensures that the network is designed and provisioned to meet performance objectives for network demands at minimum cost. Network planning ensures that switching and transport capacity is planned and deployed in advance of forecasted traffic growth. Figure 22.1 illustrates real-time traffic management, capacity management, and network planning as three interacting feedback loops around the network. The input driving the network is a noisy traffic load, consisting of predictable average demand components added to unknown forecast error and load variation components. The load variation components have different time constants ranging from instantaneous variations, hour-to-hour variations, day-to-day variations, and week-to-week or seasonal variations. Accordingly, the time constants of the feedback controls are matched to the load variations, and thereby regulate the service provided by the network through capacity and routing adjustments. Real-time traffic management provides monitoring of network performance through collection and display of real-time traffic and performance data, and allows traffic management controls and maximum flow routing designs to be introduced when circumstances warrant. Capacity management plans, schedules, and provisions needed capacity over a time horizon of several months to one year or more, based on minimum cost network design.

22.1.1 Classification of dynamic routing networks

Connection routing methods are used for establishment of a bearer path for a given service request or session flow, and include fixed routing, time-dependent routing, state-dependent routing, and event-dependent routing methods. In addition, networks can be either *flat* or *hierarchical*.

Hierarchical routing is employed in all types of networks, including packet-based and time-division-multiplex (TDM)-based networks. In both cases, there is a hierarchical relationship among various sub-networks, which depending on the switching technology can be called *regions, areas, peer groups,* or *domains*. For example, Internet Protocol (IP) networks may use open shortest path first (OSPF) (Moy, 1998) routing for multi-area hierarchical routing. Asynchronous transfer mode (ATM) networks may use private network-network interface (PNNI) (ATM Forum Technical Committee, April 2002) routing for multi-peer-group routing. Essentially all TDM-based network routing topologies use hierarchical routing, as exemplified by the five-level, multiple region hierarchical network that existed within the former Bell System (Ash, 1998). Flat network routing is exemplified by IP-based border gateway protocol (BGP) (Rekhter and Li, 1995) routing, where multiple sub-networks called *autonomous systems* appear as non-hierarchical or flat-network routing domains to the BGP routing protocol.

Fixed routing (FR) is an important routing method employed in all types of networks, including packet- and TDM-based networks. With FR protocols, routing tables remain fixed, that is, they do not change with time, or with network state. Hierarchical FR networks are widely deployed in TDM-based voice networks worldwide.

Time dependent routing (TDR) methods are a type of DR in which the routing tables are altered at a fixed point in time during the day or week. TDR routing tables are determined on an off-line, preplanned basis and are implemented consistently over a time period. The TDR routing tables are determined considering the time variation of traffic load in the network, for example based on measured hourly load patterns. Several TDR time periods are used to divide up the hours on an average business day and weekend into contiguous routing intervals sometimes called load set periods. Typically, the TDR routing tables are designed to take advantage of noncoincidence of busy hours among the traffic loads.

In state dependent routing (SDR), illustrated in Figure 22.2, the routing tables are altered automatically according to the state of the network. For a given SDR method, the routing table rules are implemented to determine the path choices in response to changing network status, and are used over a relatively short time period. Information on network status may be collected at a central routing processor or distributed to nodes in the network. The information exchange may be performed on a periodic or on-demand basis. SDR methods use the principle of routing connections on the best available path based on network state information. For example, in the least loaded routing method, the residual capacity of candidate paths is calculated, and the path having the largest residual capacity is selected for the connection. Various levels of link occupancy can be used to define link load states, such as lightly-loaded, heavily-loaded, or bandwidth-not-available states. In general, SDR methods calculate a path cost for each connection request based on various factors such as the load-state, distance, delay, transport cost, etc. of the links in the network.

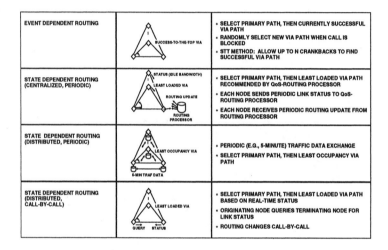

Figure 22.2   Dynamic routing methods

In SDR, the routing tables are designed on-line, in real time, by the originating node (ON) or a central routing processor through the use of network status and topology information obtained through information exchange with other nodes and/or a centralized routing processor. There are various implementations of SDR distinguished by

- whether the computation of the routing tables is distributed among the network nodes or centralized and done in a centralized routing processor, and
- whether the computation of the routing tables is done periodically or connection by connection.

This leads to three different implementations of SDR (see Figure 22.2):

- Centralized periodic SDR (CP-SDR) – here the centralized routing processor obtains link status and traffic status information from the various nodes on a periodic basis (e.g., every 10 seconds) and performs a computation of the optimal routing table on a periodic basis. To determine the optimal routing table, the routing processor executes a particular routing table optimization procedure such as least-loaded routing and transmits the routing tables to the network nodes on a periodic basis (e.g., every 10 seconds). Typically, if the shortest path is busy (e.g., bandwidth is unavailable on one or more links), the second path is selected from the list of feasible paths on the basis of having the greatest level of idle bandwidth at the time.

- Distributed periodic SDR (DP-SDR) – here each node in the SDR network obtains link status and traffic status information from all the other nodes on a periodic basis (e.g., every 5 minutes) and performs a computation of the optimal routing table on a periodic basis (e.g., every 5 minutes). Flooding is a common technique for distributing the status and traffic data, however other techniques with less overhead are also available, such as a query-for-status method. To determine the optimal routing table, the ON executes a particular routing table optimization procedure such as least-loaded routing.

- Distributed connection-by-connection SDR (DC-SDR) – here an ON in the SDR network obtains link status and traffic status information from the destination node (DN), and perhaps from selected via nodes (VNs), on a connection by connection basis and performs a computation of the optimal routing table for each connection. Typically, the ON first tries the primary path and if it is not available finds an optimal alternate path by querying the DN and perhaps several VNs through query-for-status network signaling for the load status of all links connected on the alternate paths to the DN. To determine the optimal routing table, the ON executes a particular routing table optimization procedure such as least-loaded routing.

In event dependent routing (EDR), the routing tables are updated locally on the basis of whether connections succeed or fail on a given path choice. In the EDR learning approaches, the path last tried, which is also successful, is tried again until blocked, at which time another path is selected at random and tried on the next connection request. EDR path choices can also be changed with time in accordance with changes in traffic load patterns. Success-to-the-top (STT) EDR path selection, illustrated in Figure 22.2, is a decentralized, on-line path selection method with update based on random routing. STT-EDR uses a simplified decentralized learning method to achieve flexible adaptive routing. The primary path path-p is used first if sufficient resources are available, and if not a currently successful alternate path path-s is used until it is blocked (i.e., sufficient resources are not available, such as bandwidth not available on one or more links). In the case that path-s is blocked, a new alternate path path-n is selected at random as the alternate path choice for the next connection request overflow from the primary path. In the EDR learning approaches, the current alternate path choice can be updated randomly, cyclically (round-robin), or by some other means, and may be maintained as long as a connection can be established successfully on the path. Hence the routing table is constructed with the information determined during connection setup, and no additional information is required by the ON.

There are features commonly applied in all the connection routing methods. With TDR, SDR, and EDR, dynamically activated bandwidth reservation is typically used under congestion conditions to protect traffic on the primary path. Crankback signaling may be used when an alternate path is blocked at a VN, and the connection request advances to a new path choice. Many path choices can be tried by a given connection request before the request is blocked. Paths in the routing table may consist of the direct link, a 2-link path through a single VN, or a multiple-link path through multiple VNs. Paths in the routing table may be subject to allowed load state restrictions on each link. For either SDR or EDR, as in TDR, the alternate path choices for a connection request may be changed in a time-dependent manner considering the time-variation of the traffic load.

22.1.2 DR benefits & comparison of DR & FR networks

DR brings benefits to customers in terms of new service flexibility and improved service quality and reliability, at reduced cost. DR, as shown through examples in Ash (1998), achieves the following improvements over FR:

- Essentially zero traffic loss performance on normal business days and weekends:

- Improved performance under network overloads and failures;

- Increased revenue and network throughput through maximum-flow/revenue routing design;

- Lower capital costs through minimum cost network design and voice/data integration;

- Lower expense costs through centralization and automation of network management/design functions; and

- Overall higher quality service for customers.

Peak days such as Christmas, Mother's Day, and Thanksgiving, in addition to unpredictable events such as earthquakes, hurricanes, and failures, are a test for DR, because the traffic loads and patterns of traffic deviate severely from normal business day traffic for which the network is designed. Peak day call blocking results show that the performance improvements of TDR over FR, and SDR over TDR, are dramatic. As illustrated in Figure 22.3, in the case of Thanksgiving Day traffic overload, as routing evolved in the AT&T TDM-based voice network from FR to TDR to DC-SDR, average network blocking, which is about 34% for FR, is down to 3% with TDR, and then down to 0.4% with DC-SDR, which is nearly a factor of 100 improvement. The average blocking for Christmas Day traffic overload is down from 58% for the FR network, to 12% for the TDR network, to 6% for the DC-SDR network, which is almost a factor of 10 improvement. Also, there is less network capacity relative to demand in the DC-SDR network versus the FR network, because of the efficiencies of DR network design, and there is more traffic load as the network evolved from FR to TDR to DC-SDR.

DYNAMIC ROUTING NETWORKS 579

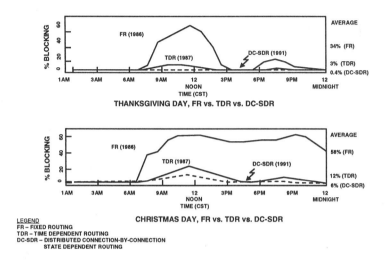

Figure 22.3 Dynamic routing performance with peak day traffic loads

DR provides a self-healing network capability to ensure a network-wide path selection and immediate adaptation to failure. As illustrated in Figure 22.4, a fiber cut near Nashville (Tennessee, US) severed 67,000 trunks in the DC-SDR network, and after several minutes dynamic transport restoration restores 37,000 trunks in the DC-SDR network. Call blocking then returned to zero even though there are 30,000 trunks still out of service, since DC-SDR is able to find paths on which to complete traffic. Hence, DR in combination with dynamic transport restoration provides a self-healing network capability, and even though the cable is repaired two hours after the cable cut, degradation of service is minimal. DR provides priority routing for selected customers and services, which permits priority calls to be routed in preference to other calls, and blocking of the priority services is essentially zero throughout the whole event.

This improved network performance provides additional service revenues as formerly blocked calls are completed, and improves service quality to the customer. DR lowers capital costs through improved network design, in the range of 10 to 30% of FR network cost over a range of network configurations and applications. Results show that voice and data services integration enables more efficient and robust network performance with independent traffic control for each voice or data class-of-service, provides efficient sharing of integrated transport network capacity, and implements an integrated class-of-service routing feature for extending DR to emerging services. DR lowers operations expense through centralized and automated network management/design.

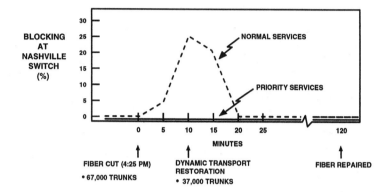

Figure 22.4 Dynamic routing performance with fiber cut

Various DR optimization methods are now described and illustrated in the remainder of the chapter. These methods include discrete event simulation, shortest path design, linear programming (LP), link capacity optimization, and multihour network design. These methods are applied to the following optimization problems:

- DR protocol design, as described in Section 22.2,
- minimum-cost DR network capacity design, as described in Section 22.3, and
- maximum-flow DR design, as described in Section 22.4.

Excellent references are available which cover network design and optimization techniques, for example, see Pioro and Medhi (2004) for comprehensive coverage of a wide range of methods and other chapters in this book for coverage of relevant optimization techniques.

22.2 DR PROTOCOL DESIGN & OPTIMIZATION

22.2.1 Problem statement

Find an optimal DR protocol for arbitrary traffic loads and network topology, which yields minimum-cost capacity design and maximum-flow/revenue.

22.2.2 Optimization methods

Many techniques are available for DR protocol optimization, and this book describes many of these methods in detail. In this chapter, we emphasize the use of shortest path routing design and discrete event simulation models for DR protocol design and optimization.

22.2.2.1 Shortest path routing design.

Some routing methods such as hierarchical FR limit path choices and provide inefficient design. If we choose paths based on cost, and relax constraints such as a hierarchical network structure, that is, use a flat network structure, a more efficient network results. Figure 22.5 illustrates the selection of shortest paths between two network nodes, SNDG and BRHM. Longer paths, such as SNDG-SNBO-ATLN-BRHM, which did arise in the 5-level hierarchical FR network through hierarchical path selection, are less efficient than shortest path selection, such as SNDG-PHNX-BRHM, SNDG TCSN-BRHM, or SNDG-MTGM-BRHM. Routing on the least costly, most direct, or shortest paths is often more efficient than routing over longer hierarchical paths.

——— LONGER, LESS EFFICIENT PATHS
—·—·— SHORTEST PATHS

Figure 22.5 Shortest path routing design

The Bellman-Ford method (Bellman, 1957) or Dijkstra's method (Dijkstra, 1959), for example, are often used for shortest path optimization. To illustrate the Bellman-Ford algorithm, let C_{ik} represent the cost of the minimum-cost path from node i to node k. If node i and node k have a direct link with link cost (administrative weight) c_{ik}, the minimum-cost path between node i and j can be obtained by solving Bellman's equation:

$$C_{ij} = \min_k c_{ik} + C_{kj} \tag{22.1}$$

One can solve for the C_{ij} by initially setting

$$C_{ij} = \begin{cases} c_{ij}, & \text{if nodes } i \text{ and } j \text{ are connected,} \\ \infty, & \text{if nodes } i \text{ and } j \text{ are not connected,} \end{cases}$$

and then iteratively solving Equation (22.1) until the C_{ij} values converge.

Shortest path optimization may use a topology database that includes, for example, the link cost and available bandwidth on each link. From the topology database, a node could determine a list of shortest paths to another node by using Bellman-Ford's or Dijkstra's algorithm based on the link costs (or administrative weights), which are communicated to all nodes within the routing domain through the routing protocol, as explained later in this section. These link costs may be set, for example, to $[1 + \varepsilon * d]$, where ε is a factor giving a relatively smaller weight to the distance d in comparison to the hop count. Alternatively, the available bandwidth on the link can be used to determine the link cost, as illustrated later in this section, or some combination of link distance, available bandwidth, or other criteria could also be used.

22.2.2.2 Discrete event simulation models. Discrete event simulation models are a versatile method to study and optimize DR performance. Such models can accurately capture non-linear and complex network behavior. A discrete event simulation model has a simulated clock that controls the time at which events are executed. Individual events, such as the arrival of voice call requests or flow connection requests, are put on an event list, and at the simulated time of their arrival they are routed on the simulated links in the model using the exact routing logic of the routing method being studied. Typically, Poisson arrivals are used to model originating traffic connections, together with an exponential holding time distribution having a given mean. However, any arrival distribution or holding time distribution could be simulated as well.

For a typical routing logic, a connection is first routed on the direct simulated link (or shortest path if the direct link does not exist), and if the direct link/shortest path is busy, the connection is routed on the next path in the routing table. If any of the simulated links on the alternate path is busy, the connection is then routed on the next path in the routing table. In the simulation, if a second or higher link in a path is busy, this condition results in a simulated crankback signaling message that notifies the originating node that the path is blocked and alternate paths might be tried, if allowed by the particular routing algorithm. If a connection finds an idle path, the simulated link capacity is made busy, the connection holding time is determined according to the exponential holding time distribution (other distributions can easily be modeled in simulations), and the end time of the connection is noted in the event list for a later disconnect event at the simulated time the connection is disconnected. At the simulated disconnect time, the simulated link capacity occupied by the connection is made idle once again, and the connections-in-progress counter for the affected node pair is decremented. For some DR logic, bandwidth reservation is applied in selecting capacity on each link.

As described later in this section, complex queuing behavior also can be readily modeled in the simulation for packet-based networks. Hence, the simulation model can very accurately model complex network routing logic and thereby give accurate estimates of network performance for such logic. Statistical data can readily be gathered in the simulation, such as the number of arriving calls, blocked calls, connections-in-progress, crankback events, etc. Simulation models can also be used to validate analytical models such as the ones presented in Section 22.3. However, connection-by-connection simulations can go far beyond the ability of analytical models to represent complex network behavior.

22.2.3 Design & optimization of circuit-switched voice network DR protocols

DR in circuit-switched TDM networks has been in operation since 1984, when dynamic nonhierarchical routing (DNHR, a TDR method) cut over into operation in the AT&T network. This was followed by many other DR implementations worldwide, and the current status of DR implementations is given in Table 22.1 (Ash and Chemouil, 2004). References are also provided in the table on the various DR systems for readers interested in more details.

Some general observations on TDM circuit-switching technology and DR network optimization implications are now discussed. TDM switching tends to be expensive relative to transport, which leads to highly-connected, meshed logical topologies and relatively low-capacity logical links. Highly-connected, meshed networks tend to minimize switching cost. Having many links tends to distribute the network load, lowering the load per link and the link capacity requirement. When doing minimum-cost capacity design for these meshed networks, DR protocols tend to have significant design-cost advantages over FR protocols, because the large number of alternate route choices affords DR an optimization advantage. In addition, low capacity links need to have higher levels of blocking and overflow (alternate routed) traffic in order to achieve high occupancy and run efficiently, as will be illustrated in Section 22.3. More alternate routed traffic tends to give DR an optimization advantage over FR. Finally, maximum flow routing design for meshed TDM networks tends to favor a DR solution for the same reasons: a large number of route choices, and the benefit of driving low-capacity links to high occupancy and more alternate routed traffic.

22.2.3.1 Example circuit-switched voice network DR protocol design.

In this section, we use discrete event simulation models to illustrate the design and optimization of a large-scale, state-dependent DR protocol, real-time network routing (RTNR, a DC-SDR method), implemented in the AT&T voice network in 1991, as depicted in Table 22.1. RTNR was selected after extensive study of a wide range of DR methods including extensive simulation comparisons of TDR, EDR, and SDR methods. Much of the comparative simulation analysis is reported in Ash (1998, Chapter 1). In this section, the simulation model is used to illustrate the design trade-offs of the RTNR protocol by optimizing the performance as a function of various design parameters, which include the number of link occupancy states, link state occupancy thresholds, triggering mechanism for congestion detection, and others.

Before proceeding with the example, we give a brief overview of RTNR. RTNR first selects the direct link if capacity is available, and if not available then selects the least-loaded 2-link alternate path from among all possible 2-link paths. The available bandwidth capacity or occupancy of each link is mapped into a discrete number of link states, and the link-state occupancy information signaling between network nodes to enable selection of the least-loaded 2-link path. Furthermore, a small amount of link capacity called the bandwidth-reservation level R is reserved on each link. A call on the direct link is allowed to access all capacity including the reserved capacity R.

Table 22.1 Operational dynamic routing systems.

Routing Type	Dynamic Routing Systems	Network	Start Year	End Year	Comments
TDR	DNHR Dynamical Nonhierarchical Routing (Ash et al., 1981; Ash, 1998)	AT&T US National Network	1984	1991	DNHR replaced by RTNR in 1991.
		AT&T FTS-2000 Network	1987	2002	DNHR replaced by RTNR in 2002.
	GTAI (GTAI is an Italian acronym: Management of the Traffic Transit Italcable switches) (DiBenedetto et al., 1989)	Italcable	1984	1985?	This routing mechanism was implemented between the 3 inter-continental switches operated by Italcable
SDR Centralized Periodic	DCR Dynamically Controlled Routing (Regnier and Cameron, 1990)	Stentor Canada National Network	1991	In operation	Also known as High Performance Routing (HPR)
		Bell Canada Network	1992	In operation	Consists of one DCR network local to the Toronto area and one local to the Montreal area
		Sprint National Network	1994	In operation	
		MCI US National Network	1995	In operation	
		Qwest Communications National Network	1999	In operation	
SDR Distributed Periodic	WIN Worldwide intelligent network routing (Ash et al., 1989)	Worldwide Intelligent Network	1993	In operation	WIN data is currently exchanged between AT&T/US, CHT-I/ Taiwan and Alestra/Mexico.
SDR Distributed call-by-call	RTNR Real-time network routing (Ash et al., 1991; Ash, 1998)	AT&T US National Network	1991	In operation	
		AT&T FTS-2000 Network	2002	In operation	
	RINR Real-time internetwork routing (Ash, 1998)	AT&T Global International Network	1991	In operation	

Table 22.1 Operational dynamic routing systems (continued).

EDR	STR State- and Time-dependent Routing (Mase and Yamamoto, 1990)	NTT Japan National Network	1992	2002	The deployment of STR started in 1992, but operation stopped in 2002 when the D60 switches were replaced by new switches.
	DAR Dynamic Alternative Routing (Gibbens et al., 1988)	British Telecom UK National Network	1993	?	
	STT Success-To-the-Top network routing (Ash, 1998)	AT&T US National Network	1995	1999	STT is a method used for a period of time to route calls with voice enhancement devices in the path.
	LAW Lastabehängige Automatische Wegesuche (in English, Automatic Last Choice Routing)	Deutsche Telekom National Network	1995	In operation	LAW is implemented in the Transit Network as well as Regional and International Access networks).
	AMI Acheminement Multiple Intelligent (in English, Multiple Intelligent Routing - MIR)	France Telecom Long Distance Network	1998	In operation	AMI/MIR is an EDR system with multiple overflow routes and crankback.

However, the reserve capacity R cannot be used on an alternate 2-link path. That is, if less than R units of bandwidth capacity is available on either link of a 2-link alternate path, a call cannot use that path. This bandwidth reservation mechanism favors direct one-link paths under network congestion conditions. Without reservation, nonhierarchical networks can exhibit unstable behavior in which essentially all connections are established on two-link as opposed to one-link paths, which greatly reduces network throughput and greatly increases network congestion (Akinpelu, 1984; Krupp, 1982; Nakagome and Mori, 1973).

We first study RTNR protocol performance as a function of the number of link occupancy states, including 3, 4, 5, and 6-state models, which are illustrated in Figure 22.6. An example of a six-state model is given at the top of the figure. The six states are maximum lightly loaded (LL3), medium lightly loaded (LL2), minimum lightly loaded (LL1), heavily loaded (HL), reserved (R), and busy (B). Illustrative values of the thresholds that define these six states are given in the figure. Five-state, four-state, and three-state models are defined that omit the LL3 state, the LL2 state, and the R state, respectively.

Figure 22.6 RTNR link state models

Table 22.2 illustrates the performance of the RTNR link-state models for a high-day network load pattern, based on the simulation model analysis. The table gives the average hourly blocking in hours 2, 3, 5, 8, and 15, which correspond to the two early-morning busy hours, the afternoon busy hour, the evening busy hour, and the weekend (Sunday night) busy hour, respectively. Also given at the bottom of the table is the average blocking over these five load set periods, then the average blocking over the five load set periods for the node pair with the highest average blocking, the node pair that had the 99th percentile highest average blocking, and the node pair that had the 90th percentile highest average blocking.

Table 22.2 Network blocking performance under high-day network load for 3-state, 4-state, & 6-state RTNR models (% blocking, 65-Node network)

Hour	RTNR (3 states)	RTNR (4 states)	RTNR (5 states)
2	12.19	0.47	0.22
3	22.38	0.58	0.18
5	18.90	0.44	0.24
8	0.00	0.00	0.00
15	0.00	0.31	0.08
average	12.05	0.39	0.16
maximum	51.35	23.03	11.85
99%	41.24	7.20	2.87
90%	28.53	1.29	0.07

We can see from the results of Table 22.2 that RTNR performance improves as more link states are used. The performance of the RTNR three-state model is far worse than the performance of the other RTNR models. The reason for this poor performance is due to the lack of reserved bandwidth capacity to favor direct one-link paths under network congestion conditions. As discussed above, without reservation nonhierarchical networks can exhibit unstable behavior in which essentially many connections are established on two-link as opposed to one-link paths, which as shown in Table 22.2 greatly reduces network throughput and increases network blocking. If we add the reserved link state, as in the four- and six-state models, performance of the RTNR network is greatly improved. Results for high-day conditions with the 103-node simulation model are given in Table 22.3, and comparison of the four-, five-, and six-state models shows that network performance is monotonically improved as more states are added, as would be expected. However, additional simulation studies show that having more than six states does not measurably improve performance.

Sensitivities of the results to various parameters were investigated, which included link state thresholds, the amount of reserved capacity R, the triggering mechanism for congestion detection, and others. We now discuss the results for the triggering mechanism for automatic bandwidth reservation, as one example, details of other examples can be found in Ash (1998).

Table 22.3 Network blocking performance under high-day network load for 3-state, 5-state, & 6-state RTNR models (% blocking, 103-node network)

Hour	RTNR (4 states)	RTNR (5 states)	RTNR (6 states)
2	0.16	0.03	0.02
3	1.03	0.80	0.76
5	0.82	0.50	0.45
15	0.00	0.00	0.00
average	0.54	0.36	0.34
maximum	19.25	15.36	15.02
99%	7.72	5.45	4.73
90%	1.57	0.61	0.52

Table 22.4 demonstrates the effect of an automatic bandwidth reservation triggering mechanism using a 60-second update interval and the calls blocked. The method based on calls blocked provides a weighted, real-time estimate of node-to-node and total node blocking, which is updated on a connection-by-connection basis. That is, node-to-node blocking is estimated as $b_{ik} = (1-\alpha)*r_i + \alpha*b_i - 1*k$, where b_{ik} is the real-time blocking estimate for node pair k based on connection i, α is a number between zero and one that represents the weight assigned to previous values of the blocking estimate, and r_i is defined to be 1 if connection i is blocked and 0 if it is not blocked. This real-time blocking estimate is then used to trigger bandwidth reservation R. Several values of parameter α were tested, as were variations of the definition of r_i. One variation defined r_i as the average blocking over a short, fixed time interval such as 15 seconds. However, the results of many simulations show that automatic bandwidth reservation based on the real-time blocking estimate is not as effective as that based on the 60-second update interval. The reason is that automatic bandwidth reservation based on the real-time blocking estimate is overly sensitive in triggering bandwidth reservation to turn on and too slow to remove it, and consequently bandwidth is over-reserved. Therefore, automatic bandwidth reservation with 60-second node-to-node blocking update time appears most promising.

We have presented simulation studies that illustrate the protocol design and optimization of RTNR, a representative voice network DR protocol.

22.2.4 Design & optimization of packet-switched data network DR protocols

Packet-switched data networks typically use link-state, or shortest-path-first (SPF), routing protocols. Such protocols are distributed-database protocols, and as such are classified as DP-SDR protocols as defined in Section 22.1.1. The functions of the link-state routing protocol include the following:

Table 22.4 Effect of triggering mechanism on RTNR network blocking performance (high day, % blocking)

Hour	Blocked Call Based	Time Based (60 seconds)
2	0.05	0.02
3	1.10	0.76
5	0.90	0.45
15	0.00	0.00
average	0.56	0.34
maximum	18.71	15.02
99%	8.22	4.73
90%	0.73	0.52

- Discovery of neighbors and link status;
- Synchronization of topology databases;
- Flooding of topology state information;
- Election of peer group or area leaders;
- Summarization of topology state information;
- Construction of the routing hierarchy.

The first link-state routing protocol was developed for use in the ARPANET (McQuillan et al., 1980), which formed the basis for all other link-state protocols. The OSPF (Moy, 1998) protocol was later developed for use in IP-based networks, and the PNNI (ATM Forum Technical Committee, April 2002) routing protocol for use in ATM-based networks. In a link-state routing protocol, each node maintains a link-state database describing the network topology. Each participating node has an identical database which includes each node's usable interfaces and reachable neighbors. Each node distributes its local state throughout the network by flooding, and constructs a tree of shortest paths with itself as the root. This shortest-path tree gives the route to each destination in the node.

Each node bundles its topology state information, which is reliably flooded throughout the peer group or area. Topology state parameters can include relatively static parameters such as link metrics (or administrative weights), and dynamic information, such as available link bandwidth. Reachability information consists of addresses and address prefixes which describe the destinations to which connections may be routed. Hello packets are sent periodically by each node on interconnecting links, and in this way the Hello protocol makes the two neighboring nodes known to each other. The Hello protocol runs as long as the link is operational. It can therefore act as a link failure detector when other mechanisms fail. Each node exchanges Hello packets with its immediate neighbors and thereby determines its local state information. This state information includes the node identity and peer group or area membership of the node's immediate neighbors, and the status of its links to the neighbors. Link-state protocols implement a hierarchy mechanism to ensure that the protocol scales for large networks. The hierarchy begins at the lowest level where the lowest-level nodes are organized into peer groups or areas.

Some general observations on packet-switching technology and data network optimization implications are now discussed. In contrast to circuit-switching technology, packet switching tends to have lower switching cost relative to transport, which leads to sparse logical topologies and relatively high-capacity logical links. Sparsely-connected networks tend to minimize transport cost. Having few links tends to concentrate the network load, raising the load per link and the link capacity requirement. When doing minimum-cost capacity design for these sparsely connected networks, DR protocols do not tend to have significant design-cost advantages over FR protocols, because the high capacity links have low (or zero) levels of blocking and overflow (alternate routed) traffic, which precludes the need for alternate routes, as will be illustrated in Section 22.3. The lack of alternate routed traffic tends to give DR little optimization advantage over FR for minimum cost design. That is, carrying all design load on the shortest path without alternate routing tends to minimize cost. However, under scenarios where links are overloaded, congest and give rise to alternate routed traffic, maximum flow routing design for sparse packet-switched networks tends to favor a DR solution in order to take advantage of a large number of route choices to access available link capacity.

22.2.5 Design & optimization of integrated voice/data network DR protocols

Networks are rapidly evolving to carry a multitude of voice/ISDN services and packet data services on IP-based networks, driven in part by the extremely rapid growth of IP-based data services. As we have discussed in previous sections, within networks and services supported by packet and TDM protocols have evolved various DR methods. Here we extend the discussion to integrated voice/data network applications and optimization of DR protocols. The scope of the integrated voice/data DR methods includes the establishment of connections for narrowband, wideband, and broadband multimedia services within multiservice networks and between multiservice networks. Here a multiservice network refers to one in which various class types (classes of service) share the transmission, switching, management, and other resources of the network. These class types can include constant bit rate and variable bit rate voice and data services traffic classes. There are quantitative performance requirements that the various classes of service normally are required to meet, such as end-to-end blocking, delay, and/or delay-jitter objectives. These objectives are achieved through a combination of DR, traffic management, and capacity management, as discussion in connection with Figure 22.1.

The DR methods discussed in Section 22.1.1 (Figure 22.2) are shown to be extensible to integrated voice/data DR networks, in particular, to IP-based multiprotocol label switching (MPLS) technology. In an example integrated voice/data DR protocol design and optimization we illustrate the tradeoffs between these various approaches, and show that a) aggregated per-class-type network bandwidth allocation compares favorably with per-flow allocation, and b) EDR methods perform just as well or better than SDR methods with flooding, which means that EDR path selection has potential to significantly enhance network scalability.

Vendors have yet to announce such DR capabilities in their products. Current practice often involves over-provisioning of IP networks, thereby avoiding the need for more efficient DR capabilities, but with concomitant low utilization and efficiency (Odlyzko, 2003). There is therefore an opportunity for increased profitability and performance in such networks through application of DR methods.

There is not an extensive literature on this emerging topic as yet, which is still under development. Awduche (1999) and Awduche et al. (2002) give excellent overviews of DR (called *traffic engineering* in the Internet terminology) approaches for IP-based networks, and also provides traffic engineering requirements (Awduche et al., 1999). Crawley et al. (August 1998) and Xiao and Ni (1999) provide good background and context for DR in the Internet. A few early implementations of off-line, network-management-based DR approaches have been published, such as in the Global Crossing network (Xiao et al., 2000), Level3 network (Pierantozzi and Springer, 2000), and Cable & Wireless network (Liljenstolpe, 2002). Some studies have proposed more elaborate DR approaches in IP networks (Apostolopoulos et al., 1999; Elwalid et al., 2001; Ma, 1998; X. Xiao, 2000), as well as in ATM networks (Ahmadi et al., 1992).

22.2.5.1 Example of integrated voice/data DR protocol design. In this section, we use discrete event simulation models, which accurately capture non-linear and complex behavior, to illustrate the design and optimization of a large-scale, event-dependent DR protocol, success-to-the-top event dependent routing (STT-EDR), for application to an integrated voice/data DR network based on IP/MPLS technology. First we illustrate the optimization of an integrated voice/data DR protocol from among a wide range of candidate DR methods based on extensive simulation comparisons of TDR, EDR, and SDR methods. An STT-EDR protocol is selected from this initial optimization study, and we then use the simulation model to optimize the performance of the STT-EDR protocol within the design tradeoffs, which include the level of reserved capacity, the number of allowed alternate paths, and the algorithm for increasing and decreasing the CT bandwidth allocation. A large-scale, integrated voice/data IP/MPLS network model is developed for the simulation studies.

22.2.5.2 Optimization of Integrated Voice/Data DR Protocol. Integrated voice/data DR protocols are optimized in the first study for these design tradeoffs, among others:

- Logical topology (sparse, mesh);

- Connection routing method (2-link, multilink, SDR, EDR, etc.); and

- Bandwidth allocation method (aggregated/per class-type (CT), per-flow).

Generally, the meshed logical topologies typical of TDM-based networks are optimized by 1- and 2-link routing, while the sparse logical topologies typical of packet-switched networks are optimized by multilink shortest path routing. Design and optimization of integrated voice/data DR protocols, discussed in this section, leads to these conclusions:

- EDR connection routing methods exhibit comparable design efficiencies to SDR routing methods;
- Sparse topology designs with multilink routing provide switching and transport design efficiencies in comparison to mesh designs with 2-link routing;
- Per-CT bandwidth allocation exhibits comparable design efficiencies to per-flow bandwidth allocation.

A discrete event flow optimization and IP/MPLS network model are used for the protocol optimization. Figure 22.7 illustrates an integrated voice/data DR model in which bandwidth is allocated on an aggregated basis to the individual service classes, which are denoted as class types (CTs). In the model, CTs have different priorities including high-priority, normal-priority, and best-effort priority services CTs. Bandwidth allocated to each CT is protected by bandwidth reservation methods, as needed, but otherwise shared. Each ON monitors CT bandwidth use on each CT label switched path (LSP), which is enabled by MPLS technology (Rosen et al., 2001), and determines when CT LSP bandwidth needs to be increased or decreased. In Figure 22.7, changes in CT bandwidth capacity are determined by ONs based on an overall aggregated bandwidth demand for CT capacity (not on a per-connection/per-flow demand basis). Based on the aggregated bandwidth demand, ONs make periodic discrete changes in bandwidth allocation, that is, either increase or decrease bandwidth on the LSPs constituting the CT bandwidth capacity. For example, if connection requests are made for CT LSP bandwidth that exceeds the current LSP bandwidth allocation, the ON initiates a bandwidth modification request on the appropriate LSP(s), which may entail increasing the current LSP bandwidth allocation by a discrete increment of bandwidth denoted here as delta-bandwidth (DBW). DBW, for example, could be the additional amount needed by the current connection request. In any case, DBW is a large enough bandwidth change so that modification requests are made relatively infrequently. The bandwidth admission control for each link in the path is performed based on the status of the link using the bandwidth allocation procedure described later in this section, where we further describe the role of the different parameters such as reserved bandwidth RBW shown in Figure 22.7 in the admission control procedure. Also, the ON periodically monitors LSP bandwidth use, such as once each minute, and if bandwidth use falls below the current LSP allocation the ON initiates a bandwidth modification request to decrease the LSP bandwidth allocation, for example, down to the current level of bandwidth utilization.

In making a CT bandwidth allocation modification, the ON determines the CT priority (high, normal, or best-effort), CT bandwidth-in-use, and CT bandwidth allocation thresholds. These parameters are used to determine whether network capacity can be allocated for the CT bandwidth modification request. In the simulation models, path selection takes different forms for the following cases:

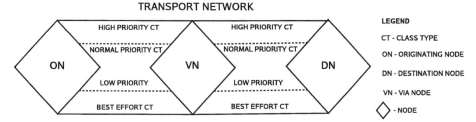

- distributed method applied on a per-class-type basis
- ON allocates bandwidth to each class type (CT) based on demand
- for CT badwidth increase request DBW
    - ON launches bandwidth modification request on primary or alternate label swicthed paths (LSPs)
    - each node in LSP decides to accept/reject DBW link-bandwidth-modification based on
        * bandwidth-in-progree (BWIP)
        * CT priority (high, normal, best-effort)
        * bandwidth allocation/constraint (BC)
        * reserved bandwidth threshold (RBW)
    - VNs send crankback/bandwidth-not-available notifiction to ON if DBW modification request rejected
        * ON can attempt alternate LSP if available

Figure 22.7 Integrated voice/data dynamic routing & bandwidth allocation

- 2-link STT-EDR – Paths are limited to 1 or 2 links, with STT-EDR path selection and a meshed network design model;

- 2-link DC-SDR – Paths are limited to 1 or 2 links, with DC-SDR path selection and a meshed network design model.

- Multilink STT-EDR - Shortest paths of an arbitrary number of links can be used, with STT-EDR path selection and a sparse network design model;

- Multilink DC-SDR – Shortest paths of an arbitrary number of links can be used, with DC-SDR path selection and a sparse network design model;

- Multilink DP-SDR – Shortest paths of an arbitrary number of links can be used, with DP-SDR path selection and a sparse network design model.

The network designs for the 2-link and multilink networks are now described. Links in the 2-link designs have fine-grained (1.536 mbps T1-level) logical transport link bandwidth allocation, and a meshed network topology results in which links exist between most (90 percent or more) of the nodes. In the 2-link designs, one and 2-link routing with crankback is used with both EDR and SDR path selection. In routing a connection with 2-link STT-EDR routing, the ON checks the equivalent bandwidth and allowed load-state threshold first on the direct path, in a manner analogous to the RTNR link-state model described in Section 22.2.1. If the direct path is unavailable, the current successful 2-link via path is tried, and if that fails then sequentially all candidate 2-link paths are tried. In routing a connection with 2-link DC-SDR, the ON checks the equivalent bandwidth and allowed load-state threshold first on the direct path, and then on the least-loaded 2-link path that meets the equivalent bandwidth and allowed load-state requirements. Each VN checks the equivalent bandwidth and allowed load-state threshold provided in the setup message, and uses crankback to the ON if the equivalent bandwidth or allowed load-state threshold are not met.

In the multilink design, high rate OC3/12/48 links provide highly aggregated link bandwidth allocation and a sparse network topology. That is, high rate OC3/12/48 links exist between relatively few (10 to 20 percent) of the node-pairs. The multilink path selection methods that are modeled include STT-EDR, DC-SDR, and DP-SDR path selection, in which crankback is used.

Figure 22.8 illustrates the operation of STT-EDR path selection and admission control combined with per-CT bandwidth allocation. ON A monitors CT bandwidth use on each CT LSP, and determines when CT LSP bandwidth needs to be increased or decreased. Based on the aggregated bandwidth demand, ON A makes periodic discrete changes in bandwidth allocation, that is, either increase or decrease bandwidth on the LSPs constituting the CT bandwidth capacity. If connection requests are made for CT LSP bandwidth that exceeds the current LSP bandwidth allocation, ON A initiates a bandwidth modification request on the appropriate LSP(s). The STT-EDR DR algorithm used is adaptive and distributed in nature and uses learning models to find good paths.

For example, in Figure 22.8 if the LSR-A to LSR-E bandwidth needs to be modified, say increased by DBW, the primary LSP-p (A-B-E) is tried first. The bandwidth admission control for each link in the path is performed based on the status of the link using the bandwidth allocation procedure described below, where we further describe the role of the reserved bandwidth RBW shown in Figure 22.8 in the admission control procedure. If the first choice LSP cannot admit the bandwidth change, node A may then try an alternate LSP. The second path tried is the last successful path LSP-p A-C-D-E (the 'STT path'), and if that path is unsuccessful another alternate path is searched out from the alternate path list. If DBW is not available on one or more links of LSP-p, then the currently successful LSP-s (A-C-D-E) is tried next. If DBW is not available on one or more links of LSP-s, then a new LSP is searched by trying additional candidate paths (not shown) until a new successful LSP-n is found or the candidate paths are exhausted. LSP-n is then marked as the currently successful path for the next time bandwidth needs to be modified. DBW, for example, can be set to the additional amount of bandwidth required by the connection request. Also, ON A periodically

DYNAMIC ROUTING NETWORKS 595

monitors LSP bandwidth use, such as once each minute, and if bandwidth use falls below the current LSP allocation the ON initiates a bandwidth modification request to decrease the LSP bandwidth allocation down to the currently used bandwidth level. In the models discussed next, the per-CT bandwidth allocation and admission control method compares favorably with the per-flow method, and STT-EDR path selection method compares favorable to the SDR methods.

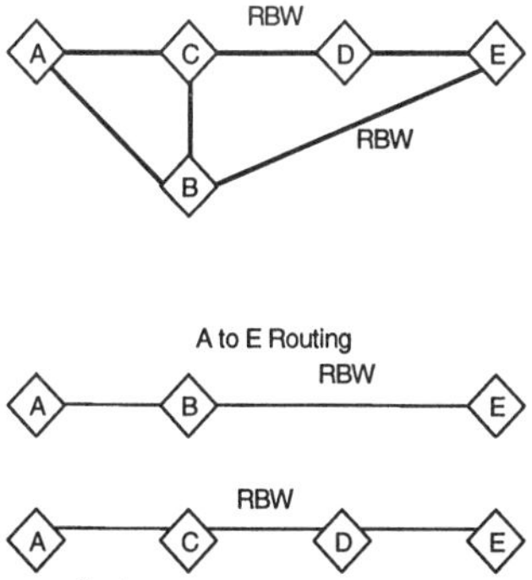

Example of STT-EDR routing method:
1. if node A to node E bandwidth needs to be modified (say increased by DBW) primary LSP-p (e.g., LSP A-B-E) is tried first
2. available bandwidth tested locally on each link in LSP-p, if bandwidth not available (e.g., link BE is in reserved RBW state & this CT is above its bandwidth allocation), crankback to node A
3. if DBW is not available on one or more links of LSP-p, then the currently successful LSP-s (e.g., LSP A-C-D-E) is tried next
4. if DBW is not available on one or more links of LSP-s, then a new LSP is searched by trying additional candidate paths until a new successful LSP-n is found or the candidate paths are exhausted
5. LSP-n is then marked as the currently successful path for the next time bandwidth needs to be modified

Figure 22.8 STT-EDR path selection & per-CT bandwidth allocation

In the model of DP-SDR, the status updates with link-status flooding occur every 10 seconds. Note that the multilink DP-SDR performance results should also be comparable to the performance of multilink CP-SDR, in which status updates and path selection updates are made every 10 seconds, respectively, to and from a DR processor. With the SDR methods, the available link bandwidth information in the topology database is used to generate the shortest, least-congested, paths. In routing a connection, the ON checks the equivalent bandwidth and allowed load-state threshold first on the first choice path, then on current successful alternate path (EDR) or least loaded shortest path (SDR), and then sequentially on all other candidate alternate paths. Each VN checks the equivalent bandwidth and allowed load-state threshold provided in the setup message, and uses crankback to the ON if the equivalent bandwidth or allowed load-state threshold are not met.

Based on a 135-node model, performance comparisons are presented in Table 22.5 for the various integrated voice/data DR methods, including 2-link and multilink EDR and SDR approaches, and a baseline case of no DR or bandwidth allocation methods applied. Table 22.5 gives performance results for a six-times overload on a single network node at Oakbrook (Illinois, US). The numbers in the table indicate the percent of traffic lost (blocked) in the admission control, in which case all paths have insufficient capacity, plus the percent of traffic delayed in the queues. Note that on high rate links, traffic delayed in the queues will almost always be dropped/lost, as shown in the case of a "bufferless" model (Bonald et al., 2002). The performance analysis results show that the multilink options in sparse topologies perform somewhat better under overloads than the 2-link options in meshed topologies, because of greater sharing of network capacity. Under failure, the 2-link options perform better for many of the CT categories than the multilink options, because they have a richer choice of alternate routing paths and are much more highly connected than the multilink networks. Loss of a link in a sparely connected multilink network can have more serious consequences than in more highly connected logical networks. The performance results illustrate that capacity sharing of different traffic class types, when combined with DR and priority queuing, leads to efficient use of bandwidth with minimal traffic delay and loss impact, even under overload and failure scenarios.

Table 22.5 Performance comparison for various DR methods for 6X focused overload on OKBK (% lost/delayed traffic)

Class Type	Mesh 2-link STT-EDR	Mesh 2-Link DC-SDR	Sparse Multilink STT-EDR	Sparse Multilink DC-SDR	Sparse Multilink DP-SDR	No DR Methods Applied
Normal Priority Voice	5.93	2.86	0.00	0.12	0.14	10.81
High Priority Voice	0.81	1.77	0.00	0.00	0.00	8.47
Normal Priority Data	7.12	3.49	0.00	2.75	3.18	12.88
High Priority Data	0.00	0.00	0.00	0.00	0.00	0.46
Best Effort Data	14.07	14.68	12.46	12.39	12.32	9.75

The EDR and SDR path selection methods are quite comparable for the 2-link, meshed-topology network scenarios. However, the EDR path selection method performs somewhat better than the SDR options in the multilink, sparse-topology case. In addition, the DC-SDR path selection option performs somewhat better than the

DP-SDR option in the multilink case, which is a result of the 10-second old status information causing misdirected paths in some cases. Hence, it can be concluded that frequently-updated, available-link-bandwidth state information does not necessarily improve performance in all cases, and that if available-link-bandwidth state information is used, it is sometimes better that it is very recent status information.

Some of the conclusions from the performance comparisons for integrated voice/data DR protocol optimization are as follows:

- DR methods result in network performance that is always better and usually substantially better than when no DR methods are applied;

- Sparse-topology multilink-routing networks provide better overall performance under overload than meshed-topology networks, but performance under failure may favor the 2-link meshed-topology options with more alternate routing choices;

- EDR DR methods exhibit comparable or better network performance compared to SDR methods;

- State information as used by the SDR options (such as with link-state flooding) provides essentially equivalent performance to the EDR options, which typically use distributed routing with crankback and no flooding;

- Single-area flat topologies exhibit better network performance in comparison with multi-area hierarchical topologies;

- Bandwidth reservation is critical to the stable and efficient performance of DR methods in a network, and differentiated services (DiffServ) to ensure the proper operation of integrated voice/data bandwidth allocation, protection, and priority treatment.;

- Per-CT bandwidth allocation is essentially equivalent to per-flow bandwidth allocation in network performance and efficiency. Because of the much lower routing table management overhead, per-CT bandwidth allocation is preferred to per-flow allocation;

- Connection admission control (CAC) with MPLS bandwidth management and queuing priority (Blake et al., December 1998) are important for ensuring that integrated voice/data network performance objectives are met under a range of network conditions. Both mechanisms operate together to enable bandwidth allocation, protection, and priority queuing.

Regarding the final point, in an integrated voice/data DR network where high-priority, normal-priority, and best-effort traffic share the same network, under congestion the DiffServ priority-queuing mechanisms push out the best-effort priority traffic at the queues so that the normal-priority and high-priority traffic can get through on the MPLS-allocated LSP bandwidth.

Table 22.6 gives a comparison of the control overhead performance of a) DP-SDR with topology state flooding and per-flow bandwidth allocation, b) STT-EDR with per-flow bandwidth allocation, and c) STT-EDR with per-CT bandwidth allocation. The numbers in the table give the total messages of each type needed to do the indicated DR functions, including flow setup, bandwidth allocation, crankback, and topology state flooding to update the topology database. The DP-SDR method does available-link-bandwidth flooding to update the topology database while the EDR method does not. In the simulation there is a 6-times focused overload on the Oakbrook node. Clearly the DP-SDR/flooding method is consuming more message resources, particular link state advertisement (LSA) flooding messages, than the STT-EDR method. Also, the per-flow bandwidth allocation is consuming far more LSP bandwidth allocation messages than per-CT bandwidth allocation, while the traffic lost/delayed performance of the three methods is comparable.

Table 22.6 Routing table management overhead SDR/flooding/per-flow, EDR/per-flow, EDR/per-CT (6X focused overload on OKBK)

DR Function	Message Type	DP-SDR Flooding (per-flow bandwidth allocation)	DP SDR Flooding (per-flow bandwidth allocation)	STT-EDR (per-CT bandwidth allocation)
Flow Routing	Flow Setup	18,758,992	18,758,992	18,758,992
LSP Routing, Bandwidth Allocation, Queue Management	LSP Bandwidth Allocation	18,469,477	18,839,216	2,889,488
	Crankback	30,459	12,850	14,867
Topology Database Update	LSA	14,405,040		

Some of the conclusions from the comparisons of routing table management overhead are as follows:

- Per-CT bandwidth allocation is preferred to per-flow allocation because of the much lower routing table management overhead. Per-CT bandwidth allocation is essentially equivalent to per-flow bandwidth allocation in network performance and efficiency.

- EDR methods provide a large reduction in flooding overhead without loss of network throughput performance. Flooding is very resource intensive since it requires link bandwidth to carry topology state messages, processor capacity to process topology state messages, and the overhead limits autonomous system size. EDR methods therefore can help to increase network scalability.

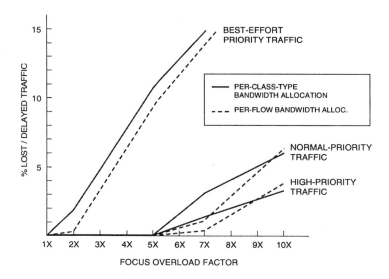

Figure 22.9 Performance under Focused Overload on CHCG Node

STT-EDR path selection, which entails the use of the crankback mechanism to search for an available path, is in sharp contrast to SDR path selection, which may entail flooding of frequently changing link state parameters such as available-link-bandwidth. With EDR path selection, the reduction in the frequency of such link-state parameter flooding allows for increased scalability. This is because link-state flooding can consume substantial processor and link resources, in terms of message processing by the processors and link bandwidth consumed by messages on the links.

Figure 22.9 compares the performance of per-CT LSP bandwidth allocation and per-flow LSP bandwidth allocation, and shows the lost or delayed traffic under a focused overload scenario on the Oakbrook node (such a scenario might occur, for example, with a radio call-in give-away offer). The size of the focused overload is varied from the normal load (1X case) to a 10 times overload of the traffic to Oakbrook (10X case). The results show that the per-flow and per-CT bandwidth allocation performance is similar; however, the improved performance of the high-priority traffic and normal-priority traffic in relation to the best-effort priority traffic is clearly evident.

In summary, integrated voice/data DR protocol design and optimization has been illustrated. These designs are based on results of discrete event analysis models, which illustrate the tradeoffs between various DR approaches and parameter selection. In the protocol design study, the per-CT bandwidth allocation method compares favorably with the per-flow method. Furthermore, we find that the fully distributed STT-EDR method of LSP bandwidth allocation performs just as well or better than the SDR methods with flooding, which means that STT-EDR path selection has potential to significantly enhance network scalability.

22.2.5.3 Optimization of STT-EDR Integrated Voice/Data DR Protocol. We now study the optimization of various design parameters for the STT-EDR algorithm. In particular, we investigate the following design tradeoffs:

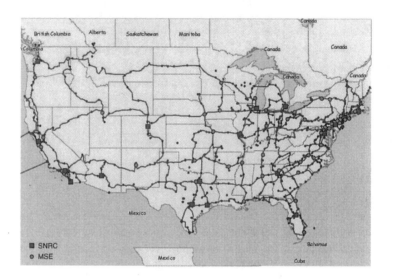

Figure 22.10 71-node model

- Level of reserved capacity: RBW_k is varied from 0 to 10% of link capacity;

- Number of allowed alternate paths: the number is varied from 2–10 paths;

- Algorithm for increasing and decreasing CT bandwidth allocation: various models are studied.

A 71-node national network model and multiservice traffic demand model are used to study various STT-EDR protocol design and optimization tradeoffs. The 71-node model is illustrated in Figure 22.10 and a 5 class-type (CT) traffic model in Table 22.7. Each voice over IP (VoIP) traffic flow for CT1 and CT2 is assumed to have a constant bit rate of 86.8 kbps, which includes a G.711 voice payload (20 ms. packets) of 160 bytes, and overhead composed of RTP/UDP/IP headers of 40 bytes, PPP with 4-byte FCS of 9 bytes, and MPLS labels of 8 bytes. Based on measured traffic, each data traffic flow for CT3, CT4, and CT5 is assumed to have a 768 kbps mean and a uniformly variability up to a maximum of 1536 kbps. Table 22.7 indicates the total traffic levels for each CT, as well as the priority of the traffic in each CT. Three levels of traffic priority – high, normal, and best-effort – are given to the CTs illustrated in Table 22.7.

Table 22.7 5 class-type (CT) traffic model.

Traffic Model	Class Types (Classes of Service)	Traffic Levels
CT1	high priority VoIP (GETS/ETS/E911)	high priority EF 0.1% of voice traffic 0.03% of total traffic
CT2	normal priority VoIP	normal priority EF 99.9% of voice traffic 29.5% of total traffic
CT3	premium private data	normal priority AF 5% of data traffic 3.5% of total traffic
CT4	premium public data	normal priority AF 45% of data traffic 31.7% of total traffic
CT5	best-effort data	best effort priority BE 50% of data traffic 35.2% of total traffic

All traffic loads are projected to 2008. Five load set periods are available for CT1 and CT2 voice traffic, which include the business day morning, afternoon, and evening busy periods, as well as the weekend morning and evening busy periods. The voice traffic is busiest at the business day morning period. A single time period is available for the data traffic used in the model. However, a separate, more aggregated set of hourly data traffic measurements was analyzed and indicates that the data traffic is evening busy both on business day and weekend. That is, voice traffic and data traffic exhibit non-coincidence, which could lead to network design efficiencies in taking advantage of the non-coincident traffic patterns.

In total, the voice traffic loads projected in the model amount to 122 Gbps in the busy hour, and the data traffic loads to 276 Gbps. That is, voice loads account for 29.5% of the total traffic load in the model. Furthermore, voice traffic varies up to 91% of the total traffic on individual links, as illustrated in Table 22.8, and up to 95% of the total traffic on individual nodes, as illustrated in Table 22.9.

Table 22.8 Traffic model – Links with > 80% voice traffic

link endpoint	link endpoint	voice traffic
ALBQNMMA	DNVRCOMA	81%
ATLNGATL	CLMASCTL	84%
CMBRMA01	SYRCNYSU	85%
DLLSTXTL	FTWOTXED	82%
GRCYNYGC	NYCMNY54	80%
HMSQNJHS	PHLAPASL	91%
HSTNTX01	NWORLAMA	81%

We considered two designs for the 71-node model, one that is sparse and another that is more meshed. These designs were constructed using the discrete-event flow optimization model described in Section 22.3. In the sparse design, there are 18 hub nodes upon which the other 53 spoke nodes are singly homed to the nearest hub node. In all, there are 86 links in the model with 53 links from spoke nodes to hub nodes, and 33 inter-hub-node links. All links are sized to a multiple of OC48 bandwidth capacity based on the traffic load projected on the link. Multilink shortest path routing is used as described above, and the network designs meet the performance objectives under the engineered load. In the meshed design, on the other hand, there are 25 hub nodes upon which the other 46 spoke nodes are multiply homed, with at least one link to the nearest hub node and any other links that are justified based on projected traffic levels. In addition, links between spoke nodes are allowed if justified based on projected traffic levels. In all, there are 122 links in the meshed design.

We now describe further details of the STT-EDR simulation model, which include the path selection algorithm, bandwidth allocation model, connection admission control procedure, priority queuing mechanism, bufferless queuing model, and capacity change algorithm.

Path selection uses a topology database that includes available bandwidth on each link. From the topology database, the originating node determines a list of shortest paths by using the Bellman-Ford algorithm. This path list is determined based on administrative weights of each link, which are communicated to all nodes within the routing domain. These administrative weights are set to $[1 + \varepsilon * d]$, where ε is a factor giving a relatively smaller weight to the distance d in comparison to the hop count. Many variations of these administrative weighting factors were studied at length, in

Table 22.9 Traffic model – Nodes with > 60% & > 80% voice traffic

	node	traffic	node	traffic
Hub nodes with > 60% voice traffic	HSTNTX01	63%	ORLDFLMA	63%
	PHNXAZMA	68%		
Spoke nodes with > 80% voice traffic	BLTMMDCH	85%	BRHMALMT	86%
	CLMASCTL	92%	DNWDGAAT	83%
	FTLDFLOV	89%	GRCYNYGC	89%
	HMSQNJHS	95%	JCVLFLCL	87%
	LSVLKYCS	82%	NWORLAMA	89%
	OMAHNENW	84%	PITBPADG	81%
	RCMDVAGR	82%	SBNDIN05	83%
	SHOKCA02	89%	SPKNWA01	85%
	SYRCNYSU	93%		

order to select the model giving the best overall performance. In addition, the available bandwidth on the link can be used to determine the link cost, as now illustrated. It should also be noted that with traffic engineering capabilities to select explicit routes, many other criteria (e.g., transport routing, transport technology, etc.) could be considered in selecting primary and alternate routes. An example of STT-EDR path selection is shown in Figure 22.8, as described earlier.

Whether or not bandwidth can be allocated to the bandwidth modification request on an LSP is determined based on the LSP link load state at each node. Three link load states are distinguished: available bandwidth (ABW), reserved-bandwidth (RBW), and bandwidth-not-available (BNA). Given a bandwidth request = DBW, link k is in the

ABW state, if $ILBW_k \geq RBW_k + DBW$,

RBW state, if $ILBW_k < RBW_k + DBW$,

BNA state, if $ILBW_k < DBW$,

where $ILBW_k$ is the idle link bandwidth on link k and RBW_k is the reservation bandwidth threshold for link k.

Management of LSP capacity uses this link state model and the allowed load state algorithm now described to determine if a bandwidth modification request can be accepted on a given LSP. In the allowed load state algorithm, a small amount of reserved bandwidth RBW_k governs the admission control on link k. Associated with each CT_c on link k are the allocated bandwidth constraints BC_{ck} to govern bandwidth allocation and protection. The reservation bandwidth on a link, RBW_k, can be accessed when a given CT_c has bandwidth-in-use $BWIP_{ck}$ below its allocated bandwidth constraint

BC_{ck}. However, if $BWIP_{ck}$ exceeds its allocated bandwidth constraint BC_{ck}, then the reservation bandwidth RBW_k cannot be accessed. In this way, bandwidth can be fully shared among CT_s if available, but is otherwise protected by bandwidth reservation methods. Therefore, bandwidth can be accessed for a bandwidth request = DBW for CT_c on a given link k based on the following rules:

For an LSP on a high priority or normal priority CT_c:

If $BWIP_{ck} \leq BC_c$: admit if $DBW \leq ILBW_k$

If $BWIP_{ck} > BC_c$: admit if $DBW \leq ILBW_k - RBW_k$; or equivalently

If $DBW \leq ILBW_{ck}$: admit the LSP.

where

$ILBW_{ck}$ is the idle link bandwidth on link k for $CT_c = ILBW_k - \delta_{0/1}(CT_{ck}) * RBW_k$, where

$$\delta_{0/1}(CT_{ck}) = \begin{cases} 0, & \text{if } BWIP_{ck} < BC_{ck} \\ 1, & \text{if } BWIP_{ck} \geq BC_{ck}. \end{cases}$$

For an LSP on a best-effort priority CT_c:

Allocated bandwidth $BC_c = 0$;

DiffServ queuing admits BE packets only if there is available link bandwidth.

In setting the bandwidth constraints for CT_{ck}, for a normal priority CT_c, the bandwidth constraints BC_{ck} on link k are set by allocating the maximum link bandwidth $LINK\_BW_k$ in proportion to the forecast or measured traffic load bandwidth $TRAF\_LOAD\_BW_{ck}$ for CT_c on link k. That is:

$PROPORTIONAL\_BW_{ck} = TRAF\_LOAD\_BW_{ck} /$
$[\sum_0^{MaxCT-1} TRAF\_LOAD\_BW_{ck}] * LINK\_BW_k$

For normal priority CT_c: $BC_{ck} = PROPORTIONAL\_BW_{ck}$;

For a high priority CT, the bandwidth constraint BC_{ck} is set to a multiple of the proportional bandwidth. That is:

For high priority CT_c:
$BC_{ck} = FACTOR * PROPORTIONAL\_BW_{ck}$,

where FACTOR is set to a multiple of the proportional bandwidth (e.g., FACTOR = 2 or 3 is typical). This results in some over-allocation ('overbooking') of the link bandwidth, and gives priority to the high priority CTs. Normally the bandwidth allocated to high priority CTs, that is, CT1 in our model, should be a relatively small fraction of the total link bandwidth, a maximum of 10–15 percent being a reasonable guideline. Given the traffic levels provided in Table 22.7, high-priority CT1 traffic could be at most 0.1% of the total link bandwidth on any given link.

As stated above, the bandwidth allocated to a best-effort priority CT_c is set to zero. That is:

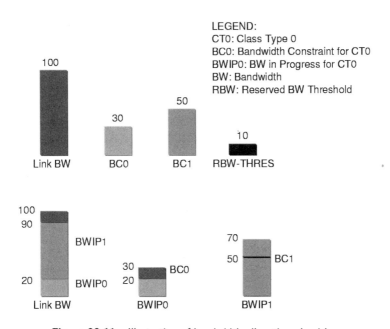

Figure 22.11 Illustration of bandwidth allocation algorithm

For best-effort priority CT_c:
$BC_{ck} = 0$.

We give a simple example of the bandwidth allocation algorithm, as illustrated in Figures 22.11 and 22.12. As shown in Figure 22.11, assume that there are two class-types: CT0 and CT1, and a particular link with

$LINK\_BW = 100$

with the allocated bandwidth constraints set as follows:

$BC0 = 30$ and $BC1 = 50$.

These bandwidth constraints are based on the forecast traffic loads, as discussed above. Either CT is allowed to exceed its bandwidth constraint BCc as long as there is at least RBW units of spare bandwidth remaining. Assume $RBW = 10$. So under overload, if

$BWIP0 = 20\ BWIP1 = 70$

Then for this loading

$ILBW = 100 - 20 - 70 = 10$

As illustrated in Figure 22.12, CT0 can take the additional bandwidth (up to 10 units) if the demand arrives, since it is below its BC value. CT1, however, can no longer increase its bandwidth on the link, since it is above its BC value and there is only RBW = 10 units of idle bandwidth left on the link. If best effort traffic is present, it can always seize whatever idle bandwidth is available on the link at the moment, but is subject to being lost at the queues in favor of the higher priority traffic.

As shown in Figure 22.12, a request arrives to increase bandwidth for CT1 by 5 units of bandwidth (i.e., DBW = 5). We need to decide whether to admit this request or not. Since for CT1

BWIP1 > BC1(70 > 50), and

DBW > ILBW − RBW(5 > 10 − 10),

the bandwidth request is rejected/blocked by the bandwidth allocation rules given above. Now let us say a request arrives to increase bandwidth for CT0 by 5 units of bandwidth (i.e., DBW = 5). We need to decide whether to admit this request or not. Since for CT0

BWIP0 < BC0(20 < 30), and

DBW < ILBW(5 < 10),

the bandwidth request is accepted by the bandwidth allocation rules given above.

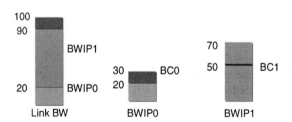

Bandwidth increase request = 5 = DBW arrives for Class Type 0 (CT0): accept for CT0 since BWIP0 < BC0 and DBW (= 5) < ILBW (= 10)

Bandwidth increase request = 5 = DBW arrives for Class Type 1 (CT1): reject for CT1 since BWIP1 > BC1 and DBW (= 5) > ILBW - RBW = 10 - 10 = 0

Figure 22.12   Illustration of bandwidth allocation algorithm

The example illustrates that with the current state of the link and the current CT loading, CT1 can no longer increase its bandwidth on the link, since it is above its BC1 value and there is only RBW = 10 units of spare bandwidth left on the link. But CT0 can take the additional bandwidth (up to 10 units) if the demand arrives, since it is below its BC0 value.

The algorithm for bandwidth additions and deletions to LSPs is as follows. The bandwidth constraint parameters are configured and do not change until reconfigured: $LINK\_BW_k$, BC_{ck}, and RBW_k. However, the reserved bandwidth variables change based on traffic: $BWIP_{ck}$, $ILBW_k$, and $ILBW_{ck}$. In the simulation, the $BWIP_{ck}$ and bandwidth allocated to each CT LSP is dynamically changed based on traffic: it is increased when the traffic demand increases and it is periodically decreased when the traffic demand decreases. Furthermore, if the CT LSP bandwidth cannot be increased on the primary path, an alternate LSP path is tried. When LSP tunnel bandwidth needs to be increased to accommodate a given connection request, the bandwidth is increased by the amount of the needed additional bandwidth, if possible. The CT allocated bandwidth quickly rises to the currently needed maximum bandwidth level, wherein no further requests are made to increase bandwidth, since departing flows leave a constant amount of available or spare bandwidth in the CT tunnel to use for new requests. CT bandwidth is reduced periodically (e.g., every 60 seconds) down to the level of the actually utilized bandwidth. Many other approaches were tried to increasing and decreasing tunnel bandwidth, such as increasing by more than currently needed bandwidth, or decreasing by only a fraction of the spare bandwidth, and these approaches are discussed later in this section.

Priority queuing for VoIP traffic and assured forwarding queuing for data traffic are maintained such that the packets are given priority according to the CT traffic priority. The queuing model assumes 3 levels of priority, high, normal, and best effort. CT1 (high priority VoIP emergency traffic) and CT2 (normal priority VoIP) use 2 priority queues with an expedited forwarding (EF) DiffServ (Blake et al., December 1998) per-hop behavior. CT1 is given a higher priority EF queue than CT2. CT3 and CT4 (premium private and premium public data, respectively) are given a normal priority assured forwarding (AF) per-hop behavior. CT5 is given a best-effort (BE) per-hop behavior, which is the lowest priority queue. All five queues have static bandwidth allocation limits applied based on the level of forecast traffic on each link, such that the bandwidth limits will not be exceeded under normal conditions, allowing for some traffic overload. Furthermore, a bufferless queuing model is assumed, in that once the bandwidth on the link is exceeded, packets are counted as lost/delayed. This bufferless model is a good approximation when the link bandwidth is large (Bonald et al., 2002), which is the case in this model in that a minimum OC48 link size is assumed.

Figure 22.13 illustrates the allocation of the classes of service to LSPs established for the five CTs, and also shows an example assignment of the EXP bits to control the DiffServ per-hop behaviors.

We now illustrate the optimization of the STT-EDR integrated voice/data model for the following parameters (note that many other parameters are investigated in ITU-T Recommendations (2002) and Ash (2003)):

- Level of reserved capacity: RBW_k is varied from 0 to 10% of link capacity;

- Number of allowed primary and alternate paths: the number is varied from 2 to 10 paths;

- Algorithm for increasing and decreasing CT bandwidth allocation: various models are studied, as described below.

Tables 22.10 and 22.11 illustrate the results for the level of reserved capacity RBW_k, as this level is varied from 0 to 10% of link capacity, for the meshed network design, with the focused overload scenario (3 times overload on Chicago) and multiple link failures scenario, respectively. The results indicate that a reservation level in the range of 3 to 5% of link capacity provides the best overall performance.

Tables 22.12 and 22.13 illustrate the results for the number of allowed primary and alternate paths, as this level is varied from 1 to 10, for the meshed network design, with the focused overload scenario (3 times overload on Chicago) and multiple link failures scenario, respectively. The results indicate that 6 or more primary and alternate paths provide the best overall performance.

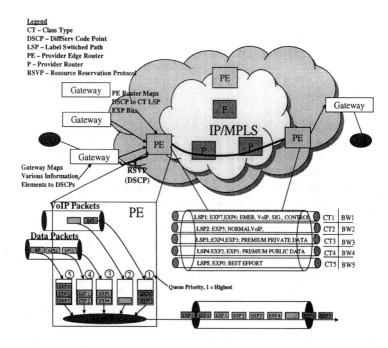

Figure 22.13 Illustration of class type bandwidth allocation & assignment of DiffServ EXP bits

Table 22.10 Simulation modeling results: Focused overload with variable level of bandwidth reservation

	STT-ERR integrated voice/data DR protocol design (reservation level – % of link bandwidth)					
Class type	0.00	0.50	1.00	3.00	5.00	10.00
Normal priority voice	**0.22**	**0.13**	**0.08**	**0.04**	**0.04**	**0.06**
GETS/ETS priority voice	**0.00**	**0.00**	**0.00**	**0.00**	**0.00**	**0.00**
Premium private data	0.01	0.01	0.01	0.00	0.00	0.00
Premium public data	0.00	0.00	0.00	0.00	0.00	0.00
Best effort data	33.15	32.86	33.10	32.44	32.80	32.47
Model assumptions	• 71-MSE simulation model, meshed design • 3X overload on CHCG • Results are in total network % lost/delayed packets. (CAC blocking is **bold face**. Queueing loss is not.)					
Observations	• Reservation level of 3–5% of link bandwidth provides best performance.					

Tables 22.14 and 22.15 illustrate the results for the algorithm for increasing and decreasing CT bandwidth allocation, for the meshed network design, with the focused overload scenario (3 times overload on Chicago) and multiple link failures scenario, respectively. In the simulation, the bandwidth allocated to each CT is dynamically changed based on traffic: it is increased as needed when the traffic demand increases above the currently allocated level and it is periodically decreased when the traffic demand decreases. Furthermore, if CT allocated bandwidth cannot be increased on a primary path, an alternate LSP path is tried. When CT bandwidth needs to be increased to accommodate a given connection request, the bandwidth is increased by a multiple (denoted as the MULTUP parameter) of the amount of the needed additional bandwidth, if possible. For example, if an increase in bandwidth to accommodate a single call is needed, then if MULTUP is set to 4, the tunnel bandwidth will be increased sufficiently to accommodate 4 additional calls. The tunnel bandwidth quickly

Table 22.11 Simulation modeling results: Multiple link failures with variable level of bandwidth reservation

	STT-ERR integrated voice/data DR protocol design (reservation level – % of link bandwidth)					
Class type	0.00	0.50	1.00	3.00	5.00	10.00
Normal priority voice	**0.08**	**0.07**	**0.06**	**0.06**	**0.06**	**0.06**
GETS/ETS priority voice	**0.00**	**0.00**	**0.00**	**0.00**	**0.00**	**0.00**
Premium private data	0.00	0.00	0.00	0.00	0.00	0.00
Premium public data	0.00	0.00	0.00	0.00	0.00	0.00
Best effort data	20.00	18.99	19.52	19.19	18.82	18.34
Model assumptions	• 71-MSE simulation model, meshed design • Multiple link failures • Results are in total network % lost/delayed packets. (CAC blocking is **bold face**. Queueing loss is not.)					
Observations	• Reservation level of > 1% of link bandwidth provides best performance.					

rises to the currently needed maximum bandwidth level, wherein no further requests are made to increase bandwidth, since departing flows leave a constant amount of available or spare bandwidth in the tunnel to use for new requests. Tunnel bandwidth is reduced periodically (e.g., every 60 seconds) by a multiplicative factor (denoted as the XDOWN parameter) times the difference between the allocated CT bandwidth and the current level of the actually utilized bandwidth (i.e., the current level of spare bandwidth). The number of times CT bandwidth is increased and decreased is also counted over a simulated hour, and recorded in the tables.

The results indicate that MULTUP = 1.0 and XDOWN = 0.5 provide the best overall performance.

Tables 22.16 and 22.17 illustrate the results for the algorithm for increasing and decreasing CT bandwidth allocation, for the meshed network design, with the focused overload scenario (3 times overload on Chicago) and multiple link failures scenario,

Table 22.12 Simulation modeling results: Focused overload with variable number of primary & alternate paths

	STT-ERR integrated voice/data DR protocol design (number of primary & alternate paths)						
Class type	1	2	3	4	5	8	10
Normal priority voice	**0.50**	**0.21**	**0.15**	**0.09**	**0.10**	**0.13**	**0.13**
GETS/ETS priority voice	**0.00**	**0.00**	**0.00**	**0.00**	**0.00**	**0.00**	**0.00**
Premium private data	1.95	0.52	0.18	0.11	0.01	0.01	0.01
Premium public data	3.74	1.85	0.67	0.41	0.00	0.00	0.00
Best effort data	33.09	31.30	32.69	33.04	33.73	32.86	32.86

Model assumptions	• 71-MSE simulation model, meshed design • 3X overload on CHCG • Results are in total network % lost/delayed packets. (CAC blocking is **bold face**. Queueing loss is not.)
Observations	• Six or more primary & alternate paths provides best performance.

respectively. In these simulations, the time period for the periodic reduction in tunnel bandwidth is varied over 15, 30, 60, 120, and 300 seconds. In each of these cases the MULTUP parameter is held at 1.0 and the XDOWN parameter is held at 0.5. The number of times CT bandwidth is increased and decreased are also counted over a simulated hour, and recorded in the tables.

The results indicate that the number of bandwidth increases and decreases is a strong function of the time period for the periodic reduction in tunnel bandwidth while the network lost/delayed traffic performance is not a strong function of this parameter. Therefore the parameter can be chosen to minimize the tunnel bandwidth changes in order to maximize scalability. However, an upper limit of 5 minutes is reasonable in order for the bandwidth allocation to react quickly to fast traffic demand changes, which can occur for example under a radio-call-in focused overload, a TV commercial break, or other often-observed traffic surges. Overall, it appears that setting this time period to 120 or 300 seconds provides good performance in terms of minimizing the bandwidth management change frequency, while still providing reasonable responsiveness to changes in bandwidth demand.

Table 22.13 Simulation modeling results: Multiple link failures with variable number of primary & alternate paths

	STT-ERR integrated voice/data DR protocol design (number of primary & alternate paths)						
Class type	1	2	3	4	5	8	10
Normal priority voice	**0.15**	**0.10**	**0.06**	**0.06**	**0.07**	**0.07**	**0.07**
GETS/ETS priority voice	**0.00**	**0.00**	**0.00**	**0.00**	**0.00**	**0.00**	**0.00**
Premium private data	5.76	2.13	0.00	0.00	0.00	0.00	0.00
Premium public data	6.85	3.22	0.00	0.00	0.00	0.00	0.00
Best effort data	21.11	22.12	18.93	19.58	18.99	18.99	18.99
Model assumptions	• 71-MSE simulation model, meshed design • Multiple link failures • Results are in total network % lost/delayed packets. (CAC blocking is **bold face**. Queueing loss is not.)						
Observations	• Four or more primary & alternate paths provides best performance.						

The optimization of the integrated voice/data DR protocol shows that DR performance can be significantly improved through the use of discrete event models. For example, we found that an STT-EDR based protocol in a sparsely connected network yields better performance than other options to wide changes of traffic patterns as well as failure scenarios. We then showed that various parameters can be optimized with discrete event models. In particular, we found that the following parameters provided the best overall performance:

- Reservation level in the range of 3 to 5% of link capacity;

- Five or more primary and alternate paths;

- CT bandwidth allocation increase factor MULTUP = 1.0 and decrease factor XDOWN = 0.5;

Table 22.14 Simulation modeling results: Focused overload with variable MULTUP & XDOWN parameters for bandwidth management

Class type	STT-ERR integrated voice/data DR protocol design (bandwidth management parameters: MULTUP, XDOWN)						
	1.,1.	2.,1.	4.,1.	1.,0.	1.,.25	1.,.5	1.,.75
Normal priority voice	**0.12**	**9.58**	**11.13**	**0.04**	**0.08**	**0.06**	**0.09**
GETS/ETS priority voice	**0.00**	**0.00**	**0.00**	**0.00**	**0.00**	**0.00**	**0.00**
Premium private data	0.01	1.00	0.97	0.00	0.01	0.01	0.02
Premium public data	0.00	2.37	2.35	0.00	0.00	0.00	0.00
Best effort data	32.86	30.59	30.09	35.38	33.86	33.37	33.80
BW increases (mil)	3.38	0.006	0.003	0.10	1.38	2.20	2.82
BW descreases (mil)	0.34	0.09	0.09	0.00	0.15	0.14	0.14

Model assumptions	• 71-MSE simulation model, meshed design • 3X overload on CHCG • Results are in total network % lost/delayed packets. (CAC blocking is **bold face**. Queueing loss is not.)
Observations	• MULTUP = 1. and XDOWN = 0.5 provides best performance.

- Time period for periodic reduction in tunnel bandwidth of 120 seconds or more provides good performance and reasonable responsiveness to changes in bandwidth demand.

22.3 MIN-COST DR NETWORK DESIGN & OPTIMIZATION METHODS

22.3.1 Problem formulation

Given traffic loads and node locations, find the minimum-cost link topology, link capacity, and DR design, subject to meeting performance constraints.

Table 22.15 Simulation modeling results: Multiple link failures with variable MULTUP & XDOWN parameters for bandwidth management

Class type	STT-ERR integrated voice/data DR protocol design (bandwidth management parameters: MULTUP, XDOWN)						
	1.,1.	2.,1.	4.,1.	1.,0.	1.,.25	1.,.5	1.,.75
Normal priority voice	**0.06**	**0.93**	**0.92**	**0.03**	**0.05**	**0.06**	**0.07**
GETS/ETS priority voice	**0.00**	**0.00**	**0.00**	**0.00**	**0.00**	**0.00**	**0.00**
Premium private data	0.00	5.71	5.66	0.14	0.00	0.00	0.00
Premium public data	0.00	6.78	6.81	0.12	0.00	0.00	0.00
Best effort data	18.98	21.05	21.17	19.15	20.22	19.84	19.41
BW increases (mil)	3.47	0.001	0.00	0.20	1.46	2.19	2.91
BW descreases (mil)	0.35	0.09	0.09	0.00	0.14	0.14	0.14

Model assumptions	• 71-MSE simulation model, meshed design • Multiple link failures • Results are in total network % lost/delayed packets. (CAC blocking is **bold face**. Queueing loss is not.)
Observations	• MULTUP = 1. and XDOWN = 0.5 provides best performance.

22.3.2 Optimization methods

In addition to discrete event simulation and shortest path design, as described in Section 22.2.2, minimum-cost network design incorporates linear programming, link capacity optimization, and multihour network design techniques. Multihour network design models are further classified as traffic-load flow optimization models, discrete-event flow optimization models, and virtual-transport flow optimization models. These optimization methods are now described. Examples are given of a large-scale min-cost DR network design using the discrete-event flow optimization model, which illustrate the versatility of these models in capturing highly complex network behavior, and allowing comparative analysis and optimization of complex network functions.

Table 22.16 Simulation modeling results: Focused overload with variable time period for periodic reduction in tunnel bandwidth

	STT-ERR integrated voice/data DR protocol design (time period for periodic reduction in tunnel bandwidth, seconds.)				
Class type	15	30	60	120	300
Normal priority voice	**0.04**	**0.04**	**0.05**	**0.06**	**0.06**
GETS/ETS priority voice	**0.00**	**0.00**	**0.00**	**0.00**	**0.00**
Premium private data	0.00	0.01	0.02	0.00	0.00
Premium public data	0.00	0.00	0.00	0.02	0.00
Best effort data	32.38	32.66	33.76	33.47	33.65
BW increases (mil)	4.25	3.02	2.14	1.49	0.92
BW descreases (mil)	0.54	0.28	0.15	0.08	0.03
Model assumptions	• 71-MSE simulation model, meshed design • 3X overload on CHCG • Results are in total network % lost/delayed packets. (CAC blocking is **bold face**. Queueing loss is not.)				
Observations	• Time period for periodic change in tunnel bandwidth of 120 seconds or more provides best performance.				

22.3.2.1 Linear programming. Linear programming (LP) techniques are well known and covered elsewhere in this book. Several standard methods are used to solve LP problems, including the simplex method (Dantzig, 1963), Karmarkar algorithm (Karmarkar, 1984; Karmarkar and Ramakrishnan, 1988), and other techniques. DR design and optimization models often incorporate LPs within a larger model, and typical LP problems embedded in these larger models include minimum-cost or maximum-flow multicommodity flow optimization problems. Examples of such

Table 22.17 Simulation modeling results: Multiple link failures with variable time period for periodic reduction in tunnel bandwidth

	STT-ERR integrated voice/data DR protocol design (time period for periodic reduction in tunnel bandwidth, seconds.)				
Class type	15	30	60	120	300
Normal priority voice	**0.06**	**0.06**	**0.05**	**0.05**	**0.04**
GETS/ETS priority voice	**0.00**	**0.00**	**0.00**	**0.00**	**0.00**
Premium private data	0.00	0.00	0.00	0.00	0.00
Premium public data	0.00	0.00	0.00	0.00	0.00
Best effort data	19.67	18.23	18.42	18.78	0.00
BW increases (mil)	4.16	3.07	2.21	1.58	1.00
BW descreases (mil)	0.50	0.27	0.14	0.07	0.03
Model assumptions	• 71-MSE simulation model, meshed design • Multiple link failures • Results are in total network % lost/delayed packets. (CAC blocking is **bold face**. Queueing loss is not.)				
Observations	• Time period for periodic change in tunnel bandwidth of 120 seconds or more provides best performance.				

minimum-cost and maximum-flow optimization models incorporating multicommodity flow LPs are given in Sections 22.3 and 22.4, respectively.

22.3.2.2 Link capacity optimization. As illustrated in Figure 22.14, link capacity optimization requires a tradeoff of the bandwidth demand carried on the link and bandwidth demand that must route on alternate paths. High link occupancy implies more efficient capacity utilization, however high occupancy leads to link congestion and the resulting need for some traffic not to be routed on the direct link but on alternate paths. Alternate paths may entail longer, less efficient paths. A good balance can be struck between link capacity design and alternate path utilization. For example, consider Figure 22.14 which illustrates a network where traffic is offered on link A-B connecting node A and node B.

Some of the bandwidth demand can be carried on link A-B, however, when the capacity of link A-B is exceeded, some of the bandwidth demand must be carried on alternate paths or be lost. The objective is to determine the direct A-B link capacity and alternate routing path flow such that all the traffic is carried at minimum cost. A simple optimization procedure is used to determine the best proportion of bandwidth demand to carry on the direct A-B link and how much bandwidth demand to alternate route to other paths in the network. As the direct link capacity is increased, the direct link cost increases while the alternate path cost decreases as more direct capacity is added, because the overflow bandwidth load decreases and therefore, the cost of carrying the overflow load decreases. An optimum, or minimum-cost condition is achieved when the direct A-B link capacity is increased to the point where the cost per incremental unit of bandwidth capacity to carry traffic on the direct link is just equal to the cost per unit of bandwidth capacity to carry traffic on the alternate network (Truitt, 1954). This is a design principle used in many design models, be they sparse or meshed networks, DR or FR networks.

22.3.2.3 Multihour network design.
DR design improves network utilization relative to FR design because FR cannot respond as efficiently to traffic load variations that arise from business/residential network use, time zones, seasonal variations, and other causes. DR design increases network utilization efficiency by varying routing tables in accordance with traffic patterns and designing capacity accordingly. A simple illustration of this principle is shown in Figure 22.15, where there is afternoon peak load demand between nodes A and B, but a morning peak load demand between nodes A and C and nodes C and B. Here, a simple DR design is to provide capacity only between nodes A and C and nodes C and B but no capacity between nodes A and B. Then the A-C and C-B morning peak loads route directly over this capacity in the morning, and the A-B afternoon peak load uses this same capacity by routing this traffic on the A-C-B path in the afternoon. A fixed routing network design provides capacity for the peak period for each node pair and thus provides capacity between nodes A and B, as well as between nodes A and C and nodes C and B.

The effect of multihour network design is illustrated by a national intercity network design model illustrated in Figure 22.16.

Here it is shown that about 20 percent of the network's first cost can be attributed to designing for time-varying loads. As illustrated in Figure 22.16, the 17 hourly networks are obtained by using each hourly load, and ignoring the other hourly loads, to size a network that perfectly matches that hour's load. The 17 hourly networks show that three network busy periods are visible: where we see morning, afternoon, and evening busy periods, and the noon-hour drop in load and the early-evening drop as the business day ends and residential calling begins in the evening. The hourly

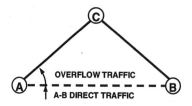

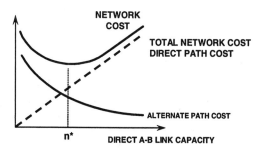

Figure 22.14 Link capacity optimization

network curve separates the capacity provided in the multihour network design into two components: Below the curve is the capacity needed in each hour to meet the load; above the curve is the capacity that is available but is not needed in that hour. This additional capacity exceeds 20 percent of the total network capacity through all hours of the day, which represents the multihour capacity cost. This gap represents the capacity of the network to meet noncoincident loads.

We now discuss the three types of multihour network design models: traffic load flow optimization models, discrete event flow optimization models, and virtual transport flow optimization models, and illustrate how they are applied to DR designs. For each model, we discuss steps that include initialization, routing design, capacity design, and parameter update.

22.3.2.4 Traffic Load Flow Optimization:. Traffic load flow optimization (TLFO) models use analytical expressions to represent often non-linear network behavior and to capture the routing of traffic loads over the network model. Approximations/heuristics are often used in such models, and the solution can involve a large-scale, multi-commodity flow LP optimization problem. TLFO methods are ap-

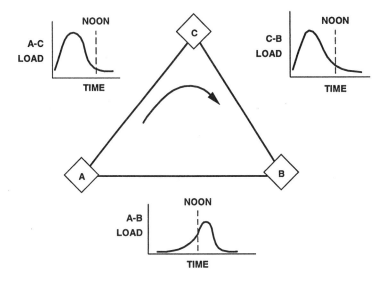

Figure 22.15 Multihour network design

proximate but give good design cost, with some uncertainty in meeting performance objectives due to approximations.

TLFO models are used for DR network design to optimize the routing of traffic flows and the associated link capacities. Such models typically solve mathematical equations that describe the routing of traffic flows analytically and, for DR network design, often solve LP flow optimization models. Various types of TLFO models are distinguished as to how flow is assigned to links, paths, and routes. For example, TLFO models can be path-based, in which traffic flow is assigned to individual paths, or route-based, in which traffic flow is assigned to routes. TLFO models do network design based on shortest path selection and LP traffic flow optimization. A TLFO model is illustrated in Figure 22.17.

The Routing Design step finds the shortest paths between nodes in the network, combines them into candidate routes, and uses the LP flow optimization model in the Capacity Design step to assign traffic flow to the candidate routes. For example, shortest path routing optimization is used in the Routing Design step to give connections access to paths in order of cost, such that connections access all direct path capacity between nodes prior to attempting more expensive overflow paths. Routes are constructed with specific path selection rules. Route-TLFO models construct routes from the selected paths and estimate the relative volume of traffic carried on each link of the route (Ash, 1998). The LP flow optimization model in the Capacity Design step then strives to minimize network design cost by sharing link capacity to the greatest

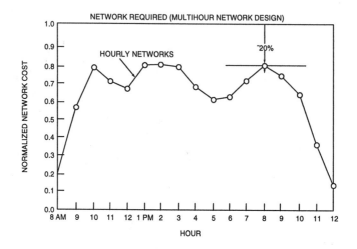

Figure 22.16 Hourly versus multihour network design

extent possible with the variation of loads in the network. This is done by equalizing the loads on links throughout the busy periods on the network, such that each link is used to the maximum extent possible in all time periods.

The Capacity Design step then takes the routing design and solves a fixed-point traffic flow model to determine the capacity of each link in the network. This model determines the flow on each link, and sizes the link to meet the performance level design objectives used in the routing design step. Once the links have been sized, the cost of the network is evaluated and compared to the last iteration. If the network cost is still decreasing, the update module:

- Computes the slope of the capacity versus load curve on each link, which reflects the incremental link cost, and updates the link *length* using this incremental cost as a weighting factor;

- Recomputes a new estimate of the optimal link utilization and overflow probability using the method described above in Section 22.3.2.2.

The new link lengths and utilization levels are fed to the routing design, which again constructs route choices from the shortest paths, and so on. Minimizing incremental network costs helps convert a nonlinear optimization problem to an LP optimization problem. Yaged Jr. (1971), Yaged (1973), and Knepley (1973) take advantage of this approach in their network design models. This favors large efficient links, which

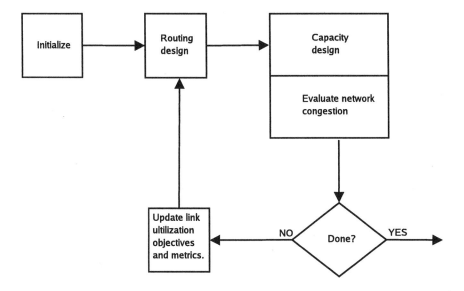

Figure 22.17 Traffic load flow optimization (TLFO) model

carry traffic at higher utilization efficiency than smaller links. Selecting an efficient utilization (blocking/delay) level on each link in the network is basic to the TLFO model. The link capacity optimization model described in Section 22.3.2.2 is used in the TLFO model to optimally divide the load between the direct link and the overflow network.

Detailed examples of TLFO model implementation are given in Ash (1998), which also illustrate the operational use of TLFO models in the on-going design of AT&T's deployment of TDR (DNHR) network technology.

22.3.2.5 Discrete Event Flow Optimization:. Discrete event flow optimization (DEFO) models are used for DR network design, and optimize the routing of discrete event flows, as measured in units of individual connection requests, and the associated link capacities. A network simulation model is incorporated directly into the optimization problem. Non-linear and complex behavior can be modeled accurately, and meeting performance objectives can be assured. However, DEFO methods can be computationally intensive. Examples are given for the optimal DR network design within a large-scale network model, including aggregated per-CT design versus per-flow design, and separate voice and data network design versus integrated voice/data network design.

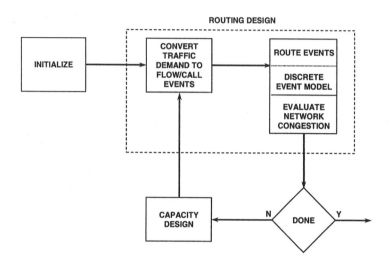

Figure 22.18 Discrete event flow optimization (DEFO) model

Figure 22.18 illustrates steps of the DEFO model. The event generator converts traffic demands to discrete connection-request events. The discrete event model provides routing logic according to the particular routing method and routes the connection-request events according to the routing table logic. DEFO models use simulation models for path selection and routing table management to route discrete-event demands on the link capacities, and the link capacities are then optimized to meet the required flow.

We generate initial link capacity requirements based on the traffic load matrix input to the model. In a meshed network design, an initial node-termination capacity is estimated based on a maximum design occupancy in the node busy hour (e.g., 0.93), and the total network occupancy (total traffic demand/total link capacity) in the network busy hour is adjusted to fall within a range (e.g., 0.84 to 0.89). Network performance is evaluated as an output of the discrete event simulation model, and any needed link capacity adjustments are determined. Capacity is allocated to individual links in accordance with the Kruithof allocation method (Kruithof, 1937), which distributes link capacity in proportion to the overall demand between nodes.

Kruithof's technique is used to estimate the node-to-node requirements p_{ij} from the originating node i to the terminating node j under the condition that the total node link capacity requirements may be established by adding the entries in the matrix $P = [p_{ij}]$. Assume that a matrix $Q = [q_{ij}]$, representing the node-to-node link capacity requirements for a previous iteration, is known. Also, the total link capacity requirements b_i at each node i and the total link capacity requirements d_j at each node j are estimated as follows:

$$b_i = \frac{a_i}{\gamma}$$

$$d_j = \frac{a_j}{\gamma}$$

where a_i is the total traffic at node i, a_j is the total traffic at node j, and γ is the average traffic carrying capacity per unit of bandwidth (e.g., $\gamma = 0.8$ network-wide is a typical approximation), or node design occupancy, as given previously. The terms p_{ij} can be obtained as follows:

$$f_i = \frac{b_i}{\sum_j q_{ij}}$$

$$f_j = \frac{d_j}{\sum_i q_{ij}}$$

$$E_{ij} = \frac{f_i + f_j}{2}$$

$$p_{ij} = q_{ij} E_{ij}$$

After the above equations are solved iteratively, the converged steady state values of p_{ij} are obtained.

In a sparse network design, the initial network topology is generated as follows. Candidate links are initially established between all the nodes, that is, the nodes are fully interconnected by candidate links which can be later eliminated in the DEFO model. Traffic is then routed on shortest paths, where the shortest primary and alternate paths are determined by the Bellman Ford algorithm. Links are then eliminated if less than a threshold level (e.g., 0.67) of a transport capacity module (e.g., OC48) is required based on the routed link traffic demand. Traffic is then routed again over this reduced link topology, and links sized to an objective occupancy (e.g., 0.8) based on the routed link traffic demand. Finally, the link capacities are rounded up to the nearest module size to yield the initial link topology.

The DEFO model can generate connection-request events according to a Poisson arrival distribution and exponential holding times, or with more general arrival streams and arbitrary holding time distributions, because such models can readily be implemented in the discrete routing table simulation model. Connection-request events are generated in accordance with the traffic load matrix input to the model. These events are routed on the selected path according to the routing table rules, as modeled by the routing table simulation, which determines the selected path for each connection event and flows the event onto the network capacity.

The output from the routing design is the fraction of traffic lost and delayed in each time period. From this traffic performance, the capacity design determines the new link capacity requirements of each node and each link to meet the design performance level. From the estimate of lost and delayed traffic at each node in each time period, an occupancy calculation determines additional link capacity requirements for an updated link capacity estimate. Such a link capacity determination is made based on the amount of blocked/delayed traffic. In the sparse design mode, the total blocked/delayed traffic Δa is estimated at each link, and an estimated link capacity increase ΔT is calculated by the relationship

$$\Delta T = \frac{\Delta a}{\gamma}$$

where γ is the average traffic-carrying capacity per unit of bandwidth.

In the meshed network design, the ΔT is calculated for each node, and this capacity increment is distributed to each link according to the Kruithof estimation method described above. The Kruithof allocation method distributes link capacity in proportion to the overall demand between nodes and in accordance with link cost, so that overall network cost is minimized.

Once the links have been resized, the network is reevaluated to see if the performance objectives are met, and if not, another iteration of the model is performed.

We evaluate in the model the confidence interval of the engineered blocking/delay. For this analysis, we evaluate the binomial distribution for the 90th percentile confidence interval. Suppose that for a traffic load of A in which calls arrive over the designated time period of stationary traffic behavior, there are, on average, m blocked calls out of n attempts. This means that there is an average observed blocking/delay probability of

$$pl = \frac{m}{n}$$

where, for example, $pl = .01$ for a 1 percent average blocking/delay probability. Now, we want to find the value of the 90th percentile blocking/delay probability p such that

$$E(n,m,p) = \sum_{r=m}^{n} C_n^r p^r q^{n-r} \geq 0.90$$

where

$$C_n^r = \frac{n!}{(n-r)!r!}$$

is the binomial coefficient, and

$$q = 1 - p.$$

Then the value p represents the 90th percentile blocking/delay probability confidence interval. That is, there is a 90 percent chance that the observed blocking/delay will be less than or equal to the value p. Methods given in Weintraub (1963) are used to numerically evaluate the above expressions.

As an example application of the above method to the DEFO model, suppose that network traffic is such that one million calls arrive in a single busy-hour period, and we wish to design the network to achieve one percent average blocking/delay or less. If the network is designed in the DEFO model to yield at most .00995 probability of blocking/delay, that is, at most 9950 calls are blocked out of 1 million calls in the DEFO model, then we can be more than 90 percent sure that the network has a maximum blocking/delay probability of .01. For a specific node pair where 2000 calls arrive in a single busy-hour period, suppose we wish to design the node pair to achieve 1 percent average blocking/delay probability or less. If the network capacity is designed in the DEFO model to yield at most .0075 probability of blocking/delay for the node pair, that is, at most 15 calls are blocked out of 2000 calls in the DEFO model, then we can be more than 90 percent sure that the node pair has a maximum blocking/delay probability of .01. These methods are used to ensure that the blocking/delay probability design objectives are met, taking into consideration the sampling errors of the discrete event model.

The greatest advantage of the DEFO model is its ability to capture very complex routing behavior through the equivalent of a simulation model provided in software in the routing design module. By this means, very complex routing networks have been designed by the model, which include all of the routing methods discussed in Section 22.2 to include TDR, SDR, and EDR methods. Complex traffic processes, such as self-similar traffic, can also be modeled with DEFO methods. A flow diagram of the DEFO model, in which multilink STT-EDR logical blocks described in Section 22.2.3 are implemented, is illustrated in Figure 22.19. The DEFO model is general enough to include all DR models yet to be determined.

22.3.2.6 Examples of DR Design using DEFO Models:. Many network design and optimization tradeoffs are examined with DEFO models, including:

- Centralized routing table control versus distributed control;

- Off-line (e.g. preplanned TDR-based) routing table control versus on-line routing table control (e.g. SDR- or EDR-based);

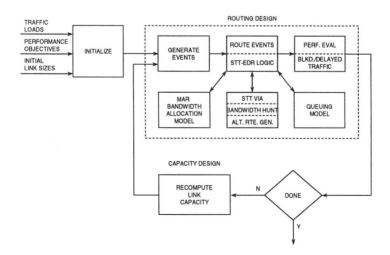

Figure 22.19 DEFO model with multilink STT-EDR

- Per-flow bandwidth allocation versus aggregated per-CT bandwidth allocation;
- Integrated voice/data design versus separate voice and data design;
- Sparse logical topology versus meshed logical topology;
- FR versus TDR versus SDR versus EDR path selection;
- Multilink path selection versus two-link path selection;
- Path selection using local status information versus global status information;
- Status dissemination using status flooding, versus distributed query for status, versus centralized status in a DR processor.

Here we illustrate per-flow bandwidth allocation versus aggregated per-CT bandwidth allocation, and integrated voice/data design versus separate voice and data design. These examples well illustrate the versatility of the DEFO model in capturing highly complex network behavior, and allowing comparative analysis and optimization of complex network functions.

Aggregated Per-CT vs. Per-Flow Network Design

Here we illustrate the use of the DEFO model to design for a per-flow integrated voice/data STT-EDR network design and a per-CT network design, and to provide comparisons of these designs. The per-flow and per-CT DEFO designs for a 135-node model are summarized in Table 22.18.

Table 22.18 Design comparison of aggregated per-CT & per-flow bandwidth allocation – multilink STT-EDR connection routing – sparse single-area flat topology (135-node network model; DEFO design model)

Network design parameters	Aggregated per-CT bandwidth allocation design	Per-flow bandwidth allocation design
Number of Links	981	982
Termination capacity (equiv. DS0s, mill.)	16.5	16.4
Transport capacity (equiv. DS0-miles, mill.)	9365.2	9285.3
Termination cost ($ mill.)	148.1	137.7
Transport cost ($ mill.)	1158.1	1129.5
Total cost ($ mill.)	1306.2	1267.2

We see from the above results that the per-CT design compared to the per-flow design yields the following results:

- The per-flow design has 0.996 of the total termination capacity of the per-CT design;
- The per-flow design has 0.991 of the total transport capacity of the per-CT design;
- The per-flow design has 0.970 of the total network cost of the per-CT design.

These results indicate that the per-CT design and per-flow design are quite comparable in terms of capacity requirements and design cost. In ITU-T Recommendations (2002) we show that the performance of these two designs is also quite comparable under a range of network scenarios.

Integrated vs. Separate Voice/ISDN & Data Network Designs:

The comparative DEFO designs for separate and integrated network designs under multilink STT-EDR, per-flow routing are given in Table 22.19 for the following cases:

- Voice-only traffic;
- Data-only traffic;
- Integrated voice/data design.

Table 22.19 Comparison of voice-only design, data-only design, & integrated voice/data design – multilink STT EDR connection routing; per-flow bandwidth allocation – sparse single-area flat topology (135-node network model; DEFO design model)

Network design parameters	Voice-only design	Data-only design	Integrated voice/data design
Number of links	254	796	982
Termination capacity (equiv. DS0s, mill.)	3.31	14.2	16.4
Transport capacity (equiv. DS0-miles, mill.)	1824.5	7816.9	9285.3
Termination cost ($ mill.)	34.2	112.8	137.7
Transport cost ($ mill.)	290.4	900.9	1129.5
Total cost ($ mill.)	324.7	1013.8	1267.2

We see from the above results that the separate voice and data designs compared to the integrated design yields the following results:

- The integrated design has 0.937 of the total termination capacity as the separate voice and data designs;

- The integrated design has 0.963 of the total transport capacity as the separate voice and data designs;

- The integrated design has 0.947 of the total cost as the separate voice and data designs.

These results indicate that the integrated design is somewhat more efficient in design owing to the economy-of-scale of the higher-capacity network elements, as reflected in the cost model.

Virtual Transport Flow Optimization

Virtual transport flow optimization (VTFO) models convert traffic load demands to virtual transport (VT) bandwidth demands at the outset, and a large-scale, multi-commodity flow LP optimization problem solved for the transport capacity. VTFO methods can provide good design cost, and in some formulations reasonable accuracy in meeting performance objectives. VTFO models optimize the routing of transport flows, as measured in units of transport bandwidth demand and the associated link capacities. For application to min-cost network design, VTFO models use mathematical equations to convert traffic demands to transport capacity demands, and the transport flow is then routed and optimized. Figure 22.20 illustrates the VTFO model steps. This model typically assumes an underlying traffic routing structure. For example, the traffic routing model can be FR or DR.

Node-to-node traffic demands are converted to node-to-node VT demands by using the link capacity optimization procedure given in Section 22.3.2.2. This approach is used to optimally divide the load between the shortest path and the overflow network, to obtain an equivalent VT demand, by time period. We first calculate the direct path capacity according to the optimal sizing procedure which yields an efficient direct path utilization level and the direct path bandwidth demands. The alternate path bandwidth demand is readily determined according to the overflow from the direct path by assuming a constant occupancy level γ (e.g., .8) for the overflow bandwidth demand.

A multicommodity LP transport flow optimization problem is then solved in the Capacity Design Step for the link capacity sizing, which routes node-to-node transport load demands by time period on the optimal, least-cost paths and sizes the links to meet the design level of flow. Once the links have been sized, the performance of the network is evaluated, and if the performance objectives are not met, further modification is made to capacity requirements.

Detailed examples of VTFO model implementation are given in Ash (1998).

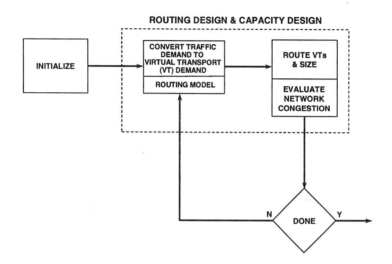

Figure 22.20 Comparison of voice-only design, data-only design, & integrated voice/data design – Multilink STT EDR connection routing; Per-flow bandwidth allocation – sparse single-area flat topology – (135-node network model; DEFO design model)

22.4 MAX-FLOW/REVENUE DR DESIGN & OPTIMIZATION METHODS

22.4.1 Problem formulation

Given traffic loads and network topology/capacity, find the maximum-flow/revenue DR design.

22.4.1.1 Optimization methods. Maximum-flow network design incorporates discrete event simulation and shortest path design, as described in Section 22.2.2, as well as LP, link capacity optimization, and multihour network design techniques, as described in Section 22.3.2. Furthermore, multihour network design models include TLFO, DEFO, and VTFO methods, as described in Section 22.3.2.

The design of the DR protocols for SDR- and EDR-based methods beneficially uses DEFO models, as illustrated by the examples given in Section 22.2, and this approach inherently maximizes flow across a wide range of network conditions including overload and failure scenarios. Max flow optimization, therefore, is one of the optimization criteria in the DR protocol design in such cases.

Other very interesting max-flow optimization algorithms have been formulated which allow the link costs (administrative weights) to be optimized such that a maximum flow design results when shortest path routing is employed (Fortz and Thorup,

2000; Fortz et al., 2002). These papers illustrate how OSPF network administrative weights can be optimized for max-flow routing; heuristics are given in the cited papers since the optimization problem is NP-hard.

In the next section we give an example of max-flow design using a VTFO model, which involves the solution of a multi-commodity LP flow optimization problem.

22.4.2 Example of VTFO max-flow design

In VTFO max-flow design, traffic load demands are converted to transport bandwidth demands at the outset, and a large-scale, multi-commodity flow LP optimization problem is solved. A max-flow design example is described here for a DC-SDR network. The model achieves significant improvement in both the average network blocking and node-pair blocking distribution when the network is in a congested state, such as under peak-day loads (e.g., Christmas Day traffic loads).

In the DC-SDR algorithm assumed in the max-flow DR design, there is a prespecified set of alternate paths for each node pair in the network and these alternate paths are segmented into two sets. When a connection request enters the DR network, the originating node first tries to set up the connection on the direct path, and if the direct primary path is not available, then the connection can be routed through one of the first set of prespecified paths selected according to its idle bandwidth status. The candidate paths in the first set are obtained from the solution of a max-flow LP optimization problem. If the connection cannot be routed on the direct path or one of the first set of paths, then the search for alternate paths is extended into the second set of via paths which are the remaining alternate paths not included in the first set of via paths. Simulation results presented below demonstrate that this network max-flow optimization reduces both the average network blocking and node-pair blocking distribution when the traffic demands are on-average higher and have a different distribution as compared to the design-level traffic demands.

In the network max-flow optimization model, the network capacity and offered loads are given, and a throughput maximization problem is solved for the first set of candidate paths for each node pair in the network. We first convert the traffic demands into virtual transport (VT) capacity demands and then optimally flow the VT demands when solving the LP problem to maximize total network throughput:

$$\max \sum_{k=1,\ldots,K; j=1,\ldots,J_k} y_{jk}$$

$$\text{s.t.} \sum_{k=1,\ldots,K; j=1,\ldots,J_k} P_{jkl} * y_{jk} \leq a_l$$

$$\sum_{j=1,\ldots,J_k} y_{jk} \leq VT_k$$

where

J_k = number of allowed paths (including the direct path) for node pair k,

$P_{jkl} = 1$ if path j of node pair k routes over link l; 0 otherwise,

y_{jk} = flow assigned to path j of node pair k,

a_l = existing transport capacity on link l available for the given CT,

VT_k = total VT demand for node pair k.

In this model, links l are numbered 1 to L, paths j are numbered 1 to J_k, and node pairs k are numbered 1 to K. The first set of constraints requires that the sum of the flows on all the paths containing link l cannot exceed the total capacity of link l. The second set of constraints requires that the sum of flows on all the paths connecting node pair k cannot exceed the VT demand of node pair k. The Karmarkar algorithm (Karmarkar, 1984; Karmarkar and Ramakrishnan, 1988) is used to solve the LP max-flow model.

In a mesh network, the total number of links is almost the same as the total number of node pairs (K). Thus, the number of constraints in the LP is approximately 2 times the number of node pairs, or $2K$. The number of variables is the same as the number of paths, which is K times the average number of paths per node pair. Assuming each alternate path has two links only, the total number of variables is of the order K^3, which is the dominating factor for the convergence speed of the Karmarkar algorithm. On the other hand, the constraint matrix of the LP problem is very sparse, because each path belongs to only one node pair, and it contains at most two links. Thus, the particular variant of the Karmarkar algorithm given in Karmarkar and Ramakrishnan (1988) is suitable for solving the LP problem. However, the very large number of variables in this problem affects the convergence speed of the Karmarkar algorithm. We found that a way of reducing the computation time is to run the Karmarkar algorithm up to a certain level of convergence and then to delete those paths with zero or very minimal path flows.

A significant number of paths can be eliminated in this way. We then use the same Karmarkar algorithm method to solve the problem with the reduced sets of paths. The shortening of the path lists not only reduces the total computation time but also helps to

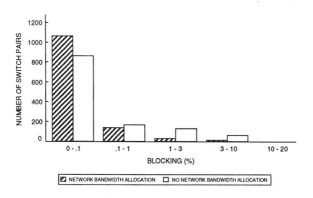

Figure 22.21 Example of VTFO max-flow design

force the flows of VT capacity demands to be concentrated on fewer paths. When we have obtained the optimal solution to the smaller problem, we then select the candidate paths as the paths that have flows exceeding a threshold in the LP solution.

We use simulation to test the max-flow solution for paths that maximize throughput. Results are shown in Figure 22.21 based on a 50-node network model. The network is first designed for the given traffic demands, using the VTFO model described in Section 22.2. We then solve the max-flow optimization problem with each node-to-node load increased by 10 percent. The first-choice via paths are then obtained from the LP solution described above. In the simulation, the traffic loads are increased randomly by a factor uniformly varying between 8 and 12 percent. We find that the network average blocking of the network max-flow design is 0.27 percent, versus 0.5 percent without network max-flow design, which is a substantial improvement. The blocking distribution is shown in Figure 22.21, which shows a clear gain in the number of low-blocking node pairs by the DR network max-flow design. We have thus seen that the VTFO max-flow DR optimization model achieves significant improvement in both the average network blocking and the node-pair blocking distribution when the network is in a congested state.

Bibliography

H. Ahmadi, J. Chen, and R. Guerin. Dynamic routing and call control in high-speed integrated networks. In *Proc. 13th International Teletraffic Congress, ITC 13*, pages 397–403, Copenhagen, 1992.

J. M. Akinpelu. The overload performance of engineered networks with nonhierarchical and hierarchical routing. *Bell System Technical Journal*, 63:1261–1281, 1984.

G. Apostolopoulos, R. Guerin, S. Kamat, A. Orda, and S. K. Tripathi. Intra-domain QoS routing in IP networks: A feasibility and cost/benefit analysis. *IEEE Network*, 13(5):42–54, 1999.

G. R. Ash. *Dynamic Routing in Telecommunications Networks*. McGraw-Hill, 1998.

G. R. Ash. Performance evaluation of QoS-routing methods for IP-based multiservice networks. *Computer Communications*, 26(8):817–833, May 2003.

G. R. Ash, R. H. Cardwell, and R. P. Murray. Design and optimization of networks with dynamic routing. *Bell System Technical Journal*, 60(8):1787–1820, 1981.

G. R. Ash and P. Chemouil. 20 years of dynamic routing in circuit-switched networks: Looking backward to the future. *IEEE Global Communications Newsletter*, pages 1–4, October 2004. Insert of IEEE Communication Magazine, October 2004 issue.

G. R. Ash, P. Chemouil, A. N. Kashper, S. S. Katz, and Y. Watanabe. Robust design and planning of a worldwide intelligent network. *IEEE Journal on Selected Areas in Communications*, 7(8), 1989.

G. R. Ash, J.-S. Chen, A. E. Frey, and B. D. Huang. Real-time network routing in a dynamic class-of-service network. In *Proc. 13th International Teletraffic Congress, ITC 13*, Copenhagen, 1991.

ATM Forum Technical Committee. Private network-network interface specification version 1.1 (pnni 1.1), April 2002. [af-pnni-0055.002].

D. Awduche. MPLS and traffic engineering in IP networks. *IEEE Communications Magazine*, 37(12):42–47, December 1999.

D. Awduche, A. Chiu, A. Elwalid, I. Widjaja, and X. Xiao. Overview & principles of internet traffic engineering. *RFC 3272*, May 2002. Available online on May 15, 2005: http://www.faqs.org/rfcs/rfc3272.html.

D. Awduche, J. Malcolm, J. Agogbua, M. O'Dell, and J. McManus. Requirements for traffic engineering over MPLS. *RFC 2702*, September 1999. Available online on May 15, 2005: http://www.faqs.org/rfcs/rfc2702.html.

R. E. Bellman. *Dynamic Programming*. Princeton University Press, 1957.

S. Blake, D. Black, M. Carlson, E. Davies, Z. Wang, and W. Weiss. An architecture for differentiated services. *RFC 2475*, December 1998. Available online on May 15, 2005: http://www.faqs.org/rfcs/rfc2475.html.

T. Bonald, S. Oueslanti-Boulahia, and J. W. Roberts. IP traffic and QoS control: The need for a flow-aware architecture. In *World Telecommunications Congress*, Paris, France, September 2002.

E. Crawley, R. Nair, B. Rajagopalan, and H. Sandick. A framework for QoS-based routing in the internet. *RFC 2386*, August 1998. Available online on May 15, 2005: http://www.faqs.org/rfcs/rfc2386.html.

G. B. Dantzig. *Linear Programming and Extensions*. Princeton University Press, 1963.

A. DiBenedetto, P. LaNave, and C. Sisto. Dynamic routing of the italcable telephone traffic: Experience and perspectives. In *Proceedings Globecom 1989*, Dallas, TX, 1989.

E. W. Dijkstra. A note on two problems in connection with graphs. *Numerical Mathematics*, 1:269–271, 1959.

A. Elwalid, C. Jin, S. Low, and I. Widjaja. MATE: MPLS adaptive traffic engineering. In *Proceedings INFOCOM'01*, April 2001.

B. Fortz, J. Rexford, and M. Thorup. Traffic engineering with traditional IP routing protocols. *IEEE Communications Magazine*, 40(10):118–124, October 2002.

B. Fortz and M. Thorup. Internet traffic engineering by optimizing OSPF weights. In *Proceedings of IEEE INFOCOM'00*, March 2000.

R. J. Gibbens, F. P. Kelly, and P.B. Key. Dynamic alternative routing: Modeling and behavior. In *Proc. 12th Int'l Teletraffic Congress, ITC 12*, Turin, Italy, 1988.

ITU-T Recommendations. QoS routing & related traffic engineering methods for multiservice TDM. In *ATM & IP-based networks*, May 2002.

N. K. Karmarkar. A new polynomial time algorithm for linear programming. *Combinatorica*, 4(4):373–395, 1984.

N. K. Karmarkar and K. G. Ramakrishnan. Implementation and computational results of the karmarkar algorithm for linear programming, using an iterative method for computing projections. In *13th International Symposium on Mathematical Programming*, Tokyo, Japan, 1988.

J. E. Knepley. Minimum cost design for circuit switched networks. Technical Report 36–73, Defense Communications Engineering Center, System Engineering Facility, Reston, Virginia, July 1973.

J. Kruithof. Telefoonverkeersrekening. *De Ingenieur*, 52(8):E15–E25, February 1937.

R. S. Krupp. Stabilization of alternate routing networks. In *IEEE International Communications Conference*, Philadelphia, Pennsylvania, 1982.

C. Liljenstolpe. An approach to IP network traffic engineering (Cable & Wireless), 2002. Available online on May 15, 2005: http://www.nanog.org/mtg-0202/ppt/lil/.

Q. Ma. *Quality-of-Service Routing in Integrated Services Networks*. PhD thesis, Computer Science Department, Carnegie Mellon University, Pittsburgh, PA, January 1998.

A. Mase and H. Yamamoto. Advanced traffic control methods for network management. *Communications Magazine*, 28(10):82–88, 1990.

J. McQuillan, I. Richter, and E. Rosen. The new routing algorithm for the ARPANET. *IEEE Transactions on Communications*, 28(5):711–719, May 1980.

J. Moy. Ospf version 2. *RFC 2328*, April 1998. Available online on May 15, 2005: http://www.faqs.org/rfcs/rfc2328.html.

Y. Nakagome and H. Mori. Flexible routing in the global communication network. In *Proc. 7th International Teletraffic Congress, ITC7*, Stockholm, Sweden, 1973.

A. Odlyzko. Data networks are lightly utilized, and will stay that way. *The Review of Network Economics*, 2(3):210–237, September 2003.

C. Pierantozzi and V. Springer. Level3 mpls protocol architecture, 2000. Internet Draft. Available online on May 15, 2005: https://datatracker.ietf.org/public/idindex.cgi?command=id_detail&id=58%93.

M. Pioro and D. Medhi. *Routing, Flow, and Capacity Design in Communication and Computer Networks*. Morgan Kaufmann, 2004.

J. Regnier and W. H. Cameron. State-dependent dynamic traffic management for telephone networks. *IEEE Communications Magazine*, 28(10):42–53, October 1990.

Y. Rekhter and T. Li. A border gateway protocol 4 (BGP-4), March 1995. Available online on May 15, 2005: http://ietf.org/rfc/rfc1771.txt.

E. Rosen, A. Viswanathan, and R. Callon. Multiprotocol label switching architecture. *RFC 3031*, January 2001. Available online on May 15, 2005: http://www.faqs.org/rfcs/rfc3031.html.

C. J. Truitt. Traffic engineering techniques for determining trunk requirements in alternate routed networks. *Bell System Technical Journal*, 31(2):277–304, March 1954.

S. Weintraub. *Tables of Cumulative Binomial Probability Distribution for Small Values of p*. Collier-Macmillan Limited, 1963.

X. X. Xiao. *Providing Quality of Service in the Internet*. PhD thesis, Michigan State University, 2000.

X. Xiao, A. Hannan, B. Bailey, and L. Ni. Traffic engineering with MPLS in the internet. *IEEE Network Magazine*, pages 28–33, March 2000.

X. Xiao and L. M. Ni. Internet QoS: A big picture. *IEEE Network*, 13(2):8–18, March–April 1999.

B. Yaged. Minimum cost design for circuit switched networks. *Networks*, 3:193–224, 1973.

B. Yaged Jr. *Long Range Planning for Communications Networks*. PhD thesis, Polytechnic Institute of Brooklyn, 1971.

23 ILP FORMULATIONS FOR THE ROUTING AND WAVELENGTH ASSIGNMENT PROBLEM: SYMMETRIC SYSTEMS

Brigitte Jaumard[1], Christophe Meyer[1], and Babacar Thiongane[1]

[1]Department of Computer Science and
Operations Research
Université de Montréal
Montréal H3C 3J7, Canada
jaumard@iro.umontreal.ca
christop@crt.umontreal.ca
babacar.thiongane@gerad.ca

Abstract: Different integer linear programming (ILP) formulations have been proposed for the routing and wavelength assignment problem in WDM optical networks, mainly for asymmetrical systems, more than for symmetrical systems, under different objectives. We propose a synthesis of the mathematical models for symmetrical systems, with a unified and simplified notation for four widely used objectives. As for asymmetrical traffic models (Jaumard, Meyer, and Thiongane, 2004), we show that all formulations, both link and path formulations, are equivalent in terms of the upper bound value provided by the optimal solution of their linear programming relaxation, although their number of variables and constraints differ. We propose an experimental comparison of the best linear relaxation bounds with the optimal ILP solutions whenever it is possible, for several network and traffic instances.

Keywords: Routing, wavelength assignment, RWA problem, mathematical programming formulation, relaxations, bounds.

23.1 INTRODUCTION

Wavelength-Division-Multiplexing (WDM) and more recently Dense WDM (DWDM) offer the capability of building large networks with Gb/s or Tb/s network traffic. This allows to answer the increase in voice, video or data demand following the reduced cost in bandwidth due to the deregulation and the breaking up of the telephone industry, see, e.g., Ramaswami and Sivarajan (2002) and Zheng and Mouftah (2004). WDM or DWDM networks correspond to *wavelength-routing* networks, where the network ensures communication through lightpaths. A lightpath is an optical end-to-end connection carried over a wavelength on each intermediate link along the route from the source to the destination node. No two lightpaths going through the same fiber link can share the same wavelength: this is the so-called *clash* constraint. Given a set of connection requests and a set of available wavelengths, the Routing and Wavelength Assignment (RWA) problem aims at defining a set of lightpaths so as to accommodate all or most of the connections, while optimizing a given objective function. Several objectives have been considered in the literature:

- Minimizing the blocking rate (or equivalently maximizing the number of accepted connections): see, e.g., Ramaswami and Sivarajan (1995), Zang et al. (2000), Chen and Banerjee (1996), Krishnaswamy and Sivarajan (2001a), and Kumar and Kumar (2002). This objective is most relevant when there is not enough transport capacity, i.e., enough available wavelengths, to accommodate all connection requests.

- Minimizing the number of used wavelengths: see Lee et al. (2002) and Margara and Simon (2000). It is usually assumed in that case that all connections can be established given the available wavelengths and the objective is to use the smallest number of them.

- Minimizing the maximum number of wavelengths going through a single fiber: see, e.g., Wauters and Demeester (1996) and Zang et al. (2000). In this case, it is again assumed that all connections can be granted.

We also introduce a new objective function which consists in minimizing the network load as defined by the fraction of the number of wavelengths used on the overall set of fiber links in the network.

Two main traffic models can be found in the literature corresponding to symmetrical or asymmetrical systems. Asymmetrical systems are associated with asymmetrical traffic assumptions and can be used, e.g., for network architectures where the transport capacities can be adapted to asymmetrical bandwidth requirements, for instance bidirectional WDM systems with more wavelengths in one direction or unidirectional WDM systems with more wavelength setups in one direction. On the contrary, symmetrical systems are associated with symmetrical traffic assumptions (bandwidth requirements are similar on the downstream and upstream directions) with "bidirectional" fibers often implemented by a pair of unidirectional fibers, one in each direction, or for systems with full-duplex wavelengths, see, e.g, Cisco Systems (2001).

Another distinction, the single-hop versus the multi-hop networks, depends on whether the optical network has some *wavelength conversion* capability or not, and on the electronic equipments at the nodes: see, e.g., (Mukherjee, 1992a;b). In multi-hop optical networks, the data can go successively through several so called *virtual wavelength paths* (VWP) (i.e., the lightpath is decomposed in several segments), each associated with a different wavelength, whereas in single-hop optical network, the same wavelength is used throughout the whole path from source to destination, with no electrical conversion at any of the nodes.

In this chapter, we restrict our attention to exact ILP formulations for the RWA problem for single-hop symmetrical systems, without wavelength converters. The reader is referred to Leonardi et al. (2000) and Dutta and Rouskas (2000) for surveys on multi-hop RWA and to Jaumard et al. (2004a) for asymmetrical systems. We consider the static RWA problem, also called the wavelength dimensioning problem, i.e., we assume, as in most practical situations today, that the network dimensioning is designed to support a certain, fixed set of connection requests where the grooming, if any, has been considered at an earlier stage. Recall that the grooming is the packing of different low speed traffic streams into the higher speed streams corresponding to the capacity of the transport signals.

We propose a unified presentation of various integer linear programming (ILP) formulations including all the classical ILP already proposed in the literature, i.e., those with a continuous relaxation that can be solved using the simplex algorithm. The ILP formulations that include decomposition features and require column generation type algorithms are reviewed in a subsequent paper (Jaumard et al., 2005) where they are also compared with the classical ILP formulations of this chapter. We then show that despite the large number of possible ILP formulations, they are almost all equivalent with respect to the bounds provided by their linear relaxation. Computational and comparative results conclude the chapter; we show that exact solution can be obtained for small to medium traffic and network size instances.

23.2 STATEMENT OF THE RWA PROBLEM

Let us consider an optical network represented by a multigraph $G = (V, E)$ with node set $V = \{v_1, v_2, \ldots, v_n\}$ where each node is associated with a node of the physical network, and edge set $E = \{e_1, e_2, \ldots, e_m\}$ where each edge is associated with a fiber and a link, i.e., a fiber link, of the physical network. See Figure 23.1 for an example of a multigraph representing a multifiber optical network.

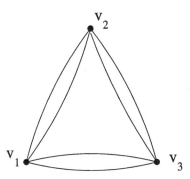

Figure 23.1 Example of a multifiber network with two fibers per link

We denote by $\omega(v_i)$ the set of adjacent edges at node v_i, for $v_i \in V$. The traffic is defined by a set K of connection requests. For each $k \in K$, let s_k and d_k denote the endpoints (i.e., the source and the destination) of the connection. The set of available wavelengths is denoted by $\Lambda = \{\lambda_1, \lambda_2, \ldots, \lambda_W\}$ with $W = |\Lambda|$.

As mentioned in the introduction, a bidirectional fiber link is often implemented by a pair of unidirectional fiber links, one for each direction. This leads to consider the directed graph $G' = (V,A)$ where A defined as follows: to each edge $e = \{v_i, v_j\} \in E$, we associate two opposite arcs $a(e) = (v_i, v_j)$ and $\bar{a}(e) = (v_j, v_i)$. We denote by $\omega^+(v_i)$ the set of outgoing arcs at node v_i and by $\omega^-(v_i)$ the set of incoming arcs for $v_i \in V$.

For each connection $k \in K$, we denote by $\mathcal{P}_k$ the set of all elementary (no loop) paths between s_k and d_k, or from s_k to d_k depending on the context.

23.3 BASIC ILP FORMULATIONS FOR THE RWA PROBLEM

All formulations presented in this section are associated with different sets of variables. The set of variables are defined in Section 23.3.1. In Section 23.3.2, we describe the different objectives that will be considered throughout the chapter, while the constraints are presented in Sections 23.3.3 – 23.3.6.

23.3.1 Variables

We define below the sets of variables that will be used in the classical ILP formulations for the RWA problem in the first half of the chapter.

The first two sets of variables deal with the selection of a connection or a wavelength:

$$x_k = \begin{cases} 1 & \text{if connection } k \text{ is accepted} \\ 0 & \text{otherwise} \end{cases} \qquad k \in K,$$

$$x^\lambda = \begin{cases} 1 & \text{if wavelength } \lambda \text{ is used} \\ 0 & \text{otherwise} \end{cases} \qquad \lambda \in \Lambda.$$

The next set of variables corresponds to decision variables associated with the control of the wavelength assignment to a connection:

$$x_k^\lambda = \begin{cases} 1 & \text{if wavelength } \lambda \text{ supports connection } k \\ 0 & \text{otherwise} \end{cases} \qquad \lambda \in \Lambda, k \in K.$$

Finally, the last set of variables is used for the lightpath settings.

$$x_{ke}^\lambda = \begin{cases} 1 & \text{if wavelength } \lambda \text{ supports connection } k \\ & \text{on the (undirected) fiber link } e \\ 0 & \text{otherwise} \end{cases} \quad \lambda \in \Lambda, k \in K, e \in E$$

$$x_{ka}^\lambda = \begin{cases} 1 & \text{if wavelength } \lambda \text{ supports connection } k \\ & \text{on the (directed) fiber link } a \\ 0 & \text{otherwise} \end{cases} \quad \lambda \in \Lambda, k \in K, a \in A$$

$$x_{kp}^\lambda = \begin{cases} 1 & \text{if wavelength } \lambda \text{ supports connection } k \\ & \text{on path } p \\ 0 & \text{otherwise} \end{cases} \quad \lambda \in \Lambda, k \in K, p \in \mathcal{P}_k.$$

Note that variables x_{ka}^λ and x_{kp}^λ will not be used in all formulations, see Section 23.3.5 for the details.

23.3.2 Objective functions

We consider four different objective functions. The first three of them have already been studied in the literature while the fourth one appears to be new, at least in the single-hop context.

23.3.2.1 Minimizing the blocking rate. Probably the most studied, the first objective corresponds to the minimization of the number of denied sessions:

$$\min \; z_{\text{BLOCKING}} \quad \text{where} \quad z_{\text{BLOCKING}} = \frac{|K| - \sum_{k \in K} x_k}{|K|}.$$

If we assume that all wavelengths are available, we should add the following constraints:

$$x^\lambda = 1 \quad \lambda \in \Lambda.$$

The blocking rate objective was considered by several authors, see Chen and Banerjee (1996), Krishnaswamy and Sivarajan (2001a), Ramaswami and Sivarajan (1995), Zang et al. (2000), although for asymmetrical systems.

Note that when solving the mathematical programs with the objective z_{BLOCKING}, we can consider only minimizing $-\sum_{k \in K} x_k$ without modifying the optimization process or the optimal vectors.

23.3.2.2 Minimizing the number of used wavelengths. A second much studied objective corresponds to the minimization of the number of used wavelengths. It can be written

$$\min \; z_{\#\lambda} \quad \text{where} \quad z_{\#\lambda} = \sum_{\lambda \in \Lambda} x^\lambda.$$

It is usually assumed that the wavelength set Λ is big enough so that all connection requests can be accepted. Therefore, we add the following constraints:

$$x_k = 1, \quad k \in K$$
$$x^\lambda \geq x_k^\lambda, \quad \lambda \in \Lambda, k \in K.$$

Several authors have derived theoretical lower bounds on the number of required wavelengths on special structured networks, see, e.g., Margara and Simon (2000). Zang et al. (2000) suggest to determine the minimum number of wavelengths using a binary search: fix a number of wavelengths and solve the RWA problem with the objective of minimizing the maximum number of wavelengths on a fiber link (see Section 23.3.2.3). If the problem is infeasible, increase the number of wavelengths; otherwise decrease this number. In a comparative study of the RWA problem with or without wavelength conversion, Wauters and Demeester (1996) consider the $z_{\#\lambda}$ objective and develop a heuristic combining an ILP formulation with the shortest path routes for each pair of source and destination nodes. Lee et al. (2002) also propose an ILP to solve the RWA problem with the $z_{\#\lambda}$ objective. However, their formulation requires column generation type algorithms. It is reviewed in Jaumard et al. (2005) together with other decomposition type formulations and compared with classical ILP formulations.

23.3.2.3 Minimizing the congestion. A third objective concerns the minimization of the congestion, that is expressed through the minimization of the maximum number of wavelengths on a given fiber link. It has been studied by Zang et al. (2000).

The objective can be written

$$\min \quad z_{\text{CONGESTION}}$$

with the addition of the following constraints:

$$z_{\text{CONGESTION}} \geq \sum_{k \in K} \sum_{\lambda \in \Lambda} x_{ke}^\lambda, \quad e \in E$$
$$x_k = 1 \quad k \in K.$$

One underlying motivation is to distribute the lightpaths on the links in the most possible uniform manner, with the aim of leaving some room for additional future connections.

23.3.2.4 Minimizing the network load. Let us assume again that there is enough wavelengths so that we can accept all connection requests. Even if fiber links can each support up to W wavelengths, it is not always necessary to light on all of them, in which case, some electronic equipment can be saved at the nodes. The fourth objective deals with the network load defined as the ratio of used wavelengths over the overall potential number of wavelengths (i.e, mW). It can be written:

$$\min \quad z_{\text{LOAD}} \quad \text{where} \quad z_{\text{LOAD}} = \sum_{k \in K} \sum_{e \in E} \sum_{\lambda \in \Lambda} \frac{x_{ke}^\lambda}{mW}$$

with the addition of the following constraints:

$$x_k = 1 \quad k \in K$$
$$x^\lambda = 1 \quad \lambda \in \Lambda.$$

This last objective was considered in the context of multi-hop optical networks, see, e.g., Banerjee and Mukherjee (2000). Again, without modifying the optimization process, the constant term mW can be omitted when solving a mathematical program with z_{LOAD} for the objective function.

23.3.3 Wavelength assignment constraints

The following set of constraints express that a unique wavelength is assigned to an accepted connection:

$$\sum_{\lambda \in \Lambda} x_k^\lambda = x_k \quad k \in K. \tag{23.1}$$

23.3.4 Clash constraints

The clash constraints express that no two lightpaths going through the same fiber link can use the same wavelength:

$$\sum_{k \in K} x_{ke}^\lambda \leq x^\lambda \quad e \in E, \ \lambda \in \Lambda. \tag{23.2}$$

23.3.5 Network constraints

The network constraints express that if a connection k is accepted, there must be a path between its source s_k and destination d_k. We need to express those constraints for each $\lambda \in \Lambda$ since the same wavelength must be used all along the source/destination path (this corresponds to the single hop assumption). We study three possible mathematical formulations of the network constraints. The first formulation, called the EDGE formulation and described in Section 23.3.5.1, is defined on the undirected graph G. The second formulation, called the PATH formulation and described in Section 23.3.5.2, is also derived on the undirected graph G, but has some interpretation with the directed graph G' (Section 23.3.5.3). This last interpretation leads to a third formulation, called the ARC formulation, and is described in Section 23.3.5.4.

23.3.5.1 EDGE network formulation. The EDGE formulation is directly derived on the undirected multigraph G and is defined by the following sets of equalities and inequalities:

$$\sum_{e \in \omega(v_i)} x_{ke}^\lambda = x_k^\lambda \quad \lambda \in \Lambda, k \in K, v_i \in \{s_k, d_k\} \tag{23.3}$$

$$\sum_{e \in \omega(v_i)} x_{ke}^\lambda \leq 2 \quad \lambda \in \Lambda, k \in K, v_i \in V \setminus \{s_k, d_k\} \tag{23.4}$$

$$\sum_{e' \in \omega(v_i), e' \neq e} x_{ke'}^\lambda \geq x_{ke}^\lambda \quad \lambda \in \Lambda, k \in K, e \in \omega(v_i), v_i \in V \setminus \{s_k, d_k\}. \tag{23.5}$$

A weaker version of this formulation that uses an additional family of variables was proposed in Jaumard et al. (2004b). Constraints (23.4) and (23.5) express that, at each node, there are either two incident edges or none, except at the endpoint nodes s_k and d_k. Constraint (23.3) express that, if $x_k^\lambda = 1$, exactly one edge incident to s_k and one edge incident to d_k are selected. Hence, all together, constraints (23.3)–(23.5) control the definition of a path between s_k and d_k associated with wavelength λ if x_k^λ is equal to 1. Note that there may also exist isolated dummy loops that are disconnected from the path. However these loops, as it will be seen later, do not affect the optimal values of the various objective functions that we consider. If $x_k^\lambda = 0$, isolated dummy loops that are not going through s_k and d_k, may also exist for wavelength λ. Again this is not a difficulty. However, in this last case, these loops can be avoided by replacing (23.4) by the stronger inequality:

$$\sum_{e \in \omega(v_i)} x_{ke}^\lambda \leq 2 x_k^\lambda \qquad \lambda \in \Lambda, \quad k \in K, \quad v_i \in V \setminus \{s_k, d_k\}. \tag{23.6}$$

In the subsequent sections, we will assume that the EDGE network formulation is defined by (23.3), (23.5) and (23.6).

23.3.5.2 PATH network formulation.

For each path p we define

$$b_{ep} = \begin{cases} 1 & \text{if edge } e \text{ belongs to the path } p \\ 0 & \text{otherwise.} \end{cases}$$

The PATH network constraints are defined by the following sets of equations:

$$\sum_{p \in \mathcal{P}_k} x_{kp}^\lambda = x_k^\lambda \qquad \lambda \in \Lambda, \quad k \in K \tag{23.7}$$

$$x_{ke}^\lambda = \sum_{p \in \mathcal{P}_k} b_{ep} x_{kp}^\lambda \qquad \lambda \in \Lambda, \quad k \in K, \quad e \in E. \tag{23.8}$$

Constraints (23.7) ensure the existence of exactly one lightpath with the wavelength λ selected for connection k. Constraints (23.8) link the variables x_{kp}^λ to the variables x_{ke}^λ that are used to express the clash constraints. These equalities may be used to eliminate the variables x_{ke}^λ if necessary.

23.3.5.3 A second thought at the path network formulation.

We will now show that the path formulation presented in the previous section and developed using the undirected graph G can also be interpreted using the directed graph G'.

In a directed graph setting, a connection $k \in K$ between s_k and d_k can be viewed as being made of two elementary connections: an upstream connection k_{UP} from s_k to d_k and a downstream connection k_{DOWN} from d_k to s_k. Denote by K_{UP} and K_{DOWN} the set of upstream and downstream flows respectively, and by $\mathcal{P}_{k_{\text{UP}}}$ and $\mathcal{P}_{k_{\text{DOWN}}}$ the set of (directed) paths for the k_{UP} upstream and the k_{DOWN} downstream flows. Paths in $\mathcal{P}_{k_{\text{UP}}}$ and $\mathcal{P}_{k_{\text{DOWN}}}$ are assumed to be elementary, i.e., they do not go more than one time through a given node. Given a path p, we denote by $\overline{p}$ its reverse path deduced by changing the direction of all its arcs. The network constraints can then be written:

$$\sum_{p \in \mathcal{P}_{k_{\text{UP}}}} x^\lambda_{k_{\text{UP}} p} = x^\lambda_{k_{\text{UP}}} \qquad \lambda \in \Lambda, k \in K \qquad (23.9)$$

$$\sum_{p \in \mathcal{P}_{k_{\text{DOWN}}}} x^\lambda_{k_{\text{DOWN}} p} = x^\lambda_{k_{\text{DOWN}}} \qquad \lambda \in \Lambda, k \in K \qquad (23.10)$$

$$x^\lambda_{k_{\text{UP}} a} = \sum_{p \in \mathcal{P}_{k_{\text{UP}}}} b_{ap} x^\lambda_{k_{\text{UP}} p} \qquad \lambda \in \Lambda, k \in K, a \in A \qquad (23.11)$$

$$x^\lambda_{k_{\text{DOWN}} a} = \sum_{p \in \mathcal{P}_{k_{\text{DOWN}}}} b_{ap} x^\lambda_{k_{\text{DOWN}} p} \qquad \lambda \in \Lambda, k \in K, a \in A \qquad (23.12)$$

where b_{ap} is equal to 1 if arc a belongs to the (directed) path p, and to 0 otherwise. Moreover, the wavelength selection constraint (23.1) and the clash constraint (23.2) are changed for:

$$\sum_{\lambda \in \Lambda} x^\lambda_{k_{\text{UP}}} = x_{k_{\text{UP}}} \qquad \lambda \in \Lambda, \quad k \in K \qquad (23.13)$$

$$\sum_{\lambda \in \Lambda} x^\lambda_{k_{\text{DOWN}}} = x_{k_{\text{DOWN}}} \qquad \lambda \in \Lambda, \quad k \in K \qquad (23.14)$$

$$\sum_{k \in K} \left(x^\lambda_{k_{\text{UP}} a} + x^\lambda_{k_{\text{DOWN}} a} \right) \leq x^\lambda \qquad a \in A, \quad \lambda \in \Lambda. \qquad (23.15)$$

Clearly, for any connection k, either both upstream and downstream connections must be denied, or they must be both accepted. In this last case, the paths used by the upstream and downstream connections must be the same except for the direction; moreover both connections must be assigned the same wavelength. This can be modeled by adding the constraint

$$x^\lambda_{k_{\text{DOWN}} p} = x^\lambda_{k_{\text{UP}} \bar{p}} \qquad \lambda \in \Lambda, \quad k \in K, \quad p \in \mathcal{P}_{k_{\text{DOWN}}}. \qquad (23.16)$$

We now use these equations to eliminate all variables related to downstream connection. Using (23.12) and (23.16), we deduce

$$x^\lambda_{k_{\text{DOWN}} a} = \sum_{p \in \mathcal{P}_{k_{\text{DOWN}}}} b_{ap} x^\lambda_{k_{\text{UP}} \bar{p}} = \sum_{p \in \mathcal{P}_{k_{\text{UP}}}} b_{a\bar{p}} x^\lambda_{k_{\text{UP}} p} \qquad \lambda \in \Lambda, k \in K, a \in A. \qquad (23.17)$$

Substituting (23.16) in (23.10), we see that $x^\lambda_{k_{\text{DOWN}}} = x^\lambda_{k_{\text{UP}}}$ for every $\lambda \in \Lambda$ and $k \in K$. Using (23.13)–(23.14), this in turn implies that $x_{k_{\text{DOWN}}} = x_{k_{\text{UP}}}$ for $k \in K$. Since $b_{a\bar{p}} = b_{\bar{a}p}$ and using (23.11) and (23.17), (23.15) becomes

$$\sum_{k \in K} \left(x^\lambda_{k_{\text{UP}} a} + x^\lambda_{k_{\text{UP}} \bar{a}} \right) \leq x^\lambda \qquad a \in A, \quad \lambda \in \Lambda. \qquad (23.18)$$

After elimination of the variables related to the downstream connections, the new wavelength selection and network constraints are:

$$\sum_{\lambda \in \Lambda} x_{k_{\mathrm{UP}}}^{\lambda} = x_{k_{\mathrm{UP}}} \qquad k \in K \qquad (23.19)$$

$$\sum_{p \in \mathcal{P}_{k_{\mathrm{UP}}}} x_{k_{\mathrm{UP}}p}^{\lambda} = x_{k_{\mathrm{UP}}}^{\lambda} \qquad \lambda \in \Lambda, k \in K \qquad (23.20)$$

$$x_{k_{\mathrm{UP}}a}^{\lambda} = \sum_{p \in \mathcal{P}_{k_{\mathrm{UP}}}} b_{ap} x_{k_{\mathrm{UP}}p}^{\lambda} \qquad \lambda \in \Lambda, k \in K, a \in A \qquad (23.21)$$

$$x_{ke}^{\lambda} = x_{k_{\mathrm{UP}}a(e)}^{\lambda} + x_{k_{\mathrm{UP}}\bar{a}(e)}^{\lambda} \qquad \lambda \in \Lambda, k \in K, e \in E. \qquad (23.22)$$

Note that constraint (23.22) allows to use the clash constraint (23.2) instead of (23.18).

By definition of the sets $\mathcal{P}_k$, a path cannot use simultaneously an arc and its opposite. Hence we can set $b_{ep} = b_{a(e)p} + b_{\bar{a}(e)p}$ for $e \in E$ and $p \in \mathcal{P}_{k_{\mathrm{UP}}}$. Using (23.21) and (23.22), we get

$$x_{ke}^{\lambda} = \sum_{p \in \mathcal{P}_{k_{\mathrm{UP}}}} \left(b_{a(e)p} + b_{\bar{a}(e)p}\right) x_{k_{\mathrm{UP}}p}^{\lambda} = \sum_{p \in \mathcal{P}_{k_{\mathrm{UP}}}} b_{ep} x_{k_{\mathrm{UP}}p}^{\lambda} \qquad k \in K, e \in E, \lambda \in \Lambda.$$

Identifying K_{UP} with K, and $\mathcal{P}_{k_{\mathrm{UP}}}$ with $\mathcal{P}_k$, we see that this network formulation coincides with the path network formulation on the undirected graph G described in Section 23.3.5.2.

23.3.5.4 ARC network formulation. It is well-known that one way to ensure the existence of a path between two nodes in a directed graph is to push one unit of flow at the source node and to express flow conservation at every node except for the destination node. Starting from the directed graph interpretation of the path formulation, we obtain the following ARC formulation:

$$\sum_{a \in \omega^+(s_k)} x_{ka}^{\lambda} - \sum_{a \in \omega^-(s_k)} x_{ka}^{\lambda} = x_k^{\lambda} \qquad \lambda \in \Lambda, k \in K \qquad (23.23)$$

$$\sum_{a \in \omega^+(v_i)} x_{ka}^{\lambda} - \sum_{a \in \omega^-(v_i)} x_{ka}^{\lambda} = 0 \qquad \lambda \in \Lambda, k \in K, v_i \in V \setminus \{s_k, d_k\} \qquad (23.24)$$

$$x_{ke}^{\lambda} = x_{ka(e)}^{\lambda} + x_{k\bar{a}(e)}^{\lambda} \qquad \lambda \in \Lambda, k \in K, e \in E. \qquad (23.25)$$

The equivalent of constraint (23.23) for node d_k is implied by constraints (23.23) and (23.24):

$$\sum_{a \in \omega^+(d_k)} x_{ka}^{\lambda} - \sum_{a \in \omega^-(d_k)} x_{ka}^{\lambda} = -x_k^{\lambda} \qquad \lambda \in \Lambda, k \in K. \qquad (23.26)$$

The following constraints may be added

$$x_{ka}^{\lambda} = 0, \qquad \lambda \in \Lambda, \quad k \in K, \quad a \in \omega^-(s_k) \qquad (23.27)$$

$$x_{ka}^{\lambda} = 0, \qquad \lambda \in \Lambda, \quad k \in K, \quad a \in \omega^+(d_k) \qquad (23.28)$$

in order to prevent the existence of dummy loops going through s_k or d_k. Although these constraints are not necessary to derive exact formulations for the RWA problem, they offer the advantage of reducing the number of variables x_{ka}^{λ}. They also allow the elimination of the variables x_k^{λ}: see Section 23.5.1.

23.3.6 Network collapse constraints

The following constraints force the variables x_{ke}^λ that do not correspond to the selected wavelength to be set to 0:

$$x_{ke}^\lambda \leq x_k^\lambda \quad \lambda \in \Lambda, \quad k \in K, \quad e \in E. \tag{23.29}$$

These constraints may or may not be added to the formulation. When added, they allow to collapse the W sets of network constraints into only one: see Section 23.5.2. See also Section 23.5.1 for formulations without those constraints.

A more compact formulation for (23.29) is the following:

$$\sum_{e \in \omega(v_i)} x_{ke}^\lambda \leq 2 x_k^\lambda \quad \lambda \in \Lambda, \quad k \in K, \quad v_i \in V. \tag{23.30}$$

Note that these correspond to constraints (23.6) of the EDGE formulation. Observe also that if (23.29) and (23.30) are equivalent when considering the optimal integer solution, it is no more the case when considering the optimal continuous solution.

When the ARC network formulation is used, a constraint stronger than (23.30) is:

$$\sum_{a \in \omega^+(v_i)} x_{ka}^\lambda \leq x_k^\lambda \quad \lambda \in \Lambda, \quad k \in K, \quad v_i \in V. \tag{23.31}$$

Using (23.24) and (23.25), it is indeed easy to see that constraints (23.31) dominate (23.30), even for the continuous relaxation. They also dominate the constraints

$$x_{ka}^\lambda \leq x_k^\lambda, \quad \lambda \in \Lambda, \quad k \in K, \quad a \in A$$

which can be viewed as the equivalent constraints of (23.29) in the directed graph setting.

Finally note that due to its nature, the PATH formulation does not benefit from constraints (23.29). Indeed, these constraints are implied by (23.7)–(23.8).

We will show in Section 23.4 that the addition of these constraints does not change the value of the continuous relaxation.

23.3.7 ILP formulations for the RWA problem

We denote by RWA($\mathcal{F}$, OBJ) a valid ILP formulation for the RWA problem. The first parameter $\mathcal{F}$ identifies the network formulation used, $\mathcal{F} \in \{\text{EDGE}, \text{PATH}, \text{ARC}\}$, while the second parameter OBJ determines the objective, OBJ $\in \{\text{BLOCKING}, \#\lambda, \text{CONGESTION}, \text{LOAD}\}$. Note that these two parameters do not uniquely determine a formulation. When it is not necessary to mention a particular parameter, we will omit it and use the notations RWA(., .), RWA(., OBJ) or RWA($\mathcal{F}$, .).

A formulation for the RWA problem is defined by constraints (23.1), (23.2), one of constraint sets associated with one of the three network formulations, and one objective function among those described in Section 23.3.2 together with its accompanying constraints. Observe that potential dummy loops are not a difficulty since they can be removed without changing the value of the objective function for the objectives that are considered in this study. The network collapse constraints presented in Section 23.3.6 can optionally be added to a formulation.

23.4 LOWER BOUNDS DEDUCED FROM THE CONTINUOUS RELAXATIONS FOR THE RWA PROBLEM

The lower bound deduced from the different continuous relaxations are obtained by relaxing the integrality constraint on the binary variables. This leads to several different lower bound values, depending, among others, on the network formulation that is used and whether we include or not the network collapse constraints.

We denote by EDGE$^+$ the network formulation EDGE augmented by the network collapse constraints (23.29), and by ARC$^+$ the network formulation ARC augmented by the network collapse constraints (23.29) and (23.31). For a fixed $k \in K$ and $\lambda \in \Lambda$, and a fixed value for x_k^λ, we denote by $\Omega^{\text{LP}}(\mathcal{F}, x_k^\lambda)$ the feasible set for variables $x_{ke}^\lambda \in [0,1]$ with the network formulation $\mathcal{F}$. We summarize below the network constraints for each network formulation.

EDGE formulation:

$$\sum_{e \in \omega(v_i)} x_{ke}^\lambda \leq 2x_k^\lambda \qquad k \in K, v_i \in V \setminus \{s_k, d_k\}, \lambda \in \Lambda \qquad (23.32)$$

$$\sum_{e' \in \omega(v_i), e' \neq e} x_{ke'}^\lambda \geq x_{ke}^\lambda \qquad k \in K, e \in \omega(v_i), v_i \in V \setminus \{s_k, d_k\}, \lambda \in \Lambda \qquad (23.33)$$

$$x_k^\lambda = \sum_{e \in \omega(v_i)} x_{ke}^\lambda \qquad k \in K, \quad v_i \in \{s_k, d_k\}, \quad \lambda \in \Lambda. \qquad (23.34)$$

PATH formulation (undirected):

$$x_k^\lambda = \sum_{p \in \mathcal{P}_k} x_{kp}^\lambda \qquad k \in K, \quad \lambda \in \Lambda \qquad (23.35)$$

$$x_{ke}^\lambda = \sum_{p \in \mathcal{P}_k} b_{ep} x_{kp}^\lambda \qquad k \in K, e \in E, \lambda \in \Lambda \qquad (23.36)$$

$$x_{kp}^\lambda \geq 0 \qquad p \in \mathcal{P}_k, \quad k \in K, \quad \lambda \in \Lambda. \qquad (23.37)$$

PATH formulation (directed):

$$x_k^\lambda = \sum_{p \in \mathcal{P}_k} x_{kp}^\lambda \qquad k \in K, \quad \lambda \in \Lambda \qquad (23.38)$$

$$x_{ke}^\lambda = x_{ka(e)}^\lambda + x_{k\bar{a}(e)}^\lambda, \qquad k \in K, e \in E, \lambda \in \Lambda \qquad (23.39)$$

$$x_{ka}^\lambda = \sum_{p \in \mathcal{P}_k} b_{ap} x_{kp}^\lambda \qquad k \in K, a \in A, \lambda \in \Lambda \qquad (23.40)$$

$$x_{kp}^\lambda \geq 0 \qquad p \in \mathcal{P}_k, \quad k \in K, \quad \lambda \in \Lambda. \qquad (23.41)$$

ARC formulation:

$$\sum_{a\in\omega^+(v_i)} x_{ka}^\lambda = \sum_{a\in\omega^-(v_i)} x_{ka}^\lambda \qquad k\in K, v_i\in V\setminus\{s_k,d_k\}, \lambda\in\Lambda \qquad (23.42)$$

$$\sum_{a\in\omega^+(s_k)} x_{ka}^\lambda - \sum_{a\in\omega^-(s_k)} x_{ka}^\lambda = x_k^\lambda \qquad k\in K, \lambda\in\Lambda \qquad (23.43)$$

$$\sum_{a\in\omega^+(d_k)} x_{ka}^\lambda - \sum_{a\in\omega^-(d_k)} x_{ka}^\lambda = -x_k^\lambda \qquad k\in K, \lambda\in\Lambda \qquad (23.44)$$

$$x_{ke}^\lambda = x_{ka(e)}^\lambda + x_{k\bar{a}(e)}^\lambda \qquad k\in K, e\in E, \lambda\in\Lambda \qquad (23.45)$$

$$(\ x_{ka}^\lambda = 0 \qquad k\in K, a\in\omega^-(s_k), \lambda\in\Lambda\ ) \qquad (23.46)$$

$$(\ x_{ka}^\lambda = 0 \qquad k\in K, a\in\omega^+(d_k), \lambda\in\Lambda\ ) \qquad (23.47)$$

$$x_{ka}^\lambda \geq 0 \qquad k\in K,\ a\in A,\ \lambda\in\Lambda. \qquad (23.48)$$

Let us now compare the feasible domains of the continuous relaxations of the various network formulations.

Lemma 23.1

$$\Omega^{LP}(\text{PATH}, x_k^\lambda) \subseteq \Omega^{LP}(\text{ARC}^+, x_k^\lambda) \subseteq \Omega^{LP}(\text{ARC}, x_k^\lambda)$$
$$\Omega^{LP}(\text{PATH}, x_k^\lambda) \subseteq \Omega^{LP}(\text{EDGE}^+, x_k^\lambda) \subseteq \Omega^{LP}(\text{EDGE}, x_k^\lambda).$$

Proof. The inclusions $\Omega^{LP}(\text{ARC}^+, x_k^\lambda) \subseteq \Omega^{LP}(\text{ARC}, x_k^\lambda)$ and $\Omega^{LP}(\text{EDGE}^+, x_k^\lambda) \subseteq \Omega^{LP}(\text{EDGE}, x_k^\lambda)$ are obvious.

Let us show the inclusion $\Omega^{LP}(\text{PATH}, x_k^\lambda) \subseteq \Omega^{LP}(\text{EDGE}^+, x_k^\lambda)$. By definition of a path $p\in\mathcal{P}_k$, the vector b_{ep} satisfies:

$$b_{ep} \leq 1 \qquad e\in E \qquad (23.49)$$

$$\sum_{e\in\omega(v_i)} b_{ep} \leq 2, \qquad v_i\in V\setminus\{s_k,d_k\} \qquad (23.50)$$

$$\sum_{e\in\omega(v_i)} b_{ep} = 1, \qquad v_i\in\{s_k,d_k\} \qquad (23.51)$$

$$\sum_{e'\in\omega(v_i), e'\neq e} b_{e'p} \geq b_{ep}, \qquad e\in\omega(v_i), v_i\in V\setminus\{s_k,d_k\}. \qquad (23.52)$$

Multiplying these inequalities by x_{kp}^λ, summing over $p\in\mathcal{P}_k$ and using (23.35) and (23.36), we show that $\Omega^{LP}(\text{PATH}, x_k^\lambda) \subseteq \Omega^{LP}(\text{EDGE}^+, x_k^\lambda)$.

A similar reasoning can be made for inclusion $\Omega^{LP}(\text{PATH}, x_k^\lambda) \subseteq \Omega^{LP}(\text{ARC}^+, x_k^\lambda)$. The vector $b_{ap}(a\in A)$ satisfies $b_{a(e)p} + b_{\bar{a}(e)p} \leq 1$ for $p\in\mathcal{P}_k$, (23.31), (23.42)–(23.44) and (23.46)–(23.48) with x_{ka}^λ replaced by b_{ap} and x_k^λ replaced by 1 for each $p\in\mathcal{P}_k$. Therefore a positive combination with weights x_{kp}^λ satisfies (23.29), (23.31), (23.42)–(23.44) and (23.46)–(23.48), which shows the inclusion $\Omega^{LP}(\text{PATH}, x_k^\lambda)$

$$\subseteq \Omega^{LP}(\text{ARC}^+, x_k^\lambda).$$

We denote by $\underline{z}^{\text{RWA}(\mathcal{F},\text{OBJ})}$ the LP relaxation value of the formulation $\text{RWA}(\mathcal{F},\text{OBJ})$.

Proposition 23.1 *For all* $\text{OBJ} \in \{\text{BLOCKING}, \#\lambda, \text{CONGESTION}, \text{LOAD}\}$,

$$\underline{z}^{\text{RWA}(\text{PATH},\text{OBJ})} \geq \underline{z}^{\text{RWA}(\text{ARC}^+,\text{OBJ})} \geq \underline{z}^{\text{RWA}(\text{ARC},\text{OBJ})} \tag{23.53}$$

$$\underline{z}^{\text{RWA}(\text{PATH},\text{OBJ})} \geq \underline{z}^{\text{RWA}(\text{EDGE}^+,\text{OBJ})} \geq \underline{z}^{\text{RWA}(\text{EDGE},\text{OBJ})}. \tag{23.54}$$

Proof. By definition, $\underline{z}^{\text{RWA}(\mathcal{F},\text{OBJ})}$ is the minimum of the OBJ objective function over the set of constraints:

$$\sum_{\lambda \in \Lambda} x_k^\lambda = x_k \qquad k \in K$$

$$\sum_{k \in K} x_{ke}^\lambda \leq 1 \qquad e \in E, \ \lambda \in \Lambda$$

$$\left(x_{ke}^\lambda\right) \in \Omega^{LP}\left(\mathcal{F}, x_k^\lambda\right) \qquad k \in K, \ \lambda \in \Lambda$$

$$\left(x_k, x^\lambda, x_k^\lambda, x_{ke}^\lambda\right) \in \Omega^{LP}(\text{OBJ})$$

$$x_k^\lambda \geq 0 \qquad k \in K, \ \lambda \in \Lambda,$$

where $\Omega^{LP}(\text{OBJ})$ is the polyhedral set defined by the constraints associated with the OBJ objective function. Let $(\hat{x}_k, \hat{x}^\lambda, \hat{x}_k^\lambda, \hat{x}_{ke}^\lambda)$ be an optimal solution of the continuous relaxation of $\text{RWA}(\text{PATH},\text{OBJ})$. Since $\Omega^{LP}(\text{PATH}, \hat{x}_k^\lambda) \subseteq \Omega^{LP}(\text{ARC}^+, \hat{x}_k^\lambda)$ for $k \in K, \lambda \in \Lambda$ by Lemma 23.1, $(\hat{x}_k, \hat{x}^\lambda, \hat{x}_k^\lambda, \hat{x}_{ke}^\lambda)$ is a feasible solution of the continuous relaxation of $\text{RWA}(\text{ARC}^+,\text{OBJ})$. Moreover, its objective value is $\underline{z}^{\text{RWA}(\text{PATH},\text{OBJ})}$. This leads to the inequality $\underline{z}^{\text{RWA}(\text{ARC}^+,\text{OBJ})} \leq \underline{z}^{\text{RWA}(\text{PATH},\text{OBJ})}$. It is then easy to see that the three other inequalities hold, using a similar reasoning. □

Up to now we did not make any assumption on $\mathcal{P}_k$. If $\mathcal{P}_k$ is the set of all possible paths from s_k to d_k, we can reinforce the result of Proposition 23.1:

Proposition 23.2 *If $\mathcal{P}_k$ is the set of all paths between s_k and d_k, then for all* $\text{OBJ} \in \{\text{BLOCKING}, \#\lambda, \text{CONGESTION}, \text{LOAD}\}$,

$$\underline{z}^{\text{RWA}(\text{PATH},\text{OBJ})} = \underline{z}^{\text{RWA}(\text{ARC}^+,\text{OBJ})} = \underline{z}^{\text{RWA}(\text{ARC},\text{OBJ})}.$$

Proof. It is enough to show that $\underline{z}^{\text{RWA}(\text{PATH},\text{OBJ})} = \underline{z}^{\text{RWA}(\text{ARC},\text{OBJ})}$. Indeed, by the sandwich principle applied on inequality (23.53) of Proposition 23.1, the remaining equality will follow straightforwardly.

It is well known (see, e.g., (Ahuja et al., 1993, p.80–81)), that for fixed $(\bar{k}, \bar{\lambda}) \in K \times \Lambda$, the set of $x_{\bar{k}a}^{\bar{\lambda}}$ satisfying (23.42)–(23.44) and (23.48) coincides with the set of $x_{\bar{k}a}^{\bar{\lambda}}$ satisfying

$$x_{ka}^{\bar{\lambda}} = \sum_{p \in \mathcal{P}_k} b_{ap} x_{kp}^{\bar{\lambda}} + \sum_{c \in C} b_{ac} z_{kc}^{\bar{\lambda}} \qquad a \in A \qquad (23.55)$$

$$\sum_{p \in \mathcal{P}_k} x_{kp}^{\bar{\lambda}} = x_k^{\bar{\lambda}} \qquad (23.56)$$

$$x_{kp}^{\bar{\lambda}} \geq 0 \qquad p \in \mathcal{P}_k \qquad (23.57)$$

$$z_{kc}^{\bar{\lambda}} \geq 0 \qquad c \in C \qquad (23.58)$$

where C is the set of cycles in the directed graph G', and b_{ac} is equal to 1 if arc a belongs to cycle c, 0 otherwise. Clearly, the result remains true if we add the constraint $x_{ka(e)}^{\bar{\lambda}} + x_{k\bar{a}(e)}^{\bar{\lambda}} \leq 1$ for $e \in E$ both in (23.42)–(23.45), (23.48) and in (23.55)–(23.58). So, let $(\hat{x}_k, \hat{x}^\lambda, \hat{x}_k^\lambda, \hat{x}_{ke}^\lambda)$ be an optimal solution of the continuous relaxation of RWA(ARC, OBJ). There exist $\hat{x}_{kp}^\lambda$ and $\hat{z}_{kc}^\lambda$ such that (23.55)–(23.58) holds for all $(\bar{k}, \bar{\lambda}) \in K \times \Lambda$. Setting to 0 the variables z_{kc}^λ in the definition of $\hat{x}_{ka}^\lambda$, let us define

$$\tilde{x}_k^\lambda = \hat{x}_k^\lambda \qquad k \in K, \quad \lambda \in \Lambda$$
$$\tilde{x}_k = \hat{x}_k \qquad k \in K$$
$$\tilde{x}^\lambda = \hat{x}^\lambda \qquad \lambda \in \Lambda$$
$$\tilde{x}_{ka}^\lambda = \sum_{p \in \mathcal{P}_k} b_{ap} \hat{x}_{kp}^\lambda \qquad a \in A, \quad k \in K, \quad \lambda \in \Lambda$$
$$\tilde{x}_{ke}^\lambda = \tilde{x}_{ka(e)}^\lambda + \tilde{x}_{k\bar{a}(e)}^\lambda \qquad e \in E, \quad k \in K, \quad \lambda \in \Lambda.$$

Clearly the vector $(\tilde{x}_k, \tilde{x}^\lambda, \tilde{x}_k^\lambda, \tilde{x}_{ke}^\lambda)$ defines a feasible solution of the continuous relaxation of RWA(PATH, OBJ). Observe that since $\tilde{x}_{ke}^\lambda \leq \hat{x}_{ke}^\lambda$ for $k \in K, \lambda \in \Lambda, e \in E$, the objective value of the new solution cannot be worse than the one of $(\hat{x}_k, \hat{x}^\lambda, \hat{x}_k^\lambda, \hat{x}_{ke}^\lambda)$ for the objective functions that we consider. This leads to $\underline{z}^{\text{RWA(PATH,OBJ)}} \leq \underline{z}^{\text{RWA(ARC,OBJ)}}$.

If constraints (23.46) or (23.47) were considered in the ARC formulation, observe that these constraints would force z_{kc}^λ to be 0 for the cycles c going through s_k or d_k. Hence the proof remains valid for this variation of the ARC network formulation. □

Proposition 23.3 *For all* OBJ $\in$ {BLOCKING, #λ, CONGESTION, LOAD},

$$\underline{z}^{\text{RWA(EDGE}^+,\text{OBJ})} = \underline{z}^{\text{RWA(EDGE,OBJ)}}.$$

Proof. Because of Proposition 23.1, it suffices to show that $\underline{z}^{\text{RWA(EDGE}^+,\text{OBJ})} \leq \underline{z}^{\text{RWA(EDGE,OBJ)}}$. Let $(\hat{x}_k, \hat{x}^\lambda, \hat{x}_k^\lambda, \hat{x}_{ke}^\lambda)$ be an optimal solution of the continuous relaxation of
RWA(EDGE, OBJ). Define

$$\tilde{x}_{ke}^\lambda = \min\{\hat{x}_{ke}^\lambda, \hat{x}_k^\lambda\} \qquad \lambda \in \Lambda, k \in K, e \in E.$$

Let us show that $(\hat{x}_k, \hat{x}^\lambda, \hat{x}_k^\lambda, \tilde{x}_{ke}^\lambda)$ is a feasible solution to the continuous relaxation of RWA(EDGE$^+$, OBJ). By construction, $(\hat{x}_k, \hat{x}^\lambda, \hat{x}_k^\lambda, \tilde{x}_{ke}^\lambda)$ satisfies the network collapse constraints (23.29). Inequalities (23.32) are satisfied due to the fact that $\tilde{x}_{ke}^\lambda \leq \hat{x}_{ke}^\lambda$. Consider now an inequality (23.33) for some fixed $k \in K$, $e \in \omega(v_i)$, $v_i \in V \setminus \{s_k, d_k\}$, $\lambda \in \Lambda$. If $\tilde{x}_{ke'}^\lambda = \hat{x}_{ke'}^\lambda$ for all $e' \in \omega(v_i), e' \neq e$, we have

$$\sum_{e' \in \omega(v_i), e' \neq e} \tilde{x}_{ke'}^\lambda = \sum_{e' \in \omega(v_i), e' \neq e} \hat{x}_{ke'}^\lambda \geq \hat{x}_{ke}^\lambda \geq \tilde{x}_{ke}^\lambda.$$

If $\tilde{x}_{ke'}^\lambda \neq \hat{x}_{ke'}^\lambda$ for at least one $e' \in \omega(v_i), e' \neq e$, then

$$\sum_{e' \in \omega(v_i), e' \neq e} \tilde{x}_{ke'}^\lambda \geq \hat{x}_k^\lambda \geq \tilde{x}_{ke}^\lambda.$$

In both cases, the inequality (23.33) is satisfied. Constraints (23.34) are satisfied by observing that $\hat{x}_{ke}^\lambda$ cannot be greater than $\hat{x}_k^\lambda$ for $e \in \omega(s_k) \cup \omega(d_k)$ due to the nonnegativity of x_{ke}^λ, hence $\tilde{x}_{ke}^\lambda = \hat{x}_{ke}^\lambda$ for those e. Finally the wavelength assignment constraints (23.1) are satisfied because the variables involved in them did not change their value, and the clash constraints (23.2) are satisfied because $\tilde{x}_{ke}^\lambda \leq \hat{x}_{ke}^\lambda$. Therefore $(\hat{x}_k, \hat{x}^\lambda, \hat{x}_k^\lambda, \tilde{x}_{ke}^\lambda)$ is a feasible solution for the continuous relaxation of RWA(EDGE$^+$, OBJ). Moreover its objective value is $\underline{z}^{\text{RWA}(\text{EDGE},\text{OBJ})}$, hence the desired inequality

$$\underline{z}^{\text{RWA}(\text{EDGE}^+,\text{OBJ})} \leq \underline{z}^{\text{RWA}(\text{EDGE},\text{OBJ})}.$$

□

Contrary to what is happening for the ARC formulation, we do not have the equality $\underline{z}^{\text{RWA}(\text{EDGE}^+,\text{OBJ})} = \underline{z}^{\text{RWA}(\text{PATH},\text{OBJ})}$. This is illustrated on the following example for the objective of minimizing the blocking rate.

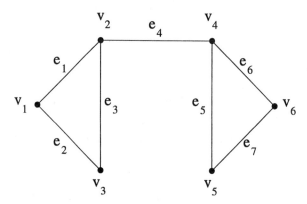

Figure 23.2 Network illustrating $\underline{z}^{\text{RWA}(\text{EDGE}^+,\text{BLOCKING})} \neq \underline{z}^{\text{RWA}(\text{PATH},\text{BLOCKING})}$

Example 23.1 *Consider the network depicted in Figure 23.2. There are 2 connections: $(s_{k_1}, d_{k_1}) = (v_1, v_6)$ and $(s_{k_2}, d_{k_2}) = (v_2, v_4)$, and only 1 wavelength is available. An optimal solution of the continuous relaxation of* RWA(EDGE$^+$, BLOCKING) *is given by $x_1 = x_2 = x_1^{\lambda_1} = x_2^{\lambda_1} = 1 = x_{2e_4}^{\lambda_1}$, $x_{1e_1}^{\lambda_1} = x_{1e_2}^{\lambda_1} = x_{1e_3}^{\lambda_1} = x_{1e_5}^{\lambda_1} = x_{1e_6}^{\lambda_1} =*

$x_{1e_7}^{\lambda_1} = \frac{1}{2}$, *with the other variables being equal to* 0. *The corresponding objective value* $\underline{z}^{\text{RWA}(\text{EDGE}^+,\text{BLOCKING})}$ *is equal to* 0. *On the other hand, the optimal value* $\underline{z}^{\text{RWA}(\text{PATH},\text{BLOCKING})}$ *of the continuous relaxation of* RWA(PATH, BLOCKING) *cannot be less than* $1/2$ *because all paths connecting* v_1 *and* v_6 *and all paths connecting* v_2 *and* v_4 *go through the same edge* e_4.

Similar examples can be easily built for the other objectives.

In the sequel, we will assume that $\mathcal{P}_k$ is equal to the set of all paths between s_k and d_k. In view of the results of this section, we denote by $\underline{z}_{\text{OBJ}}^{\text{ARC}}$ the common lower bound obtained by solving the linear relaxation of the RWA(PATH, OBJ) and RWA(ARC, OBJ) formulations, and by $\underline{z}_{\text{OBJ}}^{\text{EDGE}}$, the lower bound obtained by solving the linear relaxation of the RWA(EDGE, OBJ) formulation.

23.5 AGGREGATION WITH RESPECT TO λ

It is possible to replace some constraints by their sum while still keeping an exact formulation for the RWA problem. This transformation is called *aggregation*. In this section, we consider aggregation with respect to λ, i.e., we sum together constraints of the same nature that are related to different wavelengths. In Sections 23.5.1 and 23.5.2, we propose 2 classes of formulations that make use of some aggregations with respect to λ. In Section 23.5.3 we show that the value of the LP lower bound does not deteriorate under aggregation. Finally in Section 23.5.4 we consider an ILP lower bound, that is obtained by exhaustively aggregating with respect to λ the integer formulations for the RWA problem.

23.5.1 A first class of aggregated ILPs

It is possible to aggregate constraints (23.3), (23.7) or (23.23) (provided that (23.27) are present) of the network formulations with respect to λ, i.e., to replace them by their sum on λ. Indeed, if $x_k = 1$, exactly one x_{ke}^{λ}, x_{kp}^{λ} or x_{ka}^{λ} (depending on the network formulation used) will be equal to 1 for an edge (arc) incident to s_k, and the other network constraints will guarantee that this edge (arc) is continued by a path toward d_k with the same wavelength. If the EDGE formulation is used to model the network, it is also possible to aggregate constraints (23.6).

Using constraints (23.1), it is then possible to eliminate the variables x_k^{λ}.

Other variables may also be eliminated, sometimes at the price of adding new constraints to fulfill the bound constraints of the eliminated variables.

As an example of aggregated ILP of this first class, we propose an illustration on the RWA(ARC, BLOCKING) that is described below.

$$\min \quad z_{\text{BLOCKING}}$$

subject to

$$
\begin{cases}
\sum_{k \in K} \left(x^\lambda_{ka(e)} + x^\lambda_{k\bar{a}(e)} \right) \leq 1 & e \in E, \; \lambda \in \Lambda \\
\sum_{a \in \omega^+(v_i)} x^\lambda_{ka} = \sum_{a \in \omega^-(v_i)} x^\lambda_{ka} & \lambda \in \Lambda, \; k \in K, \; v_i \in V \setminus \{s_k, d_k\} \\
\sum_{\lambda \in \Lambda} \sum_{a \in \omega^+(s_k)} x^\lambda_{ka} = x_k & k \in K \\
x^\lambda_{ka} = 0 & \lambda \in \Lambda, \; k \in K, \; a \in \omega^-(s_k) \\
x_k, x^\lambda_{ka} \in \{0, 1\} & \lambda \in \Lambda, \; k \in K, \; a \in A.
\end{cases}
$$

23.5.2 A second class of aggregated ILPs

Observe that all the RWA(.,.) of the first class of aggregated ILPs have exactly W sets of network constraints, one for each wavelength. It is possible to reduce this number to one set of network constraints by adding the network collapse constraints (23.29), (23.30) or (23.31). These constraints allow to replace the W sets of network constraints by only one, expressed in terms of $\sum_{\lambda \in \Lambda} x^\lambda_{ke}$, $\sum_{\lambda \in \Lambda} x^\lambda_{kp}$ or $\sum_{\lambda \in \Lambda} x^\lambda_{ka}$ depending on the network formulation used. It is also possible to eliminate the variables x^λ_k. We illustrate this for RWA(ARC, BLOCKING). We first obtain:

$$\min \quad z_{\text{BLOCKING}}$$

subject to:

$$
\begin{cases}
\sum_{\lambda \in \Lambda} x^\lambda_k = x_k & k \in K \\
\sum_{k \in K} \left(x^\lambda_{ka(e)} + x^\lambda_{k\bar{a}(e)} \right) \leq 1 & e \in E, \; \lambda \in \Lambda \\
\sum_{a \in \omega^+(s_k)} x^\lambda_{ka} = x^\lambda_k & \lambda \in \Lambda, \; k \in K \\
\sum_{\lambda \in \Lambda} \sum_{a \in \omega^+(s_k)} x^\lambda_{ka} = x_k & k \in K \\
x^\lambda_{ka} = 0 & \lambda \in \Lambda, \; k \in K, \; a \in \omega^-(s_k) \\
\sum_{\lambda \in \Lambda} \sum_{a \in \omega^+(v_i)} x^\lambda_{ka} = \sum_{\lambda \in \Lambda} \sum_{a \in \omega^-(v_i)} x^\lambda_{ka} & k \in K, \; v_i \in V \setminus \{s_k, d_k\} \\
\sum_{a \in \omega^+(v_i)} x^\lambda_{ka} \leq x^\lambda_k & \lambda \in \Lambda, \; k \in K, \; v_i \in V \setminus \{s_k, d_k\} \\
x_k, x^\lambda_k, x^\lambda_{ka} \in \{0, 1\} & \lambda \in \Lambda, \; k \in K, \; a \in A.
\end{cases}
$$

Eliminating the variables x^λ_k now yield:

$$\min \quad z_{\text{BLOCKING}}$$

subject to

$$\begin{cases} \sum_{k \in K} \left(x^\lambda_{ka(e)} + x^\lambda_{k\bar{a}(e)} \right) \leq 1 & e \in E, \ \lambda \in \Lambda \\ \sum_{\lambda \in \Lambda} \sum_{a \in \omega^+(s_k)} x^\lambda_{ka} = x_k & k \in K \\ x^\lambda_{ka} = 0 & \lambda \in \Lambda, k \in K, a \in \omega^-(s_k) \\ \sum_{\lambda \in \Lambda} \sum_{a \in \omega^+(v_i)} x^\lambda_{ka} = \sum_{\lambda \in \Lambda} \sum_{a \in \omega^-(v_i)} x^\lambda_{ka} & k \in K, v_i \in V \setminus \{s_k, d_k\} \\ \sum_{a \in \omega^+(v_i)} x^\lambda_{ka} \leq \sum_{a \in \omega^+(s_k)} x^\lambda_{ka} & \lambda \in \Lambda, k \in K, v_i \in V \setminus \{s_k, d_k\} \\ x_k, x^\lambda_{ka} \in \{0, 1\} & \lambda \in \Lambda, k \in K, a \in A. \end{cases}$$

This formulation has a higher number of constraints than the formulation proposed in Section 23.5.1 but its set of feasible integer solutions is smaller due to the elimination of the dummy loops corresponding to an unused wavelength. When solving exactly the formulation, this reduction of the feasibility set may become advantageous.

23.5.3 LP lower bounds

Although we can obtain quite different formulations by using aggregation as illustrated in the previous subsections, we show that the LP lower bound obtained by solving the LP relaxation of these formulations are all equal, for a given network formulation.

Let $\text{ILP}(\mathcal{F}, \text{OBJ})$ be the straightforward ILP formulation for the RWA problem, i.e., without any aggregation with respect to λ, and $\text{ILP\_AGGR}(\mathcal{F}, \text{OBJ})$ be a formulation obtained from $\text{ILP}(\mathcal{F}, \text{OBJ})$ by performing some aggregation with respect to λ. Denote by

$$\underline{z}^{\text{ILP}(\mathcal{F},\text{OBJ})} \text{ and } \underline{z}^{\text{ILP\_AGGR}(\mathcal{F},\text{OBJ})}$$

the optimal values of the continuous relaxation of these formulations. In this section, we show that

$$\underline{z}^{\text{ILP}(\mathcal{F},\text{OBJ})} = \underline{z}^{\text{ILP\_AGGR}(\mathcal{F},\text{OBJ})},$$

i.e., that aggregation has no effect on the optimal value of the continuous relaxation. To this effect, we use the following result:

Proposition 23.4 *Consider a LP of the form*

$$\min \ \sum_{r \in R} c_r \sum_{t \in T} y^t_r + \sum_{s \in S} d_s y_s \tag{23.59}$$

subject to

$$\sum_{r \in R} \alpha_{rG} y^t_r + \sum_{r \in R} \beta_{rG} \sum_{t' \in T} y^{t'}_r + \sum_{s \in S} \gamma_{sG} y_s \leq b_G, \quad t \in T, \ G \in \mathcal{G} \tag{23.60}$$

$$y^t_r \geq 0 \quad t \in T, \ r \in R \tag{23.61}$$

$$y_s \geq 0 \quad s \in S \tag{23.62}$$

for some sets R, S and T, and coefficients $\alpha, \beta \in R^{|R|} \times R^{|\mathcal{G}|}$, $\gamma \in R^{|S|} \times R^{\mathcal{G}}$, $c \in R^{|R|}$ and $d \in R^{|S|}$.

Let LP_AGGR be the linear program obtained by replacing a subset of the constraint groups, i.e., (23.60) for $G \in \mathcal{G}' \subseteq \mathcal{G}$, by their sum constraints. The set of constraints of LP_AGGR is therefore

$$\begin{cases} \sum_{r \in R} \alpha_{rG} \sum_{t \in T} y_r^t + |T| \sum_{r \in R} \beta_{rG} \sum_{t \in T} y_r^t + |T| \sum_{s \in S} \gamma_{sG} y_s \leq b_G |T| & G \in \mathcal{G}' \\ \text{constraints } (23.60\text{-}G) & G \in \mathcal{G} \setminus \mathcal{G}' \\ \text{constraints } (23.61) - (23.62). \end{cases}$$

Then the optimal values of the two linear programs are equal, i.e.,

$$z^{LP} = z^{LP\_AGGR}.$$

Proof. Since LP_AGGR is a relaxation of LP, we easily deduce that $z^{LP\_AGGR} \leq z^{LP}$. Now let $\hat{y}$ be an optimal solution of LP_AGGR. We define

$$\tilde{y}_r^t = \frac{\sum_{t \in T} \hat{y}_r^t}{|T|} \qquad r \in R, \quad t \in T$$

$$\tilde{y}_s = \hat{y}_s \qquad s \in S.$$

Let us show that $\tilde{y}$ is a feasible solution to LP. To do that, let us consider successively each constraint of LP. If a constraint (ctr) is of the form (23.60), 2 cases are possible:

- The constraint (ctr), as well as the $(|T| - 1)$ others of the same group, are appearing as such in LP_AGGR, i.e., we have

$$\sum_{r \in R} \alpha_{rG} \hat{y}_r^t + \sum_{r \in R} \beta_{rG} \sum_{t \in T} \hat{y}_r^t + \sum_{s \in S} \gamma_{sG} \hat{y}_s \leq b_G \qquad t \in T$$

for some $G \in \mathcal{G}$. Summing these inequalities with weight $\frac{1}{|T|}$ and using the definition of $\tilde{y}$, we obtain

$$\sum_{r \in R} \alpha_{rG} \tilde{y}_r^t + \sum_{r \in R} \beta_{rG} \sum_{t \in T} \tilde{y}_r^t + \sum_{s \in S} \gamma_{sG} \tilde{y}_s \leq b_G \qquad (23.63)$$

which shows that the constraint (crt) is satisfied by $\tilde{y}$.

- The constraint (ctr), as well as the $(|T|-1)$ others of the same group, are appearing in LP_AGGR in an aggregated sum constraint. Hence (23.63) is satisfied. Therefore $\tilde{y}$ satisfies the constraint (ctr).

We have therefore showed that $\tilde{y}$ is a feasible solution to LP. Considering the objective, we see that this solution has the same value than $\hat{y}$. Hence $z^{LP} \leq z^{LP\_AGGR}$.
□

Note that the following constraints can be viewed as a particular case of constraints (23.60)

$$\sum_{r \in R} \beta_{rT} \sum_{t' \in T} y_r^{t'} + \sum_{s \in S} \gamma_{sT} y_s \leq b_T$$

for which the Proposition 23.4 also holds. If the constraint group is of the above form, it is appearing as such in *LP_AGGR*.

Corollary 23.1 *With the above definitions, we have*

$$\underline{z}^{\text{ILP}(\mathcal{F},\text{OBJ})} = \underline{z}^{\text{ILP\_AGGR}(\mathcal{F},\text{OBJ})}$$

for $(\mathcal{F},\text{OBJ}) \in \{\text{EDGE},\text{PATH},\text{ARC}\} \times \{\text{BLOCKING},\#\lambda,\text{CONGESTION},\text{LOAD}\}$.

Proof. Let $T = \Lambda$. For a given λ, the variables $x_k^\lambda, x_{ke}^\lambda, x_{kp}^\lambda, x^\lambda, \ldots$, numbered consecutively, yield the variables y_r^t and a set R. Similarly the variables x_k, numbered consecutively, yield the variables y_s and the set S. The result follows then from Proposition 23.4 with LP and LP_AGGR being respectively the continuous relaxation of ILP$(\mathcal{F},\text{OBJ})$ and of ILP_AGGR$(\mathcal{F},\text{OBJ})$. $\square$

The result of Corollary 23.1 was proved in (Ramaswami and Sivarajan, 1995) in the particular case of RWA(PATH, BLOCKING). This result can be explained by the wavelength symmetry: as the wavelengths are interchangeable, given an optimal solution of the RWA problem or of one of its continuous relaxation, one can derive a large number of equivalent solutions using any permutation of the wavelengths. Moreover, if we consider continuous relaxations, any convex combination of all these equivalent solutions define a solution as well.

23.5.4 ILP lower bounds

Consider the ILP formulation obtained by performing the maximum possible aggregation with respect to λ. All constraints are then expressed in terms of the variables x_k,

$$x = \sum_{\lambda \in \Lambda} x^\lambda \text{ and } x_{ke} = \sum_{\lambda \in \Lambda} x_{ke}^\lambda$$

$$x_{kp} = \sum_{\lambda \in \Lambda} x_{kp}^\lambda \text{ or } x_{ka} = \sum_{\lambda \in \Lambda} x_{ka}^\lambda$$

depending on the network formulation and the objective that are considered. In particular, constraints (23.2) become:

$$\sum_{k \in K} x_{ke} \leq W \qquad e \in E. \tag{23.64}$$

We denote the resulting ILP formulation by ILP$_\lambda(\mathcal{F},\text{OBJ})$. Clearly, for a fixed objective OBJ, all formulations ILP$_\lambda(.,\text{OBJ})$ express the RWA problem with the wavelength continuity constraints replaced by the weaker constraints asking for no more than W paths going through the same fiber link (this problem is sometimes referred to as the circuit-switched routing problem, see, e.g., Ramaswami and Sivarajan (1995)). Hence

$$\underline{z}^{\text{ILP}_\lambda(\text{EDGE},\text{OBJ})} = \underline{z}^{\text{ILP}_\lambda(\text{PATH},\text{OBJ})} = \underline{z}^{\text{ILP}_\lambda(\text{ARC},\text{OBJ})}.$$

This common value defines a lower bound on the optimal value of the RWA problem.

We next provide an example for each objective to illustrate a gap between the optimal value of ILP$_\lambda(.,\text{OBJ})$ and the optimal value of the corresponding RWA problem.

Example 23.2 *Minimizing the blocking rate. Consider the network of Figure 23.3. We have three connections: $(s_1, d_1) = (v_1, v_2)$, $(s_2, d_2) = (v_2, v_3)$ and $(s_3, d_3) = (v_1, v_3)$ and 2 wavelengths. The optimal value of the* $\text{ILP}_\lambda(., \text{BLOCKING})$ *formulation is equal to 0, while the optimal value of the* $\text{RWA}(., \text{BLOCKING})$ *formulation is equal to 1/3.*

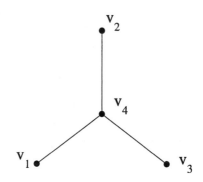

Figure 23.3 Network for Example 23.2

Example 23.3 *Minimizing the number of wavelengths. Let the network be the 5 node ring $(v_1, v_2, v_3, v_4, v_5, v_1)$. Assume that there are 5 connections: $\{v_1, v_3\}$, $\{v_2, v_4\}$, $\{v_3, v_5\}$, $\{v_4, v_1\}$ and $\{v_5, v_2\}$ and 3 wavelengths. The optimal value of* $\text{RWA}(., \#\lambda)$ *is equal to 3, while the optimal value of* $\text{ILP}_\lambda(., \#\lambda)$ *is equal to 2.*

Example 23.4 *Minimizing the congestion. Consider the network of Figure 23.4. Assume that 3 wavelengths are available, and that the following 7 connections have to be established: $k_1 = v_5 \leftrightsquigarrow v_8$, $k_2 = v_1 \leftrightsquigarrow v_5$, $k_3 = v_1 \leftrightsquigarrow v_8$ and $k_4 = k_5 = k_6 = k_7 = v_3 \leftrightsquigarrow v_4$. An optimal solution to* $\text{ILP}_\lambda(., \text{CONGESTION})$ *is obtained with the following paths: $p_1 = (v_5, v_6, v_7, v_8)$, $p_2 = (v_1, v_2, v_3, v_6, v_5)$, $p_3 = (v_1, v_2, v_4, v_7, v_8)$, $p_4 = p_5 = (v_3, v_4)$, $p_6 = (v_3, v_2, v_4)$ and $p_7 = (v_3, v_6, v_7, v_4)$. Note that no edge belongs to more than 2 paths, hence the value of this solution is 2. Note that it is not possible to assign wavelengths to these paths with only 3 available wavelengths.*
A feasible solution to $\text{RWA}(., \text{CONGESTION})$ *is obtained by replacing path p_7 by $p'_7 = (v_3, v_4)$. A wavelength assignment is then obtained by assigning wavelength λ_1 to p_1, p_4 and p_6, wavelength λ_2 to p_2, p_5 and wavelength λ_3 to p_3 and p'_7. The value of this solution is 3.*
It remains to show that there is no feasible solution with value 2. Assume that there exists a feasible solution with value 2. Observe first that, in such a solution, at least one connection between v_3 and v_4, say k_4, must consider the path $p_4 = (v_3, v_6, v_7, v_4)$. Indeed, there are 4 such connections and at most one can go through v_2 because of connections k_2 and k_3, and at most 2 can follow the direct link (v_3, v_4). Then the only possibility for p_1 is (v_5, v_6, v_7, v_8). This in turn forces p_2 to contain the subpath (v_3, v_6, v_5) and p_3 the subpath (v_4, v_7, v_8). Note that p_1, p_2 and p_3 must use 3 different

wavelengths. It follows that each wavelength is used on one edge of p_4, hence p_4 cannot be assigned a wavelength. Therefore the optimal value of RWA(., CONGESTION) is 3.

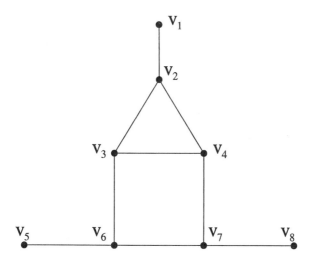

Figure 23.4 Network for Example 23.4

Example 23.5 *Minimizing the network load. Consider the network of Figure 23.5. We assume the same set of connections than in Example 3, i.e., $K = \{v_1 \leftrightsquigarrow v_3, v_2 \leftrightsquigarrow$*

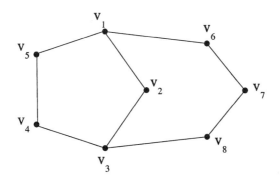

Figure 23.5 Network for Example 23.5

$v_4, v_3 \leftrightsquigarrow v_5, v_4 \leftrightsquigarrow v_1, v_5 \leftrightsquigarrow v_2\}$ but this time we assume that there are only 2 wavelengths. An optimal solution of RWA(., LOAD) *is made of paths $p_1 = \{v_1, v_6, v_7, v_8, v_3\}$ with λ_1, $p_2 = \{v_2, v_3, v_4\}$ with λ_2, $p_3 = \{v_3, v_4, v_5\}$ with λ_1, $p_4 = \{v_4, v_5, v_1\}$ with λ_2 and finally $p_5 = \{v_5, v_1, v_2\}$ with λ_1. The objective value is $2/3$ as $m = 9$ and $W = 2$. In contrast the optimal value of* ILP_λ(., LOAD) *is $5/9$.*

23.6 AGGREGATION WITH RESPECT TO PAIRS OF SOURCE AND DESTINATION

In the previous formulations, the number of variables and constraints was proportional to the total number $|K|$ of connections. In the formulations considered in this section, we group together the connections with the same endpoints.

23.6.1 Aggregated formulations

23.6.1.1 Variables. Let $\mathcal{SD} = \{(v_i, v_j) \in V \times V : T_{ij} \geq 1\}$, and denote by $\mathcal{P}_{sd}$ the set of paths between v_s and v_d.

The first two sets of variables deal with some number of connections:
x_{sd} = number of accepted connections between v_s and v_d ($0 \leq x_{sd} \leq T_{sd}$) for all $(v_s, v_d) \in \mathcal{SD}$, and
x_{sd}^λ = number of connections between v_s and v_d supported by wavelength λ ($0 \leq x_{sd}^\lambda \leq T_{sd}$) for all $(v_s, v_d) \in \mathcal{SD}, \lambda \in \Lambda$.

The next three sets of variables correspond to decision variables:

$$x_{sde}^\lambda = \begin{cases} 1 & \text{if wavelength } \lambda \text{ supports a connection between } v_s \text{ and } v_d \text{ on the (undirected)} \\ & \text{fiber link } e \\ 0 & \text{otherwise} \end{cases}$$

for any $(v_s, v_d) \in \mathcal{SD}, e \in E, \lambda \in \Lambda$.

$$x_{sda}^\lambda = \begin{cases} 1 & \text{if wavelength } \lambda \text{ supports a connection between } v_s \text{ and } v_d \text{ on the (directed)} \\ & \text{fiber link } a \\ 0 & \text{otherwise} \end{cases}$$

for any $(v_s, v_d) \in \mathcal{SD}, a \in A, \lambda \in \Lambda$.

$$x_p^\lambda = \begin{cases} 1 & \text{if lightpath } (p, \lambda) \text{ is established} \\ 0 & \text{otherwise} \end{cases}$$

for any $\lambda \in \Lambda, p \in \bigcup_{(v_s,v_d) \in \mathcal{SD}} \mathcal{P}_{sd}$.

23.6.1.2 Objectives. Using the variables defined in the previous section, the different objective functions can be rewritten:

$$z_{\text{BLOCKING}} = - \sum_{(v_s,v_d) \in \mathcal{SD}} x_{sd}$$

with the additional set of constraints

$$x^\lambda = 1 \qquad \lambda \in \Lambda.$$

$$z_{\#\lambda} = \sum_{\lambda \in \Lambda} x^\lambda$$

with the additional set of constraints

$$x_{sd} = T_{sd} \qquad (v_s, v_d) \in \mathcal{SD}.$$

FORMULATIONS FOR THE RWA PROBLEM

$z_{\text{CONGESTION}}$
with the three additional sets of constraints

$$z_{\text{CONGESTION}} \geq \sum_{(v_s,v_d)\in S\mathcal{D}} \sum_{\lambda \in \Lambda} x^\lambda_{sde} \quad e \in E \quad (23.65)$$

$$x^\lambda = 1 \qquad \lambda \in \Lambda$$
$$x_{sd} = T_{sd} \qquad (v_s, v_d) \in S\mathcal{D}.$$

$$z_{\text{LOAD}} = \sum_{(v_s,v_d)\in S\mathcal{D}} \sum_{e\in E} \sum_{\lambda \in \Lambda} x^\lambda_{sde}$$
with the two additional sets of constraints

$$x^\lambda = 1 \qquad \lambda \in \Lambda$$
$$x_{sd} = T_{sd} \qquad (v_s, v_d) \in S\mathcal{D}.$$

23.6.1.3 Network constraints. Observe first that formulation EDGE cannot be used with aggregation with respect to source and destination nodes. Indeed, consider the network graph as illustrated in Figure 23.6. Assume that there are 2 connections between v_s and v_d: nothing prevent a possible dummy loop at each node instead of two paths linking these two nodes.

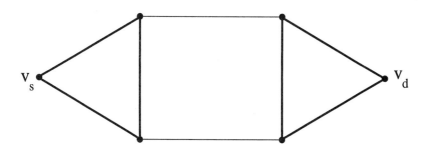

Figure 23.6 Network formulation EDGE cannot be aggregated with respect to source and destination

For the PATH formulation, the aggregated constraints are

$$\sum_{p\in \mathcal{P}_{sd}} x^\lambda_p = x^\lambda_{sd} \qquad \lambda \in \Lambda, \ (v_s,v_d) \in S\mathcal{D} \qquad (23.66)$$

$$x^\lambda_{sde} = \sum_{p\in \mathcal{P}_{sd}} b_{ep} x^\lambda_p \qquad \lambda \in \Lambda, \ (v_s,v_d) \in S\mathcal{D}, \ e \in E. \qquad (23.67)$$

For the ARC formulation, they can be written

$$\sum_{a \in \omega^+(v_s)} x^\lambda_{sda} - \sum_{a \in \omega^-(v_s)} x^\lambda_{sda} = x^\lambda_{sd} \qquad \lambda \in \Lambda, (v_s, v_d) \in S\mathcal{D} \qquad (23.68)$$

$$\sum_{a \in \omega^+(v_i)} x^\lambda_{sda} = \sum_{a \in \omega^-(v_i)} x^\lambda_{sda} \qquad \lambda \in \Lambda, (v_s, v_d) \in S\mathcal{D},$$
$$v_i \in V \setminus \{v_s, v_d\} \qquad (23.69)$$

$$x^\lambda_{sde} = x^\lambda_{sda(e)} + x^\lambda_{sd\bar{a}(e)} \qquad \lambda \in \Lambda, (v_s, v_d) \in S\mathcal{D}, e \in E \qquad (23.70)$$

with the optional loop removal constraints:

$$x^\lambda_{sda} = 0, \qquad \lambda \in \Lambda, \ (v_s, v_d) \in S\mathcal{D}, \ a \in \omega^-(v_s) \qquad (23.71)$$
$$x^\lambda_{sda} = 0, \qquad \lambda \in \Lambda, \ (v_s, v_d) \in S\mathcal{D}, \ a \in \omega^+(v_d). \qquad (23.72)$$

23.6.1.4 Other constraints. The wavelength assignment constraints and the clash constraints are respectively:

$$\sum_{\lambda \in \Lambda} x^\lambda_{sd} = x_{sd} \qquad (v_s, v_d) \in S\mathcal{D} \qquad (23.73)$$

$$\sum_{(v_s, v_d) \in S\mathcal{D}} x^\lambda_{sde} \leq x^\lambda \qquad e \in E, \ \lambda \in \Lambda. \qquad (23.74)$$

The network collapse constraints loss some interest because they cannot anymore result in the collapse of the network constraints. They are:

$$x^\lambda_{sde} \leq x^\lambda_{sd} \qquad \lambda \in \Lambda, \ (v_s, v_d) \in S\mathcal{D}, \ e \in E. \qquad (23.75)$$

23.6.2 Additional aggregation with respect to λ

The possibility of additional aggregation with respect to λ is limited. It is not anymore possible to collapse the W sets of network constraints as it was done in Section 23.5.2. The reason is that several connections with the same source and destination node pairs can use different wavelengths. It is still possible however to aggregate constraints (23.66), or (23.68) provided that (23.71) are present, with respect to λ.

23.6.3 LP lower bounds

The results of Proposition 23.2 and Corollary 23.1 extend to the aggregation with respect to source and destination and to the combination of aggregation with respect to source and destination and aggregation with respect to λ. In particular, the LP lower bound is equal to the optimal value of the LP relaxation of the RWA formulation fully aggregated with respect to the pairs of source and destination and to the wavelengths.

Proposition 23.5 *Denote by* $\text{RWA}_{S\mathcal{D}}(\mathcal{F}, \text{OBJ})$ *a formulation for the RWA problem that is aggregated with respect to source and destination, with possibly some amounts of aggregation with respect to λ. Then, for all* $(\mathcal{F}, \text{OBJ}) \in \{\text{PATH}, \text{ARC}\} \times \{\text{BLOCKING}, \#\lambda, \text{CONGESTION}, \text{LOAD}\}$*, we have*

$$z^{\text{RWA}_{S\mathcal{D}}(\mathcal{F}, \text{OBJ})} = z^{\text{PATH}}_{\text{OBJ}}.$$

Proof. Let us perform the aggregations sequentially. We denote by LP^0 the linear program prior to any aggregation, by LP^ℓ the linear program after aggregation with respect to the first ℓ pairs of source and destination, and by $LP^{|\mathcal{SD}|+1}$ the linear program obtained by performing some aggregation with respect to λ on $LP^{|\mathcal{SD}|}$. LP^1 is obtained from LP^0 by aggregating with respect to the first pair of source and destination (s^1, d^1). We use again Proposition 23.4. Let us choose $T = \{k \in K : (s_k, d_k) = (s^1, d^1)\}$. For a given $k \in T$, the variables x_k^λ, x_{ke}^λ, x_{kp}^λ, x_k, ..., numbered consecutively, yield the variables y_r^t and a set R. Similarly the variables x_k^λ, x_{ke}^λ, x_{kp}^λ, x_k, ... for $k \notin T$ and the variables x^λ, numbered consecutively, yield the variables y_s and the set S. Applying Proposition 23.4, we get that $z^{LP^1} = z^{LP^0}$.

LP^2 is obtained from LP^1 by aggregating with respect to the second pair of source and destination nodes. We choose $T = \{k \in K : (s_k, d_k) = (s^2, d^2)\}$. For a given $k \in T$, the variables x_k^λ, x_{ke}^λ, x_{kp}^λ, x_k, ..., numbered consecutively, yield the variables y_r^t and a (new) set R. Similarly the variables x_k^λ, x_{ke}^λ, x_{kp}^λ, x_k, ... for $k \notin T$ and the variables x^λ, numbered consecutively, yield the variables y_s and the (new) set S. Using Proposition 23.4, we show that $z^{LP^2} = z^{LP^1}$. Iterating this process, we get $z^{LP^{|\mathcal{SD}|}} = z^{LP^{|\mathcal{SD}|-1}} = \cdots = z^{LP^1} = z^{LP^0}$.

$LP^{|\mathcal{SD}|+1}$ is deduced from $LP^{|\mathcal{SD}|}$ after some aggregation with respect to λ. Applying Proposition 23.4 with $T = \Lambda$, we show that $z^{LP^{|\mathcal{SD}|+1}} = z^{LP^{|\mathcal{SD}|}}$, i.e., that $z^{LP^{|\mathcal{SD}|+1}} = z^{LP^0}$.

□

23.7 AGGREGATION WITH RESPECT TO SOURCES

A further step into aggregation is to aggregate with respect to sources (we assume here an orientation of the traffic following the interpretation of Section 23.3.5.3). For the same reason than mentioned in Section 23.6.1.3, it is not possible to use the network formulation EDGE with this type of aggregation. Moreover such an aggregation brings nothing new for the PATH network formulation as a path must be identified by its source and destination nodes. Therefore, in the following of this section, we assume that the ARC network formulation is used. Aggregation with respect to sources was already studied in Coudert and Rivano (2002) in the context of the RWA with wavelength translation and by Krishnaswamy and Sivarajan (2001b) for the multi-hop RWA problem.

23.7.1 Aggregated formulations

Recall that $\mathcal{SD} = \{(v_i, v_j) \in V \times V : T_{ij} \geq 1\}$. Let $D_s = \{v_d \in V : T_{sd} \geq 1\}$ be the set of destination nodes for the connections originating at v_s.

23.7.1.1 Variables. The variables x_{sd} and x_{sd}^λ have the same definition than in Section 23.6.1.1. Instead of variables x_{sda}^λ, we consider

$$x_{sa}^\lambda = \begin{cases} 1 & \text{if wavelength } \lambda \text{ supports a connection originating at } v_s \text{ on fiber link } a \\ 0 & \text{otherwise.} \end{cases}$$

23.7.1.2 Objectives.
Compared with the aggregation studied in Section 23.6, the only changes for the mathematical expressions of the objective functions are with respect to constraints (23.65) which are now written:

$$z_{\text{CONGESTION}} \geq \sum_{v_s \in V} \sum_{\lambda \in \Lambda} x_{sa}^{\lambda} \qquad a \in A$$

and in the expression of the objective of minimizing the network load:

$$\min z_{\text{LOAD}} \qquad \text{where} \qquad z_{\text{LOAD}} = \sum_{v_s \in V} \sum_{a \in A} \sum_{\lambda \in \Lambda} x_{sa}^{\lambda}.$$

23.7.1.3 ARC network constraints.
When aggregating with respect to the sources, the ARC network constraints are rewritten:

$$\sum_{a \in \omega^+(v_s)} x_{sa}^{\lambda} - \sum_{a \in \omega^-(v_s)} x_{sa}^{\lambda} = \sum_{v_d \in D_s} x_{sd}^{\lambda} \qquad \lambda \in \Lambda, v_s \in V \qquad (23.76)$$

$$\sum_{a \in \omega^-(v_d)} x_{sa}^{\lambda} - \sum_{a \in \omega^+(v_d)} x_{sa}^{\lambda} = x_{sd}^{\lambda} \qquad \lambda \in \Lambda, v_d \in D_s, v_s \in V \qquad (23.77)$$

$$\sum_{a \in \omega^+(v_i)} x_{sa}^{\lambda} = \sum_{a \in \omega^-(v_i)} x_{sa}^{\lambda} \qquad \lambda \in \Lambda, v_s \in V, v_i \in V \setminus (D_s \cup \{v_s\}), \qquad (23.78)$$

where (23.76)–(23.78) are the flow conservation constraints. Note that constraints (23.76) are here redundant as they can be written as the sum of (23.77)–(23.78).

23.7.1.4 Other constraints.
The wavelength assignment constraints are unchanged with respect to aggregation by source and destination node pairs:

$$\sum_{\lambda \in \Lambda} x_{sd}^{\lambda} = x_{sd} \qquad (v_s, v_d) \in S\mathcal{D}. \qquad (23.79)$$

The clash constraints are now written:

$$\sum_{v_s \in V} \left(x_{sa(e)}^{\lambda} + x_{s\bar{a}(e)}^{\lambda} \right) \leq x^{\lambda} \qquad e \in E, \quad \lambda \in \Lambda. \qquad (23.80)$$

23.7.2 Additional aggregation with respect to λ

Note first that we cannot anymore impose that no flow is going out the destination nodes as expressed by (23.72). Indeed, the flow may have to go through a first destination node of v_s for a given connection in order to reach another destination node of v_s for another connection.

We aggregate (23.77) with respect to λ, and using (23.79) to eliminate the x_{sd}^{λ} variables, we obtain

$$\sum_{\lambda \in \Lambda} \left(\sum_{a \in \omega^-(v_d)} x_{sa}^{\lambda} - \sum_{a \in \omega^+(v_d)} x_{sd} \right) = x_{sd} \qquad (v_s, v_d) \in S\mathcal{D}. \qquad (23.81)$$

In order to guarantee that no term inside the sum over λ becomes negative, we need to add the following constraints:

$$\sum_{a \in \omega^-(v_d)} x_{sa}^\lambda - \sum_{a \in \omega^+(v_d)} x_{sa}^\lambda \geq 0 \qquad \lambda \in \Lambda, \ (v_s, v_d) \in \mathcal{SD}. \tag{23.82}$$

The resulting new formulation has the same number of constraints, but significantly less variables.

23.7.3 Examples of formulations

We give the formulation corresponding to the additional aggregation with respect to λ for each objective function.

23.7.3.1 Minimizing the blocking rate.
After eliminating the variables x_{sd}, we obtain

$$\min \ z_{\text{BLOCKING}} \qquad \text{where} \quad z_{\text{BLOCKING}} = - \sum_{(v_s, v_d) \in \mathcal{SD}} \sum_{\lambda \in \Lambda} \left(\sum_{a \in \omega^-(v_d)} x_{sa}^\lambda - \sum_{a \in \omega^+(v_d)} x_{sa}^\lambda \right)$$

subject to:

$$\begin{cases} \sum_{a \in \omega^+(v_i)} x_{sa}^\lambda = \sum_{a \in \omega^-(v_i)} x_{sa}^\lambda & \lambda \in \Lambda, \ v_s \in V, \ v_i \in V \setminus (D_s \cup \{v_s\}) \\ \sum_{\lambda \in \Lambda} \left(\sum_{a \in \omega^-(v_d)} x_{sa}^\lambda - \sum_{a \in \omega^+(v_d)} x_{sa}^\lambda \right) \leq T_{sd} & (v_s, v_d) \in \mathcal{SD} \\ \sum_{a \in \omega^-(v_d)} x_{sa}^\lambda - \sum_{a \in \omega^+(v_d)} x_{sa}^\lambda \geq 0 & \lambda \in \Lambda, \ (v_s, v_d) \in \mathcal{SD} \\ \sum_{v_s \in V} \left(x_{sa(e)}^\lambda + x_{s\bar{a}(e)}^\lambda \right) \leq 1 & e \in E, \ \lambda \in \Lambda \\ x_{sa}^\lambda \in \{0, 1\} & \lambda \in \Lambda, \ v_s \in V, \ a \in A. \end{cases}$$

23.7.3.2 Minimizing the number of wavelengths.

$$\min \ z_{\#\lambda} \qquad \text{where} \quad z_{\#\lambda} = \sum_{\lambda \in \Lambda} x^\lambda$$

subject to

$$\begin{cases} \sum_{a \in \omega^+(v_i)} x_{sa}^\lambda = \sum_{a \in \omega^-(v_i)} x_{sa}^\lambda & \lambda \in \Lambda, \ v_s \in V, \ v_i \in V \setminus (D_s \cup \{v_s\}) \\ \sum_{\lambda \in \Lambda} \left(\sum_{a \in \omega^-(v_d)} x_{sa}^\lambda - \sum_{a \in \omega^+(v_d)} x_{sa}^\lambda \right) = T_{sd} & (v_s, v_d) \in \mathcal{SD} \\ \sum_{a \in \omega^-(v_d)} x_{sa}^\lambda - \sum_{a \in \omega^+(v_d)} x_{sa}^\lambda \geq 0 & \lambda \in \Lambda, \ (v_s, v_d) \in \mathcal{SD} \\ \sum_{v_s \in V} \left(x_{sa(e)}^\lambda + x_{s\bar{a}(e)}^\lambda \right) \leq x^\lambda & e \in E, \ \lambda \in \Lambda \\ x_{sa}^\lambda, x^\lambda \in \{0, 1\} & \lambda \in \Lambda, \ v_s \in V, \ a \in A. \end{cases}$$

23.7.3.3 Minimizing the congestion.

$$\min \quad z_{\text{CONGESTION}}$$

subject to

$$\begin{cases} \sum_{a \in \omega^+(v_i)} x_{sa}^\lambda = \sum_{a \in \omega^-(v_i)} x_{sa}^\lambda & \lambda \in \Lambda, v_s \in V, v_i \in V \setminus (D_s \cup \{v_s\}) \\ \sum_{\lambda \in \Lambda} \left(\sum_{a \in \omega^-(v_d)} x_{sa}^\lambda - \sum_{a \in \omega^+(v_d)} x_{sa}^\lambda \right) = T_{sd} & (v_s, v_d) \in \mathcal{SD} \\ \sum_{a \in \omega^-(v_d)} x_{sa}^\lambda - \sum_{a \in \omega^+(v_d)} x_{sa}^\lambda \geq 0 & \lambda \in \Lambda, (v_s, v_d) \in \mathcal{SD} \\ \sum_{v_s \in V} (x_{sa(e)}^\lambda + x_{s\bar{a}(e)}^\lambda) \leq 1 & e \in E, \quad \lambda \in \Lambda \\ z_{\text{CONGESTION}} \geq \sum_{\lambda \in \Lambda} \sum_{v_s \in V} (x_{sa(e)}^\lambda + x_{s\bar{a}(e)}^\lambda) & e \in E \\ x_{sa}^\lambda \in \{0,1\} & \lambda \in \Lambda, v_s \in V, a \in A. \end{cases}$$

23.7.3.4 Minimizing the network load.

$$\min \quad z_{\text{LOAD}} \qquad \text{where} \qquad z_{\text{LOAD}} = \sum_{\lambda \in \Lambda} \sum_{v_s \in V} \sum_{a \in A} x_{sa}^\lambda$$

subject to:

$$\begin{cases} \sum_{a \in \omega^+(v_i)} x_{sa}^\lambda = \sum_{a \in \omega^-(v_i)} x_{sa}^\lambda & \lambda \in \Lambda, v_s \in V, v_i \in V \setminus (D_s \cup \{v_s\}) \\ \sum_{\lambda \in \Lambda} \left(\sum_{a \in \omega^-(v_d)} x_{sa}^\lambda - \sum_{a \in \omega^+(v_d)} x_{sa}^\lambda \right) = T_{sd} & (v_s, v_d) \in \mathcal{SD} \\ \sum_{a \in \omega^-(v_d)} x_{sa}^\lambda - \sum_{a \in \omega^+(v_d)} x_{sa}^\lambda \geq 0 & \lambda \in \Lambda, (v_s, v_d) \in \mathcal{SD} \\ \sum_{v_s \in V} \left(x_{sa(e)}^\lambda + x_{s\bar{a}(e)}^\lambda \right) \leq 1 & e \in E, \quad \lambda \in \Lambda \\ x_{sa}^\lambda \in \{0,1\} & \lambda \in \Lambda, v_s \in V, a \in A. \end{cases}$$

23.7.4 LP lower bounds

Let us now examine the LP lower bounds.

Proposition 23.6 *Denote by* $\text{RWA}_S(\text{ARC}, \text{OBJ})$ *a formulation of the RWA problem with aggregation with respect to sources and possibly some additional aggregation with respect to* λ. *Then we have*

$$\underline{z}^{\text{RWA}_S(\text{ARC}, \text{OBJ})} = \underline{z}^{\text{PATH}}_{\text{OBJ}}.$$

Proof. Note that it is not possible to use Proposition 23.4 because the constraints

$$x_{sd} \leq T_{sd} \qquad (v_s, v_d) \in \mathcal{SD}$$

do not satisfy the assumptions of the Proposition as T_{sd} vary from one destination to the next, for $v_d \in D_s$. Instead we use again the result of Ahuja et al. (1993) that was already used in the proof of Proposition 23.2.

Assume first that there is no aggregation with respect to λ in $\text{RWA}_S(\text{ARC}, \text{OBJ})$. Because of the aggregation with respect to sources, we easily get $\underline{z}^{\text{RWA}_S(\text{ARC},\text{OBJ})} \leq \underline{z}^{\text{PATH}}_{\text{OBJ}}$. To show the reverse inequality, let $(\hat{x}_{sd}, \hat{x}^{\lambda}, \hat{x}^{\lambda}_{sd}, \hat{x}^{\lambda}_{sa})$ be an optimal solution of the continuous relaxation of $\text{RWA}_S(\text{ARC}, \text{OBJ})$. Following Ahuja et al. (1993, p.80-81), for each $v_s \in V$ and $\lambda \in \Lambda$, there exist $\hat{x}^{\lambda}_{sp}$ and $\hat{z}^{\lambda}_{sc}$ such that

$$\hat{x}^{\lambda}_{sa} = \sum_{p \in \bigcup_{v_d \in D_s} \mathcal{P}_{sd}} b_{ap} \hat{x}^{\lambda}_{sp} + \sum_{c \in C} c_{ap} \hat{z}^{\lambda}_{sc} \qquad a \in A$$

$$\hat{x}^{\lambda}_{sp} \geq 0 \qquad \qquad p \in \bigcup_{v_d \in D_s} \mathcal{P}_{sd}$$

$$\hat{z}^{\lambda}_{sc} \geq 0 \qquad \qquad c \in C.$$

Because of (23.77), we must have

$$\sum_{p \in \mathcal{P}_{sd}} \hat{x}^{\lambda}_{sp} = \hat{x}^{\lambda}_{sd}, \qquad v_d \in D_s.$$

Let us define

$$\tilde{x}^{\lambda}_{sa} = \sum_{p \in \bigcup_{v_d \in D_s} \mathcal{P}_{sd}} b_{ap} \hat{x}^{\lambda}_{sp} \qquad a \in A.$$

The vector $(\hat{x}_{sd}, \hat{x}^{\lambda}, \hat{x}^{\lambda}_{sd}, \tilde{x}^{\lambda}_{sa})$ is a feasible solution to $\text{RWA}_{S\mathcal{D}}(\text{PATH}, \text{OBJ})$ with an objective value equal to $\underline{z}^{\text{RWA}_S(\text{ARC},\text{OBJ})}$. Hence $\underline{z}^{\text{RWA}_{S\mathcal{D}}(\text{PATH},\text{OBJ})} \leq \underline{z}^{\text{RWA}_S(\text{ARC},\text{OBJ})}$. But $\underline{z}^{\text{RWA}_{S\mathcal{D}}(\text{PATH},\text{OBJ})} = \underline{z}^{\text{PATH}}_{\text{OBJ}}$ by Proposition 23.5.

If there is some further aggregation with respect to λ, we use Proposition 23.4 to show that the optimal value of the continuous relaxation with aggregation with respect to λ is equal to the optimal value of the continuous relaxation without aggregation with respect to λ. □

23.7.5 Reduction of the Number of Variables and Constraints by Carefully Choosing the Sources

Recall that by the interpretation of Section 23.3.5.3, the symmetrical RWA problem with connections between s_k and d_k, $k \in K$ can be interpreted as an asymmetrical problem where the connections are from s_k to d_k, $k \in K$ in a directed graph obtained from the undirected graph by replacing each edge by two opposite arcs. The only difference with the RWA problem for asymmetrical systems lies in the clash constraints where we consider simultaneously the two arcs that are associated with a given edge in the undirected graph. Now observe that in this interpretation, we are free to choose the direction of the connections, i.e., for a given connection $k \in K$, to choose which of its endpoint plays the role of the source or the destination. Choosing carefully the direction may lead to a reduction of the number of variables and constraints. In general the problem of finding the orientation of the connections that minimizes the number of

Table 23.1 Comparison of lower bounds and optimal value – NSF network – T_1 – 191 connection requests

# wavelengths	$z^{\text{EDGE}}_{\text{DENIED}}$	$z^{\text{ARC}}_{\text{DENIED}}$	$z^{\text{ILP}_\lambda}(\text{ARC},\text{DENIED})$	z^{*}_{DENIED}
10	73.5	75.5	76	76
12	57.8	61.0	62	62
14	42.5	48.0	48	48
16	28.6	38.0	38	38
18	15.9	30.0	30	30
20	4.3	22.0	22	22

# wavelengths	$z^{\text{EDGE}}_{\text{LOAD}}$	$z^{\text{ARC}}_{\text{LOAD}}$	$z^{\text{ILP}_\lambda}(\text{ARC},\text{LOAD})$	z^{*}_{LOAD}
26	416.0	421.0	421	421
28	415.0	416.0	416	416
30	414.0	414.0	414	414
32	414.0	414.0	414	414

variables and constraints is NP-hard. Indeed it is equivalent to find a minimum dominating set in the graph G'' with vertex set V and such that there is an edge between two vertices if and only if there is at least one connection between the two corresponding nodes. The Minimum Dominating Set problem is known to be NP-hard, see Garey and Johnson (1979).

23.8 CONCLUDING REMARKS ON THE VARIOUS MATHEMATICAL FORMULATIONS

Ramaswami and Sivarajan (1995) present a variant of the formulation RWA$_{S\mathcal{D}}$(PATH, BLOCKING) where the equality (23.74) is replaced by a $\geq$-inequality. It is easy to see that for this particular objective function this inequality will always be satisfied at equality at the optimum, even for the LP relaxation. Hence our results still apply. Note that the formulations using aggregation with respect to source and destination (Section 23.6) or with respect to the sources (Section 23.7) require less variables and constraints than the formulations without these aggregations (Sections 23.3-23.5). However the former formulations may not be suited when some additional constraints are added. For example, when hop constraints limiting the number of links in a lightpath are considered, we have no choice but to use a formulation at the connection level (see Jaumard et al. (2004a)).

Table 23.2 Comparison of lower bounds and optimal value – NSF3 network – T_1 – 191 connection requests

# wavelengths	$z^{\text{EDGE}}_{\text{DENIED}}$	$z^{\text{ARC}}_{\text{DENIED}}$	$z^{\text{ILP}_\lambda}(\text{ARC,DENIED})$	z^*_{DENIED}	
10	47.5	113.0	113	113	
12	26.5	104.0	104	104	
14	15.0	95.0	95	95	
16	10.0	86.0	86	86	
18	8.0	78.5	79	80	(gap = 1.88 %)
20	6.0	71.5	72	74	(gap = 3.38 %)
22	4.0	64.5	65	68	(gap = 5.15 %)
24	2.0	58.0	58	62	(gap = 6.45 %)
26	0.0	53.0	53	56	(gap = 5.36 %)
28	0.0	49.0	49	50	(gap = 2.00 %)

# wavelengths	$z^{\text{EDGE}}_{\text{LOAD}}$	$z^{\text{ARC}}_{\text{LOAD}}$	$z^{\text{ILP}_\lambda}(\text{ARC,LOAD})$	z^*_{LOAD}	
66	382.0	632.0	632	634	(gap = 0.32 %)
68	382.0	632.0	632	632	
70	382.0	632.0	632	632	

Completely different ILP formulations were proposed by Lee et al. (2002) and Ramaswami and Sivarajan (1995). These ILP formulations have an exponential number of variables, each variable corresponding to a set of paths that can be assigned the same wavelength.

23.9 COMPUTATIONAL EXPERIENCE

We compare the values of the lower bounds discussed in the previous sections together with the optimal ILP value, denoted by z^*, or if not available with the best known solution, on six network and traffic instances. All continuous and integer programs were solved using the CPLEX-MIP package (version 8.1) (CPLEX, 2003), on a Pentium 4 Linux Dell machine.

We consider four optical networks, two of them widely used in the literature, the NSF and the EON networks in addition to the Brazil network (Noronha and Ribeiro, 2005) and a subnet (30 nodes, 41 links) of the 30N60S network available from the web site of Grover's book (Grover, 2005; 2004). The NSF network is a network with 14 nodes and 21 optical links, with a maximum of 4 links per node, and is, e.g.

Table 23.3 Comparison of lower bounds and optimal value – EON network – T_2 – 270 connection requests

# wavelengths	$z^{\text{EDGE}}_{\leq\text{DENIED}}$	$z^{\text{ARC}}_{\leq\text{DENIED}}$	$z^{\text{ILP}_\lambda}(\text{ARC,DENIED})$	z^*_{DENIED}
10	78.0	94.0	94	94
12	51.6	75.5	76	76
14	35.0	57.5	58	58
16	22.0	45.0	45	45
18	13.0	33.0	33	33
20	9.0	21.0	21	21

# wavelengths	$z^{\text{EDGE}}_{\leq\text{LOAD}}$	$z^{\text{ARC}}_{\leq\text{LOAD}}$	$z^{\text{ILP}_\lambda}(\text{ARC,LOAD})$	z^*_{LOAD}
28	613.5	651.0	651	651
30	613.5	651.0	651	651
32	613.5	651.0	651	651

described in Krishnaswamy and Sivarajan (2001b). We also consider a subnet of the NSF network, denoted by NSF3 below, where we have removed 6 links, making the network almost a tree for some experiment purposes. The EON network is described in, e.g., O'Mahony et al. (1995); it is a network with 20 nodes and 39 optical links, with a maximum of 7 links per node. The Brazil network contains 27 nodes and 70 optical links with a maximum of 10 links per node. The last network corresponds to a subnet (30 nodes, 41 links) of the 30N60S network available from the web page of Grover's book (Grover, 2005; 2004) and used in some experiments in Doucette et al. (2003).

We used six non-uniform traffic matrices. Two of them, matrices T^1 and T^2 come from Krishnaswamy (1998). They correspond to asymmetrical systems with 268 connections for the NSF instance, and 374 for the EON one. We modify those matrices and use $\max\{T_{sd}, T_{ds}\}$ for the number of connections between a pair of source and destination nodes (v_s, v_d). The resulting symmetrical matrices lead to 191 connections for the NSF instance and to 270 for the EON one. We randomly generated a traffic matrix T^3 for the Brazil network with a random number of connections between 0 and 4 for each pair of source and destination nodes, and two traffic matrices T_4 and T_5 for the last network with a random number of connections between 0 and 3 for T_4, 0 or 1 for T_5 for each node pair. All network and traffic instances are available from Jaumard (2005).

Table 23.4 Comparison of lower bounds and optimal value – BRAZIL network – T_3 – 954 connection requests

# wavelengths	$z_{\text{DENIED}}^{\text{EDGE}}$	$z_{\text{DENIED}}^{\text{ARC}}$	$z^{\text{ILP}_\lambda}(\text{ARC,DENIED})$	z_{DENIED}^{*}
10	520.2	530.8	531	531
12	460.3	477.8	478	478
14	< 414.0	433.1	434	434
16	< 1 day	392.5	393	393
18	< 1 day	354.0	354	354

# wavelengths	$z_{\text{LOAD}}^{\text{EDGE}}$	$z_{\text{LOAD}}^{\text{ARC}}$	$z^{\text{ILP}_\lambda}(\text{ARC,LOAD})$	z_{LOAD}^{*}
80	> 1 day	2653.0	2653	2653
82	> 1 day	2649.0	2649	2649
84	> 1 day	2645.0	2645	2645

The comparison of the various lower bounds proposed in the previous sections is summarized in Tables 23.1 to 23.6 for two objectives, the minimization of the blocking rate and the minimization of the network load using the ARC network formulation with an aggregation with respect to the sources, i.e., $\text{RWA}_S(\text{ARC},\text{DENIED})$ and $\text{RWA}_S(\text{ARC},\text{LOAD})$. Note that instead of expressing the results with z_{BLOCKING}, we did it with z_{DENIED} the number of denied connections. As for the RWA problem with asymmetrical traffic matrices, we observe that, except for some rare cases (e.g., NSF network with 12 wavelengths), the lower bounds deduced from the optimal solutions of the continuous relaxations are equal to the optimal integer values. However, most of the time, the optimal vectors of the continuous relaxations are non integer.

However, we know that the gap can be high, as least for some particular network topology. For instance, consider the star network with four nodes v_1, v_2, v_3 and v_4 as illustrated in Figure 23.3. Assume that $W = 2M$ wavelengths are available on each fiber, and the symmetrical traffic matrix is defined by M requested connections between the pair of nodes (v_1, v_2), (v_1, v_3) and (v_2, v_3). Clearly, $z_{\text{DENIED}}^{*} = M$ while $z_{\text{DENIED}}^{\text{ARC}} = 0$. This leads to a gap of M between the optimal value of the continuous relaxation and the optimal integer value, illustrating the fact that the gap can be quite large in some particular cases. Note that if we add an edge between v_2 and v_3 then the gap is then equal to 0. This observation suggests that if the graph density is small, the gap may be larger between the optimal value of continuous relaxation value and the optimal integer value of the RWA problem, as least with the objective of maximizing the number

Table 23.5 Comparison of lower bounds and optimal value – 30N41S network – T4 – 899 connection requests

# wavelengths	$z_{\text{DENIED}}^{\text{EDGE}}$	$z_{\text{DENIED}}^{\text{ARC}}$	$z^{\text{ILP}_\lambda}(\text{ARC,DENIED})$	z_{DENIED}^*
10	667.5	672.0	673	673
12	637.5	647.0	647	647
14	< 621.0	622.8	623	623
16	< 587.0	600.9	601	601
18	< 577.0	580.8	581	581
# wavelengths	$z_{\text{LOAD}}^{\text{EDGE}}$	$z_{\text{LOAD}}^{\text{ARC}}$	$z^{\text{ILP}_\lambda}(\text{ARC,LOAD})$	z_{LOAD}^*
230	> 1 day	3643.0	3643	> 1 day
232	> 1 day	3639.0	3639	> 1 day
234	> 1 day	3637.0	3637	> 1 day

of accepted connections, i.e., with the blocking rate objective. This observation motivated the experiments of the NSF3 network. Although we were not able to point out some gap values, we notice that the RWA problem was much more difficult to solve in practice than on network with a larger density. Indeed, for all considered instances, we were unable to reach the optimal ILP solution, or at least to prove that the best integer solution that was generated was indeed optimal.

About the comparisons between the optimal values of the continuous relaxations that are not proved to be equal, we observe that the EDGE formulation usually leads to weaker bounds than the ARC formulation, and that it requires more computing time for its optimal solution. All programs were stopped after one day of computing, and we indicate in the Tables 23.1 to 23.6 the best value we were able to reach within that computing limit. Note that, for some instances, we were unable to get even a suboptimal value.

For some instances, we were unable to obtain the optimal ILP solution. It is mainly due to the large number of symmetrical solutions, i.e., solutions that can be deduced from each other with a permutation on the wavelengths, for the mathematical formulations considered in that paper. In a subsequent paper (Jaumard et al., 2005), we are studying decomposition methods in order to overcome that difficulty using column generation formulations that allow the elimination of symmetrical solutions.

Table 23.6 Comparison of lower bounds and optimal value – 30N41S network – T5 – 107 connection requests

# wavelengths	z^{EDGE}_{DENIED}	z^{ARC}_{DENIED}	$z^{ILP_\lambda(ARC,DENIED)}$	z^*_{DENIED}
10	8.0	39.0	39	39
12	0.0	35.0	35	35
14	0.0	31.0	31	31
16	0.0	27.0	27	27

# wavelengths	z^{EDGE}_{LOAD}	z^{ARC}_{LOAD}	$z^{ILP_\lambda(ARC,LOAD)}$	z^*_{LOAD}
30	214.0	461.0	461	461
32	214.0	459.0	459	459

Bibliography

R. K. Ahuja, T. L. Magnanti, and J. B. Orlin. *Network flows: theory, algorithms, and applications*. Prentice Hall, 1993.

D. Banerjee and B. Mukherjee. Wavelength-routed optical networks: linear formulation, resource budgeting tradeoffs, and a reconfiguration study. *IEEE/ACM Transactions on Networking*, 8(5):598–607, 2000.

C. Chen and S. Banerjee. A new model for optimal routing and wavelength assignment in wavelength division multiplexed optical networks. In *INFOCOM'96, Proceedings IEEE*, volume 1, pages 164–171, 1996.

Cisco Systems. Comparing Metro WDM systems: Unidirectional vs. bidirectional implementations, 2001.

D. Coudert and H. Rivano. Lightpath assignment for multifibers WDM networks with wavelength translators. In *IEEE Globecom*, pages 2686–2690, Taiwan, November 2002. OPNT-01-5.

CPLEX. *Using the CPLEX^{TM} Callable Library (Version 8.1)*. CPLEX Optimization Inc., 2003.

J. Doucette, M. Clouqueur, and W.D. Grover. On the availability and capacity requirements of shared backup path-protected mesh networks. *Optical Networks Magazine*, pages 29–44, November/December 2003.

R. Dutta and G.N. Rouskas. A survey of virtual topology design algorithms for wavelength routed optical networks. *Optical Networks Magazine*, 1(1):73–89, January 2000.

M.R. Garey and D.S. Johnson. *Computers and Intractability, A Guide to the Theory of NP-completeness*. W.H. Freeman, San Francisco, CA, 1979.

W.D. Grover. *Mesh-Based Survivable Networks*. Prentice Hall, 2004.

W.D. Grover. www.ee.ualberta.ca/~grover/, 2005.

B. Jaumard. Network and traffic data sets for optical network optimization. www.iro.umontreal.ca/~jaumard, 2005.

B. Jaumard, C. Meyer, and B. Thiongane. Comparison of ILP formulations for the RWA problem. Submitted for Publication, 2004a.

B. Jaumard, C. Meyer, and B. Thiongane. On column generation formulations for the RWA problem. In *INOC - International Network Optimization Conference*, pages B1.52–B1.59, Lisbon, Portugal, March 20-23 2005.

B. Jaumard, C. Meyer, B. Thiongane, and X. Yu. ILP formulations and optimal solutions for the RWA problem. In *IEEE GLOBECOM*, volume 3, pages 1918–1924, Nov.-Dec. 2004b.

R.M. Krishnaswamy. *Algorithms for Routing, Wavelength Assignment and Topology Design in Optical Networks*. PhD thesis, Dept. of Electrical Commun. Eng., Indian Institute of Science, Bangalore, India, 1998.

R.M. Krishnaswamy and K.N. Sivarajan. Algorithms for routing and wavelength assignment based on solutions of LP-relaxation. *IEEE Communications Letters*, 5(10):435–437, October 2001a.

R.M. Krishnaswamy and K.N. Sivarajan. Design of logical topologies: A linear formulation for wavelength routed optical networks with no wavelength changers. *IEEE/ACM Transactions on Networking*, 9(2):184–198, April 2001b.

M.S. Kumar and P.S. Kumar. Static lightpath establishment in WDM networks - new ILP formulations and heuristic algorithms. *Computer Communications*, 25:109–114, 2002.

K. Lee, K.C. Kang, T. Lee, and S. Park. An optimization approach to routing and wavelength assignment in WDM all-optical mesh networks without wavelength conversion. *ETRI Journal*, 24(2):131–141, April 2002.

E. Leonardi, M. Mellia, and M.A. Marsan. Algorithms for the logical topology design in WDM all-optical networks. *Optical Networks Magazine*, 1:35–46, January 2000.

L. Margara and J. Simon. Wavelength assignment problem on all-optical networks with *k* fibres per link. In *Automata, Languages and Programming. 27th International Colloquium, ICALP 2000*, volume 1853 of *Lectures Notes in Computer Science*, pages 768–779, 2000.

B. Mukherjee. WDM-based local lightwave networks. I. single-hop systems. *Networks*, 6(3):12–27, May 1992a.

B. Mukherjee. WDM-based local lightwave networks. II. multi-hop systems. *Networks*, 6(4):20–32, July 1992b.

T.F. Noronha and C.C. Ribeiro. Routing and wavelength assignment by partition coloring. *European Journal of Operational Research*, 2005. to appear.

M.J. O'Mahony, D. Simeonidu, A. Yu, and J. Zhou. The design of the european optical network. *Journal of Ligthwave Technology*, 13(5):817–828, 1995.

R. Ramaswami and K.N. Sivarajan. Routing and wavelength assignment in all-optical networks. *IEEE/ACM Transactions on Networking*, 5(3):489–501, October 1995.

R. Ramaswami and K.N. Sivarajan. *Optical Networks - A Practical Perspective*. Morgan Kaufmann, 2nd edition edition, 2002.

N. Wauters and P. Demeester. Design of the optical path layer in multiwavelength cross-connected networks. *IEEE Journal on Selected Areas in Communications*, 14 (5):881–892, June 1996.

H. Zang, J. P. Jue, and B. Mukherjee. A review of routing and wavelength assignment approaches for wavelength-routed optical WDM networks. *Optical Networks Magazine*, pages 47–60, January 2000.

J. Zheng and H. Mouftah. *Optical WDM Networks*. Wiley Interscience, 2004.

24 ROUTE OPTIMIZATION IN IP NETWORKS

Jennifer Rexford[1]

[1] Department of Computer Science
Princeton University
Princeton, NJ 08544 USA
jrex@cs.princeton.edu

Abstract: The performance and reliability of the Internet depend, in large part, on the operation of the underlying routing protocols. Today's IP routing protocols compute paths based on the network topology and configuration parameters, without regard to the current traffic load on the routers and links. The responsibility for adapting the paths to the prevailing traffic falls to the network operators and management systems. This chapter discusses the modeling and computational challenges of optimizing the tunable parameters, starting with conventional intradomain routing protocols that compute shortest paths as the sum of configurable link weights. Then, we consider the problem of optimizing the interdomain routing policies that control the flow of traffic from one network to another. Optimization based on local search has proven quite effective in grappling with the complexity of the routing protocols and the diversity of the performance objectives, and tools based on local search are in wide use in today's large IP networks.

Keywords: Internet, routing, traffic engineering, local search.

24.1 INTRODUCTION

Over the past fifteen years, the Internet has become a critical part of the world's communications infrastructure. The Internet consists of tens of thousands of domains or *Autonomous Systems* (ASes)—portions of the infrastructure that are each administered by an single institution such as a university, corporation, or Internet Service Provider (ISP). A message sent by one computer typically traverses multiple ASes before reaching its destination, making communication performance depend on the flow of traffic within and between the ASes along the path. In this chapter, we consider the role of optimization in controlling the routing of traffic through the Internet. We explain how the complexity of both the performance objectives and the network routing protocols lead to optimization problems that are not analytically tractable, forcing the use of efficient search techniques to explore a large space of tunable parameters.

24.1.1 Internet architecture

Messages sent by one computer (say, a Web client) to another computer (say, a Web server) are divided into individual datagrams or *packets* that flow through the network independently. The decision to base the Internet Protocol (IP) on packet switching has its roots in the early days of the ARPAnet in the late 1960s (Leiner et al., 1997; Clark, 1988). In contrast to the traditional *circuit switching* approach of the telephone network, packet switching acknowledges that data communication is often bursty, with users and applications alternating between periods of high network activity and relative silence. Packet switching has the allure of allowing the links in the network to multiplex traffic across multiple pairs of senders and receivers. While one sender-receiver pair is inactive, another can capitalize on the unclaimed bandwidth by exchanging traffic at a higher rate. However, this flexibility comes at a cost. If too many hosts exchange packets at the same time, the aggregate traffic may overwhelm the link bandwidth, leading to delays in transmitting the data and, ultimately, to lost packets when the buffer feeding the link overflows.

The Internet, then, has a relatively simple *best effort* service model with no guarantee that a data packet reaches its ultimate destination in a timely fashion. Although many applications (such as electronic mail) can tolerate *delay* in receiving the data, *missing* information is often unacceptable. Rather than building reliable, in-order data delivery into IP, the end hosts bear the responsibility of retransmitting lost packets and reconstructing the ordered stream of data at the receiver. These functions are performed by the Transmission Control Protocol (TCP), implemented in the operating system on the end-host computers (Stevens, 1994). In addition, the TCP sender determines the rate of data transmission by monitoring the success (or failure) of sending packets to the receiver. When packets are lost, the sending host decreases the transmission rate to help alleviate the apparent congestion; when packets are successfully delivered, the sending host optimistically increases the sending rate to capitalize on the available bandwidth along the path to the receiver. This decentralized *congestion control* scheme ensures a form of fair sharing of the bandwidth of the links amongst the competing pairs of senders and receivers.

However, transport protocols like TCP do *not* ensure that the network operates *efficiently*. For example, one link may be heavily congested while other links in the network remain lightly loaded. Or, a voice-over-IP (VoIP) call may traverse a long path with high propagation delay when a low-latency path is available. The responsibility of selecting the path a packet follows through the network falls to the routing protocols implemented by the individual routers in the network. Rather than using hard-wired tables to forward the packets, the routers exchange control messages with each other to compute the paths through the network in a distributed fashion. The distributed approach allows a collection of routers to adapt automatically to changes in the network topology. This makes IP networks robust in the presence of link and router failures, and easily accommodates the deployment of new equipment as the network grows. However, the routing protocols deployed in most IP networks do *not* incorpo-

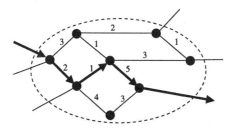

Figure 24.1 Shortest path routing based on integer link weights

rate information about network load and performance into the selection of the paths. Left to their own devices, the routers continue to forward packets over heavily-loaded links, leading to packet loss that forces the end hosts to decrease their sending rates in response.

Early forays into delay-based routing protocols led to concerns about the stability of adaptive routing in packet-switched networks. Under load-sensitive routing, the forwarding of data packets would take place on the same small timescale as the delivery of information about the prevailing load on the individual links. Initial experiences in the ARPAnet suggested that acting on out-of-date information about link load causes routers to direct traffic to seemingly underutilized links, leading to congestion that ultimately forces the routers to switch to other paths. Despite some proposals that try to prevent these kinds of oscillation (Khanna and Zinky, 1989), the conventional IP routing protocols adapt only to changes in the network topology and router configuration. While research and standards work continued (and still continues) on more flexible routing protocols (Chen and Nahrstedt, 1998; Crawley et al., 1998), practitioners needed an effective way to influence the flow of traffic through their networks. Fortunately, the IP routing protocols have various tunable parameters that the network operators can adjust to change the paths the routers use to forward data packets. For example, the operator can tune the integer link weights the routers use to compute shortest paths for carrying the data traffic, as shown in Figure 24.1. However, identifying a good setting of these parameters in a large network is challenging for a human operator, setting the stage for the application of optimization techniques.

24.1.2 Optimization of IP routing

As an alternative to having the distributed routing protocols adapt to the prevailing traffic, the network operators or automated management systems can modify the configuration of the "static" parameters that drive the operation of the routing protocols (Fortz et al., 2002). This leads to the "control loop" shown in Figure 24.2, where measurements of the operational network serve as inputs to a "what if" model that captures the effects of changes to the tunable parameters. The measurements capture the offered load on the network as well as the current topology, whereas the model predicts the outcome of the distributed path-selection process running on the routers for a given setting of the configurable parameters. This allows the use of optimization techniques for identifying parameter settings that satisfy the network's performance goals. Once

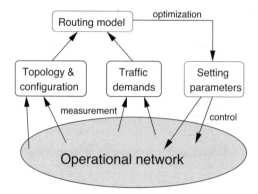

Figure 24.2 Key components of the route optimization framework

good values have been identified, the human operator or automated system can contact the routers to change the configuration of the tunable parameters. Then, the routers collectively act on the new parameter values to compute new paths for forwarding data packets through the network, as predicted by the model.

The approach in Figure 24.2 has several advantages over extending the routing protocols to adapt automatically to the traffic. First, a network can continue to use the conventional IP routing protocols, rather than deploying or enabling new (potentially complex) protocols that adapt automatically to the prevailing traffic. Second, the network avoids the protocol oscillations that can arise when routers react to locally-constructed (potentially out-of-date) views of the prevailing traffic. Instead, routing changes are carefully planned based on an accurate, network-wide view of the traffic and topology. Third, the routers do not need to maintain statistics about the load on the individual links, disseminate this information throughout the network, or recompute paths as this information changes; avoiding these operations reduces the bandwidth and computational demands on the routers. Fourth, and perhaps most importantly, the selection of routing parameters can depend on a wide variety of performance and reliability constraints that may be difficult to incorporate in a distributed routing protocol. These constraints, and the optimization techniques for satisfying them, can continue to evolve over time.

Despite these compelling advantages, the resulting optimization problem is difficult to solve because of the complexity of the underlying routing protocols and the diversity of the network objectives. The conventional IP routing protocols were not designed with optimization in mind, and the optimization of the tunable parameters is computationally difficult even for the simplest of metrics (e.g., such as minimizing the maximum link utilization) (Fortz and Thorup, 2000). The parameter space is quite large, and small changes in the value of a parameter can lead to significant differences in the flow of traffic through the network. In the core the Internet, routing changes in a single AS may affect how traffic leaves that network and enters the next AS along the path to the destination (Feldmann et al., 2000). This introduces additional complexity in modeling the effects of changes to the configurable parameters. Finally, the optimization problem needs to consider numerous objectives, such as link load, propagation delay, and hop count. More complex metrics come in to play as well (Fortz and Thorup, 2002), such as the desire to limit the total traffic directed to each neighboring AS or the frequency of configuration changes as the traffic fluctuates over time. The optimization process may also need to account for the effects of equipment failures and planned maintenance on the suitability of the parameter settings; in addition, multiple routers or links may fail or recover together if they have shared risks, such as a common power supply or optical amplifier.

These practical challenges make it difficult (and frequently impossible) to derive analytical solutions to the optimization problem. Instead, solving the optimization problem depends on having efficient models of the routing protocols and effective ways to explore large parts of the parameter space. To reduce the computational overhead, the "what if" models should not *simulate* the detailed operation of the routing protocols over time as the routers exchange hypothetical control messages. Instead, the models should determine the *outcome* of the routing protocols—the paths the routers would ultimately select for carrying the traffic. In addition, the optimization process should not require recomputing these outcomes from scratch for each candidate setting of the tunable parameters; instead, *incremental algorithms* should be used to minimize the computational overhead for exploring the search space. Finally, experimental results on real input data should be used to drive heuristics that *limit the search space*, in terms of the number of parameters and the range of their values.

In the next section, we explore these principles in the context of optimizing routing inside a single AS, based on conventional shortest-path routing protocols with tunable edge weights. This problem has been studied widely over the past several years and the solutions have been incorporated in management tools used in many operational IP networks (Feldmann et al., 2000; Cariden MATE Framework, 2005; OpNet SP Guru, 2005). Next, we describe how to extend the routing model to accurately capture the operation of large backbone networks that have multiple egress points for directing traffic toward each destination. Then, we explore the role of optimization in tuning the interdomain routing policies that control how traffic flows between ASes. In the conclusion, we briefly discuss recent research work on designing distributed routing protocols that are easier to tune as well as centralized approaches for directly computing routes on behalf of the operational routers.

24.2 SHORTEST-PATH ROUTING IN A SINGLE NETWORK

Many Autonomous Systems (ASes) run routing protocols such as Open Shortest Path First (OSPF) (Moy, 1998) or Intermediate System-Intermediate System (IS-IS) (Callon, 1990) that compute shortest paths based on configurable link weights. In this section, we formulate an optimization problem for selecting the link weights based on the network topology, traffic matrix, and an objective function. On the surface, tuning a single integer weight on each link may not seem flexible enough to satisfy diverse performance objectives. Yet, experimental results suggest that it is possible to achieve near-optimal routing for real topologies and traffic matrices. Although the optimization problem is NP-hard, local search techniques are surprisingly effective in finding

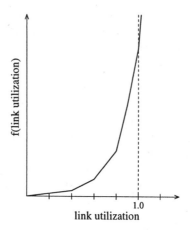

Figure 24.3 Convex function $f()$ of link utilization u_ℓ/c_ℓ

good solutions and easily support diverse performance objectives and operational constraints.

24.2.1 Formulating the optimization problem

In conventional intradomain routing protocols, each router is configured with a static integer weight on each of its outgoing links, as shown in Figure 24.1. The routers flood the link weights throughout the network and compute shortest paths as the sum of the weights. Each router uses this information to construct a table that drives the forwarding of each IP packet to the next hop in its path to the destination. These protocols view the network inside an AS as a graph $G(R, L)$ where each router is a node $r \in R$ and each directed edge is a link $\ell \in L$ between two routers. Each unidirectional link has a fixed capacity c_ℓ, as well as a configurable weight w_ℓ. The outcome of the shortest-path computation can be represented as the proportion $P_{i,j,\ell}$ of the traffic from router i to router j that traverses the link ℓ. In the simplest case, the network has a single shortest path from i to j, resulting in $P_{i,j,\ell} = 1$ for all links ℓ along the path, and $P_{i,j,\ell} = 0$ for the remaining links. More generally, the network may split traffic along multiple shortest paths, resulting in $P_{i,j,\ell} \in [0, 1]$. For example, an OSPF or IS-IS router typically splits traffic evenly along one or more outgoing links along shortest paths to the destination.

The selection of the link weights $\{w_\ell\}$ should depend on the offered traffic, as captured by a matrix $M_{i,j}$ that represents the rate of traffic entering at router i that is destined to router j. The traffic matrix can be computed based on traffic measurements (Feldmann et al., 2001; Medina et al., 2002; Zhang et al., 2003) or may represent explicit subscriptions or reservations from users. Given the traffic matrix $M_{i,j}$ and link weights w_ℓ, the volume of traffic on each link ℓ is $u_\ell = \sum_{i,j} M_{i,j} P_{i,j,\ell}$—the proportion of traffic that traverses link ℓ summed over all pairs of routers. An *objective function* can quantify the "goodness" of a particular setting of the link weights. For example, minimizing the simple objective function $\max_\ell(u_\ell/c_\ell)$ would result in a set of link weights that minimizes the worst-case utilization over all links in the network. Although minimizing the maximum utilization is a natural and intuitive objective, this metric is sensitive to individual bottleneck links that may be difficult to avoid under any routing solution. Instead, the optimization could consider a network-wide objective of minimizing $\sum_\ell f(u_\ell/c_\ell)$ for a convex function $f()$ that penalizes solutions that have heavily-loaded links, as shown in Figure 24.3.

Even for these simple objective functions, the problem of optimizing the link weights $\{w_\ell\}$ is NP-hard (Lorenz et al., 2001; Fortz and Thorup, 2004; Juttner et al., 2000). The computational challenge arises because of the limited control over the splitting of traffic over multiple shortest path routes. If a router could split traffic over multiple outgoing links in *arbitrary* proportions, the optimization problem would be much simpler—allowing the use of multicommodity-flow solutions (Wang et al., 2001). However, the routers themselves do not compute paths based on any information about the traffic matrix or the link loads, leaving them no basis for determining the appropriate "splits" of traffic over the shortest paths. In addition to the NP-hard optimization problem, the limitations on path splitting make it difficult for an optimal setting of the link weights (if known) to approximate the efficiency of a multicommodity-flow solution with the same topology and traffic matrix. In fact, the performance gap, as measured by the objective function, between an optimal setting of the link weights and a multicommodity-flow solution can be arbitrarily high in the worst case (Fortz and Thorup, 2000; Lorenz et al., 2001). Fortunately, real network topologies and traffic matrices have much smaller performance gaps between the two kinds of routing (Fortz and Thorup, 2000); in practice, the performance differences are nearly indistinguishable.

24.2.2 Using search algorithms

The computational complexity of the optimization problem and the desire to support diverse objective functions limit the options for how to solve the optimization problem. Local-search techniques (Aarts and Lenstra, 1997) are an attractive way to find a good setting of the link weights subject to an arbitrary objective function. Each iteration of the search evaluates a candidate solution (i.e., an assignment of the weights w_ℓ) and sets the stage for exploring a neighborhood of solutions by changing one, or a few, link weights. Although the search could focus exclusively on neighbors that reduce the value of the objective function, this approach runs the risk of falling in to a local minimum that is much worse than the optimal solution. Instead, the search should consider steps that move away from better solutions, while taking care to avoid repeated evaluation of the same setting of the link weights. For example, tabu search and simulated annealing can generate non-improving moves, and hashing the solutions as they are evaluated can prevent the algorithm for considering the same scenario twice (Fortz and Thorup, 2004).

Each iteration of the local search requires computing the all-pairs shortest paths— the shortest path(s) between each pair of routers—for a single setting of the link weights. In the worst case, this requires running Dijkstra's shortest-path algorithm on the weighted graph to determine the routes $P_{i,j,\ell}$, computing the total traffic $\sum_{i,j} M_{i,j} P_{i,j,\ell}$ on each link ℓ, and evaluating the objective function. For a large network with $|R|$ nodes and $|L|$ edges, the computational overhead can be quite high. Fortunately, the local-search algorithm does not need to compute these quantities from scratch in each iteration. For example, if each iteration of the local search considers a change to a single link weight, incremental graph algorithms can be used to quickly determine the *changes* in the paths, link loads, and objective function. The search can often start with a reasonable setting of the link weights taken from the existing network; often, changing one or two link weights is sufficient to achieve near-optimal performance. In addition, selecting link weights from a small number of distinct values is an effective way to substantially reduce the search space without much reduction in the quality of the solution.

Studies of various search techniques have shown that it is possible to find near-optimal settings of the link weights on graphs with hundreds of nodes in a reasonable amount of time (Fortz and Thorup, 2000; Ericsson et al., 2002; Buriol et al., 2002). In addition, a good setting of the link weights performs almost as well as the theoretical upper bound from solving the multi-commodity flow problem, which assumes greater flexibility in routing the traffic than the shortest-path routing algorithms permit. In practice, the performance gap between a good weight setting and the multi-commodity flow solution is especially small for real network topologies, which typically have a relatively low diameter and a narrow range of link capacities. Although performing the local search is computationally inexpensive for small graphs, identifying good weight settings in larger graphs may argue for an alternate approach. A promising alternative is a two-phase algorithm where the first phase allocates each traffic matrix element to a single path and the second phase tries to find a setting of the link weights that can achieve these paths (Pioro et al., 2000).

24.2.3 Incorporating practical constraints

Using local search to optimize the link weights provides enormous flexibility in addressing the many practical constraints on setting the link weights in operational networks. In particular, minimizing the number and frequency of weight changes is very important to avoid disrupting the network. When a link weight changes, the new information is flooded throughout the network and the routers recompute their shortest paths. During this convergence period, the routers in the network do not have a consistent view of the shortest-path routes for some destinations; in the meantime, packets may be lost or experience long delays. Although recent work has brought the convergence times down considerably (Alaettinoglu et al., 2000), delays of several seconds are not uncommon. As such, operators do not change the link weights unless the current routing configuration is causing performance problems.

Robustness to variations in the traffic matrix: The setting of the link weights should not be sensitive to small variations in the traffic matrix M. Existing techniques for measuring or inferring the traffic matrix (Feldmann et al., 2001; Medina et al., 2002; Zhang et al., 2003) have limitations on their accuracy, and the traffic matrix itself fluctuates over time. The simplest way to allow for uncertainty would be to increase the elements $M_{i,j}$ of the traffic matrix by some target amount. More generally, the local search algorithm could evaluate a candidate setting of the link weights over a *set* of traffic matrices, favoring a solution that performs well for each traffic matrix over one that performs best for one at the expense of the others. This approach is extremely effective for selecting link weights that accommodate diurnal cycles in the traffic, allowing network operators to have a single assignment of link weights for both daytime and nighttime traffic (Fortz and Thorup, 2002). In fact, recent work (Applegate and Cohen, 2003) shows that it is possible to find routing solutions that perform well *independently* of the traffic matrix, or across an extremely wide range of traffic demands.

Surviving equipment failures and planned maintenance: Ideally, the assignment of the link weights would be robust to common network disruptions, such as single-link failures. When a link fails, the information is flooded throughout the network and the routers compute new shortest paths over the remaining edges. Preventing congestion after a failure could require the local search algorithm to evaluate a candidate setting of the link weights under each failure scenario (Nucci et al., 2003). However, evaluating the weights under all failures is computationally prohibitive in large network settings. Fortunately, it is possible to identify and evaluate a much smaller set of critical scenarios (Fortz and Thorup, 2003). In practice, link weights that perform well on the original network topology continue to perform well after most single-link failures. However, for a few failure scenarios, the link weights may need to change to avoid congestion. Fortunately, changing one or two link weights is typically sufficient to alleviate the congestion. As a proactive measure, the necessary weight changes could be computed in advance of any link failure and stored by the network management system. These same precomputed weight changes are extremely useful when network operators must intentionally "fail" a link in order to fix or upgrade the equipment; in this case, the weight changes can be made in advance, before disabling the equipment for maintenance.

Supporting diverse constraints on path preferences: When there are multiple shortest paths between two routers, the routers along the paths split the traffic over multiple outgoing links—a practice known as equal-cost multi-path routing. Rather than alternating between these links at the packet level, routers typically attempt to forward packets for the same source-destination pair along a single path; this reduces the likelihood that packets from the same TCP connection arrive out-of-order at the receiver. Load-balancing is typically achieved by performing a hash function on the source and destination IP addresses of each packet; the value of the hash function determines which outgoing link should carry the packet. As a result, ties introduce uncertainty in the exact distribution of the traffic over the links; with two outgoing links, the traffic does not necessarily divide exactly in half. The local search can easily account for the effects of uncertainty by penalizing solutions that have ties (e.g., by setting $P_{i,j,\ell}$ to 0.6 rather than 0.5, implicitly assuming that *both* links must carry 60% of the original traffic load). On the other hand, having multiple shortest paths is advantageous in some settings. In particular, if one of the two links fails, the other shortest path still remains—allowing the network to converge more quickly, causing fewer packet losses and delays (Sridharan et al., 2003). The local search algorithm can easily bias toward solutions with a larger number of ties by including the number of ties in the objective function.

More generally, the use of local search allows the objective function itself to change over time as different metrics reign in importance. For example, for best-effort Internet traffic, minimizing the maximum link utilization (or the sum of $f()$ over all links) is a very effective way to maximize TCP throughput. Yet, the situation is completely different when interactive applications such as Voice-over-IP (VoIP) and video games enter the picture. For these applications, keeping propagation delay low (e.g., below 100 msec) is crucial, and disruptions during routing protocol convergence must be avoided by reducing the frequency of routing changes and by picking solutions that have multiple shortest paths. The underlying machinery of local search is extremely flexible for incorporating complex metrics and changing their importance in evaluating candidate settings to the link weights.

24.3 EARLY-EXIT ROUTING IN LARGE BACKBONE NETWORKS

The vast majority of Internet traffic must traverse multiple Autonomous Systems (ASes) on the way to the destination. Whereas the previous section focused on a single network in isolation, this section considers the additional challenges faced by transit ASes (such as large Internet Service Providers) that connect to other networks in the Internet. Optimizing the link weights in these networks requires a more complex model that captures the effects of *early-exit* routing, as well as optimization metrics that favor solutions that avoid shifting traffic from one exit point to another.

24.3.1 Destinations with multiple egress points

Although many networks are largely self-contained, transit providers connect to numerous other networks at multiple geographic locations. For example, in Figure 24.4, AS A allows a customer network (such as a company or university campus) connected to router i to reach destinations (such as Web servers) in ASes B and C. Large transit ASes often *peer* with each other in multiple locations to improve reliability and performance. Increasingly, customers often connect to their providers at multiple locations for improved reliability. Each connection consists of two routers—one in each domain—that exchange reachability information using the Border Gateway Protocol (BGP). In essence, BGP is the glue that holds the disparate parts of the Internet together. For example, a router in AS B would advertise a path for destination d_B to the adjacent router in AS A. Since the two networks peer in two locations, the routers in AS A have two possible *egress points* (the links to AS B at routers j and k) for directing traffic toward destination d_B.

At a high level, the two interdomain paths are largely equivalent—for example, both paths traverse the same number of ASes (and, in fact, the same AS) en route to the destination. When a router has two "equally good" interdomain routes to a destination, the BGP routing decision depends on the cost of the *intra*domain path to each egress router. For example, router i would direct traffic destined to d_B via router j, since the intradomain path of cost 10 to router j is shorter than the path of cost 20 to router k. The common practice of *early-exit* or *hot-potato* routing tries to minimize the use of internal network resources by shuttling the traffic to a neighboring AS as early as possible. In fact, AS B would perform early-exit routing in the reverse

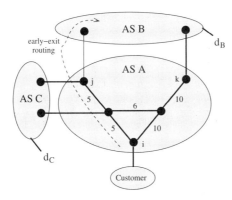

Figure 24.4 AS A with connections to AS B and AS C

direction, for the traffic from d_B to the customer in AS A's network, leading to the asymmetric routing where traffic from the customer leaves AS A via router j and traffic to the customer enters via router k. Given the many destinations spread throughout the Internet, a transit AS would have to choose between multiple egress points for a large portion of the destinations, making it crucial to capture the effects of early-exit routing on the flow of traffic.

24.3.2 Modeling the effects of early-exit routing

Optimizing the intradomain link weights requires several extensions to the model presented earlier in Section 24.2 (Feldmann et al., 2000; Teixeira et al., 2004b):

- **Egress links connecting to neighboring domains:** The graph $G(R,L)$ needs to include the external links that connect to routers in neighboring ASes. The objective function should consider the traffic load u_ℓ on these links when evaluating candidate settings of the weights on the internal links.

- **Set of egress links for reaching each destination d:** The graph can include a logical node to represent the set of egress links E_d available for reaching each destination d. For example, the logical node representing d_B would have logical links to the two border routers in AS B. Since these logical links are not real, the objective function does not need to consider the traffic load imparted on these links.

- **Offered traffic from ingress points to destinations:** Since the setting of the link weights affects how a router selects an egress point, the traffic matrix is not an appropriate representation of the offered traffic. Instead, the traffic demand $v_{i,d}$ of load from ingress point i to destination d is the correct abstraction (Feldmann et al., 2001).

- **Selection of the closest egress point:** Evaluating a candidate setting of the link weights requires comparing the intradomain path costs from ingress i to each egress point in E_d. This comparison is necessary to identify the "closest" egress point $o_{i,d}$ for ingress point i; the traffic ultimately flows over the shortest path to the closest egress point.

In the extended model, the load u_ℓ on each external link ℓ is the sum over all traffic demands $v_{i,d}$ that have a closest egress point $o_{i,d} = \ell$. For the internal links ℓ, the traffic load comes from the traffic $v_{i,d} P_{i,j,\ell}$ such that the closest egress point $o_{i,d}$ is incident to router j.

On the surface, the extended model seems much more complicated than the model in Section 24.2. For example, the routers in large backbone networks typically have routes for more than 150,000 distinct destinations (blocks of contiguous IP addresses). In addition, a network with n edge routers could conceivably have 2^n unique sets of egress routers, and even more sets of egress links. In practice, though, many destinations have the same set of egress routers, and the number of distinct egress sets is much smaller than 2^n (Feamster et al., 2003). For example, a neighboring AS typically advertises comparable BGP routing information over all of its peering links, resulting in a single egress set for of these destinations. In essence, the optimization process can treat the group of related destinations as a single destination d with one larger traffic demand $v_{i,d}$ per ingress point. This reduces the complexity of evaluating a candidate setting of the link weights. In addition, as in Section 24.2, the size of the search space can be reduced substantially by exploring incremental changes to the link weights, a small set of unique weight values, and so forth.

24.3.3 Avoiding changes in the chosen egress point

In a transit network with early-exit routing, a change in the link weights can affect where traffic leaves the network. In Figure 24.4, ingress router i initially directs traffic via egress point j with the smaller intradomain path cost of 10. However, a link failure could increase the cost to reach j from 10 to 21, making egress point k the closer egress point with a path cost of 20. Hence, when the failure occurs, router i would shift all traffic destined to d from egress point j to egress point k. In practice, small internal changes, due to a link failure or a weight change, can cause a router to shift the routes for tens of thousands of destinations at once (Teixeira et al., 2004a). This, in turn, can dramatically affect the flow of traffic in the network, and the offered load on the neighboring domain, resulting in serious congestion. To avoid these undesirable disruptions, these effects should be taken into account when designing the topology or planning changes to the network.

For example, in Figure 24.4 the router i selects the egress point j (with a path cost of 10) over egress point k (with a path cost of 20) to direct traffic toward d_B. However, if the left link from router i needs to be disabled temporarily for maintenance purposes, the path cost to j would increase to 21 ($= 10 + 6 + 5$), making k the closer egress point. This would cause i to change its routing decision and direct traffic via router k. The routing change can be avoided by changing the weight of the middle link from 6 to 4 before the maintenance activity; this ensures that the path to i has cost 19—smaller than the cost to k. The goal of avoiding unnecessary early-exit routing changes leads to a new optimization problem to identify changes to the existing link weights that would allow a link (or router) to be removed from the network with the minimal disruption to the existing traffic. More generally, network operators can use models of early-exit routing to identify particular links and routers which would cause large disruptions when they fail (Teixeira et al., 2004b). Optimizing the link weights to minimize the likelihood of such disruptions across a set of failure scenarios is useful for avoiding the performance problems that arise when unexpected early-exit routing changes occur.

24.4 INTERDOMAIN ROUTING POLICIES

The flow of traffic through a transit network depends on the BGP routes advertised by neighboring ASes, as well as the local policies configured on the individual routers. In this section, we consider how tuning the BGP policies influences the forwarding of data traffic. Next, we describe how to extend the routing model from the previous section to capture the role of routing policies. Then, we discuss approaches to exploring the very large search space of BGP policy configurations, as well as fundamental limitations on the ability of models to predict the load on the links in an AS.

24.4.1 BGP policies influencing the path-selection process

In the simplest case, the routers in an AS select BGP routes with the shortest AS path, breaking ties based on the proximity of the egress points. More generally, a router can be configured to apply a policy that assigns a *local preference* to a route, to select one route over another with a shorter AS path or closer egress point. Today's routers provide an extremely flexible "programming language" for specifying rules for assigning the local-preference attribute. For example, the policies may differentiate between routes learned from different neighboring ASes, based on the commercial relationship. In transit networks, the common practice is to assign a higher preference to BGP routes learned from customers than to routes learned from upstream providers, to ensure that data traffic traverses neighbors that are paying customers, even if the path through the provider is shorter (Huston, 1999; Gao and Rexford, 2001), as shown in Figure 24.5. Similarly, a network may assign a lower local preference to BGP routes learned over low-bandwidth links that exist only to provide backup service (Gao et al., 2001), to ensure that data traffic traverses the high-bandwidth primary links except when they have failed.

In addition to assigning preferences based on the relationship with the neighboring AS, operators configure policies to influence the load on the network links (Feamster et al., 2003; Quoitin et al., 2004; Ye and Kalyanaraman, 2003). For example, suppose an AS learns BGP routes to a destination from two upstream providers. By assigning a lower local preference to one route, the AS decides to direct traffic via the route learned from the other provider. Careful assignment of local preference over the range of destinations is helpful in balancing the load on the two links. In some cases, one provider might charge a higher price for sending traffic, or generally have poorer performance. To reduce financial cost or improve performance, the AS may tend to prefer routes through the other provider, up to the point where the link becomes too heavily loaded. More generally, an AS may connect to a single AS in multiple geographic locations. If one of the links to the neighboring AS becomes congested, the operators

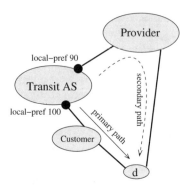

Figure 24.5 Transit AS assigning higher local preference to the route through a customer

may reconfigure the adjacent router to assign a lower local preference value to some of the BGP routes learned at this location. This ensures that the routers in the AS direct the traffic for these destinations through the other, lightly-loaded links to the same AS.

24.4.2 Modeling the effects on the flow of traffic

Capturing the effects of routing policies requires extensions to the model presented earlier in Section 24.3 (Feamster et al., 2004b). By changing how the local-preference attribute is set, tuning the BGP policies affects how a router at the periphery of the network selects a best route. This, in turn, affects the egress set for each destination address block (or *prefix*). Modeling the influence of BGP policies on the flow of traffic requires:

- **BGP-learned routes from neighbor ASes:** The input data for the computation is the set of BGP-learned routes from neighboring ASes. In practice, the BGP-learned routes can be captured by "dumping" the BGP routing table at each router, or monitoring the BGP routes as they are advertised by the neighbors.

- **Specification of routing policies:** A BGP routing policy is a sequence of clauses, where each clause specifies a set of routes (e.g., based on the destination prefix or the elements in the AS path) and assigns a local-preference value to the matching routes. For example, a policy might assign a local preference of 100 to all routes with AS "1234" as the second hop, and 90 to all others.

- **Model of BGP path selection:** Computing the egress set requires applying the policies to the BGP routes learned from neighboring demands, and then selecting the "best" of the modified paths. In particular, the model should select routes with the highest local preference and, among those, the routes with the shortest AS path length. Ultimately, each router would select the route with the "closest" egress point.

- **Offered traffic from ingress points to destination:** As in Section 24.3, estimating the flow of traffic requires overlaying the traffic demands $v_{i,d}$ on top of the paths through the AS to reach the chosen egress point for each ingress point and destination prefix.

With an accurate view of the BGP-learned routes and the traffic demands, an optimization tool can conduct a local search over possible changes to the routing policies and evaluate the influence on the flow of traffic through the network.

24.4.3 Limiting the search space

The significant flexibility in specifying BGP routing policies, coupled with the large number of destination prefixes, makes the search space extremely large—far too big to explore exhaustively. Fortunately, heuristics are helpful for limiting the overhead of the optimization process (Feamster et al., 2003):

- **Exploring incremental policy changes:** As in Section 24.2, the local search can explore incremental changes to the router configuration, rather than considering all possible policies from scratch. For example, the optimization could, on noting that one edge link is congested, focus on modifying the policy in this one location to assign a smaller local preference to the BGP routes for some destinations. This approach not only limits the search space but also reduces the overhead of evaluating the effects of the configuration change, since the optimization tool need only consider the traffic demands that *move* to new egress points.

- **Focusing only on popular destination prefixes:** A typical IP backbone network has BGP routes for more than 150,000 destination address blocks. In practice, the vast majority of these destinations contribute receive only a small fraction of the traffic (Fang and Peterson, 1999). Changing the local-preference values for a small handful of popular destination prefixes would move a significant amount of traffic from one egress point to the other, while minimizing the number of routing changes and reducing the local-search overhead. In addition, these popular destinations tend to have more stable candidate BGP routes (Rexford et al., 2002), making it likely that the new BGP policy configuration continues to be effective in the future.

- **Focusing on groups of destination address blocks:** In practice, a network has the same (or similar) BGP routing options for multiple destination prefixes. For example, a large customer may have a dozen blocks of addresses that are advertised exactly the same way, at the same locations in the network. Specifying routing policies that match on the *group* of destinations (say, by matching on common aspects of their AS paths) reduces the search space, and also makes the routing in the network less sensitive to small changes in other route attributes. In addition, groups of destinations tends to have a more stable, aggregate traffic volume, making the predictions of the traffic load in the network more accurate.

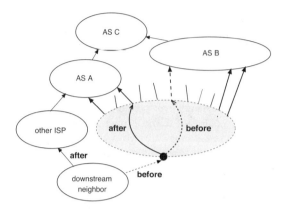

Figure 24.6 Change in neighbor's behavior upon receiving a new BGP route

A combination of these three techniques substantially reduces the overhead of exploring the search space and tends to favor robust solutions.

However, in some cases, the impact of a BGP policy change may be hard to predict because of side effects on neighboring domains (Feamster et al., 2003). In the example in Figure 24.6, ASes A and B both advertise paths to destinations in AS C. Initially, there are five "best" routes—two via AS A and three via AS B. Routers on the west coast route via AS A and routers on the east coast route via AS B. Suppose that the leftmost link to AS B is congested (as illustrated by the dashed line), and the BGP policy for this egress point is modified to assign a lower local preference to routes originating from AS C. After this change, some routers might switch from a route via the leftmost link to AS B to a route via the rightmost link to AS A. These routers would advertise the new best path to downstream neighbors. Depending on the neighbor's routing policies, the new advertisement might cause the neighbor to select a different next-hop AS (i.e., another ISP) for reaching this prefix. This could result in an unpredictable decrease in the volume of traffic entering the domain at this router. Similarly, the routing change could trigger an increase in traffic if other neighbors preferred the (A, C) route over the (B, C) route. These kinds of side effects make it difficult to predict how the change in routing policy would affect the volume of traffic on the network links.

To prevent the effects of routing changes from propagating to neighboring domains, the optimization tool should, whenever possible, achieve the load-balancing goals by adjusting routing policies for destinations for which every potential best route has the *same AS path*. In practice, an AS would have many such destinations. For example, a destination residing in AS B would be reachable through three egress points to AS B that offer the same one-hop AS path. Changing the routing policy for one (or more) of those destination prefixes would move traffic away from the dotted link to one of the other links to AS B, without changing the AS path seen by the downstream neighbor. Depending on the BGP implementation, the downstream AS might not even receive a new BGP advertisement, since the AS path has not changed. This allows the network to alleviate congestion on a link by moving traffic to other, lightly-loaded links, without changing the BGP routing information sent to neighboring domains.

24.5 CONCLUSION

IP routing protocols have tunable parameters that operators can set to control the flow of traffic through their networks. Optimization based on local-search techniques plays an important role in adapting these parameters to the prevailing network conditions. In this chapter, we considered three variants of the optimization problem, with increasing complexity:

1. When each destination connects to the network at a single location, the optimization problem consists of setting the link weights that drive how the routers direct traffic on shortest paths. The inputs to the optimization problem are the traffic matrix and the capacitated network topology.

2. When some destinations are reachable via multiple egress points, the optimization problem becomes more complicated. Instead of the traffic matrix, the offered load is represented as a set of traffic demands—the volume of traffic entering at a certain router and traveling to a particular destination. The optimization problem must also consider the set of egress points for each destination prefix.

3. In addition to selecting the link weights, the optimization problem can consider changes to the BGP policies that determine the set of egress points for each destination prefix. To determine the egress set, the optimization must consider how the configurable routing policies assign the local-preference attribute of the BGP routes learned from neighboring domains.

For each optimization problem, local-search techniques are effective in finding good solutions that support diverse performance objectives and operational constraints.

Still, the optimization problems are computationally challenging, often requiring heuristics for limiting the size of the search space. Arguably, the underlying routing protocols were not designed with optimization in mind. Just as an understanding of the routing protocols leads to interesting optimization problems, a deeper understanding of optimization techniques could lead to better routing architectures. Initial forays into redesigning IP routing with optimization in mind include:

- **Routing protocols with simpler optimization problems:** Optimizing the link weights in a shortest-path routing protocol is an NP-hard problem, as discussed in Section 24.2. The work in (Fong et al., 2001) considers a simple variant of OSPF and IS-IS routing, where the routers forward traffic on paths in *inverse proportion* to the sum of the weights, rather than sending all traffic on shortest paths. This small change leads to an optimization problem that can be solved in polynomial time for simple objective functions, as well as a protocol that has smaller reactions to small changes in the path costs.

- **Explicit negotiation between neighboring domains:** When a configuration change causes a router to direct traffic to a different egress point, the neighboring domain starts receiving traffic at a different ingress point. Rather than make configuration changes independently, neighboring ASes could coordinate their activities to find mutually beneficial ways to direct the traffic that traverses both domains (Mahajan et al., 2005; Winick et al., 2002). However, this approach depends on having an effective way to satisfy the objectives of both ASes, while ensuring that neither domain has an incentive to provide misleading information to the other.

- **Logically-centralized control over path selection:** The routing protocols in today's IP networks were designed, first and foremost, to be implemented in a distributed fashion. More recently, the increasing capabilities of computers makes it possible to select the paths for a large collection of routers in a separate platform with a network-wide view of the topology and traffic (Feamster et al., 2004a; Bonaventure et al., 2004; Farrel et al., 2005). Rather than emulating today's distributed protocols, these platforms could define new frameworks for computing paths in a logically-centralized fashion to satisfy network engineering goals directly.

Designing new IP routing architectures that are more amenable to optimization is a promising avenue for future research.

Bibliography

E.H.L. Aarts and J.K. Lenstra. *Local Search in Combinatorial Optimization*. Wiley-Interscience, 1997.

C. Alaettinoglu, V. Jacobson, and H. Yu. Towards milli-second IGP convergence. Expired Internet Draft, draft-alaettinoglu-isis-convergence-00.txt, 2000. http://www.packetdesign.com/Documents/convergence.pdf.

D. Applegate and E. Cohen. Making intra-domain routing robust to changing and uncertain traffic demands: Understanding fundamental tradeoffs. In *Proc. ACM SIGCOMM*, August 2003.

O. Bonaventure, S. Uhlig, and B. Quoitin. The Case for More Versatile BGP Route Reflectors. Internet Draft draft-bonaventure-bgp-route-reflectors-00.txt, July 2004.

L. S. Buriol, M. G. C. Resende, C. C. Ribeiro, and M. Thorup. A memetic algorithm for OSPF routing. In *Proc. INFORMS Telecom*, pages 187–188, 2002.

R. Callon. Use of OSI IS-IS for Routing in TCP/IP and Dual Environments. Request For Comments 1195, IETF, December 1990.

Cariden MATE Framework, 2005. http://www.cariden.com/products/.

S. Chen and K. Nahrstedt. An overview of quality of service routing for next-generation high-speed networks: Problems and solutions. *IEEE Network Magazine*, pages 64–79, November/December 1998.

D. D. Clark. The design philosophy of the DARPA Internet protocols. In *Proc. ACM SIGCOMM*, pages 106–114, August 1988.

E. Crawley, R. Nair, B. Rajagopalan, and H. Sandick. A framework for QoS-based routing in the Internet. Request For Comments 2386, IETF, August 1998.

M. Ericsson, M.G.C. Resende, and P.M. Pardalos. A genetic algorithm for the weight setting problem in OSPF routing. *J. Combinatorial Optimization*, 6(3):299–333, 2002.

W. Fang and L. Peterson. Inter-AS traffic patterns and their implications. In *Proc. IEEE Global Internet*, December 1999.

A. Farrel, J.-Ph. Vasseur, and J. Ash. Path computation element (PCE) architecture. Internet Draft draft-ietf-pce-architecture-00.txt, March 2005.

N. Feamster, H. Balakrishnan, J. Rexford, A. Shaikh, and J. van der Merwe. The case for separating routing from routers. In *Proc. ACM SIGCOMM Workshop on Future Directions in Network Architecture*, August 2004a.

N. Feamster, J. Borkenhagen, and J. Rexford. Guidelines for interdomain traffic engineering. *ACM SIGCOMM Computer Communication Review*, 33(5), October 2003.

N. Feamster, J. Winick, and J. Rexford. A model of BGP routing for network engineering. In *Proc. ACM SIGMETRICS*, June 2004b.

A. Feldmann, A. Greenberg, C. Lund, N. Reingold, and J. Rexford. NetScope: Traffic engineering for IP networks. *IEEE Network Magazine*, pages 11–19, March 2000.

A. Feldmann, A. Greenberg, C. Lund, N. Reingold, J. Rexford, and F. True. Deriving traffic demands for operational IP networks: Methodology and experience. *IEEE/ACM Trans. Networking*, 9(3), June 2001.

J. Fong, A. Gilbert, S. Kannan, and M. Strauss. Better alternatives to OSPF routing. In *Proc. Workshop on Approximation and Randomized Algorithms in Communication Networks*, 2001.

B. Fortz, J. Rexford, and M. Thorup. Traffic engineering with traditional IP routing protocols. *IEEE Communication Magazine*, October 2002.

B. Fortz and M. Thorup. Internet traffic engineering by optimizing OSPF weights. In *Proc. IEEE INFOCOM*, March 2000.

B. Fortz and M. Thorup. Optimizing OSPF/IS-IS weights in a changing world. *IEEE Journal on Selected Areas in Communications*, 20(4):756–767, May 2002.

B. Fortz and M. Thorup. Robust optimization of OSPF/IS-IS weights. In *Proc. International Network Optimization Conference*, pages 225–230, October 2003.

B. Fortz and M. Thorup. Increasing Internet capacity using local search. *Computational Optimization and Applications*, 29(1):13–48, 2004.

L. Gao, T. G. Griffin, and J. Rexford. Inherently safe backup routing with BGP. In *Proc. IEEE INFOCOM*, April 2001.

L. Gao and J. Rexford. Stable Internet routing without global coordination. *IEEE/ACM Trans. Networking*, 9(6):681–692, December 2001.

G. Huston. Interconnection, peering, and settlements. In *Proc. INET*, June 1999.

A. Juttner, A. Szentesi, J. Harmatos, M. Pioro, and P. Gajowniczek. On solvability of an OSPF routing problem. In *Proc. Nordic Teletraffic Seminar*, 2000.

A. Khanna and J. Zinky. The revised ARPANET routing metric. In *Proc. ACM SIGCOMM*, pages 45–56, September 1989.

B. M. Leiner, V. G. Cerf, D. D. Clark, R. E. Kahn, L. Kleinrock, D. C. Lynch, J. Postel, L. G. Roberts, and S. S. Wolff. The Past and Future History of the Internet. *Communications of the ACM*, 40(2):102–108, February 1997.

D. H. Lorenz, A. Orda, D. Raz, and Y. Shavitt. How good can IP routing be? Technical Report 2001-17, DIMACS, May 2001.

R. Mahajan, D. Wetherall, and T. Anderson. Negotiation-based routing between neighboring ISPs. In *Proc. USENIX Symposium on Networked Systems Design and Implementation*, May 2005.

A. Medina, N. Taft, K. Salamatian, S. Bhattacharyya, and C. Diot. Traffic matrix estimation: Existing techniques compared and new directions. In *Proc. ACM SIGCOMM*, August 2002.

J. Moy. OSPF Version 2. Request For Comments 2328, IETF, April 1998.

A. Nucci, B. Schroeder, S. Bhattacharyya, N. Taft, and C. Diot. IGP link weight assignment for transient link failures. In *Proc. International Teletraffic Congress*, August 2003.

OpNet SP Guru, 2005. http://www.opnet.com/products/spguru/home.html.

M. Pioro, A. Szentesi, J. Harmatos, A. Juttner, P. Gajowniczek, and S. Kozdrowski. On OSPF related network optimisation problems. In *Proc. IFIP ATM IP*, July 2000.

B. Quoitin, S. Uhlig, C. Pelsser, L. Swinnen, and O. Bonaventure. Interdomain traffic engineering with BGP. *IEEE Communication Magazine*, May 2004.

J. Rexford, J. Wang, Z. Xiao, and Y. Zhang. BGP routing stability of popular destinations. In *Proc. Internet Measurement Workshop*, 2002.

A. Sridharan, S. B. Moon, and C. Diot. On the correlation between route dynamics and routing loops. In *Proc. Internet Measurement Conference*, October 2003.

W. Richard Stevens. *TCP/IP Illustrated, Volume 1: The Protocols*. Addison-Wesley, January 1994. ISBN 0201633469.

R. Teixeira, A. Shaikh, T. Griffin, and J. Rexford. Dynamics of hot-potato routing in IP networks. In *Proc. ACM SIGMETRICS*, June 2004a.

R. Teixeira, A. Shaikh, T. Griffin, and G. Voelker. Network sensitivity to hot-potato disruptions. In *Proc. ACM SIGCOMM*, September 2004b.

Z. Wang, Y. Wang, and L. Zhang. Internet traffic engineering without full mesh overlaying. In *Proc. IEEE INFOCOM*, April 2001.

J. Winick, S. Jamin, and J. Rexford. Traffic engineering between neighboring domains. http://www.cs.princeton.edu/~jrex/papers/interAS.pdf, July 2002.

T. Ye and S. Kalyanaraman. A recursive random search algorithm for large-scale network parameter configuration. In *Proc. ACM SIGMETRICS*, June 2003.

Y. Zhang, M. Roughan, N. Duffield, and A. Greenberg. Fast, accurate computation of large-scale IP traffic matrices from link loads. In *Proc. ACM SIGMETRICS*, June 2003.

25 OPTIMIZATION PROBLEMS IN MULTICAST TREE CONSTRUCTION

Carlos A.S. Oliveira[1], Panos M. Pardalos[2], and Mauricio G.C. Resende[3]

[1]School of Industrial Engineering and Management
Oklahoma State University
Stillwater, OK 74078 USA
coliv@okstate.edu

[2]Center for Applied Optimization
Department of Industrial and Systems Engineering
University of Florida
Gainesville, FL 32611 USA
pardalos@ufl.edu

[3]Internet & Network Systems Research Center
AT&T Labs Research
Shannon Laboratory
Florham Park, NJ 07932 USA
mgcr@research.att.com

Abstract: Multicasting is a technique for data routing in networks that allows multiple destinations to be addressed simultaneously. The implementation of multicasting requires, however, the solution of difficult combinatorial optimization problems. In this chapter, we discuss combinatorial issues occurring in the implementation of multicast routing, including multicast tree construction, minimization of the total message delay, center-based routing, and multicast message packing. Optimization methods for these problems are discussed and the corresponding literature reviewed. Mathematical programming as well as graph models for these problems are discussed.

Keywords: Multicasting, data routing, networks, combinatorial optimization, multicast tree construction, delay minimization, cache placement, center-based multicast routing, and multicast packing.

25.1 INTRODUCTION

Routing is a fundamental task in network systems in general. Any computer network has as one of its main functions to send information (usually in the form of packets) between clients and servers. Techniques used for routing in traditional network systems can be classified in the following way:

- *Unicast routing*: under this routing paradigm, the objective is to transport data packets between single sources and destinations. This is the simplest routing method, and therefore is also the most frequently used. The idea of single source-destination transport of information has been explored in countless applications, and allowed the development of protocols such as ftp, http, which are used in the implementation of important Internet applications.

- *Broadcast routing*: in the broadcast paradigm of routing, a network node can send packets to all nodes directly connected to it. Broadcasting is popular in mass distribution media such as radio and TV, but it has also applications in computer networks. For example, broadcasting is generally used in packet switching networks for distributing data in the local segment, under network implementations such as Ethernet, which use a bus topology. In the case of Ethernet networks, client nodes are addressed as part of a network segment using broadcast messages (in an IP implementation, this happens when IP addresses ending in 255 are used).

Both paradigms for packet routing, although being useful, are not adequate when information needs to be sent to a relatively large group of users, that are geographically separated and have similar interest on a piece of data. This situation has lead to the development of network protocols capable of sending information to selected destinations (Deering and Cheriton, 1990). Such protocols are called *multicast routing* protocols. Different implementations have been proposed in the last 20 years, with varying levels of success. However, all multicast routing implementations share similar requirements, in terms of the distributed combinatorial optimization problems that must be solved.

In this chapter, we discuss some of the most important combinatorial optimization problems occurring in multicast routing systems. We provide formulations for these problems and discuss some of the techniques that have been used to solve them in practice, as well as remaining open problems in this area.

25.1.1 Multicast network concepts

A *multicast group* is a set of network clients and servers interested in sharing a specific set of data. A multicast protocol has the objective of connecting members of the multicast group in an optimal way, by reducing the amount of bandwidth necessary but also considering other issues such as communication delays and reliability.

Multicast systems have in common the fact that the total amount of resources used by the network becomes large as the number of members in the multicast group increases. This makes it difficult to provide services at desirable levels, unless efficient algorithms are used to reduce resource consumption.

The delivery of data to a selected number of users organized as a multicast group is called *multicast routing*. This concept was introduced by Wall (1980) in order to allow the implementation of network services that require addressing specific sets of users. Despite having been proposed back in the 1980s, implementations of the multicast concept have not been brought into practical working until the the 1990s. Even so, multicast routing is not available by default in most routers, since it has the potential to slow down network servers. This shows the need for algorithms allowing efficient implementation of multicast routing concepts.

Multicast systems can be classified according to the types of groups involved. A multicast group can be *sparse* when the number of elements in the group is small compared to the total size of the network. If this is not the case, then the group is called a *pervasive group* (Waitzman et al., 1988). Another classification of multicast groups is according to the duration of the required connection and the possibility of changes in the group membership. A *static group* is one that is fixed at its creation time, and cannot be changed afterwards. A *dynamic group*, on the other hand, is allowed to change by adding or removing nodes.

Several protocols have been proposed to implement the multicast network concept. Among the most important techniques we cite the following:

- DVMRP (Deering and Cheriton, 1990; Waitzman et al., 1988) (Distance-Vector Multicast Routing Protocol) is a protocol used to integrate multicast routing to RIP (interior gateway protocol). It allows tunneling of multicast traffic over standard routers.

- The MOSPF (Moy, 1994a) is an extension of the traditional OSPF (Moy, 1994b; Thomas II, 1998) (Open Shortest Path First) to handle multicast applications. Thus, it is able to integrate into existing routers.

- PIM (Deering et al., 1996) (protocol independent multicast) is a protocol that tries to explore sparsity in multicast groups. It is also designed to be independent of the underlying unicast routing protocol.

- Core-based trees (Ballardie et al., 1993) are a technique to allow easier creation of multicast trees, by the use of a core that can be reused by different multicast groups. It is a protocol that can be classified among other shared tree technologies (Chiang et al., 1998; Wei and Estrin, 1994).

- MBONE (Eriksson, 1994): The Internet multicast backbone (MBONE) was introduced as a implementation of multicasting capable of sending live streams of video and sound using an existing Internet connection. In fact, one of the first applications of the MBONE was the transmission of IETF (Internet Engineering Task Force) meetings.

In unicast routing systems, routes are computed using classical algorithms such as Bellman (1957) and Dijkstra (1959). Information about best routes are stored in tables in the routers' main memory. These algorithms have polynomial complexity, generally in the order of $O(n^3)$. Algorithms for routing in multicast networks, on the other hand, cannot provide exact solutions in polynomial time, since the underlying problems are NP-hard, as seen in the next sections. Some of these problems have bee studied by many researchers in the last years, starting with Dalal and Metcalfe (1978).

Several documents discussing the design and implementation of current multicast protocols are available on the Internet. For example, the Internet task force draft (Semeria and Maufer, 1996) on multicast technologies is freely available. For papers and surveys on general aspects and algorithms related to multicast systems, see Du and Pardalos (1993a), Wan et al. (1998), Pardalos et al. (1993), Pardalos and Khoury (1995), Paul and Raghavan (2002), and Salama et al. (1997b).

In this chapter, we review algorithmic and modeling strategies used to solve problems in multicast routing. Among the problems discussed, are the multicast tree construction, delay minimization, cache placement problems, center-based multicast routing, and multicast packing.

25.1.2 Applications

Several applications can benefit of the use of multicast routing schemes. We list some possible scenarios, occurring in areas such as corporate environments, education, entertainment, and real-time communication.

A possible scenario of application of multicast in corporate environments is the implementation of financial data delivery. In this application, users are interested in reliable and up to date information about stocks, bonds, and several financial services provided by the stock market. The advantage of multicast routing in this scenario is the possibility of reaching users that are registered to receive the information, with the group membership status being updated automatically, independent of the data source. This type of application requires reliable message delivery mechanisms, which is one of the main areas of research in multicast systems (IBM, 2005).

A second application that has received a lot of attention is video-conferencing (Eriksson, 1994; Sabri and Prasada, 1985). The implementation of video-conferencing requires the use of fast protocols that can deliver information to a large number of users quickly. Multicast is an important technology for video-conferencing due to several reasons, including the savings in network bandwidth, the possibility of dynamically managing the set of users receiving the information, and the use of fast mechanisms for sharing the costs of connection. In fact, one of the first practical uses of a multicast protocol was the Internet multicast of sessions of IETF meetings.

Software delivery (Han and Shahmehri, 2000) is another example of application that can benefit of the use of multicast protocols. In particular, distribution of software in a company requires a large amount of bandwidth, the deliveries are quite predictable, and there is a strong motivation for allowing the computational burden to be shared among the set of computers being updated.

Group-ware collaboration (Chockler et al., 1996; Ellis et al., 1991) requires that a set of users keep data (such as calendars, working files, sets of archives in revision systems) in a large number of locations, where it can be accessed by workers involved in a collaborative task. Group-ware requires the use of multicast protocols since it is very frequent that files need to be synchronized in this type of applications. Each set of files can be viewed as participating in a multicast group, and data is updated among elements of the group whenever necessary.

Finally, several other applications could benefit from the use of multicast protocols, including the delivery of multimedia content on the Internet (Pasquale et al., 1998), real-time video streaming (Jia et al., 1997; Kompella et al., 1996), and even networked games (Park and Park, 1997).

25.1.3 Notation

We use standard notation for graphs. A network is modeled as a graph $G = (V, E)$, where the set of nodes V represent servers, clients, or intermediate nodes, and the set of edges E represent links between nodes in V. A *multicast group* M is a set of source nodes S (also called servers), which store data that must be transfered to clients, together with a set of destinations D. All problems in multicast routing require that elements of the multicast group $M = S \cup D$ be connected in some specified way. In particular, we consider more commonly that S is composed of a single source node s.

The following functions defined on graph elements will be used throughout the paper. In particular, let $d(v)$, for $v \in V(G)$ represent the degree of a node, i.e., the number of nodes adjacent to v. Similarly, let $\delta(v)$ and $\Delta(v)$ be the minimum and maximum value of $d(v)$, for $v \in V$, respectively. A path between two nodes $a, b \in V$ is denoted by $\mathcal{P}_{ab}$. The length of $\mathcal{P}_{ab}$ is the number of edges in the path.

25.1.4 Chapter organization

This chapter is organized as follows. In Section 25.2 we describe the multicast tree routing problem (MRP) which has the objective of finding an optimal tree linking the elements of the multicast group. This is a fundamental problem in multicast routing, and several models and algorithms have been proposed for its solution. In Section 25.3 we discuss the approaches to solve the MRP based on Steiner tree algorithms; we will see different ways of modeling the problem, including the use of center nodes and rings to improve the resulting solution. The need for network designs that support multicast in an efficient way lead to the formulation of the multicast packing problem, which will be discussed on Section 25.4. In Section 25.5, other problems related to multicast routing will be discussed, such as the point-to-point connection problem, and the cache placement problem. Finally, in Section 25.6 we provide some concluding remarks.

25.2 MINIMUM MULTICAST TREE ROUTING PROBLEM

25.2.1 Problem definition

One of the most difficult problems in multicast networks consists of finding routing paths linking the source (or set of sources) to the set of destinations, while simultaneously minimizing some cost function. A way of achieving this objective is computing a tree connecting the source and destination nodes. Thus, we are frequently interested in finding trees connecting sources and destinations with low total cost. The resulting problem is called the *multicast routing tree problem* (MRP).

A mathematical formulation of the MRP based on graphs can be given as follows. Let $G = (V, E)$ be a graph representing the network, $s \in V$ the source node, and $D \subset V$

the set of destinations. We assume that each node $v \in D$ must receive a copy of the information stored in s, and therefore $D \cup s$ form a multicast group. The set E of edges has an associated function $w : E \to \mathbf{R}_+$ returning the cost of using edge e. Then, the objective of the MRP is to find a tree $T \subseteq E$ connecting the source node s to each of the destinations $d \in D$, such that the total cost $\sum_{e \in T} w(e)$ is minimum.

The multicast tree problem has been studied by many researchers (Aguilar et al., 1986; Berry, 1990; Bauer, 1996; Ballardie et al., 1993; Baldi et al., 1997; Du and Pardalos, 1993a; Pardalos and Du, 1998; Frank et al., 1985; Feng and Yum, 1999; Im et al., 1997; Jiang, 1992; Wan et al., 1998; Pardalos et al., 2000) using strategies that range from simple heuristics to approximation algorithms. However, no exact polynomial algorithm is known for the problem in general, since it is NP-hard, as shown in the next theorem.

Theorem 25.1 *The multicast routing problem is NP-hard.*

Proof. By reduction from the Steiner problem on graphs. In the Steiner problem, we are given a graph $G = (V, E)$, a cost function $w : E \to \mathbf{R}_+$, a set of required nodes R, and a set of Steiner nodes $V \setminus R$. The objective is to find a tree T connecting all nodes in R at minimum cost.

The reduction is simple: take one of the nodes $v \in R$ to become the source, and make $D = R \setminus \{v\}$. Clearly, an optimal solution to this instance of the MRP instance will be also optimal for the Steiner problem. An optimal solution to the MRP can also be found given the optimum solution for the Steiner tree instance, thus the multicast routing problem in NP-hard. □

When applying the MRP to specific problems, a number of additional constraints arise. The two most important additional constraints for the MRP are the following.

- *Delay constraints.* In applications that require real time coordination, such as video-conferencing and other multimedia systems, it is very important that the total delay in the routing of packets be as small as possible. In practice, it is required that the delivery of data take less than some delay threshold.

- *Reliability constraints.* In some multicast systems, an important concern is that data be delivered with some guarantee of reliability. This is done usually by requiring that additional capacity be reserved on each arc of the network, so that in the case of failure in some edge the data can be diverted through other paths in the network.

25.2.2 Flooding and reverse path-forwarding algorithms

The simplest way of sending packets to multiple nodes in a network is using the technique called *flooding*, which consists of recursively sending data to all neighbors of a node until the destinations are reached. The algorithm is shown in Algorithm 25.1. Note that this algorithm requires that each node keep track of packages that were sent through it during the process, otherwise a packet can loop indefinitely through the network. Keeping this extra information limits the usefulness of flooding, since the number of all packages that can be reliably sent is bounded by the available memory on any routing machine. There are also inevitable bandwidth losses in this process since many non-required edges are saturated with data. Finally, there is no confirmation that a package reached the required destinations.

```
procedure Flooding-Algorithm
begin
    Receive packet p from node u;
    if destination(p) = v then
    |   PacketReceived;
    else
        if packet was not previously processed then
        |   Send packet p to all nodes in N(v) \ {u};
    end
end
```
Algorithm 25.1: Flooding algorithm for node v

To improve the performance of the flooding algorithm, the *reverse path-forwarding algorithm* (RPFA) was proposed by Dalal and Metcalfe (1978). In the RPFA, each node is responsible for finding an edge that is in the path to the destination. In this way, traffic is not accepted from all nodes, but just from a selected number of neighbors. The edge from which traffic is accepted, when routing to a particular node, is called the *parent link*. A simple algorithm to determine the parent link for a node s is the following: let e be the first edge from which a package having origin s was received. From now on, if a package is received from an edge that is not considered to be the parent link for the given source node, then drop the package. Otherwise, receive the package and broadcast it to all other neighbor nodes. Improvements to this basic algorithm are discussed by Semeria and Maufer (1996).

25.2.3 An integer programming formulation

In this section, we discuss an integer programming model for a generalized multicast routing problem, originally proposed by Noronha and Tobagi (1994). The model is for a version of the MRP in which there are costs and delays for each link, and a set $\{1,..,T\}$ of multicast groups. Each group i has its own source s_i, a set of n_i destinations $d_{i1} \ldots, d_{in_i}$, a maximum delay Δ_i, and a bandwidth request r_i. There is also a matrix $B^i \in R^{n \times n_i}$ for each group $i \in \{1, \ldots, T\}$ of source-destination requirements. The value of B^i_{jk} is 1 if $j = s$, -1 if $j = d_{jk}$, and zero otherwise. The node-edge incidence matrix is represented by the matrix $A \in Z^{n \times m}$.

The considered network has n nodes and m edges. The vectors $W \in R^m$, $D \in R^m$ and $C \in R^m$ give respectively the costs, delays, and capacities for each link in the network. The variables in the formulation are $X^1, \ldots, X^T$ (where each X^i is an $m \times n_i$ matrix), and $Y^1, \ldots, Y^T$ (where each Y^i is a vector of m elements), and $M \in R^T$. The variable $X^i_{jk} = 1$ if and only if link j is used by group i to reach destination d_{ik}. Similarly, variable $Y^i_j = 1$ if and only if link j is used by multicast group i. Also, variable M_i represents the delay incurred by multicast group i in the current solution.

In the following formulation, the objectives of minimizing total cost and maximum delay are considered. However, the constant values β_c and β_d represent the relative weight given to the minimization of the cost and to the minimization of the delays, respectively. Using the variables shown, the integer programming formulation is given by:

$$\min \quad \sum_{i=1}^{T} r_i \beta_c CY^i + \beta_d M_i \tag{25.1}$$

subject to

$$AX^i = B^i \quad \text{for } i \in \{1,\ldots,T\} \tag{25.2}$$

$$X_{jk}^i \leq Y_j^i \leq 1$$
$$\text{for} \quad i \in \{1,\ldots,T\}, j \in \{1,\ldots,n\}, k \in \{1,\ldots,n_i\} \tag{25.3}$$

$$M_i \geq \sum_{j=1}^{k} D_j X_{jk}^i \quad \text{for } i \in \{1,\ldots,T\}, k \in \{1,\ldots,n_i\} \tag{25.4}$$

$$M_i \leq L_i \quad \text{for } i \in \{1,\ldots,T\} \tag{25.5}$$

$$\sum_{i=1}^{T} r_i Y^i \leq C \tag{25.6}$$

$$X_{jk}^i, Y_j^i \in \{0,1\}, \quad \text{for } 1 \leq i \leq T, 1 \leq j \leq K, 1 \leq k \leq n_i. \tag{25.7}$$

The constraints in the above integer program have the following meaning. Constraint (25.2) is the flow conservation constraint for each of the multicast groups. Constraint (25.3) determines that an edge must be selected when it is used by any multicast tree. Constraints (25.4) and (25.5) determine the value of the delay, since it must be greater than the sum of all delays in the current multicast group and less than the maximum acceptable delay L_i. Finally, constraint (25.6) says that each edge i can carry a value which is at most the capacity C_i. This is a very general formulation, and clearly cannot be solved exactly in polynomial time because of the integrality constraints (25.7).

This formulation was used by Noronha and Tobagi (1994) to derive exact algorithms for the general problem. Initially, a decomposition technique was used to divide the constraint matrix into smaller parts, where each part could be more easily solved. This part of the algorithm can be executed using some standard mathematical programming decomposition techniques, as shown e.g., in Bazaraa et al. (1990). Then, a branch-and-bound algorithm was proposed to solve the resulting problem. The lower bounding procedure is important in this branch-and-bound implementation, since it can benefit from the efficiency gains of the decomposition method, resulting in improved computation time.

25.3 STEINER TREE PROBLEMS AND MULTICAST ROUTING

25.3.1 Steiner tree algorithms

Due to the similarity of the MRP and Steiner tree problems, one of the simplest and more effective techniques for solving the MRP involve careful modification of algorithms for Steiner tree (Bauer and Varma, 1997; Du et al., 2001; Du and Pardalos, 1993b; Hwang and Richards, 1992; Hwang et al., 1992; Kou et al., 1981; Takahashi and Matsuyama, 1980; Winter, 1987; Winter and Smith, 1992; Pardalos and Khoury, 1995; 1996; Pardalos et al., 1993), according to the necessary extra constraints. As an example of application of this solution scheme, we introduce the algorithm for the Steiner tree problem described in Algorithm 25.2, known as the KMB algorithm (Kou et al., 1981).

procedure *KMB-Algorithm*
begin
   Let $K = (V, E)$ be a complete graph with $|D| + 1$ nodes;
   foreach $(u, v) \in E(K)$ **do**
      | $w(u, v) \leftarrow d(u, v)$, where d is the minimum distance from u to v in G;
   end
   Let T be a minimum spanning tree on K;
   Let T' be the union of all shortest paths $v \rightsquigarrow w$ in G, s.t. $(v, w) \in T$;
   Let T'' be a minimum spanning tree T';
   return T''
end

Algorithm 25.2: The KMB algorithm for Steiner tree.

The KMB is an approximation algorithm which constructs a solution for the Steiner tree problem where the required nodes are elements of $D \cup s$ and the nodes in the required set are connected using the union of shortest paths. It is easy to prove the following result:

Theorem 25.2 *The KMB algorithm returns a solution within a factor 2 of the optimum.*

The KMB heuristic shown above provides a good scheme for heuristic development for the MRP, since it gives an initial guarantee of approximate optimality. Moreover, studies on the quality of solutions computed by the KMB heuristic (Wall, 1980) showed that the algorithm performs very well in problems occurring in practice, exceeding by a large difference the factor 2 guarantee described in Theorem 25.2. Similarly, Doar and Leslie (1993) have shown that this heuristic usually achieves at most 5% of difference to the optimal solution value, for many instances of large size.

The main challenge in using techniques based on KMB to solve the multicast routing problem consists of distributing the computational effort among all nodes involved in the multicast group. For this reason, many variants of the KMB algorithm have been proposed to solve the MRP, as described in the next section, to overcome the limitations of using a sequential, deterministic algorithm (Bharath-Kumar and Jaffe, 1983; Wall, 1982; Waxman, 1988; Wi and Choi, 1995).

An alternative algorithm to solve the Steiner tree problem was proposed by Takahashi and Matsuyama (1980), and is known as the *greedy incremental tree algorithm* (GIT). This heuristic is similar to the Prim's and Dijkstra's algorithm in the sense that it starts from a single node, and at each iteration it adds the edge that is closest to the current tree and which does not form a cycle. These steps are repeated until a tree connecting all required nodes is found. The GIT algorithm has also been adapted to the MRP and employed by some other researchers as an initial step in the construction of multicast trees (Carlberg and Crowcroft, 1997; Li and Mohapatra, 2003).

25.3.2 Steiner tree-based algorithms for the MRP with delay constraints

In this section, we describe some of the algorithms that have been proposed for the multicast routing problem (Im et al., 1997; Kompella et al., 1992; 1993b;a; Jia, 1998; Sriram et al., 1998). Most of these algorithms share the following features:

- The Steiner tree model, as described in the previous section, is assumed;

- They are distributed versions of some of the well known heuristics for the Steiner tree problem;

- The distributed algorithms do not provide any guarantee of optimality;

- Unlike metaheuristic-based heuristics, these algorithms do not provide a guarantee of *local optimality*, which means that they do not check any neighborhood of the given solution for improvements that would locally improve the considered solution.

Im et al. (1997) proposed a distributed algorithm for the delay constrained multicast routing problem. In this distributed algorithm, the objective is to reduce the computational time for constructing the routing tree. The strategy used consists of creating the tree in only one round of data exchange, thus avoiding excessive message exchanges, which are the main time constraint on distributed algorithms. The basic steps in the algorithm can be described as follows. Initially, messages are sent to nodes in the set $T = D \cup s$ composed of sources and destinations, with the objective of asking each node $v \in T$ to compute the node $w \in T \setminus v$ that is closest to v. Then, each node resends the messages to its children nodes, until the delay information has been computed by all nodes, and the results are sent back to the elements of T. When a node is selected to be part of the tree, it receives an ADD message and sends back an ADD_ACK message. The algorithm stops when all nodes is T have been added to the multicast tree. The following theorem has been proved in Im et al. (1997):

Theorem 25.3 *The described algorithm produces a feasible multicast tree, and if there exists a feasible solution to the MRP, then the algorithm returns a feasible solution.*

Kompella et al. (1992) provided an algorithm based on a distributed version of the KMB heuristic. The basic modification included in the KMB algorithm is an added step, which guarantees that the constructed tree is feasible for the delay constraints. This works as follows: initially, a closure graph K is created from the original graph G, as in the original KMB algorithm. Then, at each step an edge of K is added to the tree. The edges are added, however, according to a function $g(e)$, for $e = (u, v) \in E(K)$ such that

$$g(u,v) = \frac{c(u,v)}{\mathcal{D} - \mathcal{D}(u,v) - \mathcal{D}(s,u)},$$

where $\mathcal{D}$ is the maximum delay acceptable in the MRP instance, $\mathcal{D}(u,v)$ is the delay incurred in the path from u to v, and similarly, $\mathcal{D}(s,u)$ is the delay incurred in the path from s to u.

The described algorithm runs in $O(2^{\log \mathcal{D}})$, where $\mathcal{D}$ is the maximum delay acceptable in the MRP instance. This is not polynomial, since $\mathcal{D}$ can be very large. However, in practice the value of D is small, and the algorithm behaves in most cases as a polynomial algorithm. The following interesting results provides a bound on the worst case performance of the described procedure.

Theorem 25.4 *The worst case of the algorithm described above is $l|D|/(l+|D|)$, for a given value l corresponding to edge costs. Thus, the worst case performance is asymptotically equal to $|D|$.*

25.3.2.1 Sparsity considerations and the MRP. Sparsity in the definition of multicast network instances is a major factor that must be considered in the design of new algorithms for the MRP. This is necessary since sparse instances can usually be handled by faster algorithms. As an example of application of this strategy, the heuristic proposed by Chung et al. (1997) was designed to provide solutions to the delay constrained MRP, when the instances are sparse. To do this, the algorithm employs a Lagrangian relaxation technique, implicitly used on an algorithm that was initially proposed by Blokh and Gutin (1996) (the algorithm is from now on called the BG heuristic). The BG heuristic solves a minimization problem with a constraint by recasting it as an unconstrained problem with Lagrangian multipliers. The multipliers are used in this context to force feasibility of the resulting solution.

The algorithm proposed in Chung et al. (1997) tries to adapt the BG heuristic to the MRP. To do this, it uses a well known algorithm for the Steiner problem (Takahashi and Matsuyama, 1980) as the generator for an initial solution. Then, it applies the scheme used by the BG heuristic to improve the initial solution, until a feasible solution for the MRP is found. According to empirical results in Chung et al. (1997), the heuristic provides results for the MRP that have on average 11% larger objective cost than the optimum solution.

25.3.2.2 Multicast routing as an on-line problem. A frequent problem that occurs when solving the MRP consists of dealing with reconfigurations of routes, when inclusions and exclusion of members of a group are allowed (Aguilar et al., 1986; Waxman, 1988). In this case, although the initial tree may have been optimal according to some criterion, it is easy to see that after several insertions and deletions this may not be an optimal tree, and in fact it can be far from optimality. Thus, a major challenge on dynamic multicast applications is to allow for periodic reconfiguration of routes, with the objective of avoiding excessive degradation of solution quality.

To solve this major problem, some algorithms have been presented in the last few years (Kheong et al., 2001; Waxman, 1988). A technique that was proposed by Waxman (1988) consists of maintaining caches with precomputed sub-trees that connect subsets of the destinations. Such "routing caches" can be used in two ways:

- First, caches can be used to store information about frequent combinations of nodes in a multicast group. In this sense, insertions and deletion of nodes can be used to "learn" new configurations of destinations, and how they are best satisfied, storing the resulting information in a routing tree database.

- Second, the cached information can be used to speed up the creation of multicast routing trees in the case of dynamic changes. A cache can also be used to quickly reconstruct routing trees, starting from parts of existing tree, which are known to be able to connect nodes that are currently part of the multicast group.

Waxman (1988) was one of the first researchers to consider heuristics for cost minimization taking in consideration the dynamics of inclusion and exclusion of members in a multicast group. The algorithm by Waxman (1988) presents one of the possible techniques to manage this database of routing subtrees. To do this, it provides a mathematical framework for updating trees formed by the dynamic inclusion and exclusion of nodes. He proposed a random graph model, where the probability of adding an edge between two nodes depends on the Euclidean distance between them. This probability decreases with distance between nodes. Using this model, a dynamic update rule is derived, which gives an improved way of constructing routes.

Other examples of algorithmic techniques for the on-line MRP are the following. The algorithm by Hong et al. (1998), which is in most aspects similar to Chung et al. (1997), is an implementation of dynamic resource management, and uses a strategy capable of handling additions and removals of elements to an existing multicast group. The algorithm is again based on the KMB algorithm for the Steiner tree problem. To reduce the complexity of the problem, the authors employed a Lagrangian relaxation technique.

Feng and Yum (1999) also devised a heuristic for the online version of the MRP, with the main goal of allowing easy insertion of new nodes in the multicast group. Similarly to the TM heuristic for the Steiner tree problem (Takahashi and Matsuyama, 1980), the proposed algorithm uses a strategy in which, at each step, it takes a nonconnected destination with minimum cost and tries to add this destination to the current solution (this is similar to the way the Prim's algorithm computes a solution to the minimum spanning tree problem).

Sriram et al. (1999) proposed new algorithms for the on-line, delay constrained minimum cost multicast routing which tries to maintain a fixed quality of service by specifying minimum delays. The algorithm is able to adapt the routing tree to changes in membership due to inclusions and exclusions of users.

Finally, a technique that has been explored in some algorithms for the online version of the MRP consists of using information available from unicast protocols. By doing this, the algorithm has two advantages:

- To simplify the main algorithm since it is not necessary to consider the lower level task of finding a single path — the objective becomes to integrate unicast paths in an optimal way;

- To reduce the computational time of the algorithm, by reusing information that was previously computed and left available by the unicast protocol.

An example of the use of this technique is given by Baoxian et al. (2000), who proposed a heuristic where paths are found by requesting this information to the underlying unicast protocol.

25.3.3 Distributed algorithms

Multicast routing is a task that depends essentially on the use of distributed algorithms. This is true since the MRP is defined on a distributed environment, where clients and servers are able to perform computation. This available computational power can be used to reduce the time complexity of the employed algorithms, through careful parallel implementation. A number of papers have focused on implementation of better distributed strategies for MRP problem, as discussed below (Jia, 1998; Chen et al., 1993).

A good example of distributed algorithm for the MRP is provided by Kompella et al. (1993a). The proposed algorithm was targeted at applications where audio and video need to be delivered over a network, and where delay restrictions play an important role. In the first algorithm presented in Kompella et al. (1993a), the Bellman-Ford algorithm (Bellman, 1957) (which is used to compute the shortest path between two nodes in a network) is extended with the objective of finding a minimum delay tree from the source to each destination. A second algorithm was also proposed, with similar techniques, but this time using a strategy similar to the Prim's algorithm for minimum spanning tree construction.

Chen et al. (1993) have further studied distributed algorithms to the MRP. The authors proposed a heuristic that is similar to the algorithm described by Kompella et al. (1993b) (and also to the technique used in the KMB heuristic for Steiner tree). The main difference in this case, however, is that a distributed algorithm has been employed to find a minimum spanning tree, which must be computed twice during the execution of the heuristic. The method used to find the MST is based on the distributed algorithm proposed by Gallager et al. (1983). The use of this distributed algorithm allowed a considerable reduction on computational time, providing an efficient implementation.

Shaikh and Shin (1997) have presented a distributed algorithm where the focus is to reduce the complexity of distributed versions of heuristics for the the delay constrained Steiner problem. In their paper, the authors try to adapt the general model of Prim's and Dijkstra's algorithms to the harder task of creating a multicast routing tree. The main addition done to the structure of Dijkstra's algorithm is a method to distinguish between destinations and non destination nodes, as presented in Algorithm 25.3. Here, function $I_D : V \to \{0,1\}$ is an indicator function, such that if $v \in D$, then $I_D(v) = 0$, with $I_D(v) = 1$ otherwise. The objective of this indicator function is to remove the costs of adding a destination to the current tree, making it easier to construct a tree linking the source to destinations.

25.3.3.1 Algorithms for sparse groups. An important case of the MRP occurs when the number of sources and destinations is small compared to the whole network. This is the typical case for large instances, where just a few nodes will participate in a group, at each moment. To handle sparsity in the MRP, Sriram et al. (1998)

procedure *Modified-Dijkstra*
Input: $G(V,E), s$
begin
    for $v \in V$ **do** $d[v] \leftarrow \infty$;
    $d[s] \leftarrow 0$;
    $S \leftarrow 0$;
    $Q \leftarrow V$ /* Q is a queue */ ;
    repeat
        $v \leftarrow get\_min(Q)$;
        $Q \leftarrow Q \setminus \{v\}$;
        $S \leftarrow S \cup \{v\}$;
        foreach $u \in N(v)$ **do**
            if $u \notin S$ and $d[u] > d[v]I_D[v] + w(u,v)$ **then**
                $d[u] \leftarrow d[v]I_D[v] + w(u,v)$;
        end
    until Q is empty;
end

Algorithm 25.3: Modification of Dijkstra's algorithm for multicast routing, as proposed by Shaikh and Shin (1997).

proposed a distributed algorithm which tries to explore properties of sparse instances. The algorithm uses the unicast routing algorithms that already exist in the underlying network. The algorithm is composed of two phases: in the first one, paths are computed from the source to each destination, using the underlying unicast protocol. This phase of the algorithm can be performed in a distributed way, since the unicast protocols are also distributed. In the second phase, the algorithm has to apply some heuristic rules to define what segments of the returned paths will be used as part of the final routing tree. Some rules have bee proposed on Sriram et al. (1998), depending on the configuration of the possible intersections between the given paths.

A problem that arises when multicast groups are allowed to have dynamic membership is that a considerable amount of time is spent in the process of connection configuration. Jia (1998) proposed a distributed algorithm which addresses this question. The main assumptions used by the algorithm are the following:

- *Minimizing computational effort.* Most of the computational effort should be spent by nodes that participate in the multicast group (source and destination), other nodes should not be heavily penalized.

- *Decentralized management of routing information.* Routing information must be distributed, and every node should be capable of collecting is own routing information. This allows the algorithm to operate independently of any central node that stores the global state of the system.

- *Minimizing number of messages.* The algorithm should use the least possible amount of messages, and each message should have a small size.

Using these principles, the algorithm applies the following steps. Starting from the source node s, all the shortest paths from the source to the destinations are computed. To do this, it is possible to use information available from the unicast protocol. The first step consists of linking s to the destination $v \in D$ that is closest to s. Then, in the following iterations, shortest paths are computed from the subtree T that has been constructed in the previous steps; the node $v \in D$ that is closest to one of the nodes $w \in T$ is selected, and the path $\mathcal{P}$ connecting them is added to the tree. Finally, redundant parts of the union $T \cup \mathcal{P}$ are removed, in such a way that delay constraints are satisfied.

25.3.4 Other restrictions: Nonsymmetric links, degrees, and delay variation

An interesting feature of real networks, which is not mentioned in most research papers, is that links are, in general, nonsymmetric. The capacity in one direction can be different from the capacity in the other direction, for example, due to congestion problems in some links. Ramanathan (1996) considered this kind of restriction, and proposed an approximation algorithm with constant worst case guarantee. The resulting algorithm has also the nice characteristic of being parameterizable, and therefore it allows the trading of execution time for accuracy.

If the number of links from any node in the network is required to be a fixed value, then we have the *degree-constrained* version of the multicast routing problem. As an example where this is an important feature we can cite ATM networks. In this type of application, the number of out-connections must have a fixed limit, dictated by the type of routers used in the system (Zhong et al., 1993). A combinatorial algorithm for the MRP with maximum degree constraints was provided by Bauer and Varma (1995) and Bauer (1996). In such algorithms, the basic strategy of selecting Steiner trees with the required nodes is modified in a way such that only trees with limited number of connections per node are allowed.

Another restriction that is frequently disregarded is most algorithms is related to the so called *delay variation*. The delay variation is defined as the difference between the minimum and maximum delays between source and destinations, as defined by a specific routing tree. In some applications, such as video-conferencing, virtual reality, and other collaborative environments, it is interesting that the delay variation stay within a specific range. For example, it may be desirable that all nodes receive the same information at about the same time, to improve synchrony among participants of the multicast group. The delay variation constraint was considered by Rouskas and Baldine (1996), who proposed an algorithm for its minimization.

25.3.5 Center based routing

An important technique for routing tree construction is the so called *center based algorithms* (Salama et al., 1996; Wall, 1982). The objective of center based algorithms is to find a tree rooted on some node of the original graph G, with a specific centrality property. As an example, we can consider using as root a node r that minimizes the maximum distance to all other nodes in G. This idea has been pursued in some algorithms for the MRP, and is implemented in protocols such as CBT (core-based tree) (Ballardie et al., 1993).

An example of application of center based algorithms occurs when a routing tree must be shared among several multicast groups (a technique that is called shared tree construction). In this case, it is interesting to use a central node that is as close as possible from all other sources and destinations.

Another example of center based routing usage is in finding the topological center of a set of nodes. The *topological center* is the node that is closest to all members of the involved multicast groups, and is known to be an NP-hard problem (Ballardie et al., 1993). In other words, the objective is to find a node $v \in V$ that is closest to any other node in the network, i.e., the node v minimizing $\max_{u \in V} d(v,u)$. A routing tree rooted at v is then constructed and used throughout the multicast session – the idea is that such tree will be more "balanced" and therefore will need less changes as nodes are added or removed from the original multicast group(s).

To avoid the complexity problems involved with the concept of topological center, other centrality measures have been used in practical protocol implementations. Examples of alternative centrality measures are *core points* (Ballardie et al., 1993) and *rendez-vous* points (Deering et al., 1994). A comparison of different methods for computing a center for a routing tree is provided by Calvert et al. (1995).

25.3.6 Delay constrained minimum spanning tree problem

Finally, an alternative solution method for the computation of routes in multicast systems consists of constructing a spanning tree problem with delay constraints. This problem, similarly to the shortest path problem with constraints, is NP-hard (Garey and Johnson, 1979). Therefore, heuristics are used in practice to provide solutions in polynomial time. An example of heuristic solution was proposed by Chow (1991), using the strategy of combining different routes into one single routing tree. Another heuristic algorithm for this problem was provided by Salama et al. (1997a). This heuristic is a modified version of the Prim's algorithm, implemented using distributed computation techniques. A comparison of the algorithms for this problem can also be found in Salama et al. (1997b).

25.4 THE MULTICAST PACKING PROBLEM

Capacity planning for the necessary bandwidth in a multicast network is another important problem that occurs during the implementation phase of multicast systems. The main requirement when considering this problem is that the network links must have enough capacity to satisfy all multicast groups sharing the network. Similarly, when the maximum capacity is limited, it must be shared by all groups according to their bandwidth requirements. These capacity constraints are modeled in what is called the *multicast packing problem*. This problem has attracted considerable attention from researchers in the area of network design in the past few years (Wang et al., 2002; Priwan et al., 1995; Chen et al., 1998).

In this section, we present a formulation for the multicast packing problem, along with some of the computational techniques that have been applied for its solution.

OPTIMIZATION IN MULTICAST ROUTING 717

Given a graph $G = (V,E)$ representing a network, the congestion λ_e on edge $e \in E(G)$ is defined as the sum of all load imposed by the multicast groups using e. With this notation, the maximum congestion λ is defined as the maximum congestion λ_e, taken over all edges $e \in E$. If we assume that there are K multicast groups, and each group k generates an amount t^k of traffic, an integer program for the multicast packing problem is given by

$$\min \lambda \qquad (25.8)$$

subject to

$$\sum_{i=1}^{K} t^k x_e^k \leq \lambda \quad \text{for all } e \in E \qquad (25.9)$$

$$x_e^k \in \{0,1\}^{|E|} \quad \text{fox } i \in \{1,\ldots,K\}, \qquad (25.10)$$

where variable x_e^k is equal to one if and only if the edge e is being used by multicast group k. Note that the variables x_e^k must be themselves determined using an integer programming formulation, as the one presented in the previous section.

The multicast packing problem formulated above is NP-hard, and as such it can only be approximately solved by polynomial time algorithms. Heuristic strategies for the problem have been proposed by Wang et al. (2002). The authors discuss two main algorithms to set up multiple multicast groups, and formalized this as a packing problem.

The first heuristic uses a Steiner tree-based strategy — solutions are created in a greedy way, with edges being selected to the Steiner tree if they have enough capacity to satisfy the requirements of the multiple multicast groups. The proposed algorithm has been shown to run in time $O(kn^4 + k^2mn^2)$, where k is the number of multicast groups, n the number of nodes, and m the number of links.

The second heuristic is based on the idea of computing the cut-sets between pairs of elements of the multicast group. By determining the cut-sets for each such pair of nodes, one can compute the edges that need to be part of a routing tree. The proposed algorithm repeats the tree computation for each multicast group, using each time the *residual graph* (i.e., the graph representing the remaining capacity at a specific step).

25.4.1 Alternative formulations for the multicast packing problem

Alternative formulations for packing problems in multicast networks were presented by Chen et al. (1998), using two integer programming models. Initially, suppose we are given a graph $G = (V,E)$ and a set of costs w_e, for each edge $e \in E(G)$. In their first integer formulation of the multicast packing problem, Chen et al. (1998) used a binary 0-1 variable x_e for each edge $e \in E$,. The variable x_e is equal to one if and only if the edge e is selected, with respective cost w_e. Then the integer formulation for the tree version of the problem is given by

$$\min \sum_{e \in E} w_e x_e \qquad (25.11)$$

subject to

$$\sum_{e \in \delta(S)} x_e \geq 1 \quad \text{for all } S \subset V \text{ such that } m_1 \in S \text{ and } M \not\subset S \qquad (25.12)$$

$$x \in \{0,1\}^{|E|}, \qquad (25.13)$$

where M is the set of nodes participating in a multicast group, node m_1 is an arbitrary member of M, and $\delta(S)$ represents the edges leaving the set $S \subset V$. This is basically the same formulation used for the Steiner tree problems, and can be solved using similar techniques, including branch-and-cut algorithms based on the known cuts for the Steiner problem.

In a second IP formulation to the multicast packing problem, the variables x_e, for $e \in E(G)$, are similar to the ones introduced above. The idea, however, is to use a *ring-based* version of the problem. In the ring-based multicast packing, the routes are defined as a ring (instead of a tree) linking the elements of the multicast group. The ring-based configuration is preferred in some situations, since it provides better reliability in the case of failures of a single link. The IP formulation for this version of the problem is given by

$$\min \sum_{e \in E} w_e x_e \qquad (25.14)$$

subject to

$$\sum_{e \in \delta(v)} x_e = 2 \quad \text{for all } v \in M \qquad (25.15)$$

$$\sum_{e \in \delta(v)} x_e \leq 2 \quad \text{for all } v \in V \setminus M \qquad (25.16)$$

$$\sum_{e \in \delta(S)} x_e \geq 2 \quad \text{for all } S \subset V, u \in S, \text{ and } M \not\subset S \qquad (25.17)$$

$$x \in \{0,1\}^{|E|}. \qquad (25.18)$$

Here, u is an arbitrary (but fixed) element of M. The integer solution of this formulation gives a ring passing through all nodes participating in group M. To solve this problem, one can also employ a branch-and-cut algorithm. In Chen et al. (1998), some valid inequalities for this problem have been used to implement such a branch-and-cut.

25.4.2 The multicast network dimensioning problem

Another interesting problem occurs when we consider the design of a new network, intended to support a specific multicast demand. This is called the *multicast network dimensioning problem*, and it has been studied in some recent papers (Prytz, 2002; Forsgren and Prytz, 2002; Prytz and Forsgren, 2002).

The problem consists of determining the topology (i.e., which edges need to be selected) and the corresponding capacity of the edges, such that a specific multicast service can be deployed in the resulting network. Much of the work related to this problem has used mathematical programming techniques to provide exact and approximate solutions.

The following formulation was derived by Forsgren and Prytz (2002). In that paper, the technique used has been the Lagrangian relaxation applied to an integer programming model. We assume in this formulation a network represented by the graph $G = (V,E)$ and costs w_e, for $e \in E(G)$. Moreover, we assume there are T multicast groups. The model uses variables $x_e^k \in \{0,1\}$, for $k \in \{1,\ldots,T\}$, and $e \in E$, which indicate if edge e is used by group k. There are also variables $z_e^l \in \{0,1\}$, for $l \in \{1,\ldots,L\}$, and $e \in E$ (where L is the highest possible capacity level), which determine if the capacity level of edge e is equal to l. Now, let d_k, for $k \in \{1,\ldots,T\}$, be the bandwidth demanded by group k. Also, let c_e^l, for $l \in \{1,\ldots,L\}$, and $e \in E$, be the capacity available for edge e at the level l; and w_e^l, for $l \in \{1,\ldots,L\}$, and $e \in E$, be the cost of using edge e at the capacity level l. Assume that $b \in Z^n$ is the demand vector, and $A \in R^{n \times m}$ is the node-edge incidence matrix. We can now state the multicast network dimensioning problem using the following integer program

$$\min \quad \sum_{e \in E} \sum_{l=1}^{L} w_e^l z_e^l \qquad (25.19)$$

$$\text{subject to} \quad \sum_{k=1}^{T} d^k x_e^k \leq \sum_{l=1}^{L} c_e^l z_e^l \quad \text{for all } e \in E \qquad (25.20)$$

$$\sum_{l \in L} z_e^l \leq 1 \quad \text{for all } e \in E \qquad (25.21)$$

$$Ax = b \qquad (25.22)$$

$$x, z \text{ integral.} \qquad (25.23)$$

In this integer program, constraint (25.20) ensures that the bandwidth used on each edges is at most the available capacity. constraint (25.21) selects just one capacity level for each edge. Finally, constraint (25.22) enforces the flow conservation in the resulting solution. Instances of the problem proposed above have been solved using a branch-and-cut algorithm. Valid inequalities proved in Forsgren and Prytz (2002) have been used to derive a class of cuts, which were then implemented in the branch-and-cut framework.

25.5 OTHER PROBLEMS IN MULTICAST ROUTING

25.5.1 The point-to-point connection problem

The point-to-point connection problem (PPCP) is a generalization of the Steiner problem that can be used to model multicast systems with multiple sources and destinations. In the PPCP, we are given a graph $G = (V,E)$ with cost function $w : E \to R$ on the edges, a set of sources $S \subset V$, and a set of destination $D \subset V$ such that $|V| = |D|$ and $V \cap D = \emptyset$. The objective of the problem is to find a forest (a set of o trees) $F \subseteq E$ connecting sources to destinations, such that there is a bijective mapping $\phi : V \to V$ between elements of S and elements of D, and F has minimum cost. If the map ϕ is fixed in advance, then this is known as the *fixed-destinations version* of the PPCP.

The PPCP can be thought as a general version of the MRP, where multiple sources are allowed. The requirement that the number of sources be equal to the number of destinations can be satisfied by any PPCP instance with the addition of "dummy" sources, without loss of generality. Thus, solving the PPCP is important when the multicast structure is distributed, and sources are replicated over the network.

The PPCP has been proved to be NP-hard (Li et al., 1992), considering its directed and undirected versions, and even for the fixed-destinations version. For the especial case with $p = 2$ (where p is the number of destinations), an algorithm with time complexity $O(n^5)$ was also proposed. Later, Natu (1995) improved on this case of the problem, with a dynamic programming algorithm with time complexity $O(mn + n^2 \log n)$. The same authors proposed an algorithm for the case $p = 3$ with time complexity $O(n^{11})$.

In terms of approximation complexity, it is known that the PPCP can be approximated within a constant ratio. In their seminal work, Goemans and Williamson (1995) presented approximation algorithms for a large class of forest constrained problems, including the PPCP and the Steiner problem. The approximation algorithm, when applied to the PPCP, runs in $O(n^2 \log n)$ time, and gives its results within a factor $2 - 1/p$ of the optimal solution.

Given the inherent complexity of the PPCP, finding optimal solutions requires the use of clever enumeration schemes that try to avoid the search through all possible solutions. A branch-and-bound algorithm along these lines was proposed by Meneses et al. (2004). To use such an approach, one needs to state the problem as an integer program. Let x_e be a binary variable equal to 1 if and only if the edge $e \in E(G)$ is selected to be part of the solution. Then, we can write the PPCP as the following problem:

$$\min \sum_{e \in E(G)} c_e x_e$$
$$\text{subject to} \sum_{e \in \delta(A)} x_e \geq 1 \quad \text{for all } A \subset V \text{ with } |A \cap S| \neq |A \cap D| \quad (25.24)$$
$$x_e \in \{0, 1\}.$$

An important property of this formulation is summarized in the the following theorem.

Theorem 25.5 *The inequality*

$$\sum_{e \in \delta(A)} x_e \geq 1 \quad \text{for all } A \subset V \text{ with } |A \cap S| \neq |A \cap D|$$

is a valid inequality, i.e., it must be satisfied by any feasible solutions $x \in \{0,1\}^E$ for the PPCP.

The valid inequality (25.24) can be used to provide a branch-and-cut algorithm for the PPCP. The basic strategy consists of adding only a small number of inequalities of the type (25.24), and then check if the solution of the resulting formulation is optimal. We can check if a solution does not violate any inequalities of the type (25.24) using the following algorithm: given a solution x to the integer program above, create a network with underlying graph $G = (V, E)$ and capacities $cap(e) = x_e$, for $e \in E(G)$. For each combination $v_s \in S$, $v_t \in D$ of source-destination pairs, run a maximum flow algorithm from v_s to v_t. If the maximum flow between any two source-destination pair has value less than one, this means that one of the inequalities (25.24) has not been satisfied. It is then easy to identify such inequality, since it corresponds to the minimum cut corresponding to the maximum flow computed previously.

Heuristic approaches for the PPCP have been proposed by Correa et al. (2003) and Gomes et al. (1998). The technique used for solving the problem relies on the cooperation of multiple algorithms, also called agents, that manipulate a population of feasible solutions for the problem. The technique, called *asynchronous teams* (A-Teams) combines several agents to optimize an objective function over the set of feasible solutions for a combinatorial problem (Talukdar and de Souza, 1990). An asynchronous teams strategy for the PPCP was proposed in Gomes et al. (1998), with results close to the optimal for most of the tested instances. The asynchronous nature of the A-Teams metaheuristic makes it very natural to use parallel and distributed algorithms to implement it. A parallel implementation of the A-Teams for the PPCP was proposed by Correa et al. (2003), showing a clear improvement of solution values with increasing number of processors.

25.5.2 Streaming cache placement problems in multicast routing

Although multicasting is an important operation in modern networks, most of the current routers using the TCP/IP protocol do not support multicasting by default. Therefore, the economical implementation of multicast systems is a practical issue, and one of the challenges in the design of multicast systems.

One of the models for the design of economic multicast networks is the *cache placement problem* (CPP). The main issue addressed by the CPP is that of restricting the number of replication nodes, also known as *cache nodes*. Notice however, that due to capacity constraints, a minimum number of cache nodes is required to simultaneously transfer data from the source to all destinations. The objective of the CPP is to find the minimum number of nodes necessary to satisfy all demand, given the capacity constraints.

Using a graph model to describe the CPP, we are given a graph $G = (V,E)$ with capacities c_e on each edge $e \in E$, a source node s and a set of destinations $D \subset V(G)$. Then, the objective of the CPP consists of finding a set R of replication nodes, also known as *cache nodes*, such that all destinations receive the required data, and all capacity constraints are satisfied. To better describe the requirements of the CPP, we discuss the following integer programming model of the problem. Let the integer variables y, x, b, and w be described as

$$y_e = \begin{cases} 1 & \text{if edge } e \text{ is in the spanning tree } T \\ 0 & \text{otherwise,} \end{cases}$$

$$x_i = \begin{cases} 1 & \text{if node } i \neq s \text{ is a cache node} \\ 0 & \text{otherwise} \end{cases}$$

$$b_i \in \{-1,\ldots,|V|-1\} \quad \text{the flow surplus for node } i \in V$$

$$w_e \in \{0,\ldots,|V|\} \quad \text{the amount of flow in edge } e \in E.$$

Where A is the node-arc incidence matrix corresponding to the graph G. The problem can now be stated as

$$\min \sum_{i=1}^{|V|} x_i \qquad (25.25)$$

subject to

$$Aw = b \qquad (25.26)$$
$$\sum_{i \in V} b_i = 0 \qquad (25.27)$$
$$b_s \geq 1 \quad \text{for source } s \qquad (25.28)$$
$$x_i - 1 \leq b_i \leq x_i(|V| - 1) - 1 \quad \text{for } i \in D \qquad (25.29)$$
$$x_i \leq b_i \leq x_i(|V| - 1) \quad \text{for } i \in V - (D \cup \{s\}) \qquad (25.30)$$
$$\sum_{e \in E} y_e = |V| - 1 \qquad (25.31)$$
$$\sum_{e \in G(H)} y_e \leq |H| - 1 \quad \text{for all } H \subset V \qquad (25.32)$$
$$0 \leq w_e \leq c_e y_e \quad \text{for } e \in E \qquad (25.33)$$
$$x \in \{0,1\}^{|V|}, \quad y \in \{0,1\}^{|E|} \qquad (25.34)$$
$$b \in Z, \quad w \in Z^+, \qquad (25.35)$$

where $G(H)$ is the subgraph induced by the nodes in H.

Recall that in the CPP, as in the MRP, a feasible solution is a tree, rooted at the source node s, connecting s to the destinations. Another version of the CPP arises when we relax the constraints (25.31) and (25.32), which define the solution as a tree on G. In this case, we have a problem where the objective is to send data from source to destinations using any feasible flow, not restricted to a tree. The resulting problem is called the *flow version* of the CPP. Notice that this makes sense in the CPP because only a small part of the nodes will be elements of R, and other nodes can receive flow according to multiple configurations that are not necessarily a tree.

Despite the importance of the CPP for the economical design of multicast networks, only recently it has been studied; an initial result shows that the problem is NP-hard, as stated in the following theorem (Mao et al., 2003; Oliveira et al., 2003; Mao et al., 2003):

Theorem 25.6 *The CPP in the tree-version ((25.25) to (25.35)) is NP-hard. Moreover the result is true for the flow version of the CPP, as well as the cases where the underlying graph is directed or undirected.*

Possible techniques for solving the CPP include heuristic, and enumerative methods. Heuristic algorithms are important for this problem, since it is difficult to find approximation algorithms with good approximation guarantee. In fact, it has been shown in Oliveira and Pardalos (2005), using a reduction from SET COVER (Feige, 1998), that the problem cannot be approximated by a factor better than $\log |D|$, unless NP is easy to solve:

Theorem 25.7 *If there is some $\varepsilon > 0$ such that a polynomial time algorithm A can approximate CPP within $(1-\varepsilon)\log k$, where $k = |D|$, then $NP \subset TIME(n^{O(\log \log n)})$.*

The theorem above shows that no algorithm can guarantee a good solution for the CPP problem in general. Nonetheless, heuristics proposed in Oliveira and Pardalos (2005) were demonstrated to return near optimal solutions to the CPP for most instances. It is an open problem to find algorithms for the CPP that can match the approximation lower bound given above — while the techniques developed for SET COVER appear to be a good starting point, they do not seem be directly applicable in solving the CPP.

25.6 CONCLUDING REMARKS

In this chapter, we discussed several optimization problems occurring in multicast networks. The rich combinatorial structure of multicasting makes it possible to provide several perspectives of the problem, according to the different and sometimes unrelated objectives that must be optimized, as well as the various constraints that must be satisfied by practical systems.

Most of the problems concerning multicast routing present interesting challenges for researchers working in combinatorial optimization. First, the problems in this area are of practical interest, since most network applications could benefit from good algorithms for the discussed problems. Second, many of these problems have not been fully explored using tools of optimization. As shown in this chapter, most algorithms are only concerned with the generation of feasible solutions, with some additional requirements used to maintain a minimum quality of service. It remains an open question how the solutions returned by the discussed algorithms can be improved by the application of optimization techniques, such as local search optimization, metaheuristics, branch-and-cut/branch-and-price, and approximation algorithms. Therefore, we believe that important advances can be made in the near future by the application of optimization techniques to problems in multicast routing.

Acknowledgments

This work was partially supported by NSF and US Air Force grants.

Bibliography

L. Aguilar, J.J. Garcia-Luna-Aceves, D. Moran, E.J. Graighill, and R. Brungardt. Architecture for a multimedia teleconferencing system. In *Proceedings of the ACM SIGCOMM*, pages 126–136, Baltimore, Maryland, 1986. Association for Computing Machinery.

M. Baldi, Y. Ofek, and B. Yener. Adaptive real-time group multicast. In *Proceedings of IEEE INFOCOM'97*, page 683, 1997.

A. Ballardie, P. Francis, and J. Crowcroft. Core-based trees (CBT) – An architecture for scalable inter-domain multicast routing. *Computer Communication Review*, 23(4):85–95, 1993.

Z. Baoxian, L. Yue, and C. Changjia. An efficient delay-constrained multicast routing algorithm. In *International Conference on Communication Technologies (ICCT 2000)*, page S07.2, 2000.

F. Bauer. *Multicast routing in point-to-point networks under constraints*. PhD thesis, University of California, Santa Cruz, 1996.

F. Bauer and A. Varma. Degree-constrained multicasting in point-to-point networks. In *Proceedings IEEE INFOCOM '95, The Conference on Computer Communications*, pages 369–376, 1995.

F. Bauer and A. Varma. ARIES: A rearrangeable inexpensive edge-based on-line Steiner algorithm. *IEEE Journal of Selected Areas in Communications*, 15(3):382–397, 1997.

M. Bazaraa, J. Jarvis, and H. Sherali. *Linear Programming and Network Flows*. John Wiley and Sons, 2nd edition, 1990.

R. Bellman. *Dynamic Programming*. Princeton University Press, Princeton, NJ, 1957.

L.T.M. Berry. Graph theoretic models for multicast communications. *Computer Networks and ISDN Systems*, 20(1):95–99, 1990.

K. Bharath-Kumar and J.M. Jaffe. Routing to multiple destinations in computer networks. *IEEE Transactions on Communications*, 31(3):343–351, 1983.

D. Blokh and G. Gutin. An approximate algorithm for combinatorial optimization problems with two parameters. *Australasian J. Combin.*, 14:157–164, 1996.

K.L. Calvert, E.W. Zegura, and M.J. Donahoo. Core selection methods for multicast routing. In *IEEE ICCCN'95*, pages 638–642, Las Vegas, Nevada, 1995. IEEE.

K. Carlberg and J. Crowcroft. Building shared trees using a one-to-many joining mechanism. *ACM Computer Communication Review*, 27(1):5–11, January 1997.

G. Chen, M. Houle, and M. Kuo. The Steiner problem in distributed computing systems. *Information Sciences*, 74(1):73–96, 1993.

S. Chen, O. Günlük, and B. Yener. Optimal packing of group multicastings. In *Proc. IEEE INFOCOM'98*, pages 980–987, 1998.

C. Chiang, M. Gerla, and L. Zhang. Adaptive shared tree multicast in mobile wireless networks. In *Proceedings of GLOBECOM'98*, pages 1817–1822, 1998.

G.V. Chockler, N. Huleihel, I. Keidar, and D. Dolev. Multimedia multicast transport service for groupware. In *TINA Conference on the Convergence of Telecommunications and Distributed Computing Technologies*, pages 43–54, 1996.

C.H. Chow. On multicast path finding algorithms. In *Proc. IEEE INFOCOMM'91*, pages 1274–1283, 1991.

S.-J. Chung, S.-P. Hong, and H.-S. Huh. A fast multicast routing algorithm for delay-sensitive applications. In *IEEE GLOBECOM'97*, pages 1898–1902, 1997.

R. Correa, F. Gomes, C.A.S. Oliveira, and P.M. Pardalos. A parallel implementation of an asynchronous team to the point-to-point connection problem. *Parallel Computing*, 29(4):447–466, 2003.

Y. Dalal and R. Metcalfe. Reverse path forwarding of broadcast packets. *Communications of the ACM*, 21(12), 1978.

S. Deering and D. Cheriton. Multicast routing in datagram internetworks and extended LANs. *ACM Transactions on Computer Systems*, 8(2):85–110, 1990.

S. Deering, D. Estrin, D. Farinacci, V. Jacobson, C.-G. Liu, and L. Wei. An architecture for wide-area multicast routing. *Computer Communication Review*, 24(4): 126–135, 1994.

S. Deering, D.L. Estrin, D. Farinacci, V. Jacobson, C.-G. Liu, and L. Wei. The PIM architecture for wide-area multicast routing. *IEEE/ACM Transactions on Networking*, 4(2):153–162, 1996.

E.W. Dijkstra. A note on two problems in connexion with graphs. *Numer. Math.*, 1: 269–271, 1959.

M. Doar and I. Leslie. How bad is naive multicast routing. In *Proceedings of the IEEE INFOCOM*, pages 82–89, Los Alamitos, Calif, USA, 1993. IEEE Comput. Soc. Press.

D.-Z. Du, B. Lu, H. Ngo, and P.M. Pardalos. Steiner tree problems. In C.A. Floudas and P.M. Pardalos, editors, *Encyclopedia of Optimization*, volume 5, pages 227–290. Kluwer Academic Publishers, 2001.

D.-Z. Du and P.M. Pardalos, editors. *Network Optimization Problems: Algorithms, Complexity and Applications*. World Scientific, 1993a.

D.-Z. Du and P.M. Pardalos. Subset interconnection designs: Generalizations of spanning trees and Steiner trees. In *Network Optimization Problems*, pages 111–124. World Scientific, 1993b.

C.A. Ellis, S.J. Gibbs, and G.L. Rein. Groupware: Some issues and experiences. *Commun. ACM*, 34(1):39–58, 1991.

H. Eriksson. MBONE – the multicast backbone. *Communications of ACM*, 37(8): 54–60, 1994.

U. Feige. A threshold of $\ln n$ for approximating set cover. *Journal of the ACM*, 45(4): 634–652, 1998.

G. Feng and T. Peter Yum. Efficient multicast routing with delay constraints. *International Journal of Communication Systems*, 12:181–195, 1999.

A. Forsgren and M. Prytz. Dimensioning multicast-enabled communications networks. *Networks*, 39:216–231, 2002.

A.J. Frank, L.D. Wittie, and A.J. Bernstein. Multicast communication on network computers. *IEEE Software*, 2(3):49–61, 1985.

R.G. Gallager, P.A. Humblet, and P.M. Spira. A distributed algorithm for minimum-weight spanning trees. *ACM Trans. Programming Languages and Systems*, 5(1): 66–77, 1983.

M.R. Garey and D.S. Johnson. *Computers and Intractability: A Guide to the Theory of NP-completeness*. W.H. Freeman and Company, 1979.

M.X. Goemans and D.P. Williamson. A General Approximation Technique for Constrained Forest Problems. *SIAM J. Comp.*, 24(2):296–317, 1995.

F.C. Gomes, C.N. Meneses, A.R.G. Lima, and C.A.S. Oliveira. Asynchronous organizations for solving the point-to-point connection problem. In *Proc. of the Intl. Conference on Multiagents Systems (ICMAS)*, pages 144–149. IEEE Computer Society, 1998.

L. Han and N. Shahmehri. Secure multicast software delivery. In *IEEE 9th International Workshops on Enabling Technologies: Infrastructure for Collaborative Enterprises (WET ICE'00)*, pages 207–212, 2000.

S. Hong, H. Lee, and B.H. Park. An efficient multicast routing algorithm for delay-sensitive applications with dynamic membership. In *Proceedings of IEEE INFOCOM'98*, pages 1433–1440, 1998.

F. Hwang and D. Richards. Steiner tree problems. *Networks*, 22:55–89, 1992.

F. Hwang, D.S. Richards, and P. Winter. *The Steiner tree problem*, volume 53 of *Annals of Discrete Mathematics*. North-Holland, 1992.

IBM. Reliable multicast messaging, 2005. Available at http://www.haifa.il.ibm.com/projects/software/rmsdk.

Y. Im, Y. Lee, S. Wi, and Y. Choi. A delay constrained distributed multicast routing algorithm. *Computer Communications*, 20(1):60–66, 1997.

X. Jia. A distributed algorithm of delay-bounded multicast routing for multimedia applications in wide area networks. *IEEE/ACM Transactions on Networking*, 6(6): 828–837, 1998.

X. Jia, N. Pissinou, and K. Makki. A real-time multicast routing algorithm for multimedia applications. *Computer Commun. J.*, 20(12):1098–1106, 1997.

X. Jiang. Routing broadband multicast streams. *Computer Communications*, 15(1): 45–51, 1992.

Chee Kheong, David Siew, and Gang Feng. Efficient setup for multicast connections using tree-caching. In *Proceedings IEEE INFOCOM 2001*, pages 249–258, 2001.

V. Kompella, J. Pasquale, and G. Polyzos. Multicasting for multimedia applications. In *Proceedings of IEEE INFOCOM'92*, pages 2078–2085, 1992.

V. Kompella, J. Pasquale, and G. Polyzos. Two distributed algorithms for the constrained Steiner tree problem. In *Proceedings of the Second International Conference on Computer Communications and Networking (ICCCN'93)*, pages 343–349, 1993a.

V.P. Kompella, J.C. Pasquale, and G.C. Polyzos. Multicast routing for multimedia communication. *IEEE/ACM Trans. Networking*, 1(3):286–292, 1993b.

V.P. Kompella, J.C. Pasquale, and G.C. Polyzos. Optimal multicast routing with quality of service constraints. *Journal of Network and Systems Management*, 4(2):107–131, 1996.

L. Kou, G. Markowsky, and L. Berman. A fast algorithm for Steiner trees. *Acta Informatica*, 15:141–145, 1981.

C. Li, S. McCormick, and D. Simchi-Levi. The point-to-point delivery and connection problems: Complexity and algorithms. *Discrete Applied Math.*, 36:267–292, 1992.

Z. Li and P. Mohapatra. QMBF: A QoS-aware multicast routing protocol. *Computer Communications Journal*, 26(6):611–621, 2003.

Z.M. Mao, D. Johnson, O. Spatscheck, J E. van der Merwe, and J. Wang. Efficient and robust streaming provisioning in VPNs. In *Proceedings of the WWW2003*, Budapest, Hungary, May 2003.

C.N. Meneses, C.A.S. Oliveira, and P.M. Pardalos. A branch-and-cut to the point-to-point connection problem on multicast networks. In F. Giannessi and A. Maugeri, editors, *Variational Analysis and Applications*, page 1084. Kluwer Academic Publishers, Dordrecht, 2004.

J. Moy. Multicast extensions to OSPF, RFC 1584 – IETF network working group. On-line document: http://www.ietf.org/, 1994a.

J. Moy. OSPF version 2, RFC 1583 – IETF network working group. On-line document: http://www.ietf.org/, 1994b.

M.G. Natu. *Network Loading and Connection Problems*. PhD thesis, Operations Research Dept., North Carolina State University, 1995.

C. Noronha and F. Tobagi. Optimum routing of multicast streams. In *IEEE INFO-COM'94*, pages 865–873, 1994.

C.A.S. Oliveira and P.M. Pardalos. Construction algorithms and approximation bounds for the streaming cache placement problems in multicast networks. *Cybernetics and systems Analysis*, 41, 2005. To appear.

C.A.S. Oliveira, P.M. Pardalos, O.A. Prokopyev, and M.G.C. Resende. Streaming cache placement problems: Complexity and algorithms. Technical report, University of Florida, 2003.

P.M. Pardalos and D.-Z. Du, editors. *Network Design: Connectivity and Facilities Location*, volume 40 of *DIMACS Series on Discrete Mathematics and Theoretical Computer Science*. American Mathematical Society, 1998.

P.M. Pardalos, F. Hsu, and S. Rajasekaran, editors. *Mobile Networks and Computing*, volume 52 of *DIMACS Series on Discrete Mathematics and Theoretical Computer Science*. American Mathematical Society, 2000.

P.M. Pardalos and B. Khoury. An exact branch and bound algorithm for the Steiner problem in graphs. In D.-Z. Du & M. Li, editor, *Proceedings of COCOON'95*, volume 959 of *Lecture Notes in Computer Science*, pages 582–590. Springer-Verlag, Xi'an, China, 1995.

P.M. Pardalos and B. Khoury. A heuristic for the Steiner problem on graphs. *Comp. Opt. & Appl.*, 6:5–14, 1996.

P.M. Pardalos, B.N. Khoury, and D.-Z. Du. A test problem generator for the steiner problem in graphs. *ACM Transactions on Mathematical Software*, 19(4):509–522, 1993.

J. Park and C. Park. Development of a multi-user & multimedia game engine based on TCP/IP. In *Proceedings of IEEE Pacific Rim Conference on Communications Computers and Signal Processing*, pages 101–104, Victoria, B.C. Canada, 1997.

J. Pasquale, G. Polyzos, and G. Xylomenos. The multimedia multicasting problem. *ACM Multimedia Systems Journal*, 6(1):43–59, 1998.

P. Paul and S.V. Raghavan. Survey of multicast routing algorithms and protocols. In *Proceedings of the Fifteenth International Conference on Computer Communication (ICCC 2002)*, pages 902–926, 2002.

V. Priwan, H. Aida, and T. Saito. The multicast tree based routing for the complete broadcast multipoint-to-multipoint communications. *IEICE Transactions on Communications*, E78-B(5):720–728, 1995.

M. Prytz. *On optimization in Design of Telecommunications Networks with Multicast and Unicast Traffic*. PhD thesis, Dept. of Mathematics, Royal Institute of Technology, Stockholm, Sweden, 2002.

M. Prytz and A. Forsgren. Dimensioning of a multicast network that uses shortest path routing distribution trees. Technical Report TRITA-MAT-2002-OS1, Department of Mathematics, Royal Institute of Technology, Stockholm, Sweden, 2002.

S. Ramanathan. Multicast tree generation in networks with asymmetric links. *IEEE/ACM Trans. Networking*, 4(4):558–568, 1996.

G.N. Rouskas and I. Baldine. Multicast routing with end-to-end delay and delay variation constraints. In *IEEE INFOCOM'96*, pages 353–360, 1996.

S. Sabri and B. Prasada. Video conferencing systems. *Proc. of the IEEE*, 73(4):671–688, 1985.

H.F. Salama, D.S. Reeves, and Y. Viniotis. Shared multicast trees and the center selection problem: A survey. Technical Report TR-96/27, Dept. of Electrical and Computer Engineering, NCSU, 1996.

H.F. Salama, D.S. Reeves, and Y. Viniotis. The delay-constrained minimum spanning tree problem. In *2nd IEEE Symposium on Computers and Communications (ISCC'97)*, pages 699–704. IEEE Computer Society, 1997a.

H.F. Salama, D.S. Reeves, and Y. Viniotis. Evaluation of multicast routing algorithms for real-time communication on high-speed networks. *IEEE Journal on Selected Areas In Communications*, 15(3):332–345, 1997b.

C. Semeria and T. Maufer. Introduction to IP multicast routing. Internet draft (IETF), 1996.

A. Shaikh and K.G. Shin. Destination-driven routing for low-cost multicast. *IEEE Journal of Selected Areas in Communications*, 15(3):373–381, 1997.

R. Sriram, G. Manimaran, and C. Siva Ram Murthy. Algorithms for delay-constrained low-cost multicast tree construction. *Computer Communications*, 21(18):1693–1706, 1998.

R. Sriram, G. Manimaran, and C. Siva Ram Murthy. A rearrangeable algorithm for the construction of delay-constrained dynamic multicast trees. *IEEE/ACM Transactions on Networking*, 7(4):514–529, 1999.

H. Takahashi and A. Matsuyama. An approximate solution for the Steiner problem in graphs. *Mathematica Japonica*, 24(6):573–577, 1980.

S.N. Talukdar and P.S. de Souza. Asynchronous Teams. In *Second SIAM Conf. on Linear Algebra: Signals, Systems, and Control, San Francisco*, 1990.

T.M. Thomas II. *OSPF network design solutions*. Cisco Systems, 1998.

D. Waitzman, C. Partridge, and S. Deering. Distance Vector Multicast Routing Protocol, RFC 1075 – IETF Network Working Group. On-line document: http://www.ietf.org/, 1988.

D. Wall. *Mechanisms for Broadcast and selective broadcast*. PhD thesis, Stanford University, 1980.

D.W. Wall. *Mechanisms for broadcast and selective broadcast*. PhD thesis, Computer Science Department, Stanford University, 1982.

P.-J. Wan, D.-Z. Du, and Panos M. Pardalos, editors. *Multichannel Optical Networks: Theory and Practice*, volume 46 of *DIMACS Series on Discrete Mathematics and Theoretical Computer Science*. American Mathematical Society, 1998.

C.-F. Wang, C.-T. Liang, and R.-. Jan. Heuristic algorithms for packing of multiple-group multicasting. *Computers & Operations Research*, 29(7):905–924, 2002.

B.M. Waxman. Routing of multipoint connections. *IEEE Journal on Selected Areas in Communications*, 6(9):1617–1622, 1988.

L. Wei and D. Estrin. The trade-offs of multicast trees and algorithms, 1994.

S. Wi and Y. Choi. A delay-constrained distributed multicast routing algorithm. In *Proceeding of the Twelfth International Conference on Computer Communication (ICCC'95)*, pages 883–838, 1995.

P. Winter. Steiner problem in networks: A survey. *Networks*, 17:129–167, 1987.

P. Winter and J.M. Smith. Path-distance heuristics for the Steiner problem in undirected networks. *Algorithmica*, pages 309–327, 1992.

W. De Zhong, Y. Onozato, and J. Kaniyil. A copy network with shared buffers for large-scale multicast ATM switching. *IEEE/ACM Transactions on Networking*, 1(2):157–165, 1993.

IV Reliability, restoration, and grooming

26 NETWORK RELIABILITY OPTIMIZATION

Abdullah Konak[1] and Alice E. Smith[2]

[1] Information Sciences and Technology
Penn State Berks
Reading, PA 19610 USA
konak@psu.edu

[2] Industrial and Systems Engineering
Auburn University
Auburn University, AL 36849 USA
aesmith@eng.auburn.edu

Abstract: This chapter presents design of reliable networks. The exact calculation of any general network reliability measure is NP-hard. Therefore, network designers have been reluctant to use reliability as a design criterion. However, reliability is becoming an important concern to provide continuous service quality to network customers. The chapter discusses various network reliability measures and efficient techniques to evaluate them. Two genetic algorithms are presented to demonstrate how these techniques to estimate and compute network reliability can be incorporated within an optimization algorithm. Computational experiments show that the proposed approaches significantly reduce computational effort without compromising design quality.

Keywords: Network reliability, network resilience, network design, network survivability.

26.1 INTRODUCTION

While planning a telecommunication network, several competing interests such as cost, throughput, performance, connectivity requirements, and reliability must be considered. Among them, reliability has become an important concern in recent decades. Many new telecommunication technologies such as fiber-optic cables and high capacity switches have provided economical benefits by means of capacity concentration (Ball et al., 1995). As a result, telecommunication networks tend to be sparser compared to networks based on traditional copper cables (Balakrishnan et al., 1998). A high capacity sparse network, however, is vulnerable to component (links and nodes) failures. Even a single component failure can significantly disturb the service quality of a network or leave many customers disconnected. Therefore, as the dependence on telecommunication networks increases and network topologies become sparser, network reliability becomes an important concern while designing new networks.

In the most general form, network reliability describes the ability of a network to continue network services in the case of component failures. This chapter focuses on designing reliable networks. The most challenging aspect of this problem is computing a reliability measure of a network. The exact calculation of most reliability measures is NP-hard (Ball, 1980). Therefore, an overwhelming body of research on network reliability has focused on developing efficient techniques to evaluate network reliability, including exact methods, theoretical bounds, and simulation. However, work on optimal reliable network design did not fully exploit or implement these efficient techniques in an optimization framework.

The chapter is organized as follows: Section 26.2 presents network reliability modeling and major reliability measures. Efficient evaluation of network reliability is very important for optimal network design. Therefore, the methods to evaluate network reliability measures are given in Section 26.3. Existing work on network reliability optimization is summarized in Section 26.4. Sections 26.5 presents a genetic algorithm (GA) to design reliable networks with an emphasis on demonstrating use of efficient reliability evaluation in a search algorithm. Section 26.6 presents a bi-objective GA to design resilient networks.

26.2 NETWORK RELIABILITY MODELING

Telecommunication networks consist of imperfect components. Both the links and the nodes of a network are subject to failure. Failure mechanisms of network components, especially those of links, have not been well defined in the literature (Ball et al., 1995). Nonetheless, failure rates can be derived from historical data. A telecommunication network with unreliable components is usually modeled as an undirected probabilistic network $G = (E, V)$ with node set V and arc set E such that each arc can be in either of two states: operative or failed, with associated probabilities p_{ij} and $1 - p_{ij}$, respectively. In this model, nodes represent telecommunication devices such as routers, switching stations, and computers, and arcs represent links connecting these devices. Although p_{ij} can be interpreted differently, it is usually defined as the probability that arc (i, j) is in the operative state at a random point in time. The common assumptions of this model are:

- Arc failures are independent;
- Nodes are perfectly reliable;
- No repair is allowed.

Hence, the probability of observing a particular state of the network is as follows:

$$Pr\{\mathbf{X}\} = \prod_{(i,j) \in E} [1 - p_{ij} + x_{(i,j)}(2p_{ij} - 1)] \qquad (26.1)$$

where $x_{(i,j)} = 1$ if arc (i, j) is operative in state $\mathbf{X}$, $x_{(i,j)} = 0$ otherwise.

The assumptions of independent component failures and perfectly reliable nodes have been criticized as being impractical. Although in practice component failures are often observed together, the assumption of independent component failures is very important for computational tractability. Assuming perfectly reliable nodes is not very appropriate for telecommunication networks, either. However, a probabilistic network with unreliable nodes can be transformed to a probabilistic network with perfectly reliable nodes, or node failures can be incorporated into reliability calculations in some cases (Colbourn, 1987). Therefore, most network reliability analysis and optimization papers assume perfectly reliable nodes. In Section 26.6, we relax the assumption of perfectly reliable nodes and use a reliability measure in which node failures are considered.

26.3 NETWORK RELIABILITY MEASURES

Generally, three major types of network reliability measures are considered in the network reliability literature: connectivity, resilience, and performability measures. The majority of the research on network reliability focuses on connectivity measures since the primary function of a telecommunication network is to provide connectivity. With respect to connectivity, network reliability is expressed as the probability that a specified set of nodes (T) of the network are connected at a random point in time. This probability can be computed as follows:

$$R = E[\Phi(\mathbf{X})] = \sum_{\mathbf{X} \in S} Pr\{\mathbf{X}\}\Phi(\mathbf{X}) \qquad (26.2)$$

where S is the state space of all possible network states, and $\Phi(\mathbf{X})$ is a structure function defined as follows:

$$\Phi(\mathbf{X}) = \begin{cases} 1, & \text{if all nodes in } T \text{ are connected in state } \mathbf{X}; \\ 0, & \text{otherwise.} \end{cases} \qquad (26.3)$$

The primary network reliability measures for undirected probabilistic networks are as follows

- *Two-terminal.* Probability that a selected node pair, a source node s, and a sink node t, are connected (i.e., $T = \{s,t\}$).

- *All-terminal.* Probability that every node can communicate with every other node in network G (i.e., $T = V$).

- *K-terminal.* Given a node set $K \subset V$, probability that every node in K can communicate with every other node in K (i.e., $T \subset V$).

A two-terminal measure is used when the communication between two specified nodes of a network is critical (e.g., two metropolitan areas, a file and a web server on different sites). The all-terminal measure is frequently used for the backbone level of packet switched networks since if a path fails in these networks, traffic can be rerouted around alternative paths as long as the network is globally connected. A K-terminal measure is used when connectivity of a subset of network nodes is concern (e.g., virtual local area networks and distributed computing applications). The reliability measures defined above are strongly related with each other. A method that can be applied to one can also be applied to the others. Exact calculation of these reliability measures is NP-hard for general networks (Ball, 1980).

Network resilience measures, in fact, are special cases of connectivity measures. The connectivity measures given above assume that a network is not operational if the desired connectivity is lost. In practice, however, a network continues to serve the remaining connected components even though one or more nodes have become disconnected. To evaluate the ability of a network coping with catastrophic failures and recovering network services in disconnected states, several network resilience measures have been proposed such as the probability that all operative node pairs can communicate and the expected fraction of node pairs communicating (Ball, 1979). Unlike the binary structure function in the connectivity based reliability measures, the structure function is multi-model in resilience measures, which usually makes their calculation harder than reliability measures.

Performance metrics such as response time and throughput are commonly used in design and analysis of telecommunication networks. Network performability measures are concerned with the performance of a network in various network states. Given a performance metric of a network (Ω), performability measures are usually expressed in three cases:

- $Pr\{\Omega \leq \alpha\}$;

- $E[\Omega]$;

- $E[\Omega \mid $ at most k components fail$]$.

The last group of the performability measures is primarily used to reduce the computational complexity by using a k value of usually one or two. The justification for this assumption is that network components are usually so reliable that it is adequate in practice to consider states with no and single failure only since these states cover most of the state-space probability (Sanso et al., 1992).

26.4 EVALUATION OF RELIABILITY

26.4.1 Exact Methods

Methods for exact reliability calculation are two-fold: cut/path set enumeration methods and state enumeration methods (Ball et al., 1995). Cut/path set methods require enumerating all cut/path sets of a network. For example, given all path sets of a network, $P_1, ..., P_h$, let E_i be the event that all arcs in P_i are operational, then network reliability is calculated as

$$R(G) = Pr\{E_1 \cup E_2 \cup ... \cup E_h\} \tag{26.4}$$

Unfortunately, equation (26.4) cannot be calculated easily since the E_i's are not mutually exclusive events and the number of path sets of a network is exponential in the number of arcs. In fact, a naive implementation of equation (26.4) leads to a doubly exponential time algorithm (Ball et al., 1995). For example, the backtracking algorithm (Ball and Slyke, 1977), which is based on generating cut sets, could be used in Deeter and Smith (1998) to design networks with only 5 nodes. The domination theorem developed by Satyanarayana and Prabhakar (1978) provides substantial improvement in the computation of equation (26.4). The most efficient algorithms based on cut/path sets are the algorithms proposed by Ball and Nemhauser (1979) and Provan and Ball (1984), which can compute reliability in polynomial time in the number of minimal path and cut sets.

The most basic state based method is complete state enumeration, requiring the generation of all 2^m states of a network with m arcs. Applying reliability preserving network reductions and network decomposition techniques can significantly reduce the computational effort in state space enumeration. In network reductions, a network G is reduced to a network G' with fewer nodes and/or arcs such that

$$R(G) = \lambda R(G') \qquad (26.5)$$

where λ is a reliability-preserving multiplicative constant depending on reductions. In network decomposition, a network is partitioned into two or more subnetworks; then, the reliabilities of these subnetworks are calculated and used to obtain the reliability of the original network. This technique is successfully used for two-terminal reliability in directed networks (deMercado et al., 1976; Singh and Ghosh, 1994; Hagstrom, 1984). A powerful network decomposition technique is to pivot the reliability expression on the state of an individual arc, which is also known as factoring (Colbourn, 1987; Page and Perry, 1991; Resende, 1986; 1988; Satyanarayana and Chang, 1983). Conditioning on the state of arc (i, j), the reliability expression for $R(G)$ is given by

$$R(G) = R(G \cdot (i,j))p_{ij} + R(G - (i,j))(1 - p_{ij}) \qquad (26.6)$$

where $G \cdot (i, j)$ is a network obtained from G by merging nodes i and j into a new node and connecting each arc incident to them to this new node, and $G - (i, j)$ is a network obtained by deleting arc (i, j) from G. If this decomposition is repeatedly applied with proper arc selection for pivoting and network reductions, significant improvements can be achieved in reliability computation since only relevant states of the network are considered (Satyanarayana and Chang, 1983). In reality, any technique to compute reliability requires exponential time in the worst case. For undirected networks, factoring with network reductions provides the best possible time (see (Ball et al., 1995) for a comprehensive discussion). In Section 26.5, we used a factoring algorithm to calculate the all-terminal measure.

26.4.2 Bounds

Because of the intractability of exact reliability calculation, theoretical bounds on reliability were used as a substitute for actual reliability in several network reliability papers (Jan, 1993; AboElFotoh and Al-Sumait, 2001; Dengiz et al., 1997a;b). Bounds for approximating network reliability can be considered in three groups:

- Bounds based on the reliability polynomial (Slyke and Frank, 1972; Ball and Provan, 1983);

- Bounds based on arc packing by cut or path sets (Lomonosov and Polesskii, 1972; Aboelfotoh and Colbourn, 1989; Brecht and Colbourn, 1988);

- Bounds based on the most probable states (Li and Silvester, 1984; Lam and Li, 1986; Yang and Kubat, 1989).

The first group of bounds depends on counting a small fraction of operational network states to approximate reliability. An important shortcoming of this approach is that it is applicable only to networks with identical arc reliabilities. The second group of bounds considers only a fraction of all possible cut/path sets of a network to obtain a bound. These bounds can be used for networks with arbitrary arc reliabilities. However, they are computationally demanding since they require generating cut/path sets or an effective arc packing of a network. In addition, these bounds can be used only for connectivity based reliability measures with a binary structure function.

The most probable state bounds are based on the observation that the reliabilities of individual components are usually so high that only a small fraction of all possible states will cover the majority of the probability. Although the number of the possible network states is enormous, most of them have extremely low probabilities of occurrence; hence, they can be ignored. The most probable states method requires enumerating k most probable states of a network, $\mathbf{X}^1, \mathbf{X}^2,...,\mathbf{X}^k$, such that $Pr\{\mathbf{X}^1\} \geqslant Pr\{\mathbf{X}^2\} \geqslant,..., \geqslant Pr\{\mathbf{X}^k\}$ and $Pr\{\mathbf{X}^k\} \geqslant Pr\{\mathbf{X}^i\}$ for all remaining states i of the network. Upper and lower bounds on reliability based on the k most probable states are given as follows:

$$R_U(G) = \sum_{i=1}^{k} \Phi(\mathbf{X}^i) Pr\{\mathbf{X}^i\} + (1 - \sum_{i=1}^{k} Pr\{\mathbf{X}^i\})$$

$$R_L(G) = \sum_{i=1}^{k} \Phi(\mathbf{X}^i) Pr\{\mathbf{X}^i\}$$

The tightness of these bounds depends on the number of the states considered. Several methods have been proposed to efficiently enumerate the k most probable states of a network (Li and Silvester, 1984; Lam and Li, 1986; Yang and Kubat, 1989).

26.4.3 Simulation and other Estimation Techniques

Simulation has been a major alternative to estimate new reliability, especially for large networks, due to the intractability of the exact calculation of network reliability and the absence of tight bounds. The application of simulation to estimate new reliability and performability measures is outwardly straightforward. For example, the procedure for the Crude Monte Carlo (CMC) method, which is based on sampling of network states, is given as follows:

Step 1. Set $R = 0$

Step 2. For $k = 1 \ldots K$ do following steps

a. For each $(i,j) \in E$ generate a random number U, then if $p_{ij} \leq U$ then $x_{(i,j)} = 1$ else $x_{(i,j)} = 0$

b. $R = R + \Phi(\mathbf{X})$

Step 3. $\widehat{R}(G) = R/K$

Although the CMC simulation is easy to implement, reliability estimation using simulation is computationally very expensive. A major concern is that networks and components are usually so reliable that large numbers of network states are required to be sampled in order to observe a few network failures and obtain an accurate estimate of reliability. This significantly increases the cost of accurate estimations especially for highly reliable networks. An accurate estimate is particularly important if simulation is to be used within an optimization algorithm to compare alternative designs. Several alternative approaches such as dagger sampling (Kumamoto et al., 1980), stratified sampling (Slyke and Frank, 1972), the Markov model (Mazumdar et al., 1999), the sequential construction/destruction methods (Easton and Wong, 1980; Fishman, 1986a), importance sampling using bounds (Fishman, 1986b), graph evolution method (Elperin et al., 1991) and sampling based on failures sets (Kumamoto et al., 1977) have been proposed in the literature to improve efficiency and effectiveness of simulation in estimating network reliability.

As a new approach to estimate network reliability, Srivaree-ratana et al. (2002) used artificial neural networks to estimate all-terminal and two-terminal reliability.

26.5 RELIABLE NETWORK DESIGN PROBLEM

26.5.1 Problem Formulation

In the most general form, the optimization problem is to find a network topology maximizing the reliability for a given cost constraint or minimizing the design cost for a given reliability constraint. We refer to these two problems as P1 and P2, respectively. As a part of the overall network design problem, however, P2 is a more common problem. In addition to the cost or reliability constraint, there might be some other constraints restricting topologies such as survivability constraints (e.g., node degrees, node or link connectivity requirement) or performance constraints (e.g., network diameter).

Decision variable z_{ij} represents the type of the arc installed between nodes i and j from a discrete set of arc types $\{0,1,...,L\}$ where L is the most reliable arc type and $z_{ij} = 0$ means that no arc is installed on (i, j). Then, the design cost is:

$$C(Z) = \sum_{i=1}^{n} \sum_{j=i+1}^{n} c_{ij}(z_{ij}) \quad (26.7)$$

where $c_{ij}(l)$ is the cost of installing arc type l between nodes i and j. This cost usually depends on the distance between node pairs and may include fixed and variable costs such as cabling, installation costs, and right of way costs. In many formulations of the problem, arc types are ignored and only binary decision variables z_{ij} are used.

The problem formulated in this chapter is to minimize the design cost subject to a given minimum all-terminal reliability requirement R_{min} and a 2-node connectivity constraint. A network is 2-node connected if at least 2 nodes must be removed in order to disconnect the remaining nodes. For fiber-optic networks, 2-node/link connectivity is frequently used to ensure a minimum level of survivability (Monma and Shallcross, 1989).

26.5.2 Solution Approaches

Alternative methods have been proposed in the literature to solve the network design problem considering reliability. Due to the difficultly of the problem and calculating reliability measures, however, few exact methods were developed. Aggarwal et al. (1982a;b) proposed an approach based on enumeration of the spanning trees of the possible network topology. However, this approach is only applicable to very small networks. Jan et al. (1993) developed a branch-and-bound algorithm to minimize the design cost subject to a minimum all-terminal reliability constraint. In this approach, the problem was sequentially solved by considering solutions with only $n-1$, n, $n+1$,...,$n(n-1)/2$ arcs. An upper bound on reliability was used to efficiently eliminate infeasible branches, and a lower bound was used to fathom unpromising subproblems. The approach was tested for networks up to 12 nodes. However, this approach is only applicable to networks with identical arc failure probabilities. In another paper, Jan (1993) investigated the topological features of networks that may lead to high reliability.

Belovich (1995) developed a construction heuristic to enhance the reliability of an existing network by adding the most promising arcs to the network. The reliability metric used in this study was the probability that all packets arrive at their destination, which is estimated by a linear-time upper bound.

Kumar et al. (1995a;b) developed a GA to design and extend existing networks considering various objectives such as reliability, delay, and average nodal distance. Genetic Algorithm

Dengiz et al. (1997a;b) presented a GA approach for the network design problem to maximize all-terminal reliability under a cost constraint. An upper bound on reliability was used to evaluate the fitness of candidate solutions, and simulation was used to estimate the reliability of the best solution. In their GA, the crossover operator was supported with a network repair algorithm to make sure that offspring had a minimum node degree of two. Due to the limitation of the upper bound used in their GA, only arcs with identical reliability were assumed. Deeter and Smith (1998) also proposed a GA to solve both P1 and P2 with different arc types and objective functions (two-terminal and all-terminal reliability). In their method, network reliability was exactly calculated for small size networks and CMC simulation was used for larger networks.

AboElFotoh and Al-Sumait (2001) developed a neural network (NN) approach to solve the problem. In this approach, each arc was represented by a neuron. If an arc was selected in a solution, the corresponding neuron's output became one (i.e., the neuron fired), otherwise zero. The NN aimed to minimize an energy function composed of three terms: the cost of a solution, reliability, and a penalty term for higher reliability than required. To evaluate the reliability term in the energy function, an upper bound was used when arc reliabilities were high or a lower bound was used when arc reliabilities were low. Srivaree-ratana et al. (2002) used a NN approach to estimate all-terminal reliability within a simulated annealing algorithm to evaluate candidate solutions.

26.5.3 A Genetic Algorithm to Design Reliable Networks

In this section, a GA is introduced to solve the formulated problem. The main difference between the GA herein and previous approaches is that the GA uses a combination of techniques for efficient evaluation of all-terminal reliability so that large problems with practical importance can be solved. Unlike some of the previous approaches, the reliability evaluation procedure of the GA is efficient for general types of networks. In addition, the GA employs advanced operators to deal with the 2-node connectivity survivability constraint.

26.5.3.1 Problem Encoding. A node-to-node adjacency matrix representation with arc types is used to represent solutions. In this representation, a network is stored in an $n \times n$ matrix, $Z = \{z_{ij}\}$, such that $z_{ij} = l$ if arc type l is used between nodes i and j, $l = 0,...,L$ where type 1 denotes the least reliable, type L denotes the most reliable arc type, and type 0 means that no arc exists between nodes i and j. If arc types are not considered, then $L = 1$ with $z_{ij} = 1$ denoting that arc (i,j) is selected in the solution, otherwise $z_{ij} = 0$.

26.5.3.2 Crossover Operator. The GA's crossover operator is basic uniform crossover with an efficient repair algorithm for 2-node-connectivity as follows:

Step 1. Randomly select two parents $X = \{x_{ij}\}$ and $Y = \{y_{ij}\}$
Step 2. Set $z_{ij} = \lfloor 1 - U \rfloor x_{ij} + \lceil U \rceil y_{ij}$ for $i = 1,...,n$, $j = i+1,...,n$
Step 3. **If** $Z \subseteq X$ or $Z \subseteq Y$ **Then** return Z and stop.
Step 4. For each node $v = 1,...,n$, perform the following steps.

> Step 4.1 Delete node v from offspring Z, and let s be the node with smallest index in $V \setminus v$.
> Step 4.2 $S(v) = \{s\}$, LIST=$\{s\}$, $\overline{S}(v) = V \setminus v$
> Step 4.3 Select node i from the end of LIST.
> Step 4.4 **If** there exists an arc (i,j) such that $j \in \overline{S}(v)$ **Then** $\overline{S}(v) = \overline{S}(v) \setminus j$, $S(v) = S(v) \cup j$, and LIST=LIST $\cup j$ **Else** LIST=LIST $\setminus i$
> Step 4.5 **If** LIST$\neq \emptyset$ **Then** go to Step 4.3
> Step 4.6 **If** $\overline{S}(v) = \emptyset$ **Then** stop since Z is 2-node connected with respect to v.
> Step 4.7 Find the minimum distance arc (i,j) such that $i \in S(v)$, $j \in \overline{S}(v)$, and $x_{ij} + y_{ij} \geq 1$, and set $z_{ij} = INT(1,L)$.
> Step 4.8 $\overline{S}(v) = \overline{S}(v) \setminus j$, $S(v) = S(v) \cup j$, and LIST=LIST $\cup j$
> Step 4.9 Go to Step 4.3

Step 5. Return Z

26.5.3.3 Mutation Operators. The GA has several mutation operators perturbing network topologies without disturbing 2-node connectivity. To achieve this, the mutation operators create a solution from an existing solution by changing the cycles of the existing solution. This approach was used by Monma and Shallcross (1989) to find minimum cost 2-node and 2-arc connected network topologies. The mutation operators are given below.
Procedure: One-Link-Exchange

Step 1. Find a random cycle C of at least four nodes.
Step 2. Randomly choose four nodes a, b, c, and d of C such that $z_{ab} \geq 1$, $(a,b) \notin C$, and $z_{cd} = 0$.
Step 3. Set $z_{cd} = z_{ab}$ and $z_{ab} = 0$.

Procedure: Two-Link-Exchange

Step 1. Find a random cycle C of at least four nodes.
Step 2. On cycle C, randomly determine two arcs (a,b) and (c,d) such that $z_{ab} \geq 1$, $z_{cd} \geq 1$, $z_{ac} = 0$, and $z_{bd} = 0$.
Step 3. Set $z_{ac} = z_{ab}$, $z_{bd} = z_{cd}$, $z_{ab} = 0$, and $z_{cd} = 0$.

Procedure: Three-Link-Exchange

Step 1. Find a random cycle C of at least four nodes.
Step 2. On cycle C, randomly choose three arcs (a,b), (c,d), and (e,f) such that $z_{ad} = 0$, $z_{be} = 0$, and $z_{cf} = 0$.
Step 3. Set $z_{ad} = z_{ab}$, $z_{be} = z_{ef}$, $z_{cf} = z_{cd}$, $z_{ab} = 0$, $z_{cd} = 0$, and $z_{ef} = 0$.

Procedure: Add-Two-and-Remove-One-Arcs

Step 1. Find a random cycle C of at least four nodes.
Step 2. Randomly choose an arc (a,b) on cycle C such that $z_{ac} = 0$ and $z_{bd} = 0$.
Step 3. Set $z_{ac} = \min(1, z_{ab} - 1)$, $z_{bc} = \min(1, z_{ab} - 1)$, and $z_{ab} = 0$.

Procedure: Add-One-and-Remove-Two-Arcs

Step 1. Find two random adjacent cycles C' and C'' with a common arc (a,b).
Step 2. Remove arc $(a,c) \in C'$ and $(b,d) \in C''$, i.e. $z_{ac} = 0$ and $z_{bd} = 0$.
Step 3. Set $z_{cd} = \max(L, z_{ac} + 1, z_{bd} + 1)$.

Procedure: Delete-an-Arc

Step 1. Find a random cycle C of at least four nodes.
Step 2. Randomly choose two nodes a and b such that $z_{ab} \geq 1$ and $(a,b) \notin C$.
Step 3. Set $z_{ab} = 0$

Procedure: Add-an-Arc

Step 1. Randomly choose two nodes a and b such that $z_{ab} = 0$
Step 2. Set $z_{ab} = INT(1, L)$

Procedure: Change-Arc-Type

Step 1. Randomly choose an arc (a,b) such that $z_{ab} \geq 1$
Step 2. Change type of arc (a,b) to another type.

26.5.3.4 Fitness Function. The mutation and crossover operators of the GA always produce feasible network topologies with respect to the connectivity requirement. However, a candidate solution may violate the reliability constraint. Infeasible solutions with respect to reliability are penalized using a penalty function as follows

$$f(Z,t) = C(Z) + C_{\min}(t)\theta(t)\left(\frac{\max(0, R_{\min} - R(Z))}{R_{\min}}\right) \quad (26.8)$$

where $C_{\min}(t)$ is the cost of the cheapest solution in the population, and $\theta(t)$ is an adaptive penalty factor, which is updated at each generation t as follows

$$\theta(t) = \theta(t-1)(0.5 + \text{fraction of infeasible solutions}) \quad (26.9)$$

Without requiring any parameter setting, this adaptive penalty function is a simplified version of the adaptive penalty function given by Coit and Smith (1996) and Coit et al. (1996). For the first generation, $\theta(1) = 1$, and then in the following generations it is increased if the population includes more infeasible solutions than feasible ones, or it is decreased if the otherwise is true.

26.5.3.5 Calculation of All-terminal Reliability. The reliability evaluation subroutines of the GA include an upper bound for quick assessment of all-terminal reliability, simulation (the Sequential Construction Method (Easton and Wong, 1980)), and an exact method based on the factoring procedure given by Page and Perry (1991) with minor modifications. The overall objective in the reliability evaluation is to minimize rigorous analysis of reliability without losing accuracy. The details of the reliability evaluation is given below.

When a new solution Z is produced, first an upper bound ($R_U(Z)$) on the reliability is calculated as follows:

$$R_U(Z) = 1 - \sum_{i=1}^{n}\left(\left(\prod_{k=1}^{n}(1-p_{ik})\right)\prod_{j=1}^{i-1}\left(1 - \frac{\prod_{k=1}^{n}(1-p_{jk})}{1-p_{ij}}\right)\right) \quad (26.10)$$

where $p_{ij} = 0$ if $z_{ij} = 0$. This bound can be used with arbitrary arc reliabilities, and in addition, it is computationally very efficient since only cut sets separating individual nodes, which can be identified easily, are considered. If $R_U(Z) < R_{\min}$, solution Z is infeasible; therefore, $R(Z)$ is not evaluated further by simulation or factoring, and $R(Z) = R_U(Z)$ is used in the fitness calculation.

If the infeasibility of solution Z is not determined by the upper bound, then simulation or factoring is used to evaluate its reliability. However, if solution Z is not promising, meaning that it has a higher cost than the cost of the *Best Feasible Solution* found so far in the search, a rigorous analysis of reliability is not required. Therefore, the reliability of a non-promising solution is estimated by simulation using a very low number of replications. For a promising solution, factoring or simulation with a high number of replications is used depending on the size and density of the solution. For problems with 10 or less nodes, factoring is used regardless of the density of solutions. For problems with larger than 10 nodes, if $m \leq 1.5n$, factoring is used; otherwise, simulation is used to estimate the reliability.

When simulation is used, the estimated reliability is a random variable. Therefore, to ensure the feasibility of a promising solution Z with $100(1-\alpha)$ percent confidence, the reliability constraint is modified as follows

$$\widehat{R}(G) - z_\alpha \sigma_{\widehat{R}(Z)} \geq R_{\min} \quad (26.11)$$

where $\widehat{R}(G)$ is the estimated all-terminal reliability, and $\sigma_{\widehat{R}(Z)}$ is the standard deviation of the estimation.

26.5.3.6 Overall Algorithm. The GA has a dynamic population size (μ), which is randomly and uniformly selected between μ_{min} and μ_{max} in each generation. In a generation, each solution participates in crossover exactly two times with randomly selected solutions. The mutation rate (MR) determines the probability of solutions being mutated in each generation. In mutation, one of the mutation operators is selected randomly and uniformly. Iterations continue until no new solution improving the Best Feasible Solution is found in the last gn_{max} solutions evaluated or a maximum number of g_{max} new solutions are evaluated. The details of the GA's overall procedure is given as follows:

Step 1. $t = 1$, $\theta(t) = 1$ $\mu(t) = INT(\mu_{min}, \mu_{max})$
Step 2. Generate $\mu(t)$ initial solutions.
Step 3. **Crossover:** Do the following steps for $i = 1, ..., \mu(t) - 1$:

Step 3.1 Crossover the ith and $(i+1)^{th}$ solutions.
Step 3.2 Evaluate and add offspring to the end of the population. Update the Best Feasible Solution if necessary.

Step 4. **Mutation:** Do the following steps for $i = 1, ..., \mu(t)$:

Step 4.1 Generate a random number U. If $U \leq MR$ then randomly and uniformly select a mutation operator, and mutate solution the ith solution.
Step 4.2 Evaluate the mutated solution, update the Best Feasible Solution if necessary, and replace the original solution with the mutated one.

Step 5. Stop if $t \geq g_{max}$ or the Best Feasible Solution has not been updated in last gn_{max} solutions evaluated.
Step 6. Update $\theta(t)$, and calculate the fitness of the population.
Step 7. Sort the population in the descending order of the fitness.
Step 8. $t = t + 1$ and $\mu(t) = \min(\mu(t-1), INT(\mu_{min}, \mu_{max}))$
Step 9. Shuffle the first $\mu(t)$ solutions of the population, and delete the rest while making sure that the Best Feasible Solution stays in the population. Go to Step 3.

26.5.3.7 Computational Experiments. To test the performance of the GA, several problems from the literature are studied. The first set of problems includes 8, 9, 10, 15, 20, and 25-node test problems taken from Dengiz et al. (1997a). These problems have identical arc reliabilities (i.e., no arc choice is available); therefore, the decision variables are binary variables indicating whether to include an arc in a solution or not. In Dengiz et al. (1997a), a GA is used to solve the problems, and a branch-and-bound algorithm is also implemented to find optimal solutions for small problems. The second set of problems is taken from Deeter and Smith (1998), and includes a 10-node problem with three arc reliability choices of .70, .80, and .90 and a 19-node problem with three arc reliability choices of .96, .975, and .99.

For each problem instance, the GA was run 10 times using a different random number seed in each run. The computational results given in Tables 26.1 and 26.3

Table 26.1 Computational results for the problems with identical arc reliability.

n	p	R_{min}	Solutions Search	CPU per Solution	Breakdown of the Computational Effort in Percents			
					Factoring	Simulation $K = 200$	Simulation $K = 20{,}000$	Upper Bound
8	0.90	0.90	11059.0	0.003	4.67	95.33	0.00	0.00
8	0.90	0.95	11229.2	0.003	14.97	75.76	0.00	9.26
8	0.90	0.99	11440.9	0.003	0.12	30.05	0.00	69.83
8	0.95	0.90	10649.7	0.004	0.09	99.90	0.00	0.00
8	0.95	0.95	10526.0	0.004	0.22	99.78	0.00	0.00
8	0.95	0.99	10705.6	0.004	0.25	53.88	0.00	45.86
9	0.90	0.90	11729.9	0.004	4.15	95.84	0.00	0.00
9	0.90	0.95	11705.3	0.003	7.92	64.21	0.00	27.87
9	0.90	0.99	13936.3	0.002	0.19	32.22	0.00	67.58
9	0.95	0.90	12926.0	0.003	0.12	99.88	0.00	0.00
9	0.95	0.95	11438.4	0.004	0.25	99.74	0.00	0.00
9	0.95	0.99	13057.7	0.003	0.81	49.76	0.00	49.42
10	0.90	0.90	11078.1	0.003	1.93	98.06	0.00	0.00
10	0.90	0.95	12549.4	0.003	8.78	57.12	0.00	34.10
10	0.90	0.99	14008.4	0.003	0.20	32.08	0.00	67.72
10	0.95	0.90	11710.4	0.004	0.17	99.83	0.00	0.00
10	0.95	0.95	11564.5	0.004	0.24	99.76	0.00	0.00
10	0.95	0.99	13517.0	0.003	2.86	47.29	0.00	49.85
15	0.90	0.90	22951.1	0.004	36.48	56.64	0.03	6.84
15	0.90	0.95	21954.7	0.004	15.13	43.16	0.18	41.52
15	0.90	0.99	21604.2	0.005	0.00	41.45	0.96	57.60
15	0.95	0.90	16217.2	0.004	0.77	99.20	0.02	0.00
15	0.95	0.95	15167.7	0.004	9.10	90.87	0.03	0.00
15	0.95	0.99	19444.3	0.004	0.90	39.16	0.13	59.79
20	0.90	0.90	31640.3	0.004	32.21	50.12	0.04	17.62
20	0.90	0.95	26509.1	0.008	1.49	43.30	1.31	53.89
20	0.90	0.99	33876.6	0.011	0.00	42.20	2.37	55.43
20	0.95	0.90	22477.0	0.005	6.37	93.61	0.02	0.00
20	0.95	0.95	24829.2	0.005	19.14	80.84	0.02	0.00
20	0.95	0.99	26770.1	0.005	1.30	41.76	0.33	56.60
25	0.90	0.90	42792.1	0.012	31.08	44.59	0.14	24.19
25	0.90	0.95	42863.8	0.036	3.72	43.66	4.15	48.47
25	0.90	0.99	43971.7	0.001	0.00	37.85	1.30	60.85
25	0.95	0.90	31845.4	0.006	7.61	92.36	0.02	0.00
25	0.95	0.95	32155.1	0.006	24.84	75.11	0.03	0.01
25	0.95	0.99	37142.4	0.019	2.38	43.58	1.77	52.27

represent the averaged values over ten runs. The parameters of the GA used in all runs are as follows: $\mu_{min} = 25$, $\mu_{max} = 50$, $MR = 0.1$, $gn_{max} = 10{,}000$, and $g_{max} = 100{,}000$. In simulation, $K = 20{,}000$ is used for promising solutions and $K = 200$ for non-promising solutions.

Table 26.1 lists the computational effort to solve the problems with identical arc reliabilities. In the table, the total number of solutions evaluated is broken down into groups with respect to the reliability evaluation method used. For example, for the 8-node problem with $p = 0.90$ and $R_{min} = .99$, a total of 11,440.9 solutions were evaluated on the average over 10 runs, and factoring was used for .13 percent, simulation with 200 replications for 30.05 percent, and only the upper bound for 69.83 percent to evaluate the solutions. For most cases, factoring or simulation with $K = 20,000$ was used to evaluate a very small fraction of solutions searched. When $R_{min} = .99$, the upper bound was especially effective in identifying infeasible solutions, reducing the need for factoring or simulation. However, the upper bound did not provide useful information when R_{min} and p were low. The highest percent of rigorous evaluation occurred when $R_{min} = .90$. For these cases, however, the CPU time per solution did not increase significantly or even decreased in some cases. The main reason for this is that $R_{min} = .90$ can be achieved with sparse networks with a few arcs, and the reliability of these networks can be computed by factoring using only several pivots and network reductions. Although more solutions were exactly evaluated for these cases, each evaluation took significantly less time, in turn, improving overall CPU time per solution.

Table 26.2 summarizes the results for the problems with identical arc reliabilities. In this table, the cost and reliability of the Best Feasible Solution found, the average best cost over 10 runs, the difference between the cost of the best and the worst solutions found in 10 runs, and the p-value for Best Feasible Solution are given. Here, p-value is the probability that the reliability of the Best Feasible Solution is higher than R_{min}. A p-value of one indicates that all-terminal reliability was exactly computed. In most cases, the GA found the previously reported optimal solutions or improved upon the previous best results. As seen in the table, the GA was very robust over random number seeds. In many cases, the same solution was found in 10 runs or the best and worst solutions were very close. For most cases, the reliability of the Best Feasible Solution was exactly calculated by factoring.

Table 26.3 lists the computational results for the problems with arc choices. Compared with the computational results of Set I, a higher percent of solutions were rigorously evaluated. However, allowing multiple arc choices did not complicate the reliability calculation as reasonable CPU times were observed. Similarly, for high R_{min} values, the upper bound alone was quite useful in identifying infeasible solutions. As seen in Table 26.4, the GA improved upon previous results.

26.6 RESILIENT NETWORK DESIGN PROBLEM

26.6.1 Problem Definition

In Section 26.3, network resilience measures were briefly introduced. In this section, a network resilience measure, called traffic efficiency (T), is used to design resilient networks. The traffic efficiency of a network G is given as

$$T(G) = \frac{1}{\gamma} \sum_{\mathbf{X} \in S} \left(\sum_{i=1}^{n} \sum_{j=i+1}^{n} \tau_{ij}(\mathbf{X}) t_{ij} \right) P\{\mathbf{X}\} \qquad (26.12)$$

Table 26.2 Results for the problems with identical arc reliabilities.

			Best Solution Reported	Results of the Genetic Algorithm				
n	p	R_{min}		Best Cost	Mean Cost	Range	Best Reliability	p-Value
8	0.90	0.90	203*	208.00	208.00	0	0.931722	1.0
8	0.90	0.95	247*	247.00	247.00	0	0.961377	1.0
8	0.90	0.99	-	321.00	321.00	0	0.990744	1.0
8	0.95	0.90	-	173.00	173.00	0	0.942755	1.0
8	0.95	0.95	179*	184.00	184.00	0	0.974181	1.0
8	0.95	0.99	-	247.00	247.00	0	0.991377	1.0
9	0.90	0.90	239*	239.00	239.00	0	0.906564	1.0
9	0.90	0.95	286*	286.00	287.50	15	0.956670	1.0
9	0.90	0.99	-	401.00	401.80	1	0.990750	1.0
9	0.95	0.90	-	204.00	204.50	5	0.928789	1.0
9	0.95	0.95	209*	209.00	209.00	0	0.966935	1.0
9	0.95	0.99	-	286.00	289.00	15	0.990752	1.0
10	0.90	0.90	154*	154.00	154.00	0	0.905014	1.0
10	0.90	0.95	197*	197.00	197.20	2	0.951644	1.0
10	0.90	0.99	-	283.00	283.40	1	0.990793	1.0
10	0.95	0.90	-	136.00	136.00	0	0.961130	1.0
10	0.95	0.95	136*	136.00	136.00	0	0.961130	1.0
10	0.95	0.99	-	206.00	207.80	3	0.990621	1.0
15	0.90	0.90	-	225.00	227.00	4	0.901397	1.0
15	0.90	0.95	317	262.00	263.80	6	0.953308	1.0
15	0.90	0.99	-	373.00	377.00	11	0.990589	0.984
15	0.95	0.90	-	170.00	178.00	29	0.913171	1.0
15	0.95	0.95	-	196.00	198.70	12	0.950277	1.0
15	0.95	0.99	-	262.00	264.70	11	0.990486	1.0
20	0.90	0.90	-	192.00	209.80	40	0.900449	1.0
20	0.90	0.95	-	249.00	257.10	21	0.951573	0.964
20	0.90	0.99	-	357.00	373.60	30	0.990564	0.954
20	0.95	0.90	-	147.00	154.20	16	0.911795	1.0
20	0.95	0.95	926	160.00	165.30	18	0.950021	1.0
20	0.95	0.99	-	248.00	254.80	17	0.990022	1.0
25	0.90	0.90	-	322.00	338.00	41	0.901312	1.0
25	0.90	0.95	-	391.00	405.40	32	0.950441	1.0
25	0.90	0.99	-	518.00	533.10	28	0.991245	0.9998
25	0.95	0.90	1606	247.00	258.60	31	0.903600	1.0
25	0.95	0.95	-	271.00	277.90	20	0.951735	1.0
25	0.95	0.99	-	390.00	407.90	43	0.990752	1.0

* The optimal solution reported in (Dengiz et al., 1997a).

Table 26.3 Computational results for the test problems with different arc reliabilities.

				Breakdown of the Computational Effort in Percent			
n	R_{min}	Solutions Search	CPU per Solution	Factoring	Simulation $K = 200$	Simulation $K = 20,000$	Upper Bound
10	0.900	23468.5	0.00267	36.08	44.85	0.00	19.07
10	0.950	20083.9	0.00267	23.70	41.74	0.00	34.55
10	0.990	24737.9	0.00284	7.97	39.08	0.00	52.94
19	0.900	30123.3	0.00498	3.44	96.54	0.01	0.00
19	0.950	32827.8	0.00501	14.04	85.93	0.02	0.00
19	0.990	34049.4	0.01729	37.08	50.20	4.92	7.80
19	0.995	37834.1	0.02836	13.24	44.59	9.13	33.04

Table 26.4 Results for the test problems with different arc reliabilities.

			Results of the Genetic Algorithm				
n	R_{min}	Best Solution Reported	Best Cost	Mean Cost	Range	Best Reliability	p-Value
10	0.900	-	3792.92	3868.18	166	0.902018	1.0
10	0.950	5661.32	4403.93	4560.89	318	0.950242	1.0
10	0.990	-	5843.50	5936.91	253	0.990014	1.0
19	0.900	-	1292390.00	1428689.60	291536	0.902125	1.0
19	0.950	-	1348283.00	1406466.00	262051	0.952263	1.0
19	0.990	7694708.00	1619322.00	1665539.90	160733	0.990015	0.969
19	0.995	-	1805600.00	1854732.50	103731	0.995006	1.0

where $\tau_{ij}(\mathbf{X}) = 1$ if nodes i and j are connected in state $\mathbf{X}$, $\tau_{ij}(\mathbf{X}) = 0$ if they are not, t_{ij} is the two-way traffic average demand between nodes i and j, and $\gamma = \sum_{i<j\leq n} t_{ij}$. By assigning proper values to t_{ij}, $T(G)$ can represent different network resilience measures. For example, if $t_{ij} = 1$ for each node pair i and j, $T(G)$ gives the expected fraction of node pairs communicating.

There are several important differences between network resilience and connectivity based reliability measures such as all-terminal reliability. First, all-terminal reliability implicitly ignores the effect of node failures in the reliability function since all nodes must be operational as the minimum requirement. The probability that all nodes are operational (i.e., the product of individual node reliabilities) can be incorporated into the reliability function as a constant term after which the reliability analysis is carried out without node failures (Colbourn, 1987). In other words, node failures have the same effect on all-terminal reliability independent of network topology. Therefore, the majority of the research on the reliable network design problem assumes perfectly reliable nodes. On the other hand, node failures must be taken into account while calculating a network resilience measure since a network's service availability in disconnect states is also of interest. While modeling telecommunication networks as probabilistic graphs, in fact, nodes represent complex processing units that are more likely to fail than highly reliable links (cables or microwaves) represented by arcs. Therefore, the assumption of perfectly reliable nodes does not represent reality in telecommunication networks.

Another drawback of all-terminal reliability is that disconnectivity of all nodes is considered equally in the reliability function, meaning that disconnectivity of a node with a large amount of incoming and outgoing traffic and a node with a small amount of traffic have equal weights in the reliability function. In both cases, the network service level is zero (i.e., the network is disconnected). Because of these reasons, network resilience measures are more versatile than pure connectivity based measures such as all-terminal reliability in modeling connectivity of telecommunication networks.

26.6.2 A Bi-objective Genetic Algorithm to Design Resilient Networks

In this section, the resilient network problem is formulated as a bi-objective problem with a 2-node connectivity constraint. Multiple conflicting objectives are common in most real-world telecommunication network design problems. In fact, the overall task of designing networks involves several phases, in which many conflicting objectives are considered. Choosing a network topology is the first step of the network design process. Candidate topologies are refined in the detailed design phases based on the specifications and available technology. Therefore, in topological design, cost and resilience (or reliability) are not hard constraints but competing objectives. The objective in this section is to introduce an approach to aid network designers in choosing a final network design by providing the trade-off curve of cost and network resilience in terms of a set of network topologies that are not inferior to each other with respect to cost or resilience.

26.6.2.1 Bi-objective optimization.
If the objectives under consideration conflict among each other, optimizing with respect to a single objective often results in unacceptable results with respect to the other objectives. Therefore, a perfect multi-objective solution that simultaneously optimizes each objective function is almost impossible. A reasonable solution to a multi-objective problem is to investigate a set of solutions, each of which satisfies the objectives at an acceptable level without being dominated by any other solution. The ultimate goal of a multi-objective optimization is to identify the Pareto optimal set, which is the set of feasible solutions which are not dominated by any other feasible solutions in the solution space.

In our case, a solution X is said to dominate another solution Y if and only if at least one of the following conditions is satisfied.

(i) $C(X) \leq C(Y)$ and $T(X) > T(Y)$
(ii) $C(X) < C(Y)$ and $T(X) \geq T(Y)$.

A feasible solution X is said to be Pareto optimal if it is not dominated by any other feasible solution Y in the solution space. A true Pareto optimal solution cannot be improved with respect to any objective without worsening at least one other objective.

The goal of a multi-objective algorithm is to find the set of all non-dominated solutions in the solution space, called the Pareto set. The representation of Pareto set in the objective space is called Pareto front. Depending on the problem, the Pareto set may be very large, making it very expensive or impossible to investigate fully. In many cases, therefore, investigating a set of solutions uniformly approximating the true Pareto front is preferable.

There are alternative approaches to multi-objective optimization such as, weighted sums of objectives, alternating objectives, and Pareto ranking. In the recent decade, GA has become a popular heuristic tool for multi-objective optimization problems. This popularity can be attributed to its multi-solution approach and its capability to exploit the similarities of Pareto optimal solutions in order to generate new solutions by means of crossover. Interested readers may refer to the comprehensive survey papers by Fonseca and Fleming (1995), Deb (1999), and Van Veldhuizen and Lamont (2000).

26.6.2.2 Overall Optimization Algorithm.
The bi-objective GA uses the same problem encoding scheme, crossover operator, and mutation operators as the single objective GA given in the previous section. Generally, a multi-objective GA has two sets of solutions, the population where general GA operations are applied and the elitist list storing all non-dominated solutions found so far during the search. However, the bi-objective GA does not use an elitist list and its population is made of only non-dominated solutions found so far in the search.

The overall procedure of the bi-objective GA is given below. In each iteration, a single solution is generated by either crossover or mutation, which is selected randomly and uniformly. In mutation, one of the mutation operators is also randomly and uniformly selected. Iterations continue until a maximum number of new solutions (g_{max}) is reached. The overall procedure of the bi-objective GA is as follows:

Step 1. Generate μ_0 nondominated initial solutions. $t = 1$.

Step 2. **If** $U \leq 0.5$, **Then** go to Step 3 **Else** Step 4.
Step 3. Crossover:

   Step 3.1 Randomly select two solutions X and Y, and crossover them to produce new solution Z.
   Step 3.2 Evaluate solution Z, and update the population if necessary. Go to Step 5.

Step 4. Mutation:

   Step 4.1 Randomly select a solution X, and randomly and uniformly choose a mutation operator, and mutate solution X to generate new solution Z.
   Step 4.2 Evaluate solution Z, and update the population if necessary. Go to Step 5.

Step 5. Stop if $t \geq g_{max}$.
Step 6. **If** $mod(t, K_{step}) = 0$, **Then** perform K additional replications for each solution in the population and update estimations using (26.13) and (26.14). Check the population one more time with the new estimates and remove any dominated solutions.
Step 7. Set $t = t + 1$. Go to Step 3.

26.6.2.3 Calculation of Traffic Efficiency and Evaluation of Solutions.
The exact calculation of traffic efficiency, which requires examining 2^{n+m} network states, is intractable. In a sense, the exact calculation of $T(G)$ is more difficult than the exact calculation of $R(G)$ since node failures must be considered. In addition, enumeration schemes such as factoring (Page and Perry, 1991) and domination theory (Satyanarayana and Chang, 1983), which exploit the binary nature of the reliability structure function to improve performance of enumeration, cannot be used with the same efficiency. The most probable state bounds are the only bounds applicable to estimate $T(G)$. However, these bounds were very inconsistent in our initial experiments, meaning that the upper bound did not correlate well with the actual $T(G)$. Therefore, a simulation based on Sequential Construction Sampling (Easton and Wong, 1980) is used to estimate $T(G)$. The details of the simulation procedure will not be included herein.

Estimating traffic efficiency is a very computationally intense operation. To reduce the computational effort at the beginning of the search, $T(G)$ is estimated using a low number of simulation replications assuming that solutions at the beginning are inferior. As the search progresses, the number of replications is gradually increased to the maximum replication number K_{max}, which is done in K_{step} steps by performing K_{max}/K_{step} additional replications for each solution in the population in every new g_{max}/K_{step} solutions evaluated.

This process requires updating the estimate for each solution in the current population, which is performed as follows. Let $T(Z, K_1)$ and $\sigma^2_{T(Z,K_1)}$ be estimated traffic efficiency and the variance of the estimation, respectively, using total K_1 replications. After K_2 additional replications, the new estimate can calculated as follows:

$$T(Z, K_1 + K_2) = \frac{K_1 \times T(Z, K_1) + K_2 \times T(Z, K_2)}{K_1 + K_2} \tag{26.13}$$

Table 26.5 The coordinates of the nodes for the 10- and 20-node problems

node	1	2	3	4	5	6	7	8	9	10
x	26	38	93	74	86	60	26	44	54	52
y	5	86	64	8	61	10	70	70	71	36

node	11	12	13	14	15	16	17	18	19	20
x	57	13	32	93	7	33	54	49	99	50
y	28	68	9	56	42	52	13	3	56	27

with the following variance

$$\sigma^2_{T(Z,K_1+K_2)} = \frac{K_1^2 \times \sigma^2_{T(Z,K_1)} + K_2^2 \times \sigma^2_{T(Z,K_2)}}{(K_1+K_2)^2} \tag{26.14}$$

26.6.2.4 Computational Experiments. A 10-node and a 20-node problem were used to demonstrate the effectiveness of the bi-objective GA. The x and y coordinates of the nodes for both problems are given in Table 26.5. The 10-node problem uses the first 10 nodes of the 20-node problem. The cost of each arc is equal to the Euclidean distance between its two-end nodes. For both problems, all arcs and nodes have reliabilities of .97 and .99, respectively. The parameters of the bi-objective GA in all runs were as follows: $g_{max} = 100,000$, and for simulation, $K_{max} = 40,000$ and $K_{step} = 10$.

Figure 26.1 shows the Pareto front found for the 20-node problem. As seen in the figure, the final Pareto front is diverse, and it significantly improved upon the initial one. A similar result was obtained for the 10-node problem. Figure 26.2 illustrates the network designs found at the opposing ends and in the middle of the Pareto front for the 20-node problem. The least resilient and cheapest solution is made of two cycles; a similar cycle based structure was also observed at the low resilience end of the Pareto front for the 10-node problem. The most resilient and expensive solution found is not fully dense although in theory the fully dense network is the most resilient network. However, after a level of density, adding arcs improves resilience only at a negligible level since node failures becomes main reason for disconnectivity. Therefore, at the high-cost/high-resilience end of the Pareto front, solutions did not reach full-density. Another interesting observation is that the nodes have almost a uniform number of links (3 or 4). When the nodes are subject to failure, over connecting a node really does not really improve resilience. These examples provide convincing evidence that the bi-objective GA is capable of identifying a wide spectrum of solutions for large problems in a single run.

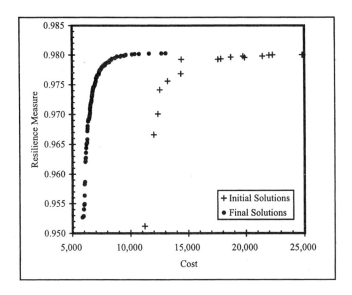

Figure 26.1 The Pareto Front found for the 20-node problem.

26.7 CONCLUSIONS

Network reliability has become important criteria because of recent technological innovations that encourage sparse networks and high traffic. Computing and estimating a network reliability measure is a difficult task. This chapter presented two GAs to solve the reliable and resilient network design problem considering a survivability constraint. The main focus of the algorithms was to efficiently evaluate the reliability of candidate solutions. State-of-the-art techniques are introduced and used to evaluate reliability, making it possible to solve large problems in relatively short time, even by exactly computing reliability.

756   HANDBOOK OF OPTIMIZATION IN TELECOMMUNICATIONS

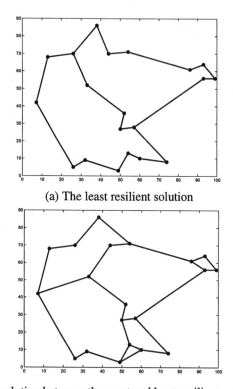

(a) The least resilient solution

(b) A solution between the most and least resilient solutions

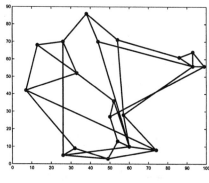

(c) The most resilient solution

Figure 26.2 Sample solutions found for the 20-node problem.

Bibliography

H.M. Aboelfotoh and C.J. Colbourn. Series-parallel bounds for the two-terminal reliability problem. *ORSA Journal on Computing*, 1(4):209–22, 1989.

H.M.F. AboElFotoh and L.S. Al-Sumait. A neural approach to topological optimization of communication networks, with reliability constraints. *IEEE Transactions on Reliability*, 50(4):397–408, 2001.

K.K. Aggarwal, Y.C. Chopra, and J.S. Bajwa. Topological layout of links for optimising the overall reliability in a computer communication system. *Microelectronics and Reliability*, 22(3):347–51, 1982a.

K.K. Aggarwal, Y.C. Chopra, and J.S. Bajwa. Topological layout of links for optimizing the s-t reliability in a computer communication system. *Microelectronics and Reliability*, 22(3):341–5, 1982b.

A. Balakrishnan, T.L. Magnanti, and P. Mirchandani. Designing hierarchical survivable networks. *Operations Research*, 46(1):116–36, 1998.

M.O. Ball. Computing network reliability. *Operations Research*, 27:823–38, 1979.

M.O. Ball. Complexity of network reliability computations. *Networks*, 10(2):153–65, 1980.

M.O. Ball, C.J. Colbourn, and J.S. Provan. *Network reliability*, volume 7 of *Handbooks in Operations Research and Management Science*, pages 673–672. Elsevier Science B.V., Amsterdam, 1995.

M.O. Ball and G.L. Nemhauser. Matroids and a reliability analysis problem. *Mathematics of Operations Research*, 4(2):132–43, 1979.

M.O. Ball and J.S. Provan. Calculating bounds on reachability and connectedness in stochastic networks. *Networks*, 13(2):253–78, 1983.

M.O. Ball and R. Van Slyke. Backtracking algorithms for network reliability analysis. *Discrete Applied Mathematics*, 1:49–64, 1977.

S.G. Belovich. A design technique for reliable networks under a nonuniform traffic distribution. *IEEE Transactions on Reliability*, 44(3):377–87, 1995.

T.B. Brecht and C.J. Colbourn. Lower bounds on two-terminal network reliability. *Discrete Applied Mathematics*, 21(3):185–98, 1988.

D.W. Coit and A.E. Smith. Penalty guided genetic search for reliability design optimization. *Computers and Industrial Engineering*, 30(4):895–904, 1996.

D.W. Coit, A.E. Smith, and D.M. Tate. Adaptive penalty methods for genetic optimization of constrained combinatorial problems. *INFORMS Journal on Computing*, 8(2):173–82, 1996.

C.J. Colbourn. *Combinatorics of network reliability*. Oxford, New York, NY, USA, 1987.

D.L. Deeter and A.E. Smith. Economic design of reliable networks. *IIE Transactions*, 30(12):1161–74, 1998.

J. deMercado, N. Spyratos, and B.A. Bowen. A method for calculation of network reliability. *IEEE Transactions on Reliability*, R25(2):71–6, 1976.

B. Dengiz, F. Altiparmak, and A.E. Smith. Efficient optimization of all-terminal reliable networks, using an evolutionary approach. *IEEE Transactions on Reliability*, 46(1):18–26, 1997a.

B. Dengiz, F. Altiparmak, and A.E. Smith. Local search genetic algorithm for optimal design of reliable networks. *IEEE Transactions on Evolutionary Computation*, 1(3):179–88, 1997b.

M.C. Easton and C.K. Wong. Sequential destruction method for monte carlo evaluation of system reliability. *IEEE Transactions on Reliability*, R-29(1):27–32, 1980.

T. Elperin, I. Gertsbakh, and M. Lomonosov. Estimation of network reliability using graph evolution models. *IEEE Transactions on Reliability*, 40(5):572–81, 1991.

G.S. Fishman. A comparison of four monte carlo methods for estimating the probability of s-t connectedness. *IEEE Transactions on Reliability*, R-35(2):145–55, 1986a.

G.S. Fishman. A monte carlo sampling plan for estimating network reliability. *Operations Research*, 34(4):581–94, 1986b.

J.N. Hagstrom. Using the decomposition tree for directed-network reliability computation. *IEEE Transactions on Reliability*, R-33(5):390–5, 1984.

Rong-Hong Jan. Design of reliable networks. *Computers and Operations Research*, 20(1):25–34, 1993.

Rong-Hong Jan, Fung-Jen Hwang, and Sheng-Tzong Chen. Topological optimization of a communication network subject to a reliability constraint. *IEEE Transactions on Reliability*, 42(1):63–70, 1993.

H. Kumamoto, K. Tanaka, and K. Inoue. Efficient evaluation of system reliability by monte carlo method. *IEEE Transactions on Reliability*, R-26(5):311–15, 1977.

H. Kumamoto, K. Tanaka, K. Inoue, and E.J. Henley. Dagger-sampling monte carlo for system unavailability evaluation. *IEEE Transactions on Reliability*, R-29(2): 122–5, 1980.

A. Kumar, R.M. Pathak, and Y.P. Gupta. Genetic-algorithm-based reliability optimization for computer network expansion. *IEEE Transactions on Reliability*, 44 (1):63–72, 1995a.

A. Kumar, R.M. Pathak, Y.P. Gupta, and H.R. Parsaei. A genetic algorithm for distributed system topology design. *Computers and Industrial Engineering*, 28(3): 659–70, 1995b.

Y.F. Lam and V.O.K. Li. An improved algorithm for performance analysis of networks with unreliable components. *IEEE Transactions on Communications*, COM-34(5): 496–7, 1986.

V.O.K. Li and J.A. Silvester. Performance analysis of networks with unreliable components. *IEEE Transactions on Communications*, COM-32(10):1105–10, 1984.

M.V. Lomonosov and V.P. Polesskii. Lower bound of network reliability. *Problemy Peredachi Informatsii*, 8(2):118–23, 1972.

M. Mazumdar, D.W. Coit, and K. McBride. A highly efficient monte carlo method for assessment of system reliability based on a markov model. *American Journal of Mathematical and Management Sciences*, 19(1-2):115–33, 1999.

C.L. Monma and D.F. Shallcross. Methods for designing communications networks with certain two-connected survivability constraints. *Operations Research*, 37(4): 531–41, 1989.

L.B. Page and J.E. Perry. A note on computing environments and network reliability. *Microelectronics and Reliability*, 31(1):185–6, 1991.

J.S. Provan and M.O. Ball. Computing network reliability in time polynomial in the number of cuts. *Operations Research*, 32(3):516–26, 1984.

L.I.P. Resende. Implementation of a factoring algorithm for reliability evaluation of undirected networks. *IEEE Transactions on Reliability*, 37(5):462–8, 1988.

M.G.C. Resende. A program for reliability evaluation of undirected networks via polygon-to-chain reductions. *IEEE Transactions on Reliability*, R-35(1):24–9, 1986.

B. Sanso, M. Gendreau, and F. Soumis. An algorithm for network dimensioning under reliability considerations. *Annals of Operations Research*, 36(1-4):263–74, 1992.

A. Satyanarayana and M.K. Chang. Network reliability and the factoring theorem. *Networks*, 13(1):107–20, 1983.

A. Satyanarayana and A. Prabhakar. New topological formula and rapid algorithm for reliability analysis of complex networks. *IEEE Transactions on Reliability*, R27(2): 82–100, 1978.

B. Singh and S.K. Ghosh. Network reliability evaluation by decomposition. *Microelectronics and Reliability*, 34(5):925–7, 1994.

R. Van Slyke and H. Frank. Network reliability analysis. i. *Networks*, 1(3):279–90, 1972.

C. Srivaree-ratana, A. Konak, and A.E. Smith. Estimation of all-terminal network reliability using an artificial neural network. *Computers and Operations Research*, 29(7):849–68, 2002.

C.-L. Yang and P. Kubat. Efficient computation of most probably states for communication networks with multimode components. *IEEE Transactions on Communications*, 37(5):535–8, 1989.

27 STOCHASTIC OPTIMIZATION IN TELECOMMUNICATIONS

Alexei A. Gaivoronski[1]

[1] Department of Industrial Economics and Technology Management
Norwegian University of Science and Technology
Trondheim, Norway
Alexei.Gaivoronski@iot.ntnu.no

Abstract: We survey different optimization problems under uncertainty which arise in telecommunications. Three levels of decisions are distinguished: design of structural elements of telecommunication networks, top level design of telecommunication networks, and design of optimal policies of telecommunication enterprise. Examples of typical problems from each level show that the stochastic programming paradigm is a powerful approach for solving telecommunication design problems.

Keywords: Stochastic optimization, telecommunications, network design, performance analysis, models of competition.

27.1 INTRODUCTION

Telecommunications have a long tradition of application of advanced mathematical modeling methods. Besides being a consumer of mathematical modeling, telecommunications provided a motivation for development of areas of applied mathematics. Important chapters of the theory of random processes have their roots in the work of telecommunication engineers. So far, this mutual influence was mainly limited to the queueing theory and the theory of Markov processes, but now new decision problems arise which require the application of optimization methods. The recent trends in telecommunications have led to considerable increase in the level of uncertainty which became persistent and multifaceted. The decision support methodologies which provide adequate treatment of uncertainty are becoming particularly relevant for telecommunications. Stochastic optimization is the methodology of choice for optimal decision support under uncertainty, see Ermoliev and Wets (1988), Kall and Wallace (1994), and Birge and Louveaux (1997). This chapter provides a survey of applications of stochastic programming for solving design and decision problems in telecommunications.

Stochastic optimization is important for a large variety of such problems. We start by defining a classification which will serve as a roadmap for the exposition. This classification is made using the scale of the decision relative to the whole telecommunication environment which defines the nature of decision itself. Besides, different types of uncertainty come into play at different levels. We distinguish three scale levels: technological, network, and enterprise. The technological level corresponds to the smallest scale and the enterprise level to the largest and the most aggregated scale.

The *technological level* deals with design of different elements of telecommunication networks, including switches, routers, multiplexers. Uncertainty on this level is a salient feature of communication requests and flows in the network. Besides, it can arise due to equipment failures. The key decisions are the engineering decisions which define the design for blueprints of these elements. Such blueprints depend on a number of parameters which should be chosen from the point of view of performance and quality of service. Traditionally, performance evaluation of the elements of telecommunication networks was the domain of queueing theory (Medhi, 1991). To be successful the methods of this theory require a specific probabilistic description of the stochastic processes which govern the behavior of communication flows. Usually such description is lacking for the new data services, and when it exists it does not satisfy requirements of the queueing theory. Stochastic optimization may help to obtain the performance estimates in the cases when more traditional methods are difficult to apply. See Section 27.2 for one such example.

Network level problems deal with design and planning of different kinds of networks which may differ by scale and by technology involved. These can be access networks, local area networks, fixed or mobile networks. The decisions involve the placement of processing and link capacities provided by a given technology in a given geographic area with the aim to satisfy aggregated demand for telecommunication services from different user groups. Decisions are often of dynamic nature and include several time periods. The main uncertainty here is due to the demand for telecommunication services. Due to the quantitative and qualitative explosion of such services, this kind of uncertainty increased considerably during the 1990s. There are important additional sources of uncertainty connected with possible network failures and future technology development. Stochastic programming methods provide an added value of identifying the robust network design which within reasonable bounds will accommodate the future demand variations. This is particularly true for stochastic programming problems with recourse and multiperiod stochastic programming problems which provide intelligent means for mediation between different and often conflicting scenarios of the future. While traditional design approach is centered around minimization of the network costs under technological and quality of service constraints, systematic application of stochastic programming techniques includes incorporation of modern tools from corporate finance like evaluation of real options. Comprehensive models which include pricing decisions and binary variables provide a motivation for further development of this methodology. Section 27.3 contains several examples of stochastic optimization models for design problems at the network level. For related examples see Bonatti et al. (1994), Sen et al. (1994), Fantauzzi et al. (1997), Tomasgard et al. (1998), Dempster and Medova (2001), Yen et al. (2001), and Andrade et al. (2002).

Finally, the *enterprise level* is the highest level of aggregation and looks at the telecommunication enterprise as a member of a larger industrial environment which includes other industrial actors and different consumer types. Decisions involve selection of the range of services which the enterprise will provide to the market, strategic investment decisions, pricing policy. Market acceptance of services, innovation process and actions of competition constitute the sources of uncertainty which are not present at the lower levels. Telecommunication and, more generally, information industry differs in important ways from traditional industries due to the rapid pace of innovation. This leads to the absence of perfect markets and to fundamental nonstationarity which makes it difficult to apply traditional microeconomic approaches based on equilibrium. Stochastic programming models enriched with selected notions of game theory can provide more adequate decision recommendations here. We outline one such model in Section 27.4.

There is no rigid boundary between various levels since decisions taken at each level influence decisions on other levels. Further exposition is organized along the lines of this classification. Each section contains examples of stochastic optimization models which illustrate the typical problems of each level. A summary concludes.

27.2 TECHNOLOGICAL LEVEL

The central problem is to find the design parameters of a piece of telecommunication equipment which will assure a given level of performance for a specified class of traffic patterns. We describe this problem on a general level and then provide a specific example which deals with access design of high speed data network.

In the vast majority of practically interesting cases, the performance is measured by the functional

$$F(x,H) = \mathbb{E}_H f(x,\xi) = \int f(x,\xi) dH(\xi) \qquad (27.1)$$

where x is the vector of design parameters, ξ denotes the values of a stochastic process defined on an appropriate probability space which describes the interaction of traffic with device and H is the stationary distribution of this process. The function $f(x,\xi)$ describes the performance of the equipment for a given traffic value ξ and a given value of design parameters x. $\mathbb{E}_H$ is the expected value with respect to H. The values of the performance measure F should belong to the set Φ of admissible values which describes requirements for the grade of service. This set is usually defined by bounds ϕ^- and ϕ^+. In this case parameters x should satisfy

$$\phi^- \leq F(x,H) \leq \phi^+. \qquad (27.2)$$

Usually equipment should work satisfactorily for a sufficiently wide range of admissible traffic patterns which are defined by a set of traffic parameters. Therefore, instead of a single distribution H in (27.1) we have a whole set of distributions G which is defined indirectly by traffic parameters of interest.

Design problem. Find the values of design parameters x for which the values of the performance functional $F(x,H)$ belong to an admissible set Φ for all $H \in G$.

The final design is selected among solutions of this problem by considering additional criteria which may have economic or manufacturing nature.

The function $f(x,\xi)$ from (27.1) is usually known and has a simple analytical structure. The major difficulty here is constituted by the distribution H. The reason is that the traffic is described in terms of characteristics of individual nonhomogeneous traffic sources. Even if description of a single source is relatively simple, the composite traffic consisting of a large amount of such sources can be complex. Complexity increases due to nontrivial interaction of traffic with device. Even the description of a single source can be difficult to obtain, especially in the case of new services which generate traffic with partly unknown properties. Therefore a characterization of the distribution H and set G can be extremely difficult.

Traditionally, performance analysis developed two complementary approaches for confronting this difficulty: analytical approach and simulation. The *analytical approach* requires that the traffic sources are described by a Markov chain. This is a serious limitation because the number of states in such a Markov chain can be high for realistic sources. After this, the whole system consisting of the traffic and device is described by a Markov chain and its stationary distribution H is computed numerically. This enables a computation of the performance functional from (27.1) by direct integration which reduces to summation. The main problem of this approach is that the resulting Markov chain is very often so large that the computation of its stationary distribution is far beyond the reach of modern computers. Approximations are necessary for a majority of the problems of interest, which may undermine the relevance of results. The analytic approach is often supplemented by a *simulation approach*. It consists of direct simulation of the interaction between the traffic and device and allows a considerably more realistic representation of the whole system. The problem with this approach is that the simulation times necessary for obtaining the estimates of stationary performance can be prohibitively long. This is especially true for the case when design requirements are expressed in terms of packet loss which is a popular performance measure for the modern data networks. Both these techniques have the common drawback that they are applicable for a given traffic pattern, while design should be valid for the whole range of traffic parameters.

Stochastic optimization can enhance both analytic and simulation approaches to performance analysis by addressing the problem of development of guaranteed estimates for performance of telecommunication systems. Such estimates involve computation of $F^+(x)$ and $F^-(x)$ such that

$$F^-(x) \leq F(x,H) \leq F^+(x) \qquad (27.3)$$

for all $H \in G$. When performance requirements are described by (27.2), one can select x from

$$\phi^- \leq F^-(x),\ F^+(x) \leq \phi^+,$$

which will guarantee satisfaction of performance requirements for all traffic patterns of interest. The bounds $F^+(x)$ and $F^-(x)$ can be obtained from the solution of

$$F^-(x) = \inf_{H \in G} F(x,H),\ F^+(x) = \sup_{H \in G} F(x,H). \qquad (27.4)$$

These problems can be classified as belonging to a special class of stochastic optimization problems, namely optimization problems in the space of probability measures (see Karlin and Studden (1966), Kemperman (1968), Dupačová (1978), Ermoliev et al. (1986), and Gaivoronski (1986)). One may object that to solve such problems should be even more difficult than to compute the value of $F(x,H)$ for a given H, a difficult problem by itself as argued above. In many cases this is not true. Firstly, the function $F(x,H)$ can be often approximated by simpler functions $F^+(x,H)$ and $F^-(x,H)$ satisfying

$$F(x,H) \leq F^+(x,H), \ F^-(x,H) \leq F(x,H)$$

for all $H \in G$. These functions can be used in (27.4) instead of $F(x,H)$ which will make these problems simpler. Even more important, the measures which solve the problems (27.4) often have a very special structure which can be obtained from analysis of function $F(x,H)$ and set G without solving the problem itself (Whitt, 1984; Ermoliev et al., 1986). Numerical methods have been developed which exploit this structure (Gaivoronski, 1986) and simplify the solution of (27.4).

27.2.1 Access engineering of broadband multiservice network

We illustrate this general approach using a specific example of access engineering of broadband multiservice network taken from Bonatti and Gaivoronski (1994b). We consider a high speed data network with data packets of fixed length, for example a network based on Asynchronous Transfer Mode (ATM) architecture (see De Prycker (1995)). The central part of an access network consists of a server (multiplexer) with one output link and L input links. Data packets arrive from input links and are sent by the server to the network, maybe staying some time in the buffer of limited capacity x_0. If some packet finds the buffer full upon arrival it is discarded. Since all the packets are of the same length, so are the service times. Therefore, it is natural to consider the system operating in discrete time with the time interval being equal to the service time of one packet. Denoting by $l(t)$ the buffer contents at time t, by $\xi_i(t)$ the number of packets which arrive from the source i at time t and by $\xi(t)$ the total number of packets arrived at time t we have the following relation which describes the dynamics of the buffer contents:

$$l(t+1) = \max\{0, \min\{x_0, l(t) + \xi(t) - 1\}\}, \ \xi(t) = \sum_{i=1}^{L} \xi_i(t),$$

where $\xi_i(t)$ is 0 or 1. Each source i at the input of a server generates a packet arrival process with distribution H_i. This distribution defines the probability that a packet arrives at time t conditioned on the history of packet arrivals and is described by a vector of parameters $a : H_i = H_i(a)$. Many distributions may correspond to a given value of the vector a. Each source belongs to one of N classes where each class corresponds to a given service. In terms of arrival distribution H_i each class is characterized by a set A_j in the space of parameters, such that if the source i belongs to the traffic class j it means that $H_i \in G_j$ where

$$G_j = \{H(a) \mid a \in A_j\}.$$

Examples of traffic classes and corresponding parameter sets will be given later. Denote by x_j the number of sources which belong to class j. Then $\sum_{j=1}^{N} x_j = L$. I_j is a subset of $\{1,...,L\}$ which indexes the sources belonging to the traffic class G_j. The maximal admissible number of sources of each class $x_j, j = 1 : N$ together with the buffer length x_0 constitute the vector x of design parameters which should be chosen so that the access system satisfies given performance requirements which are expressed in terms of the admissible upper bound ϕ^+ on the packet loss probability $F(x,H)$:

$$F(x,H) = \frac{\mathbb{E} \max\{0, \xi(t) - l(t) - 1\}}{\mathbb{E}\xi(t)} \leq \phi^+ \quad (27.5)$$

where $H = H(z;x) = \mathbb{P}\{\xi(t) \leq z\}$ denotes the stationary distribution of $\xi(t)$. Usually this bound is chosen between $10^{-10} - 10^{-9}$.

This and similar systems constitute an important part of telecommunication networks and considerable effort was dedicated to its study (see, for example, Massey (1994)). The Markov chain analysis of this system starts by assuming that the packet arrivals from a single source are independent. This is a very serious assumption because the packets in ATM networks packets of the standard size. Therefore the traffic from a single source consists of bursts which correspond to each communication request. Even then, the resulting Markov chain contains at least $\prod_{i=0}^{N}(x_i + 1)$ states which can be a very large number. The number of states explodes further if more realistic assumptions about the traffic are taken. A simulation approach also encounters difficulties because estimation of probabilities of the order of 10^{-10} requires long simulation times.

This problem can be treated by optimization over probability measures. We start by deriving a bound on the packet loss probability from (27.5).

Proposition 27.1 (Bonatti and Gaivoronski, 1994b) *Suppose that $\xi_i(t)$ are the stationary ergodic stochastic processes for all i and the length of the buffer x_0 is not smaller than the total number of sources L. Then*

$$F(x,H) \leq F^+(x,H) = \frac{\int \max\{0, z-1\} dH(z;x)}{\int z dH(z;x)} = \frac{\mathbb{E}_H \max\{0, \zeta - 1\}}{\mathbb{E}_H \zeta} \quad (27.6)$$

where ζ is a random variable distributed according to H and $\mathbb{E}_H$ is expectation with respect to H.

The arrival processes from each source usually are assumed to be independent. Then the upper bound $F^+(x)$ on the cell loss probability can be obtained by solution of

$$F^+(x) = \max_{H_i \in G_j \, \forall i \in I_j} F^+(x,H) =$$

$$\max_{H_i \in G_j \, \forall i \in I_j} \frac{1}{B(x,H)} \int \cdots \int \max\left\{0, \sum_{i=1}^{L} z_i - 1\right\} \prod_{i=1}^{L} dH_i(z_i) \quad (27.7)$$

where

$$B(x,H) = \sum_{i=1}^{L} \int z_i dH_i(z_i). \quad (27.8)$$

To advance further it is necessary to specify the sets G_j which define the traffic classes. One common way of doing so is to put bounds on the moments of the distributions belonging to this set together with the bounds on the support of these distributions. In this case

$$G_j = G_j(a_j) = \left\{ H \mid \int_0^{a_{0j}} dH(z) = 1, \int_0^{a_{0j}} \psi_{rj}(z)dH(z) \leq a_{rj}, r = 1:R \right\} \quad (27.9)$$

where $a_j = (a_{0j}, a_{1j}, ..., a_{Rj})$ and $\psi_{rj}(z)$ are known functions. From the point of view of access design this definition has an important meaning. The bound on support a_{0j} is the peak bandwidth of the source from the class j measured in fractions of the output bandwidth of the server. Let $\psi_{1j}(z) = z$ then a_{1j} will be the average bandwidth of a source from the class j. The properties of (27.7) with the sets G_j defined by (27.9) are well understood. Its solution has a special structure which was exploited for development of numerical methods in Ermoliev et al. (1986) and Gaivoronski (1986). Sometimes it is possible to obtain an explicit solution as in the important case when $G_j(a_j)$ is defined by the values of peak and average bandwidth only:

Theorem 27.1 *(Bonatti and Gaivoronski, 1994b) Suppose that the traffic classes G_j are defined by the peak bandwidth a_{0j} and the average bandwidth a_{1j}. Then, among the sources which yield the largest packet loss probability always exist those with cell arrival distribution concentrated in two points $(0, a_{0j})$ with weights $(1 - p_j, p_j)$, $p_j = a_{1j}/a_{0j}$. The tight upper bound for the cell loss probability is*

$$F^+(x) = \frac{1}{\sum_{j=1}^N a_{1j}x_j} \sum_{\substack{0 \leq k_j \leq x_j \\ 1 \leq j \leq N}} \max\left\{0, \sum_{j=1}^N a_{0j}k_j - 1\right\} \prod_{j=1}^N \frac{x_j!}{k_j!(x_j - k_j)!} p_j^{k_j}(1 - p_j)^{x_j - k_j}.$$

(27.10)

Design decision can be taken by finding feasible solutions of

$$F^+(x) \leq \phi^+$$

where ϕ^+ is in Theorem 27.1. Designs obtained by stochastic optimization have the following advantage compared to designs obtained by Markov chain modeling or simulations. It will assure the required quality of service for all sources with specified average and peak bandwidth. In contrast, the design obtained through Markov chain modeling will be valid only for a much narrower class of sources which in addition generate packets with independently distributed arrival times. Simulations can assure required quality of service only for a finite set of sources which were selected for simulation experiments.

Stochastic optimization can be applied to other design problems. Often it is important to exploit carefully the special structure of each particular case to obtain approximations of performance measure similar to (27.6). It is also possible to address the problem of this approximation from a more general point of view by developing approximations of complex random processes by simpler ones. For the case of Markov chains such approximations which yield guaranteed bounds for performance measures were developed in Bonatti and Gaivoronski (1994a).

27.3 NETWORK LEVEL

The objective of the network level is to develop a design of the telecommunication network with a given capability to provide a set of services to a population of end users. Results of technological design are used as the inputs to the network design. This design should serve different and often conflicting purposes, e.g. satisfaction of demand, maintenance of a given service quality, cost effectiveness. Important decisions to take at this level include placement and dimensioning of processing nodes and transmission links. These decisions are affected by service pricing because it affects the quantity of demand to be satisfied. There exists considerable literature dedicated to the optimal design of networks in deterministic case (Ahuja et al., 1993; Bertsekas, 1998; Fortz et al., 2000; Mitra et al., 2001).

At the age of big state monopolies, with largely immutable services and a highly predictable environment, the prevailing paradigm was the minimization of network costs under constraints on quality of service and demand satisfaction. Although this paradigm remains important, it is clearly insufficient for the highly mutable and uncertain environment of today. A robust network which can accommodate within reasonable bounds future market changes is more valuable then maybe cheaper network designed for today conditions and maybe for one specific future scenario. For this reason profit, service pricing, evaluation of investment opportunities under uncertainty become increasingly important in the network design models. This is where stochastic programming models have a competitive edge compared to deterministic models because the capability to mediate between scenarios of uncertain future is explicitly embedded into them. This capability is very important because uncertainty of different kinds is one of the defining features of the modern telecommunications. On the network level the main source of uncertainty is unpredictable user response to introduction of new services which results in highly uncertain demand variations. To this one can add uncertainty due to technological innovation and uncertainty related to possible failures.

Another important feature of network design is the dynamic character of decisions. The network development projects have a time dimension and an important decision is how to distribute the investment over time. New information about the market will become available and the possibility to react on this information should be included in the decision models. Stochastic programming with recourse provides adequate tools for doing so. It also facilitates the incorporation of adaptation policies specifically designed to allow the network to react flexibly to the changing environment. Such policies are essential for robust network design. Another important issue is the correct evaluation of flexibilities present in the network investment projects. Examples of such flexibilities or real options are option to expand, option to upgrade technology, option to alter usage. Consideration of these options can drastically change the overall evaluation of the network expansion project. For example a project which is unprofitable from the first glance can reveal hidden profit opportunities.

Further exposition is organized in the form of examples which illustrate the general ideas. Section 27.3.1 presents a series of decision models for planning of Internet based information service starting from a simple traditional deterministic cost minimization model to a two period stochastic programming model for profit maximization which can be used for evaluation of real options embedded in this project. The problem of design of access network described in Section 27.2 is considered in Section 27.3.2 on the level of network design. This is an interesting example which shows interplay between both levels. Sections 27.3.3 and 27.3.4 show examples how technology influences network design models in the case of backbone networks. Network design which takes into account possible failures is considered in Section 27.3.5. Finally, Section 27.3.6 is dedicated to design of mobile networks in the situation of leveling out of demand.

27.3.1   Planning of Internet based information service

We consider here the problem of deployment of an Internet based information service on some territory which can be a country or a region. We assume that the network itself exists already and the decision consists in deployment of servers at the nodes of this network and assignment of demand generated in different geographical locations to these servers. The service provider on behalf of which the problem is solved can be the network owner, but can also be a virtual service provider which does not possess its own network and leases network from some network owner. Decisions to consider include phased introduction of service which can take the shape of Phase 1 deployment followed by Phase 2 deployment contingent on the market reaction. In addition, decisions include pricing of service.

Among various aspects of the problem one should consider geographical dimension, uncertainty of demand and costs, cost structure which includes fixed and variable costs, competition and substitution between services, relations between different market actors, e.g. network providers and service providers. The decision to go ahead with the project depends on the project profitability which in its turn depends on various options embedded in it, e.g. option to expand, to abandon, to upgrade technology. It is advisable to start the model development from the simple case which includes only some of the relevant features and to expand the model stepwise. We present here three such model development steps.

27.3.1.1   Single period deterministic cost minimization model. We start by considering only one decision period and full knowledge about demand and other uncertainties. Although these assumptions are highly unrealistic, the resulting model sets the stage for more realistic models. In this setting we assume that the deployment program has to satisfy the known demand fully. Since the service price is fixed, the revenue becomes fixed. For this reason the only way one can influence the profit is by minimizing the costs.

Notation:

- $i = 1 : n$ – index for regions which constitute a territory. User population exists in each region which generates demand.
- $j = 1 : m$ – index for possible server locations.
- y_j – binary variable which takes the value 1 if decision to place a server at location j is taken and 0 otherwise.

- x_{ij} – amount of demand from region i served by server placed in location j.
- f_j – fixed cost for setting up a server in location j.
- c_{ij} – cost for serving of one unit of demand from region i by server at location j.
- d_i – demand generated at region i.
- g_j – capacity of server placed at location j.

Model. Find the server deployment program $y = (y_1,...,y_m)$ and assignment of user groups to servers $x = \{x_{ij}\}, i = 1 : n, j = 1 : m$ as solution of

$$\min_{x,y} \sum_{j=1}^{m} f_j y_j + \sum_{j=1}^{m} \sum_{i=1}^{n} c_{ij} x_{ij} \qquad (27.11)$$

$$\sum_{j=1}^{m} x_{ij} \geq d_i, \ i = 1 : n \qquad (27.12)$$

$$\sum_{i=1}^{n} x_{ij} \leq g_j y_j, \ j = 1 : m \qquad (27.13)$$

where $y_j \in \{0,1\}$ and $x_{ij} \geq 0$. Here the first term in (27.11) represents the fixed cost of deployment of servers while the second term represents the variable costs for serving demand. The constraints (27.12) are imposed in order to obtain full demand satisfaction, while constraints (27.13) are the capacity constraints. This is a well known facility location model and it will serve as a starting point for development of a stochastic programming model with different scenarios of the future demand and a larger number of deployment phases.

27.3.1.2 Two period stochastic cost minimization model..
Two deployment phases are considered: present Phase 1 with known demand and future Phase 2 with uncertain demand which is described by a finite number of scenarios. Each scenario is described by demand values in different regions during Phase 2 and the probability of this scenario. The Phase 2 decisions include additional deployment of servers and reassignment of demand to servers in response to the demand which becomes known. The model follows the framework of stochastic programming with recourse.

Additional notation:

- $r = 1 : R$ – index for demand scenarios.
- d_i^r – demand generated by region i under scenario r.
- p^r – probability of scenario r.
- z_j^r – binary variable which takes the value 1 if under scenario r the decision to place a server at location j is taken and 0 otherwise.

- x^r_{ij} – amount of demand from region i served by a server placed in location j under scenario r.

- α – coefficient for discounting of the Phase 2 costs to the present.

Each scenario is characterized by a pair (d^r, p^r) where $d^r = (d^r_1,...,d^r_n)$.
Model. Find the Phase 1 server deployment program $y = (y_1,...,y_m)$, and assignment of user groups to servers $x = \{x_{ij}\}, i=1:n, j=1:m$ as solution of

$$\min_{x,y} \sum_{j=1}^{m} f_j y_j + \sum_{j=1}^{m}\sum_{i=1}^{n} c_{ij}x_{ij} + \alpha \sum_{r=1}^{R} p^r Q(r,y) \quad (27.14)$$

subject to (27.12)-(27.13). The third term in (27.14) represents discounted costs of the Phase 2 deployment averaged over scenarios. The cost associated with scenario r is $Q(r,y)$ and it depends on the Phase 1 deployment decision y. These costs are obtained from solution of the *recourse* problem for each scenario r:

$$Q(r,y) = \min_{x^r, z^r} \sum_{j=1}^{m} f_j z^r_j + \sum_{j=1}^{m}\sum_{i=1}^{n} c_{ij} x^r_{ij} \quad (27.15)$$

$$\sum_{j=1}^{m} x^r_{ij} \geq d^r_i, \ i=1:n \quad (27.16)$$

$$\sum_{i=1}^{n} x^r_{ij} \leq g_j (y_j + z^r_j), \ j=1:m \quad (27.17)$$

which is similar to (27.11)-(27.13) and chooses the Phase 2 deployment $z^r = (z^r_1,...,z^r_m)$ and new assignment of user groups to servers $x^r = \{x^r_{ij}\}, i=1:n, j=1:m$ from minimization of fixed deployment costs and variable service costs for a given scenario r.

This can be a numerically challenging problem because it contains binary variables. Still, the modern optimization technology permits to solve it for nontrivial and practically important cases. For example, we solved its deterministic equivalent to optimality using MPL modeling system powered by CPLEX and XPRESS solvers with $R = 5$, $n = m = 20$ in approximately 8 minutes on a 1133 Mhz Pentium III laptop. The deterministic equivalent for this case has 120 binary and 2400 continuous variables with 240 constraints. This time has grown to 1 hour for $R = 6$, $n = m = 25$ with the deterministic equivalent having 175 binary and 4375 continuous variables and 350 constraints. Utilization of decomposition is essential for solving the problems of larger dimensions.

27.3.1.3 Two period stochastic profit maximization model with pricing..
In a competitive deregulated environment the profit maximization is a more appropriate objective than the minimization of network costs. It becomes fundamentally different from the plain cost minimization when the pricing decisions are considered simultaneously with deployment decisions. The models become more

complicated because pricing affects demand and this dependence introduces nonlinearities. Still, meaningful analysis is feasible also in this case. We start by defining the linear demand model extending the scenario framework explained before.
Additional notation:

- h_0 – reference price for service during Phase 1.
- d_{i0} – reference demand at region i during Phase 1 which corresponds to reference price h_0.
- w_i – demand elasticity at region i during Phase 1.
- h – the price increment relative to the reference price during Phase 1.
- h_0^r – reference price for service during Phase 2 under scenario r.
- d_{i0}^r – reference demand at region i during Phase 2 which corresponds to reference price h_0^r under scenario r.
- w_i^r – demand elasticity at region i during Phase 2 under scenario r.
- h^r – the price increment relative to the reference price during Phase 2 under scenario r.

Demand model. This is the crucial piece of the profit model. Consider the Phase 1 deployment. The service price equals $h_0 + h$ and the price decision consists in selecting the price increment h which may be positive or negative. Assume that the demand d_i in region i during this phase depends on the price of the service according to some function $d_i = f_i(h_0 + h)$ and in the vicinity of the point h_0 this dependence can be linearized via

$$d_i = d_{i0} - w_i h. \qquad (27.18)$$

Similar relations describe the demand behavior during Phase 2 for each of the scenarios $r = 1 : R$. Each scenario is defined in this case by a tuple $(d_{i0}^r, h_0^r, w_i^r, p^r)$ which defines the dependence of demand on price according to relation (27.18) for a given scenario r.

Decision model. Find the Phase 1 increment for the service price h, server deployment program $y = (y_1, ..., y_m)$, and assignment of user groups to servers $x = \{x_{ij}\}, i = 1 : n, j = 1 : m$ as a solution of

$$\max_{h,x,y} W(h) - C(y,x) + \alpha \sum_{r=1}^{R} p^r P(r,y) \qquad (27.19)$$

subject to

$$w_i h + \sum_{j=1}^{m} x_{ij} \geq d_{i0}, \ i = 1 : n \qquad (27.20)$$

and constraint (27.13). Here $W(h)$ is the revenue during Phase 1:

$$W(h) = \sum_{i=1}^{n} (h + h_i)(d_0 - w_i h) \qquad (27.21)$$

and $C(y,x)$ are the costs during Phase 1 defined according to (27.11). The third term in (27.19) represents the profits during Phase 2 averaged over scenarios and discounted to the present where $P(r,y)$ is the profit during Phase 2 under scenario r. It is taken as the optimal value of the following *recourse* problem:

$$P(r,y) = \max_{h^r, x^r, z^r} W(r,h^r) - C(r,z^r,x^r) \tag{27.22}$$

$$w_i^r h^r + \sum_{j=1}^m x_{ij}^r \geq d_{i0}^r, \quad i=1:n \tag{27.23}$$

subject to additional constraints (27.17). Here $W(r,h^r)$ is the revenue during Phase 2 under scenario r obtained similarly to (27.21) and $C(r,z^r,x^r)$ are the costs during Phase 2 under scenario r taken from (27.15).

There is one important feature of this model which was not present in the models (27.11)-(27.13) and (27.14)-(27.17). While (27.14)-(27.17) can be transformed to a mixed integer LP by considering the deterministic equivalent, no such transformation is possible for the model (27.19)-(27.23). This is because the revenues $W(r,h)$ and $W(r,h^r)$ depend nonlinearly on the decision variables h and h^r. Even in the simplest case of the linear demand model (27.18) this dependence is quadratic. Therefore specialized numerical techniques should be employed in this case with decomposition approaches being the most promising.

Evaluation of investment opportunities, real options. One of the most important utilizations of model (27.19)-(27.23) is the evaluation of profitability of the investment project which consists in the deployment of the new service. Recent developments in the corporate finance showed importance of evaluation of real options for correct evaluation of industrial projects (Trigeorgis, 1996). While for more traditional industries direct evaluation techniques can be similar to evaluation of financial options, for innovative industries with unique projects such approaches are difficult to apply. Stochastic programming models can represent a valid alternative for real option evaluation. Let us utilize model (27.19)-(27.23) for this purpose. In particular, let us evaluate options to expand, to upgrade technology, to abandon or to convert a part of the infrastructure.

Option to expand (wait and see option). This option is already embedded in the model (27.19)-(27.23) which contains the possibility to add additional servers during Phase 2 contingent on market reaction. The value of this option can be computed as follows. Denote by P^* the optimal value of the model (27.19)-(27.23). This is the value of the project with an option to expand. The value $\hat{P}$ of the same project without the option to expand is obtained by solving the same model with binary variables z^r fixed to zero for all scenarios. Clearly, $\hat{P} \leq P^*$. The value of the option is the difference $P^* - \hat{P}$.

Option to upgrade technology. This is a valuable option because it can dramatically change the project evaluation. The most known example is the GSM mobile network whose development started when the technology to make mobile phones small enough was not yet available. In order to evaluate this option it is necessary to have a closer look at the ways the technology development can affect various components of the model (27.19)-(27.23). For example, the technology development can lead to decreasing of fixed costs for server installation and/or increase in the possible server capacities during Phase 2. In this case it is necessary to introduce these features into definition of scenarios. The notation is:

- f_j^r – fixed cost for setting up server in location j under scenario r.
- g_i^r – capacity of server placed at location i under scenario r.

The model changes as follows. The part (27.19)-(27.20) remains the same because it describes Phase 1 to be implemented with known technology. The part (27.22)-(27.23) has a modified capacity constraint which substitutes (27.17):

$$\sum_{i=1}^{n} x_{ij}^r \leq g_j y_j + g_j^r z_j^r, \ j = 1 : m. \tag{27.24}$$

Besides, the cost term $C(r, z^r, x^r)$ from (27.22) is

$$C(r, z^r, x^r) = \min_{x^r, z^r} \sum_{j=1}^{m} f_j^r z_j^r + \sum_{j=1}^{m} \sum_{i=1}^{n} c_{ij} x_{ij}^r \tag{27.25}$$

The model (27.19)-(27.23) is solved with modification (27.24)-(27.25) which will give the value P^{**} of the project with an option to upgrade technology. This value is compared with the value of the project P^* without such an option and the difference $P^{**} - P^*$ will give the option value.

Option to abandon. This is a valuable option when the market reaction is uncertain. If demand does not catch up it is reasonable to cut maintenance costs in the regions where demand is weak and possibly recover part of the fixed costs by selling or leasing the server infrastructure. The additional notation is:

- b_j^r – maintenance costs for server at location j during Phase 2 under scenario r.
- β_j^r – fraction of fixed costs which can be recovered by abandonment of server at location j under scenario r.
- u_j^r – binary variable which equals 1 if server at location j is abandoned during scenario r.

The model changes as follows. Again, the part (27.19)-(27.20) which refers to the Phase 1 remains the same. The part (27.22)-(27.23) has a modified capacity constraint which substitutes (27.17):

$$\sum_{i=1}^{n} x_{ij}^r \leq g_j \left(y_j + z_j^r - u_j^r \right), \ j = 1 : m, \ r = 1 : R \tag{27.26}$$

and additional abandonment constraints

$$u_j^r \leq y_j, \ j = 1 : m, \ r = 1 : R. \tag{27.27}$$

The revenue term $W(r, h^r)$ and the cost term $C(r, z^r, x^r)$ from (27.19) are

$$W(r, h^r) = \sum_{i=1}^{n} (h_0^r + h^r)(d_{i0}^r - w_i^r h^r) + \sum_{j=1}^{m} \beta_j^r f_j u_j^r \tag{27.28}$$

$$C(r, z^r, x^r) = \sum_{j=1}^{m} \left(f_j z_j^r + b_j^r \left(y_j + z_j^r - u_j^r \right) \right) + \sum_{i=1}^{n} c_{ij} x_{ij}^r \right). \quad (27.29)$$

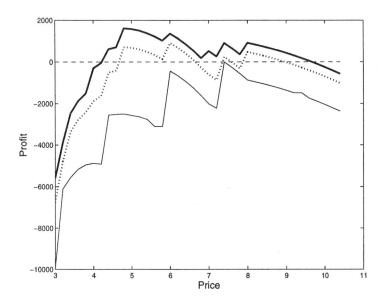

Figure 27.1 Evaluation of real options in the case of service introduction

The model (27.19)-(27.23) is solved with the modification (27.26)-(27.29) which will yield the value P^{++} of the project with an option to abandon infrastructure. This value is compared with the value of the project P^+ obtained by solving the same model with variables u_j^r set to zero which corresponds to evaluation without the option to abandon. The difference $P^{++} - P^+$ is the option value.

We now provide an example of such an option evaluation depicted in Figure 27.1. This figure shows the dependence of the project value on the service price $h_0 + h$ for the case when the Phase 2 service prices were fixed to the Phase 1 prices, i.e. $h_0^r \equiv h_0, h^r \equiv h$. Three alternatives are shown in this figure. The first alternative is depicted by a thin curve and describes the dependence of project value on price for the case when no option to expand and no option to upgrade technology are considered during Phase 2. The second alternative allows an option to expand, but not an option to upgrade technology and is depicted by a dotted curve. The third alternative shown with a thick line allows both options during Phase 2.

First of all, one notices the jumps on the curves due to the discrete character of the decisions. The objective in all three cases is full demand satisfaction. A small increase in price leads to a small decrease in demand which can make a given server redundant with a corresponding stepwise decrease in fixed costs. Another observation confirms the added value of flexibility which options provide. The value of the project without option is not positive even for the best choice of service price. The project becomes profitable when the option to expand is allowed. There are two regions of profitability with respect to the service price. The first corresponds to an aggressively low service price designed to stimulate large demand and the second corresponds to a less aggressive behavior with high prices and smaller demand. These profitability regions expand when an additional option to upgrade technology is considered. In the absence of options the model recommends defensive behavior with high pricing, while flexibility embedded in options allows to stimulate demand more aggressively with lower prices.

27.3.2 Design of access network

We consider the topic of Section 27.2.1 from the network level. The model from the technological level allowed us to understand which user population can have a guaranteed quality of service from an access server with given technological characteristics. On the basis of this information the network level design should answer the following questions. Given the present demand and future projections for a given region, what type of servers and how many of them a network operator should choose? Given a description of the typical user populations with given demand patterns, what kind of access servers should the equipment manufacturer produce? In this section we describe a stochastic programming model which helps to answer these questions (Bonatti et al., 1994).

We focus on the core of the network access design which is the design of the first statistical multiplexer stage. The model will support decisions on the number of multiplexers, the bandwidth they carry, a recommended composition of the traffic they serve. Such a model with two time periods: "the present" and "the future"incorporates some robustness against unexpected traffic evolution, otherwise the design can become obsolete less or more rapidly and considerable problems due to the system reconfigurations can arise. The present demand is relatively well known. The information available at present about the future demand is available in the form of several demand scenarios. The decision about access design should be made at the "present" only on the basis of scenario probabilities, without knowledge about which of the demand scenarios will materialize. When the demand will become known in the future some corrective action will be taken with the aim to assure the required grade of service. The objective is to select a design which provides a cost effective solution both from the point of view of the "present" costs (design implementation) and the "future" costs (design correction).

Evolution of traffic patterns.

- Traffic generated by different users are statistically independent.

- Traffic generated by any of the users belong to one of the well defined traffic classes whose characteristics are known. These classes are defined by simple parameters like maximal and average bandwidth. Initially there are N traffic classes $v_i, i = 1 : N$, in the future there are M traffic classes $\mu_j, j = 1 : M$. Some future classes may coincide with the past classes, while some others may be completely new.

- Users, presently belonging to class v_i can pass in the future to class μ_j, new users can appear in class μ_j.

- Present traffic classes are known; for the future traffic classes there are K different scenarios $\vartheta_k, k = 1 : K$ described as triples

$$\vartheta_k = \left\{X_0^k, \alpha_{ij}^k, p^k\right\}, k = 1 : K, i = 0 : N, j = 1 : M, \ \sum_{j=1}^{M} \alpha_{ij}^k = 1, \sum_{j=1}^{M} p^k = 1$$

where X_0^k is the number of new users under scenario ϑ_k, $\alpha_{ij}^k \geq 0$ for $i = 1 : N$ is the portion of the users of class v_i which pass to class μ_j under scenario ϑ_k and for $i = 0$ it is the portion of the new users which belong to the class μ_j, $p^k \geq 0$ is the probability of scenario ϑ_k.

Denoting the number of users which belong to the present traffic class v_i by $X_i, i = 1 : N$ and the number of users which belong to the future traffic class μ_j under scenario ϑ_k by Y_j^k we obtain the following relation between these two quantities:

$$\sum_{i=1}^{N} \alpha_{ij}^k X_i + \alpha_{0j}^k X_0^k = Y_j^k, \ j = 1 : M. \tag{27.30}$$

Decision variables. We consider the case when all multiplexers from the first statistical multiplexing stage handle a fixed mixture of sources. The more general case which includes multiplexers of different types can be handled similarly. Thus, we have to define the initial number n_1 of multiplexers to install, the number of multiplexers n_2^k to add later in the future when the demand scenario ϑ_k becomes known, the bandwidth a of one multiplexer and the mix of sources served by an arbitrary multiplexer at present and in the future. x_i is the number of users of class v_i supported by one multiplexer at present and y_j^k is the number of users of class μ_j supported by one multiplexer in the future under scenario ϑ_k. The vectors $x = (x_1, ..., x_N)$ and $y^k = (y_1^k, ..., y_M^k)$ describe the source mix. These decision parameters satisfy the quality constraints of two types. The first one states that all users should be covered, now and in the future:

$$n_1 x_i \geq X_i, i = 1 : N \tag{27.31}$$

$$\left(n_1 + n_2^k\right) y_j^k \geq Y_j^k, \ j = 1 : M, k = 1 : K \tag{27.32}$$

where Y_j^k is defined in (27.30). The second group of constraints should assure that each multiplexer provides the required grade of service. The quality of service (QoS) is characterized by the known function $f(a, z)$ where a is the multiplexer bandwidth and z is the source mix supported by a given multiplexer. There is an admissible bound γ on the grade of service. This yields the following representation for the QoS constraints for the present and the future:

$$f(a, x) \leq \gamma \tag{27.33}$$

$$f(a, y^r) \leq \gamma, \ r = 1 : R. \tag{27.34}$$

The specific expression for the function $f(a,z)$ or the algorithm for its computation is provided by the technological level design where an example was described in Section 27.2.1. If we take the packet loss as the measure of the quality of service we can use expression (27.10) from that section.

Costs. We take into account the costs of initial installation, additional installation, connection of new users, reconnection of old users.

- cost of initial installation and connection of users:

$$n_1 C_{11} + C_{21} \sum_{i=1}^{N} X_i$$

where C_{11} is the fixed cost for initial installation of one multiplexer and C_{21} is the initial cost for connecting one user;

- cost of additional installation and connection of users:

$$n_2 C_{12} + C_{22} X_0^k$$

where C_{12} is the fixed cost for additional installation of one multiplexer in the future and C_{21} is the cost for connecting one user in the future;

- cost of reconnecting the users in the future:

$$C_3 n_1 \sum_{j=1}^{M} \max \left\{ 0, \sum_{i=1}^{N} \alpha_{ij}^k x_i - y_i^k \right\}$$

where C_3 is the cost of reconnecting of one user in the future. The future costs C_{12}, C_{22} and C_3 may be only partially known and may differ between different scenarios.

Decision hierarchy. Different decision variables are defined by two coordinated optimization problems. They correspond to different time scales and different levels of knowledge about demand. Initial installed capacity n_1, multiplexer bandwidth a and initial source mix x are decided at the present, when the actual future demand scenario is not known. This is the *long term planning problem*. Additional installed capacity n_2^k and the final source mix y^k supported by one multiplexer are decided in the future when the demand scenario is known. This is the decision correction problem or *recourse problem* in stochastic programming terminology. The optimal value of this problem enters in the expression for the total costs of the long term planning problem. This recourse problem is defined as follows.

For scenario k, multiplexer bandwidth a, initial installed capacity n_1 and initial traffic mix x find (n_2^k, y^k) which solve

$$Q(k,a,n_1,x) = \min_{(n_2^k, y^k)} C_{12} n_2 + C_{22} X_0^k + C_3 n_1 \sum_{j=1}^{M} \max \left\{ 0, \sum_{i=1}^{N} \alpha_{ij}^k x_i - y_i^k \right\} \quad (27.35)$$

subject to constraints (27.32),(27.34).

The stochastic programming problem with recourse for long term planning is to find (a, n_1, x) which minimize the present costs and discounted future costs averaged among demand scenarios:

$$\min_{a,n_1,x} C_{11}n_1 + \alpha \sum_{k=1}^{K} p^k Q(k,a,n_1,x) \qquad (27.36)$$

subject to constraints (27.31),(27.33) where α is a discount coefficient.

It possesses features which place it apart from the vast majority of such problems found in the literature. First of all, constraints (27.33)–(27.34) are nonlinear. Moreover, while variables a, x, y^k can be considered to be continuous, the variables n_1 and n_2 are substantially discrete. Therefore the usual solution approaches based on application of large scale linear programming to the deterministic equivalent or even Benders decomposition are not applicable here. A version of stochastic random search worked well on this problem.

27.3.3 Design of backbone connection oriented network

The objective of the network design is to make a decision about the placement of the network elements in a given geographical area. The network is represented as the collection of nodes with processing capabilities which are connected by links with transmission capacities. Design involves taking decisions concerning the placement of nodes of different type and processing capability and placement of links with different capacities. The objective of design is the satisfaction of communication demand between different nodes under given requirements about the quality of service and taking into account profit, costs and other considerations. Telecommunication networks have a hierarchical structure and backbone networks are the top level of this hierarchy responsible for carrying large demand quantities between geographically distributed nodes where demand is collected with the help of local networks.

The design of backbone network involves considerable investments and the resulting network should be robust enough to accommodate unpredictable demand variations during the time horizon of a few years. The flexibility required for the adequate reaction to the changing demand patterns is provided by the network management policies. In the case of connection oriented networks such policies may include construction of a logical network on the top of the physical network by reserving transportation capacity between different nodes for different virtual paths in the network. While the change in the physical network requires considerable investment and time, the change in the logical network can be performed relatively fast and cheap following the change of the demand pattern. The design of the physical network should take into account this possibility and the stochastic programming approach provides the necessary modeling tools for doing so. We describe one such model for the case of connection oriented networks where the transportation and processing capacity for connection between any given pair of nodes should be reserved before the actual communication can take place. Examples of such networks are traditional telephone networks, broadband ATM networks (De Prycker, 1995), mobile networks.

Topology of the physical network.

- n – number of nodes in the network. The nodes are indexed by integer numbers $i = 1 : n$.
- x_{ij} – link capacity between nodes i and j.
- u_i – processing capacity of node i.
- x – vector of all link capacities.
- u – vector of all processing capacities.
- x_{ij}^- – lower bound on the transmission capacity between nodes i and j. Nonzero x_{ij}^- means that the link between nodes i and j exists already and the objective of design is to identify the necessary network expansion.
- x_{ij}^+ – upper bound on the transmission capacity between nodes i and j.
- u_i^+ – upper bound on processing capacity at node i.
- u_i^- – lower bound on processing capacity at node i.

Demand scenarios.

- $q = 1, ..., m$ – index for demand scenarios.
- d_{ij}^q – communication demand between nodes i and j under scenario q.
- z_{ij}^q – amount of demand between nodes i and j which is not served under scenario q.
- g^q – amount of the link capacity required for transportation of one demand unit by an arbitrary link under scenario q.
- h^q – amount of communication flow through a node served by one unit of processing capacity at this node under scenario q. The communication flow is measured by the amount of transmission capacity necessary to carry this flow.
- p_q – probability of scenario q.

The dependence of g^q and h^q on the demand scenario allows different service development possibilities.

Topology of the logical network. Capacities of the links in the logical network depend on the specific demand scenario because this network should be adapted to demand.

- I_{ij} – the set of admissible paths between nodes i and j.
- m_{ij} – number of different paths in the set I_{ij}.

- b_{ijr} – path number r from the set I_{ij}, $r = 1 : m_{ij}$. It is represented as a sequence of pairs of nodes where each pair identifies the link which belongs to the path b_{ijr}.

$$b_{ijr} = ((i, j_{r1}), (j_{r1}, j_{r2}), ..., (j_{rt_r}, j))$$

where t_r is the number of hops in the path b_{ijr}.

- y_{ijr}^q – capacity of path b_{ijr} between nodes i and j under scenario q.

Costs. The decision paradigm of the minimization of the network costs under constraints on quality of service and demand satisfaction is adopted here. Different objectives like maximization of revenue or maximization of profit can be considered.

- c_{ij}^v – variable cost of installation of one unit of link capacity between nodes i and j.

- c_{ij}^f – fixed cost of installation of link capacity between nodes i and j.

- c_i^v – variable cost of installation of one unit of processing capacity in node i.

- c_i^f – fixed cost of installation of processing capacity in node i.

- e_{ij}^q – opportunity cost of not meeting one unit of demand between nodes i and j under scenario q.

Decision structure. There are two coordinated decision problems here which fit the paradigm of stochastic programming problems with recourse. The decision to construct the physical network is taken now. Besides decisions about the specific values of transmission capacities x_{ij} and processing capacities u_i, this decision involves also the logical decisions to install capacities or not. Due to the presence of transaction costs these decisions should be modeled by binary variables.

- v_{ij} – binary variable which equals 1 if the link capacity between nodes i and j is increased above the already existing level x_{ij}^- and zero otherwise.

- w_i – binary variable which equals 1 if the processing capacity at node i is increased above the already existing level u_i^- and zero otherwise.

The design of the physical network is taken before the actual demand patterns become known only on the basis of the information about demand scenarios. Therefore the objective is to minimize the current costs of network installation and discounted future costs of the network adaptation to demand, averaged among demand scenarios. The design problem is to find v_{ij}, w_i, x_{ij}, u_i for all i, j that solve

$$\min_{\substack{v_{ij}, w_i, \\ x_{ij}, u_i}} \sum_{(i,j)} \left(c_{ij}^f v_{ij} + c_{ij}^v x_{ij} \right) + \sum_i \left(c_i^f w_i + c_i^v u_i \right) + \alpha \sum_{q=1}^m p_q Q(q, x, u) \quad (27.37)$$

$$x_{ij}^- \leq x_{ij} \leq x_{ij}^- + \left(x_{ij}^+ - x_{ij}^- \right) v_{ij}, \forall (i, j) \quad (27.38)$$

$$u_i^- \le u_i \le u_i^- + \left(u_i^+ - u_i^-\right) w_i, \; i = 1:n \qquad (27.39)$$

The objective function (27.37) includes fixed and variable costs for installation of processing capacities at nodes, transmission capacities at links and averaged costs of the network adaptation to demand discounted with the discount coefficient α. Constraints (27.38),(27.39) impose bounds on capacities and connect logical and continuous decision variables.

The cost $Q(q,x,u)$ of the network adaptation to a given demand pattern q is obtained by solving the *recourse problem* which is the design problem of the logical network for this demand pattern. This problem will be solved repeatedly during the life time of the physical network as a new demand scenario emerges.

Given capacities x_{ij}, u_i of the physical network and demand scenario q find the link capacities y_{ijr}^q of the logical network by solving

$$Q(q,x,u) = \min_{y_{ijr}^q, z_{ij}^q} \sum_{i,j=1}^{n} e_{ij}^q z_{ij}^q \qquad (27.40)$$

$$\sum_{r \in I_{ij}} y_{ijr}^q + g^q z_{ij}^q = g^q d_{ij}^q, \; \forall (i,j) \qquad (27.41)$$

$$\sum_{\substack{i,j \\ r \in I_{ij}, (k,l) \in b_{ijr}}} y_{ijr}^q \le x_{kl}, \; \forall (k,l) \qquad (27.42)$$

$$\sum_{\substack{i,j \\ r \in I_{ij}, l, (k,l) \in b_{ijr}, (l,k) \in b_{ijr}}} y_{ijr}^q \le h^q u_k, \; k = 1:n \qquad (27.43)$$

$$z_{ij}^q \ge 0, \; y_{ijr}^q \ge 0. \qquad (27.44)$$

The objective function (27.40) represents adaptation costs which consist of the opportunity costs of not meeting demand. The costs of reconfiguring the logical network are assumed to be either negligible or not dependent on the decision variables. Constraint (27.41) connects the link capacities of the logical network with the served demand. In the left hand side of constraint (27.42) we have the total communication capacity required by the logical network from the link between nodes k and l which should not exceed the physical capacity x_{kl}. The left hand side of constraint (27.43) represents the sum of all ingoing and outgoing communication flow at node k measured by reserved transmission capacity. It is assumed that the required processing capacity at node k is proportional to this flow.

Solving the problem (27.37)-(27.39) is via formulating its deterministic equivalent and solving the resulting mixed integer LP. Decomposition techniques may be obligatory to process problems of realistic dimensions. Possible simplifications include approximation of fixed costs by variable costs which allows to dispense with binary variables. When the bottleneck is represented by either processing capacities or transmission capacities the problem can be simplified by considering only the bottleneck capacities. Different variants of problem (27.37)-(27.39) were considered in Fantauzzi et al. (1997) where also the results of numerical experiments were reported.

27.3.4 Design of backbone connectionless network

Connectionless networks represent an important class of telecom networks where no capacity reservation is needed in order to establish communication between different nodes. An important example is the Internet network. The communication between nodes consists of the flow of data packets which are routed with the help of routing tables and routing algorithms implemented at network nodes. Therefore the logical network can be represented in the form of multicommodity network flow where each commodity corresponds to a given pair of nodes. The design objectives and the design paradigm remain the same as in the case of the connection oriented network, the main difference being in the formulation of the network adaptation problem (27.40)-(27.44).

The *topology of logical network* is represented by y^q_{ijlk} which is the communication flow between nodes i and j which passes through link between nodes l and k under scenario q. The problem of design of the physical network (27.37)-(27.39) remains the same, while in the network adaptation problem (27.40)-(27.44) constraints (27.41)-(27.44) are substituted by the following constraints:

$$\sum_{l:l\neq k} y^q_{ijkl} - \sum_{l:l\neq k} y^q_{ijlk} = 0, \forall (i,j), \forall k: k \neq i, j, \qquad (27.45)$$

$$\sum_{l:l\neq i} y^q_{ijil} - \sum_{l:l\neq i} y^q_{ijli} + g^q z^q_{ij} = g^q d^q_{ij}, \; \forall (i,j) \qquad (27.46)$$

$$\sum_{l:l\neq j} y^q_{ijjl} - \sum_{l:l\neq j} y^q_{ijlj} - g^q z^q_{ij} = -g^q d^q_{ij}, \; \forall (i,j) \qquad (27.47)$$

$$\sum_{(i,j)} y^q_{ijkl} + \sum_{(i,j)} y^q_{ijlk} \leq x_{kl}, \; \forall (k,l) \qquad (27.48)$$

$$\sum_{(i,j),l:l\neq k} y^q_{ijkl} + \sum_{(i,j),l:l\neq k} y^q_{ijlk} \leq h^q u_k, \; k=1:n \qquad (27.49)$$

$$z^q_{ij} \geq 0, \; y^q_{ijkl} \geq 0. \qquad (27.50)$$

Constraints (27.45)-(27.47) are the network flow continuity constraints. In the left hand side of constraint (27.48) we have the total communication capacity required by the logical network from the link between nodes k and l which should not exceed the physical capacity x_{kl}. The left hand side of constraint (27.49) represents the sum of all ingoing and outgoing communication flow at node k. It is assumed that the required processing capacity at node k is proportional to this flow.

27.3.5 Incorporating reliability considerations

An important source of uncertainty inherent in the telecommunication networks is presented by possible failures of links and nodes which can occur due to a variety of reasons, see Gavish and Neuman (1992). Therefore reliability and dependability of networks is a serious design issue. We show that the reliability considerations can be naturally incorporated into the stochastic programming modeling approach. We show how to extend the models of sections 27.3.3 and 27.3.4 to obtain a reliable network design.

The key is to represent the possible link and node failures in the form of failure scenarios which are similar to demand scenarios described in the previous sections. Such scenarios can be independent from demand scenarios or can be combined with them. Demand scenario q is described by the following quantities.

- α_{ij}^q – portion of the link capacity between nodes i and j which remains operational under failure scenario $q, 0 \leq \alpha_{ij}^q \leq 1$.

- β_k^q – portion of the processing capacity at node k which remains operational under failure scenario $q, 0 \leq \beta_k^q \leq 1$.

- p_q – probability of scenario q.

Among all the failure scenarios there will always be one scenario $q = 1$ corresponding to normal operation when $\alpha_{ij}^q = 1, \beta_k^q = 1$ for all links (i, j) and all nodes k. All other scenarios will correspond usually to the failure of one given link or node because in the vast majority of practical situations the simultaneous failure of several links or nodes is unlikely.

The models from sections 27.3.3 and 27.3.4 remain the same the only difference being a slight modification of capacity constraints. For connectionless network constraints (27.48),(27.49) are substituted by

$$\sum_{(i,j)} y_{ijkl}^q + \sum_{(i,j)} y_{ijlk}^q \leq \alpha_{kl}^q x_{kl}, \ \forall (k,l) \tag{27.51}$$

$$\sum_{(i,j),l:l \neq k} y_{ijkl}^q + \sum_{(i,j),l:l \neq k} y_{ijlk}^q \leq h^q \beta_k^q u_k, \ k = 1:n. \tag{27.52}$$

For connection oriented networks similar modifications should be made to constraints (27.42),(27.43).

The discussion of this and the previous two sections by presenting the Figure 27.2 which shows dependence of the optimal network costs on the opportunity costs of not meeting demand. The specific model for which these costs were calculated was the design of a connectionless reliable network. The opportunity costs $e_{ij}^q \equiv e$ are the same for all links and all scenarios.

The thin curve on the Figure 27.2 represents the dependence of the total network costs on the unit opportunity costs e. These costs equal the optimal solution of the problem (27.37)-(27.39) where $Q(q,x,u)$ is the optimal solution of the problem (27.40) with constraints (27.45)-(27.47),(27.50),(27.51),(27.52). These costs grow linearly for low opportunity costs e, after some threshold they start to grow more slowly and after another threshold they stop to grow which means that all demand is satisfied after that threshold. The dependence of the total costs on opportunity costs is concave. The thick curve represents the corresponding dependence of the pure network component of costs, namely the sum of the first two terms in (27.37). It shows some interesting phenomena which can be observed also in the common practice of the network development. When the unit opportunity costs of not meeting demand are low, no network is built at all and all the costs consist of penalties for not meeting demand. After the unit opportunity cost reach some threshold the network is built which satisfies a large portion of demand and this network remains unchanged with further growth

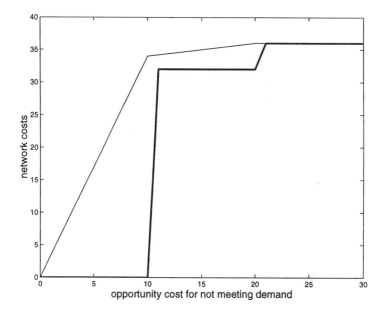

Figure 27.2 Dependence of the optimal network costs on the opportunity costs of not meeting demand

of opportunity costs until another threshold is reached. After that second threshold the network is built which satisfies all demand. An important observation is that the network is upgraded not incrementally, but in a few large steps which corresponds to the industrial practice of the telecom network deployment. Stochastic programming models coupled with the estimation of opportunity costs can help to decide about the timing of decisions to upgrade the network.

27.3.6 Planning of capacity expansion of mobile networks

The recent few years were characterized by exponential growth of mobile networks driven by exponential growth of demand. The market was capable of absorbing practically any capacity which mobile operators were capable to offer. The things are changing now with the market for voice services being close to saturation in many countries while the market for the future broadband wireless services is uncertain. The mobile operators are becoming more attentive to the optimal network planning. The objective is to avoid potential losses which may be caused from one side by deployment of excessive capacity and from the other side by deterioration of quality of service and the consequent loss of customers who may turn to competition. Stochastic programming models can provide the network designers with balanced and robust network expansion alternatives which take into account uncertainties in demand development.

In an example the objective is to develop a plan for the expansion of mobile network based on the demand forecast. The time horizon of the network expansion is 1-2 years while the plan itself is revised periodically, for example each quarter. The expansion plan consists of two components: establishing new channels (TRX) in existing cells and establishing new cells. Establishing of new channels is a relatively minor matter compared to establishing of a new cell both from the point of view of expenditure and time. Establishing of a cell is a complex process which may take up to six months of time and considerable amount of resources. It consists of the following steps.

- Preparatory period during which the necessary agreements are made, it takes approximately half the time and a few percent of resources.

- Building of a cell which takes one third of the time and two fifth of resources.

- Putting the cell into operation which takes one sixth of the time and three fifths of resources.

It is possible to suspend the establishment of a cell after each of these steps and resume it again after some time. This contains a source of additional flexibility which is possible to exploit in order to diminish the reaction time on demand changes and avoid unnecessary investment. In particular, it is possible to have pools of semi-finished cells on different levels of preparedness and to invest into further establishment only when demand requires this. The model which we present below helps to exploit such a possibility. The network expansion process is extended over several time periods $t = 1, 2, ..., T$ which in this setting can have the length of one month, while T may take the values between 6 and 18.

Demand description. It is the key input in the network expansion model. As in the previous sections, demand forecasts are represented in the form of scenarios. It is assumed that from times $t = 1$ to t_1 demand d_t^0 is known. After this demand is uncertain and this uncertainty is described via scenarios $k = 1 : K$ which describe both demand development uncertainty and seasonal variations and variations due to holidays/special events, etc.

- d_t^k – demand forecast at time t under scenario k, where $k = 0$ for

- $t = 1 : t_1$ and $k = 1 : K$ for $t = t_1 + 1, ..., T$;

- v_t^k – demand which will not be satisfied at time t under scenario k;

- c_v – opportunity cost of not meeting one unit of demand;

- p_k – probability/frequency of scenario $k, k = 1 : K$;

Network description.

- a_t – amount of capacity necessary to satisfy one unit of demand at time t, it can be variable due to introduction of new services;

- C – capacity of one cell;

- D – capacity of one TRX;

- Δ_z – time necessary for preparation of cell agreement;
- c_z – cost of preparation of cell agreement;
- Δ_y – time necessary for building of a cell;
- c_y – cost of building of a cell;
- Δ_x – time necessary for activation of a cell;
- c_x – cost of activation of a cell;
- Δ_w – time necessary for setting up of a new TRX;
- c_w – cost of setting up of a new TRX;
- α_t – coefficient used to discount costs at time t to the present.

Network dimension and dimensioning decisions. During time horizon $t = 1,...,t_1$ they are described by the following quantities which refer to the beginning of period t and which depend on demand scenario k where $k = 0$ for $t = 1 : t_1$ and $k = 1 : K$ for $t = t_1 + 1,...,T$.

- X_t^k – amount of cells in the working condition;
- Y_t^k – amount of built, but not activated cells;
- Z_t^k – amount of cell agreements;
- W_t^k – amount of TRX;
- x_t^k – amount of new cells to be activated; for $t < 1$ this is a model input resulting from previous decisions;
- y_t^k – amount of new cells to be built;
- z_t^k – amount of new cell agreements to be prepared;
- w_t^k – amount of new TRX to be set up.
- $Z_{\max}$ – maximal amount of cell agreements to initiate at any given time period.

Development of the network capacity during time. The following equations describe the development of fully operational cells, built cells, cell agreements and TRX over time.

$$X_t^k = X_{t-1}^k + x_{t-\Delta_x}^k \qquad (27.53)$$

$$0 \leq x_t^k \leq Y_t^k \qquad (27.54)$$

$$Y_t^k = Y_{t-1}^k + y_{t-\Delta_y}^k - x_{t-1}^k \qquad (27.55)$$

$$0 \leq y_t^k \leq Z_t^k \qquad (27.56)$$

$$Z_t^k = Z_{t-1}^k + z_{t-\Delta_z}^k - y_{t-1}^k \qquad (27.57)$$

$$0 \leq z_t^k \leq Z_{\max} \qquad (27.58)$$

$$W_t^k = W_{t-1}^k + w_{t-\Delta_w}^k \qquad (27.59)$$

$$w_t^k \geq 0 \qquad (27.60)$$

where $k = 0$ for the time horizon $t = 1 : t_1$ when demand is assumed to be known and $k = 1 : K$ for the time horizon $t = t_1 + 1 : T$ when demand is uncertain. The following relations connect these time horizons.

$$X_{t_1}^k = X_{t_1}^0, \; Y_{t_1}^k = Y_{t_1}^0, \; Z_{t_1}^k = Z_{t_1}^0, \; W_{t_1}^k = W_{t_1}^0, \; k = 1 : K \qquad (27.61)$$

$$x_{t-\Delta_x}^k = x_{t-\Delta_x}^0, t_1 \leq t \leq t_1 + \Delta_x, k = 1 : K \qquad (27.62)$$

$$y_{t-\Delta_y}^k = y_{t-\Delta_y}^0, t_1 \leq t \leq t_1 + \Delta_y, k = 1 : K \qquad (27.63)$$

$$z_{t-\Delta_z}^k = z_{t-\Delta_z}^0, t_1 \leq t \leq t_1 + \Delta_z, k = 1 : K \qquad (27.64)$$

$$w_{t-\Delta_w}^k = w_{t-\Delta_w}^0, t_1 \leq t \leq t_1 + \Delta_w, k = 1 : K. \qquad (27.65)$$

Relation between capacity and satisfied demand:

$$DW_t^k + a_t v_t^k \geq a_t d_t^k, \; v_t^k \geq 0 \qquad (27.66)$$

$$D\left(W_t^k + w_t^k\right) \leq CX_t^k. \qquad (27.67)$$

Costs. Two types of costs are considered: the costs of network expansion and opportunity costs of not meeting demand. Additional costs could be considered, e.g. maintenance of cells and cell agreements.

Capacity expansion decision. This decision is taken at the beginning of time period $t = 1$ from the point of view of minimization of total costs which include costs of network expansion and opportunity costs of not meeting demand during time horizon $t = 1 : t_1$ plus costs of further network expansion contingent on demand scenarios during horizon $t = t_1 + 1 : T$ averaged over scenarios and discounted to time $t = 1$. More formally, find $x_t^0, y_t^0, z_t^0, w_t^0, X_t^0, Y_t^0, Z_t^0, W_t^0, v_t^0, t = 1 : t_1$ from the solution of

$$\min_{\substack{x_t^0, y_t^0, z_t^0, w_t^0, \\ X_t^0, Y_t^0, Z_t^0, W_t^0, v_t^0}} \sum_{t=1}^{t_1} \alpha_t \left(c_x x_t^0 + c_y y_t^0 + c_z z_t^0 + c_w w_t^0\right) + \sum_{t=1}^{t_1} \alpha_t c_v v_t^0 +$$

$$\sum_{k=1}^{K} p_k Q^k(x^0, y^0, z^0, w^0, X^0, Y^0, Z^0, W^0) \qquad (27.68)$$

subject to (27.53)-(27.60),(27.66),(27.67) where $k = 0$. The arguments of the function $Q^k(\cdot)$ represent the vectors with components indexed by time, for example

$x^0 = (x_1^0, x_2^0, ..., x_{t_1}^0)$. The function $Q^k(\cdot)$ represents the optimal costs of further network expansion during time horizon $t = t_1 + 1 : T$ for a given demand scenario k. In stochastic programming terminology this is a *recourse problem*

$$Q^k(x^0, y^0, z^0, w^0, X^0, Y^0, Z^0, W^0) = \min_{\substack{x_t^k, y_t^k, z_t^k, w_t^k, \\ X_t^k, Y_t^k, Z_t^k, W_t^k, v_t^k}} \sum_{t=t_1+1}^{T} \alpha_t \left(c_x x_t^k + c_y y_t^k + c_z z_t^k + c_w w_t^k \right) + \sum_{t=t_1+1}^{T} \alpha_t c_v v_t^k \quad (27.69)$$

subject to constraints (27.53)-(27.67). Problems (27.68)-(27.69) can be transformed into their deterministic equivalent and solved by linear programming software. We utilized the capabilities present in Excel spreadsheet for its solution.

The purpose of the model described above is to provide advice about aggregated decision of the investment into network expansion. It includes aggregated network characteristics and capacity constraints (27.67) which describe the capacity of the whole network. The inherently integer variables, like the number of cells or the number of TRX were substituted by their continuous approximations. In a more detailed model constraints (27.67) should be considered for a group of similar cells, or even for a given cell.

27.4 ENTERPRISE LEVEL

The solutions of the network design problems considered in the previous section depend on a number of external parameters which were considered to be fixed. Some of these parameters derive from decisions which can be taken only on the strategic level through consideration of the whole environment in which the enterprise operates. Examples of such parameters are pricing of services, the total amount of resources allocated for investment and other quantities which characterize the strategy of the enterprise. The importance of this *enterprise decision level* has grown considerably during the recent years due to the increase in complexity of the telecommunications environment due to the convergence process with computer industry and content provision, deregulation and technological development. It is characterized by a growing uncertainty whose main sources are unpredictable market response to the new technologies and services, decreasing life cycles of products, actions of competition. Stochastic programming models represent a natural methodology for decision support on this level. However, they should be enhanced by selected ideas from the game theory in order to be capable of reflecting uncertainty which stems from actions of other decision makers.

27.4.1 Network operators and virtual providers: Service pricing

The model developed refers to the situation represented on Figure 27.3. The environment consists of the Network Operator (NO), Virtual Service Provider (VNO) and a population of customers. Both NO and VNO provide a service to customers which can decide to subscribe to this service, change provider or discontinue to use the service altogether. Service providers decide the price for their service. The Network

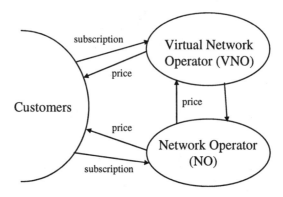

Figure 27.3 Competition between Network Operator and Virtual Network Operator

Operator possesses the network which is necessary for the service provision. The Virtual Network Operator does not have a network and in order to provide the service he has to lease the necessary network capacity from the Network Operator. Therefore the VNO has to decide how much capacity to lease and the NO decides the price to charge for the network capacity. Regulatory bodies may impose bounds on the leasing prices. Besides, the VNO may provide additional value to a service which may lead to a market expansion which, in its turn, may make the existing network capacity inadequate. Therefore the network operator may face the necessity to invest in the network expansion.

The important question is what should be the bounds imposed by regulatory authorities on the price for leasing of the network capacity. From one side they should not be too high in order to permit the VNO to compete with the NO on the service provision. From the other side they should not be too low in order to permit the Network Operator to recover investment expenditures. We develop a stochastic optimization model which helps to answer this question. We take the point of view of the Network Operator which pursues the objective of maximizing his profit by making decisions about service prices, leasing prices and investment in the network. His decision model should contain the submodels which describe reactions of customers and the VNO to his policies.

Consider two decision periods $t = 1, 2$. Decisions are taken and implemented at the beginning of each period. During the rest of periods the market and competition reactions are observed which influence revenues and profits.

Decisions of NO and VNO. They are denoted by vectors y_t and z_t for NO and VNO respectively, $t = 1, 2$ where $y_1 = (y_{11}, y_{12}, y_{13})$, $y_2 = (y_{21}, ..., y_{25})$, $z_t = (z_{t1}, z_{t2})$. Components of these vectors have the following meaning:

- y_{t1}, z_{t1} – prices charged for one unit of service by NO and VNO respectively during Period t.

- y_{t2} – the price charged by NO for one unit of leased capacity.
- y_{t3} – the maximal amount of capacity that NO is willing to lease.
- y_{24} – the amount of new capacity added to the network at the beginning of Period 2.
- y_{25} – the binary variable which equals 1 if decision is taken to expand the network capacity and zero otherwise.
- z_{t2} – the amount of capacity that VNO decides to lease.

Profit model of Network Operator for Period 1. The Network Operator takes the decision about prices at the beginning of the Period 1 without knowing precisely the reaction of customers and competition. His objective is to maximize the average profit taking into account the possibilities of profit and investment during the second period. The profit of NO is the difference between average revenue and costs during Period 1 with added discounted averaged profit from the Period 2. In order to describe it formally we need the following notations where we shall associate NO with index $i=1$ and VNO with index $i=2$.

- ω_t – vector of parameters which are uncertain for NO at the beginning of Period t. These parameters describe the market and competition reaction and will be defined more precisely when the market and competition models will be defined. It is enough that they have a known probabilistic description either in the form of continuous probability distributions or in the form of scenarios with given probabilities.
- d_{ti} – demand for service provided by operator i during Period t measured by the network capacity required for its satisfaction.
- a – network capacity owned by NO and available for service provision at the beginning of Period 1.
- e_{ti} – cost of serving a unit of demand for operator i during Period t.
- g_{ti} – opportunity cost of not meeting one unit of demand for operator i during Period t.
- h_{ti} – cost of maintenance of one unit of capacity for operator i during Period t, it is owned capacity for NO and leased capacity for VNO.
- b_f – fixed costs associated with network expansion.
- b_v – variable costs per unit of capacity associated with network expansion.
- α – coefficient for discounting the second period profit to the beginning of Period 1.

The profit of the NO is

$$F_{11}(y_1,z_1,d_{11}) = (y_{11} - e_{11})\min\{d_{11}, a - z_{12}\} + y_{12}z_{12} -$$
$$g_{11}\max\{0, d_{11} - a + z_{12}\} - ah_{11} + \alpha \mathbb{E}_{\omega_1,\omega_2} Q(y_1, z_1, \omega_1, \omega_2) \quad (27.70)$$

where $Q(y_1, z_1, \omega_1, \omega_2)$ is the profit of the NO during Period 2, it depends on the decisions of both NO and VNO during Period 1 and on uncertain parameters ω_1, ω_2. The profit of the NO during Period 1 is described by the first four terms in (27.70). The first term represents the profit due to provision of service to customers. The second term represents the profit due to leasing of capacity to the VNO. The third term represents the opportunity costs of not meeting the demand for service provision while the fourth term does not depend on decision variables and represents the variable network maintenance cost.

An important feature of this profit model which distinguishes it from the models of the previous sections is the presence of two unknown components: decisions z_1 of the VNO and service demand d_{11}. To make this model useful for decision making the NO has to predict both these quantities. Such predictions are the scope of the *market model* and *competition model* which the NO should have. The market model provides the prediction of demand $d_{1i} = d_{1i}(y_1, z_1, \omega_1)$ for the service provided by both operators as function of their pricing decisions and uncertain parameters. The competition model provides the prediction of decisions $z_1 = z_1(y_1, \omega_1)$ of the VNO as function of decisions of the NO and uncertain parameters. Demand predictions $d_{12} = d_{12}(y_1, z_1, \omega_1)$ for the service provided by the VNO are used for making this prediction. Substituting these predictions into (27.70) we obtain the profit expression which depends only on the decisions of the NO and on uncertain parameters ω_1 with known probabilistic description. The policy recommendation y_1 for the NO can be obtained by finding the values of y_1 which yield the highest mean profit. This leads to the following stochastic optimization problem:

$$\max_{y_1} \mathbb{E}_{\omega_1} F_{11}(y_1, z_1(y_1, \omega_1), d_{11}(y_1, z_1(y_1, \omega_1), \omega_1)) \quad (27.71)$$

$$y_{11}^- \leq y_{11} \leq y_{11}^+ \quad (27.72)$$

$$y_{12}^- \leq y_{12} \leq y_{12}^+ \quad (27.73)$$

$$0 \leq y_{13} \leq a \quad (27.74)$$

where $y_{11}^-, y_{11}^+, y_{12}^-, y_{12}^+$ are price constraints imposed by regulatory and other considerations. We now give examples of the market model and the competition model.

Market model. The simplest such model ignores dependence of demand on prices and considers a finite number of demand scenarios with given probabilities. This is, however, an inadequate description of demand. A step towards more realistic demand representation is in the linear autoregressive model:

$$d_{t1} = \max\{0, d_{t-1,1} + d_{t1}^0 + r_{t1}(y_{t-1,1} - y_{t1}) + q_t(z_{t1} - y_{t1})\} \quad (27.75)$$

$$d_{t2} = \max\{0, d_{t-1,2} + d_{t2}^0 + r_{t2}(z_{t-1,1} - z_{t1}) + q_t(y_{t1} - z_{t1})\} \quad (27.76)$$

where $t = 1, 2$. Here d_{0i} is demand for the service of operator i prior to the beginning of Period 1, it is assumed to be known; y_{01} and z_{01} are some initial reference service prices for both operators, d_{ti}^0 is the component of the demand change for operator i during Period t which is not related to the price changes; r_{ti} is an additional demand obtained/lost due to a unit change of the price of operator i and q_t is the flow of demand between the operators caused by a unit price difference between them. Parameters d_{ti}^0, r_{ti}, q_t are not known with certainty to the NO and constitute part of the vector ω_t of unknown parameters. We have used the linear model due to its simplicity, but nonlinear models are also possible.

Competition model. This model summarizes the knowledge which the NO has about the objectives of the VNO. In the simplest case it is assumed that the VNO at time period t wants to maximize his expected current profit $F_{t2}(y_t, z_t, d_{t2})$ which has the structure similar to the profit of NO:

$$F_{t2}(y_t, z_t, d_{t2}) = (z_{t1} - e_{t2}) \min\{d_{t2}, z_{t2}\} - (y_{t2} + h_{t2}) z_{t2} - g_{t2} \max\{0, d_{t2} - z_{t2}\} \quad (27.77)$$

where the first term represents the profit derived from the service provision, the second one reflects expenditures due to the network leasing and the third one reflects opportunity costs for not meeting demand. Parameters e_{t2}, h_{t2} and g_{t2} are uncertain for the NO and represent another part of components of the vector of uncertain parameters ω_t. The dependence of demand d_{t2} on prices of both operators is obtained from the market model (27.75)-(27.76) and substituted into (27.77). After this the prediction $z_t(y_t, \omega_t)$ of response of the VNO to decisions of the NO is obtained as solution of

$$\min_{z_t} F_{t2}(y_t, z_t, d_{t2}(y_t, z_t, \omega_t)) \quad (27.78)$$

$$z_{t1}^- \leq z_{t1} \leq z_{t1}^+ \quad (27.79)$$

$$0 \leq z_{t2} \leq y_{t3} \quad (27.80)$$

where z_{t1}^-, z_{t1}^+ are price bounds imposed by regulation and other considerations.

Profit model of Network Operator for Period 2. This model reflects the future profits $Q(y_1, z_1, \omega_1, \omega_2)$ in the total profit of the NO. This will make the first period decision more forward looking and capable to facilitate an eventual adaptation to changing market circumstances by investment in the network expansion. The profit of the NO during Period 2 is

$$F_{21}(y_2, z_2, d_{21}, \omega_2) = (y_{21} - e_{21}) \min\{d_{21}, a - z_{22} + y_{24}\} + y_{22} z_{22} - g_{21} \max\{0, d_{21} - a + z_{22} - y_{24}\} - h_{21}(a + y_{24}) - b_v y_{24} - b_f y_{25} \quad (27.81)$$

where the network expansion occurs at the beginning of the Period 2 and the total network capacity available for service provision during Period 2 is $a + y_{24}$. This expression for the profit is similar to (27.70) and contains two new last terms which reflect variable and fixed costs related to the expansion of network. Parameters $e_{21}, g_{21}, h_{21}, b_v, b_f$ can be uncertain for the NO at the beginning of Period 1 and constitute additional components of the vector ω_2, others being the components similar to those of ω_1. Prediction $d_{21} = d_{21}(y_2, z_2, \omega_2)$ of demand is obtained from the demand model (27.75)-(27.76) and prediction $z_2(y_2, \omega_2)$ of competition response is obtained from the competition model (27.78)-(27.80). These predictions depend also on y_1, z_1, ω_1 through autoregressive relations (27.75)-(27.76), but we omitted this dependence to simplify notations. The value of the future profit is obtained by solving

$$Q(y_1, z_1, \omega_1, \omega_2) = \min_{y_2} F_{21}(y_2, z_2(y_2, \omega_2), d_{21}(y_2, z_2, \omega_2), \omega_2) \tag{27.82}$$

$$y_{21}^- \leq y_{21} \leq y_{21}^+ \tag{27.83}$$

$$y_{22}^- \leq y_{22} \leq y_{22}^+ \tag{27.84}$$

$$z_{12} \leq y_{23} \leq a + y_{24} \tag{27.85}$$

$$0 \leq y_{24} \leq M y_{25} \tag{27.86}$$

where (27.83)-(27.85) are similar to constraints (27.72)-(27.74). Constraint (27.85) reflects the assumption that the NO can not offer to VNO less capacity during Period 2 than the amount offered by him during Period 1. Constraint (27.86) will force the amount of the network extension to zero if the decision not to expand the network was taken. Otherwise it will limit the network expansion to the maximal admissible level M.

The problem (27.71)-(27.74) is an extension of the classical stochastic programming problem with recourse to the case when the part of the uncertainty is due to the actions of other decision makers. This adds a new level of complexity in the form of prediction problem (27.78)-(27.80). The problem (27.82)-(27.86) is an extension of the classical recourse problem with the new feature added by prediction problem (27.78)-(27.80). This additional complexity makes the problem far more challenging than traditional linear stochastic problems with recourse. The linearity structure is never present here. However, some general approaches still can be used, notably the transformation of the problem into its deterministic equivalent by representing the uncertain parameters ω_1, ω_2 through a finite number of scenarios. Nonlinear programming software can be used for solution of such deterministic equivalent. Another promising approach is the stochastic quasigradient methods (Gaivoronski, 1988).

We finish this section by presenting on Figure 27.4 a typical example of a computation of the profit function $F_1(y_1) = \mathbb{E}_{\omega_1} F_{11}(y_1, z_1(y_1, \omega_1), d_{11}(y_1, z_1(y_1, \omega_1), \omega_1))$ of the Network Operator performed by Matlab 6.1 with Optimization Toolbox.

On this table the vertical axis marked by $F1(y)$ shows the values of the profit function of the NO computed according to (27.70). The two horizontal axes marked by $y1$ and $y2$ show the values of the service price and leasing price respectively of the NO. The figure shows the complex nature of the profit function which exhibits different patterns in different regions of the price space. This space can be divided into four regions. In the first region both service price and leasing price are moderate which results in the pattern where both operators have positive share in service market. In the second region the service price is high while the leasing price is moderate. In this region the NO has no customers and gets all his profit from leasing the capacity to the VNO which monopolizes the service market. The opposite picture can be observed

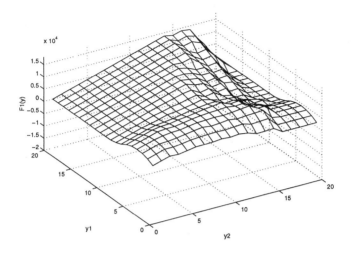

Figure 27.4 Profit function of Network Operator

in the third region where the leasing price is high while the service price is moderate. In this region the NO becomes a monopolist in service provision and the VNO is squeezed out from the market. Finally, in the fourth region where both prices are high the market does not take off at all because the high service price discourages the customers from subscribing to the service of the NO, while the high leasing price prevents the VNO from offering the service at an attractive price. The good news which can be derived from this and similar examples (Audestad et al., 2002) is that despite its complex nature the profit function has a distinctive structure and within each of the regions its behavior is close to concave. This circumstance can be exploited in numerical methods.

27.5 SUMMARY

In this chapter we gave a survey of different applications of stochastic optimization to telecommunications. Some presented models are new, while others are close to those found in the literature. Stochastic programming is a methodology of choice for support of complex network design decisions in the presence of uncertainty. In telecommunications uncertainty is present on all levels of network design, starting from the level of technology, through the level of network design to the enterprise level where the top level strategic decisions are taken. Moreover, due to the current changes in the industry environment the adequate treatment of uncertainty is becoming paramount for taking competitive design and investment decisions. As different examples in this chapter show, accumulated modeling and computational experience in solving stochastic optimization problems represent a solid background for deployment of stochastic programming applications in telecommunications. At the same time more work is needed in development of methodology, in particular in nonlinear and mixed integer stochastic programming models.

27.6 ACKNOWLEDGMENT

Thanks are due to Dr. Mario Bonatti of Italtel and Dr. Jan-Arild Audestad of Telenor for useful discussions which helped to shape some of the ideas in this chapter.

Bibliography

R.K. Ahuja, T. L. Magnanti, and J. B. Orlin. *Network flows: Theory, algorithms, and applications*. Prentice-Hall, New Jersey, 1993.

R. Andrade, A. Lisser, N. Maculan, and G. Plateau. Planning network design under uncertainty with fixed charge, 2002. Submitted for publication.

J.A. Audestad, A. Gaivoronski, and A. Werner. Modeling market uncertainty and competition in telecommunication environment: Network providers and virtual operators. *Telektronikk*, 97(4):46–64, 2002.

D.P. Bertsekas. *Network optimization: Continuous and discrete models*. Athena Scientific, Fitchburg, MA, 1998.

J.R. Birge and F. Louveaux. *Introduction to stochastic programming*. Springer, New York, 1997.

M. Bonatti and A. Gaivoronski. Guaranteed approximation of Markov chains with applications to multiplexer engineering in ATM networks. *Annals of Operations Research*, 49:111–136, 1994a.

M. Bonatti and A. A. Gaivoronski. Worst case analysis of ATM sources with application to access engineering of broadband multiservice networks. In J. Labetoulle and J. W. Roberts, editors, *Proceedings of 14th International Teletraffic Congress*, pages 559–570. Elsevier, 1994b.

M.A. Bonatti, A. Gaivoronski, P. Lemonche, and P. Polese. Summary of some traffic engineering studies carried out within RACE project R1044. *European Transactions on Telecommunications*, 5:207–218, 1994.

M. De Prycker. *Asynchronous transfer mode, Solution for broadband ISDN*. Prentice Hall, New Jersey, 1995.

M. Dempster and E. Medova. Evolving system architectures for multimedia network design. *Annals of Operations Research*, 104:163–180, 2001.

J. Dupačová. Minimax approach to stochastic linear programming and the moment problem. Recent results. *ZAMM*, 58:T466–T467, 1978.

Y. Ermoliev, A. Gaivoronski, and C. Nedeva. Stochastic programming problems with incomplete information on distribution functions. *SIAM J.Control and Optimization*, 23(5):697–716, 1986.

Y. Ermoliev and R. J.-B. Wets, editors. *Numerical techniques for stochastic optimization*. Springer Verlag, Berlin, 1988.

F. Fantauzzi, A. A. Gaivoronski, and E. Messina. Decomposition methods for network optimization problems in the presence of uncertainty. In P.M. Pardalos, D. Hearn, and W. Hager, editors, *Network optimization*, volume 450 of *Lecture Notes in Economics and Mathematical Systems*, pages 234–248. Springer, Berlin, 1997.

B. Fortz, M. Labbé, and F. Maffioli. Solving the two-connected network with bounded meshes problem. *Operations Research*, 48(6):866–877, 2000.

A. Gaivoronski. Linearization methods for optimization of functionals which depend on probability measures. *Mathematical Programming Study*, 28:157–181, 1986.

A. A. Gaivoronski. Implementation of stochastic quasigradient methods. In Y. Ermoliev and R. J.-B. Wets, editors, *Numerical techniques for stochastic Optimization*, pages 313–352. Springer Verlag, Berlin, 1988.

B. Gavish and I. Neuman. Routing in a network with unreliable components. *IEEE Transactions on Telecommunications*, 40(7):1248–1258, 1992.

P. Kall and S. Wallace. *Stochastic programming*. John Wiley and Sons, New York, 1994.

S. Karlin and W. J. Studden. *Tchebycheff systems: With applications in analysis and statistics*. Interscience Publishers, New York, 1966.

J. H. B. Kemperman. The general moment problem: A geometric approach. *Annals of Mathematical Statistics*, 39:93–122, 1968.

W. A. Massey. The fundamental role of teletraffic in the evolution of telecommunications networks. In J. Labetoulle and J. W. Roberts, editors, *Proceedings of the 14th International Teletraffic Congress*, volume 1A, pages 145–154, Amsterdam, 1994. Elsevier.

J. Medhi. *Stochastic models in queueing theory*. Academic Press, Boston, 1991.

D. Mitra, K. G. Ramakrishnan, and Q. Wang. Combined economic modeling and traffic engineering: Joint optimization of pricing and routing in multi-service networks. In *Proceedings of 17th International Teletraffic Congress, Salvador, Brazil*, Amsterdam, 2001. Elsevier.

S. Sen, R. D. Doverspike, and S. Cosares. Network planning with random demand. *Journal of Telecommunications Systems*, 3:11–30, 1994.

A. Tomasgard, J. Audestad, S. Dye, L. Stougie, M. V. der Vlerk, and S. Wallace. Modelling aspects of distributed processing in telecommunications networks. *Annals of Operations Research*, 82:161–184, 1998.

L. Trigeorgis. *Real options: Managerial flexibility and strategy in resource allocation.* MIT Press, Cambridge, MA, 1996.

W. Whitt. On approximation for queues, I: Extremal distributions. *AT&T Bell Laboratories Technical Journal*, 63(1):115–138, 1984.

J. Yen, A. Schaefer, and C. Smith. A stochastic SONET network design problem. In *9th International Conference on Stochastic Programming*, Berlin, August 25–31 2001.

28 NETWORK RESTORATION

Deep Medhi[1]

[1]Department of Computer Science & Electrical Engineering
University of Missouri–Kansas City
Kansas City, MO 64110 USA
dmedhi@umkc.edu

Abstract: An important problem in the design and deployment of communication networks is the issue of network restoration to address for various types of failures. In this chapter, we consider a variety of networks and networking technologies and argue that networks can be broadly classified as either traffic networks or transport networks. We then present optimization models for network protection for link failures and discuss how they fit into this classification. We then discuss the process of network restoration and interaction between the traffic and the transport network. Finally, we also discuss situations and failures for which restoration is difficult to model—an area that requires further exploration.

Keywords: IP networks, circuit-switched networks, optical networks, traffic and transport networks, multi-layer networks, network optimization modeling, network protection design.

28.1 INTRODUCTION

Network restoration refers to the ability of a network to restore or recover from a failure. The difficulty with this subject is that there are diverse issues such as how capable the network is to do restoration (and the capability the network has), to what extent the network can do it (fully or partially), what type of failure it is, the cost of network restoration and protection. To complicate matter, we have different types of networks as well as networks of networks; furthermore, technological changes are happening almost every day.

Often, we also tend to look at network restoration from the point of view of a specific technology, for example, how do we restore an IP network, or how do we restore a SONET network. However, as capabilities change, the technological distinction and functionalities have made it difficult to see what to do and where to do in terms of restoration. We also need to understand the role of protection (especially pre-planned) in network restoration. In this chapter, our interest is primarily about network restoration from a link failure for different types of networks. We start with a simple example about considering a failure in a network and then discuss different technologies. We then argue that most networks can be broadly classified either as traffic networks or transport networks. After that, we present a representative set of optimization models that can be applicable in a traffic or transport network setting for protection and restoration design. This is followed by considering a sample of different networks and discussing possible approaches for network restoration. Finally, we discuss different types of failures faced by a network; in particular, we discuss the issue of failure propagation (usually from one network to another) and how that can impact recovery from a failure.

28.2 A SIMPLE ILLUSTRATION

Consider a three-node network (Figure 28.1) where we have average traffic demand of 20 units between each pair of nodes. The nodes are connected by links, each with capacity of 150 units.

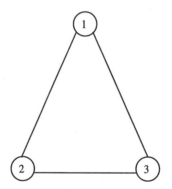

Figure 28.1 Three-node network

Suppose a link fails in the network, say it is the link 2-3. In this case, the traffic between nodes 2 and 3 which normally would use the link 2-3 could be routed via path 2-1-3. This however depends on whether the network has the ability to switch (re-route) traffic from one route to another route. In other words, whether node 2, on knowing that the link 2-3 has failed, can switch traffic quickly to path 2-1-3. If we assume that the network has this capability, then the traffic would be rerouted without any perceived effect since there is plenty of bandwidth (capacity) available on links 1-2 and 1-3 to carry the traffic between 2 and 3. Thus, in this case, the link 2-3 is not necessarily needed to be restored (at least immediately) and the failure may not be perceived in the network at the user level.

Consider now for a moment that dynamic rerouting is not available in this network, i.e., the network has only static routes where routes are only direct-link routes. In this case, with the link failure of 2-3, the traffic between 2-3 cannot be restored. The only way to address this case is to bring the link 2-3 up (somehow). These two situations illustrate two important points. If the network has enough bandwidth[1] and the network has dynamic routing capability, then traffic restoration is almost immediate and not perceived without requiring the link to be physically repaired. On the other hand, if the network does not have dynamic routing capability, i.e., it only has static routing, then this is not possible.[2] These two observations often hold in most communication networks, irrespective of the actual technology as long as the networks property described (dynamic routing or static routing) is available.

Consider next a slightly different traffic loading scenario. That is the traffic demand between nodes 2 and 3 is 140 units while all the other values remain the same as before. In this case, under the normal scenario, the network has the capacity to carry all traffic; however, this is not so if any of the links fails. If the goal of the network is to provide 'full' restoration, then the only way to do this would be to have restoration (or protection) capacity in the network ahead of time so that when the failure occurs, the network can bank on this protection capacity to route the traffic.

There are several aspects that the simple example discussed so far does not cover. For example, how does the above hold when the network is large? Are there other factors that play a role (or roles)? How about the actual technology of a network? And so on. In order to delve more into these aspects, we need to understand certain properties of different networking technologies.

28.3 TECHNOLOGY OVERVIEW

We can see from the illustration in the previous section that a closely intertwined topic with restoration is network routing. This also helps us to define the domain of our work here: we consider here network restoration for networks where routing (either dynamic or static) is applicable. In other words, local area networks based on Ethernet/Fast Ethernet, more specifically "networks" that are set up as spanning trees (or bus) would not be covered here; instead, our work here is primarily applicable to wide-area networks. However, we do clarify that our work is also applicable to access networks that are not simply spanning tree-based or bus-type networks; for example, campus or enterprise (access) networks that have backbone networks consisting of multiple routers and switches; this would fall under the domain we are discussing here. Also, our coverage in this chapter is for wired networks; that is, wireless networks restoration is outside the scope of this chapter. In this backdrop, we now review a few key technologies and networks.

28.3.1 IP network

Internet is based on the IP networking technology. Contents for applications such as web, email are broken down into small chunks (called packets or IP datagrams) at the end computers which are routed through the IP network. An IP network consists

---

[1] While bandwidth is not always plenty in real life, this is not unusual in a network with the drop in bandwidth capacity cost, especially within a geographical proximity such as a campus or enterprise network.
[2] See Section 28.4 for a modification of this statement when considered in a two-layer network architecture framework.

of a set of nodes, commonly referred to as routers. In reality, a large IP network is really a networks of networks, as a packet traversing from one computer to another computer (server) goes from an IP network run by an Internet service provider (ISP)[3] to another IP network run by another ISP, and then to another, until the packet reaches the destination computer. IP networks can be broken into two main components: access networks, and core networks consisting of multiple ISPs; in either case, the entire Internet is a collection of autonomous systems (AS) where an autonomous system is connected to one or more autonomous systems and where each autonomous system consists of multiple routers that are maintained by the ISP in charge of it.[4] Each ISP runs an intra-domain routing protocol for routing traffic within its network (autonomous system) or traffic that transitions through its network (autonomous system); the most common intra-domain routing protocols are OSPF and IS-IS[5]. Another protocol called BGP is used for the purpose of inter-domain routing that connects different autonomous systems managed by various ISPs. A pictorial view of intra-domain and inter-domain relation is shown in Figure 28.2.

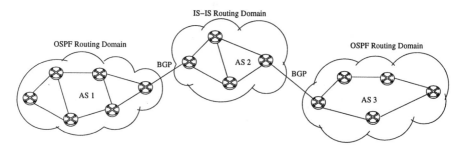

Figure 28.2 IP network routing architecture

Both OSPF and IS-IS are link-state protocols that are similar in spirit and use Dijkstra's shortest-path routing; the shortest path is determined based on the "state" information, e.g., cost of links in the network. It is important to note that the routing protocol does not define the metric value for the link; that is, the provider is left to decide on its own the link cost metric whether it is to be based just on hop counts, or delay, or some other measure. Presumably, an ISP uses a metric based on network goals such as minimizing average delay or minimizing congestion on any link. It may be noted that while a provider may chose the link metric, it may not necessarily want to update the values of the link metric frequently or periodically, unless there is a failure of a link in the network. When a failure occurs, the cost of the failed link is set to infinity and this information is distributed as link-state advertisement to every router in the network.

---

[3] Our definition of an ISP is broad: a campus network is also considered to be an ISP.
[4] While multiple autonomous systems can possibly fall under the jurisdiction of a single ISP (for example, due to merger of providers), we assume here that each ISP covers an autonomous system.
[5] See appendix for the list of acronyms.

The inter-domain routing protocol BGP works on the notion of exchange of reachability information. For the purpose of routing determination, it considers each autonomous system as if it is a mega-node, and the routing is determined based on minimum number of mega-nodes visited.

28.3.2 Circuit-switched voice telephone network

The voice service[6] is primarily provided through a circuit-switched network where a dedicated circuit with a fixed amount of bandwidth (64Kbps) is set up for the duration of a voice call and is teared down when the call is over. Different telephone service providers (TSP) play different roles in carrying a call. A call originates in an access TSP, a "Local-exchange carrier (LEC)", where the call starts from a switching node called the end office (or the central office). If the call is destined for another user in the same LEC, then the call is routed within its network. When a call starts from one LEC and is destined to terminate in another LEC, then the call typically goes through an inter-exchange carrier (IXC) before entering the destination LEC (see Figure 28.3).

Figure 28.3 Telephone network architecture

In most cases, LECs use a two level fixed hierarchical routing architecture with call overflow from the lower level to the upper level (see Figure 28.4a). An IXC can either deploy fixed hierarchical routing or dynamic routing (Figures 28.4b and 28.4c); if the network does provide dynamic routing, then the part of the call that falls within its network uses at-most two links for routing between the ingress switch and the egress switch within its network.[7] A link between two telephone switches is referred to as trunkgroups, or inter-machine trunks. Unless a call terminates in a different country, there is usually at most one IXC involved between the access LECs. For a call going from one country to another country, the call may go through the second country's inter-exchange provider or equivalent before reaching the destination address in an another access carrier. In many countries, both the access service and the inter-exchange/long-distance service are provided by the same provider; regardless, a fixed hierarchical routing is commonly deployed in access LECs.

---

[6]Carrying voice traffic on Internet that uses VoIP technology is gaining popularity; this will not be discussed in this chapter.
[7]In other words, at the level of end-offices, a call traverses through multiple links that includes at most two links when dynamic routing is used in the IXC part for this call (Ash, 1997).

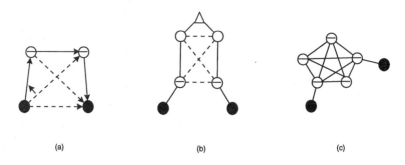

Figure 28.4 Telephone network routing: (a) Local exchange carrier: Two level fixed hierarchical routing, (b) Inter-exchange carrier: hierarchical, (c) Inter-exchange Carrier: fully-connected dynamic routing

28.3.3 MPLS network

Multi-protocol Label Switching (MPLS) is a relatively recent networking technology that is capable of carrying multiple protocols in the same data networks which may be of different types and may have different quality-of-service requirements. Certainly, the IP service is a primary candidate for MPLS networks, but other services such as voice or Ethernet services can also be handled directly by the MPLS technology.

MPLS provides a functionality called 'labels' so that different protocols/services (with differing quality-of-service requirements) can be carried separately without inter-mixing them (or some of them may be combined, if needed); the ingress and egress nodes in an MPLS network provide the translation functionality from an external protocol for use in an MPLS network. Nodes in an MPLS are called label-switched routers (LSR). Between LSRs, labeled-switched paths (LSP) can be set up. Label-Distribution protocol (LDP) is used for communicating label information for LSPs. The actual decision of how many or what labels to set for different services and usage of LSP functionality is left to the MPLS network provider.

There are several different usages of MPLS technology. An MPLS network can be deployed where LSPs are set up on a permanent or a semi-permanent basis. For example, LSPs can be set up to carry IP traffic; in this case, this LSP will act as a logical link between two IP network routers for the IP network.

MPLS also provides the capability to do constraint-based routing. This means that a service which may have different "constraint" requirements such as bandwidth requirements, or delay requirement may be taken into account in routing decision. This may be on a per "call" basis for a particular service. In that sense, this function has similarity to the circuit-switched voice service, with the important distinction that each call can be of differing bandwidth requirement (instead of just a fixed 64Kbps, in the case of voice circuit-switching) and may also have other requirements such as inter-packet delay, and so on.

There is another usage of MPLS that may be possible. In this case, the LSPs are to be setup and teared down on an on-demand basis based on customer requests, rather than being of permanent or semi-permanent nature. That is, demand requests arrive randomly requesting to set up LSPs with specific bandwidth; this usage of MPLS is essentially similar to the circuit-switched voice service.

A pre-cursor to MPLS technology is ATM (Asynchronous Transfer Mode) where packets are of fixed length called cells. ATM provides virtual path (VP) services which are similar to the notion of label-switched paths.

28.3.4 Telecommunications cross-connect network

The role of the telecommunication cross-connect network (Rey, 1983) has been that of a bearer network to provide permanent or semi-permanent capacity for 'logical' services such as circuit-switched voice, or for private line service between geographically separated locations of large companies, or more recently, for IP networks. For example, the actual physical routes for trunkgroups for the circuit-switched voice network or a link in the IP network is provided in the cross-connect network as bulk capacity service (such as, 24 voice circuits as T1). The nodes in this networks are cross-connects, or digital cross-connects (DCS) systems; they can be thought of as 'slow' switches since they do not address on-demand real time connection requests. In the literature, such networks are also referred to as transmission networks, transmission facilities networks, facilities networks, or simply as transport networks.

Through DCS functionally, a logical trunkgroup between two switches, that are in far apart places such as Los Angeles and New York (which are two extremes of continental US), can be physically provisioned through cross-connects and transmission network links. To address the economy of scale, such bulk capacities are multiplexed up at grooming nodes, such as from T1 (1.54Mbps, 24 voice circuits) to T3 (45 Mbps, 28 T1s); thus, within the transport network, a hierarchy of rates that feeds from one to another is available.

28.3.5 Optical network

The first generation of optical networks are essentially an extension of the digital cross-connect network where links are over fiber-optic cables. With optical networks, many new layers of rate hierarchy have been defined going from OC-3 (155.52Mbps) to OC-192 (9.9 Gbps) using SONET/ SDH standard.

An early implementation of SONET is in the form of rings which are self-healing, in that it can address a single cut failure very quickly, usually in less than 40 ms. A transport network encompassing a large geographical network can be formed by putting together a collection of interconnecting SONET self-healing rings (see Figure 28.5); it is important to ensure that two rings are not connected at just one point to avoid single point failure which can result in losing both the rings.

Over the past decade, the use of optical cross-connects (OXCs) to form mesh transport networks has gained popularity. Recently, there has been interest in providing on-demand optical level services at high data rates (OC-3 or higher) that can be requested much like circuit-switched voice on demand and can be set up and teared down almost instantaneously through signaling functionality.

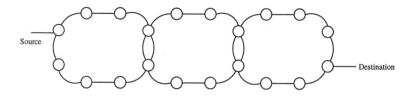

Figure 28.5 Network of SONET rings

In recent years, wave-division multiplexing (WDM) concept is gaining momentum where multiple light-waves can be multiplexed on a fiber, thereby increasing capacity without needing to add new fibers in the ground. With WDM, a node may or may not have converter; the role of converter is to switch from one frequency to another (if it is available).

28.3.6 Network hierarchy

Traditionally, network hierarchy consists of the telephone network over the underlying transmission facility made of digital cross-connect networks. Over time, the traditional transmission facility has been replaced by SONET/SDH networks. Similarly, IP networks are provided over SONET network using Packet over SONET (PoS) technology to create IP network links. Thus, an architecture where circuit-switched voice and IP services are offered over the SONET network can be envisioned as shown in Figure 28.6. With the advent of WDM, SONET networks are strong candidates for replacement with WDM technology.[8]

It is worth mentioning that network hierarchy can consist of more than two layers; for example, MPLS can be put in between the IP and the SONET layer. In reality, below the SONET layer, there is a duct layer where the actual fiber is deployed. When we consider one network on top of another network, the term "overlay network" is also used to indicate the network at the top layer.

28.3.7 Distinguishing between traffic and transport networks

Based on the technology discussion so far along with the functionalities available or emerging, we make an important distinction on how to view networks; this view can be broadly classified as *traffic networks* and *transport networks*. We argue that this classification is more powerful than taking a purely technological point of view.

Our use of these two terminologies is as follows: a network is a traffic network if the service request are on-demand (stochastic) and requests are switched or routed on a per-unit basis and are of short duration. A routed entity in the traffic network would

---

[8]Note that this is only in theory; in practice, replacing an existing (live) network with another technology is not very easy—in the end, multiple technologies are in place at one point of time when a new technology is being rolled in while another is not yet fully rolled out. This makes network restoration complicated in practice due to co-existence of multiple transport technologies.

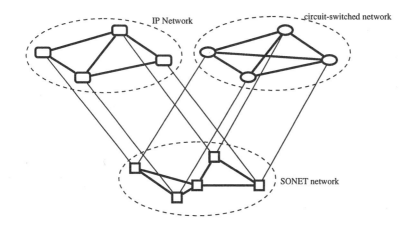

Figure 28.6 Multi-layer architecture

be referred to as *packets* or *calls*. A network is considered to be a transport network if requests are scheduled to be activated on a periodic basis and such requests when routed in the network are set up on a permanent or a semi-permanent basis. A "routed" entity in the transport network would be referred to as *circuits*. From the point of view of network restoration, the one for the traffic network can be called *traffic restoration* since traffic is rerouted whereas the one for the transport restoration can be called *link restoration* or *path restoration* (whether dedicated or shared) since either a transport link is restored or an end-to-end path is restored.

Clearly, IP networks and circuit-switched voice networks are traffic networks. Similarly, digital cross-connect networks are clearly transport networks. On the other hand, MPLS and optical networks cannot be labeled directly in either category; instead, we need to look at the specific functionality or service provided by such technology based on which we can classify them into these two categories.

Consider MPLS first. MPLS can be used for setting up semi-permanent or permanent label-switched tunnels—if it is used this way, then this particular MPLS network can be classified as a transport network. If MPLS tunnels are set up on demand (virtually with minimal set-up delay), are of short duration and can be teared down quickly, then the behavior of this MPLS network is closer to the circuit-switched voice network and hence, in this case, this MPLS network can be thought of as a traffic network. Similarly, depending on how constraint-based routing is used for services provided in an MPLS network (i.e., on demand or semi-permanent/permanent basis), the proper categorization under either traffic or transport network can be done.

Now consider optical networks. An optical network based on WDM or SONET/SDH technology is a transport network. If the optical network uses optical cross-connects (OXCs), then this would fall under the transport network. On the other hand, an on-demand high-rate optical service with optical switching capability that can be set up for short duration and then can be teared down (with minimal set-up/tear-down delay) would fit into services in a traffic network setting.

From the discussion about MPLS and optical networks, it is important to recognize that instead of purely looking at them as technologies, the service functionality provided by them can help us see whether to put each one under the traffic network or the transport network classification. There are several important benefits of this broad classification that has implications for network protection and restoration as well:

- In traffic networks, there is no distinction between working and backup capacity for network protection and restoration, while in transport networks there is often a distinction between working paths/capacity and backup paths. A corollary of this observation is that under no failure, the excess capacity provided in the traffic network serves as additional revenue-bearing capacity for the provider.

- In traffic networks, the restoration would mean being able to restore traffic in general (perhaps at a degraded performance) while for the transport network, restoration would mean whether we have the actual capacity to restore lost capacity and ability to set up paths.

- In a traffic network, the restoration would be on traffic basis while in the transport network, the restoration can be on a bulk capacity basis.

- An effected unit of traffic in a traffic network may not necessarily be recoverable when a failure occurs; for example, consider a packet in transit in an IP network, or an existing call in the circuit-switched voice network—they could both be dropped if a failure occurs[9]. In the case of transport network, the restoration is done for affected circuits.

- There are certain failures inherent to a traffic network that cannot be perceived by the underlying transport network. For example, a line card failure at an IP network router would not be recognized by the underlying transport network. Reverse is not necessarily true. That is, it is indeed true that a self-healing SONET ring can restore a failed link without the traffic network above it perceiving it; on the other hand, in a cross-connect network where there is a delay in restoration time, the failure will be perceived by the traffic network above it.

- In a multi-layer network (say, three layers), if we have more than two layers of technology, then only the top layer is the traffic network, and the second and the bottom layer would be (cascaded) transport networks. For example, consider IP over MPLS over SONET. In this case, MPLS would be set up as permanent tunnels[10] for IP network links and serves the function of a (virtual) transport network. It is neither possible nor desirable to setup MPLS in this case

---

[9]Certainly, this depends on whether the failure is perceived by the traffic network; for example, a SONET self-healing ring can restore a failed link without an IP network or circuit-switched voice network perceiving it.

[10]Note that, from a technology point of view, ATM virtual paths can be set up instead of MPLS tunnels in the middle layer in this architecture.

as a traffic network; such usage of MPLS would mean quick setup/tear down of tunnels (and associated bandwidth) that would cause the IP network routing to oscillate undesirably. An important corollary is that a transport network, especially an intermediate one, can still be a logical network.

- Network design models can be of different types depending on whether this is addressed for a traffic network or a transport network (see Section 28.5).

While the traffic network and the transport network classification has several benefits, we want to emphasize that this is a broad classification. It cannot capture how to do or what to do for all types of failure, for example when a failure propagates from one network to another, or certain intricacies between different types of traffic networks (or, for different types of transport networks). In the rest of the discussion, we also bring out some of the limitations of this classification as and when applicable.

28.4 REVISITING THE THREE-NODE EXAMPLE

First consider the three-node network again; this time as an IP network with three routers, where each router pairs have 20 Mbps of traffic on average and the link speed is 155 Mbps. With OSPF protocol, when link 2-3 fails, the routers will use link state update. Through this process, node 2 can find out that path 2-1-3 exists. In a data network such as the IP network, an important goal is that the delay is minimized, or utilization is kept as low as possible[11]. For example, ISPs often try to keep utilization lower than 50% on average in the normal operating condition. In this case, with re-routing, still the link utilization is at $(20+20)/155 = 25.81\%$, and thus, the failure may not be perceived by the user.

We now discuss the case if we were to consider a circuit-switched voice network consisting of three switches; assume that each node pair has 20 Erlangs of offered traffic and the link capacity is 150 voice trunks. If the circuit-switched network has dynamic routing,[12] then this is not different than the IP network with dynamic routing since there is plenty of bandwidth on the alternate path to route traffic. In the case of the telephone network, another possible case is when, of the three nodes, one is at a higher level of the hierarchy than the other two (e.g. two are end office switches, while the third one is used for alternate routing.). In this case, if nodes 2 and 3 are at the lower level, and node 1 is at a higher level, then for link 2-3 failure, the network can do alternate routing via node 1. On the other hand, if nodes 1 and 2 are at the lower level and node 3 is at the higher level, then for link 2-3 failure the telephone network cannot do any alternate routing (thus, relying on the underlying transport network for recovery). Finally, if three switches are at the top level of a telephone switching hierarchy without dynamic routing, this would be similar to the case of the IP network

---

[11]There is a strong relation between utilization and delay; for example, if we assume M/M/1 queueing model, the average delay, T, is related to utilization ρ by the relation $T = \frac{1}{\mu(1-\rho)}$ where μ is the effective bandwidth of the link.

[12]In the case of telephone network, this usually means that switches in the network are at the same level of hierarchy and employ dynamic routing.

with static routing. In other words, the routing capability does make a difference in a traffic network's ability to restore from a failure.

Let us consider the case of static routing in the traffic network (be it IP or circuit-switched voice), that is, only direct link route allowed for each pair of nodes. In this case, when traffic link 2-3 fails, then traffic between nodes 2 and 3 cannot communicate. On the other hand, recall from previous section that links for IP networks are set up over a transport network. Now suppose that there is a SONET ring that connects traffic link 2-3 as shown in Figure 28.7a. Thus, in this case, a traffic link failure is really a transport link that is affected in the SONET network; for example, the traffic link failure 2-3 is really due to the failure of transport link 2-4. Here, in the transport network itself, transport link 2-4 will be restored by routing around the SONET ring immediately without the traffic (IP or circuit switched voice) network perceiving it. Based on this observation, we now modify one of the observations we discussed earlier in Section 28.2: in a traffic network with static routing, restoration may be possible either partially or fully depending on the capability of the underlying transport network.

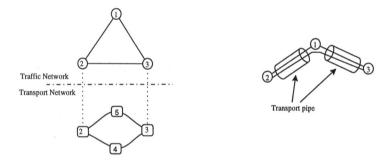

Figure 28.7 (a) Transport level diversity through ring, (b) no transport level diversity

Another important variation to consider is the following: when the perceived diversity in the traffic network is not really diverse when viewed from the transport network. For example, consider again the three node IP network with dynamic routing. Thus, for a link failure, we assume that the traffic can be routed over the alternate path. Now suppose the underlying transport capacity for the IP network is over the same fiber as shown in Figure 28.7b. In this case, if a fiber cable cut occurs, one router in the IP network will be completely isolated from the other two. This illustrates that a single fiber link (or transport link) failure can actually translate to multiple (two in this case) link failures in the IP network. The following important observations come out of this aspect:

- With the dynamic routing capability in the traffic network, it can restore services in its network more effectively only if the underlying transport network has diversity.

- Developing a design model for network restoration based on a single-link failure assumption may not always be realistic if the network under consideration is especially a traffic network.

- Logical diversity in the traffic network can be misleading.

Finally, the notion of static routes for the traffic network is similar to the idea of permanent static circuits being setup in the transport network with cross-connects. Thus, for a transport network with permanent cross-connect circuits, back up circuits on diverse paths needs to be set up for recovery. However, for a logical transport network (intermediate layer in a multi-layer network, e.g., MPLS between IP and SONET), the underlying physical transport network must provide physical diversity even if backup paths are set up in the intermediate transport network.

28.5 NETWORK PROTECTION

No matter where it lies in the network hierarchy, the notion of traffic and transport network provides us a good way to address network protection in terms of optimization model development. In this section, we present formulations for certain specific scenarios. The reader is directed to Pióro and Medhi (2004, Chapters 9 and 12) for a wide variety of different optimization models that can be used for network protection design. We also present a short overview on algorithmic approaches.

28.5.1 Traffic network protection

From the discussion earlier, it is clear that networks such as IP networks, circuit-switched voice networks, on-demand MPLS network, or on-demand optical networks, all fall under the traffic network as per our broad classification.

Assume that h_d is the average traffic volume for demand identifier d ($d = 1, 2, \ldots, D$) between two nodes i and j in a traffic network. Let $p = 1, 2, \ldots, P_d$ be the possible paths for demand d to carry this demand volume h_d. Let x_{dp} be the unknown (to be determined) for flow allocated to path p for demand d. Note that since we are considering a traffic network, the flow on a path here reflects the average amount of traffic to be allocated to this path; secondly, due to traffic nature, we can assume the flow variables to take continuous values. We denote the links in the network by $e = 1, 2, \ldots, E$. Due to path p and link e, we also introduce the indicator δ_{edp} which is set to 1 if path p for demand d uses link e; 0, otherwise. Let ξ_e be the cost of adding capacity of modular unit M on link e, and let variable y_e denote incremental number of capacity units needed on link e (given that the link already has c_e units). Then, the basic capacity design problem can be written as the following mixed-integer linear programming multicommodity flow model:

$$\text{minimize}_{x,y} \quad \sum_{e=1}^{E} \xi_e y_e$$
$$\text{subject to} \quad \sum_{p=1}^{P_d} x_{dp} = h_d, \quad d = 1, 2, \ldots, D \tag{28.1}$$
$$\sum_{d=1}^{D} \sum_{p=1}^{P_d} \delta_{edp} x_{dp} \leq M(c_e + y_e), \quad e = 1, 2, \ldots, E$$
$$x_{dp} \geq 0,$$
$$y_e \geq 0 \text{ and integer.}$$

In the above model, the first set of constraints, called demand constraints, is to indicate that demand volume is to be satisfied by the path flow variables. The second set of constraints, called capacity constraints, is to determine how much capacity is needed on each link to carry traffic taking that link. Finally, the objective function minimizes the cost of incremental capacity y_e.

It may be noted that this multicommodity flow model is based on the link-path representation which suits communication networks well since there are often restrictions on the length of paths and also due to any restriction imposed by network engineers based on their knowledge of an operational network. For example, for dynamic routing circuit-switched networks, a path is restricted to at most two links.

In order to address failures, we consider S states $s = 1, 2, \ldots, S$ and consider the normal state as $s = 0$. We also introduce the parameter α_{es} to denote 0 if the link e is affected due to failure s and 1 if it is not affected.[13] Let h_{ds} be the traffic volume to be carried in state s for demand d. Note that $s = 0$ denotes the normal state of the network; thus, we have $h_{d0} = h_d, d = 1, 2, \ldots, D$. Extending x_{dp}, for each state s we indicate flow variables as x_{dps} for path p and demand identifier d. Then the protection capacity design problem to address for any failure s can be written as follows:

$$\text{minimize}_{x,y} \quad \sum_{e=1}^{E} \xi_e y_e$$
$$\text{subject to} \quad \sum_{p=1}^{P_d} x_{dps} = h_{ds}, \quad \begin{array}{l} d = 1,2,\ldots,D \\ s = 0,1,\ldots,S \end{array} \tag{28.2}$$
$$\sum_{d=1}^{D} \sum_{p=1}^{P_d} \delta_{edp} x_{dps} \leq \alpha_{es} M(c_e + y_e), \quad \begin{array}{l} e = 1,2,\ldots,E \\ s = 0,1,\ldots,S \end{array}$$
$$x_{dps} \geq 0,$$
$$y_e \geq 0 \text{ and integer.}$$

In the above model, the first set of constraints, called demand constraints, is to indicate that demand volume is to be satisfied by the path flow variables for each

---
[13] Parameter α_{es} can be allowed to take a value between 0 and 1 f a link is partially available; for example, see Pióro and Medhi (2004).

state s. The second set of constraints, called capacity constraints, is to determine how much capacity is needed on each link to carry traffic taking that link for each failure state s. The goal is to minimize the cost that addresses any link failure. The above model is quite powerful due to the following reasons:

- The failure states can be either all single link failures (each considered independently), or multiple link failures, or one can consider only a subset of key links that is of interest to a network provider. Depending on the failure which is indicated by s, the corresponding α_{es} can be set to zero, and thus, a variety of failures including multi-link failures can be captured as well.

- Since in a failure state s, we have $\alpha_{es} = 0$, the right hand side of the capacity constraints is forced to be zero. Consequently, this forces the associated x_{dps} to be zero. Furthermore, this says that for each demand pair we do not necessarily need to consider a separate set of paths for each failure state from the normal states (thus, we have kept the number of candidate paths for demand d at P_d, instead of needing to introduce a new P_{ds}).

- Traffic volume can be appropriately captured depending on specific types of traffic network considered; accordingly, modular value M can be adjusted. We discuss below two cases:

    – For an intra-domain IP network running the OSPF/IS-IS protocol, h_d represents traffic in Gbps for average packet traffic from router at node i to router at node j; then, h_{ds} is an appropriately adjusted volume in state s. The quantity, M, is a combination of two factors: acceptable utilization times the capacity module (such as OC-12). Note that multicommodity flow x_{dps} on path p would be positive in state s only if path p is the shortest path as imposed by the link metric (weight) system $w_{es}, e = 1, 2, \ldots, E$ for state $s = 0, 1, \ldots, S$. Thus, in the above formulation, we need to replace x_{dps} by $x_{dps}(w_{es})$ to properly reflect the induced dependency on flow due to the link weight system. It is important to point out that protection design model (28.2), with explicitly writing $x_{dps}(w_{es})$ (instead of just flow x_{dps}), is no longer a mathematical programming formulation. On the other hand, the solution to the dual of the original problem (28.2) is related to the optimal link weight system; see Ahuja et al. (1993), Pióro and Medhi (2004), Pióro et al. (2002), and Wang et al. (2001). Certainly, this relationship between dual solution and the original optimal solution holds also for the basic design case of OSPF networks (i.e., no failure).

    – For a circuit-switched voice network with dynamic routing,[14] h_d can be approximated as follows: $h_d = \mathcal{E}^{-1}(a_d, B)$, where $\mathcal{E}^{-1}$ is the inverse Erlang-

---

[14]Network design formulation for dynamic routing circuit switched networks is more accurately captured through non-linear optimization formulation; for example, see Girard (1990). With the advent and availability of high-valued modular capacity units as well as drop in per unit cost of network capacity, a mixed-integer linear programming model such as (28.2) adequately captures network's need in term of capacity.

B formula[15] applied to traffic load a_d Erlangs for an acceptable blocking level (grade-of-service) B. For each state s, traffic load h_{ds} is an appropriately adjusted volume due to possibly different acceptable blocking level for each state s. In this case, M would be the capacity in terms of modular trunk units (such as 24 voice trunks/T1).

28.5.2 Transport network protection

An important problem in transport network protection is to provide a back-up path for every path with positive demand flow. The idea is that since the back-up path exists, the demand can take the back-up path automatically in case of a failure,[16] and the restoration is instantaneous. For this illustration, we assume that we are addressing protection design for a traffic network link. It is important to understand that bandwidth required for a traffic link is routed on a transport network on a permanent or semi-permanent basis (as circuits); thus, from the perspective of the transport network, this protection can be thought of as an end-to-end or path protection.

Consider now Figure 28.8. In this network, if the demand between 1 and 2 is routed on path 1-3-4-2, then a complete link and node-disjoint back-up path can be setup on 1-6-2; alternately, if the demand between 1 and 2 is routed on 1-3-5-2, then 1-6-2 can also be the back-up path. We can see that when we combine a pair of main path and the back-up path that are disjoint, we can form a cycle; for example, in this case, for demand between 1 and 2, we have two cycles: 1-3-4-2-6-1, and 1-3-5-2-6-1. Either of them is a candidate to be chosen for this demand pair—which one to chose would certainly depend on the network goal. Before we discuss the network goal, we want to point out the following: i) for a network to provide disjoint back ups, we assume that the topology allows us to find at least a pair of primary and back-up paths – otherwise, this is not possible (imagine a linear network), ii) there are already algorithms available for finding a pair of disjoint paths (for example, see Suurballe (1974) and Suurballe and Tarjan (1984)) – such an algorithm can be used for generating a set of pre-processed candidate cycles for a demand pair to be used in the protection design phase.

In order to use the notion of cycle, in a link-path multicommodity formulation, we make an important observation: a path can be replaced by a cycle since a cycle can be thought of as a series of links much like a path (Medhi, 1991). Thus, for demand identifier e connecting two end points of a traffic link, we can consider a set of candidate cycles in the transport network identified as $q = 1, 2, \ldots, Q_e$. Note again that the output capacity of the traffic network is now the demand volume for the transport network. Thus, a traffic network link is now a demand pair for the transport network (hence, we chose to use the notation e for the same entity to show this connection). A transport network link would be identified by g where $g = 1, 2, \ldots, G$. We use γ_{geq} as the indicator that takes the value 1 if cycle q for demand e takes the transport link

---

[15]That is, we need to find the smallest h such that $\mathcal{E}(a,h) \leq B$, where $\mathcal{E}(a,h) = \frac{a^h/h!}{\sum_{k=0}^{h} a^k/k!}$.

[16]For example, this is applicable to an MPLS network deployed as a virtual transport network with the fast-reroute option.

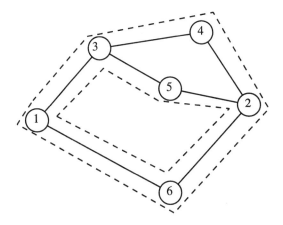

Figure 28.8 Network with multiple disjoint paths forming cycles

g; 0, otherwise. Since, we are providing back-up paths in the transport network,[17] we want the demand volume for a demand e to be assigned to just one cycle out of a set of possible cycles; we use the binary variable u_{eq} for making this decision.

Our goal is to minimize the cost of transport network capacity ($z_g, g = 1, 2, \ldots, G$) for providing the primary-back up paths for demand volume given as $\hat{c}_e, e = 1, 2, \ldots, E$; this can be formulated as follows:

$$\begin{aligned}
\text{minimize}_{u,z} \quad & \sum_{g=1}^{G} \eta_g z_g \\
\text{subject to} \quad & \sum_{q=1}^{Q_e} u_{eq} = 1, \quad e = 1, 2, \ldots, E \\
& \sum_{e=1}^{E} \sum_{q=1}^{Q_e} \gamma_{geq} \hat{c}_e u_{eq} \leq N z_g, \quad g = 1, 2, \ldots, G \\
& u_{eq} \text{ binary} \\
& z_g \geq 0 \text{ and integer.}
\end{aligned} \quad (28.3)$$

In the above formulation, η_g is the unit cost of transport network capacity on link g in modular units of N. Here, the first set of constraints indicated that only a single cycle is to be determined for each demand pair e. For the cycles chosen, the load on links needs to be less than the capacity which is indicated in the second set of constraints. Note that this is an integer linear multicommodity flow model.

In many cases, available capacity on a transport network link is known due to the planning cycle for such capacity determination being on a different time window than the planning cycle for the traffic network. In this case, the goal of a protection de-

---

[17]If completely disjoint back-up path is not achievable, then a possible approach is to provide demand splitting on diverse paths in the transport network for a demand arising for a traffic network link.

sign problem can be to maximize residual capacity[18] (so that future demand can take advantage of the residual capacity) given that $\widehat{C}_g$ is the currently available capacity in transport network link g.

$$\text{maximize}_u \quad \sum_{g=1}^{G} \left(\widehat{C}_g - \sum_{e=1}^{E} \sum_{q=1}^{Q_e} \gamma_{geq} \hat{c}_e u_{eq} \right)$$

$$\text{subject to} \quad \sum_{q=1}^{Q_e} u_{eq} = 1, \qquad e = 1, 2, \ldots, E \qquad (28.4)$$

$$\sum_{e=1}^{E} \sum_{q=1}^{Q_e} \gamma_{geq} \hat{c}_e u_{eq} \leq \widehat{C}_g, \qquad g = 1, 2, \ldots, G$$

$$u_{eq} \text{ binary.}$$

While it may not be obvious to the reader from either (28.3) or (28.4), we do point out that for most transport network design (protection or without protection) cases, the flow variables usually take integer[19] values since the flows for transport networks is to route permanent or semi-permanent circuits. This is another subtle and important distinction between traffic and transport networks and their impact on network modeling formulation. Certainly, this does not rule out the possibility that an integer programming model can be relaxed to linear programming counter part for ease of solving a model towards generating a set of usable solution.

Models (28.3) and (28.4) illustrate just one way to address for network protection, namely through pre-configuration of stand-by backup paths. It is fairly easy to see that such an approach comes at the price of high protection capacity while achieving quick restoration through back-up. Depending on the functional capability of the transport network, protection capacity can be sharable that would lead to reduction in capacity required. Often, however, this reduction in shared capacity comes at the price of the actual restoration time being more pronounced.

So far, we have not clearly stated what should be the demand volume $\hat{c}_e$ for the transport network. There are two possibilities in light of the design models presented for the traffic network. If $\hat{c}_e$ is set to be the total capacity from model (28.2), i.e., $\hat{c}_e = M(c_e + y_e^*), e = 1, 2, \ldots, E$ (where y_e^* is the solution of (28.2)), and if the back-up capacity in the transport network is designed using (28.3), then it is very clear that we are double-dipping on spare capacity for network protection—once in the traffic network and then again in the transport network! Unfortunately, this double-dipping is not always avoidable since in many instances traffic and transport networks are provided by different network providers; thus, when a traffic network provider requests capacity $\hat{c}_e$ from the transport network provider, the transport network provider has no way to know whether the quantity, $\hat{c}_e$, requested by the traffic network has already built-in protection capacity or not (recall that in the traffic network, there is no distinction between working and spare protection capacity).

---

[18] Formulation (28.4) can be easily transformed to a minimization problem.

[19] Obviously, relaxation of flow variables to be continuous to solve the overall model can be done to arrive at a good approximate solution. However, this is done primarily to address the issue of solvability of a model.

If both the traffic network and the transport network are provided by the same network provider, then the provider can coordinate its own network design steps to avoid double-dipping on capacity. Such a model is discussed in the next section.

28.5.3 Multi-layer model

In this section, we present a two-layer model that considers a traffic network and a transport network together. As we have discussed so far, we assume that the back-up path method described above in Section 28.5.2 is used for protection in the transport network part along with normal capacity design for the traffic network part. In this model, we assume that the available capacity in the transport network is given and our design goal is to minimize the total cost due to capacity for the traffic network and routing (primary and backup) in the transport network. The formulation is presented below:

$$
\begin{aligned}
\text{minimize}_{x,u,y} \quad & \sum_{e=1}^{E} \xi_e y_e + \sum_{e=1}^{E} \sum_{q=1}^{Q_e} \zeta_{eq} u_{eq} \\
\text{subject to} \quad & \sum_{p=1}^{P_d} x_{dp} = h_d, & d = 1,2,\ldots,D \\
& \sum_{d=1}^{D} \sum_{p=1}^{P_d} \delta_{edp} x_{dp} \leq M(c_e + y_e), & e = 1,2,\ldots,E \\
& \sum_{q=1}^{Q_e} u_{eq} = 1, & e = 1,2,\ldots,E \\
& \sum_{e=1}^{E} \sum_{q=1}^{Q_e} \gamma_{geq} M y_e u_{eq} \leq \widehat{C}_g, & g = 1,2,\ldots,G \\
& x_{dp} \geq 0, \\
& u_{eq} \text{ binary} \\
& y_e \geq 0 \text{ and integer.}
\end{aligned}
\quad (28.5)
$$

The common notations are the same as used earlier in Sections 28.5.1 and 28.5.2 except for a few. This model combines (28.1) and (28.3) in an integrated model. In this model, the unit routing cost for a pair of primary and backup paths is giving by ζ_{eq}, and the transport network protection is shown only for the incremental capacity of the traffic network capacity y_e. Thus, this model captures the fact that often network providers do not want to re-arrange traffic network capacity already installed from a previous planning cycle, as well as how that capacity is provisioned in the transport network There is however a different problem with the above model: this is a mixed integer *non-linear* programming model due to the product term $y_e u_{eq}$. On the other hand, this non-linearity can be avoided if a new shadow flow variable v_{eq} associated with the binary variable for selection of cycle u_{eq} in the transport network is introduced. We also need an artificially high positive number H to capture the relation between v_{eq} and u_{eq}; as a guideline, H needs to be more than $\max_{e=1,2,\ldots,E} \{My_e\}$. A model equivalent to (28.5) is shown below:

$$\begin{aligned}
&\text{minimize}_{x,u,v,y} && \sum_{e=1}^{E} \xi_e y_e + \sum_{e=1}^{E} \sum_{q=1}^{Q_e} \zeta_{eq} u_{eq} \\
&\text{subject to} && \sum_{p=1}^{P_d} x_{dp} = h_d, && d = 1,2,\ldots,D \\
& && \sum_{d=1}^{D} \sum_{p=1}^{P_d} \delta_{edp} x_{dp} \leq M(c_e + y_e), && e = 1,2,\ldots,E \\
& && \sum_{q=1}^{Q_e} u_{eq} = 1, && e = 1,2,\ldots,E \\
& && \sum_{q=1}^{Q_e} v_{eq} = M y_e, && e = 1,2,\ldots,E && (28.6)\\
& && v_{eq} \leq H u_{eq}, && e = 1,2,\ldots,E \\
& && && q = 1,2,\ldots,Q_e \\
& && \sum_{e=1}^{E} \sum_{q=1}^{Q_e} \gamma_{geq} v_{eq} \leq \widehat{C}_g, && g = 1,2,\ldots,G \\
& && x_{dp} \geq 0, \\
& && u_{eq}\ \text{binary} \\
& && v_{eq} \geq 0, \\
& && y_e \geq 0\ \text{and integer}.
\end{aligned}$$

Note that for each e, only one u_{eq} takes the value 1—the rest are zero; this forces the associated v_{eq} to be zero except for the one corresponding to the specific u_{eq} that takes the value 1. In the above model, capacity constraints in the transport network are linear constraints.

Through the above integrated model, we have shown the following: i) indeed, traffic and transport network protection design can be modeled through an integrated optimization model, ii) depending on the requirement in the transport network protection, it is possible to arrive at models with non-linear product terms; on the other hand, through the introduction of addition variables such product terms can be avoided.

28.5.4 Algorithmic approaches

In the previous subsections, we have presented a set of optimization models to address the protection and restoration design for traffic and transport networks. Approaches for solving different models cover a vast area in itself; we now briefly discuss algorithmic approaches.

The optimization models presented so far are multicommodity flow models, specifically mixed-integer linear programming models for the traffic network, and integer linear programming models for the transport network; also, design models for multi-layer protection design can be cast as mixed-integer linear programming models. Furthermore, we have commented earlier that often link-path formulation suits telecommunication network modeling well since network designers and engineers can use their prior knowledge of the network and network properties to determine a set of pre-processed candidate paths using k-shortest paths methods (and pruning paths not wanted), or determining pairs of disjoint-paths for protection.[20]

Despite having pre-generated candidate paths, such formulations are hard and time consuming to solve for large-size networks, especially if a commercial mixed-integer programming solver is used. Fortunately, these models are multicommodity flow models with special structures; for example, requirement for routing can be decoupled on a per commodity (i.e., demand) basis, flow and capacity variables are coupled only on certain type of constraints; for a multi-layer model, there is also a connection due to one network's capacity being related to another network's demand.

Based on our experience over the years, we have found that Lagrangian relaxation-based dual subgradient (LRBDS) approach can be a very effective method for solving many such problems for very large networks. The main idea behind LRBDS is to consider the original problem in the dual space by relaxing a set of constraints of the original problem so that it allows us to decouple the problem into different "like-minded" variables (e.g., flow variables, capacity variables); such relaxation leads to generating a set of subproblems which are not only decoupled, but can be solved very efficiently. On the other hand, such subproblems depend on knowing the dual solution. Since we would not know the dual solution to begin with, we need to iterate through the dual space. Furthermore, the dual problem is continuous, but non-differentiable at many points; thus, the iterative dual approach is based on obtaining a subgradient (instead of gradient) of the dual function at each dual iteration to compute a direction. Such a subgradient is readily available from solving a series of subproblems. Thus, the entire process iterates back and forth between the dual multiplier update and solving large number of subproblems efficiently. The interested reader can find more details about how LRBDS can be applied to different network design problems in Pióro and Medhi (2004).

We do want to point out that when the traffic network is the IP network with shortest path routing, then the goal is to determine link weights—we have commented earlier that this is not a direct mathematical programming formulation. For such a design problem, several approaches have been proposed such as local search heuristic (Fortz and Thorup, 2000; 2004; Yuan, 2003), genetic algorithm (Ericsson et al., 2002; Buriol et al., 2005), and a variant of the LRBDS approach (Pióro and Medhi, 2004; Srivastava et al., 2005).

28.5.5 Remarks

Although from a modeling point of view integrated protection model as described in Section 28.5.3 is possible, a network provider may still decide to do completely decoupled design for its traffic and transport networks for practical reasons: i) the network management and provisioning system needed to implement such a model can be cumbersome, and often a hindrance, ii) the transport network may provide transport

---

[20] See references such as Lawler (1976) and Pióro and Medhi (2004) for generating k shortest paths, and Suurballe (1974) and Suurballe and Tarjan (1984) for generating disjoint-path pairs.

function for many different upper layer 'logical' transport networks,[21] each of which may also fall under different planning cycles, iii) traffic and transport network design are done by different business units (partly because of item-ii) for large telecommunications network providers.

We conclude this section by noting that the traffic and transport network design models presented here are to give some flavor of models applicable. The reader is directed to Pióro and Medhi (2004, Chapters 9 and 11) for an extensive variety of optimization models and approaches.

28.6 APPROACH TO RESTORATION

In this section, we present approaches for restoration in different networks. By no means, this list is exhaustive. Our attempt here is to make the reader aware of the necessity to take different approaches to network restoration. It is important to understand that for this part we assume that protection capacity (whether partial or full) is already provided through the protection design models presented in the previous section.

28.6.1 Large-scale intra-domain IP network

In the three-node IP network discussed earlier in Section 28.4, any link-state update can be performed quickly. When an intra-domain IP network is very large (say 100 nodes or more), then communicating about a link failure can take certain amount of time as this is being disseminated throughout the network to every router. Once this information is received, then each router employs shortest path routing (by setting the affected link with infinite cost) to compute next hop for shortest paths. There is an important issue to consider here. Most large IP networks are traffic engineered for optimal movement of packets through the network (Fortz and Thorup, 2000; 2004); often, such traffic engineering requires considering traffic snapshot and capacity of links to determine an optimal (or, near optimal) set of link weights for links of the IP network so that congestion is minimized in the network. Such weight determination is computed off-line and disseminated back to each router in the network. When a link failure occurs, affected traffic will be re-routed on the newly computed shortest path (based on old link weights except for the affected link) which may result in significant congestion on some links, thereby affecting the delay perceived by users. This suggests that determination of a *new* optimal link weight system after a failure could conceivably minimize such congestion.

While theoretically the idea of determining new link weights after each specific failure is desirable, we need to understand the basic steps in the restoration process which can be summarized as follows:

failure occurs → *indicate failure to all nodes* → *determine new weight systems* → *compute shortest path* → *update routing table*

---

[21] Consider a telecommunications transport network provider that provides its own traffic network service, e.g., for voice telephony, IP networks, and virtual private network services for automated teller machine networks for banking institutions; each can be considered as a mid-level logical transport network.

Furthermore, in order to determine the new weight system, network measurements to get a snapshot of traffic demand in the network may be necessary, followed by running an optimization algorithm for determining new link weights; i.e., in reality, the new weight determination step involves two sub-steps. We can now see that the actual time to restoration starts adding up as we consider various steps involved. Finally, we need to understand something about types of failure. It has been reported that 50% of failures last for < 1 minute, and 80% of failures last for < 10 minutes for a large inter-domain IP network (Nucci et al., 2003); such short-lived failures are referred to as transient failures.

Given the observation about transient failures, there is another important issue to consider; even if a new link weight system under a specific failure might be optimal, the network may be required to revert back to the old weight system (that existed prior to the failure) for most of the transient failure scenarios—this itself can cause undesirable oscillatory behavior in the network, not to mention additional impact on computational time to obtain them for a live network. The approach proposed in Nucci et al. (2003) and Yuan (2003) (also discussed in Pióro and Medhi (2004, Chapter 7)) is elegant: instead of trying to compute a new weight system *every time* after a failure, determine *ahead* of time a robust weight system that works for *both* normal network condition as well as under likely transient failures; in other words, after a transient failure, it is not necessary to determine new traffic snapshot[22] and no new weight computation is necessary; instead, the failure of the link is communicated through the link state advertisement—each router can then compute a new shortest path preferably only for affected destinations using the robust link weights already determined except for using infinite cost for the affected link; this itself reduces computation time in each router instead of computing complete routing tables to all destinations.

We now briefly describe the off-line modeling approach for computation of link metric values that can work for both under the normal condition and transient failures. Consider that we are interested in the goal of minimizing maximum link utilization; now, minimizing maximum link utilization under no failure and transient failures can be given different priority by constructing an objective function that gives weights to each scenario (no failure or transient failures); this then can be formulated as a composite optimization problem (see Nucci et al. (2003); also, see Pióro and Medhi (2004, Chapter 7) for a different formulation). Another approach would be to consider no failure and failures as a bi-criteria optimization problem for which Pareto optimal solution can be obtained (see Yuan (2003)). Thus, off-line optimization model to understand how to determine link metric that works for no failure and transient failures is an important usage of optimization models in networking. Finally, we point the reader to the recent work (Iannaccone et al., 2004) that discusses feasibility of restoration in IP networks taking into account practical issues.

---

[22]Determining traffic demand in an IP network can be time consuming, see Feldmann et al. (2001).

28.6.2 Large dynamic routing circuit-switched networks

Most implementations of dynamic routing for circuit-switched voice networks (usually deployed by inter-exchange TSPs) have two important features: the network is almost fully-connected with the length of a call path (within its network) consisting of maximum two-links[23] and multiple possible paths are cached (instead of just the shortest path) for each pair of source-destination switches. Note that paths are updated periodically or almost on-demand; exchanges of link status information is usually done through an out-of-band network. The path list is kept in a pre-defined priority order, for example from most available bandwidth to the least available bandwidth. When a call arrives and cannot find bandwidth on the first path, it can try the second path in the list, and so on.[24]

To be able to compute path ordering quickly, most dynamic routing implementations such as real-time network routing (RTNR) and dynamically controlled routing (DCR) use heuristic algorithms based on the status of link bandwidth availability (see Ash (1997), Ash and Chemouil (2004), and Girard (1990)). This operation is followed during the normal network condition; if failure occurs, status of the failed link is updated to non-availability. Since multiple paths are cashed for each source-destination pair and assuming a failure does not affect all paths in cache, the path set need not be updated immediately. Rather, for a new call that arrives, a path with the affected link would be attempted (without success), and depending on crankback capability, the call will then attempt the next path available in the list (until all paths in the cache are exhausted). Thus, this functionality provides the traffic restoration capability in the network. The actual performance of the network and its adaptability would depend on built-in traffic restoration capacity in the network.

Since a dynamic routing circuit-switched network is almost fully-connected, a lower layer transport network link failure is likely to affect multiple links in the traffic network; this is an important point to note. Thus, trunk diversity along with restoration capacity design is needed for graceful recovery from a failure (Medhi, 1994).

28.6.3 Transport network restoration

If the transport network is a SONET/SDH ring-based network, then the self-healing property of such a ring can address any single link failure. As we have discussed earlier, a large transport network is often made up of a series of rings where two adjacent rings are usually interconnected in two different points to reduce the impact of a node failure on two rings. It is commonly understood that the overall capacity needed in a SONET ring network is much higher than if the entire network is set up as a mesh transport network. On the other hand, the benefit of immediate restoration possible with a SONET/SDH ring can be of paramount importance to a network provider; this

---

[23]This length limitation was originally done to reduce complexity of switching software along with the realization that the marginal gain was negligible when additional paths with more than two-link paths are used if the network grade-of-service were to be maintained below 1% call blocking; see Ash et al. (1981).

[24]Due to a feature called trunk reservation or state protection, a call may not be allowed on a link even if there is available capacity; see Girard (1990).

benefit may have much higher weight than the cost of the capacity over provisioned through a series of rings. Thus, it is up to the service provider to decide on this tradeoff between the high cost ring architecture and the restoration speed, and to make the best strategic decision.

Now consider a mesh transport network. With technological advantages, increasingly more intelligence is being pushed to a transport network node than it was possible before. Thus, for the rest of the discussion in this section, we will consider only mesh transport networks in general and provide specific network cases as applicable, for example, when an MPLS network serves as a transport network.

Mesh transport network restoration clearly depends on availability of protection capacity, as well as ability to compute back-up paths. If a transport network is designed without any hot-standby back-up paths (either for path or link restoration), then after a failure some algorithm is required to be solved to determine restoration paths. Determination of restoration of transport paths after a failure in the path restoration framework can be modeled as a multicommodity flow problem (Iraschko et al., 1998; Medhi and Khurana, 1995)). However, solving a multicommodity flow model in its entirety after a failure can be time consuming; thus, a heuristic approach is desirable to arrive quickly at an acceptable solution. Arriving at such a solution, however, depends on whether the protection capacity in the network is tight. A general sense is that if the protection capacity is plenty, then determining such a solution through a heuristic is usually less time consuming.

In many transport networks, a back-up path is pre-provisioned corresponding to a deployed primary path. We have already discussed earlier protection capacity design for such an environment. For an MPLS network that provides transport network services, a functionality called "MPLS fast-reroute" provides signaling capacity for restoration from a failure by switching from a primary path to a backup path quickly. In addition, automatic protection switching (APS) concept has been around for sometime to address for quick restoration in a transport network for link restoration.

It may be noted that the concept of p-cycle (Grover, 2004) which can be implemented at the signaling level of a mesh optimal network for restoration can be a very attractive method that brings the advantage of a SONET-ring like restoration functionality in a mesh network environment; for details, see Grover and Stametaelakis (2000) and Grover (2004).

The above discussion is to provide a glimpse into the vast area of transport network protection. For additional details on transport network protection, see two recent books: Grover (2004) and Vasseur et al. (2004).

28.6.4 Combined traffic and transport network restoration

Usually, large telecommunication network service providers operate both traffic networks such as circuit-switched voice and IP networks along with underlying optical transport networks. Thus, in such a situation, the question of combined or integrated protection arises. As we have discussion earlier in the context of protection capacity design model, such combined effort can lead to less capacity required for providing protection.

From the point of view of actual restoration, it is helpful to understand the impact of a severe transport network failure and staggered restoration in the transport network for overall recovery of the traffic network. To illustrate this, we consider a specific traffic/transport network scenario[25] where the traffic network is fully-connected with dynamic call routing capability (with multiple routes cached)—this traffic network is provided over a mesh transport network and the capacity required (including protection capacity) for the traffic network is diversely routed in the transport network. In this example, a single link failure in the transport network affected multiple links in the traffic network (almost 40% of the total number of traffic network links); since the logical traffic network is fully-connected, a traffic link impacted means the traffic that uses the direct link path is always impacted.

To show the severity of a transport network failure, we report impact on the affected[26] node pairs in the traffic network and where we assume that the transport capacity is not fully recoverable from a failure. We found that dynamic routing can still find available capacity to reroute calls and bring down the blocking significantly (from being nearly 100% blocking) due to built-in transport network diversity. Through staggered recovery in the transport network, if we bring up the affected traffic links in a staggered manner,[27] blocking gradually improves. This is illustrated in Figure 28.9. Specifically, we have indicated two lines to illustrate recovery; the solid line represents when the transport network path restoration is recovered without any knowledge of the current traffic volume in the traffic network, whereas the dashed line represents prioritized transport network path restoration allocation that is aided by the knowledge of the current traffic volume in the traffic network along with dynamic routing capability—it is important to note that when the transport network restoration is over (in this case, capacity only partially recoverable due to assumption on the lack of enough protection capacity in the transport network) the traffic network is not fully recovered in the former case, but is fully recovered in the latter case.

Finally, integrated restoration, especially, IP over optical networks, and its implication on traffic engineering is a growing area of interest; for example, see Fumagalli and Valcarenghi (2000), Lee and Mukherjee (2004), Medhi and Tipper (2000), and Strand (2001).

28.6.5 Lessons learned

Eventually, the question arises about what can we learn in regard to network restoration for traffic and transport networks when done independently, and also in regard to integrated traffic and transport network restoration. We also need to understand how the speed of restoration impacts what we may do in terms of restoration. Incidentally, Figure 28.9 allows us to understand several different dimensions about network restoration, especially when we interpret that the x-axis represents unit of time without being specific about what the actual value is.

---

[25] See Medhi and Khurana (1995) for details.
[26] Instead of all node pairs, we consider only the affected node pairs to show the "worst" case scenario.
[27] Note that from the transport network point of view, a traffic link restoration is equivalent to a path restoration in the transport network.

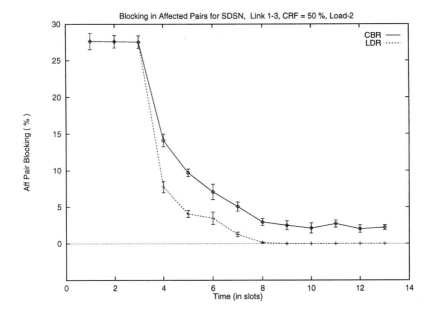

Figure 28.9 Impact on performance due to staggered restoration

- If the transport network consists of self-healing rings with restoration capability in milliseconds (e.g., SONET/SDH rings), then the failure is not perceived in the traffic network; in other words, the intermediate steps shown in the plot in terms of time is instantaneous, and may not trigger any specific traffic network level recovery.

- If transport network capacity is not recoverable, then at least some level of transport diversity should be provided in the routed circuits for traffic network capacity; this can provide some flexibility for the traffic network to address traffic restoration through re-routing. Secondly, traffic restoration protection capacity may be addressed in the traffic network; with this, the end state in the plot can be arrived at almost instantaneously. This situation also essentially applies for traffic network providers who do not own the underlying transport network, i.e., leases capacity from a transport network provider.

- Traffic network restoration can be fully recoverable without requiring fully recovery in the transport network *if* certain information exchange between the traffic network and the transport network is possible regarding a failure—for this to work smoothly, an integrated network management functionality to coordinate between two layers is needed.

- In a mesh transport network, if a dedicated back-up path is provided for every provisioned path, then depending on the speed of the switching logic to move to the back-up path, the failure may or may not be perceived in the traffic network.

- In a mesh transport network, the recovery process may have some delay due to calculation of restored paths. This graph allows us to see how the traffic network may be impacted depending on the speed of this recovery.

- Speed of restoration in the transport network can be the driver in the right combination to do between the traffic and transport network, especially in terms of restoration capacity.

- When a failure in the transport network is recognized by the traffic network (e.g., due to delay in recovery in the transport network), the traffic network starts its own restoration process by rerouting (and possibly new computation of the routing table). As the transport network starts to recover, the traffic network may start changing routing again, thus possibly causing instability in routing and traffic flow behavior during the transient period. Good understanding of protocol interactions with state information exchange is required through further study to determine an appropriate action to take.

A closely related issue with speed of restoration is the cost associated with restoration. The cost of restoration has primarily three major cost components: i) protection capacity cost, ii) switching functionality cost, iii) operational cost. In Section 28.5, we have shown how optimization models can be used for determining protection capacity cost. In many instances, switching functionality cost can be modeled in these optimization models since unit link cost can be thought of as summation of the unit distance-based cost and the unit switching gear cost. Secondly, the formulation of the model is independent of actual component costs for different technologies; thus, the appropriate unit cost can be used for a specific technology without fundamentally needing to change the optimization models. Operational cost is however very hard to model; often a general understanding is that more automated a network is, the less is the operational cost for a failure; on the other hand, it is hard to differentiate between day-to-day cost of normal operation of a network and what is specifically needed whenever a failure occurs. Finally, there is another important cost we need to be aware of which can be called "opportunity" cost: if a provider has a network down often or for a long period of time, then customers may chose to move to other providers—this cost is difficult to model solely within the framework of an optimization model for protection. It may be noted that there can be other costs as well for failures not considered here; for example, redundancy in power to address for a power failure.

28.7 TYPES OF FAILURE AND FAILURE PROPAGATION

Our discussion here has primarily centered around link failures in networks. In many cases, the result is also applicable to node failures when we consider that a node failure in abstract means that all the links connected to the failed node is not available any more. At the same time, it is more appropriate to consider node failure in a network as a separate design problem. For example, large IP networks (Iannaccone et al., 2004) use redundant routers at a point-of-presence (PoP) to address for any router failure.

In general, it is important to realize that a network may have many different types of failures. There are no recent study available in regard to sources of failure in the public switched telephone network; the last reported study was by Kuhn (1997). Regarding sources of failure in IP networks, the first one was by Labovitz et al. (1999), while there has been a series of recent works by researchers at Sprint (Bhattacharyya and Iannaccone, 2003; Iannaccone et al., 2004; Markopoulou et al., 2004). It is worth noting that IP network availability is a growing area of interest (see also Callon (2003) and Kaplan (2003)).

The study for the PSTN was based on reporting required to be submitted to Federal Communication Commission (FCC) by a provider when the provider has a failure that lasts for more than 30 minutes affecting at least 30,000 customers;[28] thus, these failures can be considered major failures and do not include short transient failures that the network has managed to heal based on capabilities such as routing. It was reported that errors due to cable cuts constitute about one-fourth of total errors while affecting the network 14% of the time in terms of downtime minutes. Interestingly, while failures due to overload were only 6% in terms of occurrence, it contributed to more than two-fifth in terms of downtime minutes. Hardware failure that included loss of power supply accounted for 19% of the time being only 6% of the total downtime-minutes. It was also reported that software failure was not a big factor. Regardless, Kuhn stated that during the period studied, the PSTN network was very reliable, being up 99.999% of the time; he attributed this reliability due to software reliability and dynamic routing capability of the network to re-route around a failure.

The report by Labovitz et al. on the source of failure in the Internet was based on the trouble ticket log of a regional Internet service provider (which leases the transport capacity from telecommunication providers). Certainly, this report cannot be generalized to the entire Internet or other ISPs; on the other hand, this report can serve primarily as indicators of the type of failures than the actual values. Nevertheless, specific values can provide some guideline on what to focus on in terms of reliability objective. The report stated that 16% of the failures were due to maintenance while another 16% were due to power outage. However, they also reported that 15% of failures were due to the underlying transport network (e.g., fiber cut) which the ISPs have little control over. Based on this report, it can be inferred that a traffic network provider such as an ISP cannot completely rely on the underlying transport provider to be reliable; thus, an ISP must engineer traffic restoration capacity in its network and request trunk diversity from its transport network provider so that with robust link metric setting (see Section 28.6.1), it can circumvent most transport network failures.

The series of reports from the Sprint research group is the most recent study available about failure characterization and restoration of an operational IP network based on measurements over a six-month window. It may be noted that in Sprint's case,

---

[28] A limitation of such reporting is that it cannot capture the benefit of protection capacity already deployed in the network which may has resulted in the ability to provide services gracefully despite some failures.

the optical transport network is also provided by Sprint (but by a different organizational/business unit). They made the following important remark: "failures are part of everyday operation and affect a variety of links" (Markopoulou et al., 2004). They reported that 20% of failures occurred during scheduled maintenance activities, although the maintenance window was only for 5% of the time every week. Of the rest of the failures, 30% were shared failures (16.5% were router-related while 11.4% optical related which affected multiple IP network links), and the rest were individual link failures. They also reported that 50% of the failures were of up to a minute duration, 81% were of up to ten minute duration (which they classify as "transient" failures); only 10% of the failures were longer than 45 minutes duration. By clever use of forward path re-computation of shortest paths (after a failure) and adjustment of various timers, they were able to bring down the recovery time to less than one second (Iannaccone et al., 2004) for failures "seen" by the IP network. It is worthwhile to note that Sprint deploys networks of SONET rings in most of its optical transport network; despite this feature, it is clear that there are many failure events still seen by the IP network (which uses this transport network), and thus, the development of design and restoration mechanisms independently for the traffic and the transport network is evidently important.

Finally, we conclude this section with an example of failure propagation somewhat typical to Internet routing architecture. Recall that Internet routing architecture is a federation of autonomous systems connected by BGP. Through BGP, reachability information is communicated about a network connected to the Internet where a network is identified by an IP-prefix.[29] When a BGP router[30] faces a problem (for example, due to router CPU overload[31]), keep-alive messages (which are used for indicating that two neighboring BGP routers are in communication) may not be communicated on a timely basis—this can make one router think that the other router is not available any more, much like as if the link between the routers has gone down. Each router can then generate a series of messages to indicate about non-reachability which can cascade from one autonomous system to another one; due to the path vector protocol nature of the BGP protocol and to avoid route looping problem, finding another path through other autonomous systems can take a long time (sometimes, in the order of 15 minutes). A second problem is that when the CPU overload subsides and both the routers determine that they can talk to each other again (that means the link is up again)—this can cause another storm of activities in regard to re-updating reachability information thus causing an instable behavior. For a specific example of CPU overload, consider the impact of a recent virus on the routing infrastructure. Routers usually cache forwarding path for commonly routed IP addresses to provide fast lookup and fast packet routing. During the Code Red and Nimda virus attack (Cowie and Ogielski, 2001), instances of the virus started random IP port scanning; this resulted

---

[29] A network identified through IP-prefix is based on higher order bits of the IP address along with a subnet mask; such a network should not be confused with an autonomous system.
[30] A BGP router is also referred to as a BGP speaker.
[31] See Labovitz et al. (1998); while router vendors have made progress on how to handle keepalive messages during CPU overload, this is a good example to understand the basics of Internet routing instability.

in cache misses as well as router error messages[32] leading to extensive CPU overload in routers, thus causing the instable behavior just described.

28.8 SUMMARY

Network restoration is a critical area to understand in the design and deployment of communication networks. In this chapter, we have considered a variety of networks and networking technologies to argue that classification of networks into two broad categories, traffic network and transport network, provides us with a systemic way to consider network protection design models and associated restoration techniques to address for a network failure. In particular, we have shown how this classification impacts the development of appropriate optimization models.

In general, existence of multiple layers (to address for different purpose) makes network protection and restoration a difficult problem. On either extremes are the scenarios where the restoration is done only in the traffic network or is only in the transport network. Certainly, a combination of strategies give the best option to address for many different types of failures which also depends on the capability available in a network and the associated cost of doing it. Finally, we point out that besides a link (or a node) failure, there are other types of failures that a network regularly faces for which a provider needs to develop an appropriate restoration strategy based on its reliability objective. Many of these problems are difficult to model since there are conflicting interests and strategies involved while ensuring that network instability can be avoided—this is an area that requires further exploration.

Appendix: Acronyms

APS := Automatic Protection Switching
ATM := Asynchronous Transfer Mode
AS := Autonomous System
BGP := Border Gateway Protocol
DCR: = Dynamically Controlled Routing
DCS := Digital Cross-Connect System
FCC := Federal Communication Commission
ICMP := Internet Control Message Protocol
IS-IS := Intermediate System-to-Intermediate System
ISP := Internet Service Provider
IXC := Inter-exchange Carrier
LDP := Label Distribution Protocol
LEC := Local-exchange Carrier
LSP := label Switched Path
LSR := Label Switched Router
MPLS := Multi-Protocol Label Switching
OC := Optical Carrier

---

[32]through ICMP error messages.

OXC := Optical Cross-connects
OSPF := Open Shortest Path First
PoS := Packet over SONET
PSTN := Public Switched Telephone Network
RTNR := Real-Time Network Routing
SDH := Synchronous Digital Hierarchy
SONET := Synchronous Optical Network
TSP := Telephone Service Provider
VP := Virtual Path
WDM := Wave Division Multiplexing

Acknowledgments

This work in some sense is a reflection of my interest to understand network protection and restoration for more than fifteen years for many different types of networks. During this period, I have benefited from funding provided by DARPA (Agreement no. F30602-97-1-0257), NSF (grant no. NCR-9506652), the University of Missouri Research Board, and Sprint Corporation. Needless to say, my interest in this area initially piqued during my employment at the AT&T Bell Laboratories (1987–1989).

I have personally benefited from communication/discussion with several people over the years in regard to traffic and transport networks, and network protection and restoration; for that, I like to thank Jerry Ash, Supratik Bhattacharyya, Robert Doverspike, Wayne Grover, Hadriel Kaplan, Karen Medhi, Jim Pearce, Michał Pióro, John Strand, and David Tipper. Finally, I like to thank Balaji Krithikaivasan for drawing several figures included in this chapter, and Amit Sinha for carefully reading a draft of this chapter.

Bibliography

R. K. Ahuja, T. L. Magnanti, and J. B. Orlin. *Network Flows: Theory, Algorithms, and Applications*. Prentice Hall, 1993.

G. R. Ash. *Dynamic Routing in Telecommunication Networks*. McGraw-Hill, 1997.

G. R. Ash, R. H. Cardwell, and R. P. Murray. Design and optimization of networks with dynamic routing. *Bell System Technical Journal*, 60:1787–1820, 1981.

G. R. Ash and P. Chemouil. 20 years of dynamic routing in telephone networks: Looking backward to the future. *IEEE Global Communications Newsletter*, pages 1–4, October 2004. (appears as insert in IEEE Communication Magazine, October 2004 issue).

S. Bhattacharyya and G. Iannaccone. Availability and survivability in IP networks. Tutorial presented at the 11th IEEE International Conference on Network Protocols (ICNP), November 2003.

L.S. Buriol, M.G.C. Resende, C.C. Ribeiro, and M. Thorup. A hybrid genetic algorithm for the weight setting problem in OSPF/IS-IS routing. *Networks*, 2005. In press.

R. Callon. Availability & security for IP data networks. Tutorial presented at the 4th International Workshop on Design of Reliable Communication Networks (DRCN), Banff, Canada, October 2003.

J. Cowie and A. Ogielski. Global routing instabilities during Code Red 2 and Nimda worm propagation. Presentation at NANOG23 Meeting, Oakland, CA, October 2001.

M. Ericsson, M. G. C. Resende, and P. M. Pardalos. A genetic algorithm for the weight setting problem in OSPF routing. *Journal of Combinatorial Optimization*, 6(3):229–333, 2002.

A. Feldmann, A. Greenberg, C. Lund, N. Reingold, J. Rexford, and F. True. Deriving traffic demands for operational IP networks: methodology and experience. *IEEE/ACM Transactions on Networking*, 9:265–279, 2001.

B. Fortz and M. Thorup. Internet traffic engineering by optimizing OSPF weights. In *Proc. 19th IEEE Conference on Computer Communications (INFOCOM'2000)*, pages 519–528, 2000.

B. Fortz and M. Thorup. Increasing Internet capacity using local search. *Computational Optimization and Applications*, 29(1):13–48, 2004. Preliminary short version of this paper published as "Internet Traffic Engineering by Optimizing OSPF weights," in Proc. 19th IEEE Conf. on Computer Communications (INFOCOM 2000).

A. Fumagalli and L. Valcarenghi. IP restoration vs. WDM protection: Is there an optimal choice? *IEEE Network*, 14(6):34–41, November 2000.

A. Girard. *Routing and Dimensioning in Circuit-Switched Networks*. Addison-Wesley, Reading, MA, 1990.

W. Grover and D. Stametaelakis. Bridging the ring-mesh dichotomy with p-cycles. In *Proc. Design of Reliable Communication Networks (DRCN'2000), Munich*, pages 92–104, 2000.

W. D. Grover. *Mesh-based Survivable Networks: Options and Strategies for Optical, MPLS, SONET and ATM Networking*. Prentice Hall, 2004.

G. Iannaccone, C.-N. Chuah, S. Bhattacharyya, and C. Diot. Feasibility of IP restoration in a tier-1 backbone. *IEEE Network*, 18(2):13–19, March-April 2004.

R. R. Iraschko, M. MacGregor, and W. D. Grover. Optimal capacity placement for path restoration in STM or ATM mesh survivable networks. *IEEE/ACM Trans. on Networking*, 6:325–336, 1998.

H. Kaplan. Resilient IP network design. Tutorial presented at the 3th IEEE Workshop on IP Operations & Management (IPOM), Kansas City, Missouri, USA, October 2003.

D. R. Kuhn. Sources of failure in public switched telephone network. *IEEE Computer*, 30(4):31–36, April 1997.

C. Labovitz, A. Ahuja, and F. Jahanian. Experimental study of Internet stability and wide-area network failures. In *Proceedings of Twenty-Ninth Annual International Symposium on Fault-Tolerant Computing (FTCS99)*, pages 278–285, Madison, Wisconsin, June 1999.

C. Labovitz, G. R. Malan, and F. Jahanian. Internet routing instability. *IEEE/ACM Transactions on Networking*, 6:515–528, 1998.

E. L. Lawler. *Combinatorial Optimization: Networks and Matroids*. Holt, Rinehart, and Winston, 1976.

Y. Lee and B. Mukherjee. Traffic engineering in next-generation optical networks. *IEEE Communications Surveys*, 6(3):16–33, 2004.

A. Markopoulou, G. Iannaccone, S. Bhattacharyya, C.-N. Chuah, and C. Diot. Characterization of failures in an IP backbone. In *Proc. of 23rd IEEE Conference on Computer Communication (INFOCOM'2004)*, pages 2307–2317, Hong Kong, March 2004.

D. Medhi. Diverse routing for survivability in a fiber-based sparse network. In *Proc. IEEE International Conference on Communication (ICC'91)*, pages 672–676, Denver, Colorado, June 1991.

D. Medhi. A unified approach to network survivability for teletraffic networks: Models, algorithms and analysis. *IEEE Trans. on Communications*, 42:534–548, 1994.

D. Medhi and R. Khurana. Optimization and performance of network restoration schemes for wide-area teletraffic networks. *Journal of Network and Systems Management*, 3(3):265–294, 1995.

D. Medhi and D. Tipper. Multi-layered network survivability – models, analysis, architecture, framework and implementation: An overview. In *Proc. DARPA Information Survivability Conference and Exposition (DISCEX'2000)*, volume I, pages 173–186, Hilton Head Island, South Carolina, USA, January 2000.

A. Nucci, B. Schroeder, S. Bhattacharyya, N. Taft, and C. Diot. IGP link weight assignment for transient link failures. In *Proc. 18th International Teletraffic Congress (ITC18)*, pages 321–330, Berlin, Germany, September 2003.

M. Pióro and D. Medhi. *Routing, Flow and Capacity Design in Communication and Computer Networks*. Morgan Kaufmann Publishers, 2004.

M. Pióro, A. Szentesi, J. Harmatos, A. Jüttner, P. Gajowniczek, and S. Kozdrowski. On OSPF related network optimization problems. *Performance Evaluation*, 48: 201–223, 2002.

R. F. Rey. *(ed.) Engineering and Operations in the Bell System – 2nd Edition*. AT&T Bell Laboratories, Murray Hill, New Jersey, 1983.

S. Srivastava, G. Agrawal, M. Pióro, and D. Medhi. Determining link weight system under various objectives for OSPF networks using a Lagrangian relaxation-based approach. *IEEE eTransactions on Network and Service Management*, 2005. In press.

J. Strand. Converging protection and restoration strategies of the IP and optical layers to support the survival of IP services. MPLS Summit, January 2001.

J. W. Suurballe. Disjoint paths in a network. *Networks*, 4:125–145, 1974.

J. W. Suurballe and R. E. Tarjan. A quick method for finding shortest pairs of disjoint paths. *Networks*, 14:325–336, 1984.

J.-P. Vasseur, M. Pickavet, and P. Demeester. *Network Recovery: Protection and Restoration of Optical, SONET-SDH, IP, and MPLS*. Morgan Kaufmann Publishers, 2004.

Y. Wang, Z. Wang, and L. Zhang. Internet traffic engineering without full mesh overlaying. In *Proc. 20th IEEE Conference on Computer Communications (IN-FOCOM'2001)*, pages 565–571, New York, USA, 2001.

D. Yuan. A bi-criteria optimization approach for robust OSPF routing. In *Proc. IEEE Workshop on IP Operations and Management (IPOM'2003)*, pages 91–98, Kansas City, USA, October 2003.

29 TELECOMMUNICATIONS NETWORK GROOMING

Richard S. Barr[1], M. Scott Kingsley[2], and Raymond A. Patterson[3]

[1]Department of Engineering Management,
Information, and Systems
Southern Methodist University
Dallas, TX 75205 USA
barr@engr.smu.edu

[2]OptionTel, LLC
Dallas, TX 75252 USA
scottkingsleyemail@yahoo.com

[3]School of Business University of Alberta
Edmonton, Alberta T6G 2R6 Canada
Ray.Patterson@ualberta.ca

Abstract: Grooming has emerged as an active area of research within the operations research and telecommunications fields and concerns the optimization of network transmissions that span multiple distinct transmission channels, protocols, or technologies. This chapter explores the meaning of grooming, the technical context in which it can be applied, and example situations. A new taxonomy captures key aspects of grooming problems and is used to summarize over 50 key publications on this important traffic-engineering and optimization problem class.

Keywords: Grooming, network design, aggregation, channel assignment, multiplexing, bundling.

29.1 INTRODUCTION

Network grooming is an industry term that has been adopted in the academic literature and applied to a variety of optimization problems. The grooming of a network to optimize utilization of its traffic-carrying capacity can have a significant impact on its profitability, reliability, and availability.

This chapter explores the meaning of grooming within various technological contexts and provides illustrative examples. A new grooming taxonomy, PACER, is introduced and key publications in the area are summarized for quick reference.

29.2 WHAT IS GROOMING?

Grooming has come to encompass a variety of meanings within the telecommunications industry and literature. The following examples reflect this diversity.

- Cinkler (2003) defines grooming as combining traffic streams to carry a more data and distinguishes between end-to-end (sub-rate) and intermediate (core) grooming. He further examines traffic grooming and λ, or wavelength, grooming and defines hierarchical grooming as the combination of both.

- Dutta and Rouskas (2002b) define grooming in WDM (wavelength-division-multiplexing) networks as "techniques used to combine low-speed traffic streams onto high-speed wavelengths in order to minimize the network-wide cost in terms of line terminating equipment and/or electronic switching."

- Weston-Dawkes and Baroni (2002) talk about grooming architectures, in the context of mesh networks, using optical switches (OSs) and optical cross connects (OCXs) as "a strategy for the placement of intermediate grooming sites, routing of traffic, and rules for how often traffic is groomed as it traverses the network."

- Zhu et al. (2003b) investigate "next-generation optical grooming switches" and their impact on network throughput and network resource efficiency. They define traffic grooming as "a procedure of efficiently multiplexing/demultiplexing and switching low-speed traffic streams onto / from high-capacity bandwidth trunks in order to improve bandwidth utilization, optimize network throughput, and minimize network cost." They further assert that "traffic grooming is an extremely important issue for next-generation optical WDM networks to cost-effectively perform end-to-end automatic provisioning."

- Zhu and Mukherjee (2003) define traffic grooming in WDM optical networks as the bundling of "low-speed traffic streams onto high-capacity optical channels" and concur with Barr and Patterson (2001) that "grooming is a term used to describe the optimization of capacity utilization in transport systems by means of cross-connections of conversions between different transport systems or layers within the same system."

Thus, a consensus is forming that telecommunications grooming is the optimization of network transmissions that span multiple distinct transmission channels or methods. Grooming can occur within multiple layers of the same technology or between technologies. Grooming can be performed when signals are bundled for extended-distance transmission and when cross-connection equipment converts signals between different wavelengths, channels, or time slots.

Hence, grooming is more than assigning time-slots or optimizing traffic routing (Bennett, 2002). Grooming is complex routing, and often implicitly assumes bundling or multiple capacities or multiple layers of transmission. In an abstract sense, grooming is a complex multicommodity network flow problem with multiple transmission layers, each having its own set of constraints related to hops, distance traveled, speed of travel, capacity, etc.

Network grooming is often described in the context of a particular technology, such as SONET, WDM rings, and WDM mesh networks (Zhu and Mukherjee, 2002). The essential ingredient for a telecom network transport problem to be called "grooming" is that there are multiple layers of transport within the system. The tell-tale signs of multiple layers are cross-connections or conversions between different transport systems or layers within the same system, which may involve time-slot or frequency conversion equipment to increase a network's efficiency and effective capacity.

The results of grooming efforts include:

- Changing circuits' channels and time-slot assignments,

- Eliminating wavelength-continuity and distinct-channel assignment constraints on some or all circuits,

- Improving capacity utilization,

- Increasing the number of utilizable routing possibilities, and

- Simplifying the problem by decomposing it into easier to solve subproblems (Barr and Patterson, 2001).

29.3 A TAXONOMY OF GROOMING PROBLEMS

Grooming of telecommunications networks, then, involves optimizing an interrelated set of functions or problems, and each definition above focuses on some subset. These signal-routing and traffic-engineering functions are summarized in the following *PACER taxonomy* of grooming activities:

- *Packing* or grouping lower-speed signal units into higher-speed transport units. Examples: aggregating a set of T-1 demands into ATM cells at an access node and combining a set of ATM cells into a SONET frame.

- *Assigning* of demand units to transmission channels (e.g., time-slots, frequencies, wavelengths) within a given transport layer. Examples: assigning demands to SONET time-slots (TSA) and assigning WDM lightpaths to specific wavelengths (or λ) on each span of a given mesh or ring network.

- *Converting* signals between channels in the same transport layer. Examples: employing Time-Slot Interchange (TSI) within a SONET ADM to reshuffle time slots of transitioning traffic and using an optical crossconnect (OXC) to change lightpaths' wavelengths at a transitioning node or to switch both time-slot and wavelength for a given signal.

- *Extracting/inserting* lower-speed signal units to/from transitioning higher-speed units. For example, using an add/drop multiplexer (ADM) or B-DCS to terminate a lower-rate (sub-wavelength) SONET demand.

- *Routing* demand units between their origins and destinations. Examples: determining the OD path that each OC-3 or DS-0 demand will follow and creating a set of lightpaths in an optical network for a given demand matrix.

Grooming research addresses different PACER subsets, as detailed in Section 29.8, where the taxonomy is used to categorize the various models and approaches. In all cases, however, grooming involves multiplexing or bundling functions.

29.4 MULTIPLEXING AND BUNDLING

Multiplexing and *bundling* methods combine multiple streams into composite streams that travel at higher speeds and with higher capacities (Doverspike, 1991). Multiplexing simultaneously transmits different messages over a communication network by partitioning the available bandwidth or other resource.

Telecommunication networks employ three types of multiplexing, each of which partitions one resource into a distinct set: space-division, frequency-division, and time-division multiplexing (Stern and Bala, 1999). The partitioning of physical space to increase transmission bandwidth is called *space-division multiplexing* (SDM); examples include the bundling of multiple fibers into a cable or using multiple fibers within a network link. Partitioning the frequency spectrum into independent channels is called *frequency-division multiplexing* (FDM) and examples are wavelength-division multiplexing, which enables a given fiber to carry traffic on many distinct wavelengths by dividing the optical spectrum into wavebands with embedded channels. The division of bandwidth time into repeated time-slots of fixed length is called *time-division multiplexing* (TDM), which allows wavelength sharing of non-overlapping time slots. Any combination of these multiplexing and bundling approaches can be utilized.

29.5 ROUTING AND CHANNEL ASSIGNMENT FOR LIGHTPATHS

Per Barr and Patterson (2001), routing and channel assignment in lightpaths often take the form of a multicommodity network flow problem with multiple layers of routing demand from point-to-point. Each commodity represents origin-destination connections (O-Ds) that are transported over the network, just as is done in the multicommodity flow problem. Without conversion equipment, lightpaths must be assigned to distinct channels or wavelengths referred to as a *distinct channel assignment* (DCA). Similar considerations and constraints apply to other types of bandwidth partitioning in FDM, TDM, and λ-channels in waveband-routed networks (Barr and Patterson, 2001). Day and Ester (1997) and Mukherjee (1997) illustrate the difficulty of routing and channel assignment due to *wavelength-continuity* and DCA constraints, which lead to channel conflict and contention resulting in misallocation of bandwidth capacity and limitations on wavelength reuse. Grooming in this context attempts to better utilize the wavelength capacities.

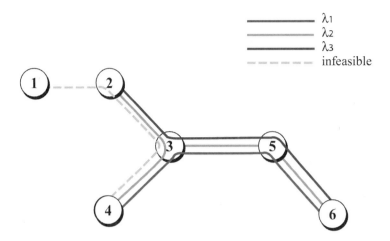

Figure 29.1 Unroutable 1-4 traffic due to wavelength-continuity constraints (Barr and Patterson, 2001; Betts, 1998)

The hypothetical optical network model shown in Figure 29.1 demonstrates the potential impact of the wavelength continuity and DCA constraints. There is a unit demand between origin-destination node pairs 1-4, 2-4, 4-6, 3-5, 6-5, and 2-6. Each span has an capacity of three units. The tree topology creates unique routings for all demands and there is sufficient aggregate span capacity to accommodate all O-D circuits. However, not all of the demands can be assigned to one of the three λ-channels available on each link. Since O-D 1-4 cannot be accommodated, the routing and channel-assignment problem—and the overall design—is infeasible. One solution is to add one unit of capacity on links $(1,2)$, $(2,3)$, and $(3,4)$ to accommodate this circuit, giving an overall capacity utilization of $13/18 = 72.2\%$.

The addition of grooming equipment to this example can enable the routing of all traffic without increasing link capacities. As shown in Figure 29.2, if wavelength conversion is introduced at node 3, all demands can be routing on existing wavelengths. Such grooming equipment eliminates lightpath 1-4's wavelength conflict with demand 4-6 on link $(3,4)$ by switching the connection's frequency from λ_2 to λ_1 at node 3. This increases the usable capacity of the network, permits λ_2 to be used on every network link, avoids the need for a fourth wavelength on three spans, and yields a 13/15 or 86.7% capacity utilization (an improvement of 14.5%). Cost figures are needed to evaluate the tradeoff between the costs of additional grooming equipment and the added wavelengths (Barr and Patterson, 2001).

29.6 GROOMING WITHIN LAYERED TECHNOLOGIES

Gilder (2000) received much attention for proclaiming that "bandwidth is free." While this statement may have been true regarding the potential bandwidth available in installed fiber-optic networks., to be used such bandwidth must be "lit" by expensive equipment and managed by systems and people—all hardly free.

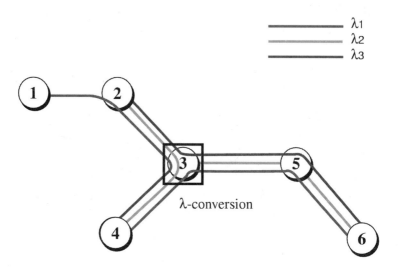

Figure 29.2 RCA feasibility achieved through λ-conversion grooming at node 3 (Barr and Patterson, 2001; Betts, 1998)

Moreover, today's telecommunications industry is complicated, capital-intensive, constantly changing, and highly competitive. With high profit margins being rare and brief, economies of scale must be achieved for a service provider to survive (Aidarous and Plevyak, 1994). Unfortunately, the provider's longevity increases this challenge, as it must also accommodate its older "legacy" equipment and technologies. The process of designing networks for adequate but not excessive capacity and assigning traffic on actual existing capacity is an important and challenging necessity.

29.6.1 Network planning and traffic engineering

Two important categories of telecommunications network design are capacity planning and traffic engineering. Network *capacity planning* determines a physical network's transmission capacity, based on historical traffic levels and anticipated future demand. Primary considerations include cost, quality of service, reliability, survivability, and availability. The time between planning and availability of such capacity is often years (Sharma, 1997).

On the other hand, *traffic engineering* determines how to place actual demands on the installed capacity. The time frame for this activity is much shorter, involves traffic routing, and has the goal of an on-demand service network.

Grooming is an extension to traffic engineering whereby traffic is assigned to capacity within the network to minimize the resources (e.g., capacity, facilities) of the network in reaction to changes that occur. Grooming is primarily a local operation performed at network nodes. Although routing of traffic is not grooming, it heavily influences what and where grooming can occur. Therefore, when demand routing is considered, grooming becomes a system-level problem.

Considering that millions of traffic demands can be quickly created, these processes are daunting tasks and are often performed inefficiently. Increasingly, the success of service providers is dependent on how quickly and effectively they are accomplished.

To illustrate the necessity of grooming, and how it is accomplished, a review of conceptual network modeling is presented . These concepts are applied to specific technologies to illustrate grooming at different levels of telecommunications networks.

29.6.2 Telecommunications network models

Telecommunications technologies and systems use layered models to structure and group required functions, enabling the modular design of networks and equipment. Each layer of a model has specific required tasks and functions, and performs services for the adjacent layers. Complications arise from the variety of competing models and layers' duplication of responsibility in practice. Discussion of the most prevalent models follow.

- **OSI.** One of the first layered models developed is the International Standards Organization Open Systems Interconnection model (ISO OSI). As shown in Figure 29.6.2, it has seven layers, with the highest being the Application Layer and the lowest the Physical Layer. Like most models, each layer has a client-server relationship with layers above and below it.

  The four upper layers ensure that information is delivered in correct and understandable form, end-to-end (Spragins, 1994) The lower layers provide transparent connections between users and operate primarily on a hop-by-hop basis on individual links between network nodes. This model in modern communications in many ways is most theoretical since even though many of the protocols that were defined were developed and are used are not as prevalent as those used in the current Internet.

- **TCP/IP.** The TCP/IP model developed by the Internet Engineering Task Force (IETF) uses five layers, the most recognizable of which is the Internet Protocol (IP). The designation "IP network" may sound like the only component is the IP, but the IP packets require the services of other layers, such as the Transmission Control Protocol (TCP) above it and a method of transport (such as T-1 or SONET) below it on a physical medium such as copper or fiber. Hence, an IP network requires many functions provided by other layers to be useful (Tanenbaum, 1996). Figure 29.6.2 shows the correspondences between the TCP/IP and OSI models' layers.

- **IEEE.** Other organizations have also developed important widely used protocols such as the International Electrical and Electronic Engineers (IEEE), which developed popular local-area-network protocols, including Ethernet.

To make an end-to-end connection in the network requires specific functions be performed at each of the major layers and there are many combinations that can be used. The layers can be viewed as a funnel whereby traffic of different types (voice,

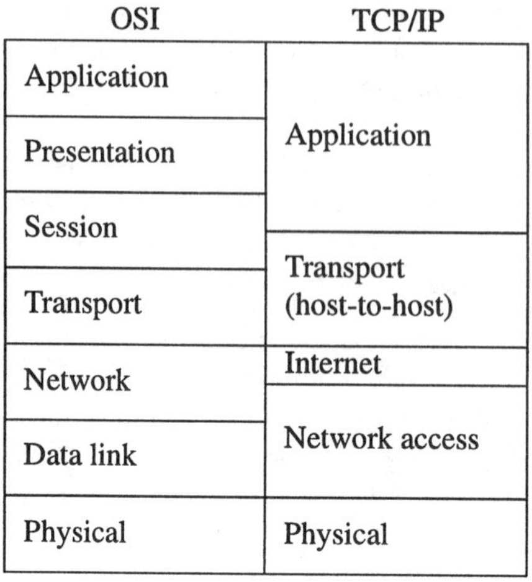

OSI	TCP/IP
Application	Application
Presentation	
Session	Transport (host-to-host)
Transport	
Network	Internet
Data link	Network access
Physical	Physical

Figure 29.3 Comparison of OSI and TCP/IP architectures (Stallings, 2002)

data, and video) and formats from many different users is channeled into a stream of bits for transport to other locations in the network.

An important element is *encapsulation*, whereby the user data is packaged with overhead information, such as origin and destination addresses, that the layer is responsible for providing (Schwartz, 1987). This header information is often combined with similar formations from other users via multiplexing, as detailed in section 29.4.

In most models, the information (e.g., voice, data, video) starts at the top of the layer "stack." Each layer attaches its overhead and passes the information to the next layer, which encapsulates it again with more overhead and instructions to be performed, possibly multiplexing it again with streams using the same or different formats. While network design and grooming would be simpler if these encapsulation and multiplexing functions were performed by one type of equipment, that is seldom the case. In the next section, a few examples further illustrate the concept of layering and multiplexing.

29.6.3 Layering examples

A company's end user sends an e-mail from his computer. The e-mail is formatted using higher-level protocols with overhead information describing its format, then further encapsulated with the TCP and IP containing addressing and other information. The TCP/IP combination is then transported across the company's local-area network, encapsulated within an Ethernet frame. Emails leaving the company are switched outside the company using a router. To be transported, these packets are multiplexed with others into some type of protocol and medium. Although they could be transported using Ethernet, they more commonly use a T-1- or SONET-formatted stream that encapsulates all the packets with additional overhead information before delivery to the Internet service provider (ISP).

The ISP receives many types of information (voice, video and data) from many sources. It may use an ATM system (described below) to create virtual circuits that combine traffic arriving from many locations with traffic with similar destinations. his process involves reversing much of the encapsulation to retrieve, for example, the IP packets from a T-1 frame, to determine their destinations. Once ATM then groups packets with similar destinations into cells for transport using T-1 or SONET. The switching of the ATM cells in the network is also decided on a per-node basis. Similarly, this also cannot be done without extracting ATM cells from any encapsulation attached by subsequent layers. Therefore, there are many stages of multiplexing, transporting, de-multiplexing, switching, re-multiplexing, and transporting before each email reaches its destination.

Note that each layer affects system efficiency by introducing complexity, delay, and overhead. In addition, each layer imposes its own set of constraints on the size, structure, and amount of information that it can carry.

An analogy for this process is a postal system, wherein a letter is placed in an envelope with a destination and return address. When it reaches the post office it is combined with other letters and packages of differing sizes, delivery locations, and requirements. These are groomed by sorting and combining into larger boxes and containers to be sent to different locations. They may be transported by truck for regular delivery or sent via airplane for overnight service. There may be intermediate destinations, at which containers must be opened, the letters and packages resorted for delivery, placed in other containers, and forwarded on. The transport units (boxes, containers, trucks, and airplanes) may be completely full or nearly empty depending on the number of letters and packages involved. With such a system, it is possible that a box or airplane could carry only one letter—an enormous waste of transport capacity—or be given more than it could deliver, thus causing message delays or losses.

These same principles apply to capacity planning and traffic engineering in telecommunications networks. Hence, determining the locations of network equipment (post offices), the protocols (boxes and packages), and the transport types (trucks and airplanes) is vitally important. Even if all the capacity were perfectly utilized, the efficiency can be less than 50% due to the overhead required at the various layers. Strategies include: optimizing capacity utilization through grooming, reducing the number of layers, process automation, and increasing transport speed.

29.7 GROOMING SPECIFICS BY TECHNOLOGY

This section describes grooming from the perspective of key transport technologies, addresses the major areas to be considered, and highlights the complexity of the planning, routing and grooming problem. It also sets the stage for discussion of work in progress to overcome these problems. Key to the understanding of the grooming problem is hidden within multiplexed systems and label switching.

29.7.1 Multiplexing and label switching

Telecommunications relies on multiplexing to combine many lower-speed signals onto higher-speed lines. TDM time-division multiplexing converts analog signals (such as voice phone calls) to digital form and mixed together in a single digital bit stream of separate fixed-length channels for each input analog signal. Two methods are prevalent: T-1/T-3 systems, based on the North American Digital Hierarchy, and the newer SONET standard, which converts the bit streams to light waves for transmission over distance. WDM combines many wavelengths by assigning them to different frequencies for simultaneous transmission on fiber optic strands. Used primarily in data networks, statistical multiplexing breaks input data streams into variable-sized containers (such as IP packets and ATM cells) for transmission and routing. These systems operate on a "first-come, first-served" basis. They are usually oversubscribed in that they assume that not everyone will be sending at the same time. If they did, the network would saturate. The network would buffer, or temporarily save, a limited amount of data but beyond that limit the data would be lost. Determining how much capacity is required as not to have too much (and therefore excess expense) or too little (resulting in lost data) capacity is determined using statistical calculation methods of queuing theory (Betsekas and Gallager, 1992).

Modern networks require the use of multiple types of multiplexing to move different types of information (voice, video, and data) across the network, creating a hierarchical, multi-commodity scenario. The placement of data within individual layers as well as nesting them together creates a hierarchical multi-commodity-flow grooming problem. Although routing is not specifically grooming, it does have a significant impact on the problem since it dictates what can be groomed into or from.

An improvement on conventional multiplexing has been the introduction of label switching. Using this technique, many of the layers of a network can be masked by assigning labels with additional information that simplifies the multiplexing and routing introduced in other layers. Label switching was introduced in asynchronous transfer mode and improved in multiprotocol label-switching networks. The following section gives more detailed descriptions of multiplexing and label switching by specific technology, to provide a context for network grooming.

29.7.2 PDH and DS-1/DS-3 multiplexing

Before the advent of the modern Internet data, the public communication network was intended primarily for transferring analog voice calls. To do the same for digital data requires the use of modems that convert the data to analog and transfer it, usually on dedicated, wired connections between endpoints. Since analog phone calls require several wires, a method of multiplexing calls onto a few wires was needed to reduce the number of cables required between telephone switches. The most important method converts 24 voice calls to digital and multiplexed them together for transmission on two copper wires.

This system is called Digital Signal Level 1 (DS-1) and operates at a speed of 1.544 Mbps; when applied to copper wire it is called T-1. Higher digital signal levels were developed to form the North American Plesiochronous Digital Hierarchy (PDH), along with a similar system in Europe (Bellamy, 1991; Grover, 2004). Originally used only in service-provider transport networks and large private networks for voice calls, small businesses now use T-1s for voice calls and Internet data.

Conventional TDM formats (DS-1, DS-3) are point-to-point connections with equipment only at the endpoints. because the T-1 or T-3 channels cannot not be dynamically accessed, to drop or add a channel between the endpoints requires the entire stream to be de-multiplexed, the channels extracted and inserted, and the new channel stream re-multiplexed before continuing. This results in added equipment and expense.

Grooming such networks with digital cross-connect switch (DCS) systems improves this situation. The DCS is a large machine placed at high-traffic locations to connect to large numbers of T-1s or T-3s. The equipment de-multiplexes and "cross-connects" or switches individual channels from incoming connections, the re-multiplexes back into outgoing T-1s and T-3s.

The DCS is an important grooming location in networks. It also made possible the mesh network architecture. However, the mesh also resulted in "backhauling," which occurs when a channel has to go to a hub location where it is taken out of the group it is in (say a T-1) and placed into another group (again maybe another T-1) that is going to another location before it can go to the location it is intended. This is analogous to a trip from Dallas to New York but having to go to Houston for a connecting flight.

29.7.3 Synchronous optical network

The advent of fiber optics and light waves greatly expanded the ability to transfer large amounts of data at high speeds. The Synchronous Optical Network (SONET) was developed to take advantage of this capability. Initially, SONET equipment initially had T-1 low-speed inputs which were time-division multiplexed, or mapped, into a SONET frame. Part of the SONET frame is reserved for extensive overhead information to allow monitoring and to identify where the T-1s were mapped into the SONET frame since they can be placed there at different starting locations. However, the size of the areas where T-1s could be mapped was rigid.

SONET has seven sections called Virtual Tributaries (VTs) that can contain four T-1s, each within a single Synchronous Transfer Signal Level 1 (STS-1) which operates at 54Mbps. Many STS-1s can be further multiplexed together to create higher-level STSs (Goralski, 1997). For instance, an STS-3 contains three STS-1s. The STS-1 is the basic rate of SONET and is electrical because SONET machines work internally on digital pulses. However, before the electrical STS-1 signals can be transmitted on fiber they must be converted to a light wave. At this point it is an Optical Channel to create an OC-3. Typical optical SONET rates are OC-3, OC-12, and OC-48.

An advantage of SONET is that it is synchronous, or precisely timed. This allows individual channels to be accessed and manipulated without breaking down the SONET frame. An initial problem with SONET, though, was that it used fixed-sized VTs were often a mismatch to formats other than T-1, resulting insufficient or unused capacity in the SONET frame when lower-speed signals were mapped. Fortunately SONET evolved to include enhanced mappings of other formats (such as ATM) by cleverly working around the fixed boundaries. The first was virtual concatenation

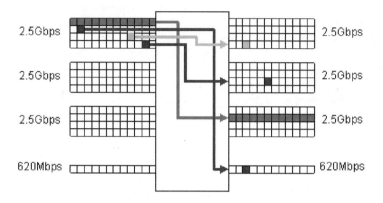

Figure 29.4 Digital cross-connect switch with STS-1 grooming granularity (Bennett, 2002)

whereby VTs could be connected to carry higher-speed inputs. The Generic Framing Procedure defines how to more efficiently map and groom a wide variety of lower-speed signals, including ATM and Ethernet, into a SONET frame. The Link Capacity Adjustment Scheme defines how to dynamically adjust the capacity on SONET links dynamically.

SONET also uses DCSs that allowed for regional grooming (see Figure 29.4) but also introduced a new type of equipment called the add-drop multiplexer (ADM). The ADM allowed adding or dropping lower-speed signals at intermediate locations along a route with less equipment than previously required and by function also became a grooming point.

It also introduced an important new architecture based on closed-loop rings which, when a failure occurred on the ring, would be "self-healing" in that only affected individual channels or the entire SONET frame will be re-rerouted around the side of the ring that is unaffected. However, the addition of protection and restoration capability as well as complexities introduced by the ring architecture resulted in much more complex planning, routing, and grooming problems.

One drawback of TDM networks is they are not historically routed adaptively like data networks are. In other words, IP and ATM networks have functionality built in whereby they route packets and cells "on the fly." These data routes can be quickly changed as needed. In contrast, TDM networks are "nailed up" or difficult to reconfigure. Channel placement and grooming in data networks is automatic. In contrast, it often takes weeks or months to provide a TDM path. Although many TDM systems such as SONET DCSs can and do select paths and groom channels for lower level connections in the network, the methods are often proprietary and require human intervention.

29.7.4 (Dense) Wavelength-division multiplexing

Fiber-optic systems originally carried only one wavelength of light per fiber-optic strand. However, the available spectrum is typically large enough to potentially carry thousands of wavelengths. Therefore, much research was focused on closely packing wavelengths while avoiding interference. The multiplexing, or combining many wavelengths within a single fiber is called *wavelength-division multiplexing* or WDM. When the wavelengths are tightly placed frequency-wise it is termed *dense wavelength-division multiplexing* or DWDM (Kartalopoulos, 2000).

Low-speed inputs to DWDM systems are usually SONET, such as sixteen OC-48s. The DWDM system assigns a different wavelength or frequency to each of the OC-48s and all sixteen wavelengths would be simultaneously transmitted on the fiber for an accumulated throughput from the WDM system of OC-768. In a point-to-point configuration this provides a tremendous gain in capacity while reducing the amount of expensive fiber that is needed.

However, DWDM introduces traffic-engineering complexities. For example, once a wavelength's path has been assigned in the network, other demand paths are blocked from using that frequency on those links. What this means is the wavelength must be available on each link from the origination to the intended destination. It may be such that the wavelength is already being used in one or more of the most directly connected links of the network that are needed to complete the connection. Therefore, the wavelength that needs to terminate at the destination location must be routed on other fibers that have that wavelength available or must converted to another available wavelength on the most direct path.

This led to the introduction of the *wavelength cross-connect* (WXC), which functions much the same as DCSs described above. The WXC can switch, or groom, wavelengths from one fiber onto another at hub locations. Since the wavelengths-in-use problem may still exist, wavelength converters were developed to convert one wavelength frequency to another. The problem in wavelength grooming is routing and assigning wavelengths in the network while minimizing the high conversion-equipment expense (Ramaswami and Sivarajan, 2002).

29.7.5 Internet protocol (statistical multiplexing)

IP networks use statistically multiplexed packets to transfer data. They are basically on "first come, first served" basis and no guarantees are provided that information sent is actually received. IP packets are adaptively routed, meaning that each node in the packet path determines what the packet's next "hop," or routing location, will be. Each packet must be analyzed and the next hop determined.

Little grooming is accomplished in IP networks even though packets can be prioritized by assigned cost parameters on links. Therefore, certain links in the network may be heavily used while others operate well below their capacities. The tremendous growth of IP-based Internet traffic and the migration to other technologies such as Voice-over-IP has shown that much more functionality and control of large IP streams (and hence grooming) is needed. This is accomplished through label-switching, as described in the next sections.

29.7.6 Asynchronous transfer mode

Asynchronous Transfer Mode (ATM) is a technology and related protocols that accepts multiple streams of data in many different formats, converts them to packets and statistically multiplexing them together into *cells*. Placing the different streams of information within an ATM cell is accomplished using *virtual circuits* whereby permanent or semi-permanent logical connections are set up in the network which are used as long as they are needed and then automatically torn down. In other words, it is a "soft" connection (rather than hard-wired) through the network for transferring multiple formats simultaneously.

To avoid individual packet inspection for routing, ATM is based on *label switching*, where cells with a common destination are assigned a label that the ATM switch uses to reference a routing table to determine the outgoing port and associated link on which the cells will be transferred. ATM uses encapsulation of input data streams (such as IP) and provides the overhead with labels to communicate the information that it provides.

ATM uses two labels: the Virtual Circuit Connection (VCC), which identifies the connection endpoints and the Virtual Path Connection (VPC), which identifies a bundle of VCCs with the same endpoints. It is at the VPC level where grooming is accomplished (Grover, 2004). When an ATM switch receives cells, it looks at the cells' VPC/VCCs as well as incoming streams from other ATM switches, combines those with common endpoints (labels) and assigns new labels to new cells that are going to other switches, a technique referred to as *label swapping*. ATM is "virtual" in that streams going to many locations are aggregated using labels and share common links while being switched to other locations when the commonality diverges.

29.7.7 Multiprotocol label switching

IP network routing and assignment is simple, effective, but not efficient. Hence ATM's label-switching technique was adapted to IP networks via *Multiprotocol Label Switching* (MPLS) whereby packets with common characteristics could be grouped and routed together without inspecting individual packet. MPLS can groom upper-layer traffic streams from many sources and formats into similar label-switched paths (LSPs). Functionally, MPLS simply appends labels to the existing IP Packets without fully encapsulating them.

MPLS networks use *Label Edge Routers* (LERs) and *Label-Switched Routers* (LSRs). IP packets enter the network via a LER where, like ATM, they are multiplexed and assigned to *Label-Switched Paths* (LSPs) based on *Forwarding Equivalency Classes* of packets that share the same destination. This is a important step because it effectively masks the requirement to analyze the layers below in the core of the MPLS network requiring this information only be necessary at the network edges where lower-level streams ingress and egress. The labels and associated data are then transferred to LSRs in the network core where other MPLS streams are also moving across the network. Much like ATM, the MPLS LSR selects labels with common destinations and, unlike ATM, can stack labels rather than just change them. This enables

LSPs with common destinations from that point to be assigned another common label over the others underneath.

MPLambdaS is an extension to MPLS for wavelength routing and grooming, in which each wavelength is assigned a label. By encoding wavelengths with labels, all the MPLS functionality can be attributed at the wavelength level. This protocol also enables dynamic, on-demand set up and tear down of wavelength routes in real or near-real time.

29.7.8 Generalized multiprotocol label switching

The IETF has proposed the *Generalized Multiprotocol Label Switching* (GMPLS) protocol as a framework for automating and simplifying the routing and grooming problem. The intent is that, for example, when a lower-level T-1 demand arrives it will automatically trigger its optimal placement in a new or existing SONET ST-1 and further into a new or existing STS-1 and wavelength. The hierarchical structure and the ability to stack labels allows for the extension of label switching across multiple layers of network functionality. Ideally this would result in automatic network routing and grooming at all levels. However, as in MPLS, the routing and many other important functions are not defined within GMPLS, leaving additional work to be completed in this area.

GMPLS defines five layers whereby labels and paths could be assigned: (1) the Packet Switching Capable layer, for IP, ATM, MPLS and similar streams; (2) the TDM-Capable layer, for older and SONET TDM systems components; (3) the Lambda (wavelength) Switching Capable layer, for wavelengths in WDM equipment and MPLambdaS-type systems; (4) the Waveband Switching Capable layer, for grouping and assignment of multiple wavelengths; and (5) the Fiber Switching Capable layer, for assigning wavelengths to groups of fibers.

29.7.9 Assembling multiple technologies

Figure 29.5 shows a hierarchy of transport technologies found in current telecommunications networks. The nesting effect denotes possible encapsulations (Cinkler, 2003); the directed arrows indicate typical technology up-conversions available from today's equipment manufacturers (Grover, 2004) and the associated multiplexing technique.

Many are the opportunities for grooming in such multi-tiered systems. The next section summarizes key grooming research efforts using the various technologies explored above.

29.8 OPTIMIZATION-BASED GROOMING

Grooming has emerged as an active area of research within the operations research and telecommunications fields. Grooming problems can usually be represented as mixed-integer linear-programming and graph problems, are typically NP-hard, and must be solved with heuristic techniques for instances of any significant size (Chu and Modiano, 1998; Gerstel et al., 2000). In the past, practitioners have typically dealt with these problems by partitioning the problem into sub-problems that can be more readily

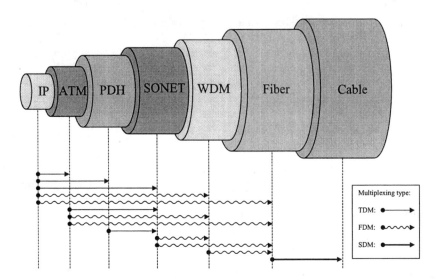

Figure 29.5 Optical transport layers, with typical up-conversion options. (Cinkler, 2003; Grover, 2004)

solved. The partitioning point is typically at the switching (or cross-connection) point. When these problems are not dealt with holistically, the solution is almost always suboptimal. However, these problems tend to be so complex that the only realistic way to solve them has been to consider the sub-problems separately. Recent advances in the areas of solution techniques, commercial software, and computer processing capabilities enable us to grapple with these extremely large and difficult problems in a holistic manner (Cosares et al., 1995; Reingold, 1999).

While the basic problems and concepts of grooming telecommunications networks are known, the underlying combinatorial problems are computationally daunting. Effective algorithms and realistic models are starting to emerge and this section summarizes the current literature on grooming techniques, from an optimization perspective.

From the growing number of research publications on grooming techniques, Tables 29.8 and 29.1 summarize the problem assumptions and the optimization approach(es) reported in 50 key papers. Table 29.8 cites grooming research on static problems (with deterministic traffic assumptions), sorted by type of technology, and Table 29.1 reports on dynamic grooming methods for restructuring and re-grooming networks in response to changes in demand patters, and lists additional survey articles on grooming. These tables contain the following information for each source:

- **Citation.** Bibliographic reference.

- **Technology.** The assumed network technology—PDH, SONET, WDM, GM-PLS, IP, waveband switching (WBS), and IP over optical (IPOO)—and structure (mesh, tree, ring).

- **S/D.** Traffic/demand assumptions: static (S), dynamic (D), or ordered arrival sequence (O).

- **PACER.** The grooming functions addressed in the model presented, denoted as $xx\ldots$, where x is a letter in the PACER acronym. For example, the classic paper by Doverspike (1991) on static bundling algorithms is classified PAER, since it addresses the packing, assignment, extraction/insertion, and routing functions, but not conversion between channels.

- **Model/algorithm.** The type of grooming model, solution algorithm, or theory presented, where

    - ILP: integer or mixed-integer linear programming model (possibly solved with a generic optimizer),
    - ACO: ant colony optimization (Dorigo and Di Caro, 1999),
    - GP: GRASP heuristic (Feo and Resende, 1995),
    - HEU: specialized heuristic,
    - LB: theoretical lower or upper bound,
    - MS: multi-start heuristic,
    - SP: shortest path,
    - SS: scatter-search heuristic (Glover, 1999), and
    - TS: tabu-search heuristic (Glover and Laguna, 1997),

- **Objective/criteria.** Optimization objective function or quality evaluation criteria used, where

    - Min: minimize;
    - $: cost or cost of;
    - A: number of add/drop multiplexers;
    - B: blocking probability;
    - D: total distance, delay, or hops in routings;
    - F: fiber cables;
    - L: number of wavelengths, channels, or time slots;
    - O: number of optical cross-connects;
    - P: number of paths;
    - p: count of equipment ports;
    - R: total routed traffic;
    - T: count of transponders or repeaters;
    - U: capacity utilization;
    - X: number of conversions; and
    - Δ: number of changes to existing network.

854 HANDBOOK OF OPTIMIZATION IN TELECOMMUNICATIONS

Table 29.1 Static grooming references, by technology and network structure.

Static-grooming: Citation	Technology	S/D	PACER	Model/algorithm	Objective/criteria	Computation?
Doverspike (1991)	PDH, mesh	S	PAER	HEU, IP	Min $	y
Betts (1998)	PDH, mesh tree	S	PACE	ILP	Min $, L	y
Carpenter et al. (1997)	SONET rings	S	AR	ILP	Min L	n
Chow and Lin (2004)	SONET rings	S	PER	LB, AA	Min A	n
Hu (2001a)	WDM ring	S	ER	Mixed ILP	Min A	y
Hu (2001b)	WDM ring	S	ER	ILP	Min A	y
Dutta and Rouskas (2002a)	WDM rings	S	PACER	ILP, LB	(other)	y
Chiu and Modiano (2000)	WDM SONET rings	S	AE	HEU	Min A	n
Wan et al. (2000)	WDM SONET rings	S	PER	AA	Min $ A	y
Zhang and Qiao (2000)	WDM SONET rings	S	PAER	HEU	Min A,L	y
Battiti and Brunato (2001)	WDM SONET rings	S	PAER	TS	Min A	y
Ghafouri-Shiraz et al. (2001)	WDM SONET rings	S	E	HEU	Min A	y
Wang et al. (2001)	WDM SONET rings	S	PACE	ILP, HEU	Min A	y
Bermond and Ceroi (2003)	WDM SONET rings	S	AE	LB	Min A	n
Zhang and Ramamurthy (2003)	WDM SONET rings	S	E	ILP, HEU, TS	Min A,...	y
Liu and Tobagi (2004)	WDM SONET rings	S	AE	NLP, ILP	Min $ A	y
Birman and Kershenbaum (1995)	WDM mesh	S	AR	HEU	Min B	y
Fang and Somani (2003)	WDM mesh	S	PACE	ILP, Cplex	Min L	y
Zhemin and Hamdi (2003)	WDM mesh	S	PAER	LP, ILP, HEU	Min $ PL	y
Cavendish et al. (2004)	WDM mesh	S	ACR	HEU	Min LX	n
Houle et al. (2004)	WDM mesh	S	R	TS	Min $	y
Hu and Leida (2004)	WDM mesh	S	PAER	ILP, Cplex	Min L	y
Kennington and Olinick (2004)	WDM mesh	S	PACER	ILP, TS	Min $	y
Zymolka and Koster (2004)	WDM mesh	S	ACR	ILP	Min O	y
Melián et al. (2005)	WDM mesh	S	PECR	ILP, SS+TS+MS	Min $ PLF	y
Prathombutr et al. (2005)	WDM mesh	S	PAER	HEU	Max t, Min RT	y
Strand et al. (2001)	WDM mesh	SO	ACR	HEU,kSP	Min L, LD	y
Cox, Jr. and Sanchez (2001)	WDM SONET mesh	S	PACER	GA,GP	Min $ FO	y
Lardies et al. (2001)	WDM SONET mesh	S	PACER	ILP	Min $ P	y
Zhu and Mukherjee (2002)	WDM SONET mesh	S	PACER	ILP, HEU	Max t	y
Dutta and Rouskas (2002b)	WDM ring, mesh	S	PACER	ILP	Min P	n
Brunato and Battiti (2003)	WDM ring, mesh	S	PAR	GP	Min O	y
Cao et al. (2003b)	WDM WBS mesh	S	ACER	ILP, HEU	Min $	y
Kuri et al. (2004)	WDM WBS mesh	S	CER	ILP, TS	Min $	y
Parthiban et al. (2003)	GMPLS, IP, WDM	S	PE	HEU	Min O	n
Chigan et al. (2003)	IPOO	S	PACER	ILP, Cplex	Min $ pL	y

Table 29.2 Dynamic grooming references, by technology and network structure, and grooming survey articles.

Dynamic-grooming: Citation	Technology	S/D	PACER	Model/algorithm	Objective/criteria	Computation?
Gerstel et al. (2000)	WDM rings	SD	PAE	na	Min $ ALT	n
Berry and Modiano (2000)	WDM SONET rings	D	PAER	Lemma, algo	Min $ A	n
Garlick and Barr (2002)	WDM mesh	D	AR	ACO	Min B	y
Grover and Doucette (2002)	WDM mesh	SD	PACER	ILP	Min $	y
Zhu et al. (2003a)	WDM mesh	SD	PACER	HEU, SP	Min L, D	y
Thiagarajan and Somani (2001)	WDM mesh	D	PACE	Rules	Fairness	y
Sreenath et al. (2001)	WDM mesh	D	AER	HEU	Min D, Δ	y
Ou et al. (2003)	WDM mesh	D	PCER	HEU	Min U	y
Zhu et al. (2003b)	WDM mesh	D	PACER	HEU	B, U	y
Zhu et al. (2003c)	WDM mesh	D	AC	SP	Min B	y
Comellas et al. (2003)	GMPLS, IP, WDM	SD	AER	Rule, sim	Min B	y

Surveys: Citation	Technology	S/D	PACER	Model/algorithm	Objective/criteria	Computation?
Chan et al. (1998)	WDM ring, mesh	–	–	na	na	n
Modiano and Lin (2001)	SONET, Mesh, IP	S	PACER	na	na	n
Weston-Dawkes and Baroni (2002)	Mesh	S	–	na	na	n
Cinkler (2003)	WDM	SD	–	na	na	n
Cao et al. (2003a)	Waveband	SD	–	na	na	n
Zhu and Mukherjee (2003)	WDM SONET, mesh	S	–	na	na	n

- **Computation?** Were the results of computational experimentation reported [per Barr et al. (1995)]?

Clearly, static grooming has received the earlier and greater attention by researchers and, as evidenced in Table 29.8, has been studied over a wide range of technologies. The introduction of WDM highlighted the value of this problem class and energized work in this area. Computationally, general-purpose optimization methods can only address relatively small problem instances of the integer-programming grooming formulations, hence a variety of heuristic approaches have been developed to meet the need for solutions to problems of more realistic dimensions.

The stochastic and temporal nature of demand has motivated the study of dynamic traffic engineering methods, as reflected in Table 29.1. Since the frequency of significant shifts in demand affects algorithmic options, techniques range from on-line algorithms (for real-time grooming) to integer linear programs (for reconfiguring networks experiencing more fundamental demand shifts.)

For additional viewpoints on telecommunications grooming, Table 29.1 also cites six excellent survey articles on the subject. Each addresses different aspects of this important and varied category of network optimization.

29.9 SUMMARY AND CONCLUSIONS

Telecommunications network grooming is an active area of research spanning multiple disciplines and technologies. The concepts and solution methods developed for telecommunications grooming problems apply to many other industries where network transportation is optimized over multiple distinct transport methods or capacities. The authors have limited their discussion to telecommunication-specific instances of grooming.

The term "grooming" has come to encompass a variety of meanings in the telecommunications literature. A consensus definition of telecommunications network grooming is developed in this paper. Grooming is usually examined in the context of a specific technology. The technical contexts in which telecommunications network grooming can be applied are also discussed.

A new taxonomy called PACER is developed to capture five key aspects of telecommunication network grooming problems: Packing, Assigning, Converting, Extracting/inserting, and Routing. Key telecommunications network grooming papers are analyzed using PACER, along with the type of technology addressed, the model or algorithmic solution approach, the objectives used by the authors, whether or not computations were used by the authors, and the traffic demand assumptions in the paper. This analysis of the literature provides a good sense of the type and nature of research of research being conducted in the area of telecommunications network grooming.

Bibliography

S. Aidarous and T. Plevyak, editors. *Telecommunications Network Management into the 21st Century*. IEEE Press, New York, 1994.

R. Barr, B. Golden, J. Kelly, M. Resende, and W. Stewart. Designing and reporting on computational experiments with heuristic methods. *Journal of Heuristics*, 1(1): 9–32, 1995.

R. Barr and R. Patterson. Grooming telecommunications networks. *Optical Networks Magazine*, 2(3):20–23, May/June 2001.

R. Battiti and M. Brunato. Reactive search for traffic grooming in WDM networks. In S. Palazzo, editor, *Evolutionary Trends of the Internet*, pages 56–66. Springer, 2001.

J. Bellamy. *Digital Telephony*. John Wiley, New York, second edition, 1991.

G. Bennett. Tutorial on grooming switches. *LightReading*, February 4 2002. www.lightreading.com.

J.-C. Bermond and S. Ceroi. Minimizing SONET ADMs in unidirectional WDM rings with grooming ratio 3. *Networks*, 41(2):83–86, 2003.

R. Berry and E. Modiano. Reducing electronic multiplexing costs in SONET/WDM rings with dynamically changing traffic. *IEEE Journal on Selected Areas in Communications*, 18(10):1961–1971, Oct. 2000.

D. Betsekas and R. Gallager. *Data Networks*. Prentice Hall, Englewood Cliffs, NJ, second edition, 1992.

D. Betts. *Optimization Models for Telecommunications Networks Planning and Design: Channel Assignment and Equipment Selection*. praxis, Southern Methodist University, School of Engineering, Dallas, Texas, 1998.

A. Birman and A. Kershenbaum. Routing and wavelength assignment methods in single-hop all-optical networks with blocking. In *INFOCOM '95: Proceedings of* the Fourteenth Annual Joint Conference of the IEEE Computer and Communication Societies (Vol. 2)-Volume, page 431, Washington, DC, USA, 1995. IEEE Computer Society.

M Brunato and R. Battiti. A multistart randomized greedy algorithm for traffic grooming on mesh logical topologies. In A. Bianco and F. Neri, editors, *Next Generation Optical Network Design and Modelling*. Kluwer, 2003.

X. Cao, V. Anand, and C. Qiao. Waveband switching in optical networks. *IEEE Communications Magazine*, 41(4):105 – 112, April 2003a.

X. Cao, V. Anand, Y. Xiong, and C. Qiao. A study of waveband switching with multilayer multigranular optical cross-connects. *IEEE Journal on Selected Areas in Communications*, 21(7):1081 – 1095, Sept. 2003b.

T. Carpenter, S. Cosares, and I. Saniee. Demand routing and slotting on ring networks. DIMACS Technical Report 97-02, Rutgers University, New Brunswick, NJ, 1997.

D. Cavendish, A. Kolarov, and B. Sengupta. Minimizing the number of wavelength conversions in WDM networks with hybrid optical cross-connects. In *Proceedings of the Seventh INFORMS Telecommunications Conference*, Boca Raton, FL, March 2004. INFORMS.

V.W.S. Chan, K.L. Hall, E. Modiano, and K.A. Rauschenbach. Architectures and technologies for high-speed optical data networks. *IEEE Journal of Lightwave Technology*, 16(12):2146 – 2168, Dec. 1998.

C. Chigan, G.W. Atkinson, and R. Nagarajan. Cost effectiveness of joint multilayer protection in packet-over-optical networks. *IEEE Journal of Lightwave Technology*, 21(11):2694 – 2704, Nov. 2003.

A.L. Chiu and E.H. Modiano. Traffic grooming algorithms for reducing electronic multiplexing costs in WDM ring networks. *IEEE Journal of Lightwave Technology*, 18(1):2 – 12, Jan. 2000.

T. Y. Chow and P. J. Lin. The ring grooming problem. *Networks*, 44(3):194–202, 2004.

A. L. Chu and E. H. Modiano. Reducing electronic multiplexing costs in unidirectional SONET/WDM ring networks via efficient traffic grooming. In *Proceedings of GLOBECOM 1998*, pages 322–327. IEEE, 1998.

T. Cinkler. Traffic and λ grooming. *IEEE Network*, 17(2):16 – 21, March-April 2003.

J. Comellas, R. Martinez, J. Prat, V. Sales, and G. Junyent. Integrated IP/WDM routing in GMPLS-based optical networks. *IEEE Network*, 17(2):22 – 27, March-April 2003.

S. Cosares, D. Deutsch, I. Saniee, and O. Wasem. SONET toolkit: A decision support system for designing robust and cost-effective fiber-optic networks. *Interfaces*, 25(1):20–40, 1995.

BIBLIOGRAPHY 859

L. A. Cox, Jr. and J. R. Sanchez. Cost savings from optimized packing and grooming of optical circuits: Mesh versus ring comparisons. *Optical Networks Magazine*, 2 (3):72–90, May/June 2001.

M. Day and G. Ester. Mesh and ring architectures for the optical network layer. In *Proceedings of NFOEC 97*, 1997.

M. Dorigo and G. Di Caro. The ant colony optimization meta-heuristic. In David Corne, Marco Dorigo, and Fred Glover, editors, *New Ideas in Optimization*, pages 11–32. McGraw-Hill, London, 1999.

R. D. Doverspike. Algorithms for multiplex bundling in a telecommunications network. *Operations Research*, 39(6):925–944, 1991.

R. Dutta and G.N. Rouskas. On optimal traffic grooming in WDM rings. *IEEE Journal on Selected Areas in Communications*, 20(1):110 – 121, Jan. 2002a.

R. Dutta and G.N. Rouskas. Traffic grooming in WDM networks: past and future. *IEEE Network*, 16(6):46 – 56, Nov.-Dec. 2002b.

J. Fang and A.K. Somani. Enabling subwavelength level traffic grooming in survivable WDM optical network design. In *Proceedings IEEE Global Telecommunications Conference (GLOBECOM 2003)*, volume 5, pages 2761 – 2766. IEEE, Dec. 2003.

T.A. Feo and M.G.C. Resende. Greedy randomized adaptive search procedures. *Journal of Global Optimization*, 6:109–133, 1995.

R. Garlick and R. Barr. Dynamic wavelength routing in wdm networks via ant colony optimization. In M. Dorigo, G. Di Caro, and M. Sampels, editors, *Ant Algorithms*, volume 2463 of *Lecture Notes in Computer Science*, pages 250–255. Springer, 2002.

O. Gerstel, R. Ramaswami, and G. Sasaki. Cost effective traffic grooming in WDM rings. *IEEE/ACM Transactions on Networking*, 8(5):618–630, 2000.

H. Ghafouri-Shiraz, Guangyu Zhu, and Yuan Fei. Effective wavelength assignment algorithms for optimizing design costs in SONET/WDM rings. *IEEE Journal of Lightwave Technology*, 19(10):1427 – 1439, Oct. 2001.

G. Gilder. *TELECOSM: How Infinite Bandwidth will Revolutionize Our World*. Free Press, New York, 2000.

F. Glover. Scatter search and path relinking. In M. Dorigo D. Corne and F. Glover, editors, *New ideas in optimization*, pages 297–316. McGraw-Hill Ltd., UK, Maidenhead, UK, England, 1999.

F. Glover and M. Laguna. *Tabu Search*. Kluwer Academic Publishers, Boston, 1997.

W. Goralski. *SONET: A Guide to Synchronous Optical Network*. McGraw-Hill, New York, 1997.

W. D. Grover. *Mesh-Based Survivable Networks*. Prentice Hall PTR, Upper Saddle River, NJ, 2004.

W.D. Grover and J. Doucette. Design of a meta-mesh of chain subnetworks: enhancing the attractiveness of mesh-restorable WDM networking on low connectivity graphs. *IEEE Journal on Selected Areas in Communications*, 20(1):47 – 61, Jan. 2002.

A. Houle, B. Jaumard, and Y. Solari. Dimensioning wdm optical networks with minimum mspp configuration. In *Proceedings of the Seventh INFORMS Telecommunications Conference*, Boca Raton, FL, March 2004. INFORMS.

J. Q. Hu. Optimal traffic grooming for wavelength-division-multiplexing rings with all-to-all uniform traffic. *Journal of Optical Networking*, 1(1):32–42, 2001a.

J. Q. Hu. Traffic grooming in wavelength-division-multiplexing ring networks: a linear programming solution. *Journal of Optical Networking*, 1(11):397–408, 2001b.

J. Q. Hu and B. Leida. Traffic grooming, routing, and wavelength assignment in optical WDM mesh networks. In *Proceedings of IEEE INFOCOM 2004*, 2004.

S. Kartalopoulos. *Introduction to DWDM Technology*. IEEE Press, Bellingham, WA, 2000.

J. L. Kennington and E. V. Olinick. Wavelength translation in WDM networks: optimization models and solution procedures. *INFORMS Journal on Computing*, 16(2):174–187, 2004.

J. Kuri, N. Puech, and M. Gagnaire. Assessing the economic benefit of introducing multi-granularity switching cross-connects in optical transport networks. Technical Report TR-ENST-2004C003, ENST Paris, Paris, July 2004.

A. Lardies, R. Gupta, and R. Patterson. Traffic grooming in a multi-layer network. *Optical Networks Magazine*, 2(3):91–99, May/June 2001.

H. Liu and F. Tobagi. A novel efficient technique for traffic grooming in WDM SONET with multiple line speeds. In *ICC '04*, 2004.

B. Melián, M. Laguna, and J. Moreno-Pérez. Minimizing the cost of placing and sizing wavelength division multiplexing and optical crossconnect equipment in a telecommunications network. *Networks*, 45(4):199–209, July 2005.

E. Modiano and P. Lin. Traffic grooming in WDM networks. *IEEE Communications Magazine*, 39(7):124 – 129, July 2001.

B. Mukherjee. *Optical Communication Networks*. McGraw-Hill, New York, 1997.

C. Ou, K. Zhu, H. Zang, L.H. Sahasrabuddhe, and B. Mukherjee. Traffic grooming for survivable WDM networks - shared protection. *IEEE Journal on Selected Areas in Communications*, 21(9):1367 – 1383, Nov. 2003.

R. Parthiban, R.S. Tucker, and C. Leckie. Waveband grooming and IP aggregation in optical networks. *IEEE Journal of Lightwave Technology*, 21(11):2476 – 2488, Nov. 2003.

Passakon Prathombutr, Jerry Stach, and E. K. Park. An algorithm for traffic grooming in wdm optical mesh networks with multiple objectives. *Telecommunication Systems*, 28(3-4):369–386, 2005.

R. Ramaswami and K. N. Sivarajan. *Optical Networks: A Practical Perspective*. Morgan Kaufmann, San Francisco, second edition, 2002.

N. Reingold. Telcordia's SONET design tool. presentation at INFORMS Telecommunications Conference, March 1999.

M. Schwartz. *Telecommunication Networks: Protocols, Modeling and Analysis*, volume Reading, MA. Addison-Wesley, 1987.

R. Sharma. *Network Design Using EcoNets*. International Thomson Publishing, Boston, 1997.

J. Spragins. *Telecommunications: Protocols and Design*. Addison-Wesley, Reading, MA, 1994.

N. Sreenath, B. H. Gurucharan, G. Mohan, and C. Siva Ram Murthy. A two-stage approach for virtual topology reconfiguration of WDM optical networks. *Optical Networks Magazine*, 2(3):58–71, May/June 2001.

W. Stallings. *High-Speed Networks and Internets: Performance and Quality of Service*. Prentice Hall, Upper Saddle River, NJ, second edition, 2002.

T. Stern and K. Bala. *Multiwavelength Optical Networks: A Layered Approach*. Addison-Wesley, Reading, MA, 1999.

J. Strand, R. Doverspike, and G. Li. Importance of wavelength conversion in an optical network. *Optical Networks Magazine*, 2(3):33–44, May/June 2001.

A. Tanenbaum. *Computer Networks*. Prentice Hall PTR, Upper Saddle River, NJ, third edition, 1996.

S. Thiagarajan and A.K. Somani. A capacity correlation model for WDM networks with constrained grooming capabilities. In *ICC 2001*, volume 5, pages 1592 – 1596. IEEE International Conference on, June 2001.

P.-J. Wan, G. Calinescu, L. Liu, and O. Frieder. Grooming of arbitrary traffic in SONET/WDM BLSRs. *IEEE Journal on Selected Areas in Communications*, 18 (10):1995 – 2003, Oct. 2000.

J. Wang, W. Cho, V. R. Vemuri, and B. Mukherjee. Improved approaches for cost-effective traffic grooming in WDM ring networks: ILP formulations and single-hop and multihop connections. *Lightwave Technology, Journal of*, 19(11):1645 – 1653, Nov. 2001.

J. Weston-Dawkes and S. Baroni. Mesh network grooming and restoration optimized for optical bypass. In *Technical Proceedings of the National Fiber Optic Engineers Conference (NDOEF 2002)*, pages 1438–1449, 2002.

S. Zhang and B. Ramamurthy. Dynamic traffic grooming algorithms for reconfigurable SONET over WDM networks. *IEEE Journal on Selected Areas in Communications*, 21(7):1165 – 1172, Sept. 2003.

X. Zhang and C. Qiao. An effective and comprehensive approach for traffic grooming and wavelength assignment in SONET/WDM rings. *IEEE/ACM Transactions on Networking*, 8(5):608 – 617, Oct. 2000.

D. Zhemin and M. Hamdi. Traffic grooming in optical WDM mesh networks using the blocking island paradigm. *Optical Networks Magazine*, 4(6):7–13, Nov/Dec 2003.

Hongyue Zhu, Hui Zang, Keyao Zhu, and Biswanath Mukherjee. A novel generic graph model for traffic grooming in heterogeneous wdm mesh networks. *IEEE/ACM Trans. Netw.*, 11(2):285–299, 2003a.

K. Zhu and B. Mukherjee. Traffic grooming in an optical WDM mesh network. *IEEE Journal on Selected Areas in Communications*, 20(1):122 – 133, Jan. 2002.

K. Zhu and B. Mukherjee. A review of traffic grooming in WDM optical networks: architectures and challenges. *Optical Networks Magazine*, 4(2):55–64, March/April 2003.

K. Zhu, H. Zang, and B. Mukherjee. A comprehensive study on next-generation optical grooming switches. *IEEE Journal on Selected Areas in Communications*, 21 (7):1173 – 1186, Sept. 2003b.

K. Zhu, H. Zhu, and B. Mukherjee. Traffic engineering in multigranularity heterogeneous optical WDM mesh networks through dynamic traffic grooming. *IEEE Network*, 17(2):8 – 15, March-April 2003c.

A. Zymolka and A. Koster. Design of survivable optical networks by demand-wise shared protection. In *Proceedings of the Seventh INFORMS Telecommunications Conference*, Boca Raton, FL, March 2004. INFORMS.

V Wireless

30 GRAPH DOMINATION, COLORING AND CLIQUES IN TELECOMMUNICATIONS

Balabhaskar Balasundaram[1] and Sergiy Butenko[1]

[1] Department of Industrial Engineering
Texas A&M University, College Station, TX 77843 USA
baski@tamu.edu
butenko@tamu.edu

Abstract: This chapter aims to provide a detailed survey of existing graph models and algorithms for important problems that arise in different areas of wireless telecommunication. In particular, applications of graph optimization problems such as minimum dominating set, minimum vertex coloring and maximum clique in multihop wireless networks are discussed. Different forms of graph domination have been used extensively to model clustering in wireless ad hoc networks. Graph coloring problems and their variants have been used to model channel assignment and scheduling type problems in wireless networks. Cliques are used to derive bounds on chromatic number, and are used in models of traffic flow, resource allocation, interference, etc. In this chapter we survey the solution methods proposed in the literature for these problems and some recent theoretical results that are relevant to this area of research in wireless networks.

Keywords: Dominating sets, independent sets, cliques, coloring, wireless networks.

30.1 INTRODUCTION

The telecommunication industry has always been a host for a wide variety of optimization problems. In recent years, extensive research has led to the rapid development of numerous mobile computing devices that are diverse and technologically intensive. This has in turn, posed a variety of challenges to the scientific and engineering research community to provide the necessary algorithms and protocols to utilize these systems effectively and at times, overcome their drawbacks. Wireless networks such as satellite networks, radio networks, sensor networks, cellular networks, ad hoc networks and other mobile networks have become predominant because of the flexibility they offer. The applications of these networks include military communications, emergency systems and disaster recovery, e-commerce, etc. In this chapter, we survey only research that applies graph coloring, domination and clique problems in models of these networks. As with any modeling situation, these models capture only the basic aspects of the problem while several technical aspects are not captured in order to maintain the simplicity of the model. In many cases, even these "simple" models cannot be solved optimally for practical purposes and hence effort is made in this chapter to identify heuristics, approximation algorithms and complexity results in this field of research.

The chapter is organized as follows, Section 30.2 provides the notations and definitions of all the graph theoretic concepts used throughout this chapter. In Section 30.3, we restrict our attention to the applications of graph domination problems to Mobile Ad Hoc Networks (MANET). Section 30.4 surveys the application of graph coloring and its variants that have been used to model channel assignment and scheduling type problems in wireless networks such as Packet Radio Networks and Cellular Networks. Cliques are most often not used in isolation for modeling, but as clusters in clustering problems or for deriving bounds for the graph coloring problems. Clique models are also utilized in traffic flow models, link interference models, etc. In Section 30.5, we deal with clique and independent set models in telecommunication in detail. Finally, we conclude with a summary in Section 30.6.

30.2 NOTATIONS AND DEFINITIONS

Let $G = (V,E)$ be a simple loopless undirected graph with vertex set $V = \{1,\ldots,n\}$ and edge set $E \subseteq V \times V$. Let n and m denote the number of vertices and edges, respectively. The *adjacency matrix* of the graph, A_G is the $n \times n$ matrix in which entry $a_{i,j} = 1$ if $(i,j) \in E$, and is 0 otherwise. For $i \in V$, let $N(i)$ denote the set of vertices adjacent to i called the *open neighborhood* and let $N[i] = N(i) \cup \{i\}$ denote the *closed neighborhood*. Let $deg_G(i) = |N(i)|$, denote the degree of vertex i in G. Denote by Δ and δ the maximum and minimum degrees respectively. The *closed neighborhood* of a set S is defined as $N[S] = \cup_{i \in S} N[i]$ and the *open neighborhood* of a set is defined as $N(S) = N[S] \setminus S$. The *complement graph* of G is the graph $\bar{G} = (V, \bar{E})$, where $\bar{E}$ is the complement of E. For a subset $W \subseteq V$ let $G(W)$ denote the subgraph induced by W on G which is obtained by deleting the set of vertices $V \setminus W$ and the incident edges from G. Also, by $d_G(i,j)$, denote the number of edges on the shortest path between vertices i and j in the graph G.

A dominating set $D \subseteq V$ is a set of vertices such that every vertex in the graph is either in this set or has a neighbor in this set. The minimum cardinality of a dominating set is called the *domination number*, denoted by $\gamma(G)$. A *connected dominating set* (CDS) is one in which the subgraph induced by the dominating set is connected. The *connected domination number* $\gamma_c(G)$ is the cardinality of the minimum connected dominating set (MCDS). The subgraph weakly induced by $D \subseteq V$ is the graph $G^w = (N[D], E^w)$, where E^w contains every edge in E that has at least one end point in D. The set D is a *weakly-connected dominating set* (WCDS) of G if D is dominating and G^w is connected. The minimum size of a WCDS is the *weakly connected domination number* of a graph, denoted by $\gamma_w(G)$. For a detailed description of notations and definitions relating to graph domination, refer (Hedetniemi and Laskar,

1997). A set of vertices $I \subseteq V$ is called an independent set if for every $i, j \in I$, $(i, j) \notin E$, i.e. the graph $G(I)$ is edgeless. An independent set is *maximal* if it is not a subset of any larger independent set, and *maximum* if there are no larger independent sets in the graph. Note that a maximal independent set is also a minimal independent dominating set. The maximum cardinality of an independent set of G is called the *independence number* (or *stability number*) of the graph G and is denoted by $\alpha(G)$.

Generalizing the neighborhood definitions above to a distance k-neighborhood, leads to k-*domination* and k-*independence*. A subset $D \subseteq V$, is said to be a k-*dominating set* if every vertex not in D is reachable in no more than k steps from some vertex in D. Note that this is different from the k-domination defined in (Cockayne et al., 1985). Vertices in $I \subseteq V$ are said to form a k-*independent set* if the length of the shortest path between any two vertices in the set is at least $k+1$. The size of the largest k-independent set in the graph is called the k-*independence number*, $\alpha_k(G)$. Note that $\alpha_1(G)$ is the same as $\alpha(G)$. $\gamma(G), \gamma_c(G)$ and $\alpha_k(G)$ are related as follows (Duchet and Meyniel, 1982).

$$\gamma(G) \leq \gamma_c(G) \leq 3\gamma(G) - 2$$

$$\gamma_c(G) \leq 2\alpha(G) - 1$$

$$\alpha_2(G) \leq \gamma(G) \leq \alpha(G)$$

The duality between $2k$-independence and k-domination for special classes of graphs is studied in (Chang and Nemhauser, 1984). The NP-completeness of the decision version of these problems is also established here for bipartite and chordal graphs.

A *clique* C is a subset of V such that an edge exists between every pair vertices in C, i.e. subgraph $G(C)$ induced by C on G is complete. The maximum clique problem is to find a clique of maximum cardinality. The *clique number* $\omega(G)$ is the cardinality of a maximum clique in G. Note that $\omega(G) = \alpha(\bar{G})$. A proper coloring of a graph is one in which every vertex is colored such that no two vertices of the same color are adjacent. A graph is said to be k-*colorable* if it admits a proper coloring with k colors. Vertices of the same color are referred to as a *color class* and they induce an independent set. The *chromatic number* of the graph, denoted by $\chi(G)$ is the minimum number of colors required to properly color G. Note that for any graph G, $\omega(G) \leq \chi(G)$, as different colors are required to color the vertices of a clique. Analogously, in the *edge coloring problem* two edges that are incident at the same vertex have to be colored differently. The minimum number of colors required to color all the edges satisfying this condition is referred to as the *edge chromatic number* or the *chromatic index* denoted by $\chi'(G)$.

Computation of $\gamma(G)$, $\gamma_c(G)$, $\alpha(G)$, $\omega(G)$ and $\chi(G)$ for general graphs is difficult as the minimum dominating set, minimum connected dominating set, maximum independent set, maximum clique problems and the minimum vertex coloring problems are well known NP-hard problems (Garey and Johnson, 1979). Minimum edge coloring has also been shown to be NP-hard (Hoyler, 1981) and hence finding $\chi'(G)$ is also difficult. Most of these problems remain NP-hard even on *Unit-Disk Graphs* (UDG) that are frequently used to model wireless networks (Clark et al., 1990). Simple heuristics for these problems on UDG can be found in (Marathe et al., 1995). It is also important to note that in these problems, the UDG model is given to us, as the problem of recognizing a UDG is also NP-complete (Breu and Kirkpatrick, 1993).

30.3 GRAPH DOMINATION IN WIRELESS NETWORKS

MANET refers to distributed, wireless, multihop networks that function without using any infrastructure such as a base station or access points for communication. These are complex, dynamic systems owing to the mobility of nodes and hence lead to ad hoc network topologies.

Historically, ad hoc networks were utilized for military applications. Survival in a battlefield entails mobile wireless communication systems that can be used to coordinate and control operations on the battlefield in a distributed fashion without relying on centralized control stations that are prone to failure. But with the development of technologies such as Bluetooth and wireless Internet, MANET have found applications in sensor networks, emergency services such as disaster recovery systems, in business environments, for conferencing and home networking, etc. In a recent work (Chlamtac et al., 2003), a comprehensive study of issues relevant to ad hoc networks and research activities in enabling technologies, networking protocols and services, etc., has been undertaken. Several books have been written recently (Perkins, 2001; Stojmenovic, 2002; Cheng et al., 2003a; Basagni et al., 2004) that provide information on research and development activities, issues and applications, with theory and relevant background material. Wireless ad hoc networks are networks which exhibit dynamic changes in their network topology. Clustering introduces a hierarchy that is other wise absent in these ad-hoc networks. Existence of a hierarchy facilitates routing of information through the network. Efficient resource management, routing and better throughput performance can be achieved through adaptive clustering of these mobile nodes.

Given the *connectivity graph* (also known as *communication graph, network graph*, etc.), $G^0 = (V^0, E^0)$, where the vertices represent the nodes in the network and the edges represent the communication links between pairs of nodes in the network, the *clustering problem* is to find subsets (not necessarily disjoint) $\{V_1^0, \ldots, V_k^0\}$ of V^0 such that $V^0 = \bigcup_{i=1}^{k} V_i^0$. Each subset is a cluster and it is required that the diameter of the subgraph induced by each subset is small. One vertex in each subset is distinguished as the *cluster-head* and the restriction on diameter ensures that every vertex in that subset is reachable in few steps from the cluster-head. After clustering, we can abstract the connectivity graph to a graph $G^1 = (V^1, E^1)$ as follows: there exists a vertex $v_i^1 \in V^1$ for every subset V_i^0 and there exists an edge between v_i^1, v_j^1 if and only if there exist $x^0 \in V_i^0$ and $y^0 \in V_j^0$ such that $(x^0, y^0) \in E^0$. We can recursively cluster the abstracted network to obtain a multi-level hierarchy.

The concept of graph domination has been frequently used for clustering. Once a dominating set is found, every vertex in it becomes a cluster-head and clusters are formed by the closed neighborhood of each vertex in the dominating set. The problem of finding a MCDS, which is frequently used in clustering is not only NP-hard, but also difficult to approximate (Lund and Yannakakis, 1994; Guha and Khuller, 1998) on general graphs. Hence several heuristics and approximation algorithms when the problem is restricted to certain special graph classes have been proposed to tackle this problem. For characterization and complexity classification of domination-type problems refer (Telle, 1994). Another problem that is equivalent to finding a MCDS is finding a spanning tree of the graph with the maximum number of leaves which is also naturally NP-hard (Garey and Johnson, 1979). The extremal aspects of these problems, bounds and heuristics are presented in (Caro et al., 2000). In a recent unpublished work (Saxena, 2003), the author conducts a detailed study of polyhedral aspects of graph domination. This paper provides integer programming (IP) formulations of the dominating set problem and studies the characteristics of the associated polytope. Valid inequalities are provided and the relationships between the dominating set polytope, spanning tree polytope and matching polytope are studied. The following is the integer programming formulation for finding the domination number:

$$\gamma(G) = \min\{\mathbf{1}^T x : Ax \geq \mathbf{1}, x \in \{0,1\}^n\} \qquad (30.1)$$

Where $A = A_G + I$, A_G is the adjacency matrix of the graph, I is the $n \times n$ identity matrix and $\mathbf{1}$ is a $n \times 1$ vector of ones. Also recently, a fully distributed approximation algorithm has been developed, based on the linear programming (LP) relaxation of (30.1) in (Kuhn and Wattenhofer, 2003).

One of the earliest and simplest clustering algorithm for ad-hoc networks is the *linked cluster algorithm* found in (Baker and Ephremides, 1981) which is a distributed algorithm that produces a dominating set by scanning vertices in the decreasing order of their indices (also known as vertex IDs). They also suggest using some vertices outside the dominating set as "gateway nodes" in order to establish a backbone. In other words, they provide procedures that add more vertices to the dominating set that was originally found in order to obtain a connected dominating set. Similar algorithms based on indices and vertex degree are also proposed in (Gerla and Tsai, 1995), where an independent set of cluster-heads is found. It was also found that the vertex ID based algorithms are more robust in the face of changing network topology due to the mobility of nodes. However, these algorithms sometimes failed to produce dominating sets and some nodes were left without cluster-heads. This was resolved in a distributed linear-time *adaptive clustering algorithm* presented in (Lin and Gerla, 1991). A more recent randomized distributed approximation algorithm that has time complexity $O(\log n \log \Delta)$ with high probability, for finding a dominating set with size no larger than $O(\log \Delta)$ of the optimal in expectation and $O(\log n \log \Delta)$ of the optimal with high probability can be found in (Jia et al., 2002).

In (Krishna et al., 1997), the authors define a *k-cluster* to be a subset of nodes that are mutually reachable by a path of length at most k. Since they deal only with 1-clusters (cliques) any further characterization was not necessary. Note that this definition allows 2 vertices to be in a cluster if the shortest path between them is of length at most k, even if the path uses vertices that are not in the cluster. Interestingly, this concept has been used in studying "cohesion" in social networks (also known as *acquaintance networks*) where it is referred to as *k-cliques*. An introduction to relevant definitions and basic properties of other clustering concepts such as *k-clubs* and *k-clans* that are used in studying social networks can be found in (Mokken, 1979). Recently in (Balasundaram et al., 2005) these models were studied in the context of biological networks where some of the drawbacks in the definitions of these models were removed and NP-completeness results on arbitrary graphs as well as diameter-bounded graphs were presented. In addition, integer programming formulations for these problems were provided along with basic polyhedral properties of the 2-club problem. Note that, in social networks, where clustering is often accomplished by 2-clubs, the vertex set is partitioned into subsets (clusters) whose induced subgraphs have diameter at most two. Hence every vertex in the cluster is no more than two hops away from every other vertex. In other words, for every two vertices in the cluster there either exists an edge between them or a common neighbor *inside* the cluster. This essentially generalizes the notion of using a closed neighborhood to form a cluster by allowing subgraphs other than *stars* and thereby possibly avoiding the use of a centralized cluster-head for intra-cluster communication which could lead to congestion in the cluster. The *k-clustering* problem, as defined in (Deogun et al., 1997), is to find a partition of the vertex set of the graph such that the number of partite sets is a minimum and each partite set induces a subgraph whose diameter is bounded by k (k-clusters). This problem has been shown to be NP-hard and also hard to approximate in (Deogun et al., 1997). The authors also present a polynomial-time approximation algorithm with a constant worst-case ratio of 3 for special classes of graphs that contain a *dominating diametral path* (DDP). The shortest path between two vertices in a graph is a DDP if its length is equal to the diameter of the graph and its vertices form a dominating set of the graph. Recently, a distributed polynomial-time approximation algorithm for this problem on unit disk graphs was presented in (Fernandess and Malkhi, 2002).

In (Basagni, 1999) a *distributed clustering algorithm* and a *distributed mobility-adaptive clustering algorithm* are presented that employ maximal weighted independent sets for clustering. A more recent distributed algorithm for clustering using maximal weighted independent sets can be found in (Basagni, 2001). Usually, the weights are assigned to reflect the node's mobility, i.e. higher the mobility, lower the weights. This produces cluster-heads that are less mobile. In a similar approach, (Chatterjee et al., 2002) presents a *weighted clustering algorithm* that finds a dominating set based on weights assigned to every node. The weights are based on several system parameters like degree, distance, mobility and battery power.

Several distributed algorithms have been proposed (Sivakumar et al., 1998; Das and Bharghavan, 1997; Das et al., 1997) that are based on the algorithms presented in (Guha and Khuller, 1998) for MCDS. In (Guha and Khuller, 1998), the authors present two polynomial-time algorithms that have approximation ratios of $2H(\Delta)+2$ and $H(\Delta)+2$, where H is the Harmonic function given by $H(\Delta) = \sum_{i=1}^{\Delta} \frac{1}{i} \leq \ln \Delta + 1$. An algorithm with a time complexity of $O(\Delta^2)$ and a message complexity of $O(n\Delta)$ can be found in (Wu and Li, 2001). Although they do not have any performance bounds, the authors find their algorithms to be more efficient compared to (Das et al., 1997) based on simulation results. An approximation algorithm for the MCDS problem on UDG with an approximation ratio of at most 8 is proposed in (Cardei et al., 2002). This algorithm finds a maximal independent set and connects it using a Steiner tree in $O(n)$ time. A similar 8-approximate algorithm that runs in linear time is presented in (Butenko et al., 2003). Here again, a maximal independent set is found in the first phase and is connected using a tree in the second phase. Both these algorithms have a message complexity of $O(n\Delta)$. Other approximation algorithms for MCDS in UDG have been proposed in (Alzoubi et al., 2002a; Wan et al., 2004) that have time complexity of $O(n)$, message complexity of $O(n\log n)$ and an approximation factor of 8. A lower bound of $\Omega(n \log n)$ on message complexity was also established in (Wan et al., 2004) showing their algorithm to be message optimal. The same authors also present an algorithm in (Alzoubi et al., 2002b) with linear time and message complexity where each message is $O(\log n)$ bits long. A heuristic for MCDS on a restricted class of graphs that do not contain cliques as minors is presented in (Wan et al., 2003). The approximation ratio depends on the clique size that's forbidden and for planar graphs in particular, this heuristic produces a connected dominating set of size at most $15\alpha_2(G) - 5$. Recently, a polynomial time approximation scheme for the MCDS problem on UDG have been proposed in (Cheng et al., 2003b). A heuristic approach to the MCDS problem that is different from other existing constructive algorithms has recently been proposed in (Butenko et al., 2004a). The authors present a $O(mn)$ time complexity, vertex elimination type algorithm that starts with the entire vertex set as the solution and proceeds by recursively removing a vertex of minimum degree, at the same time ensuring connectedness and domination property (thereby having a feasible solution at all stages of the algorithm). Two distributed algorithms for finding connected dominating sets in unit disk graphs, that break the linear-time barrier have been proposed very recently in (Parthasarathy and Gandhi, 2004). These algorithms are based on the distance-2 coloring of the ad hoc network.

The use of WCDS has also been proposed recently for clustering ad hoc networks (Chen and Liestman, 2002). The problem of finding a WCDS has been shown to be NP-hard in (Dunbar et al., 1997), where a detailed study of this problem is carried out. Sharp upper and lower bounds on $\gamma_w(G)$ and its relation to other domination parameters are also provided in the same paper. Approximation algorithms based on algorithms presented in (Guha and Khuller, 1998) have been proposed recently in (Chen and Liestman, 2002) that have an approximation ratio of $O(\ln \Delta)$. A *zonal algorithm* for finding weakly-connected dominating sets is presented in (Chen and Liestman, 2003). Distributed polylogarithmic-time approximation algorithms for finding connected and weakly-connected dominating sets has recently been proposed in (Dubhashi et al., 2003) with an approximation ratio of $O(\log \Delta)$. In (Alzoubi et al., 2003b), two linear time algorithms are presented, one of which has an approximation ratio of 5 and a message complexity of $O(n \log n)$ and the other has a higher approximation ratio but a linear message complexity. Randomized greedy algorithms for finding WCDS in regular graphs is presented in (Duckworth and Mans, 2003).

A different approach to clustering involves the use of distance k-neighborhood, where all nodes that are at a distance of no more than k from the cluster-head are included in the cluster. In other words, the set of cluster heads forms a *k-dominating set*. The problem of finding a minimum k-dominating set in UDG is shown to be NP-hard in (Amis et al., 2000). The authors also propose a heuristic to solve this problem. Such clustering algorithms, based primarily on degree (and vertex IDs for tie-breaking) can be found in (Nocetti et al., 2003). A k-dominating set is said to be small if it has no more than $\max\{1, \lfloor \frac{n}{k+1} \rfloor\}$ vertices in it. Distributed algorithms of time complexity $O(k \log^* n)$ for finding small k-dominating sets can be found in (Kutten and Peleg, 1998; Penso and Barbosa, 2004), where $\log^* n = \min\{i : \log^{(i)} n \leq 2\}$ and $\log^{(i)} n = \log \log^{(i-1)} n$ with $\log^{(1)} n = \log n$.

Other approaches include mobility-based clustering designed to capture common mobility characteristics among nodes (An and Papavassiliou, 2001; Sivavakeesar and Pavlou, 2002), access-based clustering reported in (Hou and Tsai, 2001) and a recent clustering approach proposed in (Bannerjee and Khuller, 2001) where explicit constraints on the size, connectivity and overlap of clusters is enforced. The notion of dominating sets is extended to dominating and absorbent sets in directed graphs in (Wu, 2002) and algorithms are provided that are shown to be effective by simulations. In (Huang et al., 2004), the authors introduce a *Mobile Piercing Set Problem* which is similar to the dominating set problem on UDG. They provide distributed approximation algorithms for this problem on UDG, as well as non-uniform disks (disks of different radii). Besides clustering, graph domination has been used to model other problems that arise in ad hoc networks as well. Broadcasting algorithms that utilize dominating sets can be found in (Stojmenovic et al., 2002; Wu et al., 2003; Ingelrest et al., 2004).

In this section, we have tried to capture the research activity in ad hoc networks with regards to graph domination. A literature review and references to some recent developments in clustering algorithms for wireless networks have been presented. Recent papers that deal with theoretical and algorithmic complexity aspects of graph domination have also been referred. Extensive study of domination in graphs has been carried out in (Haynes et al., 1998a;b). These books cover the basics of graph domination, as well as advanced topics and recent research activities in this field. A comprehensive bibliography is also provided. A recent survey that focuses on graph domination and clustering in ad hoc networks can be found in (Chen et al., 2005) and provides more details regarding other technical aspects of this problem as well. Also, for a recent detailed study of impacts of clustering in ad hoc networks and a comparison of different mechanisms used based on various metrics, refer (Hossain et al., 2004).

30.4 GRAPH COLORING IN WIRELESS NETWORKS

The graph coloring problem on general graphs is a well known NP-complete problem and it is hard to approximate (Garey and Johnson, 1979; Lund and Yannakakis, 1994). Even on UDG, this problem is found to be NP-hard (Clark et al., 1990; Gräf et al., 1998) and a 3-approximation algorithm for coloring UDG is known (Gräf et al., 1998). However, the maximum clique problem can be solved in polynomial time on UDG (Clark et al., 1990) and often heuristics for the maximum clique problem are utilized to derive lower bounds on the chromatic number of the graph.

The graph coloring problem is often used to model channel assignment type problems such as code assignment, frequency assignment and time-slot assignment in wireless networks. Protocols such as FDMA (Frequency Division Multiple Access), TDMA (Time Division Multiple Access), CDMA (Code Division Multiple Access), etc., are designed to limit the number of users sharing the channel simultaneously. Despite significant differences from practical implementation point of view, FDMA protocol is not very different from a TDMA protocol from an algorithmic point of view, as the same conditions are necessary for the transmissions to be collision-free. In the frequency domain however, the total bandwidth is divided into frequency bands, which are equivalent to the time-slots that are assigned in the time domain. So, the heuristics available for a TDMA framework apply for FDMA as well. In fact, most papers surveyed do not restrict the application of their models and heuristics to the TDMA networks alone, but also to the FDMA networks. However, some papers specifically address the frequency assignment problem with different models that are variants of the basic graph coloring problem. These will be dealt with separately towards the end.

In multihop wireless networks, two nodes that are not directly connected, communicate by sending information packets through intermediate nodes. Under these circumstances, *collision* is said to have occurred if any node receives packets from more than one node at the same time. Such collisions lead to a wastage of resources in terms of delay and bandwidth used to retransmit the same information. Two types of collisions, *primary or direct collisions* and *secondary or hidden collisions* are modeled using graph coloring. The former are collisions that occur between nodes that are within the transmission range of each other and the latter are collisions that occur at a node due to simultaneous transmissions from nodes that are themselves not in the hearing range of each other. Scheduling problems that arise in a TDMA framework are *link or singlecast scheduling* and *broadcast scheduling*. The fundamental difference being, in link scheduling, the transmission from a node is intended for one node only whereas in broadcast scheduling, the transmission from one node is intended for all the nodes in its neighborhood. Note that both these types are special cases of *multicast scheduling* where a node's transmission is destined for a subset of stations in its neighborhood. In this section, we look at applications of graph coloring problem and its variants to channel assignment problems (time-slot/frequency/code) and scheduling problems in wireless networks.

The greedy coloring algorithm is a simple and very popular heuristic for graph coloring. This algorithm proceeds by ordering the vertices and coloring them in that order with the smallest color (the colors are usually identified by an integer) not yet assigned to any of its neighbors that appeared before in the ordering. Since, each vertex can have at most Δ neighbors before it in the ordering, this algorithm will require no more than $\Delta + 1$ colors to color the graph and thus we have $\chi(G) \leq \Delta + 1$. As we will see very soon, this heuristic is used often in developing algorithms for scheduling problems in wireless networks. Note that this bound is sharp for complete graphs and odd cycles. It is sharp in fact only for these graphs as it has been established in (Brooks, 1941) that $\chi(G) \leq \Delta$ unless the graph is complete or an odd cycle. It should be noted that there always exists an ordering for which greedy algorithm is optimal, i.e. the vertices can be ordered in the non-decreasing order of colors assigned to them in an optimal coloring and the greedy algorithm will produce optimal coloring in this case.

Note that the link scheduling problem can be modeled as an edge coloring problem and broadcast scheduling problem as vertex coloring problem. Let $G = (V, E)$ represent the connectivity graph of the wireless network. Construct $G^2 = (V, E^2)$ where $(i, j) \in E^2 \Leftrightarrow d_G(i, j) \leq 2$. Graph coloring on G and G^2 model direct collisions and both direct and hidden collisions, respectively. Note that coloring G^2 is the same as the *distance 2-coloring* of G. In distance k-coloring, two vertices i, j can have the same color if and only if $d_G(i, j) \geq k + 1$. The distance k-coloring problem was shown to be NP-complete for any fixed $k \geq 2$ in (McCormick, 1983) in the context of approximating sparse Hessian matrices. An $O(\sqrt{n})$-approximation for distance 2-coloring was also presented in the same paper.

The graph G^2 is called the square of G. In general, the kth power of a graph G, denoted by G^k, is a graph $G^k = (V, E^k)$ where $(i,j) \in E^k \Leftrightarrow d_G(i,j) \leq k$. Note that it is a well known result (Appel and Haken, 1977; Appel et al., 1977; Robertson et al., 1997) that for a (loopless) planar graph G, $\chi(G) \leq 4$. Let Δ denote the maximum degree in G which is planar, some recent results of interest are $\chi(G^2) \leq 2\Delta + 25$ proven in (van den Heuvel and McGuinness, 2003) and $\chi(G^2) \leq 3\Delta + 9$ proven in (Jendrol' and Skupień, 2001). It can also be easily seen that $\chi(G^2) \geq \omega(G^2) \geq \Delta + 1$ as a vertex with the maximum degree in G, along with its neighbors would form clique in G^2. For general powers of graphs, a recent work (Agnarsson and Halldórsson, 2003) derives bounds on the *inductiveness* of G^k when G is planar, where inductiveness of a graph G is defined as $\max_{S \subseteq V(G)} \{\min_{v \in S} deg_{G(S)}(v)\}$. This can be used to derive bounds on the chromatic number and for developing algorithms as it leads to ordering of vertices for the greedy coloring algorithm.

If the transmission radius is not the same for every node, directed graphs need to be used as bidirectional nature of communications cannot be assumed. Then the UDG has to be generalized to a digraph $G = (V, A)$ that has vertices representing the radio stations and a directed edge from u to v if and only if v can receive u's transmission. The problem of broadcast scheduling reduces to a vertex coloring problem on graph G. Two vertices u and v can be colored the same if and only if $u \to v \notin A$ and $v \to u \notin A$ and there doesn't exist a w such that $u \to w \in A$ and $v \to w \in A$, thereby avoiding primary and secondary collisions. The link scheduling problem reduces to an edge coloring problem where two edges $u \to v$ and $x \to w$ can be assigned the same colors if and only if u, v, x, w are all distinct, the edges $u \to w$ and $x \to v$ do not exist.

The NP-hardness of various forms of the link and broadcast scheduling problems has been long since established. In (Even et al., 1984), complexity results of network testing problems are established, some of which are close to scheduling problems that are under consideration. Link scheduling problems to avoid primary and secondary collisions in packet radio networks are modeled as digraphs in (Arikan, 1984), where the problem of deciding whether a given origin-to-destination message rates are achievable via any arbitrary protocol is shown to be NP-complete. However, the link scheduling to meet link demands and end-to-end demands was shown to be polynomial time solvable in a spread spectrum framework where secondary collisions are tolerated in (Hajek and Sasaki, 1988). This algorithm was later improved in (Ogier, 1986). Some algorithms for link scheduling can be found in (Chlamtac and Lerner, 1985; Hajek and Sasaki, 1988). Similar to feasibility results obtained in (Arikan, 1984) for link scheduling, the feasibility problem for broadcast scheduling, where secondary collisions are permitted, is shown to be NP-complete and several heuristic algorithms for the problem are presented in (Bonuccelli and Leonardi, 1997). The problem of finding the largest broadcasting set, i.e. the largest set of vertices that can broadcast simultaneously (the largest 2-independent set on arbitrary undirected graphs) is shown to be NP-hard in (Ephremides and Truong, 1990; Ramaswami and Parhi, 1989). In addition, NP-hardness of minimum length schedule for broadcast scheduling (distance 2-coloring on arbitrary undirected graphs) is proven in (Ramaswami and Parhi, 1989).

Centralized and distributed algorithms based on the greedy coloring heuristic for the minimum length scheduling problem are presented in (Ramaswami and Parhi, 1989) as well. Distributed algorithms for broadcast scheduling can also be found in (Cidon and Sidi, 1989; Ephremides and Truong, 1990). The NP-hardness of distance-2-coloring of planar graphs, as well as approximation algorithms for the planar and general graphs are presented in (Lloyd and Ramanathan, 1992). A *phase allocation* algorithm for channel sharing in wireless multihop networks is presented in (Chlamtac and Kutten, 1985b) and a *distributed link allocation algorithm* is presented in (Chlamtac and Lerner, 1985) . The time-slot assignment problem is modeled on a digraph and NP-hardness of minimizing the maximum time and the average time (over all nodes) needed for the broadcasted message to reach all nodes are established and a distributed heuristic based on the greedy coloring algorithm are presented in (Chlamtac and Kutten, 1985a). A distributed version of this algorithm is used to solve the time slot assignment problem in (Chlamtac and Pinter, 1987). The digraph model is used in (Ramanathan, 1993) to study scheduling problems. Given the hardness of these problems in general graphs, the complexity aspect of these problems restricted to trees and planar graphs is studied in (Ramanathan, 1993). Although the problems are polynomial for trees, they remain NP-hard for planar graphs. Approximation algorithms for general graphs based on the thickness of the graphs are also presented. The minimum number of planar graphs into which a given graph can be partitioned is called the thickness of the graph (θ). An $O(\theta^2)$ approximation algorithm for link scheduling that runs in $O(mn \log n + n\theta^2 \Delta)$ time and an $O(\theta)$ approximation algorithm for broadcast scheduling with time complexity $O(n\theta\Delta)$ are presented in (Ramanathan, 1993). Similarly in (Sen and Huson, 1997), the digraph models (referred to as *point graphs*) are used and the NP-completeness of distance 2-coloring of a much restricted class of planar point graphs is established. Approximation algorithms for special cases are also provided. Recently, distance 2-coloring on undirected graphs has been studied in (Krumke et al., 2001) where approximation algorithms for the distance 2-coloring problem on special graphs used to model packet radio networks are presented and a 2-approximate algorithm for this problem on bounded degree planar graphs is also presented. A simulated annealing type heuristic is applied to the broadcast scheduling problem in (Wang and Ansari, 1997) and is found to be effective. A two-phased algorithm based on sequential vertex coloring is presented in (Yeo et al., 2002) where the length of the schedule is minimized in phase one and utilization is maximized in phase two. In a similar two phase heuristic is presented in (Salcedo-Sanz et al., 2003) where the first phase uses a Hopfield neural networks (Hopfield, 1982; Hopfield and Tank, 1985) for minimizing schedule length and genetic algorithms(similar to (Watanabe et al., 1998)) to maximize slot utilization. A Greedy Randomized Adaptive Search Procedure (GRASP) (Feo and Resende, 1995; Resende and Ribeiro, 2003; 2005) is developed for the broadcast scheduling problem in (Butenko et al., 2004b). GRASP is a two-phase meta-heuristic that has been used quite successfully in the recent past for various combinatorial optimization problems (Festa and Resende, 2001). In a recent paper (Commander et al., 2004) a comparative study of these four metaheuristic approaches is performed and all the heuristics are found to be competitive, with GRASP finding shorter schedules than the other heuristics in all the randomly generated test cases. In (Laguna and Martí, 2001), GRASP for coloring sparse graphs is presented with an extensive survey of literature on heuristics applied to coloring problems.

The frequency assignment problem has been researched extensively, especially because of its application in cellular networks, and the trade-offs between bandwidth usage and system interference studied. As suggested in (Hale, 1980), besides minimizing the length of the schedule, referred to as *order* in this context, it is also important to minimize *span*, which is the difference between the largest and the smallest frequencies assigned. The author also establishes some fundamental results in this respect. This work is further extended in (Cozzens and Roberts, 1982) where T-colorings of graphs and multigraphs are used to model the channel assignment problem. In T-colorings, given a set of non-negative integers T, a T-coloring on graph G 'colors' every vertex by assigning an integer from T to every vertex such that the difference of colors on any two adjacent vertices does not fall in T. A recent survey of the frequency assignment problem, detailing several technical aspects and a host of available solution methods and heuristics including graph coloring heuristics are presented in (Murphey et al., 1999). Another variation that is also applicable to frequency assignment problems is called $L(p,q)-labeling$. A labeling of a graph $\varphi: V \longrightarrow \{0,\ldots,k\}$, for given integers $p,q,k \geq 0$, is called an $L(p,q)-labeling$ if it satisfies:

$$|\varphi(u) - \varphi(v)| \geq p \ \forall u,v \ such \ that \ d_G(u,v) = 1$$

$$|\varphi(u) - \varphi(v)| \geq q \ \forall u,v \ such \ that \ d_G(u,v) = 2$$

The $p,q-span$ of a graph G denoted by $\lambda(G;p,q)$ is the minimum k for which an $L(p,q)-labeling$ exists. Note that $\lambda(G;1,0) = \chi(G)$. If G is planar, then $\lambda(G;p,q) \leq (4q-2)\Delta + 10p + 38q - 24$ (van den Heuvel and McGuinness, 2003). Other results of interest and relevant references can be found in (van den Heuvel and McGuinness, 2003).

CDMA protocols are widely used in cellular networks, wireless local area networks and wireless ad hoc networks. CDMA protocols provide higher capacity, flexibility, scalability, reliability and security than conventional FDMA and TDMA (Wan, 2004). They enable proper channel sharing by the assignment of *orthogonal codes* which are nothing but pseudo-random binary codes. The data that is transmitted is duplicated and multiplied by these codes before transmission and the entire spectrum is used to communicate. The number of duplicates is equal to the length of the code word and is known as the *spreading factor*. The longer the code, more robust is the communication but at a lower rate as rate is defined as inverse of code length.

When all the codes have a fixed code length as in conventional CDMA, it is called OFSF-CDMA (for Orthogonal Fixed Spread Factor - CDMA). In this case, the code assignment problem is combinatorially identical to the previously seen TDMA and FDMA assignment problems. Naturally, the heuristics and algorithms for this problem have also been developed in a similar way. Centralized and distributed heuristics for the hidden interference problem are presented in (Bertossi and Bonuccelli, 1995). This work is further extended in (Battiti et al., 1999) where a *saturation degree code assignment heuristic* is presented for the graph coloring problem which is based on (Brélaz, 1979). Here, the vertex with the highest number of colors in its neighborhood is colored first.

However, recently variable length codes are being used in a CDMA framework, referred to as the OVSF-CDMA (orthogonal variable spreading factor - CDMA). Here the codes are represented by a binary tree structure. Under this situation, the primary and hidden collisions need to be redefined as well as the graph model used to capture this problem. This is done using *prefix free vertex coloring*. The readers are referred to (Wan, 2004; Minn and Siu, 2000) for details regarding definitions, results, heuristics and assignment schemes used for solving these problems.

In (Ramanathan, 1999), a unified framework is presented for studying the assignment problems and algorithms for TDMA/FDMA/CDMA channel assignments are presented based on this framework. For a discussion on the various technical aspects of channel assignment problem and for a survey of assignment schemes, refer (Katzela and Naghshineh, 1996).

30.5 CLIQUES IN WIRELESS NETWORKS

The maximum clique problem is a well known NP-hard problem (Garey and Johnson, 1979) that is also hard to approximate (Håstad, 1999). The maximum independent set problem is closely related to the maximum clique problem since a set C is a clique in G if and only if it is an independent set in its complement, $\bar{G}$. Hence, we will discuss the applications of both these graph models in telecommunications in this section. An extensive survey of the maximum clique problem can be found in (Bomze et al., 1999). This contains pointers to several fundamental formulations and results besides presenting a host of available heuristics and solution techniques.

As a maximal independent set is also a dominating set, it is often used in clustering wireless networks. In fact, some of the earliest clustering algorithms find maximal independent sets and use them as dominating sets for clustering. Maximal independent sets are specially favored in clustering as they provide cluster-heads with desirable domination and independence properties. Effective algorithms are also known for finding maximal independent sets in general graphs. Probabilistic parallel algorithms for the problem are presented in (Luby, 1986; Alon et al., 1986; Goldberg and Spencer, 1989). As mentioned in Section 30.3, in (Basagni, 2001) the author presents a distributed algorithm for finding maximal weighted independent sets in graphs. It is also shown using results from the theory of *random graphs*, that the average time complexity of this algorithm is $O(\log_b n)$, where b indicates the probability of having an edge between any pair of nodes in the network. Some fundamental properties of this problem on UDG and algorithms for clustering based on the construction of maximal independent sets are presented in (Alzoubi et al., 2003a). More recently a polylogarithmic algorithm has been proposed in (Moscibroda and Wattenhofer, 2004) for solving maximal independent set problem on UDG.

Another modeling approach is to partition the graph into clusters where each cluster is a clique. This has the advantage that control in each cluster is not centralized and each node is one hop away from every other node in the cluster. The problem of partitioning a graph into cliques is NP-complete (Garey and Johnson, 1979). However, recent results in (Krishnamachari et al., 2003) regarding phase transition in the complexity of this problem with transmission radius on random graphs provide some useful insights into the practical issues involved in solving this problem. Similar results are also obtained for channel assignment problems, i.e. graph coloring and distance-2 coloring.

As the maximum clique problem is polynomial time solvable in UDG, exact algorithms or efficient heuristics are often used to find $\omega(G)$ or a lower bound on $\omega(G)$ which is used as lower bound for the chromatic number of the graph. As mentioned before in Section 30.3, this approach is used in (Battiti et al., 1999) to study the bounds and scaling properties of the code assignment problem in CDMA framework. They use a Reactive Local Search (RLS) heuristic for maximum clique problem presented in (Battiti and Protasi, 2001) for deriving bounds on the chromatic number using the best clique size found. RLS heuristic is a $O(\max\{n,m\})$ running time heuristic that has been shown to produce good computational results on the DIMACS benchmark problems (DIMACS, 1995). It complements local search by a *feedback* scheme used to control *prohibition-based diversification*. Prohibition refers to the fact that some local neighbors are forbidden from being visited in order to explore new regions of the global solution space. Feedback is an internal way of maintaining history of solutions visited in order to automate tuning of the algorithm as it is solving a particular instance. These principles are similar to Glover's *Tabu Search* algorithm (Glover and Laguna, 1997; Glover, 1989; 1990).

The maximum independent set problem on a connectivity graph, yields the maximum broadcasting set in a TDMA framework as presented in (Ramaswami and Parhi, 1989) or an activation set in a multihop radio network as suggested in (Tassiulas and Ephremides, 1992). Fundamentally, independent set models are used to identify mutually non-conflicting nodes in the network, with respect to some shared resource. In addition, if secondary conflicts such as hidden collisions defined in Section 30.3 are to be avoided, then the problem reduces to finding the largest 2-independent set. The maximum independent set is inherently a hard problem to solve. Another dimension is added to the complexity of this problem in the present context where this problem has to be solved often to keep up with the dynamic network topology. This is suggested by the NP-hardness of the problem of adapting a maximum broadcasting set to a simple topology change of adding nodes to the network (Vuong and Huynh, 1999). But this is not surprising given that the difficulty in solving this and other such problems arises from the fact that given a vertex it is hard to determine whether or not it belongs to a maximum independent set, under nontrivial circumstances.

In modeling interference between wireless links in a network, a *conflict graph* (also referred to as *contention graph, interference graph,* etc.) is often used. A conflict graph $G^c = (V^c, E^c)$ has a vertex for every edge in the connectivity graph ($V^c = E$). An edge exists between two vertices in G^c if the corresponding links in G interfere. Usually, two links in G are said to interfere if their mid-points lie within the interference radius of each other.

The authors of (Gupta and Walrand, 2004) utilize maximal cliques to model interference between the links (edges) in the connectivity graph ($G = (V, E)$) of a wireless network. A clique in G^c is set of links such that at most one link from that set can be active. In other words, no two links in a clique can be active simultaneously if link interference has to be avoided. The problem that is addressed in this work is to find, given a link $l \in V^c$, all maximal cliques that contain this link. The *scanning disk heuristic* proposed in this paper is a fully distributed heuristic to find all maximal cliques containing a given vertex in UDG. Simulation results indicate that the heuristic is fast on the random UDG tested.

In (Puri, 2002) the problem of maximizing traffic flow in a fixed wired and wireless network is studied. The wireless links can interfere and with these wireless links as vertices, a conflict graph is constructed. Time division multiplexing is assumed and a schedule dictates during which time slot each link is active. Objective is to maximize revenue which is generated according to the amount of flow between every origin-destination pair in the wireless network subject to capacity constraints on individual links. The problem of finding a feasible schedule that maximizes revenue is shown to be NP-hard. A linear programming formulation that utilizes a clique-constraint for every clique in the graph is used to obtain an upper bound on the optimum revenue and a heuristic is presented for finding good feasible schedules. Alternately, use of independent sets is also suggested and corresponding formulations are presented. In a similar approach, (Montemanni et al., 2001) uses clique constraints in order to strengthen the linear programming relaxation of the integer formulation of the frequency assignment problem to obtain good bounds on the optimum. In (Kalvenes et al., 2005), a similar approach is used for a different IP formulation. In this paper, a revenue maximization model is proposed that utilizes clique cuts that are added successively by solving maximal clique problems on the interference graph. Elaborate preprocessing and other techniques are proposed to improve the effectiveness of the LP relaxation in obtaining good solutions in reasonable amounts of time. Computational results are found to be very encouraging with this approach. The problem of resource allocation in a wireless multihop ad hoc networks is studied in (Xue et al., 2003). The objective is to maximize the aggregated utility over the network subject to bandwidth availability to competing multihop flows. A pricing approach is used to decide resource allocation. They propose the use of a link contention graph and shadow prices are associated with maximal cliques in this graph. A distributed algorithm is presented that determines shadow prices such that aggregated utility of all flows is maximized. The mathematical program is a non-linear model that is handled using Lagrangian methods and the associated Lagrange multipliers give rise to shadow price associated with the corresponding maximal clique. More recently in (Fang and Bensaou, 2004), a game theoretic model has been developed for fair-sharing of bandwidth in wireless ad-hoc networks. A primal problem which is a constrained maximization problem is developed and Lagrangian relaxation and duality theory are used to determine the game formulation of the problem. Algorithms for solving these problems are also provided.

Before concluding this section, drifting away from wireless networks, we look at an application of clique model to a different telecommunication problem. This model is interesting for several important reasons. The graph under consideration does not "directly" come from a physical network, it is very different from the applications we restricted our attention to in this section and the graphs under consideration are massive. The graphs we are dealing with are called *Call graphs* whose vertices are telephone numbers, and two vertices are connected by an edge if a call was made from one number to another. Such massive graphs representing telecommunications traffic data are presented in (Abello et al., 1999). A call graph based on one 20-

day period had 290 million vertices and 4 billion edges. The analyzed one-day call graph had 53,767,087 vertices and over 170 millions of edges. This graph appeared to have 3,667,448 connected components, most of them tiny; only 302,468 (or 8%) components had more than 3 vertices. A giant connected component with 44,989,297 vertices was computed. It was observed that the existence of a giant component resembles a behavior suggested by the random graphs theory of Erdös and Rényi, but by the pattern of connections the call graph obviously does not fit into this theory. The maximum clique problem and problem of finding large quasi-cliques with prespecified edge density were considered in this giant component. These problems were attacked using a greedy randomized adaptive search procedure (GRASP) (Feo and Resende, 1995; 1994). To make application of optimization algorithms in the considered large component possible, the authors use some suitable graph decomposition techniques employing external memory algorithms. 100,000 GRASP iterations were needed, taking 10 parallel processors about one and a half days to finish. Of the 100,000 cliques generated, 14,141 appeared to be distinct, although many of them had vertices in common. The authors suggested that the graph contains no clique of a size greater than 32. Finally, large quasi-cliques with different density parameters were computed for the giant connected component. Approaches to detect quasi-cliques are detailed in a recent work presented in (Abello et al., 2002).

30.6 CONCLUSION

In this chapter, we have attempted to provide a detailed survey of the use of graph models in telecommunications. Given the vast amount of literature generated in this line of research, it is almost impossible to present a comprehensive survey of all available literature. Hence we restrict our attention to wireless networks and identify research that elucidates the idea behind selected graph models. Our main objective was to strike a balance between the breadth of literature covered and the depth to which individual works are presented. Although we restrict our attention mostly to wireless networks, the principles and the modeling criteria are often the same wherever these graph optimization problems are used. Besides providing the models and solution methods that exist in telecommunication literature, we also present recent theoretical developments in graph theory and complexity theory with regards to these graph models that are of interest and relevance.

Acknowledgments

The authors would like to thank the editors, P.M. Pardalos and M.G.C. Resende for their encouragement and patience.

Bibliography

J. Abello, P.M. Pardalos, and M.G.C. Resende. On maximum clique problems in very large graphs. In J. Abello and J. Vitter, editors, *External Memory Algorithms*, volume 50 of *DIMACS Series on Discrete Mathematics and Theoretical Computer Science*, pages 119–130. American Mathematical Society, 1999.

J. Abello, M.G.C. Resende, and S. Sudarsky. Massive quasi-clique detection. *Lecture Notes in Computer Science*, 2286:598–612, 2002.

G. Agnarsson and M. M. Halldórsson. Coloring powers of planar graphs. *SIAM Journal on Discrete Mathematics*, 16(4):651–662, 2003.

N. Alon, L. Babai, and A. Itai. A fast and simple randomized parallel algorithm for the maximal independent set problem. *Journal of Algorithms*, 7:567–583, 1986.

K. M. Alzoubi, P.-J. Wan, and O. Frieder. Distributed heuristics for connected dominating sets in wireless ad hoc networks. *Journal of Communications and Networks*, 4:22–29, 2002a.

K. M. Alzoubi, P.-J. Wan, and O. Frieder. Message-optimal connected dominating sets in mobile ad hoc networks. In *Proceedings of the Third ACM International Symposium on Mobile Ad Hoc Networking and Computing*, pages 157–164, 2002b.

K. M. Alzoubi, P.-J. Wan, and O. Frieder. A simple parallel algorithm for the maximal independent set problem. *International Journal of Foundations of Computer Science*, 14(2):287–303, 2003a.

K. M. Alzoubi, P.-J. Wan, and O. Frieder. Weakly-connected dominating sets and sparse spanners in wireless ad hoc networks. In *Proceedings of the 23rd International Conference on Distributed Computing Systems*, page 96. IEEE Computer Society, 2003b.

A. Amis, R. Prakash, T. Vuong, and D. Huynh. Max-min d-cluster formation in wireless ad hoc networks. In *Proceedings of IEEE INFOCOM*, 2000.

B. An and S. Papavassiliou. A mobility-based clustering approach to support mobility management and multicast routing in mobile ad-hoc wireless networks. *Intl. J. Network Management*, 11(6):387–395, 2001.

K. Appel and W. Haken. Every planar map is four colourable. part I: Discharging. *Illinois Journal of Mathematics*, 21:429–490, 1977.

K. Appel, W. Haken, and J. Koch. Every planar map is four colourable. part II: Reducibility. *Illinois Journal of Mathematics*, 21:491–567, 1977.

E. Arikan. Some complexity results about packet radio networks. *IEEE Transactions on Information Theory*, IT-30:910–918, 1984.

D. J. Baker and A. Ephremides. The architectural organization of a mobile radio network via a distributed algorithm. *IEEE Trans. on Communications*, COM-29 (11):1694–1701, November 1981.

B. Balasundaram, S. Butenko, and S. Trukhanov. Novel approaches for analyzing biological networks. *Journal of Combinatorial Optimization*, 10:23–39, 2005.

S. Bannerjee and S. Khuller. A clustering scheme for hierarchical control in wireless networks. In *Proceedings of IEEE INFOCOM*, pages 1–12, 2001.

S. Basagni. Distributed clustering for ad hoc networks. In *Proceedings of the 1999 International Symposium on Parallel Architectures, Algorithms and Networks (ISPAN '99)*, page 310, 1999.

S. Basagni. Finding a maximal weighted independent set in wireless networks. *Telecommunication Systems*, 18(1-3):155–168, 2001.

S. Basagni, M. Conti, S. Giordano, and I. Stojmenovic, editors. *Mobile Ad Hoc Networking*. Wiley-IEEE Press, 2004.

R. Battiti, A. A. Bertossi, and M. A. Bonuccelli. Assigning codes in wireless networks: bounds and scaling properties. *Wireless Networks*, 5:195–209, 1999.

R. Battiti and M. Protasi. Reactive local search for the maximum clique problem. *Algorithmica*, 29:610–637, 2001.

A. A. Bertossi and M. A. Bonuccelli. Code assignment for hidden terminal interference avoidance in multihop packet radio networks. *IEEE/ACM Trans. Netw.*, 3(4): 441–449, 1995.

I. M. Bomze, M. Budinich, P. M. Pardalos, and M. Pelillo. The maximum clique problem. In D.-Z. Du and P. M. Pardalos, editors, *Handbook of Combinatorial Optimization*, pages 1–74. Kluwer Academic Publishers, Dordrecht, The Netherlands, 1999.

M. A. Bonuccelli and S. Leonardi. On scheduling variable length broadcasts in wireless networks. *Telecommunication Systems*, 8:211–227, 1997.

D. Brélaz. New methods to color the vertices of a graph. *Communications of the ACM*, 22(4):251–256, 1979.

H. Breu and D. G. Kirkpatrick. Unit disk graph recognition is NP-hard. Technical Report 93-27, Department of Computer Science, University of British Columbia, 1993.

R. L. Brooks. On coloring the nodes of a network. *Proc. Cambridge Philos. Soc.*, 37: 194–197, 1941.

S. Butenko, X. Cheng, D.-Z. Du, and P. Pardalos. On the construction of virtual backbone for ad hoc wireless networks. In S. Butenko, R. Murphey, and P.M Pardalos, editors, *Cooperative Control: Models, Applications and Algorithms*, pages 43–54. Kluwer Academic Publisher, 2003.

S. Butenko, X. Cheng, C.A.S Oliveira, and P.M. Pardalos. A new heuristic for the minimum connected dominating set problem on ad hoc wireless networks. In R. Murphey and P.M Pardalos, editors, *Cooperative Control and Optimization*, pages 61–73. Kluwer Academic Publisher, 2004a.

S. Butenko, C. W. Commander, and P. M. Pardalos. A greedy randomized adaptive search procedure for the broadcast scheduling problem. *Submitted to Journal of Combinatorial Optimization*, 2004b.

M. Cardei, X. Cheng, X. Cheng, and D. Z. Du. Connected domination in multihop ad hoc wireless networks. In H. J. Caulfield, S. H. Chen, H. D. Cheng, R. J. Duro, V. Honavar, E. E. Kerre, M. Lu, M. G. Romay, T. K. Shih, D. V., P. P. Wang, and Y. Yang, editors, *Proceedings of the 6th Joint Conference on Information Science*, pages 251–255. JCIS / Association for Intelligent Machinery, Inc., 2002.

Y. Caro, D. B. West, and R. Yuster. Connected domination and spanning trees with many leaves. *SIAM J. Discret. Math.*, 13(2):202–211, 2000.

G. J. Chang and G. L. Nemhauser. The k-domination and k-stability problems on sunfree chordal graphs. *SIAM Journal on Algebraic and Discrete Methods*, 5:332–345, 1984.

M. Chatterjee, S. Das, and D. Turgut. WCA: A weighted clustering algorithm for mobile ad hoc networks. *Journal of Cluster Computing*, 5:193–204, 2002.

Y. P. Chen and A. L. Liestman. Approximating minimum size weakly connected dominating sets for clustering mobile ad hoc networks. In *Proceedings of the Third ACM International Symposium on Mobile Ad Hoc Networking and Computing*, pages 165–172, 2002.

Y. P. Chen and A. L. Liestman. A zonal algorithm for clustering ad hoc networks. *International Journal of Foundations of Computer Science*, 14(2):305–322, 2003.

Y. P. Chen, A. L. Liestman, and J. Liu. Clustering algorithms for ad hoc wireless networks. In Y. Pan and Y. Xiao, editors, *Ad Hoc and Sensor Networks*, volume 2 of *Wireless Networks and Mobile Computing*, chapter 7. Nova Science Publishers, 2005.

X. Cheng, X. Huang, and D.-Z. Du, editors. *Ad Hoc Wireless Networking*. Kluwer Academic Publisher, The Netherlands, 2003a.

X. Cheng, X. Huang, D. Li, W. Wu, and D. Z. Du. A polynomial-time approximation scheme for the minimum-connected dominating set in ad hoc wireless networks. *Networks*, 42(4):202–208, 2003b.

I. Chlamtac, M. Conti, and J. J.-N. Liu. Mobile ad hoc networking: imperatives and challenges. *Ad Hoc Networks*, 1:13–64, 2003.

I. Chlamtac and S. Kutten. On broadcasting in radio networks - problem analysis and protocol design. *IEEE Transactions on Communications*, 33(12):1240–1246, 1985a.

I. Chlamtac and S. Kutten. A spatial reuse TDMA/FDMA for mobile multi-hop radio networks. In *Proceedings of the IEEE INFOCOM*, 1985b.

I. Chlamtac and A. Lerner. A link allocation protocol for mobile multihop networks. In *Proceedings of the IEEE Globecom*, 1985.

I. Chlamtac and S. S. Pinter. Distributed nodes organization algorithm for channel access in a multihop dynamic radio network. *IEEE Transactions on Computers*, 36 (6):728–737, 1987.

I. Cidon and M. Sidi. Distributed assignment algorithms for multihop packet radio networks. *IEEE Trans. Comput.*, 38(10):1353–1361, 1989.

B. Clark, C. Colbourn, and D. Johnson. Unit disk graphs. *Discrete Mathematics*, 86: 165–177, 1990.

E. J. Cockayne, B. Gamble, and B. Shepherd. An upper bound for the k-domination number of a graph. *Journal of Graph Theory*, 9(4):533–534, 1985.

C. W. Commander, S. Butenko, and P. M. Pardalos. On the performance of heuristics for broadcast scheduling. In D. Grundel, R. Murphey, and P. Pardalos, editors, *Theory and Algorithms for Cooperative Systems*, pages 63–82. World Scientific, 2004.

M. B. Cozzens and F. S. Roberts. T-colorings of graphs and the channel assignment problem. *Congressus Numerantium*, 35:191–208, 1982.

B. Das and V. Bharghavan. Routing in ad-hoc networks using minimum connected dominating sets. In *IEEE International Conference on Communications (ICC '97)*, pages 376–380, 1997.

B. Das, R. Sivakumar, and V. Bharghavan. Routing in ad-hoc networks using a virtual backbone. In *Proceedings of the International Conference on Computers and Communication Networks (IC3N)*, pages 1–20, 1997.

J.S. Deogun, D. Kratsch, and G. Steiner. An approximation algorithm for clustering graphs with dominating diametral path. *Information Processing Letters*, 61:121–127, 1997.

DIMACS. Cliques, Coloring, and Satisfiability: Second DIMACS Implementation Challenge. http://dimacs.rutgers.edu/Challenges/, 1995. Accessed November 2004.

D. Dubhashi, A. Mei, A. Panconesi, J. Radhakrishnan, and A. Srinivasan. Fast distributed algorithms for (weakly) connected dominating sets and linear-size skeletons. In *Proceedings of the Fourteenth Annual ACM-SIAM Symposium on Discrete Algorithms*, pages 717–724, 2003.

P. Duchet and H. Meyniel. On Hadwiger's number and stability numbers. *Annals of Discrete Mathematics*, 13:71–74, 1982.

W. Duckworth and B. Mans. Randomized algorithms for finding small weakly-connected dominating sets of regular graphs. In R. Petreschi, G. Persiano, and R. Silvestri, editors, *Proceedings of the Fifth Conference on Algorithms and Complexity*, pages 83–95, 2003.

J. E. Dunbar, J.W. Grossman, J. H. Hattingh, S. T. Hedetniemi, and A. A. McRae. On weakly connected domination in graphs. *Discrete Mathematics*, 167–168:261–269, 1997.

A. Ephremides and T. V. Truong. Scheduling broadcasts in multihop radio networks. *IEEE Transactions on Communications*, 38:456–461, 1990.

S. Even, O. Goldreich, S. Moran, and P. Tong. On the NP-completeness of certain network testing problems. *Networks*, 14:1–24, 1984.

Z. Fang and B. Bensaou. Fair bandwidth sharing algorithms based on game theory frameworks for wireless ad-hoc networks. In *Proceedings of the IEEE Infocom*, 2004.

T. A. Feo and M. G. C. Resende. A greedy randomized adaptive search procedure for maximum independent set. *Operations Research*, 42:860–878, 1994.

T. A. Feo and M. G. C. Resende. Greedy randomized adaptive search procedures. *Journal of Global Optimization*, 6:109–133, 1995.

Y. Fernandess and D. Malkhi. K-clustering in wireless ad hoc networks. In *Proceedings of the Second ACM International Workshop on Principles of Mobile Computing*, pages 31–37, 2002.

P. Festa and M.G.C. Resende. GRASP: An annotated bibliography. In P. Hansen and C.C. Ribeiro, editors, *Essays and Surveys on Metaheuristics*, pages 325–367. Kluwer Academic Publishers, 2001.

M. R. Garey and D. S. Johnson. *Computers and Intractability: A Guide to the Theory of NP-completeness*. W.H. Freeman and Company, New York, 1979.

M. Gerla and J. T. C. Tsai. Multicluster, mobile, multimedia radio network. *Wireless Networks*, 1(3):255–265, 1995.

F. Glover. Tabu search - part I. *ORSA J. Comput.*, 1:190–260, 1989.

F. Glover. Tabu search - part II. *ORSA J. Comput.*, 2:4–32, 1990.

F. Glover and M. Laguna. *Tabu Search*. Kluwer Academic Publishers, Dordrecht, The Netherlands, 1997.

M. Goldberg and T. Spencer. A new parallel algorithm for the maximal independent set problem. *SIAM Journal on Computing*, 18(2):419–427, 1989.

A. Gräf, M. Stumpf, and G. Weißenfels. On coloring unit disk graphs. *Algorithmica*, 20(3):277–293, 1998.

S. Guha and S. Khuller. Approximation algorithms for connected dominating sets. *Algorithmica*, 20:374–387, 1998.

R. Gupta and J. Walrand. Approximating maximal cliques in ad-hoc networks. In *Proceedings of the PIMRC 2004*, September 2004.

B. Hajek and G. Sasaki. Link scheduling in polynomial time. *IEEE Transactions on Information Theory*, 34:910–917, 1988.

W. K. Hale. Frequency assignment: theory and applications. *Proceedings of the IEEE*, 68(12):1497–1514, 1980.

J. Håstad. Clique is hard to approximate within $n^{1-\varepsilon}$. *Acta Mathematica*, 182:105–142, 1999.

T. W. Haynes, S. T. Hedetniemi, and P.J.Slater. *Fundamentals of Domination in Graphs*. Marcel Dekker Inc., 1998a.

T. W. Haynes, S. T. Hedetniemi, and P. J. Slater, editors. *Domination in Graphs: Advanced Topics*. Marcel Dekker Inc., 1998b.

S. T. Hedetniemi and R. C. Laskar. Bibliography on domination in graphs and some basic definitions of domination parameters. *Discrete Mathematics*, 86(1-3):257–277, 1997.

J. J. Hopfield. Neural networks and physical systems with emergent collective computational abilities. *Proceedings of the National Academy of Sciences*, 79:2554–2558, 1982.

J. J. Hopfield and D. W. Tank. "Neural" computation of decisions in optimization problems. *Biological Cybernetics*, 52:141–152, 1985.

E. Hossain, R. Palit, and P. Thulasiraman. Clustering in mobile wireless ad hoc networks: issues and approaches. In *Wireless Communications Systems and Networks*, pages 383–424. Plenum Publishing Corporation, New York, 2004.

T. C. Hou and T. J. Tsai. An access-based clustering protocol for multihop wireless ad hoc networks. *IEEE Journal on Selected Areas in Commnications*, 19(7):1201–1210, 2001.

I. Hoyler. The NP-completeness of edge-coloring. *SIAM Journal on Computing*, 10:718–720, 1981.

H. Huang, A. W. Richa, and M. Segal. Approximation algorithms for the mobile piercing set problem with applications to clustering in ad-hoc networks. *Mobile Networks and Applications*, 9(2):151–161, 2004.

F. Ingelrest, D. Simplot-Ryl, and I. Stojmenovic. A dominating sets and target radius based localized activity scheduling and minimum energy broadcast protocol for ad hoc and sensor networks. In *Proceedings of the 3rd IFIP Mediterranean Ad Hoc Networking Workshop (MED-HOC-NET)*, pages 351–359, 2004.

S. Jendrol' and Z. Skupień. Local structures in plane maps and distance colorings. *Discrete Mathematics*, 236(1-3):167–177, 2001.

L. Jia, R. Rajaraman, and T. Suel. An efficient distributed algorithm for constructing small dominating sets. *Distributed Computing*, 15(4):193–205, 2002.

J. Kalvenes, J. Kennington, and E. V. Olinick. Hierarchical cellular network design with channel allocation. *To appear in European Journal of Operational Research*, 160(1):3–18, 2005.

I. Katzela and M. Naghshineh. Channel assignment schemes for cellular mobile telecommunications: A comprehensive survey. *IEEE Personal Communications*, pages 10–31, 1996.

P. Krishna, N. Vaidya, M. Chatterjee, and D. Pradhan. A cluster-based approach for routing in dynamic networks. In *ACM SIGCOMM Computer Communication Review*, pages 49–65, 1997.

B. Krishnamachari, S. Wicker, R. Béjar, and C. Fernández. On the complexity of distributed self-configuration in wireless networks. *Telecommunication Systems*, 22(1-4):33–59, 2003.

S. O. Krumke, M. V. Marathe, and S. S. Ravi. Models and approximation algorithms for channel assignment in radio networks. *Wireless Networks*, 7(6):575–584, 2001.

F. Kuhn and R. Wattenhofer. Constant-time distributed dominating set approximation. In *Proceedings of the Twenty Second Annual Symposium on Principles of Distributed Computing*, pages 25–32. ACM Press, 2003.

S. Kutten and D. Peleg. Fast distributed construction of small k-dominating sets and applications. *J. Algorithms*, 28(1):40–66, 1998.

M. Laguna and R. Martí. A GRASP for coloring sparse graphs. *Computational Optimization and Applications*, 19(2):165–178, 2001.

C. R. Lin and M. Gerla. Adaptive clustering for mobile wireless networks. *IEEE Journal of Selected Areas in Communications*, 15(7):1265–1275, 1991.

E. L. Lloyd and S. Ramanathan. On the complexity of distance-2 coloring. In W. W. Koczkodaj, P. E. Lauer, and A. A. Toptsis, editors, *Computing and Information - ICCI'92, Fourth International Conference on Computing and Information, Toronto, Ontario, Canada, May 28-30, 1992, Proceedings*, pages 71–74. IEEE Computer Society, 1992.

M. Luby. A simple parallel algorithm for the maximal independent set problem. *SIAM Journal on Computing*, 15(4):1036–1055, 1986.

C. Lund and M. Yannakakis. On the hardness of approximating minimization problems. *Journal of the ACM*, 41(5):960–981, 1994.

M. V. Marathe, H. Breu, H. B. Hunt III, S. S. Ravi, and D. J. Rosenkrantz. Simple heuristics for unit disk graphs. *Networks*, 25:59–68, 1995.

S. T. McCormick. Optimal approximation of sparse hessians and its equivalence to a graph coloring problem. *Mathematical Programming*, 26:153–171, 1983.

T. Minn and K.-Y. Siu. Dynamic assignment of orthogonal variable spreading factor codes in W-CDMA. *IEEE Journal on Selected Areas in Communications*, 18(8): 1429–1440, 2000.

R. J. Mokken. Cliques, clubs and clans. *Quality and Quantity*, 13:161–173, 1979.

R. Montemanni, D. H. Smith, and S. M. Allen. Lower bounds for fixed spectrum frequency assignment. *Annals of Operations Research*, 107:237–250, 2001.

T. Moscibroda and R. Wattenhofer. Efficient computation of maximal independent sets in unstructured multi-hop radio networks. In *Proceedings of the First International Conference on Mobile Ad-hoc and Sensor Systems (MASS)*, 2004.

R. A. Murphey, P. M. Pardalos, and M. G. C. Resende. Frequency assignment problems. In D.-Z Du and P.M. Pardalos, editors, *Handbook of Combinatorial Optimization*. Kluwer Academic Publishers, 1999.

F. Nocetti, J. Gonzalez, and I. Stojmenovic. Connectivity-based k-hop clustering in wireless networks. *Telecommunication Systems*, 22(1-4):205–220, 2003.

R. Ogier. A decomposition method for optimal link scheduling. In *Proceedings of the 24th Allerton Conference*, pages 822–823, 1986.

S. Parthasarathy and R. Gandhi. Fast distributed well connected dominating sets for ad hoc networks. Technical Report CS-TR-4559, University of Maryland, Computer Science Department, 2004.

L. D. Penso and V. C. Barbosa. A distributed algorithm to find k-dominating sets. *Discrete Applied Mathematics*, 141:243–253, 2004.

C. E. Perkins, editor. *Ad Hoc Networking*. Addison-Wesley, 2001.

A. Puri. Optimizing traffic flow in fixed wireless networks. In *Proceedings of the Wireless Communications and Networking Conference, WCNC 2002*, volume 2, pages 904–907, 2002.

S. Ramanathan. Scheduling algorithms for multihop radio networks. *IEEE/ACM Transactions on Networking*, 1(2):166–172, 1993.

S. Ramanathan. A unified framework and algorithm for channel assignment in wireless networks. *Wireless Networks*, 5(2):81–94, 1999.

R. Ramaswami and K. K. Parhi. Distributed scheduling of broadcasts in a radio network. In *Proceedings of the INFOCOM*, 1989.

M.G.C. Resende and C.C. Ribeiro. Greedy randomized adaptive search procedures. In F. Glover and G. Kochenberger, editors, *Handbook of Metaheuristics*. Kluwer Academic Publishers, 2003.

M.G.C. Resende and C.C. Ribeiro. GRASP with path-relinking: Recent advances and applications. In T. Ibaraki, K. Nonobe, and M. Yagiura, editors, *Metaheuristics: Progress as Real Problem Solvers*. Kluwer Academic Publishers, 2005.

N. Robertson, D. Sanders, P. Seymour, and R. Thomas. The four-color theorem. *Journal of Combinatorial Theory*, 70(1):2–44, 1997.

S. Salcedo-Sanz, C. Buso no Calzón, and A. R. Figueiral-Vidal. Mixed neural-genetic algorithm for the broadcast scheduling problem. *IEEE Transactions on Wireless Communications*, 2(2):277–283, 2003.

A. Saxena. Polyhedral studies in domination graph theory (I). http://littlehurt.gsia.cmu.edu/gsiadoc/WP/2003-E80.pdf, 2003. Accessed November 2004.

A. Sen and M. L. Huson. A new model for scheduling packet radio networks. *Wireless Networks*, 3:71–82, 1997.

R. Sivakumar, B. Das, and V. Bharghavan. Spine routing in ad hoc networks. *Cluster Computing*, 1(2):237–248, 1998.

S. Sivavakeesar and G. Pavlou. A prediction-based clustering algorithm to achieve quality of service in mulithop ad hoc networks. In *Proceedings of the London Communications Symposium (LCS), London, UK*, pages 17–20, 2002.

I. Stojmenovic, editor. *Handbook of Wireless Networks and Mobile Computing*. Wiley InterScience, 2002.

I. Stojmenovic, M. Seddigh, and J. Zunic. Dominating sets and neighbor elimination-based broadcasting algorithms in wireless networks. *IEEE Transactions on Parallel and Distributed Systems*, 13(1):14–25, 2002.

L. Tassiulas and A. Ephremides. Stability properties of constrained queueing systems and scheduling policies for maximum throughput in multihop radio networks. *IEEE Transactions on Automatic Control*, 37(12):1936–1948, 1992.

J. A. Telle. Complexity of domination-type problems in graphs. *Nordic J. of Computing*, 1(1):157–171, 1994.

J. van den Heuvel and S. McGuinness. Coloring the square of a planar graph. *Journal of Graph Theory*, 42(2):110–124, 2003.

T. H. P. Vuong and D. T. Huynh. Adapting broadcasting sets to topology changes in packet radio networks. In *Proceedings of the Eight International Conference on Computer Communications and Networks*, pages 263–268, 1999.

P.-J. Wan. Lecture notes on wireless networking: OVSF-CDMA code assignment in wireless ad hoc networks. http://www.cs.iit.edu/wan/lecture10.pdf, 2004. Accessed October 2004.

P.-J. Wan, K. M. Alzoubi, and O. Frieder. A simple heuristic for minimum connected dominating set in graphs. *International Journal of Foundations of Computer Science*, 14(2):323–333, 2003.

P.-J. Wan, K. M. Alzoubi, and O. Frieder. Distributed construction of connected dominating set in wireless ad hoc networks. *Mobile Networks and Applications*, 9(2): 141–149, 2004.

G. Wang and N. Ansari. Optimal broadcast scheduling in packet radio networks using mean field annealing. *IEEE Journal on Sleceted Areas in Communications*, 15(2): 250–260, 1997.

Y. Watanabe, N. Mizuguchi, and Y. Fujii. Solving optimization problems by using a hopfield neural network and genetic algorithm combination. *Syst. Comput. Japan*, 29(10):68–73, 1998.

J. Wu. Extended dominating-set-based routing in ad hoc wireless networks with unidirectional links. *IEEE Transactions on Parallel and Distributed Computing*, 22 (1-4):327–340, 2002.

J. Wu and H. Li. A dominating-set-based routing scheme in ad hoc wireless networks. *Telecommunication Systems*, 18(1-3):13–36, 2001.

J. Wu, B. Wu, and I. Stojmenovic. Power-aware broadcasting and activity scheduling in ad hoc wireless networks using connected dominating sets. *Wireless Communications and Mobile Computing*, 4(1):425–438, 2003.

Y. Xue, B. Li, and K. Nahrstedt. Price-based resource allocation in wireless ad-hoc networks. In *Proc. IWQoS 2003*, 2003.

J. Yeo, H. Lee, and S. Kim. An efficient broadcast scheduling algorithm for TDMA ad hoc networks. *Computers and Operations Research*, 29:1793–1806, 2002.

31 OPTIMIZATION IN WIRELESS NETWORKS

Manki Min[1] and Altannar Chinchuluun[1]

[1]Department of Industrial and Systems Engineering
University of Florida
Gainesville, FL 32611 USA
mkmin@ufl.edu
altannar@ufl.edu

Abstract: Wireless ad hoc networks consist of autonomous nodes and require each node's cooperation in communications. Since the network environment does not assume any infrastructure, communication tasks are performed in an ad-hoc fashion and many well-established protocols for wired networks are not applicable in such a network. In this chapter, we survey several combinatorial optimization. problems in wireless ad hoc networks and discuss applications of the problems. To improve the solution quality, the intrinsic natures of wireless communications should be considered in designing algorithms.

Keywords: Combinatorial optimization, wireless ad-hoc network, Steiner tree problem, minimum connected dominating set problem, minimum broadcast cover problem.

31.1 INTRODUCTION

A wireless ad hoc network is an autonomous or self-organizing system consisting of mobile hosts connected by a shared, scarce wireless channel. It has attracted more and more attention because it can be quickly and widely deployed. Some of example applications of wireless ad hoc networks include emergency search-and-rescue operations, data acquisition operations in inhospitable environments, and decision making in the battlefield.

Important features of a wireless ad hoc network are:

- Resource limitation;
- Multi-hop communication;
- Dynamic topology; and

- Lack of infrastructure.

Due to physical limitation of wireless links and resources, a mobile host communicates indirectly with other hosts in a multi-hop fashion. In such a scenario, each host should cooperate with each other by relaying packets from the source host to the destination host. This means that mobile hosts should act as routers by themselves and ad hoc routing schemes are required for unicasting, multicasting, and broadcasting. In a wireless ad hoc network, the mobility of hosts and physical link failures lead to dynamic topology. Due to multiple access, background noise, and interference from other transmissions, an active link between two hosts may easily become invalid. Thus, the communication link is unreliable and retransmission is quite often necessary for reliable services. The unreliability of communication links leads to frequent topology or route updates which should also be performed cooperatively by mobile hosts. In this chapter, we focus on optimization problems in wireless ad hoc networks where no infrastructure is assumed and, hence, the optimization problems become more challenging.

31.1.1 Inefficiency of ad hoc communications

Existing ad hoc routing protocols can be categorized as *table-driven (proactive)* and *demand-driven (reactive)* (Royer and Toh, 1999). Topology (in proactive routing protocols) and route (in reactive routing protocols) are updated and maintained by *flooding* request packets. Network-wide flooding may cause *Unreliable flooding* (Sinha et al., 2001) and *Broadcast storm problem* (Tseng et al., 2002).

Sinha et al. (2001) studied the inefficiency of local broadcasts. Their experimental results show that even in moderately sparse graphs a specific broadcast message is expected to be reachable by only about 80% of nodes in the network. The reachability of a single broadcast message decreases as the network load increases. Furthermore, their experiments show the degradation of data delivery rate of local broadcasts, which confirms the inefficiency of local broadcasts.

The broadcast storm problem is caused by network-wide flooding, which is common in ad hoc routing for topology or route updates, and denotes the inefficiency of network-wide flooding. More specifically, it may result in excessive *redundancy, contention*, and *collision* which cause high protocol overhead and interference with other ongoing communication traffic. Tseng et al. (2002) provided theoretical analysis on redundancy, contention, and collision of a single packet that is relayed by flooding mechanism. Based on their analysis, a rebroadcast can, on average, provide only 41% additional coverage beyond the coverage by the previous transmission. This redundancy becomes even worse when more than one rebroadcast happen; when a host rebroadcasts a packet more than three times, the expected additional coverage is below 5%. The expected probability of contention between two hosts is about 59% and the probability that n hosts experience contention increases over 80% as $n \geq 6$. Collisions that are caused by various reasons may result in even more inefficient rebroadcasting.

31.1.2 Improving wireless ad hoc communications

Unlike unicasting communications where the shortest path between the source node and the destination node is easily computed, multicasting and broadcasting communications have many optimization problem which are often NP-hard. When the objective is to minimize the number of relaying hosts in a multicasting session, the problem can be modeled as a *Steiner tree problem*. The problem of minimizing the number of relaying hosts in a broadcasting session can be modeled as a *minimum connected dominating set problem*. When we consider the power, or energy to be consumed for a broadcasting session, we can model the problem as *minimum broadcast cover problem* (Cagalj et al., 2002). Many heuristics or approximation schemes have been proposed for these NP-hard problems. We will discuss some of major heuristics or approximation schemes for the combinatorial problems and briefly review some variations of problems.

31.1.3 Chapter organization

The rest of this chapter is organized as follows. The main problems are defined and major approaches for each problem are reviewed in Section 31.2. In Section 31.3, applications of the problems and some variations of the problems are introduced. Section 31.4 concludes this chapter.

31.2 MAIN PROBLEMS

31.2.1 Steiner tree problem

The Steiner tree problem in graphs is a very useful technique in multicast routing (Wall, 1980). The Steiner tree problem in a graph can be described as follows:

Given a graph $G = (V, E)$, and a set of terminal nodes $T \subseteq V$, find a minimum cost connected subgraph of G containing all nodes in T.

Due to interference, fades and hidden stations, wireless communication may be not bidirectional. In this case, communication networks are usually modeled using directed graphs. A version of the Steiner tree problem in directed graphs is called the Steiner arborescence problem which can be described as follows:

Given a directed graph $G = (V, E)$, and a set of terminal nodes $T \subseteq V$, find a minimum cost Steiner arborescence which is a tree spanning T directed away from a root $r \in T$.

The connected subgraph may contain points in $V \setminus T$ auxiliary to the set T of terminal nodes. These auxiliary nodes are called *Steiner points*. If $V = T$, the Steiner tree problem becomes the well known *minimum spanning tree problem*. The *shortest path problem* can also be seen as a special case ($|T| = 2$) of the Steiner tree problem. We note that, if all the edge costs are positive, then the resulting subgraph is a tree (called a Steiner tree) of which leaves consist of a subgraph of the terminal nodes. In the sequel, only connected graphs with positive edge costs are assumed. The Steiner tree problem is one of the classic NP-hard problems (Garey and Johnson, 1979) and has been extensively studied. For more comprehensive surveys of the problem, the reader is referred to Hwang et al. (1992) and Winter (1987).

There are several exact algorithms available for the Steiner tree problem. However, they are suitable only for small-size problems since the problem is NP-hard. One of the earliest approach is the *spanning tree enumeration algorithm* by Hakimi (1971). The main idea of the algorithm is as follows:

Let the set of Steiner points be S. Then, of course, any Steiner tree must be a minimum spanning tree for the subgraph of G induced by $T \cup S$. However, the set S is not known so far. Thus, one of the possibilities is to try all possible sets S.

One of the main approaches for optimization problems is a dynamic programming technique. The idea of a dynamic programming approach for the Steiner tree problem is to combine optimal solutions of smaller sub-problems in such a way that the combined solution becomes optimal for the bigger problem.

Other algorithms for solving the Steiner tree problem are discussed in the next sections. For more comprehensive surveys of exact algorithms, the reader is referred to Maculan (1987) and Winter (1987).

31.2.1.1 Integer programming formulations. Many combinatorial optimization problems can be formulated as mixed integer programming problems. Here, we present integer programming formulations of the Steiner tree problem. Before presenting, we need some definitions.

Let c_e be the cost of edge $e \in E$. Given a graph $G = (V, E)$ and a set R of vertices, $\delta(R)$ represents the cut induced by $R \subseteq V$, that is, the set of edges with exactly one endpoint in R.

Aneja (1980) has shown that the Steiner tree problem is equivalent to the following integer program:

$$\min \quad \sum_{e \in E} c_e x_e \qquad (31.1)$$
$$\text{s.t.} \quad x(\delta(R)) \geq 1 \quad \forall\, R \subseteq V,$$
$$R \cap T \neq \emptyset,$$
$$(V \setminus R) \cap T \neq \emptyset,$$
$$x_e \in \{0,1\} \; \forall\, e \in E,$$

where $x(F) = \sum_{e \in F} x_e$ for $F \subseteq E$.

The above integer programming problem is actually a set covering problem. Moreover, the set $\{e \in E \mid x_e = 1\}$ corresponding to an optimal solution x of the integer program constitutes an optimal Steiner tree. Indeed, $x \in \{0,1\}^{|E|}$ such that $\{x_e = 1 \mid e \in F\}$ and $\{x_e = 0 \mid e \in E \setminus F\}$, where F is a tree which spans T, is a feasible solution to the integer program. On the other hand, a subgraph, which consists of the set $\{e \in E \mid x_e = 1\}$, is connected and spans T.

In the formulation, there could be exponential number of constraints, however, the linear relaxation of the set covering problem can be solved in polynomial time using the ellipsoid algorithm. Several algorithms for solving the integer program have been proposed including a cutting plane algorithm by Aneja (1980), branch and cut algorithms by Chopra and Rao (1994a;b) and Koch and Martin (1998). Koch and Martin

(1998) used the bidirected version (Bahiense et al., 2003; Goemans and Myung, 1993) of the integer programming formulation, and reduction techniques of Duin and Volgenant (1989). For the branching rule, they used strong branching by Bixby (1996) while using the heuristic by Takahashi and Matsuyama (1980) for the cutting phase.

Another basic integer programming formulation was introduced by Beasley (1984). Let $r \in T$ and $T_r = T \setminus \{r\}$. Note that a Steiner tree can be seen as a minimum cost subgraph of G such that there is a path between r and any other node in T in the subgraph. Based on this fact, Beasley presented the following integer programming problem and found two lower bounds for the Steiner tree problem using different Lagrangian relaxations of the integer program.

$$\min \sum_{(i,j) \in E} c_{ij} x_{ij} \tag{31.2}$$

$$\text{s.t.} \quad x_{ij} \geq y_{ijk} + y_{jik} \ \forall (i,j) \in E, \ k \in T_r, \tag{31.3}$$

$$\sum_{(r,i) \in E} y_{rik} \geq 1 \ \forall k \in T_r, \tag{31.4}$$

$$\sum_{(i,k) \in E} y_{ikk} \geq 1 \ \forall k \in T_r, \tag{31.5}$$

$$\sum_{(i,j) \in E} y_{ijk} - \sum_{(h,i) \in E} y_{hik} \geq 0 \ \forall i \in V \setminus \{r,k\}, \tag{31.6}$$

$$x_{ij}, y_{ijk} \in \{0,1\} \ \forall (i,j) \in E, \ k \in T_r, \tag{31.7}$$

where $y_{ijk} = 1$ if edge $(i,j) \in E$ is on the unique directed path from r to $k \in T_r$ in the Steiner tree, and $y_{ijk} = 0$ otherwise. Constraint (31.3) implies that an edge is on some path in the Steiner tree if that edge is in the Steiner tree. Constraints (31.4)-(31.6) express that there exists a path from r to k containing $(i,j) \in E$ while constraint (31.7) is the integrality constraint.

The mixed integer program below for the Steiner arborescence problem was introduced by Wong (1984), and frequently referred to as the multicommodity flow formulation. Let the root r be the source, and let $K = \{k \in V \mid k \in T_r\}$ be the set of commodities. For a node $i \in V$, $I(i) = \{j \in V \mid \overrightarrow{(j,i)} \in E\}$ and $O(i) = \{j \in V \mid \overrightarrow{(i,j)} \in E\}$. Then the Steiner arborescence problem can be formulated as follows:

$$\min \sum_{\overrightarrow{(i,j)} \in E} c_{ij} x_{ij} \tag{31.8}$$

$$\text{s.t.} \quad \sum_{j \in O(i)} y_{ijk} - \sum_{j \in I(i)} y_{jik} = \begin{cases} 1 & i = r, \ k \in T_r \\ 0 & i \in V \setminus \{r,k\}, \ k \in T_r \\ -1 & i = k, \ k \in T_r \end{cases} \tag{31.9}$$

$$0 \leq y_{ijk} \leq x_{ij}, \ \overrightarrow{(i,j)} \in E \text{ and } k \in T_r \tag{31.10}$$

$$x_{ij} \in \{0,1\}, \ \overrightarrow{(i,j)} \in E, \tag{31.11}$$

where y_{ijk} indicates the flow amount of commodity k going through the edge $\overrightarrow{(i,j)} \in E$; and $x_{ij} = 1$ if $\overrightarrow{(i,j)} \in E$ is in the solution, and $x_{ij} = 0$ otherwise. Constraint (31.9) ensures the existence of a unit flow between the source node and every other terminal while constraint (31.10) allows a nonzero flow of commodity k through arc (i, j) if it is in the Steiner arborescence.

Wong (1984) introduced a dual ascent method for its linear programming relaxation. Relaxations for the above mixed integer program can be used for the Steiner tree problem since the Steiner tree problem can be seen as a bidirected Steiner arborescence problem. Recently, Bahiense et al. (2003) proposed an algorithm to obtain an approximate optimal solution for the Steiner tree problem based on a Lagrangian relaxation of the above mixed integer program. An extension of the subgradient algorithm is used to solve a Lagrangian dual problem to obtain lower bounds and to estimate primal solutions.

For more integer programming formulations for the Steiner tree problem and their comparisons, the reader is referred to Claus and Maculan (1983), Goemans and Myung (1993), Koch and Martin (1998), and Gouveia and Magnanti (2003).

31.2.1.2 Major heuristics. Since the Steiner tree problem is NP-hard, it is of practical importance to obtain heuristic methods which find trees whose costs are close to optimal. As we discussed earlier, there are many heuristic algorithms for the Steiner tree problem based on linear programming relaxations of the previous integer programming problems. Many heuristic algorithms of reasonable complexity have also been proposed for the Steiner tree problem based on the other characteristics of the problem. In this section, we discuss some of these major heuristics for solving the problem which use well known polynomial time shortest path algorithms and minimum spanning tree algorithms. One well known heuristic, KMB (Kou et al., 1981), can be described as follows:

Step 1. Construct the new complete graph $C = (T, E_C)$ such that, for every edge $(i, j) \in E_C$, the distance of (i, j) is equal to the distance of the shortest path from i to j in G.

Step 2. Find a minimum spanning tree T_C of C.

Step 3. Replace each edge (i, j) in T_C by a shortest path from i to j in G. Let the resulting subgraph of G be G'.

Step 4. Find a minimum spanning tree T' of G'.

Step 5. If there is a leaf in T' which is not in T, then go to the next step. Otherwise, T' is an approximate optimal Steiner tree and terminate the algorithm.

Step 6. Remove the leaf in T' which is not in T from T' and return to Step 5.

One of the characteristics which define the efficiency of any heuristic algorithm is the gap between a solution found by the algorithm and a global optimal solution of the problem. Kou et al. (1981) proved the following theorem regarding the approximation factor of their algorithm.

Theorem 31.1 *The KMB heuristic produces a Steiner tree with approximation factor of* $2(1 - 1/|T|)$.

They have also constructed a graph such that their algorithm has an exact performance of $2(1-1/|T|)$. However, Doar and Leslie (1993) have used the algorithm in problems occurring in real networks and reported that the total cost of trees generated using KMB is on the average only 5% worse than the cost of the optimal minimum Steiner tree.

Next, we present a heuristic algorithm by Takahashi and Matsuyama (1980). For a subset W of V and a vertex i in $V \setminus W$, let $P(W,i)$ be a path whose cost is minimum among all shortest paths from vertices in W to vertex i. Let us denote its cost by $cost(W,i)$. At each step in the algorithm, a tree containing a subset of T is constructed and a new vertex in T with some vertices in $V \setminus T$ are added to the tree constructed at the previous step.

Step 1. Choose a vertex, $v_1 \in T$, and set $T_1 = (V_1, v_1)$, where $V_1 = \{v_1\}$ and $E_1 = \emptyset$.

Step k. Find a vertex, v_k, in $S \setminus V_{k-1}$ such that
$$cost(V_{k-1}, v_k) = \min_{v_i \in T \setminus V_{k-1}} cost(V_{k-1}, v_i).$$
Let $T_k = (V_k, E_k)$ be the tree constructed by adding $P(V_{k-1}, v_k)$ to T_{k-1}.

Let $V = n$ and $E = m$. Since any shortest $P(V_{k-1}, k)$ can be found in $O(n^2)$ using Dijkstra's algorithm (Dijkstra, 1959), the worst case complexity of the above algorithm is $O(kn^2)$, however, this may be improved by using Dial's implementation (see, for example, Ahuja et al. (1993)).

Theorem 31.2 *For all k ($2 \leq k \leq n-1$), the following inequality is true.*

$$\frac{\text{The cost of } T_k}{\text{The cost of a Steiner tree}} \leq 2\left(1 - \frac{1}{k}\right).$$

Moreover if $k = n$, then the algorithm finds an optimal tree.

They have also shown that this is the best evaluation of their algorithm by constructing a graph achieving the upper bound.

Recently, Gendreau et al. (1998) proposed a heuristic algorithm based on tabu search and they reported that their algorithms outperformed algorithms by Esbensen (1995) and Voss and Gutenschwager (1998).

For more comprehensive surveys of heuristic methods for the Steiner tree problem, the reader is referred to Winter (1987), Hwang and Richards (1992) and Voss (1992).

31.2.2 Minimum connected dominating set problem

31.2.2.1 Problem definition.
The Minimum Connected Dominating Set problem, which is NP-hard, is one of the classical combinatorial optimization problems in graph theory.

Given an undirected graph $G = (V,E)$ with the set V of vertices and the set E of edges, a *dominating set* (DS) D of G is a subset of V such that any vertex in $V \setminus D$ is adjacent to at least one vertex in D. A *connected* dominating set (CDS) D of G is a dominating set such that the subgraph induced by D in G is connected. A CDS with minimum cardinality among all CDSs is called a *minimum* connected dominating set (MCDS).

The problem of finding an MCDS in G is equivalent to the problem of finding a maximum leaf spanning tree in G. A spanning tree in G is a connected spanning subgraph of G such that it contains no cycle. All non-leaf nodes in the spanning tree form an MCDS.

An *independent set* (IS) I of a graph $G = (V,E)$ is a subset of V such that no two vertices in I are joined by an edge in E. An independent set I is *maximal* if $I \setminus \{v\}$ is not an independent set for all $v \in V \setminus I$. According to this definition, it can be easily shown that a maximal independent set is a DS.

We again note that the MCDS problem is NP-hard (Garey and Johnson, 1979). However, there are several efficient polynomial time approximation algorithms for solving the MCDS problem (Guha and Khuller, 1998; Marathe et al., 1995). Guha and Khuller (1998) introduced a polynomial time approximation algorithm for finding MCDS with approximation factor of $H(\Delta) + 2$, where Δ is the maximum degree of the graph and $H(k) = 1 + 1/2 + ... + 1/k$ is the harmonic function. . Their approach first constructs a DS using a greedy heuristic. Then a Steiner tree algorithm is used to connect the nodes of the dominating set.

31.2.2.2 The minimum connected dominating set problem in wireless ad hoc networks. Ad hoc networks can be modeled using unit-disc graphs. For a graph G, if length of any edge in E is less than or equal to 1, then G is called a *unit-disk* graph. One of the main problems in ad hoc wireless networks is to operate routing protocols over a virtual backbone. The virtual backbone structure for ad hoc wireless network has attracted much attention in the recent years. However, it is quite difficult to solve the minimum connected dominating set problem in unit-disk graphs. Clark et al. (1990) have shown that computing an MCDS in a unit-disk graph is NP-hard. Here, we discuss some recent heuristics for finding approximate MCDSs.

Das et al. (1997) proposed a distributed algorithm, which is essentially a distributed version of the algorithm by Guha and Khuller (1998) for computing an MCDS. The algorithm first finds a DS, and then constructs a spanning forest of which each tree component is a union of stars centered at the nodes in the DS. In the final stage, the spanning forest is expanded to a spanning tree by connecting the forests. All internal nodes in the tree form a CDS. They also used a clustering technique to deal with large networks. In this case, the algorithm is applied to each cluster instead of the whole network. The algorithm guarantees an approximation factor of $3H(\Delta)$. The message complexity and the time complexity of the algorithm is $O(n^2)$, where n is the number of nodes in the unit-disk graph. Wan et al. (2004a) later showed that the approximation factor of the algorithm was between $(\log \Delta)/2 - 1/2$ and $3H(\Delta)$.

Wu and Li (1999) proposed an algorithm for finding MCDS with approximation factor of $n/2$. The algorithm first finds a CDS such that it includes all intermediate nodes of any shortest path between nodes in the dominating set, and then prunes some redundant nodes from the CDS in order to find better CDS. Their work was initially presented without any performance analysis and Wan et al. (2004a) later showed its performances. The algorithm has $O(m)$ message complexity and $O(\Delta^3)$ time complexity, where m is the number of edges in the unit-disk graph. A similar procedure was given by Butenko et al. (2003). The algorithm starts with a CDS. Then, at each step, a vertex is selected from the current set using a greedy method and either remove it from the set or include it in the final solution. It has time complexity equal to $O(n\log^3 n)$ and message complexity equal to $O(nm + n^2 \log^3 n)$.

Wan et al. (2004a) have proven the following theorem about the relation between the size of MIS of unit-disk graph G and the size of any MCDS.

Theorem 31.3 *The size of any IS of G is at most $4 \cdot opt + 1$, where opt is the size of any MCDS.*

They propose a distributed algorithm for finding an approximate MCDS in unit-disk graphs and prove that their algorithm finds a connected dominating set with at most $8 \cdot opt - 2$ vertices using the above theorem. The algorithm consists of two phases. First, it computes a MIS; then it uses a dominating tree to connect all vertices in the MIS. In the first phase, a MIS is constructed such that the distance between any pair of its complementary subsets is exactly two hops. In the second phase, a dominating tree is constructed to connect the nodes in the MIS. Moreover the internal nodes of the dominating tree form a nontrivial CDS. The distributed algorithm has an approximation factor of at most 8, $O(n)$ time complexity, and $O(n\log n)$ message complexity.

Cardei et al. (2002) have proposed an 8-approximate algorithm. Their algorithm also has two phases. First, it computes a MIS and then Steiner tree is used to connect all vertices in the MIS.

Other approximation algorithms are given in Marathe et al. (1995), Butenko et al. (2002), and Cheng et al. (2003b).

31.2.3 Minimum broadcast cover problem

31.2.3.1 Problem definition.
Cagalj et al. (2002) defined the *Minimum Broadcast Cover (MBC)* problem and showed its NP-hardness. The problem MBC is described as follows: Given a directed graph with power level for each node and cost for each edge, determine whether there is a power assignment for each node so that any node has a path from the source node in the induced graph using at most constant amount of total power. More formally,

---

Instance: A directed graph $G = (V, E)$, a set P consisting of all power levels at which a node can transmit, edge costs $c_{ij} : E(G) \to \mathbb{R}_+$, a source node $r \in V$, an assignment operation $p_i : V(G) \to P$, and some constant $B \in \mathbb{R}_+$.

Question: Is there a node power assignment vector $A = [p_1, p_2, ..., p_{|V|}]$ such that it induces the directed graph $G' = (V, E')$, where $E' = \{(i,j) \in E : c_{ij} \leq p_i\}$, in which there is a path from r to any node in V (all nodes are covered), and such that $\sum_{i \in V} p_i \leq B$?

MBC is proven to be NP-complete by reduction from *the Set Cover problem*. Note that MBC deals with a graph where edges and their costs are arbitrarily chosen. It is also useful to consider two-dimensional Euclidean metric space since transmission power is proportional to the Euclidean distance between two communicating hosts. The geometric version of MBC, *Geometric Minimum Broadcast Cover (GMBC)* problem (Cagalj et al., 2002), is described as follows: Given a set of nodes in two-dimensional Euclidean metric space with power level for each node and cost for each edge, determine whether there is a power assignment for each node so that any node has a path from the source node in the induced graph using at most a constant amount of total power. More formally,

---

Instance: A set of nodes V in the plane, a set P consisting of all power levels at which a node can transmit, a constant $k \in \mathbb{R}_+$, costs of edges $c_{ij} = k d_{ij}^\alpha$ where d_{ij} is Euclidean distance between i and j, a real constant $\alpha \in [2..5]$, a source node $r \in V$, an assignment operation $p_i : V(G) \to P$, and some constant $B \in \mathbb{R}_+$.

Question: Is there a node power assignment vector $A = [p_1, p_2, ..., p_{|V|}]$ such that it induces the directed graph $G' = (V, E')$, where $E' = \{(i, j) \in E : c_{ij} \leq p_i\}$, in which there is a path from r to any node in V (all nodes are covered), and such that $\sum_{i \in V} p_i \leq B$?

GMBC is proven to be NP-complete by reduction from *the Planar 3-SAT problem*. Note that GMBC is equivalent to asking whether there exists a broadcast tree that requires the total cost at most B in Euclidean metric spaces.

31.2.3.2 Heuristics. In wireless networks, the network connectivity depends on the transmission power. The received signal power varies as $r^{-\alpha}$, where r is the range and α is a parameter that typically takes on a value between 2 and 4, depending on the characteristics of the communication medium. Based on this model, the transmitted power p_{ij} required for a link between two nodes i and j separated by distance r is proportional to r^α. Ignoring the constant coefficient, we get $p_{ij} = r^\alpha$.

Wieselthier et al. (2002) proposed three heuristics: BLU (Broadcast Least-Unicast-cost algorithm), BLiMST (Broadcast Link-based MST algorithm), and BIP (Broadcast Incremental Power algorithm). They focused on the difference between wired and wireless networks. In wired networks, "link-based" approaches such as a minimum-cost spanning tree can be used for a broadcast tree. On the other hand, wireless networks have the "wireless multicast advantage" (WMA) property which enables a single transmission (broadcast in nature) to be reachable by every node within the transmission range from the source node . Hence, the total power to reach a set of nodes from a source node is the maximum power to reach any node in the set. This observation leads to a "node-based" approach. BLU and BLiMST are link-based approaches which are based on conventional networking technologies and BIP is node-based approach adopting the WMA property.

BLU uses shortest unicast paths to construct broadcast trees by superposing the paths to individual destinations. This algorithm assumes that an underlying unicast algorithm, such as the Bellman-Ford or Dijkstra algorithm, provides minimum distance paths from the source node to every destination. However, the resulting broadcast tree can be inefficient, or require more total power, since while some nodes can receive a transmission without help of relay nodes, BLU may create unnecessary links for those redundant relays.

BLiMST is based on the use of the standard MST algorithm where link cost is assigned for each pair of nodes. Again the WMA property is ignored in this algorithm. Hence, the resulting tree may be inefficient in some cases.

The inefficiency of the previous two algorithms comes from the link-based approach. To improve energy-efficiency, BIP constructs a tree rooted at the source node by adding nodes with *minimum additional cost*. This algorithm is based on Prim's algorithm with iterative modifications of the link costs. The link costs will be updated at each step as follows:

$$P'_{ij} = P_{ij} - P(i)$$

where P_{ij} is the link-based cost of a transmission between node i and node j and $P(i)$ is the power level at which node i is already transmitting. This approach does not guarantee minimum-cost spanning trees but experimental results show improved performance over a wide range of examples (Wieselthier et al., 2002).

All three algorithms (BLU, BLiMST, and BIP), are centralized algorithms and their time complexities are $O(N^2)$, $O(N^3)$, and $O(N^3)$, respectively. Here, N is the number of nodes in the network. The performance can be slightly improved by the "sweep" procedure (Wieselthier et al., 2002) which eliminates unnecessary transmissions with additional $O(N^2)$ time complexity. The authors also described how those algorithms can be applied to multicast routing and this will be reviewed in Section 31.3.

Wan et al. (2002) proved the approximation ratios of BLU, BLiMST, and BIP. In addition to the three algorithms, they proposed a variation of BIP based on Chvatal's algorithm (Chvatal, 1979) for the Set Cover problem. The variation is named Broadcast Average Incremental Power, BAIP for short. In BAIP, a set of new nodes with *minimal average incremental cost* is added to the spanning tree. Minimal average incremental cost for a set of new nodes is defined to be the average (over the set) of minimum additional power increase to reach the nodes in the set. Despite BAIP's similarity to BIP, its approximation ratio is quite different from that of BIP. Here, the approximation ratio of a heuristic is defined to be the maximum ratio of the energy necessary to broadcast a message based on the tree resulting from the heuristic to the least power required by any spanning tree. They used geometric arguments to analyze the lower and upper bounds of approximation ratios of the four algorithms (see Table 31.2.3.2).

Another heuristic based on the WMA property is proposed in Cagalj et al. (2002). The heuristic *Embedded Wireless Multicast Advantage*, EWMA for short, is based on the link-based MST but iteratively modifies the tree by excluding some transmitting nodes so that the total power decreases. EWMA consists of two phases. In the first phase, it constructs an MST, and in the second phase, it incrementally builds a broadcast tree starting from a single (source) node tree. For the selection of transmitting nodes, *gain* is used as a metric. The gain of a node v is defined as the decrease in the total power of a broadcast tree by excluding some of the transmitting nodes in the MST and instead increasing the transmission power of v. EWMA also has 12 as its upper bound of approximation ratio, but the lower bound is not clear. Its time complexity is

Table 31.1 Comparison of BLU, BAIP, BLiMST, BIP, and EWMA

	BLU	BAIP	BLiMST	BIP	EWMA
lower bound	$\frac{n}{2}$	$\frac{4n}{\ln n - o(1)}$	6	$\frac{13}{3}$	–
upper bound	–	–	12	12	12
complexity	$O(N^2)$	–	$O(N^3)$	$O(N^3)$	$O(d^4)m^2$

n is the number of receiving nodes
N is the number of nodes in the network
d is the maximum degree, m is the number of transmitting nodes

$O(d^4)m^2$, where d is the maximum degree and m is the number of transmitting nodes in the tree.

31.3 APPLICATIONS AND VARIATIONS

31.3.1 Multicasting

Traditional computer network applications involve communication between two computers and techniques for this kind of networks are often referred to as unicasting. In recent years, a number of important emerging applications of computer networks, such as video conferencing, corporate broadcasts, distance learning, and distribution of software, stock quotes, and news require simultaneous communication between groups of computers rather than just two computers. Multicasting is a technique for the efficient distribution of some specific data from one or more sources to a group of destinations that need to receive the data. Distributing data to all computers in the network is referred to as broadcasting. In multicast routing, the goal is to find a way to distribute data while minimizing the usage of resources such as connection costs, time and bandwidth. Multicast routing is one of the most important problems in wireless networks. Telecommunication networks are usually described as weighted graphs and this gives us an opportunity to use optimization approaches in graph theory to handle multicast routing problems. In graph representations, nodes in a graph represent hosts, and edges represent network link while weights represent a measure of the utilization of that link's resources. In addition, there are capacity and delay for each link. The capacity for a link represents the maximum amount of data allowed in the link while the delay for a link represent the total time between delivery and arrival of a data packet in the link. Multicast routing algorithms can be classified into two categories: shortest path algorithms and Steiner tree algorithms.

One way to handle multicast routing is to use shortest path problems to minimize the cost of each path from the source node to each multicast member. Shortest path problems can be solved in polynomial time using well known shortest path algorithms such as Bellman-Ford's algorithm (Bellman, 1958) and Dijkstra's algorithm (Dijkstra, 1959). However, some models of multicast routing consider delay constraints on each destination of the multicast group such that the sum of the delays from the source node to each destination is limited to some point. Those constraints make the shortest path problem difficult as NP-hard (Garey and Johnson, 1979). Several heuristic algorithms have been proposed for finding good approximate solutions of the problem (Deering and Cheriton, 1990) and (Sun and Langendoerfer, 1995).

The objective of the Steiner tree problem in multicast routing is to minimize the total cost of multicast tree. In this case, the set of destinations in the Steiner tree problem represent the multicast group. The Steiner tree problem is well known to be NP-hard (Garey and Johnson, 1979).

31.3.2 Virtual backbone

In a wireless ad hoc network, there are no fixed infrastructure and central administration. This deficiency exacerbate the inefficiency of ad hoc communications. To alleviate these problems, *virtual backbone-based routing* strategies have been introduced (Das and Bharghavan, 1997; Sivakumar et al., 1998; Wu and Li, 1999; Sinha et al., 2001; Stojmenovic et al., 2001). The most important benefit of virtual backbone-based routing is the dramatic reduction of protocol overhead; thus it greatly improves the network throughput. This is achieved by propagating control packets inside the virtual backbone, not the whole network. Other benefits include the support of broadcast/multicast traffic and the propagation of "link quality" information for quality-of-service (QoS) routing (Sivakumar et al., 1999).

The concept of a virtual backbone was initially studied for mobile packet radio networks (Ephremides et al., 1987; Gerla and Tsai, 1995). Those initial approaches aimed at finding a feasible interconnected set of *clusters* covering the entire node population without the objective of optimizing the size of the backbone. Ephremides et al. (1987); Gerla and Tsai (1995) proposed two algorithms for clustering : Lowest-ID Cluster Algorithm and Highest-Connectivity Cluster Algorithm. In the formation of clusters and clusterheads, the node with the lowest ID or highest connectivity is selected to be a clusterhead. Gerla and Tsai (1995) gave experiments about number of cluster changes upon a single host movement and the result shows that the lowest-ID algorithm requires less cluster changes. Both algorithms have the following two interesting properties:

- Clusterheads are not directly linked each other.

- Any two nodes in the same cluster are at most two-hops away, since the clusterhead is directly linked to every other node in the cluster.

The above two properties imply that the set of clusterhead nodes form an independent set and this idea has been commonly used in an approximation of the minimum connected dominating sets with the benefit of a constant upper bound of the backbone size.

31.3.2.1 Virtual backbone applications.
The main application of virtual backbones in wireless ad hoc networks is virtual backbone-based routing (Das and Bharghavan, 1997; Sivakumar et al., 1999; Wu and Li, 1999; Sinha et al., 2001). Virtual backbones can also be used for broadcasting (Stojmenovic et al., 2001) or mobility management (Liang and Haas, 2000).

Das and Bharghavan (1997) discussed a virtual backbone scheme with a goal of *optimizing the size of the virtual backbone* and introduced its application to routing. Their virtual backbone structure aims at supporting unicast, multicast, fault-tolerant routing in wireless ad-hoc networks. The virtual backbone is constructed from an approximation of minimum connected dominating set, MCDS for short. The backbone structure will change upon node movement and the hosts in the backbone are used only for computing and updating routes. In order to accommodate the dynamic nature of the virtual backbone, their approach splits the routing problem into two levels: find and update the virtual backbone, and then find and update routes. The route computation is a standard shortest-paths algorithm.

Low communication overhead for route computation is critical in wireless ad hoc networks since routes are computed frequently due to the dynamic topology. Also local information based route computation is more adaptable to dynamic environments. Wu and Li (1999) discussed a virtual backbone based routing as a solution of these requirements. They proposed a simple and message-efficient mechanism ("marking process" and 2 rules to reduce the size) to construct a virtual backbone, which is a connected dominating set, and the nodes in the backbone are called *gateway*. Though their routing scheme is based on the shortest path scheme, they showed that the shortest path between any two nodes should consist of gateway nodes only. Using this property, the shortest path route can be computed only among gateway nodes.

Sivakumar et al. (1999) proposed CEDAR, a core-extraction distributed ad hoc routing algorithm, aiming at *quality-of-service (QoS) routing* in wireless ad hoc network environments. Their objective is to compute unicast routes satisfying a minimum bandwidth requirement from the source to the destination. Due to dynamic topology and unreliable transmissions, their goal is to provide routes that are highly likely to satisfy the bandwidth requirement of a route. CEDAR has three components: core extraction, link state propagation, and route computation. The set of core nodes is obtained by approximating minimum dominating sets using local computation and local state. The core nodes (the dominators) maintain the local topology of its domain and compute routes on behalf of the nodes in its domain. For QoS purpose the bandwidth availability information of stable high bandwidth links is propagated to core nodes, while information about dynamic or low bandwidth links is kept local. Routes are computed iteratively, starting from core path, which consists of virtual links between core nodes, between the dominator of the source and the dominator of the destination. Using the core path, routes are computed by finding successive partial routes from the source to the furthest possible node in the core path satisfying the bandwidth requirement where the intermediate destination becomes the new source for the next partial route.

Virtual backbones can be used to *enhance the existing routing schemes* and a recent experimental result is found in (Sinha et al., 2001). The authors present extensive simulation results on the effect of applying existing routing schemes on the virtual backbone infrastructure. They use an adaptation of the core infrastructure (extracted by CEDAR algorithm) as a virtual backbone and routes are computed by applying existing routing schemes over the backbone. They chose DSR and AODV and carried out performance evaluation of the routing schemes over the backbone. The experimental results show that by restricting the exchange of routing control packet to backbone nodes, the performance of AODV and DSR is improved. Average number of messages per route request is significantly reduced to about 10% for both schemes. The protocol overhead decreases up to 50% for AODV scheme. For DSR the overhead does not decrease and this is due to DSR's aggressive caching policy. But in a highly dynamic scenarios the cached information easily becomes stale, hence the result does not imply that virtual backbones are not suitable for DSR.

Stojmenovic et al. (2001) discussed about an application of virtual backbones to *broadcasting*. They proposed two schemes; backbone broadcasting scheme and neighbor elimination scheme. In backbone broadcasting scheme, rebroadcasting is limited to the backbone nodes. Compared to flooding, this scheme requires significantly less retransmissions. Neighbor elimination scheme further reduces the number of retransmissions by removing redundant rebroadcasts. A backbone node will rebroadcast the received message only when it has a neighbor that may need the message. In order to compute the redundancy, it is required that each node knows the exact location of all its neighbors or knows its 2-hop neighbors list. Every non-backbone node registers itself to one of its backbone neighbors and backbone nodes decide whether there is a registered neighbor that will not receive the rebroadcast message from other backbone nodes. They conducted experiments on various broadcasting schemes and the result shows that backbone broadcasting performs the best and with neighbor elimination added, the performance is further improved.

Another application of virtual backbone to *mobility management* is discussed in (Liang and Haas, 2000). The authors proposed ad-hoc mobility management scheme utilizing location databases. These location databases are distributed on virtual backbone nodes. The virtual backbone is constructed using a greedy algorithm for Minimum Set-Covering problem, hence the backbone nodes need not be connected via 1-hop links, but the connectivity between the backbone nodes are maintained as multi-hop paths. Routing is performed on the flat network structure, contrary to the (hierarchical) backbone routing, which involves non-backbone nodes for more balanced loading on the nodes and the links. The virtual backbone nodes provide location information from the databases they hold. The location information helps localizing the routing procedure and managing nodal mobility. Since the location information should be kept up-to-date the virtual backbone nodes need to maintain the inter-connection among themselves.

31.3.2.2 Reliability of virtual backbones. Among many characteristics of wireless networks, nodal mobility is one of important features. When nodes move, any sparse structure is subject to lose connectivity very soon. In the construction of virtual backbones for wireless networks, the nodal mobility should be considered so that the resulting structure, virtual backbone, becomes more robust and at the same time requires less maintenance cost.

Min et al. (2004) proposed a virtual backbone scheme which is more reliable and hence requires less maintenance need for mobile ad hoc networks. The authors considered two factors: stability and coverage of nodes. Their scheme also generates a constant approximation ratio (at most 8) virtual backbone. Moreover, by constructing the backbone using nodes with more coverage and more stability, the lifetime of the resulting backbone increases significantly. The message and time complexities are $O(\Delta n)$ and $O(n)$, respectively and the extensive experimental results show that the resulting backbone is more robust than the backbones generated from other existing backbone schemes. Against the smallest size backbones, the lifetime of connectivity and coverage increased by 148% and 490% on the average under various mobility scenarios. Even against the equivalent size backbones, the lifetime increased by 88% and 32% on the average.

In terms of maintenance cost, they measured potential maintenance need which essentially counts the number of disconnected backbone and non-backbone nodes. For a backbone node, when it loses a link to its neighbor that was originally in the backbone, maintenance may be required to repair the damaged connectivity. For a non-backbone node, when it loses all the links to its neighbors that were originally in the backbone, maintenance may be required to repair the damaged coverage. In this way they measured the potential maintenance needs for several virtual backbone schemes and the simulation results show significant improvements for the potential maintenance needs.

One more interesting point is that they interleaved the conventional two phases (MIS construction and interconnection) so that after new MIS nodes are selected, those nodes select intermediate nodes which interconnect them to existing MIS nodes. This integration of phases leads to a faster construction of virtual backbone, which is preferable in highly dynamic environments.

31.3.3 Energy-efficient broadcasting

The mobile devices in a wireless ad hoc network operate on their batteries. As a result, energy-efficiency is one of the most important factors in designing such networks. For unicast communications, it is best to transmit at the lowest possible power level to save transmission energy. Throughout this chapter, we use power and energy interchangeably. In case of broadcast communications, it may not be the best strategy to use the lowest possible power level. Unlike wired networks, wireless networks have "node-based" nature of communications (Wieselthier et al., 2002), which arises from broadcast nature of wireless channel. Hence using higher power level than minimum, a single transmission can reach more hosts. As a result, the overall energy consumption can be reduced. Computing the power-optimal broadcast tree is NP-hard and several heuristics to compute energy-efficient broadcast trees are discussed in Section 31.2.3.

31.3.3.1 Energy-efficient multicasting. As discussed in Section 31.2.1, minimum-cost multicast tree problem can be modeled as an NP-hard Steiner tree

Table 31.2 Approximation ratios of MLU, MLiMST, MIP, SPF and MIPF

	MLU	MLiMST	MIP	SPF	MIPF
lower bound	$\frac{n}{2}$	$n-1$	$n-2-o(1)$	-	$\frac{13}{3}$
upper bound	-	-	-	$2c$	$2c$

n is the number of receiving nodes

c is a constant between 6 and 12

problem. In this subsection, the application of energy-efficient broadcast tree heuristics to energy-efficient multicast tree construction is discussed.

Wieselthier et al. (2002) proposed the applications of three broadcast tree heuristics to multicast tree construction; Multicast Incremental Power (MIP) algorithm, Multicast Least-Unicast-cost (MLU) algorithm, and Multicast Link-based MST (MLiMST) algorithm. In MIP, first a broadcast tree is constructed by BIP and then to obtain the multicast tree, redundant transmissions are pruned from the tree. A transmission is redundant if all the members in the multicast group can receive the multicast message without the transmission. Similarly, MLiMST prunes the broadcast tree obtained from BLiMST to get a multicast tree. The redundancy of a transmission is defined the same as MIP. MLU is almost identical to BLU, the only difference is that MLU finds unicast paths from the source to only the destinations in the multicast group. The multicast tree is formed by superposition of the unicast paths.

Experimental results are provided in (Wieselthier et al., 2002) in which MIP, MLU and MLiMST (BIP, BLU and BLiMST for broadcast tree, respectively) are simulated. Performance metric is the total power of the multicast (or broadcast) tree. Since a multicast tree is constructed as a subtree of the corresponding broadcast tree, multicast tree heuristics can be expanded to generate a broadcast tree by setting the entire hosts as the multicast group. When the multicast group size is small, MLU performs the best. This is a natural result since MLU computes optimal paths to individual destinations. However when the multicast group size becomes large, the performance of MLU degrades while the performance of MLiMST and MIP improves. Especially MIP outperforms both MLiMST and MLU for large multicast group. This improvement comes from the fact that MIP successfully exploit the "node-based" nature of wireless (broadcast) communications.

Wan et al. (2004b) proposed two heuristics, SPF (Shortest-Path First) and MIPF (Minimum Incremental Path First), for minimum power multicast tree problem based on BLiMST and BIP, respectively. They showed none of MIP, MLU and MLiMST has constant approximation ratio and their two heuristics have constant approximation ratios. Instead of pruning branches in the broadcast tree, both SPF and MIPF iteratively add paths from the source node to the required destinations. In SPF, a tree T rooted at the source node is maintained during the procedure. Starting from a single (source) node, T is iteratively grown by one path from the source node to a destination not in T yet. The path is found by first collapsing the entire tree T into a temporary node and then finding all the shortest-paths from the temporary node to all the destinations not in T. A path with the least total power is added. MIPF is based on BIP and the execution is similar to *SPF* with the exception that MIPF uses incremental power instead of total power as the metric of a tree. Starting from a single (source) node, the path from a node in T to a destination not in T yet with the least incremental power is added iteratively. The incremental power of a path is defined as the total power minus the transmission power of the starting node in the path. MIPF first collapses the entire tree T into a temporary node and then finds all the shortest-paths from the temporary node to all the destinations not in T yet. Among those paths, the one with the least incremental power is added.

31.3.3.2 Balanced power consumption.
While the most research on energy-efficient broadcasting focuses on optimizing the total power in the broadcast tree, Cheng et al. (2003a) proposed a heuristic aiming at *balancing the power consumption* in the broadcast tree. Wireless ad hoc networks require every host's cooperation, for example without any router each host has to exchange messages to find out routes. In this situation, if some hosts are completely drained of power, the network may be partitioned or may not operate normally. Hence the authors identified a slightly different energy-efficient broadcast tree problem where the objective is to balance the power consumption. In the paper, the network lifetime is defined as the first time that any host is completely drained of power. And the problem is to find a broadcast tree which minimizes the maximum power consumption of hosts.

Cheng et al. (2003a) found an interesting property that MST has the minimum longest edge among all spanning trees and based on the property MLE (Minimum Longest Edge) algorithm was proposed. MLE consists of two phases: MST construction phase and the redundancy elimination phase. Every relaying (transmitting) host is represented as a nonleaf node in the MST. And the transmission range of a relaying host is represented as the length of the longest edge between the node and a leaf node. Among the nonleaf nodes, and hence their corresponding transmissions, there may be unnecessary transmissions; for example, if a node A's children are all within the transmission range of A's parent, then A does not need to transmit. In this way the MST from the first phase is reformed into another spanning tree after the second phase. Note that the reformed tree preserves the minimum longest edge property. The second phase of MLE also implicitly makes use of WMA property discussed in Section 31.2.3.2. The experimental results show that the maximum transmitting power of MLE is decreased by 15% against BIP on average.

31.4 CONCLUSIONS

In this chapter, we reviewed three combinatorial optimization problems with their applications to wireless ad hoc networking. These problems can be used to model the wireless communication problems, especially for multicasting and broadcasting. The heuristics and approximation schemes discussed in this chapter provide good guidelines for wireless ad hoc communications, but not perfect solutions. In order to improve the quality of solutions, the environment-intrinsic natures, for example WMA (Wireless Multicast Advantage) for minimum total energy broadcasting, should be taken into account when designing a heuristic since those features are the one that complicates the problems. Each of the main problems discussed in this chapter also have variations, some introduced in this chapter, which are meaningful from different view points. Wireless ad hoc networking is truly one of the most challenging and evolving research areas in both theory and practice and still more thorough theoretical approach of optimization is required.

Bibliography

R.K. Ahuja, T.L. Magnanti, and J.B. Orlin. *Network Flows: Theory, Algorithms, and Applications*. Prentice Hall, Englewood Cliffs, New Jersey, 1993.

Y.P. Aneja. An integer linear programming approach to the steiner problem in graphs. *Networks*, 10:167–178, 1980.

L. Bahiense, F. Barahona, and O. Porto. Solving steiner tree problems in graphs with lagrangian relaxation. *Journal of Combinatorial Optimization*, 7:259–282, 2003.

J.E. Beasley. An algorithm for the steiner problem in graphs. *Networks*, 14:147–159, 1984.

R.E. Bellman. On a routing problem. *Quarterly of Applied Mathematics*, 16:87–90, 1958.

R. Bixby, 1996. Personal Communication.

S. Butenko, X. Cheng, D.-Z. Du, and P.M. Pardalos. On the construction of virtual backbone for ad hoc wireless network. In S. Butenko, R. Murphey, and P.M. Pardalos, editors, *Cooperative Control: Models, Applications and Algorithms*, pages 43–54. Kluwer Academic Publishers, 2002.

S. Butenko, X. Cheng, C.A.S. Oliveira, and P.M. Pardalos. A new heuristic for the minimum connected dominating set problem on ad hoc wireless networks. In S. Butenko, R. Murphey, and P.M. Pardalos, editors, *Recent Developments in Cooperative Control and Optimization*, pages 61–73. Kluwer Academic Publishers, 2003.

M. Cagalj, J.-P. Hubaux, and C. Enz. Minimum-energy broadcast in all-wireless networks: Np-completeness and distribution issues. In *Proceedings of the international conference on Mobile computing and networking*, pages 172–182, 2002.

M. Cardei, X. Cheng, X. Cheng, and D.-Z. Du. Connected domination in multihop ad hoc wireless networks. In *Proceedings of the International Conference on Computer Science and Informatics*, pages 251–255, 2002.

M.X. Cheng, J. Sun, M. Min, and D.-Z. Du. Energy-efficient broadcast and multicast routing in ad hoc wireless networks. In *Proceedings of the IEEE International Conference on Performance, Computing, and Communications Conference*, pages 87–94, 2003a.

X. Cheng, X. Huang, D. Li, and D.-Z. Du. A polynomial-time approximation scheme for the minimum-connected dominating set in ad-hoc wireless networks. *Networks*, 42:202–208, 2003b.

S. Chopra and M.R. Rao. The steiner tree problem i: Formulations, compositions and extension of facets. *Mathematical Programming: Series A*, 64:209–229, 1994a.

S. Chopra and M.R. Rao. The steiner tree problem ii: Properties and classes of facets. *Mathematical Programming: Series A*, 64:231–246, 1994b.

V. Chvatal. A greedy heuristic for the set-covering problem. *Mathematics of Operations Research*, 4:233–235, 1979.

B.N. Clark, C.J. Colbourn, and D.S. Johnson. Unit disk graphs. *Discrete Mathematics*, 86:165–177, 1990.

A. Claus and N. Maculan. Une nouvelle formulation du problème de steiner sur un graphe. Technical Report 280, Centre de Recerche sur les Transports, Université de Montréal, 1983.

B. Das and V. Bharghavan. Routing in ad-hoc networks using minimum connected dominating sets. In *Proceedings of the IEEE International Conference on Communications*, pages 376–380, 1997.

B. Das, R. Sivakumar, and V. Bharghavan. Routing in ad-hoc networks using a spine. In *International Conference on Computers and Communications Networks '97*, Las Vegas, NV, September 1997.

S. Deering and D. Cheriton. Multicast routing in datagram internetworks and extended lans. *ACM Transactions on Computer Systems*, 8:85–111, 1990.

E.W. Dijkstra. A note on two problems in connexion with graphs. *Numerical Mathematics*, 1:262–271, 1959.

M. Doar and I. Leslie. How bad is naive multicast routing. In *Proceedings of the Annual Joint Conference of the IEEE Computer and Communications Societies*, pages 82–89, 1993.

C. Duin and A. Volgenant. Reduction tests for the steiner problem in graphs. *Operations Research Letters*, 19:549–567, 1989.

A. Ephremides, J.E. Wieselthier, and D.J. Baker. A design concept of reliable mobile radio networks with frequency hopping signaling. In *Proceedings of the IEEE*, pages 56–73, 1987.

H. Esbensen. Computing near-optimal solutions to the steiner problem in a graph using a genetic algorithm. *Networks*, 26:173–185, 1995.

M.R. Garey and D.S. Johnson. *Computers and Intractibility: A Guide to the Theory of NP-completeness*. W. H. Freeman, San Francisco, 1979.

M. Gendreau, J.-F. Larochelle, and B. Sansò. A tabu search heuristic for the steiner tree problem. *Networks*, 34:162–172, 1998.

M. Gerla and J.T.C. Tsai. Multicluster, mobile, multimedia radio network. *Wireless Networks*, 1:225–265, 1995.

M.X. Goemans and Y.S. Myung. A catalog of steiner tree formulations. *Networks*, 23:19–28, 1993.

L. Gouveia and T.L. Magnanti. Network flow models for designing diameter-constrained minimum-spanning and steiner trees. *Networks*, 41:159–173, 2003.

S. Guha and S. Khuller. Approximation algorithms for connected dominating sets. *Algorithmica*, 20:374–387, 1998.

S.L. Hakimi. Steiner's problem in graphs and its implications. *Networks*, 1:113–133, 1971.

F. Hwang and D. Richards. Steiner tree problems. *Networks*, 22:55–89, 1992.

F. Hwang, D. Richards, and P. Winter. *The Steiner Tree Problem*, volume 53 of *Annals of Discrete Mathematics*. Elsevier, Amsterdam: North-Holland, 1992.

T. Koch and A. Martin. Solving steiner tree problems in graphs to optimality. *Networks*, 32:207–232, 1998.

L. Kou, G. Markowsky, and L. Berman. A fast algorithm for steiner trees. *Acta Informatica*, 15:141–145, 1981.

B. Liang and Z.J. Haas. Virtual backbone generation and maintenance in ad hoc network mobility management. In *Proceedings of the Annual Joint Conference of the IEEE Computer and Communications Societies*, pages 1293–1302, 2000.

N. Maculan. The steiner problem in graphs. *Annals of Discrete Mathematics*, 31: 185–212, 1987.

M.V. Marathe, H. Breu, H.B. Hunt III, S.S. Ravi., and D.J. Rosenkrantz. Simple heuristics for unit graphs. *Networks*, 25:59–68, 1995.

M. Min, F. Wang, D.-Z. Du, and P.M. Pardalos. A reliable virtual backbone scheme in mobile ad-hoc networks. In *Proceedings of the IEEE International Conference on Mobile Ad Hoc and Sensor Systems*, pages 60–69, 2004.

E.M. Royer and C.-K. Toh. A review of current routing protocols for ad hoc mobile wireless networks. *IEEE Personal Communications Magazines*, pages 46–55, April 1999.

P. Sinha, R. Sivakumar, and V. Bharghavan. Enhancing ad hoc routing with dynamic virtual infrastructure. In *Proceedings of the Annual Joint Conference of the IEEE Computer and Communications Societies*, pages 1763–1772, 2001.

R. Sivakumar, B. Das, and V. Bharghavan. The clade vertebrata: Spines and routing in ad hoc networks. In *Proceedings of the IEEE Symposium on Computers and Communications*, pages 599–605, 1998.

R. Sivakumar, P. Sinha, and V. Bharghavan. Cedar: A core-extraction distributed ad hoc routing algorithm. *IEEE Journal on Selected Areas in Communications*, 17: 1454–1465, 1999.

I. Stojmenovic, M. Seddigh, and J. Zunic. Dominating sets and neighbor elimination-based broadcasting algorithms in wireless networks. *IEEE Transactions of Parallel and Distributed Systems*, 12:14–25, 2001.

Q. Sun and H. Langendoerfer. Efficient multicast routing for delay-sensitive applications. In *Proceedings of the Second Workshop on Protocols for Multimedia Systems(PROMS'95)*, pages 452–458, 1995.

H. Takahashi and A. Matsuyama. An approximate solution for the steiner problem for graphs. *Mathematica Japonica*, 24:573–577, 1980.

Y.-C. Tseng, , S.-Y. Ni, Y.-S. Chen, and J.-P. Sheu. The broadcast storm problem in a mobile ad hoc network. *Wireless Networks*, 8:153–167, 2002.

S. Voss. Steiner's problem in graphs: Heuristic methods. *Discrete Applied Mathematics*, 40:45–72, 1992.

S. Voss and K. Gutenschwager. A chunking based genetic algorithm for the steiner tree problem in graphs. In P.M. Pardalos and D.-Z. Du, editors, *Network Design: Connectivity and Facilities Location*, volume 40 of *DIMACS Series on Discrete Mathematics and Theoretical Computer Science*, pages 335–355. American Mathematical Society, 1998.

D. Wall. *Mechanisms for Broadcast and Selective Broadcast*. PhD thesis, Stanford University, 1980.

P.-J. Wan, K. M. Alzoubi, and O. Frieder. Distributed construction of connected dominating set in wireless ad hoc networks. *Mobile Networks and Applications*, 9: 141–149, 2004a.

P.-J. Wan, G. Calinescu, X.-Y. Li, and O. Frieder. Minimum-energy broadcasting in static ad hoc wireless networks. *Wireless Networks*, 8:607–617, 2002.

P.-J. Wan, G. Calinescu, and C.-W. Yi. Minimum-power multicast routing in static ad hoc wireless networks. *IEEE/ACM Transactions on Networking*, 12:507–514, 2004b.

J.E. Wieselthier, G.D. Nguyen, and A. Ephremides. Energy-efficient broadcast and multicast trees in wireless networks. *Mobile Networks and Applications*, 7:481–492, 2002.

P. Winter. Steiner problems in networks: Survey. *Networks*, 17:129–167, 1987.

R.T. Wong. A dual ascent approach for steiner tree problems on a directed graph. *Mathematical Programming*, 28:271–287, 1984.

J. Wu and H. Li. On calculating connected dominating set for efficient routing in ad hoc wireless networks. In *Proceedings of the international workshop on Discrete algorithms and methods for mobile computing and communications*, pages 7–14, 1999.

32 OPTIMIZATION PROBLEMS AND MODELS FOR PLANNING CELLULAR NETWORKS

Edoardo Amaldi[1], Antonio Capone[1],
Federico Malucelli[1], and Carlo Mannino[2]

[1]Dipartimento di Elettronica e Informazione
Politecnico di Milano
20133 Milano, Italy
amaldi@elet.polimi.it
capone@elet.polimi.it
malucell@elet.polimi.it

[2]Dipartimento di Informatica e Sistemistica
Università di Roma "La Sapienza"
00185 Roma, Italy
mannino@dis.uniroma1.it

Abstract: During the last decade the tremendous success of mobile phone systems has triggered considerable technological advances as well as the investigation of mathematical models and optimization algorithms to support planning and management decisions. In this chapter, we give an overview of some of the most significant optimization problems arising in planning second and third generation cellular networks, we describe the main corresponding mathematical models, and we briefly mention some of the computational approaches that have been devised to tackle them. For second generation systems (GSM), the planning problem can be subdivided into two distinct subproblems: coverage planning, in which the antennas are located so as to maximize service coverage, and capacity planning, in which frequencies are assigned to the antennas so as to maximize a measure of the overall quality of the received signals. For third generation systems (UMTS) network planning is even more challenging, since, due to the peculiarities of the radio interface, coverage and capacity issues must be simultaneously addressed.

Keywords: Wireless communications, cellular networks, coverage, capacity, location problems, frequency assignment.

32.1 INTRODUCTION

Wireless communications, and in particular mobile phone systems, have rapidly pervaded everyday life. This success has triggered considerable technological advances as well as the investigation of mathematical models and optimization algorithms to support planning and management decisions. The main contribution of optimization in this field is to improve the way the scarce resources (e.g., transmission band, antennas) are used, and to enhance the service quality (e.g., bandwidth, transmission delay).

When planning and managing a cellular system, a number of aspects must be considered, including traffic estimation, signal propagation, antenna positioning, capacity allocation, transmission scheduling, power and interference control. Most of these aspects gives rise to interesting and challenging optimization problems which must account for the peculiarities of the specific network technology.

In this chapter, we summarize the most significant optimization problems arising in planning a cellular network, we describe the related mathematical models and mention some of the computational approaches that have been devised to tackle them. After recalling the main technological features and the most relevant telecommunication aspects (Section 32.2), we focus on second generation cellular systems (Sections 32.3 and 32.4). Since the corresponding planning problem is very challenging computationally, it is usually decomposed into two distinct phases: coverage planning (e.g., antennas location and transmission power selection) and capacity planning. Interference clearly plays a crucial role in the latter phase, in which frequencies have to be assigned to transmitters, and a wide class of mathematical models and of elegant solution methods have been proposed. Due to the peculiarities of air interface of third generation cellular systems, a two-phase approach is no longer appropriate: coverage planning and capacity planning have to be simultaneously addressed (Section 32.5). Different optimization models and algorithms are thus under investigation.

Although the management of cellular networks, in particular third generation ones, gives also rise to interesting optimization problems such as code assignment and packet scheduling, we do not consider this class of problems here and we refer for instance to Minn and Siu (2000), Agnetis et al. (2003), and Dell'Amico et al. (2004), and the references therein.

32.2 CELLULAR TECHNOLOGIES

Mobile phone systems provide telecommunication services by means of a set of base stations (BSs) which can handle radio connections with mobile stations (MSs) within their service area (Walke, 2001). Such an area, called *cell*, is the set of points in which the intensity of the signal received from the BS under consideration is higher than that received from the other BSs. The received power level depends on the transmitted power, on the attenuation effects of signal propagation from source to destination (path loss due to distance, multi-path effect, shadowing due to obstacles, etc.) as well as on the antenna characteristics and configuration parameters such as maximum emission power, height, orientation and diagram (Parsons, 1996). As a result, cells can have different shapes and sizes depending on BSs location and configuration parameters as well as on propagation.

When users move in the service area crossing cell boundaries, service continuity is guaranteed by handover procedures. During handovers, a connection is usually switched from a BS to a new one (hard-handover). In some cases, simultaneous connections with two or more BSs can be used to improve efficiency.

In order to allow many simultaneous connections between BSs and MSs, the radio band available for transmissions is divided into radio channels by means of a multiple access technique. In most of second generation systems (such as GSM and DAMPS) the radio band is first divided into carriers at different frequencies using FDMA (Frequency Division Multiple Access) and then on each carrier a few radio channels are created using TDMA (Time Division Multiple Access) (Walke, 2001). With bidirectional connections, a pair of channels on different carriers is used for transmissions from BS to MS (downlink) and from MS to BS (uplink), according to the FDD (Frequency Division Duplexing) scheme. BSs can use multiple frequencies by means of a set of transceivers (TRX).

Unfortunately, the number of radio channels obtained in this way (several hundreds in second generation systems) are not enough to serve the large population of mobile service users. In order to increase the capacity of the system the radio channels must be reused in different cells. This generates interference that can affect the quality of the received signals. However, due to the capture effect, if the ratio between the received power and the interference (sum of received powers from interfering transmissions), referred to as SIR (Signal-to-Interference Ratio), is greater than a capture threshold, SIR_{min}, the signal can be correctly decoded.

In order to guarantee that such a condition is satisfied during all system operations the assignment of radio channels to BSs must be carefully planned. Obviously, the denser the channel reuse, the higher the number of channels available per cell. Therefore, the channel assignment determines system capacity. Since usually BSs of second generation systems are not synchronized, radio channels within the same carrier cannot be assigned independently and only carriers are considered by the reuse scheme. For these reasons the process of assigning channels to cells is usually referred to as capacity planning or frequency planning. Although the main source of interference derives from transmissions on the same frequency (carrier), transmissions on adjacent frequencies may also cause interference due to partial spectrum overlap and should be taken into account.

Planning a mobile system involves selecting the locations in which to install the BSs, setting their configuration parameters (emission power, antenna height, tilt, azimuth, etc.), and assigning frequencies so as to cover the service area and to guarantee enough capacity to each cell. Due to the problem complexity, a two-phase approach is commonly adopted for second generation systems. First coverage is planned so as to guarantee that a sufficient signal level is received in the whole service area from at least one BS. Then available frequencies are assigned to BSs considering SIR constraints and capacity requirements.

Second generation cellular systems were devised mainly for the phone and low rate data services. With third generation systems new multimedia and high speed data services have been introduced. These systems, as UMTS (Holma and Toskala, 2000) and CDMA2000 (Karim et al., 2002), are based on W-CDMA (Wideband Code Division Multiple Access) and prior to transmission, signals are spread over a wide band by using special codes. Spreading codes used for signals transmitted by the same station (e.g., a BS in the downlink) are mutually orthogonal, while those used for signals emitted by different stations (base or mobile) can be considered as pseudo-random. In an ideal environment, the de-spreading process performed at the receiving end can completely avoid the interference of orthogonal signals and reduce that of the others by the spreading factor (SF), which is the ratio between the spread signal rate and the user rate. In wireless environments, due to multipath propagation, the interference among orthogonal signals cannot be completely avoided and the SIR is given by:

$$SIR = SF \frac{P_{received}}{\alpha I_{in} + I_{out} + \eta}, \quad (32.1)$$

where $P_{received}$ is the received power of the signal, I_{in} is the total interference due to the signals transmitted by the same BS (intra-cell interference), I_{out} that due to signals of the other BSs (inter-cell interference), α is the orthogonality loss factor ($0 \leq \alpha \leq 1$), and η the thermal noise power. In the uplink case, no orthogonality must be accounted for and $\alpha = 1$, while in the downlink usually $\alpha \ll 1$.

The SIR level of each connection depends on the received powers of the relevant signal and of the interfering signals. These, in turn, depend on the emitted powers and the attenuation of the radio links between the sources and destinations. A power control (PC) mechanism is in charge of dynamically adjusting the emitted power according to the propagation conditions so as to reduce interference and guarantee quality. With a SIR-based PC mechanism each emitted power is adjusted through a closed-loop control procedure so that the SIR of the corresponding connection is equal to a target value SIR_{tar}, with $SIR_{tar} \geq SIR_{min}$ (Grandhi et al., 1995).

For third generation systems, a two-phase planning approach is not appropriate because in CDMA systems the bandwidth is shared by all transmissions and no frequency assignment is strictly required. The network capacity depends on the actual interference levels which determine the achievable SIR values. As these values depend in turn on traffic distribution, as well as on BSs location and configuration, coverage and capacity must be jointly planned.

32.3 COVERAGE PLANNING

The general *Coverage Problem* can be described as follows. Given an area where the service has to be guaranteed, determine where to locate the BSs and select their configurations so that each point (or each user) in the service area receives a sufficiently high signal. Since the cost associated to each BS may depend on its location and configuration, a typical goal is that of minimizing the total antenna installation cost while guaranteeing service coverage.

In the literature, the coverage problem has been addressed according to two main types of approaches. In the first one, the problem is considered from a continuous optimization point of view. A specified number of k BSs can be installed in any location of the space to be covered, possibly avoiding some forbidden areas, and antenna coordinates are the continuous variables of the problem. Sometimes also other parameters, such as transmission powers and antenna orientations, can be considered as variables. The crucial element of this type of approach is the propagation prediction model used to estimate the signal intensity in each point of the coverage area. The coverage area is usually subdivided into a grid of pixels, and for each pixel the amount of traffic is assumed to be known. The signal path loss from transmitter j to the center of pixel i is estimated according to a function $g_i(x_j, y_j, z_j)$ that depends on the transmitter coordinates x_j, y_j, z_j, the distance and the obstacles between the transmitter and the pixel. In the literature, many prediction models have been proposed, from the simple Okumura-Hata formulas (Hata, 1980) to the more sophisticated ray tracing techniques (Parsons, 1996). The objective function of the coverage problem is usually a combination of average and maximum-minimum signal intensity in each pixel or other measures of Quality of Service. If this objective function is denoted by $f(\mathbf{x}, \mathbf{y}, \mathbf{z})$, where $\mathbf{x}, \mathbf{y}, \mathbf{z}$ are the vector coordinates of the k BSs, the coverage problem is simply stated as follows:

$$\max \quad f(\mathbf{x}, \mathbf{y}, \mathbf{z})$$
$$0 \leq x_j \leq h_1, 0 \leq y_j \leq h_2, 0 \leq z_j \leq h_3 \quad j = 1, \ldots, k,$$

where the coverage area is the hyper-rectangle with sides h_1, h_2 and h_3.

Although these problems have simple box constraints, the very involved path loss functions, which cannot always be defined analytically, make them beyond the reach of classical location theory methods (Francis et al., 1992). Global optimization techniques were thus adapted to tackle them, as for example in Sherali et al. (1996) where an indoor optimal location problem is considered.

The alternative approach to the coverage problem is based on discrete mathematical programming models. A set of *test points* (TPs) representing the users are identified in the service area. Each TP can be considered as a traffic centroid where a given amount of traffic (usually expressed in Erlang) is requested (Tutschku et al., 1996). Instead of allowing the location of BSs in any position, a set of *candidate sites* (CSs) where BSs can be installed is identified. Even though parameters such as maximum emission power, antenna height, tilt and azimuth are inherently continuous, the antenna configurations can be discretized by only considering a subset of possible values. Since we can evaluate (or even measure in the field) the signal propagation between any pair of TP and CS for a BS with any given antenna configuration, the subset of TPs covered by a sufficiently strong signal is assumed to be known for a BS installed in any CS and with any possible configuration. The coverage problem then amounts to an extension of the classical minimum cost set covering problem, as discussed for instance in Mathar and Niessen (2000).

Let $S = \{1, \ldots, m\}$ denote the set of CSs. For each $j \in S$, let the set K_j index all the possible configurations of the BS that can be installed in CS j. Since the installation cost may vary with the BS configuration (e.g., its maximum emission power, or the antenna diagram), an installation cost c_{jk} is associated with each pair of CS j and BS configuration k, $j \in S$, $k \in K_j$. Let $I = \{1, \ldots, n\}$ denote the set of test points.

The propagation information is summarized in the attenuation matrix G. Let g_{ijk}, $0 < g_{ijk} \leq 1$, be the attenuation factor of the radio link between test point i, $i \in I$, and a BS installed in j, $j \in S$, with configuration $k \in K_j$.

From the attenuation matrix G, we can derive a 0-1 incidence matrix containing the coverage information that is needed to describe the BS location and configuration problem. The coefficients for each triple TP i, BS j and configuration k are defined as follows:

$$a_{ijk} = \begin{cases} 1 & \text{if the signal of a BS installed in CS } j \text{ with configuration } k \text{ is} \\ & \text{sufficient to cover TP } i \\ 0 & \text{otherwise.} \end{cases}$$

Introducing the following binary variable for every pair of candidate site j and BS configuration k:

$$y_{jk} = \begin{cases} 1 & \text{if a BS with configuration } k \text{ is installed in CS } j \\ 0 & \text{otherwise,} \end{cases}$$

the problem of covering all the test points at minimum cost can be formulated as:

$$\min \quad \sum_{j \in S} \sum_{k \in K_j} c_{jk} y_{jk}$$

$$\sum_{j \in S} \sum_{k \in K_j} a_{ijk} y_{jk} \geq 1 \quad \forall i \in I \tag{32.2}$$

$$\sum_{k \in K_j} y_{jk} \leq 1 \quad \forall j \in S \tag{32.3}$$

$$y_{jk} \in \{0,1\} \quad \forall j \in S, \forall k \in K_j. \tag{32.4}$$

Constraints (32.2) ensure that all TPs are within the service range of at least one BS, and constraints (32.3) state that in each CS at most one configuration is selected for the base station.

This problem can be solved by adapting the algorithms for set covering, see Ceria et al. (1997). In practice, however, the covering requirement is often a "soft constraint" and the problem actually involves a trade-off between coverage and installation cost. In this case, constraints (32.2) are modified by introducing for each $i \in I$, an explicit variable z_i which is equal to 1 if TP i is covered and 0 otherwise. The resulting model, which falls within the class of maximum coverage problems, is then:

$$\max \quad \lambda \sum_{i \in I} z_i - \sum_{j \in S} \sum_{k \in K_j} c_{jk} y_{jk}$$

$$\sum_{j \in S} \sum_{k \in K_j} a_{ijk} y_{jk} \geq z_i \quad \forall i \in I \tag{32.5}$$

$$\sum_{k \in K_j} y_{jk} \leq 1 \quad \forall j \in S \tag{32.6}$$

$$y_{jk} \in \{0,1\} \quad \forall j \in S, \forall k \in K_j \tag{32.7}$$

$$z_i \in \{0,1\} \quad \forall i \in I, \tag{32.8}$$

where $\lambda > 0$ is a suitable trade-off parameter which allows to express both objectives in homogeneous economic terms. This problem can be efficiently solved by using, for instance, GRASP heuristics (Resende, 1998).

Note that these two discrete models do not account for the interference between cells or the overlaps between them, which are very important to deal with handover, i.e., the possibility for a moving user to remain connected to the network while moving from one cell to another. In Amaldi et al. (2005b), for instance, the classical set covering and maximum coverage problems are extended to consider overlaps in the case of Wireless Local Area Network (WLAN) design by introducing suitable non linear objective functions.

The influence of BS locations on the "shape" of the cells can be captured by introducing variables that explicitly assign test points to base stations. These binary variables are defined for every pair of TP i and CS j such that there exist at least one configuration of the BS in CS j that allows them to communicate:

$$x_{ij} = \begin{cases} 1 & \text{if TP } i \text{ is assigned to BS } j \\ 0 & \text{otherwise.} \end{cases}$$

If $K(i, j)$ denotes the set of the available configurations for the BS in CS j that allow the connection with TP i, the formulation of the full coverage problem becomes:

$$\min \quad \sum_{j \in S} \sum_{k \in K_j} c_{jk} y_{jk}$$

$$\sum_{j \in S} x_{ij} = 1 \quad \forall i \in I \tag{32.9}$$

$$\sum_{k \in K_j} y_{jk} \leq 1 \quad \forall j \in S \tag{32.10}$$

$$x_{ij} \leq \sum_{k \in K(i,j)} y_{jk} \quad \forall i \in I, \forall j \in S : K(i,j) \neq \emptyset \tag{32.11}$$

$$y_{jk} \in \{0,1\} \quad \forall j \in S, \forall k \in K_j \tag{32.12}$$

$$x_{ij} \in \{0,1\} \quad \forall i \in I, \forall j \in S : K(i,j) \neq \emptyset. \tag{32.13}$$

The crucial constraints of the above model are (32.11) stating that a TP can be assigned to a BS only if the configuration of this BS allows that connection. In order to account for the maximum coverage variant, the equality constraints (32.9) expressing full coverage can be transformed into inequalities ($\leq$) and a suitable term proportional to the number of connected TPs can be added to the objective function. Note that, in this case, a cell is defined by the set of TPs assigned to it and hence is not predefined by the incidence matrix, as in the models based on set covering.

This basic model can be amended by adding constraints related with the actual "shape" of the resulting cells. Some authors proposed a quality measure of a cell C given by

$$\frac{\sqrt{area(C)}}{boundary(C)}$$

(Zimmerman et al., 2000). Since this quantity, which is maximized when the cell is circular, is difficult to deal with in a mathematical model and in a solution algorithm, other rules based on connectivity are usually preferred. For instance, each TP is assigned to the "closest" (in terms of signal strength) activated BS. One way to express this constraint for a given TP i is to consider all the pairs of BSs and configurations that would allow connection with i and sort them in decreasing order of signal strength. Let $\{(j_1,k_1),(j_2,k_2),\ldots,(j_L,k_L)\}$ be the ordered set of BS-configuration pairs, the constraints enforcing the assignment of TP i to the closest activated BS are:

$$y_{j_\ell k_\ell} + \sum_{h=\ell+1}^{L} x_{ij_h} \leq 1 \quad \ell = 1,\ldots,L-1. \tag{32.14}$$

According to the above constraints, if a BS is activated in configuration ℓ, then TP i cannot be connected to a less convenient BS. In some settings, including second generation systems, capacity constraints can also be introduced so as to limit the number of TPs assigned to the same BS (Mathar and Niessen, 2000).

These location-allocation models can be solved efficiently with known exact and heuristic methods. See, for instance, Ghosh and Harche (1993).

32.4 CAPACITY PLANNING

In second generation systems, after the coverage planning phase, available carriers (frequencies) must be assigned to BSs in order to provide them with enough capacity to serve traffic demands. Frequencies are identified by integers (denoting their relative position in the spectrum) in the set $F = \{1, 2, \ldots, f_{max}\}$. To efficiently exploit available radio spectrum, frequencies are reused in the network. However, frequency reuse may deteriorate the received signal quality. The level of such a deterioration depends on the *SIR* and can be somehow controlled by a suitable assignment of transmission frequencies.

The *Frequency Assignment Problem* (FAP) is the problem of assigning a frequency to each transmitter of a wireless network so that (a measure of) the quality of the received signals is maximized. Depending on spectrum size, objectives and specific technological constraints, the FAP may assume very different forms.

It is worth noting that the FAP is probably the telecommunication application which has attracted the largest attention in the Operations Research literature, both for its practical relevance and for its immediate relation to classical combinatorial optimization problems. This wide production has been analyzed and organized in a number of surveys and books (Aardal et al., 2003; Eisenblätter et al., 2002; Jaumard et al., 1999; Leese and Hurley, 2002; Murphey et al., 1999; Roberts, 1991). In this section, we give an overview of the most significant contributions to the models and algorithms for the FAP and provide a historical perspective.

In the 1970s, frequencies were licensed by governments in units; since operators had to pay for each single frequency, they tried to minimize the total number of frequencies required by non-interfering configurations. It was soon understood (Metzger, 1970) that this corresponds to solving a suitable version of the well-known graph coloring problem, or some generalization of it. This immediate correspondence is obtained by associating a graph $G = (V,E)$ with network R, defining V to be the set of antennas (TRXs) of R, and by letting $\{i,j\} \in E$ if and only if TRX i and TRX j interfere. Any coloring of the vertices of G (i.e., assignment of colors such that adjacent vertices have different colors) is then an assignment of frequencies to R such that no mutual interfering TRXs receive the same frequency. A minimum cardinality coloring of G is a minimum cardinality non-interfering frequency assignment of R. Early solution approaches to the graph coloring model of the FAP were proposed in Metzger (1970) and Zoellner and Beall (1977): both papers discuss simple greedy heuristics.

The graph coloring model assumes that distinct frequencies do not interfere: this is not always the case. In general, a frequency h interferes with all frequencies $g \in [h-\delta, h+\delta]$ where, δ depends upon channel bandwidth, type of transmission and power of signals. To overcome this drawback, in the early 80's a number of generalizations of the graph coloring problem were proposed (Gamst and Rave, 1982; Hale, 1980).

In the new offspring of works an instance of FAP is represented by a complete, undirected, weighted graph $G = (V, E, \delta)$, where $\delta \in Z_+^{|E|}$ is the *distance vector*, and δ_{uv} is the (minimum) admissible distance (in channel units) between a frequency f_u assigned to u and a frequency f_v assigned to v. The problem of defining a free-interference plan becomes now that of finding an assignment f such that $|f_v - f_u| \geq \delta_{uv}$ for all $\{u, v\} \in E$ and the difference between the largest and the smallest frequency, denoted by $Span(f)$, is minimized. The Span of a minimum Span assignment of $G = (V, E, \delta)$ is a graph invariant denoted by $Span(G)$. Clearly, when $\delta_{uv} \in \{0, 1\}$, then $Span(G) = \chi(G)$ (the minimum cardinality of a coloring of G). This version of FAP, called MS-FAP, has been widely addressed in the literature; most of solution approaches are heuristic methods, ranging from the simple generalizations of classical graph coloring heuristics (Costa, 1993; Borndörfer et al., 1998) like DSATUR (Brélaz, 1979) to specific implementations of local search such as, for instance, simulated annealing (Costa, 1993), genetic algorithm (Valenzuela et al., 1998), and tabu search (Hao and Perrier, 1999).

It was soon remarked (Box, 1978) that, given an assignment f, one can build one (or more) total ordering $\sigma(V)$ on the vertices V by letting $\sigma(u) < \sigma(v)$ whenever $f_u < f_v$ (ties are broken arbitrarily). Similarly, given an ordering $\sigma(V) = (u_1, \ldots, u_n)$, one can immediately associate an assignment $f \in \{1, \ldots, f_{max}\}^{|V|}$ by letting $f_{u_1} = 1$ and

$$f_{u_j} = max_{i<j} f_{u_i} + \delta_{u_i u_j}$$

for $j = 2, \ldots, n$ (with f_{max} large enough). This observation led to the definition of a number of models and algorithms based on the correspondence between orderings of V and acyclic orientations of the edges of G. A nice relation between frequency assignments and Hamiltonian paths of G was first pointed out by Raychaudhuri (1994). First observe that, with any ordering $\sigma(V) = (u_1, \ldots, u_n)$, a Hamiltonian path

$$P(\sigma) = \{(u_1, u_2), (u_2, u_3), \ldots, (u_{n-1}, u_n)\}$$

of G is uniquely associated (where G is a complete graph and $\delta_{uv} \geq 0$ for $\{u, v\} \in E$). If δ_{uv} is interpreted as the length of $\{u, v\} \in E$, then the length $\delta(H^*)$ of a minimum length Hamiltonian path H^* of G is a lower bound on $Span(G)$. In fact, let f^* be an optimum assignment and let $\sigma^* = (u_1^*, \ldots, u_n^*)$ be one possible corresponding ordering. Then

$$f_{u_j^*}^* = max_{i<j} f_{u_i^*}^* + \delta_{u_i^* u_j^*} \geq f_{u_{j-1}^*}^* + \delta_{u_{j-1}^* u_j^*}$$

for $j = 2, \ldots, n$. Finally

$$Span(f^*) = f^*_{u_n} - f^*_{u_1} = \sum_{j=2}^{n} f^*_{u_j} - f^*_{u_{j-1}} \geq \sum_{j=2}^{n} \delta_{u^*_{j-1} u^*_j} = \delta(P(\sigma^*)) \geq \delta(H^*).$$

In order to satisfy an increasing traffic demand, the number of TRXs installed in a same BS had to be increased; in fact, for practical instances, the number of frequencies to be assigned to a BS ranges from 1 to several units (up to ten or more). In the graph model introduced so far, every TRX is represented by a vertex v of G. However, as for their interferential behavior, TRXs belonging to the same BS are indistinguishable. This yields to a more compact representation $G = (V, E, \delta, m)$, where each vertex v of G corresponds to a BS, while $m \in R^{|V|}$ is a *multiplicity* vector with m_v denoting, for each $v \in V$, the number of frequencies to be assigned to v. The FAP is then the problem of assigning m_v frequencies to every vertex of G so that (i) *every* frequency f_v assigned to v and *every* frequency f_u assigned to u satisfy $|f_v - f_u| \geq \delta_{uv}$ and (ii) the difference between the largest and the smallest frequencies assigned (*Span*) is minimized. This version of the FAP was very popular up to the 1990s; the most famous set of benchmark instances, the Philadelphia instances (FAP website, 2000), are actually instances of this problem. Whilst the majority of solution methods use demand multiplicity in a straightforward way by simply splitting each (BS) vertex v into m_v "twin" vertices (the TRXs), a few models (and algorithms) account for it explicitly, see e.g. Janssen and Wentzell (2000) and Jaumard et al. (2002). The introduction of multiplicity led to a natural extension of the classical Hamiltonian paths to the more general *m*-walks, i.e., walks "passing" precisely m_v times through every vertex $v \in V$ (a Hamiltonian path is an *m*-walk with $m = 1_{|V|}$). In fact, one can show (Avenali et al., 2002) that the length of a minimum length *m*-walk is a lower bound on the span of any (multiple) frequency assignment. These observations led to the definition of suitable integer linear programming (ILP) formulations, with variables associated with edges and walks of the interference graph. These formulations can be exploited to produce lower bounds (Allen et al., 1999; Janssen and Wentzell, 2000) or to provide the basis for an effective Branch-and-Cut solution algorithm (Avenali et al., 2002).

In the late 1980s and in the 1990s the number of subscribers to GSM operators grew to be very large and the available band rapidly became inadequate to allow for interference-free frequency plans: in addition to this, frequencies were now sold by national regulators in blocks rather than in single units. The objective of planning shifted then from minimizing the number of frequencies to that of maximizing the quality of service, which in turn corresponds to minimizing (a measure of) the overall interference of the network. This last objective gives rise to the so called *Minimum Interference Frequency Assignment Problem* (MI-FAP) which can be viewed as a generalization of the well-known *max k-cut* problem on edge-weighted graphs. Here, rather than making use of an intermediate graph-based representation of this problem (*interference graph*), we prefer to refer to a standard 0-1 linear programming formulation.

The basic version of the MI-FAP takes only into account *pairwise* interference, i.e., the interference occurring between a couple of interfering TRXs. Interference is measured as the number of unsatisfied requests of connection. Specifically, if v and w are potentially interfering TRXs and f, g two available frequencies (not necessarily distinct), then we associate a penalty p_{vwfg} to represent the interference (cost) generated when v is assigned to f and w is assigned to g. Then the problem becomes that of finding a frequency assignment which minimizes the sum of the penalty costs.

In order to describe a 0-1 linear program for MI-FAP, we introduce a binary variable x_{vf} for every vertex v and available frequency $f \in F$:

$$x_{vf} = \begin{cases} 1 & \text{if frequency } f \in F \text{ is assigned to vertex } v \in V \\ 0 & \text{otherwise.} \end{cases}$$

Since it is easy to see that the contribution to the objective value of the interference between v and w can be expressed as $\sum_{f,g \in F} p_{vwfg} x_{vf} x_{wg}$, the objective function can be written as

$$\min \sum_{\{v,w\} \in E} \sum_{f,g \in F} p_{vwfg} x_{vf} x_{wg}. \tag{32.15}$$

In order to linearize the quadratic terms $x_{vf} x_{wg}$, we define the variables $z_{vwfg} = x_{vf} x_{wg}$ for all $v, w \in V$ and all $f, g \in F$, i.e.,

$$z_{vwfg} = \begin{cases} 1 & \text{if } x_{vf} \cdot x_{wg} = 1 \\ 0 & \text{otherwise.} \end{cases}$$

To enforce z_{vwfg} to be one when $x_{vf} = x_{wg} = 1$, we add the following constraints to the formulation:

$$x_{vf} + x_{wg} \leq 1 + z_{vwfg} \qquad \forall \{v,w\} \in E, \forall f, g \in F. \tag{32.16}$$

By substitution, the quadratic form of the objective function (32.15) becomes the following linear expression:

$$\min \sum_{\{v,w\} \in E} \sum_{f,g \in F} p_{vwfg} z_{vwfg}.$$

Finally, the requirement that $m(v)$ frequencies have to be assigned to each vertex v is modeled by the following *multiplicity constraints*:

$$\sum_{f \in F} x_{vf} = m(v) \qquad \forall v \in V.$$

If only co-channel interference is involved, i.e., $p_{vwfg} = 0$ holds whenever $f \neq g$, then MI-FAP reduces to the *max k-cut problem*: given an edge-weighted graph $G = (V, E, \delta)$, find a partition of V into k classes so that the sum of the weights of crossing edges is maximized. We can solve our special instance of MI-FAP by letting $k = |F|$, solve the max k-cut problem on G and then assign to every vertex in a same class the same frequency from F, while assigning different frequencies to vertices belonging to different classes. Several algorithms exploit this natural correspondence. A special mention deserves the innovative approach proposed by (Eisenblätter, 2002) to compute strong upper bounds for MI-FAP by solving a semidefinite programming relaxation of a suitable ILP formulation of the corresponding *max k−cut* problem.

928   HANDBOOK OF OPTIMIZATION IN TELECOMMUNICATIONS

The MI-FAP model is certainly the version of FAP mostly addressed in the recent literature, mainly for its direct applicability to the solution of large practical instances of GSM network planning problems. Thus, a huge number of solution approaches, both exact and approximate, have been proposed in the last years. Among heuristic approaches, the most successful appear to be variants of Simulated Annealing as reported in the FAPweb benchmark instances section of (FAP website, 2000). Exact methods for MI-FAP include implicit enumeration as well as polyhedral approaches. In particular, the Branch-and-Cut proposed in Aardal et al. (1996) makes use of the ILP formulation above described, which can be strengthen by adding clique inequalities derived from the packing constraints (32.16).

The MI-FAP 0-1 model introduced so far can be exploited as a paradigm for several (constraint or objective based) variations. Indeed, the standard model for the FAP displays in practice a very rich speciation in order to adapt to a multitude of different technological or quality requirements and operator objectives. Two remarkable variants to the standard MI-FAP accounts for the *cumulative interference*, i.e., the (total) noise generated by multiple interfering transmitters on the service area of a target one (Fischetti et al., 2000; Capone and Trubian, 1999). In Capone and Trubian (1999), the interference generated by the BSs is evaluated on the points of a spatial grid and a quality threshold on the *SIR* is considered. Since the resulting problem formulation is quite different from the MI-FAP 0-1 model, in the following we describe the approach adopted in Fischetti et al. (2000) where only the total noise generated on the area covered by each TRX is considered.

If we denote by I_{vwfg} the noise generated by w on v when $f \in F$ is assigned to v and $g \in F$ is assigned to w, then the total noise produced by all neighboring vertices $N(v) = \{w : \{v,w\} \in E\}$ on a frequency f for $v \in V$ is simply computed as

$$\sum_{w \in N(v)} \sum_{g \in F} I_{vwfg} x_{wg}.$$

In order to control cumulative interference, we define for each $v \in V$ a "local" threshold L_v, which is the maximum interference acceptable for v. Then the following constraints are introduced:

$$\sum_{w \in N(v)} \sum_{g \in F} I_{vwfg} x_{wg} x_{vf} \leq L_v \qquad \forall v \in V, \forall f \in F.$$

These are non-linear constraints, but again we can easily linearize them by introducing suitable 0-1 variables. The resulting ILP model is the basis of a Branch-and-Cut algorithm exploited in Fischetti et al. (2000) for the solution of large real-life instances.

32.5 JOINT COVERAGE AND CAPACITY PLANNING

As mentioned in Section 32.2, while the problem of planning a second generation cellular system can be subdivided into a coverage and a capacity planning subproblems, such a two-phase approach is not appropriate for third generation systems that are based on a CDMA radio access scheme. Since the signal quality depends on all the communications in the systems, the critical issues of radio planning and coverage optimization must be tackled jointly.

Due to the many issues that can affect system performance and the huge costs of the service licenses that service operators have to face, there is an acute need for planning tools that help designing, expanding, and configuring third generation systems, like UMTS, in an efficient way. In practice, optimizing BS configuration can often be more critical than BS location since the set of candidate sites may be very limited due to authority constraints on new antenna installation and on electromagnetic pollution in urban areas (Cappelli and Tarricone, 2002).

In the *UMTS network planning problem*, given a set $S = \{1,\ldots,m\}$ of candidate sites, a set $I = \{1,\ldots,n\}$ of test points with the corresponding required number u_i of active connections for each TP i (which is a function of the traffic demand) and the propagation matrix G (providing channel attenuation between CSs and TPs), one has to select a subset of CSs where to install BSs together with their configurations so as to optimize an appropriate objective function which takes into account traffic coverage and installation costs. Due to the peculiarities of the CDMA radio access scheme, the *SIR* constraints and power limits must also be considered.

Most of the early work on planning CDMA systems still relies on classical coverage models and does not appropriately account for interference. For instance, in Calégari et al. (1997) a simple model based on the minimum dominating set problem is considered, while in Lee and Kang (2000) the traffic capacity is also taken into account and the resulting classical capacitated facility location problem is tackled with a Tabu Search algorithm. In Chamaret et al. (1997) a different approach is adopted: in a graph with nodes corresponding to the CSs and edges to pairs of CSs whose BSs would have coverage areas with too much overlap, a maximum independent set of vertices is searched for. In Galota et al. (2001) a simplified model for locating BSs in CDMA-based UMTS networks is considered and a polynomial time approximation scheme is presented. Although intra-cell interference is taken into account, the interference among BSs (inter-cell one) is neglected.

Akl et al. (2001) address the problem of locating a prescribed number k of BSs and of optimizing their configurations in the uplink direction, assuming a power-based PC. The model, which includes pilot signals but assumes that BS locations are continuous decision variables, is tacked by Branch-and-Bound. In Mathar and Schmeink (2001) a discrete mathematical programming model is given for locating up to k BSs, considering the downlink direction and with simplified *SIR*-constraints. A Branch-and-Bound algorithm is used to tackle instances with up to 1100 TPs but very small values of k, namely $k < 8$. Pointers to other work on planning third generation cellular systems can be found in Amaldi et al. (2003a), Amaldi et al. (2005a), and Eisenblätter et al. (2003).

The mixed integer programming (MIP) model we present is based on Amaldi et al. (2005a). The subproblem of locating BSs for the uplink direction is investigated in Amaldi et al. (2003a) with power-based as well as *SIR*-based PC mechanisms. The extensions with BS configurations and with downlink communications are addressed in (Amaldi et al., 2002) and, respectively, in (Amaldi et al., 2003b).

The overall model takes into account signal-quality constraints in both uplink and downlink directions, and assumes a *SIR*-based power control mechanism. BS configurations include, among others, maximum emission power, antenna height, antenna tilt, and sector orientation if directive antennas are co-located in the same CS. Since these BS characteristics have an impact on the attenuation matrix G, g_{ijk}, with $0 < g_{ijk} \leq 1$, denotes the gain factor of the radio link between TP i and a BS installed in CS j with configuration k, where $i \in I$, $j \in S$, $k \in K_j$.

Soft-handover is not considered explicitly, that is, each TP is assigned to at most one BS. Since soft-handover tends to increase *SIR* values, we may be too conservative but this can be compensated for by decreasing the value of SIR_{min}. For the sake of simplicity, we also assume that the number of connections assigned to any BS does not exceed the number of available spreading codes. In uplink, there is a very large number of nonorthogonal codes, while in downlink, where there are at most *SF* orthogonal codes, standard cardinality constraints can be added to the model.

A mixed integer programming formulation of the UMTS network planning problem involves the location variables y_{jk}, with j in S and k in K_j, as well as the assignment variables:

$$x_{ijk} = \begin{cases} 1 & \text{if TP } i \text{ is assigned to a BS in CS } j \text{ with configuration } k \\ 0 & \text{otherwise,} \end{cases}$$

with i in I, j in S and k in K_j.

If we assume a *SIR*-based power control mechanism, we also need for every TP $i \in I$ the continuous variables p_i^{up} to indicate the power emitted by the mobile in TP i towards the BS it is assigned to (uplink direction) and p_i^{dw} to indicate the power received at each TP i from the BS it is assigned to (downlink direction).

To aim at a trade-off between maximizing the total traffic covered and minimizing the total installation costs, the objective function is:

$$\max \lambda \sum_{i \in I} \sum_{j \in J} \sum_{k \in K_j} x_{ijk} - \sum_{j \in J} \sum_{k \in K_j} c_{jk} y_{jk}, \tag{32.17}$$

where $\lambda > 0$ is a trade-off parameter between the two contrasting objectives. Alternatively a budget can be imposed on the total installation costs.

The first three groups of constraints of the formulation ensure the coherence of the location and assignment variables. Every TP i can be assigned to at most one BS:

$$\sum_{j \in S} \sum_{k \in K_j} x_{ijk} \leq 1 \quad \forall i \in I \tag{32.18}$$

and at most one BS configuration can be selected for CS j:

$$\sum_{k \in K_j} y_{jk} \leq 1 \quad \forall j \in S. \tag{32.19}$$

If for a given CS j we have $y_{jk} = 0$ for all $k \in K_j$, no BS is installed in CS j. A TP i can be assigned to a CS j only if a BS with some configuration k, $k \in K_j$, has been installed in j:

$$\sum_{k \in K_j} x_{ijk} \leq \sum_{k \in K_j} y_{jk} \quad \forall i \in I, j \in S. \tag{32.20}$$

In uplink, the power emitted by any mobile terminal at TP i cannot exceed a maximum power P_i^{max-up}:

$$0 \leq p_i^{up} \leq P_i^{max-up} \quad \forall i \in I. \tag{32.21}$$

In downlink, besides the limit on the total power emitted by each BS j, we also consider an upper bound on the power used for every connection:

$$0 \leq p_i^{dw} \leq \sum_{j \in S} \sum_{k \in K_j} P^{max-dw} g_{ijk} x_{ijk} \quad \forall i \in I, \tag{32.22}$$

where P^{max-dw} denotes the maximum power per connection. According to constraints (32.18) at most one of the x_{ijk} variables in (32.22) is equal to 1 and the right-hand side amounts to the maximum power available in downlink. Thus BSs cannot use too much of their power for transmission towards mobiles with bad propagation conditions (Holma and Toskala, 2000).

Let us now turn to the *SIR* constraints, which express the signal quality requirements, and consider first the uplink direction. Given the *SIR*-based PC mechanism, for each triple of BS $j \in S$ with configuration $k \in K_j$ and TP $i \in I$ we have the uplink *SIR* constraint:

$$\frac{p_i^{up} g_{ijk} x_{ijk}}{\sum_{h \in I} u_h p_h^{up} g_{hjk} \sum_{t \in S} \sum_{l \in K_t} x_{htl} - p_i^{up} g_{ijk} + \eta_j^{bs}} = SIR_{tar} x_{ijk}, \tag{32.23}$$

where η_j^{bs} denotes the thermal noise at BS j. For any single connection between a TP i and a BS installed in CS j with configuration k, the numerator of the left-hand-side term corresponds to the power of the relevant signal arriving from TP h at CS j with BS configuration k while the denominator amounts to the total interference due to all other active connections in the system. The triple summation term expresses the total power received at the BS in j with configuration k from all TPs h that are served. Indeed, $p_h^{up} g_{hjk}$ indicates the power received at the BS j from TP h and, according to (32.18), $\sum_{t \in S} \sum_{l \in K_t} x_{htl}$ is equal to 1 if and only if TP h is assigned to a BS, namely is served. The total interference, is the obtained by just subtracting the received power of the relevant signal.

In downlink, for each triple of TP $i \in I$ and BS $j \in S$ with configuration $k \in K_j$, we have the *SIR* constraint:

$$\frac{p_i^{dw} x_{ijk}}{\alpha \left(\sum_{h \in I} u_h g_{ijk} \frac{p_h^{dw}}{g_{hjk}} x_{hjk} - p_i^{dw} \right) + \sum_{\substack{l \in S \\ l \neq j}} \sum_{z \in K_l} \sum_{h \in I} u_h g_{ilz} \frac{p_h^{dw}}{g_{hlz}} x_{hlz} + \eta_i^{mt}}$$

$$= SIR_{tar} x_{ijk}, \tag{32.24}$$

where η_i^{mt} denotes the thermal noise of mobile terminal at TP i. For any single connection between a BS located in CS j with configuration k and a TP i, the numerator of the left-hand-side term corresponds to the power of the relevant signal received at TP i from the BS j (definition of p_i^{dw}) and the denominator amounts to the total interference due to all other active connections in the system. The interpretation of (32.24) is similar to that of (32.23) except for the orthogonality loss factor α in the *SIR* formula (32.1), which is strictly smaller than 1 in downlink.

Thus, constraints (32.23) and (32.24) ensure that if a connection is active between a TP i and a BS j with configuration k (i.e., $x_{ijk} = 1$) then the corresponding *SIR* value is equal to SIR_{tar}.

Note that the resulting model is a mixed integer program with nonlinear *SIR* constraints since they contain products of assignment variables (x_{ijk} and y_{jk}) and power variables (p_i^{up} and p_i^{dw}).

If we assume a power-based PC mechanism instead of a *SIR*-based one, the model can be substantially simplified (Amaldi et al., 2005a). Indeed, the powers p_i^{up} emitted from any TP i in uplink and the powers p_i^{dw} received at any TP i from the BS they are assigned to, are no longer variable. Since all emitted powers are adjusted so as to guarantee a received power of P_{tar}, p_i^{up} and p_i^{dw}, for all $i \in I$, just depend on the value of P_{tar} and on the propagation factor of the corresponding radio links. To obtain the simplified model, we take into account that:

$$p_i^{up} = \sum_{j \in S} \sum_{k \in K_j} \frac{P_{tar}}{g_{ijk}} x_{ijk} \qquad p_i^{dw} = P_{tar}, \qquad (32.25)$$

and in the *SIR* constraints we require that the left-hand-side terms of equations (32.23) and (32.24) are greater or equal to $SIR_{min} \, x_{ijk}$.

The resulting *SIR* constraints, which have the general form

$$\frac{P_{tar}}{(\alpha I_{in} + I_{out} + \eta)} \geq SIR_{min} \, x_{ijk}, \qquad (32.26)$$

where P_{tar} is a constant, and I_{in} and I_{out} are linear functions in the x and y variables, can be linearized as follows:

$$(\alpha I_{in} + I_{out} + \eta) \leq \frac{1}{SIR_{min}} + M_{ijk}(1 - x_{ijk}) \qquad (32.27)$$

for large enough values of the constants M_{ijk}.

The simplified model for locating BSs with fixed configurations considering the uplink direction and a power-based PC mechanism is described in Chapter 20. It is worth pointing out that even the linearization of this simplified model yields integer linear programs that are very challenging computationally. In practice, even small instances of the overall problem with both uplink and downlink directions and *SIR*-based PC mechanism are beyond the reach of state-of-the-art exact methods.

Heuristics for the overall model must also face very high computational requirements. Indeed, just computing the transmitted powers corresponding to any given set of active BSs (**y**) and given assignment of TPs to these active BSs (**x**) involves finding a solution of a large linear system consisting of the *SIR* equations (32.23) and (32.24) while satisfying the power limit constraints (32.21) and (32.22). Fortunately this critical computation, which has to be carried out after every variable modification, can be speeded-up substantially by adapting (Amaldi et al., 2005a) a recently proposed iterative method (Berg, 2002). An alternative method to compute power levels and determine the feasibility of a given assignment of TPs to BSs is proposed in Catrein et al. (2004).

A Tabu Search procedure, which starts from an initial solution obtained with GRASP and explores the solution space through suitable switches of the y variable values, provides good quality solutions of relevant-size instances (with up to 200 CSs and 1000 TPs) in reasonable computing time (Amaldi et al., 2005a;c). A two-stage approach can be adopted: solutions of a simpler model, which considers a power-based PC mechanism and only the uplink direction, are exploited as good initial solutions for the overall uplink and downlink model with a SIR-based PC mechanism. Since in the uplink direction the model with power-based PC is a quite accurate approximation of the model with SIR-based PC, the insight gained from solving the former model help drastically reducing the computing times for tackling the overall model without significantly affecting the solution quality. However, the computational requirements to tackle larger instances grow very fast with the number of discretization values of the configuration parameters.

The above MIP model and algorithm can be easily extended to include pilot signals and to deal explicitly with directive antennas, that is BSs with several sectors (Amaldi et al., 2005a;c).

For additional modeling aspects such as hand-over and the limited number of codes the reader is referred to Eisenblätter et al. (2003) and to the other references related to the Momentum IST-EU project (IST-EC website, 1999). The overall model described in Eisenblätter et al. (2003) considers both network cost and service quality in the objective function and includes uplink and downlink SIR constraints, SIR-based power control, pilot signal, antenna configurations and a limited number of codes assigned to each connection. Since more detailed models are less tractable even with heuristics, the huge resulting mathematical programs are beyond the reach of state-of-the-art MIP solvers except for very small-size instances. For realistic data scenarios see for instance IST-EC website (1999).

The actual challenge is to find a reasonable trade-off between an accurate description of the UMTS network planning problem and a computationally tractable model. This is a critical issue if we want to consider a representative set of traffic scenarios (snapshots).

Eisenblätter and Geerdes (2005) make a step in this direction by trying to derive reliable average transmitted powers from average traffic intensities instead of individual snapshots. By considering average coupling matrices aimed at capturing the essential coverage and cell coupling properties of the radio network, the network design problem is casted in terms of designing a "good" average coupling matrix.

32.6 CONCLUDING REMARKS

In this chapter, we gave an overview of the most relevant optimization problems arising in planning cellular systems and we described some optimization approaches that have been developed to tackle them. These problems have stimulated interesting lines of research which not only improved the way these wireless systems are planned but also led to significant advances in discrete optimization, as for instance in the case of frequency assignment. The peculiarities and increasing complexity of new telecommunication systems shifts attention towards new interesting challenges. A relevant challenge is the development of an integrated and computationally efficient approach to coverage and capacity planning for third and fourth generation cellular systems. Extensions of the classical set covering and maximum coverage problems which explicitly account for interference (Amaldi et al., 2005b) and signal quality, deserve special attention and are leading to new research threads.

Acknowledgments

The authors were partially supported through grant 2003-014039 "Optimization, simulation and complexity in the design and management of telecommunication networks" from Ministero dell'Istruzione, dell'Università e della Ricerca (MIUR), Italy.

Bibliography

K. Aardal, S.P.M. van Hoesel, A. Koster, C. Mannino, and A. Sassano. Models and solution techniques for frequency assignment problems. *4OR*, 1(4):261–317, 2003.

K.I. Aardal, A. Hipolito, C.P.M. van Hoesel, and B. Jansen. A branch-and-cut algorithm for the frequency assignment problem. Technical Report 96/011, Maastricht University, 1996.

A. Agnetis, G. Brogi, G. Ciaschetti, P. Detti, and G. Giambene. Optimal packet scheduling in UTRA-TDD. *IEEE Communication Letters*, 52(2):229–242, 2003.

R.G. Akl, M.V. Hegde, M. Naraghi-Pour, and P.S. Min. Multicell CDMA network design. *IEEE Trans. on Vehicular Technology*, 50(3):711–722, 2001.

S. M. Allen, D. H. Smith, and S. Hurley. Lower bounding techniques for frequency assignment. *Discrete Mathematics*, 197/198:41–52, 1999.

E. Amaldi, P. Belotti, A. Capone, and F. Malucelli. Optimizing base station location and configuration in UMTS networks. *Annals of Operations Research*, 2005a. To appear, preliminary version in Proceedings of the 1st International Network Optimization Conference (INOC), Evry/Paris, October, 2003.

E. Amaldi, S. Bosio, F. Malucelli, and D. Yuan. On a new class of set covering problems arising in WLAN design. In *Proceedings of the 2nd International Network Optimization Conference (INOC)*, volume A, pages 470–477, 2005b.

E. Amaldi, A. Capone, and F. Malucelli. Planning UMTS base station location: Optimization models with power control and algorithms. *IEEE Transactions on Wireless Communications*, 2(5):939–952, 2003a.

E. Amaldi, A. Capone, and F. Malucelli. Radio planning and coverage optimization of 3G cellular networks. *Wireless Networks*, 2005c. Submitted for publication.

E. Amaldi, A. Capone, F. Malucelli, and F. Signori. UMTS radio planning: Optimizing base station configuration. In *Proceedings of IEEE VTC Fall 2002*, volume 2, pages 768–772, 2002.

E. Amaldi, A. Capone, F. Malucelli, and F. Signori. Optimization models and algorithms for downlink UMTS radio planning. In *Proceedings of IEEE Wireless Communications and Networking Conference (WCNC'03)*, volume 2, pages 827–831, 2003b.

A. Avenali, C. Mannino, and A. Sassano. Minimizing the span of d-walks to compute optimum frequency assignments. *Mathematical Programming*, 91(2):357–374, 2002.

M. Berg. Radio resource management in bunched personal communication systems. Technical report, Kungl Tekniska Hogskolan (KTH), Stockholm, Sweden, 2002.

R. Borndörfer, A. Eisenblätter, M. Grötschel, and A. Martin. Frequency assignment in cellular phone networks. *Annals of Operations Research*, 76:73–93, 1998.

F. Box. A heuristic technique for assigning frequencies to mobile radio nets. *IEEE Transactions on Vehicular Technology*, 27:57–74, 1978.

D. Brélaz. New methods to color the vertices of a graph. *Communications of the ACM*, 22:251–256, 1979.

P. Calégari, F. Guidec, P. Kuonen, and D. Wagner. Genetic approach to radio network optimization for mobile systems. In *Proceedings of IEEE VTC 97*, pages 755–759, 1997.

A. Capone and M. Trubian. Channel assignment problem in cellular systems: A new model and tabu search algorithm. *IEEE Trans. on Vehicular Technology*, 48(4): 1252–1260, 1999.

M. Cappelli and L. Tarricone. A bioelectromagnetic overview of the universal mobile telecommunication system (UMTS). In *Proceedings of IEEE MTT-S International Microwave Symposium Digest*, volume 3, pages 2265–2268, 2002.

D. Catrein, L. Imhof, and R. Mathar. Power control, capacity, and duality of up- and downlink in cellular CDMA systems. *IEEE Transactions on Communications*, 52 (10):1777–1785, 2004.

S. Ceria, P. Nobili, and A. Sassano. Set covering problem. In M. Dell'Amico, F. Maffioli, and S. Martello, editors, *Annotated Bibliographies in Combinatorial Optimization*, chapter 23, pages 415–428. John Wiley and Sons, 1997.

B. Chamaret, S. Josselin, P. Kuonen, M. Pizarroso, B. Salas-Manzanedo, S. Ubeda, and D. Wagner. Radio network optimization with maximum independent set search. In *Proceedings of IEEE VTC 97*, pages 770–774, 1997.

D. Costa. On the use of some known methods for t-colourings of graphs. *Annals of Operations Research*, 41:343–358, 1993.

M. Dell'Amico, M. Merani, and F. Maffioli. A tree partitioning dynamic policy for OVSF codes assignment in wideband CDMA. *IEEE Transactions in Wireless Communications*, 3(4):1013–1017, 2004.

A. Eisenblätter. The semidefinite relaxation of the k-partition polytope is strong. In *Integer Programming and Combinatorial Optimization: 9th International IPCO Conference*, volume 2337 of *Lecture Notes in Computer Science*, pages 273–290. Springer-Verlag, 2002.

A. Eisenblätter, A. Fugenschuh, T. Koch, A. Koster, A. Martin, T. Pfender, O. Wegel, and R. Wessäly. Modelling feasible network configurations for UMTS. In *Telecommunications network design and management (eds. G. Anandalingam and S. Raghavan)*, pages 1–22. Kluwer Publishers, 2003.

A. Eisenblätter and Hans-Florian Geerdes. A novel view on cell coverage and coupling for UMTS radio network evaluation and design. In *Proceedings of the 2nd International Network Optimization Conference (INOC)*, volume B, pages 307–314, 2005.

A. Eisenblätter, M. Grötschel, and A. Koster. Frequency assignment and ramifications of coloring. *Discussiones Mathematicae Graph Theory*, 22:51–88, 2002.

FAP website. A website devoted to frequency assignment, 2000. http://fap.zib.de, accessed April 20,2005.

M. Fischetti, C. Lepschy, G. Minerva, G. Romanin-Jacur, and E. Toto. Frequency assignment in mobile radio systems using branch-and-cut techniques. *European Journal of Operational Research*, 123:241–255, 2000.

R.L. Francis, L.F. McGinnis Jr., and J.A. White. *Facility Layout and Location: An Analytical Approach*. Prentice-Hall, Englewood Cliffs, NJ, 1992.

M. Galota, C. Glasser, S. Reith, and H. Vollmer. A polynomial-time approximation scheme for base station positioning in UMTS networks. In *Proceedings of the 5th International Workshop on Discrete Algorithms and Methods for Mobile Computing and Communications (DIAL-M)*, pages 52–59. ACM, 2001.

A. Gamst and W. Rave. On frequency assignment in mobile automatic telephone systems. In *Proceedings of GLOBECOM'82*, pages 309–315. IEEE Press, 1982.

A. Ghosh and F. Harche. Location-allocation models in the private sector: progress, problems and prospects. *Location Science*, 1:81–106, 1993.

S.A. Grandhi, J. Zander, and R. Yates. Constrained power control. *Journal of Wireless Personal Communications*, 1(4):257–270, 1995.

W.K. Hale. Frequency assignment: Theory and applications. *Proceedings of the IEEE*, 68:1497–1514, 1980.

J-K. Hao and L. Perrier. Tabu search for the frequency assignment problem in cellular radio networks. Technical report, LGI2P, EMA-EERIE, Parc Scientifique Georges Besse, Nimes, France, 1999.

M. Hata. Empirical formula for propagation loss in land mobile radio service. *IEEE Trans. on Vehicular Technology*, 29:317–325, 1980.

H. Holma and A. Toskala. *WCDMA for UMTS*. John Wiley and Sons, 2000.

IST-EC website. IST-EC MOMENTUM project: Models and simulations for network planning and control of UMTS, 1999. http://momentum.zib.de, accessed April 20,2005.

J. Janssen and T. Wentzell. Lower bounds from tile covers for the channel assignment problem. Technical Report G-2000-09, GERAD, HEC, Montreal, Canada, 2000.

B. Jaumard, O. Marcotte, and C. Meyer. Mathematical models and exact methods for channel assignment in cellular networks. *in B. Sans and P. Soriano, editors, Telecommunications Network Planning, Norwell, MA: Kluwer*, pages 239–256, 1999.

B. Jaumard, O. Marcotte, C. Meyer, and T. Vovor. Comparison of column generation models for channel assignment in cellular networks. *Discrete Applied Mathematics*, 118:299–322, 2002.

M. R. Karim, M. Sarraf, and V. B. Lawrence. *W-CDMA and CDMA2000 for 3G Mobile Networks*. McGraw Hill Professional, 2002.

C.Y. Lee and H.G. Kang. Cell planning with capacity expansion in mobile communications: A tabu search approach. *IEEE Trans. on Vehicular Technology*, 49(5): 1678–1690, 2000.

R. Leese and S. Hurley. *Methods and Algorithms for Radio Channel Assignment*. Oxford Lecture Series in Mathematics and its Applications. Oxford University Press, Oxford, United Kingdom, 2002.

R. Mathar and T. Niessen. Optimum positioning of base stations for cellular radio networks. *Wireless Networks*, 6(4):421–428, 2000.

R. Mathar and M. Schmeink. Optimal base station positioning and channel assignment for 3G mobile networks by integer programming. *Annals of Operations Research*, 107:225–236, 2001.

B.H. Metzger. Spectrum management technique, 1970. Presentation at 38th National ORSA meeting (Detroit, MI).

T. Minn and K.-Y. Siu. Dynamic assignment of orthogonal variable-spreading-factor codes in W-CDMA. *IEEE Journal on Selected Areas in Communications*, 18(8): 1429–1440, 2000.

R.A. Murphey, P.M. Pardalos, and M.G.C. Resende. Frequency assignment problems. In D.-Z. Du and P.M. Pardalos, editors, *Handbook of combinatorial optimization*, volume Supplement Volume A, pages 60–67. Kluwer Academic Publishers, 1999.

D. Parsons. *The mobile radio propagation channel*. John Wiley and Sons, 1996.

A. Raychaudhuri. Further results on t-coloring and frequency assignment problems. *SIAM Journal on Discrete Mathematics*, 7:605–613, 1994.

M.G.C. Resende. Computing approximate solutions of the maximum covering problem with GRASP. *Journal of Heuristics*, 4:161–171, 1998.

F.S. Roberts. t-colorings of graphs: Recent results and open problems. *Discrete Mathematics*, 93:229–245, 1991.

H.D. Sherali, C.M. Pendyala, and T.S. Rappaport. Optimal location of transmitters for micro-cellular radio communication system design. *IEEE Journal on Selected Areas in Telecommunications*, 14(4):662–673, 1996.

K. Tutschku, N. Gerlich, and P. Tran-Gia. An integrated approach to cellular network planning. In *Proceedings of the 7th International Network Planning Symposium - Networks 96*, 1996.

C. Valenzuela, S. Hurley, and D.H. Smith. A permutation based genetic algorithm for minimum span frequency assignment. In *Parallel Problem Solving from Nature – PPSN V: 5th International Conference*, volume 1498 of *Lecture Notes in Computer Science*, pages 907–916. Springer Verlag, 1998.

B. H. Walke. *Mobile Radio Networks*. John Wiley and Sons, 2001.

J. Zimmerman, R. Höns, and H. Mühlenbein. The antenna placement problem for mobile radio networks: An evolutionary approach. In *8th International Conference on Telecommunications Systems*, pages 358–366, 2000.

J.A. Zoellner and C.L. Beall. A breakthrough in spectrum conserving frequency assignment technology. *IEEE Transactions on Electromagnetic Compatiblity*, 19: 313–319, 1977.

33 LOAD BALANCING IN CELLULAR WIRELESS NETWORKS

Sem Borst[1], Georg Hampel[1], Iraj Saniee[1], and Phil Whiting[1]

[1]Bell Laboratories, Lucent Technologies
Murray Hill, NJ 07974 USA
sem@research.bell-labs.com
ghampel@research.bell-labs.com
iis@research.bell-labs.com
pwhiting@research.bell-labs.com

Abstract: We present a linear-programming approach for dynamic load balancing in CDMA networks. The linear program characterizes the minimum achievable base station load for a given configuration of mobiles at each time interval, and gives a useful benchmark for the potential gains from optimizing the power assignment. The solution of the linear program also offers valuable insight to the qualitative properties of the optimal power allocation. In particular, the structure of the optimal assignment reflects the critical notion that power allocation should not just be based on signal strength values but also on shadow prices which arise from load considerations. We develop a dual-ascent scheme for solving the linear program in a (mostly) distributed fashion with low communication overhead. Extensive numerical experiments demonstrate that there is scope for significant gains from balancing base station loads in typical scenarios.
Keywords: CDMA networks, cellular networks, uplink power control, Yates-Huang-Hanly fixed point, load balancing, min-max optimization, dual linear program, shadow pricing, distributed algorithm, stochastic approximation.

33.1 INTRODUCTION

A fundamental characteristic of wireless networks is the occurrence of vast load variations, both across space and time. The spatial variations manifest themselves in a broad spectrum of traffic densities, ranging from hot spots and dense urban networks to sparser networks in suburban areas and remote rural regions with scattered coverage. Similarly, the temporal fluctuations span a wide range of time scales. On a quite

Figure 33.1 Cell breathing as a dynamic mechanism to balance load across multiple cells. (Each dot represents a small cluster of mobiles.)

slow time scale of months to years, the traffic volume tends to increase due to network expansion, subscriber growth, and the increase in usage per subscriber. On a time scale of hours to days, the traffic load shows steep variations according to day-of week and hour-of-day patterns. On a minute-to-hour level, the number of mobiles varies considerably because of the intrinsic randomness in call arrival epochs and holding times. Even for a fixed population of mobiles, the power requirements may hugely vary on a second-to-minute time scale as a result of mobility in the presence of slow fading and distance-related attenuation. Finally, on a fast sub-second time scale, the power requirements wildly fluctuate due to fast fading, even for nearly stationary mobiles.

It is worth observing that some of the variations in space and time are intertwined. For example, as the load shifts across a network over the course of a day, not only does the total call volume vary drastically, but also does the spatial distribution.

The large variations in load call for some sort of network adaptation in order to optimize the performance. On the slowest time scale of months to years, such adaptation may be accomplished by splitting cells or adding carriers so as to deal with static load imbalances. In faster time scales, a further degree of adaptation may be achieved by cell breathing, i.e., shrinking heavily-loaded cells and expanding lightly-loaded cells so as to balance the load, as depicted in Figure 33.1.

While the concept of cell breathing is well-established (Goto et al., 2002; Hanly, 1995; Kim and Kang, 2002; Yates and Huang, 1995), the actual mechanics of it are not as well-developed. One potential mechanism is adjustment of antenna parameters such as azimuth or tilt. Although powerful, this approach is constrained to operate on an infrequent basis as physical antenna adjustment is mostly a manual, and thus an expensive and relatively slow, process. As a result, only systematic load imbalances can be addressed, and it is not easy to respond to temporal load variations. In addition, such a static, off-line approach relies on knowledge of the traffic profile and the propagation environment, which are both intrinsically uncertain.

An alternative mechanism for cell breathing is via reassignment of mobiles to base stations and a corresponding rearrangement of transmit powers in response to load variations. This allows the adaptation to occur on a much faster time scale based on direct observation or measurements of the actual network state. Such a dynamic, on-line, approach opens up the possibility to react to fluctuations on a faster time scale with a larger dynamic range, and thus achieve bigger performance gains. Besides the faster dynamics, a further advantage of measurement-based approaches is that they avoid the need for inherently inaccurate estimates of traffic statistics and propagation conditions. One potential drawback of cell breathing, however, is that in fast time scale cellular re-arrangements, intermittent coverage holes could be created thus resulting in high admission blocking and low quality of service.

Another mechanism for measurement-based rapid reconfiguration that goes beyond cell breathing involves (i) keeping the statically optimized cell boundaries (through use of one set of pilot tones[1]), and (ii) using measurement-based re-allocations (via a second set of pilot tones) for rapid and dynamic reaction to local mobile load variations. The advantage of this approach is that a statically optimized configuration ensures pre-assigned coverage, while the dynamic reconfiguration via re-allocations could help optimize some performance metric of the cellular structure in a fast dynamic setting. A key issue that arises in the above context concerns the minimum achievable 'load' for a given mobile configuration. In this chapter, we will specifically interpret 'load' to mean the maximum amplifier load across a given collection of base stations, and formulate a linear program (LP) which provides a lower bound for the base station load in terms of the path loss coefficients, SNIR (Signal-to-Noise-and-Interference-ratio) requirements and other possible system parameters.

The solution of the linear program gives a quantitative sense of the scope for improvement from optimizing the power assignment. In addition, it furnishes a detailed understanding of the structural properties of optimal power allocations. Specifically, the form of the optimal arrangement corroborates the important realization that power allocation should not only take into account signal strength measures but also load conditions. Also, optimal power allocation involves a modest number of soft hand-off legs. An additional important property of the optimal power allocation is that the optimal active sets of serving base stations can be indicated through pilot tones.

We present a dual-ascent scheme for solving the linear program which is largely distributed in nature and requires limited exchange of information among base stations. The numerical experiments that we conduct suggest that there is a potential for substantial improvements from balancing the base station loads in typical circumstances. We observe that the the linear program, as presented, renders a lower bound for a particular 'snapshot', and does not explicitly consider the evolution of the system over time. This evolution is addressed by solving the linear program repeatedly over time while taking into account the changing conditions (path losses) of the mobiles. To what degree the lower bound can actually be approached in a dynamic setting thus strongly depends on a variety of factors, in particular the extent to which existing control algorithms can be modified to accommodate the LP optimization. Moreover, enhancing or even just measuring the performance of wireless networks is a complex multi-faceted problem, as it involves consideration of multiple interacting resources as well as multiple, often conflicting, objectives. Thus, the ultimate performance enhancement from balancing base station loads remains a subject of further investigation.

The remainder of this chapter is organized as follows. In Section 33.2 we present a detailed model description and formulate a linear program characterizing the min-

---

[1]Pilot tone here refers to an auxiliary signal broadcast by the base station for the purpose of base station identification and acquisition of the downlink spreading code.

imum achievable (maximum) base station load for a given configuration of mobiles. We develop a dual-ascent scheme for solving the linear program and establish its convergence properties in Sections 33.3, and 33.4. In Section 33.5 we consider a toy example with just two cells in order to illuminate some of the key insights in the simplest possible setting. In Section 33.6 we discuss numerical experiments that we performed to validate the results and assess the potential gains from dynamic load balancing. In Section 33.7 we sketch a possible implementation of the linear-programming approach in a dynamic setting. We conclude by examining several extensions of the LP model in Section 33.8.

33.2 LINEAR-PROGRAMMING FORMULATION

33.2.1 Model description

We consider a system with M mobile users served by C base stations. Denote by α_{mc} the (forward link) path loss from base station c to mobile m. Let $\gamma > 0$ be the Signal-to-Noise-and-Interference-Ratio (SNIR) requirement and let $\eta > 0$ be the thermal noise. For convenience, we assume the values of γ and η to be identical for all users, but the results easily extend to the case where the various users have heterogeneous SINR (Signal to Interference plus Noise Ratio) requirements or experience different thermal noise levels. Denote by P_{mc} the amount of transmit power assigned by base station n to mobile m. Define

$$L_c := \sum_{m=1}^{M} P_{mc}$$

as the load at base station c.

We wish to minimize the maximum load

$$L := \max_{c=1,\ldots,C} L_c$$

across all base stations while satisfying the SINR requirements of all users. The minimum achievable load will be denoted by L^*, when it exists. A corresponding (not necessarily unique) optimal power assignment will be denoted by $P^* = (P^*_{mc})$. We will first focus on the 'constant-interference' case where the interference experienced by all mobiles is assumed to be independent of the actual transmit powers assigned by the base station, e.g., maximum possible interference. In that case, the SINR requirements reduce to constraints on the total received signal strength. Next, we will investigate the 'variable-interference' case where the interference experienced by the various users *does* depend on the transmit powers selected by the base stations.

33.2.2 Constant-interference case

We first examine the case where the interference experienced by mobile m is assumed to be some fixed value I_m, independent of the actual transmit powers assigned by the base stations. In that case, a lower bound for the maximum base station load may be obtained from the following linear program:

$$\min \quad L \tag{33.1}$$

$$\text{subject to:} \quad L \geq L_c = \sum_{m=1}^{M} P_{mc} \quad c = 1, \ldots, C \tag{33.2}$$

$$\sum_{c=1}^{C} \alpha_{mc} P_{mc} \geq \zeta_m \quad m = 1, \ldots, M \tag{33.3}$$

$$P_{mc} \geq 0 \quad m = 1, \ldots, M, c = 1, \ldots, C,$$

with $\zeta_m := \gamma(\eta + I_m)$.

Remark 33.1 *The above linear program corresponds to the continuous relaxation of the 'unrelated' parallel machine scheduling problem (Lenstra et al., 1987). Specifically, the problem amounts to minimizing the make span (maximum completion time) of M jobs on C parallel machines, interpreting ζ_m as the size of job m, α_{mc} as the processing speed of job m when run on machine c (so ζ_m/α_{mc} is the processing requirement of job m when run on machine c), and P_{mc} as the processing time assigned by machine m to job c.*

For later use, we also state the dual version of the linear program (33.1)–(33.3):

$$\max \quad \sum_{m=1}^{M} \zeta_m \mu_m \tag{33.4}$$

$$\text{subject to} \quad \sum_{c=1}^{C} \lambda_c \leq 1 \tag{33.5}$$

$$\alpha_{mc} \mu_m \leq \lambda_c \quad m = 1, \ldots, M, c = 1, \ldots, C \tag{33.6}$$

$$\lambda_c, \mu_m \geq 0 \quad m = 1, \ldots, M, c = 1, \ldots, C,$$

with λ_c and μ_m representing the dual variables or shadow prices associated with the constraints (33.2) and (33.3), respectively. We note that the above dual problem lends itself more readily to a distributed solution as explained in Remark 33.2 and detailed in Section 33.3.

Note that the primal problem will have a feasible solution (and the dual problem will have a finite solution) as long as

$$\min_{c=1,\ldots,C} \alpha_{mc} > 0$$

for all $m = 1, \ldots, M$. Further observe that $L^* > L_c^*$ cannot occur unless $\alpha_{mc} = 0$ for all m with $\mu_m^* > 0$. This in fact follows from the complementary slackness conditions, which imply that $L^* > L_c^*$ forces $\lambda_c^* = 0$. In particular, when $\alpha_{mc} > 0$ for all $m = 1, \ldots, M$, $c = 1, \ldots, C$, we must have $L_c^* = L$ for all $c = 1, \ldots, C$.

Since optimality demands

$$\sum_{c=1}^{C} \lambda_c^* = 1 \quad \text{and} \quad \mu_m^* = \min_{c=1,\ldots,C} \lambda_c^*/\alpha_{mc},$$

the latter variables may be eliminated, and the dual problem may be more succinctly cast as:

$$\max \quad \sum_{m=1}^{M} \zeta_m \min_{c=1,\ldots,C} \lambda_c / \alpha_{mc} \quad (33.7)$$

$$\text{subject to:} \quad \sum_{c=1}^{C} \lambda_c = 1 \quad (33.8)$$

$$\lambda_c \geq 0 \quad c = 1, \ldots, C.$$

Remark 33.2 *The dual problem has a natural interpretation as a greedy assignment problem in terms of the shadow prices, λ_cs. To see this, consider (4)-(6) or (7)-(8). Each mobile needs to get (one unit of) power from one (or more) of the base stations. The price of power at base station c is λ_c and the cost of power that mobile m pays for, when acquiring power from base station c, is $\beta_{mc} := \zeta_m \lambda_c / \alpha_{mc}$. There are two 'opposing' players. Player 1, the minimizer (the set of mobiles), aims to assign mobiles to base stations so as to minimize the total payment for the given prices. The cost minimization can be performed simply by selecting for each mobile the cheapest supplier (or suppliers when there are ties) to satisfy its demand, i.e.,*

$$c_m^* := \arg \min_{c=1,\ldots,C} \lambda_c / \alpha_{mc}.$$

Player 2, the maximizer (the set of base stations), aims to set prices so as to maximize the total payments, given any payment minimization strategy of Player 1. This interpretation leads to a distributed dual solution scheme whereby prices are set by the base stations, then mobiles determine their payment minimizing suppliers independently of each other but resulting in specific payment (load) to each base station. This leads to the recomputation of the shadow prices by the base stations, followed by reassignment of mobiles, which is repeated until a fixed point is reached.

In the spirit of the above max-min interpretation, the dual problem may equivalently be stated as:

$$\max \quad V(\lambda_1, \ldots, \lambda_C) \quad (33.9)$$

$$\text{subject to:} \quad \sum_{c=1}^{C} \lambda_c = 1 \quad (33.10)$$

$$\lambda_c \geq 0 \quad c = 1, \ldots, C,$$

where

$$V(\lambda_1, \ldots, \lambda_C) := \gamma \sum_{m=1}^{M} (\eta + I_m) \min_{c=1,\ldots,C} \lambda_c / \alpha_{mc}.$$

The latter quantity represents the minimum value of a weighted combination of the base station loads and may be obtained as the optimum value of the following linear program:

$$\min \quad \sum_{c=1}^{C} \lambda_c L_c \qquad (33.11)$$

$$\text{subject to:} \quad L_c = \sum_{m=1}^{M} P_{mc} \qquad c = 1,\ldots,C \qquad (33.12)$$

$$\sum_{c=1}^{C} \alpha_{mc} P_{mc} \geq \zeta_m \qquad m = 1,\ldots,M \qquad (33.13)$$

$$P_{mc} \geq 0 \qquad m = 1,\ldots,M, c = 1,\ldots,C.$$

In conclusion, the problem of minimizing the maximum base station load is equivalent to finding a set of coefficients (prices) for which the minimum weighted sum of the base station loads is maximized. We will revisit this interpretation in Section 33.3.

Remark 33.3 *An interesting related problem is the maximization of the minimum (weighted) throughput across all users in a data system (Bejerano and Bhatia, 2004). Denote by $H_m(Q_{m1},\ldots,Q_{mC})$ the (weighted) throughput received by mobile m as a function of the transmit powers (or time shares) $Q_{m1},\ldots,Q_{mC}$ assigned to it. The throughput optimization problem may then be formulated as:*

$$\max \quad T$$

$$\text{subject to:} \quad T \leq T_m = H_m(Q_{m1},\ldots,Q_{mC}) \qquad c = 1,\ldots,C \qquad (33.14)$$

$$\sum_{c=1}^{C} Q_{mc} \leq 1 \qquad m = 1,\ldots,M \qquad (33.15)$$

$$Q_{mc} \geq 0 \qquad m = 1,\ldots,M, c = 1,\ldots,C.$$

Now suppose that the (weighted) throughput obtained by mobile m only depends on the transmit powers (or time shares) $Q_{m1},\ldots,Q_{mC}$ through a weighted linear combination, i.e.,

$$H_m(Q_{m1},\ldots,Q_{mC}) = H\left(\sum_{c=1}^{C} w_{mc} Q_{mc}\right)$$

with $H(\cdot)$ some increasing function common to all users. A specific example of the above form is

$$H_m(Q_{m1},\ldots,Q_{mC}) = w_m \sum_{c=1}^{C} R_{mc} Q_{mc},$$

with w_m the weight factor associated with user m and R_{mc} the throughput received by user m when served by base station c at full power all the time. Then the maximum common throughput is $H(T^*)$, where T^* represents the optimum value of the above problem with the constraint (33.15) replaced by

$$T \leq \sum_{c=1}^{C} w_{mc} Q_{mc}. \qquad (33.16)$$

The above problem is then equivalent to (33.1)–(33.3) with $\alpha_{mc} \equiv w_{mc}$ in the sense that the optimal solutions are related as $Q^*_{mc} = P^*_{mc}/L^*$. Note Q^*_{mc} as defined above is indeed a feasible solution of (33.15), (33.16) with objective value $1/L^*$. Conversely, for any feasible solution Q_{mc} of (33.15), (33.16), $P_{mc} = Q_{mc}/T$ is a feasible solution of (33.1)–(33.3) with objective value $1/T$. Hence, the optimal values of the two problems must be equal.

33.2.3 Variable-interference case

We now turn to the case where the interference incurred by the various mobiles depends on the transmit powers selected by the base stations. The maximum base station load is then bounded from below by the optimal value of the following linear program:

$$\min \quad L$$

$$\text{subject to:} \quad L \geq L_c = \sum_{m=1}^{M} P_{mc} \qquad c = 1, \ldots, C \qquad (33.17)$$

$$\sum_{c=1}^{C} \alpha_{mc} P_{mc} \geq \gamma \left[\eta + \sum_{c=1}^{C} \alpha_{mc} L_c \right] \qquad m = 1, \ldots, M \quad (33.18)$$

$$P_{mc} \geq 0 \qquad m = 1, \ldots, M, c = 1, \ldots, C.$$

In the above formulation we still neglected minimum and maximum power constraints nor did we account for orthogonality factors. We will refine the model to include these features in Section 33.8.

For later use, we also formulate the dual version of the linear program (33.16)–(33.18):

$$\max \quad \gamma \sum_{m=1}^{M} \mu_m \qquad (33.19)$$

$$\text{subject to:} \quad \sum_{c=1}^{C} \lambda_c \leq 1 \qquad (33.20)$$

$$\alpha_{mc}\mu_m \leq \gamma \sum_{m'=1}^{M} \alpha_{m'c}\mu_{m'} + \lambda_c, \quad m = 1, \ldots, M, \qquad (33.21)$$
$$c = 1, \ldots, C,$$

$$\lambda_c, \mu_m \geq 0, m = 1, \ldots, M, c = 1, \ldots, C, \qquad (33.22)$$

with λ_c and μ_m representing the dual variables associated with the constraints (33.17) and (33.18), respectively.

Since dual optimality entails $\sum_{c=1}^{C} \lambda_c = 1$ and

$$\mu_m = \min_{c=1,\ldots,C} \left[\gamma \sum_{m'=1}^{M} \alpha_{m'c}\mu_{m'} + \lambda_c \right] / \alpha_{mc}, \qquad (33.23)$$

or equivalently,

$$\max_{c=1,\ldots,C} \mu_m \alpha_{mc} / \left[\sum_{m'=1}^{M} \alpha_{m'c} \mu_{m'} + \lambda_c / \gamma \right] = \gamma, \qquad (33.24)$$

(33.20) can be imposed with equality while (33.21) may be replaced by either of the latter two constraints.

Note that the primal problem may not have a feasible solution (and the dual problem may have an infinite solution) in general, even when $\alpha_{mc} > 0$ for all $m = 1,\ldots,M$, $c = 1,\ldots,C$. For example, no feasible primal solution exists when $\gamma > 1$. A necessary and sufficient condition for the existence of a feasible solution may be expressed in terms of the Perron-Frobenius eigenvalue of a matrix with relative path loss coefficients, see for instance Foschini and Milzanic (1993).

When the optimal solution of the dual exists and is finite we denote an optimum λ with λ^*. As in the constant-interference case, the dual problem may equivalently be phrased as:

$$\max \quad V(\lambda_1,\ldots,\lambda_C) \qquad (33.25)$$

$$\text{subject to:} \quad \sum_{c=1}^{C} \lambda_c = 1 \qquad (33.26)$$

$$\lambda_c \geq 0 \qquad c = 1,\ldots,C,$$

where $V(\lambda_1,\ldots,\lambda_C)$ represents the optimum value of the following linear program:

$$\max \quad \gamma \sum_{m=1}^{M} \mu_m \qquad (33.27)$$

$$\text{subject to:} \quad \alpha_{mc} \mu_m \leq \gamma \sum_{m'=1}^{M} \alpha_{m'c} \mu_{m'} + \lambda_c, \quad m = 1,\ldots,M, c = 1,\ldots,C \qquad (33.28)$$

$$\mu_m \geq 0 \qquad m = 1,\ldots,M.$$

As noted earlier, the constraints (33.28) may equivalently be replaced by either (33.23) or (33.24).

In order to see that $V(\lambda_1,\ldots,\lambda_C)$ may be interpreted as the minimum value of a weighted combination of the base station loads, simply note that the above problem is the dual version of the following linear program:

$$\min \quad \sum_{c=1}^{C} \lambda_c L_c \qquad (33.29)$$

$$\text{subject to:} \quad L_c = \sum_{m=1}^{M} P_{mc} \qquad c = 1,\ldots,C \qquad (33.30)$$

$$\sum_{c=1}^{C} \alpha_{mc} P_{mc} \geq \gamma \left[\eta + \sum_{c=1}^{C} \alpha_{mc} L_c \right] \qquad m = 1,\ldots,M \qquad (33.31)$$

$$P_{mc} \geq 0 \qquad m = 1,\ldots,M, c = 1,\ldots,C.$$

Thus, as before, the problem of minimizing the maximum base station load is equivalent to determining a set of coefficients for which the minimum weighted sum of the base station loads is maximized.

Finally, let us consider the related problem of *minimizing*

$$\sum_{m=1}^{M} \mu_m$$

subject to the constraints (33.24). Assuming reciprocity between downlink and uplink path losses, this may be interpreted an *uplink* power control problem: minimize the sum of the transmission powers used by the mobiles while meeting the SINR requirement γ at the base stations with noise level λ_c/γ at base station c. The latter problem may be solved using the Yates-Huang-Hanly algorithm (Hanly, 1995; Yates, 1995). Obviously, the optimal solution to this minimization problem will produce a smaller objective value than the optimal solution to the original maximization problem (33.19)–(33.21). In addition, for given values of λ_c, the optimum value is obtained by solving the fixed point for μ_m using (33.24). This may be verified by observing that a solution to the maximization problem that satisfies the constraints (33.24) meets the stopping criterion of the Yates-Huang-Hanly algorithm for uplink power control, and must hence be an optimal solution to the maximization problem as well. Of course, this is merely an indirect manifestation of the fact that the solution to the equations (33.24), if it exists, is unique. In conclusion, the optimal dual variables μ_m^* in (33.27)–(33.28) may be computed as the minimum uplink transmit powers using the Yates-Huang-Hanly algorithm. A similar relationship between downlink and uplink power control problems has been described in Rashid-Farrokhi et al. (1997).

33.2.4 Observations

We conclude the section with some additional remarks.

- Both in the constant-interference case and the variable-interference case, we observed that the problem of minimizing

$$L = \max_{c=1,\ldots,C} L_c$$

is equivalent to finding a vector $\lambda = (\lambda_1, \ldots, \lambda_C)$ with

$$\sum_{c=1}^{C} \lambda_c \leq 1$$

that maximizes

$$V(\lambda_1, \ldots, \lambda_C) = \min \sum_{c=1}^{C} \lambda_c L_c.$$

This may also be deduced directly. Note that

$$L \geq \sum_{c=1}^{C} \lambda_c L \geq \sum_{c=1}^{C} \lambda_c L_c.$$

The reverse inequality holds for the optimal L^* and $\lambda^* = (\lambda_1^*,\ldots,\lambda_C^*)$ since

$$\sum_{c=1}^{C} \lambda_c^* = 1$$

while the complementary slackness conditions imply $\lambda_c^*(L^* - L_c^*) = 0$.

- It follows from linear-programming theory that both in the constant-interference case and the variable-interference case there exists an optimal solution with at most $M + C - 1$ non-zero variables. In other words, there are only $C - 1$ additional 'legs' beyond the minimum number of legs that is required to connect all M users. When $M > C$ as will typically be the case, this amounts to an extremely limited degree of 'soft hand-off'. Observe however that the soft hand-off here purely serves to balance the base station loads. In practice, a further rationale for maintaining soft hand-off, is to achieve a certain extent of macro diversity and robustness in the presence of fading, which is not captured by the above linear program. We will discuss this issue further in Section 33.8.

- Linear-programming theory also implies that any feasible solution to the dual problem provides a lower bound to the value of the primal problem. Thus, both in the constant-interference case and in the variable-interference case, the value of $V(\lambda_1,\ldots,\lambda_C)$ provides a lower bound to the value of the primal problem for any given vector λ with

$$\sum_{c=1}^{C} \lambda_c = 1.$$

This is especially useful as the value of $V(\lambda_1,\ldots,\lambda_C)$ is fairly easy to obtain for any given vector λ, a fact that will be leveraged by the dual-ascent scheme described in the next section.

- The complementary slackness conditions state that in the constant-interference case the optimal primal and dual solutions satisfy the following relations:

$$P_{mc}^*(\lambda_c^* - \alpha_{mc}\mu_m^*) = 0$$

for all $m = 1,\ldots,M$, $c = 1,\ldots,C$. In other words, mobile m is only assigned power by base station c ($P_{mc}^* > 0$) when $\lambda_c^* = \alpha_{mc}\mu_m^*$. Since

$$\mu_m^* = \min_{c=1,\ldots,C} \lambda_c^*/\alpha_{mc},$$

we find that mobile m only receives power from base station(s) p with

$$\lambda_p^*/\alpha_{mp} = \min_{c=1,\ldots,C} \lambda_c^*/\alpha_{mc}.$$

Thus, mobiles should only connect to the base station(s) with the strongest received pilot strength, where the pilot power at base station c is set inversely proportional to λ_c^*. In other words, the optimal so-called 'active set' of base stations can be configured via suitably chosen pilot signals. It is worth emphasizing that 'pilot' here refers to an auxiliary tone broadcast by the base stations for the purpose of power allocation, which should be properly coordinated with, and also separated from, the default pilot signal that determines coverage.

In the variable-interference case the optimal primal and dual solutions satisfy the following slightly more complicated constraints:

$$P_{mc}^*(\lambda_c^* + \gamma \sum_{m'=1}^{M} \alpha_{m'c}\mu_{m'}^* - \alpha_{mc}\mu_m^*) = 0,$$

for all $m = 1,\ldots,M$, $c = 1,\ldots,C$. Because

$$\mu_m^* = \min_{c=1,\ldots,C}\left[\gamma \sum_{m'=1}^{M} \alpha_{m'c}\mu_{m'}^* + \lambda_c^*\right]/\alpha_{mc},$$

we conclude that mobile m only receives power from base station(s) p with

$$\left[\gamma \sum_{m'=1}^{M} \alpha_{m'p}\mu_{m'}^* + \lambda_p^*\right]/\alpha_{mp} = \min_{c=1,\ldots,C}\left[\gamma \sum_{m'=1}^{M} \alpha_{m'c}\mu_{m'}^* + \lambda_c^*\right]/\alpha_{mc}.$$

Thus, mobiles should only connect to the base station(s) with the strongest received pilot strength, where the pilot power of base station n is set inversely proportional to

$$\gamma \sum_{m'=1}^{M} \alpha_{m'c}\mu_{m'}^* + \lambda_c^*.$$

As in the previous case, the optimal active set can be determined through adequately selected pilot signals.

33.3 DUAL-ASCENT SCHEME

In principle, each of the linear programs presented in the previous section may be readily solved using standard routines. However, these algorithms generally involve centralized computation and require global knowledge of all path loss coefficients. In this section we specify an Arrow-Hurwicz type dual-ascent scheme (Hurwicz et al., 1958) which, while slower to converge, is mostly distributed in nature, and only involves a limited exchange of information among cells.

33.3.1 Algorithm description

1. Initialize $\lambda = (\lambda_1,\ldots,\lambda_C)$, e.g., $\lambda_c^{(0)} = 1/C$ for all $c = 1,\ldots,C$.

2. For given $\lambda = (\lambda_1,\ldots,\lambda_C)$, find transmit powers P_{mc} that minimize
$V(\lambda_1,\ldots,\lambda_C) = \sum_{c=1}^{C} \lambda_c L_c.$

(a) In the constant-interference case, this may be done by assigning each of the individual users m to the 'cheapest' base station

$$c_m^* := \arg\min_{c=1,\ldots,C} \lambda_c/\alpha_{mc}.$$

Thus, $V(\lambda_1,\ldots,\lambda_C) = \sum_{m=1}^{M} \beta_{mc}$ with $\beta_{mc} := \zeta_m \min_{c=1,\ldots,C} \lambda_c/\alpha_{mc}$.

(b) In the variable-interference case, this may be done in the following manner.

   i. Solve the 'uplink power control problem' for SNIR requirements γ and noise levels λ_c/γ using the Yates-Huang-Hanly fixed-point algorithm (Hanly, 1995; Yates, 1995):

   A. Initialize $\mu_m^{(0)} = 0$ for all $m = 1,\ldots,M$;
   B. Update $\mu_m^{(p)}$ as

   $$\mu_m^{(p+1)} := \min_{c=1,\ldots,C} \gamma \left[\sum_{m'=1}^{M} \alpha_{m'c}\mu_{m'}^{(p)} + \lambda_c/\gamma\right]/\alpha_{mc};$$

   C. Repeat step (B) until some convergence/stopping criterion is satisfied.

   ii. Let $\mu_m^{(P)}$ be the final solution from the above procedure and let

   $$c_m^* := \arg\max_{c=1,\ldots,C} \gamma \left[\sum_{m'=1}^{M} \alpha_{m'c}\mu_{m'}^{(P)} + \lambda_c/\gamma\right]/\alpha_{mc}\mu_m^{(P)},$$

   ties being broken arbitrarily. Then solve the following set of linear equations through repeated substitution (the constrained 'downlink power control problem').

   $$\alpha_{mc_m^*}P_{mc_m^*} = \gamma\left[\eta + \sum_{c=1}^{C} \alpha_{mc}L_c\right],$$

3. Let $L_c(\lambda^{(i)})$ be the optimal load at base station n for given

$$\lambda^{(i)} = (\lambda_1^{(i)},\ldots,\lambda_C^{(i)})$$

as determined in step 2. Update $\lambda_c^{(i)}$ as

$$\lambda_c^{(i+1)} := \lambda_c^{(i)} + \rho_i(L_c(\lambda^{(i)}) - \bar{L}(\lambda^{(i)})),$$

with $\bar{L}(\lambda^{(i)}) := \sum_{c=1}^{C} L_c(\lambda^{(i)})/C$ denoting the average base station load. To ensure that $\lambda_c^{(i+1)} > 0$ for all $c = 1,\ldots,C$, truncate the update step if needed.

4. Let $P_{mc}(\lambda^{(i)})$ be the optimal transmit power assigned by base station c to mobile m for given $\lambda^{(i)} = (\lambda_1^{(i)}, \ldots, \lambda_C^{(i)})$ as determined in step 2. Update $P_{mc}^{(i)}$ as

$$P_{mc}^{(i+1)} = (1 - \sigma_i)P_{mc}^{(i)} + \sigma_i P_{mc}(\lambda^{(i)}),$$

with $\sigma_i := \rho_i / \sum_{j=0}^{i} \rho_j$. (In particular, $\sigma_0 = 1$, so that $P_{mc}^{(1)} = P_{mc}(\lambda^{(0)})$.)

5. Repeat the above steps until some convergence/stopping criterion is satisfied.

33.3.2 Observations

- To guarantee convergence, it is required that at step (3) above,

$$\lim_{i \to \infty} \rho_i = 0$$

and

$$\sum_{i=0}^{\infty} \rho_i = \infty.$$

For example, one may take $\rho_i = \rho i^{-1/2 + \varepsilon}$ for positive constants ε, ρ.

- In order to interpret step (2), recall that problem (33.29)–(33.31) is the dual version of (33.27)–(33.28) as noted earlier. Furthermore, the optimal solution of the latter problem may be computed as the minimum transmit powers for the corresponding uplink power control problem using the Yates-Huang-Hanly algorithm. Hence, the optimal transmit powers P_{mc}^* and the minimum uplink transmit powers μ_m^* for given values of λ_c must satisfy the complementary slackness conditions: if

$$\alpha_{mc}\mu_m < \gamma \sum_{m'=1}^{M} \alpha_{m'c} P_{m'c} + \lambda_c,$$

then $P_{mc} = 0$.

- Note that step (4) is only required in order to obtain the optimal transmit powers P_{mc}^*, and is not needed for finding the optimal $\lambda^* \equiv (\lambda_1^*, \ldots, \lambda_C^*)$.

- Alternatively, the optimal transmit powers P_{mc}^* may be obtained by using the optimal $\lambda^* = (\lambda_1^*, \ldots, \lambda_C^*)$ and L^* as follows. Let μ_m^* be the minimum uplink powers corresponding to $\lambda^* = (\lambda_1^*, \ldots, \lambda_C^*)$ satisfying

$$\mu_m^* = \min_{c=1,\ldots,C} \gamma \left[\sum_{m'=1}^{M} \alpha_{m'c}\mu_{m'}^* + \lambda_c/\gamma \right] / \alpha_{mc},$$

which may be determined as described in step (2) above. Let

$$C_m^* := \arg \max_{c=1,\ldots,C} \left[\sum_{m'=1}^{M} \alpha_{m'c}\mu_{m'}^* + \lambda_c^*/\gamma \right] / \alpha_{mc}\mu_m^*.$$

Now solve the constrained downlink power control problem with the mobile to base station assignment determined by C_m^* and the total power fixed at level L^*, i.e., solve the set of linear equations

$$\sum_{c \in C_m^*} \alpha_{mc} P_{mc} = \gamma \left[\eta + \sum_{c=1}^{C} \alpha_{mc} L^* \right].$$

The latter set of linear equations may be solved in a greedy manner because the incidence matrix $I_{mc}^* := I_{\{c \in C_m^*\}}$ has a tree structure.

- As mentioned earlier, the dual-ascent scheme is mostly distributed in nature and only requires limited communication among base stations. Specifically, for a given shadow cost vector $\lambda = (\lambda_1, \ldots, \lambda_C)$, the load vector $L = (L_1, \ldots, L_c)$ in step (2) can be obtained by each of the mobiles independently updating its power. For a given load vector $L = (L_1, \ldots, L_C)$, the update in the shadow cost vector $\lambda = (\lambda_1, \ldots, \lambda_C)$ can be determined by exchanging the average load value only.

33.4 CONVERGENCE RESULTS

Let P^* be the optimal solution of the primal problem, and let λ^* be the optimal solution of the dual problem. In order to avoid technicalities, we assume that the optimal shadow cost vector λ^* and the optimal power assignment vector P^* are unique.

Theorem 33.1 *For any starting vector $\lambda^{(0)}$, $\lambda^{(i)} \to \lambda^*$ as $i \to \infty$.*

It is noteworthy that, although the shadow cost vector $\lambda^{(i)}$ converges to λ^*, that does not imply that the power assignment vector $P(\lambda^{(i)})$ and load vector $L(\lambda^{(i)})$ converge to P^* and L^*, or even converge at all. Note that even when $\lambda^{(i)}$ approaches λ^* arbitrarily close, the direction $H(\lambda^{(i)}) = L(\lambda^{(i)}) - \bar{L}(\lambda^{(i)})e$ of the update step does not vanish, where e is the all 1s vector. In fact, the vectors $P(\lambda^{(i)})$ and $L(\lambda^{(i)})$ correspond to simplex allocations where the users that are supposed to be in soft hand-off, are fully allocated to one of the supporting base stations. Since the soft hand-off users typically have huge path losses (low path gains), the load vectors associated with these simplex allocations tend to be far from balanced. This causes persistent 'bouncing' of the shadow cost vector $\lambda^{(i)}$ around the optimal point λ^*, and the reason why $\lambda^{(i)}$ finally converges, is only because the size ρ_c of the update step tends to zero. However, the smoothed power assignment vector $\bar{P}^{(i)}$ *does* converge as the next corollary shows.

Corollary 33.1 $\bar{P}^{(i)} \to P^*$ as $i \to \infty$.

The proofs of Theorem 33.1 and Corollary 33.1 may be found in the appendix of this chapter.

33.5 TWO-CELL EXAMPLE

In order to illustrate some of the key observations in the simplest possible context, we now consider a toy network with just two cells. To simplify the exposition, we focus on the constant-interference case as studied in Subsection 33.2.1. While the optimal power allocation is easy to determine in this case, the structural properties yield some crucial qualitative insights that apply in a broader sense. Without loss of generality we assume that the users are indexed according to the ratio of path losses to the two base stations, i.e.,

$$\frac{\alpha_{11}}{\alpha_{12}} \geq \frac{\alpha_{21}}{\alpha_{22}} \geq \cdots \geq \frac{\alpha_{M1}}{\alpha_{M2}}.$$

Denote by

$$L_1^{(m)} := \sum_{p=1}^{m} \frac{\zeta_p}{\alpha_{p1}}$$

the aggregate load of base station 1 when serving users $1,\ldots,m$ in simplex mode. Similarly, denote by

$$L_2^{(m)} := \sum_{p=m}^{M} \frac{\zeta_p}{\alpha_{p2}}$$

the total load of base station 2 when fully supporting users $m,\ldots,N$. Define $m^* := \min\{m : L_1^{(m)} > L_2^{(m+1)}\}$, so that $L_1^{(m-1)} \leq L_2^{(m)}$. Note that the value of m^* can easily be found through a simple linear search. The optimal power assignment is given by:

$$P_{m1}^* = \begin{cases} \zeta_m/\alpha_{m1} & m = 1,\ldots,m^*-1 \\ (\zeta_{m^*} - \alpha_{m^*2}(L_1^{(m^*-1)} - L_2^{(m^*+1)}))/(\alpha_{m^*1} + \alpha_{m^*2}) & m = m^* \\ 0 & m = m^*+1,\ldots,M \end{cases},$$

$$P_{m2}^* = \begin{cases} 0 & m = 1,\ldots,m^*-1 \\ (\zeta_{m^*} - \alpha_{m^*1}(L_2^{(m^*+1)} - L_1^{(m^*-1)}))/(\alpha_{m^*1} + \alpha_{m^*2}) & m = m^* \\ \zeta_m/\alpha_{m2} & m = m^*+1,\ldots,M \end{cases},$$

so that the minimum total load is:

$$L^* = \frac{\zeta_{m^*} + \alpha_{m^*1} L_1^{(m^*-1)} + \alpha_{m^*2} L_2^{(m^*+1)}}{\alpha_{m^*1} + \alpha_{m^*2}}.$$

Observe that the total number of legs is $M+1$ as noted earlier.

The optimal dual variables are:

$$\lambda_c^* = \frac{\alpha_{m^*c}}{\alpha_{m^*1} + \alpha_{m^*2}},$$

so that

$$\mu_m^* = \begin{cases} \frac{\alpha_{m^*1}}{\alpha_{m1}(\alpha_{m^*1}+\alpha_{m^*2})} & m = 1,\ldots,m^*-1 \\ \frac{1}{\alpha_{m^*1}+\alpha_{m^*2}} & m = m^* \\ \frac{\alpha_{m^*2}}{\alpha_{m2}(\alpha_{m^*1}+\alpha_{m^*2})} & m = m^*+1,\ldots,M \end{cases}$$

Note that the ratio of the dual variables λ_c is simply the ratio of the path losses α_{m^*c} to the two base stations of the user that is in 'soft hand-off' i.e., receives power from both base stations.

It is also interesting to examine the behavior of

$$V(\lambda_1,\lambda_2) = \min \lambda_1 L_1 + \lambda_2 L_2 = \sum_{m=1}^{M} \zeta_m \min_{c=1,\ldots,C} \lambda_c/\alpha_{mc}.$$

Define $\phi_0 = 1$, $\phi_{M+1} = 0$, and $\phi_m := \alpha_{m1}/(\alpha_{m1}+\alpha_{m2})$ for all $m = 1,\ldots,M$. It may easily be shown that $V(\lambda_1,\lambda_2)$ is a continuous piece-wise linear function with knee points at $V(\phi_m, 1-\phi_m) = \phi_m L_1^{(m-1)} + \zeta_m/(\alpha_{m1}+\alpha_{m2}) + (1-\phi_m)L_2^{(m+1)}$, $m=1,\ldots,M$, and $V(\lambda, 1-\lambda) = \lambda L_1^{(m-1)} + (1-\lambda)L_2^{(m)}$ for $\lambda \in (\phi_m, \phi_{m-1})$, $m = 0,\ldots,M$. Thus, $dV(\lambda, 1-\lambda)/d\lambda = L_1^{(m-1)} - L_2^{(m)}$ for $\lambda \in (\phi_m, \phi_{m-1})$, $m = 0,\ldots,M$. In particular, at the optimal point $(\lambda_1^*, \lambda_2^*) = (\phi_m^*, 1-\phi_m^*)$, the left derivative is $L_1^{(m^*)} - L_2^{(m^*+1)} > 0$ and the right derivative is $L_1^{(m^*-1)} - L_2^{(m^*)} \leq 0$. Hence, the sensitivity of $V(\lambda_1, \lambda_2)$ with respect to (λ_1, λ_2) around the optimal point depends on the degree of imbalance in the two simplex allocations where the user that is to be in soft hand-off, is either fully allocated to base station 1 or fully supported by base station 2.

As shown above, the optimal power allocation in a two-cell network is trivial to determine. It is instructive however to investigate the behavior of generic solution algorithms in this simple scenario, in particular the simplex algorithm and the dual-ascent scheme described in Section 33.3, as it unveils some of the typical phenomena that occur in a general setting as well. It is worth emphasizing that the 'simplex algorithm' here refers to a standard method for solving linear programs, and does not directly relate to the power allocation being simplex. (Interestingly enough however, the so-called 'basic solutions' produced by the simplex algorithm actually have at most $M+C-1$ non-zero variables, and thus correspond to allocations that are largely simplex.)

In order to describe the operation of the simplex algorithm, we first convert the inequality constraints (33.2), (33.3) into equality constraints by introducing positive slack variables in (33.2) and observing that (33.3) may be assumed to hold with equality, yielding:

$$\min \quad L$$

subject to:
$$L = \sum_{m=1}^{M} P_{mc} + u_c \quad c = 1,\ldots,C \qquad (33.32)$$

$$\sum_{c=1}^{C} \alpha_{mc} P_{mc} = \zeta_m \quad m = 1,\ldots,M \qquad (33.33)$$

$$u_c, P_{mc} \geq 0 \quad m = 1,\ldots,M, c = 1,\ldots,C.$$

It may be shown that there exists an optimal solution that is 'basic', i.e., has at most $M+C-1$ non-zero variables. As an initial basic solution, we may for example take the allocation where each user m is assigned to the strongest received base station $c_m^* := \arg\max_{c=1,\ldots,C} \alpha_{mc}$:

$$P_{mc} = \begin{cases} \frac{\zeta_m}{\alpha_{mc}} & c = c_m^* \\ 0 & c \neq c_{m^*} \end{cases},$$

so that
$$L = \max_{c=1,\ldots,C} L_c,$$

$u_c = L - L_c$, with
$$L_c = \sum_{m: c_m^* = n} P_{mc}.$$

By solving the linear equality constraints, the $M + C - 1$ non-zero basic variables may now be expressed in terms of the non-basic variables as follows:

$$P_{mc_m^*} = \frac{\zeta_m - \sum_{p \neq c_m^*} \alpha_{mp} P_{mp}}{\alpha_{mc_m^*}},$$

$L = M_{c^*}$, $u_c = M_{c^*} - M_c$, with
$$c^* := \arg \max_{c=1,\ldots,C} L_c$$

and
$$M_c := \sum_{m: c_m^* = c} \frac{\zeta_m - \sum_{p \neq c} \alpha_{mp} P_{mp}}{\alpha_{mc}} + \sum_{m: c_m^* \neq c} P_{mc}.$$

Now observe that the value of L may be reduced by increasing $P_{mc'}$ for some user m with $c_m^* = c^*$ and $c' \neq c^*$, which amounts to reducing the power assigned to a user by the most heavily loaded base station and increasing the power assigned to that user by a more lightly loaded base station so as to maintain a feasible solution. Note that $P_{mc'}$ may be increased until either P_{mc^*} becomes zero or $u_{c'}$ becomes zero. In the former case, user m has been entirely transferred from base station c^* to base station c'. In the latter case, the total load at base station c' has reached the same level as that at base station c^*. In general, several users can be shifted until at some point the loads equalize when some user m^* has partly been transferred. At that point, we have for $m \neq m^*$,

$$P_{mc_m^*} = \frac{\zeta_m - \sum_{p \neq c_m^*} \alpha_{mp} P_{mp}}{\alpha_{mc_m^*}},$$

and
$$P_{m^*c^*} = \frac{\alpha_{m^*c'}}{\alpha_{m^*c^*} + \alpha_{m^*c'}} \left[\frac{\zeta_{m^*}}{\alpha_{m^*c'}} + M_{c^*} - M_{c'} \right],$$

$$P_{m^*c'} = \frac{\alpha_{m^*c^*}}{\alpha_{m^*c^*} + \alpha_{m^*c'}} \left[\frac{\zeta_{m^*}}{\alpha_{m^*c^*}} + M_{c'} - M_{c^*} \right],$$

and
$$L = \frac{1}{\alpha_{m^*c^*} + \alpha_{m^*c'}} [\zeta_{m^*} + \alpha_{m^*c^*} M_{c^*} + \alpha_{m^*c'} M_{c'}].$$

It may now be possible to further reduce the value of L by increasing $P_{m'c'}$ for some user m' with $P_{m'c^*} > 0$, and
$$\frac{\alpha_{m'c'}}{\alpha_{m'c^*}} > \frac{\alpha_{m^*c'}}{\alpha_{m^*c^*}},$$

until at some point there is no such user left. In the case of two cells, we then have obtained a solution with $L_1 = L_2$ and the property that if $P_{m1} = 0$, then $P_{p1} = 0$ for all $p = m+1, \ldots, M$, which corresponds to the optimal power allocation characterized above.

We now turn to the behavior of the dual-ascent scheme in the case of two cells. Since (λ_1, λ_2) converges to the optimal point $(\lambda_1^*, \lambda_2^*) = (\phi_{m^*}, 1 - \phi_{m^*})$, we know that from some point on the ratio λ_1/λ_2 will be constrained to the interval

$$\left(\frac{\alpha_{m^*-1,1}}{\alpha_{m^*-1,2}}, \frac{\alpha_{m^*+1,1}}{\alpha_{m^*+1,2}} \right).$$

From that point on, the load vectors minimizing $\lambda_1 L_1 + \lambda_2 L_2$ will either be

$$(L_1^{(m^*-1)}, L_2^{(m^*)})$$

or

$$(L_1^{(m^*)}, L_2^{(m^*+1)}).$$

The above description indicates that the simplex algorithm will take at most M iterations to terminate in the case of two cells, while the dual-ascent scheme may need a large number of steps in order for the shadow cost vector to converge and an even larger number of iterations for the power allocation vector to settle down. However, the update steps made by the simplex algorithm require centralized knowledge of all the path loss coefficients, whereas the updates of the dual-ascent scheme can be performed in a distributed way.

33.6 NUMERICAL EXPERIMENTS

We now discuss some of the numerical experiments that we conducted to illustrate the results and evaluate the potential gains from dynamic load balancing. We first present some simple examples to illuminate the key concepts of the linear-programming (LP) approach. Next, we summarize the results of some benchmarking experiments comparing the perfectly balanced power allocation provided by the LP with dynamic simulations of a 3G CDMA system and the results obtained using the dual-ascent (DA) scheme described in Section 33.3.

33.6.1 Examples

We first consider a toy example with $C = 2$ base stations and $M = 4$ users. The path loss values are given in Table 33.1. The SNIR requirement is set to $\gamma = 0.0005$ and the thermal noise is assumed to be $\eta = 0.02$. The linear program (33.16)–(33.18) was solved using AMPL (Gay et al., 1993). The shadow prices associated with the constraints (33.17) were found to be $(\lambda_1^*, \lambda_2^*) = (0.669435, 0.330565)$. The dual variables μ_m^* corresponding to the SNIR requirements are listed in Table 33.2.

Using the optimal shadow price vector $(\lambda_1^*, \lambda_2^*)$, we also ran the Yates-Huang-Hanly algorithm as specified in step 2.a of the dual-ascent scheme. The resulting power values are shown in Table 33.2 as well, and are seen to be in close agreement with the values obtained from the LP using AMPL.

Table 33.1 Path gain values

User	Base 1	Base 2
0	0.0010	0.0005
1	0.0050	0.0001
2	0.0020	0.0100
3	0.0007	0.0010

Table 33.2 Optimal dual variables and results from Yates-Huang-Hanly algorithm

User	μ	YH	Base
1	703.928	703.92746	1
2	140.786	140.78549	1
3	35.1964	35.196442	2
4	351.964	351.96442	2

Table 33.3 Achieved uplink SNIR's at optimum in dual problem

User	Base 1	Base 2
1	0.020000	0.020000
2	0.020000	0.000800
3	0.002000	0.020000
4	0.007000	0.020000

The achieved uplink SNIR's for the optimal values μ_m^* are displayed in Table 33.3. The table shows that the SNIR requirement of the first user is met at either base station. Complementary slackness thus implies that the first user is in soft hand-off and may receive power from both base stations, while all other users only receive power from one of the two base stations.

In Figure 33.2 we graph the value of $V(\lambda_1, 1-\lambda_1)$ as function of λ_1 obtained using the Yates-Huang-Hanly algorithm. Observe that the maximum value is attained around $\hat{\lambda}_1 \approx 0.67$, yielding $\hat{\lambda}_2 \approx 0.33$, in agreement with the results from the LP.

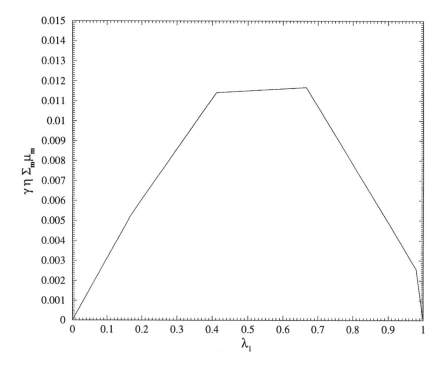

Figure 33.2 Value of $V(\lambda_1, 1-\lambda_1)$ as function of λ_1

We also ran the entire dual-ascent scheme, which was found to converge, although rather slowly. The results are presented in Table 33.4, and are seen to be in close agreement with those obtained using the LP. Observe that the value of

$$V(\hat{\lambda}_1, \hat{\lambda}_2) = \gamma \eta \sum_{m=1}^{M} \hat{\mu}_m$$

is about half the total load $\hat{L}_1 + \hat{L}_2$ as it should be since $\hat{L}_1 \approx \hat{L}_2$.

We now turn to a network with a rectangular grid of $N = 16$ base stations and $M = 32$ users which are distributed uniformly at random. The path loss values are computed using a standard power-law model.

As before, the linear program (33.16)–(33.18) was solved using AMPL. The shadow prices associated with the constraints (33.17) are listed in Table 33.5, and range from about 0.006 to 0.26. The dual variables corresponding to the SNIR requirements (33.17) are displayed in Table 33.6.

As before, we ran the Yates-Huang-Hanly algorithm using the optimal shadow price vector. The resulting power values are shown in Table 33.7 as well, and are seen to be

Table 33.4 Results from dual-ascent method

Quantity	Value
$\hat{\lambda}_1$	0.6687
$\hat{\lambda}_2$	0.3313
$V(\hat{\lambda}_1, \hat{\lambda}_2)$	0.01231
$\hat{L}_1 + \hat{L}_2$	0.023

Table 33.5 Optimal shadow prices

Base	λ	Base	λ
1	0.0430290	9	0.107572
2	0.0100268	10	0.260223
3	0.0500479	11	0.0140151
4	0.0413243	12	0.0294555
5	0.0219057	13	0.0559462
6	0.0149269	14	0.164543
7	0.0058589	15	0.0483616
8	0.0123052	16	0.120459

in close agreement with the results from the LP. The table also specifies the serving base stations for the various users as inferred from the μ_m^*, ties being broken arbitrarily.

Finally, we ran the entire dual-ascent scheme which again was found to converge (slowly) to the optimal values obtained using the LP.

33.6.2 Comparisons with 3G CDMA simulations

To quantify the potential gains from dynamic load balancing, we performed extensive benchmarking experiments comparing the perfectly balanced power allocation provided by the LP with dynamic simulations of a 3G CDMA system. We also included the results from the DA scheme in the comparisons, so as to assess the quality and speed of convergence.

Experimental set-up. The dynamic simulations are time-driven with a granularity of 500 ms. Since the LP approach basically considers a 'snapshot' of the system in time, we take 400 such 'samples' (i.e., ~ 3.5 minutes) from the simulation runs with the instantaneous user configuration and path loss values. For each of these

Table 33.6 Optimal dual variables

User	μ	User	μ
1	20.53930	17	25.41597
2	0.58952	18	61.89996
3	0.98299	19	126.47312
4	0.33666	20	141.32139
5	14.18345	21	7.23224
6	3.98482	22	0.46208
7	0.02404	23	17.79327
8	21.39627	24	3.70228
9	16.95331	25	50.37689
10	0.54712	26	0.63100
11	1.42241	27	40.81531
12	6.93493	28	57.50224
13	1.64843	29	6.52227
14	0.35836	30	5.09796
15	0.62783	31	49.37766
16	5.39319	32	71.58780

samples, we compare the maximum load across the base stations in the simulation experiments with the minimum achievable maximum load as determined by the LP. Since the LP implicitly assumes that the power allocation can be perfectly adapted every 500 ms, it provides an estimate for the maximum gains possible through load balancing. We then average the percentage gains obtained by the LP over the 400 samples. We make similar comparisons between the simulation results and the numbers produced by the DA scheme with a finite number of iterations.

Assumptions. We consider a network consisting of four base stations located at the corner points of a square coverage region. Calls originate according to a spatially uniform Poisson process and have exponential holding times. Users may roam around but do not leave the coverage area.

We assume that the downlink power load at the base stations are the only limiting resources. The power control algorithms and soft hand-off rules are simplified versions of the IS-95B standards. In particular, all base stations in the active set provide the same transmit power. In order to determine the received SINR, we assume that the users perform maximum-ratio combining, thus summing up the SINR values over all the legs in the active set, and we ignore channel orthogonality.

Table 33.7 Results from Yates-Huang-Hanly algorithm

User	μ	Base	User	μ	Base
1	20.494373	1	17	25.389114	9
2	0.589278	1	18	61.847153	13
3	0.982130	2	19	126.423298	6
4	0.336366	2	20	140.944499	10
5	14.184783	3	21	7.238571	11
6	3.985199	3	22	0.462488	11
7	0.024029	4	23	17.741493	12
8	21.418477	4	24	3.702045	12
9	16.940444	5	25	50.282238	13
10	0.547092	5	26	0.630836	13
11	1.422946	6	27	40.755045	14
12	6.934611	6	28	56.963515	2
13	1.652000	7	29	6.518343	15
14	0.359398	7	30	5.095496	15
15	0.628528	8	31	49.304174	16
16	5.401335	8	32	71.680437	3

The propagation characteristics are governed by a standard power-law path loss model, possibly superimposed with log-normal shadow fading. Given the time granularity of the simulations, we do not consider Rayleigh fading. Within the above basic framework, we consider four scenarios with an increasing degree of reality:

- *Scenario A*: No shadow fading. The size of the active set is limited to 1 (no soft hand-off), so that each user will be connected to the strongest received base station for the entire duration of the call.

- *Scenario B*: Log-normal shadow fading with a standard deviation of 8 dB, a coherence time of 10 s, and an inter-cell correlation factor of 0.5. The size of the active set is still limited to 1 (no soft hand-off). As a result of the temporal fluctuations, users may now be handed off between base stations, but the hand-off is delayed by one time step.

- *Scenario C*: In addition to the shadow fading, soft hand-off is introduced with a maximum size of the active set of 4.

- *Scenario D*: In addition, base stations can only drop out from the active set with a time delay of 5 s.

Table 33.8 Comparison of LP and DA results with 3G CDMA simulations

	Percentage gain via LP			Percentage gain via DA		
Scenario	Average	Min	Max	Average	Min	Max
A0	22.09	3.40	48.42	20.87	1.99	48.04
B0	52.47	6.95	95.55	51.56	5.25	95.50
C0	45.82	0.30	99.36	44.86	0.00	98.72
D0	60.42	16.05	98.67	59.58	13.50	98.64
A3	30.62	1.07	92.79	29.14	0.00	92.75
B3	52.36	0.51	96.82	51.23	6.70	96.75
C3	47.74	5.19	97.81	46.51	1.86	97.80
D3	62.52	13.82	97.87	61.54	7.84	97.80
A6	33.76	0.86	96.20	31.62	0.00	96.02
B6	52.36	3.27	97.02	50.92	0.00	96.86
C6	49.43	5.40	96.73	47.88	0.00	96.69
D6	61.99	20.78	97.79	60.70	12.93	97.70

Discussion of results. For each of the above four scenarios, we considered three different SINR requirements, corresponding to three different transmission rates. These are identified as cases 0 (lowest data rate), 3 (twice data rate of case 0, SINR increase 3db), and case 6 (four times the original data rate, SINR increase 6 db). In order to keep the offered load constant, we reduced the call arrival rate proportionally. Thus, we have 12 scenarios in total, corresponding to a total of 4,800 samples. The number of users in each sample varied roughly in the range between 80 and 100, so that the number of variables in the linear program approximately ranged from 320 to 400.

The linear programs were solved using Cplex (version 6.0) (ILOG, Inc., 1998) which took a fraction of a second per sample. In addition to the LP, we also included the DA results in the comparisons. In order to assess the speed of convergence, the DA scheme was run for only 100 iterations, with parameters $\rho_0 = 0.05$ and $\delta = 0.5$. Performing 100 iterations also took less than a second per sample. The results are summarized in Table 33.8.

Inspecting the results, we observe that the percentage gains from the LP range from less than 1% to more than 99% in the various scenarios, with an average of about 48%. Note that a 99% gain means that the load in the 3G system is 100 times larger than the theoretical minimum load identified by the LP! Also, the gains from the DA scheme are nearly as large as from the perfect solution provided by the LP. It is worth emphasizing that all these results are for a spatially uniform traffic distribution, where the gains are entirely due to stochastic variations in the user population and temporal fluctuation as a result of shadow fading. When the traffic distribution is non-uniform, the gains are likely to be larger.

We note that the percentage gains are consistently larger for the scenarios with higher data rates, although not by too much. This suggests that load balancing produces bigger gains when there are fewer users each with larger power requirements. Comparison between scenarios A and B indicates that the gains increase as a result of shadow fading, as expected. Further, observe from scenarios B and C that the gains slightly decrease when soft hand-off is introduced, which may be somewhat surprising at first sight. While soft hand-off will increase the total power provided to a given user and thus the total transmit power, it may also help to reduce the *maximum* base station power, which in turn may help decrease the necessary transmit power because of reduced interference. Finally, introducing a time delay in dropping base stations from the active set increases the gains.

33.7 IMPLEMENTATION CONSIDERATIONS

As noted earlier, the linear program gives a lower bound for a particular 'snapshot', and does not explicitly consider the evolution of the system over time. In reality, however, not only the path losses of the various users vary over time as a result of fading and mobility, but also the user population itself will evolve over time due to call arrivals and departures. In fact, it is exactly these temporal variations that create the need for some sort of dynamic network adjustment in order to optimize the system performance. We now describe a possible implementation of the linear programming approach in such a context. We will first outline how the optimal solution of the linear program may be tracked over time, without necessarily actively intervening in the operation of the system. Later we will also discuss how the solution may be acted upon so as to bring the actual power allocation closer to that determined by the linear program through active intervention.

We assume that the above procedure is organized in some time-slotted fashion, and thus view the various parameters and variables involved as functions of time. Accordingly, we attach a time index to the relevant quantities introduced before, e.g., $M(t)$, $\alpha_{mc}(t)$, $\beta_{mc}(t)$, $P_{mc}(t)$, $L_c(t)$, $L(t)$, $\lambda_m(t)$, $\mu_m(t)$. While the variable $\alpha_{mc}(t)$ is used as the path loss in computation cycle t, its value may actually be obtained as a smoothed average over a longer time period. In fact, since the power requirement is inversely proportional to the path loss, it may be desirable to keep track of the average of the reciprocal of the path loss, rather than the path loss itself. The exact duration of the smoothing interval remains to be specified, but a suitable value might be on the order of a second, so as to average across fast fading, but track slower variations due to shadowing and mobility.

We assume that in each computation cycle only a limited number of I_1 updates of the shadow cost vector λ can be performed, and that for each update of λ only I_2 iterations in the uplink power control problem and I_3 iterations in the downlink power control problem can be executed. Obviously, the values of I_1, I_2, I_3 are highly interdependent, and also depend on the length of the computation cycles. These are all design parameters that remain to be determined.

In view of the temporal variations and the limited number of iterations within each computation cycle, we cannot expect the algorithm to fully converge to the optimal solution. In fact, it may not be necessary or even desirable for the procedure to perfectly follow occasional short-term fluctuations but rather track slower longer-term changes. Also, the step sizes in the updates of the shadow cost vector and the power assignment vector should not be reduced, but either held fixed or reinitialized at the start of each computation cycle.

Based on the above considerations, the dual-ascent scheme described in the previous section may be adapted as follows.

33.7.1 Algorithm description

In each computation cycle t, execute the following steps.

1. Initialize $\lambda(t) = (\lambda_1(t), \ldots, \lambda_C(t))$ as $\lambda_c^{(0)}(t) = \lambda_c^{(L_1)}(t-1)$.

2. For given $\lambda(t) = (\lambda_1(t), \ldots, \lambda_C(t))$, find transmit powers $P_{mc}(t)$ that minimize
$$V(\lambda_1(t), \ldots, \lambda_C(t)) = \sum_{c=1}^{C} \lambda_c(t) L_c(t).$$
In the constant-interference case, this may be done by assigning each of the individual users m to the 'cheapest' base station $c_m^*(t) := \arg\min_{c=1,\ldots,C} \lambda_c(t)/\alpha_{mc}(t)$.
Thus, $V(\lambda_1(t), \ldots, \lambda_C(t)) = \sum_{m=1}^{M(t)} \beta_{mc}(t)$ with $\beta_{mc}(t) := \zeta_m \min_{c=1,\ldots,C} \lambda_c(t)/\alpha_{mc}(t)$.

In the variable-interference case, this may be done in the following manner.

(a) Solve the 'uplink power control problem' for SINR requirements γ and noise levels $\lambda_c(t)/\gamma$ using the Yates-Huang-Hanly algorithm (Hanly, 1995; Yates, 1995):

   i. Initialize $\mu_m^{(0)}(t) = \mu_m^{(L_2)}(t-1)$ for all $m = 1, \ldots, M(t)$ (or initialize to zero for new users);

   ii. Update $\mu_m^{(p)}(t)$ as
   $$\mu_m^{(p+1)}(t) := \min_{c=1,\ldots,C} \gamma \left[\sum_{m'=1}^{M(t)} \alpha_{m'c}(t) \mu_{m'}^{(p)}(t) + \lambda_c(t)/\gamma \right] / \alpha_{mc}(t).$$

   iii. Repeat step (ii) I_2 times.

(b) Let $\mu_m^{(L_2)}(t)$ be the final solution from the above procedure and let
$$c_m^*(t) := \arg\max_{c=1,\ldots,C} \gamma \left[\sum_{m'=1}^{M(t)} \alpha_{m'c}(t) \mu_{m'}^{(P)}(t) + \lambda_c(t)/\gamma \right] / \alpha_{mc}(t) \mu_m^{(P)}(t),$$
ties being broken arbitrarily. Then solve the constrained 'downlink power control problem' with the mobile to base station assignment determined

by $c_m^*(t)$, i.e., solve the set of linear equations

$$\alpha_{mc_m^*}(t)P_{mc_m^*}(t) = \gamma\left[\eta + \sum_{c=1}^{C}\alpha_{mc}(t)L_c(t)\right],$$

which may be done through repeated substitution (I_3 times).

3. Let $L_c(\lambda^{(i)}(t))$ be the optimal load at base station n for given

$$\lambda^{(i)}(t) = (\lambda_1^{(i)}(t),\ldots,\lambda_C^{(i)}(t))$$

as determined in step 1. Update $\lambda_c^{(i)}(t)$ as

$$\lambda_c^{(i+1)}(t) := \lambda_c^{(i)}(t) + \rho_i(L_c(\lambda^{(i)}(t)) - \bar{L}(\lambda^{(i)}(t))),$$

with $\bar{L}(\lambda^{(i)}(t)) := \sum_{c=1}^{C} L_c(\lambda^{(i)}(t))/C$ denoting the average base station load. To ensure that $\lambda_c^{(i+1)}(t) > 0$ for all $c = 1,\ldots,C$, truncate the update step if needed.

4. Let $P_{mc}(\lambda^{(i)}(t))$ be the optimal transmit power assigned by base station c to mobile m for given $\lambda^{(i)}(t) = (\lambda_1^{(i)}(t),\ldots,\lambda_C^{(i)}(t))$ as determined in step 2. Update $P_{mc}^{(i)}(t)$ as

$$P_{mc}^{(i+1)}(t) = (1-\sigma)P_{mc}^{(i)}(t) + \sigma P_{mc}(\lambda^{(i)}(t)),$$

with the convention that $P_{mc}^{(0)}(t) = P_{mc}^{(L_1)}(t-1)$.

5. Repeat the above steps I_1 times.

The above description shows how the optimal solution of the linear program may be tracked over time, without necessarily actively intervening in the operation of the system. We now discuss how the solution may be acted upon so as to bring the actual power allocation closer to that determined by the linear program through active intervention.

One possible approach is to use the optimal solution to determine the 'active sets', but not touch the actual power control mechanisms. As noted earlier, in the optimal power allocation users should only connect to base station(s) with the strongest received pilot strength, where the pilot transmit power of base station c is set inversely proportional to κ_c^*, where $\kappa_c^* = \lambda_c^*$ and

$$\kappa_c^* = \gamma \sum_{m'=1}^{M} \alpha_{m'c}\mu_m^* + \lambda_c^*$$

in the constant-interference and variable-interference case, respectively. Thus, the normal pilot signals can be replaced by or augmented with such additional pilot tones. For protection against fading, however, mobiles should not only connect to the strongest base station(s), but also to base stations which are within a certain margin from the strongest one. Depending on how that margin is set, there will be a trade-off between the closeness to the optimal solution and the degree of robustness against fading effects.

A more radical approach is to not only use the optimal solution to determine the active set, but also replace the existing power control mechanisms. In particular, not only the pilot transmit power could be set inversely proportional to κ_c^*, but all transmit powers on the traffic channels as well. A detailed investigation of such mechanisms is beyond the scope of this chapter.

33.8 EXTENSIONS

We now extend the linear programming formulation to capture features such as minimum and maximum power constraints and orthogonality factors. While power restrictions may be added in a trivial fashion, it turns out that the inclusion of orthogonality coefficients is not as straightforward since the linear program only considers aggregate interference and cannot easily be modified to distinguish between the various contributing sources. In addition, we show how the model may further be endowed with robustness constraints.

33.8.1 Constant-interference case

We first revisit the constant-interference case, where the notion of orthogonality has little meaning. The linear program may be augmented with minimum and maximum power constraints as follows:

$$\min \quad L$$

$$\text{subject to:} \quad L \geq L_c = \sum_{m=1}^{M} P_{mc}, \quad c = 1, \ldots, C \quad (33.34)$$

$$\sum_{c=1}^{C} \alpha_{mc} P_{mc} \geq \zeta_m, \quad m = 1, \ldots, M \quad (33.35)$$

$$P^{\min} \leq P_{mc} \leq P^{\max}, \quad m = 1, \ldots, M, c = 1, \ldots, C. \quad (33.36)$$

The dual version of the linear program reads as follows:

$$\max \quad \sum_{m=1}^{M} \zeta_m \mu_m + P^{\min} \sum_{m=1}^{M} \sum_{c=1}^{C} v_{mc} - P^{\max} \sum_{m=1}^{M} \sum_{c=1}^{C} \xi_{mc} \quad (33.37)$$

$$\text{subject to:} \quad \sum_{c=1}^{C} \lambda_c \leq 1 \quad (33.38)$$

$$\alpha_{mc} \mu_m \leq \lambda_c - v_{mc} + \xi_{mc}, \quad m = 1, \ldots, M, c = 1, \ldots, C \quad (33.39)$$

$$\lambda_c, \mu_m, v_{mc} \geq 0, \xi_{mc} \geq 0, \quad m = 1, \ldots, M, c = 1, \ldots, C, \quad (33.40)$$

with λ_c, μ_m, v_{mc} and ξ_{mc} representing the dual variables associated with the constraints (33.34), (33.35) and (33.36), respectively.

33.8.2 Variable-interference case

We now proceed with the variable-interference case. In case of orthogonality, the SINR requirement for mobile m becomes:

$$\sum_{c=1}^{C} \frac{\alpha_{mc} P_{mc}}{\eta + \alpha_{mc}(\theta(L_c - P_{mc}) + \delta P_{mc}) + \sum_{c' \neq c} \alpha_{mc'} L_{c'}} \geq \gamma,$$

with δ denoting a self-interference factor ($\delta = 0$ means no self-interference) and θ representing an orthogonality coefficient ($\theta = 0$ means perfect orthogonality). Unfortunately, the above constraint is no longer linear. Since the linear program does not explicitly consider cell membership, it is hard to give meaning to the distinction between intra-cell interference and out-of-cell interference. Hence, we will replace the above constraint by:

$$\frac{\sum_{c=1}^{C} \alpha_{mc} P_{mc}}{\eta + \sum_{c'=1}^{C} \alpha_{mc}(\theta(L_c - P_{mc}) + \delta P_{mc})} \geq \gamma,$$

which may be further simplified to

$$\sum_{c=1}^{C} \alpha_{mc} P_{mc} \geq \zeta \left[\eta + \theta \sum_{c=1}^{C} \alpha_{mc} L_c \right],$$

with $\zeta := \gamma/(1 - \gamma(\delta - \theta))$. Note that the latter constraint is 'optimistic' in the sense that the orthogonality coefficient is applied to the total interference, and not just the intra-cell interference. Further observe the similarity in form with (33.13).

With the above simplifications, the linear program may be generalized to account for minimum and maximum power restrictions and orthogonality coefficients as follows:

$$\min \quad L$$

subject to:
$$L \geq L_c = \sum_{m=1}^{M} P_{mc} \quad c = 1, \ldots, C \quad (33.41)$$

$$\sum_{c=1}^{C} \alpha_{mc} P_{mc} \geq \zeta \left[\eta + \theta \sum_{c=1}^{C} \alpha_{mc} L_c \right], \quad m = 1, \ldots, M \quad (33.42)$$

$$P^{\min} \leq P_{mc} \leq P^{\max}, \quad m = 1, \ldots, M, c = 1, \ldots, C. \quad (33.43)$$

The dual version of the linear program reads as follows:

$$\max \quad \zeta\eta \sum_{m=1}^{M} \mu_m + P^{\min} \sum_{m=1}^{M}\sum_{c=1}^{C} \nu_{mc} - P^{\min} \sum_{m=1}^{M}\sum_{c=1}^{C} \nu_{mc} \quad (33.44)$$

$$\text{subject to:} \quad \sum_{c=1}^{C} \lambda_c \leq 1 \quad (33.45)$$

$$\alpha_{mc}\mu_m \leq \zeta\theta \sum_{m'=1}^{M} \alpha_{m'c}\mu_{m'} + \lambda_c - \nu_{mc} + \xi_{mc}, \quad (33.46)$$

$$m = 1,\ldots,M, c = 1,\ldots,C,$$

$$\lambda_c, \mu_m, \nu_{mc}, \xi_{mc} \geq 0, \quad m = 1,\ldots,M, c = 1,\ldots,C,$$

with λ_c, μ_m, ν_{mc} and ξ_{mc} representing the dual variables associated with the constraints (33.41), (33.42) and (33.43), respectively.

33.8.3 Robustness

As observed in Subsection 33.2.4, the optimal power allocation only contains a minimum degree of soft hand-off for the sole purpose of balancing the base station loads. In practice, a further rationale for maintaining soft hand-off is to achieve a certain extent of macro diversity and robustness in the presence of fading.

A possible approach to include the latter considerations in the linear program, is to add a constraint that users should continue to receive a fraction $\kappa > 0$ of the SINR requirement in case the link to any of the base stations were to fail (mathematically, in case the corresponding path loss value were to drop to zero). In the constant-interference case, such a constraint may be cast as

$$\sum_{c \neq q} \alpha_{mc} P_{mc} \geq \kappa \zeta_m$$

for all $q = 1,\ldots,C$. Note that the above constraint will force users to be supplied with power from at least two base stations. This may in fact not be necessary for all the users, and it may be sufficient to impose the above constraint only for a subset of the users who are potential soft hand-off candidates. In order to have that flexibility, we allow the value of κ to be user-specific, with the possibility of it being zero for some users.

Further observe that the above constraint, in conjunction with the aggregate SNIR requirement (33.3), may be replaced by

$$\alpha_{mc} P_{mc} \leq (1 - \kappa_m)\zeta_m \quad (33.47)$$

for all $c = 1,\ldots,C$. Thus, no single base station may contribute more than a fraction $1 - \kappa_m$ of the SNIR requirement of user m.

Adding the constraint (33.47) to the linear program (33.1)–(33.3), the dual version reads as follows:

$$\max \quad \sum_{m=1}^{M} \zeta_m \left(\mu_m - (1 - \kappa_m) \sum_{c=1}^{C} \nu_{mc} \right) \quad (33.48)$$

$$\text{subject to:} \quad \sum_{c=1}^{C} \lambda_c \leq 1 \quad (33.49)$$

$$\alpha_{mc}(\mu_m - \nu_{mc}) \leq \lambda_c, \quad m = 1, \ldots, M, c = 1, \ldots, C \quad (33.50)$$

$$\lambda_c, \mu_m, \nu_{mc} \geq 0, \quad m = 1, \ldots, M, c = 1, \ldots, C,$$

with ν_{mc} the dual variables associated with the constraints (33.47). Observe that optimality implies $\nu_{mc}^* = \max\{\mu_m^* - \lambda_c^*/\alpha_{mc}, 0\}$.

In the variable-interference case, the additional robustness constraint may be cast as:

$$\alpha_{mc} P_{mc} \leq (1 - \kappa_m) \gamma \left[\eta + \sum_{c=1}^{C} \alpha_{mc} L_c \right] \quad (33.51)$$

for all $c = 1, \ldots, C$.

The dual version of the linear program then reads as follows:

$$\max \quad \eta \sum_{m=1}^{M} \left(\mu_m - (1 - \kappa_m) \sum_{c=1}^{C} \nu_{mc} \right) \quad (33.52)$$

$$\text{subject to:} \quad \sum_{c=1}^{C} \lambda_c \leq 1 \quad (33.53)$$

$$\alpha_{mc}(\mu_m - \nu_{mc}) \leq \gamma \sum_{m'=1}^{M} \alpha_{m'c}(\mu_{m'} - (1 - \kappa_{m'})\nu_{m'c}) + \lambda_c, \quad (33.54)$$

$$m = 1, \ldots, M, c = 1, \ldots, C$$

$$\lambda_c, \mu_m, \nu_{mc} \geq 0, \quad m = 1, \ldots, M, c = 1, \ldots, C, \quad (33.55)$$

with ν_{mc} the dual variables associated with the constraints (33.51).

We refer to Bertsimas and Sim (2004) for general notions of robust optimization in the presence of uncertain information.

Acknowledgment. The authors are grateful to John Graybeal and Alan Weiss for interesting discussions.

Appendix: Proof of convergence of dual-ascent scheme

In this appendix, we establish the convergence of the dual-ascent scheme described in Section 33.3. We first prove an auxiliary lemma. For compactness, define $H(\lambda) = L(\lambda) - \bar{L}(\lambda)e$, with e denoting the N-dimensional vector of all ones. Also recall that λ^*, P^* are supposed unique, see the beginning of Section 33.5.

Lemma 33.1 *For any* $\lambda \neq \lambda^*$,

$$(\lambda - \lambda^*)^T H(\lambda) = \sum_{c=1}^{C} (\lambda_c - \lambda_c^*)(L_c(\lambda) - \bar{L}(\lambda)) < 0.$$

Proof. As mentioned in Section 33.2, the optimal shadow vector λ^* maximizes $V(\lambda) = \sum_{c=1}^{C} \lambda_c L_c(\lambda)$ over all $\lambda = (\lambda_1, \ldots, \lambda_C)$. Since λ^* is supposed to be unique,

$$\sum_{c=1}^{C} \lambda_c L_c(\lambda) < \sum_{c=1}^{C} \lambda_c^* L_c(\lambda^*)$$

for all $\lambda \neq \lambda^*$.

Since $L(\lambda^*)$ by definition minimizes $V(\lambda^*)$,

$$\sum_{c=1}^{C} \lambda_c^* L_c(\lambda^*) \leq \sum_{c=1}^{C} \lambda_c^* L_c(\lambda).$$

Combining these two inequalities and using that $\sum_{c=1}^{C} \lambda_c = \sum_{c=1}^{C} \lambda_c^* = 1$, the statement follows. □

We proceed to prove a second auxiliary lemma. For any $\lambda = (\lambda_1, \ldots, \lambda_C)$, denote by $D(\lambda) := \|\lambda - \lambda^*\|^2 = \sum_{c=1}^{C} (\lambda_c - \lambda_c^*)^2$ the squared Euclidean distance between λ and λ^*. The idea of the convergence proof will be to show that $D(\lambda^{(i+1)}) \leq \max\{D(\lambda^{(i)}) - \rho_i K(\varepsilon_i), \varepsilon_i\}$, with $\varepsilon_i \downarrow 0$ as $i \to \infty$. Thus, $D(\lambda^{(i)})$ monotonically decreases with i whenever it exceeds some value ε_i and otherwise remains below ε_i. This then forces $D(\lambda^{(i)}) \downarrow 0$ and thus $\lambda^{(i)} \to \lambda^*$ as $i \to \infty$.

Lemma 33.2 *For any i,*

$$D(\lambda^{(i+1)}) - D(\lambda^{(i)}) \leq \rho_i \left[2(\lambda^{(i)} - \lambda^*)^T H(\lambda^{(i)}) + \rho_i \|H(\lambda^{(i)})\|^2 \right].$$

Proof. By definition,

$$D(\lambda^{(i+1)}) = D(\lambda^{(i)} + \rho_i H(\lambda^{(i)})) =$$
$$D(\lambda^{(i)}) + 2\rho_i (\lambda^{(i)} - \lambda^*)^T H(\lambda^{(i)}) + \rho_i^2 \|H(\lambda^{(i)})\|^2. \square$$

We have now gathered the necessary ingredients for the proof of the main convergence result.

Proof of Theorem 33.1. Using the above two lemmas, we find that when $D(\lambda^{(i)}) \leq \varepsilon/2$,

$$D(\lambda^{(i+1)}) < \varepsilon/2 + \rho_i^2 \|H(\lambda)\|^2,$$

so that $D(\lambda^{(i+1)}) < \varepsilon$ as long as

$$\rho_i \leq \min_{\lambda: D(\lambda) \leq \varepsilon/2} \frac{\sqrt{\varepsilon/2}}{||H(\lambda^{(i)})||}.$$

Also, when $D(\lambda^{(i)}) \geq \varepsilon/2$, we have

$$D(\lambda^{(i+1)}) < D(\lambda^{(i)}) - \rho_i K(\varepsilon),$$

with

$$K(\varepsilon) := \min_{\lambda: D(\lambda) \geq \varepsilon/2} -(\lambda - \lambda^*)^T H(\lambda),$$

provided

$$\rho_i \leq \min_{\lambda: D(\lambda) \geq \varepsilon/2} \frac{-(\lambda - \lambda^*)^T H(\lambda)}{||H(\lambda)||^2}.$$

Thus, we obtain that $D(\lambda^{(i+1)}) \leq \max\{D(\lambda^{(i)}) - \rho_i K(\varepsilon), \varepsilon\}$ whenever $\rho_i \leq \rho_\varepsilon$, with

$$\rho_\varepsilon := \min\left\{\min_{\lambda: D(\lambda) \leq \varepsilon/2} \frac{\sqrt{\varepsilon/2}}{||H(\lambda)||}, \min_{\lambda: D(\lambda) \geq \varepsilon/2} \frac{(\lambda - \lambda^*)^T H(\lambda)}{||H(\lambda)||^2}\right\}.$$

Noting that $\rho_i \downarrow 0$ as $i \to \infty$ and $\sum_{i=0}^{\infty} \rho_i = \infty$, we conclude $D(\lambda^{(i)}) \to 0$, and thus $\lambda^{(i)} \to \lambda^*$ as $i \to \infty$. $\square$

Proof of Corollary 33.1. By definition, the transmit powers P^* minimize $\sum_{c=1}^{C} \lambda_c^* L_c$. In particular,

$$\sum_{c=1}^{C} \lambda_c^* L_c(\lambda^{(i)}) \geq \sum_{c=1}^{C} \lambda_c^* L^* = L^*,$$

which gives

$$\liminf_i \sum_{c=1}^{C} \lambda_c^* L_c(\lambda^{(i)}) \geq L^*.$$

Also, the transmit powers $P(\lambda^{(i)})$ minimize

$$\sum_{c=1}^{C} \lambda_c^{(i)} L_c,$$

so

$$\sum_{c=1}^{C} \lambda_c^{(i)} L_c(\lambda^{(i)}) \leq \sum_{c=1}^{C} \lambda_c^{(i)} L^* = L^*.$$

Since $\lim_{i \to \infty} \lambda^{(i)} = \lambda^*$, then

$$\limsup_i \sum_{c=1}^{C} \lambda_c^* L_c(\lambda^{(i)}) \leq L^*$$

We thus conclude that
$$\sum_{c=1}^{C} \lambda_c^* L_c(\lambda^{(i)}) \to L^*$$
as $i \to \infty$. Denote
$$L_c^{(i)} = \sum_{m=1}^{M} P_{mc}^{(i)}.$$
Since
$$\sum_{j=0}^{i} \rho_j P_{mc}^{(i)} = \sum_{j=0}^{i} \rho_j P_{mc}(\lambda^{(j)}),$$
with $\sum_{j=0}^{\infty} \rho_j = \infty$, it follows that $\sum_{c=1}^{C} \lambda_c^* L_c^{(i)} \to L^*$ as $i \to \infty$ by summing the above equation over m.

Now observe that
$$\begin{aligned}
\lambda_c^{(i+1)} - \lambda_c^{(0)} &= \sum_{j=0}^{i} \rho_j (L_c(\lambda^{(j)}) - \bar{L}(\lambda^{(j)})) \\
&= (L_c^{(i+1)} - \bar{L}_c^{(i+1)}) \sum_{j=0}^{i} \rho_j,
\end{aligned}$$
which implies that
$$\lim_{i \to \infty} (L_c^{(i)} - \bar{L}_c^{(i)}) = 0$$
as
$$\sum_{j=0}^{\infty} \rho_j = \infty.$$
Combining the above two results, we obtain that
$$\lim_{i \to \infty} L_c^{(i)} \to L^*,$$
which means that any limit point of the sequence $P^{(i)}$ must be an optimal power assignment vector. Since the sequence $P^{(i)}$ is bounded and the optimal power assignment vector P^* is assumed to be unique, it follows that $\lim_{i \to \infty} P^{(i)} = P^*$. □

Bibliography

Y. Bejerano and R.S. Bhatia. Mifi: Managed WiFi for QoS assurance, fairness and high throughput of current IEEE 802.11 networks with multiple access points, 2004. Preprint.

D. Bertsimas and M. Sim. The price of robustness. *Operations Research*, 52:35–53, 2004.

G.J. Foschini and Z. Milzanic. A simple distributed autonomous power control algorithm and its convergence. *IEEE Trans. Veh. Techn.*, 40:641–646, 1993.

D. Gay, R. Fourer, and B. Kernighan. *AMPL (A Modeling Language for Mathematical Programming)*. Scientific Press, San Francisco, 1993.

K. Goto, T. Suzuki, and T. Hattori. Cell size adaptation in W-CDMA cellular system. In *Proc. of Veh. Tech. Conf.*, Birmingham, Alabama, 2002.

S.V. Hanly. An algorithm for combined cell-site selection and power control to maximize cellular spread spectrum capacity. *IEEE J. Sel. Areas Commun.*, 13(7):1332–1340, 1995.

L. Hurwicz, K. Arrow, and H. Uzawa. *Studies in linear and non-linear programming*. Stanford University Press, 1958.

ILOG, Inc. Cplex division, 1998.

W.-I. Kim and C.-S. Kang. A new traffic-load shedding scheme in the WCDMA mobile communication systems. *Proc. of Veh. Tech. Conf.*, 4:2405–2409, 2002.

J.K. Lenstra, D. Shmoys, and E. Tardos. Approximation algorithms for scheduling unrelated parallel machines. In *Proc. 28th IEEE FOCS*, 1987.

F. Rashid-Farrokhi, K.J.R. Liu, and L. Tassiulas. Downlink power control and base station assignment. *IEEE Commun. Lett.*, 1:102–104, 1997.

R.D. Yates. A framework for uplink power control in cellular radio systems. *IEEE J. Sel. Areas Commun.*, 13(7):1341–1347, 1995.

R.D. Yates and C.-Y. Huang. Integrated power control and base station assignment. *IEEE Trans. Veh. Techn.*, 44:638–644, 1995.

VI The web and beyond

34 OPTIMIZATION ISSUES IN WEB SEARCH ENGINES

Zhen Liu[1] and Philippe Nain[2]

[1] IBM Research
Hawthorne, NY 10532, USA
zhenl@us.ibm.com

[2] INRIA
B.P. 93, 06902, Sophia Antipolis Cedex, France
Philippe.Nain@inria.fr

Abstract: Crawlers are deployed by a Web search engine for collecting information from different Web servers in order to maintain the currency of its data base of Web pages. We present studies on the optimization of Web search engines from different perspectives. We first investigate the number of crawlers to be used by a search engine so as to maximize the currency of the data base without putting an unnecessary load on the network. Both the static setting, where crawlers are always active, and the dynamic setting where, crawlers may be activated/deactivated as a function of the state of the system, are addressed. We then consider the optimal scheduling of the visits of these crawlers to the Web pages assuming these pages are modified at different rates. Finally, we briefly discuss some other optimization issues of Web search engines, including page ranking and system optimization.

Keywords: Web search engines, web crawlers, scheduling, optimal control, queues; Markov decision process.

34.1 INTRODUCTION

The role of World Wide Web as a major information publishing and retrieving mechanism on the Internet is now predominant and continues to grow extremely fast. The amount of information on the Web has long since become too large for manually browsing through any significant portion of its hypertext structure. As a consequence, a number of Web search engines have been developed in the last decade: starting from the pioneering search engines such as Alta Vista, Lycos, Infoseek, Magellan, Excite, to the most successful ones such as Yahoo and Google.

Search engines have become an indispensable utility for Internet users. According to a recent Pew Foundation Internet and Project (January 2005), "Search engines are highly popular among Internet users. Searching the Internet is one of the earliest activities people try when they first start using the Internet, and most users quickly feel comfortable with the act of searching. Users paint a very rosy picture of their online search experiences.", and as of January 2005, "84% of Internet users have used search engines. On any given day, 56% of those online use search engines."

Thus, technologies that enhance Web search engines are of high practical interest. These search engines consist of indexing engines for constructing a data base of Web pages, and in many cases **crawlers** for bringing information to the indexing engine. To maintain currency and completeness of the data base, crawlers periodically make recursive traversals of the Web's hypertext structure by accessing pages, then the pages referenced by these pages, and so on. In the literature one finds other colorful terms for crawler, such as wanderer, robot or spider, and the notion of a crawler being 'routed to' or 'visiting' a page. This chapter keeps with the 'crawler' and 'accessing' terminology throughout.

Traditionally, crawlers visit and index the Web pages until the data base reaches certain size. Periodically, this process is repeated through the rebuilding of a brand new data collection in replacement of the old one. Alternatively, the data base can be refreshed or updated incrementally. Such an operational mode is sometimes referred to as incremental crawler, see e.g. Cho and Garcia-Molina (2000b). Throughout this chapter, we consider the latter mode, i.e. the incremental crawler, although most analyzes apply to the former as well.

Due to the critical role that these crawlers play in the Web search engines, the optimization issues are topics of a number of research papers. In this chapter we present some of these research problems. Rather than providing comprehensive, but high-level, discussions, we present detailed solutions to some of the technical problems.

More precisely, Section 34.2 considers both the issues of optimizing the number of the crawlers to be deployed when all crawlers are always active (static setting – Section 34.2.1), and of finding an optimal decision rule for the case where crawlers may be activated/deactivated as a function of the state of the system (dynamic setting – Section 34.2.2). Performance of static and dynamic policies are compared in Section 34.2.3. The optimal scheduling of the page visits of these crawlers is studied in Section 34.3. Finally, we provide pointers to some other issues such as page ranking and system optimization (Section 34.4).

A word on the notation in use: $\lfloor x \rfloor$ (respectively $\lceil x \rceil$) denotes the largest (respectively smallest) integer less (respectively greater) than or equal to x. Also for any mappings f and g, the relation $f(x) \stackrel{x}{\sim} g(x)$ is understood as $\lim_{x \to \infty} f(x)/g(x) = 1$.

34.2 OPTIMIZING THE NUMBER OF CRAWLERS

We first address in Section 34.2.1 the situation where crawlers are always active, regardless of the state of the system, and we determine the optimal number of crawlers to be deployed. Then, we move in Section 34.2.2 to the situation where crawlers may be activated/deactivated as a function of the state of the system, and we find an optimal

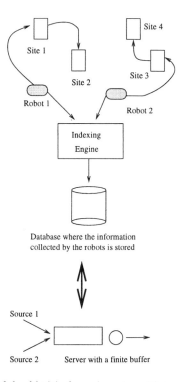

Figure 34.1 Model of search engine with two crawlers

decision rule for the number of active crawlers at any time. In both settings the cost function is a weighted sum of the starvation probability and loss rate.

The results presented in this section are based on the work of Talim et al. (2001b) and Talim et al. (2001a). Practical issues of deploying parallel crawlers are discussed in Cho and Garcia-Molina (2002).

34.2.1 The static setting

The search engine is modeled as a single server finite capacity queue. The system capacity is $K \geq 2$ (including the position in the server), see Figure 34.1.

There are $N \geq 1$ crawlers: each crawler brings new pages to the queue according to a Poisson process with rate $\lambda > 0$. These N Poisson processes are assumed to be mutually independent and independent of the indexing (service) times. Hence, new pages are generated according to a Poisson process with intensity λN. An incoming page finding a full queue is lost. Indexing times are assumed to be independent and identically random variables with common distribution $F(x)$. Let $1/\mu$ be the expected indexing time.

The search engine is therefore modeled as the well known M/G/1/K queue (see e.g. Cohen (1982, Chapter III.6)). In this notation we define the cost function as the weighted sum of two terms:

- the fraction of time that the system is empty, hereafter referred to as the *starvation probability*;
- the expected number of times when an arriving crawler finds a full system per unit time, hereafter referred to as the *loss rate*.

Let X (resp. X^*) be the stationary queue-length at arbitrary epochs (resp. stationary queue-length at arrival epochs) in a M/G/1/K queue with arrival rate λN and service rate μ.

With $\rho := N\lambda/\mu > 0$ and for $\gamma > 0$ the cost function is then defined as

$$C(\rho,\gamma,K) := \gamma \operatorname{Prob}(X=0) + \lambda N \operatorname{Prob}(X^*=K) \tag{34.1}$$

with $\operatorname{Prob}(X=0)$ and $\lambda N \operatorname{Prob}(X^*=K)$ the starvation probability and the loss rate, respectively. Since $\operatorname{Prob}(X^*=i) = \operatorname{Prob}(X=i)$ for $i=0,1,\ldots,K$ from the PASTA property Wolff (1982), (34.1) rewrites as

$$C(\rho,\gamma,K) = \gamma \operatorname{Prob}(X=0) + \rho\mu \operatorname{Prob}(X=K) \tag{34.2}$$

where λN in (34.1) has been replaced by $\rho\mu$.

Throughout Section 34.2.1 we will assume that indexing times are exponentially distributed. The general case where the indexing times are arbitrarily distributed is more involved, due to the lack of closed-form expressions for the M/G/1/K queue, and is discussed in Talim et al. (2001b).

34.2.1.1 The M/M/1/K search engine model.

We assume that the indexing times are exponentially distributed, namely, $F(x) = 1 - \exp(-\mu x)$. In other words, we model the search engine as an M/M/1/K queue.

In the M/M/1/K queue with traffic intensity ρ the stationary queue-length probabilities at arbitrary epochs are given by Kleinrock (1975):

$$\operatorname{Prob}(X=i) = \frac{1-\rho}{1-\rho^{K+1}} \rho^i \tag{34.3}$$

for $i = 0,1,\ldots,K$. Therefore,

$$C(\rho,\gamma,K) = \frac{(1-\rho)(\gamma+\mu\rho^{K+1})}{1-\rho^{K+1}}. \tag{34.4}$$

In particular, $C(\rho,\gamma,K) = (\gamma+\mu)/(K+1)$ when $\rho = 1$.

Lemma 34.1 shows the existence of a unique minimum for $C(\rho,\gamma,K)$ considered as a function of ρ. The proof is provided in Talim et al. (2001b).

Lemma 34.1 *For any $\gamma > 0$, $K \geq 2$, the mapping $\rho \to C(\rho,\gamma,K)$ has a unique minimum in $[0,\infty)$, to be denoted $\rho(\gamma,K)$. Furthermore, $0 < \rho(\gamma,K) < 1$ if $\gamma < \gamma(K)$, $\rho(\gamma,K) = 1$ if $\gamma = \gamma(K)$ and $\rho(\gamma,K) > 1$ if $\gamma > \gamma(K)$, with $\gamma(K) := \mu(K+2)/K$.*

We now return to the original problem, namely the computation of the number N of crawlers that minimizes the cost function $C(\rho, \gamma, K)$ with $\rho = \lambda N/\mu$. The answer is found in the next result which is a direct corollary of Lemma 34.1.

Proposition 34.1 *For any $\gamma > 0$, $K \geq 2$, let $N(\gamma, K)$ be the optimal number of crawlers to use.*
Then,

$$N(\gamma, K) = \arg\min_n C(n\lambda/\mu, \gamma, K) \qquad (34.5)$$

with $n \in \{\lfloor \rho(\gamma,K)\mu/\lambda \rfloor, \lceil \rho(\gamma,K)\mu/\lambda \rceil\}$. Furthermore, $N(\gamma,K) \leq \lceil \mu/\lambda \rceil$ if $\gamma < \gamma(K)$, $N(\gamma,K) \in \{\lfloor \mu/\lambda \rfloor, \lceil \mu/\lambda \rceil\}$ if $\gamma = \gamma(K)$, and $N(\gamma,K) \geq \lfloor \mu/\lambda \rfloor$ if $\gamma > \gamma(K)$.

In the next section we investigate the impact of the parameter γ on the optimal number of crawlers.

34.2.1.2 Impact of γ on the optimal number of crawlers. Recall that the parameter γ is a positive constant that allows us to stress either the probability of starvation or the loss rate. Part of the impact of γ on $\rho(\gamma, K)$, and therefore on $N(\gamma, K)$, the optimal number of crawlers, is captured in the following result.

Proposition 34.2 *For any $K \geq 2$, the mapping $\gamma \to \rho(\gamma, K)$ is nondecreasing in $(0, \infty)$, with $\lim_{\gamma \to \infty} \rho(\gamma, K) = \infty$.*

Proof. Pick two constants $0 < \gamma_1 < \gamma_2$ and define

$$\begin{aligned} \Delta(\rho, \gamma_1, \gamma_2, K) &:= C(\rho, \gamma_2, K) - C(\rho, \gamma_1, K) \\ &= \frac{1-\rho}{1-\rho^{K+1}}(\gamma_2 - \gamma_1). \end{aligned}$$

We assume that $\rho(\gamma_2, K) < \rho(\gamma_1, K)$ and show that this yields a contradiction. □
Under the condition $\gamma_1 < \gamma_2$ the mapping $\rho \to \Delta(\rho, \gamma_1, \gamma_2, K)$ is decreasing in $[0, \infty)$. Therefore,

$$\begin{aligned} 0 &< \Delta(\rho(\gamma_2,K), \gamma_1, \gamma_2, K) - \Delta(\rho(\gamma_1,K), \gamma_1, \gamma_2, K) \\ &= [C(\rho(\gamma_2,K), \gamma_2, K) - C(\rho(\gamma_1,K), \gamma_2, K)] \\ &\quad + [C(\rho(\gamma_1,K), \gamma_1, K) - C(\rho(\gamma_2,K), \gamma_1, K)] \\ &\leq 0, \qquad (34.6) \end{aligned}$$

which contradicts the fact that $\rho \to \Delta(\rho, \gamma_1, \gamma_2, K)$ is decreasing in $[0, \infty)$. Therefore $\rho(\gamma_2, K) \geq \rho(\gamma_1, K)$ and the mapping $\gamma \to \rho(\gamma, K)$ is nondecreasing in $[0, \infty)$. We may then define $L := \lim_{\gamma \to \infty} \rho(\gamma, K)$.
From the identity $\partial C(\rho, \gamma, K)/\partial \rho = 0$ for $\rho = \rho(\gamma, K)$ (see Lemma 34.1) we obtain

$$\begin{aligned} 0 &= \mu\rho(\gamma,K)^{2(K+1)} - (\gamma K + \mu(K+2))\rho(\gamma,K)^{K+1} \\ &\quad + (K+1)(\mu+\gamma)\rho(\gamma,K)^K - \gamma. \qquad (34.7) \end{aligned}$$

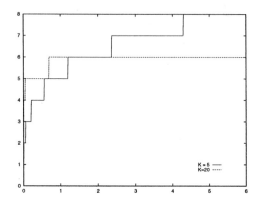

Figure 34.2 $\gamma \to N(\gamma, K)$ for $1/\lambda = 0.6$ and $\mu = 1$

Assume that $L < \infty$. Letting $\gamma \to \infty$ in (34.7) yields

$$\left(KL^{K+1} - (K+1)L^K + 1\right) \lim_{\gamma \to \infty} \gamma$$
$$= \mu L^K \left(L^{K+2} - (K+2)\right)L + (K+1)). \qquad (34.8)$$

Since $L > 1$ (we have shown in Lemma 34.1 that $\rho(\gamma, K) > 1$ for $\gamma > \mu(K+2)/K$) it is easily seen that $KL^{K+1} - (K+1)L^K + 1 > 0$, which implies that the l.h.s. of (34.8) is infinite whereas the r.h.s. is finite. Therefore, (34.8) cannot hold if $L < \infty$ and $\lim_{\gamma \to \infty} \rho(\gamma, K) = \infty$. This concludes the proof. □

Proposition 34.2 has a simple physical interpretation. As the parameter γ increases the probability of starvation becomes the main quantity to minimize. Hence, the minimization is done by increasing the arrival rate or, equivalently, by increasing the number of crawlers, as shown in Proposition 34.2. Figure 34.2 provides two numerical examples illustrating the monotonicity of the optimal number of crawlers as a function of γ.

34.2.1.3 Impact of K on the optimal number of crawlers. In this section we examine the behavior of $\rho(\gamma, K)$ as a function of K.

The following results hold (see Talim et al. (2001b)):

Proposition 34.3

(a) *If $0 < \gamma \leq \mu$ then the mapping $K \to \rho(\gamma, K)$ is nondecreasing in $[2, \infty)$;*

(b) *If $\gamma > \mu$ then there exists an integer $K_0 \geq \lfloor 2u/(\gamma - \lambda) \rfloor$ such that the mapping $K \to \rho(\gamma, K)$ is nondecreasing in $[2, K_0 - 1]$ and non-increasing in $[K_0, \infty)$.*

The next proposition examines the limiting behavior of $\rho(\gamma, K)$ as K increases to infinity.

Proposition 34.4 *For any $\gamma > 0$,*

$$\lim_{K \to \infty} \rho(\gamma, K) = 1. \tag{34.9}$$

Proof. Let $M := \lim_{K \to \infty} \rho(\gamma, K)$, where the existence of the limit follows from Proposition 34.3.

Letting now $K \to \infty$ in (34.7) we see that the r.h.s. converges to $-\gamma$ if $M < 1$ and converges to infinity if $M > 1$, thereby showing that necessarily $M = 1$, which concludes the proof. □

Proposition 34.4 shows that the optimal arrival rate converges to the service capacity when the buffer size increases to infinity.

The limiting result (34.9) can be used to derive an approximation for the optimal number of crawlers to be deployed when K is large. Indeed, the relation

$$\lim_{K \to \infty} N(\gamma, K) = \lim_{K \to \infty} \arg \min_{n \in \{\lfloor \mu/\lambda \rfloor, \lceil \mu/\lambda \rceil\}} C(\lambda n/\mu, \gamma, K), \tag{34.10}$$

which follows from (34.5), suggests the following approximation, for large K

$$N(\gamma, K) \overset{K}{\sim} \begin{cases} \lceil \mu/\lambda \rceil & \text{if } C(\rho_+, \gamma, \infty) \leq C(\rho_-, \gamma, \infty) \\ \lfloor \mu/\lambda \rfloor & \text{if } C(\rho_+, \gamma, \infty) > C(\rho_-, \gamma, \infty) \end{cases} \tag{34.11}$$

with the notation

$$C(\rho, \gamma, \infty) := \lim_{K \to \infty} C(\rho, \gamma, K), \ \rho_+ := (\lambda/\mu) \lceil \mu/\lambda \rceil \text{ and } \rho_- := (\lambda/\mu) \lfloor \mu/\lambda \rfloor.$$

Since $C(\rho, \gamma, \infty) = \gamma(1-\rho)$ for $\rho \leq 1$ and $C(\rho, \gamma, \infty) = -\mu(1-\rho)$ for $\rho \geq 1$ from (34.4), we may rewrite (34.11) as

$$N(\gamma, K) \overset{K}{\sim} \begin{cases} \lceil \mu/\lambda \rceil & \text{if } -\mu(1-\rho_+) \leq \gamma(1-\rho_-) \\ \lfloor \mu/\lambda \rfloor & \text{if } -\mu(1-\rho_+) > \gamma(1-\rho_-). \end{cases} \tag{34.12}$$

The mapping $K \to \rho(\gamma, K)$ is displayed in Figure 34.3 for $\gamma < \mu$ and in Figure 34.4 for $\gamma > \mu$. Table 34.1 gives $N(\gamma, K)$ for different values of K and compare these values with the approximation (34.12) (last column in Table 34.1). The approximation (34.12) appears to be fairly sensitive to model parameters; however, in all but one case (34.12) lies within 10% of the exact value as soon as $K \geq 10$. We also observe that the quality of the approximation increases when γ increases (within 10% of the exact value for $\gamma = 2$ for all $K \geq 2$).

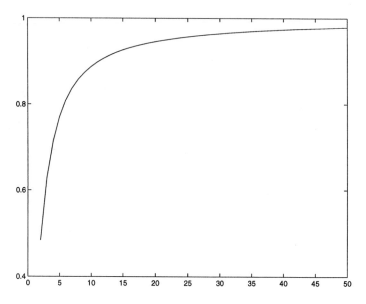

Figure 34.3 $K \to \rho(\gamma, K)$ for $\gamma = 0.5$ and $\mu = 1$

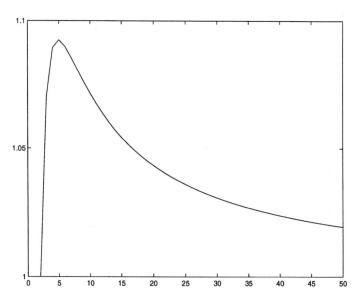

Figure 34.4 $K \to \rho(\gamma, K)$ for $\gamma = 2$ and $\mu = 1$

Table 34.1 $K \to N(\gamma, K)$ for $\lambda = 0.01$ and $\mu = 1$

$K:$	2	3	4	5	10	20	30	40	50	∞
$\gamma = 0.1$	20	34	44	52	72	85	89	92	94	100
$\gamma = 0.5$	48	63	71	77	89	95	96	97	98	100
$\gamma = 1.2$	77	88	93	96	100	101	101	101	100	100
$\gamma = 1.5$	86	96	100	102	103	102	102	101	101	100
$\gamma = 2$	100	107	109	109	107	104	103	102	102	100

34.2.2 The dynamic setting

In this section we assume that the number of active crawlers may vary in time according to the backlog in the queue and to the number of crawlers already active. To address this situation we will cast our model into the Markov Decision Process (MDP) framework (Bertsekas, 1987; Puterman, 1994; Ross, 1983).

The indexing engine is again modeled as a finite-capacity single-server queue. Service times still constitute independent random variables with common negative exponential distribution (with mean $1/\mu$) and the buffer may accommodate at most $K \geq 2$ customers, including the one in service, if any. There are N available crawlers and each of these crawlers, when activated, brings pages to the server according to a Poisson process with rate λ. We assume that these N Poisson processes are mutually independent and further independent of the service time process.

The new feature in this section is that the number of active crawlers may be modified at any arrival and at any departure epoch. When an arrival occurs, the incoming crawler is deactivated at once; the controller may then decide to keep it idle or to reactivate it. When a departure occurs the controller may either decide to activate one additional crawler, if any available, or to do nothing (i.e. the number of active crawlers is not modified).

The objective is to find a policy (to be defined) that minimizes a weighted sum of the stationary starvation probability and the loss rate.

We now introduce the MDP setting in which we will solve this optimization problem. Since the time between transitions is variable we will use the uniformization method (Bertsekas, 1987, Sec. 6.7).

At the n-th decision epoch t_n the state of the MDP is represented by the triple $x_n = (q_n, r_n, s_n) \in \{0, 1, \ldots, K\} \times \{0, 1, \ldots, N\} \times \{0, 1, 2\}$, with q_n and r_n the queue-length and the number of active crawlers just *before* the n-th decision epoch, respectively, and s_n the type (arrival, departure, fictitious – see below) of the n-th decision epoch.

The successive decision epochs $\{t_n, n \geq 1\}$ are the jump times of a Poisson process with intensity $\nu := \lambda N + \mu$, independent of the service time process. In this setting, the n-th decision epoch t_n corresponds to an arrival in the original system with probability $\lambda r_n / \nu$ (in which case $s_n = 1$), to a departure with probability μ/ν provided that $q_n > 0$ ($s_n = 0$) and to a fictitious event with the complementary probability $((N - r_n)\lambda + \mu)/\nu$ ($s_n = 2$).

Let $a_n \in \{0, 1\}$ be the action chosen at time t_n. We assume that $a_n = 1$ if the decision is made to activate one additional crawler, if any available, and $a_n = 0$ if the decision is made to keep unchanged the number of active crawlers. By convention we assume that $a_n = 0$ if the n-th decision epoch corresponds to a fictitious event ($s_n = 2$).

From the above definitions we see that states of the form $(\bullet, 0, 1)$ and $(0, \bullet, 0)$ are not feasible, as an arrival cannot occur if all crawlers are inactive and a departure cannot occur if the queue is empty, respectively. Therefore, the state-space for this MDP is

$$\{(q, r, s), 0 \leq q \leq K, 0 \leq r \leq N, s = 0, 1, 2\}$$
$$- \{(0, r, 0), (q, 0, 1), 0 \leq q \leq K, 0 \leq r \leq N\}.$$

However, this set contains one absorbing state, the "fictitious" state $(0, 0, 2)$. To remove this undesirable state we will only consider policies (see formal definition below) that always choose action $a = 1$ when the system is in state $(1, 0, 0)$ so that $(0, 0, 2)$ can never be reached. This is not a severe restriction since a policy that never activates crawlers when the system is empty is of no interest. In conclusion, the state space for this MDP is

$$\mathbf{X} := \{(q, r, s), 0 \leq q \leq K, 0 \leq r \leq N, s = 0, 1, 2\}$$
$$- \{(0, 0, 2), (0, r, 0), (q, 0, 1), 0 \leq q \leq K, 0 \leq r \leq N\}$$

and the set $\mathbf{A}_x$ of allowed actions when the system is in state $x = (q, r, s) \in \mathbf{X}$ is given by

$$\mathbf{A}_x = \begin{cases} \{0\} & \text{if } s = 2 \\ \{1\} & \text{if } (q, r, s) = (1, 0, 0) \\ \{0, 1\} & \text{otherwise.} \end{cases}$$

To complete the definition of the MDP we need to introduce the one-step cost c and the one-step transition probabilities p. Given that the process is in state $x = (q, r, s)$ and that action a is made, the one-step cost is defined as

$$c(x) = \gamma \mathbf{1}(q = 0) + \nu \mathbf{1}(q = K, s = 1), \qquad (34.13)$$

independent of a. We will show later on in this section that this choice for the one-step cost will allow us to address, and subsequently to solve, the optimization problem at hand.

For $x \in \mathbf{X}$, the one-step transition probabilities $p_{x,x'}(a)$ are given by

$$p_{x,x'}(a) = \begin{cases} \dfrac{\mu}{\nu}\mathbf{1}(q>1) & \text{if } x' = (q-1, \min\{r+a, N\}, 0) \\[4pt] \dfrac{\lambda r}{\nu} & \text{if } x' = (q-1, \min\{r+a, N\}, 1) \\[4pt] 1 - \dfrac{\mu\mathbf{1}(q>1) - \lambda r}{\nu} & \text{if } x' = (q-1, \min\{r+a, N\}, 2) \end{cases} \quad (34.14)$$

if $s = 0$, $a = 0, 1$;

$$p_{x,x'}(a) = \begin{cases} \dfrac{\mu}{\nu} & \text{if } x' = (\min\{q+1, K\}, r+a-1, 0) \\[4pt] \dfrac{\lambda(r+a-1)}{\nu} & \text{if } x' = (\min\{q+1, K\}, r+a-1, 1) \\[4pt] 1 - \dfrac{\mu + \lambda(r+a-1)}{\nu} & \text{if } x' = (\min\{q+1, N\}, r+a-1, 2) \end{cases} \quad (34.15)$$

if $s = 1$, $a = 0, 1$;

$$p_{x,x'}(0) = \begin{cases} \dfrac{\mu}{\nu}\mathbf{1}(q>0) & \text{if } x' = (q, r, 0) \\[4pt] \dfrac{\lambda r}{\nu} & \text{if } x' = (q, r, 1) \\[4pt] 1 - \dfrac{\mu\mathbf{1}(q>0) + \lambda r}{\nu} & \text{if } x' = (q, r, 2) \end{cases} \quad (34.16)$$

if $s = 2$. All other transition probabilities are equal to 0.

Without loss of generality we will only consider *pure stationary* policies since it is known that nothing can be gained by considering more general policies (Puterman, 1994, Ch. 8-9). Recall that in the MDP setting a policy π is pure stationary if, at any decision epoch, the action chosen is a non-randomized and time-homogeneous mapping of the current state (Bertsekas, 1987; Puterman, 1994; Ross, 1983). We define an *admissible* stationary policy as any mapping $\pi : \mathbf{X} \to \{0, 1\}$ such that $\pi(x) \in \mathbf{A}_x$.

For later use introduce $P(\pi) := [p_{x,x'}(\pi(x))]_{(x,x') \in \mathbf{X} \times \mathbf{X}}$, the transition probability matrix under the stationary policy π.

Let $\mathcal{P}$ be the class of all admissible stationary policies. For any policy $\pi \in \mathcal{P}$ introduce the long-run expected average cost per unit time

$$W_\pi(x) = \lim_{n \to \infty} \frac{1}{n} \mathbf{E}_\pi \left[\sum_{i=1}^n c(x_i) \,\Big|\, x_1 = x \right], \qquad x \in \mathbf{X}. \quad (34.17)$$

The existence of the limit in (34.17) is a consequence of the fact that π is stationary and $\mathbf{X}$ is countable (Puterman, 1994, Proposition 8.1.1).

We shall say that a policy $\pi^* \in \mathcal{P}$ is average cost optimal if

$$W_{\pi^*}(x) = \inf_{\pi \in \mathcal{P}} W_\pi(x) \quad \forall x \in \mathbf{X}. \tag{34.18}$$

In order to use results from MDP theory for average cost models we first need to determine to which class (recurrent, unichain, multichain, communicating, etc.) the current MDP belongs to. Consider the following example: $N = 2$ and let π be any stationary policy that selects action 1 in states $(\bullet, r, 1)$ for $r \in \{1,2\}$ and in state $(1,0,0)$, and action 0 otherwise. It is easily seen that this policy induces a MDP with two recurrent classes $(\mathbf{X} \cap \{(\bullet,1,\bullet)\}$ and $\mathbf{X} \cap \{(\bullet,2,\bullet)\})$ and a set of transient states $(\mathbf{X} \cap \{\bullet,0,\bullet\})$. We therefore conclude from this example that the MDP $\{x_n, n \geq 1\}$ is *multichain* (Puterman, 1994, p. 348).

An MDP is *communicating* (Puterman, 1994, p.348) if, for every pair of states $(x,x') \in \mathbf{X} \times \mathbf{X}$, there exists a stationary policy π such that x' is accessible from x, that is, if there exists $n \geq 1$ such that $P^n_{x,x'}(\pi) > 0$, where $P^n_{x,x'}(\pi)$ is the (x,x')-entry of the matrix $P^n(\pi)$.

Lemma 34.2 *The MDP $(x_n, n \geq 1)$ is communicating.*

The proof of Lemma 34.2 is given in (Talim et al., 2001a). The next result follows from Lemma 34.2 and Proposition 4 in (Bertsekas, 1987, Sec. 7.1):

Proposition 34.5 *There exists a scalar θ and a mapping $h : \mathbf{X} \to \mathbf{R}$ such that, for all $x \in \mathbf{X}$,*

$$\theta + h(x) = c(x) + \min_{a \in A_x} \sum_{x' \in \mathbf{X}} p_{x,x'}(a) h(x') \tag{34.19}$$

with $\theta = \inf_{\pi \in \mathcal{P}} W_\pi(x)$ for all $x \in \mathbf{X}$, while if $\pi^*(x)$ attains the minimum in (34.19) for each $x \in \mathbf{X}$, then the stationary policy π^* is optimal.

The optimal average cost θ and the optimal policy π^* in Proposition 34.17 can be computed by using the following recursive scheme, known as the relative value iteration algorithm.

Proposition 34.6 *Let $\hat{x}$ be a fixed state in $\mathbf{X}$ and $0 < \tau < 1$ be a fixed number. For $k \geq 0$, $x \in \mathbf{X}$, define the mappings $(h_k, k \geq 0)$ as*

$$h_{k+1}(x) = (1-\tau) h_k(x) + \tau (T(h_k)(x) - T(h_k)(\hat{x}))$$

with

$$T(h_k)(x) := c(x) + \min_{a \in A_x} \sum_{x' \in \mathbf{X}} p_{x,x'}(a) h_k(x'),$$

where $h_0(\hat{x}) = 0$ but otherwise h_0 is arbitrary.
Then, the limit $h(x) = \lim_{k \to \infty} h_k(x)$ exists for each $x \in \mathbf{X}$, $\theta = \tau T(h)(\hat{x})$, and the optimal action $\pi^*(x)$ in state x is given by $\pi^*(x) \in \text{argmin}_{a \in A_x} \sum_{x' \in \mathbf{X}} p_{x,x'}(a) h(x')$. □

Proof. Since the MDP is communicating (cf. Lemma 34.2) the proof follows from Puterman (1994, Sec. 8.5,9.5.3) (see also Bertsekas (1987, Prop. 4, p. 313)). □

We now return to our initial objective, namely, minimizing a weighted sum of the stationary starvation probability and the loss rate. To see why the solution to this problem is given by the solution to the MDP problem formulated in this section, it suffices to show that the average cost (34.17) is a weighted sum of the stationary starvation probability and the loss rate. It should be clear, however, that this result cannot hold for policies that induce an average cost (34.17) that depends on the initial state x as, by definition, the stationary starvation probability and the loss rate are independent of the initial state. We will therefore restrict ourselves to the class $\mathcal{P}_0 \subset \mathcal{P}$ of policies that generate a constant average cost, namely, $\mathcal{P}_0 = \{\pi \in \mathcal{P} : W_\pi(x) = W_\pi(x'), \forall x \in \mathbf{X}\}$.

The set $\mathcal{P}_0$ is non-empty as it is well-known that it contains, among others, all unichain policies (Puterman, 1994, Proposition 8.2.1). Among such policies is the *static* policy π_N that always maintain N crawlers active, namely, $\pi_N(x) = 1$ for all $x = (\bullet, \bullet, s) \in \mathbf{X}$ with $s = 0, 1$ and $\pi_N(x) = 0$ for all $x = (\bullet, \bullet, 2) \in \mathbf{X}$.

We may also note that reducing the search for an optimal policy to policies in $\mathcal{P}_0$ does not yield any loss of generality as it is also known that there always exits an optimal policy with constant average cost in the case of communicating MDP's (Puterman, 1994, Proposition 8.3.2).

Fix $\pi \in \mathcal{P}_0$. Introducing (34.13) into (34.17) yields $W_\pi(x) = \gamma S_\pi(x) + L_\pi(x)$ with

$$S_\pi(x) = \lim_{n \to \infty} \frac{1}{n} \mathbf{E}_\pi \left[\sum_{i=1}^n \mathbf{1}(q_i = 0) \,|\, x_1 = x \right]$$

$$L_\pi(x) = \nu \lim_{n \to \infty} \frac{1}{n} \mathbf{E}_\pi \left[\sum_{i=1}^n \mathbf{1}(q_i = K, s_i = 1) \,|\, x_1 = x \right].$$

In the following we will drop the argument x in $S_\pi(x)$ and $L_\pi(x)$ since these quantities do not depend on x from the definition of $\mathcal{P}_0$.

Let us now interpret S_π and L_π. S_π is the stationary probability that the system is empty at decision epochs. Since the decision epochs form a Poisson process, we may conclude from the PASTA property (Wolff, 1982) that S_π is also equal to the stationary probability that the system is empty at *arbitrary epoch* with is nothing but the stationary starvation probability.

Let us now consider L_π. Recall that $\{t_n, n \geq 1\}$, the successive decision instants, is a Poisson process with intensity ν and assume without loss of generality that $t_1 = 0$ a.s. Define $A(t)$ as the total number of customers that have arrived to the queue up to time t, including customers which have been lost, and let $Q(t)$ be the queue length at time t. We assume that the sample paths of the processes $\{A(t), t \geq 0\}$ and $\{Q(t), t \geq 0\}$ are right-continuous with left limit. With these definitions and the identity $\mathbf{E}[t_n] = n/\nu$ we may rewrite L_π as

$$L_\pi = \lim_{n \to \infty} \frac{\mathbf{E}_\pi \left[\int_0^{t_n} \mathbf{1}(Q(t-) = K) \, dA(t) \right]}{\mathbf{E}[t_n]}.$$

In other words, we have shown that L_π is the ratio, as n tends to infinity, of the expected number of losses during the first n decision epochs over the expected occurrence time of the n-th decision epoch.

The interpretation of L_π as a *loss rate* now follows from the identity

$$\begin{aligned} L_\pi &= \lim_{n\to\infty} \frac{\mathbf{E}_\pi\left[\int_0^{t_n} \mathbf{1}(Q(t-)=K)\,dA(t)\right]}{\mathbf{E}[t_n]} \\ &= \lim_{T\to\infty} \frac{1}{T}\mathbf{E}_\pi\left[\int_0^T \mathbf{1}(Q(t-)=K)\,dA(t)\right], \quad \forall \pi \in \mathcal{P}_0, \quad (34.20)\end{aligned}$$

upon noticing that the latter quantity represents the mean number of losses per unit time or the loss rate. The second identity in (34.20) is a direct consequence of the theory of renewal reward processes (Ross, 1983, Theorem 7.5) and of the definition of the set $\mathcal{P}_0$.

In summary, we have shown that for any policy π in $\mathcal{P}_0$ the average cost is

$$W_\pi = \gamma S_\pi + L_\pi,$$

with S_π the starvation probability and L_π the loss rate. □

The optimal policy has been computed for different values of the model parameters. Figures 34.5-34.7 display the optimal policy for $N=16$, $K=5$, $\lambda=0.1$, $\mu=1$ and for different values of γ ($\gamma < \gamma(K) = 1.4$, $\gamma = \gamma(K)$ and $\gamma > \gamma(K)$). The results were obtained by running the value iteration algorithm given in Proposition 34.6 with the stopping criterion $\max_{x \in \mathbf{X}} |(h_{k+1}(x) - h_k(x))/h_k(x)| < 10^{-5}$ (254, 255 and 256 iterations were needed to compute the optimal policy displayed in Figures 34.5, 34.6 and 34.7, respectively). We see from these figures that the optimal policy is a *monotone switching curve*, namely, there exist two monotone (decreasing here) integer mappings $f_s : \{0,1,\ldots,N\} \to \{0,1,2,\ldots\}$, $s \in \{0,1\}$, such that $\pi^\star(x) = \mathbf{1}(f_s(r) \geq q)$ for all $x = (q,r,s) \in \mathbf{X}$ with $s = 0,1$ (we must also have $f_0(0) \geq 1$ so that $\pi^\star(1,0,0) = 1$ as required). We conjecture that the optimal policy always exhibits such a structure but we have not able been to prove it.

OPTIMIZATION IN WEB SEARCH ENGINES 995

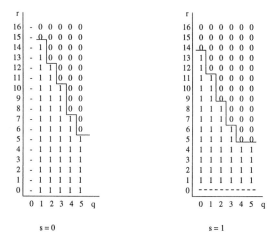

Figure 34.5 Optimal policy ($\gamma = 1$, Cost $= 0.20907$)

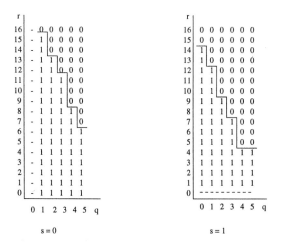

Figure 34.6 Optimal policy ($\gamma = 1.4$, Cost $= 0.25924$)

996 HANDBOOK OF OPTIMIZATION IN TELECOMMUNICATIONS

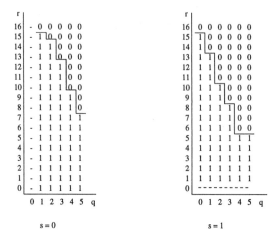

Figure 34.7 Optimal policy ($\gamma = 2$, Cost $= 0.32211$)

34.2.3 Static versus dynamic policies

In this section we compare static and dynamic policies in the case where the indexing times are exponentially distributed. The results are reported in Tables 34.2 and 34.3. Throughout the experiments $\mu = 1$. For different sets of parameters λ, K, γ, we first computed the optimal number of crawlers N_s (given by Proposition 34.1) and the average cost C_s (given in (34.4)) in the static setting.

Then, for each set of parameters λ, K, γ, we set the value of the number of available crawlers N to N_s and determined, via the relative value iteration algorithm given in Proposition 34.6 (with $\tau = 0.99999$ – the closer τ is from 1 the faster the algorithm converges), the optimal average cost C_d (given in (34.18)) as well as the minimum ($N_{\min}$) and the expected ($\overline{N}$) number of crawlers activated by the optimal dynamic policy. These results can be found in Table 34.2.

We stopped the numerical procedure when the relative error between two consecutive iterates was (uniformly) less than 10^{-5}. The number of iterations (N_{iter}) and the relative improvement ($100\% \times (C_s - C_d)/C_d$) are also reported in Table 34.2.

Last, we computed the overall optimal dynamic policy by removing the restriction on the number of available crawlers. The optimal average cost C_d as well as the minimum ($N_{\min}$), expected ($\overline{N}$) and maximum ($N_{\max}$) number crawlers used by the overall optimal dynamic policy are given in Table 34.3.

We observe that substantial gains may be achieved by dynamically controlling the activity of the crawlers. When the number of available crawlers is set to N_s (Table 34.2) the relative improvement w.r.t. to the optimal static policy ranges from 4% to 103% for the considered model parameters; when the restriction on the number of available crawlers is removed then the improvement ranges from 6% to 3226%! The gain appears to be an increasing function of the queue size K and of the arrival rate λ.

Table 34.2 Static vs. dynamic policies (with $\mu = 1$ and $\tau = 0.99999$)

λ	K	γ	Static Approach		Dynamic Approach				Rel. Impr.
			C_s	N_s	C_d	N_{min}	$\tilde{N}$	N_{iter}	
0.01	5	0.4	0.17541	73	0.16804	57	70.3	1634	4%
-	-	1.4	0.40000	100	0.38336	86	95.1	1911	4%
-	-	2.4	0.53834	114	0.51746	101	108.4	2051	4%
0.01	10	0.4	0.10207	86	0.09062	60	82.6	1794	13%
-	-	1.2	0.20000	100	0.17534	77	94.1	1939	14%
-	-	2.4	0.28347	110	0.24798	88	102.3	2039	14%
0.01	15	0.4	0.07177	91	0.05891	58	87.7	1860	22%
-	-	1.13	0.13313	100	0.10720	70	94.5	1953	24%
-	-	2.4	0.19192	107	0.15342	78	99.7	2024	25%
0.05	5	0.4	0.17578	15	0.15127	7	13.8	338	16%
-	-	1.4	0.40000	20	0.34733	12	17.7	391	15%
-	-	2.4	0.53841	23	0.46583	15	20.2	422	16%
0.05	10	0.4	0.10220	17	0.08308	5	16.2	369	23%
-	-	1.2	0.20000	20	0.14955	8	18.2	402	34%
-	-	2.4	0.28347	22	0.20541	10	19.4	423	38%
0.05	15	0.4	0.07184	18	0.05514	4	17.4	401	30%
-	-	1.13	0.13313	20	0.09117	6	18.7	426	46%
-	-	2.4	0.19372	21	0.13895	8	19.3	438	39%
0.1	5	0.4	0.17600	7	0.15239	1	6.5	167	15%
-	-	1.4	0.40000	10	0.32198	4	8.6	200	24%
-	-	2.4	0.54067	11	0.44989	5	9.3	211	20%
0.1	10	0.4	0.10403	9	0.06838	0	8.4	204	52%
-	-	1.2	0.20000	10	0.13854	2	9.0	218	44%
-	-	2.4	0.28347	11	0.18585	3	9.6	227	53%
0.1	15	0.4	0.07184	9	0.05326	0	8.7	312	35%
-	-	1.13	0.13313	10	0.08538	1	9.3	359	56%
-	-	2.4	0.19458	11	0.09606	1	9.7	376	103%

Table 34.3 Static vs. dynamic policies (with $\mu = 1$ and $\tau = 0.99999$)

λ	K	γ	Static Approach C_s	N_s	Dynamic Approach C_d	N_{min}	$\bar{N}$	N_{max}	Rel. Impr.
0.01	5	0.4	0.17541	73	0.16595	58	74.8	82	6%
-	-	1.4	0.40000	100	0.37886	88	99.9	115	6%
-	-	2.4	0.53834	114	0.51179	103	113.0	133	5%
0.01	10	0.4	0.10207	86	0.08124	62	89.3	105	27%
-	-	1.2	0.20000	100	0.15876	78	99.6	123	26%
-	-	2.4	0.28347	110	0.22777	89	107.1	137	24%
0.01	15	0.4	0.07177	91	0.04236	59	94.9	118	69%
-	-	1.13	0.13313	100	0.07812	71	99.8	131	70%
-	-	2.4	0.19192	107	0.11493	79	103.6	143	67%
0.05	5	0.4	0.17578	15	0.13770	7	15.9	20	28%
-	-	1.4	0.40000	20	0.31712	13	19.8	27	26%
-	-	2.4	0.53841	23	0.43292	16	21.9	32	24%
0.05	10	0.4	0.10220	17	0.04128	5	19.0	29	148%
-	-	1.2	0.20000	20	0.12020	8	19.9	33	66%
-	-	2.4	0.28347	22	0.20541	10	20.6	36	38%
0.05	15	0.4	0.07184	18	0.00969	2	19.8	35	641%
-	-	1.13	0.13313	20	0.01818	4	20.0	38	632%
-	-	2.4	0.19372	21	0.02782	6	20.1	41	596%
0.1	5	0.4	0.17600	7	0.11097	2	8.2	12	59%
-	-	1.4	0.40000	10	0.25924	4	9.8	16	54%
-	-	2.4	0.54067	11	0.35805	6	10.7	18	51%
0.1	10	0.4	0.10403	9	0.01937	0	9.7	18	437%
-	-	1.2	0.20000	10	0.03887	1	9.9	20	415%
-	-	2.4	0.28347	11	0.05894	2	10.1	22	381%
0.1	15	0.4	0.07184	9	0.00188	0	10.0	24	3721%
-	-	1.13	0.13313	10	0.00368	0	10.0	25	3518%
-	-	2.4	0.19458	11	0.00585	0	10.0	27	3226%

34.3 OPTIMAL SCHEDULING OF THE CRAWLERS

We now turn to the problems of scheduling a crawler that maintains the currency of existing pages in search-engine data bases. For sake of arguments, we assume that the set of Web pages is fixed. However, as we shall see, our results can be promoted as heuristics that acquire new pages and drop old pages over time. A specific objective will be to find crawler schedules that minimize the *obsolescence* of the data base in some useful sense. For example, assume there are N Web pages, labeled $1, 2, \ldots, N$, which are to be accessed repeatedly by a crawler, the duration of each access being an independent sample from a given distribution. Assume also that the contents of page i are modified at times that follow a Poisson process with parameter μ_i. A page is considered up-to-date by the indexing engine from the time it is accessed by the crawler until the next time it is modified, at which point it becomes out-of-date until the crawler's next access. Let r_i be the fraction of time page i spends out-of-date. The problem is to find relative page-access frequencies and a sequencing policy that realizes these frequencies such that the objective function $C = \sum_{1 \leq i \leq N} c_i r_i$, is minimized, where the c_i are given weights. Under simplifying but plausible assumptions on the weights, page access times, and the class of allowed policies, we obtain explicit solutions to this problem.

From a theoretical point of view, our problem is closely related to those multiple-queue single-server systems usually called *polling systems* in the queueing literature. Indeed, the crawler can be considered as the server and the pages as the stations in the polling system. The durations of consecutive page accesses correspond to switch-over times and the page modifications correspond to customer arrivals. The service times in this polling system are zero. Our two-stage approach of optimizing crawler schedules (determining access frequencies and then finding a schedule that realizes them) is similar to the approach in Borst (1994), Borst et al. (1994) and Boxma et al. (1993) of optimizing visit sequences in polling systems.

An extensive literature exists on the analysis and control of polling systems. The interested reader is referred to the book of Takagi (1986) for general references; the special issue of the journal *Queueing Systems*, Vol. 11 (1992) on polling models and the recent thesis of Borst (1994) can be consulted for more recent developments. In particular, the polling systems with zero service times were motivated by communication networks such as *teletext* and *videotex* where pages of information are to be broadcast to terminals connected to a computer network (Ammar and Wong, 1987; Dykeman et al., 1986; Liu and Nain, 1992). However, the problem here has not been analyzed. Indeed, in the usual analysis of polling systems with unbounded buffers, interest centers on mean waiting times and mean queue lengths, whereas in our problem, the performance measure of interest, viz., the obsolescence time, corresponds to the maximum waiting time of a customer during a visit cycle of the server. An alternative view of our model identifies it with a polling loss system having unit buffers, in which our obsolescence time becomes the waiting time. With this point of view, our model has potential use in maintenance applications.

The next subsection is devoted to a precise formulation of our model, and a review of some useful concepts in stochastic ordering theory. Section 34.3.2 begins by proving two properties of crawler scheduling policies: *(i)* expected obsolescence times increase as the page-access time increases in the increasing-convex-ordering sense, and *(ii)*, by Schur-convexity results, accesses to any given page should be as evenly spaced as possible. We then derive a tight lower bound on the cost function C assuming that the weights c_i are proportional to the μ_i. These results yield a formula for optimal access frequencies. Our techniques can be extended to general c_i, but explicit formulas are not attainable in general.

To motivate the assumption on weights, note that a useful choice for the c_i is the customer page-access frequency, for in this case the total cost can be regarded as a customer total error rate. The special case where the customer access frequency c_i is proportional to the page-change rate μ is reasonable under this interpretation - the greater the interest (access frequency), the greater the frequency of page modification.

Sections 34.3.3 and 34.3.4 deal with the problem of sequencing page accesses optimally, or near optimally, so as to realize a given set of access frequencies. This material is prefaced by a discussion at the end of Section 34.3.2 which relates our scheduling problem to those that come under the heading of generalized round-robin or template-driven scheduling.

In Section 34.3.3, we introduce randomized page accessing, where each access is determined by an independent and identically distributed (i.i.d.) sample from a distribution $\{f_i\}$. We show how to find that choice for this distribution which minimizes C. In Section 34.3.4, we develop a policy that performs well when N is large. It is based on work of Itai and Rosberg (1984) (in an entirely different setting) and yields a cost within 5% of optimal.

Results presented in this section are based on the work of Coffman Jr. et al. (1998). A related study was conducted in Cho and Garcia-Molina (2000a).

34.3.1 Preliminaries

Let $\{X_k\}$ be the sequence of durations of consecutive page accesses by the crawler, each X_k being distributed independently as a random variable X. For scheduling policy π, let $\pi_n \in \{1, 2, \ldots, N\}$ be the scheduling decision for the n-th access, i.e., the index of the n-th page to be accessed by the crawler under π. Define the *inter-access distance* $d_j^i(\pi) = n_j^i(\pi) - n_{j-1}^i(\pi)$, where $n_j^i(\pi)$ is the index of the j-th access of page i, i.e., $n_j^i(\pi) = \inf\{n > n_{j-1}^i(\pi) \mid \pi_n = i\}$, and where $n_0^i(\pi) \equiv 0$. Let $X_j^i = X_j^i(\pi)$ be the j-th *inter-access time* of page i, i.e., the time between the $(j-1)$-st and j-th page-i access completion times. We have $X_j^i = \sum_{k=n_{j-1}^i+1}^{n_j^i} X_k$, so the random variables X_j^i are mutually independent. Note that, if page access times X_k are exponentially distributed, then X_j^i has an Erlang distribution of d_j^i stages.

Hereafter, except in definitions, the policy π will normally be omitted from our notation; in such cases, the policy will always be clear in context.

Let $Z_j^i = Z_j^i(\pi)$ be the time that page i is out-of-date during the j-th inter-access time of page i. Let $m_n^i = m_n^i(\pi)$ be the number of accesses of page i among the first n accesses: $m_n^i = \sum_{k=1}^n \mathbf{1}\{\pi_k = i\}$, where $\mathbf{1}\{\cdot\}$ is the indicator function. Hereafter, we consider only stationary scheduling policies in the sense that, for each such policy, the limit

$$f_i = f_i(\pi) = \lim_{n \to \infty} \frac{m_n^i}{n} \qquad (34.21)$$

exists and is strictly positive for all i, $1 \le i \le N$. We call f_i the access frequency of page i. We also require that the limits $\lim_{n \to \infty} \sum_{j=1}^n Z_j^i / n$ and $\lim_{n \to \infty} \sum_{j=1}^n E[Z_j^i]/n$

exist and be equal. These last assumptions hold under fairly mild conditions, e.g., when the sequence $\{d_j^i(\pi)\}_j$ is stationary and ergodic (cf. Kingman (1968)).

The *obsolescence rate* $r_i = r_i(\pi)$ of page i is the limiting fraction of time that page i is out of date; precisely, it is defined as

$$r_i = \lim_{n \to \infty} \frac{\sum_{j=1}^{m_n^i} Z_j^i}{\sum_{j=1}^{m_n^i} X_j^i} = \frac{\lim_{n \to \infty} \frac{\sum_{j=1}^{m_n^i} Z_j^i}{n}}{\lim_{n \to \infty} \frac{\sum_{j=1}^{m_n^i} X_j^i}{n}} = \frac{1}{E[X]} \cdot \lim_{n \to \infty} \frac{\sum_{j=1}^{m_n^i} E[Z_j^i]}{n}. \qquad (34.22)$$

In particular, when policy π is cyclic with cycle length K, i.e., when $\pi_{nK+k} = \pi_{(n-1)K+k}$ for all $1 \leq k \leq K$ and all $n = 1, 2, \ldots$, then

$$r_i = \frac{1}{KE[X]} \sum_{j=1}^{m_K^i} E[Z_j^i], \qquad (34.23)$$

where m_K^i is the number of page-i accesses during a cycle. The cost function to be minimized is the weighted sum of the obsolescence rates:

$$C = C(\pi) = \sum_{i=1}^{N} c_i \, r_i, \qquad (34.24)$$

where c_i are given positive real numbers and the minimization is to be over all stationary scheduling policies.

A few basics in stochastic ordering conclude this section. For two m-dimensional real vectors $\mathbf{x}$ and $\mathbf{y}$, $\mathbf{x}$ majorizes $\mathbf{y}$, written $\mathbf{x} \succ \mathbf{y}$, if $\sum_{i=1}^k x_{[i]} \geq \sum_{i=1}^k y_{[i]}$, for $k = 1, \ldots, m-1$ and $\sum_{i=1}^m x_{[i]} = \sum_{i=1}^m y_{[i]}$, where $x_{[i]}$ is the ith largest component of $\mathbf{x}$. Intuitively, $\mathbf{y}$ is better balanced than $\mathbf{x}$. A function h is said to be *Schur-convex* if $h(\mathbf{x}) \geq h(\mathbf{y})$ whenever $\mathbf{x} \succ \mathbf{y}$. See Marshall and Olkin (1979) for more details about this and related properties.

A random variable Y_1 is said to be no greater than a random variable Y_2 in the convex ordering sense, denoted $Y_1 \leq_{cx} Y_2$, if $E[h(Y_1)] \leq E[h(Y_2)]$ for all convex functions h, provided the expectations exist. If in this definition 'convex' is replaced everywhere by 'increasing and convex,' then we write $Y_1 \leq_{icx} Y_2$. As is easily verified, $Y_1 \leq_{cx} Y_2$ implies that Y_1 has the same mean but smaller variance than Y_2. It is also easy to see that $Y_1 \leq_{cx} Y_2$ implies $Y_1 \leq_{icx} Y_2$. See Stoyan (1933) for equivalent definitions and further properties.

34.3.2 Schur convexity and a lower bound

Recall that a page is considered out-of-date from the time it is modified until the next time it is accessed by the crawler. Thus, if page i is not modified during its j-th inter-access interval, then the obsolescence time is $Z_j^i = 0$. Otherwise, Z_j^i is the time that elapses from the first moment page i is modified during its j-th inter-access interval until the end of that interval. Recall also that the modification (or mutation) epochs of page i follow a Poisson process with parameter μ_i. By the memoryless property of the

Poisson process, the time that elapses from the beginning of page i's j-th inter-access interval to the first subsequent mutation has an exponential distribution with parameter μ_i. Let $R_1^i, R_2^i, \ldots$ be an i.i.d. sequence of such random variables, so that

$$Z_j^i \stackrel{d}{=} (X_j^i - R_j^i)^+, \qquad (34.25)$$

where x^+ denotes $\max(x, 0)$ and $\stackrel{d}{=}$ denotes equality in distribution.

As an immediate consequence, we obtain

Proposition 34.7 *If the page access time is decreased in the increasing convex ordering sense, then the obsolescence rate is decreased for all pages under any scheduling policy.*

Proof. Let $\{X_k'\}$ be a sequence of access times distributed independently as X', and define $\{X''_j\}_j$ and $\{Z''_j\}_j$ as for X_k. Assume that $X' \leq_{icx} X$. Then,

$$Z''^i_j \stackrel{d}{=} \left(X''^i_j - R_j^i\right)^+ \leq_{icx} \left(X_j^i - R_j^i\right)^+ \stackrel{d}{=} Z_j^i,$$

and so $E[Z''^i_j] \leq E[Z_j^i]$. Thus,

$$r_i' = \frac{1}{E[X']} \cdot \lim_{n \to \infty} \frac{\sum_{j=1}^{m_n^i} E[Z''^i_j]}{n} \leq \frac{1}{E[X]} \cdot \lim_{n \to \infty} \frac{\sum_{j=1}^{m_n^i} E[Z_j^i]}{n} = r_i,$$

as desired. $\square$

Returning to our main problem, where the distribution of page access times is assumed given, we now show that the obsolescence rate is a Schur convex function of the vector of inter-access distances. For this, we need the following calculation which will also be useful for later results. Define $h_i = E[e^{-\mu_i X}]$, the Laplace transform of X evaluated at μ_i.

Lemma 34.3 *For any page i,*

$$E[Z_j^i] = d_j^i E[X] - \frac{1}{\mu_i}\left(1 - h_i^{d_j^i}\right).$$

Proof. Let G_j^i be the probability distribution of X_j^i. We have from (34.25) that

$$\begin{aligned}
E[Z_j^i] &= \int_0^\infty P(Z_j^i > z)\,dz \\
&= \int_0^\infty P(X_j^i - R_j^i > z)\,dz \\
&= \int_0^\infty \int_z^\infty \left(1 - e^{-\mu_i(x-z)}\right) \cdot G_j^i(dx)\,dz \\
&= \int_0^\infty \int_0^x \left(1 - e^{-\mu_i(x-z)}\right) dz\, G_j^i(dx) \\
&= \int_0^\infty \left(x - \frac{1 - e^{-\mu_i x}}{\mu_i}\right) G_j^i(dx)
\end{aligned}$$

which yields the lemma. □

We can conclude from the above proof that the result of Proposition 34.7 still holds when the increasing convex ordering is replaced by the weaker Laplace-transform ordering (see Stoyan (1933)). It follows from a result of Schur (cf. Marshall and Olkin (1979, Proposition 3.C.1, page 64)) and Lemma 34.3 that

Proposition 34.8 *For any fixed number n of page-i accesses, the expected total obsolescence time of page i, $\sum_{i=1}^{n} E[Z_j^i]$ is a Schur convex function of the distances d_j^i, $j = 1, \ldots, n$.*

Thus, in order to minimize the expected obsolescence time, the accesses to any particular page should be as evenly spaced as possible.

An algorithm that computes a schedule of the crawler that implements a given set of access frequencies in the sense of (34.21) is called an *accessing* policy. In these terms, the scheduling policies proposed in this paper consist of two stages; the first computes a set of access frequencies $\{f_i\}$ and the second is an accessing policy that implements $\{f_i\}$. The even-spacing objective of accessing policies yields a lower bound, as follows.

Proposition 34.9 *The obsolescence rate under any accessing policy implementing the access frequencies $\{f_i\}$ satisfies for each i,*

$$r_i \geq \frac{1}{E[X]} \left(E[X] - \frac{f_i}{\mu_i} + \frac{f_i}{\mu_i} h_i^{1/f_i} \right).$$

Proof.

$$r_i = \frac{1}{E[X]} \cdot \lim_{n \to \infty} \frac{\sum_{j=1}^{m_n^i} E[Z_j^i]}{n}$$

$$= \frac{1}{E[X]} \cdot \lim_{n \to \infty} \frac{1}{n} \sum_{j=1}^{m_n^i} \left(d_j^i E[X] - \frac{1}{\mu_i} \left(1 - h_i^{d_j^i} \right) \right)$$

$$= \frac{1}{E[X]} \cdot \lim_{n \to \infty} \frac{1}{n} \left(nE[X] - \frac{m_n^i}{\mu_i} + \frac{1}{\mu_i} \sum_{j=1}^{m_n^i} h_i^{d_j^i} \right)$$

$$= \frac{1}{E[X]} \left(E[X] - \frac{f_i}{\mu_i} + \frac{1}{\mu_i} \lim_{n \to \infty} \frac{1}{n} \sum_{j=1}^{m_n^i} h_i^{d_j^i} \right)$$

$$\geq \frac{1}{E[X]} \left(E[X] - \frac{f_i}{\mu_i} + \frac{1}{\mu_i} \lim_{n \to \infty} \frac{m_n^i}{n} h_i^{n/m_n^i} \right)$$

$$= \frac{1}{E[X]} \left(E[X] - \frac{f_i}{\mu_i} + \frac{f_i}{\mu_i} h_i^{1/f_i} \right),$$

where the inequality comes from the Schur convexity of $\sum_{j=1}^{m_n^i} h_i^{d_j^i}$ in the d_j^i's (cf. Proposition 34.8). □

The above lower bound can be achieved only in special cases. For instance, if the frequencies are all equal, the policy that accesses pages $1,2,\ldots,N$ cyclically yields this optimal obsolescence rate. Another example where we can find a feasible accessing policy achieving the lower bound is when the frequencies are of the form $f_i = 1/2^{k_i}$, where k_i is an integer for every i. We return to the general case after considering the cost-minimization theorem. The proof of the following theorem gives a solution technique applicable to general weights c_i and shows that the technique leads to explicit results in an interesting special case.

Proposition 34.10 *Assume that the weights in the cost function are proportional to the mutation rates of the pages, i.e., $c_i = c_0 \mu_i$ for all $i = 1,2,\ldots,N$. Then for any scheduling policy,*

$$C = c_0 \cdot \sum_{i=1}^{N} \mu_i r_i \geq c_0 \left(\mu - \frac{1}{E[X]} + \frac{1}{E[X]} \prod_{i=1}^{N} h_i \right) > 0, \qquad (34.26)$$

where $\mu = \sum_{i=1}^{N} \mu_i$.

Proof. For the moment, let the c_i be general. Following Proposition 34.9, we have $C \geq C^*$, where C^* is the solution to the following optimization problem:

$$C^* = \min \sum_{i=1}^{N} c_i \left(1 - \frac{1}{E[X]\mu_i} x_i + \frac{1}{E[X]\mu_i} x_i h_i^{1/x_i} \right) \qquad (34.27)$$

subject to $x_i \geq 0$ and

$$\sum_{i=1}^{N} x_i = 1.$$

To solve the above problem, we use Lagrange multipliers and define

$$\mathcal{L}(x_1,\ldots,x_N,\lambda) = \sum_{i=1}^{N} c_i \left(1 - \frac{1}{E[X]\mu_i} x_i + \frac{1}{E[X]\mu_i} x_i h_i^{1/x_i} \right)$$
$$+ \lambda \left(\sum_{i=1}^{N} x_i - 1 \right).$$

By the convexity of the function

$$\sum_{i=1}^{N} c_i \left(1 - \frac{1}{\mu_i E[X]} x_i + \frac{1}{\mu_i E[X]} x_i h_i^{1/x_i} \right)$$

in the vector $(x_1,\ldots,x_N)$, the solution satisfies the necessary and sufficient conditions:

$$\frac{\partial \mathcal{L}}{\partial x_i} = -\frac{c_i}{\mu_i E[X]} \left(1 - h_i^{1/x_i} + \frac{\ln h_i}{x_i} h_i^{1/x_i} \right) + \lambda = 0 \qquad (34.28)$$

$$\frac{\partial \mathcal{L}}{\partial \lambda} = \sum_{i=1}^{N} x_i - 1 = 0. \qquad (34.29)$$

Observe that $h_i < 1$, so that $h_i^{1/x_i} < 1$. One can easily check that the function $1 - y + y \ln y$ is strictly decreasing in y for $y < 1$. Thus, under the assumption that c_i is proportional to μ_i, we conclude from (34.28) that all h_i^{1/x_i} are identical and that the minimum is achieved, by (34.29), when

$$x_i = \frac{\ln h_i}{\sum_{i=1}^{N} \ln h_i} = \frac{\ln(h_i)^{-1}}{\sum_{i=1}^{N} \ln(h_i)^{-1}}. \tag{34.30}$$

This solution is positive so it is also the solution to the minimization problem in (34.27). Hence,

$$\begin{aligned} C^* &= \sum_{i=1}^{N} c_0 \mu_i - \frac{c_0}{E[X]} \sum_{i=1}^{N} x_i + \frac{c_0}{E[X]} \sum_{i=1}^{N} x_i h_i^{1/x_i} \\ &= c_0 \left(\mu - \frac{1}{E[X]} + \frac{1}{E[X]} \exp\left\{ \sum_{i=1}^{N} \ln h_i \right\} \right) \\ &= c_0 \left(\mu - \frac{1}{E[X]} + \frac{1}{E[X]} \prod_{i=1}^{N} h_i \right). \end{aligned}$$

Note that

$$\prod_{i=1}^{N} h_i = E\left[\exp\left\{ -\sum_{i=1}^{N} \mu_i X_i \right\} \right] > 1 - E\left[\sum_{i=1}^{N} \mu_i X_i \right] = 1 - \mu E[X],$$

so

$$\mu - \frac{1}{E[X]} + \frac{1}{E[X]} \prod_{i=1}^{N} h_i > 0,$$

and the proof is complete. $\square$

When the weights in the cost function are not proportional to the mutation rates of pages, one can still use Lagrange multipliers to solve the optimization problem. As noted earlier, however, we do not have closed-form solutions in general.

It is also worthwhile noticing that the optimal access frequencies (cf. (34.30)) in the above lower bound are not necessarily proportional to the page mutation rates μ_i, a fact that has emerged in the context of other polling systems (see, e.g., Borst et al. (1994) and Boxma et al. (1993)). Rather, they are proportional to $\ln(h_i)^{-1} = \ln\left(E[e^{-\mu_i X}]\right)^{-1}$. Proportionality to the μ_i occurs only when X is a constant. Note also that the magnitude of the difference between $\mu_i E[X]$ and $\ln\left(E[e^{-\mu_i X}]\right)^{-1}$ is large if $Var(X)$ is large (or X is large in the convex ordering sense).

To summarize, the results of this section show that, if the weights in the cost function are proportional to the mutation rates of the pages, then an accessing policy that comes close to the lower bound in Proposition 34.9 with the f_i nearly proportional to the $\ln(h_i)^{-1}$ will come close to minimizing C.

Finding good accessing policies that realize a given set of access frequencies is the subject of the next two sections. In Section 5, we develop an optimal randomized accessing policy, and in Section 6, we adapt the well-studied golden-ratio policy to our problem, primarily as a candidate for good asymptotic performance; we will see that this policy gives an obsolescence rate within 5% of the lower bound, in the limit of large N.

We remark that this problem is closely related to the design and analysis of polling/splitting sequences in the context of queueing (and in particular, communication) systems (Andrews et al., 1997; Arian and Levy, 1992; Borst et al., 1994; Boxma et al., 1994; 1993), where algorithms are described as template driven or generalized round robin. With future research in mind, we note that these studies suggest other approaches worth investigating, e.g., extensions of the mathematical programming techniques in Borst et al. (1994) and the algorithms (Arian and Levy, 1992) derived from Hajek's results on regular binary sequences (Hajek, 1985). Although the latter lack the established performance bounds of the golden-ratio policy, simulations in the earlier queueing models show they are superior algorithms. Thus, they make promising candidates for our page-accessing model.

34.3.3 Randomized accessing and its optimal solution

Let $f_1, f_2, \ldots, f_N$ be given access frequencies. According to the randomized scheduling policy, at each decision point, the crawler chooses to access page i with probability f_i; the decision is made independently of all previous decisions. One can easily see that $\{d_j^i\}_j$, $\{X_j^i\}_j$ and $\{Z_j^i\}_j$ are three sequences of i.i.d. random variables for all i. Moreover, d_j^i has a geometric distribution: $P(d_j^i = n) = f_i(1-f_i)^{n-1}$. Thus, we have

Lemma 34.4 *For given frequencies $f_1, f_2, \ldots, f_N$,*

$$r_i = \frac{1}{E[X]} \left(E[X] - \frac{f_i}{\mu_i} + \frac{f_i}{\mu_i} \cdot \frac{f_i h_i}{1 - h_i + f_i h_i} \right).$$

Proof. As d_j^i has a geometric distribution, we obtain

$$E[X_j^i] = \sum_{n=1}^{\infty} f_i(1-f_i)^{n-1} nE[X] = \frac{E[X]}{f_i},$$

and by Lemma 34.3 we have,

$$\begin{aligned} E[Z_j^i] &= \sum_{n=1}^{\infty} f_i(1-f_i)^{n-1} \left(nE[X] - \frac{1}{\mu_i} + \frac{1}{\mu_i} h_i^n \right) \\ &= \frac{E[X]}{f_i} - \frac{1}{\mu_i} + \frac{1}{\mu_i} \frac{f_i h_i}{1 - h_i + f_i h_i}, \end{aligned}$$

so elementary renewal theory and (34.22) imply

$$r_i(\rho) = \frac{E[Z_j^i(\rho)]}{E[X_j^i(\rho)]} = \frac{1}{E[X]} \left(E[X] - \frac{f_i}{\mu_i} + \frac{f_i}{\mu_i} \cdot \frac{f_i h_i}{1 - h_i + f_i h_i} \right).$$

□

It is interesting to compare the lower bound of Proposition 34.9 with the obsolescence rate of the randomized policy. One can see that when f_i is small (close to 0) or large (close to 1), the difference between r_i and the lower bound tends to 0. More precisely, this difference is

$$\frac{h_i}{\mu_i E[X]} f_i^2 + o((f_i)^2)$$

when f_i goes to 0, and is

$$\frac{h_i}{\mu_i E[X]}(1 + \ln h_i)(f_i - 1) + o(1 - f_i)$$

when f_i goes to 1.

We now consider the problem of finding the optimal access frequencies under the randomized policy. First, we have the following lower bound over all frequencies.

Proposition 34.11 *Assume that the weights in the cost function are proportional to the mutation rates of the pages, i.e., $c_i = c_0 \mu_i$ for all $i = 1, 2, \ldots, N$. Then*

$$C = c_0 \cdot \sum_{i=1}^{N} \mu_i r_i \geq c_0 \left(\mu - \frac{1}{E[X]} \cdot \frac{\sum_{i=1}^{N}(h_i^{-1} - 1)}{1 + \sum_{i=1}^{N}(h_i^{-1} - 1)} \right). \quad (34.31)$$

Moreover, this lower bound is achieved when the access frequencies are proportional to $h_i^{-1} - 1$.

The proof uses again Lagrange multipliers and can be found in Coffman Jr. et al. (1998). Note that if the weights in the cost function are not proportional to the mutation rates of the pages, the bound is still valid, see discussions in Coffman Jr. et al. (1998), provided

$$\min_{1 \leq i \leq N} \sqrt{\frac{c_i}{\mu_i}} \geq \frac{\sum_{i=1}^{N} \sqrt{\frac{c_i}{\mu_i}} (h_i^{-1} - 1)}{1 + \sum_{i=1}^{N}(h_i^{-1} - 1)}.$$

34.3.4 Asymptotic optimality and the golden ratio policy

In this section we consider the asymptotic large-N behavior of scheduling policies. A similar study was carried out by Itai and Rosberg (1984) in the context of the control of a multiple-access channel. Some of the results here are analogous to theirs.

We define asymptotically optimal policies with respect to the lower bound in Proposition 34.10. Hence, we assume throughout this section that the weights in the cost function are proportional to the mutation rates of the pages, i.e., $c_i = c_0 \mu_i$ for all $i = 1, 2, \ldots, N$. We say that a policy π is asymptotically optimal if

$$\lim_{N \to \infty} C(\pi) - C^* = 0.$$

Note first that if the total mutation rate μ tends to zero, then all *cyclic* policies are asymptotically optimal. Indeed, consider an arbitrary cyclic policy with cycle length

K. It follows from (34.23) and Lemma 34.3 that

$$\begin{aligned} C &= c_0 \cdot \sum_{i=1}^{N} \mu_i r_i \\ &= \frac{c_0}{KE[X]} \cdot \sum_{i=1}^{N} \mu_i \sum_{j=1}^{m_K^i} \left\{ d_j^i E[X] - \frac{1}{\mu_i} \left(1 - h_i^{d_j^i} \right) \right\} \\ &= c_0 \mu - \frac{c_0}{KE[X]} \cdot \sum_{i=1}^{N} \sum_{j=1}^{m_K^i} \left(1 - (1 - d_j^i \mu_i E[X] + O(\mu_i^2)) \right) \\ &= c_0 \mu - \frac{c_0}{KE[X]} \cdot \sum_{i=1}^{N} \sum_{j=1}^{m_K^i} \left(d_j^i \mu_i E[X] + O(\mu_i^2) \right) \\ &= O(\mu^2), \end{aligned}$$

so if $\mu \to 0$, then $C \to 0$.

Thus, we assume that when $N \to \infty$, the total mutation rate μ, as well as the expected access time $E[X]$, is fixed. However, for any i, $1 \le i \le N$, we have $\mu_i \to 0$ when $N \to \infty$. Under such assumptions, the lower bound C^* in Proposition 34.10 becomes

$$\begin{aligned} \lim_{N \to \infty} C^* &= c_0 \mu - \frac{c_0}{E[X]} \left(1 - \lim_{N \to \infty} \prod_{i=1}^{N} h_i \right) \\ &= c_0 \left(\mu - \frac{1}{E[X]} + \frac{1}{E[X]} e^{-\mu E[X]} \right), \end{aligned} \qquad (34.32)$$

where we used the facts that $E[e^{-\mu_i X}] = e^{-\mu_i E[X]} + o(\mu_i^\alpha)$ and that $\sum_{i=1}^{N} \mu_i^\alpha \to 0$ for all $1 < \alpha < 2$.

Now consider the following cyclic scheduling policy, called the Golden Ratio policy, and studied in Itai and Rosberg (1984) for the control of a multiple-access channel. The policy is defined in terms of the Fibonacci numbers

$$F_k = \frac{\phi^k - (1-\phi)^k}{\sqrt{5}}, \quad k = 0, 1, \ldots,$$

where $\phi = (\sqrt{5}+1)/2$, and where $\phi^{-1} = (\sqrt{5}-1)/2 \simeq 0.6180339887$ is the golden ratio.

For any fixed k, let $\gamma(k,N)$ denote the golden ratio policy with F_k the cycle length, and N the total number of pages. It is assumed that $F_k \ge N$. Let $M_{k,N}^i$ be the number of page-i accesses in each cycle of $\gamma(k,N)$; these numbers satisfy

$$\lfloor f_i F_k \rfloor \le M_{k,N}^i \le \lceil f_i F_k \rceil$$

and $\sum_{i=1}^{N} M_{k,N}^i = F_k$, where f_i are the optimal access frequencies given by (34.30)

$$f_i = \frac{\ln h_i}{\sum_{i=1}^{N} \ln h_i}.$$

Thus,
$$\lim_{k \to \infty} \frac{M^i_{k,N}}{F_k} = f_i.$$

Let $\text{frac}(y) = y - \lfloor y \rfloor$ be the fractional part of y. Define the set $A_k = \{\text{frac}(j\phi^{-1}) \mid j = 0, 1, \ldots, F_k - 1\}$. The s-th access of the crawler is identified with the s-th smallest point of A_k. In the golden ratio policy $\gamma(k, N)$, the points

$$\left\{ \text{frac}(j\phi^{-1}) \mid \sum_{m=1}^{i-1} M^m_{k,N} \leq j < \sum_{m=1}^{i} M^m_{k,N} \right\}$$

correspond to the accesses of page i. As an example, let us suppose $N = 4$ and $f_1 = 2/13$, $f_2 = 3/13$, $f_3 = 3/13$ and $f_4 = 5/13$. Let $k = 8$ so that $F_k = 13$. Then, the golden ratio policy $\gamma(k, N)$ defines the access sequence $\{4, 2, 4, 1, 3, 4, 2, 4, 1, 3, 4, 2, 3\}$.

Thus, again from (34.23) and Lemma 34.3,

$$C(\gamma(k,N)) = c_0 \mu - \frac{c_0}{F_k E[X]} \sum_{i=1}^{N} \sum_{m=1}^{M^i_{k,N}} \left(1 - h_i^{d^i_m}\right),$$

where the inter-access distance $d^i_m \in \{F_{j_i}, F_{j_i+1}, F_{j_i+2}\}$, where $j_i = \lceil \ln_\phi f_i \rceil$ (cf. Itai and Rosberg (1984)). Moreover, it can be shown by mimicking the proofs in Itai and Rosberg (1984) that

$$C(\gamma(N)) := \lim_{k \to \infty} C(\gamma(k, N))$$
$$= c_0\mu - \frac{c_0}{E[X]} \left\{ 1 - \sum_{i=1}^{N} \left[\left(f_i - \phi^{-j_i}\right) h_i^{F_{j_i}} \right.\right.$$
$$\left.\left. + \left(f_i - \phi^{-j_i-1}\right) h_i^{F_{j_i+1}} - \left(f_i - \phi^{-j_i+1}\right) h_i^{F_{j_i+2}} \right] \right\}.$$

Proposition 34.12 *Assume for all i that $\mu_i \to 0$ as $N \to \infty$ and that $\sum_{i=1}^{N} \mu_i = \mu > 0$. Then,*

$$\limsup_{N \to \infty} C(\gamma(N)) \leq c_0 \left\{ \mu - \frac{1}{E[X]} + \frac{1 - \phi^{-1}}{E[X]} e^{-\frac{\mu\phi}{\sqrt{5}} E[X]} + \frac{\phi^{-1}}{E[X]} e^{-\frac{\mu\phi^2}{\sqrt{5}} E[X]} \right\}. \quad (34.33)$$

Proof. By mimicking the proof of Theorem 5.3 in Itai and Rosberg (1984), we can show that

$$C(\gamma(N)) \leq c_0\mu - \frac{c_0}{E[X]} \left\{ 1 - \sum_{i=1}^{N} f_i \left[(1 - \phi^{-1}) t^\phi_{i,N} + \phi^{-1} t^{\phi^2}_{i,N} \right] \right\},$$

where $t_{i,N} = h_i^{1/(f_i\sqrt{5})}$. Note that when $\mu_i \to 0$, $h_i = e^{-\mu_i E[X]} + o(\mu_i)$ so that $f_i = \mu_i/\mu + o(\mu_i)$. These imply that $t_{i,N} \to e^{-\mu E[X]/\sqrt{5}}$ when $N \to \infty$. Hence, by noting that $\sum_{i=1}^{N} f_i = 1$, we obtain

$$\limsup_{N\to\infty} C(\gamma(N)) \leq c_0\mu - \frac{c_0}{E[X]}$$
$$\times \left\{ 1 - \left(1 - \phi^{-1}\right) e^{-\frac{\mu\phi E[X]}{\sqrt{5}}} - \phi^{-1} e^{-\frac{\mu\phi^2 E[X]}{\sqrt{5}}} \right\}.$$

□

Finally, we compare the right-hand side of (34.33) with (34.32), and obtain the following result. The detailed proof can be found in Coffman Jr. et al. (1998).

O Assume for all i that $\mu_i \to 0$ as $N \to \infty$ and that $\sum_{i=1}^{N} \mu_i = \mu > 0$. Then,

$$\limsup_{N\to\infty} \frac{C(\gamma(N))}{C^*} \leq \frac{2\phi^2}{5} = \frac{\sqrt{5}+3}{5} < 1.05. \qquad (34.34)$$

34.4 OTHER OPTIMIZATION PROBLEMS

In addition to the problems addressed above, there are a number of other optimization issues that need investigations. Among the most important ones are the page ranking and system implementations.

34.4.1 Page ranking

Searching contents on the Web without knowing specific URLs is typically through querying search engines using key words. There can be thousands or even millions of Web pages containing the key words of a query. It is therefore crucial for search engines to rank the pages in such a way that the most relevant pages are presented to the users. Search engines have developed various methods to this end. The most influential work so far in this area is that of Brin and Page (1998), the founders of Google search engine, who developed the PageRank technique. A key element in this ranking system is the Markovian representation of the Web. The ranking of a Web page is related to the stationary distribution of the state corresponding to this Web page. There are ramifications since the publication of this seminal paper, as exemplified by the work of Kamvar et al. (2003), which consists in accelerating the computation of PageRank through a novel algorithm, with the reported performance improvement of 25–300%.

Another common line of thoughts is the use of learning algorithms. In Chen et al. (2000) (and a number of other papers by the same authors), building search engines using learning techniques was explored and implemented. Such approach allows on-line learning of and adaptation to the user behaviors.

Recently, more and more Web pages are generated dynamically. It poses a big problem to the search engines from indexing (and thus ranking) perspective as such dynamic pages are invisible to (or, more accurately, unvisited by) these search engines. Some preliminary work on this issue can be found in Mukherjee (2003), where a probabilistic model together with a ranking algorithm are proposed.

34.4.2 Implementation issues

Web search engines after all are computer systems. The efficiency of such systems depends quite a lot on the ways they are implemented. It is therefore very important to consider practical issues.

Developing a crawler to crawl the Web looks simple: fetch a Web page using a URL; parse it to extract all referenced URLs, and for those URLs that are not yet seen before, recursively visit these pages. However, due to the big number of available Web pages, these tasks have to be carried out very efficiently. One research effort in this regard is reported in Broder et al. (2003), where the authors propose to use main memory cache to cache the visited pages so as to speed up the operations which determine whether the URLs are previously visited or not.

In the previous sections we provided theoretical investigations on the number of crawlers to be deployed, either statically or dynamically. While the deployment of such parallel crawlers allows search engines to scale up, there is need of coordinating the page visits of these crawlers in order to avoid page visit overlap. Cho and Garcia-Molina (2002) investigate such issues and propose and evaluate in particular tradeoffs between coordination overhead and overlapping degree.

Distributed implementation of Web crawlers increases scalability and resiliency. Boldi et al. (2002) present such an implementation. They use consistent hashing which allows a complete decentralized coordination which yields linear scalability.

34.5 CONCLUSIONS

In this chapter, we have discussed various optimization issues arising in Web search engines. There are still a lot of challenging optimization problems, from both research and system development perspectives. The interested reader is referred to

http://searchenginewatch.com/

for more up-to-date discussions on search engines. Additional information can be found in reports Huang (2000) and Boswell (2003).

Acknowledgments

Technical results presented in Sections 34.2 and 34.3 were initially obtained in Talim et al. (2001a), Talim et al. (2001b) and Coffman Jr. et al. (1998). We would like to acknowledge the contributions of the co-authors of these papers, namely, Ed. G. Coffman Jr., J. Talim and R. Weber.

Bibliography

M. H. Ammar and J. W. Wong. On the optimality of cyclic transmission in teletext systems. *IEEE Transactions on Communications*, COM-35(1):68–73, January 1987.

M. Andrews, A. Fernandez, M. Harchol-Balter, T. Leighton, and L. Zhang. Template algorithm for one-hop packet routing. Technical report, Laboratory for Computer Science, Massachusetts Institute of Technology, Cambridge, MA 02139, 1997.

Y. Arian and Y. Levy. Algorithms for generalized round robin routing. *Operations Research Letters*, 12:313–319, 1992.

D. P. Bertsekas. *Dynamic Programming. Deterministic and Stochastic Models.* Prentice-Hall, Inc., Englewood Cliffs, 1987.

P. Boldi, B. Codenotti, M. Santini, and S. Vigna. Ubicrawler: A scalable fully distributed web crawler. In *Proc. AusWeb02. The Eighth Australian World Wide Web Conference*, 2002.

S. Borst. *Polling Systems*. PhD thesis, CWI, The Netherlands, 1994.

S. Borst, O. J. Boxma, J. H. A. Harink, and G. B. Huitema. Optimization of fixed time polling schemes. *Telecommunications Systems*, 3:31–59, 1994.

Dustin Boswell. Distributed high-performance web crawlers: A survey of the state-of-the art, 2003.

O. J. Boxma, H. Levy, and J. A. Weststrate. Efficient visit orders for polling systems. *Performance Evaluation*, 18:103–123, 1993.

O. J. Boxma, H. Levy, and J. A. Weststrate. Efficient visit frequencies for polling tables: Minimization of waiting cost. *Queueing Systems (QUESTA)*, 1994.

S. Brin and L. Page. The anatomy of a large-scale hypertextual Web search engine, 1998.

Andrei Z. Broder, Marc Najork, and Janet Wiener. Efficient url caching for world wide web crawling. In *Proc. 12th WWW Conference (WWWC'03)*. ACM Press, 2003.

Z. Chen, X. Meng, B. Zhu, and R. H. Fowler. Websail: From on-line learning to Web search. In *Web Information Systems Engineering*, pages 206–213, 2000.

J. Cho and H. Garcia-Molina. Synchronizing a database to improve freshness. In *Proc. 2000 ACM SIGMOD Int. Conf. on Management of data (SIGMOD '00)*, pages 117–128, New York, NY, USA, 2000a. ACM Press.

J. Cho and H. Garcia-Molina. Parallel crawlers. In *Proc. 11th WWW Conference (WWWC'02)*. ACM Press, 2002.

Junghoo Cho and Hector Garcia-Molina. The evolution of the web and implications for an incremental crawler. In *Proceedings of the Twenty-sixth International Conference on Very Large Databases*, 2000b.

E. G. Coffman Jr., Z. Liu, and R. Weber. Optimal robot scheduling for Web search engines. *Journal of Scheduling*, 1:15–29, 1998.

J. W. Cohen. *The Single Server Queue*. North-Holland Publishing Company, 1982.

H. D. Dykeman, M. H. Ammar, and J. W. Wong. Scheduling algorithms for videotex systems under broadcast delivery. In *Proc. of Int. Communication Conf. (ICC'86)*, pages 1847–1851, 1986.

B. Hajek. Extremal splittings of point processes. *Mathematics of Operations Research*, 10:543–556, 1985.

L. Huang. A survey on web information retrieval technologies, 2000.

Pew Internet and American Life Project. Search engine users, January 2005.

A. Itai and Z. Rosberg. A golden ratio control policy for a multiple-access channel. *IEEE Transactions on Automatic Control*, AC-29(8):712–718, August 1984.

S. Kamvar, T. Haveliwala, C. Manning, and G. Golub. Extrapolation methods for accelerating pagerank computations. In *Proc. 12th WWW Conference (WWWC'03)*. ACM Press, 2003.

J. F. C. Kingman. The ergodic theory of subadditive stochastic processes. *Journal of the Royal Statistical Society. Ser. B*, 30:499–510, 1968.

L. Kleinrock. *Queueing Systems, Vol. I*. Wiley & Sons, New York, 1975.

Z. Liu and P. Nain. Optimal scheduling in some multi-queue single-server systems. *IEEE Transactions on Automatic Control*, AC-37(2):247–252, February 1992.

A. W. Marshall and I. Olkin. *Inequalities: Theory of Majorization and its Applications*. Academic Press, New York, 1979.

S. Mukherjee. A probabilistic model for optimal searching of the deep Web, 2003.

M. L. Puterman. *Markov Decision Processes*. Wiley, New York, 1994.

S. M. Ross. *Introduction to Stochastic Dynamic Programming*. Academic Press, New York, 1983.

D. Stoyan. *Comparison Methods for Queues and Other Stochastic Models*. J. Wiley & Sons, 1933. English translation (D. J. Daley editor).

H. Takagi. *Analysis of Polling Systems*. MIT Press, 1986.

J. Talim, Z. Liu, P. Nain, and E. G. Coffman Jr. Controlling the robots of Web search engines. In *Proc. ACM Sigmetrics - Performance 2001 Conf.*, volume 29 of *Performance Evaluation Review*, pages 236–244, June 2001a.

J. Talim, Z. Liu, P. Nain, and E. G. Coffman Jr. Optimizing the number of robots in Web search engines. *Telecommunication Systems*, 17(1,2):243–264, 2001b.

R. L. Wolff. Poisson arrivals see time averages. *Operations Research*, 30:223–231, 1982.

35 OPTIMIZATION IN E-COMMERCE

Markos Kourgiantakis[1],
Iraklis Mandalianos[2], Athanasios Migdalas[2], and Panos M. Pardalos[3]

[1]Economics Department
University of Crete
Rethymno 74100 Greece
mkourg@ermis.soc.uoc.gr

[2]Production Engineering & Management Department
Technical University of Crete
Chania 73100 Greece
irman@verenike.ergasya.tuc.gr
migdalas@ergasya.tuc.gr

[3]Department of Industrial and Systems Engineering
University of Florida
Gainesville, FL 32611 USA
pardalos@ufl.edu

Abstract: This chapter explores the relative literature of and provides a reference tool for optimization and game theory in e-commerce. Moreover, in order to satisfy a broad range of readers, a brief overview is given in several e-business issues. Beside its horizontal character, this chapter presents some of the most up to date applications of operations research, game theory, and mathematical programming modeling techniques to four basic issues in e-business: e-supply chain, e-auctions, data mining and the "cost and pricing" problem in the Internet. Several opportunities for future research are also presented.

Keywords: Operational research, mathematical programming, game theory, e-business, e-supply chain, e-auctions, data mining, cost and price setting.

35.1 INTRODUCTION

With no doubt, the revolution in Information and Communication Technology (ICT) changed the way people conduct business. Managers, market experts, and researchers are predicting the huge potential of the latest networking technologies in business. They forecast that the size of on-line trading revenues in the next few years will vary from a few hundred billion to a few trillion dollars for some economic sectors. Today, the "e" before words like commerce, business, marketing, etc., indicates a philosophy that must be followed by companies and organizations that want to keep their competitiveness and efficiency. The phenomenon of e-business is still quite new also to academics and researchers. It is a clear, fertile field for research. Economists, marketeers, operations researchers, computer scientists, engineers of every kind, and many other categories of scientists investigate the impact of ICT in business, develop improved information systems and techniques, and propose new business methodologies and frameworks. In other words, they are looking for the "optimum" in e-commerce. As the title declares, this chapter deals with current optimization issues in e-commerce. The objective of the chapter is to explore the relative literature and to provide a reference tool for those who are interested in optimization in e-commerce. The purpose is twofold. On one hand, beginning from an elementary level and giving a brief state-of-the art overview, this work examines horizontally several issues to satisfy a broad range of readers. The goal is to trigger as many thoughts as possible and to indicate a wide range of trends in e-business literature. On the other hand, this work refers some of the most up to date applications of operations research, game theory and mathematical modeling techniques to problems arising in e-commerce. For this reason, the chapter progressively focuses in four important issues, in which the most application of operations research in e-commerce are met. These issues are: e-supply chain, e-auctions, data mining, and the "cost and pricing" problem. The structure of the chapter is as follows. In Section 35.2, the most basic terms of e-commerce are described. Sections 35.3 and 35.4 analyze the notions of e-supply chain and e-auctions respectively, and current research issues on these topics are presented. The topic of "cost and pricing" in the Internet is shortly described in Section 35.5. Section 35.6 refers to data mining, its role in e-commerce, and some operations research applications are given. Finally, Section 35.7 concludes this chapter and several suggestions for future research are made.

35.2 BASICS IN E-COMMERCE

To understand the dynamics and the optimization issues of e-commerce, it is first necessary to understand the nature of e-commerce and how it differs from traditional commerce.

35.2.1 Definitions

Many definitions of e-commerce can be found in the literature. A very broad but clear definition of e-commerce is the definition provided by Shaw and Strader (1997): "E-commerce is a modern business methodology that addresses the needs of organizations, merchants, and consumers to cut costs while improving the quality of goods and services, and increasing the speed of service delivery." Also, according to Kalakota and Whinston (1997), e-commerce includes buying and selling of information, products and services via computer networks.

The United States Bureau of Census (2000) makes a distinction between the terms electronic business (e-business) and e-commerce. It states that "e-business is any process that a business organization conducts over a computer-mediated network" and "e-commerce is any transaction completed over a computer-mediated network that involves the transfer of ownership or rights to use goods or services". Nowadays, the term e-business includes every section of electronic activity, such as e-government, e-health, e-marketing, etc.

E-commerce transactions are usually classified according to the partners involved. The main partners consist of: consumers (C), business (B), and government (G). Thus, six practical combinations are made possible:

- Business to consumer (B2C)
- Business to business (B2B)
- Business to government (B2G)
- Consumer to consumer (C2C)
- Consumer to government (C2G)
- Government to Government (G2G)

From these types of e-commerce transactions, B2B and B2C are the most investigated in the literature. B2B commerce includes a broad range of inter-organization transactions, including wholesale trade as well as company purchases of services, resources, technology, manufactured parts and components, capital equipment, and some types of financial transactions between organizations, such as insurance, commercial credits, etc. (Lucking-Reiley and Spulber, 2001). B2C is the e-commerce application with which individual consumers can interact with companies to learn about and buy products and/or services.

35.2.2 Technology of e-commerce

Although the technology of e-commerce includes a variety of systems, from fax to information systems, the explosion of e-commerce during the early 1990s has been strongly associated with the rapid diffusion of the Internet and World Wide Web, which has made conducting business over the Internet much cheaper and easier.

The Internet can be simply described as the 'network of networks', linking together otherwise independent computer networks, like Local Area Networks (LANs), Wide Area Networks (WANs), and individual end-users. Technically, Internet Service Providers (ISPs) provide the link between end-users and the Internet infrastructure. The flow of information across interconnected computers is made possible by protocols like Transfer Control Protocol/Internet Protocol (TCP/IP), HyperText Transfer Protocol (HTTP), HyperText Mark-up Language (HTML), etc.

The World Wide Web (the web or WWW) is the most popular part of the Internet. The WWW uses HTML language and makes the publishing of information on the Internet simple. Web-pages are single HTML documents containing not only text, but images, sound, video, and more.

35.2.3 E-commerce business models

There are several kinds of business conducting models in the Internet today, that is, there are several methods of doing business online. The absence of a unique generally accepted term for e-business model creates a little confusion in e-commerce literature (e.g. see Lambert (2003) and Patteli (2002)). Different definitions and taxonomies of e-business models based on different criteria, such as buyer/supplier type, financial aspects, degree of innovation, value proportion, etc., have been proposed in the literature (e.g. see Timmers (1998), Tapscott et al. (2000), and Elliot (2002)). Empirically, there are two well-known kinds of e-business models: e-shops and e-marketplaces. E-shops refer to corporate web sites that usually offer marketing information (e.g. provide company brochure and products offers) and transaction mechanisms. This model relates mainly to B2C e-commerce (e.g. Amazon.com). Although it could not facilitate all the benefits e-commerce and the Internet offer, an e-shop can be potentially the visible part of a more extensive electronic commerce system, integrated throughout the company and whose objectives are improved quality, reduced time-to-market, and access to new markets Timmers (2000).

The major components of e-commerce are electronic marketplaces (e-marketplaces or e-markets). An e-market is an inter-organizational information system that allows at least the participating buyers and sellers to exchange information about prices and product offerings, and to perform transactions between them. It must be noted that the term *at least* in the previous broad definition means that web sites containing only information for the firms and their products, without the ability for electronic ordering and/or payment, are not e-marketplaces but can be considered as the first step in an e-business strategy.

The model of e-markets is suitable for C2C, B2C, and B2B e-commerce. However, the dominating kind is B2B e-marketplaces because of their very high liquidity, that is, the ability to bring together a large number of buyers/sellers and generate a high volume of transactions. Modern B2B e-marketplaces not only provide an electronic, or on-line, method to facilitate transactions between buyers and sellers, but also offer services that support business activities throughout the entire order fulfillment process.

It is obvious that e-marketplaces offer much more value-added perspective in contrast to e-shops. Usually, firms in their e-commerce experience pass from simpler to more advanced and sophisticate business models. The progressive behavior of firms in e-business is described in the literature with a kind of "ladder" models (Willcocks and Sauer, 2000). According to such stage models, small and medium enterprises (SMEs) first develop their web presence before moving to stage two where they attend transactions. At the next stages, firms redesign business processes and integrate their supply chain operations. Hence, the last stages lead to integrated and collaborative e-commerce.

35.2.4 Benefits and trends

The Internet, and new technologies in general, are considered to be powerful marketing, transaction and coordination paradigms which offer significant benefits both to B2C and B2B e-commerce. Theoretically, e-business can give to any kind of firm and commercial organization the following four basic sources of competitive advantage (Fraser et al., 2000):

- Reduction in intermediation costs associated with wholesale and retail activities.

- Ability to decrease costs associated with purchasing by curbing the time and effort involved in supply and logistics operations.

- Improved information selection and processing which result in improved management of the supply chain.

- Prospect of expanding market share and/or developing new markets by decreasing the cost of selecting and processing information concerning the needs and the wants of existing and potential customers.

Despite all these benefits and the optimistic forecasts by the United Nations Conference on Trade and Development (2003, 2004), the evolution of e-business had been predicted more accelerated than its present situation in global level. The rapid diffusion of e-commerce was held back not only due to the ".coms" crash during the year 2001, but also because of two correlated parameters: the so-called "digital divide", which refers to uneven diffusion of ICTs and e-business between developed and developing countries, and the small rate of e-commerce adoption by small and medium enterprises (SMEs). Micro firms and SMEs globally play a vital role within many economies and therefore the growth of e-commerce depends on the introduction of SMEs in digital economy. Many studies report that SMEs, in contrast to their larger competitors, are lagging behind in the adoption of new technology and e-business (Lewis and Cockrill, 1999; Windrum and de Berranger, 2003; Stansfield and Grant, 2003; Kourgiantakis et al., 2003; Scupola, 2002; Drew, 2003; Zhu et al., 2003). The general rule is that the smaller the company, the lower the degree of e-business adoption.

The scientific community actively tries to solve existing problems in the implementation of e-business practices and to promote the information society. Researchers propose new theoretical frameworks and paradigms in marketing (Constantinides, 2002; Varadarajan and Yadav, 2002), in electronic operations management (Silveira, 2003), in firm strategy (Porter, 2001), in economics (Kauffman and Walden, 2001), and in the horizontal issues of e-business, like e-government, e-banking, e-learning, etc. (Mandalianos, 2005; Simpson, 2002; Schweizer, 2004).

The following sections are devoted to four important issues in e-commerce, where the majority of operations research applications, game theory, and mathematical programming modeling techniques and applications are found. These issues are: e-supply chain, e-auctions, data mining, and the "cost and pricing" problem in the Internet.

35.3 OPTIMIZATION IN E-SUPPLY CHAIN

35.3.1 Basics in Supply Chain Management

Supply chains exist in every industry and their management is a complicated task because it requires many mechanisms, such as demand forecasting, customer satisfaction, purchasing and communication among supply chain partners, inventory and distribution management (Strader et al., 1999). While a large number of supply chain management definitions exists in the literature, a modern definition is given by Chopra and Meindl (2001): "Supply chain management *(SCM)* involves the management of flows between and among stages in a supply chain to maximize total profitability."

In this definition, SCM is not restricted to the management of material or product flows from the manufacturers to final consumers, but the term "flows" refers to three kinds of flows: product, money, and information flows. Moreover, the term "total profitability" emphasizes the need of added value, as well as of reduced costs, to and for any member in the entire supply chain.

The field of supply chain management has grown at a rapid pace in recent years, basically because of the extended use of new ICT practices and new management models.

35.3.2 The impact of e-commerce in Supply Chain

The major contributions of e-commerce systems to and practices in SCMs concern the optimization of information flows across the chains. The topic is referred to in the literature under the names of e-supply chain, e-procurement, or e-logistics. The Internet and other relative technologies can improve significantly the coordination within and among the trading partners through fast, accurate, and effective information sharing and communication. Lancioni et al. (2000) investigate the impact of the Internet in several fields of supply chain management and conclude that the Internet is "indeed a powerful tool for logistics managers to monitor operations and reduce costs". Chopra and Mieghem (2000) synopsize the major direct and indirect benefits that e-commerce can provide to logistics procedures (see Table 35.1).

35.3.3 Basic types of systems

The evolution in supply chain started with the development and the adoption of Interorganizational Information Systems (IOSs). Williamson et al. (2004) define an IOS as a "collection of IT resources, including communications networks, hardware IT applications, standards for data transmission, and human skills and experiences." According to Shore (2001), the evolution of IOS among firms in supply chain could be separated into four phases. With the exception of the first phase, in which ICT do not contribute significantly, each of the other phases are strongly related to the development of the following systems and technology:

- *Electronic Data Interchange (EDI):* EDIs are systems that allow the electronic transmission of business data, like orders and invoices, in a standard format among business partners. EDI systems are related not only with the "digitalization" of supply chains, but also with the development of entire B2B e-commerce. Traditionally, EDI systems have been one-to-one technology, usually between one big enterprise and its partners separately. That is why the first EDI systems were used in dedicated (closed) and expensive Value-Added-Networks (VANs) and their standards and communications protocols were different. Through the time, the need for flexible interorganizational cooperations and globalization lead to the creation of Internet-EDIs or Open-EDIs. Having

Table 35.1 Opportunities from e-supply chain (Chopra and Mieghem, 2000)

Revenue Opportunities	Cost Opportunities
Direct sales: • Increased margin from eliminating intermediaries Product information: • Flexibility on price and promotions • Wider product portfolio offering Time to market: • Faster time to market Negotiating prices and contract terms: • Price and service customization • Downward price pressure due to increased competition Order placement and tracking: • Access at any time from any place Fulfillment: • Increased availability by aggregating chain information • Shorter response time • Increased choice of delivery options Payment: • Efficient funds transfer may improve cash flow	Facility costs: • *Site costs*: eliminate intermediaries or retail and distribution sites • *Processing costs*: customer participation, smoothed capacity requirements Inventory costs: • Reduced cycle stock (geographic centralization) • Reduced safety stock (statistical aggregation) • Postponing product differentiation to after other placement Transportation costs: • Inbound • Outbound Information sharing improves supply coordination: • Reduced bullwhip effect • Shared planning and forecasting

the ability to be connected through the Internet and to be much cheaper than traditional EDIs, these systems are more suitable for SMEs and allow them to participate in EDI networks (Vlachos, 2004; Williamson et al., 2004). For the impact of EDI systems in SCM, see, e.g. Lee and Lim (2005).

- *Enterprise Resource Planning (ERP):* ERP are multi-module, business management systems for planning, scheduling, and monitoring various business functions in an organization, such as accounting, human resources, inventory management and production planning. Historically, the implementation of ERPs helped businesses to apply advanced logistics techniques, such as Manufacturing Resource Planning (MRP) and Just-In-Time (JIT). In contrast to EDI networks, the integration of several ERPs could offer full collaboration among business partners, while this kind of integration allow the diffusion and the exploitation of much more strategic information in the supply chain. For this reason, traditional ERPs became web-based ERPs in order to be used through the Internet. There are several studies that analyze the role of ERPs in SCM and reveal the usage rate of known ERPs, such as SAP and Baan (Akkerman et al., 2003; Mabert et al., 2003)

- *Internet platforms:* The evolution of the Internet and the development of web technologies, such as XML and SQL databases, allow participants in supply chain to integrate completely their information resources and therefore to accelerate the decision-making on SCM processes (Williamson et al., 2004). The matter was and remains how all different EDI, ERP, and other management systems can be homogenized and work effectively through the Internet. The solution to this problem is given by the Internet platforms, the so-called electronic marketplaces. Nowadays, the majority of B2B e-marketplaces have the appropriate mechanisms to connect several different business systems of participants. Moreover, modern B2B e-marketplaces give the ability to suppliers and buyers to enjoy on-line supply chain services like product design, supply chain planning, optimization, and fulfillment processes. In other words, e-marketplaces act also like Application Service Providers (ASP) for SCM procedures, satisfying in this manner SMEs that have not the ability to maintain their own SCM systems. Of course, the level of integration that an Internet platform can offer to supply chain members depends exclusively the e-commerce adaptability and philosophy of the members. A recent study by Eng (2004) reveals that effective participation in e-marketplaces requires companies to integrate all their internal and external supply chain activities and to proceed not only to automate transaction-based activities, but also to strategic supply chain activities. Skjott-Larsen et al. (2003) propose that different types of buyer-supplier relationships require different types of electronic marketplaces.

35.3.4 Research issues in e-supply chain

Modern research issues in e-supply chain concern problems in the implementation of ICT in SCM, and derivation of models that support the optimization of procedures among the partners in an e-chain. At macroscopic level, the diffusion of IT for SCM purposes, especially to SMEs, is associated with the diffusion of e-business in general. Key complementary factors such as liberalization of telecommunications, government promotion of e-commerce and ICT, specific legislative framework, higher degree of trade openness, and financial enhancement of e-business initiatives, have been indicated as major policy issues by many researchers (Gibbs et al., 2003; Pohjola, 2002).

At the firm level, research topics can be summarized in three correlated categories: the proper implementation and use of ICT in SCM; the improvement of SCM systems and networks; and the optimum coordination through the supply chain. For the first category, a very good and analytical review is given by Gunasekaran and Ngai (2004). They conclude their survey with some major comments that are simultaneously topics for further investigation on information systems in SCM. We list several such topics below:

- The need for alignment between information models and supply chain models.

- The need for flexible SCM systems that accommodate on one hand the individual organizational characteristics and, on the other, the strategic objectives of the entire supply chain.

- The need for changes in traditional business operations that will lead to full exploitation of IT benefits. In general, business process re-engineering (BPR) is considered as one of the basic requirements for successful e-commerce in the literature.

- The need for new metrics in information systems performance and suitability in SCM.

35.3.4.1 Network and IT design for e-supply chain systems.

The improvement of SCM systems and network design, requires the combination of operations research techniques and mathematical programming algorithms with the latest technologies. For example, using meta-data and an inference algorithm, Kuechler Jr. et al. (2001) propose an interoperability model for workflow management systems that can support optimized, extended e-commerce transactions, usually in cooperation with EDI or ERP systems. This kind of systems is often used in electronic transaction in private networks or in e-marketplaces in order to improve online trading. For the coordination of supply chain units, Gerber et al. (2003) present an information and trading network, that is, constituted from a set of holonic [1] agents, each for every member in the chain (producers, retailers, logistics providers, etc.).

Based on agent technology (i.e. the combination of heterogeneous systems for advanced functionality and more automated processes), Keskinocak et al. (2001) have

---

[1] The term "holon" (from the Greek word "holos" for "whole") was originally introduced by Arthur Koestler, and was defined as a self-similar or fractal structure that is stable, coherent and that consists of several holons as sub-structures and is itself a part of a greater whole (Schillo and Fisher, 2003). Schillo and Fisher define as holonic agents the agents consisting of sub-agents with the same inherent structure. In a holonic multiagent system, holonic agents coordinate and control the activities of their sub-agents and join other holonic agents.

built an agent-based decision-support system that supports decision-makers finding "good" matches between demand and supply and responding to opportunities in a timely manner. They introduce two major kind of agents: search agents, that take demand queries of buyers and search a set of web sites for offers that (approximately) match these queries, and matching agents, that find optimal matches between the supplies and demands. To use optimization algorithms (e.g. greedy and network-flow based matching algorithms) in these agents, the developers define demand and supply items as attribute-value pairs. Buyers and suppliers specify their preferences by giving values to the attribute pairs and agents help them to decide with the provision of a set of good "matching" alternatives.

35.3.4.2 e-Supply Chain interactions and coordination. The optimal interactions and coordination among the participants in traditional and electronic supply chains are actively investigated by the research community. The analysis of supply chain coordination and interaction, often referred also as supply chain optimization, is based on many scientific fields, such as economics, game theory, supply chain contracts, and network theory. There are several reviews in the literature of supply chain optimization and many of them examine in different degrees the impact of IT and e-business in this field (Simchi-Levi et al., 2004; Geunes and Pardalos, 2003; Geunes et al., 2002; Cachon, 2001; Tsay et al., 1999). We mention here a few examples.

Shee et al. (2000) analyze the supply-demand interaction between buyers and suppliers as a fuzzy bi-level (i.e., Stackelberg leader-follower) multiple objectives programming (BLMOP) problem that is based on the objectives of individual party and resource constraints. Letting $\{q_i^L\}_{i=1}^{\ell}$ be the objectives of the leader and $\{q_i^F\}_{i=1}^{k}$ the objectives of the follower with respect to quality, $\{f_i^L\}_{i=1}^{m}$ be the objectives of the leader and $\{f_i^F\}_{i=1}^{n}$ the objectives of the follower with respect to cost, $\mathbf{x}$ the variables controlled by the leader and $\mathbf{y}$ the variables controlled by the follower, the problem can be stated as follows:

$$\text{(Leader)} \quad \max_{\mathbf{x} \geq 0} \quad \{q_i^L(\mathbf{x},\mathbf{y}^*)\}_{i=1}^{\ell}$$

$$\min_{\mathbf{x} \geq 0} \quad \{f_i^L(\mathbf{x},\mathbf{y}^*)\}_{i=1}^{k}$$

$$\text{where} \quad \mathbf{y}^* \text{ solves}$$

$$\text{(Follower)} \quad \max_{\mathbf{y} \geq 0} \quad \{q_i^F(\mathbf{x},\mathbf{y})\}_{i=1}^{m}$$

$$\min_{\mathbf{y} \geq 0} \quad \{f_i^F(\mathbf{x},\mathbf{y})\}_{i=1}^{n}$$

$$\text{s.t. } \mathbf{A}\mathbf{x} + \mathbf{B}\mathbf{y} \leq \mathbf{b}$$

Mingzhou and Wu (2001) focus on supplier competition and different information patterns in a two-supplier, one-buyer setting but they also include market intermediaries, who are electronic intermediaries (ISPs, e-marketplace owners, etc.) in the case of e-commerce. Taking several schemes with complete and asymmetric information and calculating the optimum price and quantities for each player and the whole system, it can be shown that different forms of market intermediation and the extent of information asymmetry change the nature of buyer-supplier interaction in general, two-part contract auction achieves better system efficiency and buyer profitability. The wholesale price auction is preferable to the catalog auction, while the preference for market intermediary is more complex and depends on the strategy and the selection between transaction-based commissions or membership fees. The formulated optimization problems are simple enough to be solved in closed form for all studied cases.

Netessine and Rudi (2004) present a model for a drop-shipping supply chain on the Internet. In drop-shipping, the wholesaler takes inventory risk and performs fulfillment, while the retailers can focus exclusively on customer acquisition. Concerning customer acquisition and inventory costs, Netessine and Rudi (2004) examine three distinct drop-shipping models in a multi-period environment: one with a powerful wholesaler (as the Stackelberg leader), one with a powerful retailer, and one with a wholesaler and a retailer having equal power. They find optimal inventory and customer acquisition spending in each model, and they compare these results with the "integrated" and "traditional" supply channels. Among their results is that drop-shipping introduces a conflict between marketing and operations functions that results in inefficiencies in the form of simultaneous understocking and spending too little on customer acquisition and that none of the mechanisms described in the literature on channel coordination (except for those that allow side payments) are able to induce an optimal system behavior in the presence of customer acquisition expenses.

Ovalle and Marquez (2003) conduct an assessment study with system dynamics for the effectiveness of using e-collaboration tools in the supply chain. In their work, the term "e-collaboration tools" refers to information sharing types (product, customer demand and inventory information). Their approach involves a three step supply chain model. At the first step there is a lack of collaboration among the commercial partners. At the second step, called "Collaborative Forecasting," there is such a collaborative partnership offering the spillover of the end customer demand along the chain. The last stage, the "Collaborative Planning," asserts that Internet and related technologies allow the supply chain members to gain access to additional information that they do not control (e.g. others inventory status) and use it in their planning process. Defining variables for information, material and cash flows, the above conceptual model is transformed into a mathematical model with simple and heuristic equations and the relationships among variables are detected. Finally, they validate the behavior patterns of the model's variables using operational data (lead times, delay times, etc.) from a real four partners supply chain in the PC components business. Their analysis by means of system dynamics reveals that when e-collaboration tools are implemented in a supply chain, all the processes are more agile, the costs are more favorable, and the end customer's service is the best.

In the same spirit, Disney et al. (2004) analyze the impact of four e-business scenarios on supply chain dynamics by investigating the bullwhip effect. Bullwhip is a measure of a poorly performing supply chain and it is caused by demand uncertainties that lead to inventory buffers through the chain. Disney et al. (2004) analyze this effect using two different approaches and comparing them to a traditional supply chain. The first approach is based on an analysis of the results of a management simulator, the Beer Game, while the second approach is based on a quantitative z-transform analysis. For the above scenarios, the z-transform analysis indicates that there is an expectation that the innovative e-commerce technologies will outperform the alternative strategies, while the Beer Game results have indicated that e-commerce technology may add a degree of complexity to human decision-making. In other words, the adoption of e-commerce solutions in supply chain, which allow information sharing and offers a great opportunity for greater efficiency, creates often a complicate environment with too much information and too many calculations to manage.

Nagurney et al. (2005) develop a supply chain network model for both physical and electronic transactions. The model addresses the optimization of decision-making in a three-tiered supply chain network, that is, manufacturers, distributors and retailers are involved. Its most innovative characteristic is that manufacturers and distributors (the supply side) are multicriteria decision-makers who are concerned with both profit maximization and risk minimization. The demand side risk is represented by the uncertainty included in random demands from retailers. Let i denote a manufacturer, j a distributor, and k a retailer. Let further $\mathbf{q}_{i1} = [q_{ij}]$ denote the shipments from manufacturer i to all distributors j, $\mathbf{q}_{i2} = [q_{ik}]$ denote the direct shipments from manufacturer i to all retailers k, $\mathbf{q}_1 = [\mathbf{q}_{i1}]$ the shipments between all pairs of manufacturers and distributors, and $\mathbf{q}_2 = [\mathbf{q}_{i2}]$ the shipments between all pairs of manufacturers and retailers. Then the problem of manufacturer i is to maximize her profit p_i, which is a function of $\mathbf{q}_1$ and $\mathbf{q}_2$, that is, it depends on the shipments of the other manufacturers as well, and to minimize the risk r_i, which again depends on the actions of the other manufacturers:

$$\text{(Manufacturer } i\text{)} \quad \max_{\mathbf{q}_{i1},\mathbf{q}_{i2}} \quad p_i(\mathbf{q}_1,\mathbf{q}_2)$$
$$\min_{\mathbf{q}_{i1},\mathbf{q}_{i2}} \quad r_i(\mathbf{q}_1,\mathbf{q}_2)$$
$$\text{s.t.} \quad \mathbf{q}_{i1} \geq 0, \ \mathbf{q}_{i2} \geq 0$$

Define next $\mathbf{q}_{1j} = [q_{ij}]$ as the vector of shipments from all manufacturers i to distributor j, and the vector $\mathbf{q}_{j1} = [q_{jk}]$ of shipments from distributor j to all retailers k. Furthermore, let $\mathbf{q}_3 = [\mathbf{q}_{j1}]$ be the vector of shipments between all distributors and all retailers. Then retailer j is concerned with the maximization of the difference between her revenues and her handling cost and payouts to the manufacturers, as well as with the minimization of the risk:

$$\text{(Distributor } j\text{)} \quad \max_{\mathbf{q}_{j1},\mathbf{q}_{1j}} \quad d_j(\mathbf{q}_1,\mathbf{q}_3)$$
$$\min_{\mathbf{q}_{j1},\mathbf{q}_{1j}} \quad r_j(\mathbf{q}_1,\mathbf{q}_3)$$
$$\text{s.t.} \quad \mathbf{e}_{j1}^T \mathbf{q}_{j1} \leq \mathbf{e}_{1j}^T \mathbf{q}_{1j}$$
$$\mathbf{q}_{j1} \geq 0, \ \mathbf{q}_{1j} \geq 0,$$

where $\mathbf{e}_{j1}$ and $\mathbf{e}_{1j}$ are vectors of ones imposing a constraint which states that the retailers cannot purchase more from distributor j than she is holding in stock. Note that the objective functions of distributor j is again depending on the actions of others. Finally, retailer k is maximizing her expected profit, which is the difference between the expected revenues and the total expected cost which involves expected penalties, handling cost and payouts to manufacturers and distributors:

$$(\text{Retailer } k) \quad \max_{\mathbf{q}_{1k},\mathbf{q}_{2k}} \quad E[p_k(\mathbf{q}_2,\mathbf{q}_3)]$$
$$\text{s.t.} \quad \mathbf{q}_{1k} \geq \mathbf{0}, \ \mathbf{q}_{2k} \geq \mathbf{0}$$

Finding the optimal conditions for the various participants in supply chain, Nagurney et al. (2005) formulate the equilibrium conditions in terms finite-dimensional variational inequalities.

35.4 OPTIMIZATION IN E-AUCTIONS

35.4.1 E-auctions

An auction is a well-known and widely used market mechanism by which buyers and sellers make bids and place offers. In contrast to other transaction mechanisms, auctions are dynamic procedures where bidding prices are not fixed but change over time according to current demand and supply needs. Auctions and bidding, as transaction mechanisms, have been investigated from the point of view of economics in depth by, for example Klemperer (1999), McAfee and McMillan (1987), and Milgrom (1989).

Today, ICT technology provides the opportunity to firms to use this mechanism electronically, mainly through the Internet. Electronic auctions (e-auctions or online auctions) have received much attention recently, especially for B2C and C2C trading. However, B2B auctions have at least as much potential, especially where spare capacity and surplus stocks are concerned.

E-auctions have several differences relative to traditional auctions. According to Klein (1997), these auctions can work much better in contrast to traditional auctions as coordination mechanisms, as a social mechanism to determine a price, as efficient allocation mechanisms and as highly visible distribution mechanisms. Dans (2002) states that a set of new auction models based on variations of the original ones emerge in the Internet, and that these new models are enabled by the very own characteristics of the Internet, such as less friction, universal availability and reach, rich information, audience tracking, etc. In general, the benefits provided by e-auctions can be synopsized to the reduction of transaction and search costs and to the improved convenience, both geographic and temporal, for bidders (Lucking-Reiley, 2000). Concerning disadvantages of e-auctions in contrast to traditional auctions, Turban (1997) and Lucking-Reiley (2000) indicate the inability of bidders to inspect goods before bidding and the possible cases of fraud. However, these are two potential problems that characterize every mechanism in e-commerce.

There are many types of e-auctions. Table 35.2 provides a greater classification of e-auctions according to several criteria. From this classification, the most popular types of auctions are briefly presented here. First of all, auctions can be distinguished as forward and reverse auctions. In general, forward are the auctions in which one supplier offers a product and many buyers compete for this, while reverse auctions refer to auctions with one buyer who asks for a product and many sellers that bid for this request. The most common kinds of auctions are:

- The English auction, which belongs to open auctions in which auctioneers (buyers or sellers) set a minimum price for the offered (demanded) product(s) and bids are raised incrementally. The bidder with the highest bid wins the auction. The English auction is also known as ascending auction.

- The Dutch auction, in which auctioneers continuously lower the price until a bidder takes the item at the current price. The Dutch auction is also denoted as descending auction, because of the decreasing prices during the auction.

- Sealed-bid auctions, contrary to open auctions, mean that bids are submitted sealed (i.e. bidders do not know what other bidders offer) and only once by each bidder. There are several kinds of sealed-bid auctions, but the most common are the first and second price auctions. In the first-price sealed-bid auction, the bidder with the highest bid wins the auction for the price of her bid, while in the case of second price auction the bidder with the highest bid wins but for the second highest offered price. Second-price auction is also known as Vickrey auction.

Finally, in the Internet, multi-unit (i.e., bids for more than one unit of a good), combinatorial (for bundles of goods), and multi-dimensional or multi-attribute (i.e., bidding also for non-price attributes, such as quality, delivery time, etc.) auctions are met (Teich et al., 2004).

35.4.2 Optimized e-auctions

The phenomenon of e-auctions has been investigated systematically since late 1990s. Several works with empirical evidence on the evolution of e-auctions and covering many topics in B2C and B2B, such as seller and bidder behavior, auction design, fraud, and reputation in e-auctions, etc., have been reported (Anwar et al., 2004; Bapna et al., 2000; Lucking-Reiley, 2000). An excellent review and research framework for such empirical and theoretical investigations of e-auctions is given by Wood (2004).

Operational research and game theory applications on online auctions focus mainly on the improvement of e-auction systems and to the optimization of auction mechanism. Modern trends in e-auction information systems is the use of intelligent agents that help auctioneers and bidders to data analysis and decision making (Gregg and Walczak, 2004). The need for more sophisticated agents leads researchers to develop agents that use advanced techniques. For example, Cai and Wurman (2005) propose trading agents that model multi-unit sealed bid auctions as a sequential-incomplete information game and use Monte Carlo sampling to generate heuristic bidding policies. In the same logic, Wang et al. (2004) propose a parallel and autonomous agents based system for B2C auctions that deploys negotiation models based on fuzzy evaluation criteria. Beyond the software agents that are completely programmed to some specific negotiation strategies, learning agents also appear in the literature (Chandrashekar et al., 2005). These kinds of agents usually use genetic algorithms and learn from previous actions of bidders in order to help decision makers in the future. A recent example of these systems is proposed by Bandyopadhyaya et al. (2004), in which agents

Table 35.2 Classification of auctions (Teich et al., 2004)

Characteristic	Range
1. Number of items of a certain good	One to many
2. Number of goods auctioned	One-round vs. progressive
3. Nature of goods	One to many
4. Attributes	Homogeneous to heterogeneous
5. Type of auction	One to many
6. Nature of auction	Reverse vs. forward
7. English vs. Dutch auction	Ascending, descending price
8. Participation	By invitation vs. open
9. Use of agents	Agent mediated vs. manual mode
10. Price paid by winner	First price vs. second price vs. nth price
11. Price discrimination	Yes, no
12. Constraints exist	Implicitly, explicitly
13. Follow-up negotiation	Yes, no
14. Value function elicitation	Yes, no
15. Nature of bids	Open-cry vs. semi-sealed vs. sealed
16. Bid vector	1, 2, or n-dimensional
17. Bids divisible	Yes, no
18. Bundle bids allowed	Yes, no

use a reinforcement learning algorithm to change their pricing strategy over time in B2B reverse auctions.

With respect to auction mechanisms, there is a growing number of papers that investigate various e-auction formats. Having a satisfactory economic investigation of simple type auctions (Riley and Samuelson, 1981), researchers' interest shifts to more complicate mechanisms, such as multi-units and combinatorial auctions (Klemperer, 1999; Narahari and Dayama, 2005). Xia et al. (2005), for instance, manage to solve a general combinatorial double auction problem using integer programming and transforming this problem to multi-dimensional knapsack problem. Another recent example is due to Bourbeau et al. (2005), who develops an advanced design model for multi-units auctions using linear and non linear programming.

A detailed and comprehensive review for models concerning e-auction mechanisms is given by Chandrashekar et al. (2005). Chen et al. (2002) investigate multi-unit Vickrey auctions and introduce three models that beside bid prices incorporate transportation and production costs into auctions. In their first model (referred to as T auction) the buyer submits a fixed consumption vector, $\mathbf{q}$, and the auctioneer decides the quantities each supply location will provide to each demand center by minimizing the entire system cost. For a given consumption vector $\mathbf{q}$, the auctioneer minimizes the sum of the accepted bids and the transportation costs. To state the optimization problem corresponding to T auction, let N be the number of production facilities, K the number of suppliers, and M the number of buyers. Then, under T auction, the auctioneer's problem is

$$\text{(T auction)} \quad \min \quad \sum_{k=1}^{K} f_k(\mathbf{x}_k) + \sum_{n=1}^{N} \sum_{m=1}^{M} c_{nm} y_{nm}$$

$$\text{s.t.} \quad \sum_{n=1}^{N} y_{nm} = q_m, \ m = 1, \ldots, M$$

$$\sum_{m=1}^{M} y_{nm} = x_n, \ n = 1, \ldots, N$$

$$y_{nm} \geq 0, \ m = 1, \ldots, M, \ n = 1, \ldots, N,$$

where $\mathbf{x} = [\mathbf{x}_k] = [x_n]$, that is x_n is the production quantity at facility n and $\mathbf{x}_k$ is the production vector for the facilities owned by supplier k, $f_k(\mathbf{x}_k)$ is the production cost of supplier k for producing $\mathbf{x}_k$, c_{nm} is the per unit transportation cost from facility n to buyer m, y_{nm} is the shipment from facility n to buyer m, and the vector $\mathbf{q} = [q_m]$ gives the demand q_m by each buyer m.

35.5 COST AND PRICE OPTIMIZATION

Internet price determination is a multidimensional optimization problem. The challenge for government and industry is not only to develop effective Internet pricing mechanisms, but to do so without losing the benefits of interoperability and positive network externalities (McKnight and Bailey, 1995).

A manufacturer's decision problem for online and offline channel pricing is stated in Fruchter and Tapiero (2005) as determining a pricing strategy that enables the maximization of the manufacturer's discounted profits over a specific time horizon, given that the retailer acts as a strategic follower attempting to maximize its own profits. The manufacturer is assumed to be risk-neutral, while the competitive retailer is acting strategically. Both are attempting to maximize the present value of the profit stream over the finite horizon. The problem is formulated as a bilevel (Stackelberg) optimization problem:

(Leader) $\max_{P_{on},w} \int_0^\infty \{(p_{on} - c_{on})q + (w - c_r)(1-q)\}Ne^{-\delta t}dt$

(Follower) $\max_b \int_0^\infty bw(1-q)Ne^{-rt}dt$

s.t. $\dot{q} = -\alpha\left[q - 1 + F_\theta(1 - \frac{(1+b)w - p_{on}}{V})\right], \ q(0) = q_0,$

where p_r is the retail product price, p_{on} is the price for on-line products, w the price at which the manufacturer supplies the retailer of the physical store, c_{on} is the unit cost (on-line channel), c_r is the unit cost (off-line channel), V is the value of the product when bought through a retailer, $\theta V, 0 \leq \theta \leq 1$, is the value of the product brought on-line through virtual inspection, F_θ is the cumulative density function, b is the markup for pricing ($p_r = (1+b)w$), q_0 is an initial condition, $\alpha > 0$, and $q(t)$ is the probability for a customer to buy the product on-line. The problem is solved using a dynamic programming approach.

Gupta et al. (1995; 1999) propose, among others, the following characteristics for prices in multi-service class data communication:

- Users should be encouraged to use the network when it is less congested by shifting their demands across time;

- The impact of current load on future demand should be taken into account;

- The pricing scheme should be implemented in a completely decentralized manner, otherwise, the costs that formulate the prices may reduce the potential benefits of the pricing method;

- Prices should redistribute the load from highly loaded nodes to lightly loaded nodes, for the load management to be effective.

- Different applications and users require different quality of service (QoS,) therefore multiple priorities should be taken into account;

- The implemented pricing scheme should allow users to make decisions based on the price they pay, and service providers to provide the required QoS based on the profits they derive from pricing methods.

Optimal pricing problem for Asynchronous Transfer Mode (ATM) integrated-services networks is examined by Wang et al. (1997). The authors present a model for optimal pricing determined from the demand elasticity of each service. A three-stage procedure is implemented for solving the optimal pricing problem. Karsten et al. (1999) focus on Internet Integrated Services (IIS) and present a method that enables the comparison between reservation requests and extracts resource cost by optimizing total revenue.

35.6 OPTIMIZATION IN DATA MINING

Today's technology offers unique capabilities in storing and gathering data. Optimized use of these data sets gives enterprises the opportunity to gain a significant competitive advantage.

Data warehouse is a subject oriented, time variant, nonvolatile collection of data in support of management's decision needs. It provides tools to satisfy the information needs of users at all organizational levels – not just for complex data queries, but as general facility for getting quick, accurate, and often insightful information. A data warehouse is designed so that its users can retrieve the information they want to access using simple tools (Inmon, 1996).

Knowledge Discovery in Databases (KDD) is the process of identifying valid, potentially useful, and ultimate understandable structure in data. KDD focuses on selecting or sampling data from data warehouses, filter, transform and apply a data mining element to reveal and produce structure (Fayyad et al., 1996). The KDD process begins with data cleaning and data transforming to implement a particular data mining algorithm. Most researchers treat data mining as a synonym to knowledge discovery in databases, others define Data Mining as a step in the KDD process concerned with the algorithmic means by which patterns structures are extracted from the data under computationally acceptable efficiency limitations (Fayyad et al., 1996).

In the field of Data Mining and Knowledge Discovery in Databases, mathematical programming is used for the analysis of large data sets in order to reveal unsuspected relationships, structures, and figures. Knowledge discovery in databases is a process composed by the following steps:

- *Selection of data:* Data must be reliable with rich accurate descriptions and must contain fields that are potentially useful for data mining algorithms.

- *Preprocessing / Exploration of data:* Includes exploration and data cleaning, removal of all irrelevant text or fields. Although rich data can be proved useful, irrelevant data with different format gathered from many sources, or inaccurate data can lead to time consuming manual processes of cleaning, filtering and reformatting data.

- *Transformation:* Generally, data comes from various sources with significantly different data formats. It is of great importance to extract and normalize all the information under a certain type of data format.

- Data Mining techniques are used to reveal patterns in large datasets. A necessary property of algorithms capable of handling large growing datasets is their scalability, or linear complexity with respect to the datasets (Hegland, 2001). Different disciplines and applications reflect different goals of data mining techniques (Ramakrishnan and Grama, 1999):

- *Induction:* Locate, general, special characteristics, patterns or equations in data.
- *Compression:* Compress-reduce complexity of the data by adapting simpler concepts.
- *Querying:* Query data in better ways, extract useful information and properties.
- *Approximation:* Create models that can reveal hidden structure in data well.
- *Search:* Look for recurring patterns and restrict the space of possible patterns.

- *Interpretation / Visualization / Evaluation:* Visual representation results are much better understood by business users. Visual representations such as bar charts, line charts and heat-maps can connect a lot of information in a concise yet effective way (Kohavi et al., 2004).

35.6.1 Applications of Data Mining in E-commerce

Data mining in e-commerce has revealed important improvement avenues of long-term business relationships between collaborating parties. At the same time, it has been used to identify and resolve problems regarding web sites architecture, appearance, functionality and navigation. Major applications areas for web usage mining include personalization, system improvement, site modification, business intelligence, usage characterization (Strivastava et al., 2000).

In e-commerce, data is collected mainly by servers logs, cookies logs, session logs, user entry data, web meta data. Data mining implementations in e-commerce can help reveal significant informations about customer behavior that is essential for business. Example information that can be obtained are:

- What characterizes customers that spend most?
- What characterizes visitors that quickly become customers?
- What characterizes customer that do not buy?
- What characterizes customers that prefer special kind of promotions (promotion X over Y)?
- What characterizes customers that put a lot in the basket but do not buy?
- What characterizes customers that accept cross-sells and up-sells?
- What characterizes customers that buy with their credit card?

In e-commerce transactions are made electronically and large scale of different data can be obtained automatically by various sources. Example sources:

- Server logs are automatically created by the server upon visiting a site. The IP of the visitor, the requested URL, date, time and the duration of the visit are logged. This information are valuable but not sufficient to extract information about the behavior of the user. Public computers are used by more than one customer who share the same IP, connected to Internet by a router (or a proxy) many computers might share the same IP (which is the only one logged by the server). Ansari et al. (2001) emphasizes the need for data collection at the application server layer in order to support logging of data and meta-data.

- Cookies allow to set a special short text identification file for each customer. Setting a cookie, however requires and the permission by the customer.

- Registration and session tracking provides more descriptive and at the same time revealing information about clients behavior. Registration is a process that must be made short and simple in order to convince guests to become registered visitors and at the same time provide information that can be stored in a database for data mining.

- Even if a guest prefers not to register, entry data that are collected by search fields and the results found can reveal important patterns about customers preferences and shape new priorities for the company.

- Information in web meta data provides the topology of the site, normally stored in a site-specific index table. This information, in most cases, is represented through HTML meta tags or XML statements among other commands, text and comments.

35.6.2 Basic data mining techniques

Knowledge discovery and prediction are the two main categories of data mining problems.

35.6.2.1 Predictive Modeling. Recent developments on predictive modeling in machine learning, pattern recognition, and artificial neural networks, have contributed to meeting the challenges of data mining and ever increasing data warehouses (Hong and Weiss, 1999). The goal of predictive modeling goal is to predict attributes based on other or past attributes of the data by estimating a function g which maps points or feature vectors from an input space X to an output space Y, given only a finite sampling of the mapping $\{\mathbf{x}^i, g(\mathbf{x}^i)\}_{i=1}^{M} \subset \mathbb{R}^{n+1}$ (Fayyad et al., 1996). To create an estimator $\hat{g}$, a training set is used. The problem turns out to be a regression one if Y is continuous or numeric and a classification one if Y is discrete.

1. The goal of classification is to analyze a set of training data in order to create a model that can predict the classification of future data, or data not implemented in the training set. Machine learning literature treats in depth classification applications; see e.g. Mitchell (1997); Michie et al. (1994). Three common cases of classes are distinguished by Michie et al. (1994).

    - Classes respond to labels for different populations.
    - Classes result from a prediction problem.

- Classes are predefined by a partition of a sample space.

A classifier's performance is judged mainly by its accuracy, rather than by support and confidence. An important measure of the quality of classifiers is the misclassification rate called also Bayes risk; see e.g. Hegland (2001).

2. *Regression.* Mathematical formulations for the estimation of a true, unknown regression function function g by a linear combination of some set of predefined functions are considered in Bradley et al. (1998). It is assumed that for a true regression function, a finite number of samples $\{x_i, g(x_i)\}_{i=1}^{M} \in \mathbb{R}^{n+1}$ are available. The true function g is estimated by $\hat{g}(x) = \sum_{j=1}^{N} w_j f_j(x)$, where $\{f_1, \ldots, f_N\}$ is the set of predefined functions. By sampling these functions at the points $x_1, \ldots, x_M$, the problem of estimating the weights w_j reduces to solving the linear system of equations $Aw = b$, where the elements of A are $a_{ij} = f_j(x_i)$, and $b_i = g(x_i)$. If there is no noise in b, i.e., g is sampled precisely, the methods of frames and basis pursuit find a solution to the linear system by minimizing $w = [w_j]$ with respect to 1 or 2-norm as discussed in Daubechies (1988) and Chen et al. (1996), that is, for $q = 1, 2$, they consider the problem

$$\min_{w} \quad \|w\|_q$$
$$\text{s.t.} \quad Aw = b.$$

In the presence of noise, the linear system may not have an exact solution, in which case a least square solution is sought by solving the problem

$$\min_{w} \quad \|Aw - b\|_2.$$

In Bradley et al. (1998) the least one norm is considered in order to estimate $\hat{g}$ in the case of noises in b. A corresponding linear programming problem is formulated for that purpose:

$$\min_{w,y} \quad e^T y$$
$$\text{s.t.} \quad -y \leq Aw - b \leq y,$$

where e is a vector of ones.

35.6.2.2 Clustering.
Clustering is the division of data, with "similar" characteristics, into subsets. These groups are called clusters, the degree of similarity (or association) among members of the same cluster must be strong and weak between members of different clusters. Similarity can be revealed by a number of algorithms focusing on different aspects of the data provided.

For a given set $\mathcal{A}$ of m points in $\mathbb{R}^n$ represented by the matrix $A \in \mathbb{R}^{m \times n}$ and a number k of desired clusters, the clustering problem can be stated as follows: Find cluster centers $c_l, l = 1, \ldots, k$, in $\mathbb{R}^n$ such that the sum of the minima over $l \in \{1, 2, \ldots, k\}$ of the 1-norm distance between each point (row) $A_i, i = 1, \ldots, m$, and the cluster centers

$c_l, l = 1,...,k$, is minimized. Thus the corresponding optimization problem (c.f. the linear program of the previous subsection) is then (Mangasarian, 1997):

$$\min_{c,d} \sum_{i=1}^{m} \min_{l=1,...,k} \{e^T d_{il}\}$$
$$\text{s.t.} \quad -d_{il} \leqslant A_i^T - c_l \leqslant d_{il}, i = 1,...,m, \; l = 1,...,k,$$

where $d_{il} \in \mathbb{R}^n$ is a dummy variable which bounds the difference $A_i^T - c_l$, and $e \in \mathbb{R}^n$ is a vector of ones.

Clustering Algorithms need to have the following properties (Han and Kamber, 2001):

- Scalable in order to work on large and growing data sizes;
- Capable of handling a large number of attributes of various types and complex data;
- Capable of handling noisy data;
- Insensitive to the ordering of the data; and
- Able to represent clusters of arbitrary shapes.

AdaBoost is one of the most successful classification algorithms involving learning (Freund and Schapire, 1997). Clustering algorithms are mainly categorized in the following types:

- Partitioning methods;
- Hierarchical clustering methods;
- Association Rules;
- Density based methods;
- Model-based methods; and
- Dependency Modeling.

We briefly discuss here Partitioning methods and Hierarchical clustering methods.

Partitioning Methods: Partitioning algorithms as well as the hierarchical clustering approaches of the following subsection heavily utilize the notion of distance between points. Assuming an n-dimensional Euclidean space, the distance $d(\mathbf{x}, \mathbf{y})$ between two vectors $\mathbf{x} = [x_1, x_2, ..., x_n]^T$ and $\mathbf{y} = [y_1, y_2, ..., y_n]^T$ may be defined in several different ways, for example, the Euclidean distance (2-norm) $\|\mathbf{x} - \mathbf{y}\|_2 = \sqrt{\sum_{i=1}^{n}(x_i - y_i)^2}$, the Manhattan distance (1-norm) $\|\mathbf{x} - \mathbf{y}\|_1 = \sum_{i}^{n} |x_i - y_i|$, the maximum of dimensions (supremum norm) $\max_{i=1,...,n}\{|x_i - y_i|\}$, the Canberra distance, the Pearson correlation, the Spearman correlation, etc. Given k the number of clusters, k-means and k-medoids algorithms are the most commonly used in partition techniques.

The goal of k-Mean (Queen, 1967) is to minimize dissimilarity in the elements within each cluster while maximizing this value between elements in different clusters. Given k cluster centers $\mathbf{c}_l^t$ in iteration t, new centers $\mathbf{c}_l^{t+1}$ are computed by the following two steps:

1. *Cluster Assignment:* For $i = 1,...,m$, assign $\mathbf{x}_i$ to cluster $l(i)$ such that $\mathbf{c}_{l(i)}^t$ is nearest to $\mathbf{x}_i$ in the 2-norm distance.

2. *Cluster Center Update:* For $l = 1,...,k$ set $\mathbf{c}_l^{t+1}$ to be the median of all $\mathbf{x}_i$ assigned to $\mathbf{c}_l^t$. That is, the point $\mathbf{c}_l^{t+1}$ is a cluster center that minimizes the sum of 2-norm distances squared to all the points in cluster l.

The iterations stop when $\mathbf{c}_l^t = \mathbf{c}_l^{t+1}$, $l = 1,....,k$.

The k-Median Algorithm utilizes the 1-norm distance instead. Although both algorithms are similar when the same distance is used (Fukunaga, 1990; Jain and Dubes, 1988; Selim and Ismail, 1984), they differ both computationally and theoretically. Given k cluster centers $\mathbf{c}_1^t, \mathbf{c}_2^t, \ldots, \mathbf{c}_k^t$ at iteration t, new centers $\mathbf{c}_1^{t+1}, \ldots, \mathbf{c}_k^{t+1}$ are computed by the following two steps:

1. *Cluster Assignment:* For $i = 1,...,m$, assign $\mathbf{x}_i$ to cluster $l(i)$ such that $\mathbf{c}_{l(i)}^t$ is nearest to $\mathbf{x}_i$ in the 1-norm distance.

2. *Cluster Center Update:* For $l = 1,...,k$ set $\mathbf{c}_l^{t+1}$ to be the median of all $\mathbf{x}_i$ assigned to $\mathbf{c}_l^t$. That is, the point $\mathbf{c}_l^{t+1}$ is a cluster center that minimizes the sum of 1-norm distances to all points in cluster l.

The iterations stop when $\mathbf{c}_l^t = \mathbf{c}_l^{t+1}$, $l = 1,....,k$.

Hierarchical clustering: Agglomerative and divisive are the two basic approaches used in hierarchical partitioning (Figure 35.1). Clusters are created by a hierarchical partitioning of the data. There are several Hierarchical Clustering Methods, among them:

- *Single linkage:* In it the distances between each member of one cluster to the members of the other cluster are measured and the minimum of them is considered as the distance between clusters, that is, $\min_{\mathbf{x}_i \in S_i, \mathbf{x}_j \in S_j} d(\mathbf{x}_i, \mathbf{x}_j)$.

- *Average linkage:* The average distance of each member of one cluster to members of the other cluster is considered as the distance between clusters, that is, $\frac{1}{|S_i||S_j|} \sum_{\mathbf{x}_i \in S_i} \sum_{\mathbf{x}_j \in S_j} d(\mathbf{x}_i, \mathbf{x}_j)$.

- *Complete linkage:* In it the distance between clusters is defined as the maximum of the distances between each member of one cluster and the members of other clusters. That is, $\max_{\mathbf{x}_i \in S_i, \mathbf{x}_j \in S_j} d(\mathbf{x}_i, \mathbf{x}_j)$. This method tends to produce compact clusters.

- *Centroid method:* A centroid is defined as the center of a cluster and the distance between two clusters is equal to the distance between their centroids.

- *Ward's method (Ward, (Ward, 1963)):* Cluster membership is assessed by calculating the total sum of squared deviations from the mean of a cluster. Criterion for fusion is the smallest possible increase in the error sum of squares, i.e., $ESS = \sum_{i=1}^{n} x_i^2 - \frac{1}{n}(\sum_{i=1}^{n} x_i)^2$.

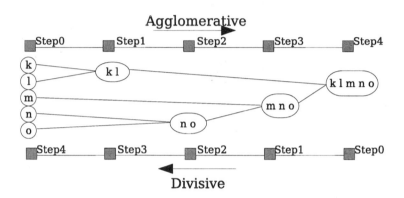

Figure 35.1 Agglomerative and Divisive approaches Steps

35.6.3 Association Rules

Association rules are used to reveal important aspects of web site implementations and point out important improvements on architecture and functionality. An association rule in then form "if a then b" is interpreted as a link between a and b such that if a is satisfied then b is likely to occur with a certain frequency and certainty. Probability is assumed to be time-invariant (Hegland, 2001) or not (Ramaswamy et al., 1998). The basic problem of allocating association rules is as follows (Agrawal et al., 1993): Let $I = \{i_1, i_2, ..., i_m\}$ be a set of binary attributes, $\mathcal{D}$ a set of transactions, $\mathcal{T}$ a database of transactions and such that if X is a subset of $\mathcal{T}$, then $\mathcal{T}$ contains X. An association rule is an implication of the form $X \Rightarrow \mathcal{Y}$, where $X \subset \mathcal{Y}$, $\mathcal{Y} \subset I$, and $X \cap \mathcal{Y} = \emptyset$. Parallel association rules algorithms are proposed by Agrawal and Shafer (1996) based on the a priori algorithm (Agrawal and Srikant, 1994).

35.7 CONCLUSIONS

This chapter explores the e-commerce literature and gives a brief survey of several issues. Wanting to reveal the role of operations research, game theory, and mathematical programming modeling in e-commerce, four basic and correlated topics were investigated: e-supply chain, e-auctions, data mining, and the "cost and pricing" problem in the Internet. The review indicates that:

- The implementation of optimization techniques in various fields of e-commerce is a necessity;

- The huge growth of e-services during the last years has been offered in a rather "confusing" way and therefor re-orientation, is a necessity for most business;

- Optimization and game theory in supply chain, data mining, cost, price, and auctions could be very useful for both enterprises and customers in their justified attempt to maximize their corresponding profits from on-line activities;

- Powerful new technology helps to implement various optimization models in e-commerce easier than ever;

- Optimal policies are key factors in e-commerce success. Game theoretic modeling optimization techniques can be very useful in achieving such goals.

Acknowledgments

Research partially supported by NSF and Dash Optimization, Inc.

Bibliography

R. Agrawal, T. Imielinski, and A. Swami. Mining association rules between sets of items in large databases. In *Proceedings of the ACM SIGMOD International Conference on Management of Database*, pages 207–216, 1993.

R. Agrawal and J. C. Shafer. Parallel mining of association rules: Design, implementation and experience. *IEEE Trans, Knowledge and Data Engineering*, 8:962–969, 1996.

R. Agrawal and R. Srikant. Fast algorithms for mining association rules. In *Proc. of the 20th VLDB Conf.*, pages 487–499, 1994.

H. Akkerman, P. Bogerd, E. Yucesan, and L. van Wassenhove. The impact of ERP on supply chain management: Exploratory findings from a European Delphi study. *European Journal of Operational Research*, 146(2):284–301, 2003.

S. Ansari, R. Kohavi, L. Mason, and Z. Zheng. Integrating E-commerce and data mining: Architecture and challenges. In *Proceedings of the IEEE International Conference on Data Mining – ICDM*, 2001.

S. Anwar, R. McMillan, and M. Zheng. Bidding behavior in competing auctions: Evidence from eBay. *European Economic Review*, 2004. In press. Available online 30 November 2004: www.sciencedirect.com.

S. Bandyopadhyaya, J. Rees, and J. M. Barron. Simulating sellers in online exchanges. *Decision Support Systems*, 2004. In press. Available online 2 October 2004: www.sciencedirect.com.

R. Bapna, P. Goes, and A. Gupta. A theoretical and empirical investigation of multi-item on-line auctions. *Information Technology and Management*, 1:1–23, 2000.

B. Bourbeau, T. G. Crainic, M. Gendreau, and J. Robert. Design for optimized multi-lateral multi-commodity markets. *European Journal of Operational Research*, 163: 503–529, 2005.

P. S. Bradley, U. M. Fayyad, and O. L. Mangasarian. Mathematical programming for data mining: Formulations and challenges. Technical report, Microsoft Press, 1998.

G. Cachon. Supply chain coordination with contracts. In S. Graves and T. de Kok, editors, *Handbooks in Operations Research and Management Science: Supply Chain Management*. North-Holland, 2001.

G. Cai and P. Wurman. Monte Carlo approximation in incomplete information sequential auction games. *Decision Support Systems*, 39:153–168, 2005.

T. S. Chandrashekar, Y. Narahari, C.H. Rosa, D. Kulkarni, and J. D. Tew. Auction based mechanisms for electronic procurement. Technical report, Electronic Enterprises Laboratory, Department of Computer Science and Automation, Indian Institute of Science, 2005.

R. R. Chen, R. Roundy, R. Q. Zhang, and G. Janakiraman. Efficient auctions for supply chain procurement. Technical report, Johnson Graduate School of Management, Cornell University, Ithaca, NY, 2002.

S. S. Chen, D. L. Donoho, and M. A. Saunders. Atomic decomposition by basis pursuit. Technical report, Dept. of Statistics, Stanford University, 1996.

S. Chopra and P. Meindl. *Supply chain management: Strategy, planning, and operation*. Prentice-Hall, 2001.

S. Chopra and J. Van Mieghem. Which e-business is right for your supply chain? *Supply Chain Management Review*, pages 32–40, July – August 2000.

E. Constantinides. The 4S web-marketing mix model. *Electronic Commerce Research and Applications*, 1:57–76, 2002.

E. Dans. Existing business models for auctions and their adaptation to electronic markets. *Journal of Electronic Commerce Research*, 3(2):23–31, 2002.

I. Daubechies. Time-frequency localization operators: A geometric phase space approach. *IEEE Transactions on Information Theory*, 34:605–612, 1988.

S. Disney, M. Naim, and A. Potter. Assessing the impact of e-business on supply chain dynamics. *International Journal of Production Economics*, 89(2):109–118, 2004.

S. Drew. Strategic uses of e-commerce by SMEs in the east of England. *European Management Journal*, 21(1):79–88, 2003.

S. Elliot, editor. *Electronic commerce: B2C strategies and models*. John Wiley, 2002.

T. Y. Eng. The role of e-marketplaces in supply chain management. *Industrial Marketing Management*, 33:97–105, 2004.

U. M. Fayyad, G. Piatetsky-Shapiro, and P. Smyth. From data mining to knowledge discovery: An overview. In U. M. Fayyad, G. Piatetsky-Shapiro, P. Smyth, and

R. Uthurusamy, editors, *Advances in Knowledge Discovery and Data Mining*, pages 1–30. AAAI/MIT Press, 1996.

J. Fraser, N. Fraser, and F. McDonald. The strategic challenge of electronic commerce. *Supply Chain Management: An International Journal*, 5(1):7–14, 2000.

Y. Freund and R. E. Schapire. A decision-theoretic generalization of on-line learning and an application to boosting. *Journal of Computer and System Sciences*, 55(1): 119–139, 1997.

G. E. Fruchter and C. Tapiero. Dynamic online and offline channel pricing for heterogeneous customers in virtual acceptances. *International Game Theory Review*, 2005. To appear.

K. Fukunaga. *Statistical pattern recognition*. Academic Press, 1990.

A. Gerber, C. Russ, and M. Klusch. Supply web co-ordination by an agent-based trading network with integrated logistics services. *Electronic Commerce research and Applications*, 2:133–146, 2003.

J. Geunes and P. Pardalos. Network optimization in supply chain management and financial engineering: An annotated bibliography. *Networks*, 42(2):66–84, 2003.

J. Geunes, P. Pardalos, and E. Romeijn, editors. *Supply chain optimization: Applications and algorithms*. Kluwer Academic Publishers, 2002.

J. Gibbs, K. Kraemer, and J. Dedrick. Environment and policy factors shaping global e-commerce diffusion: A cross-country comparison. *The Information Society*, 19: 5–18, 2003.

D. G. Gregg and S. Walczak. Auction advisor: An agent-based online-auction decision support system. *Decision Support Systems*, 2004. In press. Available online 30 November 2004: www.sciencedirect.com.

A. Gunasekaran and E. Ngai. Information systems in supply chain integration and management. *European Journal of Operational Research*, 159:269–295, 2004.

A. Gupta, D. O. Stahl, and A. B. Whinston. A priority pricing approach to manage multiservice class network in real time. *Journal of Electronic Publishing*, 1995. Available on-line: www.press.umich.edu:80/jep/econTOC.html.

A. Gupta, D. O. Stahl, and A. B. Whinston. The economics of network management. *Communications of the ACM*, 42(9):57–63, 1999.

J. Han and M. Kamber. *Data mining, concepts and techniques*. Morgan Kaufmann, 2001.

M. Hegland. Data mining techniques. *Acta Numerica*, 10:313–355, 2001.

S. J. Hong and S. M. Weiss. Advances in predictive data mining methods. In *Proceedings of First International Workshop in Machine Learning, Data Mining and Pattern Recognition*, volume 1715 of *Lecture Notes in Artificial Intelligence*, 1999.

W. H. Inmon. *Building the data warehouses*. John Wiley & Sons, 1996.

A. K. Jain and R. C. Dubes. *Algorithms for clustering data*. Prentice-Hall, 1988.

R. Kalakota and A. B. Whinston. *Electronic commerce: A manager's guide*. Addison-Wesley, 1997.

M. Karsten, J. Schmitt, L. Wolf, and R. Steinmetz. Cost and price calculation for Internet integrated services. In *Proceedings of Kommunikation in Verteilten Systemen (KiVS'99)*, 1999.

R. J. Kauffman and E. A. Walden. Economics and electronic commerce: Survey and directions for research. *International Journal of Electronic Commerce*, 5(4):5–116, 2001.

P. Keskinocak, R. Goodwin, F. Wu, R. Akkiraju, and S. Murthy. Decision support for managing an electronic supply chain. *Electronic Commerce Research*, 1:15–31, 2001.

S. Klein. Introduction to electronic auctions. *Electronic Markets*, 7(4):3–7, 1997.

P. Klemperer. Auction theory: A guide to the literature. *Journal of Economic Surveys*, 13(3):227–186, 1999.

R. Kohavi, L. Mason, R. Parekh, and Z. Zheng. Lessons and challenges from mining retail e-commerce data. *Machine Learning*, 57:83–113, 2004.

M. Kourgiantakis, N. Matsatsinis, and A. Migdalas. E-commerce: A survey for Cretan agro-food sector. In A. Nikolaidis, G. Baourakis, E. Isikli, and M. Yercan, editors, *The market for organic products in the Mediterranean region*. Cahiers Options Mediterrraneennes, 61, MAIch, 2003.

W. Kuechler Jr., V. K. Vaishnavi, and D. Kuechler. Supporting optimization of business-to-business e-commerce relationships. *Decision Support Systems*, 31:363–377, 2001.

S. Lambert. A review of the electronic commerce literature to determine the meaning of the term business model. Technical report, School of Commerce, Flinders University of South Australia, 2003. Available online 14 April 2005: www.ss.flinders.edu.au/commerce/researchpapers.

R. Lancioni, M. Smith, and T. Oliva. The role of the Internet in supply chain management. *Industrial Marketing Management*, 29:45–56, 2000.

S. Lee and G. G. Lim. The impact of partnership attributes on EDI implementation success. *Information & Management*, 42(4):503–516, 2005.

R. Lewis and A. Cockrill. Going global remaining local: The impact of e-commerce on small retail firms in Wales. *International Journal of Information Management*, 22:195–209, 1999.

D. Lucking-Reiley. Auctions on the Internet: What's being auctioned and how? *Journal of Industrial Economics*, 48(3):227–252, 2000.

D. Lucking-Reiley and D. F. Spulber. Business-to-business electronic commerce. *Journal of Economic Perspectives*, 15(1):55–68, 2001.

V. Mabert, A. Soni, and M. Venkataramanan. Enterprise resource planning: Managing the implementation process. *European Journal of Operational Research*, 146(2): 302–314, 2003.

I. Mandalianos. Prototype e-government system for municipalities. Master's thesis, Department of Production Engineering and Management, Technical University of Crete, 2005.

O. L. Mangasarian. Mathematical programming in data mining. *Data Mining and Knowledge Discovery*, 1:183–201, 1997.

R. McAfee and J. McMillan. Auctions and bidding. *Journal of Economic Literature*, 25(2):699–738, 1987.

L. W. McKnight and J. P. Bailey. An introduction to Internet economics, 1995. Presented at MIT the Workshop on Internet Economics March 1995. Available online 14 April 2005: www.press.umich.edu/jep/.

D. Michie, D. J. Spiegelhalter, and C. C. Taylor, editors. *Machine learning, neural and statistical classification*. Ellis Horwood, 1994.

P. Milgrom. Auctions and bidding: A primer. *Journal of Economic Perspectives*, 3: 3–22, 1989.

J. Mingzhou and D. Wu. Supply chain coordination in electronic markets: and contracting mechanisms. Technical report, Lehigh University, 2001.

T. M. Mitchell. *Machine Learning*. McGraw Hill, 1997.

A. Nagurney, J. Cruz, J. Dong, and D. Zhang. Supply chain networks, electronic commerce and supply side and demand side risk. *European Journal of Operational Research*, 164(1):120–142, 2005.

Y. Narahari and P. Dayama. Combinatorial auctions for electronic business. Technical report, Electronic Enterprises Laboratory, Department of Computer Science and Automation, Indian Institute of Science, 2005. Available online on 14 April 2005: http://lcm.csa.iisc.ernet.in/hari/research_reports.html.

S. Netessine and N. Rudi. Supply chain structures on the Internet and the role of marketing – Operations interaction. In D. Simchi-Levi, S. D. Wu, and Z.-M. Shen, editors, *Handbook of Quantitative Supply Chain Analysis: Modeling in the E-Business Era*. Kluwer Academic Publishers, 2004.

United States Bureau of Census. Measuring electronic business: Definitions, underlying concepts and measurement plans. Technical report, 2000. Available online on 14 April 2005: http://www.census.gov/epcd/www/ebusines.htm.

O. Ovalle and A. Marquez. The effectiveness of using e-collaboration tools in the supply chain: An assessment study with system dynamics. *Journal of Purchasing & Supply Management*, 9:151–163, 2003.

A. Patteli. A domain area report on business models, the eBusiness center. Technical report, Athens University of Economics and Business, 2002. Available online on 14 April 2005): www.eltrun.gr/whitepapers.

M. Pohjola. The New Economy: Facts, impacts and policies. *Information Economics and Policy*, 14:133–144, 2002.

M. E. Porter. Strategy and the Internet. *Harvard Business Review*, pages 63–78, March 2001.

J. Mac Queen. Some methods for classification and analysis of multivariate observations. In *Proceedings of the 5th Berkeley Symposium Math. Statist. and Probability (1965/66), Vol. I: Statistics*, pages 281–297. University of California Press, 1967.

N. Ramakrishnan and A. Y. Grama. Data mining: From serendipity to science. *Computer*, 32(8):34–37, 1999.

S. Ramaswamy, S. Mahajan, and A. Silberschatz. On the discovery of interesting patterns in association rules. In *Proceedings of 24th Intl. Conf. on Very Large Databases (VLDB'98)*, pages 368–379, 1998.

J. G. Riley and W. F. Samuelson. Optimal auctions. *The American Economic Review*, 71(3):381–392, 1981.

M. Schillo and K. Fisher. Holonic multiagent systems. *Kunstliche Intelligenz*, 4/03: 54–55, 2003.

H. Schweizer. E-learning in business. *Journal of Management Education*, 28(6):674–692, 2004.

A. Scupola. Adoption issues of business to business Internet commerce in European SMEs. In *Proceedings of 35th Hawaii International Conference on System Sciences*, 2002.

S. Z. Selim and M. A. Ismail. K-means-type algorithms: A generalized convergence theorem and characterization of local optimizality. *IEEE Transactions on Pattern Analysis and Machine Intelligence*, PAMI-6:81–87, 1984.

M. J. Shaw and M. J. Strader. Characteristics of electronics markets. *Decision Support Systems*, 21:185–198, 1997.

D. Y. Shee, T.-I. Tang, and G.-H. Gwo-Hshiung Tzenget. Modeling the supply-demand interaction in electronic commerce: A bi-level programming approach. *Journal of Electronic Commerce Research*, 1(2):79–93, 2000.

B. Shore. Information sharing in global supply chain systems. *Journal of Global Information Technology Management*, 4(3):27–50, 2001.

G. J. C. Da Silveira. Towards a framework for operations management in e-commerce. *International Journal of Operations & Production Management*, 23(2):200–212, 2003.

D. Simchi-Levi, S. D. Wu, and M. Shen, editors. *Handbook of quantitative supply chain analysis: Modeling in th e-business era*. Kluwer Academic Publishers, 2004.

J. Simpson. The impact of the Internet in banking: Observations and evidence from developed and emerging markets. *Telematics and Informatics*, 19(4):315–330, 2002.

T. Skjott-Larsen, H. Kotzab, and M. Grieger. Electronic marketplaces and supply chain relationships. *Industrial Marketing Management*, 32:199–210, 2003.

M. Stansfield and K. Grant. Barries to the take-up of electronic commerce among small-medium sized enterprises. *Informing Science*, pages 737–745, 2003.

T. Strader, F. Lin, and M. Shaw. Business-to-business electronic commerce and convergent assembly supply chain management. *Journal of Information Technology*, 14:361–373, 1999.

J. Strivastava, R. Cooley, M. Deshpande, and P. N. Tan. Web usage mining: Discovery and applications of usage patterns from web data. *ACM SIGKDD Explorations Newsletter*, 1(2), January 2000.

D. Tapscott, D. Ticoll, and A. Lowi. *Digital capital: Harnessing the power of business webs*. Harvard Business School Press, 2000.

J. E. Teich, H. Wallenius, J. Wallenius, and O. Koppius. Emerging multiple issue e-auctions. *European Journal of Operational Research*, 159:1–16, 2004.

P. Timmers. Business models for electronic markets. *Electronic Markets*, 8(2):3–8, 1998.

P. Timmers. Electronic commerce – Marketing strategies and business models, 2000. Available online on 14 April 2005: http://users.pandora.be/paul.timmers/PaperbackWebarticle.htm.

A. A. Tsay, S. Nahmias, and N. Agrawal. Modeling supply chain contract: A review. In S. Tayur, R. Ganeshan, and M. Magazine, editors, *Quantitative Models for Supply Chain Management*, pages 229–336. Kluwer Academic Publishers, 1999.

E. Turban. Auctions and bidding on the Internet: An assessment. *Electronic Markets*, 7(4):7–11, 1997.

R. Varadarajan and M. S. Yadav. Marketing strategy and the Internet: An organizing framework. *Journal of the Academy of Marketing Science*, 30(4):296–312, 2002.

I. Vlachos. Critical success factors of business to business (B2B) e-commerce solutions to supply chain management. In P. Pardalos, A. Migdalas, and B. Baourakis, editors, *Supply Chain and Finance*, pages 162–175. World Scientific, 2004.

Q. Wang, J. M. Peha, and M. A. Sirbu. Optimal pricing for integrated-services networks with guaranteed quality of service. In J. Bailey and L. McKnight, editors, *Internet Economics*. MIT Press, 1997. Available online on 14 April 2005: http://www.ece.cmu.edu/~peha/pricing.html.

Y. Wang, K.-L. Tan, and J. Ren. PumaMart: A parallel and autonomous agents based Internet marketplace. *Electronic Commerce Research and Applications*, 3:294–310, 2004.

J. H. Ward. Hierarchical grouping to optimize an objective function. *Journal of American Statistical Association*, 58(301):236–244, 1963.

L. Willcocks and C. Sauer, editors. *Moving to e-business*. Random House, 2000.

E. A. Williamson, D. K. Harrison, and M. Jordan. Information systems development within supply chain management. *International Journal of Information Management*, 24:375–385, 2004.

P. Windrum and P. de Berranger. Factors affecting the adoption of intranets and extranets by SMEs: A UK study. Technical report, Manchester Metropolitan University Business School, 2003. Available online on 14 April 2005): www.business.mmu.ac.uk.

C. A. Wood. Current and future insights from online auctions. In M. Shaw, R. Blanning, T. Strader, and A. B. Whinston, editors, *Handbook on Electronic Commerce*. Springer-Verlag, 2004.

M. Xia, J. Stallaert, and A. B. Whinston. Solving the combinatorial double auction problem. *European Journal of Operational Research*, 164:239–251, 2005.

K. Zhu, K. Kraemer, and S. Xu. Electronic business adoption by European firms: A cross-country assessment of the facilitators and inhibitors. *European Journal of Information Systems*, 12(4):251–268, 2003.

36 OPTIMIZATION ISSUES IN COMBINATORIAL AUCTIONS

Stan van Hoesel[1] and Rudolf Müller[1]

[1]Department of Quantitative Economics
Maastricht University
Maastricht, The Netherlands
S.vanHoesel@KE.unimaas.nl
R.Muller@KE.unimaas.nl

Abstract: Auctions are used more and more to sell a large variety of goods. In this chapter, it is our objective to concentrate on applications of auctions in telecommunication, which possess a part or a feature that can be optimized. Optimization methods are necessary, in particular, when auctions are used to sell or purchase goods which consist of combinations of items, and where combinations have higher or lower value than the sum of values of individual items: combinatorial auctions. In the first part, we review the theory on combinatorial auctions, starting with the various properties and mechanisms found in the literature on combinatorial auctions. Then the allocation decision is identified as the winner determination problem (WDP), which is the central subject of this chapter. The winner determination problem is formulated as an Integer Linear Program (ILP) with the structure of a set-packing problem. Therefore, complexity results, polynomial special cases, and general solution methods for the WDP are often obtained from results for the set-packing problem. In the second part of this chapter, we turn to applications from telecommunications. First, a model for bandwidth allocation in networks is discussed. The problem is translated into a formulation that has close relations to multicommodity flow and network synthesis. This guides us to alternative formulations and to solution methods. Second, the auctions of radio spectrum in the US and Europe are reviewed. The WDP of these multi-round auctions can be modeled using the XOR-of-OR bidding language, and solved by methods originally developed for set-packing.

Keywords: Auctions, combinatorial auctions, telecommunications, optimization, bandwidth allocation.

36.1 INTRODUCTION

Auctions are an ancient means to allocate efficiently a single, indivisible good among a group of potential buyers. For example, in the English auction buyers, also called bidders, overbid other buyers until there is only one buyer left who is willing to purchase the good at the current high bid. Obviously, the winning buyer values the good most. Moreover, the auction determines a Walrasian price, i.e., a price at which the market clears. The same applies to the Dutch auction, another popular auction that uses a decreasing price clock. The buyer who first stops the clock wins the good at the current price. For the Dutch auction there exists a symmetric Bayesian Nash equilibrium that guarantees again that the allocation is efficient. The same result holds for sealed-bid, first-price auctions, which are very common in public procurement, but then in form of a reverse auction. Here the potential sellers submit sealed bids to get a particular contract. The buyer chooses the cheapest contract and pays the price asked in this contract.

Two innovations in telecommunication have created a boom in applying auctions. The first has been wireless communication, in combination with the liberalization of national telecom markets. Governments had to allocate radio frequency to the telecom industry, and wanted to do so in an economically efficient way. Onwards from the beginning of the 90s of last century they used auctions to achieve this goal. The second innovation has been the Internet. For the first time, auctions could be held without auctioneer and buyers gathering at the same location. Since then billions of consumer-to-consumer auctions have been performed online. Also, online procurement auctions became feasible with the new infrastructure.

Auctions are means to design markets, whenever other means seem not to work. The key problem that is solved by an auction is information asymmetry. In the frequency assignment example, the government does not know which telecom operator can most efficiently build up a network, and thus benefit most from a frequency. In an Internet auction, the seller does not know how much somebody would be willing to pay for her used laptop, say. In a procurement auction, the buyer doesn't know which supplier is able to produce at the lowest cost, and thus to offer the lowest price. Would sellers know demand curves, they could quote a market clearing price, and tell buyers: take it or leave it. There are many markets, in which there is no such information, for example, because the product is very unique, or because the buyers are very heterogenous. For example, potential suppliers in a procurement negotiation for a construction project could have very different internal cost structures.

To overcome the asymmetry of information, an auction performs essentially a *competitive game* among the buyers. From now on we will call buyers bidders. The rules of the game have to define two things: to which allocation of goods do the actions of the bidders lead, and which price do the bidders have to pay. For example, in a second price, sealed bid auction, or *Vickrey auction* (Vickrey, 1961), the rules of the game say that every bidder submits a sealed bid. The bids are opened by the auctioneer, and he allocates the item to the bidder with the highest bid. The price the winner is going to pay is equal to the second highest bid. In a Dutch auction the only action is to stop the clock. The first bidder who takes this action wins the item, and pays the current price. The payoff, or *utility* of a bidder depends on his valuation for the goods he wins, and the prices he pays. Throughout this chapter, we assume quasi-linear utility: valuation minus price.

The prospective utility as a function of the actions taken in the game, and the actions of the other bidders, trigger the strategies of the bidders. Game theory can help us to solve the game, i.e., to find combinations of strategies which are in equilibrium. There are two characteristics of auction games that makes their solution very difficult. Firstly, the incomplete information among buyers. There is not only uncertainty about the strategies chosen by the other players, but also about their valuations for the goods (why would we need an auction otherwise?) Since information about valuations of competitors is not available one often has to use the concept of Bayesian Nash equilibrium, in which a bidder maximizes his expected utility, expectation taken with respect of a known distribution of valuations of the other bidders. Still, it is doubtful that bidders know this distribution. More attractive are therefore auctions that are robust against variations of the valuations of other bidders in the sense that they solve in a so-called ex-post equilibrium. Even better are auctions that allow for strategies that are a best response to any chosen strategies of the competitors, and any valuation that they have. Such strategies form a dominant strategy equilibrium (see, e.g., Mas-Colell et al. (1995) for a detailed treatment of these concepts).

The second complication is due to the iterative nature of the game. Unless we perform a sealed-bid auction, we will have the effect that bidders can observe other bidders' actions, and react to them. This requires to solve extensive form games with incomplete information. Most auction literature claims that this is not an issue because one can focus without loss of generality on revelation mechanisms (by invoking the revelation principle (Dasgupta and Maskin, 1979; Myerson, 1979)). In the context of auctions, revelation mechanisms are sealed-bid auctions in which bidders submit their valuations for the item(s) in a single-round, and winners and prices are determined on this information. However, sealed-bid auctions contradict not only the classical flavor of an auction, but require usually much more effort in bid preparation and communication (see Section 36.2.1).

Many other issues add to the complexity of analyzing auction games. To name two of them: asymmetries between bidders, e.g., incumbents and new entries in frequency auctions; common values, i.e., a bidder's valuation of an item depends also on signals received by other bidders. The game theoretic aspects of auctions are therefore not easy to deal with, and in this chapter we will have to leave them almost completely out of consideration. Rather we will focus on optimization aspects of auctions. These become relevant when we want to sell several items at the same time, in other words when we deal with *combinatorial auctions*.

Combinatorial auctions have been introduced as a mechanism to sell sets of items, whenever combinations of items may have values to bidders other than simply the sum of values of the individual items. This situation occurs in many settings such as airport time slots (origin and destination), railroad segments (connected subnetworks allow for a greater variety of routes), and telecom radio spectrum (licenses in geographically neighboring areas increase the value of the network). A typical example is the sale of network capacity. A customer may be interested in hiring capacity between two vertices i and j in a network. He can bid on subsets of edges in the network. If the union of two sets A and B forms a path between i and j, then the valuation $v(.)$ of the union can be higher than the valuation of the individual sets, i.e., $v(A \cup B) > v(A) + v(B)$, the sets are *complementary*. However, if each of A and B forms a path between i and j, then $v(A \cup B) < v(A) + v(B)$, the sets are *substitutable*. If in such settings an auction design does not allow for combinatorial bids, inefficient allocations can be the consequence. For example in case of complementarity, agents are exposed to the risk that they overpay because they are missing at the end some of the necessary items, e.g., the set of edges that they purchased does not contain a path connecting i and j.

What might be the simplest design for a combinatorial auction, is one auctioneer selling items from a set N to several potential buyers M, who are allowed to submit sealed bids on bundles of items. On the basis of the received bids the auctioneer must decide which bids win and which loose, under the condition that each item is sold at most once, and with the objective to maximize his revenue. This *winner determination problem* (WDP) is the main subject of this chapter. The problem has been introduced by Jackson (1976). Variants have been described in Rothkopf et al. (1998). They also provide the first results on the complexity of the problem. If in this sealed-bid design bidders have to pay their bids, they will have large incentives to bid less for subsets than they value them. There is a classic game theoretic solution to omit such strategic behavior. We change the payment rule such that bidders do not pay the bids but so-called Vickrey-Clarke-Groves (VCG) prices. This pricing rule generalizes the single-item Vickrey auction in its strategic properties: revealing the true valuation for every subset of items is a dominant strategy equilibrium—there is no way to increase own utility by lying about one's own valuation. The nice feature of this pricing rule is that the same WDP algorithm can be used to compute the VCG prices (although it has to be invoked for every winning bidder once in order to compute his prices).

The VCG auction provides thus a very attractive framework for such markets, which received nevertheless much criticism. For economic criticism we refer to Ausubel and Milgrom (2005). We will focus on computational problems and on problems related to communication complexity:

- If a bidder should submit his valuation for every subset of items, as is his dominant strategy in the VCG mechanism, he would have to communicate an exponential amount of information (in the number of items). As this is not realistic, the question arises whether in a particular situation valuations can be encoded in a compact form (e.g., additive valuations). If no compact encoding is possible, how then, does an approximate report of bids influence the quality of the allocation?

- Once bids are submitted in an appropriate bidding language, the auctioneer has to solve the WDP. In general the WDP will be equivalent to solving a set packing problem, thus NP-hard (see Section 36.2.3). The question arises in which special cases is WDP polynomially solvable, when can it be solved approximately, and how does approximate winner determination influence the strategic properties of VCG?

- Can some of the computational problems of the VCG mechanism be overcome by a multi-round design, in which bidders are free to increase bids or make bids for some subsets only in a later phase of the auction? If a multi-round approach is chosen, which information is given to the bidders about the state of the auction (e.g., high bids, the identity of bidders of high bids, dual prices from the winner determination solution), and what is the computational cost of bidders to use this information?

In Section 36.2 we describe the clearing problem for combinatorial auctions: the Winner Determination Problem. The problem is formulated and the XOR and OR bidding languages are introduced. The complexity of the problem is discussed, including polynomial cases and approximability. Finally, we describe some algorithms. In the remainder we discuss some applications of combinatorial auctions in telecommunications. In Section 36.3 we discuss the problem of bandwidth allocation in networks. The single edge, multiple bid case is modeled as an optimization problem. Then the general network single bid problem is discussed. Moreover, a reverse auction on a network with a single commodity is modeled as a shortest path problem. In Section 36.4 we discuss the well-known frequency band auctions for wireless networks as they were were formulated and applied in practice. Furthermore, we discuss a formal model for multiple round auctions introducing an extension of the XOR and OR languages.

36.2 THE WINNER DETERMINATION PROBLEM

The following setting is studied in this chapter. There is a finite set of indivisible items N to sell. There is a finite set M of buyers or bidders. A bundle S is a set of items: $S \subseteq M$. The set of subsets of M is denoted by $\mathcal{M}$. We assume that bidder i values the bundle $S \subseteq M$ by $v_i(S)$. Valuations of bidders are private information. The goal of the auction is trade: allocate the items to the bidders and agree on payments.

A combinatorial auction is a mechanism that finds an allocation and determines how much every buyer has to pay for the items allocated to him. From an economic point of view, it is meaningful to allocate the items most efficiently, i.e., such that the total value generated by the allocation is maximized. From a business point of view, one might have other objectives, in particular to maximize the revenue of the seller. As most of the combinatorial auction literature we assume that maximizing the total value is a good proxy for maximizing revenue, since we may assume that prices paid are not too far away from valuations[1].

The game theoretic problem in auctions is that the auctioneer does not know the valuations. He gets only to see bids in an auction, and can only charge prices dependent on the bids. Now, as mentioned before, VCG gives incentives to report true valuations, and computing the VCG payments can be done by the same winner determination algorithm. However, it is often impossible to report all valuations, and, as we will see, the winner determination problem is an NP-hard optimization problem. Therefore, for VCG, as well as for other combinatorial auction designs, communica-

---

[1] If they would be far away, it would also mean that competition is not strong enough in order to enforce higher prices.

tion of bids and winner determination are the main bottleneck. That is why we focus on these two issues. It will thereby not be important whether the auctioneer gets to see bids that are equal to true valuations. Therefore we make in the following no difference between bids and valuations, but assume that bids are (true or false) reports of valuations.

Communication of bids and computation of winning bids are closely related to each other, because the bids are the input of the optimization problem of winner determination. If we take the number of items and the number of bidders as a base of measuring complexity, a huge number of bids relative to the number of items makes the communication of bids hard, while winner determination becomes theoretically easy, because the number of bids is exponential in the number of items (see Sandholm (2002) for a formalization of this argument). On the other hand, a small number of bids, or a compact representation of bids, challenges winner determination since encoding length of the input is small. Therefore our discussion of the winner determination problem starts with a discussion of bidding languages.

36.2.1 Bidding languages

Our treatment is based on a formal approach to bidding languages introduced by Nisan (2000) (see also Nisan (2005)). Bidding languages formalize a semantic agreement of the meaning of syntactic combinations of atomic bids. A general assumption that we make is that there is an implicit bid $v_i(\emptyset) = 0$, and that *free-disposal* holds, i.e., that bidders value a subset S_1 of a set S_2 of items never more than the set S_2, i.e., if $S_1 \subset S_2$, then $v(S_1) \leq v(S_2)$.

The bidding languages are based on atomic bids $v_i(S)$, also called package bids, expressing the valuation of bidder i for subset S, and, by free disposal, for every superset of S. In the *OR* bidding language the atomic bids by bidder i are interpreted as: i is willing to pay for any combination of pairwise disjoint atomic bids a price equal to the sum of the bid prices. By the free-disposal assumption, i's bid for any other set S is then the maximum sum of bid prices for pairwise disjoint atomic bids contained in S. OR bids are also called *non-exclusive bundle bids*. In the XOR bidding language the set of atomic bids by bidder i is interpreted as: bidder i is willing to receive at most one of the atomic bids. By the free-disposal assumption, i's bid for any other set S becomes then the maximum price for an atomic bid contained in S. XOR bids are also called *exclusive bundle bids*. The distinction between OR and XOR has been introduced by Sandholm (2002). Nisan (2000) extended these ideas to mixtures of both OR and XOR bids, like OR-of-XOR and XOR-of-OR bids, and studied their expressiveness. An excellent survey of bidding languages can be found in Nisan (2005).

An auction design might restrict the bundles for which atomic bids may be reported. For example, in the OR bidding language one might restrict bids on subsets of at most 2. This makes sense, for example, when take-off and landing slots on airports are going to be auctioned, or in telecommunication when a bid is on a combination of a source and target node, leaving open the links over which the communication should be routed. Formally, we will assume that for every bidder i, there exists a collection of subsets $S_i \subseteq \mathcal{M}$ from which he may choose atomic bids to compose OR or XOR bids. Restrictive collections of feasible atomic bids are one of the keys to achieve computational tractable winner determination problems.

An auction design may also restrict the valuations for subsets. For example assume that the valuation has to define a submodular function, or that the bid value for a subset has to be equal to the sum of bid values for the items, minus or plus a function that depends on the number of items in the bid. Also such restrictions can be very helpful for winner determination.

Finally, bidders might submit *oracles* that compute for every subset the valuation, and which the auctioneer invokes when computing the winner or payments. Of similar flavor are proxy bids, like in ebay.com. Here a bidder submits a maximum value he is willing to pay, and the auction server creates a new bid on behalf of this bidder whenever an old high bid of him is outbid, and the maximum value is not reached yet. In iterative combinatorial auctions such proxies compute, given current prices of individual items, which bundles are most attractive for the bidder.

To summarize, bidding languages are expressions by which the auctioneer can compute for any set S the value that bidder i has for subset S. A bidding language expression from each participating bidder forms the input of the winner determination problem. Therefore, the complexity of the winner determination problem is highly dependent on the bidding language. We will now briefly survey the main results on the complexity of winner determination.

36.2.2 Problem Formulation

An allocation of the items is described by variables $x_i(S) \in \{0,1\}$. The variable $x_i(S)$ is equal to one if and only if bidder i gets bundle S. An allocation $(x_i(S)|i \in N, S \in \mathcal{S})$ is said to be feasible if it allocates no item more than once: constraints (36.2); and at most one subset to every bidder: constraints (36.3). It is optimal if the revenues (36.1) are maximum.

$$
\begin{align}
\text{ILP} \quad \max \quad & \sum_{i \in N} v_i(S) x_i(S) & & (36.1) \\
\text{s.t.} \quad & \sum_{i \in N} \sum_{S \in \mathcal{S}, j \in S} x_i(S) \leq 1 & \forall j \in M & (36.2) \\
& \sum_{S \in \mathcal{S}} x_i(S) \leq 1 & \forall i \in N & (36.3) \\
& x_i(S) \in \{0,1\} & \forall i \in N \; \forall S \in \mathcal{S} & (36.4)
\end{align}
$$

This formulation allows free disposal for the auctioneer: the auctioneer is willing to keep items. Otherwise the inequalities in (36.2) need to be replaced by equalities. If multiple copies of an item are available then one could either treat them as being different, which however increases the number of rows and columns. It is preferable to replace the right-hand side 1 of constraints (36.2) by the number of items. If bundles may contain several copies of the same type of item, one will need additional columns (for all multi-sets of items).

A well-studied version of the multi-item case is that of a single type of items, also called homogeneous multi-item auction. If the marginal valuations for additional

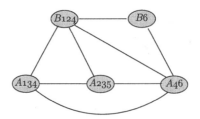

Figure 36.1 Example of the conflict graph arising in a 2-bidder situation.

items are non-increasing, WDP becomes very easy: sort the marginal valuations of all bidders, and allocate to the bidders their share in the k largest marginal valuations, if k items are on sale. If marginal valuations are not decreasing, we get knapsack or multi-knapsack problems (see Holte (2001), Kellerer et al. (2004), and Lehmann et al. (2005)). For the rest of this section we focus on the case of one copy per type of item.

The integer linear program (ILP) is independent of the bidding language, as it has a decision variable for every combination (i, S) of a bidder and a subset. It can be easily seen that a particular bidding language allows to reduce the number of variables and makes some constraints obsolete. For example, if the auctioneer receives *OR* bids, a variable for each atomic bid is sufficient if constraints (36.3) are deleted. OR bids also allow for the removal of the bidder indices in the variables, and if there are several bids for the same subset S, all but those with maximum bid value may be deleted. We use the acronym WDP_{OR} for the ILP corresponding to the OR bidding language. In the case of XOR we may not drop constraint (36.3), unless bids are super-additive, i.e., for any two bundles S_1 and S_2 with $S_1 \cap S_2 = \emptyset$ we have $v_i(S_1 \cup S_2) \geq v_i(S_1) + v_i(S_2)$. Again we need only decision variables $x_i(S)$ for subsets S for which an atomic bid has been submitted. We use the acronym WDP_{XOR} for the ILP corresponding to the XOR bidding language.

Alternatively to the ILP model above, the intersection graph or conflict graph can be used to model the constraints in the problem. The vertices of the intersection graph represent the atomic bids, possibly after deleting dominated bids. Two vertices are connected by an edge if not both can be selected, i.e., if the corresponding sets have an item in common (constraints (36.2)) or if the sets belong to the same bidder (constraints (36.3)). The winner determination problem then reduces to finding an independent set of maximum weight. In the example of figure 36.1, two bidders A and B, bid on subsets of six items numbered from 1 to 6.

The relation with the set-packing problem is obvious from this representation. Thus, it should not come as a surprise that complexity issues are also related, see the next subsection.

36.2.3 Complexity of the winner determination problem

The winner determination problem is in general NP-hard. This can be simply seen by the fact that the set-packing problem can be transformed to WDP_{OR} (Rothkopf et al.,

1998). One can refine this result to various special cases. For a recent overview, see Lehmann et al. (2005). We provide a few of the results.

Given some number k, we define the decision version of the winner determination problem as to decide whether there exists a feasible allocation of the items with total revenue greater than or equal to k. The first result makes use of the fact that the decision version of the stable set problem is NP-complete (Garey and Johnson, 1979). Given a graph $G = (V, E)$ we can construct an instance of WDP_{OR} as follows. We set $N = V$, $M = E$ and assume a bid $b_i(S) = 1$ for every $i \in N$ with $S = \{e \in E \mid i \in e\}$. Note that this is also an instance of WDP_{XOR}. It is obvious that G has a stable set of size k if and only if the instance of WDP_{OR} (or, equivalently, WDP_{XOR}) has an allocation with total bid value greater than or equal to k. This proves:

Theorem 36.1 *The decision versions of WDP_{OR} and WDP_{XOR} are NP-complete, even if we restrict to instances where every bid has a value equal to 1, every bidder submits only one bid, and every item is contained in exactly 2 bids.*

Other problems than stable set can be used, too. For example, (Rothkopf et al., 1998) uses *3-set packing* to show

Theorem 36.2 (Rothkopf et al. (1998)) *The decision version of WDP_{OR} is NP-complete even if we restrict to instances where every bid has a value equal to 1, and every bidder bids only on subsets of size of at most 3.*

Similarly, Hoesel and Müller (2001) transforms *3-dimensional matching* to WDP_{XOR} to show:

Theorem 36.3 (Hoesel and Müller (2001)) *The decision version of WDP_{XOR} is NP-complete even if we restrict to instances where every bid has a value equal to 1, and every bidder bids only on subsets of size of at most 2.*

The uncovered case here is WDP_{OR} if every bidder bids on subsets of size of at most 2. This problem reduces simply to the weighted matching problem, and is thus polynomially solvable. Other polynomial cases are either obtained by restricting the structure of bids or the valuations, i.e. the objective of (ILP).

36.2.3.1 Special structures of bundles. If the bidding language restricts atomic bids to bundles within a set $\mathcal{S} \subset \mathcal{M}$, this has direct implications for the structure of the constraint matrix of (IP). If the bids are such that the constraint matrix is either Totally unimodular, Balanced, or Perfect, then the LP-relaxation of WDP_{XOR}, say LP_{XOR}, always returns integer values on the variables, and thus the problem can be solved in polynomial time (see de Vries and Vohra (2003), Müller (2005)). For example, if all items can be arranged linearly, and $\mathcal{S}$ consists of sets of consecutive items only. In this case, WDP_{OR} becomes polynomial (Rothkopf et al., 1998), due to the consecutive-ones-property of the matrix, while WDP_{XOR} remains NP-complete (follows from Spieksma ((1999)). More generally, assume that items are one-to-one to nodes in a tree, and atomic bids are on subtrees only. Again, WDP_{OR} stays polynomial (Sandholm, 2002), since the matrix is perfect. If the conflict graph model is used, winner determination reduces to maximum weighted independent set for chordal graphs (follows from Gavril (1974)).

Further examples are summarized in Müller (2005). More generally, every positive result for the set-packing problem, or the independent set problem in graphs, provides a tractable case of the WDP with OR bids. XOR bids on the same collection of subsets S typically destroy the structure, and lead to NP-completeness. Such results naturally extend to the existence of polynomial time approximation algorithms.

36.2.3.2 Special bid values. By restricting the bid values rather than the collection of subsets of items on which bidders may bid, we may also achieve integrality of the linear programming relaxation of (ILP). We focus on XOR, because, firstly, there is no essential difference between an instance of WDP_{OR} with bids for every subset and WDP_{XOR}, and, secondly, restricted bid values have only been investigated for this model.

Recall that WDP_{XOR} has a column for every pair (i,S), $i \in N$, $S \subset M$, and we ask ourselves for which objectives does the LP-relaxation have an integral optimal solution. Sufficient for this is that the bid values satisfy the *(items are) substitutes* condition, introduced by Kelso and Crawford (1982).

In order to define this property we define for a valuation $v : 2^M \to \mathbb{R}$, and prices $p_j \in \mathbb{R}, j \in M$ the *demand correspondence* of a bidder as the collection of sets of items with the largest net utility, given prices p:

$$D(v,p) = \{S \subseteq M \mid v(S) - p(S) \geq v(T) - p(T) \text{ for all } T \subseteq S\}.$$

Here, $p(S) := \sum_{j \in S} p_j$. The substitutes condition controls the change of the demand correspondence under price changes: when increasing prices on some items, the demand for other items does not decrease.

Definition 36.1 [substitutes condition] *Given a set M, a set function $v : 2^M \to \mathbb{R}$ satisfies the substitutes condition, if for any pair of price vectors p, q with $p \leq q$, and $A \in D(v, p)$, there exists $B \in D(v, q)$ such that $\{j \in A \mid p_j = q_j\} \subseteq B$.*

Theorem 36.4 (Kelso and Crawford (1982)) *The LP-relaxation of WDP_{XOR} has an integral optimal solution if for all $i \in N$ the bid values $v_i(S)$, $S \subseteq M$ satisfy the substitutes condition.*

Economists tend to like restriction of bid values more than restrictions of bundles, because they are directly translatable into properties of preferences over bundles of items. This has inspired a lot of research on such valuations. Firstly, a direct link to discrete convexity theory has been observed independently by Danilov et al. (2001) and Murota and Tamura (2001) and Murota and Tamura (2003). Also, equivalent characterizations of the substitute condition have been given in Gul and Stacchetti (1999), Reijnierse et al. (2002), and Fujishige and Yang (2003). Furthermore, Sun and Yang (2004) have proven a generalization of Theorem 36.4, showing that a weaker condition than items are substitutes is sufficient for the LP-relaxation of WDP_{XOR} to be integral.

An interesting algorithmic consequence of the substitutes condition is the following. From Shioura (1998) it follows that if the auctioneer announces single-item prices, then valuation oracles are enough to compute a most demanded subset of items for every bidder. This can be either used in an iterative auction that is based on price-adjustments for single items (so-called clock auctions), or bidders could submit their valuation oracles, and the auctioneer could use them in a column-generation approach to solve the winner determination problem.

Another type of restricted valuations are valuations with budget constraints, and related conditions. In Penn and Tennenholtz (2000) it is shown how in this case WDP_{OR} can be modeled by b-matching. Müller (2005) shows how for the case that valuations satisfy cardinality-based subadditivity, WDP can be solved by using algorithms for min-cost-flow problems with convex costs, generalizing results by Tennenholtz (2000).

36.2.4 Solution methods

The previous section has illustrated that the winner determination problem is a combinatorial optimization problem that is closely related to set-packing, independent set and, by that, also maximum weighted clique problems. As such, any solution method developed for these problems can be used for WDP. Unfortunately, there is a limited supply in benchmark libraries stemming from practical applications, such that one has to rely on generated test-instances (Leyton-Brown and Shoham, 2005). That might be the reason why we find despite the popularity of combinatorial auctions few purely algorithmic research. An exception is Sandholm's work on WDP, summarized in Sandholm (2002). Given the problems for the bidders to communicate large sets of bids, it seems anyhow more appropriate to develop solvers of WDP for very specific settings. Applications in telecommunication come always with much special structure (see the following sections). It seems to us a fruitful direction of further research to link this special structure to the general theory from the previous section, for example, to identify cases where items are substitutes.

36.2.4.1 Economic interpretation of solution methods. In this subsection, we describe how particular solution methods can be interpreted economically and what one can gain from such interpretations. The previous section discussed two types of restrictions that made the LP relaxation of the winner determination ILP integral. In such cases the dual variables of the LP relaxation have a natural interpretation: the variables related to constraints (36.2) are item prices, the variables related to (36.3), if present, represent bidder surplus. Complementary slackness tells that at current prices, bidders should be assigned one of their surplus maximizing bundles, and bidders with a strictly positive surplus have to be guaranteed a bundle. Thus, dual optimal solutions of the LP relaxation of WDP coincide with Walrasian pricing equilibria. Bikhchandani and Ostroy (2003) have extended the primal dual interpretation to cases where the LP relaxation of IP is not integral. They provide new formulations with more variables and more constraints. The extensions lead to dual variables that assign prices to bundles rather than individual items, and eventually introduce non-anonymous prices (i.e., different bidders pay different prices for the same bundle). de Vries et al. (2003) have used these models to define an ascending price auction for general combinatorial auctions, which essentially implements the primal-dual method on one of Bikhchandani and Ostroy's models.

If one analyzes the complexity of these auctions (in terms of how many iterations of bids have to be made, how much computational effort does each round take for the bidder) it is clear that in general the auction has to be inefficient, since they solve an NP-hard optimization problem. However, for special cases this might be different. Interestingly there is not much known for special cases, except the few classical cases (homogeneous, multi-item with decreasing marginal utilities (Ausubel, 2004), the single-demand, heterogeneous item case (Demange et al., 1986), in other words, matching, the gross-substitute case (Gul and Stacchetti, 2000), and the general heterogenous item case (de Vries et al., 2003).

36.2.4.2 General approaches. A straightforward dynamic programming recursion can solve the WDP to optimality. It proceeds by computing optimal allocations for all subsets of the set of items. Let $\pi(S)$ be the optimal value for any $S \subseteq M$. It can be computed by taking the maximum of

$$v(S) \text{ and } \max\{\pi(S \setminus T) + \pi(T) : T \subset S : 0 < |T| \leq \frac{1}{2}|S|\}.$$

The order in which the subsets are treated is in increasing size of the sets S. The algorithm runs in $O(3^{|M|})$ time, see Rothkopf et al. (1998).

Based on the relation of the WDP with set-packing/covering, the methods for the latter problems could be used, such as branch-and-cut and branch-and-price. Both methods use the LP-relaxation of IP as an upper bound for the branch-and-bound tree. A detailed description of these methods is given in de Vries and Vohra (2003), extensive solver development and evaluation is reported in Sandholm (2002).

36.3 TELECOM APPLICATIONS: BANDWIDTH ALLOCATION IN NETWORKS

Most combinatorial auctions have no specific structure among the items. This is different if the resources are network related. For instance, bandwidth in telecom networks is such a resource, with limited availability on the links in the network. Customers are generally interested in point-to-point connections. Thus, the bandwidth allocated to a customer should be according to demand and should be available on the whole connection, not on just a part. Bidders, therefore, need structured bandwidth such as paths between pairs of vertices or even multiple paths in case protection is an issue.

In this section we consider models for the bandwidth allocation problem. We start with the single edge case, with complex bid structure: bids are a function of the allocated demand. Then we treat the problem on general networks, with a simpler bid structure: bids concern one demand for a point-to-point connection. This section ends with a reverse auction for a single commodity problem, where the prices of edges are determined using the VCG mechanism. The problem is elegantly solvable with shortest path methods.

The Bandwidth Allocation problem is interesting for solving congestion problems such as appearing on the Internet network, but also for pricing services that are made available on the Internet. For instance, a lot of research has been performed lately on Differentiated Services on the Internet (DiffServ environment), where a number of

QoS are available for which prices and volumes have to be determined, see Semret et al. (2000).

36.3.1 Bandwidth Allocation on an edge

Consider a single connection with a limited availability of bandwidth, say u. The bandwidth has to be divided among several users (commodities) ($k \in K$), each of which has a valuation function of the quantity of bandwidth allocated, say $b_k(q)$. Generally, $b_k(q)$ is non-decreasing in q. In Maillé and Tuffin (2004) this function is approximated by letting customers make multiple bids for different demands of capacity. Once the function is determined the optimization problem is simply the following, given that u is the capacity of bandwidth available.

$$\begin{aligned}
\max \quad & \sum_{k \in K} b_k(q_k) \\
\text{s.t.} \quad & \sum_{k \in K} q_k \leq u \\
& q_k \geq 0 \quad \forall k \in K.
\end{aligned}$$

Here the q_k ($k \in K$) represent the amounts allocated to the customers. For continuous, concave bidding functions the problem is polynomially solvable with a straightforward min-cost flow algorithm. Generally the problem is NP-hard. The most frequently arising case is the one in which a bid is generated for exactly one amount of bandwidth (the demand). Less is valued zero and more is valued equal to the value of the demand. This problem is equivalent to the knapsack problem, making the above problem NP-hard.

36.3.2 Bandwidth Allocation on a network

Let $G = (V, E)$ be an undirected network. Consider a set of commodities K, where each commodity $k \in K$ is determined by a source-terminal pair (s^k, t^k) of vertices, and a demand d^k, which is to be routed over the network from s_k to t_k. Let u_{ij} represent an upper bound on the capacity of edge $\{i, j\}$. We have to decide which demands are routed over the network, without violating capacity constraints on the edges, so as to maximize our revenues. First, let us suppose that on each commodity a bid b^k has been achieved. Then the decision to make is which bids to allocate. We introduce the following binary variables

$$x_k = \begin{cases} 1 & \text{if bid } k \text{ is selected} \\ 0 & \text{otherwise} \end{cases}$$

and flow variables f_{ij}^k which denote the flow of commodity k on edge $\{i, j\}$ in the direction from i to j. The following formulation solves the maximum revenue problem for the allocation of commodities.

$$\max \quad \sum_{k \in K} b_k x_k$$

$$\text{s.t.} \quad \sum_j f_{ij}^k - \sum_j f_{ji}^k = \begin{cases} d^k x_k & \text{if } i = s^k \\ -d^k x_k & \text{if } i = t^k \\ 0 & \text{otherwise} \end{cases} \quad \forall k \in K, \forall i \in V \quad (36.5)$$

$$\sum_{k \in K} (f_{ij}^k + f_{ji}^k) \leq u_{ij} \quad \forall \{i,j\} \in E$$

$$f_{ij}^k, f_{ji}^k \geq 0 \quad \forall k \in K, \forall \{i,j\} \in E$$

This problem has a close relation to the standard multicommodity flow problem. The only difference lies in the balance constraints (36.5) for the source-sink pair of each commodity. This formulation can easily be adapted to the problem where the flow is restricted to use a single path from source to sink, by replacing the flow variables f_{ij}^k with $d^k y_{ij}^k$, where the y_{ij}^k are binary variables. Note the difference with network synthesis problems, where design variables are introduced to the multi-commodity flow problem, instead of bid variables. The bid variables are found in the flow balance constraints (36.5), whereas the design variables are found in the capacity constraints.

The problem is NP-hard, already for one edge (see above). Thus, the problem is also NP-hard on special networks such as trees. Furthermore, there is little to no expertise with solving this problem. However, given that the problem is a similar extension of the multi-commodity flow problem as the network design problem, one can sometimes adopt the techniques that have been developed for the design problem to the allocation problem. A branch-and-cut approach will profit from additional valid inequalities. For the assignment variables cut constraints can be developed. Consider a cut in the graph, defined by $S \subset V$. Then the total available capacity must be at least as large as the allocated demand of commodities that have to be routed over the cut:

$$\sum_{k \in K: |\{s^k, t^k\} \cap S| = 1} d_k x_k \leq \sum_{\{i,j\} \in E: |\{i,j\} \cap S| = 1} u_{ij} \quad \forall S \subset V$$

36.3.2.1 Path formulation.
Instead of using flow variables on individual edges to model routing restrictions, one can also use flow variables associated with paths. Let z_p^k be the flow on a path $p \in P^k$ (where P^k is the set of all simple paths connecting the endvertices of commodity k). Let $P_{ij}^k \subseteq P^k$ denote the set of paths for commodity k that contain the edge $\{i,j\}$, then the path model reads

$$\max \quad \sum_{k \in K} \sum_{p \in P^k} b_p^k x_k$$

$$\text{s.t.} \quad \sum_{p \in P^k} z_p^k = d^k x_k \quad \forall k \in K$$

$$\sum_{k \in K} \sum_{p \in P_{ij}^k} z_p^k \leq u_{ij} \quad \forall \{i,j\} \in E$$

$$z_p^k \geq 0 \quad \forall k \in K, \forall p \in P^k, \forall \{i,j\} \in E$$

This formulation can be found in Abrache et al. (2004). Note that it gives the bidders the flexibility to bid different prices for paths connecting the end-vertices of the bidder's commodity. This may be helpful in selecting paths with, for instance, a small number of hops by simply putting low bids on larger paths. Finally, one might consider combinations of paths to bid on, for instance, if connectivity requirements play a role for the customer. Then the above formulation is remodeled as a set-packing problem again. Note, however, that the set-packing problem is defined over a huge number of items.

36.3.3 Edge Pricing in a reverse auction

The Transmission Control Protocol (TCP) on which the Internet is based, has a congestion control mechanism that uses packet-loss to determine whether the transmission rate (bandwidth) on links can be increased or should be decreased. TCP is self-regulating, so it expects individual hosts to behave in a similar manner to congestion. However, the hosts could manipulate the protocol to their advantage. Therefore, price-based mechanisms to share bandwidth have been proposed, for instance, by Shenker et al. (1996).

Hershberger and Suri (2001), consider the problem of valuation of the edges of a network, given one commodity with endvertices s and t, where all edges in the network are owned by different agents. In Hershberger and Suri (2001) the Vickrey pricing mechanism is used to determine the value of each edge. Let the network $G = (V, E)$ be given with a cost function c on the edges. The Vickrey payment or value of an edge is determined as follows. Consider the source s and sink t of the commodity. The value of an edge e is the difference in length of the shortest st-path in the network with e, and the length of the shortest st-path without e, i.e., $c_{G-e}(s,t) - c_G(s,t)$, where $c_G(s,t)$ is the length of the shortest st-path in G. Hershberger and Suri (2001) provide an efficient algorithm to compute the Vickrey payments for all edges simultaneously. The algorithm has a running time of $O(m \log n)$ in a graph with m edges and n vertices.

36.4 TELECOM APPLICATIONS: RADIO SPECTRUM ALLOCATION

One of the areas in which auctions have been applied most is the allocation of radio spectrum. From around 1990 until now, governments have been occupied in selling radio spectrum bandwidth. The 1990s were used to experiment with the ways this sale could be done most effectively, both from the viewpoint of the government in terms of revenues, and from the viewpoint of the customers, who benefit from high quality cheap wireless connections. Some experiments involved lotteries, where the winner was randomly selected and had to pay a prefixed price. Other experiments involved comparative hearings, where the winner was selected based on the plans for the bandwidth usage the company revealed. However, both types of allocation showed drawbacks on revenues and service, so the auction mechanism was soon believed to be superior to the lotteries and hearings for selecting among mutually exclusive applications for spectrum licenses.

The first auctions were all single round sealed bid auctions where one or more bundles of bandwidth were auctioned. In June 1990, New Zealand auctioned three new cellular licenses simultaneously using a sealed bid. It used a second price sealed-tender auction, which meant that the highest bidder won the license, but only paid the amount bid by the second highest bidder. This meant that one winner bid NZ $ 101 million, but only paid NZ $ 11 million. In July 1992, Greece auctioned two national GSM licenses. Participants submitted a single bid for one of the licenses and the highest bidder won the first license. The rules stated that if the second highest bidder was within 10% of the highest bid, then it had the sole right to match the highest bid. If it decided not to match the highest bid, then it and the remaining participants could participate in another round of bidding for the second license. The second highest bidder actually bid 91% of the highest bid and elected to match the highest bid, thus winning the second license. In January 1994, Columbia auctioned a second cellular license in each of three regions. The rules allowed consortia to bid in all of the regions. In a simultaneous single round auction, the highest bidder in a region won the license. The first licensee in each region was then required to pay 95% of the total amount bid by the second licensee. In August 1995, India held an auction for two GSM licenses in each of 20 regions ("circles"). The rules allowed consortia to bid in any and all of the regions. The highest bidder won the first license in each region and had to pay the amount it bid in an up-front payment and subsequent annual payments. The second highest bidder had to match the highest bidder if it wanted to receive the second license. If it declined, the right to the second license fell to the third highest bidder, which had to match the highest bidder in order to receive the license. If no bidder matched the highest bid, then the second license would be re-auctioned. In January 1996, Panama held a single round auction for a national cellular license in which the highest bidder received the license. The Panamanian authorities pre-qualified applicants based on technical, financial and business criteria before allowing them to submit a single financial bid at a public bid-revealing event. The highest bidder won the license. These data come from Spicer (1996).

The above examples represent several different designs within the category of single round sealed bid auctions. Single round sealed bid auctions generally raise less revenue, and lead to complicated bidding strategies. The design used in the United States currently is an ascending bid multiple-round auction. The US FCC now uses auctions to allocate spectrum for a wide variety of commercial purposes, from mobile telephony to broadcasting. In Europe, the success of the UK 3G auction in raising unexpectedly high revenues has encouraged many other governments to run their own auctions. As of early 2000, some eight Western European countries have auctioned their UMTS licenses. In the UK and Germany this was a great financial success with state revenues of 37.5 and 50.8 Billion Euros, respectively. The auctions in the Netherlands and Italy turned into a financial disappointment and later auctions were even canceled or had very low revenues (Switzerland had only 4 bidders for 4 licenses). Probable cause were a combination of lack of experience of the telecom companies: too high bids in the early auctions, and financial problems: economic growth decreased, and the belief that Electronic services were not as promising as expected grew. The lack of the possibility of package bidding caused even more problems: geographically close licenses can be of importance to a bidder. The absence of package bidding enforced high singleton bids by the bidders to obtain the bundles desired, and still important items were missing.

The multiple round simultaneous ascending auction was introduced by the FCC in the early 1990s. The inclusion of package bidding finally took place in 2000. In the bidding process multiple bids each consisting of a package of licenses and a bid $(S_{i,k}^r, b_{i,k}^r)$ could be handed in. Here i is the bidder, r is the round, and k is a possible package. In Günlük et al. (2004) the XOR-of-OR language is used to model the problem. The XOR-of-OR language assumes that the bids in one round are non-exclusive (no overlap) and that the bids in different rounds do exclude each other. Furthermore, the bids that are accepted consist of a package of licenses and a bid. Let $(S_{i,k}^r, b_{i,k}^r)$ be the k-th bid of bidder i done in round r. The complete set of bids done by bidder j can then be translated into the following formula.

$$\text{XOR}_{r=1}^{R} \left\{ \text{OR}_{k=1}^{l(i,r)} \{S_{i,k}^r\} \right\}$$

A natural formulation of the Winner Determination Problem in the XOR-of-OR language is as follows. An allocation of the items is described by the binary variables $x_{i,r,k} \in \{0,1\}$ and the binary variables $z_{i,r} \in \{0,1\}$:

$$x_{i,r,k} = \begin{cases} 1 & \text{if bidder } i \text{ gets the } k\text{-th bundle from round } r\text{: } S_{i,k}^r \\ 0 & \text{otherwise,} \end{cases}$$

$$z_{i,r} = \begin{cases} 1 & \text{if bidder } i \text{ gets a bid from round } r \\ 0 & \text{otherwise.} \end{cases}$$

The complete XOR-of-OR formulation then looks like the following:

$$\max \sum_{i \in N} \sum_{r \in R} \sum_{k=1}^{l(i,r)} b_{i,r,k} x_{i,r,k}$$

$$\text{s.t.} \quad \sum_{i \in N} \sum_{r \in R} \sum_{k: j \in S_{i,k}^r} x_{i,r,k} \leq u_j \quad \forall j \in N \tag{36.6}$$

$$x_{i,r,k} \leq z_{i,r} \quad \forall i \in N \; \forall r \in R \; \forall k = 1, \ldots, l(i,r) \tag{36.7}$$

$$\sum_{r \in R} z_{i,r} \leq 1 \quad \forall i \in N \tag{36.8}$$

$$x_{i,r,k} \in \{0,1\} \quad \forall i \in N \; \forall r \in R \; \forall k = 1, \ldots, l(i,r)$$

$$z_{i,r} \in \{0,1\} \quad \forall i \in N \; \forall r \in R$$

Note that a round r contains accepted bids for bidder i only if $z_{i,r} = 1$, see (36.7). Thus, the constraints (36.8) restrict the set of accepted bids to one round. Finally, the availability of an item is expressed in the constraints (36.6).

Günlük et al. (2004) concentrate on solving this model. First, another formulation with better LP-relaxation is introduced. This formulation contains a huge number of variables, each of which represents feasible combinations of sets. Note that these combinations follow directly from the restrictions of the above formulation. The combinations are modeled with single variables $y_{i,S}$, so that the problem then becomes exactly the set-packing problem of the previous section. The problem is then solved with a Branch-and-Price algorithm.

36.5 CONCLUSION

Though auction mechanisms have been worked out very well in the past years, their usage in telecom applications has been very limited. The success of radio spectrum auctions has not been followed by successes in other areas such as bandwidth allocation within networks. Nevertheless this seems a promising area of application especially for avoiding congestion. Moreover, in the ever more complicated world of different services (DiffServ) provided by telecom operators this may be a valuable tool to not only assign bandwidth to users, but simultaneously to the different QoS levels. Some first attempts have been made already, though the difficulties such as adapting protocols have prevented actual implementation of auctions.

Bibliography

J. Abrache, T.G. Crainic, and M. Gendreau. Design issues for combinatorial auctions. *4OR*, 2(1):1–33, 2004.

L. M. Ausubel and P. Milgrom. The lovely but lonely Vickrey auction. In P. Cramton, Y. Shoham, and R. Steinberg, editors, *Combinatorial Auctions*. MIT Press, 2005.

L.M. Ausubel. An efficient ascending-bid auction for multiple objects. *American Economic Review*, 94(5):1452–1475, 2004.

S. Bikhchandani and J. Ostroy. The package assignment model. *Journal of Economic Theory*, 2002:377–406, 2003.

V. Danilov, G. Koshevoy, and K. Murota. Discrete convexity and equilibria in economies with indivisible goods and money. *Mathematical Social Sciences*, 41: 251–273, 2001.

H. P. Dasgupta and E. Maskin. The implementation of social choice rules: some results on incentive compatibility. *Review of Economic Studies*, 46:185–216, 1979.

S. de Vries, J. Schummer, and R. V. Vohra. On ascending auctions for heterogeneous objects, 2003.

S. de Vries and R.R. Vohra. Combinatorial auctions: A survey. *INFORMS Journal on Computing*, 15:284–309, 2003.

G. Demange, D. Gale, and M. Sotomayor. Multi-item auctions. *Journal of Political Economy*, 94:863–872, 1986.

S. Fujishige and Z. Yang. A note on kelso and crawford's gross substitutes condition. *Mathematics of Operations Research*, 28:463–469, 2003.

M. R. Garey and D. S. Johnson. *"Computers and Intractability,"*. Freeman, San Francisco, 1979.

F. Gavril. The intersection graphs of subtrees in trees are exactly the chordal graphs. *Journal Combinatorial Theory B*, 16:47–56, 1974.

F. Gul and E. Stacchetti. Walrasian equilibrium with gross substitutes. *Journal of Economic Theory*, 87:95–124, 1999.

F. Gul and E. Stacchetti. The english auction with differentiated commodities. *Journal of Economic Theory*, 92:66–95, 2000.

O. Günlük, L. Ladányi, and S. De Vries. A branch-and-price algorithm and new test problems for spectrum auctions. *Management Science*, (to appear), 2004.

John Hershberger and Subhash Suri. Vickrey prices and shortest paths: What is an edge worth? In *Proceedings of the 42nd IEEE symposium on Foundations of Computer Science*, pages 255–262. IEEE Computer Society, 2001.

Stan van Hoesel and Rudolf Müller. Optimization in electronic markets: Examples in combinatorial auctions. *Netnomics*, 3:23–33, 2001.

R. C. Holte. Combinatorial auctions, knapsack problems, and hill-climbing search. In E. Stroulia and S. Matwin, editors, *Proc. AI'2001, the 14th Biennial Conference of the Canadian Society for Computational Studies of Intelligence*, volume 2056 of *Springer Lecture Notes in Artificial Intelligence*, pages 57–66, 2001.

C. Jackson. *Technology for spectrum markets*. PhD thesis, Department of Electrical Engineering,Massachusetts Institute of Technology,Cambridge,MA, 1976.

H. Kellerer, U. Pferchy, and D. Pirsinger. *Knapsack Problems*. Springer Verlag, 2004.

A. S. Kelso and V. P. Crawford. Job matching, coalition formation and gross substitutes. *Econometrica*, 50(6):1483–1504, 1982.

D. Lehmann, R. Müller, and T. Sandholm. The winner determination problem. In P. Cramton, Y. Shoham, and R. Steinberg, editors, *Combinatorial Auctions*. MIT Press, 2005.

K. Leyton-Brown and Y. Shoham. A test-suite for combinatorial auctions. In P. Cramton, Y. Shoham, and R. Steinberg, editors, *Combinatorial Auctions*. MIT Press, 2005.

P. Maillé and B. Tuffin. Multi-bid auctions for bandwidth allocation in communication networks. In *Proc. of IEEE INFOCOM 2004*, Hong Kong, 2004.

A. Mas-Colell, M. D. Whinsont, and J. R. Green. *Microeconomic Theory*. Oxford University Press, 1995.

R. Müller. Tractable cases of the winner determination problem. In P. Cramton, Y. Shoham, and R. Steinberg, editors, *Combinatorial Auctions*. MIT Press, 2005.

K. Murota and A. Tamura. Application of m-convex submodular flow problem to mathematical economics. In P. Eades and T. Takaoka, editors, *Algorithms and Computation, Proc. of 12th International Symposium, ISAAC 2001*, volume 2223 of *Springer Lecture Notes in Computer Science*, pages 14–25, Christchurch, New Zealand, 2001.

K. Murota and A. Tamura. New characterizations of m-convex functions and their applications to economic equilibrium models. *Discrete Applied Mathematics*, 131: 495–512, 2003.

R. Myerson. Incentive compatibility and the bargaining problem. *Econometrica*, 47: 61–74, 1979.

N. Nisan. Bidding and allocation in combinatorial auctions. Proceedings of the ACM Conference on Electronic Commerce (EC-00), available at http://www.cs.huji.ac.il/~noam/mkts.html, 2000.

N. Nisan. Bidding languages for combinatorial auctions. In P. Cramton, Y. Shoham, and R. Steinberg, editors, *Combinatorial Auctions*. MIT Press, 2005.

M. Penn and M. Tennenholtz. Constrained multi-object auctions and b-matchings. *Information Processing Letters*, 75:29–34, 2000.

H. Reijnierse, A. van Gellekom, and J.A.M. Potters. Verifying gross substitutability. *Economic Theory*, 20:767–776, 2002.

M. H. Rothkopf, A. Pekeč, and R.M. Harstad. Computationally manageable combinational auctions. *Management Science*, 44:1131–1147, 1998.

T. Sandholm. Algorithm for optimal winner determination in combinatorial auctions. *Artificial Intelligence*, 135:1–54, 2002.

N. Semret, R.R.-F. Liao, A.T. Campbell, and A.A. Lazar. Pricing, provisioning and peering: Dynamic markets for differentiated internet services and implications for network interconnections. *IEEE Journal on Selected Areas in Communication*, 18(12):2499–2513, 2000.

S. Shenker, D. Clark, D. Estrin, and S. Herzog. Pricing in computer networks: Reshaping the research agenda. *Telecommunications policy*, pages 183–201, 1996.

A. Shioura. Minimization of an m-convex function. *Discrete Applied Mathematics*, 84:215–220, 1998.

M. Spicer. International survey of spectrum assignment for cellular and PCS. Federal Communications Commission. http://wireless.fcc.gov/auctions/data/papersAndStudies.html, 1996.

F. C. R. Spieksma. On the approximability of an interval scheduling problem. *Journal of Scheduling*, 2(5):215–227, (1999.

N. Sun and Z. Yang. The max-convolution approach to equilibrium models with indivisibilities, 2004.

M. Tennenholtz. Some tractable combinatorial auctions. In *Proc. of National Conference on Artificial Intelligence (AAAI)*, pages 98–103, 2000.

W. S. Vickrey. Counterspeculation, auctions and competitive sealed tenders. *Journal of Finance*, 16:8–37, 1961.

37 SUPERNETWORKS

Anna Nagurney[1]

[1]Department of Finance & Operations Management
Isenberg School of Management
University of Massachusetts
Amherst, MA 01003 USA
nagurney@gbfin.umass.edu

Abstract: This chapter describes supernetworks as a formalism for the modeling and analysis of complex decision-making in the Information Age, which is characterized by the prominent role played by telecommunication networks coupled with other networks. The chapter traces the concept and term, whose origins lie in transportation science and computer science, and lays its foundations in the context of system-optimization versus user-optimization. The Braess paradox is recalled and its relevance to network design discussed. Multicriteria supernetworks are subsequently modeled and variational inequality formulations of the governing equilibrium conditions given, along with an explicit application to telecommuting versus commuting decision-making. Both multitiered as well as multilevel supernetworks are highlighted and a plethora of applications such as supply chains with electronic commerce, integrated social and supply chain networks, and financial networks with intermediation and electronic transactions are overviewed.

Keywords: Supernetworks, Information Age, decision-making, telecommunication networks, coupled networks, transportation science, computer science, Braess paradox, variational inequality formulations, supply chains, electronic commerce, financial networks, electronic transactions.

37.1 BACKGROUND

The Information Age with the increasing availability of new computer and communication technologies, along with the Internet, have transformed the ways in which individuals work, travel, and conduct their daily activities, with profound implications for existing and future decision-making. Indeed, the decision-making process itself has been altered due to the addition of alternatives and options which were not, heretofore, possible or even feasible. The boundaries for decision-making have been redrawn as individuals may work from home and utilize telecommuting options or purchase products from work online. Managers can now locate raw materials and other inputs from suppliers via telecommunication networks in order to maximize profits while simultaneously ensuring timely delivery of finished goods. Financing for businesses can be now obtained online. Individuals, in turn, can obtain information about products from their homes and make their purchasing decisions accordingly.

The Internet, as a telecommunications and information network, par excellence, has impacted individuals, organizations, institutions, as well as businesses and societies in a way that that no other network system has in history, due to its speed of communications and its global reach. Moreover, the intertwining of such a communication network with other network systems such as transportation networks, energy networks, as well as a variety of economic networks, including financial networks, brings new challenges and opportunities for the conceptualization, modeling, and analysis of complex decision-making.

The reality of many of today's networks, notably, transportation and telecommunication networks, include: a large-scale nature and complexity, increasing congestion, alternative behaviors of users of the networks, as well as interactions among the networks themselves. The decisions made by the users of the networks, in turn, affect not only the users themselves but others, as well, in terms of profits and costs, timeliness of deliveries, the quality of the environment, etc.

In this chapter, the foundations and applications of the theory of *supernetworks* are described. **Super** networks may be thought of as networks that are *above and beyond* existing networks, which consist of nodes, links, and flows, with nodes corresponding to locations in space, links to connections in the form of roads, cables, etc., and flows to vehicles, data, etc. Supernetworks are, first and foremost, a mathematical formalization that allows the calculation of both static and dynamic equilibria/optima of complex networks with respect to the flows, which can include information, goods, persons, and prices. The flows are associated with the relevant technology that describes the generalized costs associated with using the networks along with the associated decision-making. In addition, supernetworks integrate existing unimodal network systems by providing a structure above and beyond the component networks. Supernetworks are conceptual in scope, graphical in perspective, and, with the accompanying theory, predictive in nature.

In particular, the supernetwork framework, captures, in a unified fashion, decision-making facing a variety of economic agents (decision-makers) including consumers and producers as well as distinct intermediaries in the context of today's networked economy. The decision-making process may entail weighting trade-offs associated with the use of transportation versus telecommunication networks. The behavior of the individual agents is modeled as well as their interactions on the complex network systems with the goal of identifying the resulting flows.

For definiteness, Table 37.1 presents some basic *classical* networks and the associated nodes, links, and flows. A *classical* network is one in which the nodes correspond to physical locations in space and the links to physical connections between the nodes.

The topic of networks and their management dates to ancient times with examples including the publicly provided Roman road network and the "time of day" chariot policy, whereby chariots were banned from the ancient city of Rome at particular times of day (see Banister and Button (1993)). The formal study of networks, consisting of

Table 37.1 Examples of classical networks

Network System	Nodes	Links	Flows
Transportation			
Urban	Intersections, Homes, Places of Work	Roads	Autos
Air	Airports	Airline Routes	Planes
Rail	Railyards	Railroad Track	Trains
Communication	Computers	Cables	Messages
	Satellites	Radio	Messages
	Phone Exchanges	Cables, Microwaves	Voice, Video
Manufacturing and Logistics	Distribution Points, Processing Points	Routes Assembly Line	Parts, Products
Energy	Pumping Stations Plants	Pipelines Pipelines	Water Gas, Oil

nodes, links, and flows, in turn, involves: how to model such applications (as well as numerous other ones) as mathematical entities, how to study the models qualitatively, and how to design algorithms to solve the resulting models effectively. The study of networks is necessarily interdisciplinary in nature due to their breadth of appearance and is based on scientific techniques from applied mathematics, computer science, engineering, and economics. Network models and tools which are widely used by businesses, industries, as well as governments today (cf. Ahuja et al. (1993), Nagurney and Siokos (1997), Nagurney (1999; 2000a), Guenes and Pardalos (2003), and the references therein).

Basic examples of network problems include: *the shortest path problem*, in which one seeks to determine the most efficient path from an origin node to a destination node; *the maximum flow problem*, in which one wishes to determine the maximum flow that one can send from an origin node to a destination node, given that there are capacities on the links that cannot be exceeded, and *the minimum cost flow problem*, where there are both costs and capacities associated with the links and one must satisfy the demands at the destination nodes, given supplies at the origin nodes, at minimal total cost associated with shipping the flows, and subject to not exceeding the arc capacities. Applications of all these problems are found in telecommunications and transportation.

Table 37.2 Examples of supernetworks

Supply Chain Networks with Electronic Commerce
Financial Networks with Electronic Transactions
Telecommuting/Commuting Networks
Teleshopping versus Shopping Networks
Reverse Supply Chains with Electronic Recycling
Energy Networks/Power Grids
Dynamic Knowledge Networks
Generalized Social Networks

Supernetworks may be comprised of such networks as transportation, telecommunication, logistical and financial networks, among others. They may be *multitiered* as when they formalize the study of supply chain networks with electronic commerce and financial networks with intermediation and electronic transactions. They may also be *multilevel* as when they capture the explicit interactions of distinct networks, including logistical, financial, social, and informational as in the case of dynamic supply chains. Furthermore, decision-makers on supernetworks may be faced with multiple criteria and, hence, the study of supernetworks also includes the study of multicriteria decision-making. In Table 37.2, some examples of supernetworks are given, which highlight the telecommunication aspect. We will overview several of these later in this chapter.

In particular, the supernetwork framework allows one to formalize the alternatives available to decision-makers, to model their individual behavior, typically, characterized by particular criteria which they wish to optimize, and to, ultimately, compute the flows on the supernetwork. Hence, the concern is with human decision-making and how the supernetwork concept can be utilized to crystallize and inform in this dimension.

Below the theme of supernetworks is further elaborated upon and, in particular, the origins of the concept and the term *supernetworks* identified.

37.2 THE ORIGINS OF SUPERNETWORKS

In this part of the chapter, a discussion of the three foundational classes of networks: transportation, telecommunication, and economic and financial networks is given. Such networks have served not only as the basis for the origins of the term *supernetwork*, but, also, they arise as critical subnetworks in the applications that are relevant to decision-making in the Information Age today.

37.2.1 Transportation networks

Transportation networks are complex network systems in which the decisions of the individual travelers affect the efficiency and productivity of the entire network system.

Transportation networks, as noted in Table 37.1, come in many forms: urban networks, freight networks, airline networks, etc. The "supply" in such a network system is represented by the network topology and the underlying cost characteristics, whereas the "demand" is represented by the users of the network system, that is, the travelers.

In 1972, Dafermos (1972) demonstrated, through a formal model, how a *multi-class* traffic network could be cast into a single-class traffic network through the construction of an expanded (and *abstract*) network consisting of as many copies of the original network as there were classes. She identified the origin/destination pairs, demands, link costs, and flows on the abstract network. The applications of such networks she stated, "arise not only in street networks where vehicles of different types share the same roads (e.g., trucks and passenger cars) but also in other types of transportation networks (e. g., telephone networks)." Hence, she not only recognized that abstract networks could be used to handle multimodal transportation networks but also telecommunication networks! Moreover, she considered both user-optimizing and system-optimizing behavior, terms which she had coined with Sparrow in a paper in 1969 (see Dafermos and Sparrow (1969)). Beckmann (1967) had earlier noted the potential relevance of network equilibrium (also referred to as user-optimization) in the context of communication networks.

In 1976, Dafermos (1976) proposed an integrated traffic network equilibrium model in which one could visualize and formalize the entire transportation planning process (consisting of origin selection, or destination selection, or both, in addition to route selection, in an optimal fashion) as path choices over an appropriately constructed *abstract* network. The genesis and formal treatment of decisions more complex than route choices as *path* choices on abstract networks, that is, supernetworks, were, hence, reported as early as 1972 and 1976.

The importance and wider relevance of such abstract networks in decision-making, with a focus on transportation planning were accentuated through the term "hypernetwork" used by Sheffi (1978), and Sheffi and Daganzo (1978; 1980), which was later retermed as "supernetwork" by Sheffi (1985).

The recognition and appropriate construction of *abstract* networks was pivotal in that it allowed for the incorporation of transportation-related decisions (where as noted by Dafermos (1972), transportation applied also to communication networks) which were not based solely on route selection in a classical sense, that is, what route should one take from one's origin, say, place of residence, to one's destination, say, place of employment. Hence, abstract networks, with origins and destinations corresponding to appropriately defined nodes, links connecting nodes having associated disutilities (costs), and paths comprised of directed links connecting the origins and destinations, could capture such travel alternatives as not simply just a route but, also, the "mode" of travel, that is, for example, whether one chose to use private or public transportation. Furthermore, with the addition of not only added abstract links and paths, but abstract origin and destination nodes as well one could include the selection of such locational decisions as the origins and destinations themselves within the same decision-making framework.

For example, in order to fix ideas, in Figure 37.1, a supernetwork topology for an example of a simple mode/route choice problem is presented. In this example, it

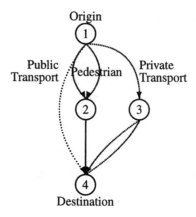

Figure 37.1 Example mode and route choice supernetwork topology

is recognized, at the outset, that the routes underlying the different modes may be distinct and, hence, rather than making copies of the network according to Dafermos (1972), the supernetwork construction is done with the path choices directly on the supernetwork itself.

In the network in Figure 37.1, decision-makers in the form of travelers seek to determine their "best" paths from the origin node 1 to the destination node 4, where a path consists of both the selection of the mode of travel as well as the route of travel. The first link, which connects node 1 to node 4, corresponds to the use of public transit, and there is only one route choice using this mode of travel. On the other hand, if one selects private transportation (typically, the automobile), one could take either of two routes: with the first route consisting of the first link joining nodes 1 and 3 and then the link joining node 3 to node 4, and the second route consisting of the second link joining nodes 1 and 3 and then onto node 4. Finally, one could choose either of two pedestrian routes to travel from node 1 to node 4, with the pedestrian routes differing by their second component links.

Additional references to supernetworks and transportation can be found in the book by Nagurney and Dong (2002b).

37.2.2 Telecommunication networks

We now turn to a discussion of the use of the term "supernetworks" in the context of telecommunication networks. Denning (1985), in the *American Scientist*, continued his discussion of the internal structure of computer networks, which had appeared in a volume of the same journal earlier that year, and emphasized how "protocol software can be built as a series of layers. Most of this structure is hidden from the users of the network." Denning then raised the question, "What should the users see?" In the article, he answered this question in the context of the then National Science Foundation's

Advanced Scientific Computing Initiative to make national supercomputer centers accessible to the entire scientific community. Denning said that such a system would be a network of networks, that is, a "supernetwork," and a powerful tool for science. Interestingly, he emphasized the importance of location-independent naming, so that if a physical location of a resource would change, none of the supporting programs or files would need to be edited or recompiled. His view of supernetworks, hence, is in concert with that of ours in that nodes do not need to correspond to locations in space and may have an abstract association.

Schubert et al. (1979) had earlier used the term in the context of knowledge representation as follows: "In the network approach to knowledge representation, concepts are represented as nodes in a network. Networks are compositional: a node in a network can be some other network, and the same subnetwork can be a subnetwork of several larger supernetworks,..."

In 1997, the Illinois Bar Association (see Illinois State Bar Association (1997)) considered the following to be an accepted definition of the Internet: "the Internet is a supernetwork of computers that links together individual computers and computer networks located at academic, commercial, government and military sites worldwide, generally by ordinary local telephone lines and long-distance transmission facilities. Communications between computers or individual networks on the Internet are achieved through the use of standard, nonproprietary protocols." The reference to the Internet as a supernetwork was also made by Fallows (1996) who stated in *The Atlantic Monthly* that "The Internet is the supernetwork that links computer networks around the world."

In his keynote address to the Internet/Telecom 95 Conference (see 95 (1995)), Mr. Vinton G. Cerf, the co-developer of the computer networking protocol, TCP/IP, used for the Internet, noted that at that time there were an estimated 23 million users of the Internet, and that vast quantities of the US Internet traffic "pass through Internet MCI's backbone." Mr. Cerf then noted that "Just a few months back, MCI rolled out a supernetwork for the National Science Foundation known as the very broadband network service or VNBS...VBNS is being used as an experimental platform for developing new national networking applications."

37.2.3 Economic and financial networks

The concept of a network in economics was implicit as early as in the classical work of Cournot (1838), who not only seems to have first explicitly stated that a competitive price is determined by the intersection of supply and demand curves, but had done so in the context of two spatially separated markets in which the cost of transporting the good between markets was considered. Pigou (1920) also studied a network system in the setting of a transportation network consisting of two routes and noted that the "system-optimized" solution was distinct from the "user-optimized" solution.

Nevertheless, the first instance of an abstract network or supernetwork in the context of economic applications, was actually due to Quesnay (1758), who, in 1758, visualized the circular flow of funds in an economy as a network. Since that very early contribution there have been numerous economic and financial models that have been constructed over abstract networks. Dafermos and Nagurney (1985), for example, identified the isomorphism between traffic network equilibrium problems and spatial price equilibrium problems, whose development had been originated by Samuelson (1952) (who, interestingly, focused on the bipartite network structure of the spatial price equilibrium problem).

Zhao (1989) (see also Zhao and Dafermos (1991) and Zhao and Nagurney (1993)) identified the general economic equilibrium problem known as Walrasian price equilibrium as a network equilibrium problem over an abstract network with very simple structure. The structure consisted of a single origin/destination pair of nodes and single links joining the two nodes. This structure was then exploited for computational purposes. A variety of abstract networks in economics were studied in the book by Nagurney (1999), which also contains extensive references to the subject.

Nagurney (2000a) used the term "supernetworks" in her essay in which she stated that "The interactions among transportation networks, telecommunication networks, as well as financial networks is creating supernetworks ...".

37.3 CHARACTERISTICS OF SUPERNETWORKS

Supernetworks are a conceptual and analytical formalism for the study of a variety of decision-making problems on networks. Hence, their characteristics include characteristics of the foundational networks. The characteristics of today's networks include: large-scale nature and complexity of network topology; congestion; alternative behavior of users of the network, which may lead to paradoxical phenomena, and the interactions among networks themselves such as in transportation versus telecommunications networks. Moreover, policies surrounding networks today may have a major impact not only economically but also socially.

Large-scale nature and complexity. Many of today's networks are characterized by both a large-scale nature and complexity of the underlying network topology. In Chicago's Regional Transportation Network, there are 12,982 nodes, 39,018 links, and 2,297,945 origin/destination (O/D) pairs (see Bar-Gera (1999)), whereas in the Southern California Association of Governments model there are 3,217 origins and/or destinations, 25,428 nodes, and 99,240 links, plus 6 distinct classes of users (cf. Wu et al. (2000).

In terms of the size of existing telecommunications networks, AT&T's domestic network had 100,000 origin/destination pairs in 2000 (cf. Resende (2000)). In AT&T's detail graph applications in which nodes are phone numbers and edges are calls, there were 300 million nodes and 4 billion edges in 1999 (cf. Abello et al. (1999)).

Congestion. Congestion is playing an increasing role in not only transportation networks but also in telecommunication networks. For example, in the case of transportation networks in the United States alone, congestion results in $100 billion in lost productivity, whereas the figure in Europe is estimated to be $150 billion. The number of cars is expected to increase by 50% by 2010 and to double by 2030 (see Nagurney (2000a)).

In terms of the Internet, according to Internet World Stats (Internet World Stats, 2005), there were over 800 million Internet users as of December 2004, with the usage growth globally between 2000-2004 being 125.2%. As individuals increasingly

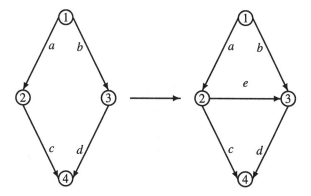

Figure 37.2 The Braess network example

access the Internet through wireless communication such as through handheld computers and cellular phones, experts fear that the heavy use of airwaves will create additional bottlenecks and congestion that could impede the further development of the technology.

System-optimization versus user-optimization. In many of today's networks, not only is congestion a characteristic feature leading to nonlinearities, but the behavior of the users of the networks themselves may be that of noncooperation. For example, in the case of urban transportation networks, travelers select their routes of travel from an origin to a destination so as to minimize their own travel cost or travel time, which although "optimal" from an individual's perspective (user-optimization) may not be optimal from a societal one (system-optimization) where one has control over the flows on the network and, in contrast, seeks to minimize the total cost in the network and, hence, the total loss of productivity. Consequently, in making any kind of policy decisions in such networks one must take into consideration the users of the particular network. Indeed, this point is vividly illustrated through a famous example known as the Braess paradox, in which it is assumed that the underlying behavioral principle is that of user-optimization. In the Braess network (cf. Braess (1968)), the addition of a new road with no change in the travel demand results in all travelers in the network incurring a higher travel cost and, hence, being worse off!

The Braess's paradox is now recalled. For easy reference, see the two networks depicted in Figure 37.2.

Example 37.1 *Braess's paradox.* *Assume a network as the first network depicted in Figure 37.2 in which there are four links: a,b,c,d; four nodes: 1, 2, 3, 4; and a single O/D pair $w_1 = (1,4)$. There are, hence, two paths available to travelers between this O/D pair: $p_1 = (a,c)$ and $p_2 = (b,d)$.*

The user link travel cost functions are given by:

$$c_a(f_a) = 10f_a, \quad c_b(f_b) = f_b + 50, \quad c_c(f_c) = f_c + 50, \quad c_d(f_d) = 10f_d.$$

Assume a fixed travel demand $d_{w_1} = 6$.

In the case of user-optimization, no traveler has any incentive to switch his path and this state is characterized by the property (for a complete description, see the next section), that all used paths connecting each O/D pair have equal and minimal travel costs (or times). Note that the user cost on a path is the sum of the user costs on the links that make up the path.

It is easy to verify that the equilibrium path flows that satisfy this condition (as well as the conservation of flow equations relating the nonnegative path flows to the travel demand) are:

$$x^*_{p_1} = 3, \quad x^*_{p_2} = 3;$$

the equilibrium link flows (incurred by the equilibrium path flows) are:

$$f^*_a = 3, \quad f^*_b = 3, \quad f^*_c = 3, \quad f^*_d = 3;$$

with associated equilibrium path travel costs:

$$C_{p_1} = c_a + c_c = 83, \quad C_{p_2} = c_b + c_d = 83.$$

Assume now that, as depicted in Figure 37.2, a new link "e," joining node 2 to node 3, is added to the original network, with user cost $c_e(f_e) = f_e + 10$. The addition of this link creates a new path $p_3 = (a, e, d)$ that is available to the travelers. Assume that the travel demand d_{w_1} remains at 6 units of flow. Note that the original flow distribution pattern $x_{p_1} = 3$ and $x_{p_2} = 3$ is no longer an equilibrium pattern, since at this level of flow, the cost on path p_3, denoted by C_{p_3}, is equal to 70. Hence, users from paths p_1 and p_2 would switch to path p_3.

The equilibrium flow pattern on the new network is:

$$x^*_{p_1} = 2, \quad x^*_{p_2} = 2, \quad x^*_{p_3} = 2;$$

with equilibrium link flows:

$$f^*_a = 4, \quad f^*_b = 2, \quad f^*_c = 2, \quad f^*_e = 2, \quad f^*_d = 4;$$

and with associated equilibrium path travel costs:

$$C_{p_1} = 92, \quad C_{p_2} = 92, \quad C_{p_3} = 92.$$

Indeed, one can verify that any reallocation of the path flows would yield a higher travel cost on a path.

Note that the travel cost increased for every user of the network from 83 to 92!

The increase in travel cost on the paths is due, in part, to the fact that in this network two links are shared by distinct paths and these links incur an increase in flow and associated cost. Hence, Braess's paradox is related to the underlying topology of the networks. It has been proven, however, that the addition of a path connecting an O/D pair that shares no links with the original O/D pair will never result in Braess's paradox for that O/D pair (cf. Dafermos and Nagurney (1984)).

In the next section, we will provide the system-optimizing solution to both of these networks. A system-optimized network never exhibits the Braess paradox.

Interestingly, as reported in the *New York Times* by Kolata (1990), the Braess paradox phenomenon has been observed in practice, in the case of New York City, when, in 1990, 42nd Street was closed for Earth Day and the traffic flow actually improved. Just to show that it is not a purely New York or US phenomena concerning drivers and their behavior an analogous situation was observed in Stuttgart where a new road was added to the downtown but the traffic flow worsened and, following complaints, the new road was torn down (see Bass (1992)).

This phenomenon is also relevant to telecommunications networks and the Braess paradox has provided one of the main linkages between transportation science and computer science. Cohen and Kelly (1990) described a paradox analogous to that of Braess in the case of a queuing network. Korilis et al. (1999), in turn, developed methods to show how resources could be added efficiently to a noncooperative network, including the Internet, so that the Braess paradox would not occur and cited the work of Dafermos and Nagurney (1984). Roughgarden (2002) further elaborated on the Braess paradox and focused on the quantification of the worst possible loss in network performance arising from noncooperative behavior (see also Roughgarden and Tardos (2002)). He also designed algorithms for the design and management of the networks so that "selfish" (a term also used by Beckmann et al. (1956)), that is, individually-optimizing, behavior, leads to a "socially desirable" outcome. He noted the importance of the work of the computer scientists, Koutsoupias and Papadimitrou (1999), who proposed the idea of bounding the inefficiency of Nash equilibria (see also Dafermos and Sparrow (1969)).

Network interactions. Clearly, one of the principal facets of the Network Economy is the interaction among the networks themselves. For example, the increasing use of electronic commerce especially in business to business transactions is changing not only the utilization and structure of the underlying logistical networks but is also revolutionizing how business itself is transacted and the structure of firms and industries. Cellular phones are being using as vehicles move dynamically over transportation networks resulting in dynamic evolutions of the topologies themselves. The unifying concept of supernetworks with associated methodologies allows one to explore the interactions among such networks as transportation networks, telecommunication networks, as well as financial networks.

37.4 DECISION-MAKING CONCEPTS

As the above discussion has revealed, networks in the Information Age are complex, typically, large-scale systems and the study of their efficient operation, often through some outside intervention, has attracted much interest from economists, computer scientists, engineers, as well as transportation scientists and operations researchers.

In particular, the underlying behavior of the users of the network system is essential in studying its operation. Importantly, Wardrop (1952) explicitly recognized alternative possible behaviors of users of transportation networks and stated two principles, which are commonly named after him:

- **First Principle:** The journey times of all routes actually used are equal, and less than those which would be experienced by a single vehicle on any unused route.

- **Second Principle:** The average journey time is minimal.

The first principle corresponds to the behavioral principle in which travelers seek to (unilaterally) determine their minimal costs of travel whereas the second principle corresponds to the behavioral principle in which the total cost in the network is minimal.

Beckmann et al. (1956) were the first to rigorously formulate these conditions mathematically, as had Samuelson (1952) in the framework of spatial price equilibrium problems in which there were, however, no congestion effects. Specifically, Beckmann, McGuire, and Winsten established the equivalence between the *traffic network equilibrium* conditions, which state that all used paths connecting an origin/destination (O/D) pair will have equal and minimal travel times (or costs) (corresponding to Wardrop's first principle), and the Kuhn-Tucker (Kuhn and W., 1951) conditions of an appropriately constructed optimization problem (cf. Bazaraa et al. (1993)), under a symmetry assumption on the underlying functions. Hence, in this case, the equilibrium link and path flows could be obtained as the solution of a mathematical programming problem. Their approach made the formulation, analysis, and subsequent computation of solutions to traffic network problems based on actual networks realizable.

Dafermos and Sparrow (1969) coined the terms *user-optimized* (U-O) and *system-optimized* (S-O) transportation networks to distinguish between two distinct situations in which, respectively, users act unilaterally, in their own self-interest, in selecting their routes, and in which users select routes according to what is optimal from a societal point of view, in that the total cost in the system is minimized. In the latter problem, marginal total costs on used paths connecting an O/D pair of nodes, rather than average costs, are equilibrated. The former problem coincides with Wardrop's first principle, and the latter with Wardrop's second principle.

The concept of "system-optimization" is also relevant to other types of "routing models" not only in transportation, but also in communications (cf. Bertsekas and Gallager (1992)), including those concerned with the routing of freight and computer messages, respectively. Dafermos and Sparrow also provided explicit computational procedures, that is, *algorithms*, to compute the solutions to such network problems in the case where the user travel cost on a link was an increasing function of the flow on the link (in order to handle congestion) and linear.

37.4.1 System-optimization versus user-optimization

The basic network models are now reviewed, under distinct assumptions as to their operation and the corresponding distinct behavior of the users of the network. The models are classical and are due to Beckmann et al. (1956) and Dafermos and Sparrow (1969). We later present more general models.

For definiteness, and for easy reference, we present the classical system-optimized network model and then the classical user-optimized network model. Although these models were first developed for transportation networks, here they are presented in the broader setting of network systems, since they are as relevant in other application settings, in particular, in telecommunication networks and, more generally, in supernetworks.

More general models are then outlined, in which the user link cost functions are no longer separable and are also asymmetric. We provide the variational inequality formulations of the governing equilibrium conditions (see Kinderlehrer and Stampacchia (1980) and Nagurney (1999)), since, in this case, the conditions can no longer be reformulated as the Kuhn-Tucker conditions of a convex optimization problem. Finally, we present the variational inequality formulations in the case of elastic demands.

37.4.1.1 The system-optimized problem. Consider a general network $G = [\mathcal{N}, \mathcal{L}]$, where $\mathcal{N}$ denotes the set of nodes, and $\mathcal{L}$ the set of directed links. Let a denote a link of the network connecting a pair of nodes, and let p denote a path consisting of a sequence of links connecting an O/D pair. In transportation networks (see also Table 37.1), nodes correspond to origins and destinations, as well as to intersections. Links, on the other hand, correspond to roads/streets in the case of urban transportation networks and to railroad segments in the case of train networks. A path in its most basic setting, thus, is a sequence of "roads" which comprise a route from an origin to a destination. In the telecommunication context, however, nodes can correspond to switches or to computers and links to telephone lines, cables, microwave links, etc. In the supernetwork setting, a path is viewed more broadly and need not be limited to a route-type decision.

Let P_ω denote the set of paths connecting the origin/destination (O/D) pair of nodes ω. Let P denote the set of all paths in the network and assume that there are J origin/destination pairs of nodes in the set Ω. Let x_p represent the flow on path p and let f_a denote the flow on link a. The path flows on the network are grouped into the column vector $x \in R_+^{n_P}$, where n_P denotes the number of paths in the network. The link flows, in turn, are grouped into the column vector $f \in R_+^n$, where n denotes the number of links in the network.

The following conservation of flow equation must hold:

$$f_a = \sum_{p \in P} x_p \delta_{ap}, \quad \forall a \in \mathcal{L}, \tag{37.1}$$

where $\delta_{ap} = 1$, if link a is contained in path p, and 0, otherwise. Expression (37.1) states that the flow on a link a is equal to the sum of all the path flows on paths p that contain (traverse) link a.

Moreover, if one lets d_ω denote the demand associated with O/D pair ω, then one must have that

$$d_\omega = \sum_{p \in P_\omega} x_p, \quad \forall \omega \in \Omega, \tag{37.2}$$

where $x_p \geq 0$, $\forall p \in P$; that is, the sum of all the path flows between an origin/destination pair ω must be equal to the given demand d_ω.

Let c_a denote the user link cost associated with traversing link a, and let C_p denote the user cost associated with traversing the path p.

Assume that the user link cost function is given by the *separable* function

$$c_a = c_a(f_a), \quad \forall a \in L, \quad (37.3)$$

where c_a is assumed to be an increasing function of the link flow f_a in order to model the effect of the link flow on the cost. The link cost functions are also assumed to be continuous and continuously differentiable.

The total cost on link a, denoted by $\hat{c}_a(f_a)$, hence, is given by:

$$\hat{c}_a(f_a) = c_a(f_a) \times f_a, \quad \forall a \in L, \quad (37.4)$$

that is, the total cost on a link is equal to the user link cost on the link times the flow on the link. Here the cost is interpreted in a general sense. From a transportation engineering perspective, however, the cost on a link is assumed to coincide with the travel time on a link. Later in this chapter, we consider generalized cost functions of the links which are constructed using weights and different criteria.

In the system-optimized problem, there exists a central controller who seeks to minimize the total cost in the network system, where the total cost is expressed as

$$\sum_{a \in L} \hat{c}_a(f_a), \quad (37.5)$$

where the total cost on a link is given by expression (37.4).

The system-optimization problem is, thus, given by:

$$\text{Minimize} \sum_{a \in L} \hat{c}_a(f_a) \quad (37.6)$$

subject to:

$$\sum_{p \in P_\omega} x_p = d_\omega, \quad \forall \omega \in \Omega, \quad (37.7)$$

$$f_a = \sum_{p \in P} x_p, \quad \forall a \in L, \quad (37.8)$$

$$x_p \geq 0, \quad \forall p \in P. \quad (37.9)$$

The constraints (37.7) and (37.8), along with (37.9), are commonly referred to in network terminology as *conservation of flow equations*. In particular, they guarantee that the flow in the network, that is, the users (whether these are travelers or computer messages, for example) do not "get lost."

The total cost on a path, denoted by $\hat{C}_p$, is the user cost on a path times the flow on a path, that is,

$$\hat{C}_p = C_p x_p, \quad \forall p \in P, \quad (37.10)$$

where the user cost on a path, C_p, is given by the sum of the user costs on the links that comprise the path, that is,

$$C_p = \sum_{a \in L} c_a(f_a) \delta_{ap}, \quad \forall a \in L. \quad (37.11)$$

In view of (37.8), one may express the cost on a path p as a function of the path flow variables and, hence, an alternative version of the above system-optimization problem can be stated in path flow variables only, where one has now the problem:

$$\text{Minimize} \sum_{p \in P} C_p(x) x_p \qquad (37.12)$$

subject to constraints (37.7) and (37.9).

System-optimality conditions. Under the above imposed assumptions on the user link cost functions, which recall are assumed to be increasing functions of the flow, the objective function in the S-O problem is convex, and the feasible set consisting of the linear constraints is also convex. Therefore, the optimality conditions, that is, the Kuhn-Tucker conditions are: For each O/D pair $\omega \in \Omega$, and each path $p \in P_\omega$, the flow pattern x (and link flow pattern f), satisfying (37.7)–(37.9) must satisfy:

$$\hat{C}'_p \begin{cases} = \mu_\omega, & \text{if } x_p > 0 \\ \geq \mu_\omega, & \text{if } x_p = 0, \end{cases} \qquad (37.13)$$

where $\hat{C}'_p$ denotes the marginal of the total cost on path p, given by:

$$\hat{C}'_p = \sum_{a \in L} \frac{\partial \hat{c}_a(f_a)}{\partial f_a} \delta_{ap}, \qquad (37.14)$$

and in (37.13) it is evaluated at the solution.

Note that in the S-O problem, according to the optimality conditions (37.13), it is the marginal of the total cost on each used path connecting an O/D pair which is equalized and minimal. Indeed, conditions (37.13) state that a system-optimized flow pattern is such that for each origin/destination pair the incurred marginals of the total costs on all used paths are equal and minimal.

We return now to the Braess network(s) in Figure 37.2. The system-optimizing solution to the first network in Figure 37.2 would be:

$$x_{p_1} = x_{p_2} = 3,$$

with marginal total path costs given by:

$$\hat{C}'_{p_1} = \hat{C}'_{p_2} = 116.$$

This would remain the system-optimizing solution, even after the addition of link e, since the marginal total cost of the new path p_3, $\hat{C}'_{p_3}$, at this feasible flow pattern (with the flow on the new path p_3 being zero) is equal to 130. Hence, in the case of a system-optimizing solution, path p_3 would not even be used and we would have that $x_{p_3} = 0$.

The addition of a new link to a network cannot increase the total cost of the network system, in the case of system-optimization, as formulated above.

37.4.1.2 The user-optimized problem.
We now describe the user-optimized network problem, also commonly referred to in the literature as the *traffic assignment* problem or the *traffic network equilibrium* problem. Again, as in the system-optimized problem, the network $G = [\mathcal{N}, L]$, the demands associated with the origin/destination pairs, as well as the user link cost functions are assumed as given. Recall that user-optimization follows Wardrop's first principle.

Network equilibrium conditions. Now, however, one seeks to determine the path flow pattern x^* (and link flow pattern f^*) which satisfies the conservation of flow equations (37.7), (37.8), and the nonnegativity assumption on the path flows (37.9), and which also satisfies the network equilibrium conditions given by the following statement.

For each O/D pair $\omega \in \Omega$ and each path $p \in P_\omega$:

$$C_p \begin{cases} = \lambda_\omega, & \text{if } x_p^* > 0 \\ \geq \lambda_\omega, & \text{if } x_p^* = 0. \end{cases} \qquad (37.15)$$

Hence, in the user-optimization problem there is no explicit optimization concept, since now users of the network system act independently, in a noncooperative manner, until they cannot improve on their situations unilaterally and, thus, an equilibrium is achieved, governed by the above equilibrium conditions. Indeed, conditions (37.15) are simply a restatement of Wardrop's first principle (Wardrop, 1952) mathematically and mean that only those paths connecting an O/D pair will be used which have equal and minimal user costs. Otherwise, a user of the network could improve upon his situation by switching to a path with lower cost. User-optimization represents decentralized decision-making, whereas system-optimization represents centralized decision-making.

In order to obtain a solution to the above problem, Beckmann et al. (1956) established that the solution to the equilibrium problem, in the case of user link cost functions (cf. (37.3)) in which the cost on a link only depends on the flow on that link could be obtained by solving the following optimization problem:

$$\text{Minimize} \quad \sum_{a \in L} \int_0^{f_a} c_a(y) dy \qquad (37.16)$$

subject to:

$$\sum_{p \in P_\omega} x_p = d_\omega, \quad \forall \omega \in \Omega, \qquad (37.17)$$

$$f_a = \sum_{p \in P} x_p \delta_{ap}, \quad \forall a \in L, \qquad (37.18)$$

$$x_p \geq 0, \quad \forall p \in P. \qquad (37.19)$$

Note that the conservation of flow equations are identical in both the user-optimized network problem (see (37.17)–(37.19)) and the system-optimized problem (see (37.7) – (37.9)). The behavior of the individual decision-makers termed "users," however, is different. Users of the network system, which generate the flow on the network now act independently, and are not controlled by a centralized controller. The relevance of these two distinct behavioral concepts to telecommunications networks is clear.

The objective function given by (37.16) is simply a device constructed to obtain a solution using general purpose convex programming algorithms. It does not possess the economic meaning of the objective function encountered in the system-optimization problem given by (37.6), equivalently, by (37.12). Of course, algorithms that fully exploit the network structure of these problems can be expected to perform more efficiently (cf. Dafermos and Sparrow (1969), Nagurney (1984; 1999)).

37.4.2 Models with asymmetric link costs

There has been much dynamic research activity in the past several decades in both the modeling and the development of methodologies to enable the formulation and computation of more general network equilibrium models, with a focus on traffic networks. Examples of general models include those that allow for multiple modes of transportation or multiple classes of users, who perceive cost on a link in an individual way. We now consider network models in which the user cost on a link is no longer dependent solely on the flow on that link. Other network models, including dynamic traffic models, can be found in Mahmassani et al. (1993), and in the books by Ran and Boyce (1996), Nagurney and Zhang (1996), Nagurney (1999), and the references therein.

We now consider user link cost functions which are of a general form, that is, in which the cost on a link may depend not only on the flow on the link but on other link flows on the network, that is,

$$c_a = c_a(f), \quad \forall a \in \mathcal{L}. \tag{37.20}$$

In the case where the symmetry condition holds, that is,

$$\frac{\partial c_a(f)}{\partial f_b} = \frac{\partial c_b(f)}{\partial f_a},$$

for all links $a, b \in \mathcal{L}$, one can still reformulate the solution to the network equilibrium problem satisfying equilibrium conditions (37.15) as the solution to an optimization problem (cf. Dafermos (1972) and the references therein), albeit, again, with an objective function that is artificial and simply a mathematical device. However, when the symmetry assumption is no longer satisfied, such an optimization reformulation no longer exists and one must appeal to *variational inequality theory*.

Indeed, it was in the problem domain of traffic network equilibrium problems that the theory of finite-dimensional variational inequalities realized its earliest success, beginning with the contributions of Smith (1979) and Dafermos (1976). For an introduction to the subject, as well as applications ranging from traffic network equilibrium problems to financial equilibrium problems, see the book by Nagurney (1999). The methodology of finite-dimensional variational inequalities has also been utilized more recently in order to develop a spectrum of supernetwork models (see Nagurney and Dong (2002b)).

The system-optimization problem, in turn, in the case of nonseparable (cf. (37.20)) user link cost functions becomes (see also (37.6)–(37.9)):

$$\text{Minimize} \sum_{a \in L} \hat{c}_a(f), \qquad (37.21)$$

subject to (37.7)–(37.9), where $\hat{c}_a(f) = c_a(f) \times f_a, \forall a \in L$.

The system-optimality conditions remain as in (37.13), but now the marginal of the total cost on a path becomes, in this more general case:

$$\hat{C}'_p = \sum_{a,b \in L} \frac{\partial \hat{c}_b(f)}{\partial f_a} \delta_{ap}, \quad \forall p \in P. \qquad (37.22)$$

Variational inequality formulations of fixed demand problems. As mentioned earlier, in the case where the user link cost functions are no longer symmetric, one cannot compute the solution to the U-O, that is, to the network equilibrium, problem using standard optimization algorithms. Such cost functions are very important from an application standpoint since they allow for asymmetric interactions on the network. For example, allowing for asymmetric cost functions permits one to handle the situation when the flow on a particular link affects the cost on another link in a different way than the cost on the particular link is affected by the flow on the other link.

Since equilibrium is such a fundamental concept in terms of supernetworks and since variational inequality theory is one of the basic ways in which to study such problems we now, for completeness, also give variational inequality formulations of the network equilibrium conditions (37.15). These formulations are presented without proof (for derivations, see Smith (1979) and Dafermos (1980), as well as Florian and Hearn (1995) and the book by Nagurney (1999)).

First, the definition of a variational inequality problem is recalled. We then give both the variational inequality formulation in path flows as well as in link flows of the network equilibrium conditions. Subsequently, in this chapter, these concepts are extended to multicriteria, multiclass network equilibrium problems.

Specifically, the variational inequality problem (finite-dimensional) is defined as follows:

Definition 37.1 *Variational inequality problem.* *The finite-dimensional variational inequality problem,* $\text{VI}(F, \mathcal{K})$, *is to determine a vector* $X^* \in \mathcal{K}$ *such that*

$$\langle F(X^*), X - X^* \rangle \geq 0, \quad \forall X \in \mathcal{K}, \qquad (37.23)$$

where F is a given continuous function from $\mathcal{K}$ *to* R^N, $\mathcal{K}$ *is a given closed convex set, and* $\langle \cdot, \cdot \rangle$ *denotes the inner product in* R^N.

Variational inequality (37.23) is referred to as being in *standard form*. Hence, for a given problem, typically an *equilibrium* problem, one must determine the function F that enters the variational inequality problem, the vector of variables X, as well as the feasible set $\mathcal{K}$.

The variational inequality problem contains, as special cases, such well-known problems as systems of equations, optimization problems, and complementarity problems. Thus, it is a powerful unifying methodology for equilibrium analysis and computation.

Theorem 37.1 *Variational inequality formulation of network equilibrium with fixed demands – Path flow version.*
A vector $x^* \in K^1$ is a network equilibrium path flow pattern, that is, it satisfies equilibrium conditions (37.15) if and only if it satisfies the variational inequality problem:

$$\sum_{\omega \in \Omega} \sum_{p \in P_\omega} C_p(x^*) \times (x - x^*) \geq 0, \quad \forall x \in K^1, \tag{37.24}$$

or, in vector form:

$$\langle C(x^*), x - x^* \rangle \geq 0, \quad \forall x \in K^1, \tag{37.25}$$

where C is the n_P-dimensional column vector of path user costs and K^1 is defined as: $K^1 \equiv \{x \geq 0, \text{ such that } (37.17) \text{ holds}\}$.

Theorem 37.2 *Variational inequality formulation of network equilibrium with fixed demands – Link flow version.*
A vector $f^* \in K^2$ is a network equilibrium link flow pattern if and only if it satisfies the variational inequality problem:

$$\sum_{a \in L} c_a(f^*) \times (f_a - f_a^*) \geq 0, \quad \forall f \in K^2, \tag{37.26}$$

or, in vector form:

$$\langle c(f^*), f - f^* \rangle \geq 0, \quad \forall f \in K^2, \tag{37.27}$$

where c is the n-dimensional column vector of link user costs and K^2 is defined as: $K^2 \equiv \{f \mid \text{there exists an } x \geq 0 \text{ and satisfying } (37.17) \text{ and } (37.18)\}$.

Note that one may put variational inequality (37.25) in standard form (37.23) by letting $F \equiv C$, $X \equiv x$, and $\mathcal{K} \equiv K^1$. Also, one may put variational inequality (37.27) in standard form where now $F \equiv c$, $X \equiv f$, and $\mathcal{K} \equiv K^2$.

Alternative variational inequality formulations of a problem are useful in devising other models, including dynamic versions, as well as for purposes of computation using different algorithms.

Variational inequality formulations of elastic demand problems. The general network equilibrium model with elastic demands due to Dafermos (1982) is now recalled. Specifically, it is assumed that now one has associated with each O/D pair ω in the network a disutility λ_ω, where here the general case is considered in which the disutility may depend upon the entire vector of demands, which are no longer fixed, but are now variables, that is,

$$\lambda_\omega = \lambda_\omega(d), \quad \forall \omega \in \Omega, \tag{37.28}$$

where d is the J-dimensional column vector of the demands.

The notation, otherwise, is as described earlier, except that here we also consider user link cost functions which are general, that is, of the form (37.20). The conservation of flow equations (see also (37.1) and (37.2)), in turn, are given by

$$f_a = \sum_{p \in P} x_p \delta_{ap}, \quad \forall a \in L, \qquad (37.29)$$

$$d_\omega = \sum_{p \in P_\omega} x_p, \quad \forall \omega \in \Omega, \qquad (37.30)$$

$$x_p \geq 0, \quad \forall p \in P. \qquad (37.31)$$

Hence, in the elastic demand case, the demands in expression (37.30) are now variables and no longer given, as was the case for the fixed demand expression in (37.2). Elastic demand models are very useful in that the demands are allowed to adjust. Hence, such models may be viewed as providing a more long-term perspective as to how decision-makers adjust given the network parameters, vis a vis the fixed demand scenario.

Network equilibrium conditions in the case of elastic demand. The network equilibrium conditions (see also (37.15)) now take on in the elastic demand case the following form: For every O/D pair $\omega \in \Omega$, and each path $p \in P_\omega$, a vector of path flows and demands (x^*, d^*) satisfying (37.30)–(37.31) (which induces a link flow pattern f^* through (37.29)) is a network equilibrium pattern if it satisfies:

$$C_p(x^*) \begin{cases} = \lambda_\omega(d^*), & \text{if } x_p^* > 0 \\ \geq \lambda_\omega(d^*), & \text{if } x_p^* = 0. \end{cases} \qquad (37.32)$$

Equilibrium conditions (37.32) state that the costs on used paths for each O/D pair are equal and minimal and equal to the disutility associated with that O/D pair. Costs on unutilized paths can exceed the disutility.

In the next two theorems, both the path flow version and the link flow version of the variational inequality formulations of the network equilibrium conditions (37.32) are presented. These are analogues of the formulations (37.24) and (37.25), and (37.26) and (37.27), respectively, for the fixed demand model.

Theorem 37.3 *Variational inequality formulation of network equilibrium with elastic demands – Path flow version.*
A vector $(x^*, d^*) \in K^3$ is a network equilibrium path flow pattern, that is, it satisfies equilibrium conditions (37.32) if and only if it satisfies the variational inequality problem:

$$\sum_{\omega \in \Omega} \sum_{p \in P_\omega} C_p(x^*) \times (x - x^*) - \sum_{\omega \in \Omega} \lambda_\omega(d^*) \times (d_\omega - d_\omega^*) \geq 0, \quad \forall (x, d) \in K^3, \qquad (37.33)$$

or, in vector form:

$$\langle C(x^*), x - x^* \rangle - \langle \lambda(d^*), d - d^* \rangle \geq 0, \quad \forall (x, d) \in K^3, \qquad (37.34)$$

where λ is the J-dimensional vector of disutilities and K^3 is defined as: $K^3 \equiv \{x \geq 0, \text{ such that } (37.30) \text{ holds}\}$.

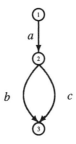

Figure 37.3 An elastic demand example

Theorem 37.4 *Variational inequality formulation of network equilibrium with elastic demands – Link flow version.*
A vector $(f^*, d^*) \in K^4$ is a network equilibrium link flow pattern if and only if it satisfies the variational inequality problem:

$$\sum_{a \in L} c_a(f^*) \times (f_a - f_a^*) - \sum_{\omega \in \Omega} \lambda_\omega(d^*) \times (d_\omega - d_\omega^*) \geq 0, \quad \forall (f,d) \in K^4, \quad (37.35)$$

or, in vector form:

$$\langle c(f^*), f - f^* \rangle - \langle \lambda(d^*), d - d^* \rangle \geq 0, \quad \forall (f,d) \in K^4, \quad (37.36)$$

where $K^4 \equiv \{(f,d),$ such that there exists an $x \geq 0$ satisfying (37.29), (37.31)\}

Note that, under the symmetry assumption on the disutility functions, that is, if $\frac{\partial \lambda_w}{\partial d_\omega} = \frac{\partial \lambda_\omega}{\partial d_w}$, for all w, ω, in addition to such an assumption on the user link cost functions (see following (37.20)), one can obtain (see Beckmann et al. (1956)) an optimization reformulation of the network equilibrium conditions (37.32), which in the case of separable user link cost functions and disutility functions is given by:

$$\text{Minimize} \quad \sum_{a \in L} \int_0^{f_a} c_a(y)dy - \sum_{\omega \in \Omega} \int_0^{d_\omega} \lambda_\omega(z)dz \quad (37.37)$$

subject to: (37.29)–(37.31).
An example of a simple elastic demand network equilibrium problem is now given.

Example 37.2 *An elastic demand network equilibrium problem.* Consider the network depicted in Figure 37.3 in which there are three nodes: 1, 2, 3; three links: a, b, c; and a single O/D pair $\omega_1 = (1,3)$. Let path $p_1 = (a,b)$ and path $p_2 = (a,c)$.
Assume that the user link cost functions are:

$$c_a(f) = 2f_a + 10, \quad c_b(f) = 7f_b + 3f_c + 11, \quad c_c(f) = 6f_c + 4f_b + 14,$$

and the disutility (or inverse demand) function is given by:

$$\lambda_{\omega_1}(d_{\omega_1}) = -2d_{\omega_1} + 104.$$

Observe that in this example, the user link cost functions are non-separable for links b and c and asymmetric and, hence, the equilibrium conditions (cf. (37.32)) cannot be reformulated as the solution to an optimization problem, but, rather, as the solution to the variational inequalities (37.33) (or (37.34)), or (37.35) (or (37.36)).

The U-O flow and demand pattern that satisfies equilibrium conditions (37.32) is: $x^*_{p_1} = 5$, $x^*_{p_2} = 4$, and $d^*_{\omega_1} = 9$, with associated link flow pattern: $f^*_a = 9$, $f^*_b = 5$, $f^*_c = 4$.

The incurred user costs on the paths are: $C_{p_1} = C_{p_2} = 86$, which is precisely the value of the disutility λ_{ω_1}. Hence, this flow and demand pattern satisfies equilibrium conditions (37.32). Indeed, both paths p_1 and p_2 are utilized and their user paths costs are equal to each other. In addition, these costs are equal to the disutility associated with the origin/destination pair that the two paths connect.

37.5 MULTICLASS, MULTICRITERIA SUPERNETWORKS

In this part of the chapter, we describe how the concept of a multicriteria supernetwork can be utilized to address decision-making in the Information Age. We then present a specific applications, in particular, telecommuting versus commuting decision-making. This section is expository. The theoretical foundations can be found in Nagurney and Dong (2002b).

The term "multicriteria" captures the multiplicity of criteria that decision-makers are often faced with in making their choices, be they regarding consumption, production, transportation, location, or investment. Criteria which are considered as part of the decision-making process may include: cost minimization, time minimization, opportunity cost minimization, profit maximization, as well as risk minimization, among others.

Indeed, the Information Age with the increasing availability of new computer and communication technologies, along with the Internet, have transformed the ways in which many individuals work, travel, and conduct their daily activities today. Moreover, the decision-making process itself has been altered through the addition of alternatives which were not, heretofore, possible or even feasible. As stated in a recent issue of *The Economist* (Economist, 2000), "The boundaries for employees are redrawn... as people work from home and shop from work."

The first publications in the area of multicriteria decision-making on networks focused on transportation networks, and were by Schneider (1968) and Quandt (1967). However, they assumed fixed travel times and travel costs. Here, in contrast, these functions (as well as any other appropriate criteria functions) are flow-dependent. The first flow-dependent such model was by Dafermos (1981), who considered an infinite number of decision-makers, rather than a finite number as is done here. Furthermore, she assumed two criteria, whereas we consider a finite number, where the number can be as large as necessary. Moreover, the modeling framework set out in this chapter can also handle elastic demands. The first general elastic demand multicriteria network equilibrium model was developed by Nagurney and Dong (2002a), who considered two criteria and fixed weights but allowed the weights to be class- and link-dependent. The models in this chapter, in contrast, allow the particular application to be handled with as many finite criteria as are relevant and retain the flexible feature of allowing the weights associated with the criteria to be both class- and link-dependent. We refer the reader to Nagurney and Dong (2002b) for additional references. We now recall the multiclass, multicriteria network equilibrium models with elastic demand and with fixed demand, respectively. Each class of decision-maker is allowed to have weights associated with the criteria which are also permitted to be link-dependent for modeling flexibility purposes. Subsequently, the governing equilibrium conditions along with the variational inequality formulations are presented.

37.5.1 The multiclass, multicriteria network equilibrium models

In this section, the multiclass, multicriteria network equilibrium models are described. The elastic demand model is presented first and then the fixed demand model. The equilibrium conditions are, subsequently, shown to satisfy finite-dimensional variational inequality problems.

Consider a general network $G = [\mathcal{N}, \mathcal{L}]$, where $\mathcal{N}$ denotes the set of nodes in the network and $\mathcal{L}$ the set of directed links. Let a denote a link of the network connecting a pair of nodes and let p denote a path, assumed to be acyclic, consisting of a sequence of links connecting an origin/destination (O/D) pair of nodes. There are n links in the network and n_P paths. Let Ω denote the set of J O/D pairs. The set of paths connecting the O/D pair ω is denoted by P_ω and the entire set of paths in the network by P.

Note that in the supernetwork framework a link may correspond to an actual physical link of transportation or to an abstract or virtual link corresponding to telecommunications. Furthermore, the supernetwork representing the problem under study can be as general as necessary and a path may consist also of a set of links corresponding to a combination of physical and virtual choices. A path, hence, in the supernetwork framework, abstracts a decision as a sequence of links or possible choices from an origin node, which represents the beginning of the decision, to the destination node, which represents its completion.

Assume that there are now k classes of decision-makers in the network with a typical class denoted by i. Let f_a^i denote the flow of class i on link a and let x_p^i denote the nonnegative flow of class i on path p. The relationship between the link flows by class and the path flows is:

$$f_a^i = \sum_{p \in P} x_p^i \delta_{ap}, \quad \forall i, \quad \forall a \in \mathcal{L}, \tag{37.38}$$

where $\delta_{ap} = 1$, if link a is contained in path p, and 0, otherwise. Hence, the flow of a class of decision-maker on a link is equal to the sum of the flows of the class on the paths that contain that link.

In addition, let f_a denote the total flow on link a, where

$$f_a = \sum_{i=1}^{k} f_a^i, \quad \forall a \in \mathcal{L}. \tag{37.39}$$

Thus, the total flow on a link is equal to the sum of the flows of all classes on that link. Group the class link flows into the kn-dimensional column vector $\tilde{f}$ with components: $\{f_1^1, \ldots, f_n^1, \ldots, f_1^k, \ldots, f_n^k\}$, the total link flows: $\{f_1, \ldots, f_n\}$ into the n-dimensional column vector f, and the class path flows into the kn_P-dimensional column vector $\tilde{x}$ with components: $\{x_{p_1}^1, \ldots, x_{p_{n_P}}^k\}$.

The demand associated with origin/destination (O/D) pair ω and class i will be denoted by d_ω^i. Group the demands into a column vector $d \in R^{kJ}$. Clearly, the demands must satisfy the following conservation of flow equations:

$$d_\omega^i = \sum_{p \in P_\omega} x_p^i, \quad \forall i, \forall \omega, \qquad (37.40)$$

that is, the demand for an O/D pair for each class is equal to the sum of the path flows of that class on the paths that join the O/D pair.

The functions associated with the links are now described. In particular, assume that there are H criteria which the decision-makers may utilize in their decision-making with a typical criterion denoted by h. Assume that C_{ha} denotes criterion h associated with link a, where

$$C_{ha} = C_{ha}(f), \quad \forall a \in L, \qquad (37.41)$$

where C_{ha} is assumed to be a continuous function.

For example, criterion 1 may be time, in which case we would have

$$C_{1a} = C_{1a}(f) = t_a(f), \quad \forall a \in L, \qquad (37.42)$$

where $t_a(f)$ denotes the time associated with traversing link a. In the case of a transportation link, one would expect the function to be higher than for a telecommunications link. Another relevant criterion may be cost, that is,

$$C_{2a} = C_{2a}(f) = c_a(f), \quad \forall a \in L, \qquad (37.43)$$

which might reflect (depending on the link a) an access cost in the case of a telecommunications link, or a transportation or shipment cost in the case of a transportation link. One can expect both time and cost to be relevant criteria in decision-making in the Information Age especially since telecommunications is at times a substitute for transportation and it is typically associated with higher speed and lower cost (cf. Mokhtarian (1990)).

In addition, another relevant criterion in evaluating decision-making in the Information Age is opportunity cost since one may expect that this cost would be high in the case of teleshopping, for example (since one cannot physically experience and evaluate the product), and lower in the case of shopping. Furthermore, in the case of telecommuting, there may be perceived to be a higher associated opportunity cost by some classes of decision-makers who may miss the socialization provided by face-to-face interactions with coworkers and colleagues. Hence, a third possible criterion may be opportunity cost, where

$$C_{3a} = C_{3a}(f) = o_a(f), \quad \forall a \in L, \qquad (37.44)$$

with $o_a(f)$ denoting the opportunity cost associated with link a. Finally, a decision-maker may wish to associate a safety cost in which case the fourth criterion may be

$$C_{4a} = C_{4a}(f) = s_a(f), \quad \forall a \in L, \qquad (37.45)$$

where $s_a(f)$ denotes a security or safety cost measure associated with link a. In the case of teleshopping, for example, decision-makers may be concerned with revealing personal or credit information, whereas in the case of transportation, commuters may view certain neighborhood roads as being dangerous.

We assume that each class of decision-maker has a potentially different perception of the tradeoffs among the criteria, which are represented by the nonnegative weights: $w^i_{1a}, \ldots, w^i_{Ha}$. Hence, w^i_{1a} denotes the weight on link a associated with criterion 1 for class i, w^i_{2a} denotes the weight associated with criterion 2 for class i, and so on. Observe that the weights are link-dependent and can incorporate specific link-dependent factors which could include for a particular class factors such as convenience and sociability. A typical weight associated with class i, link a, and criterion h is denoted by w^i_{ha}.

Nagurney and Dong (2002a) were the first to model link-dependent weights but only considered two criteria. Nagurney et al. (2000), in turn, used link-dependent weights but assumed only three criteria, in particular, travel time, travel cost, and opportunity cost in their integrated multicriteria network equilibrium models for telecommuting versus commuting.

Here, a generalized cost function is defined as follows.

Definition 37.2 Generalized link cost function. *A generalized link cost of class i associated with link a and denoted by C^i_a is given by:*

$$C^i_a = \sum_{h=1}^{H} w^i_{ha} C_{ha}, \quad \forall i, \quad \forall a \in L. \tag{37.46}$$

For example, (37.46) states that each class of decision-maker i when faced by H distinct criteria on each link a assigns his own weights $\{w^i_{ha}\}$ to the links and criteria.

In lieu of (37.39) – (37.46), one can write

$$C^i_a = C^i_a(\tilde{f}), \quad \forall i, \quad \forall a \in L, \tag{37.47}$$

and group the generalized link costs into the kn-dimensional column vector C with components: $\{C^1_1, \ldots, C^1_n, \ldots, C^k_1, \ldots, C^k_n\}$.

For example, if there are four criteria associated with decision-making and they are given by (37.42) through (37.45), then the generalized cost function on a link a as perceived by class i would have the form:

$$C^i_a = w^i_{1a} C_{1a}(\tilde{f}) + w^i_{2a} C_{2a}(\tilde{f}) + w^i_{3a} C_{3a}(\tilde{f}) + w^i_{4a} C_{4a}(\tilde{f}). \tag{37.48}$$

Let now C^i_p denote the generalized cost of class i associated with path p in the network where

$$C^i_p = \sum_{a \in L} C^i_a(\tilde{f}) \delta_{ap}, \quad \forall i, \quad \forall p. \tag{37.49}$$

Thus, the generalized cost associated with a class and a path is that class's weighted combination of the various criteria on the links that comprise the path.

Note from the structure of the criteria on the links as expressed by (37.41) and the generalized cost structure assumed for the different classes on the links according to (37.46) and (37.47), that it is explicitly being assumed that the relevant criteria are functions of the total flows on the links, where recall that the total flows (see (37.39)) correspond to the total number of decision-makers of all classes that selects a particular link. This is not unreasonable since one can expect that the greater the number of decision-makers that select a particular link (which comprises a part of a path), the greater the congestion on that link and, hence, one can expect the time of traversing the link as well as the cost to increase.

In the case of the elastic demand model, assume, as given, the inverse demand functions λ_ω^i for all classes i and all O/D pairs ω, where:

$$\lambda_\omega^i = \lambda_\omega^i(d), \quad \forall i, \quad \forall \omega, \tag{37.50}$$

and these functions are assumed to be smooth and continuous. Group the inverse demand functions into a column vector $\lambda \in R^{kJ}$.

The behavioral assumption. Assume that the decision-making involved in the particular application is repetitive in nature such as, for example, in the case of commuting versus telecommuting, or shopping versus teleshopping. The behavioral assumption that is proposed, hence, is that decision-makers select their paths so that their generalized costs are minimized.

Specifically, the behavioral assumption utilized is similar to that underlying traffic network assignment models (cf. (37.15) and (37.32)) in that it is assumed that each class of decision-maker in the network selects a path so as to minimize the generalized cost on the path, given that all other decision-makers have made their choices.

In particular, the following are the network equilibrium conditions for the problem outlined above:

Multiclass, multicriteria network equilibrium conditions for the elastic demand case. For each class i, for all O/D pairs $\omega \in \Omega$, and for all paths $p \in P_\omega$, the flow pattern $\tilde{x}^*$ is said to be *in equilibrium* if the following conditions hold:

$$C_p^i(\tilde{f}^*) \begin{cases} = \lambda_\omega^i(d^*), & \text{if } x_p^{i*} > 0 \\ \geq \lambda_\omega^i(d^*), & \text{if } x_p^{i*} = 0. \end{cases} \tag{37.51}$$

In other words, all utilized paths by a class connecting an O/D pair have equal and minimal generalized costs and the generalized cost on a used path by a class is equal to the inverse demand/ disutility for that class and the O/D pair that the path connects.

In the case of the fixed demand model, in which the demands in (37.40) are now assumed known and fixed, the multicriteria network equilibrium conditions now take the form:

Multiclass, multicriteria network equilibrium conditions for the fixed demand case. For each class i, for all O/D pairs $\omega \in \Omega$, and for all paths $p \in P_\omega$, the flow pattern $\tilde{x}^*$ is said to be *in equilibrium* if the following conditions hold:

$$C_p^i(\tilde{f}^*) \begin{cases} = \lambda_\omega^i, & \text{if } x_p^{i*} > 0 \\ \geq \lambda_\omega^i, & \text{if } x_p^{i*} = 0, \end{cases} \tag{37.52}$$

where now the λ_ω^i denotes simply an indicator representing the minimal incurred generalized path cost for class i and O/D pair ω. Equilibrium conditions (37.52) state that all used paths by a class connecting an O/D pair have equal and minimal generalized costs.

We now present the variational inequality formulations of the equilibrium conditions governing the elastic demand and the fixed demand problems, respectively, given by (37.51) and (37.52).

Theorem 37.5 *Variational inequality formulation of the elastic demand model.*
The variational inequality formulation of the multicriteria network model with elastic demand satisfying equilibrium conditions (37.51) is given by: determine $(\tilde{f}^*, d^*) \in \mathcal{K}^1$, satisfying

$$\sum_{i=1}^{k}\sum_{a\in L} C_a^i(\tilde{f}^*) \times (f_a^i - f_a^{i*}) - \sum_{i=1}^{k}\sum_{\omega\in\Omega} \lambda_\omega^i(d^*) \times (d_\omega^i - d_\omega^{i*}) \geq 0, \quad \forall (\tilde{f}, d) \in \mathcal{K}^1, \tag{37.53}$$

where $\mathcal{K}^1 \equiv \{(\tilde{f}, d) | \tilde{x} \geq 0, \text{ and } (37.38), (37.39), \text{ and } (37.40) \text{ hold}\}$; equivalently, in standard variational inequality form:

$$\langle F(X^*), X - X^* \rangle \geq 0, \quad \forall X \in \mathcal{K}, \tag{37.54}$$

where $F \equiv (C, \lambda)$, $X \equiv (\tilde{f}, d)$, and $\mathcal{K} \equiv \mathcal{K}^1$.

Hence, a flow and demand pattern satisfies equilibrium conditions (37.51) if and only if it also satisfies the variational inequality problem (37.53) or (37.54).

In the case of fixed demands, we have the following:

Theorem 37.6 *Variational inequality formulation of the fixed demand model.*
The variational inequality formulation of the fixed demand multicriteria network equilibrium model satisfying equilibrium conditions (37.52) is given by: determine $\tilde{f} \in \mathcal{K}^2$, satisfying

$$\sum_{i=1}^{k}\sum_{a\in L} C_a^i(\tilde{f}^*) \times (f_a^i - f_a^{i*}) \geq 0, \quad \forall \tilde{f} \in \mathcal{K}^2, \tag{37.55}$$

where $\mathcal{K}^2 \equiv \{\tilde{f} | \exists \tilde{x} \geq 0, \text{and satisfying } (37.38), (37.39), \text{ and } (37.40), \text{ with } d \text{ known}\}$; equivalently, in standard variational inequality form:

$$\langle F(X^*), X - X^* \rangle \geq 0, \quad \forall X \in \mathcal{K}, \tag{37.56}$$

where $F \equiv C$, $X \equiv \tilde{f}$, and $\mathcal{K} \equiv \mathcal{K}^2$.

Therefore, a flow pattern satisfies equilibrium conditions (37.52) if and only if it satisfies variational inequality (37.55) or (37.56).

Note that the above are finite-dimensional variational inequality problems. Finite-dimensional variational inequality formulations were also obtained by Nagurney (2000a) for her bicriteria fixed demand traffic network equilibrium model in which the weights were fixed and only class-dependent. Nagurney and Dong (2002a), in turn, formulated an elastic demand traffic network problem with two criteria and weights

1100   HANDBOOK OF OPTIMIZATION IN TELECOMMUNICATIONS

which were fixed but class- and link-dependent as a finite-dimensional variational inequality problem. The first use of a finite-dimensional variational inequality formulation of a multicriteria network equilibrium problem is due to Leurent (1993b) (see also, e.g., Leurent (1993a)), who, however, only allowed one of the two criteria to be flow-dependent. Moreover, although his model was an elastic demand model, the demand functions were separable and not class-dependent as above.

37.5.2 An application – Modeling telecommuting versus commuting decision-making

In this subsection, an application of the multiclass, multicriteria network equilibrium framework is presented. In particular, the fixed demand multicriteria network equilibrium model is applied to telecommuting versus commuting.

Note that, in the supernetwork framework, a link may correspond to an actual physical link of transportation or an abstract or virtual link corresponding to a telecommuting link. Furthermore, the supernetwork representing the problem under study can be as general as necessary and a path may also consist of a set of links corresponding to physical and virtual transportation choices such as would occur if a worker were to commute to a work center from which she could then telecommute.

Consider the four criteria, given by (37.42) through (37.45), and representing, respectively, travel time, travel cost, the opportunity cost, and safety cost. Consider a generalized link cost for each class given by (37.46). Thus, the generalized cost on a path as perceived by a class of traveler is given by (37.49).

The behavioral assumption is that travelers of a particular class are assumed to choose the paths associated with their origin/destination pair so that the generalized cost on that path is minimal. An equilibrium is assumed to be reached when the multicriteria network equilibrium conditions (37.52) are satisfied. Hence, only those paths connecting an O/D pair are utilized such that the generalized costs on the paths, as perceived by a class, are equal and minimal. The governing variational inequality for this problem is given by (37.55); equivalently, by (37.56).

For illustrative purposes, we now present a numerical example, which is governed by variational inequality (37.55); equivalently, (37.56). In order to compute the equilibrium flow pattern for the problem, the modified projection method was applied. (See the book by Nagurney and Dong (2002b) for complete details.)

The numerical example had the topology depicted in Figure 37.4. Links 1 through 13 are transportation links whereas links 14 and 15 are telecommunication links. The network consisted of ten nodes, fifteen links, and two O/D pairs where $\omega_1 = (1,8)$ and $\omega_2 = (2,10)$ with travel demands by class given by: $d^1_{\omega_1} = 10$, $d^1_{\omega_2} = 20$, $d^2_{\omega_1} = 10$, and $d^2_{\omega_2} = 30$. The paths connecting the O/D pairs were: for O/D pair ω_1: $p_1 = (1,2,7)$, $p_2 = (1,6,11)$, $p_3 = (5,10,11)$, $p_4 = (14)$, and for O/D pair ω_2: $p_5 = (2,3,4,9)$, $p_6 = (2,3,8,13)$, $p_7 = (2,7,12,13)$, $p_8 = (6,11,12,13)$, and $p_9 = (15)$.

The travel time functions and the travel cost functions, for this example, along with the associated weights for the two classes, are reported, respectively, in Tables 37.3 and 37.4. The opportunity cost functions and the safety cost functions for the links for

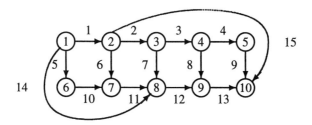

Figure 37.4 Network topology for telecommuting versus commuting example

Table 37.3 The weights and travel time functions for the links for the telecommuting example

Link a	w_{1a}^1	w_{1a}^2	$t_a(f)$
1	.25	.5	$.00005f_1^4 + 4f_1 + 2f_3 + 2$
2	.25	.5	$.00003f_2^4 + 2f_2 + f_5 + 1$
3	.4	.4	$.00005f_3^4 + f_3 + .5f_2 + 3$
4	.5	.3	$.00003f_4^4 + 7f_4 + 3f_1 + 1$
5	.4	.5	$5f_5 + 2$
6	.5	.7	$.00007f_6^4 + 3f_6 + f_9 + 4$
7	.2	.4	$4f_7 + 6$
8	.3	.3	$.00001f_8^4 + 4f_8 + 2f_{10} + 1$
9	.6	.2	$2f_9 + 8$
10	.3	.1	$.00003f_{10}^4 + 4f_{10} + f_{12} + 7$
11	.2	.4	$.00004f_{11}^4 + 6f_{11} + 2f_{13} + 2$
12	.3	.5	$.00002f_{12}^4 + 4f_{12} + 2f_5 + 1$
13	.2	.4	$.00003f_{13}^4 + 7f_{13} + 4f_{10} + 8$
14	.5	.3	$f_{14} + 2$
15	.5	.2	$f_{15} + 1$

this example, along with the associated weights for the two classes and these criteria, are reported in Table 37.5.

The generalized link cost functions were constructed according to (37.46).

Table 37.4 The weights and travel cost functions for the links for the telecommuting example

Link a	w_{2a}^1	w_{2a}^2	$c_a(f)$
1	.25	.5	$.00005 f_1^4 + 5f_1 + 1$
2	.25	.4	$.00003 f_2^4 + 4f_2 + 2f_3 + 2$
3	.4	.3	$.00005 f_3^4 + 3f_3 + f_1 + 1$
4	.5	.2	$.00003 f_4^4 + 6f_4 + 2f_6 + 4$
5	.5	.4	$4f_5 + 8$
6	.3	.6	$.00007 f_6^4 + 7f_6 + 2f_2 + 6$
7	.4	.3	$8f_7 + 7$
8	.5	.2	$.00001 f_8^4 + 7f_8 + 3f_5 + 6$
9	.2	.3	$8f_9 + 5$
10	.4	.4	$.00003 f_{10}^4 + 6f_{10} + 2f_8 + 3$
11	.7	.5	$.00004 f_{11}^4 + 4f_{11} + 3f_{10} + 4$
12	.4	.5	$.00002 f_{12}^4 + 6f_{12} + 2f_9 + 5$
13	.3	.6	$.00003 f_{13}^4 + 9f_{13} + 3f_8 + 3$
14	.2	.4	$.1 f_{14} + 1$
15	.3	.2	$.2 f_{15} + 1$

Note that the opportunity costs associated with links 14 and 15 were high since these are telecommunication links and users by choosing these links forego the opportunities associated with working and associating with colleagues from a face to face perspective. Observe, however, that the weights for class 1 associated with the opportunity costs on the telecommunication links are low (relative to those of class 2). This has the interpretation that class 1 does not weight such opportunity costs highly and may, for example, prefer to be working from the home for a variety, including familial, reasons. Also, note that class 1 weights the travel time on the telecommunication links more highly than class 2 does. Furthermore, observe that class 1 weights the safety or security cost higher than class 2.

The equilibrium multiclass link flow and total link flow patterns are reported in Table 37.6, which were induced by the equilibrium multiclass path flow pattern given in Table 37.7.

The generalized path costs were: for Class 1, O/D pair ω_1:

$$C_{p_1}^1 = 13478.4365, C_{p_2}^1 = 11001.0342, C_{p_3}^1 = 8354.5420, C_{p_4}^1 = 1025.4167,$$

Table 37.5 The weights and the opportunity cost and safety cost functions for the links for the example

Link a	w^1_{3a}	w^2_{3a}	$o_a(f)$	w_{4a}	w^2_{4a}	$s_a(f)$
1	1.	.5	$2f_1+4$	.2	.1	f_1+1
2	1.	.4	$3f_2+2$	.2	.1	f_2+2
3	1.	.7	f_3+4	.2	.1	f_3+1
4	2.	.6	f_4+2	.2	.1	f_4+2
5	1.	.5	$2f_5+1$	.2	.1	$2f_5+2$
6	2.	.7	f_6+2	.2	.1	f_6+1
7	1.	.8	f_7+3	.2	.1	f_7+1
8	1.	.6	$2f_8+1$	.2	.1	$2f_8+2$
9	2.	.9	$3f_9+2$	.2	.1	$3f_9+3$
10	1.	.8	$f_{10}+1$	.2	.1	$f_{10}+2$
11	1.	.9	$4f_{11}+3$	.2	.1	$2f_{11}+3$
12	1.	.7	$3f_{12}+2$	.2	.1	$3f_{12}+3$
13	2.	.9	$f_{13}+1$	.2	.1	$f_{13}+2$
14	.1	1.	$6f_{14}+1$	.2	.1	$.5f_{14}+.1$
15	.1	.2	$7f_{15}+4$	.2	.1	$.4f_{15}+.1$

for Class 1, O/D pair ω_2:

$$C^1_{p_5} = 45099.8047, C^1_{p_6} = 27941.5918, C^1_{p_7} = 25109.3223, C^1_{p_8} = 22631.9199,$$

$$C^1_{p_9} = 2314.7222;$$

for Class 2, O/D pair ω_1:

$$C^2_{p_1} = 15427.5996, C^2_{p_2} = 15427.2021, C^2_{p_3} = 8721.8945, C^2_{p_4} = 8721.3721,$$

and for Class 2, O/D pair ω_2:

$$C^2_{p_5} = 34924.6602, C^2_{p_6} = 34924.6094, C^2_{p_7} = 34925.3789, C^2_{p_8} = 34924.9805,$$

$$C^2_{p_9} = 41574.2617.$$

It is interesting to see the separation by classes in the equilibrium solution. Note that all members of class 1, whether residing at node 1 or node 2, were telecommuters, whereas all members of class 2 chose to commute to work. This outcome is realistic,

Table 37.6 The equilibrium link flows for the example

Link a	Class 1 - f_a^{1*}	Class 2 - f_a^{2*}	Total flow - f_a^*
1	0.000	0.0000	0.0000
2	0.0000	24.0109	24.0109
3	0.0000	22.7600	22.7600
4	0.0000	17.3356	17.3356
5	0.0000	4.6901	4.6901
6	0.0000	5.9891	5.9891
7	0.0000	1.2509	1.2509
8	0.0000	5.4244	5.4244
9	0.0000	17.3556	17.3556
10	0.0000	4.6901	4.6901
11	0.0000	10.6792	10.6792
12	0.0000	7.2400	7.2400
13	0.0000	12.6644	12.6644
14	10.0000	5.3090	15.3099
15	20.0000	0.0000	20.0000

given the weight assignments of the two classes on the opportunity costs associated with the links (as well as the weight assignments associated with the travel times). Of course, different criteria functions, as well as their numerical forms and associated weights, will lead to different equilibrium patterns.

This example demonstrates the flexibility of the modeling approach. Moreover, it allows one to conduct a variety of "what if" simulations in that, one can modify the functions and the associated weights to reflect the particular telecommuting versus commuting scenario. For example, during a downturn in the economy, the opportunity costs associated with the telecommuting links may be high, and, also, different classes may weight this criteria on such links higher, resulting in a new solution. On the other hand, highly skilled employees who are in demand may have lower weights associated with such links in regards to the opportunity costs. This framework is, hence, sufficiently general to capture a variety of realistic situations while, at the same time, allowing decision-makers to identify their specific values and preferences.

Table 37.7 The equilibrium path flows for the example

Path p	Class 1 - x_p^{1*}	Class 2 - x_p^{2*}
p_1	0.0000	0.0000
p_2	0.0000	0.0000
p_3	0.0000	4.6901
p_4	10.0000	5.3099
p_5	0.0000	17.3357
p_6	0.0000	5.4244
p_7	0.0000	1.2509
p_8	0.0000	5.9892
p_9	20.0000	0.0000

37.6 MULTITIERED AND MULTILEVEL SUPERNETWORKS

In the preceding section, the focus was on multiclass, multicriteria supernetworks and a specific application to telecommuting versus commuting decision-making was highlighted. Such a framework has also been applied to model teleshopping versus shopping decision-making (see Nagurney et al. (2001) and Nagurney et al. (2002a)). Nagurney and Dong (2003), on the other hand, proposed a supernetwork model for knowledge production in the case of multiple criteria, assuming system-optimizing behavior.

In this section, we discuss several applications in which the decision-makers are now associated with the nodes of different tiers of a supernetworks. The applications that we discuss (see also Table 37.2) are supply chain networks and financial networks with intermediation and electronic transactions. Appropriate references are noted for complete mathematical formulations and solution procedures.

37.6.1 Supply chain networks

The study of supply chain network problems through modeling, analysis, and computation is a challenging topic due to the complexity of the relationships among the various decision-makers, such as suppliers, manufacturers, distributors, and retailers as well as the practical importance of the topic for the efficient movement (and pricing) of products. The topic is multidisciplinary by nature since it involves particulars of manufacturing, transportation and logistics, retailing/marketing, as well as economics. For additional background on supply chains, see the books by Bramel and Simchi-Levi (1997) and Pardalos and Tsitsiringos (2002), which also discusses financial engineering aspects, and the volume edited by Simchi-Levi et al. (2004).

In particular, the introduction of electronic commerce has unveiled new opportunities in terms of research and practice in supply chain analysis and management since electronic commerce (e-commerce) has had an immense effect on the manner in which businesses order goods and have them transported with the major portion of e-commerce transactions being in the form of business-to-business (B2B). Estimates of B2B electronic commerce range from approximately .1 trillion dollars to 1 trillion dollars in 1998 and with forecasts reaching as high as $4.8 trillion dollars in 2003 in the United States (see Federal Highway Administration (2000) and Southworth (2000)). It has been emphasized that the principal effect of business-to-business (B2B) commerce, estimated to be 90% of all e-commerce by value and volume, is in the creation of new and more profitable supply chain networks.

In Figure 37.5, a four-tiered supply chain network is depicted (cf. Nagurney and Dong (2002b)) in which the top tier consists of suppliers of inputs into the production processes used by the manufacturing firms (the second tier), who, in turn, transform the inputs into products which are then shipped to the third tier of decision-makers, the retailers, from whom the consumers can then obtain the products. In this context, not only are physical transactions allowed but also virtual transactions, in the form of electronic transactions via the Internet to represent electronic commerce. In the supernetwork framework, both B2B and B2C can be considered, modeled, and analyzed. The decision-makers may compete independently across a given tier of nodes of the network and cooperate between tiers of nodes.

In particular, Nagurney et al. (2002d) have applied the supernetwork framework to supply chain networks with electronic commerce in order to predict product flows between tiers of decision-makers as well as the prices associated with the different tiers. They assumed that the manufacturers as well as the retailers are engaged in profit maximizing behavior whereas the consumers seek to minimize the costs associated with their purchases. The model therein determines the volumes of the products transacted electronically or physically. That work was based on the model of Nagurney et al. (2002b), which was the first supply chain network equilibrium model. It assumed decentralized decision-making and competition across a tier of decision-makers but cooperation between tiers.

As mentioned earlier, supernetworks may also be multilevel in structure. In particular, Nagurney et al. (2002c) demonstrated how supply chain networks can be depicted and studied as multilevel networks in order to identify not only the product shipments but also the financial flows as well as the informational ones. In Figure 37.6 we provide a graphic of an integrated social and supply chain networks as a multilevel supernetwork due to Wakolbinger and Nagurney (2004). In their model, they introduced flows into social networks in the form of relationship levels and allowed for transactions costs to be functions of both product transactions as well as relationship levels. The decision-making behavior assumed profit maximization, risk minimization, as well as relationship value maximization with individual associated weights for the manufacturers and the retailers.

Obviously, in the setting of supply chain networks and, in particular, in global supply chains, there may be much risk and uncertainty associated with the underlying functions. Some research along those lines has recently been undertaken (cf. Dong

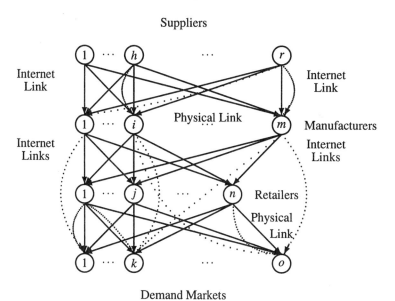

Figure 37.5 The supernetwork structure of the supply chain network with suppliers, manufacturers, retailers, and demand markets and electronic commerce

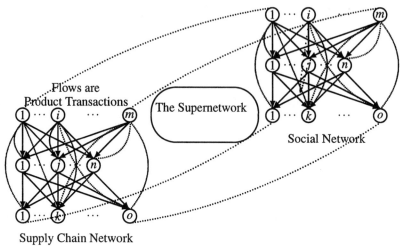

Figure 37.6 The multilevel supernetwork structure of the integrated supply chain / social network system

et al. (2004), and Nagurney et al. (2003)). Continuing efforts to include uncertainty and risk into modeling and computational efforts in a variety of supernetworks and their applications is of paramount importance given the present economic and political climate.

In addition, we emphasize that the inclusion of environmental variables and criteria is also an important topic for research and practice in the context of supply chain networks (cf. Nagurney and Toyasaki (2003)). Recently, a multitiered supply chain network equilibrium framework has been developed for reverse logistics and the recycling of electronic wastes (see Nagurney and Toyasaki (2005)).

37.6.2 Financial networks with electronic transactions

As noted earlier, financial networks have been utilized in the study of financial systems since the work of Quesnay (1758), who, in 1758, depicted the circular flow of funds in an economy as a network. His conceptualization of the funds as a network, which was abstract, is the first identifiable instance of a supernetwork.

Advances in telecommunications and, in particular, the adoption of the Internet by businesses, consumers, and financial institutions have had an enormous effect on financial services and the options available for financial transactions. Distribution channels have been transformed, new types of services and products introduced, and the role of financial intermediaries altered in the new economic networked landscape. Furthermore, the impact of such advances has not been limited to individual nations but, rather, through new linkages, has crossed national boundaries.

The topic of *electronic* finance has been a growing area of study (cf. Claessens et al. (2000), Claessens et al. (2003), Nagurney (2000b), and the references therein), due to its increasing impact on financial markets and financial intermediation, as well as related regulatory issues and governance. Of particular emphasis has been the conceptualization of the major issues involved and the role of networks is the transformations (cf. Nagurney and Dong (2002b) and the references therein).

Nevertheless, the complexity of the interactions among the distinct decision-makers involved, the supply chain aspects of the financial product accessibilities and deliveries, as well as the availability of physical as well as electronic options, and the role of intermediaries, have defied the construction of a unified, quantifiable framework in which one can assess the resulting financial flows and prices.

Here we briefly describe a supernetwork framework for the study of financial decision-making in the presence of intermediation and electronic transactions. Further details can be found in Nagurney and Ke (2001) and Nagurney and Ke (2003). The framework is sufficiently general to allow for the modeling, analysis, and computation of solutions to such problems.

The financial network model consists of: agents or decision-makers with sources of funds, financial intermediaries, as well as consumers associated with the demand markets. In the model, the sources of funds can transact directly electronically with the consumers through the Internet and can also conduct their financial transactions with the intermediaries either physically or electronically. The intermediaries, in turn, can transact with the consumers either physically in the standard manner or electronically. The depiction of the network at equilibrium is given in Figure 37.7.

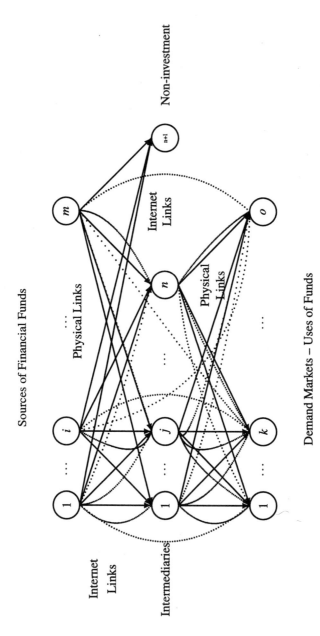

Figure 37.7 The structure of the financial network with electronic transactions

It is assumed that the agents with sources of funds as well as the financial intermediaries seek to maximize their net revenue (in the presence of transaction costs) while, at the same time, minimizing the risk associated with the financial products. The solution of the model yields the financial flows between the tiers as well as the prices. Here we also allow for the option of having the source agents not invest a part (or all) of their financial holdings. More recently, Nagurney and Cruz (2003) have demonstrated that the financial supernetwork framework can also be extended to model international financial networks with intermediation in which there are distinct agents in different countries and the financial products are available in different currencies.

Acknowledgments

The preparation of this manuscript was supported, in part by NSF Grant No. IIS 0002647, under the MKIDS program, and by two AT&T Industrial Ecology Fellowships. This support is gratefully acknowledged.

Bibliography

Telecom 95, 1995. MCI and the Internet: Dr. Vinton G. Cerf, Keynote Address, http://www.itu.int/TELECOM/wt95/pressdocs/papers/cerf.html.

J. Abello, P. M. Pardalos, and M. G. C. Resende. On maximum clique problems in very large graphs. In J. Abello and J. Vitter, editors, *External Memory Algorithms*, volume 50 of *AMS Series on Discrete Mathematics and Theoretical Computer Science*, pages 119–130. American Mathematical Society, 1999.

R. K. Ahuja, T. L. Magnanti, and J. B. Orlin. *Network Flows: Theory, Algorithms, and Applications*. Prentice-Hall, Upper Saddle River, New Jersey, 1993.

D. Banister and K. J. Button. Environmental policy and transport: An overview. In D. Banister and K. J. Button, editors, *Transport, the Environment, and Sustainable Development*, pages 130–136. E. & F.N., London, 1993.

H. Bar-Gera. Origin-based algorithms for transportation network modeling. Technical Report 103, National Institute of Statistical Sciences, Research Triangle Park, North Carolina, 1999.

T. Bass. Road to ruin. *Discover*, pages 56–61, May 1992.

M. S. Bazaraa, H. D. Sherali, and C. M. Shetty. *Nonlinear Programming: Theory and Algorithms*. John Wiley & Sons, New York, second edition, 1993.

M. J. Beckmann. On the theory of traffic flows in networks. *Traffic Quarterly*, 21: 109–116, 1967.

M. J. Beckmann, C. B. McGuire, and C. B. Winsten. *Studies in the Economics of Transportation*. Yale University Press, New Haven, Connecticut, 1956.

D. P. Bertsekas and R. Gallager. *Data Networks*. Prentice-Hall, Englewood Cliffs, New Jersey, second edition, 1992.

D. Braess. Uber ein paradoxon der verkehrsplanung. *Unternehmenforschung*, 12: 258–268, 1968.

J. Bramel and D. Simchi-Levi. *The Logic of Logistics: Theory, Algorithms, and Applications for Logistics Management.* Springer-Verlag, New York, 1997.

S. Claessens, G. Dobos, D. Klingebiel, and L. Laeven. The growing importance of networks in finance and their effects on competition. In A. Nagurney, editor, *Innovations in Financial and Economic Networks.* Edward Elgar Publishing, Cheltenham, England, 2003.

S. Claessens, T. Glaessner, and D. Klingebeiel. Electronic finance: Reshaping the financial landscape around the world. Technical report, The World Bank, Washington, DC., 2000. Financial Sector Discussion Paper No. 4.

J. E. Cohen and F. P. Kelly. A paradox of congestion on a queuing network. *Journal of Applied Probability*, 27:730–734, 1990.

A. A. Cournot. *Researches into the Mathematical Principles of the Theory of Wealth.* Macmillan, London, 1838. English Translation, 1897.

S. Dafermos. The traffic assignment problem for multimodal networks. *Transportation Science*, 6:73–87, 1972.

S. Dafermos. Integrated equilibrium flow models for transportation planning. In M. A. Florian, editor, *Traffic Equilibrium Methods*, volume 118 of *Lecture Notes in Economics and Mathematical Systems*, pages 106–118. Springer-Verlag, New York, 1976.

S. Dafermos. Traffic equilibrium and variational inequalities. *Transportation Science*, 14:42–54, 1980.

S. Dafermos. A multicriteria route-mode choice traffic equilibrium model. Technical report, Lefschetz Center for Dynamical Systems, Brown University, Providence, Rhode Island, 1981.

S. Dafermos. The general multimodal network equilibrium problem with elastic demand. *Networks*, 12:57–72, 1982.

S. Dafermos and A. Nagurney. On some traffic equilibrium theory paradoxes. *Transportation Research B*, 18:101–110, 1984.

S. Dafermos and A. Nagurney. Isomorphism between spatial price and traffic network equilibrium models. Technical Report 85-17, Lefschetz Center for Dynamical Systems, Brown University, Providence, Rhode Island, 1985.

S. C. Dafermos and F. T. Sparrow. The traffic assignment problem for a general network. *Journal of Research of the National Bureau of Standards*, 73B:91–118, 1969.

P. J. Denning. The science of computing: Supernetworks. *American Scientist*, 73: 127–129, 1985.

J. Dong, D. Zhang, and A. Nagurney. A supply chain network equilibrium model with random demands. *European Journal of Operational Research*, 156:194–212, 2004.

The Economist. Survey: E-management, November 11 2000.

J. Fallows. The Java theory. *The Atlantic Online, The Atlantic Monthly/Digital Edition*, 1996. http://www2.theatlantic.com/issues/96mar/java/java.htm.

Federal Highway Administration. E-commerce trends in the market for freight. Task 3 freight trends scans, 2000. Draft, Multimodal Freight Analysis Framework, Office of Freight Management and Operations.

M. Florian and D. Hearn. Network equilibrium models and algorithms. In M. O. Ball, T. L. Magnanti, C. L. Monma, and G. L. Nemhauser, editors, *Network Routing*, volume 8 of *Handbooks in Operations Research and Management Science*, pages 485–550. Elsevier Science, Amsterdam, The Netherlands, 1995.

J. Guenes and P. M. Pardalos. Network optimization in supply chain management and financial engineering: An annotated bibliography. *Networks*, 42:66–84, 2003.

Illinois State Bar Association. Opinion no. 96-10, Topic: Electronic communications; confidentiality of client information; advertising and solicitation, May 16 1997. http://www.chicago-law.net/cyberlaw/electric.html.

Internet World Stats. Internet usage statistics – The big picture, 2005. http://www.internetworldstats.com/stats.htm.

D. Kinderlehrer and G. Stampacchia. *An Introduction to Variational Inequalities and Their Applications*. Academic Press, New York, 1980.

G. Kolata. What if they closed 42d street and nobody noticed? *The New York Times*, December 25 1990.

Y. A. Korilis, A. A. Lazar, and A. Orda. Avoiding the Braess paradox in noncooperative networks. *Journal of Applied Probability*, 36:211–222, 1999.

E. Koutsoupias and C. H. Papadimitrou. Worst-case equilibria. In *Proceedings of the 16th Annual Symposium on the Theoretical Aspects of Computer Science*, pages 404–413, 1999.

H. W. Kuhn and Tucker A. W. Nonlinear programming. In J. Neyman, editor, *Proceedings of Second Berkeley Symposium on Mathematical Statistics and Probability*, pages 481–492, Berkeley, California, 1951. University of California Press.

F. Leurent. Cost versus time equilibrium over a network. *European Journal of Operations Research*, 71:205–221, 1993a.

F. Leurent. Modelling elastic, disaggregate demand. In J. C. Moreno Banos, B. Friedrich, M. Papageorgiou, and H. Keller, editors, *Proceedings of the Euro Working Group on Urban Traffic and Transportation*, Munich, Germany, 1993b. Technical University of Munich.

H. S. Mahmassani, S. Peeta, T. Y. Hu, and A Ziliaskopoulos. Dynamic traffic assignment with multiple user classes for real-time ATIS/ATMS applications. In *Large Urban Systems, Proceedings of the Advanced Traffic Management Conference*, pages 91–114, Washington, DC, 1993. Federal Highway Administration, U.S. Department of Transportation.

P. L. Mokhtarian. A typology of relationships between telecommunications and transportation. *Transportation Research A*, 24:231–242, 1990.

A. Nagurney. Comparative tests of multimodal traffic equilibrium methods. *Transportation Research B*, 18:469–485, 1984.

A. Nagurney. *Network Economics: A Variational Inequality Approach*. Kluwer Academic Publishers, Dordrecht, The Netherlands, second and revised edition, 1999.

A. Nagurney. Navigating the network economy. *OR/MS Today*, pages 74–75, June 2000a.

A. Nagurney. *Sustainable Transportation Networks*. Edward Elgar Publishing, Cheltenham, England, 2000b.

A. Nagurney and J. Cruz. International financial networks with intermediation: Modeling, analysis, and computations. *Computational Management Science*, 1:31–58, 2003.

A. Nagurney, J. Cruz, and D. Matsypura. Dynamics of global supply chain supernetworks. *Mathematical and Computer Modelling*, 37:963–983, 2003.

A. Nagurney and J. Dong. A multiclass, multicriteria traffic network equilibrium model with elastic demand. *Transportation Research B*, 36:445–469, 2002a.

A. Nagurney and J. Dong. *Supernetworks: Decision-Making for the Information Age*. Edward Elgar Publishing, Cheltenham, England, 2002b.

A. Nagurney and J. Dong. Management of knowledge intensive systems as supernetworks: Modeling, analysis, computations, and applications. *Mathematical and Computer Modelling*, 2003. To appear.

A. Nagurney, J. Dong, and P. L. Mokhtarian. Integrated multicriteria network equilibrium models for commuting versus telecommuting. Technical report, Isenberg School of Management, University of Massachusetts, Amherst, Massachusetts, 2000.

A. Nagurney, J. Dong, and P. L. Mokhtarian. Teleshopping versus shopping: A multicriteria network equilibrium framework. *Mathematical and Computer Modelling*, 34:783–798, 2001.

A. Nagurney, J. Dong, and P. L. Mokhtarian. Multicriteria network equilibrium modeling with variable weights for decision-making in the information age with applications to telecommuting and teleshopping. *Journal of Economic Dynamics and Control*, 26:1629–1650, 2002a.

A. Nagurney, J. Dong, and D. Zhang. A supply chain network equilibrium model. *Transportation Research E*, 38:281–303, 2002b.

A. Nagurney and K. Ke. Financial networks with intermediation. *Quantitative Finance*, 1:309–317, 2001.

A. Nagurney and K. Ke. Financial networks with electronic commerce: Modeling, analysis, and computations. *Quantitative Finance*, 3:71–87, 2003.

A. Nagurney, K. Ke, J. Cruz, K. Hancock, and F. Southworth. Dynamics of supply chains: A multilevel (logistical / informational / financial) network perspective. *Environment & Planning*, 29:795–818, 2002c.

A. Nagurney, J. Loo, J. Dong, and D. Zhang. Supply chain networks and electronic commerce: A theoretical perspective. *Netnomics*, 4:187–220, 2002d.

A. Nagurney and S. Siokos. *Financial Networks: Statics and Dynamics*. Springer-Verlag, Berlin, Germany, 1997.

A. Nagurney and F. Toyasaki. Supply chain supernetworks and environmental criteria. *Transportation Research D*, 8:185–213, 2003.

A. Nagurney and F. Toyasaki. Reverse supply chain management and electronic waste recycling: A multitiered network equilibrium framework for e-cycling. *Transportation Research E*, 41:1–28, 2005.

A. Nagurney and D. Zhang. *Projected Dynamical Systems and Variational Inequalities with Applications*. Kluwer Academic Publishers, Boston, Massachusetts, 1996.

P. M. Pardalos and V. Tsitsiringos. *Financial Engineering, Supply Chain and E-Commerce*. Kluwer Academic Publishers, Dordrecht, The Netherlands, 2002.

A. C. Pigou. *The Economics of Welfare*. Macmillan, London, England, 1920.

R. E. Quandt. A probabilistic abstract mode model. In *Studies in Travel Demand VIII*, pages 127–149. Mathematica, Inc., Princeton, New Jersey, 1967.

F. Quesnay. Tableau economique, 1758. Reproduced in facsimile with an introduction by H. Higgs by the British Economic Society, 1895.

B. Ran and D. E. Boyce. *Modeling Dynamic Transportation Networks*. Springer-Verlag, Berlin, Germany, 1996.

M. G. C. Resende. Personal communication, 2000.

T. Roughgarden. *Selfish Routing*. PhD thesis, Department of Computer Science, Cornell University, Ithaca, New York, 2002.

T. Roughgarden and E. Tardos. How bad is selfish routing? *Journal of the ACM*, 49: 236–259, 2002.

P. A. Samuelson. Spatial price equilibrium and linear programming. *American Economic Review*, 42:283–303, 1952.

M. Schneider. Access and land development. In *Urban Development Models*, volume 97, pages 164–177. Highway Research Board Special Report, 1968.

L. Schubert, R. Goebel, and N. Cercone. The structure and organization of a semantic net for comprehension and inference. In N. V. Findler, editor, *Associative Networks: Representation and Use of Knowledge by Computers*. Academic Press, New York, 1979.

Y. Sheffi. *Transportation Network Equilibrium with Discrete Choice Models*. PhD thesis, Civil Engineering Department, Massachusetts Institute of Technology, Cambridge, Massachusetts, 1978.

Y. Sheffi. *Urban Transportation Networks – Equilibrium Analysis with Mathematical Programming Methods*. Prentice-Hall, Englewood Cliffs, New Jersey, 1985.

Y. Sheffi and C. F. Daganzo. Hypernetworks and supply – Demand equilibrium obtained with disaggregate demand models. *Transportation Research Record*, 673: 113–121, 1978.

Y. Sheffi and C. F. Daganzo. Computation of equilibrium over transportation networks: The case of disaggregate demand models. *Transportation Science*, 14:155–173, 1980.

D. Simchi-Levi, S. D. Wu, and Z. J. Shen, editors. *Handbook of Quantitative Supply Chain Analysis: Modeling in the E-Business Era*. Kluwer Academic Publishers, 2004.

M. J. Smith. Existence, uniqueness, and stability of traffic equilibria. *Transportation Research B*, 13:259–304, 1979.

F. Southworth. E-commerce: Implications for freight. Technical report, Oak Ridge National Laboratory, Oak Ridge, Tennessee, 2000.

T. Wakolbinger and A. Nagurney. Dynamic supernetworks for the integration of social networks and supply chains with electronic commerce: Modeling and analysis of buyer-seller relationships with computations. *Netnomics*, 6:153–185, 2004.

J. G. Wardrop. Some theoretical aspects of road traffic research. In *Proceedings of the Institute of Civil Engineers, Part II*, pages 325–378, 1952.

J. H. Wu, M. Florian, and S. G. He. Emme/2 implementation of the SCAG-II model: Data structure, system analysis and computation, 2000. Submitted to the Southern California Association of Governments, INRO Solutions Internal Report.

L. Zhao. *Variational Inequalities in General Equilibrium: Analysis and Computation*. PhD thesis, Division of Applied Mathematics, Brown University, Providence,

Rhode Island, 1989. Also appears as: LCDS # 88-24, Lefschetz Center for Dynamical Systems, Brown University, Providence, Rhode Island, 1988.

L. Zhao and S. Dafermos. General economic equilibrium and variational inequalities. *Operations Research Letters*, 10:369–376, 1991.

L. Zhao and A. Nagurney. A network formalism for pure exchange economic equilibria. In D. Z. Du and P. M. Pardalos, editors, *Network Optimization, Problems: Algorithms, Complexity and Applications*, pages 363–386. World Scientific Press, Singapore, 1993.

Index

2-connectivity, 378
2-node connectivity, 741
3-partition inequalities, 174
Abstract network, 1077, 1079–1080
Accelerated greedy algorithm, 171
Access
  engineering, 765
  network, 273, 314, 316
  ring, 334
ACCPM, 245
Acyclic, 148
Adaptive
  memory, 108, 111
  penalty factor, 745
Add-drop multiplexer (ADM), 848
Ad hoc networks, 531
Adjacency matrix, 743
Administrative weight, 437, 449–451, 453
Advanced warmstart, 86
Agent, 1025
Aggregation, 83, 653, 660, 663
Algorithm
  accelerated greedy, 171
  analytic center cutting plane, 245
  approximation, 249, 281
  auction, 202
  augmented Lagrangian, 251
  backtracking, 739
  Bellman-Ford, 198, 581, 713
  Bellman-Ford-Moore, 202
  Benders decomposition, 43, 71, 87, 258
  Benders, 175
  BG heuristic, 711
  branch and bound, 71–73, 75–77, 84, 173, 285, 320, 444, 708, 742
  branch and cut, 71–73, 75, 84, 88, 285
  column generation, 16, 48, 72, 88, 302, 499, 405
  concave branch elimination, 444
  concave link elimination, 444
  conjugate gradient, 9
  constraint generation, 175, 178
  construction heuristic, 742
  cut/path set enumeration, 738
  Dantzig-Wolfe decomposition, 169, 245
  delayed relax and cut, 130, 134
  depth first search, 81
  D'Esopo-Pape, 199
  Dijkstra's, 401, 192, 713
  distributed, 710, 713, 952
  drop heuristic, 450
  dual ascent, 75, 943, 941, 952, 959, 972
  dual network simplex, 156
  dual simplex, 85
  duals on a tree, 152–153
  dynamic programming, 323, 328, 449, 522, 525
  exact, 738
  flows on a tree, 152
  Frank-Wolfe, 45–46, 48
  genetic, 84, 105–106, 109, 115, 117–118, 120–122, 444, 450–451, 523, 528, 742–743
  GRASP, 105–106, 109, 112, 117, 119–121, 470, 533, 853
  greedy, 104–106, 116, 118, 120, 83, 139
  greedy adaptive search, 531
  greedy incremental tree, 709
  heuristic, 78, 83, 282, 399, 437, 464, 851
  hybrid heuristics, 117
  ILP, 637
  integer linear programming (ILP), 395
  integer programming, 707, 851, 853
  interior point, 3, 78, 85, 169
  Karmarkar, 615, 631
  KMB, 709
  Lagrangean relaxation, 169, 446–447
  Lagrangian decomposition, 283, 449
  Lagrangian heuristics, 135
  Lagrangian relaxation, 52, 67, 70–72, 75, 78, 86–87, 258, 283, 320, 498, 530, 711
  linear programming, 3, 580, 614
  linear programming relaxation, 76, 282

local search, 84, 106, 109, 112, 114, 116, 118–121, 139, 451, 470, 679, 683, 685–688, 693
LP relaxation, 637
metaheuristics, 84, 104–106, 108, 117–118
modified Frank-Wolfe, 54
modified projection, 1100
multicommodity, 120, 242
multicommodity flow, 1051, 247, 278, 442, 446–447
multicommodity network flow, 839–840, 164
multilabeling, 72
multi-start heuristic, 853
Newton, 52, 54
non delayed relax and cut, 130–131
nonlinear optimization, 28
nonlinear programming, 28
on-line, 856
path relinking, 108, 113, 120, 122
Pilot method, 469
preconditioned conjugate gradient, 10
primal heuristics, 84
primal network simplex, 153
prototype shortest path, 191
proximal decomposition, 251, 258
reduced gradient method, 54
relax and cut, 130
reverse Dijkstra's, 193
scaling push-relabel, 156
scatter search, 106, 108, 117
scatter-search, 853
simplex, 78
simulated annealing, 84, 105, 110, 118–119, 121, 449, 522, 524, 528, 535–536, 685, 742
subgradient, 132, 446, 87, 132
tabu search, 107–109, 111, 117–119, 121, 471, 521, 526, 533, 685, 444
tabu-search, 853
Tarjan's, 201
threshold, 200
variable neighborhood search, 114
VNS, 106, 114, 117, 120
Allocated delay, 441
Allocation, 119, 121
All-optical, 119, 121
All-terminal, 737
reliability, 741
Amplifier, 523
placement problem, 523
Analytical approach, 764
Analytic center cutting plane method, 245
Ant colony optimization, 853
Antenna array, 19
Application delay, 453
Approximation algorithms, 249, 281
A priori efficiency, 396
Arc

capacity vector, 149
formulation, 646
unit cost vector, 149
Artificial intelligence, 449
Assignment problem, 68, 75
Association rules, 1040
Asymmetric
cost functions, 1090
interactions, 1090
Asynchronous
teams, 721
transfer mode, 765, 850
ATM, 765, 845, 850
cell, 850
networks, 715, 766
Auction, 1052
algorithms, 202
Augmented
Lagrangian method, 251
systems, 7
Availability, 254
Average
delay, 438, 440, 442, 444–446, 449
hop distance, 527
nodal degree, 396
B2B, 1019
electronic commerce, 1106
B2C, 1019, 1106
Backbone, 273
network, 314, 368–369
Backtracking algorithm, 739
Backward star, 148
Bandwidth
allocation, 1051, 439
reservation, 586–588, 604
Base station (BS), 531
Base station
locations, 532
Basic feasible solution, 152
Basis, 152
Behavioral
assumption, 1098, 1100
concepts, 1089
Bellman-Ford
algorithm, 198, 713
method, 581
Bellman-Ford-Moore algorithm, 202
Benders
decomposition, 43, 71, 87, 258
method, 175
Best
bound, 81
effort, 413, 436
BG heuristic, 711
Bi-criteria objective, 418
Bidding language, 1054
Bidirectional

INDEX 1123

line switched ring (BLSR), 391
ring, 292
Bifurcated demands, 321
Bipartition inequalities, 176
Blocking, 853
  probability, 522, 527
Blocking rate
  minimization of the, 641
Border gateway protocol (BGP), 688, 690–694
Bounds on reliability, 739
Braess
  network, 1087
  paradox, 43–44, 1081, 1083
Branch and bound, 71–73, 75–77, 84, 173, 285, 320, 444, 708
  algorithm, 742
Branch and cut, 71–73, 75, 84, 88, 285
Branching, 76, 79, 87
Broadband
  access networks, 325
  services, 325
Broadcast and select networks, 519, 523–524
Broadcast
  routing, 702
  scheduling, 873
Bronze, 413
Buffer overflow probability, 444
Bundling, 839–840
Cable networks, 329
Cache
  nodes, 721
  placement problem, 721
Call graphs, 879
Candidate sites, 921
Capacitated
  concentrator location, 529
  facility location, 320
  iterative design algorithm (CIDA), 404
  minimum spanning tree (CMST) problem, 530
  network design, 280
Capacity, 446
  allocation problems, 393
  and flow assignment (CFA) problem, 442
  and route assignment, 446
  assignment, 41, 220
  assignment and routing problem, 446–447
  assignment (CA) problem, 448
  assignment problem, 436
  constraints, 447
  costs, 441, 444
  dimensioning, 277
  expansion of mobile networks, 785
  management, 574–575, 590
  planning, 559, 842
  utilization, 853
CDMA networks, 943, 950, 953, 963, 966–967
Cell, 918

Cell-breathing phenomenon, 532
Cellular networks, 531
Center based algorithms, 715
Centralized decision-making, 1088
Channel
  assignment, 840
  assignment problems, 872
  interchange, 393
Chromatic number, 867
Circuit-switching, 583, 590
Classes
  of service, 451
  of traffic, 449, 452
Classical network, 1074
Classification, 1036
Class of traffic, 438
Class Steiner tree problem, 476
Clique, 867
  number, 867
Cluster-based topology, 534
Clustering, 1037
  problem, 868
Clusters, 1037, 445
Code assignment, 918
Code division multiple access (CDMA), 531
Coefficient reduction, 83
Coloring, 867
Color mismatch, 393
Column generation, 16, 48, 72, 88, 302, 499
Column generation algorithms, 405
Combinatorial auctions, 1051
Combined forward/reverse Dijkstra's algorithm, 193
Common pool, 430
Communication
  networks, 1077
  technologies, 1094
Commuting versus telecommuting, 1098
Competition, 1106
Competitive
  economic equilibrium, 224
  market, 224
Complementary slackness conditions, 190
Complete linear description, 498
Complexity, 1080
Complex networks, 1074
Computational experimentation, 856
Computer networks, 1078
Concave
  branch elimination procedure, 444
  capacity costs, 442
  costs, 449
  link elimination (CLE) procedure, 444
Concentrator, 316
  location, 318, 528
Conflict graph, 83, 878

Congestion, 1074, 1080–1081, 1098, 243, 246, 438, 527
   and capacity allocation cost, 221
   control, 680
   minimization of the, 642
Conjugate gradient methods, 9
Connected, 148
   dominating set, 866
   domination number, 866
Connectionless network, 783
Connectivity, 368–369, 371–377, 379
   graph, 868
   measures, 737
Conservation of flow equations, 1088, 1092
Constrained minimum Steiner tree problem, 526
Constraint generation, 175, 178
Construction heuristic, 742
Contiguity constraints, 322
Continuous capacities, 449
Converter, 521
   placement, 522
Convex
   approximation, 222
   hull, 31, 222–223
   multicommodity flow, 223, 253
Copper broadband access, 326
Core-based tree, 716
Core-based trees, 703
Core points, 716
Cost
   minimization, 1094, 213
   of capacity, 449
   of congestion, 438, 447
   of delay, 438, 446–447
Coverage problem, 920
Crew scheduling, 16
Criteria, 1096
   functions, 1104
Crossover, 743
Cumulative interference, 928
Customer concentration, 213
Custom queueing, 439, 449
Cut inequalities, 278, 505
Cut/path set enumeration, 738
Cutset, 69, 73, 717
   inequalities, 174
Cutting planes, 73
   algorithm, 284
Cutting stock problem, 88
Cycle-enumeration, 401
Cycle inequalities, 381–382
Cyclic decomposition, 55
Cyclomatic inequalities, 383
Dagger sampling, 741
Dantzig-Wolfe decomposition, 169, 245
Database correlation method (DCM), 537
Data
   mining, 1034
   warehouse, 1034
Decentralized decision-making, 1088, 1106
Decision-making, 1073–1074, 1076–1077, 1080, 1094, 1096, 1105–1106
Decomposition, 243
Dedicated protection, 293
Degradation, 79
Degree-constrained Steiner problem, 475
Delay, 437–438, 440
   allocation, 441
Delay-based routing, 681
Delay
   constraints, 706, 716
Delayed relax and cut, 130, 134
Delay
   variation, 715
Demand, 277
   matrices, 439
   node, 149
Demand target pattern matching, 423
Demand
   uncertainty, 418
Dependability, 783
Depth first search, 81
Design of backbone, 779
D'Esopo-Pape algorithm, 199
Detail graph applications, 1080
Dial's implementation, 194
Differential destination multicast (DDM) protocol, 537
Differentiated
   routing, 438
   service, 450
DiffServ, 597, 604, 607–608
Digital
   cross-connect switch (DCS), 847
   signal level 1 (DS-1), 846
   subscriber line (DSL), 316
Dijkstra, 401
Dijkstra's algorithm, 192, 713
Direct collisions, 873
Directed, 148
   formulation, 495
Directional derivative, 32
Disaggregate simplicial decomposition, 54
Discrete event flow optimization, 592
Distance labels, 188
Distinct channel assignment constraint, 840
Distributed
   algorithm, 710, 713, 952
Distributed cycle preconfiguration (DCPC), 405
Distributed
   solution, 945
Distribution channels, 1108
Diversification, 108, 111, 117–118
Dominating set, 866, 897

INDEX   1125

Domination
 number, 866
 theorem, 739
Drop heuristic, 450
DSL Access Multiplexers, 326
Dual ascent, 75, 943
 scheme, 941, 952, 959, 972
Dual failure protection, 413
Dual feasible, 156
 solution procedure, 156
Duality gap, 140
Dual
 linear program, 943–946
 network simplex procedure, 156
 simplex, 85
Duals on a tree procedure, 152–153
DVMRP, 703
DWDM, 849
Dynamic
 bandwidth allocation, 573
 grooming methods, 852
 group, 703
 programming, 323, 328, 449, 522, 525
 routing, 573, 805
 Steiner tree problem, 473
 traffic engineering, 856
Early-exit routing, 688–691
E-auctions, 1029
E-business, 1019
E-commerce, 1018, 436
Economic and financial networks, 1076
Economic equilibrium, 1080
Economics, 1080
Economy, 413
Economy of scale, 170–172, 213, 217, 395
Edge
 formulation, 643
 node, 314
Edge-path formulations, 72
Elastic
 demand model, 1092, 1098
 demand multicriteria network equilibrium
  model, 1094
 demand network equilibrium problem, 1093
Electromagnetic, 119
 pollution, 929
Electronic
 commerce, 1083, 1106
 data interchange, 1022
 transactions, 1109
Eligible cycles, 399
E-marketplaces, 1020
E-markets, 1020
Encapsulation, 844, 851
Encoding, 743
 scheme, 752
End-to-end delay, 440, 445–446, 450

Energy conserving routing, 45
Enhanced availability, 413
Enterprise
 decision level, 789
 level, 789
 network, 437, 449
 resource planning, 1024
Entropy function, 52
EON network, 669
Epsilon-optimum, 156
Equilibrium
 conditions, 1085, 1088, 1092, 1099
 flow pattern, 1082, 1100
 path flows, 1082
 patterns, 1104
Equipment failures, 762
E-supply chain, 1022
Euclidean Steiner problem, 472
Event dependent routing, 577, 591
Exact methods, 738
Expand, 401
Extreme point, 31
Facet, 371, 377–378, 380–381, 383
Facility location, 318
 model, 770
Factoring, 739
Failure, 254
 independent path protecting (FIPP), 430
 propagation, 828
Fathoming, 77
FDD, 919
FDMA, 919
Feasible flows, 149
Fiber nodes, 330
Fiber-to-the-curb (FTTC), 326
Financial
 data delivery, 704
 intermediaries, 1108–1109
 networks, 1083, 1108
 networks with intermediation, 1111
 optimization, 16
 services, 1108
 supernetwork, 1111
 transactions, 1108
Fixed charge cost, 317
Fixed demand multicriteria network equilibrium
  model, 1099–1100
Flooding, 706
Flow assignment (FA) problem, 442
Flow assignment problem, 450
Flow balance equations, 41
Flow conservation, 244
 equations, 150
Flow deviation, 249
 (FD) algorithm, 442
Flow
 formulation, 243

of funds, 1079
protecting, 393
vector, 149
Flows, 164
Flows on a tree procedure, 152
Forcers, 421, 427
Forward star, 148
Foundational networks, 1080
Frank-Wolfe algorithm, 45–46, 48
Frequency
  assignment, 17, 1052
  assignment
    problem, 924
  division multiplexing, 840
  planning, 919
Full Steiner tree, 472
Gateway, 314
General equilibrium, 552
Generalized
  cost, 1098, 1074, 1100
  cost
    function, 1086, 1097
  link cost, 1100–1101
  multiprotocol label switching, 851
  spanning tree, 476
  Steiner problem, 476–477
  upper bound, 78
General
  network equilibrium models, 1089
  ring network design problem, 479
Generate rings, 298
Generic framing procedure, 848
Genetic algorithm, 84, 105–106, 109, 115, 117–118, 120–122, 444, 450–451, 523, 528, 742–743
Geometric minimum broadcast cover problem, 900
Global minimum, 30
Global optimization, 28
Global supply chains, 1106
GMPLS, 851–852
Gold, 413
Gold-plus, 413
Goldstein-Levitin-Polyak projection algorithm, 52
Gomory cuts, 135
Grade of service problem, 479
Graph
  coloring, 925
  evolution, 741
  problems, 851
GRASP, 105–106, 109, 112, 117, 119–121, 470, 533
  heuristic, 853
Greedy, 104–106, 116, 118, 120
  adaptive search procedure (GRASP), 531
  algorithms, 83
  heuristic, 139
  incremental tree algorithm, 709

Grooming, 837–838
  research, 852
  taxonomy
    PACER, 838–839, 853
Group Steiner problem, 476
Group-ware, 704
Grow, 401
GSM, 531
Hamiltonian cycle, 397
Handover, 919
Head, 148
Heap implementations, 195
Heuristic, 78, 83, 464, 851
  algorithm, 282
  measure, 464
  methods, 399
  procedures, 437
Hidden collisions, 873
Hierarchical
  clustering, 444
  data network, 444
  design, 214
  level, 214
  network design, 280
  organization, 213
  tiers, 273
High-connectivity, 372
Hop-constrained
  path, 495
  spanning tree, 494
Hop constraints, 475
Hop-constraints
  expanded graph, 501
Hop distance, 521, 526
Hop-indexed model, 503
Hot-potato routing, 688–691
Hub location, 315, 452
Hybrid fiber coaxial (HFC), 330
Hybrid heuristics, 117
Hybridization, 117
Hypernetwork, 1077
IETF, 703
ILP, 637
Importance sampling, 741
Incidence matrix, 243
Incident, 148
Incremental algorithms, 683
Independence
  number, 867
Independent
  set, 867, 898
Information age, 1073, 1076, 1083, 1094, 1096
Insertion delay, 438, 446, 449
Integer linear programming (ILP), 395
Integer programming, 707, 851, 853
Integer programs, 67, 76
Integrated

multicriteria network equilibrium models, 1097
traffic network equilibrium model, 1077
Intensification, 108
Interactions among networks, 1074, 1080
Interdependence and intertemporal aspects, 232
Interior point methods, 3, 78, 85, 169
Intermediate system-intermediate system (IS-IS), 683–684
Internet, 1074, 1079–1080, 1083, 1094, 1106, 435, 679–680, 682, 688–689
 price determination, 1032
 Protocol (IP), 436
 telephony, 435–436
 traffic, 439
IP Network Restoration, 822
IP networks, 439, 849
IP over optical (IPOO), 852
Jitter, 438
Joint p-cycle design, 397
Jump inequalities, 506
Karmarkar algorithm, 615, 631
Karush-Kuhn-Tucker necessary conditions, 34
k-domination, 867
k-independence, 867
KMB algorithm, 709
k-mean, 1039
k-median, 1039
Knapsack, 70
 problem, 135
Knowledge
 discovery in databases, 1034
 representation, 1079
k-terminal, 737
Kuhn-Tucker constraint qualification, 34
Label
 correcting methods, 192
 setting methods, 191
 switching, 850
Lagrangean
 relaxation, 169, 446–447
Lagrange multipliers, 449
Lagrangian
 decomposition, 283, 449
 dual problem, 132
 heuristics, 135
 multipliers, 132
 relaxation, 52, 67, 70–72, 75, 78, 86–87, 258, 283, 320, 498, 530, 711
 relaxation problem, 132
λ-grooming, 838
Large-scale nature, 1080
Layered network models, 843
Leaf node, 148
Learning automata, 449
Least-hop routing, 396
Lifted cover inequalities, 73
Lightpath, 399, 520, 638, 840

Light tree continuity constraint, 526
Light trees, 526
Linear ordering, 16
Linear programming, 3, 580, 614
 relaxation, 76, 282
Link capacity, 273, 436, 445
Link dependent
 factors, 1097
 weights, 1097
Linked, 148
Link
 failure, 443
 restoration, 254, 809
 scheduling, 873
 sizing, 437
 topology, 436–437, 451
Load balancing, 941, 944, 959
Local access, 214, 368
Local area, 214
Local area network, 437
Local minimum, 29
Local optimality, 710
Local optimum, 84
Local search, 84, 106, 109, 112, 114, 116, 118–121, 139, 470, 679, 683, 685–688, 693
 heuristic, 451
Location, 117–119
 area (LA), 534
 management, 531
 problems, 518
Locations, 121
Logical implications, 83
Logistical networks, 1083
Long-term planning, 367–369
Low-connectivity, 372–373, 378
Lower bound, 648, 655, 657, 662, 666
LP relaxation, 637
MANET, 868
Marginal production costs, 224
Markov chain, 764
 analysis, 766
Markov model, 741
Mathematical programming problem, 1084
Max cut, 16
Maximum allowable delay, 441
Maximum concurrent flow, 243, 248
Maximum concurrent flows, 48
Maximum covering problem (MCP), 531
Maximum degree constraints, 715
Maximum flow problem, 1075
Maximum leaf spanning tree, 898
Maximum spanning tree preconditioner, 11
Max k-cut problem, 927
MBONE, 703
MCNF, 242
Mesh-based protocols, 538
Metaheuristics, 84, 104–106, 108, 117–118

ant colony optimization, 853
genetic algorithm, 84, 105–106, 109, 115,
    117–118, 120–122, 444, 450–451, 523, 528,
    742–743
GRASP, 105–106, 109, 112, 117, 119–121, 470,
    533
    heuristic, 853
path relinking, 108, 113, 120, 122
Pilot method, 469
scatter search, 106, 108, 117
scatter-search
    heuristic, 853
simulated annealing, 84, 105, 110, 118–119, 121,
    449, 522, 524, 528, 535–536, 685, 742
tabu search, 107–109, 111, 117–119, 121, 471,
    521, 526, 533, 685
    algorithm, 444
tabu-search
    heuristic, 853
variable neighborhood search, 114
VNS, 106, 114, 117, 120
Metric inequalities, 175, 382
M/G/1, 450
Minimum
    additional cost, 901
    broadcast cover problem, 899
    connected dominating set, 897
    cost flow problem, 1075
    cost multicommodity flow, 166, 168, 171, 279
        model, 166
    cost network flow problem, 10, 150
    cuts, 74
    degree ordering, 7
    interference frequency assignment problem, 926
    local fill-in ordering, 7
    longest edge, 908
    number of wavelength converters, 520
    spanning tree, 894, 86
    spanning tree problem, 461
    Steiner tree (MST) problem, 526
    weight arborescence problems, 71
Min-max optimization, 946–948
Mixed integer linear program, 281
Mixed integer nonlinear optimization, 28
Mixed integer programming (MIP), 317
Mixed integer program, 67
M/M/1, 445–446, 449, 452
Mobile, 118–119, 121
    and ad hoc networks (MANET), 537
    networks, 785
    switching center (MSC), 531
    wireless networks, 531
Mobility
    management scheme, 536
    management using virtual backbone, 905
Modified Frank-Wolfe algorithm, 54
Modified projection method, 1100

Modular, 395
Modular capacity, 395
MOSPF, 703
Most infeasible, 79
Most probable state, 740
Moving location areas, 536
MPLambdaS, 851
MPLS, 392, 438, 591–592, 597, 600, 806, 850
MPLS-enabled network, 439
MPLS p-cycles, 410
Multicast, 120
    group, 702
    network dimensioning problem, 718
    packing problem, 716
    routing, 472, 702
    routing tree problem, 705
    scheduling, 873
    traffic, 274
Multicasting, 525, 701, 902
Multiclass
    multicriteria network equilibrium, 1100
    multicriteria network equilibrium models, 1095
    traffic network, 1077
Multicommodity, 120, 242
Multicommodity flow, 1051, 247, 278, 442,
    446–447
    formulation, 495
    problems, 88
Multicommodity network flow, 839–840
    model, 166
    problem, 72, 521, 14
Multicommodity network flow, 164
Multicriteria
    decision-making, 1094
    network equilibrium conditions, 1098, 1100
    network equilibrium problem, 1100
    network model with elastic demand, 1099
    supernetwork, 1094
Multi-hop network, 519
Multihour network design, 580, 614, 618–620, 629
Multilabeling algorithm, 72
Multi-layer protection, 819
Multilevel
    networks, 1106
    supernetwork, 1106
Multimedia applications, 452
Multi-objective, 752
    GA, 752
    optimization, 752
Multiple access, 919
Multiplexing, 840, 846
Multiprotocol label switching, 850
Multi-QOP, 415
Multi-restorability capacity placement (p-cycle
    MRCP), 414
Multi-start heuristic, 853
Multitiered supply chain network equilibrium, 1108

INDEX 1129

Mutation, 743
Nash equilibrium, 35, 1083
Negative-cycle detection algorithms, 201
Negotiation, 695
Neighborhood, 30, 106–107, 111, 113–114, 119
NEPC, 406
Network, 148
   access design, 776
   administrator, 270
   architecture, 393
   decomposition, 739
   dependability, 783
   design, 17, 19, 41, 118–119, 214, 269, 369, 372–374, 376–377, 381, 383, 436–437, 440, 459, 495, 735
     cost, 277
   economy, 1083
   equilibrium, 1080, 1092
     conditions, 1088, 1090, 1092, 1098
     link flow pattern, 1091
     path flow pattern, 1091
     problem, 1089
   failure, 439
   flow, 185, 321
   flow problem specializations, 151
   hierarchy, 808
   level, 392
     design, 768, 776
   life extension, 45
   loading, 118
   loading problem, 280
   load
     minimization of the, 642
     template, 423
   models, 1075, 1089
   optimization, 185
   pricing, 546
   redesign, 324
   reductions, 739
   reliability, 736–737, 783
   reliability measures, 737
   re-routing, 392
   resilience, 738, 748
   substructure, 70
   synthesis, 1051
   system, 1074, 1079
   topology, 164–165
Neural network, 741–742
Newton method, 52, 54
Node, 392
Node-arc incidence matrix, 150, 165
Node costs, 167
Node-encircling p-cycle, 392
Node-encircling p-cycles, 392, 406
Node processing delay, 438, 440
Node requirement vector, 149
Node restoration, 392

Node-weighted Steiner problem, 474
Noncooperation, 1081
Noncooperative
   behavior, 1083
   network, 1083
Non delayed relax and cut, 130–131
Nonlinear optimization, 28
Nonlinear programming, 28
Nonnegative weights, 1097
Normal
   direction, 148
   equations, 7
North American Plesiochronous Digital Hierarchy, 847
NP-hard, 78
   problems, 281
NSF network, 669
Objective function, 683–685, 687–688
On-line algorithms, 856
Online Steiner tree problem, 473
Open Shortest Path First (OSPF), 683–684
Open systems interconnection model, 843
Opportunity cost, 1096–1097, 1100, 1102
   minimization, 1094
Oppositely directed, 148
Optical
   crossconnect (OXC), 839
   crossconnect, 853
   network, 637, 807
   network units, 326
Optimality conditions, 1087, 189
Optimal solution, 30
Ordered arrival sequence, 853
OSI, 843
OSPF, 121, 437, 449–450, 804
Over-subscription, 411
PACER grooming taxonomy, 839, 853
Packet
   delay, 438, 440
   distribution tree, 538
   loss probability, 766
   loss rate, 438
   switching, 590
Paging, 534
Paradoxical phenomena, 1080
Parent link, 707
Pareto optimal, 752
Partial linearization, 50
Partitioning, 78, 1038
   of the arc set, 152
Path, 148
   decomposition, 15
   formulation, 243, 644
   restoration, 254, 809
   segment, 393
Path relinking, 108, 113, 120, 122
p-cycle, 391

network design, 394
PDH, 852
Penalties, 79
Penalty function, 450, 744
Performability measures, 738
Performance, 276
  guarantee, 223, 435–436, 438
Personal computers, 436
Pervasive group, 703
Physical-layer, 392
Pilot method, 469
PIM, 703
Pivot and complement, 84
Platinum, 413
Platinum class, 413
Plesiochronous Digital Hierarchy (PDH), 846
Point-of-presence (PoP) placement problem, 530
Point-to-point connection problem, 719
Polyhedral characterization, 498
Polyhedron, 31, 371, 379
Polynomial approximation scheme, 248
Polynomial-time algorithm, 78
Polytope, 31, 376–378, 381, 383
Ports, 853
Power
  constrained networks, 45
  control, 920
  splitters, 526
Preconditioned conjugate gradient algorithm, 10
Preconditioner, 9
Predictive modeling, 1036
Preemptible, 413
Preprocessing, 78, 82, 89
Pricing, 1105
  of services, 789
  policy, 763
Primal-dual, 84
Primal feasibility, 79
Primal heuristics, 84
Primal network simplex procedure, 153
Primary network, 214
Prim's algorithm, 716
Priority
  branching, 81
  classes, 447
  queueing, 438
Private virtual circuits, 120
Prize-collecting, 120
  Steiner problem, 478
  traveling salesman problem, 302
Probabilistic network, 736
Probing, 83
Profitability of the investment, 773
Profit
  maximization, 1094, 1106
  risk, 554
Projection, 503

Project management, 185
Propagation delay, 438, 440
Protected working capacity envelope (PWCE), 420
Protection, 392
  paths, 392
Protocol, 274
Prototype shortest path algorithm, 191
Proximal decomposition, 251, 258
Pruning, 77
Pseudo-costs, 79
Pseudo costs, 81
Pseudo-shadow prices, 80
Pseudo shadow prices, 82
Public monopoly, 224
Push operation, 156
QoS, 243
Quadratic assignment problem, 521
Quality of protection (QoP), 413
Quality of service, 234, 243, 436–437
  constraints, 494
Queueing
  algorithm, 438
  delay, 438, 443, 446, 449
  theory, 762
Queuing, 582, 596–597, 602, 607
Radiolocation, 536
Randomization, 105–106, 116–117
Real options, 773, 775
Real-time decisions, 452
Re-configuration, 392
Reconfiguring networks, 856
Recourse problem, 771, 782
Rectilinear Steiner problem, 471
Reduced cost, 151, 302–303
  fixing, 78, 83, 85
Reduced gradient method, 54
Reduction techniques, 462
Redundancy, 277
Reformulation, 82
Registration, 534
Regression, 1037
Regular topology, 521
Relationship value maximization, 1106
Relax and cut, 130
Reliability, 735, 783
  constraints, 706
  estimation, 741
  evaluation, 745
Reliable
  network design, 743
  virtual backbones, 906
Rendezvous, 716
Rerouting, 254
Reserve network, 256
Resilience, 738
Resiliency, 277
Resilient networks, 751

INDEX 1131

Resource allocation, 439
Restoration, 392, 439, 453
  in dynamic call routing networks, 824
Revenue management, 557
Reverse Dijkstra's algorithm, 193
Reverse direction, 148
Reverse path-forwarding algorithm, 707
Ring, 118, 367, 369, 378–379
Ring-based multicast packing, 718
Ring-cut
  graph, 380–381
  inequalities, 380–381
Ring generation problem, 298, 301
Ring-mining, 426
Ring network
  design, 479
  design
    problem, 298, 300
Rings, 120
Risk minimization, 1094, 1106
Road traffic network, 28
Robust network, 768
  design, 762
Robust optimization, 686–687
Root node, 89
Rosen's gradient projection algorithm, 55
Rounded metric inequalities, 174
Round robin, 439
  queueing discipline, 449
Route assignment, 445
  problem, 436–437
Router location, 445, 452
  problem, 437
Routers, 439
Routing, 119–121, 219, 243, 246, 274, 437, 442,
    446, 637, 702, 840
  and wavelength assignment, 527–528
  and wavelength assignment (RWA), 521
  cache, 711
  demand, 840
  discipline, 437
  in data networks, 185
  models, 1084
  policy, 691–693
  protocols, 680–684, 695–696
  tables, 574–577, 617
RWA, 637
Safety cost, 1097, 1100
SBPP, 421
Scaling push-relabel algorithm, 156
Scatter search, 106, 108, 117
Scatter-search
  heuristic, 853
SDH, 292
SEACP, 413
Search neighborhood, 139
Secondary network, 214

Second order cone programming, 18
Security, 1097
Self-healing rings, 333, 848
Semidefinite programming, 18
Semi-homogenous, 396
Sequential construction/destruction, 741
Server deployment program, 770
Servers, 436
Service
  differentiation, 436
  introduction, 775
  section, 214
Set cover, 723
Shadow
  cost, 555
  prices, 945–946
Shared backup path protection (SBPP), 421
Shopping versus teleshopping, 1098
Shortest path, 437, 445, 451, 577, 580–582, 592,
    595, 614, 619–620, 628–629, 853
  metric, 448, 453
  problem, 70, 185, 461, 1075
  protocols, 453
  reformulation, 501
  routing protocol, 449
  tree problem, 187
Signal-to-interference ratio (SIR), 532
Silver, 413
Simple flow bounds, 150
Simplex, 78
Simplicial decomposition, 54
Simulated annealing, 84, 105, 110, 118–119, 121,
    449, 522, 524, 528, 535–536, 685, 742
Simulation, 740, 745, 764
  approach, 764
Single-commodity network flow model, 165
Single-hop network, 519
Single source / single destination shortest path
    problem, 186
SIR, 919–920
SIR-based power control, 930
Slack variables, 82
SLF and LLL algorithms, 199
Software delivery, 704
SONET, 292, 393, 847, 852
  rings, 81
Source-sink, 392
Space-division multiplexing, 840
Space of probability measures, 765
SP-Add, 401
Span costs, 396
Spanning
  subgraph, 148
  tree, 148, 323, 716
Span-protecting p-cycles, 391–392
Span-restorable mesh network, 392
Spare capacity, 243

Spare capacity allocation, 395
Spare channel, 391
Sparse group, 703
Sparsity, 711
Spatially separated markets, 1079
Spatial price equilibrium, 1084, 224
Spatial price equilibrium problem, 1080
Splitter placement problem, 527
Splitting capability, 526
SRLGS, 417
Stackelberg game, 44
Star/star, 319
Star/tree topology, 322
State dependent routing, 575
Static grooming, 856
Static grooming methods, 852
Static group, 703
Statistical
  distribution, 400
  multiplexing, 777, 849–850
Steepest-edge, 80
Steiner, 107, 120
  arborescence, 893
  arborescence problem, 893
  points, 893
  problem on graphs, 706
  tree, 70, 75, 461, 893
  tree game, 474
  tree problem, 459, 893
  tree problem in graphs, 459
  tree problem with basic sets, 476
  tree-star problem, 476
SteinLib, 462
Step-increasing cost functions, 172
Stochastic
  approximation, 972
  flows, 411
  integer programming, 72
  optimization, 28, 761, 764
  profit maximization, 771
  programming, 17, 761–762
  programming with recourse, 768, 770
  random search, 779
Straddling link algorithm (SLA), 401
Straddling spans, 391
Strategic investment decisions, 763
Stratified sampling, 741
Strong branching, 80
Sub-differential, 259
Subgradient, 132, 446
  algorithms, 87
  method, 132
  optimization, 132, 445, 447, 510
Subgradients, 259
Subgraph, 148
Subnetwork, 1079
Supernetwork, 1079, 1095, 1100, 1106

Supernetworks, 1074–1075, 1079–1080, 1083, 1108
Supernetwork
  construction, 1078
Supply chain
  analysis, 1106
  management, 1022
  network, 1105–1106
  network equilibrium model, 1106
Supply node, 149
Survivability, 243, 277, 367–369, 371–372, 374, 377–378, 439, 443, 453, 741, 755
  constraints, 177–178
Survivable network, 73–74, 331
Swapping methods, 223
Switching
  center, 214, 368
  costs, 167
Symmetric
  system, 637
Synchronous, 847
Synchronous optical network (SONET), 847
Synchronous transfer signal level 1 (STS-1), 847
System-optimality conditions, 1090
System-optimization, 1081, 1084, 1087–1088
  problem, 1086–1087, 1090
System-optimized, 1079, 1084
  problem, 1086
System optimum flow assignment, 41
Tabu search, 107–109, 111, 117–119, 121, 471, 521, 526, 533, 685
  algorithm, 444
Tabu-search
  heuristic, 853
Tail, 148
Tarjan's algorithm, 201
Taxonomy of location problems, 518
Taylor expansion, 45
TCP/IP, 1079
  model, 843
TDMA, 120, 919
TDM networks, 848
Technological level, 763
Telecommunication, 1076, 118, 121
  links, 1102
  network design, 242
  networks, 1074, 1080, 1083, 494
Telecommunications, 1108, 118–120
  link, 1096
  networks, 1080
Telecommuting, 1096
  versus commuting, 1094, 1097, 1100, 1104–1105
Teleshopping, 1096–1097
  versus shopping, 1105
TELPAK, 321
Terminal

INDEX 1133

boxes, 214
sections, 214
Theory of nonlinear optimization, 29
Threshold algorithm, 200
Time dependent routing, 575
Time division multiple access (TDMA), 531
Time division multiplexing, 840, 846
Time minimization, 1094
Time slot
    assignment (TSA), 839
    interchange (TSI), 839
Tomo-gravity, 439
Topological, 71, 75
    center, 716
    score, 396
Topology, 436, 446, 451
    capacity and flow assignment (TCFA) problem, 444
    capacity and route assignment, 446
    capacity and route assignment problem, 450
    of logical network, 783
Traffic
    assignment, 1088, 120
    assignment and routing, 36
    demand, 272
    efficiency, 748
    engineering, 679, 842
    equilibrium, 37
    load flow optimization, 618
    management, 574–575, 590
    matrices, 439
    matrix, 437, 684–687
    model, 272
    network assignment models, 1098
    network equilibrium, 1084, 1088–1089
    network protection, 813
    networks, 808
    restoration, 809
    volume, 272
Transiting, 392
Transit nodes, 293
Transmission
    bandwidth, 213
    control protocol (TCP), 680, 687–688
    facilities, 436
    rate assignment, 45
Transponders, 853
Transportation, 1076
    link, 1096
    networks, 1076, 1080, 1083–1084
    planning process, 1077
    problems, 185
Transport
    network protection, 816
    networks, 808
Transshipment node, 149
Travel cost, 1097, 1100

Traveling purchaser problem, 476
Traveling salesman problem, 89
Traveling salesman subtour problem, 301
Travel time, 1097, 1100
Tree, 148, 316, 322
Two-connected, 369, 374, 376, 378–381, 383
Two-node connectivity, 742
Two-terminal, 737
Type of service (TOS), 438
UMTS, 531–532, 537
    network planning problem, 929
Uncapacitated network design
    fixed charge, 279
Uncertainty, 761, 783
Undirected links, 151
Unicasting, 525, 902
Unicast
    routing, 702
    traffic, 274
Unidirectional ring, 293
Unit-disc graphs, 898
Universal mobile telecommunications system (UMTS), 530
Upper bound, 745
Used wavelengths (minimization of the), 641
User optimization, 1081–1082, 1088
    problem, 1088
User optimized, 1079, 1084
    network problem, 1088
Utility function, 28
Valid inequalities, 173
Variable dichotomy, 78
Variable fixing tests, 140
Variable neighborhood search, 114
Variational inequalities, 1089
Variational inequality, 1090–1091, 1099–1100
    formulation, 1085, 1090, 1092, 1099
    problem, 1090, 1099
Video
    conferencing, 435, 704
    streaming, 705
Violated inequalities, 74
Virtual
    backbone-based routing, 903
    backbone broadcasting, 905
    private networks, 435
    transport flow optimization, 614, 618, 628
    tributaries (VTs), 847
VNS, 106, 114, 117, 120
Voice over IP, 435
Volume maximization, 422
Walk, 148
Walrasian price equilibrium, 1080
Wardrop
    equilibrium, 37
    first principle, 1084, 1088
    second principle, 1084

Waveband switching, 852
Wavelength
  assignment, 393, 637
  assignment
    problem, 526
  continuity constraint, 520, 522, 840
  conversion, 521, 849
  converters, 519–520, 526
  cross-connect, 849
  cross-connects (WXC), 519
  division multiplexing, 530, 849
  division multiplexing (WDM), 519
  routed networks, 520
W-CDMA, 920

WDM, 637, 849, 852
Weakly connected
  dominating set, 866
  domination number, 866
Weights, 1097
Weight setting, 121
Wide area network (WAN), 449
Winner determination problem, 1051
Wireless
  local area network, 923
  multicast advantage, 900
  networks, 868
Working flow, 399
Working-weighted efficiency, 396
Workstations, 436
World wide web, 435